NATURAL SCIENCES IN AMERICA

NATURAL SCIENCES IN AMERICA

SELECTED WORKS OF JOEL ASAPH ALLEN

Introduction by
Keir B. Sterling

ARNO PRESS
A New York Times Company
New York, N. Y. • 1974

Reprint Edition 1974 by Arno Press Inc.

Reprinted from copies in The American
Museum of Natural History Library

NATURAL SCIENCES IN AMERICA
ISBN for complete set: 0-405-05700-8
See last pages of this volume for titles.

Manufactured in the United States of America

Publisher's Note: The articles in this compilation have been reproduced from the best available copies.

Library of Congress Cataloging in Publication Data

Allen, Joel Asaph, 1838-1921.
Selected works of Joel Asaph Allen.

(Natural sciences in America)
CONTENTS: Catalogue of the mammals of Massachusetts, with a critical revision of the species.--On the mammals and winter birds of east Florida, with an examination of certain assumed specific characters in birds, and a sketch of the bird-faunae of eastern North America.--Influence of elevation and latitude upon the distribution of species. [etc.]
1. Zoology--North America--Addresses, essays, lectures. 2. Mammals--North America--Addresses, essays, lectures. 3. Birds--North America--Addresses, essays, lectures. 4. Allen, Joel Asaph, 1838-1921.
QL151.A442 1974 598.2'97 73-17843
ISBN 0-405-05765-2

CONTENTS

Geographical Variation in North American Birds (Reprinted from *Proceedings of the Boston Society of Natural History,* Vol. XV), Boston, 1872

To What Extent Is It Profitable To Recognize Geographical Forms Among North American Birds? (Reprinted from *The Auk,* Vol. VII), New York, 1890

The Geographical Origin and Distribution of North American Birds, Considered in Relation to Faunal Areas of North America (Reprinted from *The Auk,* Vol. X), New York, 1893

Geographical Variation Among North American Mammals, Especially in Respect to Size (Reprinted from *Bulletin of the U.S. Geological and Geographical Survey of the Territories,* Vol. II), Washington, D. C., 1876

Geographical Distribution of the Mammalia, Considered in Relation to the Principal Ontological Regions of the Earth, and the Laws That Govern the Distribution of Animal Life (Reprinted from *Bulletin of the U.S. Geological and Geographical Survey of the Territories,* Vol. IV), Washington, D. C., 1878

The Geographical Distribution of North American Mammals (Reprinted from *Bulletin of the American Museum of Natural History,* Vol. IV), New York, 1892

Recent Work in North American Mammalogy (Reprinted from *Transactions of the New York Academy of Sciences,* Vol. X), New York, 1891

Recent Progress in the Study of North American Mammals (Reprinted from *Abstract of the Proceedings of the Linnaean Society of New York for the Year Ending March 27, 1894*), New York, 1894

Autobiographical Notes and a Bibliography of the Scientific Publications of Joel Asaph Allen (Reprinted from *Special Publication of the American Museum of Natural History*), New York, 1916

INTRODUCTION

Joel Asaph Allen (1838-1921) was born in Springfield, Massachusetts, to a farmer who did not particularly appreciate his son's early interest in nature, preferring that Joel eventually prepare to follow in his footsteps. Only after much pleading would the father purchase his son a shotgun when the latter was thirteen, which was soon being used to secure a great variety of birds. With no formal training or assistance whatever, young Allen showed an early scientific tendency to keep detailed records on each bird secured, though he had no access to any of the standard texts of that day. He also made efforts to draw some of his specimens, and was delighted when after several years he met a drawing teacher who happened to know something of bird study and taxidermy. Through this young man and other teachers who encouraged the lad, Allen was exposed to the entire range of the natural sciences. He began keeping detailed records of weather conditions in the vicinity of his home, and published a brief account of them in the *New England Farmer* in 1858.

Not until he was twenty did Allen enjoy any systematic schooling beyond some winter sessions at the small country schoolhouse a mile from his home. For four years (1858-1862), Allen learned much concerning his chosen vocation from Prof. Oliver Marcy, who then taught natural sciences at Wilbraham Academy in Western Massachusetts. Permitting his son to attend this institution was a major concession on Farmer Allen's part, and he insisted that his son spend his summers working in the family fields. Young Allen did so, but so strained his never robust constitution that he was subject for the rest of his life to extended bouts of illness. By now he was a dedicated collector of all kinds of specimens, ranging from birds and mammals to reptiles, fish, amphibians, insects, rocks, minerals, and the like, also venturing into plant study.

Following the completion of his studies at Wilbraham, Allen travelled to Cambridge, where he studied under the eminent Louis Agassiz for three years at Harvard, without, however, taking a degree there. He was compelled, however, to sell most of his natural history collections to the Academy in order to finance his initial work with Agassiz. Despite enforced breaks brought about by poor health, during which he did field work, Allen received excellent training at Agassiz's hands, accompanying that indefatigable scientist on an eleven month trip to Brazil from which he returned in February, 1866. Digestive problems, aggravated by the strain of the trip compelled Allen to return to his father's home for eighteen months, but he took up field collecting in the Great Lakes region in the summer of 1867, following which he returned to Cambridge to take charge of the bird and mammal collections at the Museum of Comparative Zoology under Agassiz's direction.

Though he remained in the museum's employ consistently for eighteen years, he was not made a member of its scientific staff until 1871, when he became Assistant in Ornithology, the title he retained for fourteen years. From 1871 to 1873, he also served as Lecturer in Zoology, but the years from 1868 to 1873 were largely taken up with strenuous field work in the west, and one expedition, sponsored by both the Museum and the Smithsonian Institution was twice attacked by large bands of Indians. Allen laconically noted years later that conditions for collecting on that trip "were far from ideal," and his younger colleague Frank M. Chapman remarked in summing up Allen's career that rare is the field naturalist who has run "the risk of being himself collected."

Except for a trip to Colorado with his good friend the ornithologist William Brewster in 1882, Allen never again entered the field, attacks of poor health having made that an impossibility. The Colorado trip was in fact made in part to regain a measure of physical well-being. Between 1876 and 1882 Allen researched and wrote his famous monographic works on the American Bison and North American Pinnipeds (both included in this Arno Press series), while continuing with his curatorial duties at Harvard. From time to time his strength would give out and he was forced to abandon all desk work for extended periods.

In 1885, at the age of forty-seven, Allen was invited to come to New York City to become Curator of Birds and Mammals at the American Museum of Natural History, with which he remained associated until his death in 1921. During nearly all of this thirty-seven year period, he wrote, in addition to performing his administrative functions, nearly two hundred papers on mammals and birds, based upon the museum's collections, and for more than thirty years he edited the Museum's *Memoirs* and *Bulletin*.

Probably the one outside activity which consumed most of his leisure time was the American Ornithologists' Union, of which Allen was one of three co-founders in 1883. For eight years its first president, he remained a member of its governing council until his death. He edited three A.O.U. checklists of North American Birds and was editor of *The Auk,* the Council's journal, for twenty-eight years, during which time he wrote nearly 650 pieces for it. Classification and nomenclature were among his chief interests, and from 1911 until 1921 he was a member of the International Commission on Zoological Nomenclature. In 1886 Allen received an honorary Ph.D. from the University of Indiana and in 1903 the Walker Grand Prize, awarded by the Boston Society of Natural

History. His interest in mammals was recognized in 1919 when the organizers of the newly formed American Society of Mammalogists wished to make him its president. He declined because of the infirmities of age, however, the honor going to Merriam instead.

The papers reprinted in this anthology represent a cross-section of some of Allen's major works published between 1869 and 1916, having mainly to do with taxonomy and geographic distribution. The reader's attention is drawn first to the final selection, *Autobiographical Notes and a Bibliography* (1916), which will help to place the papers selected for inclusion in their proper historical context. Allen was an extremely reticent man, however, and though the bibliography is the fullest to be found anywhere, the sketch he writes concerning his career should be supplemented by Frank M. Chapman's sensitively written memoir of his mentor and colleague (Chapman succeeded Allen as Curator of Birds at the American Museum) which appeared in *Memoirs of the National Academy of Sciences,* vol. XXI, 1927, pp. 1-20.

Allen's study of the *Mammals of Massachusetts,* published when he was thirty, is a good example of his early work in mammalogy, and of the thoroughness that characterized everything he did. His "On the Mammals and Winter Birds of East Florida," however, is justly regarded as his first major publication, since it was in this paper that he gave the subject of variation in the size and color of birds due to ecological factors greater attention than it had previously received. In part V of this paper, Allen discusses the laws governing geographic distribution of the world's fauna as well as the biogeography of North America. This was a subject to which he returned in the next two selections published in the mid 1870's, which deal with the factors controlling the distribution and genesis of species. These questions Allen and his younger colleague, C. Hart Merriam, were to address themselves to for a quarter of a century. Allen was clearly the leader in this field for a time, but he saw his ideas eclipsed because of the enthusiasm engendered by Merriam's Life Zone theory in the 1890's. He is now considered to have been closer to the mark than Merriam since many of the biogeographical units he postulated accord well with the later biome concept. In the second of these papers, "The Influence of Physical Conditions in the Genesis of Species," (1877), Allen revealed a strong Neo-Lamarckian tendency, which was to color his thinking and that of others, including Merriam, at least to the end of the 1890's.

Allen long resisted the growing trend in both ornithology and mammalogy in the 1870's and 1880's to proliferate the number of species. This tendency had come about because of improvements in collecting procedures, a growing conviction that minor geographic variations, if consistent, merited specific or subspecific recognition, and greater sophistication in field methodology. Allen's attitude is reflected in the relative conservatism displayed in his 1872 article on geographical variation in birds. Chapman has pointed out that Allen "described only 16 species and subspecies of mammals" between 1870 and 1890, whereas in the next two decades he described 415. A similar change in his thinking becomes apparent in his ornithological articles dealing with the recognition of geographic forms in birds (1890) and the geographic origin and distribution of birds (1893). Chapman has suggested that Allen gradually shifted from "lumping" to "splitting" in specific determinations, both because of his assessment of the growing collections of specimens in various collections and also because most of his colleagues gradually adopted the new approach.

The next items reveal that Allen was giving considerable thought to mammalian distribution. The first article on geographical variation in North American mammals (1876) enunciated what became known as "Allen's Rule," that the appendages of mammals became smaller the further north one found them, since it was necessary for them to preserve body heat in the arctic cold. In his "Geographical Distribution of the Mammalia," he argued that some mammals had circumpolar distribution, contrary to the views of Sclater and A. R. Wallace, and his ideas are now generally accepted. He also suggested that most species had originated in the Arctic region and moved southward, but present day biogeographers suggest that the pattern of dispersal began in tropical Africa, with the Americas having been reached via the Siberian-Alaskan land bridge. While "Geographical Distribution of the American Mammalia," (1892) was written in part to modify or rebut certain of Merriam's views, and did contain some errors, it is a sound paper in many respects, with the biological zonation Allen describes remarkably in consonance with present day ideas. Allen's two articles on developments in mammalogy (1891 and 1894) parallel the two written earlier on birds, and are an invaluable summary of the bewildering growth taking place in the field between the Civil War and the end of the nineteenth century.

Allen is a major figure in the development of American natural science, particularly ornithology, mammalogy, and biogeography, in the late nineteenth and early twentieth centuries. A careful, methodical, and productive scientist, he provided much of the basis for our present day understandings of systematics and the distribution of life, and his writings are an important part of the literature of these subjects.

February, 1974

Keir B. Sterling
Tarrytown, N. Y.

Catalogue of the Mammals of Massachusetts

No. 8. — *Catalogue of the Mammals of Massachusetts: with a Critical Revision of the Species.* By J. A. ALLEN.

THE original design of the present catalogue was simply to contribute a few data concerning the distribution of the Mammals of New England; but in order to explain certain views entertained by the writer in respect to the character of a number of currently received species, many critical notes were gradually incorporated, until finally it was thought best to extend the paper so as also to embrace a systematic revision of the species. The catalogue is based mainly on observations made by myself at Springfield. In its faunal characteristics this locality does not differ much from those parts of the State lying east of the Connecticut River generally. A few species which occur only in the western mountainous portions have been included on data afforded chiefly by the official report on the Mammals of the State by the late Dr. Ebenezer Emmons, but in part as the result of observations and inquiries of my own recently made in that section. Respecting the marine species, I have consulted Captain N. E. Atwood, of Provincetown, a gentleman well known as a reliable observer, and whose forty years' experience along our coast has rendered him very familiar with our larger marine Vertebrata. I have thus been able to add not a little to our knowledge of some of those species least known, and the most difficult to observe, of all our Mammalia. The great obligation I am under for his kind co-operation is fully evident from the valuable notes he has furnished on the Cetaceans. I am also greatly indebted to Professor E. D. Cope, of Philadelphia, to whom I transmitted the notes of Captain Atwood, for kindly identifying the species.

Less attention seems to have been paid by our naturalists to the Mammals of the State than to the Birds, or several of the other classes of our animals. This may be owing to the greater difficulty of observing and procuring the former, arising from either their scarcity or reclusive habits.

The first general scientific notice of Massachusetts Mammalia seems to have been a simply nominal catalogue by Dr. Edward Hitchcock, published in his Report on the Geology, Mineralogy, Botany, and

Zoölogy of the State (pp. 526, 527), in 1833. Forty-five species are there given, including the two Seals and three Cetaceans. To a few only are notes added respecting their relative abundance. Dr. Emmons's first Report, under the Act of the Legislature of 1837 for a Natural History Survey of the State, was published in 1838. In 1840 a second and final Report* was presented, containing the substance of the first, and considerably increased by additions. These Reports contain descriptions of all the land Mammalia then known to inhabit the State, with interesting notes on their habits and distribution, but nothing on the marine. The whole number of species given is forty-four, two of which (*Arvicola hirsuta* = *A. riparia*, and *A. albo-rufescens* = *A. riparia*, albino) were erroneously described as new. Eliminating three that have since been reduced to synonymes (*Condylura macroura*, *Sciurus niger*, *Arvicola albo-rufescens*) leaves forty-one as the number of valid species embraced in this report. The animal now known as *Hesperomys leucopus* Baird was described as *Arvicola Emmonsii* De Kay. On the whole, however, the work is remarkable for its accuracy, and, compared with those of most recent writers, for the small number of merely nominal species it contains.

The only other special treatise on our Mammals is an article by Mr. E. A. Samuels, in the Ninth Annual Report of the State Board of Agriculture,† in which thirty-nine species are described, excluding two merely nominal (a *Blarina* and *Arvicola rufidorsum*), mainly from Massachusetts specimens in the State Cabinet of Natural History; it also contains notes on their habits, and several woodcuts of the animals. Though not assuming to give all the species of the State, Mr. Samuels includes five or six described since the publication of Dr. Emmons's Report, but omits several of that author that are not uncommon in certain sections of the State, as well as all the marine species. In Audubon and Bachman's "Viviparous Quadrupeds of North America" (three volumes, 8vo, 1846 - 1853) are numerous references to Massachusetts Mammals specimens of which were frequently furnished these authors by our well-known ornithologist, Dr. T. M. Brewer, of Boston. But since the publication of Dr. Emmons's Report, no one, excepting perhaps Dr. Brewer and

* Report on the Quadrupeds of Massachusetts. By EBENEZER EMMONS, M. D. 1840. 8vo. pp. 86. This is the edition cited in the following pages.

† Agr. of Mass., 1861, pp. 137 - 191.

Mr. Samuels, has done more to increase our knowledge of their history than Mr. J. W. P. Jenks, of Middleboro'. From this locality Mr. Jenks has sent large collections of the smaller species to the Smithsonian Institution, which have been carefully worked up by Professor S. F. Baird in his invaluable Report on the Mammals of North America,* and by Dr. H. Allen in his recent excellent Monograph of the North American Bats.† In the Journal and Proceedings of the Boston Society of Natural History, among the very few notices of our Mammals, is an important paper by the Rev. John Bachman on the Mole Shrews (genus *Scalops*),‡ in which a new species (*S. Breweri*) is described from specimens from this State contributed by Dr. Brewer. In Professor Baird's Report on North American Mammals two species of *Arvicola* (*A. Breweri* and *A. rufidorsum*) are also described as new, solely from specimens from Massachusetts; the first was collected by Dr. Brewer on Muskeget Island. (On these see remarks beyond.) In February, 1863, Professor A. E. Verrill mentions, in a valuable contribution on the Shrews of New England,§ the first known occurrence of a *Neosorex* (*N. palustris*) in this State.

The more important publications on the Mammals of adjoining States, which in this connection demand a passing notice, are the Rev. J. H. Linsley's "Catalogue of the Mammalia of Connecticut,"‖ Dr. J. E. De Kay's well-known Report on the Mammals of New York, and Professor Zadoc Thompson's notes on those of Vermont.¶ Mr. Linsley's list numbers seventy-one species, embracing the marine and domesticated, and nine that are merely nominal. Removing the latter, the eight domestic, and two ("*Arvicola floridanus* Ord" and "*Phoca grœnlandica*? Mull.") of doubtful reference, leaves fifty-two as the number of valid indigenous and naturalized species (the latter being the three species of *Mus*), ten of which are marine and the remaining forty terrestrial. Two bats (*Vespertilio subulatus* Say and *Scotophilus noctivagans* = *V. noctivagans* Cooper) and one shrew (*Sorex platyrhinus*) are given in

* Pacific Railroad Reports of Expl. and Surv., VIII, 1857.

† Monograph of the Bats of North America. By H. ALLEN, M. D. Smithsonian Miscellaneous Collections, June, 1864.

‡ Proc., Vol. I, p. 40, 1841; Journ., Vol. IV, p. 46, 1842.

§ Proc. Bost. Soc. N. H., Vol. IX, 164.

‖ Am. Journ. of Science and Arts, XLIII (Oct. 1842), pp. 345-354.

¶ History of Vermont, Natural, Civil, and Statistical, etc. By ZADOC THOMPSON. 8vo. Burlington, 1842, and Appendix, 1853.

addition to the land animals described in Professor Emmons's Massachusetts report, while two of Emmons's *Cervidæ* and the Wolverine are very properly omitted.

Dr. De Kay's Report, which appeared but a few months later than Linsley's Catalogue, gives seventy-eight species as either actual or former inhabitants of the State of New York, including, in addition to the domestic and marine species of Linsley's list, five fossil species. No new ones are added, though several are described as such, and several previously well known are separated from their supposed distinct European allies and receive new names. Two species given by Linsley for Connecticut ("*Arvicola floridanus* Ord" and "*Phoca grœnlandica?* Mull.") are rightly omitted, and others, including the Opossum (*Didelphys virginiana*), added. This is a southern species which has not yet, so far as I can learn, been detected east of the Hudson. Deducting the nominal species and those of doubtful reference, nine in number, and the eight domestic and five fossil, leaves fifty-six as the number of living valid ones, forty-six being land and ten marine. This is an excess of four only, — two bats and two very small species of shrew, — excluding the marine and the extra-limital *Didelphys virginiana*, over the number given by Dr. Emmons for Massachusetts.

Professor Thompson's Natural History of Vermont, published at about the same time, contains forty-three valid species, with descriptions of them drawn up mainly from Vermont specimens, and short general accounts of their habits. It embraces but one or two species not given in Dr. Emmons's report, one of which is the common Seal (*Phoca vitulina*). A single specimen of this is reported to have been captured on the ice in Lake Champlain, and in the Appendix, published in 1853, another similar instance is recorded.

The present catalogue embraces sixty-five species, giving for the first time a probably nearly complete list of the marine, the Seals and Cetaceans. The latter are now supposed to number eighteen species. Four land species (*Scotophilus georgianus*, *Scalops Breweri*, *Neosorex palustris*, and *Arvicola pinetorum*) are also added, that are not mentioned by either Dr. Emmons or Mr. Samuels, or by either of the extra-limital authors mentioned above.

In Massachusetts, as far as Mammals and Birds are concerned, portions of two Faunæ are represented,— the Canadian and the Alleghanian;

the former occupying a large part of Berkshire and most of the western half of Franklin, Hampshire, and Hampden Counties, or those portions of the State having an elevation of and above fifteen hundred feet above the sea ; the latter the remaining and by far the larger portion.*

The recent or historic changes that have occurred in the Mammalian Fauna of the State consist mainly in the decrease in numbers of the larger species, amounting to a complete extirpation of a few of the large Carnivora and Cervidæ (*Felis concolor*, *Mustela Pennantii*, *Cervus canadensis*, *Alce malchis*, *Tarandus rangifer*), and the great reduction, almost to extinction, of several others (*Lynx canadensis*, *Lynx rufus*, *Canis lupus*, *Ursus arctos*, *Cervus virginianus*). None of these species are now anywhere common, though there is good reason to believe that several of them were once so, while a few are known to have been of very frequent occurrence. The smaller species, including most of the rodents, the bats, moles, and shrews, seem to be fully as numerous as they ever were, while it is not improbable that a few, especially the *Arvicolæ* and other field mice, and perhaps the woodchuck (*Arctomys monax*), are even increasing in numbers. The three species of exotic or eastern origin are the now almost cosmopolitan *Mus decumanus*, *M. rattus*, and *M. musculus*, which long since became annoying pests, and constitute the only additions to our feral Mammalia that have become fully naturalized.

Several of the species of this list are considered to be identical with species of the Old World, although most late writers have separated as specifically distinct all but one of our New England Mammals — the *Gulo luscus* — from their Old World relatives. Only two or three species of land Mammalia are now generally considered as common to any portions of both the Eastern and Western hemispheres.† Several

* The Canadian fauna, as represented in Massachusetts, may be characterized by the present or former occurrence among Mammalia of the following species: *Mustela Pennantii*, *M. martes*, *Gulo luscus*, *Alce malchis*, *Tarandus rangifer*, *Cervus canadensis*, *Arvicola Gapperi*, and *Erethizon dorsata*. The Alleghanian may be distinguished by the absence of the preceding and the presence of *Vulpes virginianus*, *Scalops aquaticus*, *S. Breweri*, *Sciurus cinereus*, *Arvicola pinetorum*, and *Lepus sylvaticus*, which do not occur in the Canadian fauna.

† The same is also true of the land birds, while a large proportion of those marine species that are probably really common to both sides of the Atlantic are regarded as distinct. It should be observed, however, that the separations in both classes have been made mainly by the same persons. On the other hand, the highest authorities in entomology admit many species to be common throughout the northern hemisphere, par-

others, particularly of the genera *Arvicola* and *Blarina*, currently received as valid, are here treated as merely nominal. While our reasons therefor are given somewhat fully in their proper connection, a few general remarks in further explanation seem called for here.

In the present greatly increased state of our knowledge of American mammals, not a few characters once very naturally considered of great importance in a specific diagnosis are to be regarded as far from decisive, they now being known to be dependent either upon age, season, or locality, or to be mere individual variations. A difference in size, for instance, is at present well known in mammals, as well as in birds, to almost universally accompany differences in the latitude and elevation of their respective habitats, the southern representatives of species widely diffused being very appreciably smaller than the northern. The difference between the extremes amounts not unfrequently to nearly one fourth, and occasionally even to one third, of the average size, so that, considered apart from the connecting stages afforded by representatives from the intervening districts, they might well be regarded as belonging to distinct species. It is also now well known that mammals vary geographically in respect to color, though not yet fully to what extent, and also in the character of the pelage. These latter facts have been long recognized practically in respect to the fur-bearing species, but it appears equally true of most of the others. Experienced trappers and fur-dealers readily distinguish the Mink and Sable skins of the north from those of the south, by the comparatively greater fineness, density, and length of the fur of the northern animal;* similar differences are equally evident in the pelage of the Wolves, Foxes, Lynxes, and Hares. This difference is similar to that observable between winter and summer specimens from the same locality, the northern corresponding in the character of the pelage to the winter and the southern to the summer ones. The resemblance is perhaps still more striking in regard to the

ticularly among the Hymenoptera, Neuroptera, and Coleoptera, and not a few are regarded as primitively almost cosmopolitan. The same is true in regard to plants, quite a large proportion of the species of the northern North American flora being considered identical with European and Asiatic. Hence we naturally inquire, Is there really this discrepancy in the distribution of species in the different classes of organized beings, or is it only apparent through the biased opinions of one or the other of these schools of observers?

* In the case of the Minks, those of the prairies are distinguished as readily from those inhabiting the adjoining wooded districts, the former having coarser and browner fur, the difference being sufficient to materially affect their price in the market.

clothing of the feet, species with the soles thickly furred in winter often having them sparsely so in summer, northern individuals differing in like manner from southern. The variation in this respect increases with the distance in latitude between the localities whence the specimens compared are taken.

Besides these geographical or climatic variations, we have found by a careful comparison of scores of specimens of the *same species, collected at the same locality*, that there is a much greater range of variation between individuals of the same species — the variation extending to every part — than is commonly conceded; and also that differences depending upon season,* as in the color, thickness, length, and general texture of the pelage, and others depending upon age * and sex, instead of being always recognized by authors as such, have not unfrequently been taken to indicate a constant specific diversity. From this cause there has arisen, in numerous instances, an undue increase of so-called species. *Specimens* have too often been described instead of *species*. It is not surprising that these mistakes should have happened in the earlier days of our science, when the material for study was scanty and diagnoses were commonly drawn up from stuffed skins, the authors being in total ignorance of the appearance of the animal in life; when the extent of individual variation had not been especially investigated, and it was unknown that in animals possessing a wide distribution there were marked variations accompanying wide differences in locality. But even now

* In spring, as is generally well known, mammals shed the long, thick coat wo in winter; this is replaced by a much shorter, thinner, less soft, and generally differently colored pelage. In this there is a gradual change throughout the summer, and late in fall it becomes either entirely replaced or effectually concealed by the growth of the long winter coat. The winter differs from the summer pelage not only in being longer and thicker, but generally in the different character of the hair composing it, and in the fulness of the soft under fur, as well as more or less in color. The shortness of the summer coat renders the ears of such animals as have these members very short, as the different species of *Arvicola*, *Sorex*, *Sciurus*, &c., much more conspicuous at that season than in winter, when in some of them they are nearly concealed. In young animals, too, the first pelage differs much from the succeeding, being shorter, darker, and generally more or less crisp. The general health of the animal, as no one need be told who has attentively observed domestic animals, has a marked effect upon the character of the coat, and on the time it is changed, as does also scantiness or abundance of food.

As previously stated in the text, species with the soles of the feet furred have them less densely so in summer than in winter. It is perhaps needless to advert to the fact of the existence of a temporary set of teeth in young animals, which gradually give place to a permanent one differing from the first in number and character.

but few mammalogists have come to recognize these variations as manifestations of general laws, and we are consequently scarcely surprised at the glaring inconsistencies into which even our best authorities are frequently betrayed, they at times assigning to these several variations their true character, and again, in apparently equally clear cases, considering them as indications of specific diversity. It thus happens that species are still not unfrequently based solely on differences that are but individual peculiarities, from these differences being first detected in comparing specimens from widely separated districts, whereas they are not different from variations presented by occasional specimens of the same species at any given locality. Oftener still, perhaps, species are founded on slight geographical variations, either solely or in connection with exceptional individual peculiarities, or on differences depending upon age. A remarkable instance of this latter kind seems to have occurred in our *Sorecidæ*, and especially in *Blarina*, where no less than eight at present currently received species are apparently based on one. Imperfectly understood sexual variations, associated with other differences, in some cases render the complication still greater. This occurs in the *Mustelidæ*, where the female is found to be very much smaller than the male in almost or quite all species when the sexual differences are well known. In the weasels the large amount of this difference seems to have thus far generally escaped notice, especially by American writers. As wide a range of variation, aside from the sexual, obtains in these as in their near allies, the mink and the marten. In this group, differences in size and in the relative length of the tail as compared with the body — the latter an extremely variable element — have been taken as important specific distinctions, and on these grounds alone some five species (so called) appear to have been based on two.

In respect to the differences that have been claimed to separate specifically the Old and the New World representatives of those species we in this paper consider identical, only those of very slight importance have as yet been adduced; they are only such as might be anticipated to occur when, as has repeatedly happened, the comparisons have been made between only a few specimens known to have been collected at localities widely differing in latitude, and hence in climatic conditions, and at different seasons of the year. More frequently, however, the exact origin and history of the specimens compared appears to have been wholly unknown. In no case are the differences greater, but generally less, than those presented by specimens from different localities on the same continent, where the species is admitted to be the same; sometimes not greater than is seen at the same locality. From similar unsatisfactory comparisons, and undoubtedly in part from theories of distribution, representatives from distant points in the United States of species ranging from the Atlantic to the Pacific have been described as distinct species. Not till large series of specimens from hundreds of localities have been carefully compared can all these disputed points be properly settled, through the tolerably exact determination of the influence of "locality on the individual"; and we believe that no work more important than this can at present be done.

In this connection I can hardly avoid a word or two in reference to the spirit which evidently incites many zoölogists in their researches. I refer, of course, to that eagerness for describing "new species" so patent in all their publications, — an influence highly derogatory to the advancement of scientific knowledge. It tends to divert attention from such a critical study of those species living in the naturalist's immediate vicinity as will alone acquaint him with the amount of variation a species may be expected to present.* Only by such a preparation can one be prepared to estimate properly the character and value of differences presented by specimens from remote districts, of which only a limited number of prepared examples can be examined. Almost all writers on the different classes of Vertebrata have fallen in a greater or less degree into the fault of describing species as new from either improper or insufficient material, or of founding them on characters that a critical study of numerous fresh specimens of a few well-known species would have shown were of very slight, and often of even no value as specific distinctions. The inquiry with many naturalists respecting doubtful specimens seems rarely to be whether they may not be referred to some already known species, and the points of resemblance to their nearest known ally accordingly carefully weighed against the differences, but rather are not they sufficiently different to warrant a description of them as new species? This greediness for *species nova* renders it

* In respect to Birds, I have already called attention (Memoirs Bos. Soc. Nat. Hist., Vol. I, p. 512) to the importance of collecting and comparing a very large number of specimens from the same locality, to learn the extent of the variation a species may present at the same point; it is no less essential in Mammals, where seasonal variations and those depending upon age are not always so evident.

difficult to eradicate from our systems those even but doubtfully admitted when once they have been proposed by authors high in authority, such species being ultimately accepted without having ever been scientifically established. Authors afflicted with this mania rarely reject any species of their contemporaries, but they virtually indorse the doubtful ones by adding others of their own based on similar characters. The great proportion of merely nominal species hence annually added to our lists is a detriment to science deeply to be regretted.

Perhaps the strictures contained in this article will by some be deemed too severe; they are nevertheless made, not only reluctantly and in all cases without the slightest personal feelings, but from a conviction of their necessity, and with the sole object of advancing the truth. Gladly would I have left to others the unpleasant task.

While much of the material forming the basis of this list has been, as previously stated, that of my own collecting at Springfield, I am deeply indebted to the Museum of Comparative Zoölogy for additional data, and especially for a large amount of invaluable material for the revision of the species.* It has also afforded me the opportunity of comparing American with European specimens of the species of *Mustelidæ* and *Canidæ*, and of examining specimens of most of the Mammals of North America. The very complete collection of Massachusetts mammals in the Springfield Museum of Natural History, mainly collected and prepared by Mr. C. W. Bennett, embracing as it does several unique specimens, has likewise been freely consulted, and with much profit. I have already referred to my indebtedness to Captain N. E. Atwood, of Provincetown, for notes on the Cetacea, and to Professor E. D. Cope for the identification of the species.

The names used in Dr. Emmons's Report are generally added as synonymes when different from those now adopted. A tabular comparison of the species given by Dr. Emmons from this State, by Dr. De Kay from New York, and by Mr. J. P. Linsley from Connecticut is made with those of the present list, in order to indicate their synonymy. In general only such synonymes are given, always from original examination, as are necessary to render clear the views of the writer on the

* Probably no other Natural History Museum in the world affords facilities for the investigation of the individual variation of species equal to those presented by the immense collections of New England, and especially Massachusetts, Vertebrata contained in this Institution, brought together by the Director in great part for this especial purpose.

points in question. The thorough and exhaustive manner in which this part of the subject has already been treated by Professor Baird and Dr. Allen has rendered anything further than this unnecessary.

FELIDÆ.

1. **Lynx canadensis** Raf. Canada Lynx. Rare, and generally occurring only in the more thinly settled and mountainous parts of the State. A very large one was killed in November, 1866, in the town of Ware. Reports of their capture in the towns of western Hampden, Hampshire, and Franklin Counties, as well as in Berkshire, are not very infrequent.

2. **Lynx rufus** Raf. Bay Lynx. Apparently rather more common than the preceding species, but, like this, it is generally confined to the more wooded and mountainous districts. One was taken at Ipswich a short time since, and they seem to occur at intervals in all sections of the State.

The *Felis concolor* Linn. (Panther) has probably been for some time extinct in Massachusetts, though undoubtedly once occurring here. There is a stuffed specimen in Springfield said to have been killed a year or two since in the Adirondack Mountains of New York. A few months since the writer saw another that was captured on Pine Hill, in Weathersfield, Vermont, January 31, 1867. This specimen is said to have measured seven feet from the tip of the nose to the tip of the tail, to have stood two feet nine inches high, and to have weighed one hundred twenty-two and a half pounds. It had lived for some time previously on Ascutney Mountain, a few miles from where it was captured. Very good photographs of this rare animal, taken from this specimen before it was skinned, can be obtained of Mr. J. D. Powers, of Springfield, Vermont.

Professor Thompson states, in his Natural History of Vermont (p. 37), that for some time after the settlement of that State had commenced the Panther was so common there as to be considered dangerous to travellers unless they were well armed. In his Appendix (p. 12) he states that the last one he had known to be killed in that State, and also the only one for many years, was captured in Bennington, in February, 1850.

CANIDÆ.

3. **Canis lupus** Linn. (*C. occidentalis* var. *griseo-albus* Baird.) Gray Wolf. Occasional in the sparsely populated districts of the western counties. Like the species of *Felidæ*, it has been nearly extirpated.

Authorities have differed greatly in their views respecting the identity of the American and European wolves; some, forming the majority, and among them apparently those whose opportunities for judging have been most favorable, have considered them the same, while others, and among them many who seem to have but casually examined the subject, have regarded them as distinct. Not only so, but — omitting certain varieties based on color and commonly received as merely nominal, though repeatedly raised to the rank of species — specimens from the middle and western portions of the continent have been described as specifically distinct, both from the Old World wolves and those of the eastern side of the continent.* Dr. Richardson, than whom probably no one has had better opportunities for studying American wolves, after pointing out some trivial differences in physiognomy and in the character of the pelage between the wolves of Arctic America and the Pyrenees, observes: "Notwithstanding the above enumeration of the peculiarities of the American wolf, I do not mean to assert that the differences existing between it and its European congener are sufficiently permanent to constitute them, in the eye of the naturalist, distinct species. The same kind of differences may be traced between the foxes and native races of the domestic dog of the New World and those of the Old; the former possessing finer, denser, and longer fur, and broader feet, well calculated for running on the snow.† These remarks have been elicited by a comparison of live specimens of American and Pyrenean wolves; but I have not had an opportunity of ascertaining whether the Lapland and Siberian wolves, inhabiting a similar climate with the American ones, have similar peculiarities of form, or whether they differ in physiognomy from the wolf of the south of Europe." For this reason he considered it "unadvisable to designate the northern wolf of America by a distinct specific appellation"; "the word *occidentalis*" (*Canis lupus occidentalis*), he further observes, "which I have affixed to the Linnæan name of *Canis lupus*, is to be considered as merely marking the geographical position of the peculiar race of wolf which forms the subject of this article."

Audubon and Bachman, the former having been long familiar with the American wolf in all its different varieties, unhesitatingly pronounced, after

* As *C. nubilus* Say, *C. variabilis* Maximilian, *C. gigas* Townsend, &c.

† The comparisons in this case, it should be remembered, are between specimens from localities possessing widely differing climates.

careful and extended comparisons of specimens from the two continents, the common wolves of the Old World and the New to be, in their opinion, identical. But Dr. De Kay, giving but two lines to a consideration of the subject, very summarily separates the American wolf from its Old World congener under the name of *Lupus occidentalis*. Professor Baird, after admitting the weight of authority to be in favor of the supposition of their specific identity, considers them distinct, and adopts the name of "*Canis occidentalis*" for the American species. In referring to the different varieties of the North American wolf this author says: "For the present I prefer to consider all as one species, and to assume this with good reason as distinct from some at least of the European wolves, if that continent possesses more than one." Although previously admitting the unsatisfactory character of his materials,* such a conclusion is but in accordance with his usual apparent predilection for considering American animals as distinct from their intimate affines of the Eastern continent, sometimes even where the weight of authority is by far in favor of their identity, and his own materials for an original examination of the subject are either entirely wanting or too scanty to be of much account.†

In his article on the Wolf (p. 108) Baird gives us, however, a most interesting and very valuable table of measurements of twenty-six skulls, chiefly from the Platte River, but which includes others from Sweden and Russia, as well as such remote points in North America as New York, Oregon, Texas, and Mexico. Aside from the markedly smaller size of those from the southern localities, the specimens do not appear to differ more than the same number might from either of the localities mentioned. The table shows variations in the proportion of breadth to length in the muzzle and in the whole skull, and in its relative breadth at similar points; but a careful examination of all the measurements given shows that these differences are inconstant, specimens from near the same locality differing as much or more than those from distant points. Neither are the differences greater nor different in kind from those New England specimens of the common fox (*Vulpus fulvus*), the woodchuck (*Arctomys monax*), the northern hare .*Lupus Americanus*), or the gray rabbit (*L. sylvaticus*), present, and which in some of these species are sometimes exceeded.

* "In the lack of perfect specimens of the North American wolf, I find it very difficult to throw any light upon the long-vexed questions of our species, all before me being mutilated in some way, and not allowing a satisfactory comparison with each other and with descriptions." — *N. Am. Mam.*, p. 105. After stating his conclusions in regard to the matter, however, he in a later paragraph mentions the receipt of additional specimens from the Yellowstone River.

† But one species, the *Gulo luscus*, is admitted in the Report on North American Mammals, as specifically identical with any species of the Old World. In this case a strong probability, in his estimation, of distinctness is hinted at.

The variations, particularly in point of color, presented by the species under consideration do not appear restricted to its American representatives, in the north of the Old World, the wolves, according to authors, varying from the white ones of Lapland and Siberia to the gray, pied, dusky, and even black ones of the more southern States; and here also the differences in color have been considered as indicating different species. In North America, where the wolf is quite fully known, the differences between the large white, or nearly white, races of the extreme north of the continent and the smaller dusky and rufous races of the south, in size, color, in the character of the pelage, and perhaps in other points, are so great that, without the intermediate links through which these widely differing extremes almost insensibly pass into each other, through individuals inhabiting the intervening districts, these extremes might be considered as well-marked species. At the far north, and "particularly in districts nearly destitute of wood," says Dr. Richardson, "wolves totally white are not uncommon," while grayish white is the prevailing color. The gray occupy, in general, the northern and elevated parts of the continent, including the elevated and more northern sections of the United States, and pass into the white and lighter gray wolves occupying the region farther north, and into the darker colored ones existing at the south. Southwards the color increases, tending more and more towards black and red, till in Florida * and the Gulf States dusky and black wolves predominate, and in Texas red or rufous. Yet in no portion of the continent is the color of the wolves at all uniform, the same packs generally presenting a great variety in this respect, even those of the same litter often widely differing. Dr. Richardson mentions, under his "variety *sticte*," that of five young wolves, "leaping and tumbling over each other, with all the playfulness of puppies of the domestic dog," which he thought were probably of one litter, one was "pied, another entirely black, and the rest showed the common gray colors." In speaking of the black American wolf, which forms his "variety *ater*," he says the Indians do not consider them to be even a distinct race, but report that one or more black whelps are occasionally found in a litter of a gray wolf. Audubon and Bachman, in referring to the red wolf of Texas ("*Canis lupus* Linn. var. *rufus*" of these authors), state that this variety is by no means the only one found there, "where wolves black, white, and gray are to be met with from time to time. We do not think, however," say they, "that this red wolf is an inhabitant of the more northerly prairies, or even of the lower Mississippi bottoms, and have therefore called him the Red Texan

* "The varieties, with more or less of black, continue to increase as we proceed farther to the south, and in Florida the prevailing color of the wolves is black." — AUD. & BACH., *Quad. of N. Am.*, Vol. II, p. 130. These observations of Audubon my own inquiries made during a recent journey in this State tend to confirm.

Wolf." On the Missouri we find, according to Lewis and Clark, that the wolves are chiefly yellow, as also, according to Professor Baird, on the Platte and Yellowstone (N. Am. Mam., p. 110), where they appear to gradually merge into the gray and white ones of the north. These latter evidently form the so-called varying wolf (*C. variabilis*) of Prince Maximilian,* some of which, he says, are entirely white, others yellowish white, some more mixed with gray, and others still entirely gray, in the same pack. The black wolf noted by Say on the Missouri, and which he describes as *C. nubilus*, like the gray and white ones, seems to occur everywhere, but apparently much more abundantly at the south, thus corresponding in its distribution, as in general character, with the black variety of Southern Europe, described by Linnæus, and afterwards by Cuvier, as *Canis lycaon*. This name was also applied by Dr. Harlan to the American black wolf. The red, or rufous, seems likewise southern, occurring in great abundance in Texas, and thence northward through the middle region of the continent, passing gradually through paler rufous and yellowish to the prevalent gray and grayish-white wolves of the north. Though perhaps our data are at present too few to warrant positive conclusions on the subject, the facts appear to point rather strongly to a localization of these different colors; it is nevertheless true that, as already stated, the wolves present at every locality a wide range of variation, and that neither variety of color is entirely restricted to any particular region. The gray is apparently the most widely diffused, occurring in greater or less numbers almost everywhere. † We find, however, that authors have considered these color differences as indicating not only permanent varieties, worthy of distinctive names, but even species, as is shown by a glance at the subjoined table of synonymes of the American animal. Not a few, including Audubon, Bachman, Dr. Richardson, and others, have been so inconsistent as to name and characterize as "varieties" what they at the same time admit to be either positively or probably only individual variations, occurring sometimes in the same litter with the common form. ‡

* Reise in das innere Nord-Amerika, Vol. II, 1841, p. 95. Ib., Archiv für Naturgeschichte, Vol. XXVII, 1861, p. 247.

† Dr. Coues observes, in a series of interesting papers on the "Quadrupeds of Arizona," in the American Naturalist (Vol. I, p. 288), that all the wolves seen by him in Arizona were of the grizzly or grayish-white variety, which in winter, at a distance, appear almost white.

‡ Dr. Richardson, after saying "these variations of color, however, not being attended with any differences of form, nor peculiarities of habit, I deem them to be no more characteristics of proper species, or *even permanent varieties*, than color would be in the domestic dog," proceeds at once to formally name and describe five "varieties," as though they were tangible, permanent forms, — so great apparently is the fascination to some minds of bestowing names, to be followed by their own as authority, in Natural History.

In some previous citations of the synonymes of this species, I find that Dr. Richardson has on several occasions been incorrectly quoted, first by De Kay and afterwards by Baird; his name, *Canis lupus, occidentalis*, having been rendered by them " *Canis* (*Lupus*) *occidentalis*," thus incorrectly conveying the impression that he regarded the wolf of North America as distinct from the European, and as also having placed it in a sub-genus (*Lupus*) of *Canis*. Dr. Richardson, however, expressly states that he did not regard them as distinct, and did not wish to further burden the science by imposing a new name to indicate what at most he thought might be but a geographical race.

Canis lupus.

Canis lupus LINNÆUS, Syst. Nat., I, 1767, 58.
" *mexicanus* IB., 60.
" " SHAW, Gen. Zoöl., I, 1800, 296.
" " DESMOREST, Mam., I, 1820, 199.
" " FISCHER, Syn., 1829, 183.
" " BERLANDIER, Proc. Acad. Nat. Sc. Phil., V, 1851, 157.
" *lupus, albus* SABINE, Franklin's Journ, 652.
" *lupus, griseus* IB., 654.
" *lupus, occidentalis* RICHARDSON, Faun. Bor. Amer., I, 1829, 60.
" " " var. A, *griseus*, Ib., 66.
" " " " B, *albus*, Ib., 68.
" " " " C, *sticte*, Ib., 68.
" " " " D. *nubilus*, Ib., 69.
" " " " E, *ater*, Ib., 70.
" *lupus* HARLAN, Faun. Amer., 1825, 84.
" *lupus*, var. *ater* AUDUBON and BACHMAN, Quad. N. Am., II, 1851, 126, pl. 67.
" " " *albus* IB., 156, pl. 72.
" " " *rufus* IB., 240, pl. 82.
" " EMMONS, Quad. Mass., 1838, 26; Ib., 1840, 28.
" *nubilus* SAY, Long's Exped. R. Mts., I, 1823, 168.
" " HARLAN, Faun. Amer., 84.
" *lycaon* IB., 126.
" *variabilis* MAXIMILIAN, Reise in das innere Nord Amer., II, 1841, 95.
" " IB., Arch. Naturgesch., XXVII, 1861, 247.
" *gigas* TOWNSEND, Journ. Acad. Nat. Sc. Phil. (2d series), II, 1850, 75.
" *occidentalis*, var. *griseo-albus* BAIRD, N. Am. Mam., 1857, 104, pl. 31.
" " " *nubilus* IB., 111.
" " " *mexicanus* IB., 113.
" " " *ater* IB., 113.
" " " *rufus* IB., 113.
Lupus occidentalis DE KAY, Nat. Hist. N. Y. I, i, 1842, 42, pl. 26, fig. 2.

4. **Vulpes vulgaris.** (*V. fulvus* RICH., and of most modern authors.) RED FOX. More or less common throughout the State.

The varieties called "Silver Fox," "Black Fox," and "Cross Fox," are taken at long intervals.

These so-called varieties, to which have been given such distinctive names as *Canis decussatus*, *C. argentatus*, *C. fulvus* var. *decussatus*, etc., etc., and which some authors have regarded as species and the majority as permanent "varieties," are but different degrees of melanism of the common red fox, as they sometimes all occur in the same litter of young.* They appear exactly parallel to the dusky and black varieties of marmots, which are usually considered as only variations of this character. The dusky of the preceding species (*C. lupus* Linn.) and the black form of several species of *Sciurus* are probably but the result of the same tendency more highly developed. Foxes in other countries, and particularly the European, are well known to present corresponding dusky and black variations, which have likewise been described as permanent varieties, and even as species.

Respecting the identity of the red fox of North America with that of Europe there is a diversity of opinion. Most of the old authors considered them specifically the same, while later they were almost as generally regarded as distinct. Recently their identity has been maintained by several high authorities in Europe, among whom are Giebel, Wagner, and Maximilian, and not without a fair show of reason. Professor Baird observes, that careful comparisons of the two show "appreciable differences, although the resemblance is very close in external appearances, and scarcely to be expressed except comparatively."† The

* Audubon and Bachman, in their account of the Cross Fox ("*Vulpes fulvus* Desm., var. *decussatus* Pennant"), in Quadrupeds of North America (Vol. I, pp. 52, 53), incidentally relate the following: "In the spring we induced one of our servants to dig for the young foxes that had been seen at the burrow which was known to be frequented by the Cross Fox. With an immense deal of labor and fatigue the young were dug out from the side of a hill; there were seven. Unfortunately, we were obliged to leave home, and did not return until after they had been given away and were distributed about the neighborhood. Three were said to have been black, the rest were red. The blackest of the young whelps was retained for us, and we frequently saw, at the house of a neighbor, another of the litter that was red, and differed in no respect from the common Red Fox. The older our little pet became the less it grew like the Black, and more like the Cross Fox. It was, very much to our regret, killed by a dog when about six months old, and, as far as we can now recollect, was nearly of the same color as the specimen figured in our work."

In the following autumn the female was killed: "It was nearly jet black, with the tip of the tail white. This was the female that produced the young we have just spoken of; and as some of them, as we have already said, were Cross Foxes and others Red Foxes, this has settled the question in our minds that both the Cross Fox and Black Fox are mere varieties of the Red."

† Mamm. of N. Am., p. 126.

differences in the color and texture of the fur, to which he and others have called attention, seems the most tangible difference, though not one of high value. Several specimens from different parts of Europe, in the Museum of Comparative Zoölogy, show that some of the other differences specified by Professor Baird, particularly that of the form of the tail and the greater length of its hairs in the American animal, are far from constant, there being no such differences in this respect between them and others from the United States, as has been claimed. One of the European has the tail remarkably full, the longer hairs being fully an inch longer, instead of an inch shorter, as according to authors they should be, than average American specimens. Prince Maximilian has also observed that this distinction in regard to the form of the tail is inconstant and invalid.* While, as Professor Baird remarks, European specimens can be readily separated from American, as in the case of most species commonly admitted as identical on the two continents, it does not follow necessarily that they are specifically distinct, since in very many species of animals specimens from not very remote localities can be similarly distinguished, where naturalists never question their identity. The very exact agreement in the southward distribution of the red fox in the Old World and in the New, — their southern limitation on both continents, as nearly as can be judged, coinciding with the same isotherm, — and the occurrence of the same varieties, as "cross," "black," and "silver," and in about the same relative proportion of individuals, if indicating anything, seems to point to their identity. In considering this subject it is necessary to take into account the remarkable tendency to variation presented by other members of this family in a state of nature, and the readiness with which widely distinct breeds are developed under domestication in the common dog. The European specimens to which we have referred differ considerably among themselves, these differences being in some cases greater than between some of them and the average type of the American animal. I hence do not hesitate to consider the North American red fox as identical with the common red fox of Europe, the average amount of difference being not greater than might be anticipated in specimens from so distant localities.

5. **Vulpes virginianus** De Kay. Gray Fox. Though essentially southern, this species is said by De Kay to be rather common in the southern counties of New York, and particularly on Long Island;† Audubon and Bachman give it as not uncommon in the vicinity of Albany, N. Y., but as scarce in New England, and state

* Arch. für Naturgesch., XXVII, Theil 1, p. 259.

† Zoöl. of New York, Vol. I, p. 46.

that they had not heard of it to the north of the State of Maine.* Dr. Emmons gives it as "rare in Massachusetts."† Mr. C. W. Bennett informs me that he knew of the capture of two specimens in Leominster a few winters since. The skins of this species frequently seen in our fur stores come, so far as I have learned, altogether from Eastern Virginia and the Southern Atlantic States.

MUSTELIDÆ.

6. **Mustela Pennantii** Erxl. (*M. canadensis*, Emmons Rep.; *Martes* ‡ *Pennantii* Gray.) Fisher. Probably still of rare occurrence in the Hoosac ranges. In 1840 Dr. Emmons wrote: "It is occasionally found in the vicinity of Williamstown, particularly in that range of mountains which extends northeast through Stamford, Vermont." §

This species seems to be the only one of the old Linnæan genus *Mustela* (*Martinæ* of recent authors) peculiar to the northern parts of North America, with no very near ally in the corresponding portion of the Old World.

7. **Mustela martes** Linn. (*Martes americana* Gray; "*Mustela americana* Turton" of recent American authors; *M. zibellina* Brandt.) Pine Marten. Sable. Occasional in the mountains of Berkshire County. Thirty years since Dr. Emmons mentioned it as not infrequent there, but as most common "where pine forests abound. It is, however," he says, "often found in beech woods, where it is sure of a more ready supply of food. Its nocturnal habits, and native shyness, effectually screen it from observation, even in districts where it abounds." ‖

The variations presented by the sables and martens, at single localities as

* Quad., Vol. I, p. 172.

† Rep., p. 31.

‡ Each of the three generally recognized genera of the sub-family *Martinæ* ("tribe *Mustelina*" of Gray) — *Mustela* embracing the sables and martens; *Putorius*, the minks, weasels, and ermines, and *Gulo*, the wolverine — has been recently subdivided, the sections being ranked by some as sub-genera, and by Dr. J. E. Gray as genera. In his Revision of the Genera and Species of *Mustelidæ* (Proc. London Zoöl. Soc., 1865, pp. 100–154), he restricts *Mustela* to the weasels and ermines, and *Putorius* to the polecat, while the sables and martens he places under *Martes*, and the minks under *Vison;* the distinctions, based on differences either in the dentition, form of the skull or color, are, however, very slight.

§ Rep., p. 39.

‖ Rep., p. 41.

well as in different districts, have been very perplexing, and have given rise to a considerable number of supposed species and a very great number of "varieties," the alleged distinctions between which are quite uncertain and inconstant. Some of these variations are doubtless referable to seasonal changes,* and not a few others to individual peculiarities. Dr. Gray admits six species as inhabitants of the North Old World,† several of which he divides into three to five varieties each. To a few of them only, however, does he assign separate geographical districts; in general they vary in such a way as to render the forms recognized by him as species quite intangible, the varieties forming gradations between them. Two of the three attributed to Japan (*Martes japonica* and *M. brachyura*) rest on exceedingly unsatisfactory data, while the third (*M. melanopus*) has a striking resemblance to the common form of the American species, and to varieties of both the so-called *M. abietum* and *M. zibellina* of Europe and Asia. Aside from these divisions of Dr. Gray, three principal races or forms (species of many writers) have for a long time been recognized as occurring on the Eastern continent, — the sable (*Mustela zibellina* Linn.), the pine marten (*M. martes* Linn.), and the beech marten (*M. foina* Brisson; *M. martes*, var. *fagorum* Linn.). The principal distinctions between them consist in the relative length of the tail, which varies in being sometimes longer, equal to, or shorter than the body, and in the color, which varies in general tint, and differently in the different regions of the animal, and especially on the throat, which is sometimes white, or nearly so, but more commonly yellowish or yellowish-brown; occasionally the "throat patch" is nearly obsolete. The color of the head is sometimes like that of the body, and again much lighter; the general color varies from blackish through different shades of brown to light yellowish brown and whitish. But instead of either of these differences being limited, or peculiar, to either "species," "variety," or race, or to special localities,* they are all given by Dr. Gray under the five divisions of his fifth species, — "*Mustela zibellina* Linn."; while he says of his *M. abietum*, var. *altaica*, that it is "intermediate between *M. abietum* and *M. zibellina;* but the feet are not so hairy"!* Brandt, in his Beiträge Säugtheire Russlandt, recognizes three species. The American animal (*M. americana* auct.) he considers as a yellowish or more yellowish-brown and less densely furred variety of the Asiatic sable than as a pure marten (*M. martes*), and calls it *Mustela zibellina*, var. *americana*.

Dr. Gray of course regards the American as distinct, and divides it into three varieties, — *abietinoides*, *huro*, and *leucopus*, — which seem to vary only in intensity of color, the first being "black-brown," the second

* See *postea*, pp. 165 - 167.

† Proc. Lond. Zoöl. Soc., 1865, p. 104.

"yellowish-brown," and the third much lighter than the second. The habitat of the first is given as the "Rocky Mountains"; of the second, "Fort Franklin"; that of the third is not stated, and may be supposed to be general, or at least those districts not occupied by the others. It is evident, however, that these different varieties are not local, as they occur more or less frequently at the same localities, and likewise at as distant points as the two sides of the continent. Dr. Gray refers to a series of specimens of the American pine marten in the British Museum, collected by Dr. Lord during his excursion with the Boundary Commissioners, that "vary greatly in color, from pale brown to nearly black," and have "the throat variously mottled with yellow." * Mr. Bernard R. Ross says that the farther north the skins are obtained the darker the pelage, and that on the Youkon River they strongly resemble the Siberian sable.† While in general the specimens from North America are of the white-headed or sable, rather than of the marten, type, dark-headed ones also occur, not exclusively on the western side of the continent, as some have supposed, but more or less frequently at all points.

Professor Baird has described ‡ specimens from the West Coast that do not differ essentially from others from the Adirondacks, though having the head much less white. Dr. Brandt's series of American skins from the Northwest Coast, as far south as Columbia, on the contrary, had the head very light colored, and hence resembled in this respect the generality of specimens from New York, Maine, and Nova Scotia. In other general characters he also found a close agreement with the Asiatic sable, and, as already stated, he believed them specifically identical. Dr. Gray also mentions a close resemblance in the color of the head between specimens from Russia and the Northwest Coast of America. Professor Baird, after comparing American with Swedish specimens, states that "in some respects, as in certain features of the skull and teeth, the American marten approximates to the beech marten, *M. foina*, more than to the European true marten"; and that it differs from the latter (*M. martes*) in certain proportions of the skull, in the texture and paler colors of the pelage, in the relatively longer tails of the latter, and in the extent of the naked pads of the feet. He also finds resemblances in color to the Russian *M. zibellina*, but finally concludes, after quoting Dr. Brandt's reasons for considering them identical, by saying that he is "far from admitting the identity of the American marten with the Russian sable, although it occupies a position intermediate between the latter and the *M. martes* in size,

* L. c., p. 107.

† List of Mammals, Birds, and Eggs observed in the McKenzie's River District. Nat. Hist. Rev., July, 1862, p. 272.

‡ Mam. N. Am., p. 153.

length of tail and coloration, as well as in intrinsic value of the fur. The white-headed varieties of New York are most like the sable; the darker-headed ones of the Western country like the pine marten." He is "inclined to the belief," he says still later, in an interpolated note, "that we have two species, one representing the pine martin, with dark head, the other similar to the sable, with whitish head, — both probably distinct from the corresponding Old World species, the martens at least."

In Dr. Brandt's diagnosis of the martens, the relative length of the tail is dwelt upon as an important character. In *M. zibellina* the tail without the hairs is given as one third the length of the body; in *M. martes*, one half or more than one half. Professor Baird says the tail vertebræ in *M. americana* are about one half the head and body; hence not differing much from the same proportion in *M. martes*, while quite different from the same in *M. zibellina*, which Dr. Brandt considers the *M. americana* to most resemble; while Dr. Gray observes that the tail of some of Dr. Lord's specimens from Western America is almost as short as it is in the Russian sable. A marked discrepancy is evident in these statements, explainable on the ground of the inconstancy of the distinction based on the relative length of the tail. Brandt also states that the *M. foina* differs from *M. martes* somewhat in general color (but apparently not essentially, considering the much wider differences in this respect his varieties of *M. zibellina* present among themselves), and in having the tail generally longer, with more vertebræ. Since, however, the number of tail vertebræ is far from constant in most mammals with this member considerably developed (as I have myself observed in the mice, squirrels, ermines, and foxes), this latter character must lose much of its weight till repeatedly verified. Dr. Gray says, in urging the non-identity of the American and Old World martens, that "It is curious that both Brandt and Baird seem to have overlooked the small size of the last tubercular grinder, which separates the American from the Old World pine martens"; a fact he claims to have discovered. From variations I have observed in this respect in our common *Mephitis*, it would be interesting to know whether Dr. Gray has found this difference constant in a considerable series, or whether the observation rests on a single specimen, as, in the same connection, he refers to "the skull of the American specimen we have in the Museum," in speaking on another point.

I have shown in the foregoing remarks that the martens and sables of the Old World and the New are not without close points of affinity in all essential particulars; that on both continents they present almost innumerable differences, principally in respect to color, but few of which, if any, appear to be geographical, or even constant; that on both continents the variations are similar; that the points of distinction between the supposed species are slight, and rest mainly on characters which in mammals are the most likely of all others to be variable; that authors, in their statements and opinions, are widely discrepant and often contradictory; finally, that the American animal is most closely allied to the Asiatic, grading through it into the European. At present there seems to be no middle ground between considering all as forming one circumpolar species and admitting a considerable and indefinite number, since some of the so-called "varieties" seem as strongly marked forms as some of the "species." If we must consider the American as distinct from those of the Old World, we can hardly do less, on parallel grounds, than to recognize two or more in America. It seems probable that in time the greater part will be found to be not permanent or uniformly transmissible varieties, but merely irregular individual variations; — in other words, that more than one so-called variety may be represented in the same family, as has been shown is the case in the foxes and wolves, and as is well known to occur in *Mephitis*.* The comparison of a great number of specimens from many localities will be necessary before we can consider the matter as satisfactorily settled.

Since writing the foregoing, I have met with a very valuable paper on the Fur-Bearing Animals of the Mackenzie's River District,† and another on the Martens and Weasels of Nova Scotia; ‡ I have also had an opportunity of comparing a large number of skins of the Siberian sable with an extensive series of others from Hudson's Bay. Much additional information has been derived from these sources, which tends to confirm the opinion above expressed; namely, that most of the so-called varieties and species would prove to be based on seasonal and individual variations of a single circumpolar species. The writer of the first of these papers, Mr. Bernard R. Ross, is well known from his extensive Natural History explorations in the boreal regions of this continent, and his experience of thirteen years in this district as a successful trapper entitles his statements and opinions to more than ordinary weight. He seems to have been a critical observer, and in this paper adds much to our knowledge of the fur-bearing animals of North America. His remarks on the seasonal changes in the color and character of the fur in several species are particularly valuable. The following extracts from them explain to a great extent the nature of the wide variations which, in many characters, the martens and sables everywhere exhibit.

* See *postea*, p. 173 *et seq.*

† A Popular Treatise on the Fur-bearing Animals of the Mackenzie's River District. By BERNARD ROGAN ROSS, C. T. — *Canadian Naturalist and Geologist*, Vol. VI, January, 1861, pp. 5 – 36.

‡ On the Mammals of Nova Scotia, No. III. By Dr. J. BERNARD GILPIN. — *Transact. Nova Scotia Inst. of Nat. Science*, Vol. II, Part I, pp. 8 - 16.

"It is difficult to describe," he says, "the color of the marten fur accurately. In a large heap of skins (upwards of fifty) which I have just examined minutely, there exists a great variety of shades, darkening from the rarer yellowish-white and bright orange into various shades of orange-brown, some of which are very dark. However, the general tint may with propriety be termed an orange-brown, considerably clouded with black on the back and belly, and exhibiting on the flanks and throat more of the orange tint. . . . . The ears are invariably edged with a yellowish-white, and the cheeks are generally of the same hue. The forehead is of a light brownish-gray, darkening towards the nose, *but in some specimens it is nearly as dark as the body.** The yellowish marking under the throat (considered as a specific distinction of the pine marten) is in some *well defined,* and of an *orange tint,* while in others it is *almost perfectly white.* It also varies much in extent, reaching to the forelegs on some occasions. At other times it consists merely of a *few spots, while in a third of the specimens under consideration it is* ENTIRELY WANTING." In respect to other characters he observes: "The tail is considerably less than half the length of the body *generally,* though it is sometimes longer; it is well covered and tolerably bushy. The feet are comparatively large, densely covered with short woolly fur, mingled with stiffer hairs, which prevent the naked balls from being visible in winter, *though they are distinctly so when the animal is in summer pelage.*"† Respecting the seasonal changes he says: "When casting its hair the animal has far from a pleasing appearance, as the under fur falls off, leaving a shabby covering of the long, coarser hairs, which have then assumed a rusty tint. . . . . After the fall of these long hairs, and towards the end of summer, a fine, short fur pushes up. When in this state the pelage is very pretty, and bears a strong resemblance to a dark mink in its winter coat." He further observes: "In summer, when the long hairs have fallen off, the pelage of this animal is darker than in winter. The forehead changes greatly, *becoming as deeply colored as any other part of the body, which is of an exceedingly dark brown tint* on the back, belly, and legs. The yellow throat markings are much more distinct at this season, but vary much both in color and extent, though in only our summer skins are they entirely wanting." Mr. Ross also adds, that the martens of the Mackenzie's River district "bear a greater resemblance to the sable of Eastern Siberia than to the martens of Europe, holding, as it may be with propriety said, an intermediate position."

Dr. Gilpin, in his paper on the Nova Scotian Mammals already cited, has the following remarks on the variations presented by different indi-

* The italicizing in these quotations is my own.

† This may explain the differences in the hairiness of the soles pointed out by different authors, and claimed as a distinctive character of considerable importance.

viduals at the same locality: "When we begin to study this species, we soon find a very great variety in color, not only between the summer and winter specimens, but between winter skins themselves, that are all in the highest condition. Whilst they all coincide in what may be called typical marks, such as color of legs, tail, and especially ears, all of which have a very pale but conspicuous rim or border, *they vary much in color of face*, some having black, others faces so pale as to be nearly white, and the pale faces have a lighter brown color, and the orange throat much less vivid." Of seven skins described by this gentleman in detail, two "are nearly uniform mahogany brown" from the nose to the tail; the other five, though varying somewhat among themselves, are generally lighter, with much lighter faces, and the orange spot on the throat very bright, "almost fulvous." He adds that the skins from "Newfoundland and Labrador are much finer, darker in color, and more lustrous in pelage" than those from Nova Scotia.

Through the kindness of several of the fur-dealers of Boston I have had an opportunity to make a careful comparison of scores of skins of the Siberian sable from Russia with as large a series from the Territory of Hudson's Bay. The differences between them, although through the whims of fashion producing considerable difference in the mercantile value of the skins, are really quite slight. The fur of the Hudson's Bay skins is a little coarser, and the color slightly more rufous, with much fewer of the white-tipped hairs that in the Siberian skins are sometimes sufficiently numerous to give them a slight grayish cast, and which is considered to greatly increase their value. As one of the dealers practically remarked, they differ no more than the horses raised in Pennsylvania do from those bred in Massachusetts. Some of the skins of both varieties had tails much shorter than the average, showing the unreliability of this character. In a few instances this member was distinctly tipped with white, in both the Hudson's Bay and Siberian skins.

In the light of the now well-substantiated facts of a wide range of seasonal and intergrading, inconstant individual variation, it seems to me to be beyond reasonable doubt that, as I have already stated, the martens and sables, at least all thus far described, belong to a single circumpolar species, with possibly two or more well-marked and tolerably constant continental races.

8. **Putorius vulgaris** Cuv. (*Mustela vulgaris* Linn.; *Putorius pusillus* Aud. and Bach.) Least Weasel. Rather rare. Far less numerous than the next.

9. **Putorius ermineus** Cuv. (*Mustela erminea* Linn.; *Putorius noveboracensis* De Kay; *Mustela Richardsonii* and *M. Cicognanii*

Bon.; *Putorius fuscus*, *P. agilis*, and *P. ermineus* Aud. and Bach.) COMMON WEASEL. ERMINE. Comparatively common. It varies considerably in size, like other members of this family, according to sex and age.

I have obtained specimens at Springfield, identified some years since as belonging to the three species currently admitted by American authors as inhabiting Eastern North America, — "*P. Richardsonii* Bon."; "*P. Cicognanii* Bon.," and "*P. noveboracensis* De Kay." I have not access to the specimens for re-examination, but that these forms, or so-called species, occur in Massachusetts there can be little doubt, since Professor Baird, in his Report on the Mammals of North America, cites eleven examples from Middleboro', collected by Mr. J. W. P. Jenks, of his *P. Cicognanii*, two of *P. Richardsonii* and one of *P. noveboracensis*. As indicated by the synonymy already given, I consider all these as forming but a single species, which, after careful comparison of American with European specimens, I fully believe to be identical with the ermine (*P. ermineus*) of the Old World. I also feel obliged to consider the common American weasel, after similar comparisons, as identical with the common weasel (*P. vulgaris*) of the Eastern continent.

Although three species of ermines, or stoats, have been supposed to inhabit New England, in common with Eastern North America generally, no constant character has yet been indicated by which more than a single one can be positively distinguished. In size there is an almost imperceptible gradation from the smallest specimens to the largest, and similar gradations in all other characters, not excepting the relative length of the tail to the body. This latter character and that of size have formed the two distinctions most strongly urged as specifically separating them.

Previous to 1838, all the known weasels of North America were considered as belonging to two species, identical with the *Mustela vulgaris* and *M. erminea* of the Old World. At this time Bonaparte, in his Fauna Italica, added a third, which he called *Mustela Cicognanii*. He gave of it the following short and very unsatisfactory diagnosis: "*M. rufo-cinnamomea, subtus flavo-albida; cauda corporis dimidio sub-breviori, apice nigricante*"; which contains the single tangible character of "tail rather less than half the body." In the same year, in Charlesworth's Magazine of Natural History,* he added a fourth, which he called *Mustela longicauda*. This species was based on a variety mentioned in the Fauna Boreali-Americana,† by Dr. Richardson, as differing from the common ermine in being larger and in having a longer tail. Bonaparte, in the same communication, changed the name of the ermine weasel of Rich-

* Vol. II, p. 38. † Vol. I, p. 47.

ardson's work from *M. erminea* to *M. Richardsonii*, he believing them to be distinct species, and thus separated all the larger American weasels from those of the Old World. At this point begins the uncertainty and confusion that has long existed in regard to the number of species of American weasels and their distinctive characters. But no changes were currently adopted by American authors till ten or twelve years later, when, in 1841, Audubon and Bachman, in the Proceedings and Journal of the Philadelphia Academy of Natural Sciences, described a specimen taken on Long Island, New York, as a new species, under the name of *Mustela fusca*.* In the following year Dr. De Kay, in his Report on the Mammals of New York, redescribed this specimen under the name applied to it by Audubon and Bachman, and at the same time separated the larger representatives of the ermine as a species distinct from the Old World ermine and from the supposed northern *M. Richardsonii* of Bonaparte. But this author very frankly adds: "I have never seen the true ermine in its summer dress, and only know it from Pennant's description (Arct. Zoöl., Vol. I, p. 75)." He calls the American ermine weasel *Putorius noveboracensis*, and regards it as differing generically from two other species of weasel (*M. pusilla* = *M. vulgaris* Linn, and *M. fusca* Aud. and Bach.) described by him as also inhabiting New York. In 1853, the authors of Viviparous Quadrupeds of North America, in the third volume of that work (p. 184), characterized another species as new, also from New York specimens, which they called *Putorius agilis*. In the same volume. under *P. fuscus*, they observe that whereas the number of North American weasels was believed by the older authors to be at most two, while some admitted but one, "there are now five, four of which are found in New York." If we add to the new names of Audubon and Bachman and De Kay the three bestowed on American weasels by Bonaparte, we have seven specific designations for those of Eastern North America alone; to these may be added *P. erminea* and *P. vulgaris*, Audubon and Bachman fully believing these species to be common to both continents, thus making nine.

This was the condition of the subject when Professor Baird revised the group in his Report on the Mammals of North America, in 1857. In this work eight species are admitted as inhabitants of North America. Two (*P. frenatus* and *P. xanthogenys*) are considered as exclusively southern and western in their distribution; one (*P. Kaneii*) as northwestern ("Behring's Straits and Siberia"), and three *P. Pusilla*, *P. Cicognanii*, and *P. Richardsonii*) as distributed throughout the northern parts of the continent and extending southwards into the United States. Another (*P. noveboracensis*) is regarded as ranging from Massachusetts and Northern New

* Proc., Vol. I, p. 92; Journ. Vol. VIII, 1842, p. 280.

York, west and south, to Southern Pennsylvania, Illinois, and Arkansas. The locality of still another is given as Carleton House, H. B. T., this being the variety described by Richardson as occurring at that locality, and named *Mustela longicauda* by Bonaparte. But Baird doubtfully refers to it also some long-tailed ermines from the Upper Missouri.

Concerning the Least Weasel (*P. pusillus* Aud. and Bach. of Baird's Rep.), the only queries relating to it have been principally in reference to its relationship to *P. vulgaris*, *P. pusillus* forming its principal synonyme. Bonaparte, however, doubted its occurrence in America, supposing his *P. Cicognanii* had been generally mistaken for it, as he claims he found it had been in some of the Middle States, and on his authority Dr. Godman excluded it from his American Natural History. Afterwards, however, Dr. Richardson, in the Zoölogy of Beechey's Voyage, applied to it the name of *P. Cicognanii.*

For the smaller weasels with a distinct black tip to the tail, Professor Baird retains the name of *P. Cicognanii*, referring to it the *Mustela* (afterwards *Putorius*) *fusca* of Audubon and Bachman. He gives as its distinctive character, "Length to tail, eight inches or less. Tail vertebræ, one third this length. Black of tail, two fifths its length," etc. He adds, this "species is readily distinguished from the other American weasels by the small size, and the tail, which, with the hairs, is rather less than half the body.' In a note he mentions the later reception of some hunter's skins from Nova Scotia and Labrador, among which were some that agreed very well with typical specimens from Massachusetts, while others were considerably larger, though in general preserving the same proportions. The average length of the body in the measurements of twelve specimens given by him is 8.25 inches, the largest being 10, and of tail 3.62; but between the extremes of the series there is a variation in total length of thirty-six per cent. of the average, and in the relative length of tail to the body of twelve per cent.

Putorius Richardsonii is characterized by the same author thus: "Length to tail, nine inches or less. Tail vertebræ, about half this length. Black of tail, nearly one half to one third its length," etc. "Is readily distinguished from *Putorius Cicognanii* by the longer tail, the vertebræ alone of which are fully half the length of the body, instead of requiring the entire tail to effect this proportion."* Of this "species," the measurements of two speci-

* In the account of *P. Richardsonii* in the Mammals of North America there occurs the following singular but important discrepancy, probably the result of a typographical error. In the third paragraph of page 165 it is stated, "This species, a true *Putorius* differs materially from the larger North American Weasels *in the absence of a black tip to the short tail; in this respect resembling* P. Cicognanii." But in the specific diagnosis of *P. Richardsonii* the author says: "*Black of tail nearly one half to one third its length*"; and in that of *P. Cicognanii*, "*Black of tail two fifths its length.*"

mens from Eastern Massachusetts are given, both of which, in general size, fall within the average of the twelve of *P. Cicognanii;* thus showing that "small size" fails to sufficiently distinguish the latter, and also that short tails and small size do not always go together in specimens from the same locality; the tails in these two exceed the average in the *P. Cicognanii* by about thirty per cent. of the average of the whole series. The distinction based on the relative length of the black tip seems also intangible, "two fifths" coming just between "nearly one half" and "one third." To this species he refers the *P. agilis* of Audubon and Bachman, and of course the *Mustela* (*Putorius*) *erminea* of Richardson, for which the name *Richardsonii* was substituted by Bonaparte for *erminea*. Yet the dimensions given by Richardson accord in the proportions of the tail to the body, not with Baird's diagnosis of *P. Richardsonii,* but with that of *P. Cicognanii,* the tail vertebræ being but little more than one third the body, and the hairs and vertebræ together being *less* than one half.*

Putorius noveboracensis of Baird's Report is characterized as "Length to tail about ten inches. Tail vertebræ about half this length. Black of tail about half its length," etc. It thus differs from the last only in being larger. Yet one of the three specimens of which measurements are given scarcely exceeds the size of the larger of the two specimens of *P. Richardsonii,* and falls considerably below several of the *P. Cicognanii* in length of body. One of the *P. Cicognanii* specimens even equals the average of those of *P. Richardsonii,* although *P. Cicognanii,* as previously observed, is supposed to be distinctively characterized by its small size. Some differences in the proportional length of the feet, and in the color, are mentioned as existing between this and *P. Richardsonii,* but they are evidently merely individual, and would disappear in a comparison of a large series. To this species he refers the *P. ermineus* of Audubon and Bachman and the *P. noveboracensis* of De Kay.

In comparing some of the "*noveboracensis*" specimens with a short-tailed one of the European *P. ermineus,* I am not surprised that Professor Baird found "very decided points of distinction," "notwithstanding the assurance of authors" to the contrary. The principal one mentioned, however, is the greater brevity of the tail in the European, in which the proportion of the tail to the body is about as it is in *P. Cicognanii.*

In *Putorius longicaudus* the dimensions are given as, "Length to tail about eleven inches. Tail vertebræ about half this length. Black of tail about one fourth its length," etc. The measurements given of three specimens average 10.78 inches in the length of the body, one only reaching eleven, while the tail vertebræ alone equal fully half of this length. It

* "Length of head and body, 11 inches; of tail (vertebræ), 4 inches; of tail, including fur, 5 inches." — *Faun. Bor. Am.*, Vol. I, p. 47.

differs, then, from *P. Richardsonii* only in its slightly larger size, the *proportion of length of tail to length of body being essentially the same in both.* Some smaller specimens are referred to this from the Upper Missouri, of which measurements are not given. Two of the large specimens are marked males; the sex of the other is not indicated. To this species is of course referred the long-tailed Carleton House variety mentioned by Richardson, to which, as already observed, Bonaparte gave the name *longicauda.*

From the preceding comparisons and remarks the inconstancy and the arbitrary character of the distinctions claimed as specific are fully evident. It appears that short tails by no means always accompany small size, nor long tails large size; that both occur at the same localities, as well as at points as remote from each other as the most distant localities at which the species has been found, as Hudson's Bay Territory and the Arctic Regions on the one hand and Massachusetts, Pennsylvania, and Illinois on the other; that between the "species," as characterized by Professor Baird, there is an almost insensible intergradation in all the essential characters, some of the so-called species resting on distinctions that are by no means differences (as *P. Richardsonii* and *P. longicauda; P. Cicognanii* and *P. noveboracensis*, very nearly); finally, that, contrary to the belief of this author, the short-tailed species (*P. Cicognanii* and *P. noveboracensis*) have a range to the northwards equal to that of the others, the *P. erminea* of Richardson being distinctly referable in its proportions to *P. Cicognanii.*

Although differing radically with the eminent author of the Report on the Mammals of North America in respect to the number of valid species of this group in America, — the only American zoölogist who has given it special attention, — I can but commend the candor he has exhibited in his attempt to clear up the discrepancies of former authors, and to sift the subject of its obscurities, as well as the manner in which he has presented his material.

An examination of numerous specimens from the New England and other Northern States has shown me that the variations in the relative length of the tail to the body are merely analogous to similar individual variations in the squirrels and other small mammals that have this part considerably developed, — a variation not always due merely to the lengthening or shortening of the vertebral segments, but occasionally to an increased or diminished number of the vertebræ themselves. Also, that the variation in size so noticeable in specimens from the same locality is in great part sexual, — the males in nearly all species of *Mustelidæ* being considerably larger than the females, — but in many cases to immaturity, and somewhat also to the natural individual range in this respect, which, as in their allies, the mink and marten, and in the Carnivora generally, is much greater than in some other groups. The differences in color claimed now and then as distinctive of different species are generally either such as are evidently seasonal, or such as, like those of the form and proportions of the feet, etc., would disappear in a large series. I hence feel convinced of the existence of but two species of weasels in Northeastern North America, and that these are circumpolar, identical with the *P. vulgaris* and *P. ermineus* of the Old World. These two are always distinguishable with certainty, while their representatives do not present a wider range of variation in size and other characters than is currently admitted for several of their congeners. More than this number being admitted, the whole question as to how many should be recognized, and what constitutes their distinctive characters, becomes involved in the greatest uncertainty.

Two interesting facts in respect to color in the weasels should not in this connection pass unnoticed. One is that both species generally become white in winter; apparently invariably so at the far North, and usually so as far south as Northern New England, but in Massachusetts only the larger one (*P. ermineus*) thus changes, and this not always. Still farther south such a change in *P. ermineus* occurs only occasionally, and in the extreme southern portion of its habitat not at all.* This whitening of the pelage in winter corresponds in geographical relation to the white or light gray color seen in the common wolf at the north, and the gradual darkening of its color southward. The other fact is the usual greater intensity of the yellow on the under parts in specimens from the central portions of the continent, — a variation parallel with the rufous form of the common wolf of the same region, and the comparatively more rufous tint of the pelage seen in specimens from the same district in most continentally distributed species.

Another fact in respect to size is also noteworthy, as corroborative of the general law of the larger size of the representatives of a species from the northern parts of its habitat than those from the southern. The measurements given of the length of the body by those authors who have had only southern specimens for examination is seven inches for *Putorius vulgaris*, and eight to ten inches for the corresponding measurement of *Putorius ermineus*, but Richardson, whose specimens were extremely northern, gives nine inches for the same measurement of the former, and eleven and twelve for that of the latter.†

* Respecting this seasonal change of color, compare the observations of Richardson (Fauna Boreali-Americana), Audubon and Bachman (Quadrupeds of N. Amer.), and Baird (Mam. N. Amer.).

† Professor Baird, in order to reconcile the identification of Richardson's specimens with his *P. Richardsonii*, supposes the body to have been overstretched, as he says he never saw any American ermines that would measure eleven inches before skinning;

In concluding this brief review of the American weasels I will add that, whether *P. frenatus* and *P. xanthogenys* prove ultimately distinct from each other, as they are likely to from the northern species (*P. ermineus*), I regret to feel obliged to assign the *P. Kaneii* Baird to the synonymes of *P. ermineus*, not less from my regard for its describer than for the memory of that admirable man its name is so appropriately designed to commemorate. To the same category I think must also be referred the *P. bocamelus* Bonaparte, founded on the southern race of this species in Europe (Sardinian specimens), as his *P. Cicognanii* was on a similar American race.

Since writing the above I have found that Dr. J. E. Gray, of the British Museum, has recently referred *Putorius Kaneii* Baird to *Mustela erminea* Linn., it forming his "variety 2, *Kaneii*" of this species.* To the same species he has also referred the *Putorius noveboracensis* De Kay, and the *Mustela Cicognanii* and *M. longicauda* Bonap., he calling them altogether "variety 3, *americana*," of *ermineus*. Dr. Gray adds: "Dr. Spencer Baird, in his work on the Mammals of North America, divides the stoats into six species [*P. Richardsonii*, *P. noveboracensis*, *P. longicauda*, *P. Cicognanii*, *P. ermineus*, and *P. Kaneii*], by the length of the tail and the black on the tail. . . . . When the bodies of several English stoats have been compared they show how deceptive that character is. I do not say that they may not be distinct; but if they are, there must be other characters to separate them besides the mere length of the tail." He accordingly gives as "species 2" of the stoats, *Mustela Richardsonii*, on Professor Baird's authority, and as chiefly distinguished by the upper lips and legs being "entirely brown." He adds, "I have not seen this species." He further observes: "The specimen formerly named *M. Richardsonii* [by Bonaparte ?], in the British Museum, has the hinder part of the upper lip white, but the hair is bent back and lost off the front part." In respect to the white on the upper lip, he states that English specimens sometimes have it reduced to a very narrow margin.

The American weasel (*P. pusillus* auct.) Dr. Gray likewise considers identical with the European *P. vulgaris*. But Bonaparte's *Mustela bocamela* of Southern Europe he admits as a valid species, under the section of weasels, or of species with the "back and tail uniformly colored," and extends its habitat to include North Africa (Algiers and Cairo). The correctness of this view seems highly questionable, since New England specimens of *P. ermineus* sometimes have the tip of the tail merely

forgetting apparently for the time being this law of variation which he was one of the first to recognize, and towards establishing which no one else has done so much.

* Proc. Lond. Zoöl. Soc., 1865.

dusky, the black being almost obsolete, in which condition they seem not essentially different from the figure and original description of *P. bocamela* in the Fauna Italica.

10. **Putorius lutreolus** Cuv. (*P. vison* Gapper; *Vison lutreocephala* Gray; *Mustela lutreola* Linn.) Mink. Common.

I am not prepared to admit Audubon and Bachman's Little Black Mink (*P. nigrescens*) as distinct from the above. Specimens referable to this supposed species are not of uncommon occurrence. Mr. B. R. Ross considers that the *P. nigrescens* "is nothing more than the young of the *P. vison*," * an opinion I have also long entertained.

In this species we again have an animal of questioned identity, some authors considering it the same as the European *Mustela lutreola* Linn., while others maintain its distinctness. But the differences seem very slight, and have generally been supposed to consist in the front of the upper lip being white in the European, while there is no white on that of the American; in size, proportions, and general color, no one claims that they materially differ. This single character is one of great variability in their near allies, the ermines, some having the white margin of the upper lip very broad, while in others it is very narrow and occasionally entirely obsolete. The other white markings on the mink are notoriously variable, some specimens having this color restricted to a very narrow chin patch, or even *entirely wanting*, while in others there are spots of white on the throat and between the fore legs; in still others white spots occur also along the middle of the abdomen and between the hind legs, forming an interrupted median line of white patches. I also feel confident that I have seen specimens of the American animal with a white margin to the upper lip. Experienced trappers positively assure me that such examples are of occasional occurrence.† Dr. Gray, however, gives a second character of

* Natural History Review, July, 1862, p. 273. In a later paper in the Canadian Naturalist and Geologist (Vol. VI, p. 30), Mr. Ross says the *P. nigrescens* of Audubon and Bachman are "merely common minks under three years of age." He states in another place (l. c. p. 29), "I have remarked that the color of this animal, as well as that of the otter and beaver, grows lighter as it advances in years, and that the white blotches or spots are of greater size and more distinct in the young than in the old. The color of a young mink (under three years), when killed in season, is very handsome; its color is often an almost pure black." I have myself observed a similar variation in color with age in the common black rat, and in other mammals, as well as in many birds.

† Since writing the above I find Mr. Ross says, in referring to Professor Baird's remark that the American mink *never* has the edge of the upper lip white, "I have never seen *the whole* of that part so colored, but in one specimen now on my table there is a white spot beneath the nostril."

distinction between the American and European animals, — a difference in the size of the upper tubercular tooth, — the value or constancy of which I have at present no means of determining.

Of the American animal Dr. Gray makes three "varieties." The first is dark, with unspotted throat and chest, whose habitat he gives as "Vancouver's Island"; but it also occurs in Massachusetts, Michigan, and Illinois, as I have myself observed, and probably throughout the habitat of the species. The second is characterized simply as having the "chin entirely brown," while the third is Audubon and Bachman's *P. nigrescens.* No special habitat is given for the last two. Neither of them, however, is a permanent variety. In the general color, as well as in the white spots, there is a wide variation, different specimens varying from pale brown to quite intense black. There is also an extensive variation in size, but as very large and very small individuals occur in each stage of color it is very difficult to consider any of these variations as other than individual, or such as are evidently to be referred to season, sex, or age.

Numerous supposed species of the Old World mink have also been characterized, chiefly from the warmer regions, five of which are recognized as valid by Dr. Gray. The first of these is the common *M. lutreola* of Linnæus, the habitat of which is given as "Europe." The second is the *M. siberica* of Pallas, which Dr. Gray says is paler and smaller than *M. lutreola*, with the tail relatively longer and the end paler colored, or like the back, instead of darker than the back.* He observes that it "varies greatly in the quantity of white on the chin and throat," and adds that the "*males are much smaller.*" The last statement, if true, indicates a remarkable exception to the sexual law of variation in size in this family. The habitat is given as Siberia, Himalaya, Japan, China, and Formosa. Dr. Gray's third species is the *Mustela canigula* of Hodgson, originally described from specimens from the Nepaul Hills of India. Its chief distinction seems to be an unusual amount of white on the face, chin, throat, neck, and chest, although Gray mentions as a variety a specimen with darker fur and much less white. His fourth species, *Mustela* (*Vison*) *Horsfieldii* Gray, seems not to differ particularly from the others, or from frequent American specimens, as its "variety two" is characterized as "chin brown, *edge of under lip only* white." This is likewise from India (Bootan) and Japan. The fifth, from Nepaul, the *Putorius subhemachalna* of Hodgson, differs from the preceding in being generally lighter or redder, — in other words, having less intensity of color, — with minor differences in the amount and distribution of the white. If all these species are valid, it will be seen that Southern and Eastern Asia and Japan are peculiarly rich in species of this

* The relative shade of color of the tip of the tail as compared with the back is a character too inconstant in this group to merit serious mention.

group.* But, in view of the well-known similar variations presented by our American mink, they seem to rest on very unsatisfactory distinctions, especially as the "varieties" admitted under some of them cover the differences considered as distinctive of the different species. The general paler color and somewhat smaller size of the southern forms † is paralleled by similar differences in specimens of the American animal from the southern portion of its habitat. In view of all these facts, I strongly incline to the opinion that we have here again but one circumpolar and widely dispersed species, with possibly two continental or geographical races that may be more or less easily recognized. Else, as in similar cases previously discussed, we must admit an indefinite number, subject in this respect and in their limitation to the caprice of those authors whose forte is in the description of "supposed new species." ‡

11. **Gulo luscus** Sabine. Wolverine. Dr. De Kay, in his

* It is a fact especially noteworthy that regions whose Natural History is considered as but partially explored are far richer in species (I refer more especially to mammals and birds), accepting only such as are currently allowed, than those much longer and more familiarly known. To be assured of this one needs but to compare Southern and Middle Europe with the corresponding parts of Asia, or Eastern and Northern America with Mexico and Central America, adopting as a basis for the comparison only those types or groups widely distributed. This fact is especially illustrated in the Carnivora, as the present family of *Mustelidæ* exemplifies. While distinct types appear in different regions, as some in the warmer latitudes that are not found in the colder, and *vice versa*, the martens and sables, as well as the minks, under not very different physical conditions, far outnumber in Eastern Asia alone, in reputed species, their representatives in Europe. While I would not deny the possibility of this being a fact, the intimate relationship which these several supposed species bear to each other, as well as to the European, and the unsatisfactory distinctions on which they are founded, seem to render it extremely improbable. If we extend the comparison to other groups, and to other regions, we constantly meet with cases parallel in all respects to this. This excess of species also almost always happens, in mammals, among those least known, either through their great scarcity or their nocturnal or recluse habits rendering them difficult to obtain. The explanation of this seems to be that new species are not anticipated to occur in a region that has been for a long time thoroughly explored, while specimens from imperfectly known districts, or of species in groups where the species are supposed to be difficult to distinguish, are most critically examined, and those differing slightly from others previously described — though not more, in many cases, than specimens unquestionably of the same species and obtained at the same locality frequently do — are presumed to represent undescribed species.

† See Gray's table of comparative measurements of the skulls of his several species Proc. Lond. Zoöl. Soc., 1865, p. 118.

‡ In the mink, as in the marten, it is an interesting fact that the Asiatic specimens bear a stronger resemblance to the American than the European do. According to authors, specimens not unfrequently occur in Japan and portions of Eastern Asia that are hardly distinguishable from average American ones.

Report on the Mammalia of New York, published in 1842 (p. 28), says: "Professor Emmons states that they still exist in the Hoosac Mountains, Massachusetts." But the species is not given in Emmons's Report, published two years before; it occurs, however, in Dr. Hitchcock's List, with the following note: "On Hoosac Mts.; rare.— Emmons." It is more or less common from Northern New England to the Arctic coast.

This species is remarkable for being the only one in the Mammalian Fauna of the State usually regarded as common to both the Eastern and Western Hemispheres. The existence in all together of but two or three circumpolar species of land mammals is admitted by many naturalists. It must also present an unusual constancy of character, since not only has it escaped subdivision into pseudo-species, but even no "varieties" have been generally recognized.

12. **Lutra canadensis** SABINE. (*Latax canadensis* Gray; *Lutra canadensis* and *L. destructor* Barnston.*) OTTER. Not rare; still not often captured. At Springfield I have known some half-dozen specimens taken in the last ten years.

13. **Mephitis mephitica** BAIRD. (*M. chinga* Tiedemann; *M. varians* Gray; *M. mesomelas* and *M. chinga* Maximilian.) SKUNK. Abundant. Individuals from the same locality, and even from the same litter, are very variable in color, some being almost entirely black, while others have a very large proportion of white. The amount of baldness on the soles of the feet is also very variable, independently of season or age, although this has been deemed by some naturalists, as Lichtenstein and others, as a character of great importance. Attention has been previously called to its inconstancy.†

Probably no other North American mammal is so strikingly variable in color as the common skunk; it is hence not surprising that foreign naturalists, unacquainted with the animal in life, have made of it a considerable number of supposed species. So well known is this variability to most persons at all familiar with the animal that it is all the more unexpected to find from a naturalist so justly reputed for accuracy as the author of the Report on the Mammals of North America a statement like the following: "This species varies considerably in its markings, *though individuals from the same locality are usually quite similar*."‡ Especially is this so after the

* Canadian Naturalist and Geologist, April, 1863, p. 147.

† See Dr. J. E. Gray's Review of the Mustelidæ, Proc. Lond. Zoöl. Soc., 1865, p. 147.

‡ Mam. N. Amer., p. 195.

detailed account given by Audubon and Bachman of very wide differences in color between individuals of the same litter.* The majority of the Massachusetts specimens I have seen accord very well with Professor Baird's diagnosis, the general color being black, with a narrow white streak down the face, a large white nuchal patch, and a broad white streak on each side of the back reaching commonly nearly to the tail, which is tipped with the same color. Sometimes the face streak is united with the nuchal patch, but oftener it is separated by a narrow space of black, and is occasionally absent. The dorsal streaks vary in breadth and posterior extent, generally enclosing a narrow band of black; but the latter is sometimes wanting, when they, uniting along the median line, form but one; they run nearly parallel or widely diverge posteriorly, where frequently each is deeply bifid; more frequently than otherwise they entirely cease near the loins. The nuchal patch also varies in form and extent; generally it is continuous with the dorsal streaks, but is often entirely separate from them, and is itself sometimes divided, forming two small lateral patches; its general outline is variable almost beyond description. The white on the tail is sometimes terminal and sometimes basal; now and then it is quite absent, but occasionally it preponderates over the black. The distinct terminal pencil of long white hairs in the tail, so often described, seems generally best defined in young specimens; in full-grown ones it is frequently absent. Individuals occasionally occur that are either entirely, or almost entirely, black; much more rarely others with nearly the whole of the dorsal surface white, as in a specimen in the Museum of Comparative Zoölogy, collected in Newton, Mass., by Mr. C. J. Maynard. This has the black restricted to a narrow dorsal line, a few scattering hairs in the tail and to the lower surface of the body, the white dorsal band being nearly two inches broad on the neck and seven at the loins. Mr. Maynard has another specimen, taken at the same locality, which has still more white, there being no black median line, and the white extends still lower on the sides of the body. In short, the variations in color in the skunks are almost endless, scarcely any two specimens being quite alike. It therefore seems preposterous to found species on particular styles of coloring, or on the relative proportion and distribution of white and black, as several authors have done.

Eight species were described by Lichtenstein in his monograph of the genus *Mephitis* † from Mexico and the United States alone, while from North and South America together he gave sixteen! Professor Baird recognized six in his Report, and mentions three others described by

* Quad. N. Amer., Vol. I, p. 319.

† Ueber die Gattung *Mephitis*, Afhand. Akad. Wiss., Berlin, for 1836, 1838, pp. 249–315, and 2 plates.

Lichtenstein from Mexico as probably valid and also likely to occur in the United States. Dr. Gray* has very judiciously reduced the number to five, including those of both North and South America, but he places them in what he considers three genera, — *Conepatus* (1837, nearly equal to *Thiosmus* Lichtenstein, of subsequent date), *Mephitis*, and *Spilogale*. He gives all as occurring in North America. To the first, *Conepatus nasutus* Gray (*M. nasuta* of Bennett †), he refers, and it appears to me very properly, the *M. leuconota* and *M. mesoleuca* of Lichtenstein and Baird, and numerous other species of other authors, thus greatly reducing the number previously received. He separates it, however, into four "varieties," which are based on the distribution of the colors, although they seem to be about as uncertain in extent and relative proportion in this species as in the more northern one. Of *Mephitis* proper Gray gives three species, two of which (*M. vittata* Licht. and *M. mexicana* Gray, = *M. macroura* Licht.) are from Mexico, and the other (*M. varians* Gray, = *M. mephitica* Baird, = *M. chinga* Tiedemann) is generally distributed over North America, from Great Slave Lake ‡ southwards; of *Conepatus* and *Spilogale* one each. It is highly probable, however, that Mexico is not thus pre-eminently rich in species of these animals, and that Gray's two Mexican species may be referred to the common North American one, since they rest almost solely on distinctions of color that are far from peculiar to the Mexican examples. This being true, we have three supposed genera containing a single species each, or, what seems to me more probable, the alleged differences being slight, a single genus with three species, which agree rather closely in their general style of coloring and in possessing a remarkably large range of indefinite color variation. In distribution, one (*M. mephitica*) is northern, ranging from Mexico almost to the Arctic regions, and the others (*M. mesoleuca*, = *Conepatus nasutus* Gray, and *M. bicolor*) southern, inhabiting from Mexico and the Southern States to Patagonia.

Our common species (*M. mephitica* Baird) Dr. Gray divides into five "varieties," based on the relative extent of the white dorsal streaks, which form among themselves a graduated series. The inconstant nature of the characters assigned to these as distinctive it seems to me renders them unworthy of recognition, since they not only all occur at single localities, but, as Audubon has shown, § several of them sometimes appear in the

* Proc. Lond. Zoöl. Soc., 1865, pp. 145 *et seq.*

† Ibid., 1833, p. 29.

‡ B. R. Ross, l. c., p. 273.

§ "In the winter of 1844 we caused a burrow to be opened in Rensselaer County, N. Y., which we knew contained a large family of this species. We found eleven; they were all full grown, but on examining their teeth and claws we concluded that the family was composed of a pair of old ones, with their large brood of young of the previous season. The male had a white stripe on the forehead; and from the occiput down the same litter. Most of these pseudo-varieties and others of a similar character have been described by authors as distinct species. Prince Maximilian in his latest work * still maintained the existence of two species, *M. mesomelas* Licht and *M. chinga* Tiedem., in the United States. The latter (*M. chinga*) he seems to have known only from imperfect skins brought by the Indians of the Upper Missouri from, as he presumed, the Red River of the North and the Saskatchewan. They were used by them as trappings for the legs, and were all very white, differing only in this respect from the common skunk. As specimens similarly colored occur more or less frequently throughout the United States, it seems more probable that the Indians may have selected skins of this color for the special use to which we are informed they applied them than that the skunks of any given region are generally so colored.

Without going into the synonymy of the subject in detail, I may add that for the common North American species Dr. Gray strangely adopts the specific name of *varians* (*M. varians* Gray, 1837), this name being superseded in priority by both *chinga* of Tiedemann (1808) and *americana* of Sabine (1823), as well by *mephitica* of Shaw (*Vivera mephitica*, 1792). This latter being the one first given, has very properly been adopted by Professor Baird.

URSIDÆ.

14. **Procyon lotor** STORR.† RACCOON. Formerly numerous,

whole of the back had another white stripe four inches in breadth; its tail was also white. The female had no white stripe on the forehead, but had a longitudinal stripe on each side of the back, and a very narrow one on the dorsal line; the tail was wholly black. The young differed very widely in color; we could not find two exactly alike; some were in part of the color of the male, others were more like the female, whilst the largest proportion were intermediate in their markings, and some seemed to resemble neither parent. We recollect one that had not a white hair except the tip of the tail and a minute dorsal line." — AUDUBON and BACHMAN, *Quadrupeds*, Vol. I, p. 319. See also the two young figured by these authors (Plate 42), one of which has white stripes on the back and a black tail, and the other no stripes and the end of the tail white, though both were of the same litter. I have myself met with similar variation in the same litter of young.

* Verzeichniss Nordamerikanischer Säugethiere, Archiv fur Naturg., XVII, 2, p. 218.

† *Ursus lotor* LINNÆUS, Syst. Nat., I, 1758, 48, Ib., I, 1766, 70.

Procyon lotor STORR, Bod. Meth. Anim., 1780.

" *Hernandezii* WAGLER, Isis, XXIV, 1831, 514.

" " BAIRD, N. Am. Mam., 1857, 212.

" " IB., U. S. & Mex. Bound. Surv., II, Mam., 1859, 22.

" " var. *mexicana* BAIRD, Ib., 22.

" *lotor*, var. *mexicana* ST. HILAIRE, Voyage de la Venus, Zoöl., I, 1855, 25, pl. VI.

" *nivea* GRAY, Charlesw. Mag. Nat. Hist., I, 1837, 580.

" *psora* IB., Ann. & Mag. Nat. Hist., X, 1842, 261.

and still more or less common in the mountainous and sparsely settled parts of the State.

Quite variable in color, the variations on the one hand tending strongly towards melanism and on the other towards albinism. On specimens presenting the latter kind of variation seems to have been founded the *Procyon nivea* of Gray from Texas,* as probably also the *P. psora* of Gray † from California. ‡ With the variations in the general tint the markings usually become more or less indistinct. In even what may be considered as the normal or average type the dark rings of the tail vary from four to six in number, in intensity of color, and in relative breadth to the interspaces; sometimes the dark rings are only about half the width of the intervening lighter ones, but, as I have observed to be the case in numerous specimens killed in Massachusetts, Western New York, and Florida, they often equal, and not unfrequently exceed them. The tail varies also in its form and size, as it does in the foxes and marmots, sometimes tapering considerably towards the tip, though generally but slightly. Yet these characters have been assumed by some authors to be indicative of specific differences, the *Procyon Hernandezii* of Wagler § having been founded originally on such slight variations. Professor Baird, however, has gone quite fully into a discussion of its merits as a species, ‖ but the distinctions he particularly mentions as separating it from *P. lotor* — the more tapering form of the tail, the rings of which he deems "narrower and better defined," with "the light intervals wider," and a "*nearly constant* difference in the color of the upper surface of the hind feet," which he says is darker in *P. Hernandezii* — are so slight, and based withal on characters so exceedingly liable to variation, that they can scarcely be considered as of specific value. Though apparently of somewhat larger size the relatively larger and stouter feet claimed by him to distinguish *P. Hernandezii* his measurements seem to scarcely sustain. He admits that *P. Hernandezii* bears a very close relationship to the *P. lotor*, and says that "without close comparison the differences are perhaps intangible," and that "its characteristics are more comparative than absolute." Still "an examination of a large number of North American raccoons," he affirms, "has resulted in the appreciation of certain differences which appear quite constant." They are those above specified, and, as I have already ob-

* Charlesworth's Mag. Nat. Hist., Vol. I, 1837, p. 580.

† Ann. and Mag. Nat. Hist., Vol. X, 1842, p. 267.

‡ In his recent revision of the "Ursine Animals," Dr. Gray has referred both these to the *P. lotor*. See Proc. Lond. Zoöl. Soc., 1864, p. 684.

§ Isis, XXIV, 1831, p. 514.

‖ See N. Am. Mam., p. 213, and Mex. Bound. Survey, Vol. II, Mammals, p. 22.

served, are very slight, and pertain to the most variable parts of the animal. Some of them I feel sure are but individual differences, depending mainly, especially those in respect to the form of the tail, on age or season. In respect to the black annuli, hardly two specimens can be found that do not vary more or less. In the large series of New England specimens in the Museum of Comparative Zoölogy, the variation between the extremes in this respect covers the whole range of the differences assumed to distinguish the two supposed species. The single authentic specimen of *P. Hernandezii*, labelled apparently by Professor Baird himself, that I have seen * is not appreciably different in general color from many Massachusetts specimens. The supposed differences, it seems to me, are hence reduced to the single one of absolute size, which a large number of specimens of the so-called *P. Hernandezii*, from different localities, might very considerably modify. From a comparison of authorities, as will be seen from the remarks that follow, this seems to be indeed the fact. The example of *P. Hernandezii* above referred to (No. 67, Smithsonian Cat.), from Bodega, California, is actually smaller than the average of New England specimens.

Professor Baird remarks that some of the characters of *P. Hernandezii* given by Wagler and Wiegmann, as the prevailing color of the back and sides, differed from specimens he referred to it; in other words, they were more like his *P. lotor*. St. Hilaire, in the Zoölogy of the Voyage of the Venus, † also described and figured a specimen from Mazatlan that varied similarly from *P. Hernandezii* Baird, it being smaller and colored more like *P. lotor*. Under *Procyon Hernandezii* var. *mexicana*, Baird describes a single skin brought by the Boundary Commissioners from Espia, Sonora, that he says agrees with St. Hilaire's Mazatlan specimen (already referred to), which St. Hilaire considered to differ in nothing but in intensity of color from the common *P. lotor*. Professor Baird remarks that this Espia specimen exhibits a close relationship to *P. lotor*, though readily distinguishable from it, he claims, by its "larger and more naked feet and hands." These specimens, in resembling *P. lotor* more than some others from the same region referred to *P. Hernandezii*, show still more fully the inconstancy of some of the characters on which the latter is founded. In habits the two supposed species have not been found to differ. ‡ Hence, unless the more southern *P. cancrivorus* occurs in Cali-

* Contained in the Museum of the Boston Society of Natural History, and received from the Smithsonian Institution.

† Vol. I, 1855, p. 25, pl. VI.

‡ Professor Baird observes: "According to Dr. Berlandier, the habits of this species [*P. Hernandezii*] are precisely similar to those of the common raccoon." Dr. C. B. Kennerly's notes are also of the same purport. — *Report on the Mammals of the United States and Mexican Boundary Survey*, p. 22.

fornia, as supposed by Audubon and Bachman, I see no reason why, in view of the known variability of *P. lotor* in the Eastern States and the relatively small differences only thus far pointed out between them and their Western relatives, all the raccoons of the United States thus far known should not be referred to *P. lotor*.*

15. **Ursus arctos** Linnæus. (*U. americanus* Pallas.) Black Bear. Extinct in the more thickly settled parts of the State; occasional among the mountains of the western counties.

In respect to the occurrence of this species in this State, Dr. Emmons remarked in 1840: "It is not many years since great numbers appeared there [on the Hoosac Mountain range] at once, and between twenty and thirty were taken in the course of one autumn, on the mountains in Adams and Williamstown. They are still to be found, and several have been taken every year since." (Rep., p. 24.) The local newspapers yet frequently chronicle their capture in Berkshire County.

Contrary to what was formerly supposed, bears everywhere appear to be among the most variable of mammiferous animals, not only in coloration, but in size, proportions, and in the conformation of the skull and other parts of the skeleton. Those familiar with them in life say it is rare to find two alike. A writer in the American Naturalist† has alluded to two females of the same litter, captured by him in Maine when young and raised as pets, that differed so essentially in their general build as to correspond respectively with what has been termed "ranger bear" and "hog bear," they differing also as much in disposition as in form. I am also informed by my friend Mr. C. W. Bennett that he has known two cubs of the same litter, taken in one of the Western States, that as they grew up differed very materially from each other in color, one being *black* and the other *brown*. They differed also widely in form and disposition, one being docile and playful while the other was ferocious and dangerous. The leading varieties in color of the American and European bears, as the brown and the

* This is also the view now taken by Dr. Gray, who remarks respecting *P. lotor* as follows: "This species varies rather in the tint of its colors in the different parts of North America. It is very apt to become white, and is thus the *Procyon nivea* (Gray, Mag. Nat. Hist., 1857, p. 580) from Texas; I described a specimen from California, with the tail imperfect, as *P. psora* (Ann. and Mag. Nat. Hist., 1842); and Wiegmann described two other varieties under the names of *P. brachyurus* and *P. obscurus* (Arch. III, 369). Dr Baird, in the Mammals of North America, considers *P. Hernandezii* as a species, and calls it the Black-footed Procyon, including *P. psora*, which has feet as *pale or paler than* P. lotor." — *Proc. Lond. Zoöl. Soc.*, 1864.

† Vol. I, p. 657.

black bears are now generally deemed to be but varieties and not species, though so regarded by Cuvier and the earlier naturalists generally. Great variations in the form of the skull in individuals from the same locality not unfrequently occur, aside from the differences caused by age and sex. Professor Baird mentions a skull from Saranac Lake, New York, which differs very appreciably from the ordinary type, agreeing quite nearly in some respects with the *Ursus arctos* of Europe. Concerning this specimen he remarks: "A large number of specimens from this locality may perhaps furnish a clew to this remarkable variation, which, under other circumstances, would be readily allowed as indicating a distinct species." * I some time since began to consider many of the so-called specific characters drawn from the skull as of very doubtful value, from the wide range of variation any considerable series of specimens from the same locality, and unquestionably of the same species, usually exhibit, aside from those arising from differences of age and sex. In the foxes and wolves, the common bear, the different species of *Mustelidæ*, and the larger rodents, such differences are often very considerable. On this point I find the following concurrent testimony from an author little liable to the charge of conservatism in respect to the multiplication of species or other groups.

Dr. J. E. Gray, of the British Museum, in his recent monograph of the bears, in the Proceedings of the London Zoölogical Society, † thus calls attention to the subject. "The examination of the series of skulls of bears in the [British] Museum, like the examination of the series of bones of the *Viverridæ*, has strongly impressed me with the uncertainty that must always attend the determination of fossil bones, or indeed of bones of all animals, when we have only the skulls or other bones to compare with each other. There can be no doubt that the study and comparison of the bones of the different species is very important, — that the skull and teeth afford some of the best characters for the distinction of genera and species; but few zoölogists and palæontologists have made sufficient allowance for the variations that the bones of the same species assume. In the bears, I have observed that there is often more difference in skulls of bears of the same species from the same locality than between the skulls of two undoubted species from very different habitats and with very different habits. Thus I have the skulls of some bears the habitat of which is not certainly known, which I have doubts whether they should be referred to the Thibet Bear (*U. torquatus*), or to the North American species (*U. americanus*), but I have referred them to the latter, as they were said to have come from that country. It is the same with regard to the skull of a bear that lived in the Zoölogical Gardens for years, which has the general form of the skull and the wide palate of the European bear, but the long last grinder

* N. Am. Mam., p. 227.

† 1864, p. 684.

and some other characters of the *U. ferox*. This similarity of skulls," he says, "is the more remarkable, as no two bears can be more distinct from each other than the species above named,* which have such similar skulls, showing that similar skulls do not always imply very nearly allied or doubtfully distinct species."

The bears have ever been a perplexing group, and accordingly the opinions advanced by different authors respecting the number of species vary widely. Several high authorities consider the land bears of Northern North America, Northern Asia, and Europe as forming but one, or at most two, species, among which are Blainville and Middendorff, the latter of whom, with access to a large amount of material, has especially and most minutely investigated the subject. Other authors are disposed to allow a much greater number. But, unfortunately, their statements in reference to the differences that should distinguish them are frequently contradictory. Dr. Gray recognizes eight † in his recent monograph, with numerous "varieties" and "subvarieties" of each. Of the *Ursus arctos*, or brown bear of Europe, he describes four varieties, and of the first of these eight subvarieties, to all of which he gives distinctive names. All of these are chiefly based on variations in color, the teeth, or the skull, although he states in the same paper that characters based on the latter are to a considerable extent unreliable for even the determination of species. ‡ Nilsson, in his Scandinavian Fauna, distinguishes six varieties that differ widely in color from Sweden alone. A careful study of Middendorff's tables of measurements, embracing some fifty specimens of bears from Europe and Asia, show how very extended is the range of variation in osteological and external characters, and how irregular is its nature. Schrenck has also called attention to the great variation in the size of the tuberculated grinders in the bears of Northern Asia, — a character which is unfortunately made the principal basis of Dr. Gray's specific and subspecific distinctions. Dr. Gray himself mentions that there are considerable variations in the series of skulls of American bears in the British Museum, particularly in the amount of depression in front of the orbits. His several tables of measurements of skulls that he himself refers to one

* In respect to this point I shall soon show that naturalists high in authority do not agree with Dr. Gray in regard to the great distinctness claimed by him for these species.

† It seems to me that no recent writer has been guilty of greater inconsistency than is exhibited by the author of the monograph on the *Ursidæ* above cited; for after calling attention to the variability of craniological characters, and their consequent unreliability as specific distinctions, he adopts some of those that can be readily shown to be the most trivial — even manifestly so from his own paper — as the basis of his classification of his species and varieties. So difficult is it apparently to overcome long-established habits of thought and modes of reasoning.

‡ See preceding page.

species indicate very considerable differences in the proportion of breadth to length in the entire skull, and in the relative length and slenderness of the muzzle. In consequence of such variations Dr. Gray and Professor Baird arrive at widely different conclusions in reference to the relationship of the *U. cinereus* Gray (*U. horribilis* Baird) to the *U. americanus*.

There is a strong tendency among naturalists to consider the Old World bears as all distinct from those of North America, and to recognize at least two species among the latter, — the grizzly bear (*Ursus horribilis*) of the West, and the continentally dispersed black and brown bears (*U. americanus*). Professor Baird, in his Mammals of North America, gives the probable number as five, four of which he seems to consider well founded, and thinks that there may be two others. But each of the recognized species presents so many varieties, which to a greater or less extent intergrade, that well-marked lines of distinction cannot at present be drawn. This has led a recent writer to observe, and it seems to me very justly, "If the same consolidation of species which some authors practise in plants was carried out in animals, we should have but one species [of bear] for the whole northern hemisphere." *

The present indications are that the *U. horribilis* is hardly so distinct from the common *U. americanus* as has been currently supposed; † it also presents close affinities in many respects with the *U. arctos* of Europe. Towards the north it shades into what is called the Barren Ground bear, which latter has been repeatedly referred, with more or less positiveness, by different authors to the *U. arctos* rather than to *U. americanus* or *U. horribilis*. Middendorff found the bears of Northeastern Asia equalling in size and generally resembling in other characters the *U. horribilis* (*ferox* of authors) of the Western Coast of America. The *U. americanus* also presents numerous variations in color and in other points quite parallel with similar variations in the European *U. arctos*.‡ Specimens often occur on the one continent that are strikingly like others from the other. Middendorff expressly states that the differences between *U. arctos* and *U. ferox* (*horribilis*) are not greater than occur between different specimens of *U. arctos*. Dr. Gray admits that it is only a knowledge of the locality that in some cases enables him to separate them.

* Andrew Murry, Geog. Distr. of Mammals, p. 119.

† See Professor Baird's N. Am. Mam., pp. 219 – 228.

‡ I learn from Mr. W. H. Dall, who has recently returned from a three years' exploration of Alaska, bringing with him important information relative to the natural history, geography, etc., of that country, that three kinds of bears are distinguished there; the larger and the more common being the grizzly, the second the so-called Barren Ground bear, while the third and smallest is a black bear; showing that there is found the usual variety, in point of size and color, seen on the Pacific Coast farther south.

In the Natural History Reports of the United States and Mexican Boundary Survey,* Professor Baird gives much valuable information, in addition to that contained in his Mammals of North America, respecting the bears of the Rocky Mountain and West Coast regions of the Continent. On the whole it tends to render the subject still more difficult and complex, if we recognize more than a single species in North America, as many of the different specimens described represent intermediate stages between the two commonly recognized American species. A specimen collected by Dr. Kennerly, at Los Nogales, Sonora, and others at the copper mines in New Mexico, by Mr. H. J. Clark, differ so much from the "grizzlies" of California, that Professor Baird described them as a distinct variety of the latter, — *Ursus horribilis*, var. *horriæus.* Although the leading characters are the same, this "variety" differs in being smaller, with relatively shorter ears and a longer tail, these parts being nearly equal, instead of the ears twice the length of the tail, as in California specimens; it also differs in the texture of the fur, in the arrangement of the colors, in the greater relative breadth of the skull, its narrower and slenderer muzzle and more vaulted palate, and in the shape of the teeth. While the "*horriæus*" specimens are quite distinct from either of the so-called varieties of *U. americanus*, the variation from the typical *U. horribilis* of California is towards *U. americanus; U. americanus* of the Eastern States differing from them chiefly in being smaller. In the smaller size, compared with *U. horribilis*, and the great breadth of the head, "*horriæus*" also affiliates with the *U. arctos.* The latter is usually supposed to never attain the size exhibited by many specimens of the *U. ferox* (*horribilis*); but Prince Maximilian says that this is incorrect, as he has seen Russian bears that were fully as large as the latter; and Middendorff, as already stated, remarks that the bears of Northeastern Asia are as large as those of the Pacific coast of America.†

In reference to the peculiar bears of the Sonoran region, Dr. Kennerly has observed as follows: "In regard to the bears that are found along the northern frontier of Mexico and the southern portion of New Mexico, there seems to be some confusion. In addition to the common black bear,

* Vol. II, Mammals, pp. 24 – 81.

† The facility with which the bears can cross in winter from one continent to the other by way of Behring's Strait, and the known fact that they do thus cross (I am assured of this fact by Mr. Dall), renders the close mutual resemblance of the bears of Northeastern Asia and Northwestern America a matter of no great surprise. The similar resemblance between the martens and the representatives of the other circumpolar species from these countries, which has been already pointed out, though some of them may be able to pass less readily than the bears from one continent to the other, would seem to be fully accounted for by a similar occasional migration, if any hypothetical explanation for so natural a phenomenon as the great similarity of the animals specifically identical in these slightly separated districts is required.

Ursus americanus, and the large Grizzly, *U. ferox*, there is found another, intermediate in size to these, generally of a brownish color, with the tips of the hairs often silvered, especially in the old individuals, and in appearance, except in size, is almost identical with the *U. ferox* found in such great numbers in California. Among the people of the country they are known as *brown bears;* but this term is variously modified by the most experienced hunters, as we have heard applied by them, to the same individual, the name grizzly bear, touch of the grizzly, cross between the grizzly and brown bear, and common brown bear; but on no occasion have we heard them assign any relationship between these animals and the common black bear, causing us to believe that there must be a considerable difference between this animal and the brown bear of Oregon, which is called by naturalists only a variety of the black; in fact, its size generally, would necessarily preclude such comparison, while even the very old individual falls far short of the weight and dimensions of the *U. ferox* of California, of which we could much more easily consider it a variety than of the *U. americanus*." * One of the three of Mr. H. J. Clark's specimens, however, referred by Professor Baird to the *U. cinnamomeus* Baird, was "glossy black," and the others brown.

Audubon remarks that the *U. horribilis* varies in color from nearly white through various shades of pale and dark brown to black, it being difficult to find two specimens alike. The young are generally much darker than the adult. Yellowish gray and grayish brown are common varieties, while some are of a rufous tint. This account is confirmed by Prince Maximilian's observations on the bears of the Upper Missouri.†

The specimens from New Mexico and the adjoining country southwards, which present the peculiar characters mentioned above, have usually been referred to the *U. horribilis*, as previously stated; but others that are equally perplexing, but commonly referred to the brown variety of *U. americanus*, also occur in the same region. Some of these latter differ so much from other brown bears from Oregon, also referred to *U. americanus*, that Professor Baird has considered the probability of their proving distinct species very great, and proposed to call the former, in that event, *U. amblyceps*. These Sonoran specimens differ from representatives of *U. americanus* from the Eastern States in nearly the manner that the Sonoran variety *horriœus* of the grizzly, *U. horribilis*, does from the true *U. horribilis* of California; namely, in the greater relative breadth of the head, the relatively smaller size of the molar teeth, and the irregular character of the

* United States and Mexican Bound. Surveys, Vol. II, Mammals, p. 28.

† Verzeichniss der auf seiner Reise in Nordamerika beobachteten Säugethiere, Vom Prinzen Maximilian zu Wied. Archiv für Naturgeschichte, XXVII, 1861, Theil 1, p. 203, Taf. VIII.

coloration. Professor Baird provisionally refers them to the *Ursus americanus*, var. *cinnamomeus*, of Audubon and Bachman, to which, he says, they bear the nearest resemblance.

Prince Maximilian, in his memoir "Über die Selbständigkeit der species des *Ursus ferox* Desm.," * urges strongly the distinctness of *U. horribilis* (*ferox* Maxm.) from both *U. americanus* and *U. arctos*, in which he is supported by the anatomical observations of Dr. C. Mayer, which form an appendix to his paper. Several specimens of the former, of different ages, from the Upper Missouri, are described in detail, but no differences other than those previously pointed out by other authors, are mentioned. They consider that the shorter ears and longer claws of *U. horribilis*, with certain minor osteological peculiarities, sufficiently distinguish it from *U. arctos*. These authors admit that bears from northern countries present great individual differences; yet, in reviewing Middendorff's arguments, they offset their conclusions, based on an examination of a very limited number of specimens, against those of the latter, formed from probably as careful an elaboration of many times their amount of material. The differences that have been described by authors as occurring between specimens of *U. arctos* from different parts of Europe and Asiatic Russia, or between different specimens of either *U. horribilis* or *U. americanus* from different localities on this continent, are as great as those they urge as peculiar to their so-called species.

I have not space to notice in detail each point urged as distinctive by those authors who divide the bears into a large number of species. As they mainly rest on the shape and size of the molar teeth, the relative length of the claws and the ears, and the proportions of the skull, a few further remarks on these characters may not be out of place. In Professor Baird's table of measurements of skulls of American bears, the average proportion of breadth to length in the seven specimens cited is sixty per cent, with a minimum of fifty-five per cent, and a maximum of seventy-one. Only one of the series, however, exceeds sixty. Adding four other specimens referred by Baird to "*cinnamomeus?*" the average of the eleven is fifty-nine and a half per cent; the minimum is fifty-three, and two specimens range above sixty. The proportional breadth of the skull in eight specimens of *U. horribilis* is fifty-six per cent. Between the extremes of this series (Nos. 1218 and 2037) the variation amounts to ten per cent. In his comparison of *U. horribilis* with *U. arctos*, Baird cites two of Blainville's specimens in which the same proportion is sixty-six per cent; in reference to which he adds: "This width of head far exceeding that of any well-known American species, would appear to be quite conclusive as to identity," — Professor

* Verhandlungen der Kaiserlichen Leopoldinisch-Carolinischen Akademie der Naturforschung, Band. XXVI, erste Abtheil., 1857, pp. 33 – 85, Taf. III, IV, and V.

Baird not having then received the Saranac (New York) specimen, with the breadth of the head seventy-one per cent of the length. In five skulls of the *U. arctos*, of which measurements are given by Dr. Gray, the average proportion of breadth to length is sixty-seven per cent; in five of the *U.* "*torquatus*," sixty-one; in two of *U.* "*syriacus*," sixty; in four of *U.* "*Isabellinus*," sixty. The average of these sixteen European and Asiatic skulls is thus sixty-two per cent. Gray also gives measurements of five American skulls; viz., two of *U. cinereus* (= *U. horribilis* Ord) and three of *U.* "*americanus*"; the proportional breadth of the skull in the latter is sixty-one per cent, and in the former fifty-eight. This would seem to indicate a tolerable constancy in the greater narrowness of the skull in the American bears. But from Middendorff's table of measurements of fifty-five skulls, from different parts of Russia (chiefly from Northeastern Asia), the percentage of breadth to length falls to fifty-eight and a quarter, and is hence almost precisely that of the American. The maximum breadth of skull seems to be reached in Western Europe; thence eastward to Kamschatka there is a nearer and nearer approximation in this character, as in general appearance, to the American animal.

In respect to the variability of the skull in other particulars, Dr. Gray, in referring to two skulls of *U. horribilis*, remarks that they differ very considerably; the one is much broader, with the palate wider, the nose shorter, and the orbits higher and rounder.

In comparing the teeth of the American bears with those of the European, when but a single example of each is taken, the differences *may* be considerable, so great, indeed, that if constant they might be regarded as sufficient to decide the question of the distinctness of the species; but since specimens frequently occur from the same locality that differ as much, and others from the different continents that are almost or quite indistinguishable, the unreliability of such distinctions becomes sufficiently apparent.

Variation in the size and shape of the molar teeth are found in other groups than the bears, though rarely perhaps so great. According to Professor Peters of Berlin, in the *Otariæ*, or eared Seals, the variation in this respect seems to be even somewhat greater. Most authors have heretofore looked upon the teeth of the Otariæ as affording good generic characters, but Professor Peters has found them to be so exceedingly variable that he does not consider them reliable for even specific distinctions.*

The ears, in length and form, are found to vary greatly in specimens of *U. horribilis* from different localities, especially from points on different sides of the Rocky Mountains; whether variations of this sort are found in *U. arctos*, it is difficult from the few sufficiently detailed measurements given by authors to determine. That such do occur in specimens of bears referred

* Monatsber. Ak. Wiss., Berlin, 1866, pp. 261–281 and 655–672.

to the same species by authors who separate them into numerous species, lessens the importance of distinctions based on them as separating the bears of America from those of the Old World.

The claws are well known to vary in length at localities not very remote, in the Old World and in the New. Although the differences between specimens of *U. horribilis*, which seem to have them ordinarily the most developed, and others of *U. arctos* from Western Europe, is very great, they do not appear to have the importance as specific characters assigned them by Prince Maximilian and Dr. Mayer.

Finally, in weighing the evidence in reference to the number of species of North American bears and their relationship to those of the Old World, it is evident that the comparatively small number of specimens thus far examined, either from a single region or in altogether, and the vast areas from which no specimens have been received, should be carefully considered as showing how few the data are on which any opinion must be based. The inconstancy of character presented by those from the same locality, especially in the breadth and other proportions of the skull, in the shape and relative size of the molar teeth, in color, and in size, should also be duly considered, as well as the fact that however wide the differences between specimens from distant localities are, those from intermediate ones are generally of an intermediate character. In some districts bears find an abundant supply of animal food, while in others they are more or less restricted to a vegetable diet, and that these differences must give rise to modifications in the teeth and bones of the skull is to be expected. From the wide geographical range of even the different restricted so-called species, their representatives are subjected to widely different climatic and other modifying influences. In America, the coincidence of the greatest number of individuals with the maximum development in size seen in the region occupied by the typical *U. horribilis*, as in California, and the gradual transition in the easterly portions of the Rocky Mountain district to aberrant forms of this type, some of which indisputably approach quite near the common style of *U.* "*americanus*" of the eastern portions of the United States, and at the extreme north of the continent to the *U. arctos* of the Old World, especially to the Russian type of that animal, are facts which render the separation of the bears of these several regions into well-defined species quite improbable, if not impossible. I hence see no alternative but to consider with Blainville, Middendorff, and Murry, all the bears of the Northern Hemisphere, excepting *Ursus maritimus*, as forming but a single species. Here, as in other similar cases already considered in this paper, if the opposite view be adopted, it appears inevitable that still other species than those authors have already recognized must be allowed, with numerous "sub-species," or "varieties" and "sub-varieties" of each, in order to dispose of the constantly occurring intermediate forms.

PHOCIDÆ.

16. **Phoca vitulina** Linn. (*Callocephalus vitulinus* F. Cuv.) Common Harbor Seal. Abundant. I have observed it to be particularly numerous about Ipswich, as many as eight or ten being sometimes seen at once. In June the females are accompanied by their young, then apparently about one fourth grown. Though so common, their habits seem to be little known. They are rarely captured, as when killed they sink to the bottom and are thus difficult to obtain. A fine nearly adult male, now in the Museum of Comparative Zoölogy, was obtained at Wellfleet, in June, 1868, by Mr. C. J. Maynard and the writer. The specimen having been stranded, it had just died of exhaustion when discovered, from its frantic efforts to regain the water. It had repeatedly floundered several yards up the steep sand beach.

In reply to inquiries of mine respecting our seals, Captain N. E. Atwood, of Provincetown, has kindly written me respecting this species as follows: "At Provincetown we occasionally see a straggling specimen of what we call the Harbor Seal; in the vicinity of Cape Cod it is not very common; but there are localities on our New England coast where, in summer, they are found in great numbers. In Boston Harbor, west of Rainsford Island, there is a shoal-water bay of considerable extent, in which is a small ledge of rocks that at low water rise several feet above the surface; on these rocks many hundreds of these seals may be seen at any time during the summer. If the ledge is approached, they all dive into the water and rapidly disperse, but soon return again if they perceive no danger. These seals are small, and of little value, and are hence unmolested."

17. **Cystophora cristata** Nilsson. Hooded Seal. From accounts I have received from residents along the coast of a seal of very large size observed by them, and occasionally captured, I am led to think this species is not of very unfrequent occurrence on the Massachusetts coast. Mr. C. W. Bennett informs me of one taken some years since in the Providence River, a few miles below Providence, which he saw shortly after. From his very particular account of it I cannot doubt that it was of this species. Mr. C. J. Maynard also in-

forms me that a number of specimens have been taken at Ipswich within the past few years, that have weighed from seven hundred to nine hundred pounds. It seems to be most frequent in winter, when it apparently migrates from the north.

CERVIDÆ.

18. **Cariacus virginianus** GRAY. (*Cervus virginianus* Boddært.) VIRGINIA DEER. A few still exist in Plymouth, Barnstable, and Berkshire Counties, where they have been for some time stringently protected by law. Mr. Samuels, in the report of the State Board of Agriculture of Massachusetts for 1861 (p. 189), observes: "This beautiful animal is now rare in this State, and will soon, probably, be extinct; it is found in the woods in Plymouth and Barnstable Counties, in the neighborhood of the Hoosac Mountains, and on several of the islands on the southeast coast." It has for a long time been extinct in most parts of the State. They were last seen in the vicinity of Springfield about fifty years ago.

Respecting individual variation in species of the *Cervidæ*, and especially in *Cervus* (*Cariacus*) *virginianus*, I find the following important observations by Hon. John D. Caton, in the Transactions of the Ottawa Academy of Natural Sciences.* Referring to our standard works on the American *Cervidæ*, he observes: "The superficial marks which assign to each of these species its appropriate classification are properly described. Yet this description is generally from a single specimen, while in fact individuals differ very widely, both in color and form; so much so that even among the few I have in my parks we might almost persuade ourselves we have distinct varieties. Among the fifty or sixty deer which I have, there are three distinct shades of color, which also seem to be characterized by a peculiarity of form. The lightest colored have long legs and slim bodies; they have the largest horns, do not fatten readily, and are more wild and restless than the others. The next are of a considerably darker shade; in some instances quite black along the top of the neck and down the back, and a black tail, as distinctly so as the California deer; they have frequently other black marks. I have one specimen with a distinct black line over each eye, of a triangular form, passing towards the ear; and several others in which this mark is quite visible, though not so conspicuous, giving them rather a ferocious appearance. This variety has short legs, rather short, heavy bodies, are very tame, and always the fattest in the park. The smallest variety, both in

* Part I, 1868, p. 43.

size and numbers, is of a distinct russet color, and has less white under the throat and belly than either of the former. In one specimen the white is nearly wanting beneath the lower jaw, and there is very little under the neck. They are not so wild as the first class mentioned, but are more timid than the second, and, in their disposition to fatten, seem also to be intermediate between the other two."

In regard to the Elk he also observes: "There seem to be distinct classes of the elk, which are as manifestly hereditary as those of the deer, especially so as to form and size. Of these I recognize in my parks two classes, varying in form and size, but not materially in color. . . . . One variety is larger, and has longer legs, and is much more graceful in form and carriage than the other. The largest variety seems to be the most hardy, and fattens the most readily; it is also less vicious."

The Elk or Wapati (*Cervus canadensis* Erxl.), judging from what is known of its former distribution, undoubtedly once inhabited Massachusetts. According to Professor Baird and others, it is still found in the Alleghanies in Pennsylvania.*

The Moose (*Alce malchis* Ogilby; *Cervus alces* Linn.) also formerly undoubtedly existed in Massachusetts, though it has now been long extinct here. It still occurs in Maine, as far south as the Umbagog Lakes, whence specimens have been received at the Museum of Comparative Zoölogy.

As to whether the individuals found in America are identical with those of the Old World, there is at present some discrepancy of opinion, though formerly regarded as the same. The distinctions between them are very slight, and to what extent constant is hardly known. While the Moose of Asia and Europe are considered identical, Sir John Richardson has pointed out some slight differences in the skeleton of those of the New and the Old World, which incline him to the opinion that they may be distinct species, and as such he bestowed on the American the specific name of *muswa*. Whether these distinctions are more than individual, or such as would disappear in a large series of specimens, it is perhaps impossible to say. Their distribution, however, is remarkably alike, reaching the Arctic coast on both continents, and extending southwards to about the same isotherm; on the whole I consider their identity as extremely probable, if not absolutely certain.

* Audubon states that fifty years ago a few still lived in Kentucky, near the Ohio River, and that they were not very uncommon at that time in Southern Illinois, — localities much more southern than Massachusetts. — *Quad. N. Am.*, Vol. II, p. 88.

The Woodland Caribou, or Reindeer (*Tarandus rangifer* Gray; *Rangifer caribou* Aud. and Bach.), like the preceding, probably once inhabited Western Massachusetts, judging from what is known of their earlier distribution, though probably rather as an occasional visitant from the north than as a numerous or permanent resident. It is still found occasionally in Northern Maine, whence specimens have been received on several occasions at the Museum of Comparative Zoölogy, from Mr. J. G. Rich. In March, 1863, according to Professor Verrill,* this gentleman brought seven to Boston, killed on the head waters of the Kennebec, out of a herd of about twenty, supposed by Mr. Rich to have come from regions farther north, as the caribou had been noticed there by him but once before during the previous five years. It is said to occur also in the Adirondacks of New York.†

In this species we are again met by the old question of identity with a closely allied Old World congener. Several high authorities still maintain its identity with the European and Asiatic reindeer, while others, whose opinions are equally entitled to consideration, believe them distinct. Unlike the moose, the reindeer, if forming but a single species, are apparently easily separated into several very distinct races, in some cases differing in size, but chiefly in the character of the horns. In America, the woodland caribou constitutes a southern form, and inhabits the northern wooded districts of the continent; in the Arctic Barren Grounds it is replaced by a much smaller race, but which, it is said, has larger horns. This smaller race seems to be a circumpolar one, inhabiting the similar woodless tracts of the extreme north of the Old World, and also Greenland, but differing somewhat in different districts, it is claimed, by peculiar modes of branching of the horns, especially in respect to the form of the brow antlers. Whether these differences that have been pointed out are to be considered as constantly characterizing the reindeer of these different regions is still questionable, as but comparatively few specimens appear to have been yet compared. From the great variability in the branching of the horns presented by all the different species of the *Cervidæ*, the right and left horns in the same individual, as well as the successive sets, being often most notably unlike, it seems to be a distinction of doubtful value.

In reference to the disputed question of whether there are one or several

* Proc. Bost. Soc. Nat. Hist., Vol. IX, p. 226.

† In reference to the much farther extension southward of the habitat of this species in earlier times than even two centuries ago, see the general remarks on the geographical distribution of the Massachusetts Mammals, at the close of the paper.

species of reindeer, Mr. Andrew Murray, in his valuable work on the Geographical Distribution of Mammals (p. 151), observes: "But there are several varieties; how far some of these deserve to be reckoned species, and if so, which of them, is a more difficult question. There is, first, the fossil variety; next, the Lapland reindeer, which is nearest to the fossil type; then the Siberian, which, although very close to the Lapland, differs in the character of the horns; moreover, there are two varieties in North America, and one in Greenland and Spitzbergen. I believe all these to be altered forms developed out of one stock, modified to an extent corresponding with the degree of deviation of their respective climates from the original condition of existence of that stock." Mr. Murray gives figures, copied from various authors, of the antlers of these different varieties, and mentions in detail their points of difference and resemblance; concerning which he adds: "The resemblance between them is too constant, and, as will be seen by the figures, is too considerable to be a matter of accident or coincidence."

Dr. Richardson seems to have first clearly distinguished the two varieties of American reindeer, or caribou, now so generally recognized, but of the woodland variety ("*Cervus tarandus*, var. β *sylvestris*") he claims to "know little, having," he says, "seen few of them alive or in an entire state. It is," he adds, "much larger than the Barren Ground caribou, has smaller horns, and when in good condition is vastly inferior as an article of food." The Barren Ground variety he seems to have studied with great care. Respecting the character of the horns, the peculiar form of which seems to be the chief character yet discovered by which to distinguish the different so-called varieties, he observes: "The old males have, in general, the largest and most palmated horns, while the young ones have them less branched, and more cylindrical and pointed; but this is not uniformly the case, and the variety of forms assumed by the horns of the caribou is, indeed, so great, that it is difficult to comprehend them all in a general description. Some have the branches and extremities *broadly palmated* [the italicizing is my own], and set round with finger-like points; others have them *cylindrical, and even tapering, without any palmated portion whatever.* The majority of adult males have a brow antler, in form of a broad vertical plate, running down betwixt the eyes, and hanging over the nose. In some, this horn springs from the *right* horn, in others from the *left;* in many there is a plate from *each side*, and in a considerable number it is *altogether wanting;* the plate is, in general, widest at its extremity, and is set with four or five points, which are sometimes recurved. The *main stem* of the horn also exhibits an *endless variety* in its *thickness*, *altitude*, and *curvature.*" Major Smith observes, that a "probable distinction, by which some, if not all, of the varieties of caribou may be distinguished from the reindeer of the

Old Continent, is, that their horns are always shorter, less concave, more robust, the palm narrower, and with fewer processes than those of the former," — a view that has been adopted by other writers. Respecting this assumption Dr. Richardson thus observes: "I can with confidence say, after having seen *many thousands* of the Barren Ground kind, that the horns of the old males are as much, if not more, palmated than any antlers of the European reindeer to be found in the British Museum." If attention is given to the parts of the above quotation from Dr. Richardson that I have *italicized*, it will be seen how unreliable must be any distinctions based on the horns, unless the comparisons are more extended than they thus far seem to have been. That the horns of the Barren Ground form may differ from those of the wooded districts in other points than size is quite possible, but in the several pairs of horns of the latter in the Museum of Comparative Zoölogy there is a very close resemblance to those of the Barren Ground and Greenland caribous figured by Baird and Richardson, the Northern Maine specimens much more nearly agreeing with these than with Professor Baird's figures of the Lake Superior one (No. 900), which is evidently an extreme form. The horns of the northern or Barren Ground race of the American reindeer, according to the best authorities, do not differ essentially from those of the reindeer of the corresponding districts of the Old World. Mr. Murray quotes Mr. Alfred Newton as saying, in reference to the reindeer he saw in Spitzbergen: "The average type of a good Spitzbergen head is very well represented by the first figure in the Fauna-Boreali Americana (Vol. I, p. 240) of the so-called Barren Ground caribou (*Cervus tarandus*, var. *arcticus* Richardson)"; which testimony of Mr. Newton, he states, is supported by that of Mr. Lamont.* Mr. Newton, however, says the Spitzbergen reindeer are "certainly smaller than the Lapland reindeer."

Professor Baird observes, respecting the American woodland race, that its relationship to the European reindeer is not well ascertained. "The opinion," he says, "is gaining ground that the Barren Ground reindeer is distinct, and as this species cuts it off from the Arctic Circle, it would seem most probable that it cannot be the same with the animal inhabiting the circumpolar region of the Old World." But the recorded observations seem fully to prove, as is now, indeed, currently admitted, the existence of two similar races on the Old Continent, — a northern and a southern, differing from each other nearly as do the Barren Ground and woodland varieties in North America. Hence if we allow two species of reindeer for America, why not two for the Old World? But there, where the species has been longer and is better known, competent authorities seem not to doubt their identity, and from which some even regard the American as

* Geog. Distr. of Mam., p. 155.

inseparable. I have already shown that the characters used for their separation are by no means reliable. Concerning the Greenland reindeer, Mr. Robert Brown, in a recent valuable paper on the Mammals of Greenland,* says, "that after very excellent opportunities of comparison and study," he considers "the Greenland reindeer only a climatic variety of the European species. I have, moreover," he adds, "seen specimens of reindeer horns from Greenland, which could not be distinguished from European, and *vice versa*. On the whole, however, there is a slight variation."

As I have previously remarked, I see no good reason why all may not be considered as one species, within which may be distinguished several quite well-marked geographical races.

In relation to other facts, the differences in size presented by the two races of American reindeer, the woodland and the Barren Ground, becomes extremely interesting; for, supposing them to form one species, as there seems to be little reason to doubt, the variation in this respect is directly the reverse of that ordinarily presented by individuals of the same species from localities differing considerably in latitude; the general law being an increase in size at the northward. But here there is a marked decrease. It is yet not quite exceptional, as a point is reached in the habitat of the non-migratory circumpolar species, where the rigor of the climate, and the consequent altered conditions of life, seem unfavorable to a maximum development of the animal. This is exemplified by the small stature attained by the circumpolar tribes of men, as the Esquimaux of Greenland and of the north of America, and the Laplanders of the Old World. The common wolf (*Canis lupus*) has its smaller northern form, which, in America at least, occupies the Barren Grounds and the region northwards to the Arctic coast, and which differs quite positively from its more southern relatives.†

A smaller circumpolar Arctic form of the fox has long been recognized, differing in color, in size, and in the texture of its fur from the common species (*Vulpes vulgaris* and *V. fulvus* auct.). And there is a well-known corresponding race of bears, commonly referred to the *Ursus arctos*, which in America pass almost insensibly into the more southern and larger *Ursus horribilis*. Whether this decrease in size in the extreme boreal regions

* Proc. Lond. Zoöl. Soc., 1868, Part II, p. 352.

† "Of this species (*Canis griseo-albus* Rich.) I consider that there are two varieties, one of which is of a dark color and large size, inhabiting the wooded portions of the [Mackenzie's River] District as far north as the Youcon River. The other is usually of a dirty white tint, with, in general, a dark stripe down the back, and frequents the Barren Grounds northwards to the Arctic coast. It is of smaller size than the first-mentioned variety, and lives in much larger bands; indeed, it may possibly be a distinct species." — B. R. Ross, *Nat. Hist. Rev.*, July, 1862, p. 271.

extends to other species I have not at present the means of determining, though it is hardly to be expected that it will to all, since some of them are to a considerable degree migratory, going southward in winter, as the lynxes, martens, and some others. Hence extremes of climate, whether of heat or cold, seem to unfavorably influence the development of animal life generally, a mean or temperate region being as necessary for the highest development of the lower orders of mammalia as for that of man.

Besides the marked climatic modifications in size and in other features in the species cited above, certain other variations in them may be here appropriately referred to. These, though slight, so commonly appear in a number of species inhabiting the same region as to lead one at once to suspect a common cause for such differences. Dr. Richardson * long since pointed out slight differences in the color and texture of the fur, and in the breadth of the foot, in species which he considered identical in North America and Europe, between their representatives from Northern North America and Central Europe; the former having a finer and thicker coat, and broader feet, to better adapt them to a colder climate and a more snow-covered country, as well as brighter and livelier colors. These modifications appear also, he says, in the native domestic dogs.*

Naturalists have repeatedly remarked the narrower form of the head in the moose, bear, fox, and wolf in Eastern North American specimens as compared with others from Western Europe. In the former, the absolute breadth of the skull is generally less, while there is at the same time a greater development of the facial portion. In these animals a difference in size has also been claimed to distinguish their representatives from the two continents; but, owing to the variation in size on either continent with the latitude and elevation of the locality at which they were collected, observations on this point are somewhat contradictory. The general indication, however, seems to be that the American somewhat exceed the European when both are from near the same isotherm.

I have already called attention to the fact of the same species varying in color in different portions of its habitat, as in the case of the *Canis lupus*. On both continents, this species gradually changes from nearly white (yellowish or grayish white) in the Arctic regions to very dark or "black" in the southern. Individuals of the black and cross varieties of the fox (*Vulpes vulgaris*) are most numerous on both continents towards the north; † at the south, while the general fulvous color prevails on the dorsal

* Fauna Boreali-Americana, Vol. I, p. 91.

† Mr. B. R. Ross gives the proportion of the different colors in the foxes killed in the Mackenzie River District as red $\frac{6}{15}$ths, cross $\frac{7}{15}$ths, silver $\frac{2}{15}$ths; or sixty per cent of the dark variety to forty of the red; while as far south as the United States the dark varieties probably scarcely exceed one per cent. — *Nat. Hist. Rev.*, 1862, p. 272.

surface, there is apparently a greater development of dusky on the ventral; this type forming the *Vulpes* "*melanogaster*" of the south of Europe. According to Professor Baird, the black varieties in some of the American squirrels reach their greatest numerical development in the northern portions of their habitat; * where also melanic specimens of the marmot and racoon are most frequent. On the Atlantic slope there is a noticeable tendency to a predominance of gray rather than rufous tints, while in the interior, particularly in the Mississippi Valley, and on the Plains, the reverse is the case, in at least a number of species. I have in another place † called attention to the faded appearance of the plumage of many species of birds on the Plains, in those that range across the continent; in others there is a tendency to an increase of fulvous and rufous, as is noticeable in some mammals. In the Sonoran region there is a marked inclination to pied varieties, such occurring in the weasels (*P. frenatus* and *P. xanthogenys*), skunks (*Mephitis bicolor* and also in *M. mephitica*), the bears and squirrels. The changing to white in winter of many species at the north which at the south constantly retain their summer colors, as the weasels, the Arctic fox,‡ the wolf, and some of the hares,‡ it seems to me is also to be properly classed in the category of climatic and geographical peculiarities of coloration. The prevalence of neutral mouse-gray tints in so large a proportion of the mammals of Australia, and of plumbeous and black in those of Africa, in contrast with the brighter and more varied colors of those of the other continents, is but a grander exhibition of the same kind. The hibernation of certain species in the cold regions that in the warmer are constantly active, as in the *Ursidæ* and *Vespertilionidæ*, for example, is in some respects a similar phenomenon.

There are differences in size between specimens of the same species from different localities that are not apparently explainable on the ground of difference in the latitude and altitudes of their respective places of birth. On the Mississippi prairies, for example, some species of *Muridæ*,

* North Amer. Mam., p. 244.

† Mem. Bost. Soc. Nat. Hist., Vol. I, p. 513.

‡ Concerning this point Mr. Alfred Newton observes: "I have never seen it remarked, but it is unquestionably the case, that nearly all the Icelandic examples of *Canis lagopus* are 'blue' foxes; that is to say, their winter coat is nearly the same color as their summer coat. This fact, I think, must be taken in connection with the comparatively mild climate which Iceland enjoys in winter; and if so, is analogous to the circumstance that of the Alpine Hare (*Lepus timidus* Linn., *non auct.*) always becoming white in Scandinavia, generally so in Scotland, but seldom in Ireland." (*Proc. Zoöl. Society of London*, Dec. 1864, p. 497.) Dr. Richardson also states that the Arctic fox is of a purer white on the shores of Hudson's Bay than at Bhering's Straits, where, as is well known, the climate is considerably milder. (*Faun. Bor. Amer.*, I, p. 87.)

Talpidæ, and *Soricidæ* attain an appreciably larger size than under nearly the same latitude and degree of elevation at the eastward. The same fact is also observed in the mink; while the bears of the Pacific slope are larger than from most other parts of the Continent. Whether a greater abundance of their proper food may be the cause of this, it is impossible now to determine. They are facts, however, that are worthy of careful consideration, and they are cited here simply to call to them further attention.

It may be observed, in passing, that allied species, as the fox and wolf, vary differently under the same conditions; melanism being most developed in the one at the south, and in the other at the north. It is also noteworthy that circumpolar species follow the same law in their climatal variations that obtains in the differentiation in both the fauna and flora of the northern hemisphere in passing from the north southwards. As is well known, there are many species of animals and plants at the north, where their habitats approximate, that are common to the two continents. Such species become less and less numerous to the southwards, and beyond the tropics very few occur on both the Eastern and Western continents. In like manner, specimens from towards the north of the two continents of circumpolar species that range over the north temperate regions are much nearer alike than those collected from near their southern limits of distribution.

For the following notes on the Cetaceans of the Massachusetts coast, and their local names, I am indebted, as previously stated, to Captain N. E. Atwood, of Provincetown. For the scientific names I am under obligations to Professor E. D. Cope, of Philadelphia, to whom I forwarded Captain Atwood's notes for the determination of the species. Professor Cope's identifications and remarks are distinguished by being enclosed in brackets.

BALÆNIDÆ.

19. [**Balæna cisarctica** Cope.] "Right Whale. Occasional. "This well-known species is at times taken here; in former years they were much more frequent in their visits than now. Although a straggling specimen may be seen at any time, they are generally more common during the latter part of April and the early part of May. They yield a larger amount of oil than any other species that visits our coast; besides which they have a large quantity of whalebone that finds a ready market, known as the "black whalebone" of commerce.

The skeleton of the right whale in the Museum of Comparative Zoölogy was taken here. The specimen yielded eighty barrels of oil, and the bone that was taken from its mouth was sold for $1,000."

20. [**Agaphelus gibbosus** (Erxl.) COPE.] "SCRAGG WHALE. Rare. A species of whale known by this name, nearly allied to if not identical with the right whale, is sometimes taken here. It is the opinion of many of our whalemen that they are not a distinct species, but the young right whale that lost its mother while very young and grew up without parental care, which has caused a slight modification. The most prominent feature is that in its dorsal ridge, near the tail, there are a number of small projections or bunches, having some resemblance to the teeth of a saw. It has no dorsal fin or hump on its back."

21. [**Megaptera osphyia** COPE, or another **Megaptera**.*] "HUMPBACK WHALE. This species is common on our coast, and sometimes comes into Provincetown harbor, where it is attacked and killed by our whalemen. They yield but a small quantity of oil compared with the yield of the right whale, the usual quantity being from ten to fifteen barrels. The bone in its mouth, unlike that of the right whale, is of little value and not considered worth saving. When harpooned it will run with great swiftness, and continues to do so while it is being killed. Its affection for its young seems stronger than that of any other species, as the mother will expose her own life in defence of her offspring."

22. **Eschrichtius robustus** LILJ. Professor Cope informs me that he has found a jaw of this species on the New Jersey coast; it should in all probability be enumerated in the present list.

23. [**Sibbaldius tectirostris** COPE, and probably another species.] "FINBACK WHALE. Frequent.

"This species is the most common large whale found along our coast, and is frequently seen at all times in the year. They are not har-

* Professor Cope believes that under the name of "Humpback," of Captain Atwood's list, more than one species may be embraced; and also more than one under the species called "Scragg Whale."

pooned by the whalemen, as they run so swiftly they cannot be killed. I have known a few to be killed by shooting them with a bomb lance. When they have been killed in this way in our bay they always sink to the bottom (they being not a fat whale), and remain there some few days, during which time much of the blubber is eaten off by sharks. I have known two of this species to run on shore in the night, in our harbor, and be left by the receding tide. When they were killed there appeared to be no indications of disease, and the cause of their running on the beach could not be learned. One of them yielded fourteen and the other twenty barrels of oil." In a subsequent communication Captain Atwood adds: "The finback is a species that yields only a small quantity of oil compared with its size; the blubber is thinner than in other species. The right whale killed here, of which the skeleton is in the Museum of Comparative Zoölogy, was forty-seven feet long, and yielded eighty barrels and fourteen gallons of oil; a finback since killed here was fifty-four feet long, and made only twenty barrels of oil, though a good fat whale of its kind."

24. **Sibbaldius tuberosus** Cope. A specimen at first doubtfully referred to the *S. laticeps* Gray,* by Professor Cope, but since regarded by him as a new species,† was captured in Mobjack Bay, Virginia, in May, 1866. It being a somewhat northern species, it should probably be included in the present list.

25. [**Sibbaldius borealis** Fisch.] "Sulphur-bottom Whale. Rare. "This species is said to occur on our coast. Like the finback, it has on its back a very small dorsal fin. Being very much elongated, it is a swift runner, and passes through the water with a velocity so great that the whalemen cannot kill them in the same way that they take the other species. I have never seen it dead, and know but little about it."

26. [? **Balænoptera rostratra.** I have not yet identified this one.] "Grampus. Occasional. When seen here alone, we know it by that name. It is the opinion of some of our whalemen, with whom I have conversed respecting this whale, that it is not a distinct species, but the young of the finback."

* Proc. Phil. Acad. Nat. Sci., 1866, p. 297. † Ibid., 1869, p. 16.

PHYSETERIDÆ.

27. **Physeter macrocephalus** Pander. Sperm Whale. Occasional off the coast; formerly much more frequent.

28. [**Mesoplodon sowerbiensis.**] To this species Professor Cope refers a specimen found stranded a short time since on Nantucket Island. I learn from Mr. S. C. Martin that it was called "Grampus" by the whalemen, and that its length was sixteen feet and three inches, and girth seven feet. The skull, presented by Mr. Martin to Professor Agassiz, is now in the Museum of Comparative Zoölogy, and is the specimen referred to by Professor Agassiz at the meeting of the Boston Society of Natural History, held November 6, 1867. He remarked that it was a species new to America, and that it belonged to the genus *Mesoplodon*, as characterized by Gervais, and ought to be separated from the fossil *Xiphius*, described by Cuvier.*

DELPHINIDÆ.

29. [**Orca gladiator** Sundeval.] "Killer. This species visits our bay occasionally in small schools. Their dorsal fin is several feet high when fully grown. They are at times in summer seen coming into our harbor. The horse-mackerel fears them, and will run in shore when they appear."

30. [**Globiocephalus melas** Traill. (*D. intermedius* Harlan and *G. intermedius* Gray.)] "Blackfish. Common. This well-known species sometimes come into our bay in large schools in summer and autumn. They are then attacked by a number of boats from the shore, and often driven into shoal water or on shore and hundreds killed."

31. **Hyperaodon bidens** Owen. A specimen referred by Professor Cope to this species came ashore at North Dennis in January, 1869; its skeleton, secured by Mr. J. H. Blake, is now in the Museum

* Proc. Bost. Soc. Nat. Hist., Vol. XI, p. 318.

of Comparative Zoölogy. A few weeks later Professor Cope obtained another that was stranded near Newport, R. I.

32. [**Beluga canadensis** Erxl. White Whale.] At the close of his list Captain Atwood thus mentions a species identified by Professor Cope as above: "Besides those already named, some few years ago a species was killed in our harbor and brought on shore which no one knew. I examined it, and found it to differ from all other species. Not long after it was announced in the papers that there was a white whale on exhibition at the Aquarial Gardens in Boston, that Mr. Cutting had brought alive from the river St. Lawrence; a species that had never been seen south of that river. Soon after I visited Boston and called to see it. I pronounced it to be identical with the unknown species taken at Provincetown." This undoubted occurrence of the white whale at Provincetown is the only instance of its having been found so far south that has come to my knowledge. The skeleton of the specimen exhibited at the Boston Aquarial Gardens, and referred to above by Captain Atwood, is in the Museum of Comparative Zoology. Is was presented by Mr. Cutting.

33. [? **Lagenorhynchus** sp.] "Cow Fish. Occasional.

"This species differs from the blackfish in being much smaller, and in yielding much less oil. Its blubber is thinner, and its color is a light marble. It is sometimes called white blackfish by our whalers. It is occasionally killed here, but it does not appear in large schools, like the blackfish. It is a distinct species, intermediate in size between the blackfish and the species we call porpoise (dolphin)."

34. [**Delphinus erebennus** Cope. [" Porpoise. This is not an abundant species here. They are at times in summer seen passing along the shore in large schools, going northward; in autumn they may be seen going back to the southward."

35. **Delphinus clymene** Gray. According to Professor Cope this species has been taken on the coast of New Jersey,* and it is not unlikely to occasionally visit our shores.

36. [**Phocæna americana** Agass. (or *P. brachycium* Cope; I

* Proc. Phil. Acad. Nat. Sci., 1865, p. 261.

do not know which name will stand as yet.)] "SNUFFER or PUFFING PIG (*Phocæna americana*). This is the smallest of all the species. It is very common here at all seasons, and is occasionally caught in nets set for mackerel or blue-fish." There are several skeletons in the Museum of Comparative Zoölogy.

VESPERTILIONIDÆ.

37. **Lasiurus noveboracensis** GRAY. (*Vespertilio noveboracensis* Erxl.) RED BAT. NEW YORK BAT. Common; in some sections of the State the most numerous species of the family.

This species varies greatly in color, but the difference seems to be chiefly sexual. The adult males are generally much lighter than the females. In the young the sexual variation in color seems to be often much less marked.

The only well-marked distinguishing characteristics between this species and the next, except in more highly colored specimens of the latter, is generally the black border to the ear, and the black on the lips in *L. cinereus*. In each there are the same bands of color on the hairs, distributed in the same way, — dusky, verging to black at the base, then pale yellowish brown, succeeded by darker or brighter bands of red, and tipped with whitish. In some specimens the terminal band of whitish is quite absent, particularly on the anterior part of the body, the subterminal bright red zone being thus continuous to the tips of the hairs. In other specimens the terminal band of white is developed to a great degree, so as to very much obscure the red or dark chocolate zone beneath. Such specimens often strongly approximate to what is called *L. cinereus* (*V. pruinosus* Say), where the terminal white zone reaches its maximum of development, and the subterminal russet zone its greatest intensity. I feel, in fact, far from sure that the species are distinct. In a series of about twenty Massachusetts skins, nearly all marked for sex by the collector (Mr. C. J. Maynard), all the males are of a beautiful light, bright, yellowish red, with scarcely a trace of the apical white; the females, though somewhat more variable, are universally darker, the light red of the males being replaced in these by dark russet, which is more or less obscured by the whitish tips of the fur. The alcoholic series, so far as carefully examined in reference to this point, indicates this sexual difference to be quite constant; but there are occasional exceptions.

Very little seems to be known respecting the time of copulation or the period of gestation of the bats. From Mr. J. G. Shute, of Woburn, I learn a fact in reference to this point observed by him some few years since. Soon after sunset one evening in October he observed a strange object pass him in the air, which seemed to fall to the ground not far from where he was standing. Repairing immediately to the spot he soon found it, which proved to be a pair of these bats *in coitu.* They were captured and thrown into alcohol, and thus forwarded to the Museum of Comparative Zoölogy. About the 20th of June I once found, in Northern Illinois, a number of the *Scotophilus georgianus* containing quite advanced fœtuses, usually four or five in number. Dr. C. C. Abbott says that the *V. subulatus* brings forth its litter of three to five young late in June.*

38. **Lasiurus cinereus** H. Allen. (*Vespertilio pruinosus* Say.) Hoary Bat. Not common. Probably the rarest species of the family found in the State. Though commonly given in New England lists, I have never seen it from Massachusetts. I have been able to find but two specimens in the Museum collection referable to it, and those are, unfortunately, without localities. I have often seen in local collections specimens labelled with this name, but they were only the more hoary form of the common *L. noveboracensis.* From Dr. Allen's list of specimens its range seems to be nearly that of the preceding, — throughout temperate North America at least, — as some of them are stated to have been received from Nova Scotia, Red River Settlement, Louisiana, Matamoras, New Mexico, California, &c. As already observed, I question the validity of this species.

39. **Scotophilus fuscus** H. Allen. (*Vespertilio fuscus* Pal. de Bouv.; *V. carolinensis* Geoff. St. Hil.) Carolina Bat. Common.

I not only consider the suspicion of Dr. Allen that *S. carolinensis* and *S. fuscus* "may prove to be the same" well founded, but to his list of synonymes of this species would add *Eptisecus melanops* of Rafinesque. I would remove from it the *V. gryphus* of F. Cuvier, which I consider refers to the *V. subulatus* Say.

40. **Scotophilus georgianus** H. Allen. Less common than several of the other species, but apparently not excessively rare. There are several specimens in the Museum of Comparative Zoölogy

* Geology of New Jersey, Appendix, p. 752.

from Massachusetts, and others from Maine, the latter being at present its most northern known locality. This species is believed to be now for the first time reported from the Eastern States.

It appears to me that it would have been better to have entirely ignored the synonymes considered by Dr. Allen as doubtfully referring to this species than to have adopted any of them for its designation. The *V. georgianus* of F. Cuvier seems to me to be undoubtedly referable to *V. subulatus*. If any of F. Cuvier's names are to be considered as referring to it, it seems to me it is the *V. Saleri* of the same data, though it appears highly questionable whether this also, as well as the *V. monticola* of Bachman may not be more appropriately referred to *V. subulatus*, judging from the very imperfect descriptions alone. Dr. Allen, however, has had the types of some of these for examination, and finds them to correspond with what he calls *S. georgianus*, and it is this that appears to have guided him in determining these references.

41. **Scotophilus noctivagans** H. ALLEN. (*Vespertilio noctivagans* Le Conte.) SILVERY-HAIRED BAT. Rather common.

42. **Vespertilio subulatus** SAY. LITTLE BROWN BAT. Common, especially in the Connecticut Valley. At Springfield it is *one* of the most common, if not *the* most common species.

Prior to the publication of Dr. Allen's monograph, but one species of the genus *Vespertilio*, as now restricted, had been recognized from Massachusetts, though others, based however on very doubtful characters, had been given by different authors from the Middle States. All who have critically studied the bats are well aware that they are quite variable in color and in many other characters. Thus Dr. Allen, under *Scotophilus fuscus*,* in alluding to certain variations in the form of the ear pointed out by Major Le Conte as distinguishing certain species of European authors, which Dr. Allen very properly deems to be merely nominal, observes: "While acknowledging that these differences may exist, I do not consider them constant. In a species so extensively distributed, and in a family so well known for its Protean tendencies as that to which *S. fuscus* belongs, slight and variable changes, confined entirely to the parts of the ear, are hardly sufficient data for these separations." Under *Vespertilio*† he remarks: "Owing to the fact that species of this genus have a widely spread distribution, minute differences in form and color in specimens brought from distant localities

* Monograph, p. 33.

† Ibid., p. 46.

have been made of more importance than they deserve. Species have thus sprung up, many of which have never been identified, and seem only to retard progress by a useless synonomy." We fear, however, that Dr. Allen, with all his care, and the almost unexceptionable character of his admirable Monograph, has fallen in this group into an error which he found it necessary to criticise in others. With original specimens of most of his species for examination, I am unable to convince myself, either from these or from his descriptions, that several of the species recognized or described as new by him — especially *V. lucifugus* and *V. evotis*, and also *V. affinis* — are not really referable to *V. subulatus*. Among the large lot of bats furnished by the Museum of Comparative Zoölogy for use in the preparation of his Monograph, including some two hundred specimens from different parts of North America (besides many from foreign countries), specimens of *Vespertilio* from various localities in Maine and Massachusetts were labelled by him, when returned, respectively *V. evotis*, *V. subulatus*, and *V. lucifugus*. Individuals of the same colony, and that I scarcely doubt in some cases belonged to the same litter, of what I call *V. subulatus*, vary considerably in color, and not a little in the form of the ear. Dr. Allen says: "The specimens of *V. subulatus* arrange themselves into two groups, one of which may be considered typical, the other tending in the shape of the ear to the *preceding* species [*V. evotis*]. Indeed, the changes from one species to the other is so gradual that it is difficult to assign a boundary to each. I have included under *V. subulatus* a number of specimens which have the ear higher than those from which the description has been taken, but agreeing with *V. subulatus* in other particulars." *

From a critical analysis and comparison of the tables of measurements given by him of the different species of this genus, they appear most decidedly to intergrade, no less in the size and form of the ear — the character on which their separation is mainly based — than in other points. The *V. lucifugus* has, perhaps, the best claims to be regarded as a species, but these seem to be highly equivocal. *V. evotis* is the form with the highest, and relatively the largest ear, grading in this particular into *V. subulatus*, the more common form, and this again into *V. affinis* (of which but one specimen had been received) and *V. lucifugus*, in which the ear exhibits the minimum of size. In the latter the snout is blunter, and in the first more produced, this character correlating with the narrowed and elongated or shortened and blunted ear. In other words, the *V. evotis* is the slender form, the *V. lucifugus* the robust form, *V. subulatus* coming in between the two.† They all appear to have the same geograph-

* Monograph, p. 51.

† Naturalists seem to overlook the fact that feral animals may vary in size, in general form, in physiognomy, in temperament and disposition, in the same way as different

ical distribution, and specimens of each generally occur in collections from the same localities, whenever the number of specimens received is at all large. They are sometimes found in cool weather clinging together in the same "festoons."

Each species ranges, according to Dr. Allen, from ocean to ocean, and from very far north nearly or quite to the tropics.

Prior to 1864 only five species of bats were currently reported from New England; Dr. Allen's Monograph nearly doubled the number, increasing it to nine. Only six, however, are recognized in the present catalogue, one only (*Scotophilus georgianus*) having been added to those previously well known.

In respect to the many species of bats imperfectly described by some of the earlier authors, I have little hesitancy in referring to *V. subulatus* of Say the following: —

V. lucifugus Le Conte, Cuv. An. King. (McMurtrie's ed.), 1831, p. 431.
V. Caroli Zimm., Man. de Mam., II, 1835, p. 236.
V. gryphus F. Cuv., Nouv. Ann. du Mus. d'Hist. Nat., I, 1832, p. 15.
V. Salari Ibid., p. 16.
V. crassus Ibid., p. 18.
V. georgianus Ibid., p. 16.
V. subflavus Ibid., p. 17.
V. brevirostris Pr. Maximilian, Verzeich. Beobach. Säugethiere in Nord Amer., p. 19.
V. monticola Aud. and Bach., Journ. Phil. Acad. Nat. Sc., Vol. VIII, 1842, p. 280.
V. virginianus Ibid., p. 282.
V. californicus Ibid., p. 285.
V. Leibii Ibid., p. 284.

SORICIDÆ.

43. **Neosorex palustris** Verrill.* (*Sorex palustris* Rich.;

individuals of any given nationality of men or breed of domesticated animals, in which such variations are patent to the most casual observer. In wild animals it needs only a critical comparison of many individuals of any species, concerning the identity of which there is no question, to satisfy careful investigators that it is equally the case here. It fails to be as well recognized only because it is impossible for us to be in sufficiently intimate relation with animals in a state of nature. In many instances where they are brought under the same conditions relatively for observation, as in the case of different species of *Cervidæ*, when kept in parks, it is soon detected. In this connection compare the observations of Judge Caton on "American Cervidæ" (see *antea*, p. 194).

* Notice of a *Neosorex* from Massachusetts, and of *Sorex Thompsoni* from Maine. By A. E. Verrill, Proc. Bost. Soc. Nat. Hist., Vol. IX (Oct. 1862), p. 164.

Neosorex albibarbis Cope.) MARSH SHREW. But three specimens of this species are as yet known from New England, two of which were captured by Professor E. D. Cope, at Franconia, N. H., and the other by Mr. F. W. Putnam, at Warwick, Mass. Professor Cope's specimens were swimming in a lake when first seen, about forty feet from the bank. As observed by Professor Verrill, the species of this genus are eminently adapted to an aquatic mode of life, they having large fringed feet and valvular ears.

44. **Sorex platyrhinus** LINSLEY. BROAD-NOSED SHREW. Comparatively common. I have taken a considerable number at Springfield, and Professor S. F. Baird, in his Mammals of North America (p. 26), cites nineteen examples in his list of specimens of this species from Massachusetts, eighteen of which were from Middleboro', and collected by Mr. J. W. P. Jenks.

45. **Sorex Cooperi** BACHMAN. COOPER'S SHREW. This rare species I have never seen myself from this State; Professor Baird mentions two specimens from Middleboro', received from Mr. Jenks. Professor Verrill, in his paper already cited, refers to a specimen from Danvers, in the collection of the Essex Institute, as being the only one he had seen from New England. Last winter I received it from Wayne Co., N. Y., from my friend, Mr. Charles Potwine. The specimen was captured in the daytime, while running on the snow in the woods.

46. **Sorex Forsteri** RICH. FORSTER'S SHREW. From its known range* this species is most likely to occur in Massachusetts. It has, in fact, been reported as often met with here, both in summer and in winter.†

Thompson's shrew (*Sorex Thompsoni* Baird) is also to be expected to occur in this State, it having been received by Professor Baird from Halifax, N. S., and Zanesville, Ohio, and by Professor Verrill from Maine.

* "Hudson's Bay to Carlisle, Pa." — BAIRD.

† "In the latter season they are found beneath a pile of wood or logs, and their tracks in the snow show their wanderings in search of food." — E. A. SAMUELS, *Agriculture of Mass.*, 1861, p. 142.

47. **Blarina brevicauda.** (*Sorex brevicaudus* Say, Emmons's Rep., p. 13; *Blarina talpoides* Gray). MOLE SHREW. Common. By far the most numerous species of the family.

A second species of *Blarina*, the *B. brevicauda* of Gray (*Sorex brevicaudus* Say) was formerly reported to exist in this State, Connecticut, New York, and throughout Eastern North America generally. But Professor Baird supposes it, if distinct from *B. talpoides*, to be exclusively Western; he has, however, failed to point out any differences of much weight between specimens he refers respectively to *S. brevicaudus* Say and *S. talpoides* Gapper (*B. talpoides* Gray). In his diagnosis of *B. brevicauda* he says: "Largest of all American shrews hitherto discovered (?)," and gives its dimensions as "Length, unstretched, over four inches to the root of the tail"; while he gives the "average length of head and body" of *B. talpoides* as "three and a half inches." Say gives the length of the head and body of *S. brevicaudus* as three inches and five eighths, or 3.62, which but slightly exceeds Professor Baird's average for *B. talpoides*; the two largest specimens of which he gives measurements (No. 2,078, from Massachusetts, and No. 2,146, from Illinois) slightly exceed this size. A Massachusetts specimen before me measures fully four inches, and two others exceed 3.75. Under *B. talpoides* he says, "With a large number of specimens before me, I have been more than usually perplexed in the attempt to determine the species of short-tailed shrews, as given by authors, and especially to distinguish between *S. brevicaudus* and *S. Dekayi*, of Bachman, De Kay, and others. I am satisfied that the latter species is identical with *S. talpoides* of Gapper (which indeed has priority of date), having found no essential differences between Canadian specimens and those from Massachusetts, Vermont, New York, Michigan, Wisconsin, and elsewhere. Gapper's specimen, it will be remembered, was taken in the district between York and Lake Simcoe, in Upper Canada.

"Thus far," he continues, "I have not been able to find any shrews from Massachusetts, New York, or adjoining States, possessing all the characters assigned by Bachman and De Kay. The hair of the same species varies with the season, being longer, softer, and fuller in winter; the precise shade of color is likewise not constant. The proportions of the shrews, unless taken from alcoholic or fresh specimens, vary exceedingly in the same species, according as the skin is under or over stuffed.

"For the present, therefore, I shall refer all the large shrews with short tails from the Atlantic States to the *S. talpoides*. I have, however, before me some specimens from the Upper Missouri and Iowa, which, as they differ in size from any in the East, and agree rather more closely with the *S. brevicaudus* of Say, I shall refer to this species."*

* North American Mammals, p. 41.

Under *B. brevicaudus* Professor Baird further observes: "I have found very great difficulty in identifying with any certainty the *S. brevicaudus* of Say, at least in the references to this species, as supposed to be found in the eastern portion of the continent. I have, however, I think, discovered it in some specimens of very large size from Nebraska and Iowa, localities nearer to that of the original specimen (Council Bluffs) than of any specimen yet discovered." In his list of the specimens referred to this species Professor Baird gives two from Nebraska, two from Iowa, and one each from Illinois and Wisconsin. The latter four are, however, referred with a mark of doubt. It is to be regretted that full measurements of all these specimens are not given for comparison with the excellent series of *B.* "*talpoides*"; * as the size of two out of the three given is equalled by several of the *B. talpoides*, they being respectively but 3.50 and 3.65 inches in length. In view of the generally admitted variability of this species in size, color, length of tail, &c., at single localities, and which some seventy specimens now before me from Massachusetts fully demonstrate, and the but slightly larger size of Mr. Say's single example from Council Bluffs (which forms the original of *S. brevicaudus*) than the average of our short-tailed shrews, I refer to one species, and to this of Say, all the short-tailed shrews of the Northern and the Eastern States, Canada and the adjoining Provinces, of which the more recent name (*S. talpoides*) of Gapper becomes a synonyme. Also, in view of the already known wide distribution of this species, and the law of variation in size with respect to latitude and elevation, I must also consider the *S. carolinensis* of Bachman, which only differs from the northern specimens of *S. brevicaudus* (*B. talpoides* Gray, Baird's Report) in its slightly smaller size, as merely the more southern and hence the smaller race. Indeed, in consequence of the large size allowed it by Dr. Bachman, Professor Baird is inclined to consider this name as a synonyme of *B. talpoides*, as under this species he states: "Nor do I feel quite sure that the *Sorex carolinensis* of Bachman is really anything else than a small *S. talpoides*. The measurements given by him (length three inches) agree

* There has never been a more valuable contribution to the Natural History of the Mammals and Birds of North America, or of any country, than the lists of specimens and tables of measurements published by Professor Baird in his great and invaluable works on these two classes of the North American Vertebrata, contained in Volumes VIII and IX of the Reports of the Pacific Railroad Explorations and Surveys. They show not only, to a considerable extent, the geographical range of the different species, but their variation in size and proportion at different localities, and, when the number is large from one locality, the variation at single localities. The possession of these tables and his accompanying minute descriptions is next to having in hand the specimens themselves. It is very much to be regretted that so small a proportion of our natural history descriptions have been written with this great care and minuteness of detail.

precisely with many from Massachusetts and elsewhere, and are essentially the same in proportion with those of the largest-sized specimens of *S. talpoides*." But he adds: "There is, however, a distinct species in the Southern States, considerably smaller than *S. talpoides*, to which Bachman's name may be applied." Further on he gives a diagnosis of a "*B. carolinensis*," under which he cites Bachman's "*S. carolinensis*" as a synonyme. He describes it as "size considerably less than adults of *B. talpoides*," and gives the length of head and body as "about 2.50 inches." Comparing it with *B. brevicauda*, he says it differs from that species in its considerably smaller size, proportionally smaller feet, and in having the "third and fourth lateral teeth larger in proportion to the first and second," &c. Under this head he cites four specimens, three of which are from Missouri, and the other from South Carolina. These, he says, "agree in the main very well together, and as indicating a southern species smaller than *B. talpoides* or *brevicauda*." After finally referring *S. carolinensis* of Bachman to this species, he says: "I am by no means clear, however, that the particular measurements cited by him do not belong really to a specimen of *B. talpoides*; but," he strangely adds, Dr. Bachman having given us no such intimation, "he [Dr. Bachman] undoubtedly was acquainted with a species smaller than the latter" (*S. carolinensis* Bachman). That there is a somewhat smaller race in the South is unquestionable, but its specific rank is not to me so clear. This smaller form seems to occur generally throughout the Southern States, and along the low coast border as far north as New Jersey, and even perhaps to New York, corresponding in the limits of its distribution northward with the northern boundary of the Carolinian Fauna; the larger form occupying the Northern States generally, and the highlands of the Alleghanies south to Georgia; it thus occurring throughout the whole extent of the Alleghanian Fauna, and possibly throughout the Canadian. The range of *B. brevicauda* is now carried southwards to Florida and Texas, with only such differences in size between northern and southern specimens as are admitted to occur in other unquestioned species of mammals that have the same geographical range; the difference in size being the only constant or tangible distinction yet pointed out. The difficulty experienced by Professor Baird in determining the species of the older authors, it seems to me results chiefly from two causes: first, the imperfect character of the descriptions, which are generally of single specimens only, and of skins and stuffed examples; second, the by far too great number indicated.

In this connection it is proper to notice a species of *Blarina* described as new in the Report on North American Mammals (p. 47) from a single specimen from Burlington, Vermont. This specimen, its describer says, "in external appearance perfectly resembles specimens of *B. talpoides*," but "has some remarkable peculiarities of the skull. While it has no satisfactory external characters by which to designate it," "the skull is so entirely different from all others" he had seen, he says, as "almost to make a distinct subgenus." This difference consists in its being much narrower than in other short-tailed shrews, and in the greatest interorbital constriction being placed a little in front of the middle, instead of behind it, as in the others, and in its being greater in amount. In regard to this specimen, I need only add that, in respect to its skull, and in this character alone,* whether really a distinct species or an abnormal individual variation, it still remains unique, no other like it having yet become known to naturalists.

In continuing this preliminary revision of the *Blarinæ*, we find that ten species of this strictly American genus † of the short-tailed shrews have been described, all from the United States, three of which were first characterized by Professor Baird in his North American Mammals. Seven are recognized in this work as valid; two are given as doubtful or unidentified, and one is doubtfully referred to one of the others. These are arranged in two sections, according to the number of premolars; section "A" having five, and section "B" four. Their dental formulæ are as follows: —

$$\text{Section A, } \frac{2}{2}+\frac{5-5}{2-2}+\frac{4-4}{3-3}=32; \text{ section B, } \frac{2}{2}+\frac{4-4}{2-2}+\frac{4-4}{3-3}=30.$$

A lengthy diagnosis is given of each section, but no other essential differences are pointed out, the distinctions in respect to color, &c., being, as is evident from the descriptions of the species that follow, inconstant and invalid. In section B the first premolar is said to be slightly larger than the second, and in section A to be smaller than the second. But in the de-

* That is, judging from Professor Baird's description; but from the figures of its skull (Pl. XXX, Fig. 7), it seems to have had an imperfect or abnormal dentition, the number of visible premolars being three instead of four, in the upper jaw, and one instead of two in the lower, with a naked space between them and the incisors. It is possible, however, that the first premolar in each jaw had become accidentally lost before the skull passed into the hands of the artist.

† *Sorex brevicaudus* SAY, Long's Exped., I, 1823, 164.
" *parvus* SAY, Ibid., 163.
" *talpoides* GAPPER, Zool. Journ., V, 1830, 208, Pl. VIII.
" *carolinensis* BACHMAN, Journ. Phil. Acad. Nat. Sc., VII, 1837, 366, Pl. XXIII, Fig. 1.
" *cinereus* Ibid., 373, Pl. XXIII, Fig. 3.
" *Dekayi* Ibid., 377, Pl. XXIII, Fig. 4.
" (*Brachysorex*) *Harlani*, DUVERNOY, Mag. de Zool., 1842, 40, Pl. III, Fig. 6.
Blarina angusticeps BAIRD, N. Am. Mam., 1857, 47.
" *exilipes* Ibid., 51.
" *Berlandieri* Ibid., 53.

scriptions of *B. cinerea*, *B. Berlandieri*, and *B. exilipes*, which constitute section B, it is distinctly stated that the first premolar is *smaller* than the second. Figures of the skulls of all the species of both sections are given in Pls. XXVIII and XXX, but in no case does the first premolar *appear* to be quite equal to the second. In regard to section B, there are several circumstances suggestive of its being founded on immature examples of section A, in which the dentition is incomplete.* All the species are diminutive, and vary but little in size; the teeth are generally proportionally large compared with the size of the skull, as is always the case in young animals, and other characters seem to indicate immaturity. The missing premolar is the one we should expect the animal to acquire latest.† All the species of section B come from within the admitted geographical range of the species of section A, one only (*B. Berlandieri*) possibly excepted. Unfortunately, very young specimens of shrews are extremely rare in collections, and in the large series of *Blarinæ* in the Museum of Comparative Zoölogy there are none so small as those embraced under Baird's section B. In several of the smallest of them the fifth premolar is scarcely visible, forming a minute uncolored point on the inside of the jaw. In a single specimen from Middleboro', the smallest of the lot, it is wholly wanting. I regret that I have been unable to examine any of the original types of the species of section B. Between the three supposed species of this section (*B. cinerea*, *B. exilipes*, *B. Berlandieri*) the differences (which seem to consist chiefly in color, especially between the first two) are not greater nor different from those seen in a large series of specimens from Massachusetts or other localities. The differences between the different specimens referred to either of the species are also very appreciable, and in some cases (see under *cinerea* and *exilipes* in North American Mammals) so great that their assignment was very doubtfully made. While the evidence of the existence of so many species of *Blarina* in the Eastern United States, if really of more than one, is evidently very slight, I do not claim to have fully shown that but the one exists; my design has been mainly to call attention to the great need of a thorough revision of this

* It is well known that in *Scalops aquaticus* the number of teeth in the young is less than in the adult, and this difference has resulted in discrepant statements in respect to its dentition. (See BACHMAN on the Mole Shrews of North America, in Proc. Bost. Soc. Nat. Hist. I, 40. Also, Quad. N. Amer., Vol. I, p. 92.)

† The species of *Sorex* are divided into two sections on similar characters, where small size again accompanies the lesser number of teeth. There are other circumstances that render it not improbable that we have here again a section "B," based on immature representatives of a section "A." The number of species of *Sorex* admitted for the United States, twelve or more, is probably quite too large, though undoubtedly there may be half that number.

group. What I do claim is, that there is as yet no good evidence of the existence of more than the common and widely dispersed *B. brevicauda;* that the numerous other supposed species that have been described are mainly based, in the first section, on variations in size dependent upon locality, and that there are strong indications that those of the second section rest on variations, dependent upon immaturity, of the representatives of the first; that if other species do exist, as is not of course improbable, naturalists have thus far failed to satisfactorily establish the fact. In number of species, *Blarina* thus corresponds with *Condylura*, and in distribution with *Scalops aquaticus.*

In the following comparative analysis of the diagnoses of sections A and B of *Blarina*, given in the Report on North American Mammals, some points but casually alluded to above are more fully discussed. A table of synonymes is also added.

Genus **Blarina** Gray.

List of the Species.

Section A.	Section B.
B. talpoides.	B. cinerea.
B. brevicauda.	B. exilipes.
B. carolinensis.	B. Berlandieri.
B. angusticeps.	

Diagnoses.

Color.

"Nearly uniform plumbeous on the body and tail; scarcely lighter beneath."

Exceptions. — Specimens of *B. talpoides* are mentioned as "slightly paler beneath," "fading to the belly into a still paler tint," &c.; of *B. carolinensis* as being "a little paler beneath." Massachusetts specimens of *Blarina* are generally *nearly* uniform, but many specimens occur that are considerably lighter beneath. The general color also varies from ashy and brownish through grayish plumbeous to exceedingly dark, almost black. Occasionally the hairs are so varied with light and dark as to present a hoary appearance.

"Lower parts of the body usually lighter than the upper, with *the line of demarcation* distinctly visible."

Exceptions. — *B. cinerea:* Hoary above, "somewhat resembling pepper and salt"; below, "a lighter tint of brownish gray or light ash; the *line of demarcation* in one specimen indistinct, in another more evident." *B. Berlandieri:* "In one [specimen] the prevailing tint is a chestnut brown at the tips of the hairs, with paler next to the tips, producing a slight hoariness. The under parts are a yellowish-brownish white; *the line of demarcation on the sides quite indistinct.*"

Dental Formulæ.

$$\frac{2}{2}+\frac{5-5}{2-2}+\frac{4-4}{3-3}=32 \qquad \frac{2}{2}+\frac{4-4}{2-2}+\frac{4-4}{3-3}=30$$

Incisors.

(1)* "The upper anterior incisor with the basal portion of the cutting edge formed by a nearly rectangular lobe, (2) the entire tooth forming only a single hook."

"Lower anterior incisor (1) *stout,* (2) *much curved,* (3) *with two or three lobed dentations.*" (4) "It extends back as far as the middle of the first molar." (5) "*The first and second premolars are placed above this incisor.*"

The variation presented by different specimens renders null distinctions 1 and 2, the lobe being sometimes much produced posteriorly.

(1)* "Anterior upper incisor with the basal lobe more conical and further forward than in the other section."

"Lower anterior incisor (3) *with two or three lobed serrations,* (1) *stout,* (2) *much curved,* (4) not reaching posteriorly as far as the middle of the first molar; (5) *the two first lateral teeth entirely above it.*"

On page 9, the teeth in section A are described as "nearly uncolored," — that is, brown to the base, and in section B as "bicolored," — white at the base and tipped with brown. But in *B. brevicauda,* the second type of coloration is also quite frequent.

Upper Premolars.

(1) "The first two premolars are nearly equal, (2) the second *usually* a little larger; (3) the next two much smaller; (4) the fifth very small and usually not visible externally. (5) The first four with a basal-colored point on the inner side."

(1) "The first premolar tooth slightly larger than the second. (2) *The third decidedly smaller than either,* though larger than in the other group. (5) The small cusps on the inner side of the base of the first three lateral teeth, either wanting or very small."

Exceptions. — *B. cinerea:* "The first premolar tooth is a little *smaller* than the second."

B. exilipes: "The first lateral tooth is *rather smaller* than the second," &c.

B. Berlandieri: The first lateral tooth is "rather shorter than the second." See also the figures, which so represent them. Hence this main distinction of "first premolar tooth slightly larger than the second" by no means holds.

* The numbers prefixed to the characters in the diagnoses refer to the same character in each section. Those that seem to be nearly or quite synonymous in the two sections are italicized.

Hands.

"Hand contained about two and a third times in the hind feet."

In forty-seven specimens of *B. talpoides* the proportion is 74 to 100; in three specimens of *B. brevicauda* the proportion is 72 to 100; in three of *B. carolinensis* it is also 72 to 100. The range of variation, however, in *B. talpoides* (see Baird's table) is from .55 (specimens No. 2,076, 2,080, &c.) to .80 (specimen No. 2,083).

"Feet smaller than in section **A**; the anterior contained about one and a half times in the posterior."

In four specimens of *B. cinerea* the proportion is 75 to 100; in six specimens of *B. exilipes* 68 to 100; in four of *B. Berlandieri* 66 to 100.

Before closing my remarks on this subject I should call attention to the fact of the repetition of the same character, described in slightly different language, that so constantly occurs in diagnoses of the different species of the same genus, of different genera of the same sub-family, &c., and even of characters of ordinal value in specific descriptions, in the writings of even some of the best naturalists; — to the mixing up of non-essential or irrelevant characters with, and thus obscuring, those peculiar to the group in question. Sometimes, in fact, the really essential points are omitted, the diagnosis being almost as equally applicable to several species, or to any of quite a large group, as to one. All naturalists are not, of course, equally culpable in this respect. But in general, by sifting descriptions of their generalities, they could be greatly reduced and their definiteness and accuracy proportionally increased. The labor of preparing diagnoses would of course be thus increased, but the advantages arising therefrom would be immense. I am not the first, I am happy to find, to make strictures of this character, and hope that the matter will soon receive at the hands of descriptive naturalists the consideration it merits. Neither, I should say, are these strictures introduced at this time as a special criticism upon any particular author.

Blarina brevicauda.

Sorex brevicaudus SAY, Long's Exped., I, 1823, 164.
" " HARLAN, Faun. Amer., 1825, 29.
" " GODMAN, Am. Nat. Hist., I, 1831, 79. (From Say.)
" " BACHMAN, Journ. Phil. Acad. Nat. Science, VII, 1837, 381.
" " EMMONS, Quad. Mass., 1840, 13.
" " DE KAY, N. York Fauna, I, 1842, 18.

Sorex brevicaudus LINSLEY, Am. Journ. Sc., XLIII, 1842, 346.
" " THOMPSON, Hist. Vermont, 1842, 27.
" " PLUMBER, Am. Journ. Sc., XLVI, 277.
Blarina brevicauda BAIRD, Mam. N. Am., 1837, 42, Pl. XXX, Fig. 5.
" " SAMUELS, Agr. Mass., 1861, 144.
Sorex talpoides GAPPER, Zoöl. Journ., V, 1830, 208, Pl. VIII.
Corsira (Blarina) talpoides GRAY, Proc. Lond. Zool. Sc., V, 1837, 124.
Blarina talpoides BAIRD, Mam. N. Am., 37, Pl. XXX, Fig. 6.
" " SAMUELS, Agr. Mass., 1861, 145.
" " VERRILL, Proc. Bost. Soc. Nat. Hist., IX, 1863, 172.
Sorex parvus SAY, Long's Exped., I, 164.
" " HARLAN, Faun. Am., 29.
" " BACHMAN, Journ. Phil. Ac. N. Sc., VII, 394. (From Say.)
" " DE KAY, N. Y. Fauna, I, 19.
" " LINSLEY, Am. Journ. Sc., XLIII, 346.
" " AUD. & BACH., Quad. N. Am., II, 1851, 145, Pl. LXX.
" *Dekayi* BACHMAN, Journ. Phil. Acad. Nat. Sc., VII, 377, Pl. XXIII, Fig. 4.
" " DE KAY, N. Y. Fauna, I, 17, Pl. V, Fig. 2.
" " LINSLEY, Am. Journ. Sc., XXXIX, 388, Ib. XLIII, 346.
" " AUD. & BACH., Quad. N. Am., III, 1853, 246, Pl. CL, Fig. 2.
" *cinereus** BACHMAN, Journ. Phil. Acad. Nat. Sc., VII, 373, Pl. XXIII, Fig. 3.
Blarina carolinensis BAIRD, Mam. N. Amer., 45, Pl. XXX, Fig. 8, skull.
" *angusticeps* Ib., 47, Pl. XXX, Fig. 7, skull.
" *cinerea* Ib., 48, Pl. XXX, Figs. 9 & 10, skulls. (Young.)
" *exilipes* Ib., 51, Pl. XXVIII. (Young.)
" *Berlandieri* Ib., 53, Pl. XXVIII. (Young.)

TALPIDÆ.

48. **Scalops aquaticus** FISCHER. (*Scalops canadensis* Emmons, Rep., p. 15.) COMMON MOLE. Common.

49. **Scalops Breweri** BACH. HAIRY-TAILED MOLE. Apparently rare in Massachusetts, and not numerous anywhere. The original specimen described by Dr. Bachman came from Martha's Vineyard, and was collected by Dr. L. M. Yale, and presented by Dr. T. M. Brewer

* Afterwards considered by Dr. Bachman to be the young of *S. carolinensis*. See Quad. N. Am., III, p. 344. Same as *B. cinerea* Baird.

to Dr. Bachman. There is a specimen in the Museum of Comparative Zoölogy from Warwick, and others from Upton, Maine, and Halidaysburg, Pennsylvania.

50. **Condylura cristata** DESMOREST. (*C. longicauda* Desm. and *C. macroura* Harlan of Emmons's Rep., pp. 17, 18.) STAR-NOSED MOLE. Common, but apparently more so in some parts of the State than in others. At Springfield this and *Scalops aquaticus* are about equally numerous, but in the eastern part of the State the present species seems to many times outnumber the other. From considerable variations in the length and size of the tail presented by different individuals, it was formerly incorrectly supposed that two species of *Condylura* existed in Massachusetts, and the eastern parts of the United States generally. The thickening of the tail appears to be connected with the rutting season.

SCIURIDÆ.

51. **Sciurus cinereus** LINN. (? "*S. vulpinus* Gmel.," Emmons's Rep., p. 66.) FOX SQUIRREL. Rare in most parts of the State.

52. **Sciurus carolinensis** GMELIN. ("*S. leucotis* Gapper" and "*S. niger* Linn.," Emmons's Rep., pp. 66, 67. *Macroxus* * *carolinensis* Gray.) GRAY SQUIRREL. Generally distributed, but much more common in some sections than in others, being most numerous where the forests have been least disturbed. Generally they are of the gray type, but the black variety is quite prevalent at some localities. In Wayne County, New York (on the south shore of Lake Ontario), I have found the black variety to be the most common, with every gradation between the two. All those observed that were pure glossy black seemed to be very old individuals, while the young generally presented a mixture of tawny, gray, and black, the hairs being annulated

* Dr. J. E. Gray, in his several Synopses of the Asiatic, African, and American Squirrels (Ann. and Mag. Nat. Hist., 3d Ser. Vol. XX, 1867), has recently divided the old genus *Sciurus* into several genera. *Sciurus*, as restricted by him, and *Macroxus* contain all the American species, by far the larger part of which are placed in *Macroxus*. Only the group to which *S. hudsonius* belongs, the *S. cinereus* or Northern fox squirrel, and Abert's squirrel from New Mexico (called by Gray *S.* "*Albertii*" = *S. Abertii* Woodhouse), remain in the genus *Sciurus* as restricted by Dr. Gray.

with these colors, varying in the proportion of each in almost every individual. The intensity of the black appears to increase with age.

Dr. Emmons's *S. vulpinus* seems to refer to large examples of this species rather than to the true fox squirrel (*S. cinereus* Linn.).

53. **Sciurus hudsonius** PALLAS. RED SQUIRREL. CHICKAREE. Abundant.

The variations in color, in the hairiness of the soles, the presence or absence of ear-tufts, according to the season of the year, in this and other species, have already been pointed out by Professor Baird.* The lateral dusky stripe is perhaps the most variable feature in the present animal, in many specimens it being quite absent, and in the greater portion but faintly indicated, but it is not unfrequently one of the most conspicuous features of coloration. In fall specimens, particularly around Springfield, the black lateral line is generally conspicuous, being a well-defined, quite broad black band. Specimens from Northern Maine † differ from the majority of Massachusetts specimens in possessing a relatively very much shorter tail, somewhat in general color, the back being "rusty-yellow" rather than ferruginous, and in the greater fulness and softness of the fur. The black at the end of the tail is much broader and more conspicuous. In several points these specimens thus approach *S. Richardsonii*. Specimens entirely black have been received from Mr. G. A. Boardman from near Calais, Maine. In view of the wide range of variation presented by *S. hudsonius*, the descriptions of some of its near allies, especially of *S. Fremontii* and *S. Richardsonii* of Townsend and Bachman, seem scarcely to indicate more than slight local variations of one species. The specimens of the latter thus far examined have been too few to establish any very important differences between them and *S. hudsonius*, if such exist.

Professor Baird in his admirable article on the *Sciurinæ*, or typical squirrels of the United States, was able, through the very abundant material at his disposal, to eliminate a very large proportion of the invalid species that had from time to time crept into the works of preceding authors, including many described by Bachman and other Americans as well as by foreign naturalists. The variations pointed out by him as being dependent upon season and locality are important discoveries, since such variations are also of common occurrence among other groups. Two or three species only, besides those above specified, of the twelve species of *Sciurus* admitted in the work of this author seem at all questionable. These

* N. Amer. Mam., pp. 244 and 270.

† In the Mus. Comp. Zoöl., and C. J. Maynard's collection.

are the *S. castanonotus* and *S. limitis* from the little known region of Northern Mexico and the adjoining Territories northward, whose somewhat doubtful character is particularly mentioned.

Dr. Gray, in his "Synopsis of American Squirrels," * quotes Professor Baird's remarks respecting the wide variation in color presented by individuals of the same litter, the geographical variation in size, the variations in the hairiness of the soles of the feet at different seasons and between northern and southern representatives of the same species at the same season, and also in respect to the absence or presence of the ear-tufts in different individuals of the same species from the same locality; and so far as he has followed Baird's memoir his paper is to be commended. As soon, however, as extralimital species are encountered he seems to have lost sight of all these important facts quoted by him, and takes every considerable variation in color as the basis of a species. Hence the greater part of those described by previous authors receive his approval, and some *ten or twelve*, apparently, are added as new! The whole number of American *Sciuri* is thus increased to thirty-nine species. That some of the Mexican species are as variable as those of the United States is beyond question, while it is probable that some of the still more southern ones also are. According to Dr. Gray, the number of species of Asiatic *Sciuri* is forty-nine, an improbably large number, from which the excess can only be properly eliminated by a careful observer residing where these animals live, and the elaboration of a mass of material far greater than has thus far been brought together.

54. **Pteromys volucella** Desm. Flying Squirrel. Common, but, from its nocturnal habits, not often seen.

Apparently equally mature individuals from the same locality are quite variable in size, and somewhat in other characters. One, remarkably large, collected by Mr. S. Jillson at Hudson (Mass.), corresponds very well with the *P. hudsonius* Fischer (*P. sabrinus* Rich.), which supposed species is almost unquestionably but the large northern race of *P. volucella*.

Richardson described, in the "Fauna Boreali-Americana," † a variety of his *P. sabrinus* from the Rocky Mountains, to which he gave the name *alpinus* (*P. sab.*, var. *alpinus*). Wagner, in his Supplement to Schroeber's Säugethiere, ‡ and Audubon and Bachman in their North American Quadrupeds, § afterwards raised it to the rank of a species, but apparently with insufficient reason. Professor Baird also admits *P. alpinus* as a species in

* Ann. and Mag. Nat. Hist., 1867, p. 415.
† Vol. I, p. 195, pl. 18.
‡ Vol. III. p. 230.
§ Vol. III. p. 206.

his Mammals of North America (p. 289), but remarks that, from insufficient data, he was unable to arrive at a definite conclusion as to whether it was really distinct from *P. hudsonius*. The *P. oregonensis* of Bachman seems also very doubtfully distinct from *P. volucella*, as it does not differ very appreciably from the Eastern animal. The following remarks from Audubon and Bachman's North American Quadrupeds* in respect to the number of species of North American *Pteromys* are very suggestive. "As long," they observe, "as only two species of flying squirrel were known in North America, — the present species (*P. sabrinus*) and the little *P. volucella*, — there was no difficulty in deciding on the species, but since others have been described in the far West, the task of separating and defining them has become very perplexing."

Specimens in the Museum of Comparative Zoölogy from Lake Superior, Northern Maine, New Hampshire, Massachusetts, and the Middle States, form a graduated series in size, the first-mentioned, or northern, correponding with the *P.* "*sabrinus*"; the southern, of course, with the true *P. volucella* of authors. Difference in size has been the only appreciable character that has been advanced as distinguishing them.

55. **Tamias striatus** BAIRD. (*T. americanus* Kuhl. *Sciurus striatus* Klein, Emmons's Rep., p. 68.) STRIPED SQUIRREL. CHIPMUNK. Abundant. Usually first seen abroad in spring towards the close of March, when they are readily detected by their loud clucking note.

A series of nearly fifty specimens in the Museum of Comparative Zoology, from various localities in Eastern Massachusetts, are extremely uniform in color, the variations being so slight as to be scarcely appreciable. A considerable number of others, from different localities in Maine, are generally very much lighter or paler colored. These, also, vary a good deal among themselves, chiefly, however, in the character of the stripes, which in several specimens are much less distinct than usual. In one they are quite faint and irregular, the light central one on the sides being alone well defined, and this is at one point interrupted. The difference in general tint between these Massachusetts and Maine specimens is quite marked in the rufous-colored regions of the animal, and especially on the posterior part of the back.

56. **Arctomys monax** GMELIN. WOODCHUCK. Abundant. At Springfield a number of specimens of the black variety have been taken within the last few years, and also three albinos. One of these is nearly white (pale grayish-white), and the other two are pale yellowish-brown

* Vol. III, p. 205.

or cream-colored. The latter are preserved in the Springfield Natural History Museum.

I have known of a few instances of the capture of this species in nearly midwinter. Once a specimen was taken running in the highway early in February, when the snow was a foot and a half deep. They generally leave their burrows very early in spring, often before the ground is fully thawed, but for some time after are irregular in going abroad, and are able to remain six or eight days inside their burrows without food, as they will often do when a trap is set for them. Till the season and vegetation are somewhat advanced they seem to take or require but little nourishment. Later, and especially after the birth of the young in June, they are forced in a much shorter time to leave their holes to obtain food. In fall they become very fat, and early in October generally permanently retire to their burrows, or at least go abroad then much less frequently than earlier, and apparently take very little food.

The Beaver (*Castor fiber* Linn.; *C. canadensis* Kuhl) is to be reckoned among those few animals that, in this State, have become fully exterminated.

The few differences pointed out by authors between the European and American beavers, including the distinction based on a comparison of the skulls, are too trivial, in the light of the extensive individual variations now so well known to be almost invariably presented by a large series of specimens of the same species from any given locality, to be taken as satisfactory evidence of their diversity. The weight of authority is also by far in favor of their identity.

57. **Jaculus hudsonius** Baird. (*Meriones** *hudsonius* Aud. and Bach.) Jumping Mouse. Rather common, but far from numerous.

This species has distinct cheek-pouches, — a fact I have not before seen stated.

58. **Mus decumanus** Pallas. Brown Rat. Wharf Rat. Norway Rat. Abundant in the cities and larger villages generally; rare or quite unknown in the remote farming districts.

* *Meriones*, F. Cuvier, Dents des Mam., 1825, 187; type, *Dipus americanus* Barton. Not *Meriones* Illiger, Prod., 1811.

59. **Mus rattus** LINN. BLACK RAT. Abundant in the farming districts, but rare wherever the brown rat is numerous. In the vicinity of Boston and of the larger cities generally it seems to be quite unknown. Twenty or thirty miles from the coast, and at a little distance from the large towns along the railways, it becomes numerous, and the only species there found. The brown rat is its mortal enemy. With age this species changes from black to gray, very old individuals becoming very light colored.

60. **Mus musculus** LINN. HOUSE MOUSE. Everywhere a numerous pest. Is frequent in the fields under stacks of grain as well as in houses and outbuildings.

61. **Hesperomys leucopus** LECONTE. (*H. leucopus* and *H. myoides* Baird.) WHITE-FOOTED MOUSE. DEER MOUSE. A common species of the fields and woods. In winter it (sometimes at least) retires to a warm nest in a hollow stump or log, in which in severe weather I have found five or six together in a torpid state.

No species of our *Muridæ*, excepting possibly the *Jaculus hudsonius*, presents so great variations in color with season and age as the present. The young for the first two or three months, or till nearly full-grown, are dark slate or plumbeous above, somewhat lighter below. From the casting of the winter coat in spring till late in autumn the adult differs more or less in color with almost every individual, none presenting the bright yellowish or ferruginous brown seen in winter and early spring, but every stage between it and the plumbeous hue of the young; the adult being also more or less dusky for some time after moulting. Generally there is a darker band along the back, varying in width in different specimens, and in distinctness of outline; sometimes, however, the back is uniform in color with the sides. The variation in size is also considerable between specimens apparently fully adult. The tubercles on the soles of the hind feet, on which specific distinctions are sometimes based, vary both in relative size and position. The posterior one is usually situated midway between the toes and the heel, but sometimes more posteriorly or more anteriorly. The next one is placed between this and the third, and is usually nearer to this than to the first, it being sometimes opposite to the third. The third anterior tubercle occasionally has a minute supplemental one at its outer base. But the most variable character consists in the relative length and number of the caudal vertebræ. About one fifth of the Massachusetts specimens have the tail vertebræ equal to or longer than the head and body together; occasionally a specimen is found in which the tail vertebræ alone exceed this length by one fourth to one half an inch. At least four fifths, however, have the tail shorter than the head and body, and occasionally one occurs with the tail only equal to the body alone. In these latter the proportional length of the tail vertebræ to the length of the head and body is as 68 to 100; in the other extreme, or in those with long tails, as 118 to 100. The variation between these extremes is hence about fifty per cent of the mean, — a striking example of the unreliability of this character as a specific distinction already claimed in discussing the species of *Mustelidæ*. The number of the vertebræ varies from twenty-four or twenty-five to above thirty. In regard to absolute size, the length of the head and body together, in Massachusetts specimens, rarely exceeds four inches; the average is between three and a quarter and three and a half; perhaps nearer the latter. The variation in this respect is well illustrated in Professor Baird's table of measurements of a large number of Middleboro' and other specimens of this species, given in the Mammals of North America (p. 462).

Through the seasonal and other variations in color, as well as in size and proportions, it becomes extremely difficult to distinguish the different North American species of the restricted genus *Hesperomys*, if so many species are to be recognized as have been described, similar variations apparently occurring in all the species. That several exist in the eastern part of the United States seems unquestionable, but the validity of many that have been described from this region is at the same time highly doubtful. The *H. gossypinus*, as defined by Professor Baird, would at first seem readily distinguishable by its comparatively large size, coupled with a southern habitat and its short tail; in color and proportions it closely resembles *H. leucopus*. But since in *H. cognatus* we have a form intermediate between the two and intimately allied to both, the true standing and affinities of each of the three become questionable. Some specimens of *Hesperomys* before me from Florida* differ in no essential particular from examples of *H. leucopus* in summer pelage from Massachusetts and Maine. Well-marked examples of either of the two first mentioned of these so-called species seem sufficiently distinct, but a large series of specimens is constantly presenting intermediate stages, and a large amount of variation in each of the would-be distinctive characters. A single Florida specimen of *H. Nuttallii* (*Mus aureolus* Aud. and Bach.) differs much in color from the other Florida specimens of *Hesperomys*, and from *H. leucopus*.

* In addition to the specimens collected by myself in Florida the past winter, I am indebted to Mr. C. J. Maynard for the opportunity of examining others obtained there by himself the same season.

H. michiganensis, of which I have also had fresh specimens for examination, seems as well marked as any of the group, through its small size, very short tail, and dark plumbeous color at all seasons. Other specimens collected by myself in Western Iowa, supposed from their locality to be referable to *H. sonoriensis*, differ in no way appreciably, except in being a little lighter colored, from average specimens of Massachusetts *H. leucopus*.

H. myoides, described by Baird from Canada and Vermont specimens, is positively identical with *H. leucopus*, the cheek-pouches — the only character supposed to distinctively characterize it — being probably common to all the species of the genus, as well as to *Jaculus*.* I first became aware of the existence of cheek-pouches in *H. leucopus* by capturing the animal with the pouches distended with seeds and grain; a subsequent examination of many specimens in alcohol from Berlin, Middleboro',† Springfield, and other localities in Massachusetts, and from Waterville, Norway, Bethel, Upton, and other places in Maine, has fully confirmed this discovery, as I have yet to find the first specimen without the pouches. They almost uniformly exist as described by Gapper, — that is, extending upwards to the eye and posteriorly to the ear. They are equally well marked in specimens of *H. gossypinus* and *H.* "*cognatus*," from Florida. ‡

In the large proportion of equivocal species included among the thirteen recognized in the General Report, to which one since described from In-

* See *antea*, p. 226.

† The Middleboro' specimens were collected by Mr. J. W. P. Jenks, and presented by the Smithsonian Institution to the Museum of Comparative Zoölogy, labelled "*Hesperomys leucopus*."

‡ In the Report on North American Mammals (p. 460) it is stated, "No traces of cheek-pouches can be detected" in *H. leucopus*. Under *H. myoides* the same author remarks (Ib., p. 472) that he found, much to his astonishment, decided indications of cheek-pouches in all the alcoholic specimens of that "species" he examined. "I then," he says, "investigated a considerable number of Middleboro' specimens, and in none could I detect the slightest indication of anything of the kind." "In another specimen," he says later (No. 2776), "from Watervile, New York, referable probably to the same species [*H. myoides*], I found the cheeks crammed with large seeds, and on cutting them open could see that the latter occupied a pouch of considerable size. It is possible that this specimen (immature) may not belong to *H. myoides*, if so, we must conclude that in the ability to distend the cheeks very much, even temporarily, the *H. leucopus* approaches very closely to the *H. myoides*, and this diminishes still more the propriety of placing the latter in a distinct genus. It is quite possible that others of our species may have the cheek-pouches more or less developed." It hence appears that the existence of cheek-pouches in the other species of *Hesperomys* was finally strongly suspected by the author in question. The oversight of their presence in *H. leucopus*, however, is somewhat surprising, since they are not difficult to discover in specimens preserved in alcohol, when search for them is properly made, though in specimens badly contracted by the alcohol they might quite readily escape observation.

diana by Prince Maximilian is added,* there are besides the several doubtful ones already mentioned, others equally questionable. Of those assigned to that part of the United States east of the Rocky Mountains, the *H. michiganensis*, *H. leucopus*, and *H. Nuttallii* (*aureolus* Aud. and Bach.), seem to be those best entitled to recognition, while possibly *H. gossypinus* may be also valid; but with my present knowledge of the subject, I fail to see why *H. texanus*, *H. indianus* (of Prince Maximilian), *H. sonoriensis*, *H. myoides*, and *H. cognatus*, should be thus regarded, all but the latter, and perhaps also both this and *H. gossypinus*, being apparently referable to *H. leucopus*. I do not hesitate to thus refer *H. sonoriensis*, and *H. myoides*, both of which I have examined in the fresh state, and numbers of the latter that were preserved in alcohol.

Of the Pacific Coast species, of which at least five have been described, several are intimately allied to the *H. leucopus* of the East, as well as to each other. Whether any of them are identical with *H. leucopus* is not at present, from want of sufficient material, easy to decide. Should they prove to be so, it would substantiate a more extended geographical range for *H.*

* 1. *Hesperomys leucopus* BAIRD, N. Am. Mam., 1857, 459; = *Musculus leucopus* RAFF., Amer. Monthly Mag., III, 1823, 307.

2. *Hesperomys myoides* BAIRD, N. Am. Mam., 472; = *Cricetus myoides* GAPPER, Zoöl. Journ., 1830, 204.

3. *Hesperomys indianus* MAXIMILIAN, Archiv für Naturgesch., XVIII, 1, 1862, 111.

4. *Hesperomys sonoriensis* LECONTE, Proc. Phil. Acad. Nat. Sci., VI, 1853, 413; = *H. sonoriensis* BAIRD, N. Am. Mam., 474.

5. *Hesperomys texanus* WOODHOUSE, Proc. Phil. Acad. Nat. Sci., VI, 1853, 242; = *H. texanus* BAIRD, N. Am. Mam., 464.

6. *Hesperomys Nuttallii* BAIRD, N. Am. Mam., p. 467; =? *Arvicola Nuttallii* HARLAN, Month. Amer. Journ., 1832, 446; = *Mus* (*Calomys*) *aureolus* AUD. and BACH., Jour. Phil. Acad. Nat. Sci., VIII, 1842, 302.

7. *Hesperomys cognatus* LECONTE, Proc. Phil. Acad. Nat. Sci., VII, 1855, 442; = *H. cognatus* BAIRD, N. Am. Mam., 469.

8. *Hesperomys gossypinus* LECONTE, Proc. Phil. Acad. Nat. Sci., VI, 1853, 411; = *H. gossypinus* BAIRD, N. Am. Mam., 469.

9. *Hesperomys Boylii* BAIRD, Proc. Phil. Acad., VII, 1855, 335; = Ibid., N. Am. Mam., 471.

10. *Hesperomys californicus* BAIRD, N. Am. Mam., 478; = *Mus californicus* GAMBEL, Proc. Phil. Acad. Nat. Sci., IV, 1848, 78.

11. *Hesperomys eremicus* BAIRD, N. Am. Mam., 479.

12. *Hesperomys austerus* BAIRD, Proc. Phil. Acad. Nat. Sci., VII, 1855, 336; = Ibid., N. Am. Mam., 466.

13. *Hesperomys Gambelii* BAIRD, N. Am. Mam., 464.

14. *Hesperomys michiganensis* WAGNER, Archiv für Naturgesch., 1843, 2, 51; = *Mus michiganensis* AUD. and BACH., Journ. Phil. Acad. Nat. Sci., VIII, 304; = *H. michiganensis* BAIRD, N. Am. Mam., 476; = *Mus Bairdii* HOY & KENNICOTT, Patent-Office Rep., Agr., 1856 (1857), 92.

leucopus than many of the rodents possess, particularly the smaller species, but no greater than seems to be admitted for *Jaculus hudsonius*, its somewhat near ally. The habitat of *Jaculus hudsonius*, as now commonly defined, extends from ocean to ocean, and from the Arctic regions southward through at least the Middle States and to Missouri. This, also, is a species remarkable for its variability in color, size, proportional length of the tail to the body, etc.; but in the General Report on the Mammals of North America these differences were allowed only their proper value, and several species of authors were reduced to synonymes in consequence. Had the same course been taken in respect to the genus *Hesperomys*, undoubtedly a large proportion of the nominal species now admitted would have been referred to their proper rank. There seems to be no reason why *Hesperomys leucopus* may not range as widely as *Jaculus hudsonius*, and but little to show that such is not the case.

62. **Arvicola Gapperi** VIGORS. RED-BACKED MOUSE. Apparently not very rare in some localities in the eastern part of the State. Professor Baird mentions seven specimens sent him by Mr. J. W. P. Jenks from Middleboro'.* There are also several specimens in the Museum of Comparative Zoölogy from localities near Cambridge. It has not yet been met with, however, in the vicinity of Springfield. It is apparently less southern in its distribution than the next following species.

63. **Arvicola riparius** ORD. COMMON MEADOW MOUSE. Abundant; periodically excessively so. At such times they often do great harm by destroying fruit and other trees. Apple-trees a foot in diameter are sometimes killed by being girdled by these destructive animals. They also occasionally destroy large numbers of those of smaller size, as well as of young pitch-pines (*Pinus rigida* Linn.) and other native trees. Their excessive increase is generally coincident with a series of winters during which the ground is covered with a heavy deposit of snow, which protects them from cold, and beneath which they burrow and commit their ravages. Their decrease generally occurs during a series of "open" winters, when in searching for their food they are wholly unprotected from severe cold, and the deep freezing of the ground obstructs their shallow burrows, within which they are doubtless often frozen. They frequent every variety of situa-

* N. Am. Mam., p. 521.

tion, from half-submerged meadows to the driest sandy plains. Dr. Godman, in his American Natural History,* under *Arvicola xanthognathus*, has very minutely described the habits of this species. While in meadows it forms roadways among the roots of the grass on the surface, in grain-fields it burrows beneath the surface, its habits varying with circumstances. In the latter situation the vegetation is not generally sufficiently dense to screen it, hence its more subterranean mode of life. Their nests are found containing newly born young from early in May till November. The number of litters produced by a single female in a year is probably generally not less than three, and may be more; the young of the early litters also themselves appear to have young the same season; hence the great rapidity of increase that obtains in this species.

Specimens, even from the same locality, vary considerably in size, color, the texture of the fur, and even in the shape of the skull, independently of considerable variations that result from age and season. On these variations have been erected numerous nominal species, some of which are already currently considered as synonymes of *A. riparius* Ord, and several more, doubtless, should be added to the list. Among those described from or attributed to Massachusetts which I refer to *A. riparius* are *A. hirsutus* and *A. albo-rufescens* Emmons,† *A. nasuta* Audubon and Bachman, ‡ and *A. Breweri* and *A. rufidorsum* Baird; § also, *A. rufescens* De Kay, ‖ from New York.

On Muskeget Island (a small, uninhabited, low sandy island between Nantucket and Martha's Vineyard) I recently found the so-called *A. Breweri* excessively abundant. This is the only locality from which this supposed species has been reported. They are generally much paler in color than the *A. riparius* of the interior, and though not differing from them appreciably in any other respects, they form an interesting insular race. From the peculiar character of the locality, the scattered beach-grass growing upon it affording but slight protection to these animals from the sunlight, the intensity of which is greatly heightened by the almost bare, light-colored sands, the generally bleached appearance of the Muskeget *Arvicola* might have been anticipated. Specimens occasionally occur of nearly the ordinary color, or which are undistinguishable from the lighter-colored speci-

* Vol. II, p. 66. † Report on Quad. of Mass., p. 60.

‡ Journ. Phil. Acad. Nat. Sci., Vol. VIII, p. 296; Quad. N. Am., Vol. III, p. 211, Pl. 144, Fig. 2.

§ N. Am. Mam., pp. 525, 526.

‖ N. Y. Fauna, Vol. I, p. 85, pl. XXII, Fig. 1.

mens from the interior; but most of them seem to be quite like the ones described by Professor Baird. The mice living on the extensive sand-dunes at Ipswich, under circumstances similar to those of the Muskeget mice, often present, as I have recently ascertained, the half-white appearance of the *A.* "*Breweri*."

The *A. albo-rufescens*, described by Dr. Emmons from two nearly white or cream-colored specimens procured at Williamstown, is, as first suggested by Audubon, undoubtedly but an albinic variety of *A. riparius*. Having obtained two specimens at Springfield that almost exactly accorded with Emmons's description of *A. albo-rufescens*, I was led at first to consider it a valid species. Subsequent experience convinced me that this is not its character. Two similarly colored specimens of the woodchuck (*Arctomys monax*), unquestionably albinic, have been since obtained at Springfield, which differ from the ordinary condition of that animal in the same way that these specimens of *Arvicola* do from the ordinary state of *A. riparius*. Aububon and Bachman mention similar examples that came under their nótice; in one case different stages of albinism were observed in the different individuals of the same litter. A short time since I myself received an interesting albinic example of this species from Weathersfield, Vermont, from my friend Mr. J. P. Stoughton, of which the following is a description: Beneath, except the extreme posterior part of the body, pure white; mainly white above, with a wide, rather irregular band of dusky along the back; the anterior part of the head and the cheeks dusky; posterior part of the head white, with several dusky spots; ears, thighs, and a large spot on the left shoulder, dusky, with small axillary spots of the same color; all the feet and the terminal third of the tail, white. Irides a little lighter than the natural color, but not red. Ears conspicuous; much longer than the short, soft fur. A little smaller, and rather slenderer than ordinary specimens. Apparently a mature female, taken August 18, 1868. Albinos of this species appear to be not infrequent, the capture of a litter in which all the individuals greatly resembled the parti-colored one above described having come to my knowledge since the above was written.

The single specimen from Holmes's Hole, described as *A. rufidorsum*,* which is thus far the only recognized specimen of this supposed species extant, seems to be but an unusually highly colored example of *A riparius*. At Springfield, where I have examined hundreds of specimens at different seasons of the year, the variation in color is very considerable, ranging from decidedly gray on the one extreme to as decidedly rufous chestnut-brown on the other. They are usually much grayer in March and April than they are late in the fall.

* See N. Am. Mam., p. 526, as previously cited.

The following is a partial list of the synonymes of

Arvicola riparius.

Arvicola riparius ORD, Journ. Phil. Acad. Nat. Sci., IV, 1825, 305.
" " DEKAY, N. Y. Fauna, Pt. I, 1842, 84, Pl. XXII, Fig. 2. (Young.)
" " AUD. and BACH., Quad. N. Am., III, 1854, 302.
" " KENNICOTT, Pat. Off. Rep., 1856, Agr., 1857, 304.
" " BAIRD, N. Am. Mam., 1857, 522.
" *palustris* HARLAN, Faun. Am., 1825, 126.
" *albo-rufescens* EMMONS, Quad. Mass., 1840.
" " DEKAY, N. Y. Fauna, 1842, I, 89.
" *hirsutus* EMMONS, Quad. Mass., 1840, 60.
" " DEKAY, N. Y. Fauna, I, 86.
" *oneida* Ibid., 88, Pl. XXIV, Fig. 1.
" *rufescens* Ibid., 85, Pl. XXII, Fig. 1.
" *nasuta* AUD. and BACH., Journ. Phil. Acad. Nat. Sc., VIII (2), 1842, 296.
" " Ibid., North Am. Quad., III, 1853, 211, Pl. CLXIV, Fig. 2.
" *pennsylvanica* AUD. and BACH., Quad. N. Am., I, 1849, Pl. XLV, 341.
" *rufidorsum* BAIRD, Mam. N. Am., 1857, 526.
" *Breweri* Ibid., 525.
" *xanthognathus* * GODMAN, Am. Nat. Hist., II, 1826, 65.
" " DEKAY, N. Y. Fauna, I, 1842, 90.
" " LINSLEY, Am. Jour. Sc., XLIII, 1842, 350.

64. **Arvicola pinetorum** AUD. & BACH. (*A.* [*Pitymys*] *pinetorum* Baird.) The only specimens of this species I have seen from this State are one captured at Springfield in May, 1868, by my brother, Mr. E. Allen, and one taken by myself a few weeks later. Both were taken in the same field on the "pine plains" east of the city. Audubon and Bachman, I find, speak of having received it from near Boston, from Dr. Brewer. These authors also speak of it as occurring in Connecticut, and as abundant in certain portions of Rhode Island.† Professor Baird cites it from Long Island,‡ whence Audubon and Bachman derived their first specimens of *A.* "*scalopsoides*,"§ which they afterwards very properly considered as a synonyme of *A. pinetorum.* It

* Whatever the "*A. xanthognathus*" of Leach and Richardson (Faun. Bor. Am., I, 122) may have been, the *A. xanthognathus* of Godman, DeKay, and Linsley unquestionably refers to the *A. riparius* of Ord.

† Quad. N. Am., II, p. 216.

‡ Mam. N. Am., p. 544.

§ Journ. Phil. Acad. Nat. Sci., VIII, p. 299.

being a southern species, Massachusetts is probably its northern limit Its occurrence here is comparatively rare.

65. **Fiber zibethicus** Cuv. Muskrat. Abundant. Individuals nearly black are taken occasionally.

HYSTRICIDÆ.

66. **Erethizon dorsatus** F. Cuv. (*E. dorsatus* and *E. epixanthus* Auct.) Porcupine. "Hedgehog." Occasional on the Hoosac ranges.

Professor Baird, in his description of this species,* thus observes: "Fur, dark brown; the long projecting bristly hairs dusky, with white tips; spines white, the points dusky. Nasal bones not more than one third the length of the upper surface of the skull." He adds: "I regret not to have a sufficiently perfect specimen of the common Eastern porcupine before me to furnish a satisfactory description. The differences, however, from *E. epixanthus* † are not very great, *consisting chiefly in the color of the tips of the long hairs*, and one description will answer very well for both, except where the peculiarities of each are specially indicated. The range of this species is much more limited than previously supposed, as it is replaced west of the Missouri by the *E. epixanthus*."

He thus describes *E. epixanthus*, from several good specimens: "General color dark brown, nearly black; the long hairs of the body tipped with greenish-yellow. Nasal bones nearly one half or two fifths the length of the upper surface of the skull"; which he says are not more than one third in *E. dorsatus*. Nine very fine specimens of *E. dorsatus* in the Museum of Comparative Zoölogy, from Central Maine, show that the color of the projecting bristly hairs is variable. In one they are *entirely black*, except a very few about the head, which are tipped with lighter; in another those of the back are black, while on the head, sides of the shoulders, etc., they are tipped with dull yellowish-white. Several have them of the greenish-yellow supposed to characterize exclusively *E. epixanthus*; in one or two only can they be called white, while in one these bristly hairs are almost entirely absent, being quite so on the back. The quills usually project considerably beyond the fur, but are sometimes quite concealed within it. Their color varies from white to dull yellow. Professor

* Mam. N. Am., p. 569.

† "*E. epixanthus* Brandt, Mém. Acad. de St. Petersbourg, 1835, 388, 416; Plate I (animal) and Plate IX. Figs. 1–4, skull."

Baird's detailed description of the exterior characters of *E. epixanthus* is in every respect applicable to fully one half the specimens from Maine referred to above, while none differ essentially from it. The differences referred to by him in the relative length of the nasals in the two supposed species are relatively very slight, especially as compared with the large amount of variability presented in a large series of the skulls of *Arctomys monax*, or of our common squirrels or rabbits; the difference in the proportional length of the nasals to the whole length of the skull, in five specimens of *E. epixanthus* and three of *E. dorsatus*, as given by Professor Baird, being but 4 per cent; the nasals in *E. dorsatus* being 37 per cent of the whole length of the skull, and in *E. epixanthus* 41. In No. 676 (*E.* "*dorsatus*") of Baird's table, the proportional length of the nasals to the entire skull is 39 per cent; in No. 3066, 32 per cent. In No. 822 (*E.* "*epixanthus*"), 39 per cent. In other words, the specimen in the series of *E. dorsatus* in which the nasals are longest differs less than one-third of one per cent in the proportional length of the nasals to the whole skull from the specimen with relatively the shortest nasals in the seies of the *E. epixanthus* specimens.

I am not able at this time to refer to M. Brandt's paper, but Waterhouse, in his Natural History of the Mammalia,* refers to it as follows: "Five specimens of an *Erethizon* from the West Coast of North America, in the Museum of St. Petersburg, having the exposed ends of the longest hairs of the fur of a brownish-yellow color instead of white, as the same hairs are stated to be in the *E. dorsatus*, M. Brandt is inclined to suppose there are two species of *Erethizon*, but not having specimens of the Canada animal for comparison, he is not able to satisfy himself upon this point. The specimens examined by M. Brandt are from California and Unalaska, and I may add that a similar specimen is found at Sitka, as I remember to have seen a specimen in the Leyden Museum from there agreeing with M. Brandt's description; its *spines* [not *hairs*] were most of them of a delicate yellow below the dark point." The following is Mr. Waterhouse's description of *E. epixanthus*, compiled from M. Brandt's memoir: "The longer and coarser hairs brownish-yellow at the point; spines white or yellowish at the base, and most of them brownish-black or dusky at the apex."

It hence appears that the three principal writers on the subject — Brandt, Waterhouse, and Baird — have neither of them had specimens of the two species for comparison at the time of writing; Brandt having only his five West Coast specimens, Waterhouse compiling from Brandt, and Baird's specimens coming, two from the Republican Fork, one from New Mexico, and one from California, with three or four skulls from the East.

* Vol. II, p. 442.

Dr. Brandt must have been much influenced by the difference in locality whence his specimens came, in supposing there might be two species of *Erethizon*, since the only difference he points out — that of the color of the tips of the long hairs — is one of a trivial, and, as all mammalogists must be aware, most inconstant character. The differences in the skulls discovered by Professor Baird, though so appreciable, have less weight since we know that skulls of individuals of the same species from the *same locality* not unfrequently vary as much, and in the same way. Again, according to the measurements he has given, and which are discussed above, one specimen of the one series of three is not appreciably different from a specimen of the other series of five. Hence, though having only Eastern specimens for examination, I quite confidently refer, for the reasons given above, the *E. epixanthus* Brandt to the *E. dorsatus* F. Cuvier. I am quite sure, also, that, had either Professor Baird or Dr. Brandt possessed a good series of *E. dorsatus* from Eastern North America, they could hardly have admitted the latter's doubtfully proposed species, even provisionally.

Prince Maximilian, in speaking of the porcupines of the Upper Missouri,* mentions them simply under the generic name *Erethizon*, stating that he was unable to decide whether the animal he observed should be referred to *E. dorsatus* or to *E. epixanthus*.

Dr. J. E. Gray, in the proceedings of the London Zoölogical Society,† has described a small specimen of *Erethizon* from Columbia as a new species, under the name of *E.* (*Echinoprocta*) *rufescens*, although there is nothing to indicate that it is in any way different from the young of the common *E. dorsatus*. The differences on which he has raised it to a distinct section or subgenus are only such as characterize the young or half-grown animal in *E. dorsatus*, with which also his corresponds in size.

LEPORIDÆ.

67. **Lepus americanus** Erxl. (Emmons's Rep., p. 56.) White Rabbit. Common, but generally less so than the next. Rare in the immediate vicinity of Springfield, though numerous at localities less than ten miles distant, in several directions.

68. **Sylvilagus nanus** Gray.‡ (*Lepus sylvaticus* Bach. *Lepus*

* Wiegmann's Archiv, XVIII, Theil I, p. 150.

† 1865, 121, Pl. XI; also in the Annals and Magazine of Natural History of the same year.

‡ In a recent paper entitled "Notes on the Skulls of Hares (*Leporidæ*) and Picas (*Lagomyidæ*) in the British Museum," Dr. J. E. Gray has given names to the sections of the old genus *Lepus*, first indicated by Professor Baird in his well-studied essay on this group (N. Am. Mam., pp. 572–620), and raised them to the rank of genera, thereby, of

virginianus Harlan, Emm. Rep., p. 58.) Gray Rabbit. Abundant in most parts of the State. Less common in the more elevated portions, and quite unknown in the higher ranges of the western counties.

General Synopsis and Remarks on the Geographical Distribution of the Species.

I. *Indigenous Species still existing in the State.*

1. Lynx canadensis *Raf.**
2. " rufus *Raf.**
3. Canis lupus *Linn.**
4. Vulpes vulgaris *Cuv.*
5. " virginianus *DeKay.**
6. Mustela Pennantii *Erxl.**
7. " martes *Linn.**
8. Putorius vulgaris *Linn.*
9. " ermineus *Linn.*
10. " lutreolus *Cuv.*
11. Gulo luscus *Sabine.**
12. Lutra canadensis *Sab.*
13. Mephitis mephitica *Baird.*
14. Procyon lotor *Storr.*
15. Ursus arctos *Linn.**
16. Phoca vitulina *Linn.*
17. Cystophora cristata *Nilsson.*
18. Cariacus virginianus *Gray.**
19. Balæna cisarctica *Cope.*
20 Agaphalus gibbosus *Cope.*
21. Megaptera osphyia *Cope.*
22. Eschrichtus robustus *Lilj.**
23. Sibbaldius tectirostris *Cope.*
24. " tuberosus *Cope.**
25. " borealis *Fisch.**
26. ? Balænoptera rostrata.
27. Physeter macrocephalus *Pander.**
28. Mesoplodon sowerbiensis.*
29. Orca gladiator *Sund.*
30. Globiocephalus melas *Traill.*
31. Hyperaodon bidens *Owen.**
32. Beluga canadensis *Erxl.**
33. Largenorhynchus sp.?
34. Delphinus erebennus *Cope.*
35. " clymene *Gray.**
36. Phocæna americana *Agass.*
37. Lasiurus noveboracensis *Tomes.*
38. " cinereus *H. Allen.**
39. Scotophilus fuscus *H. Allen.*
40. " noctivagans *H. Allen.*
41. " georgianus *H. Allen.*
42. Vespertilio subulatus *Say.*
43. Neosorex palustris *Verrill.**
44. Sorex platyrhinus *Linsley.*
45. Sorex Cooperi *Bach.**
46. " Forsteri *Rich.**
47. Blarina brevicauda *Baird.*
48. Scalops aquaticus *Fisch.*
49. " Breweri *Bach.**
50. Condylura cristata *Ill.*

course, introducing numerous changes in nomenclature. *Lepus* is restricted to the larger species, typically represented by *L. americanus* Erxl. and the European *L. timidus* Linn. Thirty species of the old genus *Lepus* are enumerated, but a considerable proportion appear to rest on highly questionable grounds. Dr. Gray enumerates in this paper thirty-nine species of *Leporidæ* alone, of which sixteen are North American and two South American. The characters of these groups, so far at least as they relate to the North American species, are those developed by Professor Baird in his excellent elaboration of this family.

* Species marked with the asterisk are very sparsely represented; among the Carnivora most of those thus distinguished have become nearly exterminated.

51. Sciurus cinereus *Linn.**
52. " carolinensis *Gmelin.*
53. " hudsonius *Pall.*
54. Pteromys volucella *Linn.*
55. Tamias striatus *Baird.*
56. Arctomys monax *Gmelin.*
57. Fiber zibethicus *F. Cuv.*
58. Jaculus hudsonius *Baird.*
59. Hesperomys leucopus *LeConte.*
60. Arvicola Gapperi *Vigors.*
61. " riparius *Ord.*
62. " pinetorum *LeConte.**
63. Erethizon dorsatus *F. Cuv.**
64. Lepus americanus *Erxl.*
65. Sylvilagus sylvaticus *Gray.*

II. *Extirpated Species.*

1. Felis concolor *Linn.*
2. Alce malchis *Ogl.*
3. Tarandus rangifer *Gray.*
4. Cervus canadensis *Linn.*
5. Castor fiber *Linn.*

III. *Adventitious Species.*

1. Mus decumanus *Linn.*
3. " rattus *Linn.*
3. " musculus *Linn.*

IV. *Northern Species.*

[Not occurring in this State south of the Canadian fauna (excepting *Lepus americanus*, which ranges through the Alleghanian), and hence represented only in portions of the western counties.]†

1. Mustela Pennantii.
2. " martes.
3. Gulo luscus.
4. Alce malchis.
5. Tarandus rangifer.
6. Arvicola Gapperi.
7. Erethizon dorsatus.
8. Lepus americanus.

V. *Southern Species.*

[Not occurring north of the Alleghanian Fauna, and hence unrepresented in the more elevated parts of the State, though more or less common in the other portions.]

† *Antea*, in a foot-note to page 147, *Cervus canadensis* is included among the species there mentioned as characteristic of the Canadian fauna, as formerly represented in Massachusetts. I have since found, from what is known of its earlier range, that it probably once extended over the greater part of the States lying east of the Mississippi, and undoubtedly extended along the Atlantic coast farther south even than Southern New England. There is unquestionable evidences of its existence within the last fifty years on both sides of the Ohio River near its mouth; a locality much more southern, faunally as well as geographically, than any part of New England. Hence it cannot be taken as a species the southern boundary of whose habitat marks the lower limit of the Canadian fauna, as there stated.

1. Vulpes virginianus.
2. Scalops aquaticus.
3. " Breweri.
4. Sciurus cinereus.
5. Sciurus carolinensis.
6. Arvicola pinetorum.
7. Sylvilagus sylvaticus.

VI. *Restricted to the Eastern Province.*

1. Cervus canadensis.
2. Cariacus virginianus.
3. ? Scotophilus georgianus.
4. Neosorex palustris.
5. Sorex Cooperi.
6. " Forsteri.
7. " platyrhinus.
8. Blarina brevicauda.
9. Scalops aquaticus.
10. " Breweri.
11. Condylura cristata.
12. Sciurus cinereus.
13. " carolinensis.
14. " hudsonius.
15. Tamias striatus.
16. ? Arctomys monax.
17. ? Hesperomys leucopus.
18. Arvicola Gapperi.
19. " riparius.
20. " pinetorum.
21. Lepus americanus.
22. Sylvilagus sylvaticus.

VII. *Species restricted to America, but which range over the greater portion of the Northern Continent.**

1. Felis concolor.
2. Lynx canadensis.
3. " rufus.
4. Vulpes virginianus.
5. Mustela Pennantii.
6. Mephitis mephitica.
7. Procyon lotor.
8. Vespertilio subulatus.
9. Scotophilus fuscus.
10. " noctivagans.
11. Lasiurus noveboracensis.
12. " cinereus.
13. Pteromys volucella.
14. Fiber zibethicus.
15. Jaculus hudsonius.
16. Erethizon dorsatus.

VIII. *Species that occur throughout the colder portion of the Northern Hemisphere.*

(Cetacea not included.)

1. Canis lupus.
2. Vulpes vulgaris.
3. Mustela martes.
4. Putorius erminea.
5. " vulgaris.
6. " lutreolus.
7. Gulo luscus.
8. Ursus arctos.
9. Phoca vitulina.
10. Cystophora cristata.
11. Alce malchis.
12. Tarandus rangifer.
13. Castor fiber.

* Probably *Sciurus hudsonius* and *Hesperomys leucopus* should be transferred from the preceding list to this.

IX. *Comparative Table.*

Showing what species of Dr. Emmons's Report on the Quadrupeds of Massachusetts, Dr. DeKay's on the Mammalia of the State of New York, and Mr. J. P. Linsley's Catalogue of the Mammals of Connecticut, are synonymous with those of the present list.

NOTE. — The numbers refer to the same species in each column. Domesticated and fossil species and the Cetacea are omitted.

Present List.	*Emmons's Report.*	*Linsley's Catalogue.*	*DeKay's Report.*
1. Felis concolor.	1. Felis concolor.	1. Felis concolor.	1. Felis concolor.
2. Lynx canadensis.	2. Lyncus borealis.	2. Lyncus borealis.	2. Lyncus borealis.
3. " rufus.	3. " rufus.	3. " rufus.	3. " rufus.
4. Canis lupus.	4. Canis lupus.	4. Canis lupus.	4. Lupus occidentalis.
5. Vulpes vulgaris.	5. Vulpes fulvus.	5. { " fulvus " decussatus. " argentatus. }	5. Vulpes fulvus.
6. " virginianus.	6. Vulpes virginianus.	6. " cinero-argentatus.	6. " virginianus.
7. Mustela Pennantii.	7. Mustela canadensis.	7. Mustela canadensis.	7. Mustela canadensis.
8. " martes.	8. " martes.	8. " martes.	8. " martes.
9. Putorius vulgaris.	9. Putorius vulgaris.	9. " pusilla.	9. " pusilla.
10. " ermineus.	10. " noveboracensis.	10. { " fusca. Putorius noveboracensis. }	10. { " fusca. Putorius noveboracensis. }
11. " lutreolus.	11. " vison.	11. " vison.	11. " vison.
12. Gulo luscus.	12. —— ——	12. —— ——	12. Gulo luscus.
13. Lutra canadensis.	13. Lutra canadensis.	13. Lutra canadensis.	13. Lutra canadensis.
14. Mephitis mephitica.	14. Mephitis americana.	14. Mephitis americana.	14. Mephitis americana.
15. Procyon lotor.	15. Procyon lotor.	15. Procyon lotor.	15. Procyon lotor.
16. Ursus arctos.	16. Ursus americanus.	16. Ursus americanus.	16. Ursus americanus.

Present List.	Emmons's Report.	Linsley's Catalogue.	DeKay's Report.
17. Alce malchis.	17. Cervus alces.	17. —— ——	17. Cervus alces.
18. Tarandus rangifer.	18. " tarandus.	18. —— ——	18. Rangifer tarandus.
19. Cervus canadensis.	19. —— ——	19. —— ——	19. Elaphus canadensis.
20. Cariacus virginianus.	20. " virginianus.	20. Cervus virginianus.	20. Cervus virginianus.
21. Phoca vitulina.	21. —— ——	21. { Phoca concolor. ? " grœnlandica.	21. Phoca concolor.
22. Cystophora cristata.	22. —— ——	22. Stemmatopus cristatus.	22. Stemmatopus cristatus.
23. Lasiurus noveboracensis.	23. Vespertilio noveboracensis.	23. Vespertilio noveboracensis.	23. Vespertilio noveboracensis.
24. " cinereus.	24. " pruinosus.	24. " pruinosus.	24. " pruinosus.
25. Scotophilus fuscus.	25. " carolinensis.	25. " carolinensis.	25. " carolinensis.
26. " georgianus.	26. —— ——	26. —— ——	26. —— ——
27. " noctivagans.	27. —— ——	27. " noctivagans.	27. " noctivagans.
28. Vespertilio subulatus.	28. —— ——	28. " subulatus.	28. " subulatus.
29. Neosorex palustris.	29. —— ——	29. —— ——	29. —— ——
30. Sorex platyrhinus.	30. —— ——	30. Sorex platyrhinus.	30. Otisorex platyrhinus.
31. " Forsteri.	31. —— ——	31. —— ——	31. Sorex Forsteri.
32. " Cooperi.	32. —— ——	32. —— ——	32. —— ——
33. Blarina brevicauda.	33. —— ——	33. { Sorex brevicaudus. " parvus. " DeKayi.	33. { Sorex brevicaudus. " parvus. " DeKayi. " carolinensis.
34. Scalops aquaticus.	34. Scalops canadensis.	34. Scalops canadensis.	34. Scalops aquaticus.
35. " Breweri.	35. —— ——	35. —— ——	35. —— ——
36. Condylura cristata.	36. { Condylura longicaudata. " macroura.	36. { Condylura longicaudata. " macroura.	36. Condylura cristata.
37. Sciurus cinereus.	37. Sciurus vulpinus.	37. Sciurus vulpinus.	37. Sciurus vulpinus.

38. Sciurus carolinensis.	38. Sciurus leucotis. " niger.	38. Sciurus carolinensis. " niger.	38. Sciurus leucotis. " niger.
39. " hudsonius.	39. " hudsonius.	39. " hudsonius.	39. " hudsonius.
40. Pteromys volucella.	40. Pteromys volucella.	40. Pteromys volucella. " ——?	40. Pteromys volucella.
41. Tamias striatus.	41. Sciurus striatus.	41. Sciurus striatus.	41. Sciurus striatus.
42. Arctomys monax.	42. Arctomys monax.	42. Arctomys monax.	42. Arctomys monax.
43. Castor fiber.	43. Castor fiber.	43. Castor fiber.	43. Castor fiber.
44. Fiber zibethicus.	44. Fiber zibethicus.	44. Fiber zibethicus.	44. Fiber zibethicus.
45. Jaculus hudsonius.	45. Gerbillus canadensis.	45. Gerbillus canadensis.	45. Meriones americanus.
46. Mus decumanus.	46. Mus decumanus.	46. Mus decumanus.	46. Mus decumanus.
47. " rattus.	47. " rattus.	47. " rattus. Arvicola albo-rufescens.	47. Mus rattus. " americanus.
48. " musculus.	48. " musculus.	48. Mus musculus.	48. " musculus.
49. Hesperomys leucopus.	49. Arvicola Emmonsii.	49. " agrarius.	49. " leucopus.
50. Arvicola Gapperi.	50. —— ——	50. —— ——	50. —— ——
51. " riparius.	51. Arvicola hirsutus. " albo-rufescens.	51. Arvicola riparius. " xanthognathus.	51. Arvicola riparius. " rufescens. " hirsutus. " oneida. " albo-rufescens. ? " zanthognathus.
52. " pinetorum.	52. —— ——	52. —— ——	52. —— ——
53. Erethizon dorsatus.	53. Hystrix dorsata.	53. Hystrix dorsata.	53. Hystrix hudsonica.
54. Lepus americanus.	54. Lepus americanus.	54. Lepus americanus.	54. Lepus americanus.
55. Sylvilagus sylvaticu	55. " virginianus.	55. " virginianus.	55. " nanus.
— —— ——	— —— ——	— —— ——	— Didelphys virginiana.

X. *General Summary.*

Number of	indigenous species still living in the State		*65
"	species already extirpated		5
"	adventitious species		3—
	Whole number		73
Number of	land species (including the seals)		52
"	marine species (the cetaceans)		18
"	northern species †		7
"	southern species †		8
"	species restricted to the region east of the great sterile plains		22
"	"	that range over the greater part of the continent	15
"	"	common to North America and the North Old World	13
"	"	that are numerously represented	28
"	"	that are sparsely represented	45
"	"	of Felidæ (including 1 extirpated)	3
"	"	Canidæ	3
"	"	Mustelidæ	8
"	"	Ursidæ	2
"	"	Phocidæ	2
"	"	Cervidæ (including 3 extirpated)	4
"	"	Balænidæ	8
"	"	Physeteridæ	2
"	"	Delphinidæ	8
"	"	Vespertilionidæ	6
"	"	Soricidæ	5
"	"	Talpidæ	3
"	"	Sciuridæ (including 1 extirpated)	7
"	"	Muridæ (including 3 adventitious)	9
"	"	Hystricidæ	1
"	"	Leporidæ	2
"	"	Carnivora (5 families)	18
"	"	Ruminantia (1 family)	4
"	"	Cetacea (3 families)	8
"	"	Insectivora (3 families)	14
"	"	Rodentia (4 families)	19
Number of	families represented		16

Less than one half of the indigenous species existing in the State, as indicated above in Table I, are common, and more than a third are

* Emmons gave 41; Linsley, for Connecticut, 52; DeKay, for New York, 60.

† See notes to Tables IV and V, *antea*, p. 239.

rare. The common ones, with a few exceptions (*Putorius lutreolus*, *P. ermineus*, and *Mephitis mephitica* among the carnivores, *Vespertilio subulatus* and *Lasiurus noveboracensis* among the bats), belong to the three families of rodents, — the squirrels (*Sciuridæ*), the mice (*Muridæ*), and the hares (*Leporidæ*), — and to the *Balænidæ* and *Delphinidæ*, which latter are, of course, marine. In species and families, the carnivores and rodents are about equally represented, but in individuals any one of the more common rodents outnumbers all the carnivores together. Probably a single species of *Arvicola* (*A. riparius*) alone outnumbers, when it is most abundant, all the other mammals.

The list of Extirpated Species, forming Table II, five in number, is composed entirely of such animals as, from their large size and being special objects of the chase, would be expected to earliest disappear. Two of the four species of *Cervidæ* (*Alce malchis*, *Tarandus rangifer*) have not existed in the southern half of New England since the discovery of the continent by Europeans, except in the mountains of Western Massachusetts, and there probably only as occasional migrants from the contiguous region north. They may have existed in comparatively recent times in portions of the Alleghanies, but respecting such existence we have no certain record. At a remote period they must have lived much farther south than they do now, or than they have within the last three centuries, since bones of the Caribou have been found by Professor Wyman in the Kjoekkenmœddings of Southern Maine, and teeth that he believes, but does not positively assert, belong to this species in those of Cape Cod. A positive evidence of the former much greater southward extension of the habitat of this animal is indeed already at hand, a small antler and fragments of others of the Caribou being included in the very large collection of the remains of living and extinct species of mammalia recently brought by Professor N. S. Shaler to the Museum of Comparative Zoölogy from Big Bone Lick, Kentucky.* Remains of the elk and the moose having been found in the shell-mounds of the Atlantic coast as far south as New Jersey, we have evidence that these species existed thus far south in comparatively recent times.

To the list of the "extirpated species," nine † that are now ex-

* See Professor Shaler's remarks concerning these specimens in Proc. Bost. Soc. Nat. Hist., Vol. XIII, 1869.

† *Lynx canadensis*, *L. rufus*, *Canis lupus*, *Mustela Pennantii*, *M. martes*, *Gulo luscus*, *Ursus arctos*, *Cariacus virginianus*, *Erethizon dorsatus*.

tremely rare, some of them probably being but casual visitors from Vermont or New York, must soon be added. The fisher and the wolverine may be even now extinct, and the common deer exists in the wild state only by legal protection.

The three adventitious species (see Table III), which are the most noxious of our mammalia, are intruders that, like many of the common weeds, have accompanied civilized man in his voyages till they are almost cosmopolitan in their distribution.

Table IV, composed of northern species, consists, with one exception (*Arvicola Gapperi*), also of species of large size, and such as are special objects of the chase, either for their fur or for food. They hence early disappear before the advance of civilization, and it is now almost impossible to determine in respect to some of them where was formerly their natural southern limit of distribution. At present none of them (*Lepus americanus* excepted) range below the southern boundary of the Canadian fauna, though some may have formerly extended across the next fauna south. The occurrence of *Mustela martes* and *M. Pennantii* in the Alleghanies, the latter as far south as Buncomb County, North Carolina, is well established,* but they seem to be, or to have been, — they being now apparently nearly exterminated there, — confined to the mountains, and hence also to the Canadian fauna. Yet one or both of them have occurred in a few known instances at points rather more southern, faunally, than their usual range, but apparently only during casual migrations in winter.

The *Erethizon dorsatus*, however, seems to have formerly occurred at points clearly within the Alleghanian fauna, as in Western New York,† Northern Ohio,‡ Northern Indiana, Southern Michigan, and Southern Wisconsin; § but it has disappeared in all the more thickly settled parts of the United States; east of the Mississippi it does not now occur south of the Canadian fauna.

The *Lepus americanus*, also chiefly northern in its distribution, ranges, as before stated, a little farther south than the others, and finds its southern limit near the southern boundary of the Alleghanian fauna.

* Audubon and Bachman, Quad. N. Am., Vol. I, p. 314.

† Dr. J. E. DeKay, N. Y. Fauna, Vol. I, p. 79.

‡ Wm. Case, Esq., in Audubon and Bachman's Quad. N. Am., Vol. I, p. 285.

§ R. Kennicott, Pat. Off. Rep., Agr., 1857, p. 91; I. A. Lapham, Transact. Wisc. State Agr. Soc., 1852, p. 340.

Table V, comprising those species that do not occur north of the Alleghanian fauna, embraces but one of relatively large size, — *Vulpes virginianus*, — which is also the only carnivore ; the others are two moles and four rodents. The presence of the species of this list, and the absence of those of the preceding, form the faunal differences that, among mammals, distinguish the Alleghanian from the Canadian fauna. The other thirty-three species of land mammals represented in the fauna of Massachusetts, and which are common to the other New England States, New York, the northern tier of the States westward to the Mississippi, and the greater portion of the Canadas, range widely both to the north and to the south, and some of them also to the westward, extending throughout the colder parts of the northern hemisphere, as is indicated by Tables VII and VIII.*

* In this connection a word in reference to the nature of faunæ may not be out of place, since naturalists of some eminence, but who cannot have thoroughly investigated the subject, appear to think that no faunal districts are recognizable unless there is an entire or almost an entire change in the species represented, while some altogether discard such distinctions. Such an extensive change more properly characterizes the larger divisions in geographical zoölogy, as the provinces and realms, rather than faunæ. It rarely happens that any species is restricted within the limits of a single fauna, and also rarely within those of two. There is not a single well-known species of mammal or bird but inhabits (taking the breeding range only of the latter) an area embracing two or more faunæ, and but few that do not range over more than two. The greater part extend over three, and a large proportion have a still wider distribution, as shown by Tables VII and VIII (see remarks respecting these beyond). But in going north or south from any point within the temperate zones, one observes at certain intervals (generally of about six or seven degrees of mean annual temperature) a marked change in the species, through the disappearance of some and the appearance of others; this change giving rise to well-marked differences in the general facies of the fauna at points not far distant. The habitats of species being in the main nearly coincident in their northern and southern boundaries with isothermal lines, and not with paralells of latitude; and since a number of species usually disappear at nearly the point at which a number of others first make their appearance, the limits of faunæ are thus readily defined, at least approximately. As isotherms necessarily vary with every inequality in the surface of the country, they rarely correspond, as is well known, with the parallels of latitude ; and plants and animals sharing the same apparent irregularity in their distribution, some naturalists have been led to discredit the existence of recognizable zoölogical and botanical districts, or of any definite system in the distribution of animals and plants.

Faunæ, then (the term *fauna* in its restricted sense being usually and properly employed to designate the smallest zoölogico-geographical district), it may be added, are characterized by the peculiar association of species. Generally about twenty-five per cent of those embraced in either of two adjacent faunæ are absent from the other. Rarely do adjoining faunæ differ essentially in genera, though necessarily more or less occasionally. The absence or presence of genera, sub-families, families, and even sometimes orders, more properly characterizes the higher sub-divisions, as provinces and realms.

Each of the twenty-one species mentioned in the next table (Table VI) has a comparatively restricted range, the western limit of their habitats being in most cases the eastern border of the sterile plains of the middle province. This list is composed principally of shrews, moles, and rodents; none of the first two groups and but a few of the latter ranging across the continent. The absence of carnivores from this list is its most striking feature.

Table VII embraces fifteen species that, while restricted to America, range from the Atlantic to the Pacific, and possess a correspondingly wide distribution in latitude, most of them occurring nearly throughout the northern continent. This list is composed almost exclusively of carnivores and bats, all but one of the Massachusetts species of the latter having been found in California, and at various intermediate points.

Table VIII contains thirteen species that are regarded in this paper as common to the Old Word and the New; ten of these are carnivores, and include all the New England species of that group, except those embraced in the preceding list. The geographical distribution of these species, and of the groups to which they belong, affords further evidence in favor of the supposition of the specific identity of their representatives on the two continents above assumed; each species ranging as far north on both as it seems possible for mammalian life to exist. Each has also an extended distribution southward, on each continent, some of them ranging nearly or quite to the tropics; which shows them to be fitted to exist under widely varying physical conditions. These conditions in the northern portions of their respective habitats differ much more from those of the southern portions than those of localities on the two continents ordinarily do when situated under the same isotherm. The representatives of the species in question from the eastern and western continents differ less, as has been previously stated, when the specimens compared are taken from those portions nearly contiguous, as Northwestern America and Northeastern Asia, than when they come from such widely distant points as Eastern North America and Western Europe, the nearest affinity being between those from the localities first mentioned, and the widest differences between those from the latter. The eastern and western continents, moreover, approach each other so nearly at Behring's Straits, that several of the species in question are able to pass occasionally from one to the other. It hence seems unnecessary to suppose the former existence of an Atlantic continent to explain their present distribution. It is also a noteworthy fact that no cases of close affinity among the mammals inhabiting these two continents occur in species that do not range very far to the northward, as in the *Felidæ*, for example, where the only case at all suggestive of identity, or even of close relationship, occurs between the *Lynx canadensis* of Northern North America and the *Lynx lynx* of Northern Europe; both of which species range the farthest north of any of their family, and reach the Arctic regions.

All the circumpolar species, the beaver alone excepted, pertain to the most highly organized groups found in the colder portion of the northern hemisphere, and to which belong not only all the widely ranging species of the north temperate and boreal regions, but those of this character everywhere. With three exceptions, all are carnivores. Two of the others are ruminants, and one is a rodent.

The species most highly organized in their respective families, orders, or classes are almost universally those that possess the widest geographical distribution; partial exceptions occur only in groups where the means of locomotion is specialized, or unusually developed, as in the bats among mammals. The shrews, moles, and rodents, which comprise about three fifths of the species of the North American mammals, are groups of low structural rank, and abound in species of comparatively local distribution. In this great number there are but five or six, allowing the broadest latitude in respect to the limitation of the species, that at all approach to a continental distribution, and only three as the species are usually restricted.* This is about two one-hundredths of one per cent. Only one can be regarded as identical with any Old World species. In the canivores, on the other hand, excluding sub-tropical and nominal species, the number of those that range over most of the continent reaches nearly seventy-five per cent, while fifty per cent, or one half, are identical with Old World species. In the ruminants, which rank below the carnivores, but far above the rodents and insectivores, the species having a similarly wide range on this continent, number not far from thirty per cent. Several of them are identical with Old World species. The bats, though a low group, are,

* *Castor fiber*, *Fiber zibethicus*, and *Jaculus hudsonius*. Probably the following may be added to the list of those that range across the continent: *Erethizon dorsatus*, *Sciurus hudsonius*, *Pteromys volucella*, *Hesperomys leucopus*.

from their special means of locomotion, able to range widely; but to their allies, the moles and shrews, mountain chains and arid plains prove impassable barriers.

The same laws in respect to the character of the species that among mammals have a wide distribution are equally exemplified in birds, all the wide-ranging species being of high rank, or such members of lower groups as have the power of flight unusually developed. The modification of the anterior limbs into organs of flight specially characterizing the class of birds among vertebrates, it is evident that well-developed wings are one of the elements essential to a high grade of structure; and this renders necessary the coincidence in this class of high rank with a wide geographical range. The few land-birds that embrace a large portion of the two northern continents within their respective habitats belong principally to three families, — the finches, and the hawks and owls. The first is one of the highest, if not the highest, family of the class, and the others are by no means low. The other species which have a circumpolar distribution are among the highest members of their respective families, and are rarely of a low grade. The finches thus distributed all belong to the highest genera of their family. Among the birds having a wide distribution, but which are restricted to a single continent, are the typical thrushes, another of the higher groups. The species of the short-winged genera of the Fringillidæ and Turdidæ, on the other hand, are almost invariably the most circumscribed in their habitats.* This coincidence in respect to structure and distribution is also exemplified in every sub-family, as well as family, among the water-birds; but it is not necessary to trace it further here.

Hence the view above taken in reference to the species claimed to be common to the Old World and the New is supported, not only by the

* Compare the species of *Turdus* with those of *Harporhynchus* and *Mimus;* of *Poocætes* and *Passerculus* (see observations on some of the supposed species of *Passerculus* in Mem. Bost. Soc. Nat. Hist., Vol. I, p. 515) with those of *Melospiza*, *Coturniculus*, and *Ammodromus;* or those of the sub-family *Coccothraustinæ* with those of the sub-family *Spizellinæ*. Compare, also, in the *Sylvicolidæ*, the species of *Dendrœca* with those of *Geothlypis*. Also note the very high rank of the species of *Ægiothus*, *Pinicola*, and *Plectrophanes*, and the wide extent of their habitats. Compare further, in *Falconidæ*, the species of *Falconinæ*, with their long pointed wings and compact firmly knit muscular bodies, giving unequalled powers of flight, and their extensive habitats, in several instances embracing a whole hemisphere, with the comparatively short-winged, sluggish, and clumsy species of *Buteoninæ*, of a much lower type of structure and much narrower range.

evidence already given in the special discussion of each case, but by the fact of the near approximation of their habitats, and by general principles.

The thirteen species of land mammalia common to North America and the Old World embraced in the fauna of Massachusetts comprise all thus distributed now known, except two or three very boreal ones. The faunæ of the two continents are really quite different, — not totally so, as has been claimed, — though represented largely by genera and families common to the two. These and the circumpolar species show that a close relationship exists between them, the resemblance being, in fact, far greater than between the faunæ of Southern Mexico and Canada. The difference between the faunæ of the subtropical and cold temperate zones on either continent is many times greater than between the faunæ of the temperate and boreal regions of North America and the same regions of the Old World.*

But four species have been attributed to the States adjoining Massa-

* The distribution of vegetable life in zones, differing from each other in general character and corresponding in their limitation with climatic or isothermal zones, and their similar succession at different altitudes on mountain slopes and in different latitudes at the ordinary level of the land, was partially very early recognized, but first fully demonstrated only half a century ago, by Baron Alexander von Humboldt. It was somewhat later before it was clearly shown that the same law holds in respect to the distribution of terrestrial animal life, which was done in 1845 by Professor Louis Agassiz,[1] and somewhat later still Professor Dana disclosed its presence in the distribution of marine life, in his admirable essay on the geographical distribution of the crustacea.[2] Yet most recent writers who have given attention to the geographical distribution of animals appear to have overlooked this grand fact, and hence have been led to adopt a highly artificial division of the earth's surface in respect to its primary ontological regions. While geographical botanists have so generally recognized the influence of climate, and especially of temperature, in determining the limits of distribution of plants in latitude and in altitude, zoölogists, with only a few exceptions, have very imperfectly appreciated these important influences upon the distribution of animals. While the relation of the present distribution of life to the existing means of communication between the different bodies of land and to the earlier conditions in this respect are of the highest importance in investigations of this kind, if this is the only element taken into account, as is sometimes the case, climatic influences being for the time over-

1 "Note sur la Distribution Géographique des Animaux et de l'Homme." Bulletin de la Societé des Sciences Naturelles de Neuchatel, Tom. I, 1845. See also, by the same author, a paper on the "Geographical Distribution of Animals," in the Edinburgh New Philosophical Journal, Vol. XLVI, 1850, pp. 1-25. Also his "Sketch of the Natural Provinces of the Animal World and their Relation to the different Types of Man," in Nott and Gliddon's Types of Mankind, 1854, p. lviii.

2 U. S. Expl. Exped. Reports, Crustacea, Vol. II, 1852, pp. 1451-1500.

chusetts that have not been detected in the latter. Two of them — *Didelphys virginiana* Shaw, and *Lepus glacialis* Leach, the former occurring in Southern New York, and the other attributed to Northern Maine, and known to occur in Newfoundland * — are not likely to occur here. The other two, *Sorex Thompsonii* Baird † and *Blarina augusticeps* Baird, ‡ — the latter described from a specimen taken at Burlington, Vermont, and the other reported from the same locality, from Halifax, N. S., and Maine,§ — are of a highly questionable character. What has been called *Sorex Thompsonii* (the young probably of either *S. Forsteri* or *S. Cooperi*) doubtless occurs here.

looked, the argument is one-sided, only half the truth is reached, and the general view is a distorted one.[1]

As I have already remarked above, the mutual resemblance between the faunæ and floræ of the boreal portions of North America and those of the Europeo-Asiatic continent is exceedingly great, amounting in the arctic portion, as was long since pointed out,[2] almost to identity. In the Arctic province, which occupies the woodless tracts in the extreme north of both continents, more than four fifths of the species found on the one continent occur on the other. While a few of the small number that inhabit this region are restricted to it, the larger part range much farther to the southward, the majority even over the colder part of the north temperate zone, and several throughout this zone. Besides the mutual floral and faunal resemblance between the two northern continents imparted by this wide distribution of the circumpolar species, this resemblance is increased by the large number of genera that are circumpolar, besides those that embrace the circumpolar species, and the occurrence of other forms, both specific and generic, that are closely allied. It is also true that among the forms restricted to each continent are a few family groups; yet the number of these, as of species and genera, that occur in the tropical and not in the colder temperate regions on either continent is far greater than that of those peculiar to either of the two northern continents. Consequently to apply as ontologico-geographic designations such terms as "Palæogean Creation" to the Eastern world and "Neogean Creation" to the Western, virtually implies the ignoring of the real close affinity of the life of the whole northern hemisphere at the northward, and the vast difference between that of the tropical and the cooler north temperate regions on the same continent. But a further discussion of this point is uncalled for now, and is, moreover, the more out of place here, since I shall, I trust, soon have an opportunity to treat it in detail in a more legitimate connection.

* Quad. N. Am., Vol. I, p. 248.

† N. Am. Mam., p. 34.

‡ Ibid., p. 47.

§ Proc. Bost. Soc. Nat. Hist., Vol. IX, p. 169.

1 See Murray's Geog. Distrib. of Mammals; Wallace's Malay Archipelago, etc.

2 See Agassiz's papers, cited above.

CAMBRIDGE, October, 1869.

ERRATA.

Page 144, lines 11 and 12, for *riparia* read *riparius*.
" 147, line 9 from bottom, for *dorsata* read *dorsatus*.
" 148, " 15, *for* fourth *read* sixth ; *for* third *read* fifth.
" 152, " 27, *after* species *insert* of this group.
" 155, " 32, for *fulvus* read *vulgaris*.
" 187, " 9 from bottom, *for* Murry *read* Murray.
" 191, " 6, for *Isabellinus* read *isabellinus*.
" 192, " 37, *for* Murry *read* Murray.
" 207, " 7, *for* Gray *read* Tomes.

-*On the Mammals and Winter Birds of East Floridc*

No. 3.—*On the Mammals and Winter Birds of East Florida, with an Examination of certain assumed Specific Characters in Birds, and a Sketch of the Bird-Faunæ of Eastern North America.* By J. A. ALLEN.

INTRODUCTION.

THE present paper embraces five more or less distinct parts. The first consists of introductory remarks respecting the topographical, climatic, and faunal features of that part of the peninsula of Florida usually known as East Florida. The second is an annotated list of the mammals of this region. The third is devoted to a consideration of individual, seasonal, age and geographical variation among birds, with reference to certain characters commonly assumed to be specific. The fourth contains a list of the winter birds of East Florida, with field and revisionary notes. The fifth is given to an examination of the geographical distribution of the birds and mammals (more particularly of the birds) of Eastern North America, in which is considered the number of the natural faunæ of this region, their distinctive features and their boundaries.

The enumeration of the mammals and birds, forming Parts II and IV, is based partly on my own observations and partly on notes kindly furnished me by Messrs. C. J. Maynard and G. A. Boardman. These observations may be considered as equivalent collectively to the labors of a single observer constantly in the field for at least four or five winters.

My own observations were made during a three months' exploration of the country bordering the St. John's River, between Jacksonville and Enterprise, in the winter of 1868 and 1869, under the auspices of the Museum of Comparative Zoölogy. The greater part of January was passed at Jacksonville, where I also spent the last week of March. Five weeks were also passed in the vicinity of Enterprise, and the balance of the time at various intermediate points.

Mr. Maynard's explorations were made during the same winter, mainly in portions of the country unvisited by myself, a large part of his collection coming from the Upper St. John's and Indian Rivers. He also spent several weeks at Dummitt's, twenty miles south of New Smyrna. During most of December and January he collected in the vicinity of Jacksonville, at which point one of his assistants, Mr. Charles Thurston, remained during April and a portion of May, collecting, among other things, the later arriving birds. Nearly all the birds and mammals collected by these gentlemen, and by Mr. J. F. Le-Baron, a third member of Mr. Maynard's party, have been added to the collection of the Museum of Comparative Zoölogy, and are accompanied by measurements carefully taken before skinning.

Mr. Boardman's observations were continued through three successive winters, during which he spent considerable time at the following points: St. Augustine and Fernandina on the coast, Jacksonville, Green-Cove-Springs and Enterprise on the St. John's River. Although the numerous specimens he collected at these and intermediate points were presented by him to the Smithsonian Institution, I am indebted to him for an annotated manuscript list of the species he met with. I am also indebted to the Rev. Thomas Marcy, who accompanied me on my Florida trip, for valuable assistance in collecting, and to Mr. J. E. Brundage for similar aid.

Having made use of the reports of previous visitors on the faunæ of this region, the following lists are believed to embrace all the species of mammals thus far known from East Florida, and all the birds regularly present in winter, of nearly all of which I have examined specimens from Florida. A few other birds not included in my list doubtless occasionally visit this region from the North, and others may linger here which usually pass the winter further south. In order to increase the value of the bird list as a faunal record, those species known to be resident throughout the year have been indicated by an asterisk (*), and those known only as winter visitors by an obelisk (†). The date of the first appearance of the strictly spring visitors is also noted, so far as such arrivals were observed.

The specimens on which the investigations detailed in Part III are based, as well as the revisionary notes of Parts II and IV, are mainly those of the Museum of Comparative Zoölogy, which embrace, among others, nearly a thousand specimens of birds from Florida.*

The topics discussed in Part III, namely, individual and climatic variation, necessarily involve the question of the nature of species, as well as the validity of various diagnostic characters. Many details

* I have also made use of measurements, taken by Mr. Wm. Brewster and Mr. C. J. Maynard, of hundreds of specimens not in the collection of the Museum.

in reference to these variations are given in this part, but a large proportion are recorded in the general and revisionary notes of Part IV. The conclusions arrived at, it may be here premised, are mainly the following: (1.) That the majority of nominal species originate in two principal sources of error; namely, (*a*) an imperfect knowledge of the extent and character of individual variation, and (*b*) of geographical variation. (2.) That this imperfect knowledge is mainly due to the neglect of zoölogists to study with sufficient care the common species of their respective countries, whence has arisen a faulty method of investigation and erroneous ideas respecting species and specific characters. (3.) Instead of the method at present pursued by a large school of descriptive naturalists — the analytic, or the search for differences — being the proper one, that synthesis should be duly combined with analysis, and that general principles should be sought as well as new forms, or so-called "new species" and "new genera." (4.) It is claimed that nothing is to be gained by giving binomial names to climatic or other forms, in cases where, however considerable the differences between them may be, a complete transition from the one to the other can be traced in specimens from intermediate localities, notwithstanding the plea sometimes urged that their use affords "convenient handles to facts."

In accordance with such views a partial revision of the species of certain groups is incidentally attempted in Part IV, more especially of the *Icteridæ*, the raptorial birds, and the genera *Parus*, *Turdus*, *Passerculus*, etc.

Part I.

The Topographical, Climatic, and Faunal Characteristics of East Florida.

No part of the Florida Peninsula, as is well known, is much elevated above the level of the sea, the greater portion being extremely low and large areas swampy. The surface is slightly undulating, but the higher ridges rarely attain a height of more than fifty or seventy-five feet, and the highest eminence is less than two hundred. A large part of Northern Florida, including what is usually termed East and West Florida, is covered with open pine forests, constituting the so-called "pine barrens." These barrens frequently rise into dry knolls, but they likewise embrace considerable tracts that are so low as to be more or less submerged during a portion of the year, especially in wet seasons; they are also interspersed with cypress swamps of varying extent. Such swamps usually border the St. John's on its upper course, sometimes extending back from the river for several miles. Other portions of the low grounds support a mixed forest of live-oak, water-oak, elm, bitter-nut hickory, maple, laurel, sweet gum, etc., with a more or less dense undergrowth, such forests forming the so-called "hummocks." Some portions of these forests are swampy; others are dry, and slightly elevated. The saw and dwarf palmettos (*Sabal serrulata* R. & S. and *S. Adansonii* Guerns.) frequently render the former difficult to penetrate, and extensive groves of the cabbage palm (*Chamærops palmetto* Michx.; *Sabal palmetto* R. & S.) here and there occupy the banks of the streams. At intervals in the pine barrens extensive thickets of low trees and thickly growing shrubs are met with, which are exceedingly difficult to enter, and are appropriately termed "scrubs." Each of these kinds of country, as would be naturally expected, forms the favorite haunt of certain species of birds and mammals, the grassy or open pineries being frequented by some that rarely visit the swamps and hummocks, and the latter by others that rarely visit the open pineries. The extensive savannas which occur along the upper portion of the St. John's River and elsewhere form the favorite haunts of numerous wading birds; and the numerous lakes are congenial to the swimming birds.

East Florida hence differs but little in its general character from the lower portions of Georgia and the low lands of the coast of South Carolina. The trees, especially of the hummocks and swampy forests, are usually covered with the pendant *Tillandsia usnoides*, or "Spanish moss," and the abundance of epiphytic orchids and other plants, as well as the palms, clearly indicates the subtropical and peculiar character of the climate.

From the great extent in latitude of the Florida peninsula — from 25° to 31°, or about four hundred miles — considerable differences necessarily exist between the fauna and flora of the northern and southern portions. Although the change in these features from the north southward is more or less gradual, it seems to be appreciably greater near Lake George than elsewhere. At this point so well-marked a change occurs in the vegetation as to attract the attention of unscientific observers, and a corresponding change in the fauna is readily traced. Above Lake George the general aspect of both the flora and fauna is decidedly more southern than it is below the lake. The boundary between the Floridian and Louisianian faunæ and floræ, it would hence seem, may be properly regarded as passing near this point, the portion of the State to the southward being alone properly Floridian, the northern resembling more the Louisianian type.*

As already observed, Florida, from its excessively marshy character, is pre-eminently suited to the wants of the grallatorial birds. Immense numbers of the heron tribe hence make it their permanent home, while it is the favorite winter resort of numerous species of *Grallæ* that pass the breeding season far to the northward. Ibises and egretts abound in its swamps and savannas, forming at all times, by their numbers and showy plumage, a characteristic feature of the fauna. In winter the abundance of snipe and other species of *Grallæ* and ducks render it at that season a sportsman's paradise. Florida hence attracts great numbers of sportsmen in winter, through whose reckless and often wanton waste of life the water-fowl, especially of late years, are annually decimated.

The summer bird fauna of Florida is probably not better represented in species than that of the temperate parts of the continent generally; but this State being the winter resort of numerous species of sparrows and warblers, and of those smaller land birds generally that pass the summer in much higher latitudes, its winter bird fauna, as compared with that of the Northern States, is extremely rich. In New England the number of species of birds that can be regarded as "common" in winter does not exceed fifteen,† but in Florida at that season at least five times that number can be so regarded. This, however, accords with a general law of distribution in respect to the relative number of species found at different points in latitude from the arctic zone southward, the number increasing in proportion to the decrease of the latitude, or with the increase of the mean temperature. In winter, through the southward migration of many species, the minimum number of species which in summer is characteristic of the arctic zone is carried down nearly to the Northern States, there being a marked decrease from summer to winter as far south as the warm temperate or subtropic belt; within the tropics, on the contrary, the number of species is far greater in winter than in summer, through the temporary influx of species from colder regions.

* For a further definition of the Floridian bird fauna, as distinguished from the Louisianian, see beyond, Part V.

† See American Naturalist, Vol. I, p. 47, March, 1867.

In consequence of the subtropical character of the climate of Florida certain peculiarities occur in respect to the development of vegetation at the vernal period, and in the time of breeding of the resident birds, that seem in this connection worthy of record. The mildness of the winter climate is such that the verdure of the forests is to a greater or less degree perennial, severe frosts being of rare occurrence. Some of the early flowering trees, such as the maples, ashes, and elms, begin to bloom and to gradually unfold their leaves early in January. Although the forest trees in general put forth their leaves in February, and a few have acquired their full summer dress by the 1st of March, their development is slow and irregular. I observed peach-trees in flower at the same locality (Jacksonville) in January and in April; and the flowering period of some of the forest trees is nearly as protracted. The development of vegetation is hence as great during a single week in May, in New England, as during any four weeks in February and March, in Florida.

A similar irregularity is observed in respect to the pairing and breeding of the resident birds. Some of the rapacious species, as the fish-hawk and the white-headed eagle, commence incubation in January, and, as I have been informed, occasionally in December; other members of the same species delay breeding till February or March. The great blue heron and the egretts nest in February, as do also the courlans, several of the hawks, the sandhill crane, the wood-duck and the blue-bird; the mocking-bird and other resident song-birds, in March and April.

In the Northern States the vivacity of the birds during the pairing season is as much greater than it is in Florida as is the rapidity of the development in vegetation. In spring at the North the woods, the fields, and the hedgerows are ever vocal with bird music; but in Florida no such outburst of song marks the arrival of the vernal season. The brown thrush, the blue-bird, the cat-bird, the towhee, and the various kinds of sparrows that are common in the breeding season to both New England and Florida, seem to lose at the latter locality the vivacity which characterizes them at the North, their attempts at song being listless and feeble. The songs of some are also much abbreviated, and so different from what they are at the North as to be sometimes scarcely recognizable as proceeding from the same species. Even the mocking-bird sings far less than in the Middle States, and with much less power. Such at least is the general fact as indicated by my own limited experience in Florida, which accords, I find, with that of various other observers.

In recounting the faunal peculiarities of Florida it is necessary to allude further to a few facts that will be more fully presented in the following chapters, namely, the differences which distinguish the Florida representatives of species that have a wide distribution to the northward from the northern ones. It has for some time been well known that a difference in size in birds and mammals usually accompanies differences of locality in respect to latitude and elevation. Other differences, however, are found to accompany these with considerable uniformity; namely, a relative increase in the length or general size of the bill, and an increase in the intensity of the general color of the plumage.* Florida birds, in short, usually differ considerably in these respects from their New England cospecific representatives; so much so, indeed, that in many cases the majority of ornithologists would probably regard the two forms as distinct species, though few of them have as yet been specifically separated.

Hence not only do birds of the same species living at distant points differ considerably in size, color, and other features, but also in their habits, notes, and songs. With the decrease in size to the southward there seems to be a corresponding decrease in vivacity, — a fact which accords with the general law of the distribution of the higher forms of life in the temperate latitudes. Although a few structurally high types are, from certain peculiarities of their conformation, necessarily tropical, the highest races of men, whether considered physically, intellectually, or morally, are inhabitants of a medium climate, and gradually decline in rank both to the northward and southward from this favored region, animal and vegetable life reaching, as a whole, its highest manifestation in the temperate latitudes. The excessive variety of forms within the tropics mainly results from the addition of those of comparatively low or medium grades, only a few of the exclusively tropical forms being of absolutely high rank. Generally, too, the forms to be properly regarded as temperate are represented in the tropics by only their lower members, while, conversely, many of the higher types of the tropics are really cosmopolitan.

* See Annual Report of the Mus. Comp. Zoöl., 1869, p. 16.

PART II.

List of the Mammals of East Florida, with Annotations.

FELIDÆ.

1. **Felis concolor** *Linné.* PANTHER.

Not very unfrequent in the more unsettled parts of the State. I saw several hunter's skins of it at Jacksonville, said to have been taken up the river.

2. **Lynx rufus** *Rafinesque.* BAY LYNX.

Abundant. Especially numerous on the Upper St. John's and Indian Rivers, according to Mr. Maynard and others.

CANIDÆ.

3. **Canis lupus** *Linné.* GRAY WOLF.

Canis lupus LINNÉ, Syst. Nat., I, 58, 1767. — ALLEN, Bull. Mus. Comp. Zoöl., I, 154, October, 1869

Canis lupus, occidentalis RICH., Fauna Bor. Amer., I, 60, 1829.

Canis occidentalis et var. BAIRD, Mam. N. Amer., 104, 111, 113, 1857.

Not numerous. They were described to me as being very dark colored, or black.* This account tends to confirm the statement of Audubon and Bachman in respect to this point.† After citing the comparative frequency of this form of the common wolf in Kentucky, and in several of the Southern Atlantic and Gulf States, as compared with its occurrence in regions more to the northward and westward, they observe: "The varieties with more or less of black continue to increase, as we proceed farther to the south; and in Florida the prevailing color of the wolves is black." ‡

4. **Vulpes virginianus** *Richardson.* GRAY FOX.

Canis virginianus ERXL., Syst. Reg. Anim., 567, 1777. — "SCHREBER, Säugeth., III, 361, pl. xcii, 1778."

Canis cinereo-argentatus ERXL., Syst. Reg. Anim., 567, 1778. — "SCHREBER, Säugeth., 360, pl. xcii." — GODMAN, Am. Nat. Hist., I, 280, 1826.

Canis griseus BODD., Elenchus Anim., I, 77, 1784.

* Since writing the above, I have received a letter from Mr. G. A. Boardman, of Milltown, Me., in which he also refers to the dark color of the Florida wolves.

† Quad. N. Amer., Vol. II, p. 130.

‡ Respecting the distribution of the different color races of the common wolf in North America, see my paper on the Mammals of Massachusetts, Bulletin Mus. Comp. Zoöl., Vol. I, p. 156, 1869.

Canis (Vulpes) virginianus RICH., Faun. Bor. Am., I, 96, 1829.
Vulpes virginianus DEKAY, New York Fauna, I, 45, pl. vii, fig. 2, 1842. — AUD. & BACH., Quad. N. Am., I, 162, pl. xxi, 1849.
Vulpes (Urocyon) virginianus BAIRD, Mam. N. Am., 138, 1857.

Common.

MUSTELIDÆ.

5. **Putorius lutreolus** *Cuvier.* MINK.

Mustela lutreola LINN., Syst. Nat., 66, 1766.
Putorius lutreolus CUV., Règ. Anim., I, 148, 1817. — ALLEN, Bull. Mus. Comp. Zoöl., I, 175, October, 1869.
Putorius vison GAPPER, Zoöl. Journ., V, 202, 1830.
Putorius nigrescens AUD. & BACH., Quad. N. Am., III, 104, pl. cxxiv, 1853.

"Not common." — *Boardman.* I did not meet with it. It is well known to be common, however, in the adjoining States. Audubon and Bachman speak of it as being very numerous in the rice-fields of South Carolina

6. **Lutra canadensis** *Sabine.* OTTER.

Abundant. Its fur, however, is of little value, compared with that of northern specimens, and the animal is hence not much hunted.

7. **Mephitis mephitica** *Baird.* COMMON SKUNK.

Viverra mephitica SHAW, Mus. Lever., 172, 1792. — IBID., Gen. Zoöl., I, 390, 1809.
Mephitis chinga TIEDEM., Zoöl., 362, 1808.
Mustela (Mephitis) americana DESM., Mamm., I, 186, 1820.
Mustela varians GRAY, Charlesw. Mag. Nat. Hist., I, 581, 1837.
"*Mustela mesomelas* LICHT., Darst. Säugeth., I, fig. 2." — GEOFF. ST. HIL., Voy. de la Venus, Zool., I, 133, 1855. — MAX. ZU WIED, Archiv für Naturgesch., XXVII, 218, 1861. — BAIRD, Mam. N. Am., 199, 1857.
Mephitis macroura AUD. & BACH., Quad. N. Am., III, 11, 1853.
Mephitis mephitica BAIRD, Mam. N. Amer., 195, 1857. — ALLEN, Bull. Mus. Com. Zoöl., I, 178, October, 1869.
Mephitis occidentalis BAIRD, Mam. N. Amer., 194, 1857.

Common on the Lower St. John's, but, according to Mr. Maynard, quite unknown on the Indian River.

8. **Mephitis bicolor** *Gray.* LITTLE STRIPED SKUNK.

Mephitis bicolor GRAY, Charlesw. Mag. Nat. Hist., I, 581, 1837. — BAIRD, Mam. N. Amer., 196, 1857.
Mephitis zorilla LICHT., Abhand. Akad. Wiss. Berlin, for 1836, 281, 1838. — AUD. & BACH., Quad. N. Amer., III, 276, 1854.
Mephitis interrupta LICHT., Abhand. Akad. Wiss. Berlin, for 1836, 283, 1838.

This beautiful little animal was obtained by Mr. C. J. Maynard at Captain Dummitt's, where it was said to be common in the scrub. Mr. Maynard says they are domesticated and used there as cats, the odor glands being removed when the animals are young; they become very tame and are quite efficient in destroying the mice (*Hesperomys* sp.) that infest the houses. I am not aware that this animal has been reported before from any point east of the Mississippi River. It has been recently ascertained to extend northward in the interior as far as Central Iowa.*

URSIDÆ.

9. Procyon lotor *Storr.* RACCOON.

Ursus lotor LINNÉ, Syst. Nat., 48, 1758.
Procyon lotor BAIRD, Mam. N. Amer., 209, 1857. — ALLEN, Bull. Mus. Comp. Zoöl., I, 181, October, 1869.
Procyon Hernandezii WAGLER, Isis, XXIV, 514, 1831. — BAIRD, Mam. N. Amer., 212, 1857. — IBID., U. S. and Mex. Bound. Surv., II, Mam., 22, 1859.

Exceedingly numerous.

10. Ursus arctos *Linné.* COMMON BEAR.

Ursus arctos LINNÉ, Syst. Nat., 69, 1766. — CUVIER, Règ. Anim., I. 142, 1817. — BLAINVILLE. — MIDDENDORFF, Sibirische Reise, II, ii. 1854. — GRAY, Proc. London Zoöl. Soc., 1864, 682. — ALLEN, Bull. Mus. Comp. Zoöl., I, 184, October, 1869.
Ursus americanus PALLAS, Spicelegia Zoöl., XIV, 6, 1780. — GMELIN, Syst. Nat., I, 101, 1788. — RICHARDSON, Faun. Bor. Amer., I, 14, 1829. — AUD. & BACH., Quad. N. Amer., III, 187, 1853. — MAX. ZU WIED & MAYER, Verhandl. Akad. der Naturf., XXVI, i, 33, 1857. — BAIRD, Mam. N. Amer., 225, 1857.
Ursus (Euarctos) americanus GRAY, Proc. Lond. Zoöl. Soc. 1864, 692.
Ursus horribilis ORD, "Guthrie's Geog., 2d Amer. ed., II, 291, 299, 1815." — SAY, Long's Exped., II, 53, 1823. — BAIRD, Mam. N. Amer. 219.
Ursus horribilis, var. *horriaceus*, BAIRD, U. S. & Mex. Bound. Survey, Rep., II, Mam., 24, 1859.
Ursus cinereus DESM., Mam., I, 164, 1820.
Ursus (Danis) cinereus GRAY, Proc. Lond. Zöol. Soc., 1864, 690.
Ursus ferox RICHARDSON, Faun. Bor. Amer., I, 24, 1829. — MAX. ZU WIED, Reise in das innere Nord Amer., I, 488, 1839. — MAX. ZU WIED & MAYER, Verhandl. Akad. der Naturforsch., XXVI, 39.
Ursus cinnamomeus BAIRD, U. S. & Mex. Bound. Survey Rep., II, Mam., 29.

* See H. W. Parker, in Amer. Nat, Vol. IV, 376, August, 1870.

Numerous and often troublesome, occasionally destroying swine, of which they are exceedingly fond. Judging from their tracks in the swamps, they must not only be exceedingly numerous, but some of them of enormous size. The several skins seen by me were all intensely black.*

CERVIDÆ.

11. Cariacus virginianus *Gray.* VIRGINIA DEER.

Cervus virginianus BODDÆRT, Elench. Animal., I, 136, 1784. — GMELIN, SCHREBER, DESMEREST, AUD. & BACH., BAIRD, &c.
Cariacus virginianus GRAY, Cat. of Bones in Brit. Mus., 266, 1862.

Abundant almost everywhere. Not so numerous along the Lower St. John's as in the more unsettled districts further south. As remarked by Professor Baird, the Florida deer are considerably smaller than those of the Northern States; so much so that it is a fact of common observation.

MANATIDÆ.

12. Trichechus manatus *Linné.* MANATEE.

Trichechus manatus LINNÉ, Syst. Nat., I, 34, 1758.
"*Manatus australis* TILESIUS, Jahrb. der Naturg., I, 23." — GRAY, Cat. Seals and Whales, 358, 1866. — MURRAY, Geog. Distr. Mam., 202, 1866.
Manatus amer anus DESM., Mam., 507, 1822.
Manatus latirostris HARLAN, Journ. Phil. Acad. Nat. Sci., III, 390, pl. xii, fig. 1-3, 1824. — IBID., Faun. Amer., 277, 1825.

I learn from Mr. Maynard that the manatee is still quite common in Indian River, where they are often caught in nets. They come into the river at night to feed on the mangrove bushes. Mr. Maynard did not meet with them in Mosquito Lagoon, which he traversed nearly its whole length, and he thinks they do not occur there.

The manatees of America and Africa seem to be very closely allied, and to number at most but two species. Those of the same species also appear to be exceedingly variable in their osteological characters. Dr. J. E. Gray,

* In my recent paper in this Bulletin, cited above, I have discussed the mutual relationship of the numerous supposed species of land bears of the northern hemisphere. The close affinity between the bears of Northwestern America and Northeastern Asia is especially noticed; but at that time I was not aware that Temminck, in the Fauna Japonica, had referred the large land bear of Japan to the *U. ferox* of authors, or to the so-called "grizzly bear" of Western America. This indicates the very close affinity, in this author's opinion, of the Japan and American bears.

in a valuable paper entitled "On the Species of Manatees (*Manatus*), and on the Difficulty of distinguishing such Species by Osteological Characters,"* states that he finds the African and American species are distinguished by only a single character, — the absence of the nasal bones in the African species. Concerning the individual variation in the skulls of the two species, he observes as follows: "When Cuvier had a skull of the American and one of the African Manatee, he gave eight characters by which the African skull could be known from the American. Now we have a series of skulls of each kind, we find that not one of these characters will separate the skulls of the two countries from one another. Indeed, the skulls of each kind are so variable that, after having them laid out before me for two or three days, studying them every now and then, and inducing two proficients in the study of bones, and in observing minute characters, to give me their assistance, we came to the conclusion that we believed there was no character, common to all the skulls of each kind, which could be used to separate them. As a proof of the difficulty of so doing, I may state that there was one skull in the series which had been long in the collection, and had been received without any habitat, and neither of the three could decide to which of the series this skull should be referred; and it was not until I accidentally observed the character, derived from the absence of the nasal bones in the African kind, that this question could be settled."

Having myself been struck with the variability of osteological as well as external characters in individuals of the same species, in both birds and mammals, — a matter to which I have already often called attention, and the consideration of which occupies a considerable portion of Part III of the present paper, — I can hardly refrain, in this connection, from citing further the judicious remarks of Dr. Gray on this point. "The examination," he says, "of a large series of skulls of the bears (*Ursus*) and *Paradoxuri*, shows how difficult it is to distinguish species by the study of the skulls alone. Thus, when we have a series of skulls of bears from different localities, which, from their external form and habits, are known to be distinct species, it is easy to say which is the skull of *U. thibetanus*, *U. syriacus*, *U. arctos*, *U. cinereus*, and *U. americanus*, when we have the habitat marked on each; but the true test of the power of distinguishing the one from the other is to determine to what species a skull belongs, of which we have no information as to its origin; and we have several skulls in the British Museum under these circumstances, and I cannot, with the best assistance at my command, determine to which species they ought to be referred. And it is the same with the *Paradoxuri*." "If this is the case with the skulls," he continues, "how must the difficulty of distinguishing species with certainty be increased when we have only fossil bones, which are generally more or less imperfect,

* Ann. and Mag. Nat. Hist., 3d Ser., Vol. XV, pp. 130 - 139, 1865.

to examine and compare, or of which only a limited number of examples are to be obtained and compared? They [the skulls] vary in most genera much more than was expected, before series of the skulls of each species were collected and compared."

These observations by Dr. Gray are fully confirmed by my own studies; and I hence believe that, as the number of specimens of different species increases in our museums, many species now believed to be valid will be found to rest merely on individual characters.

VESPERTILIONIDÆ.

13. **Lasiurus noveboracensis** *Gray.* RED BAT.

Vespertilio noveboracensis ERXL., Syst. Règ. Anim., 135, 1717.
Vespertilio lasiurus GMEL., Syst. Nat., 1788.
Vespertilio rubellus PAL. DE BEAUV., Cat. Peale's Mus., 1796.
? *Vespertilio cinereus* PAL. DE BEAUV., Ibid.
? *Vespertilio pruinosus* SAY, Long's Exped., 67, 1823. — RICH., Faun. Bor. Am., I, 1, 1829.
Taphozous rufus HARLAN, Faun. Amer., 23, 1825.
Lasiurus rufus GRAY, List Mam. Brit. Mus., 32, 1842.
Lasiurus noveboracensis TOMES, Proc. Lond. Zoöl. Soc., 1857, 34.
? *Lasiurus pruinosus* TOMES, Ibid., 37.
Lasiurus noveboracensis H. ALLEN, Mon. N. Am. Bats, 15, 1864. — J. A. ALLEN, Bull. Mus. Comp. Zoöl., I, 207, 1869.
? *Lasiurus cinereus* H. ALLEN, Mon. N. Am. Bats, 21.

Common. All of the several specimens obtained, both by myself and Mr. Maynard, are of a deep cherry red, with but a slight skirting of ash, and are uniformly much darker or deeper colored than any I have seen from the Northern States. All examined (nine specimens) were males.

14. **Scotophilus fuscus** *H. Allen.* CAROLINA BAT.

Vespertilio fuscus PAL. DE BEAUV., Cat. Peale's Mus., 14, 1796. — LECONTE, Proc. Phil. Acad. Nat. Sci., VII, 437, 1855.
Vespertilio carolinensis GEOFF. ST. HIL., Ann. du Mus., VIII, 193, 1806, pl. xlvii, fig. 7. — HARLAN, North Am. Jour. Geol. & Nat. Sci., I, 218, 1831 — LECONTE, Proc. Phil. Acad. Nat. Sci., VII, 437.
Vespertilio arcuatus SAY, Long's Exped., 167, 1823.
Vespertilio phaiops RAF., Amer. Month. Mag., 445, 1818.
Vespertilio ursinus TEMM., Mam., II, 234, 1835.
Scotophilus fuscus H. ALLEN, Mon. N. Am. Bats, 31, 1864.
Scotophilus carolinensis H. ALLEN, Ibid., 28.

Common. Several specimens taken.

15. **Scotophilus georgianus** *H. Allen.* GEORGIA BAT.

Scotophilus georgianus H. ALLEN, Mon. N. Am. Bats, 35, 1864, *nec. syn.* — J. A. ALLEN, Bull. Mus. Comp. Zoöl., No. 8, 1869.

This species doubtless occurs in Florida, at least in the northern part, since the capture of specimens at different localities in Georgia and at New Orleans is on record.*

16. **Nycticejus crepuscularis** *H. Allen.*

Vespertilio crepuscularis LECONTE, McMurtrie's Cuv. An. King., I, 432, 1831. IBID., Proc. Phil. Acad. Nat. Sci., VII, 438, 1855.

Nycticejus crepuscularis H. ALLEN, Mon. N. Am. Bats, 12, 1864.

A specimen collected by Mr. Maynard at Jacksonville, in January, but afterwards lost, I refer from his measurements and description of it to this species. There is also a specimen (No. 731) in the Museum of Comparative Zoölogy, collected in Florida by Mr. Chas. Belknap.

17. **Corynorhinus macrotis** *H. Allen.* BIG-EARED BAT.

Plecotus macrotis LECONTE, McMurtrie's Cuv. An. King., I, 431, 1831.

Plecotis LeContei COOPER, Ann. Lyc. Nat. Hist., IV, 12, 1837.

Synotus macrotis H. ALLEN, Mon. N. Am. Bats, 63, 1864.

Corynorhinus macrotis H. ALLEN, Proc. Phil. Acad. Nat. Sci., XVII, 173, Aug. 1865.

A specimen of this species from Micanopy, Florida, collected by Dr. Bean, is cited by Dr. Allen.† This Southern species ranges northward along the coast nearly or quite to the Middle States, it being comparatively common, according to authors, in South Carolina.

NOCTILIONIDÆ.

18. **Nyctinomus nasutus** *Tomes.*

Molossus nasutus SPIX, Sim. et Vesp. Bras., 60, pl. xxxv, fig. 7, 1823.

Nyctinomus nasutus TOMES, Proc. Lond. Zoöl. Soc., 1861, 68. — H. ALLEN, Mon. N Am Bats, 7, 1867.

This widely distributed southern species should unquestionably be included among the mammals of Florida. It has been reported from Texas, Louisiana, South Carolina, and the West Indies,‡ as well as from South America, as far south even as Buenos Ayres.§ Specimens in

* Dr. H. Allen, Monograph of North American Bats, p. 38.

† Ibid., p. 55.

‡ Ibid., p. 10.

§ Tomes, Proc. Lond. Zoöl. Soc., 1861, p. 68.

the Museum of Comparative Zoölogy from Hayti, collected by Mr. P. R. Uhler, have been identified by Dr. Harrison Allen as of this species.

19. Megadermatidæ *Sp.?*

A large species of bat was noticed by both Mr. Maynard and myself, but as it always flew very high, neither of us obtained it. It was very much larger than any other species yet described from the United States, and is doubtless a West Indian form; probably a species of *Megadermatidæ*.

SORECIDÆ.

20. Blarina brevicauda *Baird.* MOLE SHREW.

Sorex brevicaudus SAY, Long's Exped., I, 164, 1862-63.
Sorex parvus SAY, Ibid., 164.
Sorex talpoides GAPPER, Zoöl. Journ., V, 208, pl. viii, 1830.
Sorex carolinensis BACH., Journ. Phil. Acad. Nat. Sci., VII, 366, pl. xxvi, fig. 3, 1837.
Sorex cinereus BACH., Ibid., 373, fig. 3.
Sorex Dekayi BACH., Ibid., 377, fig. 4.
Corsira (Blarina) talpoides GRAY, Proc. Lond. Zoöl. Soc., V, 124, 1837.
Blarina brevicauda BAIRD, Mam. N. Am., 42, pl. xxx, fig 6, 185?.— ALLEN, Bull. Mus. Comp. Zoöl., I, 212, October, 1869.
Blarina talpoides BAIRD, Mam. N. Am., 37, pl. xxx, fig 5.
Blarina carolinensis BAIRD, Ibid., 45, pl. xxx, fig. 8.
Blarina cinerea BAIRD, Ibid., 48, pl. xxx, figs. 9 and 10, young.
Blarina exilipes BAIRD, Ibid., 51, pl. xxviii, young.
Blarina Berlandieri BAIRD, Ibid., 53, pl. xxviii, young.

A single specimen of *Blarina* from Indian River, Florida, collected by Mr. G. Wurdemann, is mentioned under "*Blarina cinerea*," by Professor Baird, as having been received at the Smithsonian Institution.* While it may be of a species distinct from *B. brevicauda*, it seems more probable that it is the young of that species, as I have elsewhere stated.† *Sorex cinereus* of Bachman,‡ which Professor Baird cites as a synonyme of his *Blarina cinerea*, Dr. Bachman subsequently regarded as the young of his *S. carolinensis*,§ which is the same as *B. talpoides et brevicauda* of recent writers.

* North American Mammals, p. 50.
† Bull. Mus. Com. Zoöl., Vol. I, No. 8, p. 212.
‡ Journ. Phil. Acad. Nat. Sci., Vol. VII, 1837, p. 373, pl. xxiii, fig. 3.
§ Quadrupeds of North America, Vol. III, p. 344.

TALPIDÆ.

21. **Scalops aquaticus** *Fischer.* SHREW MOLE.

Several specimens of this species from Indian River and Jacksonville, Florida, are mentioned by Professor Baird in his list of the specimens of this species in the Museum of the Smithsonian Institution, in his Report on North American Mammals. Mr. Boardman has also informed me that it is not uncommon there.

SCIURIDÆ.

22. **Sciurus niger** *Linné.* SOUTHERN FOX SQUIRREL.

Sciurus niger LINNÉ, Syst. Nat., I, 64, 1758.
Sciurus vulpinus GMEL., Syst. Nat., I, 147, 1788.
Sciurus vulpinus et syn. BAIRD, Mam. N. Am., 246, 1857.

Common. Confined chiefly to the pine woods. Extremely variable in general color, the variations in this respect ranging from pale yellowish gray to black. The specific name *niger* of Linné is the one which has unquestionably the priority, as observed by Professor Baird, and its applying only to a single stage of coloration, inasmuch as it is a common one, does not seem to be sufficient reason for rejecting it, since it is as applicable as any name referring to its color can be, and is not likely to seriously mislead.*

23. **Sciurus carolinensis** *Gmelin.* GRAY SQUIRREL.

Sciurus carolinensis GMEL., Syst. Nat., I, 143, 1788. — BAIRD, Mam. N. Am., 256, 1857.
" *Sciurus cinereus* SCHREBER, Säugeth., IV, 766, pl. ccxiii, 1792."
Sciurus niger GODMAN, Am. Nat. Hist., II, 133, 1826.
Sciurus leucotis GAPPER, Zoöl. Journ., V, 206, pl. xxi, 1830.

Exceedingly abundant, and generally very tame. Two of my party shot a dozen one evening in less than half an hour at Hawkinsville. They are considerably smaller than at the North, and also differ somewhat in color from northern specimens, the gray being more suffused with brownish than in the gray northern type.

The fifty or sixty specimens carefully examined were quite uniform in color and generally so in size. The yellowish-brown patch on the back usually present in the gray type of this species was of greater extent and less distinctly defined than in northern examples. No

* See Baird, North American Mammals, p. 248.

dusky or black varieties were noticed, nor could I learn that they existed here. Their voice is not so heavy as that of the northern animal, but in no other respects than in those above mentioned do they differ from it. Professor Baird has quite fully described the gradual transition from the common gray to the glossy black type of coloration seen at the North, where the dark varieties are most common.*

Measurements of Florida Specimens.

M. C. Z No.	Original No.	Sex.	Locality.	Date.	Collected by	Total Length.	Nose to Occiput	Nose to Tail.	Tail to end of Vertebræ	Tail to end of Hairs.	Length of Fore Foot.	Length of Hind Foot.
2455	203	♂	Jacksonv'l	Jan. 12	C.J.Maynard	20.00	2.50	9.00	8.00	11.00	2.00?	2.45
2454	206	♀	"	" 10	"	18.20	2.40	9.00	7.75	10.45	1.35	2.35
2453	207	♀	"	" 12	"	17.75	2.45	8.45	7.50	10.25	1.45	2.45
2040	351	♀	"	" 25	J. A. Allen	19.50	—	9.50	7.50	10.00	—	—
2041	352	♂	"	" 25	"	20.50	—	10.50	7.50	10.00	—	—
	221	♀	Dummitt's	Mar.16	C.J.Maynard	20.00	2.45	9.50	8.15	10.50	1.60	2.56
	222	♂	"	" 18	"	18.56	2.55	10.20	8.00	10.56	1.35	2.40
2054	365	♂	Welaka	Feb. 6	J. A. Allen	21.00	—	10.00	8.25	11.00	—	—
2055	366	♀	"	"	"	21.50	—	10.00	8.75	11.50	—	—
2056	367	♂	"	"	"	19.00	—	10.00	8.40	9.00	—	—
2057	368	♂	"	"	"	19.00	—	9.00	7.50	10.00	—	—
2058	369	♀	"	"	"	19.00	—	9.15	7.60	9.85	—	—
2059	370	♂	"	"	"	21.50	—	10.15	9.60	11.35	—	—
2066	377	♂	Hawkinsv'l	Mar.12	"	19.50	2.40	9.50	8.00	10.50	1.50	2.15
2067	378	♀	"	"	"	19.50	2.42	9.25	8.00	9.25	1.60	2.23
	379	♀	"	"	"	19.25	2.33	8.25	8.75	11.00	1.35	2.22
	380	♀	"	"	"	20.00	2.50	9.25	8.00	9.75	1.40	2.20
	381	♀	"	"	"	18.75	2.45	8.75	7.60	10.00	1.50	2.25
2068	382	♂	"	"	"	19.50	2.35	8.90	7.80	10.60	1.50	2.05
	384	♂	"	"	"	19.00	2.45	9.00	7.60	10.00	1.50	2.25
	385	♂	"	"	"	18.50	2.30	9.50	6.75	9.00	1.40	2.20
	386	♀	"	"	"	18.50	2.35	8.50	8.00	10.00	1.40	2.15
	387	♂	"	"	"	19.75	2.45	9.00	8.50	10.75	1.45	2.20
	388	♂	"	"	"	20.50	2.50	10.00	8.00	10.50	1.45	2.37
	389	♂	"	"	"	19.75	2.35	8.75	8.00	11.00	1.42	2.25
	390	♂	"	"	"	19.25	2.40	9.00	8.00	10.25	1.45	2.22
	391	♂	"	"	"	19.00	2.35	9.25	7.65	9.75	1.40	2.30
	392	♂	"	"	"	18.50	2.40	9.00	7.50	9.50	1.45	2.32
	393	♂	"	"	"	19.25	2.40	9.15	7.85	10.10	1.45	2.22
	394	♀	"	"	"	20.00	2.38	9.25	8.15	10.75	1.45	2.20
	395	♂	"	"	"	19.50	2.40	9.00	7.90	10.50	1.50	2.40
	396	♂	"	"	"	20.00	2.50	9.75	8.15	10.25	1.55	2.35
	397	♂	"	"	"	19.25	2.40	9.00	7.75	10.25	1.40	2.30
	398	♂	"	"	"	18.35	2.40	8.75	7.95	9.60	1.55	2.20
	399	♀	"	"	"	20.50	2.50	10.15	7.85	10.35	1.65	2.45
	400	♂	"	"	"	19.00	2.40	8.90	7.35	10.10	1.60	2.15

* N. Am. Mam., p. 259. See further on this point my remarks on this species in No. 8 of the first volume of this Bulletin, already cited.

24. **Geomys pineti** *Rafinesque.* "SALAMANDER."

Geomys pinetis RAF., Amer. Month. Mag., II, 45, 1817.
Pseudostoma floridana AUD. & BACH., Quad. N. Am., III, 242, pl. cl, fig. 1, 1853.

Common, but mainly confined to the drier portions of the pine woods. The five specimens collected by me differ very much in size, and considerably in color, some of them being plumbeous and others brownish-plumbeous; in other words, some are much darker than others. The difference in size appears to be mainly due to age. This species extends southwards at least as far as Lake Harney, and at some localities is particularly numerous, the little hillocks of earth it throws up sometimes nearly covering the ground.

MURIDÆ.

25. **Mus decumanus** *Pallas.* BROWN RAT.

Abundant at Jacksonville, but not observed by any of my party on the Upper St. John's, nor by Mr. Maynard on Indian River.

Although no other species of *Mus* was observed, it is not improbable that the common mouse (*M. musculus*) occurs in the vicinity of the towns. It was not found on the Upper St. John's (to which locality it probably has not yet extended), where the common house mice are a species of *Hesperomys*, as are also the house mice on Indian River, according to Mr. Maynard. Neither was any species of *Reithrodon* obtained. The *R. humilis*, which occurs in Georgia and South Carolina, is certainly to be expected in Northern Florida; but it has not yet to my knowledge been reported from there.

26. **Hesperomys leucopus** *Wagner.* WHITE-FOOTED MOUSE.

Mus sylvaticus, var. ERXL., Syst. Reg. An., I, 390, 1775.
Mus leucopus DESM., Mam., II, 307, 1822. — AUD. & BACH., Quad. N. Am., I, 300, pl. xlvi, 1849.
Mus agrarius GODMAN, Am. Nat. Hist., II, 1826.
Mus noveboracensis SELYS-LONGCH., Étude Micromam., 67, 1839.
Mus Emmonsii DEKAY, Emmon's Rep. Quad. Mass., 61, 1840.
Cricetus myoides GAPPER, Zoöl. Journ., 1830, 204.
Hesperomys polionotus WAGNER, Wiegm. Arch., 1843, ii, 52.
? *Hesperomys cognatus* LECONTE, Proc. Phil. Acad. Nat. Sci., VI, 442, 1852.
Hesperomys leucopus LECONTE, Ibid., 413. — BAIRD, Mam. N. Am., 459, 1857. — ALLEN, Bull. Mus. Comp. Zoöl., I, 227, October, 1869.
Hesperomys myoides BAIRD, Mam. N. Am., 472.
Hesperomys indianus MAX. ZU WIED., Arch. für Naturg., XXVIII, i, 111, 1862.

A mouse provisionally referred to this species was abundant, especially at certain localities. At my first camp, about twenty-five miles above Jacksonville (near Hibernia), an *Hesperomys* and the wood rat (*Neotoma floridana*) were excessively numerous. At evening they began scampering over the leaves, their little footsteps being heard in every direction ; at times they approached so near the camp-fire as to be distinctly seen. They ascended the bushes, and could be heard on the lower branches of the trees. Some of my party being unaccustomed to such manifestations of nocturnal life, were at first filled with apprehension as to the character of their visitors, and could scarcely be convinced that the place was not infested with poisonous snakes or other dangerous animals. Depending upon my traps for specimens, which unfortunately for me the mice avoided, I secured but two or three examples of the *Hesperomys* so abundant here. These, with several others obtained by me elsewhere, as also others obtained on Indian River by Mr. Maynard, including both young and adult, are undistinguishable from the common *H. leucopus* of the North, the young being deep plumbeous.

I observed at this place a fact in respect to the habits of the *Hesperomys* I had not previously noticed nor seen pointed out, though it was noticed in all the parts of Florida I visited. I refer to its habit of cutting off the branches and main stems of the young saplings. I at first supposed this work to be that of the wood-boring larvæ of some coleopterous insect, so nearly did the "pruning" resemble that of the so-called "oak-pruners" (*Cerambycidæ* sp.). A closer examination, however, showed that, instead of the twigs being smoothly cut, as by a boring insect working from within outwards and severing the bark last, the cutting was begun from without, and that a considerable portion of wood had been gnawed away, both the cut surfaces being highest at the middle. Marks of the teeth of these little gnawers were also generally clearly distinguishable. No traces of boring by insect larvæ could be detected near the severed point. The branches thus cut are generally of about the size of one's finger, and are usually the main stem of a young sapling. Various species of trees are thus mutilated ; but as they are usually destitute of fruit, the purpose of these animals in this work is not apparent. It is a habit that may be common to the *Hesperomys* of the North, but I have never seen it referred to. These little animals being a hundred-fold more numerous in East Florida than they generally are in the Northern States, their work would here be of course much more noticeable.

27. **Hesperomys aureolus** *Wagner*. GOLDEN MOUSE.

Arvicola Nuttalli HARLAN, Month. Amer. Journ. Geol. &. Nat. Sci., I, 446, 1842. — IBID., Med. & Phys. Researches, 55, pl. ——, 1835.
Mus (Calomys) aureolus AUD. & BACH., Journ. Phil. Acad. Nat. Sci., VIII, 302, 1842. — AUD. & BACH., Quad. N. Am., II, 305, pl. xcv, 1851.
Hesperomys aureolus WAGNER, Wieg. Archiv, 1843, ii, 51.
Hesperomys Nuttalli BAIRD, Mam. N. Am., 467, 1857.

A single specimen which I refer to this so-called species was obtained by Mr. Maynard at Dummitt's. While this example is of the size and general proportions of *H. leucopus*, it is markedly different in color, being of a bright golden yellow above, which color reaches on the outside of the legs to the feet; the under surface has also a yellowish wash. It also differs in the texture of its fur, which is remarkably soft and fine. It is a little lighter colored than Audubon and Bachman's description and figure of *H. aureolus* represent that animal to be, but the distribution of the colors is the same, the specimen in question being not orange, but bright yellowish-cinnamon. It is, however, much nearer this than to Dr. Harlan's *Arvicola Nuttalli*. The latter does not differ very appreciably, judging from Dr. Harlan's very unsatisfactory description and his wretched figure of it, which was evidently made from a badly stuffed skin. Mr. Maynard believes the specimen referred to above to be a young animal, and states that it was so regarded by the people in whose house it was caught. He further informs me that he captured another of the same color, but very much larger, which was lost. This he regards as merely the adult of the same species. His measurements show the latter to have agreed in size and proportions with the so-called *H. gossypinus*. The texture of the fur of the small specimen above referred to agrees with that of the plumbeous, immature stage of *H. leucopus*. This form, whether a valid species or not,* is now known to occur in Southeastern Pennsylvania, Southern Illinois, Georgia, Florida, Mississippi, Missouri, and at several intermediate points.

28. **Hesperomys gossypinus** *LeConte*. COTTON MOUSE.

Hesperomys gossypinus LECONTE, Proc. Phil. Acad. Nat. Sci., VI, 411, 1853. — BAIRD, Mam. N. Am., 469, 1857.

* This and the following species are only provisionally adopted. See a previous number of this Bulletin (Vol. I., No. 8, p. 227) for a fuller expression of my views as to the number of North American species of this group, and their mutual affinities.

Several specimens were obtained, corresponding in size and color with what LeConte and Baird have described under this name. It is apparently common. As I have previously stated elsewhere,* these Florida specimens have well-developed cheek-pouches.

The specimens in question are rather larger than any examples of *H. leucopus* I have seen from the Northern States, they agreeing very well in measurements with the two specimens cited by Professor Baird.† The large size of these specimens, conjoined with their southern habitat, would seem at first to clearly indicate their being distinct from *H. leucopus*, as they are at least one third larger than the average size of the latter at the North. Professor Baird in speaking of this species observes: "There is every reason to consider this mouse as specifically distinct from *H. leucopus* of the North; although skins, when much stretched (as Nos. 1105, 1112, from Middleboro', Massachusetts), of the latter, may measure as much as those recorded here, yet they are certainly actually smaller, as shown by the feet, which never attain anything of the length of .45 for the anterior and .90 for the posterior." But he is "hardly satisfied," he adds, "that this animal is different from the smaller *H. leucopus*, as the difference in size is no greater than is to be seen in a series of *Hesperomys* from more northern localities. The tail is duskier beneath than in *H. cognatus*, and the sides more rusty; otherwise I can realize only the larger size. Should both [*H. cognatus* and *H. gossypinus*] prove to be the same, the name *H. gossypinus* must of course take precedence."

As already observed, the prevailing form of the *Hesperomys* of East Florida is not essentially different from a large proportion of the *H. leucopus* of the North, either in measurements, proportions, or color, although it is unmistakably referable to the so-called *H. cognatus*, which has been supposed to replace in the Southern Atlantic and Gulf States the *H. leuçopus* of the more northern ones. If, as I have elsewhere suggested (*loc. cit.*), as Professor Baird admits may be, and as the facts seem to indicate, *H. gossypinus* is inseparable from *H. cognatus*, and the latter being most unquestionably referable to *H. leucopus*, it would seem that *H. gossypinus* must also be referred to the *H. leucopus*.

Respecting the variations in this species and the affinities of the *H. gossypinus*, Audubon and Bachman observe as follows: "That a species so widely distributed and subject to so many variations in size, length of tail, and color, should have been often described under different names is not surprising. We have ourselves often been in a state of doubt on obtaining some striking variety. The name *Hypudæus gossypinus* of our

* Bulletin Mus. Comp. Zoöl., Vol. I, No. 8, p. 229, 1869.

† Mam. N. Am., p. 469.

friend Major LeConte (see Appendix to McMurtrie's translation of Cuv. An. Kingd., Vol. I, p. 434) was intended for this species, as it is found in the Southern States. We were for several years disposed to regard it as distinct, and have, not without much hesitation, and after an examination of many hundred specimens, been induced to set it down as a variety only." These authors also remark that they are considerably larger in the Carolinas than in the Eastern States.*

29. **Hesperomys palustris** *Wagner.* RICE-FIELD MOUSE.

Mus palustris HARLAN, Am. Journ. Sci., XXXI, 386, 1837.
"*Hesperomys palustris* WAGNER, Supplem. Schreb. Saugeth., III, 543, 1843."
Hesperomys (*Oryzomys*) *palustris* BAIRD, Mam. N. Am., 482, 1857.
Arvicola oryzvіora AUD. & BACH., Quad. N. Am., III, 214, pl. cxliv, fig. 3, 1857.

No specimens of this species were obtained by either Mr. Maynard or myself. Its habitat is usually given as South Carolina and Georgia, but Audubon and Bachman state: "The late Dr. Leitner brought us a specimen obtained in the Everglades of Florida." † It in all probability occurs also in East Florida. The above-mentioned authors give it as somewhat common in the salt-marshes near Savannah and Charleston. Professor Baird has received it from Columbus and St. Simon's Island, Georgia, and Society Hill, South Carolina.

30. **Neotoma floridana** *Say & Ord.* WOOD RAT.

Mus floridanus ORD, Bull. Soc. Philom., 1818, 181. — SAY, Long's Exped., I, 54, 1823.
Arvicola floridana HARLAN, Faun. Amer., 141, 1825.
Neotoma floridana SAY & ORD, Journ. Phil. Acad. Nat. Sci., IV, ii, 352, 1825. — BAIRD, Mam. N. Am., 487, 1857.

I found this species very abundant on the Lower St. John's, especially around Jacksonville and Hibernia, but I did not meet with it above Lake George. The old residents about Hawkinsville seemed wholly unacquainted with it. Mr. Maynard also failed to meet with it on Indian River. It hence appears probable that it may not occur very frequently in the southern part of the peninsula. Professor Baird, however, has recorded a specimen from "Indian River, Fla.," collected by Dr. Wurdemann.

The present usual northward range of this species does not appear to extend beyond North Carolina; but Professor Baird, writing in 1857,‡

* Quad. N. Amer., Vol. I, pp. 301, 305.
† Ibid., Vol. III, p. 216.
‡ Mam. N. Am., p. 489.

remarks: "A few specimens of unusually large size were captured some years ago by J. G. Bell, near Piermont, on the Hudson River, but I have not heard of any in intermediate localities [New York and Society Hill, South Carolina]." Mr. George Gibbes states that he "caught a specimen, many years ago, in Massachusetts."* Audubon and Bachman remark that specimens of it have been obtained in North Carolina, and that they had "observed a few nests in the valleys of the Virginia mountains," and that they had "somewhere heard it stated that one or two had been captured as far to the north as Maryland." †

31. **Sigmodon hispidus** *Say & Ord.* COTTON RAT.

Arvicola hispidus GODMAN, Am. Nat. Hist., II, 68, 1826.
Arvicola hortensis HARLAN, Faun. Am., 138, 1825.
Arvicola ferrugineus HARLAN, Am. Journ. Sci., X, 285, 1826.
Sigmodon hispidum SAY & ORD, Journ. Phil. Acad. Nat. Sci., IV, ii, 354, pl. x, figs. 5-8, 1825. — BAIRD, N. Am. Mam., 503, 1857.
Sigmodon Berlandieri BAIRD, Proc. Phil. Acad. Nat. Sci., VII, 333, 1855. — IBID., N. Am. Mam., 504.

Abundant throughout the country along the St. John's River, and also on Indian River, whence Mr. Maynard brought fifteen specimens. They are quite a pest to the farmers, who often successfully resort to poison to reduce their numbers. By scattering grain poisoned with strychnine about their fields they are able to destroy hundreds with slight trouble. Different specimens vary considerably in color, from gray through yellowish-brown to rufous. The so-called *Sigmodon Berlandieri*, from Texas and New Mexico, seems undistinguishable from *S. hispidus*.

In its general economy, the cotton rat represents the *Arvicolæ* of the North, especially *A. riparius*.

Concerning *S. Berlandieri*, Professor Baird remarks: "This species is readily distinguishable from *S. hispidus* by the much lighter color above, where it is grayish-yellow brown instead of distinct reddish-brown; the tail is considerably longer and covered by finer annuli. The toes are shorter, and the metatarsus shorter, while the feet are nearly the same length. The claws, however, are much weaker." The tail in this species is said to be "equal to or longer than the trunk"; the "color above grayish-yellow brown, lined with black"; while *S. hispidus* is said to have the

* Nat. Hist. Wash. Terr., Zoöl., p. 128, 1860.

† Quad. N. Am., Vol. I, p. 36.

tail "less than the trunk," and "the color above reddish brown, lined with very dark brown." The specimens from Florida examined by me are mainly of the gray type, and hence like *S. Berlandieri*, but some were decidedly rufous, or like *S. hispidus*. In "Mammals of North America" measurements of specimens of the so-called *S. Berlandieri* are given, and of twelve of *S. hispidus*. In the latter the length of the tail to the length of the trunk is as 69 to 100; in the former (*S. Berlandieri*) as 63 to 100! It hence appears from Professor Baird's own measurements that the *S. Berlandieri* is far from having the tail relatively the longer. The other distinctions are based on too few specimens to have much value, since individual variations of the same character are common.

32. **Arvicola pinetorum** *LeConte.* PINE MOUSE.

Psammomys pinetorum LECONTE, Ann. N. Y. Lyceum Nat. Hist., III, 132, pl. ii, 1829.

Arvicola scalopsoides AUD. & BACH., Journ. Phil. Acad. Nat. Sci., VIII, 299, 1842.

Arvicola pinetorum AUD. & BACH., Quad. N. Am., II, 216, pl. lxxx, 1851.

Arvicola (*Pitymys*) *pinetorum* BAIRD, N. Am. Mam., 544, 1857.

Included on the authority of Audubon and Bachman, who state that they had received it from Florida, Alabama, Mississippi, and Georgia. Professor Baird also cites specimens from Georgia and Louisiana. This is the most southern of the *Arvicolæ*, and the only one, except *A. austerus*, whose habitat includes the Gulf States.

LEPORIDÆ.

33. **Lepus sylvaticus** *Bach.* GRAY RABBIT.

Lepus americanus DESM., Mam., II, 351, 1822. — HARLAN, Faun. Amer., 193, 1825.

Lepus sylvaticus BACH., Journ. Phil. Acad. Nat. Sci., VII, 1837. — WATERH., Nat. Hist. Mam., II, 116, 1848. — AUD. & BACH., Quad. N. Am., I, 173, pl. xxii, 1849. — BAIRD, Mam. N. Am., 597, 1857.

Abundant. Mr. Maynard obtained a specimen but a few weeks old, at Dummitt's, as early as the 16th of February.

34. **Lepus palustris** *Bachman.* MARSH RABBIT.

Lepus palustris BACH., Journ. Phil. Acad. Nat. Sci., VII, 194, 336, pl. xv, xvi, 1837; Ibid., VIII, 79, 1839. — AUD. & BACH., Quad. N. Am., I, 151, pl. xviii, 1849. — BAIRD, Mam. N. Am., 615, 1857. — COUES, Proc. Bost. Soc. Nat. Hist., XIII, 86, 1869.

Common, especially on the Lower St. John's.*

* See on the habits and anatomy of this species a paper by Dr. Elliott Coues, Proceed. Bost. Soc. Nat. Hist., Vol. XIII, pp. 86 - 101, June, 1869.

DIDELPHIDÆ.

35. Didelphys virginiana *Shaw.* Opossum.

Didelphys virginiana Shaw, Gen. Zoöl., I, 473, pl. cvii, 1800. — Desmarest, Harlan, Temminck, Waterhouse, Baird, and most other authors.

? "*Didelphys marsupialis* Schreb., Säugeth., III, pl. cxlv, 1778."

Didelphys californica Bennett, Proc. Lond. Zoöl. Soc., I, 40, 1833. — Also Wagner, Waterhouse, Aud. & Bach. (from Bennett). — Baird, Mam. N. Am., 233, 1857. — Baird, U. S. & Mex. Bound. Surv. Rep., II, Zoöl., 32, 1859.

Didelphys breviceps Bennett, Proc. Lond. Zoöl. Soc., I, 40, 1833. — Waterhouse, Nat. Hist. Mam., I, 477, 1846 (from Bennett ?). — Aud. & Bach., Quad. N. Am., III, 330, 1851 (from Bennett).

Didelphys pruinosus Wagner, Wiegmann's Archiv, 1842, 358. — Waterhouse, Nat. Hist. Mam., I, 477, 1846, (from Wagner).

Abundant.

This species is quite variable in its color-markings, and remarkably so in many other features, especially in the length and size of the nose, and in the size and proportions of the skull, even in specimens from the same locality.* Slight and quite inconstant differences also occur between examples from the Southern States, Texas, Mexico, and California. It would, in fact, be quite unusual if specimens of any species ranging so widely should not be found to differ somewhat at localities so widely separated. Two supposed species of North American *Didelphys* described by Mr. Bennett, as cited above, have been quoted by numerous other authors, and by them currently adopted, without apparently an examination of their merits. Professor Baird, rejecting one of them, has endeavored to separate the opossums occurring west of the Mississippi valley from those living farther eastward, designating the western one as *D. californica.* The distinctions claimed are somewhat similar to those urged as distinguishing the so-called *Procyon Hernandezii* of the western half of the continent from the *P. lotor* of the Atlantic States. They are equally slight and unsatisfactory, and at most mark but a geographical race, so intimately allied to and intergrading with the better-known eastern form that the point at which the one supplants the other is thus far undetermined. The *Didelphys breviceps* of Bennett was founded on a single specimen from California, which differed from the so-called *D. californica* only in having a relatively shorter head.

* Since writing the above I have been incidentally informed by Dr. Coues that, in preparing his memoir on the anatomy of *Didelphys virginiana* (now publishing in the Mem. of the Bost. Soc. Nat. Hist., Vol. II, Pt. I), he had occasion to examine a large number of specimens, and that he found the variation in size and proportions to amount to nearly twenty per cent.

Part III.

On Individual and Geographical Variation among Birds, considered in respect to its bearing upon the Value of certain assumed Specific Characters.

A systematic investigation of the extent and character of individual variation in birds seems not to have hitherto been attempted; in fact, few collections exist that furnish the material necessary to such a work. In occasional instances considerable differences between individuals of the same species, other than those that result from age and sex, have, however, already been pointed out, but these instances seem to have been generally, but improperly, regarded as exceptional cases.

The collection of birds in the Museum of Comparative Zoölogy now offers unusual facilities for a general investigation of this subject, most of the common species of Eastern North America being each represented by fifty to one hundred and fifty or more specimens. The greater part of them having been collected in Southern New England, and a large proportion in Eastern Massachusetts, they are the more valuable for this purpose, from their having been collected essentially from the same locality. The examination of this material has disclosed a hitherto unsuspected range of purely individual differentiation in every species thus far studied. At the same time regard has been had to the more obscure seasonal variations in color, and to the general differences that depend upon age, including such as result from senility as well as from immaturity. Local or geographical variations have likewise been carefully considered, with results that a short time since were unsuspected. These several lines of investigation have shown that in many instances what have been regarded as reliable characteristics of species have in not a few cases really little or no value; that the importance of many diagnostic features has been too highly estimated, and that consequently a careful revision of our published faunæ will be necessary for the elimination of the merely nominal species. In the following pages many of the data which have led to these conclusions will be presented.

Individual variation not only affects color and size, but the proportions of different parts, as the relative size and form of the wings, tail, bill, toes, and tarsi, including the skeleton as well as the external organs;

of the soft parts no account can as yet be given. Geographical variation has an equally universal range, but is most strikingly exhibited in the color, in size, and in the form of the bill. Individual variation will be first considered, and subsequently geographical variation. In each case each prominent phase of variation will be more or less fully described.

1. Individual Variation.

Individual Variation in Color. — In birds of whatever age, two lines of variation from the average or medium type of coloration are readily distinguishable, the variation depending essentially on differences in the depth or intensity of the general tint. On the one hand, individual variation in color results from a greater than the average amount of coloring matter in the integuments ; on the other hand, from an amount less than the average amount. The difference in this respect between the extremes of a series of fifty or one hundred specimens of any species, collected in course at a single locality, and nearly at the same season of the year, is often as great as occurs between truly distinct species. But the difference is here solely one of *intensity of color*, while in allied species there is almost always an appreciable variation in the *style of coloration.* In individual variation the differences usually extend alike to all parts of the integuments ; that is, if the plumage of the upper surface of the body is brighter or paler than usual, the same difference extends to the plumage of the lower surface of the body, and also to the bill and the feet. This is noticeable not only in species that have the color in uniform masses, differing in tint on different regions of the body, as in the robin (*Turdus migratorius*), the bluebird (*Sialia sialis*), the Maryland yellow-throat (*Geothlypis trichas*), the mocking-bird (*Mimus polyglottus*), and species generally of that type of coloration, but also in spotted birds, as in the various spotted species of *Fringillidæ*, *Turdus*, *Dendrœca*, etc., where the plumage on certain regions of the body is marked with numerous streaks and spots differing from the ground color, in which case the intensity of the color of the markings correllates in its variations with that of the ground color. Closely allied species, on the contrary, usually vary more or less, not only in respect to the ground color, but also to a greater or less degree in the style of the markings. In illustration of this point the familiar group of the small, spotted-breasted wood-thrushes of Eastern

North America — the group *Hylocichla* of Professor Baird — may be taken. Three of these species (*Turdus fuscescens*, *T. Swainsoni*, and *T. Pallasi*) are so closely related that for many years they were variously confounded with each other by almost all who wrote of them, one of them not being clearly recognized as distinct from the others till thus established by Dr. T. M. Brewer,* in 1844, and also at about the same time by Mr. J. P. Giraud,† each apparently independently of the other. Yet they are so distinct that there seems to be not the slightest excuse for again confounding them. While they all agree so closely in general size, in form, and in proportions, that a series of detailed measurements of many specimens of each species gives in the average no constant differences in any of these particulars, each differs from the other radically and constantly in style of coloration, and somewhat in general tints, in habitat, nidification, habits and song. Two of these species (*T. fuscescens* and *T. Swainsoni*) agree in the *style* of the coloration of the dorsal surface, but differ so much in the *color* of this part, that this character alone is always sufficient to separate them, while a still wider difference is seen in the color and markings of the ventral surface, a glance at this part of *T. fuscescens* being sufficient to invariably distinguish it from either of its above-named allies. The third species (*T. Pallasi*) differs markedly from both the other two in the style of coloration of the *dorsal surface*, the rump and tail being conspicuously different in color from the anterior part of the body, whilst the others exhibit no contrast of color between these regions. But the under surface of *T. Swainsoni* is so like that of *T. Pallasi* that frequently specimens cannot readily be referred to the one species rather than to the other from a view of this surface alone. This group serves as a fair general illustration of the kind of variation in color usually seen in closely allied species, but there occur occasional exceptions, where a difference in the relative proportions of different parts, or a wide difference in size, is the prominent specific distinction, the smaller species, so far as color is concerned, being a diminutive representative of the larger.

Taking the present group of *Hylocichla* (for reasons that will appear hereafter ‡) as a group illustrative also of individual variation, it is found that the differences in color in different individuals of either species

* Proc. Boston Soc. Nat. Hist., Vol. I, p. 191, July, 1844.

† Birds of Long Island, p. 91, 1843-44.

‡ See the remarks on these species in Part IV.

results from the amount of rufous pervading the plumage. Individuals of *Turdus Swainsoni* of the rufous or bright-colored type have the dorsal surface of a uniform brownish-olivaceous tint, and the sides of the head and breast strongly suffused with yellowish-brown, which tint is also traceable throughout the lower plumage, in the brighter color of the basal brownish band on the inside of the wings, and in the color of the mouth and base of the bill. In other individuals the upper plumage is of a dark olivaceous tint, without any trace of brownish, the sides of the head, neck, and breast being ashen, with often no appreciable tinge of ferruginous; specimens of this type thus differing widely in general aspect from those of the other. Between these extremes, of which examples are not unfrequent, nor confined to any particular locality or season of the year, there is every degree of intergradation, specimens intermediate between the two being by far the most frequent, and constituting the average or common form.

Turdus Pallasi and *Turdus fuscescens* present precisely similar variations. They are also seen in *Turdus mustelinus*, in *Turdus migratorius*, in *Sialia sialis*, in *Seiurus noveboracensis*, in many species of *Dendrœca*, sparrows, and other species which I have especially investigated in reference to this point, embracing examples of all the leading families of birds. The ruffed grouse (*Bonasa umbellus*), as is well known, varies in the color of the upper parts from reddish-brown to gray; the great horned owl (*Bubo virginianus*) from dusky through numerous shades of rufous and fulvous to nearly white, the fulvous suffusion so commonly present in this species varying from ferruginous on the one extreme to its complete obsolescence on the other. In such common and thoroughly known species as the robin, blue-bird, etc., the true character of these variations is recognized, but in groups where the species are not well known, and especially in specimens from partially explored regions, they are frequently regarded as of specific value, and the addition of numerous nominal species is the result.

Besides the variation in the *depth* of color already noticed, birds having the plumage varied with streaks and spots differ exceedingly in different individuals of the same species in respect to the size, shape, and number of these marks, and in the general aspect of the plumage resulting from such variations. Generally, as already stated, such differences correllate with the variations in the intensity of the ground color, the darker or more deeply colored birds being usually those with the markings largest and brightest. A wide range of variation in this respect is seen in all birds which have the breast and lower plumage marked with dark streaks and spots on a lighter ground, or that have the whole plumage streaked. In the common song sparrow (*Melospiza melodia*), the fox-colored sparrow (*Passerella iliaca*), the swamp sparrow (*Melospiza palustris*), the black and white creeper (*Mniotilta varia*), the water wagtail (*Seiurus noveboracensis*), in *Turdus fuscescens* and its allies, etc., the difference in the size of the streaks is often very considerable. In the song sparrow they vary to such an extent that in some cases * they are reduced to narrow lines; in others so enlarged as to cover the greater part of the breast and sides of the body, sometimes uniting on the middle of the breast into a nearly continuous patch. Variation in this respect is equally great in the fox-colored sparrow and in the grass finch (*Poocætes gramineus*). Massachusetts specimens of the savanna sparrow (*Passerculus savanna* auct.) also present variations exactly parallel with those of the song sparrow. Yet these differences, with other variations to be hereafter mentioned,† have been regarded, as in the case of *Passerculus savanna*, as of specific value. Similar variations in the *Hylocichla* group are very marked, as in *Turdus* (*Hylocichla*) *fuscescens* especially. In some specimens of this species the colors are on all parts not only very pale, but the markings on the breast are reduced to indistinct narrow lines; in others, in which the general color of the plumage is darker, the markings on the breast are dark, broad, and triangular. Two specimens taken in Cambridge the same day (early in May), both of which are males, exhibit these extremes. Average male specimens of the black and white creeper (*Mniotilta varia*), in which the plumage is varied with longitudinal black and white streaks, have the black streaks about a third broader than the white ones; but other specimens occur in which the white ones are equal to and even broader than the black ones; others have the black streaks so much broader than they usually are, — the white ones of course being proportionally reduced, — that the general aspect of the plumage at a short distance is nearly black. The difference between these two extremes is strikingly great. Yet similar variations, scarcely less in degree, occur in nearly all of the striped-breasted warblers.

In birds which have the ground color of certain areas of the body

* Perfectly mature specimens only are here referred to.

† See the remarks on the genus *Passerculus* in Part IV.

black spotted with white, as in some of the woodpeckers (*Picus villosus* and *P. pubescens*, for example), the white markings vary in size most notably, and sometimes in number. The white markings so common on the wings and tails of birds, as the bars formed by the white tips of the greater wing coverts, the white patch occasionally present at the base of the primary quills, or the white band crossing them, and the white patch near the end of the outer tail feathers, are also extremely liable to variation in respect to their extent and the number of feathers to which, in the same species, these markings extend. Variation in the tail markings is particularly common, as may be seen by comparing numerous specimens of almost any species of *Dendrœca*, *Junco*, *Pipilo*, of *Mimus polyglottus*, *Chordeiles popetue*, etc. In the latter species the white patch on the wing does not ordinarily encroach upon the outer vane of the first primary, and rarely upon its shaft, but in several specimens before me it covers not only the shaft of the first primary, but extends completely across its outer vane! The black subterminal bar on the upper surface of the tail of the ruffed grouse (*Bonasa umbellus*) ordinarily crosses all but the middle pair of feathers, on which there is usually no trace of this bar; in many specimens, however, it is barely traceable on them, and in others it is as distinct and perfect on the middle pair as on the others.*

The *Parula americana* presents also remarkable examples of individual color variation. The colors of the males are usually much brighter than those of the females, but cases are frequent where the sex cannot be determined by the color of the plumage. Adult males also vary greatly in the *style* of coloration. They are generally bright yellow anteriorly below, with a broad band of dusky reddish-brown across the breast, varying in tint from nearly pure chestnut to dusky reddish-brown, and even black, and also greatly in extent. In some, however, this band is partially obsolete, in which case the whole plumage is generally paler than in average specimens. More rarely large, brightly colored males are taken, even in New England, with the whole breast bright yellow, the brownish pectoral band being entirely absent. This condition, however, seems to be more frequent in specimens of *Parula* collected in Mexico, and Central and Northern South America, which on this account have been regarded as distinct from the *Parula* of the North; yet all the conditions of color seen in specimens from

* See remarks on color variations in other species in Part IV.

the North are also common to those from the South, and *vice versa.*

In species in which the female usually differs from the male in being paler colored, the pattern of coloration being the same in both sexes, females occur more or less frequently which are as brightly colored as the brightest males, and males that are paler than the generality of the females.

Variation in Color depending on Season. — A word in this connection seems necessary concerning some of the more obscure variations depending upon season and age, since it is sometimes difficult to avoid confounding these differences with those resulting from individual variation. In many species there is a marked change in the color of the plumage without a change of the plumage itself. No experienced collector can have failed to notice the much brighter and livelier tints the plumage of most song birds presents immediately after the autumnal moult, in species in which there is no marked seasonal change of color, in comparison with the faded appearance they exhibit towards the close of the breeding season. This brighter autumnal tint is particularly marked in the *Vireos*, the different species of *Empidonax*, *Sayornis*, *Contopus*, and in some of the *Sylvicolidæ*, and is clearly traceable in hundreds of other species. But almost as great a difference is seen when specimens of any species taken in spring, on its first arrival at its breeding station, are compared with those collected several weeks later, or just before the autumnal renewal of the plumage. In this case the variation results in part from an actual fading of the color, and in part from the wearing of the edges of the feathers. Seasonal differences of this character are often only readily appreciable to the experienced eye, and the failure to recognize the cause of these differences has led in many instances to their being regarded as of specific value. Especially noteworthy instances of such mistakes will be noticed later. Collectors, and even naturalists, generally place little value on faded or dull-colored specimens, so that ordinarily in collections of our native birds only fine-looking specimens are preserved. But travellers and explorers of new localities are often compelled to content themselves with any representative they may be able to get, so that the "closet" or exclusively "museum naturalist" has not usually the material necessary to furnish him with a clew to the cause of these variations.

Generally, aside from the paler tints of late-collected birds, as compared with those taken early in the season, there is a total absence of the grayish, yellowish, brownish, or rufous suffusions (the particular tint varying of course in different species) that tinges the feathers early in the season. The general aspect of the plumage at the two periods in question is thus essentially different. The common chickadee (*Parus atricapillus*) will illustrate this point, in which the brownish tint so conspicuous on the lateral portions of the ventral plumage in autumn and winter is gradually lost as spring approaches, and in summer is almost entirely wanting, especially in nesting females, which at this season have the plumage generally much more worn than the males. The savanna sparrow will also illustrate the differences resulting simply from the fading of the color during the breeding season. In spring both sexes have a greenish-yellow, superciliary stripe, varying more or less in intensity in different specimens, but rarely or never of the pale soiled-whitish so frequently met with late in the breeding season. In the large series of specimens before me collected at that season in Massachusetts, few if any have this stripe so bright as average spring specimens have it, in many it having faded to soiled white. Scores of similar cases might be cited, but the above are sufficient for illustration.

Variations in Color depending upon Age. — So well known are many of the variations depending upon age, that it seems necessary to advert to only a few of the lesser known phases. In many species there is no marked difference between old and young birds, after the moulting of the first or nestling plumage, which usually occurs in the oscine groups in a few weeks after they leave the nest. But even in these, in many cases, sufficient marks of immaturity remain for a time to enable any one acquainted with such features to recognize birds of one or two years of age from those that are older. Yearling birds of this group are often recognizable by their having more or less well-defined bars across the wings, formed of light-colored, hastate, or drop-shaped spots on the ends of the greater wing-coverts and inner secondaries, which in many genera are peculiar to yearling birds, though in other respects, so far as the plumage is concerned, they are not distinguishable from adults, — a difference which in some instances has been considered specific. Similar marks are also seen in older birds, in species that do not obtain their adult colors till later in life.

Yearling and two-year-old birds are also often distinguishable from older ones by the presence, after the spring moult, of a greater than the ordinary amount of ferruginous, ashy, or yellowish edging to the feathers, such as is often seen in the winter plumage of adult birds. In some cases such a bordering to the clothing feathers, especially those of the back, is often strictly distinctive of young birds, and is, moreover, a feature of common occurrence.

Generally speaking, several years elapse before the purity of the colors and the definiteness of outline of the markings characteristic of maturity is fully obtained, especially in highly colored species. In birds of variegated colors the contrasts of color become for a time more and more decided with each moult, and the markings better and better defined, especially in respect to the white bars of the wings and the spots on the tail common to a large number of species. The latter markings usually gradually increase in extent for a considerable period. A good illustration of this is seen in many of the gulls, particularly in the genus *Larus*. In *L. argentatus* the following gradual change with age occurs in the white markings on the tips of the primaries. At first, as ornithologists are aware, the plumage of this species is uniformly dusky, the adult colors not being acquired before the second year, and apparently frequently not before the third, there being in the breeding season usually a large proportion of individuals in the brown plumage.* But there are wide differences in the intensity of the color in different individuals in this stage of plumage, some being but slightly dusky and others extremely dark, — differences that probably result mainly from differences in age, the darker birds being probably yearling birds and the lighter ones two years old, though part of the difference is doubtless due to individual differentiation. In this stage the wings and tail are of nearly the same uniform dusky tint as the general plumage. In what may be considered as the second stage, the general color is somewhat lighter, the tail much lighter, and the primaries much darker, with a distinct paler apical margin. At a third stage the tail becomes white, the dorsal plumage begins to assume the blue tint characteristic of maturity, the primaries change from dull blackish brown to black, and a small white spot appears near the end of the inner vane of the first

* Generally the large parties that spend the summer on the coast of Massachusetts, where none of these birds now breed, consist almost wholly of birds in the brown stage of plumage. See American Naturalist, Vol. III, p. 640, 1870.

primary, separated from the white at the extreme tip by a broad space of black. A subsequent gradual increase occurs in the purity of the colors and in the extent and form of the wing markings. The complete series of the changes in the latter is as follows: At first, as previously remarked, the primaries are dull brownish black, a little darker than the general plumage, with their extreme apical margins lighter. At the next stage the three inner primaries have become much lighter, and the light border to all broader and whiter. Later the three inner primaries and the distal portions of most of the others become wholly ashy white, and the outer portion of the other primaries much blacker. The subapical dark portion of the wing now embraces only the seven outer primaries, and is of a triangular form, the first primary forming the base of the triangle. The black on the outer vane of the first primary reaches nearly to the base of the outer vane of the second, and is more and more restricted on the others, till on the sixth (or, more rarely, on the seventh) it forms only a narrow bar near the tip. In other words, the black, if present on the seventh primary, exists as a narrow transverse subapical bar, which bar increases in distal extension on the sixth, fifth, fourth, third, and second, to the first, and embraces the whole outer vane of the first primary. The basal outline of the black area being an oblique one, a much larger portion of the outer than of the inner vane of each feather is embraced in the black space. All the primaries are now terminated with a narrow white border, the first primary having also an oval white spot on the inner vane, near the end of the feather. Subsequently this spot enlarges so as to embrace a part of both vanes, the white at the tip of the feather also meantime increasing somewhat in extent, and the two being separated by a broad bar of black. Coincident with this increase in the amount of white on the first primary, a small white spot appears on the inner vane of the second primary. Subsequent increase in the extent of these white markings goes on until the white area on the second primary extends to both vanes, and the two white spots on the first primary are separated by only a narrow bar of black. Later still this bar becomes broken, through the partial union of the two white spots, and finally becomes entirely obsolete, leaving the first primary with a single continuous white apical area, an inch and a half to two inches in length. It is probable that not all individuals reach this final stage, though most doubtless do in old age. A large series of specimens of mature birds usually exhibit the gradual change above described, and indicate the inconstancy of these markings and their unreliability as specific characters. Often, as is well known, these markings in the gulls differ considerably in the two wings of the same bird.

Although the *L. argentatus* has been taken as a general illustration, the same variations with age, or in different individuals, are exhibited by most species of the genus *Larus*. Generally they are admitted to have no value as specific characters, even by those who in the case of *L. argentatus* have accorded to them this importance.

In some of the species of *Junco* and *Pipilo*, in *Mimus*, in numerous species of *Dendrœca*, in *Parula*, *Mniotilta*, etc., there is a similar increase with age in the extent of the white markings on the tail, sometimes three and sometimes four pairs of feathers being spotted or terminated with white in different specimens of the same species. In short, these variations occur in so many species that they may be looked upon as indicating a general law of variation in color depending upon age, namely, *an increase in the purity or intensity of the general color, and an increase in the size of the wing and tail markings, for a time, with age.*

After complete maturity is attained there is, however, unquestionable evidence of a decline in color, which in many cases, and especially in bright-colored species, is quite marked. So general is such a decline in other groups of the animal kingdom that a citation of evidence on this point seems wholly needless. Yet in birds, in numerous instances, it is scarcely appreciable, and doubtless is in most species too slight to be readily traced. This obscurity may result, however, more from an absence of favorable conditions for such a decline to be recognized than from its real absence. It can hardly be doubted, in fact, that a share of the color variation seen in mature birds is attributable to this cause. It is well known that young mammals in their first pelage are, as a general rule, much darker colored than the adults of the same species. At a later period the color fades more slowly, but in old age the hair often becomes more or less gray, the blanching being in some cases very marked. Nearly all birds are also darker in their nestling and immature stages of plumage than after they arrive at maturity, especially if in the adult stage the plumage is light colored; and it is more than probable, and in some cases certain, that the decline in color continues in a slight degree through life. The change of *Falco candicans* from dusky when young to nearly white when fully mature may be hardly referable wholly to the blanching of age; but the gradual obsolescence of the dusky mottlings of the snowy owl (*Nyctea nivea*), as it advances in age, seems strictly parallel to the blanching of the gray colt to a white horse. Hence a second law of variation in color in old age, namely, that of *senile decline.*

Individual Variation in General Size and in the Relative Size of Different Parts.

Individual Variation in General Size and Form. — Measurements of scores of specimens of birds of the same species and sex, collected at the same locality and season, show the existence of a large range of individual variation, both in size and in general proportions; the variation extending to every external part of the body, and implying a corresponding variation in the internal anatomy. In birds size has usually been regarded, from its comparative constancy in the same species, as an important specific character. But from the fact that specimens of closely allied species often differ but little from each other in this respect, it has been justly looked upon as being in some cases more or less unreliable; but from the great importance commonly attached to it, it is evident that such instances are usually regarded as exceptional. Individual variation in this respect having been formerly regarded as too slight to have any significance, the size of a single specimen has usually been given as that of the species to which it belonged; hence subsequent variations from it discovered in other specimens of the same species has sometimes led to the recognition of the latter as specifically distinct. Especially has this been the case when a difference in size has been associated with a wide difference of locality. The facts in the case, however, show that a variation of fifteen to twenty per cent in general size, and an equal degree of variation in the relative size of different parts, may be ordinarily expected among specimens of the same species and sex, taken at the same locality, while in some cases the variation is even greater than this. Table A (p. 198) shows to some extent the general variation in size, but it does not always give, nor even generally, the extreme differences in the size of similar parts, as the wing, tail, etc., since those averaging the largest or smallest for the four measurements given are often not those having the longest or the shortest wing, tail, or tarsus, or which measure the most or the least in length or alar extent. The extremes of variation in the size of the wing and tail is given in Tables B, C, and D.*

Table A.— Variation in General Size.

Mus Comp. Zoölogy No.	Collector's Number.	Sex.	Name.	Locality.	Date.	Collected by	Length.	Alar Extent.	Wing.	Tail.
8844	693	♂	Turdus Swainsoni	Belmont, Mass.	May 27, '68	C.J.Maynard	7.76	12.65	4 20	3.00
1520	—	—	Turdus Swainsoni	Springfield, "	May 14, '62	J. A. Allen	6.62	11.40	3 80	2.83
——	250	♂	Turdus Pallasi	Newton, "	Apr. 25, '68	C J.Maynard	7.30	12.83	3 80	3.83
——	608	♀	Turdus Pallasi	Newton, "	May 25, '68	C.J.Maynard	7.00	10 64	3.00	2.45
8830	337	♂	Turdus fuscescens	Newton, "	May 5, '68	C.J.Maynard	7.81	13.70	4 16	4.00
8834	553	♂	Turdus fuscescens	Watertown, "	May 20, '68	C.J.Maynard	7.00	11.95	3.58	3.55
——	621	♂	Dendrœca striata	Newton, "	May 27, '68	C J.Maynard	5.45	8.75	2.80	1.70
——	641	♂	Dendrœca striata	Newton, "	May 27, '68	C.J.Maynard	5.51	9.30	3.10	2.00
——	546	♂	Dend. pennsylvanica	Newton, "	May 20, '68	C.J Maynard	5.40	8 25	2.61	2.07
——	451	♂	Dend. pennsylvanica	Newton, "	May 20, '68	C J.Maynard	5.00	7.56	2 35	1.75
4674	1011	♂	Spizella pusilla	Newton, "	July 22, '68	C J.Maynard	6.00	8.32	2.62	2.60
4633	—	♂	Spizella pusilla	Waltham, "	Sept.19, '67	C J Maynard	5.06	7.62	2.00	2.62
4990	1411	♂	Sayornis fuscus	Newton, "	Sept 30, '68	C J.Maynard	7.50	11.40	3.60	3.00
——	164	♂	Sayornis fuscus	Waltham, "	Mar. 28, '68	C.J.Maynard	7.25	12.60	3 82	2.20
4819	5	♂	Sayornis fuscus	Newton, "	Oct. 9, '69	C.J.Maynard	6.51	10.32	3 20	2.85
4701	317	♂	Passerculus savanna	Newton, "	Apr. 5, '68	C.J.Maynard	5.85	9.73	3.00	2 26
5098	873	♂	Passerculus savanna	Ipswich, "	June17, '68	J. A. Allen	5.83	7.75	2.80	2.10
——	890	♀	Passerculus savanna	Ipswich, "	June17, '68	J. A. Allen	5.35	7.75	2.51	1.95
5088	851	♀	Passerculus savanna	Ipswich, "	June17, '68	J. A. Allen	5.75	9.75	2.78	2.05
——	714	♂	Vireo olivaceus	Waltham, "	May 30, '68	C J.Maynard	6.55	10 65	3.48	2.46
——	713	♂	Vireo olivaceus	Waltham, "	May 30, '68	C.J Maynard	6.25	9.88	3.22	2.10
4624	1407	♂	Chrysomitris tristis	Newton, "	Sept.28, '68	C J Maynard	5.35	9.40	3.00	2.00
4926	150	♂	Chrysomitris tristis	Newton, "	Mar 26, '68	C.J.Maynard	5.00	8.60	2.65	1.85
——	106	♂	Melospiza melodia	Newton, "	Mar. 12, '68	C J.Maynard	6.75	9.15	2.65	2.80
——	55	♂	Melospiza melodia	Newton, "	Nov. 24, '67	C.J.Maynard	6.00	8.25	2.35	2.68
1456	—	♂	Sialia sialis	Worthingt'n,"	July —, '63	C.H Hamlin	6 50	11.10	3 75	2.45
2378	—	♂	Sialia sialis	Waterville, Me.	Apr. 15, '64	W. H Niles	7.00	12.25	4 10	2.60
4371	334	♂	Mniotilla varia	Newton, Mass.	Apr. 20, '68	C.J Maynard	5 40	9 00	3 00	2 15
4376	372	♂	Mniotilla varia	Waltham, "	May 6, '68	C.J Maynard	5.35	8.25	2.87	2.05
——	887	♂	Cotyle riparia	Ipswich, "	June17, '68	J. A. Allen	5.45	11 00	4.65	2.10
5111	868	♂	Cotyle riparia	Ipswich, "	June17, '67	J. A. Allen	5 25	10.55	3.75	1.86
——	1082	♂	Passerella iliaca	Springfield, "	Mar. 22, '68	J. A Allen	7.50	11.65	3.65	2.98
——	—	♂	Passerella iliaca	Newton, "	Mar. 27, '68	C J Maynard	6.80	10.80	3.40	2 50
——	783	♂	Icterus Baltimore	Newton, "	June 6, '68	C J Maynard	8.00	12.25	4.00	3.10
4852	653	♂	Icterus Baltimore	Newton, "	May 27, '68	C.J.Maynard	7 30	11.15	3.56	2.55
——	946	♂	Sterna hirundo	Muskeget Isl	June29, '68	J. A. Allen	15.50	31.85	11.30	7.00
10485	916	♂	Sterna hirundo	Muskeget Isl.	June29, '68	J. A. Allen	14.90	29 30	10 40	5 50
10464	932	♂	Sterna arctica	Muskeget Isl.	July 2, '68	J. A. Allen	16.15	32.15	11.60	7.50
——	992	♂	Sterna arctica [mus	Muskeget Isl.	July 2, '68	J A. Allen	14.40	29.00	10.70	6.30
4009	—	♂	Pipilo erythrophthal-	Milton, Mass	——	H C. Daring	8.55	12.25	3.57	3.85
10151	—	♂	Pipilo erythrophthal.	Cambridge, "	——	L. Agassiz	7.50	10.20	3.34	3.60

As a large proportion of the specimens mentioned in some of the following tables (most of Tables A to G) were taken during the season of their migration, they may have originated at widely different localities, and thus the differences indicated may be in some measure due to geographical causes. In other cases, however, all the specimens

* The measurements given in this paper were all taken either from fresh specimens by the collector, or by myself from specimens preserved in spirits.

were taken in the breeding season; while in still other instances (Tables H to P) the species were purposely chosen from among such as find their northern limit of distribution near the locality where all were taken. Of ten species of the latter class, twenty perfect male specimens have been carefully measured,* the measurements embracing a series of eighteen to twenty distinct parts; under such circumstances the variation in general size, in length, in alar extent, in the length of the folded wing, the tail, the tarsus, the head, the bill, etc., etc., commonly ranged from twelve to eighteen per cent.

In respect to the differences in the general form of the body, two leading styles of variation from the average form may be recognized in nearly all species, namely, a relatively robust form, in which the stoutness extends to all parts, and a relatively slender form, in which the slenderness is equally general. Variations of this general and symmetrical character are remarkable only for their extent, since in such cases there are no marked discrepancies between the relative size of different parts. Contrary, however, to our usual notions of exact symmetry in animals, the unsymmetrical variations are by far the most frequent and important.

Variation in the Relative Size of Different Parts. — In specimens of average size of any given species, considerable differences exist in the relative size of different parts. In individuals of the average alar extent of their species, for example, the length of the folded wing may vary very considerably, in consequence of a difference in the length of the primary quills as compared to the length of the bones of the wing. The length of the folded wing or the alar extent may vary with reference to the whole length of the specimen, in consequence of differences in the relative length of the tail, the neck, or the body. The tarsus also varies independently of variation in the general size, as do also the toes to the tarsi, relatively short toes being found to accompany tarsi of ordinary length, and, conversely, long toes short tarsi.† The wing varies in its form in consequence of the different relative development of the primary and secondary quills.‡ The tail varies in respect to its form, especially in regard to the degree of its emargination or graduation, and, in some groups, in respect to the number of its feathers. The bill also varies greatly in size and form. The variations in these various parts will be considered separately and in detail.

* See below, Tables H to P, pp. 210–219.

† See Table F, p. 205.

‡ See Table E, p. 204.

Variation in the Length of the Folded Wing and the Tail. — The measurements given in the following table (Table B) sufficiently illustrate the variation in the length of the folded wing in fully mature specimens of the same sex and species, while Table C indicates the variation in the length of the tail, in specimens of a similar character. All the specimens, with a few exceptions, were taken within a few miles of Cambridge; the others are mainly also from Eastern Massachusetts, a few * being from a single locality in Florida. The series from which these extremes are taken embrace ordinarily not more than twenty-five or thirty specimens; with larger suites the differences would in many cases doubtless be much increased. The largest and smallest only are taken, between which, however, there is every gradation. The difference between these extremes is indicated, and also the percentage of the variation, based on the average of the two extremes. The amount of the variation in the length of the folded wing ranges, as will be seen from the table, from twelve to twenty-one per cent of the average length. In the tail the amount of variation in respect to length ranges from fourteen to twenty-three per cent. The different species vary considerably in respect to the amount of variation each presents, some being much more variable than others. It should be stated, however, that as a general rule the widest extremes, or the highest percentages of variation, occur in those species of which the greatest number of specimens has been examined. It will also be noticed that the tail usually varies more than the wings. In species with a relatively long tail the percentage of variation in the length of this member is found to be greater than in those species in which it is of medium length or short, as would have been naturally expected. In several cases the greater differences occur between females, but this may be a mere coincidence.

In this connection it may be added that the variation proves to be much less between specimens of the same species and sex when taken at a single locality in the breeding season than when taken during the period of migration. In many instances specimens of the same species may be obtained at one locality which shall represent the whole range of its geographical variation, as well as its individual variation, as in the case of those species which breed far to the North, but migrate in winter to the tropics, being thus but transient visitors to the temperate portions of the United States.

* Those of Mimus polyglottus, Cardinalis virginianus, Picus borealis.

Table B.—Individual Variation in the Length of the Folded Wing.

M. C. Z. No.	Orig. No.	Sex.		Folded Wing.	Difference.	Per cent of Variation.
10596	2510	♀	Mimus polyglottus	4.75	.75	17.0
——	2485	♀	Mimus polyglottus	4.00		
10716	1987	♂	Cardinalis virginianus	3.85	.55	14.6
——	1993	♂	Cardinalis virginianus	3.30		
——	316	♂	Passerculus savanna	2.95	.40	14.5
——	820	♂	Passerculus savanna	2.55		
8830	367	♂	Turdus fuscescens	4.16	.61	15.8
8834	556	♂	Turdus fuscescens	3.55		
4821	148	♂	Sayornis fuscus	3.87	.67	19.0
4819	5	♂	Sayornis fuscus	3.20		
9057	618	♂	Geothlypis trichas	2.56	.50	21.0
5020	703	♂	Geothlypis trichas	2.06		
4648	1389	♂	Carpodacus purpureus	3.70	.60	17.6
4655	751	♂	Carpodacus purpureus	3.10		
9696	——	♂	Pipilo erythrophthalmus	3.68	.51	14.6
1421	——	♂	Pipilo erythrophthalmus	3.17		
——	170	♂	Junco hyemalis	3.20	.45	18.0
4910	140	♂	Junco hyemalis	2.75		
1568	——	?	Tyrannus carolinensis	4.85	.68	15.0
10025	——	?	Tyrannus carolinensis	4.17		
10014	——	?	Galeoscoptes carolinensis	3.85	.60	17.0
2734	——	?	Galeoscoptes carolinensis	3.25		
786	——	♂	Icterus Baltimore	4.00	.58	16.0
1334	——	♂	Icterus Baltimore	3.42		

Table C.—Individual Variation in the Length of the Tail.

M. C. Z. No.	Orig. No.	Sex.		Tail.	Difference.	Per cent of Variation.
——	2474	♂	Mimus polyglottus	5.15	.95	20.5
10592	2372	♂	Mimus polyglottus	4.20		
——	1955	♂	Cardinalis virginianus	4.30	.90	23.4
——	2460	♂	Cardinalis virginianus	3.40		
——	317	♂	Passerculus savanna	2.26	.41	19.5
5086	846	♂	Passerculus savanna	1.85		
8830	528	♂	Turdus fuscescens	3.00	.45	14.4
8835	556	♂	Turdus fuscescens	2.55		
——	——	♀	Parus atricapillus	2.63	.48	20.0
——	——	♀	Parus atricapillus	2.15		
9056	454	♂	Geothlypis trichas	2.15	.45	23.4
5020	703	♂	Geothlypis trichas	1.70		
4651	1071	♀	Carpodacus purpureus	2.57	.52	22.5
4653	1371	♀	Carpodacus purpureus	2.05		
4614	1330	♂	Pipilo erythrophthalmus	4.00	.71	19.5
4727	415	♂	Pipilo erythrophthalmus	3.29		
——	160	♂	Junco hyemalis	2.78	.38	15.0
4917	201	♂	Junco hyemalis	2.40		
10646	1972	♀	Picus borealis	3.75	.50	14.0
10633	41	♀	Picus borealis	3.25		
1317	——	?	Tyrannus carolinensis	2.93	.61	19.0
1568	——	?	Tyrannus carolinensis	3.54		
2734	——	?	Galeoscoptes carolinensis	3.35	.75	20.0
10014	——	?	Galeoscoptes carolinensis	4.10		
1334	——	♂	Icterus Baltimore	2.70	.40	13.8
2289	——	♂	Icterus Baltimore	3.10		

Variation in the Relative Length of the Wings and Tail. — Table D illustrates the irregularity of the variation in the wings and the tail. The first column of measurements gives the length of the folded wing, and

Table D. — Individual Variation in the relative Length of the Folded Wing and Tail.

M. C. Z. No.	Original No.	Sex.		Wing.	Tail.	Diff. betw'n Wing and Tail.	Amount of Variation.
——	2429	♂	Mimus polyglottus	4.35	4.35	.00	
10590	2342	♀	Mimus polyglottus	3.25	4.35	+1.00	
——	2560	♀	Mimus polyglottus	4 15	4.35	+ .25	
——	2614	♂	Mimus polyglottus	4.40	4 90	+ .50	1.20
——	2340	?	Mimus polyglottus	4.40	4.50	+ .10	
——	2478	♀	Mimus polyglottus	4.40	4 20	— .20	
——	2374	♂	Mimus polyglottus	4 30	4.16	— .14	
8881	441	♂	Galeoscoptes carolinensis	3 60	3.60	.00	
——	1376	♀	Galeoscoptes carolinensis	3.70	3 60	— .10	.45
8879	412	♂	Galeoscoptes carolinensis	3.75	4.10	+ .35	
8841	495	♂	Turdus fuscescens	4.00	3.00	1.00	
8832	382	♂	Turdus fuscescens	4.10	4.00	.10	1.15
8835	581	♂	Turdus fuscescens	4.15	2 90	1.25	
8821	374	♂	Seiurus aurocapillus	3.00	2.00	1.00	.66
——	423	♂	Seiurus aurocapillus	3.00	2 66	1.66	
8851	322	♀	Turdus Pallasi	3.50	2 60	.90	.67
——	301	♂	Turdus Pallasi	3.43	3 17	.23	
4301	514	♂	Dendrœca æstiva	2.85	1.80	1.05	.58
——	362	♂	Dendrœca æstiva	2 45	1.98	.47	
5058	707	♂	Dendrœca striata	2.85	2.00	.85	
——	1341	♂	Dendrœca striata	3.00	1.75	1.25	
5062	734	♀	Dendrœca striata	2.45	1.98	.46	.79
——	741	♀	Dendrœca striata	2.80	1.80	1 00	
5041	665	♂	Setophaga ruticilla	2.60	2.10	— .50	.57
——	698	♂	Setophaga ruticilla	2.43	2.50	+ .07	
——	693	♀	Regulus satrapa	2.20	1.52	.68	.49
——	50	♀	Regulus satrapa	1.94	1.75	.19	
4808	711	♂	Contopus virens	3 35	2.36	.99	.54
4994	1116	♂	Contopus virens	3.15	2.70	.45	
10645	1924	♂	Picus borealis	4.80	3.32	1.48	.48
10646	1972	♀	Picus borealis	4.75	3.75	1.00	
4587	323	♂	Agelæus phœniceus	4.85	3.40	1.45	.67
4589	214	♂	Agelæus phœniceus	4.60	3.82	.78	
4654	1069	♂	Carpodacus purpureus	3 85	2.00	1.35	.64
4655	286	♂	Carpodacus purpureus	3.03	2.32	.71	
——	288	♂	Poocætes gramineus	3.55	2.41	1.14	.54
——	846	♀	Poocætes gramineus	3.10	2.50	.60	
——	881	♂	Passerculus savanna	2.75	1 85	.90	.41
——	127	♂	Passerculus savanna	2.74	2.25	.49	
——	115	♂	Passerella iliaca	3.75	2.65	1.10	.78
——	55	♀	Passerella iliaca	3.32	3.00	.32	
——	177	♂	Melospiza melodia	2.35	2.68	+ .33	.53
——	2363	♂	Melospiza melodia	2 60	2.40	— .20	
——	2369	♂	Cardinalis virginianus	3.60	3 40	— .20	.70
——	——	♂	Cardinalis virginianus	3.60	4.10	+ .50	
2293	——	♂	Dolichonyx oryzivora	3.75	2.78	.96	.32
5741	——	♂	Dolichonyx oryzivora	4 00	2.72	1.28	
10107	——	♂	Hedymeles ludoviciana	4.20	2.93	1.27	.39
9787		♂	Hedymeles ludoviciana	3.83	2.95	.88	

the second the length of the tail, of the same specimens; the third column shows the difference in length between the tail and the wing, and the fourth column the amount of the difference between the two extremes. In *Mimus polyglottus* the tail is usually one fourth to one half an inch longer than the wing; but in many specimens the wings and tail are equal, and in a small proportion the tail is *shorter* than the wing. In the seven specimens of this species cited in the table, the variation ranges from the tail being one fifth of an inch shorter than the folded wing to one inch longer. In the three specimens which agree in the length of the tail (4.35 in.), the variation in the length of the folded wing ranges from 3.25 in. to 4.25 in., or is nearly twenty-seven (26.85) per cent. The larger specimen, however, is a male, while the others are females; but between the two females the difference is over twenty-four (24.3) per cent. Similar differences have been met with in various other species, but it has not been deemed necessary to cite a larger list of examples.

Variation in the Form of the Wing. — By the form of the wing is meant its general outline when expanded, which is mainly determined by the relative length of the remiges. The form of the wing, and especially the relative length of the different primary remiges, has direct relation to the power of flight. In strong, swift-flying birds, the outer primaries are the longest, giving a narrow pointed form to the expanded wing, as in the swifts, the swallows, in *Chordeiles*, in the *Sterninæ* and in most of the *Procellaridæ*. In birds of medium powers of flight, as in most of the true finches (*Coccothraustinæ*) and *Turdinæ*, the *Tyrannidæ*, the *Sylvicolidæ*, etc., etc., the third, fourth, and fifth primaries are the longest, the wing being less pointed and broader. In species with low power of flight, as the *Troglodytidæ*, several genera of sparrows, the grouse, etc., the outer primaries are still more reduced, the wing is much more rounded and shorter, and the power of flight is in each case correspondingly less. In birds of the first class, which live almost wholly on the wing, little variation is seen in the relative length of the primaries. In those of the second and third classes, slight variations affect in less degree the particular habits of life, so that among the latter would be naturally expected the greatest range of individual variation.

Correlating with the variation in the form of the wing, as determined by the relative length of the outer primaries to the length of the inner primaries are similar variations in the relative length of the inner secondaries as compared with the outer secondaries. Relatively short inner secondaries (generally improperly called "tertiaries") hence accompany long primaries, and, conversely, long inner secondaries, short outer primaries. The particular form of the wing in any group depending upon the relative development of these several elements, they hence afford excellent generic characters; but while thus important, they are subject to a considerable range of individual variation. The form of the wing being readily determined by measurements, and easily expressed mathematically, the amount of the variation is easily measured and tabulated. In the following table (Table E) the extent and character of this variation is to some degree illustrated. In the first column of measurements is given the length of the folded wing; in the second the extent of the longest primary beyond the outer (or shortest) secondary, and in the third the extent of the longest primary beyond the inner (or longest) secondary. The fourth column gives the amount of variation in each specimen cited.

Table E.—Variation in the Form of the Wing.

M. C. Z. No.	Sex.	Species.	Length of the Wing.	Ext. of Pr. beyond Outer Sec	Ext. of Pr. beyond Inner Sec.	Amount of Variation.
2319	♂	Icterus Baltimore	3.75	.77	.90	.13
2290	♂	Icterus Baltimore	3.83	.67	.81	.14
1333	♂	Icterus Baltimore	3.64	.57	1.06	.49
1567	♂	Icterus Baltimore	3.80	.56	.92	.36
2964	♂	Icterus Baltimore	3.85	.77	1.07	.30
2299	♂	Icterus Baltimore	3.85	.87	1.12	.25
2296	♂	Dolichonyx oryzivora	3.80	.98	1.42	.44
5741	♂	Dolichonyx oryzivora	4.00	1.20	1.40	.20
119	♂	Dolichonyx oryzivora	3.82	.78	1.23	.45
9854	♂	Dolichonyx oryzivora	3.53	.98	1.14	.16
284	?	Tyrannus carolinensis	4.30	.85	1.15	.30
113	?	Tyrannus carolinensis	4.60	.90	1.45	.55
1317	?	Tyrannus carolinensis	4.25	.76	1.10	.34
4008	?	Tyrannus carolinensis	4.60	1.35	1.62	.27
10107	♂	Hedymeles ludoviciana	4.20	.90	1.05	.15
590	♂	Hedymeles ludoviciana	4.00	.90	1.25	.35
9935	♂	Hedymeles ludoviciana	4.00	.60	1.06	.40
1456	♂	Sialia sialis	3.75	1.00	1.10	.10
1945	♂	Sialia sialis	3.90	1.03	1.10	.07
338	♂	Sialia sialis	4.07	1.30	1.30	.00
5606	♂	Sialia sialis	4.05	1.25	1.40	.15
10292	♂	Sialia sialis	3.90	.95	1.15	.20
256	?	Galeoscoptes carolinensis	3.37	.55	.50	—.05
1790	?	Galeoscoptes carolinensis	3.75	.55	.70	+.15
5358	?	Galeoscoptes carolinensis	3.55	.35	.57	+.22
10014	?	Galeoscoptes carolinensis	3.85	.70	.75	+.05
2274	?	Galeoscoptes carolinensis	3.75	.67	.70	+.03

Variation in the relative Length of the Primary Quills. — From the great stress laid upon the relative length of the outer primaries by descriptive ornithologists in determining genera and species, one would be led to expect but a slight amount of variation in this respect in specimens of the same species. On the contrary, however, it is soon found, on giving special attention to this character, that a considerable amount of individual variation in this regard really exists. That the wing formula, so generally introduced of late years into specific diagnoses, is in a great degree unreliable as a specific character, is sufficiently shown by the subjoined table (Table F, p. 206) of the relative proportions of the primaries. The comparison, extended in the table to only a few species has been carried to scores of others with similar results.

In general, in species of the *Oscines* which have the second primary usually the longest, it is sometimes the first and sometimes the third that is the longest. In those which have the third ordinarily the longest, the second and third, the third and the fourth, or the second, third, and fourth are frequently equal. In those in which the first (or the second when the first is very short) is intermediate to the second and fourth or to the third and fifth, it may be equal to or longer than the second or third, or only equal to the fourth or fifth.

Variation in the Form of the Tail, and in the Number of the Rectrices. — Individual variation in the form of the tail is often quite marked. In species with the tail deeply forked, different specimens vary considerably in respect to the depth of the fork. Those with the tail rounded and much graduated differ greatly in respect to the amount the middle feathers exceed the outer ones in length. In species with a normally nearly even tail, the tail is sometimes distinctly emarginate, and sometimes as distinctly rounded in different specimens of the same species.

In regard to the number of rectrices, in those groups in which the number exceeds twelve, as in the *Rasores*, the *Lamellirostres*, etc., the number is frequently variable. The rectrices of the common ruffed grouse (*Bonasa umbellus*) are usually eighteen in number, but an examination of numerous specimens shows that the number varies from sixteen to twenty. The usual number in *Tetrao canadensis* is sixteen, but the number varies from fourteen to eighteen. In *Cupidonia cupido*, and in other species of grouse, similar variations also occur. They are also frequent in the *Anserinæ*. In *Bernicla canadensis*, for example, the usual number of rectrices is eighteen, but the number

Table F.— Variation in the relative Length of the Primaries.

M. C. Z. No	Species.	Longest.	2d in Length.	3d in Length.	4th in Length.	5th in Length.	6th in Length.	7th in Length.	8th in Length.	9th in Length.
3289	Turdus fuscescens	4	3	2	5	6	7			
6764	Turdus fuscescens	3	4	2	5	6	7			
8837	Turdus fuscescens	4 } 3 }	2	5	6	7	8			
8848	Turdus Pallasi	4	3	5	6	2	7			
5197	Turdus Pallasi	4	5	3	6	2	7			
8205	Turdus Pallasi	4	3	5	6 } 1 }	7	8			
8206	Turdus Pallasi	4	3 } 5 }	6	2	7	8			
10698	Myiarchus crinitus	3	4	2	5	6	1	7		
10699	Myiarchus crinitus	3	2	4	5	6	1	7		
8166	Myiarchus crinitus	4	3	2	5	6	1 } 7 }	8		
10700	Myiarchus crinitus	3 } 4 }	2	5	6	1	7	8		
10701	Myiarchus crinitus	3	4 } 2 }	5	6	1	7	8		
12420	Tyrannus carolinensis	2	3	1	4	5	6			
4612	Tyrannus carolinensis	2	3	4	1	5	6			
6457	Tyrannus carolinensis	2	3	1 } 4 }	5	6	7			
4816	Contopus borealis	2	1	3	4	5	6			
6665	Contopus borealis	2	3	1	4	5	6			
6938	Sayornis fuscus	3	4	2	5	6	1	7		
6932	Sayornis fuscus	3 } 4 }	2 } 5 }	6	1	7	8	9		
5364	Lophophanes bicolor	4	5	6	3	7	8	9	2	1
5248	Lophophanes bicolor	5	4	6	7	3	8	9	2	1
——	Lophophanes bicolor	5 } 4 } 6 }	7	3	8	9	2	1		
5080	Dendrœca coronata	2	3	1	4	5	6			
5176	Dendrœca coronata	3	2	4	1	5	6			
6678	Dendrœca coronata	3	2 } 4 }	1	5	6	7			
3412	Dendrœca coronata	2 } 3 }	4	1	5	6	7			
10533	Dendrœca coronata	2 } 1 } 3 } 4 }	5	6	7	8	9			
5056	Dendrœca striata	2	1	3	4	5	6			
5057	Dendrœca striata	1	2	3	4	5	6			
3390	Dendrœca striata	2	1 } 3 }	4	5	6	7			
6675	Dendrœca striata	2	3	1	4	5	6			
10958	Pinicola enucleator	3	4	2	5	1	6	7		
10963	Pinicola enucleator	3	2	4	1	5	6	7		
10960	Pinicola enucleator	3	2	1	4	5	6	7		
10962	Pinicola enucleator	3 } 2 }	4	1	5	6	7	8		
8114	Pinicola enucleator	3	4 } 2 }	1	5	6	7	8		
4843	Ampelis cedrorum	2	3	1	4	5				
4633	Ampelis cedrorum	1	2	3	4	5				
4844	Ampelis cedrorum	2	1	3	4	5				

varies from fourteen to twenty. Specimens with sixteen are tolerably frequent. Yet one of the principal characters urged as separating the *B. Hutchinsii* from the *B. canadensis* is the possession of two more feathers in the tail by the latter than the so-called *B. Hutchinsii* is assumed to have. In *Bernicla brenta* the usual number is sixteen, but in different specimens they vary from fourteen to eighteen. A greater or less amount of variation in the number of the feathers of the tail is more or less common to numerous other species of the duck tribe. An odd number is even quite frequent, one half of the tail having normally one more feather than the other.

Variation in the Relative Length of the Tarsus and Toes. — A common feature in modern generic and specific diagnoses is a statement of the ratio the length of the tarsus bears to the length of the middle toe or to the hallux, and the relative length of the hallux to the outer or inner toe, as though we had here constant structural proportions. The following table (Table G) shows that such is not the case, the varia-

Table G.—Relative Length of Tarsi and Toes.

M. C. Z. No.	Sex.	Species.	Tarsus.	Middle Toe.	Outer Toe.	Hallux.
5858	?	Galeoscoptes carolinensis	1.08	1.04	.77	.75
2273	?	Galeoscoptes carolinensis	1.15	.98	.70	.70
10356	?	Galeoscoptes carolinensis	1.00	1.00	.70	.73
5857	?	Galeoscoptes carolinensis	1.18	1.08	.70	.75
2229	?	Galeoscoptes carolinensis	1.07	.93	.68	.67
5605	♂	Sialia sialis	.77	.77	.57	.58
1456	♂	Sialia sialis	.74	.84	.62	.65
5766	♂	Sialia sialis	.83	.80	.56	.60
1883	♂	Sialia sialis	.80	.91	.77	.65
1946	♂	Sialia sialis	.80	.84	.77	.61
1881	♂	Sialia sialis	.77	.85	.56	.72
1771	♂	Pipilo erythrophthalmus	.98	.95	.73	.80
1399	♂	Pipilo erythrophthalmus	1.05	1.05	.80	.78
350	♂	Pipilo erythrophthalmus	1.05	1.12	.76	.84
1476	♂	Pipilo erythrophthalmus	1.10	1.03	.75	.78
2985	♂	Pipilo erythrophthalmus	1.13	1.00	.80	.80
9854	♂	Dolichonyx oryzivora	.98	1.17	.83	.82
5585	♂	Dolichonyx oryzivora	1.15	1.27	.98	.93
9894	♂	Dolichonyx oryzivora	1.00	1.00	.83	.81
10219	♂	Dolichonyx oryzivora	1.03	1.25	.98	.76
2320	♂	Icterus Baltimore	.83	.68	.88	.72
9793	♂	Icterus Baltimore	1.02	.85	.70	.70
1567	♂	Icterus Baltimore	.97	1.00	.75	.80
10025	?	Tyrannus carolinensis	.67	.73	.53	.54
10027	?	Tyrannus carolinensis	.80	.85	.55	.61
10028	?	Tyrannus carolinensis	.70	.87	.53	.57
5546	?	Tyrannus carolinensis	.70	.80	.60	.60

tion being as great between different specimens of the same species as between different species of the same genus, and even of different genera. The variation in the length of the toes is often due to an increase or a decrease in the length of the nail, but by no means rarely to variations in the length of the phalanges themselves. As already stated, and as appears from the table, toes of less than the average length accompany tarsi of the average or of more than the average length, and toes of more than the average length accompany tarsi of medium or less than the medium length. In compiling the above table the specimens mentioned have been selected in each case from a series of only twenty specimens of the species to which they respectively belong, and represent the longest and shortest tarsus, middle toe, outer toe, and hind toe met with in each series, and also the greatest and least amount of difference in these several elements. They are all taken from Tables H to Q (see pp. 210 – 219), which serve to show the usual range of variation, in respect to size and proportions, in ten species.*

Individual Variation in other Parts. — In addition to the instances already mentioned, individual variation of a similar character and equal extent occurs in the relative size of other parts. The length of the bill, for instance, is often compared to the length of the head, or to that of the tarsus in specific diagnoses. Table G[1] (see next page) serves to show the individual variation in respect to the proportion of length to alar extent ordinarily met with in specimens of the same species.

To show more fully, however, the exact nature and extent of what may be considered as purely individual variation, tables of detailed measurements of about twenty specimens of each of a number of species are herewith appended (Tables H to Q). Care has been taken to not only select specimens of the same sex, collected at the same locality, and as nearly as possible at the same season, but also such species as find their northern limit so near the locality at which they were taken as to obviate the complication of individual with geographical variation, which would result if the range of the species extended far to the northward of the locality in question. In general, the specimens are all from Eastern Massachusetts, and

* Icterus Baltimore, Dolichonyx oryzivorus, Pipilo erythrophthalmus, Sialia sialis, Galeoscoptes carolinensis, Pyranga rubra, Geothlypis trichas, Harporhynchus rufus, Tyrannus carolinensis, Hedymeles ludoviciana.

Table G[1] — Individual Variation in the Proportion of Length to Alar Extent.

M. C. Z. No.	Orig. No.	Sex.	Species.		
5056	668	♀	Dendrœca striata	5.45	9.70
——	777	♀	Dendrœca striata	5.50	8.68
5087	848	♂	Passerculus savanna	5.50	9.13
——	981	♂	Passerculus savanna	5.83	7.75
——	——	♂	Passerculus savanna	6.00	8.27
——	1987	♂	Cardinalis virginianus	9.00	11.50
——	2394	♀	Cardinalis virginianus	8.00	11.75
9901	——	♂	Dolichonyx oryzivorus	6.65	11.50
2295	——	♂	Dolichonyx oryzivorus	7.50	11.50
——	2340	?	Mimus polyglottus	9.60	14.25
——	2374	♀	Mimus polyglottus	9.75	14.00
——	2371	♀	Mimus polyglottus	9.80	13.00
5757	——	?	Turdus Swainsoni	7.25	12.15
2930	——	?	Turdus Swainsoni	7.75	11.20
1829	——	?	Turdus Swainsoni	6.90	11.20
307	——	?	Turdus Swainsoni	7.24	11.00
9691	——	?	Turdus Pallasi	7.00	10.50
145	——	?	Turdus Pallasi	7.00	11.40
5756	——	?	Turdus Pallasi	7.38	11.05
——	314	♂	Turdus Pallasi	7.38	12.33
——	363	♂	Turdus Pallasi	7.23	11.94
——	26	♂	Turdus Pallasi	6.80	11.28
——	367	♂	Turdus fuscescens	7.81	13.70
——	495	♂	Turdus fuscescens	7.87	11.91
——	551	♂	Turdus fuscescens	7.00	11.95
——	112	♂	Parus atricapillus	5.50	8.12
4946	268	♀	Parus atricapillus	5.00	8.60
11714	114	♀	Parus atricapillus	5.75	7.88
95	——	♂	Agelæus phœniceus	9.00	15.10
93	——	♂	Agelæus phœniceus	9.20	14.40
5723	——	♂	Agelæus phœniceus	8.45	14.45

from within a short distance of Cambridge. A very few are from Southern Maine and from the Connecticut valley at Springfield; but the general faunal character of all these localities is essentially the same.*

In addition to the measurements given in these tables, several others are sometimes taken by collectors, as the relative posterior extent of the outstretched feet and the wing, as compared with the tail. As they are, however, among the most variable of proportions, and are likewise among the most difficult measurements to take with accuracy, they have been here neglected.

* In consequence of the small size of these pages, it has been found impracticable to include the names of the localities, the date of collecting, and the name of the collector in the tables, as would have been desirable.

Table H. — Icterus Baltimore *Daud.*

M. C. Z. Number.	Sex.	Length.	Alar Extent.	Wing.	Tail.	Extent of Primaries beyond Inner Secondaries.	Extent of Primaries beyond Outer Secondaries.	Extent of Tail beyond Upper Coverts.	Extent of Tail beyond Lower Coverts.	Tarsus.	Hind Toe.	Middle Toe.	Outer Toe.	Inner Toe.	Head.	Bill. Culmen.	Bill. Commissure.	Bill. Gonys.	Bill. Height.	Bill. Width.
1332	♂	7.45	11.75	3.70	2.90	.70	.97	1.50	1.40	.96	.72	.89	.67	.65	1.62	.76	.77	.47	.32	.33
9793	♂	7.50	11.60	3.70	2.95	.76	1.07	1.67	1.33	1.02	.70	.85	.70	.66	1.58	.76	.77	.47	.38	.35
1333	♂	7.20	11.20	3.64	2.88	.57	1.06	1.52	1.40	.93	.67	.85	.70	.66	1.62	.78	.81	.50	.35	.32
2289	♂	8.00	11.90	3.85	3.10	.87	1.12	1.77	1.65	.91	.69	.95	.75	.73	1.63	.75	.77	.52	.37	.34
9884	♂	7.00	10.40	3.54	2.82	.86	.93	1.37	1.15	.87	.65	.93	.70	.69	1.60	.73	.46	.46	.33	.35
2733	♂	7.55	11.30	3.75	3.00	.82	1.03	1.82	1.18	.87	.70	.92	.70	.68	1.60	.80	.83	.48	.34	.35
2962	♂	7.50	11.50	3.72	3.00	.68	.95	1.63	1.38	.90	.70	.87	.70	.65	1.60	.77	.78	.48	.37	.35
1567	♂	7.60	11.75	3.80	3.00	.56	.92	1.53	1.40	.97	.75	1.00	.80	.78	1.63	.77	.80	.50	.32	.32
1334	♂	7.20	11.00	3.45	2.70	.60	.84	1.45	1.20	.90	.70	.90	.73	.73	1.55	.75	.77	.48	.35	.35
1566	♂	7.40	11.50	3.60	2.90	—	—	1.55	1.10	.90	.73	.86	.73	.70	1.60	.82	.82	.53	.35	.34
2993	♂	7.60	11.90	3.75	2.97	.68	.96	1.43	1.30	.91	.70	.90	.70	.67	1.58	.78	.78	.52	.34	.33
388	♂	7.40	11.00	3.68	2 78	.85	1 00	1.58	1.27	.95	.70	.88	.75	.72	1.62	.78	.77	.51	.37	.35
2320	♂	7.30	11.50	3.60	2.75	.77	.95	1.30	1.25	.83	.68	.88	.72	.68	1.50	—	—	—	—	—
2964	♂	7 75	11.90	3.85	3.00	.77	1.07	1.57	1 50	.90	.62	.82	.73	.68	1.62	.75	.76	.50	.32	.33
2290	♂	7.80	11.75	3.75	2.95	.62	.85	1.67	1.35	.86	.68	.87	.64	.61	1.60	.77	.78	.45	.34	.34
2499	♂	7.75	11.75	3.78	2.96	.90	1 03	—	—	.94	.70	.82	.68	.65	1.55	.72	.74	.44	.38	.38
2290	♂	7.55	11.75	3.83	2.95	.67	.81	1.58	1.37	.95	.72	.80	.67	.67	1.62	.82	.84	.50	.40	.38
2319	♂	7.65	11.75	3.75	2.85	.77	.90	1.54	1.40	.95	.70	.98	.70	.73	1.58	—	—	.50	.35	.33
9885	♂	7.50	11.75	3.55	3 00	.78	.95	1.87	1.65	.90	.70	.90	.75	.75	1.60	.80	.82	.47	.35	.35
2500	♂	7.65	12.00	3.83	3.00	—	—	—	—	.98	.72	.90	.75	.72	1.58	.75	.78	.50	.37	.37

Table I. — Dolichonyx oryzivorus *Swain.*

M. C. Z. Number.	Sex.	Length.	Alar Extent.	Wing.	Tail.	Extent of Primaries beyond Inner Secondaries.	Extent of Primaries beyond Outer Secondaries.	Extent of Tail beyond Upper Coverts.	Extent of Tail beyond Lower Coverts.	Tarsus.	Hind. Toe.	Middle Toe.	Outer Toe.	Inner Toe.	Head.	Bill.				
																Culmen.	Commissure.	Gonys.	Height.	Width.
2293	♂	7.50	11.50	3.75	2.68	.93	1.27	1.10	1.00	1.05	.87	1.12	.77	.80	1.32	.60	.63	.34	.36	.36
10342	♂	7.00	11.00	3.64	2.45	.88	1.20	.80	—	1.07	.92	1.17	.73	.75	1.37	.62	.65	.35	.37	.37
10343	♂	7.30	11.80	3.87	2.80	.90	1.30	.95	.90	1.12	.95	1.28	.87	.87	1.37	—	—	.33	.34	.37
5585	♂	7.35	11.55	3.80	2.66	.97	1.40	—	.83	1.15	.98	1.27	.93	.87	1.35	.62	.62	.35	.36	.36
2295	♂	7.50	11.50	3.71	2.73	1.05	1.33	.93	.90	1.00	.92	1.18	.85	.85	1.37	.62	.62	.35	.36	.36
119	♂	7.25	11.65	3.82	2.57	.78	1.23	—	.80	1.00	.83	1.12	.76	.78	1.34	.60	.60	.35	.38	.37
5743	♂	7.35	11.75	3.65	2.68	1.03	1.33	—	.80	1.10	.92	1.13	.78	.80	1.38	.60	.63	.36	.37	.40
9901	♂	6.65	11.50	3.70	2.53	.82	1.25	.95	.90	1.00	.85	1.07	.70	.76	1.35	.57	.60	.35	.35	.36
389	♂	7.20	11.20	3.55	2.50	.85	1.20	.95	.87	1.08	.86	1.10	.75	.78	1.40	.58	.62	.35	.37	.37
9854	♂	6.80	11.15	3.53	2.57	.98	1.14	1.03	.82	.98	.83	1.17	.82	.82	1.37	.60	.62	.35	.38	.38
2298	♂	7.70	11.80	3.75	2.78	1.03	1.35	1.07	.95	1.05	.90	1.12	.82	.85	1.37	.62	.64	.40	.38	.38
5741	♂	7.15	11.65	4.00	2.72	1.20	1.40	.97	.82	—	.87	1.13	.83	.82	1.36	.58	.60	.37	.38	.37
2296	♂	7.15	11.50	3.80	2.75	.98	1.42	.98	1.00	1.05	.82	1.20	.85	.88	1.33	.57	.63	.37	.38	.37
9893	♂	7.35	11.75	3.90	2.73	.98	1.37	1.15	.92	1.10	.95	1.27	.85	.85	1.36	.58	.65	.35	.38	.37
9892	♂	7.25	11.90	3.86	2.82	.97	1.35	1.10	.90	1.04	.87	1.23	.82	.83	1.33	.57	.62	.35	.37	.38
1482	♂	7.10	11.90	3.85	2.47	1.00	1.35	1.10	.95	1.05	.90	1.20	.88	.88	1.35	.60	.64	.35	.38	.37
9896	♂	7.25	12.15	3.90	2.67	1.05	1.40	1.08	.91	1.07	.85	1.17	.78	.78	1.36	.60	.60	.35	.36	.35
9894	♂	6.90	11.75	3.75	2.70	.95	1.32	.90	1.00	1.00	.83	1.00	.81	.82	1.35	.62	.64	.38	.39	.39
10219	♂	7.55	12.00	3.83	2.75	.98	1.33	1.23	1.05	1.03	.98	1.25	.76	.76	1.33	.60	.60	.37	.35	.39
5744	♂	7.45	11.35	3.85	2.82	1.08	1.43	1.00	1.05	1.10	.95	1.13	.82	.83	1.35	.60	.62	.37	.37	.39

Table J. — Pipilo erythrophthalmus *Vieillot.*

M. C. Z. Number.	Sex.	Length.	Alar Extent.	Wing.	Tail.	Extent of Primaries beyond Inner Secondaries.	Extent of Primaries beyond Outer Secondaries.	Extent of Tail beyond Upper Coverts.	Extent of Tail beyond Lower Coverts.	Tarsus	Hind. Toe.	Middle Toe.	Outer Toe.	Inner Toe.	Head.	Bill. Culmen.	Bill. Commissure.	Bill. Gonys.	Bill. Height.	Bill. Width.
350	♂	7.80	10.50	3.20	3.35	.33	.43	2.05	2.25	1.05	.76	1.12	.84	.82	1.37	.60	.67	—	.37	.37
1366	♂	8.25	10.65	3.43	3.70	.40	.50	2.25	2.15	1.03	.78	1.02	.85	.80	1.47	.56	.73	.40	.40	.37
1543	♂	8.00	11.00	3.60	3.90	.43	.48	2.35	2.27	1.05	.77	1.08	.88	.88	1.50	.60	.68	.40	.42	.34
1476	♂	7.80	10.30	3.20	3.40	.45	.45	2.38	2.18	1.10	.75	1.03	.78	.77	1.50	.60	.72	.37	.39	.35
1397	♂	8.25	10.25	3.45	3.70	.53	.58	2.60	2.13	1.12	.80	1.10	.82	.77	1.48	.53	.67	.34	.34	.38
1421	♂	8.10	10.60	3.17	3.48	.30	.20	—	1.85	1.07	.80	1.00	.78	.73	1.45	.53	.66	.40	.40	.33
349	♂	8.15	10.70	3.50	3.75	.43	.50	—	1.82	1.08	.80	1.00	.78	.77	1.46	.55	.70	.36	.38	.30
1399	♂	7.75	10.00	3.35	3.57	—	—	—	2.15	1.05	.80	1.05	.78	.77	1.46	.57	.70	.37	.35	.35
2985	♂	8.35	11.05	3.43	3.65	.35	.52	2.23	2.00	1.13	.75	1.00	.80	.80	1.50	.53	.72	.37	.35	.35
2965	♂	8.25	10.50	3.50	3.70	.45	.55	2.40	2.40	1.06	.81	1.08	.83	.78	1.49	.60	.70	.35	.37	.35
4009	♂	8.55	12.25	3.57	3.85	.47	.47	2.58	2.20	1.07	.77	1.04	.83	.78	1.40	.56	.70	.35	.35	.35
1783	♂	8.25	11.20	3.54	3.68	.45	.53	2.20	2.15	1.05	.80	1.04	.76	.68	1.50	.55	.73	.42	.35	.35
1612	♂	8.00	10.45	3.22	3.45	.30	.41	2.20	2.50	1.10	.80	1.10	.75	.72	1.50	.62	.73	.37	.32	.37
1771	♂	7.75	10.85	3.38	3.70	.35	.50	2.35	2.85	.98	.73	.95	.80	.73	1.45	.57	.68	.35	.33	.32
1784	♂	7.90	10.80	3.43	3.62	.40	.57	2.30	2.17	1.10	.78	1.08	.78	.73	1.50	.59	.72	.38	.38	.35
81	♂	8.25	11.15	3.54	3.80	.40	.45	2.53	2.30	1.08	.80	1.04	.80	.75	1.45	.55	.66	.38	.34	.35
83	♂	8.50	11.15	3.52	3.75	.42	.53	2.22	2.18	1.06	.80	1.05	.82	.75	1.50	.60	.73	.37	.41	.42
10151	♂	7.50	10.20	3.34	3.60	.46	.55	2.58	1.93	1.08	.75	1.12	.77	.75	1.45	.53	.62	.35	.37	.35
9696	♂	8.20	11.00	3.68	3.93	.50	.50	2.25	2.05	1.10	.78	1.07	.82	.82	1.50	.60	.73	.37	.42	.39
84	♂	8.15	10.25	3.30	3.70	.38	.47	—	2.06	1.02	.75	1.00	.83	.80	1.40	.60	.67	.37	.37	.35

Table K. — SIALIA SIALIS *Hald.*

M. C. Z. Number.	Sex.	Length.	Alar Extent.	Wing.	Tail.	Extent of Primaries beyond Inner Secondaries.	Extent of Primaries beyond Outer Secondaries.	Extent of Tail beyond Upper Coverts.	Extent of Tail beyond Lower Coverts.	Tarsus.	Hind Toe.	Middle Toe.	Outer Toe.	Inner Toe.	Head.	Bill. Culmen.	Bill. Commissure.	Bill. Gonys.	Bill. Height.	Bill. Width.
2331	♂	7.00	12.30	4.00	2.50	1.00	1.20	1.30	1.20	.75	.63	.87	.68	.60	1.48	.48	.77	.30	.18	.35
1945	♂	6.75	12.20	3.90	2.60	1.03	1.11	1.56	1.04	.77	.56	.85	.72	.65	1.48	.50	.70	.33	.20	.34
2378	♂	7.00	12.25	4.10	2.60	1.20	1.26	1.27	1.05	.77	.60	.85	.64	.65	1.45	.50	.76	.32	.21	.33
2479	♂	6.10	11.35	3.87	2.50	1.10	1.20	1.25	1.05	.78	.60	.87	.65	.58	1.50	.56	.77	.30	.18	.35
1946	♂	6.85	12.10	4.03	2.58	1.03	1.20	1.33	1.13	.80	.77	.84	.61	.50	1.50	.53	.80	.32	.18	.37
1883	♂	6.85	12.55	4.07	2.56	1.17	1.26	1.20	1.15	.80	.77	.91	.65	.60	1.48	.50	.80	.30	.20	.38
10292	♂	6.45	11.75	3.90	2.60	.95	1.15	—	—	.76	.85	.82	.62	.55	1.35	.49	.72	.30	.20	.38
1881	♂	6.70	11.65	3.82	2.43	.98	1.13	1.20	1.08	.77	.67	.85	.56	.50	1.50	.56	.80	.33	.20	.33
72	♂	6.80	12.00	3.87	2.33	1.03	1.12	1.17	1.03	.75	.60	.85	.68	.62	1.50	.55	.78	.33	.20	.33
10226	♂	6.75	12.40	4.00	2.77	1.15	1.25	1.50	1.08	.77	.62	.83	.63	.58	1.50	.52	.80	.34	.20	.33
10231	♂	6.75	12.10	3.90	2.50	1.10	1.30	1.37	1.23	.75	.55	.80	.60	.54	1.42	.48	.76	.27	.18	.32
1456	♂	6.50	11.10	3.75	2.45	1.00	1.10	1.15	.95	.74	.62	.84	.65	.60	1.53	.55	.80	.35	.20	.35
5605	♂	6.70	12.00	3.85	2.35	1.10	1.25	1.08	.82	.77	.57	.77	.58	.54	1.50	.47	.77	.30	.18	.35
5763	♂	6.68	12.25	4.03	2.63	1.10	1.25	1.15	.90	.78	.55	.83	.58	.53	1.53	—	—	.32	—	—
5765	♂	6.65	11.80	3.90	2.63	1.05	1.17	1.58	1.23	.76	.57	.80	.61	.56	1.50	.55	.80	.31	.17	.30
5606	♂	6.50	11.65	4.05	2.68	1.25	1.40	1.27	1.18	.78	.60	.85	.63	.60	1.47	.53	.83	.32	.17	.34
338	♂	6.40	11.10	4.07	2.63	1.30	1.30	1.25	1.07	.80	.60	.88	.68	.62	1.50	.45	.77	.30	.21	.35
5760	♂	7.00	12.25	3.85	2.50	1.15	1.30	1.20	1.05	.80	.60	.83	.60	.55	1.45	.45	.70	.29	.18	.32
5766	♂	6.80	12.00	3.85	2.48	1.00	1.20	1.20	1.12	.83	.56	.80	.60	.58	1.45	.52	.70	.32	.16	.35
77	♂	6.75	11.75	3.95	2.55	1.15	1.30	—	—	.82	.60	.87	.65	.62	1.45	.45	.77	.30	.18	.34

Table L. — GALEOSCOPTES CAROLINENSIS *Cab.*

M. C. Z. Number.	Sex.	Length.	Alar Extent.	Wing.	Tail.	Extent of Primaries beyond Inner Secondaries.	Extent of Primaries beyond Outer Secondaries.	Extent of Tail beyond Upper Coverts.	Extent of Tail beyond Lower Coverts.	Tarsus.	Hind Toe.	Middle Toe.	Outer Toe.	Inner Toe.	Head.	Bill.				
																Culmen.	Commissure.	Gonys.	Height.	Width.
5858	?	8.75	11.35	3.55	3.75	.35	.57	2.20	2.05	1.08	.75	1.04	.77	.73	1.75	.80	1.00	.45	.23	.37
1790	?	8.75	11.75	3.75	3.85	.55	.70	2.50	1.85	1.10	.70	1.06	.76	.72	1.90	.73	.98	.43	.24	.37
10132	?	7.80	10.50	3.35	3.70	.42	.55	—	—	1.10	.73	1.03	.76	.73	1.67	.65	.92	.40	.24	.38
10014	?	8.75	11.25	3.85	4.10	.70	.75	2.20	1.95	1.12	.71	1.10	.77	.74	1.80	.78	1.00	.45	.27	.38
2273	?	8.50	11.25	3.40	3.65	.48	.65	—	1.85	1.15	.70	.98	.74	.70	1.75	.74	1.00	.45	.25	.37
234	?	8.40	10.80	3.60	3.68	.45	.62	2.10	1.87	1.10	.70	.96	.75	—	1.75	.75	1.00	.43	.24	.37
257	?	9.00	11.75	3.75	4.10	.58	.67	2.40	2.05	1.10	.72	1.03	.74	.70	1.73	.78	.96	.45	.24	.37
2229	?	8.65	10.75	3.65	3.90	.50	.62	2.35	2.00	1.07	.68	.93	.67	.64	1.70	.80	.95	.47	.24	.39
2734	?	8.00	10.50	3.25	3.35	—	—	—	—	1.15	.70	1.05	.70	.65	1.65	.70	.92	.43	—	—
10356	?	8.35	11.00	3.35	3.57	.53	.65	—	—	1.00	.70	1.00	.73	.70	1.68	.72	.95	.42	.23	.36
1754	?	8.80	11.45	3.62	3.87	.65	.70	2.40	1.75	1.08	.69	1.02	.73	.71	1.70	.73	.93	.45	.23	.36
2274	?	9.00	11.85	3.75	4.00	.67	.70	2.40	1.20	1.12	.70	1.00	.73	.70	1.76	.78	.97	.50	.25	.37
256	?	8.25	11.25	3.37	3.55	.55	.50	—	1.60	1.10	.68	.97	.72	.68	1.75	.80	.97	.45	.26	.35
1638	?	8.80	11.40	3.55	3.70	.47	.60	—	—	1.06	.70	1.00	.77	.73	1.75	.75	1.00	.46	.24	.36
2281	?	8.85	11.50	3.75	3.80	—	—	—	—	1.10	.71	1.04	.74	.71	1.70	.75	.98	.47	.24	.32
5857	?	8.75	10.75	3.42	3.78	.46	.52	—	1.82	1.18	.70	1.08	.75	.72	1.75	.76	1.00	.46	.23	.37
235	?	8.62	11.20	3.50	3.63	—	—	—	—	1.05	.72	1.07	.75	.70	1.75	.73	.96	.44	.23	.35
1481	?	8.80	11.25	3.63	3.95	.53	.45	2.50	2.00	1.13	.71	1.05	.75	.72	1.75	.80	.94	.46	.21	.33
254	?	8.70	10.90	3.40	3.67	.40	.53	—	—	1.10	.66	1.00	.70	.67	1.72	.67	.97	.48	.23	.37
2986	?	8.50	10.75	3.40	3.45	.55	.60	—	—	1.05	.66	1.02	.75	.70	1.62	.63	.93	.37	.22	.37

Table M. — TYRANNUS CAROLINENSIS *Temm.*

M. C. Z. Number.	Sex.	Length.	Alar Extent.	Wing.	Tail.	Extent of Primaries beyond Inner Secondaries.	Extent of Primaries beyond Outer Secondaries.	Extent of Tail beyond Upper Coverts.	Extent of Tail beyond Lower Coverts.	Tarsus.	Hind Toe.	Middle Toe.	Outer Toe.	Inner Toe.	Head.	Bill.				
																Culmen.	Commissure.	Gonys.	Height.	Width.
1071	?	8.00	13.50	4.37	3.25	1.00	1.40	1.50	1.30	.75	.50	.80	.62	.56	1.67	.80	1.00	.44	.32	.53
10027	?	8.20	14.25	4.63	3.20	.87	1.42	1.70	1.30	.80	.55	.85	.61	.58	1.67	.80	1.02	.47	.30	.54
10026	?	7.85	14.20	4.67	3.50	1.05	1.53	1.80	1.46	.70	.55	.77	.60	.56	1.66	.77	.98	.47	.27	.45
10023	?	8.00	14.25	4.55	3.50	1.06	1.48	1.78	1.50	.70	.53	.87	.57	.53	1.58	.77	.92	.46	.28	.37
5546	?	8.15	14.15	4.63	3.43	1.09	1.56	1.62	1.36	.70	.60	.80	.60	.57	1.71	.83	1.02	.51	.27	.43
114	?	8.00	14.10	4.45	3.22	1.03	1.40	1.77	1.38	.80	.54	.77	.55	.50	1.73	.80	1.05	.50	.30	.50
112	?	8.20	13.50	4.47	3.33	.95	1.44	1.71	1.40	.70	.53	.77	.60	.56	1.60	.74	.98	.42	.30	.47
10025	?	7.00	13.00	4.17	3.15	.96	1.22	1.80	1.43	.67	.53	.73	.54	.50	1.58	.70	.97	.42	.28	.44
10376	?	7.75	13.75	4.42	3.25	.90	1.38	1.77	1.57	.70	.55	.77	.60	.53	1.68	.80	1.02	.50	.27	.47
10029	?	8.00	13.80	4.45	3.25	.90	1.30	1.73	1.73	.70	.72	.85	.62	.60	1.72	.78	1.05	.50	.27	.47
284	?	8.00	13.50	4.30	3.15	.85	1.15	1.70	1.47	.75	.67	.75	.60	.58	1.63	.67	1.00	.40	.28	.46
113	?	8.25	14.15	4.60	3.20	.90	1.45	1.55	1.20	.78	.58	.85	.70	.65	1.68	.80	1.05	.46	.27	.45
1317	?	7.40	12.50	4.25	2.93	.76	1.10	1.55	1.20	.73	.57	.78	.62	.55	1.57	.62	.90	.37	.20	.46
4008	?	8.20	14.15	4.60	3.30	1.35	1.62	1.78	1.42	.71	.57	.87	.65	.58	1.68	.82	1.04	.48	.26	.49
285	?	7.80	13.00	4.38	3.25	.83	1.15	1.77	1.45	.73	.54	.77	.64	.60	1.73	.82	1.05	.45	.29	.50
1318	?	7.85	13.25	4.20	3.25	.87	1.22	1.68	1.10	.75	.54	.78	.60	.53	1.63	.73	.95	.37	.27	.43
1758	?	8.10	13.80	4.50	3.30	.95	1.43	1.75	1.68	.70	.55	.77	.58	.55	1.68	.83	1.00	.50	.28	.46
1568	?	8.40	14.80	4.85	3.54	1.05	1.47	1.83	1.30	.76	.62	.83	.66	.60	1.70	.80	1.00	.45	.30	.45
283	?	8.65	14.25	4.58	3.47	.98	1.45	1.75	1.37	.70	.58	.82	.57	.63	1.75	.84	1.07	.48	.25	.48
4003	?	8.30	13.50	4.60	3.45	1.00	1.40	1.97	1.44	.73	.56	.77	.62	.60	1.63	.75	.90	.45	.28	.45

Table N. — Pyranga rubra *Vieillot.*

M. C. Z. Number.	Sex.	Length.	Alar Extent.	Wing.	Tail.	Extent of Primaries beyond Inner Secondaries.	Extent of Primaries beyond Outer Secondaries.	Extent of Tail beyond Upper Coverts.	Extent of Tail beyond Lower Coverts.	Tarsus.	Hind Toe.	Middle Toe.	Outer Toe.	Inner Toe.	Head.	Bill.				
																Culmen.	Commissure.	Gonys.	Height.	Width.
—	♂	6.75	11.75	3.57	2.62	.80	1.13	—	—	.77	.52	.75	.53	.45	—	.58	.73	.44	.33	.38
10300	♂	6.90	11.30	3.75	2.75	.85	1.25	1.50	1.05	.82	.60	.70	.57	.50	1.53	.58	.75	.45	.35	.37
2929	♂	7.25	11.30	3.65	2.62	.75	1.12	1.23	.85	.73	.56	.67	.53	.47	1.48	.62	.75	.45	.33	.36
9795	♂	7.25	11.30	3.70	2.55	.92	1.12	1.41	1.13	.86	.60	.80	.57	.57	1.50	.64	.76	.40	.35	.40
10119	♂	7.25	11.60	3.80	2.73	.95	1.25	1.25	1.00	.68	.57	.80	.57	.53	1.48	.65	.73	.42	.35	.40
589	♂	7.30	11.75	4.00	2.73	1.07	1.30	1.35	1.00	.72	.57	.75	.58	.52	1.52	.63	.76	.47	.37	.40
10366	♂	7.30	11.75	3.85	2.85	1.00	—	—	—	.76	.58	.76	.53	.51	1.52	.63	.75	.45	.36	.39
10388	♂	7.20	11.60	3.87	2.75	.95	1.20	1.25	1.33	.77	.53	.79	.55	.50	—	—	—	—	—	—
10133	♂	7.00	11.00	3.77	2.77	.85	1.00	1.45	.88	.78	.58	.75	.55	.53	1.48	.57	.73	.40	.35	.36
1834	♂	6.75	11.30	3.60	2.50	.85	1.17	1.12	1.12	.70	.60	.73	.56	.54	1.46	.62	.73	.45	.37	.42
190	♂	6.75	10.75	3.80	2.64	.90	1.20	1.50	1.60	.67	.55	.70	.56	.56	1.38	.64	.70	.42	.33	.38
1833	♂	6.90	11.25	3.70	2.55	.85	1.05	1.33	1.05	.73	.55	.70	.57	.52	1.50	65	.77	.43	.35	.40
1642	♂	7.00	10.65	3.85	2.75	1.07	1.23	1.50	1.10	.75	.53	.76	.55	.48	1.50	.65	.77	.45	.33	.37

Table O. — Hedymeles ludovicianus *Cab.*

M. C. Z. Number.	Sex.	Length.	Alar Extent.	Wing	Tail.	Extent of Primaries beyond Inner Secondaries.	Extent of Primaries beyond Outer Secondaries.	Extent of Tail beyond Upper Coverts.	Extent of Tail beyond Lower Coverts.	Tarsus.	Hind Toe.	Middle Toe.	Outer Toe.	Inner Toe.	Head.	Bill.				
																Culmen.	Commissure.	Gonys.	Height.	Width.
1438	♂	7.50	12.25	3.87	3.05	.80	1.17	1.25	1.00	.92	.70	.75	.75	.70	1.50	.72	.78	.43	.54	.45
590	♂	7.75	12.30	4.00	3.00	.90	1.25	1.55	1.24	.93	.70	.74	.74	.72	1.55	.77	.80	.48	.55	.47
9787	♂	7.75	11.75	3.83	2.95	.85	1.13	1.50	1.40	.90	.70	.67	.67	.62	1.53	.73	.75	.45	.50	.43
2882	♂	8.30	12.50	4.15	3.05	.93	1.15	1.45	1.15	.88	.70	.75	.75	.74	1.55	.73	.77	.45	.51	.46
9935	♂	7.55	11.90	4.00	3.02	.60	1.06	1.28	.95	.87	.67	.75	.75	.70	1.55	.70	.78	.48	.56	.45
9580	♂	8 12	12.50	4.25	3.05	1.06	1.24	1.30	1 00	.88	.72	.65	.65	.55	1.58	.75	.88	.50	.60	.50
1518	♂	7.75	11.80	3.85	2.75	.70	1.00	1.20	1.00	.80	.70	.65	.65	.60	1.54	.75	.82	.40	.55	.45
1065	♂	7.60	12.00	3.83	2.88	.82	.97	1.24	.98	.90	.60	.65	.65	.58	1.55	.70	.81	.45	.55	.43
1547	♂	7.50	12.10	3 86	2.70	.95	1.10	1.36	—	.82	.67	.63	.63	.56	1.50	.65	.75	.44	.51	.45
187	♂	7.75	11.50	3.85	2.77	.80	1.07	1.40	1.42	.82	.65	.67	.67	.63	1.52	.70	.77	.45	.53	.45
9581	♂	7.85	12.50	3.95	2.82	.77	1.00	—	1.12	.84	.67	.68	.68	.61	1.42	.66	.70	.43	.53	.42
5591	♂	7.75	11.90	4.00	3.08	.82	1.00	—	1.18	.87	.67	.68	.68	.58	1.50	.73	.82	.45	.54	.44
5592	♂	8.10	12.90	4 15	3.00	.75	1.15	1.70	1.35	.92	.65	.72	.72	.65	1.55	.70	.77	.45	.54	.43
2942	♂	7.85	12.15	3.97	2.93	.80	1 08	1.92	1.10	.90	.68	.65	.65	.60	1.50	.68	.72	.45	.55	.50
9934	♂	8.10	12.75	4.00	3.00	.83	1.05	1.40	1.22	.83	.70	.67	.67	.57	1.58	.75	.77	.46	.52	.47
1823	♂	7.15	12.00	4.05	2.85	.90	1.07	1.25	1.00	.70	.62	.78	.78	.74	—	—	—	—	—	—
10107	♂	7.65	11.75	4.20	2.93	.90	1.05	1.05	1.10	.86	.70	.73	.73	.60	1.57	.75	.73	.45	.50	.48

Table P. — GEOTHLYPIS TRICHAS *Cab.*

M. C. Z. Number.	Sex.	Length.	Alar Extent.	Wing.	Tail.	Extent of Primaries beyond Inner Secondaries.	Extent of Primaries beyond Outer Secondaries.	Extent of Tail beyond Upper Coverts.	Extent of Tail beyond Lower Coverts.	Tarsus	Hind Toe.	Middle Toe.	Outer Toe.	Inner Toe.	Head.	Bill. Culmen.	Bill. Commissure.	Bill. Gonys.	Bill. Height.	Bill. Width.
9976	♂	5 00	7.00	2.10	2.00	.35	.42	.92	—	.82	.55	.68	.52	.51	1.20	.45	.57	.31	.17	.22
358	♂	5.17	7.25	2.20	2.07	—	—	—	—	.80	.53	.75	.54	.53	1.20	.47	.53	.30	.17	.27
1808	♂	5.25	7.25	2.25	2.05	.36	.32	1.04	.75	.77	.53	.68	.50	.48	1.17	.48	.59	.30	.17	.27
2534	♂	5.15	7.20	2.12	1.88	.35	.35	—	.64	.74	.55	.75	.55	.52	1.25	.55	.67	.30	.18	.26
9775	♂	4.65	6.75	2.37	2.05	.25	.33	—	—	.78	.52	.73	.54	.53	1.13	.43	.55	.27	.15	.24
1809	♂	5 25	6.75	2.20	2.00	.26	.33	.87	1.00	.77	.52	.72	.51	.51	1.28	.48	.58	.31	.16	.27
2997	♂	5.25	7.10	2.18	1.95	.22	.35	—	.70	.78	.53	.73	.55	.53	1.22	.45	.57	.31	.17	.26
9858	♂	4.88	6.45	2.20	2.10	—	—	1.06	.82	.77	.52	.72	.55	.53	1.22	.47	.57	.34	.18	.27
7515	♂	5.37	7.25	2.27	2.10	—	—	—	—	.74	.50	.72	.53	.51	1.18	.50	.60	.32	.18	.22
330	♂	5.10	6.75	2.13	2.10	.33	.37	—	—	.72	.47	.68	.48	.48	1.22	.50	.60	.32	.18	.27
1444	♂	5.63	7.50	2.32	2.17	.35	.40	—	—	.77	.52	.75	.55	.53	1.26	.52	.65	.32	.17	.28
1971	♂	4.60	6.30	1.95	1.78	—	—	—	—	.75	.51	.62	.46	.44	1.16	.45	.54	.28	.15	—
2996	♂	5.37	7.30	2.27	2.08	.32	.38	.97	.70	.78	.54	.73	.53	.52	1.22	.48	.60	.32	.18	.25
2361	♂	4.87	6.75	2.12	1.82	—	—	—	—	.82	.44	.70	.52	.52	1.18	.43	.60	.31	.13	.26
2427	♂	4.88	6.75	2.03	1.88	—	—	—	—	.79	.50	.66	.47	.45	1.15	.43	.58	.30	.15	.28
9669	♂	5.05	6.75	2.13	2.05	—	—	—	—	.80	.48	.73	.53	.51	1.22	.48	.57	.32	.15	.22
9836	♂	5.00	6 85	2 05	1.85	—	—	—	—	.76	.51	.66	.52	.50	1.15	.48	.57	.30	.14	.21
2398	♂	5.20	7.00	2.25	2.00	—	—	—	—	.76	.50	.70	.53	.51	1.20	.49	.59	.32	.15	.23
1445	♂	5.33	6.90	2.15	2.05	—	—	—	—	.78	.48	.73	.55	.54	1.23	.48	.57	.32	.16	.24
2436	♂	5.10	6.80	2.10	1.95	—	—	—	—	.74	.52	.67	.52	.50	1.15	.44	.56	.27	.15	.24

Table Q. — Harporhynchus rufus *Cab.*

M. C. Z. Number.	Sex.	Length.	Alar Extent.	Wing.	Tail.	Extent of Primaries beyond Inner Secondaries.	Extent of Primaries beyond Outer Secondaries.	Extent of Tail beyond Upper Coverts.	Extent of Tail beyond Lower Coverts.	Tarsus.	Hind Toe.	Middle Toe.	Outer Toe.	Inner Toe.	Head.	Bill. Culmen.	Bill. Commissure.	Bill. Gonys.	Bill. Height.	Bill. Width.
1752	?	11.85	14.00	4.25	5.05	.50	.65	—	3.20	1.42	.93	1.25	.85	.80	2.30	1.13	1.25	.70	.35	.45
1625	?	10.55	12.75	3.90	4.60	.43	.60	3.15	2.95	1.27	.90	1.25	.84	.77	2.15	1.00	1.24	.60	.28	.41
237	?	11.00	12.55	3.80	4.78	.38	.56	3.13	2.75	1.33	.84	1.17	.86	.80	2.30	1.06	1.33	.76	.27	.38
10320	?	11.25	13.30	3.95	4.95	.43	.68	3.35	2.95	1.42	.87	1.22	.78	.72	2.25	1.05	1.34	.72	.29	.43
1670	♂	11.50	13.25	4.15	5.18	.47	.60	3.60	3.20	1.36	.93	1.25	.85	.80	2.18	—	1.28	.65	.27	.44
2585	?	10.85	12.90	3.95	4.86	.50	.65	—	2.65	1.25	.87	1.17	.80	.70	2.26	1.03	1.35	.68	.28	.40
1439	♂	11.10	12.80	4.00	4.93	.36	.56	3.17	2.95	1.35	.92	1.28	.90	.85	2.25	1.12	1.33	.67	.27	.43
1067	?	10.85	13.35	4.15	5.00	.32	.55	3.38	3.05	1.32	.86	1.25	.90	.78	2.30	1.06	1.35	.73	.30	.45
8887	♂	11.21	13.50	4.20	4.50	—	—	—	—	1.25	.79	1.25	.75	.73	2.25	1.05	1.35	.74	.30	—
8886	♂	11.41	13.00	4.07	4.59	—	—	—	—	1.30	.85	1.23	.82	.75	2.15	1.00	1.23	.67	.27	.40
8888	♀	11.31	13.30	4.16	5.00	—	—	—	—	1.30	.82	1.23	.85	.82	2.23	1.00	1.25	.65	—	—
8889	♀	11.30	13.00	5.00	5.10	—	—	—	—	1.20	.78	1.25	.90	.84	—	1.00	1.28	.72	.27	.40
8885	♂	11.35	13.35	4.11	5.00	—	—	—	—	1.22	.85	1.30	.87	.78	2.22	1.04	1.33	.70	.27	.37
8891	♂	11.60	13.55	4.25	5.00	—	—	—	—	1.27	.85	1.24	.87	.77	—	1.00	1.26	.67	.25	.38
8996	♂	11.50	13.25	4.10	5.00	—	—	—	—	1.35	.87	1.25	.86	.78	2.23	1.05	1.32	.72	.27	.38
8897	♂	11.75	13.35	4.32	5.30	—	—	—	—	1.24	.85	1.30	.83	.79	2.22	1.05	1.28	.73	.28	.36
8896	♂	11.60	13.00	4.15	5.20	—	—	—	—	1.35	.85	1.27	.83	.75	—	.94	1.30	.68	.25	.40

Individual Variation in the Size and Form of the Bill. — That considerable variation occurs in the size and shape of the bill, in specimens of the same sex and species living together at the same locality, is evident from a glance at some of the preceding tables of measurements. The variation in this organ is further illustrated in the accompanying plates (Plates IV – VIII), in which are given figures of the bills of several specimens of each of a number of species. Much greater differences are here shown to exist in cospecific specimens of the same sex and from the same locality than occur between those supposed to be distinct, of which comparative figures of the bills have been published with a view of demonstrating their specific diversity. In only a few groups in fact, and mainly in the long-billed *Grallæ*, is the bill generally admitted to be too variable to afford an important basis for the discrimination of species.

The principal points of variation in the form of the bill consist in variations in its general size, without corresponding variations in the general size of the individual, and in the details of its form in regard to thickness and length. There are also other variations in respect to the emargination or dentation of the terminal portion, especially in the vast group of the insectivorous species, and in the "festooning" of the bill in many of the hawks.* In respect to the size of the bill, it is a noteworthy fact that birds specifically and sexually identical vary in such a way that specimens much below the average size possess bills above the average size for their respective species, and, conversely, that specimens above the average size have bills much smaller than the average for their respective species, the general proportions of the bill in each case being essentially the same. In such cases, with the increase or decrease in length, there are corresponding differences in the thickness of the bill, both in the vertical and transverse directions. In other cases with the increase in length there is no corresponding increase in thickness, such a differentiation thus resulting in a relatively attenuated form of the bill. In other cases the bill is shortened without a corresponding decrease in its thickness, from which results a short, thick, or robust bill. The variation in thickness is again sometimes relatively greater in the vertical

* In respect to this point, see Dr. Henry Bryant's paper on "Variations in the Plumage in *Buteo borealis* auct. and *B. Harlani* Aud.?" (Proc. Boston Soc. Nat. Hist., Vol. VIII, p. 107 *et seq.*, 1861, where the variation in this feature is especially noticed.

than in the transverse direction, and sometimes the reverse, thus giving in some cases a deep, narrow bill, and in others a broad, depressed bill. In the latter case the differences are especially important, as will be more fully shown later. In regard to the tooth-like indentation near the tip of the bill in so many of the insectivorous birds, it is found that in some species which usually have it strongly developed, specimens occasionally occur with the indentation nearly or quite obsolete. Again in other cases where this feature is usually but slightly developed, some specimens have the notch at the tip of the bill exceedingly prominent. Similar variations occur in regard to the development of the so-called "festoon" of the upper mandible in the hawks, as Dr. Bryant has already sufficiently shown.

The greatest range of individual differentiation in any given organ occurs, as would be naturally expected, in those species which have that organ more than ordinarily developed, and also in species of a low grade of structure. In the long-billed *Grallæ* both these conditions exist, and it is in such genera as *Numenius*, *Gambetta*, *Limosa*, *Scolopax*, *Philohela*, and *Gallinago*, that the maximum of bill variation is seen. It is less marked in the song-birds, though in many members of this group the variation is by no means small. In the typical woodpeckers, on the other hand, which have the bill especially adapted to a peculiar function, that of digging into wood, the variation is scarcely appreciable, since any considerable variation from its usual form would seriously impair its efficiency. In the semi-frugivorous and terrestrial *Picidæ*, however, we again meet with the usual range of variation.

In the accompanying plates illustrative of variation in the bill, representatives from the higher types of the *Oscines* have mainly been chosen, several representatives from widely different families having been selected. Plate IV, figures 1 and 1*a*, 2 and 2*a*, give a view of the bills of two specimens of the common king-bird (*Tyrannus carolinensis*), from Eastern Massachusetts, which differ from each other as much as the bills of different genera sometimes do. One of them, as will be seen, is so much narrower and deeper than the other as to give very different proportions and outlines. The skulls of these two specimens vary in the same manner as do the bills, the one having a broad, flat skull, and the other a narrow, high one. Two specimens of *Myiarchus crinitus*, one of which is from South Carolina and the other from Western New York, differ as much from each other, and in nearly the same way, as do those of the king-bird. Similar and nearly as great variations occur also between different specimens of *Contopus borealis*, *C. virens*, *Empidonax minimus*, *E. flaviventris*, *Sayorius fuscus*, and in several species of the South American *Tyrannidæ*. But between these two extremes are found in other specimens nearly every possible degree of gradation.

Figures 3 and 3*a* to figures 7 and 7*a* (same plate) represent different forms of the bill in *Troglodytesaëdon*. Between these specimens there are great differences both in respect to absolute size and to general form, greater than would be deemed necessary by most ornithologists for the differentiation of species. These examples are all from Florida, and essentially from the same locality. Other specimens in the Museum come between these extremes in such a way as to show the inconstancy of all these forms. The variation in color, which is considerable in this species, does not accord with the variation in the bill, specimens exhibiting the extremes of color as often having the bills alike as otherwise, and, conversely, those with bills alike differ widely in color.

Figures 8 and 8*a* to 11 and 11*a* (same plate) indicate the variability of the bill, especially in respect to length, in Massachusetts specimens of *Seiurus noveboracensis*. The first corresponds essentially with, and unquestionably is, an example of the so-called *Seiurus ludovicianus*, which, in all probability, is but the darker colored, longer-billed southern form of *S. noveboracensis*. This species varies also remarkably in color, but the variation in color, as in the case of *Troglodytes aëdon*, and as is commonly the case in other species, does not accord with the variation in the bill, some of the long-billed specimens being in color almost undistinguishable from some of the short-billed ones, while some of those with medium bills present the extreme degrees of variation in respect to color.

Figures 12 and 12*a* to 14 and 14*a* (same plate) represent the bills of three male specimens of *Mniotilta varia* from the vicinity of Cambridge, which present as great differences as modern ornithologists would ordinarily deem sufficient, if the specimens had come from Mexico instead of from Massachusetts, to warrant their recognition as types of three distinct species. The correspondingly great variations in color in this species have already been adverted to (p. 190). The bill, however, in specimens presenting extreme forms of color variation, unfortunately for ultra-divisionists, may be either of the ordinary form or of either of the forms figured, or of any intermediate form, as exemplified by the specimens of this species in the collection of the Museum. Figure 15 and 15*a* and 16 and 16*a* (same plate) are accurate representations of the bills of two Massachusetts males of *Dendrœca striata.* The differences between these specimens, though so great, are not greater than occur in different cospecific examples of several other species of this genus contained in the Museum.

Massachusetts specimens of *Certhia familiaris* differ even more in the form of the bill than do the specimens above figured of either *Troglodytes aëdon* or *Mniotilta varia.* They also present a similar range of color variation in the plumage, and one equally at variance with the variation in the bill.

Figures 19 and 19*a*, 20 and 20*a* (same plate), show how widely two Florida specimens (both males) of *Pyranga æstiva* vary in respect to the size of the bill, the specimens in question differing but little in general size. If these figures are compared with the figures recently published of the bills of certain supposed species of *Pyranga*,* they will be found to vary more than some of the latter do, and indicate how unsatisfactory the nature of species must be when based mainly upon differences in the bill. Other cospecific specimens of *Pyranga* in the Museum exhibit great difference in the size, form, and position of the tooth-like processes of the upper mandible, and in the color of the bill, — differences that have been regarded as specific characters. The color of the bill in many species of birds, in fact, varies greatly in specimens of the same species taken at the same season, and generally in those taken at different seasons; yet it is a character that has been relied upon for the distinction of species.

Figures 1 and 1*a*, 2 and 2*a*, 4 and 4*a*, and 5 and 5*a*, Plate V, illustrate variations in the bill in Massachusetts representatives of *Ægiothus linarius.* Figures 3 and 3*a*, and 5 and 5*a*, are drawn from specimens from Arctic America, the first being an original specimen of the *Æ. fuscescens* Coues ex auct., and the other a similar specimen of the *Æ. exilipes* Coues. Figures 7 and 7*a* to 10 and 10*a*, inclusive (same plate), represent variations of the bill in male specimens of *Chrysomitris tristis*, a species allied to *Æ. linarius.* It will be seen that the two series are nearly parallel in respect to the amount and character of the variations in the bill. Figures 11 and 11*a* and 12 and 12*a* indicate similar variations in an-

* Proceed. Phil. Acad. Nat. Sci., June, 1869, pp. 130 – 133.

other allied species, the *Chrysomitris pinus*, and figures 13 and 13*a* to 15 and 15*a*, inclusive (same plate) similar variations in another species (*Curvirostra americana*), of the same sub-family. In the latter case the specimens are also all males, and all from the vicinity of Cambridge, they having been killed in fact from the same flock. In the *Ægiothus* group numerous so-called "species" have been described by different writers, six or seven of which were recognized by Dr. Coues a few years since in his monograph of that genus.* A considerable number of these species have been generally looked upon as equivocal, and the exact number in the group and their distinctive characteristics have been a matter of much uncertainty. Recently the writer above referred to has again revised the group,† and arrives at the conclusion that if more than one species exists, all the forms previously recognized by him as species are valid species. I can readily grant this alternative, being fully convinced that the genus consists of but a single known species, which has a circumpolar distribution. The alleged specific distinctions have consisted in differences in general size, in the relative size of the bill, the length of the tarsus, wing, and tail, and in color. Some of these differences are doubtless climatic and local, while others may be due to age, but the greater part I believe to be to a great degree purely individual, inasmuch as they are paralleled in allied species, whose standing has not been and cannot reasonably be questioned. But the special consideration of the variations presented by the *Ægiothi* and similar groups will be reserved till after the facts relating to geographical variation have been presented, since they can then be more appropriately discussed.

Figures 16 and 16*a* to 18 and 18*a*, inclusive (Plate V), represent the bills of three male specimens of *Passerculus savanna*, from different localities on the Atlantic coast. The specimen represented in figures 18 and 18*a*, has the bill of minimum size, being in bulk less than half that of the one represented in figures 17 and 17*a*.‡ Figure 17, it will be observed, corresponds nearly with the so-called *P. sandwichensis* § of

* A Monograph of the genus *Ægiothus*, etc., Proc. Phil. Acad. Nat. Sci., Vol. XII, p. 373, 1861, Vol. XV, p. 40, 1863.

† On variations in the plumage of the *Ægiothi*, Ibid., Vol. XXI, p. —, 1869.

‡ Other specimens received from Grinnell, Iowa, from Professor H. W. Parker, since the above was written, have bills still smaller than any of those here figured.

§ Baird's Birds of N. Amer., p. 444, 1858.

the Pacific coast, and figure 17 with the so-called *Passerculus alaudinus*,* also of the Pacific coast.

Plate VI, although designed more especially to illustrate local variation, indicates to some extent the individual variation existing in *Agelæus phœniceus.* Figures 1 and 1*a* represent the average type of the bill in this species in Massachusetts, and figures 3 and 3*a*, and 4 and 4*a*, unusually long and unusually short forms of the bill found at the same locality. Figures 2 and 2*a*, 5 and 5*a*, and 6 and 6*a*, represent a similar series from the St. John's River, Florida. All the specimens of the two series are adult males.

Plate VII represents similar variations of the bill in *Quiscalus purpureus.* Figures 1 and 1*a*, 3 and 3*a*, 4 and 4*a*, and 6 and 6*a*, represent the average and the extreme types of the bill met with in Massachusetts males. The latter also represents an inflexed type of bill, a modification seen in many species, it being especially common in the *Quiscali* and other genera having the bill of a similar form. It is unmistakably an individual peculiarity, evidently depending mainly upon age, and resulting from the upper mandible outgrowing and overhanging the lower. In *Quiscalus purpureus* such specimens are more or less frequent at probably all localities, they having been received at the Museum from Maine, Massachusetts, New Jersey, Florida, and Illinois, and I have seen them from the West Indies. It often gives rise to the name *inflexirostris*, which is found so frequently a synonyme.† The figures of the bills of four females of *Sturnella ludoviciana* (Plate VIII), from Florida, indicate the character of the bill variation exhibited by different individuals of this species at the same locality, independently of any variation attributable to sex. Figures 5 and 5*a*, and 6 and 6*a* (same plate) show that like variations occur in *Colaptes auratus*, the figures being drawn from two Massachusetts females.

Similar comparisons, with similar results, might be made with scores of other species, but the above illustrations will doubtless suffice to show that individual variation in the form of the bill is not only great, but that it exists in groups having a high grade of structure. Other groups might have been chosen in which the individual variation in the form of the bill, as already stated, is far greater than in the instances above

* Bonaparte, Comptes Rendus, Vol. XXXVII, p. 918, 1853.

† Concerning *Quiscalus inflexirostris* Swainson, see below (Part IV), under *Q. purpureus.*

cited. The *Grallæ* have already been referred to as presenting remarkable examples of bill variation. In some of the *Anatidæ*, however, it is scarcely less, whilst it is especially great among many of the *Longipennes*. Hence some authors evidently attach too high importance to the exact form of the bill in these groups.

All the illustrations referred to above have been drawn, with one or two exceptions, from fully adult specimens. One of these is a specimen of *Passerculus savanna* (Plate V, fig. 18), which is a bird of the year, killed in Labrador in August, before it had quite completed its first moult. Another is the smallest billed specimen of *Chrysomitris tristis* (Plate V, fig. 10), which is also evidently a bird of the year. The other is an autumnal specimen of *Dendrœca striata* (Plate IV, fig. 15). They all, however, would be ordinarily considered as adult in size.

Variations in the Size and Form of the Bill, Wing, etc., resulting from Age.

In the foregoing remarks on the variations in general-size, in proportions, and in the form of different parts, exclusive reference has been had to adult specimens. It is easy, however, to confound difference depending upon age with those strictly resulting from individual differentiation. The form of the bill is especially subject to variation by age in specimens that upon casual inspection would seem to be full-grown In long-billed birds the bill increases in length for several months after the bird is full-fledged, and even after it has once moulted. In short- and thick-billed birds, the bill increases considerably in thickness as well as in length after the individual seems to have acquired its adult size and proportions. As a general rule, then, "birds of the year" possess a relatively shorter and thicker bill than those fully adult, or three or four years of age. In old age an abnormal elongation of the upper mandible occasionally occurs, especially in species in which the tip of the upper mandible is decurved and projects slightly beyond the lower, as in *Corvus*, *Quiscalus*, *Vireo*, *Turdus*, *Larus*, etc. Since, however, great differences occur in the form of the bill in specimens of the same age, in birds of the year as well as in those unquestionably adult, it is sometimes difficult to determine how much of the difference in certain cases is to be considered as due to age and how much to individual variation.

The wing also varies considerably in form with age. In many of the song-birds, at least, and also in the raptorial birds, the wing becomes more pointed with the second and third moultings of the remiges. Birds of the first year hence have, even after the flight feathers are fully grown, a shorter and more rounded fore-wing, as a general rule, than birds of two or three years of age. These differences of course result from variations in the relative length of the primaries, the outer primaries being the last to acquire their ultimate proportions, as they are also the last primaries to be renewed in the annual moult. A similar change with age occurs in the form of the inner point of the wing, or that formed by the inner secondaries. These, like the primaries, are subject to a gradual increase in length for a time with each moult, they likewise being the latest of the secondaries to acquire their mature size, as they are also the last of the secondaries changed in each normal moult. Thus, through the gradual elongation of the outer primaries and the inner secondaries, a slight change is produced in the general form of the wing. It is, however, only slight, and since some young birds have as pointed wings as any of the same species which are fully adult, and some adult birds have wings as much rounded as the full-grown young, the rule is subject to many exceptions. The sexes of the same species also often differ similarly with the young and old in respect to the form of the wing. This is more especially the case in those species in which the female is much smaller and much duller colored than the male, the structural inferiority of the female to the male being thus evident in various features.

While the wing may be regarded, as already stated, as generally smaller and more rounded in the younger individuals, it not unfrequently happens that the specimens having the greatest alar extent are immature birds. This has been particularly noticed in the eagles and hawks, as well as in some of the gulls, in which it is so frequent as to have attracted the attention of numerous observers.* The feathers of the wings and tail are not only longer, but they are also broader, and hence in the expanded wing present a greater resisting surface to the air. Two explanations of this fact present themselves. First, in the cases referred to, the birds may have been born at a very northern locality, whence only the younger birds ever descend so far south. Second, the greater lack of power in the muscles of flight in the young birds, as

* See American Naturalist, Vol. III, 1869, p. 517.

compared with those fully mature, may be counterbalanced by a relatively larger supporting surface in the wings and tail. Whatever the explanation may be, the facts seem to be unquestionably as above stated.

Other variations in the plumage and in other characters depending upon age, but which are liable to be confounded with individual differentiation, might be cited, but none seem to be of sufficient importance to require a special description.

General Remarks on Individual Variation.

After the preceding remarks on this subject, I should perhaps state expressly what I regard to be the bearing of the facts above discussed, otherwise I might be understood as in a great measure discarding the majority of the characters used in the diagnoses of species and genera. Nothing, however, is further from my purpose. What I urge is simply this: that the extent of purely individual variation is far greater than has usually been recognized, and that as a result numerous strictly nominal species have found their place in our systems, from naturalists having mistaken these differences for true specific characters. Individual variation, however, is so complicated with geographical variation, that the general bearings of the whole subject will be deferred till the end of the discussion of the latter topic.

As regards the general cause of individual differences in animals, it is too evidently constitutional to allow of any other hypothesis, and akin to that seen in domestic animals, and which in man gives to each individual his unlikeness in temperament and physical structure to all other men. While individuality is so patent and so universal in the human species, and scarcely less so in domesticated animals, it is one of the most surprising facts in zoölogy that so many naturalists should have entertained the idea that there is an almost total absence of it in feral animals, and that the description of a single specimen will suffice for that of its species. Practically, however, this has been the fact, and eminently so with that large class of "species hunters," who have not inaptly been characterized as "closet naturalists"; for to this class and not to the field naturalists are we mainly indebted for the long lists of synonymes that form so vexatious a burden to zoölogical science.

Certain secondary causes that share in producing individual variation are doubtless more or less obscurely traceable. Among these are certain circumstances attending the time of hatching, as well as, of course, the vigor of the parent. Not unfrequently the first attempts of birds to rear their brood are unsuccessful, from their eggs or young being destroyed by their enemies. Persisting, however, in their efforts, it is late in the season before their brood is fledged, several sets of eggs or young having been previously destroyed. The birds of such broods are found to be smaller and paler colored than those hatched earlier in the season. In cases where several broods are reared each year, as a general rule the birds of the earlier brood seem in all respects the most perfect and vigorous. Various other causes operating during their infancy doubtless more or less affect their general size, their proportions, and colors when mature. Food has doubtless much to do with variation in color, though but few facts bearing upon this point have been yet recorded. Professor Agassiz informs me, however, that many years since, in Switzerland, he raised many *Pyrrhula vulgaris*, and found that by feeding them on the seeds of hemp the red on the breast changed to black. The well-known fact that certain brightly colored birds, as the purple finch (*Carpodacus purpureus*) and the crossbills (*Curvirostra*), change, when kept in cages, from bright red to dull olive with their first moult, and never again, or at least so long as kept in confinement, regain their original color, shows how susceptible the color of birds is to the influences of food and artificial conditions of life.

Climatic Variation.

Climatic variation involves as completely all parts of the animal as does individual variation. It is more marked, however, in some features than in others. The three most prominent phases of climatic variation in birds are the following: variation in general size, variation in the size and form of the bill, variation in color.

Climatic Variation in Size. — Variation in the size of individuals of the same species with differences in the latitude and altitude of their respective places of birth is a fact already so well known as to be quite generally recognized; hence any demonstration of such a variation is in the present connection unnecessary. A few tables of comparative measurements of New England and Florida specimens given in Part IV serve to illustrate its general character and extent. Similiar illustrations are abundantly afforded by the tables of measurements published in Professor Baird's Birds of North America,* in the text of which work frequent reference is made to the differences in size between northern and southern specimens of the same species. The same author also subsequently called attention to the subject, and explicitly announced a general law of geographical variation in size ; namely, a gradual decrease in size in individuals of the same species with the decrease in the latitude and altitude of their birth-places.†

In some species, and throughout some entire families, climatic variation is more marked than in others; generally, however, it is very appreciable, and amounts, in respect to size, not unfrequently to from twelve to twenty per cent ‡ of the average dimensions of the species.

Climatic Variation in the Bill. — The climatic variation in the size of the bill is, in general, inverse to that of the general size of the individual. In some species, as in the *Sittæ* and the typical members of the *Picidæ*, I have as yet been unable to trace an independent variation in the size of the bill to that of the body ; but in many species there is not only a marked *relative* increase in the size of the bill to the southward, but, in some, an *absolute* increase, especially in its length.

* Pacific Railroad Explorations and Surveys, Vol. IX, Birds. By Professor S. F. Baird, with the co-operation of Mr. John Cassin and Mr. George N. Lawrence. 1858. Subsequently republished under the title of "The Birds of North America," with an Atlas of one hundred plates.

† Proc. Phila. Acad. Nat. Sci., Vol. XI, p. 300, November, 1859. Also in Am. Journ. Sci. and Arts, 2d Ser., Vol. XLI, p. 190, March, 1866.

‡ Variation in size with differences in habitat is by no means confined to birds. In mammals it is well known to be as great, if not greater, than among birds. In some wide-ranging species of mammals there appears to be a double decadence in size, — a diminution to the northward, in those non-migratory species whose habitats extend into the arctic regions, as well as a diminution to the southwards of the point where in general the maximum of size is attained, — as I have elsewhere had occasion to remark. (Bull. Mus. Comp. Zoöl., Vol. I, p. 199.) But in these exceptional cases of a decline in size to the northward, the cause of such a decline must result from climatic conditions the reverse of those producing the decline at the southward, — from the excessive rigor of the arctic climate instead of from the enervating influence of warm temperate and sub-tropical latitudes.

In the case of reptiles, the larger representatives of a given species are generally found at the North, as has also been observed to be the case with the edible marine and fluviatile fishes. (I am credibly informed that this is markedly the case with the codfish and the halibut.) In some groups of crustacea and mollusca the same fact has been repeatedly observed; but in insects, as in plants, the increase in size is generally to the southward, as is especially noticeable in the diurnal Lepidoptera. In plants, however, the increase is a purely vegetative one, the northern representatives of a given species being generally far the most prolific, in proportion to the size of the plant, near the northern limit of their respective habitats.

An increase in the length of the bill is most frequent in long-billed species, while in short-billed ones the increase is in general size, without material change in its proportions. With the increased length and slenderness of the bill there is in many cases also a tendency to greater curvature.

An increase in the length of the bill is quite marked in the genera *Quiscalus*, *Agelæus*, *Geothlypis*, *Troglodytes*, *Seiurus*, *Harporhynchus*, *Galeoscoptes*, etc. *Quiscalus purpureus* and *Agelæus phœniceus* afford good illustrations of geographical variation in the size and shape of the bill. Notwithstanding that the northern specimens are the larger, the southern ones have, in the average, bills as long, though slenderer, than the northern, and occasionally even longer. These differences are shown to some extent in Plates VI and VII, where the figures of the bills of Massachusetts and Florida specimens of these species are given side by side. In Plate VI, figures 1 and 1*a* represent the bill of an average Massachusetts male *A. phœniceus*, and figures 2 and 2*a* the bill of an average Florida male of the same species. The latter, while much less thick, is fully as long as the former. Figures 4 and 4*a* represent the shortest bill of a considerable series of Massachusetts specimens, and figures 6 and 6*a* the shortest or thickest bill of a similar series of Florida specimens. Figures 3 and 3*a* give the longest bill of the Massachusetts series, and figures 5 and 5*a* the longest of the Florida series, the specimens being in each case adult males. Plate VII, figures 3 and 3*a* represent the bill in average Massachusetts males of *Quiscalus purpureus*, and figures 2 and 2*a* that of average Florida specimens, while figures 1 and 1*a*, and 4 and 4*a*, show respectively the longest and the shortest bills of a considerable series of Massachusetts specimens. Figures 5 and 5*a* are from a New Jersey specimen, and figures 6 and 6*a* from a Florida specimen, the latter showing an inflection of the upper mandible more or less frequent in the various species of *Quiscalus*. The figures, as in the previous plate, were all drawn from adult males. In each of these species the average difference in the bills of Florida and Massachusetts birds is as great as is frequently considered to be sufficient to constitute specific differentiation, and between the extremes, especially of *A. phœniceus*, even subgeneric. Yet specimens from intermediate localities present such a gradual and complete transition between the two forms as to render their specific identity unquestionable.

A similar difference between Massachusetts and Florida examples, with a gradual transition from the one to the other, through specimens from intermediate localities, is seen in *Troglodytes aëdon*, *Geothlypis trichas*, and *Seiurus noveboracensis*. In *Pipilo erythrophthalmus*, *Ortyx virginianus*, *Corvus americanus*, and *Cyanura cristata* the bill is appreciably larger in the Florida than in the northern form. In *Corvus americanus* this difference was long since noticed by Professor Baird, the larger bill of South Florida specimens having led him to recognize a variety *floridanus* of this species, based chiefly on this difference.* The same author has also referred to the larger size of the bill in Florida specimens of *Ortyx virginianus*.†

In some species individual variation is so great that it is unsafe to draw conclusions respecting geographical variation from the examination of a small number of specimens. This is notably the case in *Sturnella ludoviciana*, in which the bill varies greatly in size and form, as does the bird in general bulk, at all localities. In the average, however, Florida specimens of this species seem to have a relatively longer and slenderer bill than those from the Northern States.

As already noticed, variation in the bill is not equally marked in all species, but it occurs in too many to admit of the supposition that the numerous cases wherein it is clearly marked are exceptional, or that it does not follow a general law of geographical variation. The observations above detailed are based on specimens collected on the Atlantic coast, from New England southward to Florida, and refer exclusively to species breeding within that range. But specimens of species which breed entirely to the northward of this range, collected during their semi-annual migrations, corroborate the law already stated, namely, an increase in the size of the bill to the southward in specimens of the same species from different breeding stations. In the *Anatidæ* and *Tringæ*, which breed far to the northward and pass the winter in lower latitudes, it is noticeable that, while those which arrive first in the fall, and those which return north latest in the spring, are smaller than those that arrive later and depart earlier, they have, nevertheless, relatively larger bills. This has been especially noticed in species of *Fulix*, *Bernicla*, *Actodromas*, and *Macrorhamphus*. Professor Baird has also referred to the larger size of the bill of the southern representatives of *Lagopus albus* as compared with those from further north,

* Birds of North America, p. 568, 1858.

† Am. Journ. Sci. and Arts, 2d Ser., Vol. XLI, p. 191, 1866.

"those from Eastern Labrador and Newfoundland," he says, appearing "to have decidedly broader, stouter, and more convex bills than those from the Hudson's Bay and more northern countries."* In the writings of various authors on the birds of Southern Mexico, Central America, Southern Asia, and Northern Africa, frequent mention is incidentally made of the larger size of the bills of southern representatives of northward ranging species. Although such statements record what have been apparently regarded as only isolated facts, their frequency indicates that the increase in the size of the bill to the southward is not confined to the birds of Eastern North America, nor exclusively to those of temperate and sub-tropical countries, but that it is a general geographical law, similar to that of the variation with locality in the general bulk of the individual.

Geographical Variation in Color. — Geographical variation in color in birds may be regarded as of two kinds, which may be termed, from their different geographical relations, latitudinal variation and longitudinal variation. The first is coincident with differences in latitude, and the second with differences in longitude. Both are due, however, to climatic peculiarities, and are hence, strictly speaking, climatic. The latitudinal is perhaps at present the best known, and will be first considered.

(*a*) *Latitudinal Variation.* — In those species of North American birds whose breeding range extends over a wide range of latitude, the southern-born specimens are, as a general rule, appreciably darker or brighter, or more intensely colored, than northern-born ones of the same species; in many instances the difference being so great as to impress even the casual observer. Dark colored birds, like the *Quiscali*, *Agelæus phœniceus*, etc., become blacker towards the southern limit of their respective habitats, where those with metalic reflections have the iridescence more intense and of a darker hue, greenish and bronzy reflections changing to purple. The slaty, ferruginous, and olive tints, and the various shades of red and yellow of others, become also far more intense. In species barred transversely with dark and light colors, the dark bands, as a general rule, become broader, and the light ones narrower. Those with white spots on a black ground have the spots reduced in size and number, the smaller ones becoming obsolete. White bars on the wings and terminal white spots on the tail feathers

* Birds of N. Amer., p. 634.

are also of less extent in southern specimens. There hence results, as already observed, a generally darker aspect in the plumage of the southern representatives of wide-ranging species; the bill and the feet also usually sharing in the general accession of coloring matter in the integuments. The difference in color between the extremely northern and the extremely southern representatives of a given species is often so great that, taken in connection with other differences, as in general size and in the size and form of the bill, the two extremes might be excusably taken for distinct species, especially if viewed aside from the connecting series between the two types formed by specimens from successively intermediate points, which beyond question show their specific identity.

As in the case of climatic variation in the bill and in general size, the variation in color differs greatly in degree in different species. Climatic difference in color is particularly striking in *Agelæus phœniceus*. In the males the black is greatly intensified and more lustrous at the South, and the red on the shoulders becomes equally heightened. Instead of the *light* red shoulder-patch, bordered externally with *whitish or pale yellowish-whitish*, seen in Massachusetts specimens, the shoulder-patch in the Florida males is of a brilliant *dark* red, with a *rich cream-colored or orange-yellow* border. While the differences in the bills of the two types might in extreme cases be taken as indicative of different sub-genera, the difference in color is as great as occurs between the northeastern type of *A. phœniceus*, and either the so-called *A. tricolor* or *A. gubernator* of the Pacific slope, or between any of these *inter se*. *Quiscalus purpureus* also affords a similar example of climatic variation, as well in color as in the bill and general size. In the males the change in general tint is in the black becoming more intense at the South, and the iridescence being dark purple or bluish instead of bronzy or greenish. The change in the females is as great as that in the males. At the North their plumage is nearly lustreless brownish-black, but at the South it becomes nearly as black as that of the northern males, and has considerable iridescence, so that the northern collector, judging from color alone, would at first be likely to mistake the southern females for males.

In *Ortyx virginianus*, through the increased breadth of the transverse bars of black at the South, on the dorsal as well as on the ventral surface, the general aspect of the plumage is very much darker in Florida specimens than in New England ones. In *Sturnella ludoviciana* the yellow of the ventral surface in Florida specimens is far more intense than it is in northern ones; the slate color of *Galeoscopies carolinensis* is correspondingly darker, and the ferruginous of *Harporhynchus rufus* is much redder. In *Centurus carolinus* not only are the black transverse bars on the back broader and darker, but the red on the head and abdomen becomes more extended and lustrous. In *Picus pubescens* the white spots on the wings become smaller and fewer, with a greater tendency to black streaks on the sides of the breast, a variation in the direction of *P. Gairdneri* and *P. Harrisi*, as will be noticed at length in the remarks on *P. pubescens* and *P. villosus* in Part IV. Similar differences occur between northern and southern specimens of *Picus borealis*, which are so great as to have led Mr. Cassin to regard the southern type as specifically distinct from the northern. Similar differences to those above described occur between northern and southern specimens of *Thryothorus ludovicianus*, *Troglodytes aëdon*, *Geothlypis trichas*, *Colaptes auratus*, *Buteo lineatus*, and various other species, as will be described more in detail in Part IV.

The climatic variation in respect to the relative size of the white spaces on the rectrices and primary remiges may be illustrated by a single example. In northern specimens of *Pipilo erythrophthalmus* the terminal white spots of the tail feathers are found on the *four* outer feathers of each side; but in Florida-born ones they occur on only the *three* outer feathers on each side; and are correspondingly reduced in length. The white area on the tail of Florida specimens hence has only about the extent that would be presented in northern specimens if the outer pair of feathers were removed. The extent of the white space at the base of the primaries is correspondingly reduced in size in the southern type.

Extending the examination to northern species, it is found that similar color differences with the latitude of the birthplace are of frequent occurrence. In *Bernicla brenta* and *Bernicla canadensis* the smaller southern-born birds are, as a general rule, considerably darker than the larger northern-born ones. The same is true of *Fulix marila* and *Bucephala americana*, the so-called *Bucephala islandica* being the larger northern type of *B. americana*, in which the white markings on the wings and head occupy a somewhat larger area. It is altogether probable also that the so-called *Anser frontalis* holds a similar relation to *A. Gambeli* (= *A albifrons* ?), and the *Anser cærulescens* to the *A. hyperboreus,* though by some the former has been regarded as the young of the later. In *Larus argentatus* the southern specimens are not only smaller, with the "mantle" somewhat darker, but as a general rule the white spots at the tips of the first and second primary quills are more restricted.

The changing of the pelage to white in winter in certain northern mammals, and of the plumage in certain birds, as the ptarmigans, correlates perfectly with these geographical differences in color; and since in some species of mammals only the northern representatives change to white in winter, while the southern ones are of the same color throughout the year, this seasonal change seems evidently to come under the above-stated general law of geographical or climatic color variation, namely, a *gradual increase in color to the southward in individuals of the same species.*

A comparison of Florida birds with West India specimens of the same species shows that the difference between them in color (and, it may be added, in size and other general features) are generally not greater, and in some cases far less, especially between Cape Florida and Cuba specimens, than obtains between Florida and Massachusetts examples, and that it is of precisely the same character. West Indian specimens of course differ more from Massachusetts examples of the same species than the latter do from others from East Florida, yet by means of the South Florida specimens, which differ but slightly from the Cuba type, a gradual transition is evident from the extreme northern to the extreme southern forms. Of late many Jamaican, Porto Rican, and Cuban forms have been regarded, by many writers, as specifically distinct from their representatives in the Northern States, and in many cases they might well be so regarded, were there not a succession of intermediate forms connecting them, — a fact which seems to have been hitherto overlooked. The earlier writers considered the *Ortyx,* the *Sturnella,* the *Strix,* the *Circus,* several of the *Buteos,* etc., of the West Indies as specifically identical with the *Ortyx virginianus, Sturnella ludoviciana, Strix flammea, Circus hudsonius, Buteo borealis,* etc., of the United States, and doubtless justly, notwithstanding that the comparison of specimens reveals certain relatively slight but constant differences in color and size, and to some extent in other features.

(*b*) *Longitudinal Variation.* — In comparing the birds of the Atlantic States with specimens specifically identical from the interior of the continent, one is soon struck with the brighter colors of the latter, and especially with a tendency, in many species, to more ferruginous tints, and to melanism in others. In comparing again the birds of the Mississippi valley with those of the Pacific slope, especially that portion north of the fortieth parallel, a similar difference is also noticeable, the extremes of color variation in truly continental species being met with (especially to the northward of this parallel) at the Atlantic seaboard on the one hand, and the Pacific on the other, between which there is a gradual and, with an exception soon to be noticed, a uniform increase in intensity of color to the westward. This tendency to more ferruginous and melanic colors to the westward is especially marked in *Falco peregrinus*,* *Accipiter fuscus*, *Circus hudsonius*, *Buteo lineatus*, *Buteo borealis*, *Archibuteo lagopus*, *Hypotriorchis columbarius*, *Otus vulgaris*, and other species of *Strigidæ*, *Tetrao canadensis*, *Bonasa umbellus*, *Bernicla canadensis*, *Bernicla brenta*, *Larus argentatus*, *Parus atricapillus*, *Carpodacus purpureus*, etc., etc. The western representatives of *Melospiza melodia*, *Passerella iliaca*, *Junco hyemalis*, *Pipilo erythrophthalmus*, *Parus hudsonicus*, etc., differ mainly from their Eastern congeners in their more ferruginous or darker colors, according to the species.

While the general tendency from the East westward is thus to darker or deeper colors in specimens of the same species, and in representative species of the same genus, the rule is not without exceptions, nor is the transition quite uninterrupted. On the arid sterile plains the representatives of not a few, and probably of most, species are much lighter colored than their relatives either to the eastward or to the westward. Also at the southward on the Pacific slope there is not the tendency to deeper colors seen farther to the northward, specimens from Northwestern Texas, New Mexico, much of the Colorado basin and Lower California, being lighter than others of the same species from Northern California, Oregon, and Washington, an explanation of which will be suggested later.†

In comparing again the European representatives of circumpolar species with their representatives in Eastern North America, a difference

* For the synonymy and other remarks on these species, see Part IV.

† See below, p. 239 *et seq.*

similar to, but hardly so great as, that between the Atlantic and Pacific coast examples of indentical species is likewise seen, the American being in general several shades darker than the European. In certain cases there is also a difference in the markings, as in some of the hawks, in which in the European the transverse bars are broader and better defined, and the longitudinal ones less so than in the American. This is illustrated in *Astur palumbarius* and *A. atricapillus*, in *Accipiter nisus* and *Ac. fuscus*, etc. In many instances the only tangible differences between so-called representative American and European species consists in the darker, brighter, or intenser color of the American, the differences being oftentimes less than that between specimens of the same species from the Atlantic States and the Mississippi valley, or between those from the Mississippi valley and the Pacific coast. Not unfrequently, however, are American and European specimens so nearly alike, even of species that have rarely been considered as identical, that without a knowledge of the locality whence they came it would be impossible to confidently refer them to the one species rather than to the other.

There are also indications of various local differences in color in specimens specifically identical within the larger areas above considered, and which are in a measure exceptional to the general law of a westward increase in color. The data at hand are at present too few either to limit these exceptional areas or to indicate to what extent they are exceptional. They appear, however, to be coincident with peculiar climatic conditions, the exact nature and extent of which are likewise imperfectly known.*

Variation in the Length of the Tail and in other Characters. — At certain localities, and more especially to the southward, there are well-known instances of an increase in the length of the tail, without an appreciable modification of other parts. Marked examples of this are seen in *Icteria virens*, *Harporhynchus rufus*, and *Mimus polyglottus*, as has been pointed out by Professor Baird and other writers,† each of which species has a western long-tailed variety. The *Quiscalus macrura* is also little else than a long-tailed variety of *Q. major*. A tendency is seen to this variation in *Geothlypis trichas* at the southward,

* See on this point below, p. 239 *et seq.*

† See especially Prof. Baird in Amer. Journ. of Science and Arts, 2d Series, Vol. XLI, p. 191.

while it seems to be a marked characteristic of many of the birds of Lower California. The tendency in southern forms to an elongation of the tail seems, however, less general than the southward decrease in size and the increase in color, or the tendency to an elongation of the bill.

Among other local variations may be mentioned the white instead of a red iris in the South Florida representatives of *Pipilo erythrophthalmus*; the yellow instead of a black bill in the magpies of the coast of California; the white basal half of the feathers of the neck of the raven of Southwestern Texas and Mexico, by which it is chiefly distinguished from the common species; the greater continuation anteriorly of the superciliary stripe in the western forms of *Zonotrichia leucophrys*, by which alone it is distinguishable from the eastern form; the white frontlet of one of the western forms of the *Parus atricapillus* group, etc. There appears frequently to be also a locally greater development of the foot in western and southern forms of wide-ranging species, and occasionally an exceptional increase in general size under identical isothermes.

Causes of Climatic Variation. — The facts respecting climatic variation are at present too imperfectly known to be fully explained. There are, however, certain peculiarities of climatic variation, especially in color, coincident with certain meteorological peculiarities of the regions where they occur, that demand attention. The increase in color to the southward, especially the tendency to darker tints above shown to be so general, coincides with the increase in the intensity of the solar rays to the southward, and in the humidity of the climate. The southward increase in depth of color and in iridescence in birds specifically identical coincides also with the general increase in brilliancy of color in birds, taken as a whole, in the lower latitudes (as well as in insects and animals generally), the maximum being reached in the tropics.

The longitudinal variation, or the westward increase in color, seems to be also coincident with the increased humidity to the westward, the darker representatives of any species occurring where the annual rain-fall is greatest, and the palest where it is least. This coincidence is clearly illustrated in the birds of the United States, where the darkest representatives of a species, as a general rule, (indeed without exception so far as known to me,) come from regions of maximum annual rain-fall, and the palest from those of minimum annual rain-fall. In the Northeastern States the amount of rain is only one half to two thirds what it is in the Northwestern States, while on the Great Plains it is less than one half what it is in the Northeastern States. In the lower part of the Mississippi basin and in the Southeastern States it is much greater than to the northward under the same meridians. Within the tropics, in America and Asia at least, the humidity, as well as the intensity of the solar rays, reaches the maximum, as does the intensity of color in both birds and other animals. In Europe, as is well known, the birds from near the Scandinavian coast, where the annual rain-fall reaches forty inches, are darker than in Central Europe, where the yearly rain-fall is only half this amount. So much darker, in fact, are the Scandinavian forms, that by some writers they have been regarded as specifically distinct from their representatives in Southern Germany, the Scandinavian forms of circumpolar species being as dark as their Eastern North American allies. There is again a striking parallelism between the relative humidity of Western Europe and Eastern North America, and the relative depth of color in the representatives of circumpolar species living in these two countries, the rain-fall of the latter region being double that of the former, and the birds of darker and livelier colors. As already intimated, this coincidence is not confined to the birds of these different regions, the same correlation of livelier, brighter, deeper tints with increased humidity being also exhibited by the mammals of these various districts, the Europeo-North American species being higher colored, as a general rule, in Eastern North America than in Europe, as the western forms of the continentally distributed American species are often higher colored than the eastern.

It is a most striking fact that the birds, and even the mammals and reptiles, of the almost rainless districts of Lower California, the Gila and Colorado deserts, are almost all so much paler in color than their relatives of the better-watered neighboring districts, that many of them have been described as distinct species, and the others referred to as strongly marked varieties, they all being characterized to a greater or less degree by a faded or bleached aspect. The birds and mammals of the arid plains of the middle region of the continent exhibit also the same bleached appearance, but in a somewhat less degree.

I had long suspected that hygrometric conditions had much to do with local variations in color in individuals of the same species, but I was not a little surprised when I came to compare the known areas most prolific of dark and light local forms with rain-fall charts, — which may be assumed as indicating relatively the hygrometric conditions of different regions, — to find the distribution of the light-colored races so strictly coincident with the regions of minimum mean annual rail-fall, and the dark forms with those of maximum mean annual rain-fall, as seems to be the case.

Humidity has hence apparently far more to do with climatic variation in color than solar intensity, though the latter has undoubtedly an influence upon color. The occurrence of a light-colored race of *Arvicola riparius* on Muskeget Island and the sandy sea-beaches of the coast of Massachusetts shows clearly that the intense light caused by reflection from a sandy surface tends to the diminution rather than to an increase of color in animals, and even plants, since the foliage of the latter in arid districts so commonly assumes a dull grayish tint. The capture on Muskeget Island last season (July, 1870), by Messrs. Maynard and Brewster, of two pairs of the short-eared owl (*Otus brachyotus*) with the color of the plumage so pale as at first to suggest their being albinos, is additional evidence of the bleaching effect of strong light upon the colors of animals. Such facts render it doubtful whether the increased intensity of the light in the tropics has really much to do with the brighter colors of tropical birds and insects, and suggest that humidity alone may be the principal agent in producing this accession of color.

In regard to the cause of other climatic variations, certain other facts are naturally recalled. In the remarks on the climatic and faunal peculiarities of East Florida,* attention was called to the less degree of vivacity and energy exhibited by the southern as compared with the northern members of the same species, and the general higher physiological development of essentially extra-tropical species in the temperate portions of their habitats. Is it hence improbable that the southward deterioration in size seen in such species is directly related to the enervating influence of increased heat? And why is it that so large a proportion of the birds pre-eminently singing-birds are found in temperate latitudes?

In the increased size of the bill and tail to the southward, especially of the former, we have a fact somewhat parallel to what is not unfrequently seen in mammals. The ears, for example, of the arctic repre-

* See above, p. 166.

sentatives of species ranging to warm-temperate latitudes are smaller at the northward than at the southward, as is seen in the native dogs, the foxes, and the wolves, and in the arctic races of man. The explanation generally given of this seems possibly applicable to the beaks of birds, namely, a greater activity in the circulation of the blood in the peripheral parts of the body in the temperate latitudes.

Species, Varieties, and Geographical Races.

The foregoing remarks on individual and geographical or climatic variation necessitates a brief consideration of the character of species, varieties, and races, and the propriety of applying binomials to such forms as can be clearly shown to be connected by intergrading links with others previously known. As preparatory to what follows, it seems proper to refer briefly to the origin of the excessive synonymy with which our descriptive ornithological works are burdened.

Ornithological synonymes may be arranged, as regards their origin, under four primary heads, namely: (1) Those arising from the description of immature and adult birds of the same species for different species, (2) from authors mistaking sexual for specific differences, (3) individual variation for specific differentiation, and (4) climatic differentiation for specific. A fifth source of error, and one which has given rise to a large class of synonymes, results from a combination of the causes indicated under (3) and (4).

Synonymes arising from the first two causes mainly preceded the others in regard to the relative frequency of their occurrence, especially so far as regards the birds of this continent. During the previous century, and the first two decades of the present, our birds were mainly described by European naturalists, who had no acquaintance with them in life, and whose resources often consisted of single and imperfect specimens received from chance travellers, without any indication of their sex or age. Later they were studied by resident naturalists, by whom the mistakes of their predecessors in this respect were to a great extent corrected. The laws of sexual and age variation becoming gradually known, errors from this source were soon far less frequent than in earlier times. When at a comparatively recent date critical comparisons were made of specimens from distant localities before regarded as specifically identical, it was found that occasionally distinct species had been confounded. Such results led in the end to undue importance being attached to trivial differences, so that assumed species were frequently based solely on either individual or climatic variation, but oftener on both combined.

As the rage for describing new species increased, differences seemed alone to be sought; and so long as a given species was usually deemed sufficiently represented, even by the best ornithologists of the day, by a single pair,* the subject of individual and climatic variation was necessarily almost wholly neglected, the custom of many naturalists being to describe species from single specimens, as though all the representatives of a species were cast after an unvarying pattern. As the number of specimens of well-known species increased in our large museums, it was soon seen that some of the supposed most reliable diagnostic features were subject to considerable variation. The collections brought together from various parts of the continent by the Pacific Railroad surveying parties and from other sources, and the reports published thereon, formed the beginning of a new era in the history of the ornithology of North America, and in ornithological science. The facts thus disclosed in respect to geographical range, and individual and climatic variation, opened new fields of inquiry. Old theories and blind adherence to authorities, however, still impeded progress and led to frequent inconsistencies, which only time and further investigations could correct. Hence has gradually dawned the fact of the existence of a range of individual variation previously unsuspected, and of general laws of climatic variation, the full scope of which, as bearing upon the character of species, is yet to be determined.

Nearly half a century since it was discovered that the North American representatives of what were then commonly regarded as circumpolar species could not in all cases longer be regarded as identical with the European. Further comparisons showed that in most cases of the supposed circumpolar distribution of species, specimens from the Old World and the New could be more or less readily distinguished, yet the differences were in most cases slight, more or less inconstant, and not unfrequently due more to differences in the latitude whence the specimens came than to other causes. Yet a precedent for specific

* Not many years since amateur ornithologists were kindly informed, by one of the leaders in the science of ornithology, that his collection of the birds of a certain country, numbering over two thousand species, required for their convenient storage a space equal to only about one hundred cubic feet, the specimens averaging less than two to a species!

separation in such cases having been established by recognized authorities, it was followed till all the land-birds and a large proportion of the water-birds of the two continents were separated, in many cases, it would appear, on purely theoretical or geographical grounds.* When the comparison was carried to specimens of continentally distributed species from distant localities, differences between these were also detected, and the theory of specific diversity assumed, till the Pacific representatives of such species were separated from the Atlantic ones, and in like manner the southern from the northern, and those of particular areas, as insular, peninsular, and interior basins, from the others. In some cases such separations were of course properly made, but a high percentage of such forms are now found to intergrade through specimens from the intermediate localities.

Not a few of the species of our faunal lists have been based on, and are still only known from, single specimens, and often on differences manifestly within the range of individual variation; others represent local races, which only appear distinct when extremes alone are considered, the intermediate stages being unknown or ignored. The increase of synonymes from this fruitful source appears to have not yet culminated, a large proportion of the "new species" now annually described being but slight local differentiations of previously known specific forms, from which they often differ only in being a little smaller, a little darker or brighter colored, and in the individual peculiarities of the single specimens on which some of them are based. In many cases this process of ultra subdivision has furnished stepping-stones to later generalizations; in too many other cases it has been in its results only unmitigatedly injurious.

So large a proportion of the commonly recognized species are virtually nominal, or rest on a false basis, it is not surprising that in the reaction consequent upon a fuller knowledge of the birds of this continent, which has already commenced, the reality of species should be to some extent ignored. Whether, however, species are considered as entities or only as arbitrary inventions, convenience demands some established definition of them.

* Audubon, writing in 1838 (Orn. Biog., Vol. IV, p. 608), refers to the Prince of Musignano (by whom a large part of the circumpolar and cosmopolitan species were separated into numerous assumed species) as "having altered his notions so far as to seem desirous of proving that the same species of birds cannot exist on both the continents"; and there seems to have been good reason for the remark, only instead of ***proving*** them distinct, he in most cases merely ***assumed*** them to be so.

Not a few naturalists have hence adopted the test of intergradation, which seems a reasonable and an unobjectionable one. The question of species and of specific synonymy is thus simplified to this: that whenever two forms which have both received names are found to intergrade, the more recent name shall become a synonyme of the older. Some, however, still urge that every recognizable form, however closely allied to others, and even intergrading, should be recognized by a binomial epithet, and that whether we call them species, or varieties, or races, or simply *forms*, that such names are none the less convenient expressions for certain facts. It seems to me, however, that there are insuperable objections to this course; for however distinct the extreme geographical forms of a species may be, a vast proportion of its representatives are intermediate to them, and could never be but doubtfully referred to the one rather than to the other. Ordinarily, for instance, in the birds of the Atlantic slope, the representatives of a given species at the extreme north of its breeding range almost always differ very tangibly from its representatives at the extreme southern limit, sometimes more, sometimes less, according to the species. Those living only a little to the northward of the middle region differ less from the extreme southern type than the extreme northern type does, and those a little to the southward of the middle region differ still less from the southern type, and are quite distinguishable from the extreme northern form. In other words, in species ranging from Southern Labrador or Northern New England to Florida, of which there are numerous unquestioned instances, specimens from Southern New England differ somewhat from the more northern ones; those from Southern New England from those of Southern New Jersey and Eastern Maryland, and these latter from those of Georgia and Florida. It hence depends entirely upon individual predilection whether two, three, or four "species" or "binomial forms" shall be recognized; and in either case there is the same difficulty in disposing of the intermediate types. Again, specimens from the Mississippi valley differ more or less from their relatives from the Atlantic coast, the central plains, and the Pacific slope. Here again similar difficulties are encountered. Hence it is necessary to decide between recognizing a single binomial form, with a considerable but definite range of climatic variation, or three, or six, or nine, or even more, which cannot be rigidly defined, and between each of which will always be found a greater or less proportion of intermediate types, doubtfully referable to one of the binomial forms rather than to another. Another important objection may be urged against giving binomial names to intergrading forms. In faunal and nominal lists of the species of a large or continental area, scarcely distinguishable forms take equal rank with the most distinct congeneric species. For instance, in a list of the birds of North America, *Turdus Aliciæ* and *Turdus Swainsoni*, *Turdus Auduboni* and *Turdus Pallasi*, stand side by side with *Turdus mustelinus* and *Turdus fuscescens*, though in the former cases *Turdus Aliciæ* and *T. Auduboni* are founded at best on slight, and in the one case on inconstant individual or local differences, while in the latter no two congeneric species need be more distinct. In the one case only experts can distinguish the forms, and frequently they only by an actual comparison of specimens, and then too frequently but doubtfully, while in the other case a casual observer need not mistake them. The names alone give no clew to their real character, and are hence in a great measure meaningless when separated from the most explicit diagnoses, and whose affinities can frequently only be settled by the arbitrary criterion of locality. But it is urged that cognizance should in some way be taken of these differences; and "How can they be better recognized," it may be asked, "than in the way proposed?"

As already shown, and as I trust a large proportion of ornithologists are willing to admit, these local forms occur in accordance with recognizable laws of climatic variation, similar variations with locality occurring, to a greater or less extent, in all species having nearly the same geographical range. Eventually, then, will not the recognition of these laws be sufficient, and should not a statement of the tendencies to variation with locality, and the degree to which it is developed, be embraced in the specific diagnosis of each species as a part of its specific description? Is not this, in fact, actually essential to the proper characterization of a species? The average characters being given, a line or two would suffice for a statement of its variations, both geographical and individual. Then only in one case where now there are hundreds would there be instances of doubtful identification. Till within a very recent period, perhaps, no other course could have been pursued than that of giving binomial names to each apparently distinct form, however slightly it may have differed from others previously known. In many cases, indeed, the differences between strictly intergrading geographical forms are very great, — greater, indeed, if they were not thus serially connected, than would be deemed necessary for specific separation; and so long as the extremes only were known, no one could have regarded them otherwise than as well-defined species. But the time has already come, it seems to me, for a different and a more philosophic method, and that to further increase synonymy by giving new names to slightly different local forms of the same species is worse than useless.

It is important, in this connection, to observe that the species occurring at any point on the Atlantic coast, or on the Pacific coast, or in the Mississippi valley, or on the Great Plains, in short, at any restricted locality, have, as compared with each other, with scarcely an exception, an unequivocal character; they are based on differences that place them beyond controversy. It is not so, however, when we compare the species of distant localities with each other, whether the localities differ in latitude or longitude. In such cases we constantly meet with controverted species. At the South are species admitted as doubtfully distinct from others found farther north; at the West, those holding the same relation to others of the East; while at intermediate points either both the disputed forms occur with greater or less frequency, or there is a gradual transition of the one into the other, neither form being typically represented. This is evidently what should be expected to occur, if what has been said above in respect to climatic variation be correct, and is evidently a suggestive and important fact. Is the theory of hybridization, so often appealed to in such cases, necessary to explain these facts? and is it, in fact, true? By uniting the intergrading forms, the number of species occurring at any single locality is not essentially reduced, but such a union would considerably reduce the total number recognized, as well as the number usually assigned to the several continents, as at present not a few are repeatedly counted.

The many facts bearing upon individual and geographical variation, presented in the foregoing pages form but an imperfect exposition of the subject. They are, nevertheless, eminently suggestive of interesting results, and the conclusions above deduced I can but believe will be only the more fully confirmed by further research. Additional details are given in the general remarks embraced in Part IV, where various facts merely hinted at above are more fully presented, and an application is made in many cases of the pinciples deducible from them.

As previously stated, individual and geographical variations are in some cases difficult to distinguish. They can be satisfactorily investigated only from extensive suites of specimens taken from the same locality in the *breeding season*, and sufficiently extensive suites of this character are, with rare exceptions, still wanting. In specimens taken during migration it is difficult to determine what share of the variation is due to birthplace and what to individuality. Whilst, however, the variations noticed cannot be always traced with certainty to their origin, their bearing upon the general subject of variation within specific limits is in no way vitiated. In considering hypothetical species, it is frequently clearly evident that they are based in part upon slight and tolerably constant climatic differences, and in part and sometimes wholly upon the individual peculiarities of the single specimen upon which the original description of the species was based; in part, too, upon seasonal differences, and upon characters of immaturity. It seems to me that in the numerous closely allied species of the *Ægiothus* group, to cite a case in point, some are based in part upon one and in part upon other of these differences of a single circumpolar species. As already shown, the bill in different specimens of *Æ. linarius* varies greatly in size, yet an examination of a considerable series of specimens of several of its allies shows an amount of variation in the bill closely approximate to that seen in the specimens of the various assumed species of *Ægiothus*. Much of the variation in color seen in the flocks of *Ægiothi* that visit the Northern States in winter is due to age, yet it has been taken as characteristic of different species. These birds only visiting us in winter, those inhabiting widely distant localities in the breeding season are probably then more or less associated. The light-colored specimens are doubtless in part old or fully mature birds, or inhabitants in summer of more northern districts than the browner or more fulvous ones, a large portion of which, however, are unquestionably young birds. The short-billed ones have also relatively longer setæ at the base of the bill, which, by concealing a large portion of it, give it the appearance of being shorter than it really is. Analogy would lead us to infer that those with the shorter and more heavily clothed bills have a more northern habitat than the others.

The persistency with which nominal species when characterized by "authorities" are retained in our literature is not a little remarkable. If specimens from the original localities cannot be found to exactly fit the descriptions, the diagnosis is slightly amended to suit examples that somewhat approach them, and the name retained. In other cases the species is retained without its character being questioned, the name and the original description being copied by succeeding writers, till the species becomes traditionally accepted without its claims to recognition having been critically examined.

Another noteworthy coincidence in regard to nominal species is the fact of their most frequent occurrence in obscurely known groups, which obscurity usually results from the difficulty of obtaining specimens of the forms in question, — either from the remoteness of their habitat, their scarcity, or the peculiarities of their habits, — or from preconceived notions of the intimate relationship of the species of such groups.

Since the above was put in type, I have for the first time met with some important and timely remarks by an eminent English botanist concerning variation within specific limits in plants, which are so apropos to what has been said above in regard to individual and climatic variation in birds, and contains, moreover, such judicious strictures on various practices indulged in by botanists, and of which zoölogists are equally guilty, that a short abstract of them forms a fitting conclusion to the present paper. Says Dr. J. D. Hooker, in the introductory essay to his " Flora Novæ-Zelandiæ " (Part I, pp. xii, xiii, xv, 1853) : —

> " Some naturalists consider every minute character, if only tolerably constant or even prevalent, as of specific value; they consider two or more doubtful species to be distinct till they have been proved to be one; they limit the ranges of distribution, and regard plants from widely severed localities as almost necessarily distinct; they do not allow for the effects of local peculiarities in temperature, humidity, soil, or exposure, except they can absolutely trace the cause to the effect; and they hence attach great importance to habit, stature, color, hairiness, period of flowering, etc. These views, whether acknowledged or not, are practically carried out in many of the local floras of Europe, and by some of the most acute and observant botanists of the day; and it is difficult to overestimate the amount of synonomy and confusion which they have introduced into some of the commonest and most variable of plants. . . . . In working up incomplete floras especially, I believe it to be of the utmost importance to regard dubious species as varieties, to take enlarged views of the range of variation in species, and to weigh characters not only *per se*, but with

reference to those which prevail in the order to which the species under consideration belong; and to resist steadily the temptation to multiply names; for it is practically very difficult to expunge a species founded on an error of judgment or observation. The state of the British flora proves not only this, but further, that one such error leads to many more of the like kind; students are led to overestimate inconstant characters, to take a narrow view of the importance and end of botany, and to throw away time upon profitless discussions about the differences between infinitely variable forms of plants, of whose identity really learned botanists have no doubt whatever. There is, further, an inherent tendency in every one occupied with specialties to exaggerate the value of his materials and labors. . . . .

"To the amateur these questions are perhaps of very trifling importance, but they are of great moment to the naturalist who regards accurately defined floras as the means of investigating the great phenomena of vegetation; he has to seek the truth amid errors of observation and judgment, and the resulting chaos of synonomy which has been accumulated by thoughtless aspirants to the questionable honor of being the first to name a species. The time, however, has happily passed when it was considered to be an honor to be the namer of a plant; the botanist who has the true interests of science at heart not only feels that the thrusting of an uncalled-for synonyme into the nomenclature of science is an exposure of his own ignorance and deserves censure, but that a wider range of knowledge and a greater depth of study are required to prove those dissimilar forms to be identical, which any superficial observer can separate by words and a name."

The above remarks are as strictly applicable to zoölogy and zoölogists as they have ever been to botany and to botanists. The present state of ornithology, and the tendency the majority of ornithologists have to multiply species on improper grounds, find here a fitting rebuke.

Part IV.

*List of the Winter Birds of East Florida, with Annotations.**

TURDIDÆ.

1.† **Turdus migratorius** *Linné*. Robin.

Seen daily, sometimes in considerable flocks, till about the first of March, after which time few were observed. It was shot by me at

* An asterisk (*) prefixed to the name of a species indicates that it is a constant resident; an obelisk (†), that it is a winter visitor.

Jacksonville, April 1st, but according to general report it does not breed in the State.

In this species the females are commonly supposed to be paler colored than the males, which is undoubtedly usually the case, but specimens as brightly colored as any I ever saw proved on dissection to be *females*, and other specimens as palely colored as any I ever met with have likewise proved on dissection to be *males*. This shows the importance of determining the sex in all cases by dissection, and not from external appearances. It also indicates a wide range of variation in color in the present species, as great as is seen between typical representatives of the so-called *Turdus Swainsoni* and *T. Aliciæ*, and which is, moreover, of the same character, namely, simply a variation in intensity.

2.† **Turdus Swainsoni** *Cabanis*. OLIVE BACKED THRUSH.

Turdus minor GMELIN, Syst. Nat., I, 817, 1788; in part only. — VIEILLOT, Ois. Am. Sept., II, 7, pl. lxiii, 1807; in part only. — BONAPARTE, Geog. and Comp. List, 1838.

Turdus solitarius WILSON, Am. Orn., V, pl. xiii, fig. 2; not the text.

Turdus nanus AUDUBON, Birds of Amer., III, pl. cxlvii; * not the text. — SAMUELS, Am. Nat., II, 218, 1868

Turdus olivaceus GIRAUD, Birds of Long Island, 92, 1843-44. Not the *T. olivaceus* of Linné

Turdus Swainsonii CABANIS, "in Tschudi's Fauna Peruana, 188, 1844-46." — BAIRD, Birds N. Am., 216, 1858. — SCLATER, Cat. Am. Birds, 2, 1862. — ALLEN, Proc. Essex Inst., IV, 56, 1864. — BAIRD, Rev. Am. Birds, I, 19, 1864. — ALLEN, Mem. Bost. Soc. Nat. Hist., I, 514, 1868. — RIDGWAY, Proc. Phil Acad. Nat. Sci., XXI, 128, 1869.

Turdus Aliciæ BAIRD, Birds N. Am., 217, 1858. — COUES and PRENTISS, Smithsonian Rep., 1861, 405. — COUES, Proc. Phil. Acad. Nat. Sci., XIV, 217, 1861 — BAIRD, Rev. Am. Birds, I, 21, 1864. — RIDGWAY, Proc. Phil. Acad. Nat. Sci., XXI, 128, 1869.

Merula Wilsonii SWAINSON, Faun. Bor. Am., I, 182, 1831.

Merula olivacea BREWER, Proc. Bost. Soc. Nat. Hist., I, 191, 1844.

Rare. Given on the authority of Mr. Boardman, who writes me he obtained one specimen at Enterprise, February 18th, and another at St. Augustine, in the same month. The greater part pass the winter farther south.

* The plates in "Birds of America" are too poorly colored, as is well known, to be recognizable representations of the species whose names they bear, including all those representing wood-thrushes, they having but little resemblance to those of the folio edition. The figures of "*Turdus nanus*," *Turdus solitarius*, and *Turdus mustelinus*, might all pass for the *Turdus Swainsoni*, so far as the color of the dorsal surface is concerned.

In my "Catalogue of the Birds of Massachusetts,"* published in 1864, I first advanced the opinion that the so-called *Turdus Aliciæ* Baird was the paler form of *T. Swainsoni*. To this view other writers have taken exception. Professor Baird, in his "Review of American Birds" (p. 21), summarily disposes of the matter by presuming that I had not seen what *he* called *T. Aliciæ*. In 1868, in my "Notes on the Birds of Iowa, Illinois," etc.,† I again reviewed the subject, having in the mean time examined some twenty specimens sent out by the Smithsonian Institution to different scientific institutions, labelled respectively, "*Turdus Aliciæ*," "*Turdus Aliciæ?*" "*Turdus Aliciæ?* hybrid?" "*Turdus Swainsoni*," "*Turdus Swainsoni?*" "*Turdus Swainsoni?* hybrid?" After having examined these authentic specimens of the bird in question, and also large numbers of Massachusetts examples of what I called *Turdus Swainsoni*, — among which are a considerable number that correspond in every particular respectively with the typical, authentic specimens of "*Turdus Swainsoni*" and "*Turdus Aliciæ*" of Baird, the larger number, however, being intermediate in character between them, and agreeing with specimens sent out from the Smithsonian Institution as "*T. Swainsoni?*" "*Turdus Aliciæ?*" "*Turdus Aliciæ?* hybrid?" etc., — I state in this paper that the opinion I had previously expressed in respect to *Turdus Swainsoni* and *Turdus Aliciæ* was fully confirmed. In this paper I discussed at some length the variations presented, not only by this species, but by *Turdus Pallasi* and *Turdus fuscescens*, and the character of their supposed allies, *T. Auduboni*, *T. nanus*, and *T. ustulatus*, and their supposed respective habitats. I gave also some details in respect to the variations in general size, form of the bill, proportions of the primary quills of the wing, etc., as well as in color, and concluded that *Turdus Aliciæ* was based on simply individual variation in color, the other differences, as of size, form of bill, etc., supposed at first to characterize it, being rarely coincident with the variations in color, they occurring as frequently in the one type of coloration as in the other. *Turdus nanus* and *Turdus ustulatus* I also deemed to hold the same relationship to *T. Pallasi* and *T. fuscescens* that *T. Aliciæ* does to *T. Swainsoni*. Though described as exclusively western, I stated I had found specimens in Massachusetts that accorded with them in every particular. After having given the subject still further attention, I am but the more fully confirmed in these opinions.

Dr. Coues, thus far one of the most strenuous advocates of the validity of these nominal species, in a somewhat recent paper of his,‡ after stating

* Proceedings of the Essex Institute, Vol. IV, p. 56.

† Memoirs of the Bost. Soc. Nat. Hist., Vol. I, p. 507.

‡ "A List of the Birds of New England," Proceedings Essex Institute, Vol. V, p. 267, 1868.

that he had shown the *T. Aliciæ* to be "a very common eastern bird, having a range of habitat as extensive as, and nearly identical with, that of *T. Swainsoni*," says, in referring to my earlier remarks on this subject, that they "illustrate very fully the well-known seasonal and other variations to which *T. Swainsoni* and *T. fuscescens* are subject," and adds that I appear to have been "autoptically unacquainted" with *T. Aliciæ* at the time of writing them. In respect to this supposition of Dr. Coues, I will merely add that one of the numerous specimens considered by me to typically represent the supposed *T. Aliciæ* has been sent to the Smithsonian Institution, and pronounced by Professor Baird himself to "typically represent the *T. Aliciæ*."

The measurements given below of this species and the two following indicate the average size and the usual range of variation in this respect in these species as represented in the Atlantic States. These measurements embrace twenty-four specimens of *Turdus Swainsoni*, nearly fifty of *T. Pallasi*, and about forty of *T. fuscescens*, nearly all of which are from New England, and by far the greater part from Eastern Massachusetts.

The following is the range of variation in the series of twenty-four specimens of *T. Swainsoni*: Length, 6.62 to 7.75; alar extent, 10.75 to 12.65; wing, 3.47 to 4.30; tail, 2.40 to 3.40 (4.00 ?); tarsus, 1.02 to 1.27. The average dimensions are as follows: Length, 7.17; alar extent, 11.65; wing, 3.86; tail, 2.88; tarsus, 1.15.

Measurements of New England Specimens of TURDUS SWAINSONI.

M. C Z. No.	Collector's Number.	Sex.	Locality.	Date.	Collector.	Length.	Alar Extent.	Wing.	Tail.	Tarsus.
2877	—	—	Springfield, Mass.	May 14, '63	J. A. Allen	6.75	11.90	3.19	2.78	1.12
—	—	—	" "	May 14, '63	"	6.62	11.40	3.80	2.83	1.11
—	18	—	" "	May 27, '61	"	6.75	—	3.92	2.78	1.10
—	19	♂	" "	May 27, '61	"	6.98	—	4.30	3.40	1.22
2930	—	—	" "	May 25, '63	"	7.60	12.50	4 12	3.10	1.20
—	37	—	" "	May 25, '63	"	7.06	11.40	3 65	2.86	1.13
2940	—	—	" "	May 29, '63	"	7.10	12.00	4.00	2.83	1.13
—	44	♂	" "	May 29, '63	"	7.25	12.00	4 00	2.95	1.10
2827	—	—	" "	May 14, '63	"	7.15	11.10	3 90	2.70	1.05
1829	29	—	" "	May 30, '62	"	6 90	11.20	3.55	2 58	1.02
1830	23	—	" "	May 30, '62	"	7.75	12.20	3.95	3.07	1.17
1831	1	—	" "	May 30, '62	"	7.35	11.50	3.80	2.73	1.10
2930	—	—	" "	May 25, '63	"	7.75	12.20	3 95	3.07	1.17
2940	—	—	" "	May 29, '63	"	7.38	11.37	3.93	2.87	1.14
5757	—	—	Concord, "	——	H. Mann	7.25	12.15	4.10	2.95	1.16
8844	646	♂	Belmont, "	May 27, '68	C. J. Maynard	7.76	12.65	4.20	4.00	—
4999	1326	—	" "	Sept. 21, '68	"	7.75	—	4.05	3 07	1.20
—	17	♀	Watertown, "	Oct. 2, '69	Wm. Brewster	7.12	12.00	3.94	3.00	1.19
—	213	♀	" "	May 24, '69	"	7.12	—	4.00	2.87	1.27
282	—	—	Malden, "	——	D. Higgins	7 10	11.00	3.70	2.48	1.08
307	—	—	Norway, Maine	——	A. E. Verrill	7.24	11 00	3.47	2.73	1.14
308	—	—	" "	——	"	7.00	10.75	3.48	2.40	1.12
5963	—	—	Upton, "	——	J. G Rich	7.30	11.10	3.71	2 68	1.03
1520	—	—	Glen House, W. Mts.	——	S. H. Scudder	7.00	11.50	3.84	2.74	1.10

3.† **Turdus Pallasi** *Cabanis.* HERMIT THRUSH.

Turdus solitarius WILSON, Am. Orn., V, 95, 1812. Not the figure (pl. xliii, 2), which is of *T. Swainsoni*. Not *T. solitarius* LINNÉ. — BONAPARTE, Geog. and Comp. List, 17, 1838. — AUDUBON, Synop., 91, 1839. — IBID., Birds of Amer., III, 29, pl. cxlvi, 1841.

Turdus minor BONAPARTE, Obs. on Wilson's Nomenclature, Journ. Phil. Acad., IV, 33, 1824. — NUTTALL, Man. Am. Orn., I, 346, 1830. — AUDUBON, Orn. Biog., I, 303, pl. lviii, 1831. — IBID., V, 445, 1839. — GAMBEL, Proc. Phil. Acad. Nat. Sci., III, 113, 1846. — GIRAUD, Birds of Long Island, 90, 1843-44.

Turdus Pallasi CABANIS, Wiegm. Archiv, I, 205, 1847. — BAIRD, Birds N. Am., 212, 1858. — SCLATER, Cat. Am. Birds, 2, 1862. — BAIRD, Review Am. Birds, Part I, 14, 1864. — ALLEN, Mem. Bost. Soc. Nat. Hist., I, 514, 1868. — RIDGWAY, Proc. Phil. Acad. Nat. Sci., XXI, 128, 1869.

Turdus nanus AUDUBON, Orn. Biog., V, 201, pl. ccccxix, 1839 (*T. minor* on the plate). — IBID., Birds of Am., III, 32, 1841. — BAIRD, Birds N. Am., 213, 1858. — SCLATER, Cat. Am. Birds, 2, 1862. — BAIRD, Rev. Am. Birds, I, 15, 1864. — RIDGWAY, Proc. Phil. Acad. Nat. Sci., XXI, 129, 1869. — COOPER and BAIRD, Orn. Cal., I, 4, 1870.

Turdus Audubonii BAIRD, Rev. Am. Birds, I, 16, 1864. — RIDGWAY, Proc. Phil. Acad. Nat. Sci., XXI, 129, 1869.

Merula solitaria SWAINSON, Faun. Bor. Amer., II, 184, pl. xxxvii, 1831. — BREWER, Proc. Bost. Soc. Nat. Hist., I, 191, 1844.

Merula silens SWAINSON, Faun. Bor. Amer., II, 186, 1831. — SCLATER, Cat. Am. Birds, 2, 1862.

Common. Last seen about March 25th.

As already observed in the remarks under *Turdus Swainsoni*, I regard the *Turdus nanus* of authors as identical with *T. Pallasi*. The assumed differences are slight and inconstant, and seem to be principally individual variation in color. Although of late supposed to be exclusively western, representing on the Pacific slope the *T. Pallasi* of the Atlantic and Central States, Audubon's original specimen came from Pennsylvania, though he subsequently received it from the valley of the Columbia River. In his "Synopsis" he gives its habitat as "Columbia River. Accidental in the United States." His description of its color is identical with that he gives of *T. Pallasi* (*T. solitarius* Aud.), even the words used being almost entirely the same throughout each description. In size, however, he gives *T. nanus* as being one inch less in length and one inch less in extent than *T. Pallasi*. Since Professor Baird, in 1858, recognized the *T. nanus* as a valid species and its habitat as "Pacific coast of North America to the Rocky Mountains," and restricted the *T. Pallasi* to "Eastern North America to the Mississippi River," the validity of *T. nanus* has been generally accepted. Professor Baird himself, however, speaks of it in this work as though it was in his opinion doubtfully distinct, and observes that, "if really distinct, is so closely allied to *T. Pallasi* as to render a separation of the two exceedingly difficult." The *T. Pallasi* was formerly recognized as inhabiting California by good authorities. Dr. Gambel, in his "Remarks on the Birds of Upper California," etc.,* after stating that "the dwarf thrush of Audubon was founded upon specimens from the Atlantic States, and no doubt upon the true hermit thrush," remarks: "An examination of specimens of the *T. minor* [=*T. Pallasi*] from the Atlantic and Pacific coasts of North America shows no difference in any way, except that perhaps the western one is somewhat smaller, yet the difference is scarcely appreciable. From the measurement of many western specimens I found its length to be 6¼ inches, and the extent of wings 10½ inches; the tail, wings, and relative length of quills the same as in our eastern one, and, in fact, I think it can in no possible way be distinguished as specifically different." California specimens, however, seem to average a little smaller than New England ones, so that the *T. nanus* seems best entitled to recognition of any of the several disputed forms of this group.

The habits of *T. nanus*, as described by Dr. Cooper, are exactly like those of the *T. Pallasi* of the East, except in regard to the situation of its nest, *his account of its nest and eggs according exactly with those of T. Swainsoni*, and not at all with those of *T. Pallasi*, its nearest ally.†

The *Turdus Auduboni* of Baird, of the Rocky Mountains, I have already also referred to *T. Pallasi*, from average specimens of which it differs only in being slightly larger. My reasons for this opinion have been given with sufficient detail elsewhere. ‡

It is difficult to reconcile the account given by Wilson,§ and corroborated by Audubon,‖ of the breeding habits of this species with what is now known of the distribution in the breeding season of this group (subgenus *Hylocichla*) of thrushes. The account given by these authors of the situation and structure of the nest is applicable to only *T. Swainsoni*, which, as well as the *T. Pallasi*, is not known to breed so far south by several hundred miles as the localities they give. The only species which may probably breed there is the *T. fuscescens;* but this species does not nest on trees. To determine to which species of thrush these authors refer

* Proc. Phil. Acad. Nat. Sci., Vol. III, p. 14, October, 1844. Also Journal Phil. Acad. Nat. Sci., 2d Series, Vol. I, p. 41, 1847.

† According to Professor A. E. Verrill, the *T. Pallasi* nests on the ground, and lays "bright-blue" eggs. *Proc. Essex Inst.*, Vol. III, p. 145.

‡ Mem. Bost. Soc. Nat. Hist., Vol. I, p. 512.

§ Am. Orn., Vol. V, p. 91.

‖ Orn. Biog., Vol. I, p. 303; Birds of America, Vol. III, p. 30.

as breeding in this manner on the Lower Mississippi would solve an interesting problem.

The following table will indicate the average size of *Turdus Pallasi* in the Atlantic States. The extremes in size of forty-six specimens are as follows: Length, 6.50 and 7.65; alar extent, 10.00 and 12.25; wing, 3.30 and 3.90; tail, 2.47 and 3.17; tarsus, 1.12 and 1.33. The average dimensions of these specimens are as follows: Length, 7.04; alar extent, 11.17; wing, 3.79; tail, 2.72; tarsus, 1.15.

Measurements of Specimens of Turdus Pallasi.

M. C. Z. No.	Collector's Number.	Sex	Locality.	Date.	Collector.	Length.	Alar Extent.	Wing.	Tail.	Tarsus.
9835	—	—	Milltown, Maine	June —, '64	G. A. Boardman	6.98	——	3.50	2.68	1.17
4071	—	—	Upton, “	June —, '65	J. G. Rich	6.88	10.00	3.48	2.67	1 16
1990	—	—	Norway, “	———	A. E. Verrill	6.90	10.50	3.30	2.51	1.15
1991	—	—	“ “	———	“	7.00	10.60	3 55	2 65	1.17
5610	—	—	“ “	———	S. I. Smith	7.25	11.55	3.73	2.93	1 20
5611	—	—	“ “	———	“	6.80	10.95	3.45	2.78	1.12
312	—	—	“ “	———	“	7.15	11 50	3.70	2 92	1.15
321	—	—	“ “	———	“	7.32	11.00	3.59	2.98	1 17
1897	—	—	Waterville, “	Apr. 20, '62	C. E. Hamlin	7.20	10.77	3.66	2.93	1.07
1943	—	—	“ “	Apr. 14, '62	“	6.80	10.75	3.49	2.65	1.18
4235	—	—	“ “	Oct. 21, '63	“	6.90	10.25	3.37	2.60	1.18
4251	—	—	“ “	Oct. 24, '63	“	6.80	10.75	3.57	2.93	—
5754	—	—	Concord, Mass.	———	H. Mann	7 50	11.82	3.83	2.95	1.23
5755	—	—	“ “	———	“	7.15	10.15	3.45	2.57	1.14
4060	—	—	“ “	———	“	7.25	11.62	3.77	2.94	·1 18
2584	—	—	Woburn, “	———	J. G. Shute	6.80	10.25	3.36	2 57	1.17
2862	—	—	Springfield, “	May 8, '63	J. A. Allen	7.25	10.70	3.52	2.83	1 20
9690	1002	—	“ “	Oct. 17, '63	“	7.10	11.20	3.57	2.70	1.13
9691	1021	—	“ “	Oct. 29, 63	“	7.00	10.50	3.47	2.74	1.18
——	49	♂	Watertown, “	Dec. 10, '69	Wm. Brewster	7.00	11.63	3 63	2 63	1.33
——	58	♂	“ “	Nov. 6, '69	“	6.87	11.19	3.61	2.52	1.20
——	60	♂	“ “	Nov. 10, '69	“	7.12	11.12	3.74	2 58	1.27
——	62	♂	Belmont, “	Nov. 22, '69	“	7.12	11.50	3.80	2 71	1.20
——	363	♂	Cambridge, “	Apr. 16, '70	“	7.23	11.94	3 71	2.85	1 29
——	59	♀	Waltham, “	Oct. 26, '69	“	7.20	11.12	3 74	2.58	1.27
——	61	♀	“ “	Nov. 1, '69	“	7.06	11.37	3 72	2.59	1.25
——	62	♀	Watertown “	Oct. 26, '69	“	6.50	10.50	3 50	2 37	1.15
——	283	♀	“ “	Nov. 22, '69	“	6.56	11.00	3.50	2.58	1.16
——	353	♀	“ “	Apr. 26, '70	“	6.75	11.30	3 45	2 50	1.17
8845	8	♀	Newton, “	Oct. 12, '67	C. J. Maynard	6.83	11.00	3.32	2.65	—
——	13	♀	“ “	Oct. 12, '67	“	7.00	10.77	3.37	2.47	—
8848	250	♀	“ “	Apr. 18, '68	“	7 21	11.32	3 60	2 80	—
8852	322	♀	“ “	Apr. 25, '68	“	7.27	11.00	3.50	2.60	—
8847	358	♀	“ “	May 5, '68	“	7.00	11.30	3.50	2.73	—
——	24	♂	“ “	Oct. 19, '68	“	7.00	11.47	3.45	2.75	—
——	14	♂	“ “	Oct. 16, '68	“	7.00	11.43	3.50	2.74	—
8846	26	♂	“ “	Oct. 16, '68	“	6.80	11.28	3.57	2.75	—
8849	301	♂	“ “	Apr. 25, '70	“	7.60	11.75	3 43	3.17	—
——	314	♂	“ “	Apr. 25, '70	“	7.38	12.83	3.80	2.83	—
8853	338	♂	“ “	Apr. 28, '70	“	7.45	11.83	3.90	2.90	—
5129	—	?	Jacksonville, Fla.	Jan. 21, '68	J. A. Allen	7.40	12.25	3.85	—	—
5145	—	?	“ “	Jan. 25, '68	“	7.00	11.60	3 60	—	—
5146	—	?	“ “	Jan. 25, '68	“	6.75	11.10	3.45	—	—
5147	—	?	“ “	Jan. 25, '68	“	7.00	11.50	3.60	—	—
5197	—	?	Hibernia, “	Feb. 3, '68	“	7.65	11 87	3.65	—	—
5320	—	?	Enterprise, “	Mar. 1, '68	“	6.75	10.90	3 40	—	—

4.† **Turdus fuscescens** *Stephens.* Wilson's Thrush.

Turdus mustelinus Wilson, Am. Orn., V, 98, pl. xliii, 1812. (Not *T. mustelinus* Gmelin.)

Turdus fuscescens STEPHENS, Shaw's Gen. Zoöl., X, i, 182, 1817. — G. R. GRAY, Gen. Birds, 1849. — BAIRD, Birds N. Am., 214, 1858. — SCLATER, Cat. Am. Birds, 2, 1862. — BAIRD, Rev. Am. Birds, I, 17, 1864. — ALLEN, Mem. Bost. Soc. Nat. Hist., I, 514, 1868. — RIDGWAY, Proc. Phil. Acad. Nat. Sci., XXI, 127, 1869.

Turdus Wilsonii BONAPARTE, Obs. on Wilson's Nomenclature. — NUTTALL, Man. Am. Orn., I, 349, 1832. — AUDUBON, Orn. Biog., II, 362, pl. clxvi, 1834. Ibid., V, 446. — GIRAUD, Birds L. Island, 89, 1843-44.

Turdus ustulatus NUTTALL, Man. Am. Orn., I, (2d ed.) 400, 1840. — BAIRD, Birds N. Am., 215, 1858. — IBID., Rev. Am. Birds, I, 18, 1864. — RIDGWAY, Proc. Phil. Acad. Nat. Sci., XXI, 127, 1869. — COOPER & BAIRD, Orn Cal., I, 5, 1870.

Merula minor SWAINSON, Faun. Bor. Am., II, 179, pl. xxxvi, 1831.

Merula Wilsonii BREWER, Proc. Bost. Soc. Nat. Hist., I, 191, 1844.

Not common, the greater part passing the winter in the tropics. A few specimens were taken by Mr. Boardman at Green Cove Springs, February 20th and 22d. I did not meet with it.

The considerable variation in color exhibited by different specimens of this species have perhaps been already sufficiently adverted to. It may be added that some of the brightest colored specimens of this species proved on dissection to be females, as well, also, as some of the palest. As in *T. migratorius*, *T. Swainsoni*, etc., these variations in color do not depend entirely upon sex, age, nor season. The latter, however, doubtless has much to do with it, as has also age, as already explained; * but the variation is in the main strictly the result of individual differentiation.

Dr. Cooper says † that in habits this species is the "exact counterpart of *T. nanus*," the resemblance extending to the situation and structure of the nest, and also to the color of the eggs. In this connection it may be remarked that it is not a little remarkable that the eggs and nests of both the so-called *T. ustulatus* and *T. nanus* should so exactly coincide with those of *T. Swainsoni* (which breeds where the other species are said to), when the birds themselves are scarcely distinguishable respectively from *T. fuscescens* and *T. Pallasi*, both of which nest on the ground and lay unspotted eggs, while *T. Swainsoni* nests in trees and lays spotted eggs. The nests and eggs I have seen purporting to be those of *T. ustulatus* and *T. nanus* (and also of *T. Aliciæ*) were so closely like those of *T. Swainsoni*, — not differing more from those of this species than those of the same species usually differ, — as to at once raise the suspicion in my mind that they might all be really those of *T. Swainsoni*, and that they may have been in some accidental way wrongly identified by the collector.

* In Part III, pp 193 *et seq.*

† Ornithology of California, Vol. I, p. 5.

In the following table are given the measurements of forty specimens, some twenty-five of which were taken in Massachusetts during the breeding season. The extremes of the series are as follows: Length, 6.95 and 7.87; alar extent, 11.05 and 12.70; wing, 3.55 and 4.16; tail, 2.63 and 3.02; tarsus, 1.06 and 1.18. The average dimensions are as follows: Length, 7.38; alar extent, 11.83; wing, 3.82; tail, 2.88; tarsus, 1.13.

Measurements of Specimens of TURDUS FUSCESCENS.

M. C. Z. No.	Collector's Number.	Sex.	Locality.	Date.	Collector.	Length.	Alar Extent	Wing.	Tail.	Tarsus.
2272	—	—	Waterville, Maine	June 2, '62	C. E. Hamlin	7.12	11.25	3.90	2.87	1.10
2275	—	—	" "	June 5, '62	"	7.39	11.90	3.79	2.84	1.18
2276	—	—	" "	June 2, '62	"	7.53	11.73	3.86	2.88	1.06
2277	—	—	" "	June 2, '62	"	7.40	12.00	3.83	2.87	1.13
9607	—	—	Canton, St. Lawrence Co., N. Y.	June —, '60	J. S Foley	7.36	11.40	3.72	2.92	1.12
9608	—	—	"	June —, '60	"	7.50	11.75	3.93	3.00	1.10
9609	—	—	"	June —, '60	"	7.50	——	3.92	2.93	1.18
10382	—	—	"	June —, '60	"	7.30	11.43	3.77	2.97	1.13
10383	—	—	"	June —, '60	"	6.95	11.05	3.55	2.63	1.09
10384	—	—	"	June —, '60	"	7.45	11.75	3.69	2.80	1.12
10385	—	—	"	June —, '60	"	7.12	11.75	3.68	2.67	1.11
8830	367	♂	Newton, Mass.	May 5, '68	C. J. Maynard	7.81	12.70	4.16	3.00	—
8831	382	♂	" "	May 6, '68	"	7.75	12.55	4.10	3.00	—
8841	495	♂	" "	May 15, '68	"	7.87	11.91	4.00	3.00	—
8832	528	♂	" "	May 16, '68	"	7.70	12.45	4.00	3.00	—
8833	538	♂	Wayland, "	May 18, '68	"	7.35	11.91	3.91	2.73	—
8834	556	♂	Weston, "	May 20, '68	"	7.00	11.95	3.55	2.55	—
8835	581	♂	Newton, "	May 22, '68	"	7.41	12.50	4.15	2.90	—
8337	611	♂	" "	May 25, '68	"	7.50	12.45	3.80	2.90	—
8838	683	♂	" "	May 28, '68	"	7.50	12.30	4.15	3.17	—
8839	692	♂	" "	May 28, '68	"	7.45	12.16	3.76	2.85	—
8336	610	♀	" "	May 25, '68	"	7.30	11.36	3.55	2.60	—
8840	768	♀	" "	June 5, '68	"	7.50	12.15	3.90	2.89	—
2876	2876	—	Springfield, "	June 14, '62	J. A. Allen	7.00	11.70	3.75	2.80	1.17
1749	1749	—	" "	June 14, '62	"	7.00	11.35	3.63	2.65	1.10
1828	1828	2	" "	May 29, '62	"	7.50	11.50	3.99	2.86	1.08
1832	1832	3	" "	May 29, '62	"	7.50	11.55	3.79	2.80	1.12
——	408	—	Newton, "	May 18, '70	Wm. Brewster	7.50	12.00	3.74	2.74	1.14
2873	2873	—	Springfield "	May 12, '63	J. A. Allen	7.75	12.40	3.98	3.02	1.08
2876	2876	—	" "	May 14, '63	"	7.12	11.30	3.73	2.91	1.12
2937	2937	—	" "	May 29, '63	"	7.45	11.80	3.65	2.78	1.12
2938	2938	—	" "	May 29, '63	"	7.65	11.50	3.86	2.99	1.15
1431	—	—	Malden, "	May 22, '62	"	7.15	11.87	3.75	2.93	1.14
1432	—	—	" "	May 22, '62	"	7.52	11.80	3.93	3.00	1.15
280	—	—	" "	——	D. Higgins	7.35	11.75	3.76	2.89	1.10
281	—	—	" "	——	"	7.25	11.90	3.68	2.73	1.15
143	—	—	" "	June —, '61	"	7.30	11.72	3.72	2.83	1.12
144	—	—	" "	June —, '61	"	7.25	11.75	3.78	2.70	1.15
145	—	—	" "	June —, '61	"	7.00	11.40	3.63	2.68	1.11
146	—	—	" "	June —, '61	"	7.45	11.75	3.69	2.80	1.12

5.* **Harporhynchus rufus** *Cabanis.* BROWN THRUSH.

Very abundant. The specimens examined were smaller and much brighter colored than any I have seen from the Northern States. Commences nesting the last week in March.

6.* **Galeoscoptes carolinensis** *Cabanis.* CAT-BIRD.

Abundant. Smaller and darker colored than at the North. Some evidently remain and breed. Audubon states that none breed so far south as South Carolina, and that few remain so far north as Florida in winter; but Dr. Coues, in his "Synopsis of the Birds of South Carolina,"* gives it as abundant and resident in that State.

7.* Mimus polyglottus *Boie.* Mocking-Bird.

Common. Contrary to my anticipations, I failed to hear this bird sing during my three months' stay in Florida, except in a few instances near Jacksonville early in April, at which time they were nesting, although everywhere more or less common. It was more frequent along the borders of the forest and about clumps of bushes in the pine barrens than in the hummocks. It differed from its relatives, the brown thrush and cat-bird, in avoiding the denser thickets, which are the favorite resorts of the latter. The resemblance of the mocking-bird to the loggerhead shrike, in mode of flight and general appearance, which must strike every observer, has been properly referred to by Dr. Coues. †

Different specimens of the mocking-bird from Florida differ considerably from each other in intensity of color, some being much darker than others, and in the extent of the white on the outer tail feathers, and also in the length, thickness, and curvature of the bill. Some have the commissure but slightly curved and the tip of the bill moderately depressed; others have the commissure much arched and the tip much decurved. Several specimens before me from Cape Florida are smaller than those from the St. John's River, with longer, slenderer, and more curved bills. There seems to be as much difference between specimens from South Florida and the Middle States, as between the numerous so-called species of the West Indies, which, many of them at least, are scarcely more than local forms of the original or first-described *M. polyglottus.*

The following measurements of forty-four Florida specimens of this species indicates its usual range of variation in size and proportions. The extremes of this series are as follows: Length, 9.25 and 11.00; alar extent, 13.00 and 14.75; wing, 4.00 and 4.75; tail, 4.10 and 5.15. The average dimensions are as follows: Length, 9.91; alar extent, 13.69; wing, 4.28; tail, 4.87.

* Proc. Bost. Soc. Nat. Hist., Vol. XII, p. 113.

† "Synopsis of the Birds of South Carolina," Proc. Bost. Soc. Nat. Hist., Vol. XII, p. 113, October, 1868.

Measurements of Florida Specimens of MIMUS POLYGLOTTUS.

M. C. Z. No.	Original No.	Sex.	Locality		Date.	Collector.	Length.	Alar Extent	Wing.	Tail.
5118	——	♂	Jacksonville,	Fla.	Jan. 19, '68	J. A. Allen	9.75	14.00	4.60	
5124	——	♂	"	"	Jan. 21, '68	"	9.75	14 35	4 50	
5350	——	♂	Enterprise,	"	Nov. 4, '69	"	10.00	14.15	4 25	
——	——	♂	Hawkinsville,	"	Mar. 15, '69	"	10.60	14 75	4 50	
5415	——	—	"	"	Mar. 15, '69	"	9.85	14.00	4 40	
5395	——	—	"	"	Mar. 14, '69	"	9.85	14.00	4.30	
5355	——	♀	"	"	Mar 10, '69	"	10.12	13 00	4.26	
5185	——	♀	Hibernia,	"	Jan. 30, '69	"	10.30	14.15	4 30	
10586	1955	♀	Jacksonville,	"	Dec. 31, '68	C. J. Maynard	10.00	13 00	4 05	4.65
——	2005	♂	"	"	Jan. 10, '69	"	10.20	14.00	4.50	4.70
——	2407	♂	Dummitt's,	"	Mar. 2, '69	"	11.00	14 65	4.50	4.60
——	2614	♂	"	"	Mar 13, '69	"	10.15	14 00	4.40	4.90
10589	2341	♂	"	"	Feb. 16, '69	"	9.75	13.50	4.25	4.75
——	2449	♂	"	"	Feb. 24, '69	"	9.25	14 00	4.35	4.35
——	2613	♂	"	"	Mar. 13, '69	"	10 20	14.00	4.40	4.75
——	2374	♂	"	"	Mar. 13, '69	"	9.80	14.00	4 30	4.10
——	2370	♂	"	"	Feb. 17, '69	"	9.50	13.25	4.25	4.40
10592	2372	♂	"	"	Feb. 17, '69	"	9.75	13.75	4.20	4.50
——	2528	♂	"	"	Mar. 5, '69	"	10 30	14.00	4.45	4.95
——	2474	♂	"	"	Feb. 16, '69	"	10.15	14.30	4 60	5.15
10595	2486	♂	"	"	Feb. 16, '69	"	10.00	13.00	4.00	4.38
10596	2518	♀	"	"	Mar. 2, '69	"	9.75	13 50	4.75	4 50
——	2428	♀	"	"	Mar. 2, '69	"	10.00	13 50	4.50	4.45
——	2429	♀	"	"	Mar. 2, '69	"	10.00	13 50	4 35	4.35
——	2478	♀	"	"	Mar. 11, '69	"	10.50	13.50	4.18	4.75
——	2419	♀	"	"	Mar. 11, '69	"	10.00	13 35	4.25	4.75
10590	2340	♀	"	"	Mar. 13, '69	"	9.60	14.35	4.40	4.50
10594	2346	♀	"	"	Mar. 16, '69	"	9.75	13 60	4.25	4.35
10597	2569	♀	"	"	Mar. 11, '69	"	9.30	13.00	4.00	4.17
——	2507	♀	"	"	Mar. 11, '69	"	9.60	13 35	4 10	4 60
10587	2339	♀	"	"	Feb. 10, '69	"	9.40	13.50	4 05	4.20
10588	2333	♀	"	"	Feb. 17, '69	"	9.75	13.50	3 40	4.50
10591	2509	♀	"	"	Mar. 13, '69	"	9.50	13.25	4 20	4.30
——	2560	♀	"	"	Mar. 8, '69	"	9 50	13.20	4.15	4 35
10593	2375	♀	"	"	Feb. 17, '69	"	9.80	13.00	4 10	4.60
——	2478	♀	"	"	Feb. 16, '69	"	9.50	13 50	4 40	4.20
——	2485	♀	"	"	Feb. 16, '69	"	10.00	13.00	4.00	4.38

SAXICOLIDÆ.

8.* **Sialia sialis** *Haldemann.** BLUE-BIRD.

Common. In this species the smaller size of the Florida specimens, as compared with those from Massachusetts, is very marked, as is also the greater intensity of color.

SYLVIADÆ.

9.† **Regulus calendula** *Lichtenstein.* RUBY-CROWNED KINGLET.

Abundant. One of the most numerous of the winter birds. Chiefly confined to the swamps and hummocks.

10.† **Regulus satrapa** *Lichtenstein.* GOLDEN-CRESTED KINGLET.

Not common. A single pair was collected by Mr. Maynard at Jacksonville in January.

* *Sialia sialis* HALDEMANN, Trego's Geography of Pennsylvania, p. 77, 1843. —BAIRD, Birds of N. Am., 222, 1858. See American Naturalist, Vol. III, p. 159, 1869.

11.* **Polioptila cærulea** *Sclater.* Blue-gray Gnatcatcher.

Common. Generally seen in the same situations as *R. calendula.*

PARIDÆ.

12.* **Lophophanes bicolor** *Bonaparte.* Crested Titmouse.

Common.

13.* **Parus atricapillus** *Linné.* Black-capped Titmouse. Chickadee.

Parus atricapillus Linné, Syst. Nat., I, 341, 1766. — Wilson, Am. Orn., I, 137, 1808. — Bonaparte, Obs. Nom. Wils. Orn., Journ. Phil. Acad. Nat. Sci., IV, 254, 1825. — Rich. & Swain., Faun. Bor. Am., II, 226, 1831. — Audubon, Birds Am., II, 146, pl. cxxvi, 1841. — Cassin, Ill. Birds Cal. I, 17, 1853. — Baird, Birds N. Am., 390, 1858. — Sclater, Cat. Am. Birds, 13, 1862. — Baird, Rev. Am. Birds, I, 80, 1864.

Parus palustris Nuttall, Man. Orn., 241, 1832.

Parus carolinensis Audubon, Orn. Biog., II, 341, 1837 ; V, 474, pl. clx, 1839. — Audubon, Birds Am., II, 152, pl. cxxvii, 1841. — Cassin, Ill. Birds Cal., I, 17, 1853. — Baird, Birds N. Am., 392, 1858. — Sclater, Cat. Am. Birds, 14, 1862. — Baird, Rev. Am. Birds, I, 81, 1864.

Parus septentrionalis Harris, Proc. Phil. Acad. Nat. Sci., II, 300, 1845. — Cassin, Ill. Birds Cal., I, 17, 80, pl. xiv, 1853. — Baird, Birds N. Am., 389. — Sclater, Cat. Am. Birds, 14. — Baird, Rev. Am. Birds, I, 82.

Parus meridionalis Sclater, Proc. Lond. Zoöl. Soc., 1856, 293. — Baird, Birds N. Am., 392. — Sclater, Cat. Am. Birds, 14. — Baird, Rev. Am. Birds, I, 81.

Parus occidentalis Baird, Birds N. Am., 391, 1858. — Sclater, Cat. Am. Birds, 14, 1862. — Baird, Rev. Am. Birds, I, 81, 1864.

Pœcila atricapilla Bonap., Consp. Av., 230, 1850.

Pœcila carolinensis Bonap., Ibid.

Seen by Mr. Marcy at Jacksonville, where also specimens of it were collected by Mr. Maynard. Not observed by any of us up the river. Audubon speaks of having found it abundant in the Floridas in the winter of 1831 and 1832, and "breeding in the swamps as early as the middle of February."*

The common titmouse (*P. atricapillus*), although not more subject to geographical variation than many other birds, is one of the species in which such differences were first detected, though not recognized at the time as such. Audubon, in 1833, upon returning to Charleston, South Carolina, from a visit to the Eastern States, the British Provinces, and Labrador, noticed a considerable difference in size between the examples of this bird he met with at the North, and those of the lowlands of the

* Birds of America, Vol. II, p. 153.

Carolinas. Though no other difference was appreciable, he and his friend Bachman thought this was sufficient to warrant the description of the southern form as specifically distinct from the northern. He accordingly thus separated them in the second volume of his "Ornithological Biography." But if the black-capped titmice of the Carolinas, the lower parts of Virginia, Maryland, and Southern New Jersey are distinct from those of Massachusetts, on precisely the same grounds are those of Massachusetts distinct from those of Northern Maine. Even the titmice of Massachusetts are not just the same in winter that they are in summer, those which breed here doubtless mainly going south in winter, while their place is filled by others that spend the summer more to the northward. This at least is what the slight average difference in size between summer and winter specimens seems to indicate. But the Carolina titmouse (*P. carolinensis*) has been recognized as valid by most subsequent writers, and in accordance with the principle upon which this supposed species was admitted, several others have been added by other authors.

The titmice from the middle, elevated regions of the continent, in accordance with a general law of geographical variation among both birds and mammals, are a little larger than those of either the Mississippi valley or the Pacific coast, and have also, apparently, a relatively slightly longer tail and paler colors, — variations which occur in a number of other birds that have a similar distribution. The titmice of this region form the *Parus septentrionalis* of authors. Specimens labelled "*Parus septentrionalis*," collected near Chicago, have been received at the Museum of Comparative Zoölogy from the Chicago Academy. They do not differ, however, from numerous others collected in Massachusetts, though the true *P. septentrionalis*, or the black-capped titmice of the Rocky Mountains, does have a slightly longer tail than those from the other parts of the continent.

Those which occur on the Pacific slope of the continent, though forming the *P. occidentalis* of authors, are admittedly the same in size and general appearance as the *P. atricapillus* of the Atlantic States, this species having been introduced to the world with the following suggestive remarks: "It is rather a hazardous undertaking to add another to the list of North American black-capped and throated titmice; but if we have three good species now, instead of one, then the present is equally entitled to specific distinction with *carolinensis* and *septentrionalis*."

The *P. meridionalis* was first made known from a single specimen from Mexico, and of which very few specimens seem to have been recognized as belonging to it. The original type certainly recalls only a worn summer specimen of the common titmouse, though its darker color may be due to its southern habitat. Towards the end of the breeding season specimens of *P. atricapillus*, more especially females, have the plumage, particularly that of the lower surface of the body, much darker than in fall and winter, simply from the wearing off of the rufous and ashy extremities of the feathers, July specimens generally differing much in color from winter ones.

In respect to *P. carolinensis*, as already observed, the only difference urged as distinguishing it from *P. atricapillus* is that of its smaller size. Yet this difference is so slight that it is admitted that if *P. carolinensis* and *P. atricapillus* were "separated by a wide interval of locality, it might be a question whether it [*P. carolinensis*] might not be a variety. As, however," it is urged, "both are found together in the Middle States, and preserving together their characteristics, there will be little risk in considering them distinct." Since the larger birds are, in the main, either northern or occupy the elevated regions of the Alleghanies, the two forms must necessarily be found associated together, especially in winter, through their migrations. Unfortunately, in the work where this group has been most elaborately considered,* but two examples of each are cited, with a statement of their measurements; the two of *P. atricapillus* being from Carlisle, Pennsylvania, and the two of *P. carolinensis* from Washington, D. C. From the annexed table of measurements of *P. atricapillus* from Massachusetts and Maine, it will be seen that a few are small enough to be regarded as belonging to the *P. carolinensis*. There is, also, a larger amount of seasonal difference in the color and general character of the plumage than has been either admitted or suspected, as well as in size. No one who has previously written on this group appears yet to have compared many specimens of these supposed two species, or to have examined a sufficiently large number of either to become aware of the wide differences that exist between specimens from the same locality.

Variations similar to those assumed to specifically distinguish *P. carolinensis* from *P. atricapillus* occur in *P. hudsonicus* between specimens from localities quite distant in latitude. Dr. Bryant has already called attention to such differences in the *P. hudsonicus*, and at the same time proposed for the southern "variety" the name of "*P. hudsonicus* var. *littoralis*." Concerning this variety and the general subject in question, he remarks as follows: "The specimens of *Parus hudsonicus* from Yarmouth [Nova Scotia] and those from the Hudson Bay territory present as great, if not greater, differences in size than exist between *P. carolinensis* and *P. atricapillus*, and in color, between *P. septentrionalis* and *P. atricapillus*. I am inclined myself to consider *P. atricapillus*, *septentrionalis*, *meridionalis*, and *occidentalis* as varieties of one species; but, if they are considered as specifically distinct, there can be little question of the propriety of

* Baird's Birds of North America.

separating the Yarmouth bird from those found in the Hudson Bay territory." *

In the following table of measurements of twenty-seven specimens, all taken within ten miles of Cambridge, and all but two in December and January, the extremes of size are as follows: Length, 4.70 and 5.75, both specimens being females; alar extent, 7.50 and 8.60, both specimens being also females; wing, 2.33 and 2.63, also both females; tail, 2.15 (female) and 2.67 (male); tarsus, .62 (male) and 77 (female). The average size of these specimens is as follows: Length, 5.38; alar extent, 8.37; wing, 2.47; tail, 2.50; tarsus, .70. The females average a little smaller than the males, but the difference is only slight.

The largest specimen of the group of black-capped and black-throated titmice cited by Professor Baird † measures as follows: Length, 5.75; alar extent, 8.37; wing, 2.75; tail, 2.86 (*Parus septentrionalis*, from the Black Hills, Neb., Sm. Inst. No. 8827). A specimen of the *P. carolinensis*, cited by the same author, measures as follows: Length, 4.62; alar extent, 7.00; wing, 2.50; tarsus, .60 (Sm. Inst. No. 706, from Washington, D. C.). So far as the length of the wing and tail are concerned, specimens are fre-

Measurements of Massachusetts Specimens of PARUS ATRICAPILLUS.

M. C. Z. No.	Collector's Number.	Sex.	Locality.	Date.	Collector.	Length.	Alar Extent.	Wing.	Tail.	Tarsus.
11703	52	♂	Cambridge, Mass.	Dec. 10, '69	Wm. Brewster	5.38	8.12	2.62	2.67	.75
11704	86	♂	" "	Dec. 14, '69	"	5.25	8 12	2.55	2.37	.75
11705	87	♂	" "	Dec. 14, '69	"	5.06	7.88	2.62	2 50	.62
11708	96	♂	" "	Dec. 14, '69	"	5.00	7.50	2.44	2.30	.75
11707	94	♂	" "	Dec. 17, '69	"	5.06	7.87	2.50	2.50	—
—	99	♂	" "	Dec. 17, '69	"	5.12	8.12	2.55	2.43	.72
—	104	♂	" "	Dec. 20, '69	"	5.25	8.00	2.55	2.50	.70
11711	103	♂	" "	Dec. 20, '69	"	5.06	8.00	2.62	2.43	.75
11710	101	♂	" "	Dec. 20, '69	"	5 12	8.06	2.55	2.43	.75
—	153	♂	" "	Jan. 7, '70	"	4.94	7.80	2.43	2.25	.69
—	112	♂	" "	Dec. 24, '69	"	5 50	8 12	2.62	2.56	.69
—	206	♂	" "	Jan. 20, '70	"	5.40	8.21	2.58	2.55	.70
11706	88	♀	" "	Dec. 14, '69	"	4.94	7 50	2.43	2.31	—
—	95	♀	" "	Dec. 17, '69	"	5.00	8.00	2.50	2 37	.75
11709	97	♀	" '	Dec. 17, '69	"	5.06	7.83	2.43	2.42	.—
11712	105	♀	" "	Dec. 20, '69	"	5.19	7.75	2.50	2.25	.67
—	100	♀	" "	Dec. 20, '69	"	5 19	8.12	2.56	2.50	.75
11713	114	♀	" "	Dec. 24, '69	"	5.75	7.88	2.50	2.33	.63
—	161	♀	Watertown, "	Jan. 7, '70	"	5.45	8.00	2.63	2 64	.69
—	163	♀	" "	Jan. 7, '70	"	4.94	7.50	2.33	2 25	.69
—	179	♀	Belmont, "	Jan. 13, '70	"	4 84	7.55	2.35	2.15	.77
—	205	♀	" "	Jan. 20, '70	"	5.41	8.17	2.54	2.42	.70
—	239	♀	" "	Jan 26, '70	"	5.25	7.52	2.35	2.30	.67
—	203	♀	Arlington, "	Jan. 20, '70	"	4 70	7.75	2.45	2.40	.68
4946	268	♀	Newton, "	Apr. 21, '68	C. J. Maynard	5.00	8.60	2.58	2.30	—
5011	1216	♀	" "	Sept. 8, '68	"	5.35	7.85	2.50	2.40	.65
4945	269	♂	" "	Apr. 21, '69	"	5.00	8.00	2.41	2.35	—

* Proc. Bost. Soc. Nat. Hist., Vol. IX, p. 368, April, 1865.

† Birds of North America, p. 390.

quently taken in Massachusetts (and of which I have measurements before me) that are considerably smaller than this one from Washington, or than any given in the above table.

SITTIDÆ.

14.* **Sitta carolinensis** *Gmelin.* White-breasted Nuthatch.

Common; especially in the pineries.

15.* **Sitta pusilla** *Latham.* Brown-headed Nuthatch.

Common in the pineries; rarely seen elsewhere.

TROGLODYTIDÆ.

16.* **Troglodytes aëdon** *Vieillot.* Common Wren.

Troglodytes aëdon Vieillot, Ois. Am. Sept., II, 52, pl. cvii, 1807. — Bonaparte, Richardson & Swainson, Audubon. — Baird, Birds N. Am., 367, 1858. — Ibid., Rev. Am. Birds, I, 138, 1864. — Maynard, Naturalist's Guide, Part II, p. 95, 1870.

Troglodytes fulvus, Nuttall, Man. Am. Orn., I, 422, 1832.

Troglodytes americanus Audubon, Orn. Biog., II, 452, pl. clxxix, 1834. — Baird, Birds N. Am., 368. — Ibid., Rev. Am. Birds, I, 141.

Troglodytes Parkmani Audubon, Orn. Biog., V, 310, 1839. — Baird, Birds N. Am., 367. — Ibid., Rev. Am. Birds, I, 140.

Troglodytes sylvestris Gambel, Proc. Phil. Acad. Nat. Sci., III, 113, 1864.

Sylvia domestica Wilson, Am. Orn., I, 129, pl. viii, fig. 3, 1808.

Abundant, occurring everywhere. It keeps so closely concealed that it is difficult to shoot, except when on the wing. Both this and the Carolina wren are exceedingly quick in their movements, and if they are watching the collector when he is about to shoot at them, they are pretty sure to dodge the charge; although he finds the bushes and foliage where the bird sat riddled by the shot, he usually searches in vain for the specimen he is sure he ought to have killed. When approached in old grassy fields or pine openings, they will allow one to almost tread on them before attempting to get away, and then, instead of taking to wing, often seek to escape by running off like a mouse beneath the grass. The term "house" wren, usually applied to this bird, is decidedly a misnomer, since it frequents the fields, the thickets, and even the forest, as much as the vicinity of houses. In the wilds of Florida, where human habitations are few, there is nothing whatever in its habits to suggest this name.

The "wood wren," *Troglodytes americanus* of Audubon, I am sure is only the brighter colored form of *T. aëdon;* in size or proportions there is nothing, though the contrary has been claimed, to distinguish them. Specimens equally large and equally small occur in each state of plumage, in which the same general range of variation in proportions is presented. There is also an intergradation in color, and no observable difference in habits. Both forms were common in Florida; both also occur in New England, whence Audubon obtained the first specimen of his supposed new species. Audubon admits that it "can hardly be distinguished in description" from the house wren. The large size assumed by him as characterizing it may be readily accounted for by the fact of his obtaining his first specimens at Eastport in Maine, which is the extreme northern limit of the habitat of this species.

The following measurements of fifteen Florida specimens indicates the usual range of variation in respect to size and proportions found in specimens from the same locality. The extremes of this series are as follows: Length, 4.30 and 5.10, both specimens being females; alar extent, 6.10 and 6.95, both specimens being males; wing, 1.90 and 2.44; tail, 1.30 and 2.40; tarsus, .50 and .68; bill, .47 and .60 (.80?). The differences between these extremes, it will be noticed, are very great, considering the small size of the bird. The average dimensions are as follows: Length, 4.89; alar extent, 6.61; wing, 2.05; tail, 1.80; tarsus, .52.

Measurements of Florida Specimens of TROGLODYTES AËDON.

M. C. Z. No.	Collector's Number.	Sex.	Locality.	Date.	Collector.	Length.	Alar Extent.	Wing.	Tail.	Tarsus.	Bill.
10681	1900	♂	Jacksonville	Jan. 1, '69	C. J. Maynard	4.70	6.60	2 44	2.40	.50	.47
—	1942	♂	"	Jan. 1, '69	"	5.00	6.50	2.00	1.70	.57	.50
—	1966	♂	"	Jan. 3, '69	"	5.00	6.75	2.05	1.75	.65	.50
10682	1967	♂	"	Jan. 3, '69	"	4.75	6.75	2.05	1.95	.55	.52
—	1968	♂	"	Jan. 3, '69	"	4.50	6.50	2.05	.1.65	.61	.50
—	2790	♂	"	Mar. 20, '69	"	5.65	6.95	2.00	1.64	.54	.51
—	2576	♂	Dummitt's	Mar. 10, '69	"	5.00	6.50	2.10	2.00	.62	.50
—	4	♂	Jacksonville	Mar. 29, '69	"	4.60	6.10	2.00	1.80	.61	.60
—	2033	♀	"	Jan. 5, '69	"	5.70	6.75	2.10	1.75	.60	.80
—	1979	♀	"		"	4.30	6 50	2.00	1.30	.60	.56
—	2588	♀	Dummitt's	Mar. 11, '69	"	5.00	6.70	1.90	1.70	.65	.50
5178	—	—	Hibernia	Jan. 20, '69	J. A. Allen	5.20	6.75	2.03	1.70	.65	.60
5179	—	—	"	Jan. 20, '69	"	4.75	6.50	2.00	1.65	.67	.55
—	—	—	Hawkinsville	Mar. 10, '69	"	5.00	6.50	2.00	—	—	—
5361	—	—	"	Mar. 10, '69	"	4.87	6.75	2.00	—	.68	—

17.* **Thryothorus ludovicianus** *Bonaparte.* CAROLINA WREN.

Common. Rarely seen outside of thickets.

In few species is the difference in color between northern and southern specimens greater than in this. Florida specimens have the reddish-brown of the dorsal surface many shades deeper than Maryland ones, and the under surface strongly rufous. The tail and wings, besides being much darker, have the dark bars black, they being deep black on the tail, and consequently far more conspicuous. The crissum, however, is lighter than in the Maryland specimens, with the black bars broader. The Florida specimens have also a much longer bill, they closely agreeing in every particular with the so-called *Thryothorus Berlandieri* of Northeastern Mexico, the Florida specimens even possessing the interrupted black bars on the sides of the body said to occasionally characterize that species as distinguished from the *T. ludovicianus.* The differences between Florida and Maryland specimens of *T. ludovicianus* in the length of the bill, as well as in color, are very striking. They are paralleled, however, in *Harporhynchus rufus* and in other species. The *T. Berlandieri* hence appears to be only the smaller, darker form of *T. ludovicianus,* — the Mexican homologue of the Florida representatives of this species.

The *Thryothorus Bewickii,* from what is known of its range, doubtless occurs as a resident bird in Florida, but is probably rare there, as it generally is elsewhere.

18.† **Anorthoura hyemalis** *Rennie.* WINTER WREN.

Rare. — *Boardman.*

19.† **Cistothorus stellaris** *Cabanis.* SHORT-BILLED MARSH WREN.

Rare. Enterprise, February. — *Boardman.*

The *Telmatodytes palustris* doubtless also occurs as a winter resident.

MOTACILLIDÆ.

20.† **Anthus ludovicianus** *Lichtenstein.* TITLARK.

Common. Several were usually seen in company, but along the river I saw no large flocks. According to Mr. Maynard, however, they occurred in large flocks in the "old fields" away from the river.

SYLVICOLIDÆ.

21.† **Mniotilta varia** *Vieillot.* BLACK AND WHITE CREEPER.

Not uncommon throughout the winter, but much more numerous in March.

22.† **Parula americana** *Bonaparte.* BLUE YELLOW-BACKED WARBLER.

Occasional during the winter months, but very numerous after the 1st of March, soon after which time they were in full song.

23.† **Helminthophaga celata** *Baird.* Orange-crowned Warbler.

"Enterprise, 15th of February. Rare." — *Boardman.*

24.* **Dendrœca pinus** *Baird.* Pine Warbler.

Abundant. Is much on the ground at this season, as it sometimes is at the north in spring; on the whole, however, it is much less terrestrial in its habits than is *D. palmarum.* In full song in February.

25.† **Dendrœca palmarum** *Baird.* Yellow Redpoll Warbler.

Extremely abundant. Probably the most numerous of the winter birds in East Florida, where it is more or less common in all situations. Exceedingly terrestrial in its habits, being generally seen hopping along the ground or fallen timber. At the 1st of April they had considerably decreased in numbers, but many were at that time observed at Jacksonville.

There is some indication that the males and females, and possibly the adult and young, frequent separate districts at this season. When at Jacksonville in January I saw only males; on the Upper St. John's, in February and March, only females or immature males; but these were in excessive abundance, as were also the males at the earlier date around Jacksonville. Is it not probable that the old males either do not go quite so far south as the females and immature males, or that the species was already on its way north? As is well known, the males in the species of this family, as probably in most other birds, precede the females in their journey northward.

26.† **Dendrœca coronata** *Gray.* Yellow-crowned Warbler.

More or less common till the 1st of April, and probably some remained still later. During the last half of March they began to moult, but at the end of the month a large part were still in winter dress. The same remarks in respect to moulting apply also to *D. palmarum.*

27.* **Dendrœca dominica** *Baird.* Yellow-throated Warbler.

Seen at Jacksonville in January, but much more abundantly up the river in February and March. March 5th I found them in great numbers in the cypress and maple swamps near Lake Munroe, at which time the spring migration had commenced.

28.* **Dendrœca discolor** *Baird.* Prairie Warbler.

Abundant at Jacksonville, April 1st, and occasionally seen at earlier dates. This species is undoubtedly resident in Florida the whole year.

29.† **Seiurus aurocapillus** *Swainson.* GOLDEN-CROWNED WAGTAIL.

Not common. A few were seen in February, as well as later.

30.† **Seiurus noveboracensis** *Nuttall.* WATER WAGTAIL.

Rare. Found at Dummitt's by Mr. Maynard in Februrary.

31.* **Geothlypis trichas** *Cabanis.* MARYLAND YELLOW-THROAT.

Abundant. Though somewhat brighter colored throughout, they differ mainly from the northern type in the greater breadth of the black facial band. There is but little difference in general size, that is, so far as I have had an opportunity of observing; occasionally a Florida example has a bill considerably longer than the average in northern examples, but this does not appear to be a very constant difference between the southern and northern specimens. It would probably be more marked in specimens from South Florida.

Other species of this family were seen in March that are not to be reckoned as winter residents. Among them are the following: *Dendrœca maculosa*, *D. virens*, and *D. pennsylvanica*, *Euthlypis canadensis*, *Setophaga ruticilla*, and *Helminthophaga ruficapilla*, all of which began to appear on the Upper St. John's, near Enterprise, about the middle of March, and most of them were also seen later at lower points on the river. *Helmitherus vermivorus* and *H. Swainsoni* were taken at St. Augustine, by Mr. L. L. Thaxter, in April.

HIRUNDINIDÆ.

32.† **Tachycineta bicolor** *Cabanis.* WHITE-BELLIED SWALLOW.

More or less numerous, but observed at irregular intervals. Large flocks were seen near the St. John's River in January. It probably does not breed in Florida.

33.† **Cotyle riparia** *Boie.* BANK SWALLOW.

Not observed by either Boardman, Maynard, or myself prior to the last of March, but Mr. Audubon saw it in immense flocks "in winter," first at St. Augustine, and afterwards in other parts of the State.*

The *Stelgidopteryx serripennis* was seen about Jacksonville the first week in April, and specimens of it were obtained. Several pairs were seen flying about some bluffs a few miles below the town, apparently with the intention of selecting breeding-places.

* Birds of America, Vol. I, p. 187.

VIREONIDÆ.

34.† **Lanivireo solitarius** *Baird.* SOLITARY VIREO.

Rather common. In full song early in March.

35.* **Vireo noveboracensis** *Bonaparte.* WHITE-EYED VIREO.

Common. In full song in March.

36.† **Vireosylvia olivacea** *Bonaparte.* RED-EYED VIREO.

"A few all winter." — *Boardman.* Common after the 1st of March, on the Middle St. John's.

The Yellow-throated Vireo, *Lanivireo flavifrons*, was quite common early in March, and is undoubtedly a winter resident in South Florida.

AMPELIDÆ.

37.† **Ampelis cedrorum** *Baird.* CEDAR BIRD.

Common. Perhaps resident.

LANIIDÆ.

38.* **Collurio ludovicianus** *Baird.* LOGGERHEAD SHRIKE.

Lanius ludovicianus LINNÉ, Syst. Nat., I, 184, 1766. — BONAPARTE, NUTTALL, AUDUBON. — GAMBEL, Proc. Phil. Acad. Nat. Sci., III, 200, 1847.
Lanius garrulus BARTRAM, Travels, 289, 1791 (no description).
? *Lanius ardosiaceus* VIEILLOT, Ois. Am. Sept., I, 81, pl. li, 1807. — BONAPARTE, Obs. on Wils. Nomenc.; Journ. Phil. Acad. Nat. Sci., III. 358, 1824.
Lanius carolinensis WILSON, Am. Orn., III, 57, pl. xxii, fig. 5, 1811.
Lanius excubitoroides SWAINSON, Faun. Bor. Am., II, pl. xxxiv, 1831.
Lanius elegans SWAINSON, Ibid., 122. — NUTTALL, Man. Am. Orn., I, 2d ed., 287, 1840. — GAMBEL, Proc. Phil. Acad. Nat. Sci., I, 261, 1843.
Lanius mexicanus BREHM, Cab. Journ. für Orn., II, 145, 1854. — SCLATER, Catal. Am. Birds, 46, 1861.
Collurio ludovicianus BAIRD, Birds of N. Am., 325, 1858. — ALLEN, Amer. Nat., III, 579, 1869. — BAIRD, Rev. Am. Birds, I, 443, 1866.
Collurio excubitoroides BAIRD, Birds N. Am., 337. — BAIRD, Rev. Am. Birds, I, 445. — COOPER & BAIRD, Orn. Cal., I, 138, 1870.
Collurio elegans BAIRD, Birds N. Am., 328. — BAIRD, Rev. Am. Birds, I, 444. COOPER & BAIRD, Orn. Cal., I, 140, 1870.

Not very numerous.

I have already referred to the questionable distinctness of the so-called *C. excubitoroides* from the present species.* Further examination of the

* See a series of articles in the "American Naturalist," entitled "Notes on some of the Rarer Birds of Massachusetts," Vol. III, 1869.

subject has only confirmed me in the opinion that they are not distinct, and that in all probability the *C. elegans* of California should also be referred to the *C. ludovicianus.**

TANAGRIDÆ.

The *Pyranga æstiva* became common on the Lower St. John's April 1st to 5th, but was not observed previously. *P. rubra* was not seen at all.

A considerable number of specimens of this species (*P. æstiva*) in the Museum, from the Atlantic States, present great differences in the size of the bill in respect to vertical and lateral thickness, as well as in the position and distinctness of the "tooth" of the bill, and in the curvature of the commissure, as indicated by the accompanying figures (Plate IV, figs. 19, 20). They also vary greatly in intensity of color, both of the bill and plumage, as do different specimens of *P. rubra* from Massachusetts. Hence species based solely on such distinctions should be accepted, if at all, with great hesitancy.†

FRINGILLIDÆ.

39.† **Chrysomitris tristis** *Bonaparte.* YELLOW BIRD.

Common throughout the winter, and as numerous the first week in April as earlier.

I am sure I heard the notes of the Pine Finch (*Chrysomitris pinus*), but as I obtained no specimens of it and do not find it reported by others, I do not include it in the present list. It is not improbable that this species and the Purple Finch (*Carpodacus purpureus*) are occasional winter visitors.

* Since writing the above I have met with the following observations on this group, made by Dr. Gambel, in his "Remarks on the Birds observed in Upper California" (Proc. Phil. Acad. Nat. Sci., Vol. III, p. 200, 1847): "In the shrikes we are presented with a group of birds closely allied to each other, and undergoing such changes in plumage as renders them difficult to discriminate. Although examined with great care by Swainson in the Fauna Boreali-Americana, yet he appears to have laid too much stress upon characters subject to great variation, as size, relative length of quills, and color. . . . . The relative length of quills in the shrikes is an uncertain character, and differs very much according to age. In the young of this species the second quill is generally much shorter than the sixth, but in the adult equals and may even exceed the sixth in length; the proportion of the third, fourth and fifth to each other is also exceedingly various, and indeed in each wing of the same bird it is very common to find the proportions of the quills differing very materially. This I have found to be the case in the European and both American species [*Collurio ludovicianus* and *C. borealis*]."

† See some remarks on the "Uniformly red species of *Pyranga*," in Proceed. Phil. Acad. Nat. Sciences, p. 127, June, 1869.

40.† **Passerculus savanna** *Bonaparte.* SAVANNA SPARROW.

Emberiza sandwichensis GMELIN, Syst. Nat., I, 875, 1788.

Emberiza arctica LATHAM, Ind. Orn., I, 414, 1790.

Emberiza chrysops PALLAS, Zool. Rosso-Asiat., II, 45, pl. xlviii, fig. 2, 1811.

Fringilla savanna WILSON, Am. Orn., III, 55, pl. xxii, fig. 2, 1811.

Passerculus savanna BONAPARTE, Geog. and Comp. List., 33, 1838.—BAIRD, Birds N. Am., 442, 1858.—SCLATER, Cat. Am. Birds, 112, 1862.

Passerculus alaudinus BONAPARTE, Compte Rendu, XXXVII, 918, 1853.—BAIRD, Birds N. Am., 446, 1858.—SCLATER, Cat. Am. Birds, 112, 1862.—COUES, Proc. Phil. Acad. Nat. Sci., XVIII, 84, 1866.—COUES, Proc. Essex Inst., V, 281, 1868.—COOPER & BAIRD, Orn. Cal., I, 181, 1870.

Passerculus anthinus BONAPARTE, Compte Rendu, XXXVII, 919, 1853.—BAIRD, Birds N. Am., 445, 1858.—SCLATER, Cat. Am. Birds, 112, 1862.—COOPER & BAIRD, I, 183.

Passerculus sandwichensis BAIRD, Birds N. Am., 444, 1858.—SCLATER, Cat. Am. Birds, 112, 1862.—COOPER & BAIRD, Orn. Cal., I, 180.

Abundant, especially on the savannas, where it was the principal sparrow seen.

This species, like all the sparrows, varies considerably in color with the season of the year. Fall specimens, and especially the young of the year, have the yellow superciliary stripe very indistinctly defined, it being in numerous cases entirely obsolete. The general plumage is also much browner, with the streaks on the dorsal surface suffused and obscured with ferruginous, and those below, as in fall specimens of *Melospiza melodia*, bordered with the same tint. Different individuals also vary considerably in the breeding season, some being much grayer above than others; the superciliary line varies from bright yellow to grayish white, with the yellow either entirely wanting or limited to a slight wash on the part anterior to the eye. This grayer plumage and faded condition of the superciliary stripe is more especially seen towards the end of the breeding season. The spots below also vary so much in size as to give very different aspects to the plumage of the lower surface of the body in different specimens. In some they form little more than a narrow line along the middle of the feathers of the breast and sides of the body; in others they are quite broad, occupying relatively a much larger surface; occasionally, also, they are aggregated on the lower part of the breast, forming a large conspicuous patch, as distinct as is ever seen in *Melospiza melodia*. The general size of the bird also varies considerably, as is indicated in the accompanying table of measurements, and the bill is subject to very marked variations in respect to length, size, thickness, and slenderness, as substantiated by a series of nearly one hundred specimens now before me, including some thirty specimens taken at Ipswich, Massachusetts, in the breeding season.

These specimens are separable to some extent into several series, which may be based either upon difference in general size, the character of the bill, or upon coloration; but these several kinds of variation fail to corroborate each other. If separated upon differences in size, the two or more series thus separated embrace every combination of the other differences; and similar incongruities result when the separation is made upon differences in coloration or other characters. Yet the Massachusetts specimens present among themselves differences as well marked and of the same character as is assumed to distinguish several of the so-called species from the Pacific coast, that have been proposed and adopted by different authors.

Alexander Wilson was the first naturalist who gave any adequate description of the species in question, though the *Emberiza sandwichensis* of Gmelin unmistakably refers to this bird, and this name having been given long before that of Wilson, should, in accordance with the rule of priority, supplant Wilson's more euphonious and familiar one of *savanna*. The first supposed species recognized by modern writers after the well-known one of Wilson was the *P. alaudinus*, described by Bonaparte in 1853, in his notes on the Delattre collection,* from a specimen from California. He says it is not easily distinguished from *P. savanna*, but differs from it in being smaller, with the bill shorter and slenderer, and in wanting the yellow superciliary line.† Professor Baird redescribed it in his Birds of North America in similar language, and cites under it five specimens, which came respectively from Brownsville, Texas; Tamaulipas, Mexico; Petaluma, Cal., and Shoalwater Bay, W. T. He remarks respecting it as follows: "This species, if really distinct from *P. savanna*, differs in the rather smaller size, although the difference is not great, and in the considerably paler colors. The superciliary stripe shows a very faint trace of yellow, especially anteriorly near the bill. In some specimens, as 4342, there is none at all." Bonaparte, in his paper just cited, added another "new species" from Kodiak, Alaska, which he called *Passerculus anthinus*, and described as follows: "*Passerculus anthinus*, Bp., ex Kadiak, Am. Ross. *Simillimus præcedenti, sed rostro etiam graciliore et capite flavo induto; subtus albo-rufescens magis maculatus.*" He says it is still smaller and has the bill slenderer even than the other, and that it appears to live farther north. Professor Baird also redescribes this species, and is much more explicit in his account of it. He says: "Similar to *P. savanna*, but smaller. . . . . Breast and upper part of belly thickly spotted with sharply defined sagittate brown spots, exhibiting a tendency to aggregation on the

* Compte Rendu, Tome XXXVII, p. 918.

† "*Passerculus alaudinus*, Bp., ex Wils., mais plus petite sans jaune aux sourcils et à bec plus court et plus effilé."

middle of the belly," etc. He adds: "This species is the smallest of its group, and differs from all in the much greater amount of spotting on the under parts. The streaks, indeed, extend over the whole breast and upper part of the abdomen, instead of being mainly confined to the jugulum." It differs, he says, from *P. alaudinus* "in the strong shade of yellow on the head, the much darker tints above, and the thick crowding of larger and better defined spots beneath, with a faint tinge of reddish." He refers to it three specimens from San Francisco, Benicia, and Petaluma, California.

In 1858 Professor Baird added still another species of *Passerculus* to those previously recognized, through the redescription of the original type of this group, the *Emberiza sandwichensis* of Gmelin, based upon Latham's Sandwich Bunting * and Pennant's Unalaska Bunting.† The name Sandwich, as Professor Baird has remarked, refers not to the Sandwich Islands, but to Sandwich Sound, on the northern coast. To this species Baird judiciously refers the *Emberiza arctica* of Latham ‡ and Vigors, § and the *E. chrysops* of Pallas. ‖ Professor Baird's description of it is as follows: "Almost exactly like *P. savanna*, but half an inch larger, with much larger bill. Length, 6.12; wing, 3.00; tail, 2.55. Habitat, northwestern coast, from the Columbia River to Russian America." He also further observes: "This species is extremely similar to the *P. savanna*, and is only distinguishable by its greater size and more western locality. The tail feathers also are rather more acutely pointed. There is also a greenish-yellow shade on the top and sides of the head, brighter than is seen in *P. savanna*. The bill is considerably larger and longer, measuring .51 of an inch above instead of .44." To this is referred one specimen from "Russian America," one from Fort Steilacoom, W. T., and three from Shoalwater Bay, W. T., three of which measure as is indicated in the above-quoted description, and the other nearly three fourths of an inch less.

In respect to size, then, it appears that the so-called *P. sandwichensis* is the larger, the *P. savanna* the next in size, *P. alaudinus* the third, and *P. anthinus* the smallest. So, at least, it is claimed; but from the measurements published in Birds of North America, a female of *P. savanna* from Carlisle, Pa. (No. 780), is, with one exception (No. 4340, from Brownsville, Texas), the smallest of the specimens of this genus of which measurements are there given; two others from Pennsylvania are below the average of *P. alaudinus*. No. 10,203, from Russian America, referred to *P. sandwichensis*, is scarcely larger than an average *P. savanna*. The

* Latham's Synopsis, Vol. II, p. 202, 1783.

† Pennant's Arctic Zoology, Vol. II, Species No. 229, pp. 320, 368.

‡ Indian Ornithology, Vol. I, p. 414, 1790.

§ Zoölogy of the Blossom, p. 20, 1839.

‖ Zoographia Rosso-Asiatica, Vol. II, p. 45, pl. xlviii, fig. 1, 1811.

accompanying series of measurements shows that specimens occur in Massachusetts as large and as small as any specimens of the genus of which measurements are given by Professor Baird.

In respect to the geographical distribution of these different supposed species, it will be observed that of the three West Coast species, the larger, *P. sandwichensis*, is northern, and the others, *P. alaudinus* and *P. anthinus*, southern, which perfectly explains the difference in size that occurs between them.* In respect to *P. alaudinus* and *P. anthinus*, one is only the paler colored and the other the brighter colored form of the common savanna sparrow as represented in the Pacific States; the three supposed species together forming a series similar to what is seen when a large number of specimens of this bird from the Atlantic States are compared. In other words, the characters whereon these species are based are evidently only individual differences. The *P. alaudinus* is the form with narrow streaks and generally paler tints, or that having a minimum intensity of color; the *P. anthinus* is that with the brighter tints, or with the maximum intensity of color, the greater breadth of the streaks, and the rufous suffusion below correlating with the generally brighter tints. Aside from this normal range of variation referred to at length in Part III as obtaining in all species, there is that of season to be taken into account, as the fading of the superciliary stripe and the grayer aspect of the plumage above towards the end of the breeding season, through the natural wearing and bleaching of the plumage,† and also the rufous suffusion and greater amount of color characteristic of the renewed plumage in fall. It will be noticed that authors report the occurrence of all the western species either actually at or near the same points,‡ while *P. savanna* was not until recently supposed to occur on the Pacific slope of the continent.§ But one of the others have been announced from the plains as far east as Nebraska,‖ and from Brownsville, Texas.¶

In respect to the habits of these supposed species, there is nothing attributed to the western one that is not equally applicable to the eastern bird. Dr. Coues, it is true, says that in Southern California *P. anthinus* seemed confined to the moist salt grass and sedgy weeds of the sea-shore

* Since the above was written, Mr. Dall has given, not only *P. savanna* and *P. sandwichensis*, but also *P. alaudinus* and *P. anthinus* in his list of the birds of Alaska. (See Trans. Chicago Acad. Sciences, Vol. I, pp. 283, 284.)

† See Part III, p. 193.

‡ See Professor Baird, "Birds of North America," Dr. Coues, "Notes on the Birds of Arizona Territory," and Cooper's Ornithology of California.

§ It has recently been reported by Mr. Dall as common in Alaska.

‖ *P. alaudinus*, Sclater's Catalogue of American Birds. p. 112.

¶ *P. alaudinus*, Baird, in Birds of North America, p. 446.

itself. "When with difficulty it was flushed, its flight was," he remarks, "very rapid and irregular; and it would alight again almost immediately, and run with great celerity among the roots of the thick grasses. It was then exceedingly difficult to procure." * All of which is quite true of *P. savanna* when frequenting the salt marshes, which form its most favorite resort in Massachusetts. "*P. alaudinus*," he says, "was common two or three miles away from the coast, but on the sea-shore itself I never found one mixing with *P. anthinus*; it is a bush-and-weed rather than a grass species." *P. savanna* also frequents similar localities. Mr. Dall, under *P. anthinus*, has also accurately indicated the habits of the eastern *Passerculus*. Under *P. savanna*, however, he mentions a fact in respect to the breeding habits of this species I have never before seen mentioned as characterizing any of the *Passerculi*, namely, *its nesting in bushes*. I have met with many nests of the eastern savanna sparrow, and have always found them placed on the ground, usually in a tuft of grass.

To recur again to the series in the Museum of Comparative Zoölogy, I may add that while some of the Ipswich specimens, taken late in June, have a decidedly yellow superciliary stripe, none have it so bright as it is usually in spring specimens; in a considerable proportion it is very pale, and in Nos. 4700, 10668, etc., it is grayish-white, with no perceptible trace of yellow. No. 5099, and some others, have the spots on the breast and sides very narrow, occupying but a small share of the surface; on the other hand, in No. 5088, as also in several others of the series, the spots are so broad as to occupy more space than the enclosing white portion. In other specimens, taken at a different season of the year, the "rufous tinge" surrounding the spots referred to in the above-quoted description of *P. anthinus* is very marked. There is likewise great difference in the color of the upper surface in different specimens. In some the black central spots of the interscapularies are so broad as to give to the dorsal aspect a very dark tint; in others, taken the same day at the same locality, they are so restricted that the general aspect of this surface is gray. The bills of the different specimens vary as much in length and robustness as they are represented to do in the two extremes in this respect in the western bird. Some of the long-billed ones have the bill slender; others have it thick and stout. Occasionally one has the upper mandible projecting considerably beyond the lower, but only in cases where it is abnormally developed. A specimen from Fort Bridger, Utah (No. 11115 of the Smithsonian Catalogue), in the Museum, labelled *Passerculus alaudinus* at the Smithsonian Institution, is of this character, the upper mandible being very much abnormally developed and decurved, and projecting much beyond the lower.

* Ibis, July, 1866, 268.

In short, while not denying that there may be a slight average difference between eastern and western specimens, as I know there is between those of the Central Plains and those of the Atlantic States, I cannot allow that it is at all sufficient to substantiate a specific difference. On the contrary, I am confident that the above-named supposed species of the Pacific States are based chiefly on individual variation perfectly parallel with that seen in a large series of specimens from the Atlantic States. No one, in fact, seems to have felt very confident that any of them were distinct from the eastern *P. savanna*. Dr. Coues has even repeatedly expressed his belief that *Passerculus alaudinus* is not permanently distinct from that species. "In a large series of the latter," he says, "shot about Washington, I have found fully as great differences as I have ever detected in comparing the eastern with the western forms." *

Dr. Cooper also refers as follows to the close resemblance of the *P. alaudinus* to the *P. sandwichensis*. He says, "I think it very doubtful whether these specimens (which measure larger than the dimensions given by Baird, though otherwise agreeing) are anything more than a southern form of *P. sandwichensis*, though collected near San Diego. . . . . Baird considers it almost identical with *P. savanna* of the east, and says that *P. sandwichensis* differs from that species in its larger size. Spring specimens have the superciliary stripe more decidedly yellow, so that there only remains a more slender bill to distinguish this from *P. savanna*, and the larger size (characteristic of northern specimens generally), with darker hues, from *P. sandwichensis*." † Respecting *P. anthinus* Dr. Cooper remarks, "This species appears better marked, as compared with *P. savanna*, than the preceding [*P. alaudinus* and *P. sandwichensis*], although I am not entirely satisfied that it is different." ‡

The following measurements of twenty-six specimens (fourteen males and twelve females), all taken at Ipswich during June and July, 1868, and measured before skinning, indicates the range of individual variation presented by this species. The extremes are as follows: Length, 5.20 and 6.00, both males; alar extent, 7.61 and 9.75, both females; wing, 2.44 and 2.95; tail, 1.64 and 2.25; tarsus, .75 and .88. The average dimensions are: Length, 5.20; alar extent, 8.79; wing, 2.70; tail, 1.96; tarsus, .84. The following are the extremes of the series of measurements of the western *Passerculi*, given in Birds of North America: Length, 5.00 (*P. alaudinus* Tamaulipas, Mex.) and 6.12 (*P. sandwichensis*, Fort Steilacoom, W. T.); alar extent, 8.50 and 9.37 (same specimens); wing, 2.50 and 2.95 (same specimens); tail, 2.00 and 2.57 (same specimens). It thus appears that

* Ibis, July, 1866, p. 289.

† Ornithology of California, Vol. I, p. 182.

‡ Ibid., p. 183.

specimens taken in the breeding season in Massachusetts, overlap in two out of, the four measurements given, all the so-called western species, while specimens taken in Massachusetts at other seasons, vary still more than the specimens cited in the following table.

Measurements of Massachusetts Specimens of PASSERCULUS SAVANNA, *taken in the Breeding Season.*

M. C. Z. No.	Collector's Number.	Sex.	Locality.	Date.	Collector.	Length.	Alar Extent.	Wing.	Tail.	Tarsus.
5083	811	♂	Ipswich	June 12, '68	Allen & Maynard	5.76	8.32	2.72	2.07	—
—	820	♂	"	June 13, '68	"	5.76	9.25	2.55	2.00	.85
5086	846	♂	"	June 14, '68	"	5.65	9.15	2.75	1.85	.85
5087	848	♂	"	June 14, '68	"	5.50	9.13	2.75	2.07	.84
5089	852	♂	"	June 15, '68	"	5.40	9.10	2.65	2.00	.85
5090	853	♂	"	June 15, '68	"	5.70	9.25	2.63	2.10	.87
5094	854	♂	"	June 15, '68	"	5.70	9.25	2.64	1.95	.88
—	855	♂	"	June 15, '68	"	5.50	9.15	2.75	1.85	84
5901	856	♂	"	June 15, '68	"	5.40	9.25	2.60	1.93	.80
5092	857	♂	"	June 15, '68	"	5.20	9.37	2.95	2.05	—
5092	858	♂	"	June 15, '68	"	5.40	9.25	2.63	2.06	.83
5096	862	♂	"	June 17, '68	"	5.75	8.00	2.90	2.00	—
5098	873	♂	"	June 17, '68	"	5.83	7.75	2.80	2.10	—
—	881	♂	"	June 17, '68	"	6.00	8.27	2.74	2.25	—
5082	810	♀	"	June 12, '68	"	5.42	8.81	2.71	1.81	—
5084	819	♀	"	June 13, '68	"	5.75	8.85	2.57	1.83	.83
5084	847	♀	"	June 14, '68	"	5.54	8.55	2.70	1.90	.80
5088	851	♀	"	June 15, '68	"	5.75	9.75	2.70	2.05	.75
5094	859	♀	"	June 17, '68	"	5.45	8.90	2.65	2.05	.85
5095	860	♀	"	June 17, '68	"	5.25	8.50	2.44	1.85	—
5096	862*	♀	"	June 17, '68	"	5.70	8.90	2.70	1.90	—
5097	878	♀	"	June 17, '68	"	5.75	7.61	2.60	1.80	—
5099	877	♀	"	June 17, '68	"	5.75	8.05	2.75	2.00	—
—	890	♀	"	June 17, '68	"	5.35	9.75	2.77	1.95	—
5100	1006	♀	"	July 15, '68	C. J. Maynard	5.65	7.95	2.72	2.00	.87
—	1158	♀	"	Aug 19, '68	"	5.36	8.50	2.70	1.64	.85

41.† **Poocætes gramineus** *Baird.* GRASS FINCH.

Abundant, especially in and about the old fields. The most numerous sparrow in East Florida in winter.

42.† **Junco hyemalis** *Sclater.* SNOW BIRD.

"Common in January." — *Boardman.* Not seen by either Mr. Maynard or myself. Probably of somewhat irregular occurrence so far south.

43.† **Spizella socialis** *Bonaparte.* CHIPPING SPARROW.

Common. A large proportion of those seen were young birds.

44.* **Spizella pusilla** *Bonaparte.* FIELD SPARROW.

Common. More numerous than the preceding species (*S. socialis*). They appeared to be breeding at Jacksonville the first week in April.

The songs of the males were so different from those of the northern bird that the species was almost unrecognizable by me from its notes.

45.† **Zonotrichia albicollis** *Swainson.* WHITE-THROATED SPARROW. Generally more or less common.

46.† **Melospiza melodia** *Baird.* SONG SPARROW.

Not numerous. At least comparatively few were seen.

47.† **Melospiza palustris** *Baird.* SWAMP SPARROW.

Common, frequenting the hummocks and swamps.

48.† **Passerella iliaca** *Swainson.* FOX-COLORED SPARROW.

A single specimen was seen by Mr. G. A. Boardman at Enterprise. None were seen by Mr. Maynard or myself.

49.† **Ammodromus maritimus** *Swainson.* SEASIDE FINCH.

"Abundant at Fernandina." — *Boardman.*

50.† **Ammodromus caudacutus** *Swainson.* SHARP-TAILED FINCH.

"Abundant, with the preceding." — *Boardman.* Although I have marked as winter visitors both these species of *Ammodromus*, they may be resident.

51.† **Coturniculus Henslowi** *Bonaparte.* HENSLOW'S SPARROW.

Stated by Audubon to be abundant in winter on the grassy pine barrens of Florida.*

52.* **Peucæa æstivalis** *Baird.* PINE-WOOD SPARROW.

Fringilla æstivalis LICHTENSTEIN, Verzeich. Doubleder Zool. Mus. der königl. Univ. zu Berlin, 25, 1823.
Fringilla Bachmani AUDUBON, Orn. Biog., II, 366, pl. clxv, 1834.
Fringilla æstiva NUTTALL, Man. Orn., I, 2d ed., 568, 1846.
Peucæa Bachmani AUDUBON, Syn. Am. Birds, 112, 1839.
Peucæa æstivalis CABANIS, Mus. Hein., 132, 1850.
Zonotrichia Cassinii WOODHOUSE, Proc. Phil. Acad. Nat. Sci., 1852, 60.
Peucæa Cassinii BAIRD, Birds N. Am., 485, 1858.

Common, but confined to the pine woods.

The twenty-two specimens, collected by Mr. Maynard's party and myself, now in the Museum, present considerable differences. Several are so different in color from most of the others as to almost have the appearance of being a different species, the general color of the upper

* Birds of America, Vol. III, p. 76.

parts being rufous instead of gray. These are all females, the others being males. But the males differ greatly in color, few of our sparrows being more variable in this respect than the present species.

The following measurements of twenty-two Florida specimens indicate quite a constancy in size, much greater than in color. The extremes of this series are as follows: Length, 5.75 and 6.20; alar extent, 7.60 and 8.30; wing, 2.17 and 2.55; tail, 2.25 and 2.68. Average: Length, 5.88; alar extent, 8.99; wing, 2.40; tail, 2.49.

Measurements of Florida Specimens of PEUCÆA ÆSTIVALIS.

M. C. Z. No.	Collector's Number.	Sex.	Locality.	Date.	Collector.	Length.	Alar Extent.	Wing.	Tail.	Tarsus.
5377	5377	♂	Hawkinsville	Mar. 13, '69	J. A. Allen	6.00	8.20	2.35	2.50	.76
5396	5396	—	"	Mar. 15, '69	"	5.85	8.05	2.28	2.30	.75
5425	5425	♀	Jacksonville	Apr. 2, '69	"	5.90	8.00	2.40	2.40	.76
5426	5426	♂	"	Apr. 2, '69	"	5.90	8.30	2.55	2.50	.70
5427	5427	♂	"	Apr. 2, '69	"	5.60	7.85	2.50	2.25	.76
5428	5428	♂	"	Apr. 2, '69	"	5.80	8.20	2.40	2.50	.74
5429	5429	♂	"	Apr. 2, '69	"	5.90	7.85	2.45	2.47	.76
5430	5430	♂	"	Apr. 2, '69	"	5.60	7.85	2.40	2.50	.68
10616	18	♂	"	Apr. 3, '69	C. Thurston	5.80	8.00	2.45	2.50	.70
10617	21	♂	"	Apr. 3, '69	"	5.90	7.75	2.17	2.50	.67
—	24	♂	"	Apr. 5, '69	"	5.75	8.00	2.45	2.35	.66
10618	27	♂	"	Apr. 6, '69	"	5.75	8.00	2 50	2.55	.65
10619	28	♂	"	Apr. 6, '69	"	5.90	8.00	2.45	2.65	.62
10620	36	♀	"	Apr. 7, '69	"	5.90	8.00	2.25	2 55	.68
10624	72	♂	"	Apr. 24, '69	"	6 00	8.00	2.40	2.55	.70
10623	68	♂	"	Apr. 16, '69	"	6.00	8.00	2.40	2.56	.75
10622	67	♂	"	Apr. 16, '69	"	6.00	8.00	2.45	2.50	.74
10625	73	♂	"	Apr. 24, '69	"	6.20	8.15	2.40	2.50	.70
—	651	♂	"	Apr. 15, '69	"	6.00	8.00	2.40	2.53	.72
10621	45	♂	"	Apr. 18, '69	"	6.10	8.10	2.35	2.60	.73
—	37	♂	"	Apr. 7, '69	"	5.90	8.00	2.45	2.68	.75
10626	2934	♂	"	Apr. 13, '69	"	5.85	7.60	2.25	2.34	.70

53.* **Cardinalis virginianus** *Bonaparte.* CARDINAL BIRD.

Exceedingly numerous. Their clear, musical, loud call-note was heard everywhere, this being the most noisy bird of the forest.

None of the specimens I have seen from Florida are as large as those from the Middle States. The colors of the former are also somewhat brighter, especially in the females, in which the brownish-yellow of the lower parts is not only much deeper, but a large proportion have the breast and middle of the abdomen strongly tinged with bright red, giving a very different appearance from northern females.

The following measurements of fifty-eight specimens shows the amount of variation in size in specimens from the same locality. The females, it will be seen, average a little smaller than the males, but the sexual difference in this respect is not very great. The range of variation, which is much less in this species than in many, is as follows: In the males: Length, 7.75 to 9.10; alar extent, 11.00 to 11.78; wing, 3.50 to

3.85 ; tail, 3.40 to 4.20 ; tarsus, .62 to .80. In the females : Length, 7.50 to 8.75 ; alar extent, 10.70 to 11.75 ; wing, 3.25 to 3.85 ; tail, 3.40 to 4.10 ; tarsus, .62 to .75 Average size of the males : Length, 8.46 ; alar extent, 11.43 ; wing, 3.63 ; tail, 3.87. Average of the females : Length, 8.27 ; alar extent, 11.27 ; wing, 3.53 ; tail, 3.77.

Measurements of Florida Specimens of CARDINALIS VIRGINIANUS.

M. C. Z. No.	Collector's Number.	Sex.	Locality.	Date.	Collector.	Length.	Alar Extent.	Wing.	Tail.
5164	5164	♀	Hibernia	Jan. 30, '69	J. A Allen	8.60	11.45	3.55	4.10
5165	5165	♂	"	Jan. 30, '69	"	8 45	11 70	3.65	3 83
5166	5166	♂	"	Jan. 30, '69	"	8 45	11.50	3.60	4 10
5167	5167	♂	"	Jan. 30, '69	"	8.75	11.50	3 55	3 95
5189	5189	♀	"	Feb. 3, '69	"	8.60	11.25	3.50	3 95
5192	5192	♂	"	Feb. 3, '69	"	8 75	11.75	3.50	4.10
5193	5193	♀	"	Feb. 3, '69	"	8.45	11.35	3.60	3 83
5230	5230	—	Volusia	Feb. 12, '69	"	8 15	11.00	3.30	3.78
5311	5311	♂	Enterprise	Mar. 1, '69	"	8.75	11.60	3.60	3 90
5312	5312	♀	"	Mar. 1, '69	"	8.50	11.35	3.40	3.78
5347	5347	♂	"	Mar. 4, '69	"	9.10	11.50	3.58	4.15
—	—	♀	Hawkinsville	Mar. 13, '69	"	7.75	10.70	3 25	—
—	—	♂	Jacksonville	Mar. 31, '69	"	7.75	11.15	3.55	—
5424	5424	♂	"	Apr. 2, '69	"	8 50	11.50	3.65	3.90
—	—	♀	"	Apr. 2, '69	"	8.55	11.10	3.55	—
—	1955	♂	"	Jan. 2, '69	C. J. Maynard	9.00	11.50	3.65	4 30
—	1987	♂	"	Jan. 5, '69	"	9.00	11.50	3 85	4.05
10706	1988	♂	"	Jan. 5, '69	"	8.50	11 51	3 75	4.20
10707	1989	♂	"	Jan. 5, '69	"	8 50	11.45	3.75	3 80
—	2003	♂	"	Jan. 10, '69	"	8 05	11.60	3.75	4.15
—	2041	♂	"	Jan. 7, '69	"	8.00	11.00	3.75	4.00
—	2460	♂	Dummitt's	Feb 24, '69	"	8.00	11.25	3.50	3.40
—	2418	♂	"	Feb. 22, '69	"	8 75	11.10	3.80	3.90
—	2536	♂	"	Feb. 7, '69	"	8 60	11.00	3.65	4 10
—	2537	♂	"	Feb. 9, '69	"	8.70	11.50	3.65	4.00
—	2447	♂	"	Feb. 24, '69	"	8.00	11.56	3.60	3 60
10709	2337	♂	"	Feb 16, '69	"	8 50	11.50	3 60	3.80
—	2328	♂	"	Feb. 25, '69	"	8.25	11 50	3 50	3.80
10710	2338	♂	"	Feb. 16, '69	"	8.50	11.50	3 60	3.80
10713	2398	♂	"	Feb. 16, '69	"	8.50	11.50	3.65	3.65
—	2324	♂	"	Feb. 25, '69	"	8.60	11.50	3.60	3 50
—	2369	♂	"	Feb. 25, '69	"	8.75	11.78	3.65	3.90
—	2368	♂	"	Feb. 25, '69	"	8.40	11.50	3 60	4 10
—	2335	♂	"	Feb. 17, '69	"	8.00	11 50	3.60	3.90
—	2364	♂	"	Feb. 17, '69	"	8.50	11.50	3.60	3.60
—	2363	♂	"	Feb. 17, '69	"	8.50	11.50	3 60	4 00
—	2538	♂	"	Mar. 10, '69	"	8 15	11.18	3.60	3 40
—	2535	♂	"	Mar. 4, '69	"	8 40	11 00	3 57	3 95
—	2459	♂	"	Feb. 25, '69	"	8.50	11.50	3.75	3.95
—	2008	♀	Jacksonville	Jan. 9, '69	"	8.25	11.25	3 85	3 85
—	2042	♀	"	Jan. 6, '69	"	8.75	11.24	3 50	3.70
—	2334	♀	"	Jan. 11, '69	"	8.00	11.00	3.40	3.40
—	2579	♀	Dummitt's	Mar. 10, '69	"	7.75	10.75	3.40	3 95
—	2337	♀	"	Feb. 24, '69	"	7.50	11.00	3.30	3.50
—	2594	♀	"	Feb. 11, '69	"	8.50	11.10	3.50	3.91
—	2595	♀	"	Feb 11, '69	"	8.50	11.05	3.50	3.85
—	2334	♀	"	Feb. 24, '69	"	8.75	11.75	3 67	4.00
—	2415	♀	"	Feb. 20, '69	"	8 00	11.10	3.50	3.80
—	2394	♀	"	Feb 16, '69	"	8.00	11.75	3.30	3.55
10716	2324	♀	"	Feb 15, '69	"	8.50	11.50	3 70	4 10
—	2458	♀	"	Feb 25, '69	"	8.10	11 15	3 75	3 65
—	2336	♀	"	Feb. 17, '69	"	8 50	11.50	3.60	3 70
—	2474	♀	"	Feb. 16, '69	"	8.00	11.25	3.45	3 55
—	2475	♀	"	Feb. 26, '69	"	8.00	11 00	3.40	3 75
10117	2489	♀	"	Mar. 1, '69	"	8.05	11.15	3.64	3 40
10715	2488	♀	"	Mar 1. '69	"	8.50	11.20	3.60	4.00
10714	2427	♀	"	Mar. 4, '69	"	8.20	11.25	3.55	3 60
—	2043	♀	"	Feb 11, '69	"	8.50	11 50	3 75	3.90

54.* **Pipilo erythrophthalmus** *Vieillot.* CHEWINK.

Exceedingly numerous.

Mr. C. J. Maynard detected an interesting local race or variety of this bird at Dummitt's. Besides having the irides *yellowish-white* instead of *red*, there is less white at the base of the primaries, less skirting the secondaries, and much less on the tail. The whole bird is also smaller. The white on the tail generally extends only to the three outer pairs of feathers; in the common northern form it extends over the four outer pairs, and on the first is much more extended than in the Florida one. The tail of the common form, with the outer pair of feathers removed, would resemble, in respect to the distribution and extent of the white, that of the Florida bird. The song of this bird, as I heard it at Jacksonville in April, is quite different from that of the northern bird, it being ordinarily only about half as long, and uttered with much less spirit. As is well known, the song of the towhe, or chewink, at the north consists of two parts, nearly equal in length but otherwise quite different. In the Florida bird the last half is almost entirely omitted. According to Mr. Maynard, this variety is almost the only one occurring on Indian River, and of which he brought home some forty or more specimens. I found also one among half a dozen I shot at Jacksonville in January. In April, among a few towhes exposed in cages for sale in the market, were several of this kind. There is probably a large proportion of northern birds among the *Pipilones* of Northern Florida in winter, while probably in summer the majority are of the southern type above described, as are those of Middle and Southern Florida, doubtless, at all seasons.

Had this form been discovered ten, or even five years since, it would probably have been regarded by most ornithologists as entitled to specific rank, and not as a local race of *P. erythrophthalmus*, as it evidently is. Indeed, there are many species still on our lists that are far less entitled to rank as species than this, but which, though at first only provisionally adopted, have become traditionally established as valid species.

The two tables of measurements of specimens of this species given below, with Table J (p. 212), show the difference in size that obtains between Massachusetts and Florida specimens. The first table embraces twenty-nine specimens (nineteen males and ten females) of the white-eyed Florida type; the second table embraces sixteen specimens (ten males and six females) of the common northern type from Eastern Massachusetts; the measurements of twenty other Massachusetts males having been also already given in Table J, on p. 212. The following are the extremes of the two series. Northern type, males: Length, 7.50 and 8.80; alar extent, 10.00 and 12.25; wing, 3.17 and 3.90; tail, 3.30 and 3.93;

Measurements of Specimens of PIPILO ERYTHROPHTHALMUS *from Indian River, Florida.*

M. C. Z. No.	Collector's Number.	Sex.	Locality.	Date.	Collector.	Length.	Alar Extent.	Wing.	Tail.	Tarsus.
10722	2477	♂	Dummitt's	Feb. 26, '69	C. J. Maynard	8.10	10.25	3.25	3.60	1.00
——	2476	♂	"	Feb. 26, '69	"	8.00	10.50	3.20	3.65	.95
——	2531	♂	"	Mar. 5, '69	"	8.00	10.00	3.00	3 60	1.00
——	2530	♂	"	Mar. 5, '69	"	7.20	10.10	3.00	3.65	.90
——	2529	♂	"	Mar. 6, '69	"	8.30	10.20	3.25	3.70	1.00
——	2559	♂	"	Mar. 7, '69	"	7.90	9.85	3.42	3.50	1.02
10729	2669	♂	"	Mar. 12, '69	"	8.20	11.00	3.45	3.57	1.01
——	2592	♂	"	Mar. 11, '69	"	8.50	11.00	3.50	3.60	.87
——	2417	♂	"	Feb. 22, '69	"	8.00	10.00	3.05	3.70	.95
——	2416	♂	"	Feb. 22, '69	"	8.00	10.25	3.25	3.35	—
——	2394	♂	"	Feb. 25, '69	"	8.00	10.10	3 10	3.50	.95
——	2426	♂	"	Feb. 20, '69	"	7.78	10.00	3.00	3 25	.90
——	2514	♂	"	Mar. 2, '69	"	7.70	9.50	2.80	3.40	.80
——	2512	♂	"	Mar. 2, '69	"	7.25	10.25	3.10	3.70	.80
10724	2511	♂	"	Feb. 20, '69	"	7.75	10.50	3.20	3.50	1.09
10721	2395	♂	"	Feb. 18, '69	"	7.40	10.00	2.92	3.90	—
10729	2668	♂	"	Feb. 22, '69	"	7.40	10.75	3.00	3.45	.95
——	2044	♂	"	Feb. 17, '69	"	8.50	11.30	3.00	3 50	.95
10728	2516	♂	"	Mar. 2, '69	"	8.00	10.25	3 05	3 45	.90
——	2481	♀	"	Feb. 24, '69	"	7.50	9.48	3 00	3.60	.90
——	2483	♀	"	Feb. 20, '69	"	7.50	9.50	3.00	3.54	.90
——	2590	♀	"	Mar. 11, '69	"	7.65	10.05	3.00	3 75	.92
——	2593	♀	"	Mar. 11, '69	"	8.20	10 10	3.05	3 58	1.00
——	2591	♀	"	Mar. 11, '69	"	7.90	9.95	3.03	3 60	.92
——	2578	♀	"	Mar. 10, '69	"	7.40	9.75	3.00	3.80	.95
10727	2515	♀	"	Mar. 2, '69	"	7.50	9.75	3.10	3.05	.95
10725	2513	♀	"	Mar. 2, '69	"	7.35	10.50	3.35	3.10	.85
——	2445	♀	"	Feb. 18, '69	"	8.05	10.53	3 50	3.45	.95
10726	2514	♀	"	Mar. 2, '69	"	7.50	9 50	3 90	3.05	.95

Measurements of Specimens of PIPILO ERYTHROPHTHALMUS *from Eastern Massachusetts.*

M. C. Z. No.	Collector's Number.	Sex.	Locality.	Date.	Collector.	Length.	Alar Extent.	Wing.	Tail.	Tarsus.
——	384	♂	Newton	May 6, '68	C. J. Maynard	8.30	12 35	3.30	3.60	—
——	414	♂	Weston	May 9. '68	"	8.25	11.00	3 35	3 64	—
——	415	♂	"	May 9, '68	"	8 20	11.30	3 45	3 49	—
——	430	♂	Newton	May 11, '68	"	8 25	11:00	3.65	3.76	—
——	439	♂	Waltham	May 13, '68	"	8 45	11.30	3.51	3.85	—
——	539	♂	"	May 18, '68	"	8.35	11.23	3 46	3.55	—
——	1008	♂	Weston	July 10, '68	"	8.50	11.00	3.41	3.46	—
4616	1329	♂	Newton	Sept. 21, '68	"	8 80	11.65	3 42	3.70	1 10
——	1330	♂	"	Sept. 21, '68	"	8 51	11 41	3 51	4.00	1 00
4615	1295	♂	"	Sept. 17, '68	"	8 50	11 50	3.55	3.75	1.00
4725	496	♀	"	May 15, '68	"	7.60	10.45	3.31	3 35	—
4726	527	♀	Waltham	May 16, '68	"	8 25	11.76	3.35	3.60	—
——	——	♀	"	May 16, '68	"	8.26	10.55	3.55	3 65	—
4724	555	♀	Weston.	May 20, '68	"	8 00	10.56	3 25	3 36	—
4617	1028	♀	Newton.	July 22, '68	"	8.50	11.00	3.41	3.52	1 10
4613	1328	♀	"	Sept. 21, '68	"	8.45	11.00	3 35	3 65	1 05

tarsus, .98 and 1.13. Southern type, males: Length, 7.20 and 8.50; alar extent, 9.50 and 11.30; wing, 2.80 and 3.50; tail, 3.25 and 3.90; tarsus,

.80 and 1.09. The females in both cases average a little smaller than the males. The average dimensions of thirty northern males are as follows: Length, 8.19; alar extent, 11.32; wing, 3.43; tail, 3.66; tarsus, 1.06. Of nineteen southern males: Length, 7.88; alar extent, 9.88; wing, 3.13; tail, 3.56; tarsus, .94. The measurements given in the two preceding tables were all taken by Mr. Maynard from fresh specimens.

Other species of *Fringillidæ* that from their general distribution one naturally expects to meet with in East Florida in winter, but which, so far as I can learn, have not yet been met with there, are the Yellow-winged Sparrow (*Coturniculus passerinus*), Black-throated Bunting (*Euspiza americana*), Indigo Bird (*Cyanospiza cyanea*), and the Nonpariel (*C. ciris*). Specimens of the latter, collected at Cape Florida in winter, have been received at the Museum, and it was taken in April at Jacksonville and St. Augustine by Mr. Thurston and Mr. L. L. Thaxter.

ICTERIDÆ.

55.† **Molothrus pecoris** *Swainson.* Cow Blackbird.

Not numerous. Sometimes seen in small parties by themselves, but more frequently associating with the red-wings and grackles.

56.* **Agelæus phœniceus** *Vieillot.* Red-winged Blackbird.

Abundant. Apparently chiefly Florida born birds seen, especially in February and March. The sexes were usually in separate flocks.

The differences in respect to size and color between Florida and New England specimens usually seen in individuals of the same species from these localities are very marked in the present species, especially in respect to color. In no group, in fact, is it generally more so than in the *Icteridæ*.

In the Florida red-wings the general form is slenderer and more delicate, the bill relatively longer and more pointed, and the general color more intense and lustrous. The difference is particularly marked in the shoulder-patch, in which the red of its anterior portion is darker, approaching bright orange, and the posterior part, which in the northern bird is usually pale cream-color, whitish, or even nearly pure white, is orange-yellow, — very nearly as in the *A. gubernator* of the Pacific States. The difference in color, size, and especially in the form of the bill, is much greater than the differences existing between many currently received species of North American birds, and it is surprising that the two forms have not been specifically separated. I can only account for it on the supposition that specimens from Florida and the Gulf States have not fallen into the hands of the assiduous species hunters. As remarked in Part III (p. 234), Florida and New England specimens are as different from each other as are the so-called *Agelæus phœniceus* of the Northeastern States, the *A. tricolor* and the *A. gubernator* from each other.

Specimens of *A. phœniceus* from Louisiana I find correspond very nearly in every respect with the specimens from Florida. I have also before me one specimen from Maine with the shoulder-patch as highly colored, and with nearly as long a bill as is found in the specimens from Florida.

Plate VI shows the average form of the bill in Florida and Massachusetts specimens, and the annexed table of measurements the difference in general size. They also illustrate individual variation.

The following measurements of seventy specimens of this species from Massachusetts (forty males and thirty females), eighteen specimens from South Carolina and Florida (eleven males and seven females), and thirteen specimens from California (four males and nine females), exhibit, besides the average size and the individual variation at the same locality (especially in the case of those from Massachusetts), several interesting facts in respect to geographical variation. While the northern specimens (see the summary of these measurements given below) are somewhat larger than the southern ones, the latter have the longer head (including the bill), and also the longer bill. The height and width of the bill at the base remaining essentially the same in both, the southern ones have the bill relatively more attenuated. The difference in this respect is more striking than the measurements given seem to indicate. The California specimens closely resemble those from Florida, not only in respect to size, but in regard to the size and form of the bill, and also in respect to color; these, as well as the Florida ones, belonging to the southern type. As previously remarked, they bear a much closer resemblance to the Florida form in every respect than to that found in New England.*

The individual variation in this species seems to be very great everywhere, the variation in specimens of the same sex from the same locality being fully fifteen per cent of the average size at that locality.

* The affinities of *Agelæus gubernator* and *A. tricolor* with *A. phœniceus* are acknowledged to be exceedingly close. Professor Baird cites, in his Birds of North America, one specimen of the *A. phœniceus* from San José, California, and five from Fort Steilacoom, W. T. He also cites specimens of *A. gubernator* from Petaluma and San Francisco, Cal.; but Dr. Cooper regards this species as "limited to the interior of the State" (California), while those found along the coast, he says, clearly resemble the eastern bird. (Ornithology of California, Vol. I, p. 264.) From the close resemblance, already alluded to, of both the *A. gubernator* and *A. tricolor* to *A. phœniceus*, and their occurrence mainly in the hot valleys of California and the region more to the southward, I can scarcely doubt that these forms, especially *A. gubernator*, are the southern smaller, brighter colored, more attenuated billed western homologues of the similar eastern form from Florida and the Gulf States.

Measurements of Northern Specimens of AGELÆUS PHŒNICEUS.

M. C. Z. No.	Collector's Number.	Sex.	Locality.	Date.	Collector.	Length.	Alar Extent.	Wing.	Tail.	Head.	Bill.		
											Culmen.	Height	Width.
1274	—	♂	Vassalboro', Me.	——	Mr Becker	9 40	14.90	4.65	3.82	1.82	.87	.45	.37
9577	—	♂	Waterville, "	June 20, '64	C. E. Hamlin	9.45	14.75	4.82	3.88	1.77	.87	46	.40
391	—	♂	Malden, Mass.	1859	D. Higgins	9.35	14.75	4.67	3 75	1.76	.86	.47	.40
96	—	♂	" "	"	"	9.15	14.35	4.55	3.50	1.75	.80	.50	.37
392	—	♂	" "	"	"	9.20	14 75	4.70	3.66	1.82	.89	.47	.38
92	—	♂	" "	"	"	9.25	14.50	4.75	3.82	1.80	.92	.50	.40
93	—	♂	" "	"	"	9.20	14.40	4 50	3.45	1.85	.88	.45	.40
94	—	♂	" "	"	"	9.00	14.25	4.60	3.55	1.76	.90	.50	.38
95	—	♂	" "	"	"	9.00	15.10	4.65	3.65	1.83	95	.50	.40
393	—	♂	" "	"	"	9.00	14.55	4.65	3 35	1.82	84	45	.37
394	—	♂	" "	"	"	8.40	14.10	4.50	3.40	1.60	.75	46	.40
5727	—	♂	Concord, "	"	H. Mann	8 58	13.95	4.50	3.46	1.78	.92	.48	.40
5723	—	♂	" "	"	"	8 45	14.45	4.55	3 35	1.75	.90	.43	45
5724	—	♂	" "	"	"	8.40	14 25	4.43	3.12	1.80	.84	.45	38
5729	—	♂	" "	"	"	9.00	14.51	4.54	3.33	1.82	.85	.47	.39
5726	—	♂	" "	"	"	9.20	14.95	4.75	3.75	1.80	.87	.44	38
5720	—	♂	" "	"	"	9 05	15.00	4.86	3.86	1.84	.88	.44	.37
5721	—	♂	" "	"	"	9.70	14.88	4.68	3 80	1.77	.92	.46	.40
5722	—	♂	" "	"	"	9 00	14.50	4.65	3.65	1.74	83	.43	38
5725	—	♂	" "	"	"	9.85	14.25	4.73	3.76	1.85	.93	.43	.35
5728	—	♂	" "	"	"	9.25	14.50	4.60	3.81	1.78	.88	.44	.33
5732	—	♂	" "	"	"	9 58	14.62	4.74	3 85	1.83	.91	48	.38
10096	—	♂	Ipswich, "	June 14, '68	J. A. Allen	9.15	14.50	4.70	3.78	1.79	.92	44	.40
1674	—	♂	Springfield, "	June 26, '62	"	9.00	14.00	4.59	3.42	1.62	87	.45	.40
1675	—	♂	" "	June 26, '62	"	9.25	15 00	4.82	3.78	1.84	.93	.46	43
1782	—	♂	" "	July 12, '62	"	9 50	14.50	4.62	3 65	1.94	.97	.45	.37
1781	—	♂	" "	July 12, '62	"	9 00	14.60	4.50	3.38	1.80	.86	.43	.40
626	—	♂	Auburndale, "	Mar. 23, '57	S. Tenney	9 62	15.35	4.87	3.77	1.85	.91	.43	.40
1022	—	♂	Wenham, "	May —, '61	J. Bartlett	9.25	15.00	4.67	3.55	1.84	.89	.48	.43
——	114	♂	Newton, "	Mar. 13, '68	C. J. Maynard	8.75	14.83	4.70	3.53	—	—	—	—
——	180	♂	" "	Mar. 28, '68	"	9 38	15.00	4 75	3.60	—	—	—	—
——	214	♂	" "	Apr. 11, '68	"	8.90	14 85	4.60	3.52	—	—	—	—
——	251	♂	Weston, "	Apr. 18, '68	"	9.52	15.00	4.82	3.80	—	—	—	—
——	302	♂	Newton, "	Apr. 23, '68	"	9.00	15.00	4.66	3.65	—	—	—	—
——	323	♂	" "	Apr. 25, '68	"	9 00	15.00	4.85	3 40	—	—	—	—
——	352	♂	Weston, "	May 1, '68	"	9 16	15 10	4 80	3.70	—	—	—	—
——	371	♂	Newton, "	May 5, '68	"	9.00	15.00	5.00	3.90	—	—	—	—
——	351	♂	" "	May 1, '68	"	9.50	15 25	4 92	3.85	—	—	—	—
——	417	♂	Weston, "	May 9, '68	"	9.50	15 15	4.90	3.77	—	—	—	—
——	3047	♂	Newton, "	Mar. 23, '70	"	9.50	15.00	4.60	3.56	—	—	—	—
9843	—	♀	Milltown, Me.	——	G. A. Boardman	8.00	12.50	3.95	3.05	1.57	.73	.38	32
9844	—	♀	" "	——	"	8 00	12 50	4 08	3 10	1.54	.79	.40	.30
399	—	♀	Malden, Mass.	1859	D. Higgins	7 50	11.75	3.60	2.70	1 48	.72	.42	33
396	—	♀	" "	"	"	7.75	12.30	3 90	3.06	1 60	.82	43	.35
97	—	♀	" "	"	"	7.35	11.75	3 63	2.80	1 55	.70	.40	.37
402	—	♀	" "	"	"	8 55	13.55	4.26	3.15	1.66	.75	.44	.41
398	—	♀	" "	"	"	8.05	13.50	4.16	3.10	1.67	.77	.43	36
98	—	♀	" "	"	"	7.75	12.10	3.74	2.97	1 58	72	.37	33
403	—	♀	" "	"	"	7.42	11.55	3.87	2.73	1 62	.73	.37	35
397	—	♀	" "	"	"	7.50	11.25	3.70	2.98	1.58	76	34	33
395	—	♀	" "	"	"	7.40	12.50	4.11	2.68	1.50	70	.38	.38
99	—	♀	" "	"	"	7.45	11.50	3.75	2.90	1.55	.75	.42	.43
5730	—	♀	Concord, "	——	H. Mann	7.75	12.50	3 80	3.02	1.54	.70	.40	.33
1611	—	♀	Springfield, "	July 15, '62	J. A. Allen	7.50	12.00	3.73	2.90	1.68	.79	.40	.34
1639	—	♀	" "	June 26, '62	"	7.65	11.75	3.67	2.82	1.54	.70	.38	.37
1672	—	♀	" "	June 26, '62	"	7.85	11.82	3.75	2.89	1.60	.75	.40	.38
1673	—	♀	" "	June 26, '62	"	7.75	12 00	3.77	2.85	1.57	.78	.38	.37
1679	—	♀	" "	June 26, '62	"	8.00	12.25	3.79	3.00	1 55	.74	.39	36
1680	—	♀	" "	June 26, '62	"	8.00	12.10	3.85	2.93	1.53	.78	.38	—
——	350	♀	Weston, "	Apr. 30, '68	C. J. Maynard	7 73	12.61	3.95	2 95	—	—	—	—
——	850	♀	Ipswich, "	June 15, '68	"	7.45	13.60	4.00	3.05	—	—	—	—
——	891	♀	Essex, "	June 17, '68	"	8.00	12.35	4.00	3.00	—	—	--	—
——	893	♀	" "	June 17, '68	"	7.75	12.54	3.90	2.90	—	—	—	—
——	1093	♀	Waltham, "	Aug. —, '68	"	7.45	12.37	3.80	2.75	—	—	—	—
——	1075	♀	" "	Aug. —, '68	"	7.67	12.30	3.85	2.85	—	—	-	—
——	1096	♀	" "	Aug. —, '68	"	7.50	12 40	3.95	2.72	—	—	—	—
——	1097	♀	Newton, "	Aug. —, '68	"	7.75	12.30	3.71	2.80	—	—	—	—
——	1098	♀	Waltham, "	Aug. —, '68	"	7.50	12.00	3.85	2.65	—	—	—	—
——	1099	♀	Weston, "	Aug. —, '68	"	7.50	12.10	3.67	2.80	—	—	—	—
——	2830	♀	Newton, "	June 8, '69	"	7.50	12.00	4.00	2.90	—	—	—	—

Measurements of Southern Specimens of AGELÆUS PHŒNICEUS.

M. C. Z. No.	Coll. No.	Sex.	Locality.	Date.	Collector.	Length.	Al. Ext.	Wing.	Tail.	Head.	Bill. Cul.	Bill. Hgt.	Bill. Wid.
4126	—	♂	Charleston, S. C.	———	L. Agassiz	9.55	14.75	4.75	3.65	1.90	1.00	.45	.37
4127	—	♂	" "	———	"	8.80	14.30	4.50	3.55	1.74	.87	.47	.40
4128	—	♂	" "	———	"	9.45	14.50	4.60	3 72	1.80	.90	.50	.40
4129	—	♂	" "	———	"	9.05	13.50	4.37	3.45	1.78	.85	.43	.42
4125	—	♂	" "	———	"	9.05	14.12	4.42	3.35	1.94	.95	.46	.35
——	—	♂	Hawkinsville, Fla.	Mar. 15,'69	J. A. Allen	8.25	13.60	4.34	—	—	—	—	—
——	1928	♂	Jacksonville, "	Dec. 31,'69	C. J. Maynard	9 10	14 90	4 75	3.58	—	—	—	—
10563	1929	♂	" "	Dec. 31,'69	"	9.20	14.80	4.80	3.90	—	—	—	—
10561	2013	♂	" "	Dec. 31,'69	"	8.80	14 15	4.55	3 58	—	—	—	—
10574	2552	♂	Dummitt's, "	Mar. 8,'69	"	9 50	14.20	4.75	3.90	—	—	—	—
10573	2450	♂	" "	Feb. 24,'69	"	8.50	14.00	4.75	3.45	—	—	—	—
5153	—	♀	Hibernia, "	Jan. 30,'69	J. A. Allen	7 65	12.60	3.85	3.05	—	—	—	—
5154	—	♀	" "	Jan. 30,'69	"	7.85	12.50	3.90	3 07	—	—	—	—
5155	—	♀	" "	Jan. 30,'69	"	7.80	12.85	—	3 20	—	—	—	—
4141	—	♀	" "	Jan. 30,'69	"	8.00	12.25	3 80	3.05	—	—	—	—
5209	—	♀	Welaka, "	Feb. 8,'69	"	7.65	12.50	3 75	2 80	—	—	—	—
5208	—	♀	" "	Feb. 8,'69	"	7.50	11 85	3 63	2 75	—	—	—	—
5210	—	♀	" "	Feb. 8,'69	"	7.65	12.55	3.95	—	—	—	—	—

Measurements of California Specimens of AGELÆUS PHŒNICEUS.

M. C. Z. No.	Sex.	Locality.	Date.	Collector.	Length.	Al. Ext.	Wing.	Tail.	Tarsus.
5885	♂	San Francisco, Cal.	Winter '59-'60	A. Agassiz	8.50	14.98	4.83	3.50	1.63
5884	♂	" "	Winter '59-'60	"	8.75	15 05	4 95	3.35	1.74
566	♂	" "	Winter '59-'60	"	8.60	14 55	4.47	3 09	1.90
2188	♂	Gulf of Georgia, W.T.	Sept. —, '60	"	8.71	13.50	4.45	3.26	1.75
5889	♀	San Francisco, Cal.	Winter '59-'60	"	7 58	12 80	4.03	2.73	1.63
5893	♀	" "	Winter '59-'60	"	7.55	12.35	3.95	2.47	1.46
5887	♀	" "	Winter '59-'60	"	7.81	12.80	4.25	3.86	1.54
5890	♀	" "	Winter '59-'60	"	7 50	12.75	3 94	2.47	1.56
5886	♀	" "	Winter '59-'60	T. G. Cary	7.82	12.77	4.04	2.62	1.56
2075	♀	" "	Winter '59-'60	"	8.29	13.27	4 32	3.00	1.62
2074	♀	" "	Winter '59-'60	"	8.18	13.25	3.85	2.95	1 67
2078	♀	" "	Winter '59-'60	"	8.50	13.00	4.15	3.10	1 65
5888	♀	" "	Winter '59-'60	A. Agassiz	7.25	12.25	3 90	3.71	1.50

Summary of the above Measurements of Specimens of AGELÆUS PHŒNICEUS.

Locality.	Sex.	No. of Specim's.		Length.	Al. Ext	Wing.	Tail.	Head.	Culmen	Height.	Width.
Massachusetts	♂	40	Aver.	9.16	14.71	4 69	3.63	1.79*	.88*	.46*	.39*
	♀	28	Aver.	7.53	12.24	3.86	2 93	1.57†	.75†	.395†	.357†
South Carolina and Florida	♂	11	Aver.	9.02	14.41	4.62	3.61	1.83	.91‡	.46‡	.39‡
	♀	7	Aver.	7.73	12 44	3.83	2.99	—	—	—	—
California	♂	7	Aver.	8.64	14.52	4.67	3.30	1.75	—	—	—
	♀	9	Aver.	7.83	12.70	4.00	2.99	1 57	—	—	—
Massachusetts	♂	40	Max.	9.85	15.35	5.00	3.90	1.94*	.97*	.50*	.45*
	♂	40	Min.	8.40	13.95	4.43	3.12	1.60*	.75*	.43*	.33*
	♀	28	Max.	8 55	13.55	4.26	3.15	1.68†	.82†	.44†	.43†
	♀	28	Min.	7.35	11.25	3.63	2 65	1.48†	.70†	.37†	.30†
South Carolina and Florida	♂	11	Max.	9.55	14 90	4.80	3.90	1.94‡	1.00‡	.50‡	.40‡
	♂	11	Min.	8.25	13.60	4 34	3.35	1.74‡	.85‡	.43‡	.35‡
	♀	7	Max.	8.00	12.85	3 90	3.20	—	—	—	—
	♀	7	Min.	7.50	11.85	3.63	2.75	—	—	—	—
California	♂	7	Max.	8.75	15.05	4.95	3.50	1 90	—	—	—
	♂	7	Min.	8.50	13 50	4.45	3.09	1.63	—	—	—
	♀	9	Max.	8.50	13.27	4 32	3.86	1.67	—	—	—
	♀	9	Min.	7.25	12 25	3.85	2 47	1 46	—	—	—

* 29 specimens. † 19 specimens. ‡ 5 specimens.

57.* **Sturnella ludoviciana** *Swainson.* MEADOW LARK.

Alauda magna LINNÉ, Syst. Nat., I, 167, 1758. — WILSON, Am. Orn., III, 20, pl. xix, 1811.

Sturnus ludovicianus LINNÉ, Syst. Nat., I, 290, 1766. — BONAP., Journ. Phil. Acad. Nat. Sci., IV, 180, 1824. — NUTTALL, Man. Orn., I, 147, 1832. — AUDUBON, Orn. Biog., II, 216, 1834.

Sturnus collaris WAGLER, Syst. Avium, I, 1827.

Sturnella ludoviciana SWAINSON, Faun. Bor. Am., II, 282, 1831. — BONAP., Geog. and Comp. List, 1838. — AUDUBON, Synop. Am. Birds, 148, 1839. — CABANIS, Mus. Hein., 192, 1851. — SCLATER, Cat. Am. Birds, 139, 1862. — CASSIN, Proc. Phil. Acad. Nat. Sci., 1866, 23.

Sturnella magna SWAINSON, Phil. Mag., I, 436, 1827. — BAIRD, Birds N. Am., 535, 1858. — ALLEN, Mem. Bost. Soc. Nat. Hist., I, 496, 1868.

Sturnella collaris VIEILLOT, Analyse, 1816.

Sturnella hippocrepis WAGLER, Isis, 1832, 281. — LAWRENCE, Ann. N. York Lyceum N. Hist., VII, 266, 1860. — SCLATER, Ibis, 1861, 79. — CASSIN, Proc. Phil. Acad. Nat. Sci., 1866, 24.

Sturnella neglecta AUDUBON, Birds of Am., VII, 339, pl. cccclxxxvii, 1843. — BAIRD, Birds of N. Am., 537, 1858. — CASSIN, Proc. Phil. Acad. Nat. Sci., 1866, 23.

Sturnella mexicana SCLATER, Ibis, 1861, 79. — CASSIN, Proc. Phil. Acad. Nat. Sci., 1866, 24.

Sturnella meridionalis SCLATER, Ibis, 1861, 79. — CASSIN, Proc. Phil. Acad. Nat. Sci., 1866, 24.

Abundant. Found chiefly in the moister parts of the pineries.

Somewhat smaller than in the Northern States, but in most cases with longer and larger bills, brighter colors, and a quite different song. The latter somewhat resembles that of the western meadow lark, but is still as distinct from it in its general character as it is from that of the New England bird. The present species has a wide geographical range, throughout the greater part of which it is resident. The Alleghanian fauna forms its northern limit, from which it mostly retires during winter. To the southward it extends to Cuba and the other larger West India Islands, throughout most of Central America, and to the elevated parts of Northern South America. It ranges westward over the elevated arid plains of the middle of the continent to the Pacific. As might be expected, it is not quite uniform in its characters at all points. The main differences, however, consist merely in the lighter color of those from the plains, and the smaller size of those from the south. The former constitute the *Sturnella neglecta* of Audubon and most other writers since his time. In Cuba it is the *S. hippocrepis* of Wagler and others, and the Mexican and Guatemalan form is the *S. mexicana* of Sclater, and the South American form the *S* *meridionalis* of the same author. Yet the distinctions between them are trivial, all of these so-called species having been generally looked upon as doubtfully distinct from the *S. ludoviciana* of the United States, especially the three last named. The *S. collaris* of Vieillot has very generally been referred by subsequent writers to the *S. ludoviciana.* The main distinctive feature of the *S. neglecta* has been its song, — a very doubtful basis on which to found a species. The Florida specimens are intermediate in size and other characters between the Cuban and New England representatives of this species. As already remarked, the song of the Florida birds is as widely different from that of the New England bird as the song of the latter is from that of the western ones. Concerning the affinities of *S. neglecta* I have already remarked.* Concerning those of the other supposed species, I may well borrow the appropriate remarks of the late Mr. Cassin, who observes in respect to them, in his "Study of the *Icteridæ*,"† as follows : —

"This bird [*Sturnella ludoviciana*] is nearly related to the next four species of this genus [*S. neglecta*, *S. hippocrepis*, *S. mexicana*, *S. meridionalis*], equally in structure and in colors, and it would be difficult to describe by positive characters either species of this group, so as to insure recognition absolutely, or without comparative characters being given. . . . . No other genus or sub-genus of this family presents so many species of such uniformity of structure and similarity of color, and there are, assuredly, few such in the kingdom of birds." Under *S. neglecta* he further remarks in respect to the transition that is so apparent between it and *S ludoviciana :* "In the central regions of North America it is possible that a hybrid race between the two species may be produced, to be referred with about equal propriety to either." *S. hippocrepis*, he says, is very nearly related to *S. mexicana*, "and can scarcely be distinguished from it by any characters which seem to be reliable." He thinks it to be somewhat more distinct, however, from *S. neglecta.* Mr. Lawrence had previously remarked that the *S. hippocrepis* is somewhat smaller than *S. ludoviciana* of the United States, and that he "thinks it is specifically distinct"; although he adds, "it would be difficult to point out any reliable differences in coloration, especially of the upper plumage, as individuals even of the same species are very variable."‡ He says, further, that specimens of it from Jalapa, Mexico, differ "only in the pectoral band appearing broader in the Mexican bird, and the tertials much shorter than the primaries, but this last may not be a reliable character." In the

* See Memoirs of the Boston Soc. Nat. Hist., Vol. I, p. 494, 1868.

† Proceedings of the Phil. Acad. Nat. Sciences, 1866, p. 23.

‡ Annals of New York Lyceum of Nat. Hist., Vol. VII, 266, 1860.

following year, however, Mr. Sclater separated the Mexican bird from those of Cuba and the United States, under the name *S. mexicana*, and also the South American under the name *S. meridionalis*. Mr. Cassin says of the latter: "Very nearly related to the preceding (*S. hippocrepis*), if distinct, and I give it, at present, as a species provisionally only. . . . . The colors of the upper parts seem to be less clearly defined, and of a slightly different style and pattern from the preceding, and it may bear about the same relation to that species (*S. hippocrepis*) that *S. neglecta* does to *S. ludoviciana*. Such relation I hold to be rather probable from the specimens now at hand."

Having given the views of the describers of these several "species," I may add that I have seen examples of each, and do not question that they should all be referred to one. As is evident from the above-quoted remarks, these different species gradually pass into each other, — the *S. magna* into the *S. neglecta*, the *S. neglecta* into the *S. mexicana*, and the *S. mexicana* into the *S. hippocrepis*, which is their exact geographical relation.

In regard to the Florida specimens, as compared with New England ones, the most striking differences consist in their smaller size and much brighter colors, especially of the ventral surface.

The following tables of measurements indicate the individual and sexual differences in size, and also the difference in size between specimens from the Northern States and from Florida.

Measurements of Northern Specimens of STURNELLA LUDOVICIANA.

M. C. Z. No.	Coll. No.	Sex.	Locality.	Date.	Collector.	Length.	Al. Ext.	Wing	Tail.
4862	416	♂	Newton, Mass.	May 8, '68	C. J. Maynard	10.75	16.59	5.13	3.50
——	1100	♂	Waltham, "	Aug. 6, '68	"	10 20	16.30	4 98	2.90
4863	1134	♂	" "	Aug. 19, '68	"	10.25	15.85	4.80	2.92
——	2696	♂	Newton, "	May 15, '69	"	10 75	16 75	5.00	3.35
——	2698	♂	" "	May 15, '69	"	11.00	17.00	5.15	3.35
——	1700	♂	Waltham, "	Aug. 6, '68	"	10 20	16.30	4.98	2 90
——	4045	♂	Newton, "	Aug. 2, '69	"	11.00	17.00	5.00	3 40
——	4061	♂	" "	Aug. 2, '69	"	11.00	16.00	4.80	3.35
——	2738	♀	" "	May 15, '69	"	9.75	15.00	4.55	2.65
362	——	♀	Malden, "	——	D. Higgins	9.25	13.50	4.17	2.50
363	——	♀	" "	——	"	9.58	14.00	4.35	2.90
364	——	♂	" "	——	"	10.50	15.33	4.82	3 11
365	——	♂	" "	——	"	10.35	15.65	4.83	3.13
366	——	♂	" "	——	"	10.00	15 05	4.75	3 14
367	——	♂	" "	——	"	10.75	15 50	4.82	3 30
568	——	♀	" "	——	"	8.90	14.00	4.35	2.60
569	——	♂	" "	——	"	9.50	15.68	5.00	3.05
9764	——	♀	Evanston, Ill.	——	O. Marcy	9.25	13.92	4.15	2 82
9765	——	♀	" "	——	"	9.75	14.65	4.50	3 10
9766	——	♂	" "	——	"	10.00	15.50	4.74	3.15
2646	——	♀	Lawn Ridge, "	——	K. Butler	9.60	14.75	4.55	2.84
4042	——	♂	Concord, Mass.	——	F. C. Brown	10 25	16.00	4.77	2.83
4102	——	♀	" "	——	"	10.33	15.65	4.67	3.08

Measurements of Florida Specimens of STURNELLA LUDOVICIANA.

M. C. Z. No.	Coll. No.	Sex.	Locality.	Date.	Collector.	Length.	Al. Ext.	Wing.	Tail.
—	2067	♂	Jacksonville	Jan. 20, '69	C. J. Maynard	9.55	15.60	4.50	2.85
—	2817	♂	Dummitt's	May 15, '69	"	10.20	15.10	4.50	3.20
—	2816	♂	"	May 15, '69	"	10.00	15.15	4.60	2.95
5335	—	♂	Enterprise	Mar. 4, '69	J. A. Allen	9.75	14.75	4.50	—
5336	—	♂	"	Mar. 4, '69	"	9.85	15.20	4.40	2.89
5337	—	♂	"	Mar. 4, '69	"	9.70	14.80	4.45	2.82
5368	—	♂	Hawkinsville	Mar. 12, '69	"	9.75	15.00	4.50	3.05
5369	—	♂	"	Mar. 12, '69	"	9.50	14.75	4.25	4.83
5371	—	♂	"	Mar. 12, '69	"	10.00	15.75	4.50	3.07
—	—	♀	Volusia	Mar. 17, '69	"	8.75	13.75	4.05	—
5370	—	♀	Hawkinsville	Mar. 12, '69	"	8.90	14.15	4.10	2.65
5372	—	♀	"	Mar. 12, '69	"	9.50	14.65	4.20	2.90
5125	—	♀	Jacksonville	Jan. 19, '69	"	8.75	14.25	4.20	2.70
—	2072	♀	"	Jan. 20, '69	C. J. Maynard	8.75	14.00	4.40	2.50
—	2070	♀	"	Jan. 20, '69	"	8.50	13.55	3.90	2.55
—	2070	♀	"	Jan. 20, '69	"	8.75	13.00	4.00	2.55
—	2068	♀	"	Jan. 20, '69	"	9.25	14.75	4.50	2.80
—	2069	♀	"	Jan. 20, '69	"	8.76	14.25	4.20	2.40
—	2071	♀	"	Jan. 20, '69	"	9.00	14.00	4.30	2.80
—	2051	♀	"	Jan. 20, '69	"	9.50	14.75	4.65	2.50
—	2791	♀	"	Apr. 15, '69	"	9.05	14.00	4.10	2.88

The following is a tabulated summary of the two preceding tables: —

No. of Specimens.	Sex.	Locality.		Length.	Alar Extent.	Wing.	Tail.
15	♂	Northern States	Average	10.43	16.30	4.91	3.16
8	♀	"	Average	9.55	14.43	4.29	2.82
15	♂	"	Maximum	11.00	17.00	5.15	3.50
15	♂	"	Minimum	10.00	15.05	4.74	2.83
8	♀	"	Maximum	9.75	15.65	4.55	3.10
8	♀	"	Minimum	8.90	13.50	4.15	2.50
9	♂	Florida	Average	9.81	15.70	4.47	2.85
12	♀	"	Average	8.93	14.09	4.22	2.57
9	♂	"	Maximum	10.20	15.75	4.60	3.20
9	♂	"	Minimum	9.50	14.75	4.25	2.82
12	♀	"	Maximum	9.50	14.75	4.65	2.90
12	♀	"	Minimum	8.50	13.00	3.90	2.40

58.† **Scolecophagus ferrugineus** *Swainson.* RUSTY GRACKLE.

Abundant. Occasionally met with in large flocks.

59.* **Quiscalus purpureus** *Cassin.* PURPLE GRACKLE.

Gracula quiscula LINNÉ, Syst. Nat., I, 165, 1766. — WILSON, Am. Orn., III, 44, pl. xxi, fig. 4, 1811.

Gracula barita LINNÉ, Syst. Nat., 165, 1766. — ORD, Journ. Phil. Acad. Nat. Sci., I, 253, 1818.

Gracula purpurea BARTRAM, Travels, 289, 1791. (No description.)

? *Oriolus ludovicianus* GMELIN, Syst. Nat., 387, 1788.

Quiscalus baritus VIEILLOT, Nouv. Dict., XXVIII, 487, 1819. — BAIRD, Birds North Amer., 556, pl. xxvii, 1858. — CASSIN, Proc. Phil. Acad. Nat. Sci., 1866, 405.

Quiscalus versicolor VIEILLOT, Nouv. Dict., XXVIII 488, 1819. — BONAPARTE, SWAINSON, NUTTALL, AUDUBON, BAIRD.

Quiscalus purpureus CASSIN, Proc. Phil. Acad. Nat. Sci., 1866, 403. — RIDGWAY, Ibid., 1869, 133.

Quiscalus purpuratus SWAIN., Lardner's Cab. Cyclop., 299, 1838 (female).

? *Quiscalus lugubris* SWAIN., Lardner's Cab. Cyclop., 299, 1838. — ? CASSIN, Proc. Phil. Acad. Nat. Sci., 1866, 408.

Quiscalus inflexirostris SWAIN., Lardner's Cab. Cyclop., 300, 1838. — CASSIN, Proc. Phil. Acad. Nat. Sci., 1866, 407.

Quiscalus crassirostris SWAIN., Lardner's Cab. Cyclop., 355, 1838. — GOSSE, Birds of Jamaica, 217, 1847.

Quiscalus aglæus BAIRD, Amer. Journ. Sci. and Arts, XLI, 87, 1866. — CASSIN, Proc. Phil. Acad. Nat. Sci., 1866, 404. — RIDGWAY, Ibid., 1869, 135.

Quiscalus æneus RIDGWAY, Ibid., 134.

Quiscalus mexicanus CASSIN, Ibid., 1866, 408.

Quiscalus Gundlachii CASSIN, Ibid., 406.

Quiscalus brachypterus CASSIN, Ibid., 406.

Quiscalus niger CASSIN, Ibid., 407.

? *Quiscalus rectirostris* CASSIN, Ibid., 409.

Chalcophanes quiscalus WAGLER, Syst. Avium, 1827. — CABANIS, Mus. Hein., 197, 1851.

Chalcophanes baritus WAGLER, Syst. Avium, 1827. — CABANIS, Mus. Hein., 197, 1851

Very abundant everywhere. Flocks containing many hundreds were frequently met with.

As already remarked in Part III, few species present such marked climatic variations as the present, or better illustrate the three principal laws of geographical variation already enumerated; namely, a decrease in general size from the north southward, and at the same time an increase in the length and slenderness of the bill, and an increase in the intensity and brilliancy of the color of the plumage. Far to the north, as in Labrador, the colder parts of Canada, and Northern New England, the bill is shortest and thickest, the size of the bird at its maximum, and the colors of the plumage least brilliant, with the metallic reflections of a light tint, tending to green rather than to blue. In Southern New Jersey the change from the northern type is already considerable; even between summer specimens from Calais (Maine) and Eastern Massachusetts there is an appreciable difference. In the lowlands of South Carolina and Georgia the divergence from the northern type is still greater, and it goes on rapidly increasing in Florida, especially in South Florida, the maximum of divergence from the northern type being attained in the West Indies. In East Florida, while the general size of the bird is less than in New England, the bill is considerably longer, much slenderer and much more decurved, as is shown by the accompanying figures (Plate VII). The change in color is equally marked. Not only do the reflections become much darker at the south, but form prismatic bars across the interscapularies and the feathers of the rump, especially in the South Atlantic States. In South Florida and the West Indies these prismatic bars, in some specimens at least, seem to lose their distinctness, evidently through the continued darkening or increased intensity of the general color. The difference in size between Florida and Massachusetts specimens is considerable, especially between those from South Florida and Massachusetts. Those from the West Indies are still smaller; and in comparing specimens of these with others from Northern New England, the difference is so striking that it seems impossible at first to believe that both can belong to the same species, yet a gradual transition between the two, through the individuals inhabiting the intermediate region, fully proves it. Even between Florida and New England specimens the difference is so great that, were there no transition from one to the other, the two extremes might well be regarded as not only valid species, but as well-marked ones. Being familiar with the so-called *Quiscalus aglæus* before visiting Florida, through specimens in the Museum of Comparative Zoölogy from Cape Florida, I had no doubt that it was a species distinct from the *Q. purpureus.* But a subsequent study of these birds in Florida, and an examination of specimens from various points between Florida and Northern Maine, and also from the West Indies, has forced me to the conclusions indicated in the above table of synonymes.

The purple grackles of the Mississippi Valley have recently been separated as specifically distinct from those of the Atlantic States, under the name *Q. æneus*, *Q. purpureus* being retained for the latter. The range of *Q. purpureus* is given as "Atlantic and Gulf? States, north to Nova Scotia, west to the Alleghanies." The New England type, however, is entirely referable to the *Q. æneus*, as defined by its describer. The same writer also follows some of his predecessors in separating those of South Florida from the *Q. purpureus*, under the name of *Q. aglæus.* But Cape Florida specimens differ but little — being, in fact, scarcely distinguishable except in size — from those from the St. John's River.

Mr. Cassin, in one of his latest papers,* took the ground that each of the larger West India Islands has a distinct species of this group, peculiar to itself. That these forms, many of them evidently difficult of recognition, should be distinct species is quite contrary to general principles. These islands are generally separated by a distance of rarely more than a hundred miles; yet a near ally of these "species," the *Q. purpureus* (or *Q. æneus* as recently restricted), is admitted to range from the Gulf of Mexico

* "A Second Study of the Icteridæ," Proc. Phil. Acad. Nat. Sci., 1866, pp. 403–417.

to the arctic regions, so that those that breed farthest north make annually a journey of fully a thousand miles to reach their breeding-grounds. As I have already observed, individuals of species possessing a very northern habitat usually present a great uniformity of character, while those of species ranging farther to the southward are more variable; also that within the warm-temperate and tropical latitudes, islands but slightly separated from each other or the mainland, and peninsulas which, like Florida, are almost insular in their geographical relations, present each peculiar modifications of species ranging throughout not only all of them, but portions of the adjoining continents, which render the individuals from these different localities more or less readily distinguishable. This results partly, doubtless, from the isolation of these different districts, partly from the more sedentary habits of birds in warm countries, as compared with those of cold latitudes, and partly from the greater tendency to variation in species inhabiting tropical and sub-tropical countries.

In the subjoined tables measurements are given of thirteen males and eight females from the Northern States, and of twenty-three males and seven females from Florida, of which the following is a tabulated summary: —

No. of Specimens.	Sex.	Locality.		Length.	Alar Extent.	Wing.	Tail.
13	♂	Northern States	Average	12.63	17.73	5.66	5.30
8	♀	"	Average	11.45	15.76	4.94	4.49
13	♂	"	Maximum	13.50	18.43	6.05	6.00
13	♂	"	Minimum	12.00	17.00	5.20	4.58
8	♀	"	Maximum	12.05	16.30	5.20	4.85
8	♀	"	Minimum	10.90	15.38	4.60	4.10
23	♂	Florida	Average	12.19	16.64	5.42	5.22
7	♀	"	Average	11.12	14.86	4.75	4.55
23	♂	"	Maximum	13.00	17.80	5.75	5.50
23	♂	"	Minimum	11.00	15.25	5.00	4.55
7	♀	"	Maximum	11.75	16.75	5.00	4.77
7	♀	"	Minimum	10.25	13.75	4.50	4.45

Measurements of Northern Specimens of QUISCALUS PURPUREUS.

M. C. Z. No.	Coll. No.	Sex.	Locality.	Date.	Collector.	Length.	Alar Extent.	Wing.	Tail.
1234	—	♂	Red River, B. A.	—	S. H. Scudder	12.50	17.50	5.65	5.68
—	187	♂	Watertown, Mass.	Apr. 3, '68	C. J. Maynard	12.50	17.80	5.77	5.20
—	186	♂	" "	Apr. 3, '68	"	12.45	18.25	5.85	5.30
—	185	♂	" "	Apr. 3, '68	"	12.80	17.53	5.57	5.43
—	—	♂	" "	Mar. 28, '68	"	12.80	18.10	5.86	5.50
—	3097	♂	Ipswich, "	Aug. 28, '69	"	13.10	18.00	5.85	5.60
9768	—	♂	Evanston, Ill.	—	O. Marcy	12.40	17.75	5.62	5.07
9770	—	♂	" "	—	"	12.25	17.05	5.42	4.87
2643	—	♂	Lawn Ridge, "	—	K. Butler	12.50	17.25	5.60	5.20
1401	—	♂	" "	—	"	13.50	18.43	6.05	6.00
1871	—	♂	Springfield, Mass.	July 29, '62	J. A. Allen	12.48	17.25	5.50	5.00
1874	—	♂	" "	July 29, '62	"	12.12	17.00	5.20	4.58
2574	—	♂	" "	July 29, '62	"	12.77	17.50	5.68	5.50
1602	—	♀	" "	July 12, '62	"	11.47	15.75	5.00	4.37
1873	—	♀	" "	July 29, '62	"	10.90	15.45	4.75	4.10
9769	—	♀	Evanston, Ill.	—	O. Marcy	11.30	15.38	4.60	4.25
9767	—	♀	" "	—	"	11.48	15.67	4.95	4.40
9598	—	♀	Waterville, Me.	—	C. E. Hamlin	11.50	16.00	5.00	4.65
2284	—	♀	" "	May 3, '62	"	11.40	16.00	4.98	4.46
2271	—	♀	" "	June 9, '62	"	12.05	16.30	5.20	4.85
2501	—	♀	Lynn, Mass.	—	S. Jillson	11.50	15.50	5.05	4.85

Measurements of Florida Specimens of QUISCALUS PURPUREUS.

M. C. Z. No.	Coll. No.	Sex.	Locality.	Date.	Collector.	Length.	Alar Extent.	Wing.	Tail.
5201	—	♂	Welaka	Feb. 6, '69	J. A. Allen	12.75	17.80	5.65	5.20
5202	—	♂	"	Feb. 6, '69	"	12.40	16.87	5.45	5.00
5264	—	♂	Hawkinsville	Feb. 18, '69	"	12.80	17.15	5.50	5.37
5251	—	♂	"	Feb. 18, '69	"	11.75	17.00	5.50	4.87
5266	—	♂	"	Feb. 18, '69	"	13.00	17.60	5.55	5.40
5267	—	♂	"	Feb. 18, '69	"	11.50	16.70	5.50	—
—	—	♂	Enterprise	Feb. 21, '69	"	12.85	16.87	—	—
5345	—	♂	"	Mar. 5, '69	"	12.30	17.38	5.50	5.25
5346	—	♂	"	Mar. 5, '69	"	12.37	16.60	5.20	5.25
10604	2583	♂	Dummitt's	Mar. 9, '69	C. J. Maynard	12.40	17.50	5.55	—
—	2344	♂	"	Feb. 26, '69	"	12.25	17.50	5.75	5.50
10602	2469	♂	"	Feb. 25, '69	"	12.50	17.30	5.50	5.55
—	2470	♂	"	Feb. 25, '69	"	12.50	17.00	5.50	5.00
10603	2471	♂	"	Feb. 25, '69	"	11.75	16.75	5.75	5.55
6848	—	♂	Cape Florida	Mar. 31, '58	G. Wurdemann	11.50	15.50	5.50	5.12
6851	—	♂	"	Apr. 10, '58	"	12.00	16.25	5.25	5.00
6852	—	♂	"	Apr. 22, '58	"	11.75	16.25	5.75	5.00
—	10335*	♂	"	Mar. 31, '58	"	11.50	16.00	5.50	—
—	10336*	♂	"	Apr. 15, '58	"	11.50	15.25	5.00	—
—	10337*	♂	"	Apr. 15, '58	"	12.00	15.50	5.00	—
—	10340*	♂	"	Apr. 22, '58	"	12.00	16.50	5.12	—
—	10341*	♂	"	Apr. 9, '58	"	11.00	15.25	5.25	—
—	10342*	♂	"	May 18, '58	"	11.75	16.25	5.00	—
—	2342	♀	"	Feb. 26, '69	"	11.50	15.50	5.00	4.77
—	2344	♀	"	Feb. 26, '69	"	11.00	15.50	5.00	4.40
10601	2468	♀	"	Feb. 26, '69	"	11.50	16.00	4.50	4.60
5263	—	♀	Hawkinsville	Feb. 18, '69	J. A. Allen	11.45	15.25	4.85	4.55
6853	—	♀	Cape Florida	Apr. 22, '58	G. Wurdemann	11.00	14.50	4.50	4.45
—	10338*	♀	"	Apr. 22, '58	"	11.12	14.50	4.75	—
—	10339*	♀	"	Mar. 31, '58	"	10.25	13.75	4.75	—

The specimens from Cape Florida are considerably smaller than those from the St. John's River; but the same difference occurs in other species between specimens from these two localities. The Cape Florida specimens of *Quiscalus purpureus* differ from others from North Florida also in having a relatively longer, slenderer, and more decurved bill, but not appreciably in color.

60.* **Quiscalus major** *Vieillot.* BOAT-TAILED GRACKLE.

Abundant. Particularly numerous along the St. John's River. According to Dr. Bryant they breed about the first of April. He says that about Lake Monroe some of the birds, as late as the 6th of April, had not commenced laying, "though the majority had hatched, and the young of others were almost fledged." † He notes also their sandpiper-like habit of running along the edge of the water. At Lake Dexter I observed great numbers of them walking on the floating aquatic plants.

The females of this species present very singular variations in color. Of four specimens collected at Lake Dexter, in March, one is pale ashy-

* Smith. Inst. No. Copied from Baird's Birds of North America, p. 557.

† Proc. Bost. Soc. Nat. Hist., Vol. VII, p. 9, January, 1859.

brown below, on the throat and breast nearly white, and dull dusky-brown above; while another is deep reddish-brown below and proportionally darker above, and the others are intermediate to these.

Between the two extremes there is more difference than usually obtains between valid congeneric species. The series of twenty-four males, on the other hand, are quite uniform in color, there being only a slight difference in its intensity and in the prevailing tint of the iridescence.*

The average dimensions of the thirty-three specimens of which measurements are given below are as follows:

Length (males): 16.51; alar extent, 22.48; wing, 7.19; tail, 7.00.

Length (females): 12.95; alar extent, 17.94; wing, 5.67; tail, 5.11.

The individual variation is as follows.

Males, length, 15.50 to 16.80; alar extent, 21.10 to 23.50; wing, 6.25 to 8.35; tail, 6.25 to 7.60.

Females, length, 12.10 to 13.40; alar extent, 17.25 to 18.25; wing, 5.25 to 5.95; tail, 4.75 to 5.60.

Measurements of Florida Specimens of Quiscalus major.

M. C. Z. No.	Coll. No.	Sex.	Locality.	Date.	Collector.	Length.	Alar Extent.	Wing.	Tail.
5272	—	♂	Blue Springs	Feb. 21, '69	J. A. Allen	16.00	22.15	7.25	6.80
5252	—	♂	Enterprise	Mar. 1, '69	"	16.25	22.50	7.15	7.15
5283	—	♂	"	Feb. 22, '69	"	16.25	21.75	7.15	7.10
5332	—	♂	"	Mar. 4, '69	"	15.50	22.00	6.85	6.70
5333	—	♂	"	Mar. 4, '69	"	15.75	22.30	7.20	6.85
5334	—	♂	"	Mar. 4, '69	"	15.60	21.85	7.00	—
5407	—	♂	Hawkinsville	Mar. 15, '69	"	16.00	23.00	7.00	7.00
5243	—	♂	"	Feb. 18, '69	"	15.75	22.25	7.30	—
5408	—	♂	"	Mar. 15, '69	"	16.50	22.50	7.30	7.15
5244	—	♂	"	Feb. 18, '69	"	17.30	23.50	7.80	—
5409	—	♂	"	Mar. 15, '69	"	16.00	22.75	7.15	7.00
5410	—	♂	"	Mar. 15, '69	"	16.00	22.75	7.20	6.90
5411	—	♂	"	Mar. 15, '69	"	16.35	22.50	7.50	—
—	—	♂	"	Mar. 15, '69	"	16.50	23.25	7.25	7.40
—	2408	♂	Dummitt's	Feb. 19, '69	C. J. Maynard	16.50	23.00	7.40	7.10
10607	2405	♂	"	Feb. 19, '69	"	16.75	23.50	7.50	7.25
—	2406	♂	"	Feb. 19, '69	"	17.50	23.00	8.35	7.60
—	2345	♂	"	Mar. 17, '69	"	16.00	21.10	6.75	6.60
10610	2585	♂	"	Mar. 9, '69	"	16.90	23.00	7.70	7.50
—	2409	♂	"	Mar. 9, '69	"	16.25	22.00	7.00	7.20
—	2586	♂	"	Mar. 9, '69	"	16.75	22.00	7.00	7.00
—	2399	♂	"	Mar. 9, '69	"	15.75	22.00	6.90	7.00
—	2431	♂	"	Mar. 9, '69	"	16.00	22.25	7.25	6.75
—	2404	♂	"	Mar. 9, '69	"	16.17	20.75	6.50	6.50
—	2345	♂	"	Feb. 16, '69	"	16.50	22.30	6.25	6.25
10609	2563	♀	"	Mar. 9, '69	"	13.00	17.50	5.85	5.60
—	2343	♀	"	Mar. 9, '69	"	13.00	18.25	5.80	5.00
—	2464	♀	"	Mar. 9, '69	"	13.00	18.25	5.95	5.25
5290	—	♀	Enterprise	Feb. 25, '69	J. A. Allen	13.40	18.25	5.85	4.75
5334	—	♀	"	Mar. 4, '69	"	12.75	17.50	5.50	5.00
5412	—	♀	Lake Dexter	Mar. 23, '69	"	13.00	18.05	5.60	5.20
5413	—	♀	"	Mar. 23, '69	"	12.10	17.25	5.25	5.00
5414	—	♀	"	Mar. 23, '69	"	12.50	17.60	5.45	—

* For a very full biography of this species, see an article by Dr. Elliott Coues in the Ibis. Vol. VI, pp. 367 - 378, 1870.

The present species is hence not only remarkable for variation in size between specimens of the same sex, but especially so for its sexual variation in size, the sexual difference in this respect being greater than in any other species of insessorial bird with which I am acquainted, and it is rarely, if ever, exceeded in any group.

CORVIDÆ.

61.* Corvus americanus *Audubon.* COMMON CROW.

Corvus corone WILSON, Am. Orn., IV, 79, pl. xxv, fig. 3, 1811. — NUTTALL, Man. Orn., I, 209, 1832.

Corvus americanus AUDUBON, Orn. Biog., II, 317, 1834. — BAIRD, Birds N. Am., 566, 1858.

Corvus americanus var. *floridanus* BAIRD, Ibid., 568, 1858.

Corvus minimus GUNDLACH, Cabanis's Journal für Ornithologie, IV, 97, 1856.

Everywhere abundant.

In the average, while the general size of Florida specimens is smaller than New England ones, the bill is somewhat larger. As is well known, the crow is exceedingly variable in the size and shape of its bill even in specimens collected from the same flock. There is, however, an appreciable average difference in the size of the bill, as in general size, between northern and southern examples. This was some time since observed by Professor Baird in comparing a single specimen from the southern point of the Florida peninsula with others from the Northern States, and so strongly was he impressed by it that he thought if his Florida specimen did not represent a distinct species, it did at least a distinct variety, and as such he characterized it, calling it *Corvus americanus* var. *floridanus.* He at the same time referred to the little crow of Cuba, described by Dr. Gundlach as *Corvus minutus*, to which he said it was more nearly allied than either are to *C. americanus.* I have no examples of the latter, but from descriptions of it see no reason why it should be regarded as other than the extreme southern form of *C. americanus.*

62.* Corvus ossifragus *Wilson.* FISH CROW.

Abundant. Perhaps rather more numerous than the common crow.

63.* Cyanurus cristatus *Swainson.* BLUE JAY.

Very abundant and unsuspicious. It frequents the towns, where it seems half domestic.

The same difference occurs in this species between Florida and northern specimens in size and shape of bill as has been already pointed out in respect to *Corvus americanus*, but it is far less marked than in *Agelæus phœniceus*, *Quiscalus purpureus*, and *Sturnella ludoviciana*. The brilliancy of its colors seems not much greater than in New England specimens.

The difference in size between northern and southern specimens is as follows: Average of eighteen Massachusetts specimens (eleven males and seven females): Length, 11.71; alar extent, 16.87; wing, 5.13; tail, 4.89. Average of eleven Florida specimens (proportion of males and females nearly the same as in the previous case): Length, 10.98; alar extent, 15.11; wing, 4.75; tail, 5.00. The maxima and minima of the eleven males from Massachusetts are as follows: Length, 12.25 and 11.35; alar extent, 17.50 and 16.30; wing, 5.50 and 5.00; tail, 5.65 and 4.25.

Measurements of Specimens of CYANURA CRISTATA.

M. C. Z. No.	Coll. No.	Sex.	Locality.	Date.	Collector.	Length.	Alar Extent.	Wing.	Tail.
—	34	♂	Newton, Mass.	Oct. 25, '67	C. J. Maynard	11 62	16.30	5.32	5.06
—	90	♂	“ “	Feb. 5, '68	“	11.35	17.00	5.00	4.78
—	94	♂	“ “	Feb. 8, '68	“	12 00	17.00	5.00	5.00
—	93	♂	“ “	Feb. 8, '68	“	11.55	17.20	5.25	4.80
—	—	♂	“ “	Feb. 21, '68	“	12.00	16.80	5.00	5.00
—	687	♂	“ “	May 28, '68	“	12 16	17 00	5.45	5.40
—	1667	♂	“ “	—	“	12.25	17.25	5.65	5 65
12393	—	♂	Springfield, “	Feb. 25, '70	Irving Allen	12.00	17.20	5.15	5.15
12392	—	♂	“ “	Feb. 25, '70	“	11.50	17 00	5.30	5.10
12388	—	♂	“ “	Feb. 25, '70	“	12.00	17.00	5.00	4 25
12385	—	♂	“ “	Feb. 25, '70	“	12.00	17.50	5.50	5.40
12389	—	♀	“ “	Feb. 25, '70	“	12.00	17 00	4.40	4.45
12392	—	♀	“ “	Feb. 25, '70	“	11.00	16 50	4.33	4.80
12391	—	♀	“ “	Feb. 25, '70	“	11.00	17.00	5 25	4.75
12386	—	♀	“ “	Feb. 25, '70	“	11.50	17.00	5.50	5.15
—	33	♀	Newton, “	Oct. 25, '67	C. J. Maynard	11.40	16.32	5.30	5.30
4875	688	♀	“ “	May 28, '68	“	11.62	16.53	4.75	4.77
—	1685	♀	“ “	Nov. 4, '68	“	11.75	16.00	5.20	4.35
10733	1951	♂	Jacksonville, Fla.	Jan. 2, '69	“	11.15	16.00	5.00	5.00
10734	1973	♀	“ “	Jan. 3, '69	“	11.00	15.50	4.80	4 80
10734	1974	♀	“ “	Jan. 3, '69	“	11 00	14.75	4.00	4.80
5522	—	?	Blue Springs, “	Feb. 21, '69	J. A. Allen	10.75	15.75	4 20	—
5128	—	?	Jacksonville, “	Jan. 21, '69	“	10 75	15.50	4.70	5.12
5190	—	♂	Welaka, “	Feb 3, '69	“	10.70	15.60	5.10	5.10
—	—	♂	Enterprise, “	Mar. 1, '69	“	11.00	15 75	5.00	—
—	—	?	“ “	Mar. 4, '69	“	10.70	15.15	4.50	—
5348	—	♀	“ “	Mar 4, '69	“	11.00	16.00	5.00	5.05
5162	—	?	Hibernia, “	Jan. 30, '69	“	11.25	15 75	5.00	5.15
5163	—	?	“ “	Jan. 30, '69	“	11.50	15 50	5.00	—

64.* **Cyanocitta floridana** *Bonaparte*. FLORIDA JAY.

Corvus floridanus BARTRAM, Travels, 291, 1791. — AUDUBON, Orn. Biog., I, 444, pl. lxxxvii, 1831.

Garrulus floridanus BONAP., Am. Orn., II, 11, pl. ix, 1828.

Garrulus cœrulescens ORD, Journ. Phil. Acad. Nat. Sci., I, 347, 1818.

Garrulus californicus VIGORS, Zoöl. Beechey's Voyage, 21, pl. v, 1839.

Cyanocitta floridana BONAP., Consp. Gen. Avium, 377, 1850.

Cyanocitta superciliosa STRICKLAND, Ann. & Mag. Nat. Hist., XV, 260, 1845.

Cyanocitta californica STRICKLAND, Ibid., 342.

Cyanocitta Woodhousei BAIRD, Birds N. Am., 585, 1858.

Numerous in the scrub, but does not appear to frequent the pine woods the hummocks or swamps. I saw none along the St. John's, except at Blue Springs, but they occur in numbers a few miles back from the river.

On comparing a number of specimens of the so-called *Cyanocitta californica* with numerous others from Florida, I find, as previous writers have observed, that the differences between them are very slight, and not so great as obtain between Florida and New England specimens of *Pipilo erythrophthalmus*, *Agetæus phœniceus*, and other species where there is no reason to question their specific identity. The so-called *C. Woodhousei* is described as, and is, intermediate in character between *C. floridana* and *C. californica*. The habitat of *C. Woodhousei* is also intermediate between those of the other two, but adjoins that of *C. californica*, to which it is most nearly allied. How great the interval is between the habitats of *C. floridana* and *C. Woodhousei* I have not been able to accurately determine. Bonaparte * reported the former as being found in Louisiana and northward to Kentucky, and the latter occurs in Western Texas.

In the following measurements of twelve specimens of this species (six males and six females) the extremes are as follows: Length, 11.00 and 12.50 (both specimens being females); alar extent, 13.50 (female) and 15.00 (male); wing, 4.00 and 4.75 (both specimens females); tail, 4.25 and 5.35 (both specimens females). The average dimensions of these specimens are as follows: Length, 11.74; alar extent, 14.44; wing, 4.42; tail, 4.80. The females average slightly smaller than the males.

Measurements of Florida Specimens of Cyanocitta floridana.

M. C. Z. No.	Coll. No.	Sex.	Locality.	Date.	Collector.	Length.	Alar Extent.	Wing.	Tail.
10739	2480	♂	Dummitt's	Feb. 22, '69	C. J. Maynard	11.50	14.50	4.30	4.35
—	2377	♂	"	Feb. 22, '69	"	12.00	15.00	4.45	4.75
—	2421	♂	"	Feb. 15, '69	"	12.00	15.00	4.75	5.00
10736	2326	♂	"	Feb. 15, '69	"	12.00	14.50	4.50	4.60
—	2329	♂	"	Feb. 15, '69	"	11.50	14.25	4.50	4.25
—	2379	♀	"	Feb. 22, '69	"	11.50	14.25	4.50	5.35
10737	2328	♀	"	Feb. 15, '69	"	12.50	14.50	4.75	4.90
—	2378	♀	"	Feb. 15, '69	"	11.50	14.10	4.30	5.15
—	2375	♀	"	Feb. 15, '69	"	11.60	14.40	4.60	4.25
5271	—	♀	"	Feb. 21, '69	J. A. Allen	11.00	13.50	4.00	5.35
5272	—	♂	"	Feb. 21, '69	"	12.00	14.50	4.30	4.75
5523	—	—	"	Feb. 21, '69	"	11.75	14.80	4.20	—

TYRANIDÆ.

65.† Sayornis fuscus *Baird.* Pewee.

Abundant all winter, and a few remain till into April.

The king-bird (*Tyrannus carolinensis*), the great-crested flycatcher (*Myiarchus crinitus*), and the wood pewee (*Contopus virens*) became

* Am. Orn., Vol. II, p. 60, 1828.

common the last week in March, as also, according to Mr. Boardman, the least flycatcher (*Empidonax minimus*).

Several specimens of the gray king-bird (*Tyrannus dominicensis*) were obtained by Mr. L. L. Thaxter at St. Augustine, about the first of May.

ALCEDINIDÆ.

66.* Ceryle alcyon *Boie.* KINGFISHER.

Abundant. As shy and distrustful here as in the more thickly settled parts of the country. Begins to breed very early. Mr. Maynard saw them forming their holes in the coquina rock, in the banks of the canal connecting Indian River with Mosquito Lagoon, the first week in February.

CAPRIMULGIDÆ.

67.* Antrostomus carolinensis *Gould.* CHUCKWILL'S WIDOW.

Abundant. Not observed till about the first of March, when its notes are usually first heard. Said by Audubon to be resident; which statement is confirmed by the testimony of old residents of the State.

68.* Antrostomus vociferus *Bonaparte.* WHIPPOORWILL.

Apparently not numerous in winter. I heard it once in February, and Mr. Maynard took it at Dummitt's in the same month. The inhabitants along the St. John's agree with Audubon that this species is also a winter resident.

The night hawk (*Chordeiles popetue* * Baird) was collected at Jacksonville by Mr. Thurston as early as April 20th.

* *Caprimulgus virginianus* BRISSON, Orn., II, 477 (in part).
Caprimulgus popetue VIEILLOT, Ois. Am. Sept., I, 56, pl. liv, 1807.
Caprimulgus americanus WILSON, Am. Orn., V, 65, pl. cxl, 1812.
Caprimulgus (Chordeiles) virginianus SWAIN., Faun. Bor. Am., II, 62, 1831.
Chordeiles virginianus BON., Geog. & Comp. List, 8, 1838. — GOSSE, Birds of Jamaica, 33, 1847.
Chordeiles sapiti BONAP., Consp. Gen. Avium, I, 63, 1849. — CASSIN, Ill. N. Am. Birds, 238, 1855.
Chordeiles brasilianus LAWR., Ann. N. Y. Lyceum Nat. Hist., V, 114, 1851.
Chordeiles Henryi CASSIN, Ill. N. Am. Birds, 239. — BAIRD, Birds N. Am., 153.
Chordeiles Gundlachii LAWR., Ann. N. Y. Lyc. Nat. Hist., VI, 167, 1856.
Chordeiles texensis LAWR., Ibid., 165. — BAIRD, Birds N. Am., 154.
Chordeiles minor CABANIS, Journ. für Orn., 5, 1856.
Chordeiles popetue BAIRD, Birds N. Am., 151.
This widely distributed species presents only the usual variations in size and color

CYPSELIDÆ.

The chimney swift (*Chætura pelasgia*) arrives about the last week in March. It was common at Jacksonville, April 1st.

TROCHILIDÆ.

The ruby-throated humming-bird (*Trochilus colubris*) became common about March 1st. Some probably spend the winter in South Florida.

PICIDÆ.

69.* **Campephilus principalis** *Gray*. Ivory-billed Woodpecker.

Picus principalis Linné, Syst. Nat., I, 173, 1767.
Campephilus principalis Gray, Genera of Birds, 1840.
Campephilus Bairdii Cassin, Proc. Phil. Acad. Nat. Sci., 1863, 322. (West Indian form.)

Rather rare; at least far less numerous than most of the other species of woodpecker.

With only Florida specimens of this species before me, I am unable to give comparisons between them and specimens from other localities. According to the late Mr. Cassin, those found in Cuba differ from those of the Southern States, in being smaller, as would be expected, with very slight deviations in color-markings. He has, however, given to the Cuba race the name of *Campephilus Bairdii*, remarking that it appears to be "one of those singular insular species which have become well known to naturalists."

Measurements of Florida Specimens of Campephilus principalis.

M. C. Z. No.	Sex.	Locality.	Date.	Collector.	Length.	Alar Extent.	Wing.	Tail.
5221	♂	Volusia.	Feb. 12, '69	J. A. Allen	20.00	32.25	10.40	6.90
5222	♂	"	Feb. 12, '69	"	19.50	32.50	10.25	6.90
5229	♀	"	Feb. 12, '69	"	19.30	31.50	10.60	6.85
5354	♂	Enterprise	Mar. 5, '69	"	19.25	30.50	9.70	6.40
5399	♀	Hawkinsville	Mar. 15, '69	"	19.50	31.50	10.25	6.75

seen in other species of our birds. Yet these variations have in the present case been mistaken as indicating numerous species. The southern representatives of it are appreciably smaller than the northern, and have the white markings on the wings more restricted, — variations that have already been pointed out in this paper as occurring in numerous others similarly distributed. Those from the central arid region of the continent are also lighter in general color than those from the eastern or western portions; also a common color variation in other species. The latter type forms the so-called *Chordeiles Henryi*; the southern ones have been variously characterized as *C. sapiti*, *C. texensis*, *C. Gundlachii*, etc., as indicated in the above-cited synonymes.

70.* Hylotomus pileatus *Baird.* Pileated Woodpecker.

Abundant. Much smaller than at the north, but not otherwise appreciably different.

The average dimensions of fourteen Florida specimens (seven males and seven females) are as follows: —

Males, length, 17.48; alar extent, 28.07; wing, 9.21; tail, 6.82.

Females, length, 16.44; alar extent, 26.80; wing, 8.98; tail, 6.54.

The individual variation is as follows: —

Males, length, 17.25 to 17.75; alar extent, 27.50 to 28.50; wing, 9.00 to 9.50; tail, 6.20 to 6.75.

Females, length, 15.50 to 16.80; alar extent, 26.00 to 27.75; wing, 8.50 to 9.50; tail, 5.85 to 6.80.

Measurements of Florida Specimens of Hylotomus pileatus.

M. C. Z. No.	Coll. No.	Sex.	Locality.	Date.	Collector.	Length.	Alar Extent.	Wing.	Tail.
5118	—	♂	Hibernia	Jan. 30, '69	J. A. Allen	17.75	28.25	9.25	6.65
5203	—	♂	Welaka	Feb. 7, '69	"	17.25	28.00	9.00	6.50
5215	—	♂	"	Feb. 10, '69	"	17.50	28.50	9.25	6.75
—	—	♂	Hawkinsville	Mar. 10, '69	"	17.25	27.50	9.50	—
—	1937	♂	Jacksonville	Dec. 31, '69	C. J. Maynard	17.75	28.50	9.50	6.40
—	2076	♂	"	Jan. —, '69	"	17.25	27.75	9.00	6.20
—	2543	♂	Dummitt's	Feb. 15, '69	"	17.60	28.00	9.00	6.45
—	2334	♀	"	Mar. 11, '69	"	15.50	26.40	8.70	5.85
—	2602	♀	"	Mar. 5, '69	"	16.60	27.75	9.00	6.75
5204	—	♀	Welaka	Feb. 7, '69	J. A. Allen	16.75	26.25	8.50	6.75
5214	—	♀	"	Feb. 10, '69	"	16.35	26.75	9.15	6.60
5216	—	♀	"	Feb. 10, '69	"	16.30	27.25	9.00	6.80
5274	—	♀	Blue Springs	Feb. 21, '69	"	16 75	27 20	9 50	6.50
—	—	♀	Hawkinsville	Mar. 10, '69	"	16.80	26.00	9.00	—

71.* Picus villosus *Linné.* Hairy Woodpecker.

Picus villosus Linné, Syst. Nat., I, 175, 1767. — Forster, Philosoph. Transact., LXII, 383, 1772. — Wilson, Am. Orn., I, 150, pl. ix, fig. 3, 1808. — Audubon, Orn. Biog., V, 164, pl. ccccxvii, 1837. (Northern form.)

Picus leucomelanus Wagler, Syst. Av., No. 18, 1827. (Immature male.)

Picus Auduboni Swainson, Faun. Bor. Am., II, 306, 1831. (Immature male.) — Trudeau, Journ. Phil. Acad. Nat. Sci., 404, 1837. (Immature male). — Audubon, Orn. Biog., V, 194, 1839. (Same as the last.)

Picus Martinæ Audubon, Ibid., 181, pl. ccccxvii. (Very immature.)

Picus Phillpsii Audubon, Ibid., 186, pl. ccccxvii. (Immature.)

Picus Harrisii Audubon, Ibid., 191, same plate. (Northwestern form.) — Baird, Birds N. Am., 87.

Picus septentrionalis Nuttall, Man. Orn., I (2d Ed.), 685, 1840.

Picus rubricapillus Nuttall, Ibid., 684. (Immature male.)

Picus Cuvieri Malherbe, Mon. Picidæ, I, 85, pl. xxii, fig. 3. (Young female.)

Picus Jardinei Malherbe, Ibid., I, 85, pl. xxv, fig. 4, 5. — Cassin, Proc. Phil. Acad. Nat. Sci., 1863, 201.

Not numerous in Florida in comparison with the other species of *Picidæ.*

The difference in size between northern and southern specimens of all the species of the *Picidæ* is greater than obtains in most other families of birds. So great is it in *Picus villosus* and *Picus pubescens* that it was in these species that such variations were first noticed. This difference is well pointed out by Professor Baird in his work on the North American Birds, and fully demonstrated in his table of measurements. On this ground he distinguished three varieties of *P. villosus*, — *P. villosus major*, occupying the northern and western portions of the continent; *P. villosus medius*, occupying the Middle States; and *P. villosus. minor*, occupying the Southern States. Audubon regarded the two former as distinct species. In addition to these variations in size, my Florida specimens indicate a well-marked variation in color between the northern and extreme southern races, the Florida specimens differing from New England ones in having the white markings of relatively less extent, which gives to the plumage a considerably darker aspect. Through this variation there is an approach in the Florida examples of *P. villosus* to the so-called *P. Harrisii* of the Pacific coast and Rocky Mountain regions of the continent, and in the Florida examples of *P. pubescens* to the so-called *P. Gairdneri*, also of the middle and western regions of the continent. These, as is well known, differ respectively from *P. villosus* and *P. pubescens* almost solely in a general darker aspect, resulting simply from the relatively greater predominance of the black color of the plumage over the white markings in the western type; there being no change whatever in the general style of coloration, though some of the smaller white spots seen in the eastern are entirely obsolete in the western type. Under *Picus Gairdneri* Professor Baird thus describes these variations. "There is," he says, "the same series in specimens of *Picus Gairdneri* that were indicated under *P. Harrisii.* Thus the most northern from Washington Territory and Oregon have the under parts more brown, with faint black streaks, the white spots above smaller and less numerous. In specimens from California and farther east the white is purer, the spots more conspicuous." "The almost perfect parallelism," he further observes, "with appreciable differences between the markings of the northwestern and southeastern varieties of *Picus Harrisii* and *Gairdneri*, and their relationship to *P. villosus* and *pubescens*, is a remarkable fact in American ornithology, and may possibly indicate the necessity either of dividing the dark ones into a Pacific and Rocky Mountain series, or of considering all as variations of two species, a larger [*P. villosus*] and a smaller [*P. pubescens*], changing their character with longitudinal distribution." And he aptly adds, "Many other supposed species are involved in the same consideration." * Professor Baird in his account of these species, expressly refers to California specimens that have less white on the wings than the one form and more white than the other.† This with the color differences existing between Florida specimens and New England ones, similar in character to these, though less in degree, seems to confirm the necessity alluded to by Professor Baird of regarding the small spotted woodpeckers in question as forming only two species, — the *Picus villosus* and *Picus pubescens*, — with parallel and remarkable geographical variations. So great is the difference, however, between typical representatives of the two leading forms of each, that their discoverers, with too few specimens of each to enable them to detect the gradual passage of the one into the other, — a fact which now seems well substantiated, — were quite excusable in regarding them as distinct species. Several other supposed species, as indicated by the synonymes given above, and previously by other authors, have been based on phases of immaturity. The young of either sex often have the crown spotted with red or yellow, while the mature male alone has red on the head, and in which it is usually confined to a narrow occipital transverse band. In respect to the number, shape, position, and size of the white spots on the wings, however, there is always considerable variation in specimens from the same locality, these variations being dependent upon neither sex nor age.

Florida specimens of not only *Picus pubescens* and *P. villosus*, but of *Centurus carolinus*, *Sitta carolinensis*, and *Sitta pusilla*, often have the plumage of the lower surface of the body so much soiled and darkened by running over the blackened trees in recently burnt districts as to materially alter their appearance, so that they might almost be taken for distinct species, as previously noted by Audubon. ‡

72.* **Picus pubescens** *Linné*. DOWNY WOODPECKER.

Picus pubescens LINNÉ, Syst. Nat., I, 175, 1766. — WILSON, AUDUBON, BONAPARTE, NUTTALL, BAIRD, CASSIN, etc.

Picus (Dendrocopus) pubescens SWAINSON, Faun. Bor. Am., II, 307, 1831.

Picus (Dendrocopus) medianus SWAINSON, Ibid., 308. (Described from New Jersey specimens).

* Birds of North America, p. 91.

† In accounting for these intermediate forms, Mr. Cassin adopts the very convenient but, as it seems to me, uncalled-for and incorrect theory of hybridity, so often resorted to in similar cases. Under *Picus villosus*, he says that *P. villosus* and *P. Harrisii* probably associate in a region intermediate between the proper ranges of the two species, "and produce hybrids, which present difficulties to naturalists." Under *Picus pubescens* he makes similar remarks in respect to *P. pubescens* and *P. Gairdneri*. *Proc. Phil. Acad. Nat. Sci.*, 1863, pp. 200, 201.

‡ Orn. Biog., Vol. II, p. 82.

Picus (*Dendrocopus*) *meridionalis* SWAINSON, Ibid. (Southern race.)

Picus Gairdneri AUDUBON, Orn. Biog., V, 317, 1839. (Northwestern form.) — BAIRD, Birds N. Am., 91, 1858.

Picus meridionalis NUTTALL, Man. Orn., I, (2d Ed.) 690, 1840. (Not of Swainson).

Picus Lecontei JONES, Ann. N. York Lyc. Nat. Hist., IV, 489, pl. xviii, 1848. (Three-toed specimen.)

Picus Turati MALHERBE, Mon. Pic., I, 125, pl. xxix, fig. 5, 6. — CASSIN, Proc. Phil. Acad. Nat. Sci., 1863, 202.

Common. Much more numerous than *Picus villosus.*

The difference in size and color between northern and southern specimens has been sufficiently detailed under the previous species.

73.* **Picus borealis** *Vieillot.* RED-COCKADED WOODPECKER.

Picus borealis VIEILLOT, Ois. Am. Sept., II, 66, pl. cxxii, 1807. — CASSIN, Proc. Phil. Acad. Nat. Sci., 1863, 203.

Picus querulus WILSON, Am. Orn., II, 103, pl. xv, fig. 1, 1810. — CASSIN, Proc. Phil. Acad. Nat. Sci., 1863, 203.

Common in the pineries.

Mr. Cassin regards the Carolina and Georgia representatives of this species as specifically distinct from the Pennsylvania ones. He says that they are as distinct and as easily recognized as are *Picus villosus* and *P. Harrisii*, which he of course regards as valid species. He assigns Vieillot's

Measurements of Florida Specimens of PICUS BOREALIS.

M. C. Z. No.	Coll. No.	Sex.	Locality.	Date.	Collector.	Length.	Alar Extent.	Wing.	Tail.
10641	1919	♀	Jacksonville	Dec. 31, '68	C. J. Maynard	8.40	14.20	4.75	3.52
10642	1920	♂	"	Dec. 31, '68	"	8.30	14.20	4.76	3.62
10643	1921	♀	"	Dec. 31, '68	"	8.30	14.80	4.80	3 56
——	1922	♂	"	Dec. 31, '68	"	8.50	14.50	4.75	3.69
10644	1923	♂	"	Dec. 31, '68	"	8.20	14 45	4.75	3.39
——	1924	♂	"	Dec. 31, '68	"	8.50	15.00	4 80	3.32
10645	1925	♂	"	Dec. 31, '66	"	8 50	14.75	4.85	3.60
——	1971	♀	"	Jan. 3, '69	"	8.50	15.00	4.85	3.50
10646	1972	♂	"	Jan. 3, '69	"	8.50	14 30	4.75	3.75
10631	29	♂	"	Apr. 11, '69	"	8.00	14.75	4.90	3.45
10632	30	♀	"	Apr. 6, '69	"	8.50	15.00	4.90	3 35
10633	31	♀	"	Apr. 6, '69	"	8.30	14.90	4.85	3.35
10634	41	♂	"	Apr. 7, '69	"	8.15	14.50	4.70	3.25
10637	47	♂	"	Apr. 8, '69	"	8.60	15.15	4.87	3.40
10638	48	♂	"	Apr. 8, '69	"	8.50	15.00	4.95	3.46
10639	58	♀	"	Apr. 13, '69	"	8.50	14.10	4.75	3.59
——	49	♀	"	Apr. 8, '69	"	8.50	14.15	4.85	3.49
10640	59	♀	"	Apr. 12, '69	"	8.50	15.00	4.80	3.50
10636	44	—	"	Apr. 7, '69	"	8.30	15.00	4.80	3.60
10635	43	♀	"	Apr. 7, '69	"	8.35	14.60	4.60	3.60
——	32	♂	"	Apr. 3, '69	"	8.30	14.90	4.85	3.50
——	42	♂	"	Apr. 7, '69	"	8.20	14.70	4.75	3.29
5116	—	♂	"	Jan. 19, '69	J. A. Allen	8.50	15 20	4.40	3 30
5137	—	♂	"	Jan. 25, '69	"	8.38	14.75	4.57	3.42
5375	—	♂	Hawkinsville	Mar. 12, '69	"	8.55	14.55	4.50	3.40
5393	—	♂	"	Mar. 15, '69	"	8.50	14.50	4.45	3.20
5394	—	♂	"	Mar. 15, '69	"	8.25	14.50	4.40	3.15
5414	—	♀	Volusia	Mar. 25, '69	"	7.90	14.60	4.45	3.25

name *borealis* to the Pennsylvania type, and Wilson's name *querulus* to the more southern form. In recognizing two species of red-cockaded woodpecker in the Atlantic States, Mr. Cassin differs from all previous writers. Having only Florida specimens, a series of twenty-two, before me, I cannot state from personal observation as to how they differ from northern ones. They appear, however, to be merely a little smaller and darker.

The average size of the twenty-eight Florida specimens of which measurements are given in the foregoing table is as follows: Length, 8.34; alar extent, 14.46; wing, 4.71; tail, 3.41.

74.† **Sphyrapicus varius** *Baird.* YELLOW-BELLIED WOODPECKER.

Common.

75.* **Centurus carolinus** *Bonaparte.* RED-BELLIED WOODPECKER.

Picus carolinus LINNÉ, Syst. Nat., I, 174, 1767.
Picus griseus VIEILLOT, Ois. Am. Sept., II, 52, pl. cxvi, 1807.
Centurus carolinus BONAP., Geog. & Comp. List, 40, 1838.

Abundant. The most numerous species of its family in Florida. Specimens in the Museum from Cape Florida, taken the 8th of May by Mr. G. Wurdemann, indicate it as resident throughout Florida, though considered by Audubon and others as only a winter visitant to this and the other Gulf States.

The Florida specimens are all very much brighter colored than others before me from Maryland, Indiana, Illinois, and Michigan, the Michigan specimens being the palest. Professor Baird has remarked, in regard to a specimen from Amelia Island, Florida,* that it was not only very much smaller than northern ones, but had the white transverse bands on the back much narrower, the black ones being three times the breadth of the white ones, instead of twice, as in the northern specimens. These differences my large series from the St. John's River indicate as constant. A similar increase in the breadth of the black bands over the white ones in southern specimens as compared with northern ones, in species banded transversely, is seen in numerous other species. It is well marked in *Colaptes auratus* (where the bands are dark and light brown), in *Sphyrapicus varius*, and, as I shall show more fully subsequently, in *Ortyx virginianus*. The extent and intensity of the red on the abdomen and head, and especially its brilliancy on the head, is much greater in the Florida specimens of *C. carolinus*. In this respect there is also a well-marked difference between Cape Florida specimens and those from the St. John's River, the Cape Florida ones being much the brighter. These seem to accord in every particular with

* Birds of North Amer., p. 109.

the so-called *Centurus subelegans* of Lower California and Mexico. It is interesting to note that variations in color occur between the northern and southern representatives of *Centurus flaviventris* similar to those exhibited by northern and southern examples of *C. carolinus*. The southern forms of *C. flaviventris* were long since characterized by Wagler, Swainson, and Bonaparte as specifically distinct from the northern, under the names of *C. elegans*, *C. santacruzi*, etc., etc., which many authors still rank as species.

76.* **Melanerpes erythrocephalus**, *Swainson*. RED-HEADED WOODPECKER.

Rare in winter; said to be common in summer. I saw two only, about March 15th. Mr. Boardman also gives it as rare, while Mr. Maynard did not meet with it at all. Audubon speaks of its being very abundant in winter in Louisiana, and Dr. Coues gives it as resident in South Carolina; but it is certainly not common in winter in East Florida.

77.* **Colaptes auratus** *Swainson*. GOLDEN-WINGED WOODPECKER.

Abundant.

Considerably smaller than at the north, with the colors much more intense, and the transverse black bars on the back relatively broader. The individual variations in this species, even at the same locality, are very considerable, especially in respect to the bill. Figures 5 and 6, Plate VIII, illustrate the variation in the form and size of the bill of two specimens from Massachusetts, both of which are females.

The following summary of the subjoined tables indicates the difference in size between Massachusetts and Florida specimens, and the individual differentiation in the same respect at each locality. The sexes seem not to differ essentially in size.

No. of Specimens.	Sex.	Locality.		Length.	Alar Extent.	Wing.	Tail.
18	—	Massachusetts.	Average	12.45	19.94	6 24	4.35
11	—	Florida	Average	11.66	18.82	5.84	4.40
18	—	Massachusetts.	Maximum	13.00	20.75	6.60	4.70
18	—	"	Minimum	12.00	19.00	6.00	4.00
11	—	Florida.	Maximum	12.75	19.75	6.25	4.85
11	—	"	Minimum	10.60	17.60	5.60	4.10

While the Florida specimens are considerably smaller than the northern in three of the measurements, the tail is actually longer in the Florida birds, and hence relatively much longer. In most of the species of which comparative tables of measurements are given in the present paper, there is a decided tendency to an elongation of the tail at the southward, the tail decreasing less in length than the wing or the general size.

Measurements of Massachusetts Specimens of Colaptes auratus.

M. C. Z. No.	Coll. No.	Sex.	Locality.	Date.	Collector.	Length.	Alar Extent.	Wing.	Tail.
—	29	♂	Watertown	Oct. 14, '67	C. J. Maynard	12.50	20.15	6.00	4.15
—	232	♂	"	Mar. 17, '68	"	12.60	20.00	6.31	4.56
—	252	♀	Waltham	Apr. 18, '68	"	13.00	20.75	6.60	4.60
—	281	♂	Newton	Apr. 21, '68	"	12.30	19.60	6.00	4.24
4880	280	♂	"	Apr. 21, '68	"	12.25	19.90	6.25	4.62
—	325	♀	"	Apr. 25, '68	"	12.55	20.05	6.27	4.35
—	—	♂	"	Aug. 5, '68	"	12.67	19 80	6 10	4 00
4881	356	♂	"	May 1, '68	"	12.00	19.00	6.10	4.10
—	988	♂	"	June 2, '68	"	12 00	19.45	6.10	4.25
5460	1011	♂	Waltham	Aug. 6, '68	"	12.50	19 90	6.20	4.05
—	2902	♂	Newton	June 12, '69	"	12.50	20.50	6.40	4 30
—	4028	♀	Waltham	Aug. 22, '69	"	12.50	20.00	6.25	4.70
—	4029	♂	Newton	Aug. 22, '69	"	13.00	20.60	6.25	4 70
—	4034	♀	"	Aug. 26, '69	"	12.50	20.27	6.25	4 35
—	2913	♂	"	June 30, '69	"	12.10	20.00	6 45	4.55
—	2916	♂	"	June 22, '69	"	12.00	19.50	6.15	4.25
—	2915	♂	"	June 22, '69	"	12.50	19.50	6.25	4 20
—	2939	♂	"	May 1, '69	"	12 50	20.00	6.30	4.45

Measurements of Florida Specimens of Colaptes auratus.

M. C. Z. No.	Coll. No.	Sex.	Locality.	Date.	Collector.	Length.	Alar Extent.	Wing.	Tail.
—	2075	♀	Jacksonville	Jan. 20, '69	C. J. Maynard	10.75	18.50	5.75	4.55
—	2074	♀	"	Jan. 20, '69	"	11.00	17.60	5 50	4.50
10612	2346	♀	Dummitt's	Feb. 16, '69	"	11.75	19.50	6.25	4.10
10614	2601	♀	"	Mar. 11, '69	"	12.00	19 00	5.90	4 60
10611	2584	♀	"	Mar. 9, '69	"	10.60	17.75	5.70	4.30
10613	2542	♀	"	Mar. 5, '69	"	12.75	19.10	6.00	4.85
10610	2385	♂	"	Mar. 5, '69	"	12.00	19.20	6.00	4.25
5196	—	♂	Welaka	Feb. 5, '69	J. A. Allen	12 20	19.10	5.85	4.30
5321	—	♀	Enterprise	Mar. 1, '69	"	11.50	18.75	5.60	4.15
—	—	♀	Volusia	Mar. 25, '69	"	12.25	19.75	6.00	—
—	—	♂	"	Mar. 25, '69	"	11.50	18.75	5.65	—

Of the eight species of woodpecker mentioned above as occurring in Florida in winter, all but one or two (*Melanerpes erythrocephalus* and *Campephilus principalis*) are numerously represented. Most of them are exceedingly abundant, the woodpeckers hence forming a conspicuous element in the bird-fauna of East Florida. All of them are resident, according to Dr. Coues, in South Carolina. Audubon, however, states that two of them (*Sphyrapicus varius*, *Centurus carolinus*) do not breed south of Maryland, but Dr. Coues gives them as resident the whole year in South Carolina.

PSITTACIDÆ.

78. **Conurus carolinensis** *Bonaparte*. Carolina Parokeet.

Common. Hundreds are captured every winter on the Lower St. John's by professional bird-catchers and sent to the northern cities. Thousands of others are destroyed wantonly by sportsmen. Concerning this needless slaughter Mr. Boardman thus writes: "The little parokeet must soon be exterminated. Some of our Enterprise party would sometimes shoot forty or fifty at a few discharges, for sport, as they hover about when any are shot until the whole flock is destroyed." From its habit of feeding upon the tender maize in autumn, it is sometimes somewhat injurious to the farmer, and for this cause many are also killed. It is also more or less hunted as a game-bird. It is well known that the parokeet formerly inhabited large portions of the United States where it is now never seen, and the cause of its disappearance has been deemed a mystery. Such facts as these, however, seem to render clear what its ultimate fate must be in the United States, — extermination.

I could learn nothing from the inhabitants in regard to the time, manner, or place of breeding of this species, even old residents professing total ignorance in regard to these points.

The following table of measurements of specimens of this species serves to indicate its average size and proportions in Florida. In mature specimens the sexual difference in color and size is very slight. Neither sex acquires its adult colors before the second or third year.

The average size of the nineteen specimens (six males and thirteen females) cited below is as follows: Length, 13.10; alar extent, 21.76; wing, 7.59.

The extremes are as follows: —

Length, 12.50 and 13.60 (both specimens females); alar extent, 21.10 (female) and 22.50 (male); wing 7.00 and 7.85. These specimens seem to indicate a tolerable constancy in general size and proportions.

Measurements of Florida Specimens of CONURUS CAROLINENSIS.

M. C. Z. No.	Sex.	Locality.	Date.	Collector.	Length.	Alar Extent.	Wing.	Tail.
5205	♂	Welaka	Feb. 8, '69	J. A. Allen	13.25	22.00	7 70	6.05
5206	♂	"	Feb. 8, '69	"	13.55	22.30	7.85	6.75
5207	♀	"	Feb. 8, '69	"	12.90	21.50	7.45	6.10
5225	♀	Volusia	Feb. 12, '69	"	13.00	21.75	7.00	5.80
5226	♀	"	Feb. 12, '69	"	13.00	21.60	7 35	5.80
5227	♀	"	Feb. 12, '69	"	13.00	21.75	7 30	6.00
5228	♀	"	Feb. 12, '69	"	13 00	21.50	7.50	6.00
5294	♂	Enterprise	Feb. 25, '69	"	13.25	21.50	7 40	—
5295	♀	"	Feb. 25, '69	"	13.00	22.45	7.60	6.00
5296	♀	"	Feb. 25, '69	"	13.60	22 00	7.34	6.60
5297	♂	"	Feb. 25, '69	"	13.45	22.00	.7.50	—
——	♂	Hawkinsville	Mar. 13, '69	"	13.25	22.50	7.75	—
——	♂	"	Mar. 13, '69	"	13.15	21 25	7.50	—
——	♀	"	Mar. 13, '69	"	12.50	21 35	7.30	—
——	♀	Orange Bluffs	Mar. 24, '69	"	12.85	21 75	7.40	—
——	♀	"	Mar. 24, '69	"	13.60	22.30	7.75	—
——	♀	"	Mar. 24, '69	"	13.05	21.10	7.50	—
——	♀	"	Mar. 24, '69	"	13.25	21.30	7 50	—
——	♀	"	Mar 24, '69	"	13.25	21.50	7.55	—

VULTURIDÆ.

79.* Cathartes aura *Illiger*. Turkey Vulture.

Vultur brasiliensis Brisson, Orn., I, 468, 1760.

Vultur aura Linné, Syst. Nat., I. 122, 1767. — Vieillot, Ois. Am. Sept., I, 25, pl. 2 bis, 1807. — Wilson, Am. Orn., IX, pl. lxiv, fig. 1, 1814.

Cathartes aura Illiger, Prodromus, 283, 1811. — Bonaparte, Ann. N. Y. Lyc. Nat. Hist., II, 23, 1828. — Audubon, Orn. Biog., II, 296, pl. clii, 1835. — Bonaparte, Geog. and Comp. List, I, 1838. — D'Orbigny, Voy. dans l'Amer. Merid., IV, iii, 38, 1844. — Cassin, Proc. Phil. Acad. Nat. Sci., 1849, 159. — Bonaparte, Consp. Gen. Av., I, 9, 1850.

Vultur jota Molina, Saggio sul stor. nat. del Chile, 1782.

Cathartes ruficollis Spix, Av. Spec. Novæ, 2, 1824.

Vultur jota Molina, Sagg. sul stor. nat. del Chile, 235, 1782. — Gmelin, Syst. Nat., I, 347, 1788.

Cathartes jota Bonaparte, Consp. Gen. Av., I, 9, 1850. — Cassin, U. S. Nav. Astr. Exp., II. 172, 1855.

Cathartes septentrionalis Pr. Maximilian, Reise in das Nord-Amer., I, 162, 1839.

? *Cathartes Burrovianus* Cassin, Proc. Phil. Acad. Nat. Sciences, 1843, 212. — ? Cassin, Baird's Birds of N. Am., 6, 1858.

Abundant. Collect in large companies about the dead alligators so numerous in the St. John's River.

Both this species and the following (*Carthartes atratus*) paid us frequent visits at our camps at Enterprise and Hawkinsville, and whenever we left them they did not fail to gather up and devour the carcasses of the birds and mammals thrown away by us after skinning. We found them, in fact, rather troublesome neighbors, since on more than one occasion they proceeded, in our absence, to investigate the character of the specimens we had left in the sun to dry, and in a manner so unsatisfactory to ourselves that one of the party was frequently obliged to stay in camp to protect them while the others were away collecting.

Both this and the following species were represented as breeding late in the season, and as frequenting the palmetto swamps as well as some of the islands above Enterprise for this purpose.

The synonymy here given of the present and following species indicates clearly the confusion which several continental European authors have introduced through their descriptions of these species, to which attention has been previously called by Mr. Cassin.* While a *Vultur* (or *Cathartes*) *aura* has been described by most authors who have written of the two species in question, the name *aura* has been applied sometimes to the one and

* Proc. Phil. Acad. Nat. Sci., 1849, 159.

sometimes to the other, but when given to the true *aura* of Linné, Vieillot, and Wilson, the *atratus* of Bartram and Wilson has been cited as a synonyme, and the true *atratus* described under a new name. The name *jota* has likewise been repeatedly applied to both species by different authors, and in some cases even by the same author, as has been also the name *brasiliensis*. The description given by Linné in the twelfth edition of his Systema Naturæ, under *V. aura*, clearly refers to the *V. aura* of Wilson, of which the *V. jota* of Molina and Gmelin are synonymes; although some of Linné's synonymes may refer to the *C. atratus* of modern writers. Bonaparte, however, in both his Synopsis of the Birds of the United States and in his Geographical and Comparative List, strangely applied the name *jota* to the *atratus* of Wilson, in which he was for a time followed by other writers. By those who have regarded the South American representatives of *C. aura* as distinct from its North American ones, the name *jota* has latterly been applied to the supposed distinct South American representative of the supposed true or northern *C. aura*.

The distinctions between the so-called *C. jota* and *C. aura* seem, judging from the published accounts, to be by no means clear. Mr. Cassin, in his report on the birds of Lieutenant Gilliss's Expedition, says the *C. jota* "is apparently, or so far as can be ascertained from prepared specimens, a more slender bird, *and longer in all its measurements*. The last character is particularly applicable to its wings." * In his Illustrations of the birds of California and Texas, published the following year, he reverses this statement, and says: "The South American species [*C. jota*] *is the smaller*," and "is the more slender in all its members"; and adds: "All the specimens that we have seen have been of a more uniform clear black color." Having myself examined numerous specimens, both in Brazil and in Florida, I find the difference in the average exceedingly slight, and nearly as stated by Mr. Cassin in his later work; that is, the Brazilian are slightly smaller, and have the plumage appreciably darker.

Bonaparte, in his Conspectus, gives the *jota* of Molina as being simply smaller and with a shorter tail than *aura* of Linné. The differences are indeed very slight; they are, moreover, strictly in accordance with the well-known general laws of variation between specimens of the same species from northern and southern localities, and by no means indicate a diversity of species. Because formerly not known to occur in some of the West India Islands, it was at one time supposed by some that the habitats of the two supposed species did not meet, or that there was a region in Central and Northern South America where neither existed. As I have elsewhere stated,† this is a mistake, both this species and the *C. atratus* ranging from

* U. S Naval Astronomical Expedition, Vol. II, p. 173, 1855.

† Memoirs Bost. Soc. Nat. Hist., Vol. I, p. 500, 1868.

the middle and northern portions of the United States nearly to the southern extremity of South America; the *C. aura* also extending as much beyond the southern limit of the *C. atratus* in South America as it does to the north of it in North America.

The *Cathartes Burrovianus* of Cassin, described in 1843, from a single specimen from Mexico, is referred by Bonaparte, in his Conspectus, to *C. jota*, or to what I regard as the typical form of *C. aura*, and evidently with good reason. It differs from *C. aura* only in being smaller. I am therefore disposed to regard it as based on an unusually small specimen of that species. Though Dr. Gambel supposed he had seen it with the other species in Lower California, but two specimens seem to have been known to Mr. Cassin, one of which was from an unknown locality.

Measurements of Florida Specimens of CARTHARTES AURA.

M. C. Z. No.	Coll. No.	Sex.	Locality.	Date.	Collector.	Length.	Alar Extent.	Wing.	Tail.
5143	—	♂	Jacksonville	Jan. 25, '69	J. A. Allen	27.50	72.50	22.50	11.10
5180	—	♂	Hibernia	Feb. 1, '69	"	27.50	72.00	22.00	11.75
5187	—	♂	"		"	—	—	21.00	12.00
10746	2541	♀	Dummitt's.	Mar. 11, '69	C. J. Maynard	26.50	68.00	21.00	11.00
—	2603	♀	"	Mar. 11, '69	"	—	68.00	20.00	10.50
—	2433	♀	"	Mar. 10' '69	"	27.50	72.00	21.75	11.25

80.* **Cathartes atratus** *Swainson.* BLACK VULTURE.

? *Vultur brasiliensis* aut *mexicanus* RAY, Synop. Meth. Avium, 10, 1713.

Vultur atratus BARTRAM, Travels, 289, 1791.

Cathartes atratus SWAINSON, Faun. Bor. Am., II, 6, 1831. — AUDUBON, Synopsis, 3, 1839. — BONAPARTE, Consp. Gen. Av., I, 9, 1850. — CASSIN, Illust. Birds Cal., Texas, etc., 58, 1854. — CASSIN, Gilliss's U. S. Nav. Astr. Exp., II, 173, 1855.

Vultur jota WILSON, Am. Orn., IX, 104, pl. lxxv, fig. 2, 1814. (Not of Molina; not of Gmelin.)

Cathartes jota BONAPARTE, Ann. N. Y. Lyc. Nat. Hist., II. — AUDUBON, Orn. Biog., II, 33, 1835. — BONAPARTE, Geog. and Comp. List, I, 1838.

Vultur urubu VIEILLOT, Ois. Am. Sept., I, 53, pl. ii, 1807.

Cathartes urubu LESSON, Voy. autour du Monde, 614. — D'ORBIGNY, Voy. dans l'Amer. Merid., 1844.

Cathartes aura SPIX, Av. Spec. Novæ, 2, 1824.

Cathartes brasiliensis BONAP., Consp. Gen. Av., I, 9, 1850.

Abundant. On the whole, probably about as numerous as the preceding, but the two species occur in different proportions at different localities, and at different times at the same locality. None were seen about Jacksonville during the two weeks I spent there in January, and none were met with for some distance up the river. Above Lake

George it was generally common, and sometimes outnumbered the other species, as it did often at Hawkinsville during my stay there. The younger birds appear to be generally not so highly colored as the fully mature, nor to have the naked skin of the head and neck so rugose and corrunculated as the older. The differences in these respects are very considerable between individuals of the same flock.

A comparison of Florida specimens with Brazilian ones shows that the latter are slightly smaller than the former; in color or other general features they do not appear to differ. Most writers have regarded the South American as identical with the North American, but Mr. Cassin,* apparently on the authority of Bonaparte,† says the South American bird "is the *Vultur brasiliensis* Ray," and that "it is considerably smaller, and otherwise quite distinct." But he only refers definitely to the difference in size. The year preceding the publication of these remarks, however, he gives *C. atratus* as inhabiting Chili.‡ In speaking of the Chili specimen, he says: "A single specimen in mature plumage and excellent condition is exactly identical in size and other characters with the common species [*C. atratus*] of the southern parts of North America." He adds: "It is the only specimen presenting this similarity that we have ever seen from South America, and is larger and in other respects different from the allied *Cathartes brasiliensis*, which is an inhabitant also of that division of this continent." *C. atratus*, he says, is "not abundant in Chili, though represented to be occasionally met with in the interior"; these larger individuals referred to being doubtless the birds that inhabit the more elevated districts. Whatever Mr. Cassin's *Cathartes brasiliensis* may prove to be, it remains unquestionable that the *C. atratus* is a general inhabitant of South America, and that Bonaparte's *brasiliensis* is merely the southern type of this species. The exact parallelism of its range on the two continents as compared with that of *C. aura* has already been alluded to.

The PAINTED or SACRED VULTURE ("*Vultur sacra*"), § an apocryphal species described by Bartram ‖ as inhabiting Florida, demands in this connection a passing notice. Though not identified by any succeeding author (by some, however, it has been referred to the king vulture, *Sarcorham-*

* Illust. Birds of Cal. and Texas, p. 58, 1856.

† Conspectus Generum Avium, Tom I, p. 9, 1850.

‡ U. S. Naval Astronomical Expedition, Vol. II, p. 173, 1855.

§ Travels in Florida, etc., p. 150, 1790.

‖ *Vultur sacra* BARTRAM, Travels, pp. 150, 289, 1791. — VIEILLOT. — NUTTALL, Man. Orn. I, 42.

Sarcorhamphus sacer CASSIN, Illust. Birds of Cal. and Texas, 59, 1856.

See also BONAPARTE, Conspectus Gen. Av., I, 9.

phus papa), Bartram's account of it leads one to infer that he found it quite abundant. His description of it is given with satisfactory detail. He says it is " near the size of the turkey-buzzard, but his wings are much shorter, and consequently he falls greatly below that admirable bird in sail. I shall call this bird the painted vulture. The bill is long and straight almost to the point, where it is hooked, or bent suddenly down, and sharp; the head and neck bare of feathers nearly down to the stomach, where the feathers begin to cover the skin, and soon become long and of a soft texture, forming a ruff or tippet, in which the bird, by contracting his neck, can hide that as well as his head; the bare skin on the neck appears loose and wrinkled, which is of a deep bright yellow color, intermixed with coral red; the hinder part of the neck is nearly covered with short, stiff hair; and the skin of this part of the neck is of a dun-purple color, gradually becoming red as it approaches the yellow of the sides and fore part. The crown of the head is red; there are lobed lappets of a reddish orange color, which lay on the base of the upper mandible. But what is singular, a large portion of the stomach hangs down on the breast of the bird, in the likeness of a sack or half wallet, and seems to be a duplicature of the craw, which is naked and of a reddish flesh color; this is partly concealed by the feathers of the breast, unless when it is loaded with food (which is commonly, I believe, roasted reptiles), and then it appears prominent. The plumage of the bird is generally white or cream color, except the quill feathers of the wings, and two or three rows of the coverts, which are of a beautiful dark brown; the tail, which is large and white, is tipped with this dark brown or black; the legs and feet of a clear white; the eye is encircled with a gold-colored iris; the pupil black.

" The Creeks or Muscogulgees," he continues, " construct their royal standard of the tail feathers of this bird, which is called by a name signifying the eagle's tail; this they carry with them when they go to battle, but then it is painted with a zone of red within the brown tips, and in peaceable negotiations it is displayed new, clean, and white; this standard is held most sacred by them on all occasions, and is constructed and ornamented with great ingenuity. These birds seldom appear but when the deserts are set on fire (which happens almost every day throughout the year in some part or other, by the Indians, for the purpose of rousing up game, as also by the lightning), when they are seen at a distance soaring on the wing, gathering from every quarter, and gradually approaching the burnt plains, when they alight upon the ground yet smoking with hot embers; they gather up the roasted serpents, frogs, and lizards, filling their sacks with them. At this time a person may shoot them with pleasure, they not being willing to quit the feast, and indeed seem to brave all danger."

Mr. Cassin * refers the species described as above by Bartram to the genus *Sarcorhamphus* (*S. sacer* Cassin = *Vultur sacra* Bartram), believing it to be a valid species, and remarks that its identification "may be considered as one of the most important services to be performed in North American ornithology." It is related, Mr. Cassin continues, "to the king vulture (*S. papa*), but that species has a black tail, and in case of mistake or misprint in Bartram's description, it may be presumed, at any rate, to relate to an occurrence of that species within the United States.† There is no more interesting nor more singular problem in North American ornithology." Two years later, in Baird's Birds of North America, Mr. Cassin again refers to the subject, and says that "recent information renders it probable that this [*Vultur sacra* Bartram], or a species different from the vultures just described [*Cathartes aura*, *C. atratus*, *C. Burrovianus*], is found about Lake Okechobee in Southern Florida, where it is called king buzzard. The verification of this statement by actual specimens would be one of the most important discoveries yet to be made in North American ornithology."

Although the description of Bartram's "*Vultur sacra*" accords more nearly with the *Sarcoramphus papa* than with any other known species, I cannot avoid the conclusion that it is in the main a *purely mythical species*, notwithstanding the high reputation for veracity generally accorded to Mr. Bartram. I mainly so regard it for the reason that Florida has of late been too often traversed by naturalists, *and especially all the parts visited by Bartram*, for a bird of so striking an appearance, and so numerous as Bartram represented his *V. sacra* to be, to remain undiscovered if such a species exists there. While it nearly accords with the *S. papa* in size and general color, it is most radically different from this species, in the color of the tail, and in having a "large portion of the stomach hanging down on the breast, in the likeness of a sack or half-wallet." In the latter feature it is structurally widely different from any known American bird. It is mentioned as though it was an abundant species on, at least, the upper portion of the St. John's River, inasmuch as he speaks of large flights of them. As to the feathers of its tail being used by the Creek Indians for a royal standard, and to which feathers they give a "name signifying an eagle's tail," it seems to me more probable that they were really feathers of the white-headed eagle (*Haliaëtus leucocephalus*), since it is well known that the tail feathers of that bird are very generally used for this and similar purposes by the Indian tribes of this continent, whereas the tail feathers of so foul a bird as the vulture must in all

* Illustr. of Birds of Cuba and Texas, p. 59.

† The *S. papa*, a Central and South American species, appears to have not yet been seen north of Mexico.

probability be too ill scented to suit even the unfastidious taste of an Indian. As to Mr. Cassin's supposition that the word *white* in the description of the tail should perhaps read *black*, the context wholly forbids its probability. If thus changed the passage referred to would read, "the tail which is rather large and *black*, is tipped with this *dark brown or black!*" which makes simply an absurdity. Besides this, the tail is again mentioned in the following paragraph as being painted by the Indians, when used in their war standards, etc., "with a zone of red within the brown tips," and afterwards as being "displayed new, clean, and *white*." As to the information referred to by Mr. Cassin as having been received by him respecting a "king buzzard" existing in Southern Florida, it may be remarked that this is the name by which the caracara eagle (*Polyborus tharus* Cassin) is commonly known in Florida, and which is undoubtedly the bird of which, under the name of "king buzzard," Mr. Cassin had heard.

On the whole, it seems evident that Bartram's account of the *Vultur sacra* is a confused mixture either of pure fiction and truth, with the former largely in preponderance, or of the characters of several different species. The description would seem to have been mainly drawn from an example of *Sarcoramphus papa* that he may have somewhere met with, but with which he combined certain features of this or other species which he had only observed at a distance, and that he thus misjudged their exact character (as in respect to the strange external food-pouch) or else added them solely on popular, fabulous rumors. The flights of these birds, which he observed assembling over recently burned districts, I think must refer to the *Polyborus tharus*, which is well known to have this habit, while the tail feathers he speaks of as used by the Indians in their councils were more probably either those of the *Haliaëtus leucocephalus* or *Polyborus tharus* than of any species of vulture, since a white-tailed American vulture, I believe, is a bird thus far unknown. If the "*V. sacra*," then, is to be regarded as anything else than a myth, it should in all probability be identified with the *S. papa*, as already stated, and as was done by Bonaparte in his Conspectus.

FALCONIDÆ.

81† **Falco peregrinus** *Linné*. DUCK HAWK.

Falco peregrinus GMELIN, Syst. Nat. I, 272, 1788. — WILSON, Am. Orn., IX, 120, 1814. — BONAPARTE, Journ. Phil. Acad. Nat. Sci., 1st Ser., I, 342, 1824. — AUDUBON, Orn. Biog., I, 85, 1832; V, 365, pl. xvi. — NUTTALL, Man. Orn., I, 53, 1832.

Falco anatum BONAPARTE, Geog. and Comp. List, I, 1838. — CASSIN, Illust. Birds Cal. and Texas, 86, 1853. — CASSIN, Baird's Birds of N. Am., 7, 1858. — ALLEN, Proc. Essex Inst., IV, 153, 1865.

Falco nigriceps CASSIN, Illust. Birds of Cal., 87, 1853. — CASSIN, Baird's Birds of N. Am., 1858.

"One instance, St. Augustine, February, 1868." *Boardman.* Mr. Maynard found it rather common near Dummitt's, where he observed its peculiar manner of capturing the ducks. Also well known to occur in winter in Cuba and other of the West India Islands.

In 1838, Bonaparte, in his "Geographical and Comparative List," gave to the American peregrine or duck hawk the name *Falco anatum.* Previous to this time all writers had considered it, and it seems to me justly, as identical with the European peregrine, or *F. peregrinus,* — an opinion still held by many eminent ornithologists. Until about this date the peregrine falcon was believed to have a nearly cosmopolitan distribution, but since then the Australian and other supposed species have been separated from it on grounds that it now seems should be reconsidered. Among these supposed species is the *Falco nigriceps* of Cassin, first described in 1858, from specimens received from California and Chili. These first specimens were smaller, with the rufous color of the under parts in the young of a stronger tint than in the so-called *F. anatum,* they more resembling the African, Australian, and especially the Indian type of *F. peregrinus.* Specimens since obtained from farther north, however, fully equal those from Eastern North America, and the slight differences found to really exist between them seem to be by no means of specific value.

Formerly a difference in breeding habits was supposed to obtain between the American and European peregrines, the American peregrine being for a long time believed to breed in trees, whilst the European was well known to nest on cliffs. Recently, however, the American bird has been repeatedly found nesting in similar situations, but never yet in trees.*

82.† Falco columbarius *Linné.* PIGEON HAWK.

Falco columbarius WILSON and subsequent American writers generally.
Falco æsalon SWAINSON, Faun. Bor. Am., II, 35, pl. xxv, 1831. — NUTTALL, Man. Orn., I, 60, 1832.
Falco temerarius AUDUBON, Orn. Biog., I, 381, pl. lxxv, 1832.

"Frequent." *Boardman.*

* For an account of the breeding habits and nesting-places of the American bird in the Atlantic States, see the author's papers in Proc. Essex Inst., Vol. IV, pp. 153 - 161, and American Naturalist, Vol. III, p. 514. The past summer (1870) its eggs have been received by Mr. C. W. Bennett from Vermont. Prof. S. S. Haldeman was not only the first naturalist who made known the fact of its breeding on cliffs, but of its breeding in the United States. See Proc. Phil. Acad. Nat. Sci., Vol. I, p. 54, July, 1841.

Many of the earlier ornithological writers regarded, as is well known, a considerable proportion of the rapacious birds of North America as identical with species inhabiting the Old World. More accurate comparisons of specimens from the two continents, however, eventually revealed appreciable differences between them, and one after another of those of the American continent were regarded as specifically distinct from their Old World relatives; and now there is not one of the diurnal species that has not been separated by one author or another. The owls of the two continents, with two exceptions, have also been similarly separated. While in many of these cases there are appreciable differences that seem more or less constant, in the majority of instances there appears to be no just cause for the separation. Especially is this the case in respect to *Falco peregrinus* (as already observed), *Falco candicans*, *Archibuteo lagopus*, *Aquila chrysaëtos*, *Pandion haliaëtus*, *Otus vulgaris*, *Brachyotus palustris*, *Nyctale Tengmalmi*, and *Strix flammea*, in all of which species the American birds have been specifically separated from the European. *Buteo borealis*, *Astur atricapillus*, and *Falco columbarius* present stages of plumage that are scarcely distinguishable from certain stages of respectively *Falco æsalon*, *Buteo vulgaris*, and *Astur palumbarius*, and it is hence not strange that each of these European species have been described by many good authorities as occurring in the northern parts of North America. Certain styles of plumage presented by *Falco columbarius*, especially at northwestern localities, so strongly resemble common phases of *F. æsalon*, that one is readily puzzled to know whether to recognize the latter as also inhabiting North America, or whether, since these types imperceptibly grade into the so-called typical *F. columbarius*, all should not be regarded as forming a single species, since they differ essentially only in coloration, and never very widely. The specimens of *F. æsalon* before me (all immature) mainly differ from average specimens of *F. columbarius* of corresponding age in being less ferrugineous, the style of coloring being the same in both.

83.* Falco sparverius *Linné.* Sparrow Hawk.

Falco sparverius Linné, Syst. Nat., 128, 1766; and of subsequent writers generally.
Falco dominicensis Gmelin, Syst. Nat., I, 285, 1788.
Falco gracilis Swainson, Lardner's Cab. Cyc., 281, 1838.
Falco cinnamominus Swainson, Ibid., 281.
Falco isabellinus Swainson, Ibid., 281.
Falco sparveroides Vigors, Zoöl. Journ., III, 436, 1827.

Abundant. Breeds in March. As has been previously pointed out, though not observed by all writers, the sexes differ greatly in color, the adult females being banded transversely above, much as the young birds are.

Florida specimens are considerably smaller than New England ones, the former being intermediate in size between the latter and the West Indian and South American representatives of this species, which have been regarded as distinct species, and to which various names have been applied by different writers. Audubon observes that he found this species in the Southern States, and more especially in Florida, so much smaller than the northern birds that he was at first inclined to consider them specifically distinct, but finally felt sure they were the same. The colors, as usual in other species, are generally brighter in the more southern examples. Wide variations in the color of the plumage in this species have been long recognized, but, as Mr. Cassin has remarked, " they do not appear to be constant, nor peculiar to any locality." *

84 * **Accipiter fuscus** *Bonaparte.* SHARP-SHINNED HAWK.

Falco fuscus GMELIN, Syst. Nat., I, 280, 1788.
Accipiter fuscus BONAPARTE, Geog. and Comp. List, 5, 1838.
Astur fuscus AUDUBON, Syn., 18, 1839.
Falco dubius GMELIN, Syst. Nat., I, 281, 1788.
Falco velox WILSON, Am. Orn., V, 116, 1812.
Falco pennsylvanicus WILSON, Ibid., VI, 13, 1812.
Accipiter striatus VIEILLOT, Ois. Am. Sept., I, 42, 1807.
Accipiter fringilloides VIGORS, Zoöl. Journ., III, 434, 1827.
Accipiter pennsylvanicus RICH. & SWAIN., Faun. Bor. Am., II, 44, 1831.
Nisus Malfini LESSON, Traité d'Ornithol., I, 58, 1831.

Common. I was unable, however, to obtain specimens.

In this species, as in the hawks generally, but more especially in the group to which the present species belongs, there are wide variations in color and size, not only with age and sex, but independently of either. One of the most interesting features in the specimens before me, in respect to these variations, is the much brighter color of the several western and southwestern examples in the collection of the Museum, as compared with New England ones. In one from Cheltenham, Missouri, the color of the lower parts is nearly uniformly red; the transverse dark lines, which in adult eastern specimens usually occupy half the exposed surface of the feathers, and often more, being in this specimen almost obsolete. The tibial feathers are especially bright, while the tints are livelier throughout the plumage. Other specimens from Fort Steilacoom, received from the Smithsonian Institution, present nearly the same appearance. Although the western representatives of the present species yet await some enter-

* Illust. Birds of California and Texas, etc., p. 93.

prising divisionist to give them a distinctive name, they are interesting as indicating a rufous western race, corresponding with the *Accipiter mexicanus* form of the *A. Cooperi*, the *Falco nigripes* form of the *F peregrinus*, the *Archibuteo ferrugineus* form of the *A. lagopus*, and the western rufous forms of *Buteo borealis* and *Circus hudsonius.**

Although the *Accipiter fuscus* has always been regarded as closely related to the *Accipiter nisus* of the Old World, they have, with one or two exceptions,† been regarded by all authors as specifically distinct. The only distinctive difference between them, however, has been properly regarded as a slight difference in color, which difference is merely one of tint, the style of coloration being precisely the same in both. In the Museum of Comparative Zoölogy are several specimens of *A. nisus* from Germany and Switzerland, which represent both the adult and the young. The brown transverse markings on the lower plumage of the mature *A. nisus* are rather darker and broader than in most New England specimens of *A. fuscus;* but they still more closely resemble average New England specimens than the latter do any specimens of *A. fuscus* I have seen from the western parts of the United States. The western form of *A. fuscus*, as already stated, is brighter colored or more rufous than the eastern, while the eastern differs similarly from the European, the latter being much duller colored than the eastern form of *A. fuscus.* So closely, however, does one of the immature examples of *A. nisus* resemble several of the immature New England specimens of *A. fuscus*, that, if their origin was unknown, few ornithologists would probably consider them as otherwise than specifically identical; especially if placed in a large series composed of both eastern and western specimens of the *A. fuscus.* As I have previously remarked, the transverse markings on the lower plumage in the adult stage are broader and more regular and distinct in *A. nisus* than in *A. fuscus.* This, it may be added, is also the only difference observable between *A. palumbarius* and *A. atricapillus.* Such a coincidence of parallel differences between *Accipiter nisus* and *Accipiter fuscus*, and between *Astur palumbarius* and *A. atricapillus*, is a point of much interest to any

* For further remarks concerning the rufous western races of several of these species see the following pages.

† Prince Max zu Wied, in his "Beiträge zur Naturgeschichte von Brasilien," referred a hawk, probably of this species, of which he obtained a single immature male in Eastern Brazil, to the *Falco nisus* Linn. Respecting this species he observes: "Der Vogel dieser Beschreibung scheint von dem europäischen Sperber nicht abzuweichen. . . . . Dieser Sperber ist mir selbst in Braslien nicht vorgekommen, allein *Freireifs* hat mir ein Exemplar davon mitgetheilt, welches in der Gegend von *Camamú*, südlich von *Bahia*, geschossen wurde. So viel ich von diesem einzigen Individuo urtheilan kann, so scheint es identisch mit dem europäichen *Nisus* zu seyn; denn sowohl seine Verhältnisse als sein Gefieder stimmen vollkommen überein." Vol. III, pp. 112, 114.

one interested in geographical color variations in animals; the more so, perhaps, from the two latter species being so intimately related as to have been at one time generally regarded as identical. Yet so far as can be judged from a limited number of specimens, *Astur palumbarius* differs more from *A. atricapillus* than *Accipiter nisus* does from *Accipiter fuscus*, which latter species have never been considered as identical.*

85.* **Accipiter Cooperi** *Cassin.* COOPER'S HAWK.

Falco Cooperi BONAP., Am. Orn., II, 1, 1828.
Falco Stanleyi AUDUBON, Orn. Biog., I, 186, 1831 (young).
Astur Cooperi BONAP., Geog. and Comp. List, 5, 1838.
Accipiter Cooperi CASSIN, Illust. Birds of Cal., etc., 96, 1854.
Accipiter mexicanus SWAIN., Faun. Bor. Am., II, 45, 1831. — CASSIN, Baird's Birds N. Am., 17, 1858.
Accipiter Gundlachi LAWR., Ann. N. Y. Lyc. Nat. Hist., VII, 252, 1860.

Common.

Mr. Cassin has very properly indicated the variations in size and color commonly seen in this species in the following remarks: "Rather a difficult species to the ornithologists, on account of the great variations in its colors, and in size also. It is, in fact, unusual to find two alike in a dozen specimens."† Its relationship to *Accipiter fuscus* is of course well understood, it holding a similar relationship to that species that *Picus villosus* does to *Picus pubescens*, the essential difference between them being mainly a great difference in size. But the specific distinctness of *A. mexicanus* from it is not so clear. Being without authentic specimens of *A. mexicanus*, and having only New England specimens of *A. Cooperi*, ‡ I cannot speak confidently respecting the character and affinities of the former. According to authors, however, it seems to differ from *A. Cooperi* in being somewhat smaller and more highly colored. It is also more southern in its distribution. Hence these variations, being in accordance with the general laws of geographical variation in size and color, do not necessarily

* In this connection I wish to cite some interesting variations in color presented by Massachusetts and Maine specimens of *Astur atricapillus*. Ordinarily this species has each feather below centred with a longitudinal dark shaft-line, with several transverse broader but somewhat irregular dark ashy-brown bars on a lighter ground. Some specimens, however, as one from Maine, have the transverse bars so narrow and broken that the lower surface presents a nearly uniform, minutely mottled appearance. Another specimen (from Springfield, Mass.) represents the opposite extreme, it having the transverse bars broad, regular, and quite far apart, so that its resemblance to average specimens of *Astur palumbarius* is very close. The color in this specimen is much darker throughout than is usual in this species.

† Illustrations of Birds of California, etc., p. 93, 1854.

‡ Since the above was written, specimens have been received at the Museum from Jalapa, Mexico, from Sn. R. Montes-de-Oca.

imply a diversity of species; they only accord with what would naturally be expected to occur if *A. mexicanus* and *A. Cooperi* were known to constitute but a single species.*

Accipiter Cooperi, as is well known, is not only closely allied in general structure to *Buteo lineatus*, but also in style of coloration in both the immature and adult stages. It may be fair, then, to test the value of the distinctive characters assigned to *A. mexicanus* by what obtains as geographical variations in size and color in *Buteo lineatus*. Of this species I have fortunately a large number of specimens, including some from localities similarly separated to those whence *A. Cooperi* and *A. mexicanus* respectively come. In the case of *Buteo lineatus* there is no reason whatever to doubt that my specimens from Florida and New England are specifically identical. Yet the Florida specimens are very much brighter colored, and very much smaller; the difference in the length of the folded wing between two males, one of which is from Maine and the other from Florida, being *two and one half inches*, with corresponding differences in general measurements. This is relatively much greater than the difference in size between specimens of the so-called *A. Cooperi* and *A. mexicanus*. Similar variations in color and size to those between *A. Cooperi* and *A. mexicanus* also occur between northeastern and southwestern specimens of *A. fuscus*, the latter, as already noted under *A. fuscus*, being smaller than the former, and very much brighter colored; the difference in color between specimens from Maine and the State of Missouri being greater than is represented to occur between *A. Cooperi* and *A. mexicanus*, and of a parallel kind. In accordance with the evident inference that may be drawn from these facts, I provisionally include *A. mexicanus* among the synonymes of *A. Cooperi*. The *A. Gundlachi* of Cuba differs from the southern *A. Cooperi* in the way southern birds usually differ from the northern ones of the same species, that is, in being smaller and brighter colored, and in having the dark transverse bars on the under plumage increased in breadth at the expense of the alternating light ones.

86.* **Buteo borealis** *Bonaparte*. RED-TAILED HAWK.

Falco borealis GMELIN, Syst. Nat., I, 266, 1788. — WILSON, Am. Orn., VI, 75, pl. lii, fig. 2, 1812. — RICH. & SWAIN., Faun. Bor. Am. II, 50, 1831. — AUDUBON, Orn. Biog., I, 265, pl. II, 1832.

Buteo borealis BONAPARTE, Geog. and Comp. List, 3, 1838. — GOSSE, Birds of Jamaica, II, 1847. — LEMBEYE, Av. de la Isla de Cuba, 18, 1850. — CASSIN, Syn. N. A. Birds (Illust. Birds Cal. and Texas, etc.), 97, 1854. — BREWER, N. Am. Oölogy, 21, 1857. — CASSIN, Baird's Birds of N. Am., 25. 1858. — BRYANT, Proc. Bost. Soc. Nat. Hist., VIII, 109, 1861. — ALLEN, Memoirs Bost. Soc. Nat. Hist., I, 499, 1868.

* Bonaparte indeed long since cited *A. mexicanus* Swainson as a synonyme of *A. Cooperi*.

Falco leverianus GMELIN, Syst. Nat., I, 266, 1788. — WILSON, Am. Orn., VI, 78, pl. lii, 1812.

Falco jamaicensis GMELIN, Syst. Nat., I, 266, 1788.

Falco aquilinus BARTRAM, Travels, 290, 1791.

Falco Harlani AUDUBON, Am. Orn., I, 441, 1831.

Accipiter ruficaudus VIEILLOT, Ois. Am. Sept., I, 47, 1807.

Buteo ferrugineicaudus VIEILLOT, Ibid., 32.

Buteo fulvus VIEILLOT, Nouv. Dict. Hist. Nat., IV, 472, 1816.

Buteo americanus VIEILLOT, Ibid., 477.

Buteo vulgaris RICH. & SWAIN., Faun. Bor. Am., II, 47, pl. xxvii, 1831. — AUDUBON, Syn., 5, 1839.

Buteo buteoides NUTTALL, Man. Orn., I, 100, 1832.

Falco buteo AUDUBON, Orn. Biog., IV, 108, 1838.

Buteo Swainsoni BONAPARTE, Geog. and Comp. List, 3, 1838. — CASSIN, Illust. Birds Cal. Texas, etc., 98, 1854. — BREWER, N. Am. Oölogy, 24, 1857. — COOPER & BAIRD, Orn. Cal., I, 476, 1870.

Buteo Harlani BONAPARTE, Geog. and Comp. List, 3, 1838. — CASSIN, Illust. Birds Cal., Texas, etc., 101, 1854. — CASSIN, Baird's Birds N. Am., 14. — ? BRYANT, Proc. Bost. Soc. Nat. Hist., VIII, 115, 1861. — COOPER & BAIRD, Orn. Cal., I, 473.

Buteo montanus NUTTALL, Man. Orn. I (2d ed.), 112, 1840. — CASSIN, Baird's Birds N. Am., 26. — COUES, Proc. Phil. Acad. Nat. Sci., 1866, 43. — COOPER & BAIRD, Orn. Cal., I, 469.

Buteo Bairdii HOY., Proc. Phil. Acad. Nat. Sci., 1853, 451. — CASSIN, Baird's Birds N. Am., 21.

Buteo insignatus CASSIN, Birds Cal. and Texas, 102, pl. xxi, 1854. — CASSIN, Baird's Birds N. Am., 23. — COOPER & BAIRD, Orn. Cal., I, 474.

Buteo calurus CASSIN, Proc. Phil. Acad. Nat. Sci., 1855, 281. — CASSIN, Baird's Birds N. Am., 22. — COOPER & BAIRD, Orn. Cal., I, 471.

Not apparently uncommon, but far less numerous than the next species.

The *Buteoninæ*, or the group of hawks to which the present and the two following species belong, is well known to embrace species more variable in color than those of any other section of the *Falconidæ*, although all the members of this family are more or less remarkable for individual and other variations of plumage. The present species, however, admitting for it the wide variation in this respect herein claimed, scarcely equals the immense range of color variation well known to characterize its near ally and representative in the Old World, the *Buteo vulgaris* auct. (*Falco buteo* Linné). Six specimens of this species in the Museum from Switzerland and Germany, received under the name *Falco buteo*, vary in color as follows: One is almost entirely black; another is nearly black throughout, with obscure narrow transverse bands of ferruginous on the crissum and abdomen; another is mainly black, but varied below with bars of pale rufous and blotches of white; a fourth is also nearly black, or very dark brown, but considerably more relieved with white below than the last; a fifth is mainly white below, with longitudinal stripes of dark brown, and so nearly resembles a common immature stage of the American *Buteo borealis* that if placed together the most discriminating observer could not tell which specimen was the European or which the American one. The sixth is very light colored throughout, with only a few dusky longitudinal spots on the breast. This specimen is also not readily distinguishable from certain common phases of *B. borealis*. Another specimen of *B. vulgaris*, in the La Fresnaye collection in the Museum of the Boston Society of Natural History, is still lighter than this, being nearly uniform whitish below, and very light colored, almost white above. The latter specimen and the first-mentioned dark specimen present as great differences in color as two specimens of one species can well be conceived to exhibit.

The variations presented by the American *B. borealis* have already been fully detailed by the late Dr. Henry Bryant, in his "Remarks on the Variations of the Plumage of *Buteo borealis* auct., and *B. Harlani* Aud."* He observes that the variation in plumage of the species of *Buteo*, common in the Atlantic States, "are so slight that it is not to be wondered at that the first specimens from other parts of the country, presenting as they did such extraordinary variations in color, should have been described as distinct species. At present, however," he continues, "the number of specimens known is so large that on careful examination it seems to me necessary to adopt one of two conclusions, namely, either to increase the number of species indefinitely, or to reduce them to a much smaller number than are now supposed to exist. As the European buzzard, *Buteo vulgaris*, is well known to present the greatest variety of color, it seems to me more reasonable to adopt the last conclusion."† With the above opinions and remarks I in the main agree, but do not regard the variations presented by the *Buteo borealis* as by any means slight, even in the Atlantic States. Although instances of such excessive variation as are seen in the Central and Pacific States are apparently more rare in the Atlantic States, speci-

* Proc. Bost. Soc. Nat. Hist., Vol. VIII, p. 107, 1861.

† In respect to the variety of color in the *B. vulgaris*, Dr. Bryant makes the following quotation from Naumann's Natural History of the Birds of Germany (Vol. I, p. 347): "In the coloring of the feathers of the bird there prevails a most extraordinary difference, and one which is not often seen in other birds of prey. From the darkest uniform blackish-brown to the purest white, we find all the shades, and also both colors mixed and spotted, in such various ways that the countless transitions cannot be described; this difference is independent of age and sex." Many other European writers, it may be added, have made similar remarks in respect to its astonishing range of variation in color.

mens from Massachusetts now before me vary as follows: Some are nearly unspotted beneath, others, sparsely spotted, have the spots mainly restricted to the pectoral region; others, in which the spots are equally few, have them mainly accumulated on the abdominal region, while still others have them so numerous as to occupy the greater part of the lower surface, sometimes covering the abdomen in an almost unbroken broad band. They likewise vary in the amount of rufous tint in the plumage, in some it being very slight, while others are as strongly ferruginous as any of the California specimens (*B. montanus*) I have yet seen.

The *Buteo borealis* was first described by Latham in his "General Synopsis of Birds,"* in 1781, under the names of "cream-colored buzzard" and "American buzzard," the first name being applied to the young,† and the last to the adult stage of plumage. Pennant, in his "Arctic Zoölogy,"‡ also redescribes the immature bird as the "Leverian falcon," and to these several descriptions of Latham and Pennant, Gmelin, in his "Systema Naturæ," gave respectively the names *Falco jamaicensis*, *F. borealis*, and *F. Leverianus*. Some twenty years later the *Buteo borealis* was redescribed by Vieillot, in his "Histoire des Oiseaux de l'Amerique Septentrionale," as *Accipiter ruficaudus* and *Buteo ferrugineicaudus*, both names evidently referring to the mature or nearly mature bird; and again ten years later, in the "Nouveau Dictionnaire d'Histoire Naturelle," as *Buteo fulvus* and *B. americanus*. Audubon, in 1831, figured and described a specimen from Louisiana under the name *Falco Harlani*. This specimen, which was finally sent to the British Museum, has been regarded by Mr. G. R. Gray and others as only a very dark-colored example of *B. borealis*.§ In the same year Richardson and Swainson reported the *Buteo vulgaris*, in their "Fauna Boreali-Americana," as an inhabitant of North America, and of which they figure an immature male. As already remarked, the *B. vulgaris*, in certain stages of plumage, is not readily distinguishable from *B. borealis*, so that the mistake is a perfectly excusable one. This form, however, was for some time currently received by most writers as a species distinct from the *B. borealis*, and to which the name *B. Swainsoni* was given by Bonaparte. In 1832 Nuttall described a *Buteo buteoides*, which, though referred by Bonaparte to *B. lineatus*, and by Cassin to *B. pennsylvanicus*, seems to me to much more nearly agree with *B. borealis*. In 1840 the same writer described a *B. montanus*, which was subsequently

* Vol. I, pp. 49, 50, Nos. 30 and 31.

† Latham observes: "This beautiful specimen was sent to me from Jamaica by an intelligent friend and a good naturalist, who did not hint the least of its being a variety of the common buzzard [*Buteo vulgaris* auct.], which I should have otherwise suspected."

‡ Vol. II, p. 206, No. 101.

§ Cat. of Birds in British Museum.

referred by Bonaparte to his *B. Swainsoni*, but has since been recognized as a valid species by Cassin and other recent American authors. In 1853 Mr. P. R. Hoy described a *Buteo Bairdii*, and in 1854 Mr. Cassin added *B. insignatus*, in 1855 *B. calurus* and *B. oxypterus*, and in 1856 *B. Cooperi*. In 1861 Dr. Bryant made a revision of the group, then containing eight or nine species currently recognized by American ornithologists, and reduced the number of species to two, one of which he called *B. borealis* and the other *B. Harlani;* which latter, however, is not the *Harlani* of Cassin, and probably not the *Harlani* of Audubon.

Dr. Bryant, in the above-cited paper, describes in detail the leading variations presented by our red-tailed hawks, and the character of the numerous supposed species of this group that had then been recently described. He having at his command all the specimens of this group contained in the Museums of the Philadelphia Academy and the Smithsonian Institution, including the original types of Mr. Cassin's species, as well as the specimens in his own collection, his opportunities for investigating the subject were unusually favorable. The results of his examination of this material may be briefly stated in his own words. He says that after examining this large series of specimens, he found "that of all those belonging to *Harlani*, *insignatus*, *Swainsoni*, *Bairdii*, *oxypterus*, *borealis*, *montanus*, *calurus*, and perhaps *Cooperi*," could be "easily reduced to two very distinct groups, each of which is distinguishable by definite external characters, and in which the variations of plumage, though apparently so great, if the extremes of the series only are taken into consideration, can, it seems to me, be arranged in a series, in which the connecting of the different members may be readily traced. Of these two groups, or rather species, one, which should be called *B. borealis*, as the first described, consists of that species, *montanus*, *calurus*, *Harlani*, and probably *Cooperi*, and is characterized by a very muscular body,* stronger and longer bill, longer and more powerful tarsi, and a more rounded wing, the fourth quill generally the longest, the fifth little if any shorter than the third, and the first always longer than the eighth. The other species, to which *Harlani ?*, *insignatus*, *Swainsoni*, *Bairdii*, and *oxypterus* belong, is distinguishable by a more slender body, shorter and weaker tarsi, and a more pointed wing, the third quill generally the longest, the fifth considerably shorter than the third, and the first always longer than the eighth."

"On making the examinations which led to the conclusion above stated," he further observes, "I was struck by the small number of specimens in which all the feathers were equally developed, and when they were so, the variation in the proportions of the primaries, and of the wings and

* Stuffed skins evidently afford rather unsatisfactory data for the determination of the relative muscularity of the body.

tail, in specimens of the same variety, was much greater than I had expected to find"; a result which indicates how unreliable such features are as specific distinctions, as I have already repeatedly remarked, and also, of course, the fallacy of the belief so generally held, that they are really among the most trustworthy.* After detailing some of the instances of variation in this respect in the specimens in question, he makes the following remarks on variations in other characters: "The variation in the number and shape of the tarsal scales is considerable, as is usual in birds of this order. The development of the festoon of the lower edge of the upper mandible, *one of the principal generic characters,*† varies particularly in *B. montanus*, the series of which is the largest, *from a sharp, almost tooth-like process to an entire absence of it.*"‡

Dr. Bryant described each of the so-called species of the later authors, and generally several authentic specimens of each, showing the variations of color they present. *B. montanus* is the so-called "western red-tail," replacing, it is supposed, *B. borealis* in the western half of the continent, and differing from it in the main only in being more rufous or brighter colored. Some specimens, however, from California and Oregon are not appreciably different from others from the Atlantic States, and among them is one received at the Museum from the Smithsonian Institution labelled "*B. borealis.*" *B. calurus* differs from these in being much darker throughout, and especially below. It has, however, according to Dr. Bryant, two varieties, one of which is much darker than the other. The *B. Harlani* of Cassin, Dr. Bryant says, "resembles very closely the dark variety of *calurus*, with the exception of its tail, which resembles *montanus*." Respecting the single known specimen of *B. Cooperi*, he says there is nothing in its coloration "that would make the supposition of its being a variety of *montanus* improbable." The tail presents the greatest dissimilarity and "has very much the appearance it would have in a semi-adult of this species, if the color were partially washed out." The tarsus, though long, he says is not longer than in some specimens of *montanus*; but observes that the scutellation of the tarsus presents certain peculiarities not seen in the others, there being but *two* rows of lateral scales instead of *three or four*, and two more than the usual number of transverse scales. § In respect to these supposed species he then observes: "After

* See the remarks on this point in Part III.

† The italicizing is my own.

‡ On differences of this kind the several supposed species of the *B. borealis* group have been arranged *in different subgenera!*

§ Since writing the above I have learned from Professor Baird that he is inclined to regard this specimen as "only an *Archibuteo ferrugineus* without feathers on the tarsus; at any rate, hardly a species." It is hence omitted in Cooper and Baird's "Ornithology of California," which has just appeared.

carefully examining the birds described above, I do not see, if *Buteo borealis*, *montanus*, and *calurus* are to be considered distinct species, that we can avoid increasing the number by separating from *montanus* two species, — one the dark Steilacoom variety, and the other that from Cape St. Lucas (which, by the way, is the most distinct variety that I have seen); from *calurus*, one species, the ferruginous variety from Fort Tejon; and adding to this group one species based on the adult *Harlani* of the Academy [*Harlani* of Cassin, not of Audubon], making in all seven species of this group. I have not included in this list the young *Harlani* of the Academy, which differs as much from the adult as from any other specimen of this group; or *Cooperi*," etc. After next describing in detail *Buteo Harlani* (*B. Harlani* of Bryant, not *B. Harlani* of Cassin, nor of Audubon), and its several varieties, which form the "species" *B. insignatus*, *Swainsoni*, and *oxypterus* of Cassin and the *B. Bairdii* of Hoy and Cassin, *with several varieties under each*, some of which he clearly shows are connecting links to others, Dr. Bryant concludes his paper with the following summary: "Taking color, therefore as a sufficient ground for specific distinction, we find that we have in the red-tailed group seven species, and in the other nine, which, with the young *Harlani* of the Academy, *Cooperi*, *fuliginosus*, *albonotatus*, *lineatus*, *elegans*, and *pennsylvanicus*, give a total of twenty-three species of this genus which are found in the United States."

But Dr. Bryant by no means admits color in this group to be a specific characteristic, and, as I have already remarked, in reducing the number of species of the red-tailed hawks to two, he takes general size and the proportions of the primary quills of the wing as the basis of distinction. He has accordingly given a table of comparative measurements and proportions of the two species, in which he has arranged, as he says and doubtless supposed, the larger specimens under *B. borealis*, and the smaller under *B. Harlani*. Size and the proportions of the quills, however, it seems to me, are equally arbitrary grounds for their separation, as an examination of his tables and descriptions evidently proves. It happens that in the first, or *B. borealis* series, nearly all the specimens are fully adult, as indicated by the tail being uniformly red, with a subterminal black band, — a stage of plumage which characterizes only adult individuals. In the second, or *B. Harlani* series, but one specimen (which does not appear in the table of measurements), is described that is not evidently somewhat immature, while the greater part of them are quite so.* Respecting the so-called *Buteo Bairdii*, of which numerous specimens have been reported, some from quite eastern localities, Dr. Bryant

* They have at least the tail numerously banded, as all immature *B. borealis* do have, and their general diagnosis is that of immature birds.

remarks that a single specimen in the Museum of the Philadelphia Academy is the only one he had seen "presenting the least appearance of adult plumage." In regard to the size of the specimens of the two series, adopting the length of the folded wing as the basis of comparison, — the best element in the tables available for comparison, in this respect, — the smallest and the largest specimens, measuring 370 and 438 millimetres respectively, occur in the *B. borealis* series. The average length of wing in twenty specimens of *B. borealis* is 409 millimetres, and in fourteen * specimens of *B. Harlani* Bryant, 405. The difference of 4 millimetres is an amount too trivial to be of account, as the addition of a single specimen to either series might reverse the difference. Hence the impression possessed by Dr. Bryant of an average difference in size between the two series was evidently an erroneous one.

There hence remains but a single difference, that in respect to the form of the wing, or the relative length of the primaries, by which to distinguish the two series, which is at best one of doubtful value. My present opinion is that all the so-called species of these two groups may be safely referred to the original *Buteo borealis*, except the *B. oxypterus*, which should be undoubtedly referred to the *B. pennsylvanicus*.

87.* **Buteo lineatus** *Jardine*. RED-SHOULDERED HAWK.

Falco lineatus GMELIN, Syst. Nat., I, 268, 1788. — WILSON, Am. Orn., VI, 86, pl. liii, fig. 3, 1812. — AUDUBON, Orn. Biog., I, 296, pl. lvi, 1832.

Buteo lineatus JARDINE, Am. Orn., I, 1832. — AUDUBON, Syn., 7, 1839. — CASSIN, Baird's Birds N. Am., 28, 1858. — VERRILL, Proc. Essex Institute, III, 141, 1862.

Falco hyemalis GMELIN, Syst. Nat., I, 274, 1788. — WILSON, Am. Orn., IV, 73, 1812. — NUTTALL, Man. Orn., I, 106, 1832. — AUDUBON, Orn. Biog., V, pl. lxxi, 1832 (young).

Buteo Cooperi ALLEN, Amer. Nat., III, 518, 1869.

Circus hyemalis BONAP., Journ. Phil. Acad. Nat. Sci., 1st Ser., III, 305, 1824.

Buteo elegans CASSIN, Proc. Phil. Acad. Nat. Sci., 1855, 281. — CASSIN, Baird's Birds of N. Am 28, 1858.

Very abundant. By far the most numerous species of the family.

Generally smaller and much brighter colored than New England specimens. The dark line along the shaft of the feathers below, especially on the throat and breast, is very distinct, in this respect and in the bright colors greatly resembling the so-called *Buteo elegans* of Cassin. *B. elegans*,

* The *B oxypterus*, referred to the *B. Harlani* by Bryant, is very much smaller than any other specimen in either series, and it seems to me has decided affinities, in its small size as in other features, with the *B. pennsylvanicus*, as stated by Mr. Cassin, and it is hence excluded in my computation of the average length of the folded wing.

however, has been generally considered as the western representative of *B. lineatus*, but it differs from the latter only in being brighter colored, or in having the ferruginous of the under parts more intense. In this it resembles the western representatives of the *B. borealis*, *Archibuteo lagopus*, *Accipiter fuscus*, *Circus cyaneus*, *Falco peregrinus*, and other species of this family, the western specimens of which are ordinarily more rufous than the eastern, though in only a part of them have the eastern and western races as yet been separated as distinct species.

The considerable difference in size between specimens of this species from New England and Florida has led to the supposition that the former may be specifically distinct from the latter, or at least that they form well-marked varieties.* The following measurements, however, show that specimens occur in Florida, in winter at least, nearly as large as average-sized New England specimens. But these may have been merely winter visitors, since two of the three specimens taken in February on the St. John's River are larger than any of the others, all of which were taken later in the season. Those taken by Dr. Würdemann at Cape Florida and Indian Key are also smaller than those from the St. John's River.

Measurements of Florida Specimens of BUTEO LINEATUS.

M. C. Z. No	Sex.	Locality.	Date.	Collector.	Length.	Alar Extent.	Wing.	Tail.
5223	♂	Volusia	Feb. 12, '69	J. A. Allen	22.25	41.50	13.00	7.75
5224	♂	"	Feb. 12, '69	"	20.00	39.50	12 25	7 50
5276	♂	Blue Springs	Feb. 21, '69	"	20.00	42.00	13.00	8.40
5310	♂	Enterprise	Mar. 1, '69	"	17.65	39.15	12.25	8.00
5331	♂	"	Mar 1, '69	"	17.75	40.25	12.30	7.50
5398	♂	Hawkinsville	Mar. 15, '69	"	18.00	40.50	12.85	7.75
10744	♀	Jacksonville	Dec. 31, '68	C. J. Maynard	19 20	41.50	12.60	8.50
10743	♀	"	Jan. 11, '69	"	19.20	40.05	12.60	8.50
6899	♀	Cape Florida	Apr. 5, '58	G. Würdemann	15.75	35.75	11.00	6 75
8630†	♂	Indian Key	Aug. 31, '57	"	17.50	37.00	11.20	—
6898	—	"	Aug. 1, '58	"	15.50	34.50	10.50	7.15
8629†	—	"	Nov. 10, '57	"	17.75	40.00	12.00	—
8631†	♂	"	Aug. 31, '57	"	17.50	37.00	11.10	—

88.† **Buteo pennsylvanicus** *Bonaparte.* BROAD-WINGED HAWK.

Falco pennsylvanicus WILSON, Am. Orn., VI, 22, 1812.

Buteo pennsylvanicus BONAP., Geog. and Comp. List, 3, 1838. — AUDUBON, Syn., 6, 1839. — CASSIN, Illust. Birds Cal., Texas, etc., 100, 1854. — CASSIN, Baird's Birds N. Am., 29, 1858.

Falco latissimus WILSON, Am. Orn., VII, 22, 1812. (Later published copies.)‡

* See Prof. A. E. Verrill in Proc. Essex Institute, Vol. III, p. 141, 1862.

† Smithson. Inst., No. (Copied from Cassin in Baird's Birds of North America, p. 28.)

‡ Concerning the names *F. pennsylvanicus* and *F. latissimus* given by Wilson to this species, see Mr. Cassin's remarks, Illust. Birds of Cal., Texas, etc., p. 101.

Falco Wilsoni BONAP., Journ. Phil. Acad. Nat. Sci., III, 348, 1824.
Sparvius platypterus VIEILLOT, Encyc. Meth., III, 1273, 1823.
Buteo oxypterus CASSIN, Proc. Phil. Acad. Nat. Sci., 282, 1855. — CASSIN, Baird's Birds of N. Am., 31, 1858.

"Common." — *Boardman.* Audubon, however, gives it as rare south of the Middle States, and it is not mentioned by Dr. Coues in his list of the birds of South Carolina. There is, however, a specimen in the Museun of Comparative Zoölogy labelled as having been taken in Florida.

As previously observed, it appears to me that the *Buteo oxypterus* of Cassin, described from a single specimen taken at Fort Filmore, New Mexico, corresponds more nearly with the young of this species than with any known stage or form of *B. borealis.*

89.* Circus cyaneus *Boie.* MARSH HAWK.

Falco cyaneus LINN., Syst. Nat. I, 126, 1766. — BONAP., Am. Orn., II, 30. — AUDUBON, Orn. Biog., IV, 396, pl. ccclvi, 1838.
Circus cyaneus BOIE, Isis, 1822, 549. — AUDUBON, Synop., 19, 1839. — G. R. GRAY, Gen. of Birds, I, p. 32. — IBID., Cat. Brit. Birds, 17, 1863.
Falco hudsonius LINN., Syst. Nat., I, 128, 1766.
Falco uliginosus GMELIN, Syst. Nat., I, 278, 1788.
Circus uliginosus VIEILLOT, Ois. Am. Sept., I, 37, 1807.
Falco uliginosus WILSON, Am. Orn., VI, 67, pl. li, fig. 2, 1812.
Buteo (Circus) cyaneus ? var. ? *americanus*, RICH. and SWAIN., Faun. Bor. Am., II, 55, pl. xxix, 1831.
Circus hudsonius VIEILLOT, Ois. Am. Sept., I, 36, 1807. — CASSIN, Ill. Birds Cal., Texas, etc., 108, 1854. — BREWER, N. Am. Oöl., 42, 1857. — CASSIN, Baird's Birds N. Am., 38, 1858.
Circus variegatus VIEILLOT, Ois. Am. Sept., I, 37, 1807.
Strigiceps uliginosus BONAP., Geog. and Comp. List, 5, 1838.
Strigiceps pygargus BONAP., Ibid.

Common about the savannas.

The present species has been considered by most writers as identical with the *C. cyaneus* of the Old World. It was first separated as a distinct species by Bonaparte in 1838, in his Geographical and Comparative List. Mr. Cassin also regarding it as distinct, this opinion has been generally adopted by recent American ornithologists. They seem to be, however, quite identical.

The same variation in color between eastern and western specimens is seen in this species that has been noted in others of this family, the young western ones especially being much brighter colored than the eastern.

The great variation in plumage attending differences of age and sex in this species have given rise to numerous synonymes, of which twenty are cited by Mr. G. R. Gray in his Catalogue of British Birds.

90.* **Pandion haliaëtus** *Cuvier*. FISH HAWK. OSPREY.

Falco haliaëtus LINNÉ, Faun. Suec., 22, 1735. — WILSON, Am. Orn., V. 13, pl. xxxvii, 1812. — BONAP., Ann. N. Y. Lyc. N. Hist., II, 26, 1828. — AUDUBON, Orn. Biog., I, 415, pl. lxxxi, 1832. — NUTTALL, Man. Am. Orn., I, 78, 1832.

Pandion haliaëtus CUV., Règ. An., I, 316, 1817. — AUDUBON, Synopsis, 12, 1839. — G. R. GRAY, Cat. Brit. Birds, 5, 1863. — PELZELN, Ornithol. Brasiliens, 4, 1868. — HEUGLIN, Ornithol. Nordost-Afrika's, 54, 1869.

Falco arundinaceus GMELIN, Syst. Nat. I, 263, 1788.

Falco carolinensis GMELIN, Ibid.

Pandion carolinensis BONAP., Geog. and Comp. List, 3, 1838. — CASSIN, Illust. Birds Cal., Texas, etc., 112, 1854. — BREWER, N. Am. Oöl., 53, 1857. — CASSIN, Baird's Birds N. Am., 44, 1858.

Falco cayanensis GMELIN, Syst. Nat. I, 268, 1788.

Aquila piscatrix VIEILLOT, Ois. Am. Sept., I, 29, 1807.

Pandion fluviatilis SAVIG., Descr. de l'Egypte, Hist. Nat., I, 96, 1809.

Pandion americanus VIEILLOT, Gal. des Ois., I, 33, 1828.

Pandion indicus HODGSON, Journ. As. Soc. Bengal, 366, 1837.

Abundant everywhere; especially so around the lakes of the Upper St. John's. Commences nesting in January. At Lake Monroe I counted six nests from a single point of view. Their nests were also frequent all along the river. They generally selecting a dead tree in which to build, and often those situated in cleared fields, their nests were conspicuous objects, and could usually be seen from a long distance. Even these harmless birds do not fail to attract the fire of the numerous sportsmen who visit this region in winter, some of whom are ignorant enough to believe that when shooting them they are killing "bald eagles."

Gmelin, in his "Systema Naturæ," described the present species not only as *Falco haliaëtus*, but he gave to it also the names *F. carolinensis*, *F. arundinaceus*, and *F. cayanensis*, apparently indicating under them, however, what he regarded as varieties rather than as distinct species. For many years, however, the common fish-hawk was generally regarded as having an almost cosmopolitan distribution. Bonaparte spoke of it in 1826, in his Synopsis of the Birds of the United States,* as follows: "Inhabits almost every part of the globe near waters; much more common in North America than in Europe." Ten or twelve years later, however, he seems to

* Annals of the N. Y. Lyceum of Nat. History, Vol. II, p. 26.

have changed this opinion, since in his Geographical and Comparative List of the Birds of Europe and the United States (to which paper, by the way, we are indebted for the separation of eight of the American species of raptorial birds previously considered identical with the European,* embracing all thus separated up to the present time, except two †) he calls the American fish-hawk *Pandion carolinensis*, and gives its habitat as "America generally." Other authors have since separated the West Indian and South American as a third, the Asiatic as a fourth, and the Australian as still another. The numerous specimens in the Museum show that considerable variation obtains in color, size, and proportions among those recognized by authors as belonging to the *P. carolinensis*, much greater differences in color — the main ground on which they have been separated from the European — being presented among the Florida specimens alone than obtains in the average between Brazilian and New England specimens, or American and European. Generally the feathers of the breast are each centred with a broad longitudinal spot or stripe of brown, which spots sometimes cover the greater part of the breast; but they are often simply narrow lines, and are not unfrequently entirely wanting. Sometimes these spots are uniform dark-brown, at others suffused or broadly margined with ferruginous, and are occasionally altogether of the latter color. In reuniting the American fish-hawk with the osprey of the Old World, I but adopt the view always held by a large number of ornithologists, though by all American authors they have for the last fifteen years been commonly considered as distinct.

Measurements of Florida Specimens of PANDION HALIAËTUS.

M. C. Z. No.	Sex	Locality.	Date.	Collector.	Length.	Alar Extent.	Wing.	Tail.
5268	♂	Blue Springs	Feb. 21, '69	J. A. Allen	21.75	64.00	19.50	8.75
5298	♂	Enterprise	Feb. 25, '69	"	24.25	68 75	20.25	10.00
5330	♂	"	Mar. 4, '69	"	22.00	63.50	19.25	9.00
5356	♂	Hawkinsville	Mar. 10, '69	"	20 75	63.00	18.75	8.60
5355	♀	"	Mar. 10, '69	"	——	——	20.25	7.80
——	♀	"	Mar. 15, '69	"	24.25	66.25	19.00	——
——	♀	"	Mar. 15, '69	"	23.50	68.50	20.25	——

91. **Haliaëtus leucocephalus** *Savigny*. WHITE-HEADED EAGLE.

Falco leucocephalus GMELIN, Syst. Nat., I, 255, 1788. — WILSON, Am. Orn., IV, 89, pl. xxxvi, 1811. — AUDUBON, Orn. Biog., I, 58, pl. xxi, 1832; II, 160; V, 354, pl. cxxvi.

* *Pandion carolinensis* from *P. haliaëtus*, *Butætes* (or *Archibuteo* as now called) *Sancti-Johannis* from *B. lagopus*; *Buteo Swainsoni* from *B. vulgaris*; *Falco anatum* from *F. peregrinus*; *Astur atricapillus* from *A. palumbarius*; *Strigiceps* (*Circus* as now called) *uliginosus* from *S. pygargus* (*cyaneus* auct.); *Otus americanus* (or "*Wilsonianus*") from *O. vulgaris*; *Nyctale Richardsoni* from *N. Tengmalmi*; *Strix pratincola* from *S. flammea*.

† *Aquila chrysaëtos*, *Brachyotus palustris*.

Haliaëtus leucocephalus SAVIGNY. — BONAPARTE, Geog. and Comp. List, 3, 1838. — AUDUBON, Synop., 10, 1839. — CASSIN, Illust. Birds Cal., Texas, etc., 111, 1854. — CASSIN, Baird's Birds N. Am., 43, 1858.

Falco ossifragus WILSON, Am. Orn., VII, 16, pl. lv, 1813.

Aquila (*Haliaëtus*) *leucocephalus* RICH. & SWAIN., Faun. Bor. Am., II, 15, 1832.

Falco Washingtoni AUDUBON, Orn. Biog., I, 58, pl. xi, 1831 (plate published 1827).

Falco Washingtoniana AUDUBON, Loudon's Mag. N. Hist., I, 115, 1828.

Haliaëtus Washingtoni AUDUBON, Synop., 10, 1839. — CASSIN, Baird's Birds N. Am., 42, 1859.

Common. Breeds in January and later. Very abundant on the Upper St. John's, and especially at Lake Monroe. Saw them repeatedly dive and catch their own fish, though usually depending upon robbing the fish-hawks for them. The same fact has been reported by other observers,* although it was formerly supposed they never caught any fish themselves.

The large specimen of an eagle taken by Audubon in Kentucky, and figured and described by him as *Falco Washingtoni*, seems not to have been preserved; it is at least not known to be extant, and appears to have never been examined by any other naturalist. Audubon states that he altogether saw not "more than eight or nine" specimens, † and deemed it very rare. He does not appear, however, to have really examined but the one figured. Numerous local observers have reported it as occasional at different localities, and Mr. Cassin has doubtfully referred specimens to it taken in New Jersey. Nuttall believed the young were more or less common near Boston every winter, and considered it as "probably also indigenous to northern Europe, but confounded with the ordinary sea eagle." ‡ But, as remarked by Mr. Cassin, "No specimen precisely corresponding to Mr. Audubon's bird has been obtained since its discovery, and it has latterly been looked upon by naturalists, especially in Europe, as an unusually large specimen of the white-headed eagle." § The important point of difference between Audubon's bird and other representatives of this genus consists in the scutellation of the tarsi, which are covered in front with broad transverse scales, instead of with a great number of small irregular ones, as in other sea eagles. This, Mr. Cassin

* WILLIAM COUPER, Massachusetts Ploughman, August 26, 1870. CHARLES H. NAUMAN, on his own authority and that of Professor S. S. Haldeman, Ibid., September 24, 1870. HENRY REEKS, Can. Nat., Vol. V, No. 1, p. 43, 1870.

† Loudon's Mag. of Nat. Hist., Vol. I, p. 116, April, 1828.

‡ Mem. Am. Acad., 1st Ser., Vol. I, p. 92, 1831.

§ Illustrations of Birds of California, Texas, etc., p. 111, 1854.

has observed, is " a character quite unusual in any rapacious bird," * though I do not see that in this respect it differs essentially from *Buteo lineatus*, *B. pennsylvanicus*, or *Circus cyaneus*, etc. Its other main point of difference from the *H. leucocephalus* is its greater size. Audubon described his bird as measuring " 3 feet 7 inches in length," " 10 feet 2 inches " in extent of wings, and the folded wing " 32 inches." In this series of measurements there is no discrepancy between the different dimensions given — the proportions being exactly the same as in *H. leucocephalus* — that might lead to the suspicion of a typographical or other accidental error, as some writers have suggested there may be in respect to the alar extent. It is, then, either a valid species or a large individual of *H. leucocephalus*, or a large immature *H. albicilla*. Since known specimens of *H. leucocephalus* sometimes nearly approach the supposed *H. Washingtoni* in size, it seems not unreasonable, on the whole, to regard it as really a remarkably large example of *H. leucocephalus* in immature plumage. Audubon describes his bird as breeding within the United States, and hence it is hardly probable it could have been the arctic *H. albicilla*, which has never, so far as known to me, been observed so far south at any season of the year. In reference to its fishing habits, supposed by Audubon to distinctively characterize it, it is now well known that the *H. leucocephalus* will occasionally capture its own fish, instead of depending wholly upon robbing the fish-hawk for them.

Mr. Cassin further observes, † respecting the *H. Washingtoni*, that he believes it to be more nearly related to his *H. pelagica*, which he describes as " the largest of eagles," than to any other. In the same connection he judiciously remarks respecting the numerous apocryphal species of eagles on record as follows: " But there is no end to the accounts of strange eagles given by travellers and naturalists. Some of them may have reference to peculiar species which have in later times escaped attention, but the probability is they more frequently allude to accidental varieties, or that the authors describe from such reports as they had heard at second hand, or fell into error from insufficient personal observation." Many of these reports he alludes to in detail, including the reference by Captain Cook ‡ to a " black eagle " with a " white breast " seen by him at Kay's Island, on the northwest coast of America. A specimen of the *H. leucocephalus* in peculiar (probably albinic) plumage in the Museum of Comparative Zoölogy, taken in Eastern Massachusetts, seems to indicate that the eagle of Captain Cook may have been but an unusual stage of coloration of the common white-headed eagle. The Massachusetts specimen

* Baird's Birds of N. America, p. 42.

† Illust. Birds of Cal. and Texas, p. 36.

‡ Cook's Voyages, II, 352, 1784.

above referred to has the general color of the under parts white, with most of the feathers centred with spots of dusky brown of varying size, but with a nearly uniform dusky brown patch on the middle of the breast. The interscapulars are also mainly white, and the general plumage above, except the wings, more or less varied with the same color. The tail below is mottled with irregularly shaped specks and spots of dusky or black on a white ground, and above with white on a nearly black ground, and tipped with dusky. The appearance of the under side of the bird at a distance would be nearly uniform whitish.

Mr. Cassin having stated repeatedly that his *Haliaëtus pelagicus* (the *Aquila pelagica* Pallas *) is the largest and most powerful of all known eagles,† I was greatly surpised, in critically studying his description, to find it in every respect evidently far inferior in size to Audubon's bird of Washington, and scarcely equalling the *H. albicilla*, as described by himself; the folded wing, in fact, of his *H. pelagicus* is *one inch shorter* than the folded wing of his *H. albicilla*, *four inches shorter* than the wing of the *H. Washingtoni*, as measured by Audubon, and *two inches shorter* than the folded wing of several different Massachusetts specimens of *H. leucocephalus!* The length he gives of "a skin from Behring's Strait" — the only specimen, he says, at that time in America — is "about 3 feet 8 inches," which exceeds by *one inch only* the length of Audubon's *H. Washingtoni*, as given by Audubon, doubtless from the fresh bird. But the length given by Mr. Cassin for his *H. pelagicus* is evidently too great, as, taken in connection with the other measurements of the same specimen given by Cassin, if correct, it would indicate a bird of the most anomalous and improbable proportions. Mr. Cassin's erroneous conception of the gigantic size of his bird was doubtless formed from the length of his specimen, which if a flat or unfilled skin, as it probably was, must have measured several inches more than the natural length of the bird.‡ While I do not in the least question the sincerity of Mr. Cassin's belief in the large size of his bird, I have felt it proper to call the attention of future investiga-

* Zoographia Rosso-Asiatica, I, p. 343.

† "The bird which is the subject of our present article is the largest and most powerful of the eagles." — *Illust. Birds Cal. and Texas*, p. 32, first paragraph. "Even the famous condor of the Andes, the largest of vultures, scarcely exceeds him in size," etc. *Ibid.*, p. 32, third paragraph. "The largest of all known eagles, and nearly related to *H. Washingtoni* (Aud.). It differs from the latter as described by Audubon in being generally *larger*," etc. *Ibid.*, p. 38. "It is the largest of the eagles and appears to be related to the species immediately succeeding" (*H. Washingtoni*). *Ibid.*, p. 110.

‡ Pallas says of his *Aquila pelagica*, which Cassin makes identical with his *H. pelagicus:* "Caudæ 1′ 1″; longitudo alæ compositæ 1″, 11″, 2‴"; which dimensions do not indicate a bird larger than avarage examples of *H. leucocephalus* or *H. albicilla.*

tors of this group to this evident discrepancy of proportions in Mr. Cassin's description. An error in Mr. Cassin's figure also demands attention, which is doubtless due to an inadvertency of the artist. This consists in the scales on the front of the tarsus being arranged as Mr. Cassin says he never saw in any rapacious bird, namely, continued to the toes in broad, *unbroken transverse plates*, nearly as in Audubon's figure of the *H. Washingtoni!*

92.* **Polyborus brasiliensis** *Audubon*. CARACARA EAGLE. "KING BUZZARD."

Milvus brasiliensis RAY, Synop. Method. Av. et Pisc., 17, No. 6, 1713.
Circus brasiliensis BRISSON, Ornithologie, I, 116, No. 31, 1760.
Falco brasiliensis GMELIN, Syst. Nat., I, 262, 1788.
Falco tharus MOLINA, Sagg. sul. Storia Nat. del Chile, 1782.
Polyborus tharus CASSIN, Illust. Birds of Cal. and Texas, 113, 1856. — CASSIN, Baird's Birds N. Am., 45, 1858.
Polyborus vulgaris VIEILLOT, Nouv. Dict., V, 257, 1816. — AUDUBON, Orn. Biog., II, 350, pl. clxi (young).
Polyborus brasiliensis AUDUBON, Synop., 4, 1839. — BONAP., Consp. Gen. Av., 13, 1850.

"Frequent at Enterprise, associating with the vultures." — *Boardman*.

The swallow-tailed hawk (*Nauclerus furcatus*) became more or less common early in March. I also saw a specimen of the Mississippi kite (*Ictinia mississippiensis*) at Hawkinsville, March 15th.

STRIGIDÆ.

93. **Bubo virginianus** *Swainson*. GREAT-HORNED OWL.

Strix virginiana GMELIN, Syst. Nat., I, 287, 1788. — WILSON, NUTTALL, AUDUBON.
Strix (Bubo) virginiana SWAINSON, Faun. Bor. Am., II, 82, 1831.
Bubo virginianus BONAPARTE, Geog. and Comp. List, 6, 1838. — AUDUBON, Synop., 29, 1839. — CASSIN, Illust. Birds Cal. and Texas, 177, 1854. — CASSIN, Baird's Birds of N. Am., 1858.
Strix bubo, var. *magellanicus* GMELIN, Syst. Nat., I, 286, 1788
Strix pythaules BARTRAM, Travels, 289, 1791.
Bubo ludovicianus DAUDIN, Traité d'Orn., II, p. 210, 1800.
Bubo pinicola VIEILLOT, Ois. Am. Sept., I, 51, 1807.
Strix (Bubo) arctica SWAINSON, Faun. Bor. Am., II, 86, pl. xxx, 1831.
Bubo sub-arcticus HOY, Proc. Phil. Acad. Nat. Sci., VI, 211, 1852.

Not apparently numerous. Mr. Boardman states that he saw only a single specimen, which was killed at Enterprise. I did not observe it above Lake George, and only heard its notes a few times below. Mr. Maynard gives it as rather common about Jacksonville, and says he frequently observed it elsewhere.

Mr. Cassin has very properly remarked that different specimens of this widely distributed species vary materially in size and color, and states that after having examined a large number of specimens from many localities he believed that they were all of one species. He thought, however, that four leading varieties, which he called *atlanticus*, *pacificus*, *arcticus*, and *magellanicus*, could be distinguished. I am not disposed to regard them, however, as by any means strictly geographical, since specimens have been taken recently in Massachusetts that typically represent each of them.* While there are doubtless more or less well-marked local forms of this species, as of all other widely distributed species, many of the differences on which the different varieties have been based are probably only individual.

94.* Scops asio *Bonaparte*. Mottled Owl.

Strix asio Linné, Syst. Nat., I, 132, 1767. — Wilson, Am. Orn., V, 83, pl. xliii, fig. 1, 1812. — Audubon, Nuttall, etc.

Scops asio Bonaparte, Geog. and Comp. List, 6, 1838. — Cassin, Illust. Birds Cal. and Texas, 179, 1854. — Cassin, Baird's Birds N. Am., 51, 1858. — Allen, Amer. Nat., II, 327, 1868.

Strix nævia Gmelin, Syst. Nat., 289, 1788. — Wilson, Am. Orn., III, 16, pl. xix, fig. 1, 1812.

Bubo striatus Vieillot, Ois. Am. Sept., I, 54, pl. xxi, 1807.

Ephialtes choliba Lawrence, Ann. N. Y. Lyc. Nat. Hist., VI, 4, 1854.

Scops McCalli Cassin, Illust. Birds Cal. and Texas, 180, 1854. — Cassin, Baird's Birds N. Am., 52, 1858.

Scops Kennicotti Elliot, Proc. Phil. Acad. Nat. Sci., 1867, 69. — Ibid., Illust. Birds N. Am., pl. xi. — Baird, Trans. Chicago Acad. Sci., I, 311, pl. xxvii, 1869.

Specimens were procured by Mr. Maynard, by whom, and also by Mr. Boardman, it is reported as not unfrequent.

The remarkable differences in the color of the plumage this species presents has led many to suppose it embraced two well-marked species, the red stage being recognized as one and the gray or mottled as another. Gmelin described the red stage as *Strix asio* (which is the same as the *Strix asio* of Linné, and the *Scops carolinensis* of Brisson) and the gray stage as *Strix nævia*. Wilson redescribed these different stages as distinct species. Bonaparte was the first to regard them as identical, he believing

* See Part III, p. 189.

the differences in plumage to be the result of age.* The red he believed to be the young bird, and the mottled the adult, which opinion was also entertained by Audubon. During the last thirty years, however, they have been by some authors again regarded as distinct species; † by others ‡ the gray were regarded as the adult and the red as the young, while some have held the opinion that the difference in color was sexual. A general survey of the facts, either on record or known to me, show that the young birds are sometimes gray and sometimes red; that red young have sometimes *red* parents and sometimes *gray;* that the female is sometimes red and sometimes gray; and also that both sexes of a mated breeding pair of old birds are sometimes alike in color and sometimes different. Hence the opinion already advanced, § that this variation is dependent upon neither age nor sex, but is simply a case of irregular and somewhat remarkable individual variation of a single species, seems a well-founded one. But these different stages, though usually so different, are not always well marked, so that one is often at a loss to know whether to refer certain specimens to the red series or to the gray. In other words, specimens occur of every intermediate grade between the typically bright red stage and the typically gray stage.

I have already given my reasons for referring the *Scops McCalli* of Cassin to the common *S. asio*, of which it is merely the somewhat smaller southern type. § It is also difficult to perceive wherein the *Scops Kennicotti* Elliot, known thus far from a single specimen, differs essentially from a common phase of *S. asio.* ‖

* "Observations on the Nomenclature of Wilson's Ornithology," Journ. Phil. Acad. Nat. Sci., 1st Ser., Vol. III, p. 357, 1824. — "Synopsis of the Birds of the United States," Annals N. Y. Lyc. Nat. Hist., Vol. II, p. 36, 1828.

† MICHNER, Dr. EZRA, "A few Facts in Relation to the Identity of the Red and Mottled Owls," Journ. Phil. Acad. Nat. Sci., 1st Ser., Vol. VII, p. 53, 1834. — HOY, Dr. P. R., "Notes on the Ornithology of Wisconsin," Proc. Phil. Acad. Nat. Sci., Vol. VI, p. 306, 1853; Ibid., Transact. Wisconsin Agr. Soc., Vol. II (1852), p. 344, 1853.

‡ CABOT, Dr. S., Jr., "Observations on the Plumage of the Red and Mottled Owls," Journ. Bost. Soc. Nat. Hist., Vol. II, p. 126, 1838.

§ ALLEN, J. A., "Notes on the Red and Mottled Owls," American Naturalist, Vol. II, p. 327, 1868.

‖ Since the above was written two adult specimens of this species have been received at the Museum from Dallas, Texas, one of which is of the mottled and the other of the red type of plumage. The specimen in mottled plumage, besides being generally darker throughout than northern specimens, has also the dark markings broader and blacker. The specimen in red plumage has the red more intense than it is in specimens of the northern red type. Both the Texas specimens are a little smaller than average New England specimens.

I have seen no specimens as yet from Florida, but from Mr. Cassin having referred a specimen from Indian River, (Fla.,) provisionally to his *Scops McCalli*, they would seem to differ but little from Texas specimens, resembling them, as would be naturally expected, more than northern ones.

95.* Syrnium nebulosum *Gray*. Barred Owl.

Strix nebulosa Forster, Trans. London Philos. Soc., LXII, 386, 424, 1772. — Wilson, Am. Orn., IV, 61, pl. xxxiii, fig. 2, 1812. — Audubon, Orn. Biog., I, 242, pl. xlvi, 1832.

Syrnium nebulosum Gould, Birds of Europe, I, pl. xlvi, 1832. — Audubon, Synop., 27, 1839. — Cassin, Illustr. Birds of Cal. and Texas, 184, 1854. — Brewer, N. Am. Oöl., I, 72, 1857. — Cassin, Baird's Birds N. Am., 56, 1858.

Ulula nebulosa Bonap., Geog. and Comp. List, 7, 1838. — Bonap., Conspect. Gen. Av., I, 53, 1851.

Strix chichictli Gmelin, Syst. Nat., I, 296, 1788.

Strix acclamator Bartram, Travels, 289, 1791.

Strix fernandica Shaw, Gen. Zoöl., VII, 263, 1809.

Very abundant. The only species of owl at all common. Their ludicrous notes are heard at night everywhere, and not unfrequently during the day. At night they often startle the traveller by their strange utterances from the trees over his head.

The four Florida specimens of this species before me are several shades darker than New England specimens, one only of a considerable series of the latter being as dark as the lightest-colored Florida example. The Florida specimens are also a little smaller than the northern ones.

Measurements of Florida Specimens of Syrnium nebulosum.

M. C. Z. No.	Sex.	Locality.	Date.	Collector.	Length.	Alar Extent.	Wing.	Tail.
5241	♀	Lake Dexter	Feb. 14, '69	T. Marcy	20.00	45.75	14.00	8.75
5242	♂	"	Feb. 14, '69	"	20.00	46.25	14.00	8.75
5299	—	Enterprise	Feb. 25, '69	"	19.50	45.75	13.00	9.00
——	—	Hawkinsville	Mar. 15, '69	J. A. Allen	19.75	46.00	13.25	—

96.* Otus brachyotus *Boie*. Short-eared Owl.

Strix brachyotus Gmelin, Syst. Nat., I, 289, 1788. — Forster, Trans. Lond. Phil. Soc., LXII, 384, 1772. — Wilson, Am. Orn., IV, 64, pl. xxxiii, fig. 3, 1812. — Bonap., Ann. N. Y. Lyc. N. Hist., II, 37, 1828. — Audubon, Orn. Biog., V, 373, pl. ccccxxxii, 1835. — Rich. & Swain., Faun. Bor. Am., I, 75, 1831.

Otus brachyotus Boie, Isis, 1822, 549. — Audubon, Syn., 28, 1839. — Cassin, Illust. Birds Cal. and Texas, 182, 1854. — G. R. Gray, Gen. of Birds, I, 40. — Ibid., Cat. Brit. Birds, 27, 1863.

Otus palustris Brehm, Vög. Deutschl., I, 124.

Brachyotus palustris Bonap., Geog. and Comp. List, 7, 1838.

Brachyotus Cassini Brewer, Proc. Bost. Soc. Nat. Hist., V, 321, 1856. — Brewer, N. Am. Oöl., I, 68, 1857. — Cassin, Baird's Birds N. Am., 54, 1858.

"Quite common about marshes." — *Boardman.*

Specimens of this bird from Europe, in the Museum of Comparative Zoölogy, are not appreciably different from others from various parts of the United States. Neither do the habits of the European bird appear to differ from those of the American, as some have supposed. Dr. Richardson described its principal haunts in the Fur Countries as being "dense thickets of young pine-trees or dark entangled willow clumps, where it sits on a low branch, watching assiduously for mice." But it is now well known to more commonly frequent open fields and savannas, situations similar to those the European frequents.

An interesting state of plumage of this owl is exhibited by two pairs taken on Muskeget Island, Massachusetts, about July 1, 1870, by Messrs. C. J. Maynard and William Brewster, in which the color is so light as to almost suggest their being albinos. They are many shades lighter than the specimens of this species are from the interior, and show clearly, when taken in connection with the light race of *Arvicola riparius* (*Arvicola Breweri* Baird), also occurring on this small sandy island, the effect of the combined influence of an absence of shade and the increased light caused by reflection from the light-colored sand. The influence of similar circumstances is seen on a large scale in the birds and mammals of the Colorado desert and the arid peninsula of Lower California, and in less degree on the open arid plains of the middle region of the continent.

The long-eared owl, *Otus vulgaris* Fleming,* may be expected, from its known distribution, to also occur in Florida.

97.* Strix flammea *Linné*. Barn Owl.

Strix flammea Linné, Syst. Nat., I, 133, 1767. — Wilson, Nuttall, Audubon (Orn. Biog.), Bonaparte (Synop.).

Strix pratincola Bonap., Geog. and Comp. List, 7, 1838. — Cassin, Brewer, and recent American authors.

Strix americana Audubon, Synop., 25, 1839.

Strix perlata Bonap., Consp. Gen. Av., I, 55, 1850.

Strix furcata Temm., Pl. Col., I, 432.

A specimen was taken by Mr. Thaxter at St. Augustine. Mr.

* *Strix otus* Linné, Faun. Suec., 24, 1761.

Strix otus americana et *mexicana* Gmelin, Syst. Nat., I, 288, 1788

Strix otus Wilson, Bonap. (Synop.), Nuttall, Audubon (Orn. Biog.).

Otus vulgaris Fleming, British Animals, 60, 1828. — Audubon, Synop., 28, 1839. — G. R. Gray, Gen. Birds, I, 40; Cat. Brit. Birds, 26, 1863.

Otus Wilsonianus Lesson, Traité d'Orn., I, 110, 1831. — Cassin, Brewer, and recent American authors.

Otus americanus Bonap., Geog. and Comp. List, 7, 1838.

Maynard informs me it was said to be common, and that at Dummitt's a hollow tree was shown him in which a pair of these birds had bred for several years. Audubon also speaks of it as being common in Florida.

Respecting the numerous species of late recognized in the *Strix flammea* group of owls, Mr. Cassin has, with great propriety, remarked that naturalists have "established species on very slender characters."

As is well known, different specimens from near the same locality vary considerably in color and size, while specimens from different continents are frequently almost undistinguishable. From the considerable number of specimens I have seen from distant points, as Europe, the United States, South America, Southern Asia, the West Indies, Australia, and South Africa, I see no reason why the *Strix flammea* may not be regarded as having a nearly cosmopolitan distribution, which indeed seems to be the present opinion of several European ornithologists. Nearly the same variations in color appear to occur on each continent, the general color in specimens from near the same locality varying from yellowish rufous to pale fulvous, and the dusky spots from being large and conspicuous to nearly obsolete or entirely wanting.

COLUMBIDÆ.

98.* **Chamæpelia passerina** *Swainson.* GROUND DOVE.

Common, especially about cultivated grounds.

99.* **Zenædura carolinensis** *Bonaparte.* MOURNING DOVE.

Columba carolinensis LINNÉ, Syst. Nat., I, 286, 1766.— GMELIN, WILSON, NUTTALL, AUDUBON (Orn. Biog.).

Columba marginata LINNÉ, Syst. Nat., I, 286, 1766.

Ectopistes marginellus WOODHOUSE, Proc. Phil. Acad. Nat. Sci., Vol. VI, 104, 1852.

Zenædura carolinensis BONAPARTE, Consp. Gen. Avium, II, 84, 1854.

Zenædura marginellus BONAPARTE, Ibid., 85.

Abundant. Among its favorite resorts are the wild orange-groves, where it feeds on the seeds of the decaying fruit. Smaller than at the north, with the metallic tints much brighter and more bronzy.

MELEAGRIDÆ.

100.* **Meleagris gallopavo** *Linné.* WILD TURKEY.

Meleagris gallopavo LINNÉ, Syst. Nat., 268, 1766. — GMELIN, WILSON, BONAPARTE, AUDUBON, NUTTALL, BAIRD, etc.

Meleagris americana BARTRAM, Travels, 290, 1791.

Meleagris sylvestris VIEILL., Nouv. Dict., IX, 447.

Meleagris fera VIEILL., Galérie des Ois., II, 10, pl. x, 1824.

Meleagris mexicana GOULD, Proc. Lond. Zoöl. Soc., 1856, 61. — BAIRD, Birds N. Am., 618, 1858. — COOPER & BAIRD, Orn. Cal., I, 523, 1870.

Gallopavo sylvestris, Novæ Angliæ RAY, Synopsis, 51, 1713. — LECONTE, Proc. Phil. Acad. Nat. Sci., IX, 179, 1857.

Common and even quite numerous in those sections where it is not too much hunted. Mr. Boardman informs me that very fat male birds often weigh twenty-five to twenty-eight pounds, but that the average weight of the males is eighteen to twenty pounds, and of the females six to ten pounds.

THE ORIGIN OF THE DOMESTIC TURKEY.

Although it had been for a long time previously vaguely conjectured that the domestic turkey did not originate from the common wild turkey of North America, it was not until about 1856 that it was fully asserted that such was not its origin. In a paper read before the Zoölogical Society of London, in April, 1856, Mr. John Gould, the well-known English ornithologist, assigned this bird to the list of those domesticated animals whose origin had become involved in obscurity. He refers, however, to the fact of its known introduction into Europe from Mexico about 1524, and to the belief, shared by all naturalists from Linné up to that time, that the domesticated turkey was derived from the wild turkey of North America. He also states that, "on account of the great differences which are met with among our domestic turkeys, and the circumstance that the wild turkeys recently imported from North America not readily associating or pairing with them," he had for some years entertained the opinion that the wild turkey of the United States was not the original of the domestic turkey. He also at this time described a single specimen of a turkey from Mexico as belonging to a species distinct from the wild turkey of the United States, to which he gave the name of *Meleagris mexicana*. It differed, however, but slightly from the northern bird, mainly in having more white on the upper tail coverts. Although he claimed that it was of larger size, his measurements indicate it to be only barely above the average, and considerably smaller than the larger specimens from the Northern States. In considering it as distinct from the common wild turkey, he seems to have been greatly influenced by the locality whence his specimen came; as he states that he hardly thinks it probable that the common turkey, "authors to the contrary, notwithstanding," ranges very far into Mexico, since it is found, he says, along the southern boundary of Canada, which is nearly two thousand miles from Mexico. He deems it unlikely that a bird inhabiting "the cold regions of Canada should also be indigenous to the hotter country of Mexico, whence," he adds, "and not from North America, the turkey was originally introduced into Europe"; thus leaving it to be inferred that, in his opinion, the Mexican bird — his new species — was the ancestor of the domestic turkey. The facts in respect to the distribution of the wild turkey are briefly these: It exists in Canada only in the warmer portions of that country, and thence southward uninterruptedly throughout the table-lands of Mexico.

Dr. Henry Bryant, of Boston, in reviewing Mr. Gould's paper, a few months after its appearance, took exceptions to the views of that gentleman, and in referring to the two principal statements made by Mr. Gould, namely, that the wild and domestic turkeys were structurally different, and refused to breed together, Dr. Bryant thus observes: "How far climate and other influences may have affected the domestic variety in England I do not yet know, but with us neither of these two statements is correct. If it were not for the difference in the plumage it would be impossible in many cases to distinguish the two birds; and even with this aid it is sometimes very difficult to decide with certainty when the specimen is a female. . . . . The wild turkey breeds here with the tame variety quite as readily as could be expected; wherever the wild turkeys are numerous, it is an ordinary occurrence for the tame hen to prefer the wild gobbler to the domestic ones. I have had in my own possession wild hens that bred with the tame gobblers, — a fact much stranger than that of the wild gobbler breeding with the tame hen. But the most satisfactory proof of their specific identity is that the offspring of mixed blood is known to be hardier and more prolific than the domestic variety, — a fact which cannot be reconciled with their specific diversity." *

Dr. Bryant's facts, with those of previous writers, seem amply sufficient to settle the question as to the origin of the domestic turkey; yet a few months later Major John LeConte, who probably at that time had not seen Dr. Bryant's remarks, published a paper entitled "Observations on the Wild Turkey, or *Gallopavo sylvestris* of Ray." † In this paper he took the ground that the tame turkey could not possibly have been derived from the wild turkey of the United States. And, if what he states in support of his opinion as facts were such, they would go far towards rendering his position a tenable one, but in reality they are but baseless, dogmatic assumptions, which not only ran counter to the then generally received opinion, but were squarely opposed to unquestionable evidence already on record. Major LeConte's opinions, notwithstanding

* Proc. Bost. Soc. Nat. Hist., Vol. VI, p. 158, March, 1857.

† Proc. Phil. Acad. Nat. Sci., Vol. IX, p. 179, September, 1857.

that they were based on groundless assumptions, as an investigation of the subject fully proves, have been so generally entertained by subsequent authors, who have accepted his statements without investigating the facts for themselves, that a careful revision of the subject is now required. Major LeConte observes: "Whoever has compared the wild turkey of the United States with the domestic animal of the same genus must have observed that there existed very striking differences between them." While asserting that "these differences do not consist of slight and unimportant particularities, but in radical disagreements, which ought to remain unchangeable under all circumstances, and which form good specific characteristics," his sole point of distinction consists "in the possession by the tame bird of an enormous palear or dewlap," which he affirms, contrary to fact, is not possessed by the wild bird.* He refers also to the conviction that had long existed in his mind, that the two birds — the wild and domestic — "were really distinct species." "More than fifty years ago," he says, "when I first saw a wild turkey, I was led to conclude that one never could have been produced from the other. I have mentioned this to many ornithologists, but no one would take the trouble to investigate the matter [!]," etc. It does not appear, however, that even with him this long-standing conviction had resulted from a thorough investigation of the subject, for he gives no detailed comparison of the two, and many of his statements are not simply erroneous, but diametrically opposed to facts previously well substantiated. He refers to the early introduction of the turkey into Europe, and to the fact that it was found by the first explorers of America in both the wild and domesticated state. He alludes also to Mr. Gould's above-cited paper, remarking respecting it that he was unable to determine whether Mr. Gould's supposed new Mexican species was the same as the *M. gallopavo*, or was the original of the domestic bird. He thought, however, that the Mexican was identical with the common wild bird. He then remarks: "I have before observed that the turkey was found domesticated among the nations of Central America. *Now the bird which we have native among us has never been domesticated. All attempts to conquer its peculiar habits have failed, notwithstanding what has been said and written on the subject to the contrary.* I DEFY ANY ONE TO SHOW A TURKEY, EVEN OF THE FIRST GENERATION, PRODUCED FROM A PAIR HATCHED FROM A WILD HEN.† We have every year in our market offered for sale birds of a very dark color, and in some degree resembling the wild species; but in every instance, by the presence of the palear, the imposition can be detected at

* It is usually, however, either entirely absent in the wild bird, or present only in a rudimentary state.

† The italicizing in this extract is of course my own.

first sight, and the cheat exposed. I have known the eggs found in the woods hatched by a domestic hen, the chickens brought up carefully, and rendered so tame and familiar as to eat out of the hand, and to show considerable pleasure whenever persons with whom they were acquainted approached them. *Yet they never would associate with the domestic turkeys, studiously avoiding their company*, and in little more than a year running off to the woods, and never again returning to the haunts of their infancy. *I know*," he continues, "*that I shall be contradicted in this statement, and many quotations from authors brought forward against me. I repeat*, contrary to the assertions of many others, THAT NO ONE HAS EVER SUCCEEDED IN DOMESTICATING OUR WILD TURKEY. I speak not only from my own personal observations, but from the undivided testimony of many southern gentlemen. The turkey of our own poultry-yards, which, when young, is difficult to bring forward, it was thought might be obtained of a hardier race by a new domestication ; but every attempt has failed, nor can I find a single well-authenticated case of a mixed breed being obtained." One is certainly at a loss to know what the self-confident Major would call a well-authenticated case of a mixed breed of wild and tame turkeys, since he must have been familiar with Bonaparte's excellent account, derived mainly from notes furnished him by Mr. Audubon, of this bird given in the first volume of his continuation of Wilson's "American Ornithology." In speaking of the mixing of the wild and tame turkeys, this author remarks as follows : "This crossing often occurs in countries where wild and tame turkeys are frequent ; it is well known that they will readily approach each other ; and such is the influence of slavery upon even the turkey, that the robust inhabitant of the forest will drive his degenerate kinsfolk from their own food and from their females, being generally welcomed by the latter and by their owners, who well know the advantage of such a connection. . . . . Eggs of the wild turkey have been frequently taken from their nests and hatched under the tame hen ; the young preserve a portion of their uncivilized nature, and exhibit some knowledge of the difference between themselves and their foster-mother, roosting apart from the tame ones, and in other respects showing the force of hereditary disposition. The domesticated young, reared from the eggs of the wild turkey, are often employed as decoy birds to those in a state of nature." *

Audubon, in his account of the Canada goose, also incidentally refers to the crossing of the wild and tame turkeys, in a manner that leads us to suppose that it was to his knowledge a matter of common occurrence. He says : "The crossing of the Canada goose with the com-

* Nearly the same words are used by Audubon in his Ornithological Biography and in his Birds of America.

mon domestic species has proved as advantageous as that of the wild with the tame turkey."* He also states, "My friend, Dr. Bachman, assures me that in a state of domestication the wild turkeys, *though kept separate from tame individuals*, lose the brilliancy of their plumage in the third generation, becoming plain brown, and having here and there white feathers intermixed "†

The assertions of Major LeConte are so fully controverted by previously recorded testimony that they might have been justly ignored, had they not received, as already observed, the sanction of eminent authorities, and thus have come to be more or less currently adopted. Among the first to give them support was Professor Baird, of the Smithsonian Institution. This gentleman, in his work on the "Birds of North America," published less than two years subsequently to Major LeConte's paper, cites LeConte's opinions and statements, and partially indorses them, though he had not, he says, specimens at hand of the domestic bird for comparison with the wild one. To the data for their distinction adduced by Major LeConte, he adds a statement from Bonaparte in respect to the difference in color between the domestic and wild bird; Bonaparte observing that the wild bird never has the whitish tip to the tail which distinguishes the domestic ones. Professor Baird also adds that the flesh of the two differs in color, that of the wild bird being "much darker." He adds that, upon the whole, it is exceedingly probable that they are specifically distinct. "If the dewlap," he says, "be characteristic of a species at present only known in captivity, then, as Major LeConte remarks, it should bear the name of *M. gallopavo*, as based by Linnæus essentially upon the description by Brisson of *Gallopavo sylvestris*, in which this dewlap is particularly mentioned. In this event our wild bird will be entitled to a new name," etc. Professor Baird concludes his remarks on this subject with the following ingenious theory, which has been to some extent accepted as a probably correct one. "In conclusion," he says, "I venture to suggest the following hypothesis, which, however, is not original with myself: That there are really three species of turkey, besides the *M. ocellata*, a fourth species from Central America, entirely different from the rest. That one of them, *M. americana*, is probably peculiar to the eastern half of North America; another, *M. mexicana*, belongs to Mexico, and extends along the table-lands to the Rocky Mountains, the Gila, and the Llano Estacado; and a third is the *M. gallopavo*, or domesticated bird. That it is not at all improbable that the last was originally indigenous to some one or more of the West Indian Islands, whence it was transplanted as tamed to Mexico, and from Mexico taken to Europe about A. D. 1520.

* Birds of America, Vol. VI, p. 190.

† Ibid., Vol. V, p. 55.

Finally, that the wild turkeys were probably completely exterminated by the natives, as has been the case with equally large birds in other islands, as the dodo and solitaire.* This hypothesis," he continues, "will explain the fact of our meeting nowhere at the present day any wild turkeys resembling the domestic one.† . . . . The entire subject is one of much interest, and deserves to be investigated thoroughly. It is quite possible that a careful examination of the external form and habits of the New Mexican bird may do much to throw full light on the whole question."

It is not surprising that a theory presenting to the imagination so many attractive features, and indorsed by authority so eminent, should have been currently received, as has this, by those who have not had the opportunity, nor perhaps the desire, to examine the subject for themselves. But, if I mistake not, it has also been accepted as at least a *probably* correct hypothesis by many ornithologists.‡ I have, however, already adduced evidence from Bonaparte, Bachman, Audubon, and Bryant sufficient to show, not only the erroneous character of Major LeConte's fundamental proposition, to wit, that the wild turkey of the United States has never been and never can be domesticated, but that such an hypothesis as the one above quoted is wholly uncalled for. As the whole question of the origin of the domestic turkey and its relationship to the wild turkey of the United States turns, however, upon the fact of the domesticability or non-domesticability of the common wild turkey, it may perhaps be proper to bring forward some recent testimony respecting this disputed point.

I have myself always been more or less familiar with the domestic bird, and with the fact that breeds exist which closely resemble the wild bird, and which their owners claimed were one fourth, one half, or one eighth

* Mr. Darwin, in referring to this gratuitous theory, refers to the fact of the deterioration of the turkey within the tropics, and very properly to the "improbability of a bird having long ago become extinct in these large and luxuriant islands, or of its ever having been aboriginally an inhabitant of the lowlands of the tropics." (Animals and Plants under Domestication, Am. ed., Vol. I, p. 353, note.)

† But does it explain the frequent occurrence of domestic ones so closely resembling the wild ones as to be quite undistinguishable from them?

‡ Dr. Cooper, who considers the western wild turkey specifically distinct from the wild turkey of the east, appears to believe that the domestic turkey originated from the wild turkey of Mexico. He says: "It is well known that at the period of the Spanish discovery the native turkey was widely domesticated in Mexico, and was introduced thence first into Europe, and thence into North America. Furthermore, the native bird of Eastern North America does not occur in Mexico at all. The markings of the domestic turkey are sometimes exactly like those of the wild bird of Mexico, while they never assume the plumage of the wild *Meleagris gallopavo* of the north." (Orn. Cal., Vol. I, p. 523, 1870.) Dr. Cooper's last remark is unfortunately erroneous, since domestic birds do often occur, especially females, that cannot well be distinguished from wild northern birds.

wild blood, as the case might be, and which differed in habits in some respects from the common breeds. I have also been long conversant with the fact that in the Western States, and in those other parts of the country where the turkey exists in its native state, that the eggs of the wild bird are frequently taken and hatched under the domesticated turkey, the young carefully raised and held at high prices, they being considered as highly valuable for the purpose of improving the domestic breeds. In a recent correspondence with Mr. D. Darwin Hughes, an able ornithologist of Marshall, Michigan, I alluded to the fact that the domestication of the wild bird had been disputed, and requested him to give me any facts he might possess in reference to the subject. The facts given in the following extracts from his letters are fully corroborated by other private testimony in my possession.

Under date of October 25, 1869, he wrote me respecting the domestication of the wild bird as follows: "Here [Calhoun County, Michigan], where the wild bird is abundant, they mix freely with the tame ones, and it is a common thing to see large flocks of half-breeds; I have owned them myself. They are fond of roaming and are apt to stray; not to the woods exclusively, but also to other farms. I have known the pure wild bird, hatched from wild eggs and raised in the poultry-yard, to remain for years in the yard without being confined; but this is not usual. One fine gobbler, as beautiful a bird as I ever saw, was hatched from a wild egg and headed a flock of mixed turkeys in a barn-yard. He was tame, like the others, but easily distinguished by his wild plumage; at night the flock roosted in the yard, but this bird could not brook so low a perch, and when the flock went to roost he invariably took wing and perched on an immense forest-tree one fourth of a mile away, where he spent the night; but in the morning he always returned to the barn-yard. Such instances are not uncommon. The eggs are eagerly sought for for hatching, and in this manner, as I have before said, there is a liberal sprinkling of wild blood in domestic birds, where the wild birds are abundant. The eggs of the wild bird are found every year, and although I have offered at the rate of six to eight dollars per dozen for them, there is not one in my collection of eggs, which numbers over two hundred species, so eager are the finders of them to hatch them, the chicks selling for a large price."

In another letter, dated November 5, 1869, Mr. Hughes wrote me further concerning this subject, in which he remarks as follows: "I have already said that the wild bird has been so domesticated as to reproduce its kind in the poultry-yard, and inquiries made since my last letter show that in the more northern counties of the State such cases are quite common. I cannot agree with what is said in the ninth volume of the Pacific Railroad Reports (p. 617), that there is an appreciable difference in the color of the flesh of the wild and tame birds when cooked. There probably is some difference in color, but so little that one must have very acute powers of observation to tell the difference when brought to the table roasted. There is a difference in the color of the head, caruncles, and dewlaps, as stated by Professor Baird, but with my present means of knowledge, having no fresh specimens before me, I will not undertake to describe the differences. One thing, however, should not be forgotten; that we see the tame bird under all circumstances of passion, — in fear and when proudly strutting; in short, under all the different emotions that turkeys are heirs to, while we rarely or never see the wild turkey under such varied circumstances, but only when they are terror-stricken or dead. The head and neck in the tame bird makes rapid and surprising changes in sympathy with its emotions, and it may be so, and probably is, with the wild."

From the evidence that has now been given, it is sufficiently apparent that Major LeConte's two fundamental assumptions, — first, that the wild bird will not mix or breed with the domesticated; and, second, that the wild bird never has been and cannot be domesticated, — upon which was erected an hypothesis to explain the origin of the domesticated bird by referring it to an extinct ancestor that probably inhabited some of the West Indian Islands, are entirely groundless, and never had for their support only the negative evidence afforded by the limited experience of Major LeConte and a few of his friends.

Inasmuch as the domestic turkey was first introduced into Europe from Mexico, it may be well in this connection to inquire further into the relationship of the so-called *M. mexicana*, or Mexican turkey, to the wild turkey of the eastern part of the United States. As already stated, the *M. mexicana* was originally described by Mr. Gould from a single specimen from Mexico. This specimen differs but slightly from the common wild turkey of the eastern part of the continent. But like many other merely nominal species, this "Mexican turkey" has been since generally recognized by writers on American ornithology, doubtless mainly because its describer was deemed too eminent a naturalist to be mistaken on such a point. Its habitat has been since extended to embrace half of that portion of the continent over which the wild turkey ranges, — the entire western half of the United States; yet the point at which the habitat of the eastern species ceases and that of the western begins, no one has yet ventured to attempt to definitely indicate. It is universally conceded to be exceedingly closely allied to the *M. gallopavo*, as the latter is now defined. Though admitted provisionally as a valid species by Professor Baird in his work already cited, he says that "whether these differences can be considered as establishing a second species for the United States is a question yet to be decided." Dr. Coues, however, in his "List of the Birds of Fort Whipple, Arizona," * says he thinks there can be no doubt respecting the propriety of separating the "western turkey from the common species of the Eastern United States"; but he has given us no information as to how great the differences between them are, or in what they consist. As mentioned by Gould and by Baird, the Mexican bird differs from the eastern one only in being lighter colored, and in having, in correlation with the generally lighter color of the plumage, the terminal band of the tail, as also the tips of the tail coverts, whitish instead of pale brown, as the eastern bird usually has them. This, however, seems by no means necessarily a specific difference, it being only a slight geographical variation, not restricted to the turkey, but which runs through most species of both birds and mammals that have the same distribution; the probable cause of which variation I have already adverted to in Part III. The common eastern turkey occasionally approaches much nearer to the so-called Mexican bird than appears to be generally supposed. According to some authors, the tip of the tail in *M. gallopavo* is never whitish, but "plain chestnut, lighter than the ground color" of the tail. Yet of five specimens in the Museum of Comparative Zoölogy from one of the Western States, probably either Ohio or Michigan, two correspond with the description of the assumed typical *M. gallopavo*, two very nearly as well with that of the so-called *M. mexicana*, and one is intermediate between them. Three of them are decidedly lighter colored, and possess a lighter terminal band to the tail than they should to correspond with the true *M. gallopavo* as recently defined. I have, on the whole, no hesitancy in referring the *M. mexicana* Gould to the *M. gallopavo* Linné. The unquestionable specific identity of the domestic turkey with the wild one of the Eastern United States, though originally derived from the Mexican bird, seems further to support this view. From the great constancy of the white on the tail and its coverts in the domestic turkey, it has been thought to more resemble the western bird, or the *M. mexicana*, than the eastern. I need, however, only to recall the testimony of Dr. Bachman, already given in discussing another point, to show that it has necessarily no such significance. It will be remembered that Dr. Bachman states that he had known the wild birds of the Atlantic States, when kept entirely by themselves, to become more or less white under confinement in three generations.† Instead of this being either a "reversion" or a distinctive specific feature, it can be regarded only as the result of a diminution of the coloring matter through degeneracy, under the influences of domestication.

* Proc. Phil. Acad. Nat. Sci., Vol. XVIII, p. 93, 1866. Republished under the title of "Prodrome of a Work on the Ornithology of Arizona Territory."

† Mr. Darwin mentions a similar fact as having happened in England. (Animals and Plants under Domestication, Vol. I, p. 354).

As the whole plumage becomes lighter, those portions that are naturally lightest are those we should expect would soonest become white; and such is actually the case. Under domestication the turkey not only degenerates in size and hardiness, but is well known to soon lose much of the brilliancy of plumage that characterizes it in a state of nature. In a few generations it loses to a great extent its metallic tints, and becomes much lighter colored; the terminal band of the tail, as well as its coverts, changes to white, and in succeeding generations the cream-colored and pure white birds often seen in our poultry-yards are gradually developed.

The fact of the domestic turkey having been first introduced into Europe from Mexico, and into the United States from Europe, admits of easy explanation; since the advanced state of civilization enjoyed by the native Mexicans had enabled them to domesticate the turkey, while their more degraded neighbors of the north had accomplished nothing of the kind. The turkey having been introduced into Europe nearly a century before the establishment of permanent settlements in the northern portions of the continent, it was, of course, as naturally introduced thence into this country as were our other domesticated animals.

PERDICIDÆ.

101.* Ortyx virginianus *Bonaparte.* QUAIL.

Tetrao virginianus LINNÉ, Syst. Nat., I, 277, 1766.

Tetrao marilandicus LINNÉ, Syst. Nat., I, 277, 1766.

Ortyx borealis STEPHENS, Shaw's Zoöl., XI, 377, 1819.

Perdix (*Ortyx*) *virginiana* BONAP., Obs. on Wils. Nomen., Journ. Phil. Acad. Nat. Sci., 1st Ser., IV, 268, 1825.

Ortyx virginianus BAIRD, Birds N. Am., 640, 1838. — MARCH, "Notes on Birds of Jamaica," Proc. Phil. Acad. Nat. Sci., XV, 303, 1863.

Ortyx texanus LAWRENCE, Ann. N. Y. Lyc. Nat. Hist., VI, 1, 1853. — BAIRD, Birds N. Amer., 641, 1858.

Abundant.

The quails of Florida differ from those of the Northern States in being smaller, larger billed, and darker colored. While the difference in size is very appreciable, as is also that in respect to the size of the bill, — the bill being actually larger while there is a general decrease in the size of the individual, — the most marked dissimilarity is in the coloration, through the darker color of the Florida birds. In the latter the ground color above is rufous instead of ashen, as in northern specimens, and the transverse black markings are broader. In average northern specimens the transverse black bars on the lower surface of the body are scarcely half the breadth of the intervening white spaces; in the Florida specimens they are much more than half, and in some cases nearly equal them. In general the proportion of black in the Florida *females* is the same as that in the northern *males*. There is a similar relative increase in the extent of the black markings on the wing coverts, scapulars, and interscapulars, and on the dorsal surface generally. The black border to the white throat-patch is also broader, and extends back on the sides of the head so as usually to cover the auriculars, which in average northern specimens are dark rufous. The bill is also much darker, being generally jet black; in Massachusetts specimens it is brownish black, with the tip decidedly lighter than the other parts.

The so-called Texas quail (*Ortyx texanus* Lawr.) does not differ very greatly from either the Florida or the northern ones, it combining some of the essential characters of each, but more resembling Florida specimens than northern ones. Lawrence and Baird mention the ashen or decided gray hue on portions of the dorsal surface as distinguishing it from the *O. virginianus*, which has these parts of a "dull pinkish red." "A dull pinkish red," however, is just the color of these parts in my Florida specimens; but the Massachusetts specimens, on the contrary, are ashen, as already stated, and in this respect agree with the descriptions of the Texas form, and differ from the Florida ones in the same way that the Texas ones are said to do from those of the Atlantic coast of the Middle and Southern States. In both the Florida and Texas specimens there is a similar increase in the breadth of the black transverse markings, Lawrence describing them as being twice as broad in the Texas specimens as in the northern ones.

The *Ortyx cubanensis* of Cabanis appears scarcely to differ from the quails of Florida and Texas. D'Orbigny and Lembeye were hence doubtless correct in believing the so-called *Ortyx cubanensis* to be identical with the *O. virginianus*.

The following summary of the subjoined tables shows the difference in size that obtains between northern and southern specimens, and also in the sexes. The largest Florida specimen, it will be seen, scarcely equals the smallest northern one, when those of the same sex are compared.

No. of Specimens.	Sex.	Locality.		Length.	Alar Extent.	Wing.	Tail.
7	♂	Illinois	Average	10.18	15.44	4.47	2.82
16	♂	Florida	Average	9.46	14.16	4.22	2.52
6	♀	Illinois	Average	9.83	15.10	4.36	2.67
10	♀	Florida	Average	9.37	14.02	4.17	2.54
10	♀	"	Maximum	10.00	14.50	4.40	2.77
10	♀	"	Minimum	9.00	13.10	3.35	2.50
16	♂	"	Maximum	10.00	14.75	4.50	3.00
16	♂	"	Minimum	9.00	13.80	4.00	2.30
7	♂	Illinois	Maximum	10.50	15.60	4.60	3.00
7	♂	"	Minimum	10.00	15.00	4.37	2.55
6	♀	"	Maximum	10.25	15.50	4.50	2.85
6	♀	"	Minimum	9.50	14.50	4.25	2.45

Measurements of Florida Specimens of Ortyx virginianus.

M.C.Z. No.	Coll. No	Sex.	Locality.	Date	Collector.	Length.	Alar Extent.	Wing.	Tail.
5151	—	♂	Hibernia	Jan. 30, '69	J A. Allen	9.25	14.75	4.15	2.40
5152	—	♂	"	Jan 30, '69	"	9.25	14.10	4.00	2.60
5183	—	♂	"	Jan. 30, '69	"	9.00	14.00	4.00	2.30
5184	—	♂	"	Jan. 30, '69	"	9.25	14.25	4.10	2.50
5337	—	♂	Enterprise	Mar 4. '69	"	9.65	14.50	4.40	2.40
5336	—	♂	"	Mar. 4, '69	"	9.50	14.25	4.15	2.50
10578	1990	♂	Jacksonville	Jan. 9, '69	C. J. Maynard	9.50	13.80	4.30	2.80
10579	1990	♂	"	Jan 9, '69	"	9.35	14.15	4.45	2.53
10580	1991	♂	"	Jan. 9, '69	"	9.30	14.30	4.00	2.30
—	2547	♂	Dummitt's	Mar. 8, '69	"	10.00	14.08	4.10	2.65
—	2546	♂	"	Mar. 7, '69	"	9.30	14.05	4.25	3.00
—	2562	♂	"	Mar. 9, '69	"	9.85	13.80	4.45	2.84
10583	2472	♂	"	Feb. 24, '69	"	9.50	14.00	4.25	2.50
—	2517	♂	"	Mar. 4, '69	"	9.25	14.00	4.40	2.70
—	2561	♂	"	Mar. 9, '69	"	9.70	14.08	4.25	2.65
10581	2356	♂	"	Feb. 16, '69	"	9.70	14.50	4.50	2.70
10582	2456	♀	"	Feb. 24, '69	"	10.00	14.50	4.25	2.55
—	2795	♀	"	Feb. 16, '69	"	9.00	13.75	3.35	2.70
—	2615	♀	"	Mar. 8, '69	"	9.50	14.20	4.40	2.57
—	1993	♀	Jacksonville	Jan 9, '69	"	9.35	14.10	4.35	2.70
—	1994	♀	"	Jan. 9, '69	"	9.40	13.10	4.35	2.65
—	1995	♀	"	Jan. 9, '69	"	9.50	13.60	4.40	2.77
5182	—	♀	Hibernia	Jan. 30, '69	J. A. Allen	9.35	14.25	4.10	2.30
5338	—	♀	Enterprise	Mar. 4, '69	"	9.00	14.00	4.05	2.33
5351	—	♀	"	Mar. 5, '69	"	9.40	14.50	4.30	2.45
5352	—	♀	"	Mar. 5, '69	"	9.25	14.25	4.15	2.47

Measurements of Northern Specimens of Ortyx virginianus.

M. C. Z. No.	Sex.	Locality.	Date.	Collector.	Length	Alar Extent.	Wing.	Tail.
13096	♂	Northern Illinois	Jan. 18, '71	—	10.25	15.00	4.45	2.72
10410	♂	"	Jan. 18, '71	—	10.00	15.45	4.60	2.75
10408	♂	"	Jan. 18, '71	—	10.00	15.00	4.40	2.85
10411	♂	"	Jan. 18, '71	—	10.50	15.50	4.50	2.75
13099	♂	"	Jan. 18, '71	—	10.25	15.60	4.60	3.00
13098	♂	"	Jan. 18, '71	—	10.28	15.25	4.50	2.90
13097	♂	"	Jan. 18, '71	—	10.00	15.25	4.37	2.55
13101	♀	"	Jan. 18, '71	—	10.25	15.50	4.45	2.72
10407	♀	"	Jan. 18, '71	—	9.50	14.50	4.25	2.73
10409	♀	"	Jan. 18, '71	—	10.00	14.85	4.50	2.85
10412	♀	"	Jan. 18, '71	—	9.85	15.25	4.38	2.48
10406	♀	"	Jan. 18, '71	—	9.50	15.20	4.50	2.45
13100	♀	"	Jan. 18, '71	—	9.85	15.10	4.30	2.60

CHARADRIIDÆ.

102.† **Squartarola helvetica** *Cuvier.* Black-bellied Plover.

"Some remain on the shores of the Floridas in winter." — *Audubon.**

103.† **Charadrius virginicus** *Borck.* Golden Plover.

"St. Augustine; rare." — *Boardman.*

* Birds of America, Vol. V, p. 200.

104.* **Ægialitis vociferus** *Cassin.* KILLDEE PLOVER.

Abundant.

105.* **Ægialitis Wilsonius** *Cassin.* WILSON'S PLOVER.

Not recently reported as found in Florida in the winter months. Audubon observes: "While in the Floridas, near St. Augustine, in the months of December and January, I found this species much more abundant than any other." *

106.† **Ægialitis semipalmatus** *Cabanis.* SEMI-PALMATED PLOVER.

"Not uncommon at St. Augustine throughout the winter." — *Boardman.*

107.† **Ægialitis melodus** *Cabanis.* PIPING PLOVER.

Observed at St. Augustine in the winter months by Mr. Boardman.

HÆMATOPODIDÆ.

108.† **Hæmatopus palliatus** *Temminck.* OYSTER-CATCHER.

Given by Mr. Boardman as rare in winter at St. Augustine.

109.† **Strepsilas interpres** *Illiger.* TURNSTONE.

"Rare at St. Augustine in winter." — *Boardman.*

SCOLOPACIDÆ.

110.* **Philohela minor** *Gray.* WOODCOCK.

More or less common. Probably resident.

111. (†?) **Gallinago Wilsoni** *Bonaparte.* SNIPE.

Abundant. Probably resident. Florida specimens are darker colored and have longer bills than northern ones.

112.† **Calidris arenaria** *Illiger.* SANDERLING.

"Common at St. Augustine." — *Boardman.* "Abundant on Indian River." — *Maynard.*

113.† **Pelidna americana** *Coues.* RED-BACKED SANDPIPER.

"Common." — *Maynard. Boardman.*

114.† **Ereunetes pusillus** *Cassin.* SEMI-PALMATED SANDPIPER.

"Common." — *Maynard.*

* Ibid., p. 216.

115.† **Actodromas minutilla** *Coues.* LEAST SANDPIPER.

"Common." — *Maynard.*

116.† **Actodromas Bonapartei** *Cassin.* WHITE-RUMPED SANDPIPER.

"St. Augustine." — *Audubon.*

117.* **Symphemia semipalmata** *Hartlaub.* WILLET.

"Indian River to St. Augustine. Breeds in March." — *Maynard.*

118.† **Gambetta flavipes** *Bonaparte.* YELLOW-LEGS.

Common.

119.† **Gambetta melanoleuca** *Bonaparte.* GREATER YELLOW-LEGS.

Common.

120.* **Tringoides macularius** *Gray.* SPOTTED SANDPIPER.

Common.

121.* **Limosa fedoa** *Ord.* MARBLED GODWIT.

Common. Reported to Mr. Maynard as common all the year near St. Augustine, but where it nested was unknown to his informants.

122.† **Numenius hudsonicus** *Latham.* HUDSONIAN CURLEW.

123.† **Numenius borealis** *Latham.* ESQUIMAUX CURLEW.

I have no knowledge of the actual occurrence of these two species in East Florida, yet they apparently must occur as winter visitors. Dr. Coues gives them as winter visitors in his South Carolina list, and they are well known to range at this season southward into the tropics.

124.† **Numenius longirostris** *Wilson.* LONG-BILLED CURLEW.

"Very abundant on the coast." — *Boardman.*

Several other species of this family are well known to pass through East Florida in their migrations, and perhaps a few others are winter residents there.

RECURVIROSTRIDÆ.

125.* **Himantopus nigricollis** *Vieillot.* BLACK-NECKED STILT.

Audubon says it is found in Florida in winter.* Mr. Boardman gives it as "quite common at Enterprise after the 15th of March."

* Birds of America, Vol. VI, p. 35.

126.† **Recurvirostra americana** *Gmelin.* AVOSET.

This species must occur in Florida as a winter visitor, but I have as yet seen no specimens that were collected there.

GRUIDÆ.

127.* **Grus canadensis** *Temminck.* BROWN CRANE.

Abundant.

In 1853, in the Proceedings of the Boston Society of Natural History,* Dr. Bryant discussed at length the question of the relationship of *G americana* Ord to the *G. canadensis*, and arrived at the conclusion that while the young of the *G. americana*, or white whooping crane, might be brown like the mature *G. canadensis*, or sand-hill crane, that the two were distinct species; and this conclusion ornithologists seem to have generally adopted. I saw none of the white birds in Florida, where the brown were very numerous. In Iowa I have seen both, but only at a distance. The account given by Dr. Bryant of the breeding of the sand-hill crane in Florida is very complete and interesting. According to this author the eggs, two in number, are laid from early in February till about the middle of April.†

RALLIDÆ.

128.* **Rallus elegans** *Audubon.* MARSH HEN.

Common.

129.* **Rallus crepitans** *Gmelin.* CLAPPER RAIL.

Common.

130.† **Rallus virginianus** *Linné.* VIRGINIA RAIL.

"Common along the St. John's River." — *Boardman.*

131.† **Porzana carolina** *Vieillot.* CAROLINA RAIL.

"Common." — *Maynard.*

132.(†?) **Porzana noveboracensis** *Cassin.* YELLOW RAIL.

"Common throughout the winter along the St. John's." — *Boardman.*

133.† **Fulica americana** *Gmelin.* COOT.

Abundant. As numerous the 1st of April as during the winter.

134.* **Gallinula galeata** *Bonaparte.* FLORIDA GALLINULE.

Abundant.

* Vol. IV, p. 303.

† See also on this point the same Proceedings, Vol. VII, p. 14.

135.* **Gallinula martinica** *Latham.* PURPLE GALLINULE.

Well known as a resident bird of Florida, but not observed by either Messrs. Maynard and Boardman or myself.

ARDEIDÆ.

136.* **Demiegretta ludoviciana** *Baird.* LOUISIANA HERON.

Common.

137.* **Demiegretta Pealei** *Baird.* PEALE'S EGRET.

Several specimens of this beautiful species were brought home by Mr. Maynard from Indian River, taken at Dummitt's. This is somewhat farther north than any point from which it has been previously reported.

138.* **Garzetta candidissima** *Bonaparte.* LITTLE WHITE HERON.

Abundant. Breeds in February and March.

139.* **Herodias egretta** *Gray.* WHITE HERON.

Abundant. Breeds early in the season. At a small heronry on an islet in Lake Dexter I found several nests containing nearly fledged young, March 23d. The nests, built eight to fifteen feet above the ground, were composed of a few sticks loosely put together. Often they were placed in the tops of bushes which were thickly overgrown with woody vines. The young, when shaken from the nest, climbed through the vines, using their bills as an organ of prehension, either seizing the branches between their mandibles or hooking their bills over them, and clung so closely that it was exceedingly difficult to dislodge them.

This and the preceding species are greatly persecuted by the hunters, who sometimes destroy great numbers at their breeding places, so many of the birds being killed and the others so much alarmed, that large heronries are thus completely broken up. Some gunners make it their business to hunt them for their plumes. Some means should be devised, however, for the protection of these beautiful birds, as at their present rate of decrease their number will soon be greatly diminished.

140.* **Ardea herodias** *Linné.* GREAT BLUE HERON.

Abundant. Breeds in the retired swamps, nesting in the highest cypress-trees. It is rare that more than a single nest is seen in one tree, but often several pairs breed near each other. Young, a third grown, were met with as early as the 12th of March. This species breeds while in immature plumage, young females being found mated with adult males, and *vice versa*. The only very appreciable external sexual difference is that of size, the males, as is generally the case in this family, being much larger than the females.

141.* **Florida cærulea** *Baird.* SMALL BLUE HERON.

Common.

142.* **Ardetta exilis** *Gray.* LITTLE BITTERN.

Not common.

143.† **Botaurus lentiginosus** *Stephens.* BITTERN.

Very common at some localities.

144.* **Butorides virescens** *Bonaparte.* GREEN HERON.

Not uncommon. Smaller than northern specimens, the Florida examples being intermediate in size between those from New England and the West Indies, the latter of which are usually regarded as a distinct species, under the name of *B. brunnescens*. They also decidedly approach the West Indian type in coloration.

145.* **Nycticorax griseus** *Gray.* NIGHT HERON.

Ardea nycticorax LINNÉ, Syst. Nat., I, 235. — WILSON, AUDUBON, NUTTALL, BONAPARTE, etc.

Ardea grisea LINNÉ, Syst. Nat., I, 239, 1766.

Ardea Gardeni GMELIN, Syst. Nat., I, 644, 1788.

Nycticorax europœus STEPH., Gen. Zoöl., XI, 609, pl. xlvii.

Nycticorax americana BONAP., Geog. and Comp. List, 48, 1838.

Nycticorax Gardeni JARDINE, Notes to Wilson's Orn. — BONAP., Conspectus Gen. Avium, II, 141, 1855.

Nyctiardea Gardeni BAIRD, Birds N. Am., 678, 1858, and subsequent American authors.

I did not observe this species on the St. John's, but Mr. Maynard found it more or less common on Indian River and Mosquito Lagoon. Mr. Boardman gives it as "not rare." It is said to be resident the whole year in Florida, by Audubon.

Having compared specimens of the American night heron with others from various parts of the Old World, I see no reason for considering them specifically distinct, though so considered by all American and some European ornithologists. The differences between them are scarcely appreciable.

TANTALIDÆ.

146.* **Tantalus loculator** *Linné.* Wood Ibis. "Gannet."

Common on the Upper St. John's. In March they were undergoing their spring moult, and were consequently in poor plumage. According to Dr. Bryant, who is the first and only writer, so far as I am aware, who has minutely described their eggs and breeding habits, incubation is generally commenced by the 1st of April. Dr. Bryant visited two of their breeding places, one of which was between New Smyrna and Enterprise, in a large cypress swamp on the southern border of Lake Ashby. He estimated that a thousand pairs were breeding there.

There is a singular discrepancy in the accounts of authors in respect to the habits of this bird. Bartram mentions it as solitary in its habits, not associating in flocks. Audubon, always finding it in large flocks, calls attention to this remark of Mr. Bartram as being wholly erroneous, and regrets that his account had been so extensively copied by authors. Dr. Bryant fully corroborates Bartram's account, and censures Audubon for not remembering that birds vary in their habits at different times and places. He says he never saw it in flocks except at its breeding places, and that they usually went off and returned either singly or in pairs. I saw wood ibises more or less frequently on the Upper St. John's for four or five weeks, and only in two or three instances singly or in pairs. I almost invariably saw them in flocks, both at their feeding grounds and flying in the air, they varying in number from a dozen to a hundred. While more or less gregarious at all times, they often doubtless also separate into pairs or wander singly.

In East Florida the wood ibises are called "gannets." Under this name they were described to Audubon when he visited that country, and concerning which he remarks: "On asking the appearance of the Gannets, I was told they were large white birds, with wings black at the end, a long neck, and a large sharp bill. The description so far agreeing with that of the common gannet or solan goose, I proposed no questions respecting the legs or tail, but went off." On visiting the locality where they were said to occur, he was surprised to find the trees covered with wood ibises. He hence adds: "Now as the good people who gave the information spoke according to their knowledge, and agreeably to their custom of calling the ibises gannets, had I not gone to the pond, I might have written this day that gannets are found in the interior of the woods in the Floridas, that they alight on trees, etc., which, if *once* published, would in all probability have gone down to future times through the medium of compilers." * Numbers of similar errors have in fact crept into our natural-history literature, and after they have become well known as such to investigators, they are perpetuated for a generation or two by superficial compilers. The same may almost equally well be said in respect to nominal species.

147.* **Ibis alba** *Vieillot.* WHITE IBIS.

Abundant. Towards the end of February they were moulting and in very poor plumage. Most of the young still retained their brown dress, but in a large proportion the moulting was considerably advanced. Before the end of March it was completed, and April 1st I saw large flocks passing northward high in the air, apparently migrating.

During the winter these birds have the peculiar habit, on the Upper St. John's, of daily flying up the river at evening and down again early in the morning. They usually fly very low, passing just over the tree-tops when cutting across a bend in the river, and at other times close to the water. They are hence in easy gun-shot range from the river or its banks, and, flying in dense flocks, afford fine sport to the numerous sportsmen camping along its banks, who make great havoc among them. They breed much later in the season than the herons. Dr. Bryant states that as late as the 20th of April they had not commenced laying, and that they fly up and down Indian River in the same manner as on the St. John's.† Mr. Maynard informs me he did not meet with this bird on Mosquito Lagoon.

148. **Ibis falcinellus** *Linné.* GLOSSY IBIS.

Tantalus mexicanus ORD., Journ. Phil. Acad. Nat. Sci., I, 53, 1817.

Ibis falcinellus BONAP., Obs. on Nomencl. Wilson's Orn., Ibid., V, 70, 1825. — IBID., Am. Orn., IV, 23, pl. xxiii, 1831. — AUDUBON, Orn. Biog., IV, 608, pl. ccclxxxvij, 1838.

Ibis Ordi BONAP., Geog. and Comp. List, 1838. — BAIRD, Birds N. Amer., 685, 1858.

"Pine barrens between Lake Harney and Indian River, in the ponds, in flocks of twelve to twenty." ‡

* Birds of America, Vol. VI, p. 68.

† Proc. Bost. Soc. Nat. Hist., Vol. VII, 15.

‡ The above is a memorandum of the recent occurrence of this species in East Florida, obtained from Mr. Maynard, but whether given by him on his own authority or on that of Mr. C. H. Nauman, I am at present uncertain.

ARAMIDÆ.

149.* **Aramus giganteus** *Baird.* CRYING BIRD. "LIMPKIN."

This singular and stupid bird is at present more or less common about the grassy lakes and bayous from Lake Dexter southward. Now that Florida has become such a favorite winter resort for health-seekers, pleasure-seekers, and sportsmen, it will be surprising if it is not soon exterminated, as it seems to have almost no fear of man or the gun. They are generally seen in pairs, rarely, however, more than a few occupying the same vicinity; and when one of a party of them is shot, the others, instead of seeking safety by flight, remain and salute the intruder with their singularly discordant cries. Their excellent flesh will tend to favor their rapid extermination. They build their nests in bushes along the river and its bayous, occasionally at a considerable height, but make no effort to conceal them. At Hawkinsville I found a newly built nest, containing a single egg, March 20th, and a few days later, at Lake Dexter, I met with young nearly full grown. Hence they must breed very early, and, perhaps, somewhat irregularly. Dr. Bryant gave the first detailed account of the habits of this bird,* to which there is little to be added. He says he found it more or less common on the St. John's from Lake George to Lake Harney, but most abundant on the Wikiva Creek, which empties into the St. John's about twenty-five miles below Enterprise. This account agrees with my own experience in respect to its distribution. I did not ascend the Wikiva, but was informed that this bird was much more abundant there than on the St. John's. Dr. Bryant says that incubation usually commences in February, and that the number of eggs it lays is very large, sometimes numbering fifteen. Its popular name in Florida is "limpkin."

Possessing many features that ally them to the rails, they in other respects resemble the herons, and especially the ibises, besides having peculiar characters which mark them as a group distinct from either.

ANATIDÆ.

150.† **Anas boschas** *Linné.* MALLARD.

"Common all winter in very large flocks." — *Boardman.* Audubon speaks of their occurring in such numbers in portions of Florida, when

* Proc. Bost. Soc. Nat. Hist., Vol. VII, p. 12.

he was there in 1831, as to darken the air, and the noise of their wings, when rising from the large submerged savannas, he compares to the rumbling of thunder. Mr. Maynard also found them in vast numbers in 1869 on Indian River.

151.† **Anas obscura** *Linné.* BLACK DUCK.

"Quite common." — *Maynard.*

152.† **Dafila acuta** *Jenyns.* PINTAIL DUCK.

"St. John's River; not common." — *Boardman.* Mr. Maynard says that on Indian River he found them in immense numbers, passing over in clouds for hours together.

153.† **Nettion carolinensis** *Baird.* GREEN-WINGED TEAL.

Abundant.

154.† **Querquedula cyanoptera** *Cassin.* RED-BREASTED TEAL.

This species was found by Mr. Maynard in great numbers in the savannas of the upper part of Indian River, but unfortunately the specimens he obtained were lost. This, I believe, is the first time it has been reported from any of the Atlantic States.

155.† **Querquedula discors** *Stephens.* BLUE-WINGED TEAL.

Abundant.

156.† **Spatula clypeata** *Boie.* SHOVELLER.

"Common." — *Maynard. Boardman.*

157.† **Mareca americana** *Stephens.* BALDPATE.

"Common." — *Boardman.*

158.* **Aix sponsa** *Boie.* WOOD DUCK.

Abundant. Breeds early. Saw young March 15th.

159.† **Fulix marila** *Baird.* SCAUP DUCK.

Anas marila LINNÉ, Syst. Nat., I, 1766, 196. — WILSON, Am. Orn., VIII, 84, pl. lxix, 1814.
Fuligula marila AUD., Birds of America, VII, 355, pl. ccccxcviii, 1843.
Fuligula affinis EYTON, Mon. Anat., 157, 1838.
Fuligula mariloides VIGORS, Zoöl. Blossom, 31, 1839.
Fuligula minor GIRAUD, Birds of Long Island, 323, 1844. — BELL, Proc. Phil. Acad. Nat. Sci., I, 141, 1842.
Fulix marila et *affinis* BAIRD, Birds N. Amer., 791, 1858.

Very abundant. By far the most numerous duck on the St. John's River. Quite common at Jacksonville as late as April 1st.

The *Fulix*, or *Fuligula*, *affinis* auct. is evidently only the smaller, darker southern form of the *F. marila* auct. Most of the specimens collected in Florida were of the so-called *F. affinis* type.

160.† **Aythya americana** *Bonaparte.* RED-HEAD.

Abundant in the marshes near St. Augustine, in 1831.—*Audubon.** I find the *A. vallisneria* recorded in my notes made at Jacksonville. I saw none, however, myself, but it was reported by sportsmen to not unfrequently occur there.

161.† **Bucephala albeola** *Baird.* BUTTER-BALL.

Observed in Florida by Audubon.†

162.† **Erismatura rubida** *Bonaparte.* RUDDY DUCK.

More or less common on the Lower St. John's. Also observed by Audubon when he was on the plantation of General Hernandez, in East Florida, and "in immense flocks" about a hundred miles up the St. John's River, in February, 1832.‡ Also obtained by Mr. Maynard at Dummitt's.

163.† **Lophodytes cucullatus** *Reichenbach.* HOODED MERGANSER.

"Very abundant on the coast."—*Boardman.* "Numerous at Dummitt's."—*Maynard.* Occasional on the St. John's.

Geese are currently reported by the inhabitants to occur in winter in North Florida, but I am unable to state what species. Probably *Bernicla canadensis* and *B. brenta*, and perhaps others, are at times more or less common, since they are well known to occasionally visit Cuba.

PELECANIDÆ.

164.* **Pelecanus erythrorhynchus** *Gmelin.* WHITE PELICAN.

"Seen in large flocks near the mouth of the St. John's all winter."—*Boardman.* "Common on Indian River. Said to breed on an island near Dummitt's, and at Jupiter Inlet."—*Maynard.*

165.* **Pelecanus fuscus** *Linné.* BROWN PELICAN.

"Abundant on the coast in winter."—*Boardman.*

* Birds of America, Vol. VI, p. 312.

† Ibid., p. 370.

‡ Ibid., p. 325.

SULIDÆ.

166.† **Sula bassana** *Brisson.* Common Gannet.

"Abundant on the coast." — *Boardman.*

167.* **Sula fusca** *Linné.* Booby Gannet.

A few were seen on the coast near St. Augustine by Mr. Boardman. Mr. Maynard also observed it at Cape Canaveral

PHALACROCORACIDÆ

168.* **Graculus floridanus** *Bonaparte.* Florida Cormorant.

Common on the St. John's, and, according to Mr. Boardman, abundant on the coast.

PLOTIDÆ.

169.* **Plotus anhinga** *Linné.* Snake Bird. Water Turkey.

Abundant. Breeds in February and March, sometimes nesting in the tops of the highest trees, and sometimes quite low. Both sexes incubate.

PROCELLARIDÆ.

170.† **Oceanites oceanica** *Coues.* Wilson's Stormy Petrel.

"A few about the coast at Fernandina." — *Boardman.*

171.† **Puffinus major** *Fabricius.* Greater Shearwater.

"A few about the coast at Fernandina." — *Boardman.*

LARIDÆ.

172.† **Larus argentatus** *Brünnich.* Herring Gull.

Common. Seen up the St. John's as far as Hibernia.

On my voyage from New York to Augusta, Ga., on my way to Florida, small parties of these gulls, numbering usually six to twenty, were almost constantly hovering near the vessel. In the Bay of New York, as along the coast of New England, and doubtless along that of all the Atlantic States at this season, the birds in immature plumage far outnumbered the others; but a hundred miles from land all the gulls of this species seen were old birds, which accords with observations of mine made on other winter voyages in the North Atlantic. It hence appears that the young birds are less venturesome than the adult, and keep mainly near the land. This accords also with the well-known fact that young birds, in migratory species, do not generally attain so high latitudes in the breeding season as the fully adult. It is also highly probable that, generally, the young birds of this family do not range quite so far southward in winter as the older. The mature herring gulls, so far as I had an opportunity of observing, far outnumbered the young ones along the Carolina coast and on the St. John's River.

173.† **Larus delawarensis** *Ord.* RING-BILLED GULL.

"Not numerous." — *Boardman.*

174.* **Chrœcocephalus atricilla** *Lawrence.* LAUGHING GULL.

Common along the coast and on the Lower St. John's.

175.† **Chrœcocephalus philadelphia** *Lawrence.* BONAPARTE'S GULL.

With the preceding, and equally numerous. Also common, according to Mr. Maynard, on Indian River.

176.† **Gelochelidon anglica** *Bonaparte.* MARSH TERN.

Obtained by Mr. Maynard on Indian River.

177.* **Thalasseus regius** *Gambel.* ROYAL TERN.

"Abundant about the coast." — *Boardman. Maynard.*

178. **Sterna hirundo** *Linné.* COMMON TERN.

"Common at Dummitt's." — *Maynard.*

The following table of measurements of sixty-five specimens (forty-five males and twenty females) of this species, taken in the breeding season at Muskeget Island, Massachusetts, indicates the considerable range of individual differentiation that obtains in this species. Though so great, it does appear to be greater than occurs in *Sterna macrura*, of which I have the measurements of twenty-five specimens taken at the same locality and during the same excursion, nor is it probably greater than most of the terns and gulls present, as is evidently indicated by the great number of measurements of specimens of other species of the *Laridæ* of our coast now before me.

The average dimensions of the specimens cited in the subjoined table are as follows: —

Males: Length, 14.51; alar extent, 30.72; wing, 10.47; tail, 5.80; culmen, 1.40; tarsus, .78. *Females:* Length, 13.85; alar extent, 30.59; wing, 10.57; tail, 5.74; culmen, 1.36; tarsus, .77. The extremes of the same are as follows: —

Males: Length, 13.00 to 15.77; alar extent, 29.00 to 32.00; wing, 9.65 to 11.70; tail, 5.00 (4.81?) to 7.00; bill (culmen), 1.28 to 1.55; tarsus, .70 to .87.

Females: Length, 13.10 to 15.50; alar extent, 28.20 to 32.00; wing, 9.90 to 11.50; tail, 5.20 (4.75?) to 6.11; bill (culmen), 1.25 to 1.55; tarsus .70 to .90.

Measurements of Massachusetts Specimens of STERNA HIRUNDO, *Taken in the Breeding Season.*

M. C. Z. No.	Coll. No.	Sex.	Locality.	Date.	Collector.	Length.	Alar Extent.	Wing.	Tail.	Bill.	Tarsus.
—	814	♂	Ipswich	June 16, '68	Allen & Maynard	14 00	30 50	10.62	6.90	1.33	.73
—	807	♂	"	June 16, '68	"	14.36	29.00	10.10	6 56	1.33	.70
10477	905	♂	Wellfleet	June 26, '68	"	14 75	31.90	10.50	6.50	1.35	.80
10478	906	♂	"	June 26, '68	"	14.30	31.70	10.45	6.00	1.41	.76
10479	907	♂	"	June 26, '68	"	14.00	31 00	10.60	5.50	1 33	.74
10480	908	♂	"	June 26, '68	"	13 75	31.60	10.75	5.40	1.28	.76
—	911	♂	Muskeget Isl.	June 29, '68	"	14.75	30.50	10.50	6 00	1.30	.87
10481	913	♂	"	June 29, '68	"	14.40	29.30	10.15	6.00	1.50	.80
10485	916	♂	"	June 29, '68	"	14.80	29.30	10.40	5.50	1.45	—
10486	917	♂	"	June 29, '68	"	14.90	29.60	10.40	5.55	1.36	.85
—	920	♂	"	June 29, '68	"	14 00	30.00	11.00	5.75	1.35	.76
—	923	♂	"	June 29, '68	"	14.00	30.50	10.40	5.55	1.40	.77
10489	925	♂	"	June 29, '68	"	14.40	29.80	10.25	5.45	1 37	.81
10490	926	♂	"	June 29, '68	"	14.50	30.25	10.30	5.40	1.35	.80
—	927	♂	"	June 29, '68	"	14.60	31.00	10.50	6.00	1 40	.75
10491	928	♂	"	June 29, '68	"	14.50	30.80	10.15	5.60	1.52	.74
—	940	♂	"	June 29, '68	"	13.60	31.15	10.25	5.60	—	—
—	941	♂	"	June 30, '68	"	14.50	31.65	10.15	6.10	—	—
—	942	♂	"	June 30, '68	"	14.50	31 50	9.65	6 90	—	—
—	943	♂	"	June 30, '68	"	14.25	30.25	9.75	5.60	—	—
—	944	♂	"	June 30, '68	"	14.60	30.20	10 80	6.00	—	—
10492	945	♂	"	June 30, '68	"	14.10	30.50	10.25	5.50	—	—
—	946	♂	"	June 30, '68	"	15.50	31.85	11.30	7.00	—	.80
—	947	♂	"	June 30, '68	"	15.75	31.50	10.75	6.00	—	.80
—	949	♂	"	July 2, '68	"	13.75	29.90	10 45	5.85	—	.79
—	957	♂	"	July 2, '68	"	15.65	32.00	11.50	5.95	1.50	.75
10498	935	♂	"	July 2, '68	"	14.00	30.25	11 70	5.00	1.45	.77
10500	963	♂	"	July 2, '68	"	14.30	31.27	10.70	5.01	1.33	.77
10501	937	♂	"	July 2, '68	"	14.26	31.00	10.65	5.61	1.30	.76
—	939	♂	"	July 2, '68	"	15.60	31.60	10 85	6.70	1.30	.75
—	970	♂	"	July 2, '68	"	14.28	30.80	10.50	5.40	1 35	.81
—	971	♂	"	July 2, '68	"	14 40	31.60	10.30	5.70	1.40	.75
—	972	♂	"	July 2, '68	"	14.00	30.00	9.80	5.15	1.40	.78
—	973	♂	"	July 2, '68	"	15.00	31.00	9.90	5.80	1.50	.75
10503	975	♂	"	July 2, '68	"	15.20	30.50	10.56	5.85	1.43	.85
—	977	♂	"	July 2, '68	"	14.25	31.20	10.25	5.70	1.51	.80
10504	979	♂	"	July 2, '68	"	15 25	31.00	10.00	6 27	1.45	.81
—	930	♂	"	July 2, '68	"	14.70	30.55	10 40	5.45	1.51	.85
—	981	♂	"	July 2, '68	"	14 55	31 00	10.55	5.55	1.41	.75
—	932	♂	"	July 2, '68	"	13.00	29.00	10.30	4.81	1 35	.76
—	983	♂	"	July 2, '68	"	15.00	31.43	10.80	6.11	1 35	.75
—	993	♂	"	July 2, '68	"	14 50	31.50	10 70	5.80	1.45	.85
—	1000	♂	"	July 2, '68	"	15.77	30.00	10.50	5.75	1 55	.77
—	1002	♂	"	July 2, '68	"	14.25	31.00	10.65	5.75	1.50	.70
—	1003	♂	"	July 2, '68	"	14.00	30.30	10.35	5.60	1.41	.70
10476	904	♀	Wellfleet	June 26, '68	"	14.25	30.60	9.90	6.00	1.35	.70
10481	912	♀	Muskeget Isl.	June 29, '68	"	14.20	30.00	10.00	6.00	1.40	.76
10484	915	♀	"	June 29, '68	"	14.75	30.75	10.55	6.07	1 42	.75
10483	914	♀	"	June 29, '68	"	13.90	29.80	10.05	5.75	1.30	.75
—	918	♀	"	June 29, '68	"	13.60	28.50	10.00	5.50	1.36	.74
—	919	♀	"	June 29, '68	"	13.50	28.20	10.30	5.85	1.25	.75
—	921	♀	"	June 29, '68	"	13.60	30.25	10.25	5.65	1.26	.71
10487	922	♀	"	June 29, '68	"	13.55	30.55	10.63	5.70	1.25	.80
10488	924	♀	"	June 29, '68	"	13.10	29.50	10.50	5.20	1.35	.73
—	948	♀	"	June 30, '68	"	14.50	31.75	10.80	5.90	—	.80
10494	950	♀	"	July 2, '68	"	14.30	31.00	10.50	5.80	1 28	.80
10495	959	♀	"	July 2, '68	"	13.60	32 00	11 50	5.75	1 30	.80
—	960	♀	"	July 2, '68	"	15 50	31.75	11.25	5.41	1.43	.71
10498	965	♀	"	July 2, '68	"	13.56	30.00	10.30	5.14	1.30	.74
—	974	♀	"	July 2, '68	"	15.25	32 00	11.25	6.11	1 34	.90
—	976	♀	"	July 2, '68	"	14.20	30.20	10.26	5.55	1.31	.75
—	978	♀	"	July 2, '68	"	14.60	31.80	10.70	5.45	1.55	.80
10504	997	♀	"	July 2, '68	"	14.45	31.70	10.45	6.46	1.41	.80
10505	999	♀	"	July 2, '68	"	14.35	30.50	10.45	5.60	1.49	.80
—	1001	♀	"	July 2, '68	"	14.40	31.00	10.65	5.85	1.40	.85

179. **Sterna macrura** *Naumann.* ARCTIC TERN.

"Common at Dummitt's." — *Maynard.*

As already remarked under *Sterna hirundo* the individual variation in the present species is very great. The largest and smallest specimens in a series of twenty-five, taken at Muskeget Island in the breeding season measured as follows: —

Largest (♂): Length, 16.00; alar extent, 32.75; wing, 11.75; tail, 6.00.

Smallest (♀): Length, 14.33; alar extent, 27.52; wing, 9.85; tail, 4.26.

The maxima and minima of this series are as follows: —

Length, 14.10 and 17.00; alar extent, 27.52 and 32.75; wing, 9.85 and 11.84; tail, 4.26 and 8.25.

While the females average a very little smaller than the males, several of the females are very nearly as large as the largest males.

The *Sterna Forsteri* may also occur as a winter resident, but I have at present no evidence of its occurrence there at this season. A specimen from the "St. John's River, Florida," collected by Dr. Würdemann, is cited by Mr. Lawrence * and Dr. Coues † (Smithsonian collection No. 4928), but no information is given as to when it was collected.

180.* **Rhynchops nigra** *Linné.* BLACK SKIMMER.

Abundant on the coast, occurring in large flocks. Not observed by me on the St. John's.

COLYMBIDÆ.

181.† **Colymbus torquatus** *Brünnich.* LOON.

"A single specimen at Mandarin, on the Lower St. John's; abundant off Fernandina harbor." — *Maynard.*

The considerable number of specimens of this species in the Museum of Comparative Zoölogy show a wide range of individual variation. In a series of fifteen specimens from various localities in New England, but mainly from Massachusetts, the variation in the length of the folded wing amounts to twenty per cent of its average length in the whole series; in the length of the tarsus, to twenty-nine per cent; in the length of the outer toe, to thirty per cent; in the length of the head, to twenty-eight per cent; and in the length of the culmen to twenty-three per cent.

The form described some years since as *Colymbus Adamsi* seems to have been founded on very old specimens of the large northern race of *C. tor-*

* Baird's Birds of North America, p. 863.

† Proc. Acad. Nat. Sci. Phila., 1862, p. 547.

quatus, in which the color of the bill is unusually light, and the bill itself unusually produced.

182.† **Podiceps cornutus** *Latham.* HORNED GREBE.

"Not uncommon on the St. John's." — *Boardman.*

183.† **Podilymbus podiceps** *Lawrence.* CAROLINA GREBE.

Abundant on the St. John's.

Résumé of the preceding Tables of Measurements, with suvplemental Remarks.

The following tables present a brief summary of the measurements given in Part IV. In the first table is given the average dimensions of thirty-two species, based on specimens collected, in each case, essentially from the same locality, and generally based on twenty or more specimens, the number varying in the different species from thirteen to sixty-five specimens. In all cases where the average sexual difference in size is appreciable, the dimensions are given for each sex. In most cases very nearly all the specimens are from Eastern Massachusetts, a few being from different localities in Southern Maine, and a few from Northern Illinois. In a few species all the specimens cited are from Eastern Florida; in a few other species part of the specimens are from Southern New England and a part from Eastern Florida; but in these cases a separate average is made of those from each of the two localities. The number of the specimens on which the average is based is given in each instance.

The second table shows the range of individual variation in size in the same species, based also on the same specimens.

The third table shows the amount of geographical variation in size in specimens of the same species from northern and southern localities, these localities being generally Southern New England (Eastern Massachusetts in the main) and East Florida. Only seven species are cited, but I have traced about the same ratio of difference in a score or more of others, of which the measurements have not yet been published. Although the number of specimens compared from the two localities has in many of these cases been comparatively small, enough have been examined to show the general constancy of the variation in all the species which breed at both these localities.

It should be added that the specimens on which the generalizations given in Table III are based were not taken at the seasons likely to give the greatest differences, the northern specimens having been taken in summer and the southern ones in winter. Had *summer* Florida specimens been used instead of *winter* specimens, the differences would have been doubtless much greater, since in some cases, and especially in the cases of *Agelæus phœniceus* and *Quiscalus purpureus*, the summer home of a part at least of the Florida specimens must have been somewhat to the northward of Florida.

I. *Table showing the Average Dimensions of Thirty-two Species of American Birds, based on Measurements of Thirteen to Sixty-five Specimens of each Species.*

Species.	Locality.	No. of Spec'ns.	Sex.	Length.	Alar Extent.	Wing.	Tail.	Tarsus.
Turdus Swainsoni	Southern New England	24	—	7.17	11.65	3.86	2.88	1.15
Turdus Pallasi	" "	46	—	7.04	11.17	3.97	2.72	1.15
Turdus fuscescens	" "	40	—	7.38	11.83	3.82	2.88	1 13
Harporhynchus rufus	" "	17	—	11.29	13.09	4.15	5.00	1.31
Mimus polyglottus	Florida	37	—	9.91	13.69	4.28	4.87	—
Galeoscoptes carolinensis	Southern New England	20	—	8.60	11.16	3.53	3.76	1.10
Sialia sialis	" "	20	♂	6.80	11.93	3.94	2.55	.78
Geothlypis trichas	" "	20	♂	5.10	6.93	2.17	2.00	.77
Parus atricapillus	Eastern Massachusetts	27	—	5.38	8.37	2.47	2.50	.70
Tyrannus carolinensis	Southern New England	20	—	8.00	13.77	4.49	3 30	.73
Pyranga rubra	" "	13	♂	7.05	11.33	3 76	2.69	.75
Troglodytes aëdon	Florida	15	—	4.89	6.61	2.05	1.80	.52
Passerculus savanna	Massachusetts	26	—	5.20	8 79	2.70	1.96	.84
Peucæa æstivalis	Florida	22	—	5.88	8 99	2.40	2.49	70
Cardinalis virginianus	"	32	♂	8.46	11.43	3.63	3.87	—
	"	26	♀	8.27	11.27	3.53	3.77	—
Pipilo erythrophthalmus	Southern New England	30	♂	8.19	11.32	3 43	3.36	1.06
	Florida	19	♂	7.88	9.88	3.13	3.56	.94
Hedymeles ludoviciana	Southern New England	17	♂	7.77	12 15	3 93	2.82	.86
Icterus Baltimore	" "	20	♂	7.52	11.55	3.71	2.92	.92
Dolichonyx oryzivorus	" "	20	♂	7.24	11.67	3.78	2 67	1.00
Agelæus phœniceus	" "	40	♂	9.16	14.71	4.69	3.63	—
	" "	28	♀	7.53	12.24	3 86	2 93	—
Sturnella ludoviciana	Northern States	15	♂	10.43	16.30	4 91	3.16	—
	Florida	12	♂	9.81	15 70	4.47	2.85	—
	Northern States	8	♀	9 55	14.43	4.29	2.82	—
	Florida	9	♀	8.96	14.09	4.22	2.57	—
Quiscalus purpureus	Northern States	15	♂	10.43	16.30	4.91	3.16	—
	Florida	12	♂	9.81	15.70	4 47	2.85	—
	Northern States	8	♀	9.55	14.43	4.29	2.82	—
	Florida	9	♀	8.96	14.09	4.22	2.57	—
Quiscalus major	"	24	♂	16.51	22.48	7.19	7.00	—
	"	8	♀	12.95	17.94	5.67	5.11	—
Cyanura cristata	Massachusetts	18	—	11.71	16.87	5.13	4.89	—
	Florida	11	—	10.98	15 11	4.75	5.00	—
Cyanocitta floridana	"	12	—	11.74	14.44	4.41	4.80	—
Hylotomus pileatus	"	7	♂	17.48	28 07	9.21	6.82	—
	"	7	♀	16.44	26.80	8 98	6.54	—
Picus borealis	"	28	—	8.34	14.46	4.71	3.41	—
Colaptes auratus	Massachusetts	18	—	12.45	19.94	6.24	4.35	—
	Florida	11	—	11.66	18.82	5.84	4.40	—
Conurus carolinensis	"	19	—	13.10	21.76	7 59	—	—
Oxtyx virginianus	Illinois	7	♂	10.18	15.44	4.47	2.82	—
	Florida	16	♂	9 46	14.16	4.22	2.52	—
	Illinois	6	♀	9.83	15 10	4.36	2.67	—
	Florida	10	♀	9.37	14.02	4.17	2.54	—
Sterna hirundo	Massachusetts	45	♂	14.51	30.72	10.47	5.80	.78
	"	20	♀	13.85	30.59	10.57	5.74	.77

II. *Table showing the Range of Individual Variation in Thirty Species of American Birds, based on the Measurements of Thirteen to Sixty-five Specimens of each Species, collected at the same Locality.*

Species.	Locality.		No. of Speci'ns.	Sex	Length	Alar Extent.	Wing.	Tail	Tarsus
Turdus Swainsoni	Southern New England	Min.	24	—	6.62	10.75	3.47	2.40	1.02
	" "	Max.	24	—	7 75	12 65	4.30	3.40	1.27
Turdus Pallasi	" "	Min.	46	—	6.50	10.00	3.30	2.47	1.12
	" "	Max.	46	—	7.65	12.25	3 90	3.17	1.33
Turdus fuscescens	" "	Min.	40	—	6.95	10.05	3.55	2.63	1 06
	" "	Max.	40	—	7.87	12.70	4 16	3.02	1 18
Harporhynchus rufus	" "	Min.	17	—	10 55	12.55	3.80	4.50	1 20
	" "	Max.	17	—	11.85	14.00	4.25	5.30	1.42
Galeoscoptes carolinensis	" "	Min.	20	—	7.80	10.50	3.25	3.35	1.05
	" "	Max.	20	—	9 00	11.95	3.85	4 10	1.18
Mimus polyglottus	Florida	Min.	37	—	9.27	13.00	4.00	4.10	—
	"	Max.	37	—	11.00	14 75	4.75	5.15	—
Sialia sialis	Southern New England	Min.	20	♂	6.10	11.10	3 85	2.33	.74
	" "	Max.	20	♂	7.00	12.55	4.10	2.77	.83
Geothlypis trichas	" "	Min.	20	♂	4.60	6.30	1.95	1.78	.72
	" "	Max.	20	♂	5.63	7.50	2 37	2.10	.82
Pyranga rubra	" "	Min.	13	♂	6.75	10 65	3.57	2 55	.67
	" "	Max.	13	♂	7.30	11.75	4.00	2 85	.86
Parus atricapillus	Eastern Massachusetts	Min.	27	—	4.70	7.50	2.33	2.15	.62
	" "	Max.	27	—	5.75	8.60	2.63	2.67	.77
Troglodytes aëdon	Florida	Min.	15	—	4.30	6.10	1.90	1.30	.50
	"	Max.	15	—	5.10	6.95	2.44	2.40	.68
Passerculus savanna	Eastern Massachusetts	Min.	26	—	5.20	7.61	2.44	1.64	.75
	" "	Max.	26	—	6.00	9.75	2.95	2.25	.88
Peucæa æstivalis	Florida	Min.	22	—	5.75	7.60	2.17	2.25	—
	"	Max.	22	—	6.20	8.30	2.55	2.68	—
Cardinalis virginianus	"	Min.	32	♂	7.75	11.00	3.50	3 40	.62
	"	Max.	32	♂	9.10	11.78	3.85	4.20	.80
	"	Min.	26	♀	7.50	10.70	3.25	3 40	.62
	"	Max.	26	♀	8.75	11.75	3.85	4.10	.75
Pipilo erythrophthalmus	Southern New England	Min.	30	♂	7.50	10.00	3.17	3.30	.98
	" "	Max.	30	♂	8.80	12.25	3 90	3 93	1.13
	Florida	Min.	19	♂	7.20	9.50	2.80	3.25	.80
	"	Max.	19	♂	8.50	11.30	3.50	3.90	1.09
Hedymeles ludovicianus	Southern New England	Min.	17	♂	7.15	11.50	3 83	2.70	80
	" "	Max.	17	♂	8.30	12.90	4.25	3.08	.93
Icterus Baltimore	" "	Min.	20	♂	7.00	10.40	3.45	2 70	.83
	" "	Max.	20	♂	8.00	12.00	3.85	3.10	1.02
Dolichonyx oryzivorus	" "	Min.	20	♂	6.65	11.00	3 53	2.45	.98
	" "	Max.	20	♂	7.70	12.15	4 00	2.82	1.15
Agelæus phœniceus	Massachusetts	Min.	40	♂	8.40	13.95	4.43	2 99	—
	"	Max.	40	♂	9 85	15.35	5.00	3.90	—
	"	Min.	28	♀	7.35	11.25	4.26	2.65	—
	"	Max.	28	♀	8.55	13.55	4.43	3.15	—
Sturnella ludoviciana	"	Min.	15	♂	10.00	15.05	4.74	2.82	—
	"	Max.	15	♂	11.00	17.00	5.15	3.58	—
	Florida	Min.	12	♂	8.50	13.00	3 90	2.40	—
	"	Max.	12	♂	9.50	4.75	4 65	2.90	—
Quiscalus purpureus	Northern States	Min.	13	♂	12 00	17.00	5.20	4 58	—
	" "	Max.	13	♂	13.50	18 43	6.05	6.00	—
	Florida	Min.	23	♂	11.00	15.25	5 00	4.55	—
	"	Max.	23	♂	13.00	17.80	5 75	5.50	—
Quiscalus major	"	Min	24	♂	15.50	21.10	6 25	6 25	—
	"	Max	24	♂	16.80	23.50	8.35	7.60	—
	"	Min.	8	♀	12.10	17 25	5.25	4.75	—
	"	Max.	8	♀	13.40	18.25	5.95	5.60	—
Cyanura cristata	Massachusetts	Min	18	—	11 00	16.00	4 33	4.25	—
	"	Max.	18	—	12.25	17.50	5.65	5.65	—
	Florida	Min.	11	—	10.70	14.75	4.00	4 80	—
	"	Max.	11	—	11 25	16 00	5.00	5.15	—
Cyanocitta floridana	"	Min.	12	—	11.00	13.50	4.00	4.25	—
	"	Max.	12	—	12.50	15.00	4 75	5.35	—
Tyrannus carolinensis	Southern New England	Min.	20	—	7.00	12 50	4.17	2.93	.67
	" "	Max.	20	—	8 65	14.80	4.85	3.54	.80
Picus borealis	Florida	Min	22	—	7 90	14.10	4.40	3.15	—
	"	Max.	22	—	8.60	15.20	4.95	3.75	—

Table II. (Continued.)

Species.	Locality.		No of Speci'ns.	Sex.	Length.	Alar Extent.	Wing.	Tail.	Tarsus
Colaptes auratus	Massachusetts	Min.	18	—	12.00	19.00	6.00	4.00	—
	"	Max.	18	—	13.00	20.75	6 60	4.70	—
	Florida	Min.	11	—	10.60	17.60	5 60	4.10	—
	"	Max.	11	—	12.75	19.75	6.25	4.85	—
Conurus carolinensis	"	Min.	19	—	12.50	21 10	7.00	—	—
	"	Max.	19	—	13.60	22.50	7.85	—	—
Ortyx virginianus	Illinois	Min.	7	♂	10 00	15.00	4.37	2.55	—
	"	Max.	7	♂	10.50	15.60	4.60	3.00	—
	Florida	Min.	16	♂	9.00	13.80	4.00	2.30	—
	"	Max.	16	♂	10.00	14.75	4.50	3.00	—
	Illinois	Min.	6	♀	9.50	14.50	4.25	2.45	—
	"	Max.	6	♀	10.25	15.50	4.50	2.85	—
	Florida	Min.	10	♀	9.00	13.10	3.35	2.50	—
	"	Max.	10	♀	10.00	14 50	4.40	2.77	—
Sterna hirundo	Massachusetts	Min.	45	♂	13.00	29.00	9.65	4.81	—
	"	Max.	45	♂	15.70	32 00	11.70	7.00	—
	"	Min.	20	♀	13.10	28.20	9.90	4 75	—
	"	Max.	20	♀	15.50	32.00	11.50	6.11	—

III. *Table showing the Geographical Variation in Size in Seven Species of American Birds, between Specimens from Florida and the Northern States.*

Species.	Locality.	No. of Speci'ns.	Sex.	Length.	Alar Extent.	Wing.	Tail.	Tarsus.
Pipilo erythrophthalmus	Southern New England	32	♂	8.19	11 32	3.43	3 36	1.06
	Florida	26	♂	7.88	9.88	3.13	3.56	.94
Agelæus phœniceus	Southern New England	40	♂	9.16	14 71	4.69	3.63	—
	South Carolina & Florida	11	♂	9 02	14.41	4.62	3.61	—
Sturnella ludoviciana	Northern States	15	♂	10.43	16.30	4 91	3.16	—
	Florida	12	♂	9.81	15.70	4 47	2.85	—
	Northern States	8	♀	9.55	14.43	4.29	2.82	—
	Florida	9	♀	8 96	14.09	4.22	2.57	—
Quiscalus purpureus	Northern States	15	♂	10.43	16.30	4.91	3.16	—
	Florida	12	♂	9.81	15.70	4.47	2.85	—
	Northern States	8	♀	9.55	14.43	4.29	2.82	—
	Florida	9	♀	8.96	14 09	4.22	2.57	—
Cyanura cristata	Massachusetts	18	—	11.71	16.87	5.13	4.89	—
	Florida	11	—	10.98	15.11	4.75	5.00	—
Colaptes auratus	Massachusetts	18	—	12.45	19 94	6.24	4.35	—
	Florida	11	—	11 66	18.82	5.84	4.40	—
Ortyx virginianus	Illinois	7	♂	10 18	15 44	4.47	2.82	—
	Florida	16	♂	9.46	14.16	4.22	2.52	—
	Illinois	6	♀	9.83	15.10	4.36	2.67	—
	Florida	10	♀	9.37	14.02	4.17	2.54	—

In the tables and remarks contained in the preceding pages many facts have been given bearing upon the subject of geographical variation in birds, and especially in reference to the differences that almost universally obtain between specimens of the same species from northern and southern localities. In addition to the smaller size of the southern specimens, — a fact which has been for some time quite generally recognized, — attention has been called to the differences in color and in the form of the bill that seem almost equally constant and easy of recognition. In several species that range in the breeding season from Florida to Maine, a tendency to a relatively greater elongation of the tail in the Florida specimens has also been noticed, — a feature so well known to characterize a large proportion of the birds of Lower California, as pointed out some years since by Professor Baird, — but this variation is not so frequent as the differences in size, color, and in the length and form of the bill. As already remarked, the tail is not usually abbreviated proportionally to the general diminution in size in the southern or Florida forms of the birds of Eastern North America, and in some species it is actually longer than in the larger northern birds. As shown in the above tables, this is the case in *Pipilo erythrophthalmus*, *Cyanura cristata*, and *Colaptes auratus*, or in three species out of the seven cited in the last table.

In numerous instances the southern forms of the birds enumerated in Part IV of this paper have already been specifically separated from their northern relatives; and if the example of some previous writers was to be followed at least a dozen other similar species might still be added from among the birds of Florida. Some, indeed, might be referred to the already separated West Indian and Mexican or Central American so-called species rather than to the northern type. As already stated, I consider this almost universal similar variation of the southern representatives of species from their northern representatives to be the result of a law of gradual geographical differentiation, and that the interest of science is better subserved by simply recognizing these differences, and the law of geographical variation of which they are the result, than by giving to each newly discovered race a distinctive binomial name; and the more especially since in numerous instances there is the most indubitable proof of the gradual and almost imperceptibly minute intergradation of the extreme northern and extreme southern types, even in cases where they are the most widely diverse.

In conclusion, it may be stated that the differential diagnoses of the southern types, in cases where they have been specifically separated from the northern, and the comparisons of them made with the northern for the purpose of showing their specific distinctness, are in many cases admirable descriptions of the usual differences found to distinguish the Florida-born birds from their co-specific representatives born in the Northern States. These differences are commonly solely the following: In the southern types the size is smaller, the bill longer, and the colors generally darker; the latter resulting from the greater predominance of the black in those in which portions of the plumage are mottled with this color, and the greater breadth of the dark transverse bars, and the correspondingly diminished breadth of the alternating lighter ones. To illustrate this point more fully, I herewith append a list of some of the so-called species of American birds that have been specifically separated by different authors from their northern representatives, but which are in reality only the extreme southern forms of species previously well known, with which they were considered by the older writers to be specifically identical, the most of them having been separated within the last ten or fifteen years: —

Accipiter Gundlachi,	*separated from*	Accipiter Cooperi.
Accipiter fringilloides,	"	Accipiter fuscus.
Falco dominicensis, cinnamominus et sparveroides, etc.	"	Falco sparverius.
Mimus Gundlachi et Hillii, etc.	"	Mimus polyglottus.
Seiurus ludovicianus,	"	Seiurus noveboracensis.
Thryothorus Berlandieri,	"	Thryothorus ludovicianus.
Dendrœca Gundlachi, etc.	"	Dendrœca æstiva.
Chordeiles minor et Gundlachi, etc.	"	Chordeiles popetue.
Antrostomus cubanensis,	"	Antrostomus vociferus.
Xanthornus affinis,	"	Icterus spurius.
Sturnella hippocrepis et mexicana,	"	Sturnella ludoviciana.
Quiscalus baritus,	"	Quiscalus purpureus.
Corvus minutus,	"	Corvus americanus.
Ortyx cubanensis et texanus,	"	Ortyx virginianus.
Campephilus Bairdii,	"	Campephilus principalis.
Colaptes chrysocaulosus,	"	Colaptes auratus.
Butorides brunnescens,	"	Butorides virescens.
Actiturus longicaudus,	"	Actiturus Bartramius.
Macrorhamphus scolopaceus	"	Macrorhamphus griseus.
Charadrius tenuirostris,	"	Charadrius melodus
Larus argentatoides et Smithsonianus,	"	Larus argentatus.

In other cases the arctic forms, or the northern types, having been discovered subsequently to the southern ones, these have been described as specifically distinct from the latter. The *Bucephala islandica*, separated from the *B. americana* et *clangula*, and the *Collurio excubitoroides* from the *C. ludovicianus*, will serve to indicate the class of so-called species here referred to.

The Pacific Slope of North America furnishes a similar list of species, based on either southern or northern forms of others previously known; and the middle region of the continent its list of similar nominal species, mainly based on the desert forms of widely ranging species. In the northern half of the Old World, also, have the northern and southern geographical forms of the same species been specifically separated; but it is not my intention to call farther attention to them at present.

As already remarked, the American representatives of circumpolar species differ from the European and Asiatic principally in two ways, namely, in the apparently slightly larger size of the American, and in their somewhat brighter colors; but specific separations seem to have been based almost as frequently upon some theory of geographical distribution, or upon the individual variation of single specimens, as upon the real though slight differences that frequently obtain in such cases.

PART V.

On the Geographical Distribution of the Birds of Eastern North America, with special reference to the Number and Circumscription of the Ornithological Faunæ.

1. Introductory Remarks.

The distribution of plants and animals in circumpolar zones over the earth's surface has been long recognized; Humboldt* first making known the fact of such a natural distribution of the plants, and Agassiz,†

* Humboldt, A. von, et Bonpland, Aimé. "Essai sur la Géographie des Plantes," etc. 4to. Paris. 1805.

† Agassiz, Louis. "Essai sur la Géographie des Animaux," Revue Suisse et Chronique Littéraire, Tome VIII, pp. 441-452, 538-585, 1845. "Note sur la Distribution Géographique des Animaux et de l'Homme," Bulletin de la Société des Sciences Naturelles de Neuchatel, Tome I, pp. 162-166, 357-361, 366-369, 1845. "Sur la Distribution Géographique actuelle et le mode de l'apparition actuelle des Animaux à la surface du Globe." Ibid., Tom. 2, pp. 347-351, 1847. "Geographical Distribution of Animals," Edinburgh New Philosophical Magazine, Vol. XLVI, pp. 1-25, 1850. Ibid., Christian Examiner, Vol. XLVIII, pp. 184-204, 1850. "Sketch of the Natural Provinces of the Animal World and their Relation to the different Types of Man," Nott and Gliddon's Types of Mankind, pp. lviii-lxxxii, 1854. Also especially insisted upon in a course of unpublished Lectures delivered before the Lowell Institute, Boston, December, 1869.

Wagner,* Dana,† and others, subsequently establishing the same in regard to animals; the distribution of both plants and animals being primarily determined by the same influences. It has been further shown that these influences are mainly climatic, temperature having been justly recognized as governing the limitation, especially in latitude, of not only the species, but of faunæ and floræ. Their limitation in longitude is likewise as directly determined by climatic influences,‡ though indirectly by physical barriers, as oceans, mountain chains, and deserts. Humidity, in many instances, is scarcely a less, and in some cases a more, powerful limiting agent than temperature, plants being highly sensitive to hygrometric conditions, and their distribution intimately affects that of animals, since the existence of the latter is dependent upon the presence of the former, and their variety and numbers upon the degree of luxuriance of the vegetation. The faunal and floral zones hence coincide in their limitation in latitude with the climatic zones, but by no means necessarily with the geographical circles; isothermal lines, and not parallels of latitude, forming their boundaries. Their limits in longitude are determined by the influence geographical barriers, especially long chains of high mountains, exert upon climate.

* WAGNER, ANDREAS. "Die geographische Verbreitung der Säugethiere," Abhandlungen der baierischen Akådemie der Wissenschaften, Math. Phys. Classe, Band IV, Abth. I, pp. 1-146, 2d Abth., pp. 1-108, 3d Abth., pp. 3-114. Mit 9 Karten, 1844-1846.

† U. S. Exploring Expedition Report, Crustacea, Vol. II, pp. 1451-1500, 1852.

‡ I am aware of the diversity of opinions still prevalent among naturalists in regard to the influence climate exerts in determining the geographical distribution of species, and that many writers on this subject attribute to it only a slight importance, or altogether ignore it. The limits of these preliminary remarks will not allow of an extended comparison of the views of different authors on this point, nor a detailed consideration of the supposed objections that have been raised against the proposition above expressed. I agree with Mr. Andrew Murray in his remark, that, although "various authors have endeavored to embody the differences between the faunas and floras of the different regions of the globe into some kind of system, ... they, with one or two exceptions, have worked upon no definite principle, and the result has been a mere catalogue of regions which possessed peculiarities without distinguishing their relative importance, or their relation to each other" (Geographical Distribution of Mammals, p. 296, 4to, London, 1866), — a remark which unfortunately seems in some degree applicable to Mr. Murray's own generalizations. That temperature is a powerful limiting influence affecting the range of species, especially in respect to their northward and southward extension, is so easily demonstrable that I am surprised to see it still questioned. I have myself subjected this principle to a rigid examination in studying the distribution of the animals and plants of Eastern North America, and have been surprised at the exact coincidence I have almost constantly met with between their northern and southern

The uniform character of both the flora and the fauna throughout the arctic zone is one of the most striking onto-geographical features thus far known, and one of primary importance, especially when taken in connection with its relation to the faunæ and floræ of more southern latitudes. Not less significant is the fact that in the temperate zone there is still a prevalence of identical forms in each of the three northern continents, where the resemblance of the animals and plants of either continent to those of the others is far greater than is the resemblance of those of the temperate regions of either continent to those of the tropical portions of the same continent.

Different animals and plants, as every one recognizes, are differently limited in respect to their geographical range. A small proportion of the species are almost or quite cosmopolitan; others range over the greater part of the northern hemisphere, finding their southern limit of distribution near the tropics. A few are exclusively arctic, or range only over the arctic and cold-temperate zones. Many are limited to the temperate zone, throughout nearly the whole of which they find a congenial home. A large number can only exist within the tropics, often

limits of distribution and isothermal lines, they following them in all their numerous undulations, sweeping northward in the valleys and southward along the sides of mountain ranges. The occurrence on isolated alpine summits of species existing at a lower level only far to the northward, is of itself suggestive of the powerful influence temperature has on the distribution of animals and plants. In the northern hemisphere a northern fauna and flora everywhere extends along the mountains hundreds of miles to the southward of their respective limits in the adjoining plains and valleys. Various other causes have, of course, a greater or less influence in determining the range of species, but none other, on the land areas, humidity perhaps alone excepted, is nearly so potent. The want of conformity of isothermal lines with parallels of latitude has doubtless led to confusion in regard to this subject, since vain attempts have often been made to circumscribe the botanical and zoölogical zones by the latter.

Differences of temperature evidently explain many of the otherwise seemingly inexplicable sudden transitions in the faunæ and floræ of adjoining regions, especially in regard to the marine animals and plants, temperature forming a strong barrier to the commingling of species inhabiting the waters of opposite sides of peninsulas having a north and south trend, or such long narrow points of land as terminate the South American and African continents. Those of the one side cannot pass to the other without passing through a zone of colder water than their organization will allow them to sustain. The isotherms of the continents are widely deflected by the irregularities of the surface of the land, running nearly straight and parallel across level areas; but in mountainous districts they bend abruptly northward or southward, following along the sides of mountains instead of crossing them. In the same manner are species, and faunæ and floræ, limited, — a coincidence clearly indicative of the strong influence climates exert in determining their geographical limits.

embracing whole families, none of whose representatives are found outside of the torrid zone of a single continent. Others are again equally at home in the torrid and warm temperate zones, but which do not exist either in the arctic or cold temperate zones; others range throughout the temperate and subtorrid. Nearly an equal number, some tropical, but the greater part temperate species, range across continental areas, within which, however, they are restricted. A great number of others find their range limited in longitude to the half or the third of a continent, and others within still more circumscribed boundaries, fluviatile species being frequently confined to single river basins. Through this diversity of geographical range we have what may be termed *cosmopolitan*, *semi-cosmopolitan*, *circumpolar*, *continental*, *semi-continental*, and (relatively speaking) *restricted* species. The circumpolar and the continental are again *realm* species, the semi-continental and restricted, *province* species. Rarely is any species limited to a narrower area than that of two or three faunæ or floræ. Hence faunæ and floræ — which terms, in their restricted sense, are properly applied only to the smallest of the onto-geographical divisions — are determined by the peculiar association of species, and not by the range of a single or of a few "restricted" species; hence by their general facies. Provinces, and realms, on the other hand, may have species, and even genera and families, exclusively distinctive of them. As there are cosmopolitan, circumpolar, continental, and other kinds of species, so there must be cosmopolitan, circumpolar, continental, and other kinds of genera and families; the latter, as well as species, having each a definite or specific geographical range as distinctive of them as any biological or anatomical character may be. They are each circumscribed within definite areas, beyond which their special adaptation to their natural surroundings forbids their extension, unless aided by extraneous and unusual circumstances.

The three divisions of zones, realms (or "regions"), provinces, and faunæ and floræ,* comprise the phyto-zoölogic divisions usually recog-

* *Zone*, *realm*, *region*, *kingdom*, and *province*, are terms which have been used by different authors to designate the primary natural-history divisions of the earth's surface. In deciding as to which of these terms should be exclusively applied to these divisions, not only priority of use, but appropriateness, should of course be considered, and also the sense in which they are at present currently employed, in order to avoid, as far as possible, the confusion necessarily attending changes of nomenclature. So far as priority is concerned, zone undoubtedly has the precedence, it having been used for animals by Wagner in 1844, by Agassiz in 1845, and much earlier than this by Humboldt and others in relation to the distribution of plants. It is, however, not always a strictly convenient

nized. The boundaries of realms and provinces have often been arbitrarily fixed, inasmuch as they have been frequently limited and named in conformity to the continental areas, regardless of the fundamental law of the distribution of life in circumpolar zones.*

In addition to the law of the circumpolar distribution of life in zones, another may be recognized, namely, that *of a differentiation from the north southward*, since in passing from the northern pole to the equator we meet with a constant and accelerated divergence in the character of the animals and plants of successive regions of the continent. More or less related to the last is a third law of differentiation, namely, *a divergence of the life of given portions of continental areas from that of the corresponding portions of other continents, in proportion to the oceanic space separating such corresponding regions.* As evidence of this fact we have but to compare successively the life of the north temperate, tropical, and south temperate zones of the Western hemisphere with the life of the corresponding zones of the Eastern hemisphere; or that of Australia with the life of the other continents, as a whole; or that of tropical Asia with Africa or South America. A comparison of Africa with South America, and the faunæ and floræ of islands with those of the different continents, further corroborates this law. There is, furthermore, a correlation between the diversity

term. Realm, region, fauna and flora, and province, have been also successively used in the same sense, and also for divisions of subordinate rank, and in different ways by even the same writers. In regard to the names of the divisions of the second, third, and fourth rank, there is an equal want of uniformity in the use of the terms by which they have been designated. As being most convenient and least opposed to current usage, the following schedule of names for the primary and subordinate divisions has been adopted in the present paper: —

Realms for divisions of the first rank.

Regions for divisions of the second rank.

Provinces for divisions of the third rank.

Districts for divisions of the fourth rank.

Faunæ and *floræ* for the smallest or ultimate divisions, like the bird faunæ of Eastern North America, presently to be characterized.

Intermediate divisions to some of those above mentioned may in special cases be required; but until the necessity for them is made apparent, no names for such need be proposed.

* SCLATER, P. L. "On the general Geographical Distribution of the Members of the Class Aves," Jour. of the Proc. of the Linnæan Society, Vol. II, Zoölogy, pp. 130 - 149, 1858. The divisions proposed by this author have been quite generally adopted, but without corroboration, or apparently a critical examination of their merits.

of life in a given area and the relative temperature of that area, the number of distinct forms increasing directly with the increase in the temperature, other conditions remaining essentially unchanged. The number of distinct species and geographical races also increases directly with the increase in the diversity of the conditions of life resulting from differences of geographical configuration. Hence, faunæ and floræ cover a smaller area in the warm temperate and tropical latitudes than at the northward, and in a mountainous region than in a level region. Hence within the torrid zone, where a maximum temperature is generally associated with a highly diversified surface, species, genera, and families are the most numerous, and faunæ and floræ, as well as species, are ordinarily the most narrowly circumscribed.

In accordance with the facts stated above respecting the mode of the distribution of animals and plants over the earth's surface, and the zoölogical and botanical laws of the differentiation and mutual relations of the different regions, the following primary natural-history divisions may be recognized: I, an *Arctic Realm;* II, a *North Temperate Realm;* III, an *American* Tropical Realm;* IV, an *Indo-African Tropical Realm;* V, a *South American Temperate Realm;* VI, an *African Temperate Realm;* VII, an *Antarctic Realm;* VIII, an *Australian Realm.*

The *Arctic Realm* presents a nearly uniform character throughout its extent, and, though embracing several faunæ, is not divisible into

* The terms "Palæogean" and "Neogean," "Palæarctic" and "Nearctic," etc., like those of "Old World" and "New World," have been given with reference solely to the length of time the different land areas of the earth's surface have been known to the dominant race of mankind, and hence regardless of the zoölogical history of these different land areas. Modern science has taught us that the latest discovered continent (Australia) is peopled with the most ancient types of animals and plants now in existence, and that it is, zoölogically considered, the ancient continent. Also that North and South America are behind Europe, Asia, and Africa in their zoölogical and geological development, while they are far in advance of Australia. To apply the term "ancient" to what is really the most recent, and "modern" to what is mediæval, is evidently too great a misuse of language to be allowable in scientific nomenclature. The sciences of geographical zoölogy and geographical botany concern not merely the geographical distribution of the animals and plants now living, but also those of the past. If such descriptive terms as the above are to be employed, it is evidently important that they should be used in their legitimate sense. In the present paper it has hence been considered advisable to altogether discard these terms, since to use them properly would necessitate their adoption in a manner directly opposite to their original and generally accepted application.

provinces. Its southern boundary may be considered as the northern limit of forest-trees, or about the isotheral of 50° F.

The *North Temperate Realm* presents a more varied character, and is divisible into an *American Region* and an *Europæo-Asiatic Region*, each of which is divisible into provinces, districts, and faunæ and floræ. Its boundaries may be provisionally considered as the isotherms of 32° and 70° F.

The *American Tropical Realm*, and also the *Indo-African Tropical Realm*, may be regarded as bounded by the isotherms of 70° F. The first is far more homogeneous than the second. Though the American Tropical Realm is perhaps not divisible into distinct regions, it certainly embraces several provinces and districts, and is rich in faunæ and floræ. The Indo-African Tropical Realm may be divided into an African Region and an Indian Region, each composed of several provinces and districts, and a great number of faunæ and floræ.

The *South American Temperate Realm* embraces that part of South America south of the isotherm of 70° F.; the *African Temperate Realm* includes that part of Africa south of the same isotherm, whilst the *Antarctic Realm* is restricted to the antarctic islands.

The *Australian Realm*, embracing Australia, New Zealand, New Guinea, and their dependent islands, including those to the eastward as far as Timor and Celebes, is zoölogically as distinct from the other primary regions as it is in its geographical position. It is divisible into a Temperate and a Tropical Region, the former embracing New Zealand and the southern third of Australia. Each of these regions includes two or three well-marked provinces.

The above division of the earth's surface* avoids the arbitrary partitioning of an almost homogeneous Arctic Realm between two

* It is not within the scope of the present article to trace the subdivisions of the earth's surface in relation to the distribution of its organic life any further than to furnish illustrations of the general principles according to which it is believed animals and plants are distributed, and by which the land surface of the earth is divided. Determining the rank of the several divisions by the amount of variation from others they present, it is found, as indicated above, that the divisions of co-ordinate rank increase in number to the southward. The Arctic Realm is homogeneous to such an extent as not to admit of divisions of a higher grade than faunæ and floræ. In the Temperate Realm the animals and plants of the Eastern and Western hemispheres are, as a whole, so far different as to admit of the division of this zone into two grand divisions (divisions of the second order), with other divisions between these and the ultimate ones. In the Tropical Realm the differences in the life of the two hemispheres is so great as

implied totally distinct life regions, and also a similar division of the two slightly differentiated regions of the North Temperate Realm. For nearly all the species, and hence of course the genera and families, of the Arctic Realm, and a considerable percentage of the species, a larger proportion of the genera, and nearly all the families of the Temperate Realm, occur in the northern parts of both the so-called "Neogean" and "Palæogean Creations."* It is thus seen that the life of the North Temperate Realm differs far more from that of the Tropical Realms than the life of the Old World does from that of the New. Hence the subdivision of the earth's surface into primary ontological regions, according in area with the two primary divisions of the land, now so generally adopted, is contrary to the facts, since it wholly ignores the close resemblance of the animals and plants inhabiting the north temperate and arctic regions, and the striking differences between them and those of the intertropical zone. The recognition of a "Nearctic" as contradistinguished from a "Palæarctic Region" is almost equally arbitrary and at variance with the law of the distribution of life in circumpolar zones.†

Dana, in his map of the geographical distribution of marine animals,‡ divided the Tropical Zone into four subzones, — a North and a South Torrid and a North and a South Subtorrid; and each temper-

to require a division of the torrid zone, considered as a *climatic* zone, into two primary divisions, with subdivisions of each of several ranks. The south temperate (climatic) zone is similarly divisible, while the Australian Realm, from its isolated position and its remarkable individualization, forms a primary region, with subdivisions of various grades.

* Dr. Sclater properly observes: "It cannot be denied that the ornithology of the Palæarctic, or great temperate region of the Old World, is more easily characterized *by what it has not rather than by what it has.* There are certainly few among the groups of birds occurring in this region which do not develop themselves [to an equal or] to a greater extent elsewhere," etc. — *Journ. of Proc. Linn. Soc.*, *Zoölogy*, Vol. II, p. 137.

† I may here add that the homogeneousness of the life of the boreal regions has been recognized by a number of recent writers, among whom are Dr. L. K. Schmarda, Dr. Von Middendorff, and Professor Huxley, who have each recognized a circumpolar region. Professor Huxley has also called attention (see Proc. Zoöl. Soc. Lond., 1868, pp. 313–319) to the wide divergence of the life of the tropics from that of the north-temperate regions of even the same continents, and the individualization of Australia and its adjacent islands. He considers that the whole surface of the globe may be "primarily subdivided into two principal areæ, — a northern and a southern," for which he has proposed the names *Arctogæa* and *Notogæa.* The latter he has divided into three regions, — *Austro-Columbia* (= Neotropical Region of Sclater), *Australasia*, and *New Zealand.*

‡ Rep. on Crust. of U. S. Expl. Exped. (Vol. II), under Capt. Wilkes.

ate zone into five subzones, — a Temperate proper, a Subtemperate, a Warm Temperate, a Cold Temperate, and a Subfrigid. These zones are equally recognizable in the distribution of terrestrial life; but, owing to inequalities of its surface, they are of course less regular on the land than on the oceans.

The zones and subzones, or the Provinces and the minor phytological and zoölogical divisions of the globe, are usually not trenchantly defined. Their boundaries being determined by climatic conditions, the transition between adjacent zones, or between ontological divisions of whatever rank, is rarely abrupt; like the climatic zones, they blend more or less at their edges, their boundaries being strongly marked only in regions possessing a highly varied surface, as in mountainous districts. They are, nevertheless, easily recognizable, and can be approximately defined. Generally the dividing lines are more or less undulating, and, being determined indirectly by chains of mountains and other physical barriers, adjoining faunæ and floræ, and even adjoining provinces and realms, almost always interdigitate, and frequently enclose isolated areas of others, as will be presently shown in describing the ornithological faunæ of Eastern North America.

The boundaries of faunæ and floræ, like the range of species, are determined indirectly by elevations and depressions of the earth's surface, these variations in the altitude of the land producing varying conditions of temperature and humidity, which latter, as already stated, are the direct limiting influences of species, and of the botanical and zoölogical divisions of the globe. The permanency of their boundaries hence depends upon the constancy of the physiographic conditions of these areas, a migration of species, and of faunæ and floræ, necessarily following changes in these conditions. That such migrations have taken place is evident from the occurrence in the post-tertiary deposits of the warm temperate latitudes of the fossil remains of species found now only in the cold temperate and arctic regions, and in the tertiary strata of high latitudes of the remains of other species whose nearest allies are now found in the warm temperate and subtropical zones. These facts indicate clearly the great changes in temperature that have repeatedly occurred at given localities during the earth's history. In respect to existing animals, however, it is difficult to determine how much their known recession northward, as of the reindeer, for example, is due to climatic changes, and how much to human agency, or whether it may not be due exclusively to the latter cause.

2. The Natural Provinces of the North American Temperate Region.*

Before passing to the special subject of the present article, it will be necessary to consider briefly the North American continent as a whole. As already shown, North America embraces portions of three realms, the Arctic, the North Temperate, and the Tropical. It belongs mainly, however, to the North Temperate Realm, of which the temperate portions of North America form the Western Region. Within this Region may be recognized two Provinces, — an Eastern and a Western, — quite distinct from each other in their general features as well as in many special characteristics. The Eastern Province is characterized by the uniformity of its geographical and climatic features and by a corresponding uniformity in its faunal and floral aspects. The Western Province, on the other hand, is characterized by the diversity of its geographical and climatic features, — different areas situated under the same parallels differing greatly in these respects, — and by the number and small extent of its zoölogical and botanical areas, and its comparatively numerous restricted floræ and faunæ.

The Eastern Province † extends in the United States from the Atlantic seaboard to the vicinity of the 100th meridian, but to the northward its western boundary sweeps rapidly westward, and extends to the Rocky Mountains, whilst farther northward, where it approaches the Arctic Realm, it occupies the whole breadth of the continent. Its western border is not generally abruptly defined, and is, moreover, quite irregular, through its extension up the valleys of the numerous rivers which enter it from the westward. According to Professor Baird, its western boundary "starts on the Gulf of Mexico near the eastern border of Texas, perhaps between the Brazos and the Sabine, and follow-

* The "Districts" of the North American Region, or the ontological divisions of this region of the fourth rank, can be more conveniently characterized after the several faunæ have been defined, to which point in the paper their consideration is accordingly deferred.

† The boundaries of these two regions have been sketched with apparent accuracy by Professor Baird. See American Journal of Science and Arts, 2d Series, Vol. XLI, pp. 82-85, Jan., 1866.

ing up the direction of the former river to the approaches of the Great Desert, nearly on the meridian mentioned [the 100th], proceeds northward, forced sometimes more or less westward, especially along the Platte, sometimes eastward. It crosses the Platte between Forts Kearney and Laramie and intersects the Missouri between Fort Randall and Fort Pierre, perhaps near Fort Lookout, as it is between the first mentioned two points that in ascending the river we find the change to take place in the ornithology of the country. Soon after crossing the northern boundary of the United States the line rapidly inclines westward and extends to the Rocky Mountains." To the southeastward this region embraces the whole of the United States, except perhaps the southern portion of Florida, which is decidedly West Indian and tropical in its affinities. To the northward it embraces the whole northern and eastern portions of the continent up to the Arctic Realm.

The Western Region commences at the western border of the Eastern, and extends thence to the Pacific coast. In the United States its area is about two thirds that of the Eastern Province, but a little farther to the northward it narrows rapidly, and is finally bounded in this direction by the Alaskan mountains.* To the southward it of course merges in Mexico into the Tropical Realm, but its southern limit is not as yet well known. While its varied character renders it subdivisible into several more or less distinct longitudinal areas, each of which may be again divided transversely into numerous faunæ and floræ, many species range throughout its whole extent and give to it a certain degree of homogeneousness. This portion of North America is, however, as yet too indefinitely known, geographically and meteorologically, as well as ontologically, to admit of the exact definition of its primary and ultimate life regions.

The Eastern Province, notwithstanding its larger area, has not only a less number of ornithological faunæ than the Western, but has also a smaller number of species represented in it, as well as a smaller number exclusively restricted to it. The following list of one hundred and eight species embraces most of the birds that are exclusively restricted to the Eastern Province, and hence those that distinctively characterize this Province.

* According to Mr. W. H. Dall. See Proc. Bost. Soc. Nat. Hist., Vol. XII, p. 144, Dec., 1868.

List of Species limited in their Longitudinal Distribution to the Eastern Province of the North American Temperate Region.

1. Turdus mustelinus.
2. Sialia sialis.
3. Mniotilta varia.
4. Parula americana.
5. Prothonotaria citrea.
6. Geothlypis philadelphia.
7. Oporornis agilis.
8. Oporornis formosa.
9. Helmitherus vermivorus.
10. Helminthophaga Swainsoni.
11. Helminthophaga pinus.
12. Helminthophaga chrysoptera.
13. Helminthophaga Bachmani.
14. Helminthophaga ruficapilla.
15. Helminthophaga peregrina.
16. Seiurus aurocapillus.
17. Seiurus noveboracensis.
18. Dendrœca virens.
19 Dendrœca cærulescens.
20. Dendrœca coronata.
21. Dendrœca blackburniæ.
22. Dendrœca castanea.
23. Dendrœca pennsylvanica.
24. Dendrœca cærulea.
25 Dendrœca striata.
26. Dendrœca maculosa.
27. Dendrœca palmarum.
28. Dendrœca dominica.
29. Dendrœca discolor.
30. Perisoglossa tigrina.
31. Wilsonia mitrata.
32. Euthlypis canadensis.
33. Setophaga ruticilla.
34. Pyranga rubra.
35. Pyranga æstiva.
36. Vireosylvia olivacea.
37. Vireosylvia philadelphica.
38. Lanivireo flavifrons.
39. Vireo noveboracensis.
40. Galeoscoptes carolinensis.
41. Harporhynchus rufus.
42. Thryothorus ludovicianus.
43. Cistothorus stellaris.
44. Lophophanes bicolor.
45. Parus hudsonicus.
46. Coturniculus passerinus.
47. Coturniculus Henslowi.
48. Ammodromus caudacutus.
49. Ammodromus maritimus.
50. Zonotrichia albicollis.
51. Junco hyemalis.
52. Spizella monticola.
53. Spizella pusilla.
54. Peucæa æstivalis.
55. Passerella iliaca.
56. Euspiza americana.
57. Hedymeles ludoviciana.
58. Cyanospiza ciris.
59. Cyanospiza cyanea.
60. Cardinalis virginianus.
61. Pipilo erythrophthalmus.
62. Dolichonyx oryzivorus.
63. Icterus spurius.
64. Icterus baltimore.
65. Quiscalus purpureus.
66. Corvus ossifragus.
67. Cyanura cristata.
68. Sayornis fuscus.
69. Campephilus principalis.
70 Picus borealis.
71. Sphyrapicus varius.
72. Centurus carolinus.
73. Melanerpes erythrocephalus.
74. Colaptes auratus.
75. Coccygus americanus.
76. Coccygus erythrophthalmus.
77. Conurus carolinensis.
78. Trochilus colubris.

79. ?Chætura pelasgia.
80. Antrostomus vociferus.
81. Antrostomus carolinensis.
82. Nauclerus furcatus.
83. Ictinia mississippiensis.
84. Rosthramus sociabilis.
85. ?Tetrao canadensis.
86. Cupidonia cupido.
87. Ortyx virginianus.
88. Grus americanus.
89. Florida cærulea.
90. Ibis alba.
91. Platalea ajaja.
92. Ægialitis Wilsonius.
93. Ægialitis melodus.
94. ?Actiturus Bartramius.
95. Limosa hudsonica.
96. ?Numenius borealis.
97. Rallus crepitans.
98. Porzana jamaicensis.
99. Porzana noveboracensis.
100. Gallinula galeata.
101. Gallinula martinica.
102. Anas obscura.
103. Querquedula discors.
104. Camptolæmus labradorius.
105. Sula fiber.
106. Graculus floridanus.
107. Plotus anhinga.
108. Chrœcocephalus atricilla.

3. The Ornithological Faunæ of the Eastern Province of the North American Temperate Region.

Passing to the Eastern Province of the North American Region, the distribution of the birds will be now mainly considered. But a somewhat detailed comparison of the distribution of the representatives of this class with the distribution of the mammals and reptiles, and to some extent with the insects, mollusks, and plants, shows that the same divisions apply almost equally well to all. The distribution of plants, however, is everywhere greatly affected by the nature of the soil, as well as by humidity and temperature; and the character of the vegetation is also intimately connected with the distribution of the insects. The character of the soil, and especially the nature and amount of the mineral matter held in solution in the waters of the streams and lakes, has much to do with the relative abundance and distribution of the terrestrial and fluviatile mollusca, neither of which influences materially affects the distribution of the birds and mammals. The presence or absence of forests only, in respect to vegetation* and the soil, has much

* At the junction of the prairies with the eastern wooded districts there is quite an appreciable change in the fauna, especially in respect to the birds and mammals. The faunal differences between these regions, in respect to these two classes, result mainly through the addition of a relatively small number of strictly prairie species, the westward extension of none of the species of the Eastern Province wholly terminating at this point. The number of their representatives, however, becomes greatly reduced, and their distribution from being general and uniform is restricted to the belts of

influence on the distribution of the terrestrial vertebrates. The distribution of the fishes, the aquatic reptiles and certain groups of batrachians is, however, in great measure determined by the hydrographic basins. Hence we meet with relatively more restricted forms among the latter, as well as in insects, mollusks, and plants, than we find in either mammals or birds, the latter class being the most independent of all animals of geographical barriers.

It has been remarked that the great extent of the Eastern Province, as compared with the Western, is due to the great extent of the lowlands of Eastern North America, or of that area which has an elevation not exceeding eight hundred feet above the sea.* This is unquestionably the true reason, there being no highlands of sufficient altitude to interpose serious obstacles to the range of species. Some portions of this area, however, as the Arctic lowlands, do not belong to this region, while large portions of the country included in the Eastern Province more or less exceed that altitude. These differences of elevation are sufficient to cause the marked interdigitation of the faunæ of contiguous regions lying under the same parallels, as in the Eastern United States, where the upper portions of the Appalachian system support a Canadian or subalpine fauna and flora as far south as Georgia. Yet this elevation, in consequence of its nearly meridional trend and its lack of perfect continuity, forms a barrier to but few vertebrates except the strictly aquatic ones. If, however, the trend of the Appalachian range had been an easterly and westerly one, the influence of these highlands as a geographical barrier would have been most marked. Without the differences in altitude it affords, the faunæ and floræ of Eastern North

forest skirting the streams. At the eastern limit of the prairies, in fact, the distinctively western species begin to appear, thence westward few additional western species being met with till the edge of the great central plateau of the continent is reached, where the differentiation is further increased not only by the addition of many new forms, but by the gradual disappearance of eastern types. Whether the addition of a few prairie species be sufficient reason for recognizing a western subdivision of each of the faunæ of the Eastern Province may perhaps be thought questionable.

The forested portion of the Eastern Province also presents a lack of total uniformity between its eastern and western portions, a few species of birds occurring east of the Appalachians in the Eastern States only as stragglers, whilst they are quite common west of these highlands. About half a dozen species avoid the region circumscribed by the valleys of the St. Lawrence, Lake Champlain, the Hudson River, and the Atlantic Coast, that are found west of this area.

* See Baird, Am. Journ. of Science and Arts, 2d Series, Vol. XLI, p. 86.

America would have extended in regular and parallel zones from the Atlantic seaboard to the central plateau of the continent, whereas they now sweep far southward near the coast, and passing around the Appalachian highlands extend northward again along their western base.

In attempting to determine the number and limits of the ornithological faunæ of a large area, it is evident that the distribution of the birds in the breeding season should be taken as the basis for the investigation rather than their entire range, since during no other portion of the year can the migratory species be regarded as being at their true homes. The species numerously represented are also of far greater importance than those having but few representatives, as it is the common species which are not only the most characteristic, but those whose distribution is at present best known.

Applying the term *fauna*, when used in a special sense, to the smallest of the natural divisions in zoölogical geography, and considering faunæ to be characterized by their general facies, as determined by a peculiar assemblage of species, rather than by the restriction of a certain number of specific forms within their areas, Eastern North America may be considered as embracing seven ornithological faunæ, which occupy narrow, somewhat parallel zones or belts of varying breadth, extending from the Atlantic coast westward to the great middle plateau of the continent. The extent of each in an east and west direction is generally many times greater than what may be considered as its meridional extent. Their breadth, however, is quite unequal, not only as compared with each other, but that of the same fauna varies greatly at different points. They have their minimum breadth on the slopes of the mountains, and attain their maximum breadth on the plains. Each species having its own peculiar limits, which may or may not coincide with those of other species, it usually happens that at somewhat regular intervals, in passing either northward or southward from a given point, a greater number of species disappear at some points than at others, at which point also other species first appear. These divisional lines usually coincide with some marked physical change in the general character of the country, more especially in respect to its elevation, and form the boundaries between adjoining faunæ.* These faunal boundaries, as has been before remarked, coin-

* The first terrace of the Atlantic slope, which marks not only the transition from the tertiary deposits of the coast of the Middle and South Atlantic States to the older forma-

cide with isothermal lines. These isothermal lines, however, are not so often the yearly isotherms as those of particular seasons. While some writers have considered isocrymal lines as those having the greatest amount of limiting influence, as Dana has supposed to be the case with marine animals,* and as may be true in the case of plants, and possibly also of some terrestrial animals, † the mean temperature of the breeding season must necessarily more affect birds, especially the migratory species, than that of any other part of the year, or than the mean annual temperature. Isotherals hence most nearly coincide with the lines limiting the distribution of birds in the breeding season, and also the ornithological faunæ, since the majority of the species in the region now under consideration breed almost exclusively during the summer months, and mainly in June and July. Some breed in May, and a few of the rapacious birds in April, and even in March, but they are the exceptions to the general rule. The isotheral lines are hence adopted in the present essay in giving the boundaries of the ornithological faunæ. ‡

Owing to the imperfect state of our knowledge of the summer distribution of the birds of North America, the present attempt at a defini-

tions of the interior, as well as forming the limit of steam navigation on the rivers of the lower Atlantic States, forms also the dividing line between the faunæ of the coast and those situated next to them in the interior, although having an altitude of generally less than three hundred feet. The rise from the succeeding plateau to the more abrupt slope of the Appalachians forms likewise the boundary between the second and third tiers of faunæ in the Atlantic States. The terrace forming the northern boundary of the tertiary deposits of the Gulf States, and of the lower Mississippi Valley generally, coincides likewise with faunal boundaries, as do similar slight changes in elevation elsewhere.

* See Report on the Crustacea collected by the U. S. Expl. Expd. under the command of Captain Wilkes, Vol. II, p. 1452.

† There must, however, be many exceptions, since in cold climates many mammals and all reptiles, as well as a large proportion of the mollusca and insects, hibernate, and thus are to a great degree (especially the reptiles) beyond the influence of excessive cold. In regard to plants, also, their northward range seems to be limited more by the amount of heat in summer than by the cold of winter, particularly in the case of annuals. As soon as the sum of the heat of summer is diminished to such a degree as to be insufficient to mature the plant, or to allow it to ripen its fruit, whether an annual, a shrub or tree, it must at that point cease to propagate, and there find its polar limit.

‡ Professor A. E. Verrill states that he has found the "boundaries between the Canadian and Alleghanian Faunæ" to be "coincident with a line which shall indicate a mean temperature of 50° Fahrenheit during the months of April, May, and June." *Proc. Bost. Soc. Nat. Hist.*, Vol. XII, p. 260, May, 1866.

tion of the faunæ of this region is to be regarded as merely a provisional one, to be perfected as the required data become known. The distribution in summer of the birds of the United States, even of that portion situated east of the Mississippi River, is still too little known to afford even there entirely satisfactory data. The data are tolerably full only for the region embraced between the St. Lawrence and the Upper Lakes on the north, and the Ohio River and Virginia on the south. Much is also known, however, in regard to the summer distribution of the birds in the other Atlantic States; but in respect to the whole region of the lower Mississippi and the Gulf States, the recorded facts bearing on this subject are lamentably few.* The isothermal lines of even our best climatological charts are also obviously more or less erroneous, and are nowhere laid down with sufficient detail to meet the wants of the student of zoölogical geography. The following lists of those species which by their presence or absence determine the facies of the several faunæ of the Eastern Province are hence not only often incomplete, but will in some cases, doubtless, require more or less modification as our knowledge of the subject increases.† The facts at hand for the work herein attempted are, however, far more numerous than would at first seem probable; ‡ and doubtless the general conclusions reached in the following pages will be in the main substantiated by future investigations.

Beginning with Florida and passing northward, we meet with the following ornithological Faunæ: —

I. Floridian Fauna. As stated in Part I (p. 164), that part of Florida south of Lake George in the interior, and of Cape Canaveral

* The importance of complete and carefully annotated lists of the birds of many localities in the South Atlantic and Gulf States, and in the Mississippi Valley, is hence clearly manifest. Now that the necessity of a precise knowledge of the habitats of animals is so generally recognized, it is to be hoped that every year will add something to our knowledge in regard to these regions.

† This is especially true in respect to the Floridian, Louisianian, and Carolinian Faunæ. Over this large area I have been unable to determine satisfactorily the exact southern limit of the breeding range of any species. Their northern limit, however, is readily approximately determined.

‡ See the Appendix to Part V for a list of the special papers that have been consulted in the present connection. In addition to these papers the specimens of birds in the collection of the Museum of Comparative Zoölogy have been of great use, whilst many additional facts have been gathered from correspondents and from other sources not there indicated.

on the coast, differs quite sensibly in its general faunal and floral characteristics from that part of the State situated farther to the northward, its fauna, especially the ornithological portion, having a decidedly West Indian or tropical aspect, as has also its flora. Dr. Stimpson has recently shown that on the Gulf coast of the State the southern forms, among the marine animals, extend considerably farther north than they do on the Atlantic coast; * but whether the warm waters of the Gulf of Mexico sensibly modify the land fauna of the northern coast of the Gulf sufficiently to affect the distribution of the birds is a point I have been as yet unable to determine. As it seems probable, however, that it does not to any great degree, the Floridian Fauna may accordingly be provisionally regarded as terminating near the latitude of Lake George.

The peculiarities of the bird fauna of Southern Florida in summer is still too imperfectly known to admit of the Floridian ornithological fauna being fully characterized. The occurrence within it of the following species which do not appear to extend much, if any, to the north of it, may serve for the present to distinguish this fauna from the Louisianian.

Species limited in their Northward Range by the Floridian Fauna.

1. Vireosylvia barbatula.
2. Certhiola flaveola.
3. Icterus vulgaris.
4. Cyanocitta floridana.
5. Tyrannus dominicensis.
6. Coccygus minor.
7. Crotophaga ani.
8. " Crotophaga rugirostris."
9. Columba leucocephala.
10. Zenæda amabilis.
11. Oreopelia martinica.
12. Starnænus cyanocephalus.
13. Rostrhamus sociabilis.
14. Polyborus tharus.
15. Aramus giganteus.
16. Demiegretta Pealei.
17. Demiegretta rufa.
18. Audubonia occidentalis.
19. Ibis rubra.
20. Phœnicopterus ruber.
21. Haliplana fuliginosa.
22. Anoüs stolidus.

II. LOUISIANIAN FAUNA. The Louisianian Fauna may be provisionally considered as limited at the northward by the isotheral line of 77° F., it embracing all that part of the United States south of this line east of the Great Plains, except the Floridian Fauna. Beginning on the Atlantic coast apparently as far north as Norfolk, Virginia, it oc-

* See American Naturalist, Vol. IV, p. 536, December, 1870.

cupies a narrow belt thence southward along the coast, and in the latitude of Columbia, South Carolina, begins to expand to the westward. Farther southward its northern boundary passes to the southward of the mountains in Georgia, west of which it rises obliquely northward, and extends in a narrow point up the valley of the Mississippi as far as the mouth of the Ohio. West of the Mississippi it bends again somewhat to the southward.

The Louisianian Fauna hence embraces the coast of North Carolina, the lowlands of South Carolina and Georgia, nearly all of Alabama, all of Mississippi and Louisiana, nearly all of Arkansas, Western Tennessee, the extreme western part of Kentucky, Southern Missouri, the extreme southern part of Illinois, and a small portion of Eastern Texas. Most of the following species range throughout its entire extent, but appear farther to the northward only as stragglers. The presence of these species, and the absence of those given in the preceding list, will serve to distinguish it from the Floridian Fauna. It is similarly distinguished from the Carolinian Fauna, as will be presently shown.

Species limited in their Northward Range by the Louisianian Fauna.

1. Peucæa æstivalis.
2. Cyanospiza ciris.
3. Quiscalus major.
4. Helminthophaga Swainsoni.
5. Helminthophaga Bachmani.
6. ?Prothonotaria citrea.
7. Dendrœca dominica.
8. Sitta pusilla.
9. Antrostomus carolinensis.
10. Campephilus principalis.
11. Picus borealis.
12. Conurus carolinensis.
13. Chamæpelia passerina.
14. Cathartes atratus.
15. Ictinia mississippiensis.
16. Elanus leucurus.
17. Demiegretta ludoviciana.
18. Platalea ajaja.
19. Ibis alba.
20. Tantalus loculator.
21. Porzana jamaicensis.
22. Plotus anhinga.
23. Graculus floridanus.
24. Puffinus obscurus.

III. Carolinian Fauna. The Carolinian Fauna extends from the northern boundary of the Louisianian Fauna northward to about the isotheral line of 71° F. On the Atlantic coast this fauna includes Long Island and a small portion of Southeastern New York, which form its northern limit. In New Jersey it is restricted to the lowlands, extending westward in Southern Pennsylvania, Maryland, and Virginia to the Appalachian highlands. It embraces the middle portion of the Carolinas, and a narrow belt in Northern Georgia. West of the Appalachian highlands its northern boundary sweeps to the northeastward as far as Northern Ohio, and thence runs westward, probably along the water-shed of that State; rising somewhat to the northward in passing farther west, it crosses Michigan near the southern border of that State, and embraces a portion of Southern Wisconsin and Southern Minnesota.

The Carolinian Fauna hence embraces Long Island and Southeastern New York, the greater part of New Jersey, all of Delaware, a small portion of Southeastern Pennsylvania, the greater part of Maryland and East Virginia, all of North Carolina, except the extreme eastern and western portions, the northwestern half of South Carolina, a narrow belt of Northern Georgia south of the mountains, the eastern part of Tennessee, the larger part of West Virginia, nearly all of Ohio and Kentucky, all of Indiana, nearly all of Illinois, a narrow strip of Michigan and Wisconsin, nearly all of Iowa and the greater part of Missouri, and the eastern portions of Nebraska and Kansas. It also occupies the lower Appalachian valleys.

It is distinguished from the Louisianian Fauna by the absence of the species mentioned in the preceding list, and by the presence of those given in the list next following. The features distinguishing it from the Alleghanian will be presently given in the diagnosis of that fauna.

*Species limited in their Northward Range by the Carolinian Fauna.**

1. Cardinalis virginianus.
2. Euspiza americana.
3. Guiraca cærulea.
4. Helmitherus vermivorus.
5. Icteria virens.
6. ?Prothonotaria citrea.
7. Wilsonia mitrata.
8. ?Dendrœca cærulea.
9. Pyranga æstiva.
10. Mimus polyglottus.
11. Thryothorus ludovicianus.
12. Polioptila cærulea.
13. Lophophanes bicolor.
14. Corvus ossifragus.
15. Centurus carolinus.
16. Stelgidopteryx serripennis.
17. Nauclerus furcatus.
18. Strix flammea.
19. Cathartes aura.
20. Ægialitis Wilsonius.
21. Gallinula galeata.
22. Gallinula martinica.
23. Garzetta candidissima.
24. Herodias egretta.
25. Florida cærulea.
26. Nyctherodius violaceus.

* A few of these species occur as stragglers in the Alleghanian Fauna.

27. Rallus elegans.
28. Rallus crepitans.
29. ?Hæmatopus palliatus
30. Recurvirostra americana.
31. Himantopus nigricollis.
32. Sterna aranea.
33. Rhynchops nigra.

IV. ALLEGHANIAN FAUNA. The Alleghanian Fauna has the Carolinian for its southern boundary. Its northern boundary, from the ample data for its determination at the eastward, appears to nearly coincide with the isotheral line of 65° F. It is, however, an extremely irregular line, with abrupt and deep sinuosities. Beginning on the coast to the eastward of the Penobscot Bay, it sweeps first somewhat to the northeast, nearly or quite reaching Bangor; thence passing westward and southward, it follows the northern boundary of the lowlands through Southern Maine and Southern New Hampshire. In the Connecticut valley it rises farther to the northward, and in its southern descent skirts the eastern base of the Green Mountains, passing to the southward and westward of these highlands in Connecticut, and thence abruptly to the northward. Skirting the eastern boarder of the Champlain valley, it continues still northward to the valley of the St. Lawrence as far as Quebec; thence turning again southwestward, it passes along the northern border of the lowlands east of the Lawrentian Hills (including the valley of the Ottawa), and crosses the southern peninsula of Michigan near the forty-fifth parallel; continuing thence northwestward it passes near Fort Ripley. Reaching the valley of the Red River of the North, it turns abruptly to the northward, enclosing the lowlands around Lake Winnipeg and embracing the valley of the Saskatchewan and those of its northern and southern branches, passing westward till it meets the higher plateau forming the eastern slope of the Rocky Mountains. This may be considered as *approximately* the northern boundary of the Alleghanian Fauna; the physical, climatic, and phytozoölogical character of the interior of British North America being at present too imperfectly known to render it easy to determine definitely the northwestern limit of the Alleghanian Fauna.*

* As already stated, the mean temperature of the breeding season (May, June, and July) has been taken as limiting the breeding range of the species. But this criterion associates regions which have very different *climatic* peculiarities, when the temperature of the whole year is considered, the isotheral lines diverging more widely from the isothermal or yearly lines in the interior than on the Atlantic coast. While in the Winnipeg basin the summer heat is sufficient to ripen corn and to permit of the cultiva-

The Alleghanian Fauna hence includes all of Southern New England, except the higher parts of the Green Mountain ranges, including even the southern third of Maine and a considerable part of New Hampshire and Vermont; all of New York, except the higher portions of the Adirondacks and the southeastern extremity of that State (which belongs to the Carolinian Fauna), all the lowlands of the Canadas, as far east at least as the vicinity of Quebec; the northern border of Ohio, the greater part of Wisconsin and Minnesota (in fact, very nearly all of these two States), and the valleys of the Red River of the North, the Assinniboine, and large portions of the valleys of the Saskatchewan and its two main branches, including also the extensive lowlands surrounding Lake Winnipeg. It also embraces all the Appalachian highlands southward to Georgia, except the higher parts (which belong to the Canadian Fauna), and hence includes a large part of Pennsylvania, the greater part of the highlands of Maryland, Virginia, and the Carolinas. The isolated areas within this region belonging to the Canadian Fauna are the highlands of Northeastern New York, and the most elevated parts of Pennsylvania, the Virginias, North Carolina, and Georgia. The northwestern part of New Jersey seems also to belong to the Canadian Fauna.

The Alleghanian Fauna is characterized by the absence of those species already mentioned as finding their northern limit within the Carolinian Fauna, by the presence of those mentioned below as limited in their northward range by the Alleghanian Fauna, and by the absence of a considerable number which occur abundantly in the Canadian Fauna. It is further distinguished from the Carolinian Fauna by the occurrence within it in the breeding season of the species enumer-

tion of tobacco, the winter climate is almost arctic, ice remaining in the lakes in sheltered places till late in May. Yet in summer the Winnipeg district is frequented by birds that find their northern range limited on the Atlantic coast to Southern Maine, where the winters are much shorter and the cold far less severe than on the prairies of the Saskatchewan. The same continental character of the climate of the interior is similarly seen as far south as the prairies of the Upper Mississippi, to which the northern birds descend in winter in greater numbers and with greater regularity than in the corresponding latitudes near the Atlantic coast. A limitation of the ornithological faunæ by the distribution of the birds in winter, — in other words, by their maximum range, — would hence differ considerably from the circumscription of these faunæ based on the breeding range of the species. This remark applies, of course, not only to the present fauna (Alleghanian), but to Eastern North America as a whole, especially to that portion north of the Louisianian Fauna.

ated in the second list next subjoined, to which the present fauna forms the southern limit of their breeding range.

1 *Species limited in their Northward Range by the Alleghanian Fauna.**

1. Turdus mustelinus.
2. Sialia sialis.
3 Pyranga rubra.
4. Dendrœca discolor.
5. Lanivireo flavifrons.
6. Vireo noveboracensis.
7. Troglodytes aëdon.
8. Harporhynchus rufus.
9. Cyanospiza cyanea.
10. Pipilo erythrophthalmus.
11. Spizella pusilla.
12 Coturniculus passerinus.
13 Ammodromus caudacutus
14. Ammodromus maritimus.
15. Icterus Baltimore.
16. Icterus spurius.
17. Sturnella ludoviciana.
18. ?Antrostomus vociferus.
19. Zenædura carolinensis.
20. Cupidonia cupido.
21. Ortyx virginianus.
22. Meleagris gallopavo.
23. Ardetta exilis.
24. Rallus virginianus.
25. Chrœcocephalus atricilla.
26. Sterna paradisea.
27. Hydrochelidon fissipes.

2. *Species limited by the Alleghanian Fauna in their Southward Range in the Breeding Season.*

1. Turdus fuscescens.
2. Dendrœca pennsylvanica.
3. Parula americana.
4. Helminthophaga ruficapilla.
5. Helminthophaga chrysoptera.
6. Hirundo bicolor.
7. Lanivireo solitarius.
8. Carpodacus purpureus.
9. Melospiza palustris.
10. Passerculus savanna.
11. Dolichonyx oryzivorus.
12. Contopus borealis.
13. Empidonax minimus.
14. Empidonax flaviventris.
15. Pediœcetes phasianellus.
16. Rallus virginianus.
17. Porzana carolina.
8. Ægialitis melodus.
19. Nettion carolinensis.
20. Querquedula discors.
21. Mareca americana.
22. Anas boschas.
23. Anas obscura.
24. Mergus merganser.
25. Lophodytes cucullatus.
26. Bernicla canadensis.
27. Colymbus torquatus.
28. Podilymbus podiceps.
29. Larus argentatus.
30. Sterna macrura.

* A few of the species mentioned in this list are more or less frequent stragglers into the Canadian Fauna, but none of them seem to occur there except as irregular and infrequent visitors.

V. CANADIAN FAUNA. The next fauna to the northward of the Alleghanian is the Canadian. The southern boundary of the Canadian is hence, of course, the northern limit of the Alleghanian, which boundary has been already defined. Its northern limit coincides very nearly with the isotheral line of 57° F. The region to the northward of the Alleghanian Fauna is unfortunately too little known to permit of a very satisfactory determination of the northern boundary of either the Canadian Fauna or of the faunæ to the northward of the Canadian. On the Atlantic coast the Canadian Fauna appears to embrace the greater part of Newfoundland, nearly or quite all of Nova Scotia * and New Brunswick, Northern New England, including the crests of the Green Mountain ranges southward to Connecticut, the greater part of the province of Quebec, including the Lower St. Lawrence valley as far up as the city of Quebec, the southern slope of the Height of Land in Northern Ontario, and the highlands on both sides of Lake Superior. To the southward it also embraces as outlying islands the Adirondacks of Northern New York, and the higher crests of the Appalachians southward to Georgia. To the northward it probably extends nearly to the summit of the Height of Land, and may embrace part of the lowlands bordering the southwestern shore of Hudson's Bay. Its northern boundary hence sweeps northwestward in the interior nearly or quite to Fort Resolution, on the southern shore of Great Bear Lake.

The Canadian Fauna, as above limited, may be characterized as follows. It is distinguished from the Alleghanian Fauna by the absence of the species mentioned above as limited in their northward range by that fauna, and by the presence in the breeding season of those mentioned in the first subjoined list; from the Hudsonian Fauna by the presence of those given in the second subjoined list, and by the absence of those given in the first list under the Hudsonian Fauna. It is further distinguished by its forming the breeding haunts of a large proportion of the *Sylvicolidæ*, especially of the species of *Dendrœca*, several of which are in summer mainly restricted to it.

* Nova Scotia, zoölogically considered, presents somewhat anomalous characters. In summer a number of birds which are reported as either rare or accidental at Calais, Me., are represented as common summer residents in Nova Scotia, while other northern species breed there in numbers which do not usually breed at localities where the other species referred to are summer residents. The half-insular position of Nova Scotia is doubtless the explanation of the faunal peculiarities above mentioned.

1. *Species limited by the Canadian Fauna in their Southward Range in the Breeding Season.*

1. Turdus Pallasi.
2. Turdus Swainsoni.
3. Regulus satrapa.
4. Regulus calendula.
5. Dendrœca castanea.
6. Dendrœca Blackburniæ.
7. Dendrœca coronata.
8. Dendrœca cærulescen
9. Dendrœca striata.
10. Dendrœca palmarum.
11. Dendrœca maculosa.
12. Euthlypis canadensis.
13. Troglodytes hyemalis.
14. Parus hudsonicus.
15. Pinicola enucleator.
16. Curvirostra americana.
17. Curvirostra leucoptera.
18. Chrysomitris pinus.
19. Zonotrichia leucophrys.
20. Zonotrichia albicollis.
21. Junco hyemalis.
22. Spizella monticola.
23. Passerella iliaca.
24. Scolecophagus ferrugineus.
25. Perisoreus canadensis.
26. Picoides hirsutus.
27. Picoides arcticus.
28. Falco columbarius.
29. Astur atricapillus.
30. Surnia ulula.
31. Nyctale Tengmalmi.
32. Tetrao canadensis.
33. Calidris arenaria.
34. Macrorhamphus griseus.
35. Phalaropus Wilsoni.
36. Aythya vallisneria.
37. Aythya americana.
38. Fulix marila.
39. Fulix collaris.
40. Erismatura rubida.
41. Bucephala clangula.
42. Bucephala albeola.
43. Mergus serrator.
44. Somateria mollissima.
45. Cymochorea leucorrhea.
46. Puffinus anglorum.
47. Larus marinus.
48. Rissa tridactylus.
49. ?Chrœcocephalus philadelphia.
50. Pelecanus erythrorhynchus.
51. Sula bassana.
52. Graculus carbo.
53. Graculus dilophus.
54. Podiceps cristatus.
55. Podiceps cornutus.
56. Podiceps Holbölli.
57. Fratercula arctica.
58. Utamania torda.
59. Uria grylle.
60. Lomvia ringvia.
61. Lomvia svarbag.

2. *Species limited by the Canadian Fauna in their Northward Range.*

1. ?Turdus Pallasi.
2. Mimus carolinensis.
3. Dendrœca virens.
4. Dendrœca cærulescens.
5. Dendrœca castanea.
6. Dendrœca Blackburniæ.
7. Euthlypis canadensis.
8. Parus atricapillus.
9. Chrysomitris tristis.
10. Chrysomitris pinus.
11. Curvirostra americana.
12. Poocætes gramineus.

13. Melospiza melodia.
14. Melospiza palustris.
15. Dolichonyx oryzivorus.
16. Tyrannus carolinensis.
17. Myiarchus crinitus.
18. Contopus virens.
19. Trochilus colubris.
20. Antrostomus vociferus.
21. Sayornis fuscus.
22. Cyanura cristata.
23. Buteo lineatus.
24. Buteo pennsylvanicus.
25. Accipiter Cooperi.
26. ?Syrnium nebulosum.
27. Butorides virescens.
28. Porzana carolina.
29. Ægialitis vociferus.
30. Ægialitis melodus.
31. Philohela minor.
32. Actiturus Bartramius.
33. Aix sponsa.
34. Chaulelasmus streperus.
35. Aythya vallisneria.
36. Hydrochelidon fissipes.

VI. Hudsonian Fauna. The next ornithological fauna north of the Canadian may well be termed the Hudsonian Fauna. Its northern limit seems to nearly coincide with the isotheral line of 50° Fahrenheit, its southern limit being the isotheral of 57°, or the northern boundary of the Canadian Fauna. It will include at least the southern third of Labrador, the northern peninsula of Newfoundland, Anticosti Island, the more elevated parts of the Height of Land separating the lowlands bordering Hudson's Bay from the lowlands of the St. Lawrence and the Winnipeg district, and the basin of the Mackenzie's from Lake Athabasca to a point considerably north of Fort Simpson, extending in the Mackenzie's River valley some distance within the Arctic Circle, probably to the Arctic coast. Extending still westward, it embraces the valleys of Liard's and Peel's Rivers, and probably the valley and adjoining lowlands of the Youkon, including the greater part of that portion of the Territory of Alaska situated to the southward of the Arctic Circle. In other words, that portion of boreal America situated between the Canadian Fauna and the Barren Grounds. It is far from certain that a western or Alaskan Fauna will not have to be separated, embracing all the more temperate portions of Alaska. Although strictly western species occur here, they appear to be confined mainly to the western coast and the lower part of the Youkon valley. The faunal differences between the western shore of Alaska and the valley of Mackenzie's River become far greater when the marine species are taken into account, even if only the birds and mammalia are considered. The Pacific coasts of Alaska and Siberia have many species peculiar to the shores of those countries and to the intervening islands, constituting a distinct fauna, which may well be called the *Aleutian Fauna.* The mingling of Asiatic and American species forms its distinctive feature. There is also a slight commingling of western species in the valley of the Mackenzie's River, as there is also in the valley of the Saskatchewan. The Hudsonian Fauna doubtless embraces outlying islands in the Canadian Fauna, as the upper part of the White Mountains, and the summits of some of the higher peaks of the Adirondacks. The southern point of Greenland embraces many species common to the Hudsonian Fauna, and though Greenland belongs almost wholly to the Arctic Realm, its extreme southern portion is doubtless Hudsonian.*

The Hudsonian Fauna being coextensive northward with the limit of forest-trees, it forms the northern limit of distribution of all the species of birds whose mode of life renders them dependent upon a forest vegetation. The distinction between the Hudsonian Fauna and the Arctic Realm, as well as between the Temperate Realm and the Arctic Realm, is hence a strongly marked one, nearly one hundred species, nearly all of them land birds, finding their northern limit of distribution near the polar limit of forests, or at least within the Hudsonian Fauna.

The Hudscnian Fauna may be distinguished from the Canadian by the absence of the species given in the preceding lists and by the presence of those enumerated in the first of the lists next subjoined, and from the Arctic Realm by the presence of those given in the second list below.

1. *Species limited by the Hudsonian Fauna in their Southward Range in the Breeding Season.*

1. Anthus ludovicianus.
2. Saxicola œnanthe.
3. Ampelis garrula.
4. Ægiothus linaria.
5. Plectrophanes lapponicus.
6. Plectrophanes nivalis.
7. Plectrophanes pictus.
8. Leucosticte tephrocotis.
9. ?Picoides arcticus.
10. ?Picoides hirsutus.
11. Falco candicans.
12. Archibuteo lagopus.
13. Syrnium cinereum.
14. Nyctea nivea.
15. Lagopus albus.
16. Lagopus rupestris.

* For remarks respecting the similarity of the Fauna of Northern Labrador and Southern Greenland, see Dr. A. S. Packard, Proc. Bost. Soc. Nat. Hist., Vol. X, p. 255, 1866.

17. Charadrius virginicus.
18. Ægialitis semipalmatus.
19. Squartarola helvetica.
20. Strepsilas interpres.
21. Actodromas maculata.
22. Actodromas Bonapartei.
23. Actodromas minutilla.
24. Ereunetes pusillus.
25. Anser Gambeli.
26. Anser hyperboreus.
27. Bernicla brenta.
28. Dafila acuta.
29. Harelda glacialis.
30. Somateria spectabilis.
31. Pelionetta perspicillata.
32. Procellaria glacialis.
33. Sterna caspia.
34. Larus glaucus.
35. Colymbus septentrionalis.
36. Stercorarius parasiticus.

2. *Species limited in their Northward Range by the Hudsonian Fauna.*

1. Turdus migratorius.
2. Turdus Swainsoni.
3. Regulus calendula
4. Regulus satrapa.
5. Helminthophaga ruficapilla.
6. Helminthophaga peregrina.
7. Helminthophaga celata.
8. Seiurus aurocapillus.
9. Seiurus noveboracensis.
10. Dendrœca coronata.
11. Dendrœca striata.
12. Dendrœca æstiva.
13. Dendrœca maculosa.
14. Dendrœca palmarum.
15. Wilsonia pusilla.
16. Setophaga ruticilla.
17. Hirundo horreorum.
18. Hirundo lunifrons.
19. Hirundo bicolor.
20. Ampelis garrula.
21. Collurio borealis.
22. Vireo olivaceus.
23. Vireo gilvus.
24. Parus hudsonicus.
25. Pinicola enucleator
26. Curvirostra leucoptera.
27. Ægiothus linaria.
28. Plectrophanes nivalis.
29. Plectrophanes lapponicus.
30. Plectrophanes pictus.
31. Passerculus savanna.
32. Zonotrichia leucophrys.
33. Zonotrichia albicollis.
34. Junco hyemalis.
35. Spizella monticola.
36. Spizella socialis.
37. Melospiza Lincolnii.
38. Passerella iliaca.
39. Molothrus pecoris.
40. Agelæus phœniceus.
41. Scolecophagus ferrugineus.
42. Quiscalus purpureus.
43. Corvus corax.
44. Corvus americanus.
45. Pica caudata.
46. Perisoreus canadensis.
47. Contopus borealis.
48. Empidonax minimus.
49. Empidonax Traillii.
50. Picus villosus.
51. Picus pubescens.
52. Picoides hirsutus.
53. Picoides arcticus.
54. Sphyrapicus varius.
55. Hylotomus pileatus.
56. Colaptes auratus.
57. Chordeiles popetue.
58. Ceryle alcyon.

59. Falco candicans.
60. Falco peregrinus.
61. Falco columbarius.
62. Falco sparverius.
63. Astur atricapillus.
64. Archibuteo lagopus.
65. Buteo borealis.
66. Accipiter fuscus.
67. Circus cyaneus.
68. Bubo virginianus.
69. Otus vulgaris.
70. Otus brachyotus.
71. Ectopistes migratori
72. Tetrao canadensis.
73. Pediœcetes phasianellus.
74. Bonasa umbellus.
75. Lagopus leucurus.
76. ?Grus americanus.
77. Gallinago Wilsoni.
78. Macrorhamphus griseus.
79. Ereunetes pusillus.
80. Gambetta melanoleuca.
81. Gambetta flavipes.
82. Tringoides macularius.
83. Limosa hudsonica.
84. Porzana carolina.
85. Fulica americana.
86. ?Dafila acuta.
87. Nettion carolinensis.
88. Querquedula discors.
89. Spatula clypeata.
90. Mareca americana.
91. Fulix marila.
92. Fulix collaris.
93. Erismatura rubida.
94. Lophodytes cucullatus.
95. Graculus dilophus.
96. Pelecanus erythrorhynchus.

VII. AMERICAN ARCTIC FAUNA. The Arctic Realm may be considered as occupying that portion of the northern hemisphere north of the isotheral of 50° F. Though presenting a great uniformity of character throughout its extent, it seems to be divisible into four faunæ, — an American Arctic Fauna, an Americo-Asiatic Fauna, an Europeo-Asiatic Arctic Fauna, and an Europeo-American Arctic Fauna, the second and fourth being essentially marine. While a few species of mammals seem to be almost wholly restricted within the Arctic Realm, it contains but few resident birds, and no species of birds seem to be wholly confined to it, even in the breeding season. The following species are reported to range to the Arctic coast of North America, the most of which have been observed at Melville Island and in Greenland. The greater part are *Natatores* and *Grallæ*, the only commonly so-called land birds being two or three species of grouse, a sparrow or two, and a few hawks and owls.

Species found in the American Arctic Fauna in the Breeding Season.

1. Cotyle riparia.
2. Corvus corax.
3. Aquila chrysaëtos.
4. Haliaëtus leucocephalus.
5. Pandion haliaëtus.
6. Nyctea nivea.

7. ?Surnia ulula.
8. Tetrao canadensis.
9. Lagopus albus.
10. Lagopus rupestris.
11. Grus canadensis.
12. Botaurus lentiginosus.
13. Charadrius virginicus.
14. Ægialitis semipalmatus.
15. Strepsilas interpres.
16. Phalaropus fulicarius.
17. Calidris arenaria.
18. Tringa canutus.
19. Pelidna "americana."
20. Arquatella maritima.
21. Cygnus "americanus."
22. Anser hyperboreus.
23. Anser Gambeli.
24. Bernicla canadensis.
25. Anas boschas.
26. Bucephala albeola.
27. Bucephala clangula.
28. Histrionicus torquatus.
29. Harelda glacialis.
30. Melanetta velvetina.
31. Pelionetta perspicillata.
32. Œdemia "americana."
33. Somateria spectabilis.
34. Somateria mollissima.
35. Buphagus skau *Coues.*
36. Stercorarius pomarinus.
37. Stercorarius parasiticus.
38. Stercorarius Buffoni *Coues.*
39. Fulmarus glacialis.
40. Larus argentatus.
41. Larus glaucus.
42. Larus leucopterus.
43. Rissa tridactyla.
44. Pagophila eburnea.
45. Xema Sabini.
46. Sterna arctica.
47. ?Sula bassana.
48. Colymbus septentrionalis.
49. Colymbus arcticus.
50. Colymbus torquatus.
51. Utamania torda.
52. Fratercula glacialis.
53. Lunda cirrhata.
54. Mergulus alle.
55. Uria grylle.
56. Lomvia troile.
57. Lomvia ringvia.
58. Lomvia svarbag.

The Faunæ of the Eastern Province considered in Reference to the Distribution of the Mammals and Reptiles.

The faunæ of the Eastern Province of the North American Region above characterized from the distribution of the birds seem to be equally well marked as natural zoölogical districts by the distribution of the mammals and reptiles. About the same proportionate number of mammals are limited similarly with the birds in regard to their northward and southward distribution. The correspondence in the geographical limitation of the species of the two groups will be briefly shown by the following remarks respecting the range of the mammals.

The Arctic Realm is well known to be characterized by a few species nearly or quite restricted to it, as the polar bear (*Ursus maritimus*), the artic fox (*Vulpes lagopus*), the musk ox (*Ovibos moschatus*), the lemmings (*Myodes*), the small northern race of the caribou or reindeer, the Eskimos, etc.

The Hudsonian Fauna forms the southern limit of the polar hare (*Lepus glacialis*) and the northern limit (at least in winter) of the *Lynx canadensis*, *Mustela* "*americana*," *Mustela Pennantii*, *Putorius vulgaris*, *Putorius ermineus*, *Ursus* "*americanus*," *Ursus* "*horribilis*," *Sciurus hudsonius*, *Arctomys monax*, *Vespertilio subulatus*, and others.

The Canadian Fauna forms, at present,* the southern limit of *Mustela Pennatii*, *Mustela* "*americana*," *Gulo luscus*, *Arvicola xanthognathus*, *Erethizon dorsatus*, *Alces malchis*, *Rangifer tarandus*, etc., and the northern limit of *Felis concolor*, *Lynx rufus*, *Mephitis mephitica*, *Procyon lotor*, *Bos americanus*, *Condylura cristata*, *Blarina brevicauda*, and others.

The Alleghanian Fauna forms the southern limit of *Lynx canadensis*, *Sciurus hudsonius*, *Arvicola Gapperi*, *Jaculus hudsonius*, *Lepus americanus*, *Cervus canadensis*, *Sorex platyrhinus*, *Condylura cristata*, and doubtless of several other species; and the northern limit of *Sciurus carolinensis*, *Lepus sylvaticus*, *Arvicola pinetorum*, *Cervus virginianus*, *Scalops aquaticus*, etc.

The Carolinian Fauna forms the southern limit of *Mustela vulgaris*, *Tamias striatus*, *Arctomys monax*, and *Fiber zibethicus*; and the northern limit of *Vulpes virginianus*, *Nycticejus crepuscularis*, *Didelphys virginiana*, etc.

The Louisianian Fauna seems to form the southern limit of *Putorius vison*, *Blarina brevicauda*, *Scalops aquaticus*, and doubtless thus limits several other species, though not a small proportion of those which occur in this fauna range also into South Florida, or into the Floridian Fauna. The Louisianian Fauna limits the northward range of *Neotoma floridana*, *Reithrodon humilis*, *Sigmodon hispidus*, *Hesperomys palustris*, *Geomys pineti*, and *Lepus palustris*.

In respect to reptiles, a similar proportion of species are limited in either their northward or southward range by each fauna. Several species of batrachians range into the Hudsonian Fauna, but apparently this fauna must be the northern limit of their distribution. The Cana-

* It is probable that some of the fur-bearing species, as well as *Hystrix dorsatus*, the moose and the caribou, once ranged southward throughout the Alleghanian Fauna, and have been exterminated there by man.

dian Fauna forms the northern limit of the reptiles proper, where this class is represented by two orders only, the *Testudinata* and the *Ophidia* The *Testudinata* are there represented by only three species (*Chelydra serpentina*, *Glyptemys insculpta*, *Chrysemys picta*), and the *Ophidia* by five (*Bascanion constrictor*, *Tropidonotus sirtalis*, *Diadophis punctatus*, *Storeria occipito-maculata*). In the Alleghanian Fauna the number of species in each of these groups is more than doubled. The Carolinian Fauna forms the northern limit of the *Sauria*, of which two species (*Plestiodon fasciatus*, *Tropidolepis undulatus*) here first make their appearance, and the number of species of the other groups is still further increased, several additional generic types being added. In the Louisianian Fauna the number of species of *Sauria* becomes considerably greater, and while few of the northern species of either the true reptiles or the batrachians have disappeared, other more southern forms have been added in almost every family.

These several faunæ, it may be added, seem in general to coincide in their number and limits with the floræ of the same region. These several floræ, as thus circumscribed, form successively the northern limit of the successful cultivation of some more or less important cultivated plant. But a detailed consideration of the distribution of the vegetation of the region under consideration, in respect to the number and circumscription of the floræ, and their correspondence with the faunæ, cannot of course well be here introduced.

4. The Ornithological Districts of the North American Temperate Region.

The subdivision by Professor Dana of the tropical and temperate climatic zones of the oceanic areas into several zoölogical zones has been already alluded to as being equally applicable to the land areas. To these life zones I propose to apply the term "districts." Dana's divisions of the north temperate climatic zone correspond respectively in latitudinal extent with the several ornithological faunæ of the Eastern Province, as defined in the preceding pages. Unlike the faunæ, however, the districts extend in an east and west direction across the North American Region, each district embracing not only one of the faunæ of the Eastern Province, but also its representative fauna (or faunæ) in the Western Province. The Hudsonian Fauna corresponds in latitudinal extent with Dana's subfrigid division of the north temperate zone, and the term *Subfrigid District* may be very properly applied to that district of which this fauna forms the eastern portion. The zone corresponding with the Canadian Fauna may in like manner be termed the *Cold-temperate District;* that corresponding with the Alleghanian Fanua the *Subtemperate District;* that corresponding with the Carolinian Fauna the *Temperate District;* and that corresponding with the Louisianian Fauna the *Warm-temperate District;* the Floridian Fauna in like manner corresponding with the *Subtorrid District*, or with Dana's subtorrid zone. Each of these districts is distinguished, in contradistinction from the faunæ, by species which range across the continent, while the districts are distinguished from each other by the same kind of difference as has been shown above to characterize the several faunæ among themselves.

5. On the Geographical Range of the Species.

The preceding tables, while serving to characterize the ornithological faunæ of Eastern North America, indicate only very obscurely the range of the species. The following tables have hence been prepared in order to show more clearly the breeding range, and also the winter quarters, of those species whose distribution in the breeding season is tolerably known. For this purpose the birds occurring in the Eastern Province of the North American region have been grouped, according to their geographical distribution, into the following classes, beginning with those which have the widest breeding range: I. *Cosmopolitan Species.* II. *Circumpolar Species.* III. *Species which range across the whole breadth of the North American Temperate Region.* IV. *Species limited in longitude to the Eastern Province of this region.* The birds of the Eastern Province are further subdivided according to the range of the species in the breeding season in latitude.*

* In a preliminary notice like the present it has been found impracticable to give the authorities in detail on which the generalizations given in the following synopsis have been based. The list of papers given in the Appendix serve in a general way to indicate the principal sources from which information has been derived. It is believed, however, that the limits assigned each species will be found in the main correct, though in many cases the accessible data have been quite too few to be satisfactory. The generalizations are given, of course, as a representation of our present knowledge of the subject rather than as final. The polar and equatorial limits of the migratory range of the species varies, as is well known, more or less in different years, according to the season. It is also somewhat different on the coast from what it is in

I. Cosmopolitan Species. A large proportion of ornithologists have of late been unwilling to admit that any bird has what is usually termed a "cosmopolitan" range, while others recognize only about twenty such species, taking into account, of course, their total range. These embrace two or three species each of hawks and owls, the rest being either *Grallæ* or *Natatores*. Very few of them, however, breed within both the tropic and the polar zones; many of those which visit the shores of all lands in their migratory journeys being restricted in the breeding season to comparatively limited areas. *Pandion haliaëtus* and *Otus brachyotus* are the only examples of commonly so-called cosmopolitan species which appear to breed from the Arctic Circle southward through the tropics to the southern extremity of the southern continents. *Falco peregrinus* may form a third, but its peculiar breeding habits give it a very irregular dispersion at that season. *Strix flammea* appears to be also everywhere resident, except in the arctic and cold-temperate zones. *Cotyle riparia* and *Hirundo rufa* (including under the latter name the several slightly differing geographical races of this group, which have of late been regarded as species), seem also to be nearly cosmopolite. The list of species which are permanently cosmopolitan will hence not exceed half a dozen, and are those above enumerated.

II. Circumpolar Species. Regarding as circumpolar species only those numerously represented in both the eastern and western hemispheres, nearly one hundred species* can be included in the list of

the interior, as has been previously explained; so that an indication of only the average boreal and austral limits of the species at this season has been aimed at, and only so far as their winter range is circumscribed within the region under special consideration. The blanks in the third column of the tables hence indicate that the species winter entirely within the American Tropical Realm; those in the fourth, that the austral limit is within that realm. The few occurring in the second column of the tables indicate that the species in question also ranges southward in the breeding season into the Tropical Realm. A [?] in place of a blank indicates that the southward range of the species is supposed to be limited to the Eastern Province, but as being too vaguely known to warrant a specification of its limit in the direction indicated by the column in which the query stands.

* Dr. Richardson, in 1831 (in the "Fauna Boreali-Americana," p. xxxix), gave thirty-two species of land birds and "upwards of sixty-two species of water birds" (ninety-four in all) as "common to the Old World and the Fur Countries." A few truly circumpolar species were not included in this list, and others were included which were merely accidental visitors from one continent to the other. Since the date of that list the identity of the greater part of the species therein mentioned has been questioned by

circumpolar species. A small number of others that are properly either exclusively American or Europeo-Asiatic species occur more or less frequently as accidental visitors to the continents not embraced within their usual habitats.

one or another writer, and their representatives on the two hemispheres separated under different names. But a considerable proportion of those mentioned in the next subjoined table are still regarded as truly circumpolar by a number of leading European onithologists. Dr. Von Middendorff ("Uebersicht der Natur Nord- und Ost-Sibiriens, Theil 2, Erste Lieferung," etc.; see Newton's Ibis, April, 1870, p. 275), in 1867, gave lists of eighty-seven circumpolar species, a part of which (called "Hyperboreal Birds") are distinctive of what has been termed above the Arctic Realm, whilst many of the others range quite far southward even in summer. These lists, however, do not embrace a number of circumpolar species whose boreal limit does not extend to the districts named. A dozen or more Europeo-Asiatic species, in addition to those given below, have representatives in America so closely resembling them in habits and in geographical distribution, as well as structurally, that they have often been confounded, specimens frequently occurring on the one continent that are undistinguishable from those from the other continent.

In 1846 Professor Edm. de Selys-Longchamps, in his excellent paper entitled "Sur les Oiseaux américains admis dans la Faune européenne" (Mém. de la Soc. R. de Liège, Vol. IV, pp. 35 - 50, 1849), included thirteen species in his list of "Oiseaux terrestres communs à l'Europe et à l'Amérique," and mentions nine other terrestrial American species which he regards as "ne semblent être en réalité que des modifications climatiques de nos oiseaux européens." All but two of these, and also one or two in addition to them, have been regarded in the present paper as specifically identical. In his list of "Oiseaux aquatiques communs à l'Europe et à l'Amérique" he includes fifty-five species, and mentions thirteen others, "décrits comme espèces distinctes, ne semblent être que des races locales," three or four of which I have regarded as specifically identical. The whole number mentioned by Selys-Longchamps as common to Europe and boreal America is seventy-six, plus twenty-four "autres qui semblent n'être que des races légèrement modifiées par le climat." (See l. c., p. 48.) In the same paper he gives a list of twenty-eight American species as of accidental occurrence in Europe, eight of which are land birds, eight échassiers or waders, and twelve palmipèdes or swimmers, and also a list of twenty American species which he considers to have been improperly included among the birds of Europe, among which are *Haliaëtus leucocephalus*, *Strix nebulosa* (= *Syrnium nebulosum*), *Loxia* (= *Curvirostra*) *leucoptera*, *Struthus* (= *Junco*) *hyemalis*, and *Parus* (= *Lophophanes*) *bicolor*.

II. *List of Circumpolar Species, with Indications of their Boreal and Austral Limits.*

Species.	Boreal Limit in the Breeding Season.	Austral Limit in the Breeding Season.	Boreal Limit in Winter.	Austral Limit in Winter.
Certhia familiaris	Canad. Fauna	Allegh. Fauna	Allegh. Fauna	Carolin. Fauna
Saxicola œnanthe	Hudson. Fauna	Hudson. Fauna	—	—
Ampelis garrula	Hudson. Fauna	Hudson. Fauna	Hudson. Fauna	Canad. Fauna
Cotyle riparia	Arctic Coast	—	Louis. Fauna	—
Plectrophanes nivalis	Arctic Coast	Hudson. Fauna	Hudson. Fauna	Allegh. Fauna
Plectrophanes lapponica	Hudson. Fauna	Hudson. Fauna	—	Allegh. Fauna
Ægiothus linaria	Arctic Coast	Hudson. Fauna	Hudson. Fauna	Allegh. Fauna
Pinicola enucleator	Hudson. Fauna	Canad. Fauna	Hudson. Fauna	Allegh. Fauna
Curvirostra leucoptera	Hudson. Fauna	Canad. Fauna	Hudson. Fauna?	Canad. Fauna
Eremophila alpestris	Hudson. Fauna	Carolin. Fauna	Canad. Fauna?	Louis. Fauna
Corvus corax	Arctic Coast	Allegh. Fauna	Arctic Realm	?
Pica caudata	Hudson. Fauna	Allegh. Fauna	Canad. Fauna	?
Aquila chrysaëtos	Hudson. Fauna	Canad. Fauna	Hudson. Fauna?	Carolin. Fauna?
Archibuteo lagopus	Arctic Coast	Hudson. Fauna	Hudson. Fauna	Allegh. Fauna
Pandion haliaëtus	Arctic Coast	—	Carolin. Fauna	—
Falco candicans	Arctic Coast?	Hudson. Fauna	Canad. Fauna	Canad. Fauna
Falco peregrinus	Hudson. Fauna	—	—	—
Circus cyaneus	Arctic Coast?	Louis. Fauna?	Canad. Fauna	—
Strix flammea	Carolin. Fauna	—	Carolin. Fauna	—
Nyctale Tengmalmi	Hudson. Fauna	Canad. Fauna	Hudson. Fauna	Canad. Fauna
Syrnium cinereum	Hudson. Fauna	Hudson. Fauna	Hudson. Fauna	Canad. Fauna
Otus vulgaris	Hudson. Fauna	—	Canad. Fauna	—
Otus brachyotus	Arctic Coast	—	Hudson. Fauna?	—
Surnia ulula	Arctic Coast	Canad. Fauna	Arctic Realm	Canad. Fauna
Nyctea nivea	Arctic Coast	Hudson. Fauna?	Arctic Realm	Allegh. Fauna
Lagopus albus	Arctic Coast	Hudson. Fauna	Arctic Realm	Hudson. Fauna
Lagopus rupestris	Arctic Coast	Hudson. Fauna	Arctic Realm	Hudson. Fauna
Nycticorax griseus	Hudson. Fauna	Carolin. Fauna?	Carolin. Fauna	—
Strepsilas interpres	Arctic Coast	Hudson. Fauna	—	—
Charadrius pluvialis	Arctic Coast	Arctic Realm	Tropical Amer.	—
Squartarola helvetica	Arctic Coast	Hudson. Fauna	Carolin. Fauna	—
Hæmatopus palliatus	Arctic Coast	—	—	—
Arquatella maritima	Arctic Coast	Arctic Realm	Canad. Fauna	?
Calidris arenaria	Arctic Coast	Canad. Fauna	Allegh. Fauna	—
Tringa canutus	Arctic Coast	Arctic Realm	Carolin. Fauna	—
Ancylocheilus subarnuatus	Arctic Coast	Arctic Realm?	Carolin. Fauna?	—
Pelidna alpina	Arctic Coast	Hudson. Fauna?	Allegh. Fauna	—
Actodromas maculata	Arctic Coast	Arctic Realm	Carolin. Fauna	—
Actodromas Bonapartei	Arctic Coast	Arctic Realm	Carolin. Fauna	—
Tringites rufescens	Arctic Coast?	Hudson. Fauna	Louis. Fauna	—
Lobipes hyperboreus	Arctic Coast	Canad. Fauna	Carolin. Fauna	?
Phalaropus fulicarius	Arctic Coast	Hudson. Fauna?	Carolin. Fauna?	?
Anser hyperboreus	Arctic Coast	Hudson. Fauna	Canad. Fauna?	—
Anser albifrons	Arctic Coast	Hudson. Fauna	Allegh. Fauna	—
Bernicla brenta	Arctic Coast	Hudson. Fauna	Allegh. Fauna	Louis. Fauna
Anas boschas	Arctic Coast	Allegh. Fauna	Canad. Fauna	—
Dafila acuta	Arctic Coast	Hudson. Fauna	—	—
Spatula clypeata	Arctic Coast	Hudson. Fauna	Allegh. Fauna	—
Chaulelasmus streperus	Hudson. Fauna	Hudson. Fauna	?	Louis. Fauna
Somateria mollissima	Arctic Coast	Canad. Fauna	?	Carolin. Fauna
Somateria spectabilis	Arctic Coast	Hudson. Fauna	?	Allegh. Fanna
Bucephala clangula	Arctic Coast	Canad. Fauna	Canad. Fauna	Louis. Fauna
Histrionicus torquatus	Arctic Coast	Canad. Fauna	Hudson. Fauna?	Allegh. Fauna
Harelda glacialis	Arctic Coast	Hudson Fauna	?	Louis. Fauna
Fulix marila	Hudson. Fauna	Canad. Fauna	—	—
Pelionettaper spicillata	Arctic Coast	Hudson. Fauna	Canad. Fauna?	Louis. Fauna
Melanetta fusca	Arctic Coast	Hudson. Fauna	Canad. Fauna?	Louis. Fauna
Mergus merganser	Hudson. Fauna	Canad. Fauna	Canad. Fauna?	Louis. Fauna
Mergus serrator	Hudson. Fauna	Canad. Fauna	Canad. Fauna	Louis. Fauna
Graculus carbo	Arctic Coast	Hudson. Fauna	?	Allegh. Fauna
Sula bassana	Arctic Coast	Canad. Fauna	?	Louis. Fauna
Gelochelidon anglica	Carolin. Fauna	—	—	—
Thalasseus caspius	Arctic Coast	Hudson. Fauna	?	Carolin. Fauna
Thalasseus cantiacus	Louis. Fauna	—	—	—
Sterna hirundo	Arctic Coast?	Carolin. Fauna	Florid. Fauna	—
Sterna macrura	Arctic Coast	Allegh. Fauna	Florid. Fauna	—
Sterna paradisea	Allegh. Fauna	—	—	—
Hydrochelidon fissipes	Carolin. Fauna	—	Florid. Fauna	—

List of Circumpolar Species. (Continued.)

Species.	Boreal Limit in the Breeding Season.	Austral Limit in the Breeding Season.	Boreal Limit in Winter.	Austral Limit in Winter.
Pagophila eburnea . . .	Arctic Coast	Arctic Realm	?	Huds. Fauna
Xema Sabini	Arctic Coast	Hudson. Fauna	?	Canad. Fauna
Rissa tridactyla	Arctic Coast	Canad. Fauna	?	Allegh. Fauna
Rhodostethia rosea . . .	Arctic Coast	Arctic Realm	?	?
Larus glaucus	Arctic Coast	Hudson. Fauna	?	Canad Fauna
Larus marinus	Arctic Coast	Canad. Fauna	?	Louis. Fauna
Larus leucopterus . . .	Arctic Coast	Hudson Fauna	?	Allegh. Fauna
Larus argentatus . . .	Arctic Coast	Canad. Fauna	?	Trop Realm
Buphagus skau	Arctic Coast	Hudson. Fauna	?	Canad Fauna
Stercorarius pomarinus .	Arctic Coast	Hudson. Fauna?	?	Canad. Fauna
Stercorarius parasiticus .	Arctic Coast	Hudson. Fauna?	?	Allegh. Fauna
Stercorarius "Buffoni" . .	Arctic Coast	Hudson. Fauna?	?	Allegh. Fauna.
Thalassidroma Leachii . . .	?	Canad. Fauna	?	Allegh. Fauna?
Podiceps griseigena	Hudson. Fauna	Allegh. Fauna	?	Louis Fauna
Podiceps cornutus	Hudson. Fauna	Allegh. Fauna	?	Louis Fauna
Podiceps cristatus	Hudson. Fauna	Allegh. Fauna	?	Louis. Fauna
Podiceps auritus	Hudson. Fauna	Allegh. Fauna?	?	Louis. Fauna
Colymbus torquatus . . .	Arctic Coast	Allegh. Fauna	?	Florid. Fauna
Colymbus arcticus	Arctic Coast	Hudson. Fauna	?	Canad. Fauna
Colymbus septentrionalis . .	Arctic Coast	Canad. Fauna	?	Canad. Fauna
Fratercula arctica	Arctic Coast	Canad. Fauna	?	Allegh. Fauna
Uria grylle	Arctic Coast	Canad. Fauna	?	Carol. Fauna
Lomvia svarbag	Arctic Coast	Canad. Fauna	?	Carol. Fauna?
Lomvia troile	Arctic Coast	Canad Fauna	?	Allegh. Fauna
Mergulus alle	Arctic Coast	———	?	Allegh Fauna

Summary of the Preceding Table. — The whole number of species in the preceding list is 93. Its most striking feature is the great predominance of the water birds, less than one third of the whole being land birds. Of the 27 land birds, 7 are owls, 6 are hawks, and 5 belong to the family *Fringillidæ;* 9 species embracing all the representatives of other families. The water birds include 1 heron, 14 *Grallæ*, 17 *Anatidæ*, 19 *Laridæ*, 5 *Alcidæ*, 3 species of *Colymbus*, and 4 of *Podiceps.*

In summer 65 species are inhabitants of the Arctic coast and adjacent seas; 22 have their boreal limit near the northern border of the Hudsonian Fauna; 2 are similarly limited by the Canadian Fauna, 5 by the Alleghanian, 3 by the Carolinian, and 1 by the Louisianian; 3 are essentially tropical aquatic species.

Seven seem to be altogether restricted in the breeding season to the Arctic Realm; 36 find their austral limit during the same season near the southern border of the Hudsonian Fauna; 23 are similarly limited by the Canadian, 9 by the Alleghanian, and 3 by the Carolinian, while 10 extend nearly to or within the Tropical Realms, 4 being also inhabitants of the greater part of the southern hemisphere.

The winter quarters of the land birds of this list are the cold-temperate and middle-temperate districts of the northern hemisphere. Most of the water birds visit the warm-temperate parts of the same hemisphere; a considerable number also visit the tropics, and a few wander, at this season, over the greater part of the warmer regions of the globe.

III. Species mainly restricted in the Breeding Season to the North American Temperate Region.

1. *List of Species which breed throughout the greater Part of Temperate North America, with Indications of their Boreal and Austral Limits Distribution in the Eastern Province.*

Species.	Boreal Limit in the Breeding Season.	Austral Limit in the Breeding Season.	Boreal Limit in Winter.	Austral Limit in Winter.
Turdus migratorius	Hudson. Fauna	Carol Fauna	Carol. Fauna	——— *
Turdus fuscescens	Canad Fauna	Carol. Fauna	——— *	———
Geothlypis trichas	Hudson Fauna	Florid. Fauna	Florid. Fauna	———
Hirundo horreorum	Hudson. Fauna	Louis. Fauna	———	———
Petrochelidon lunifrons .	Hudson. Fauna	Carol. Fauna ?	———	———
Tachycineta bicolor	Hudson. Fauna	Carol. Fauna	Louis. Fauna	———
Cotyle riparia	Arctic Coast	Carol. Fauna	Florid. Fauna	———
Progne subis	Canad. Fauna	?	Louis. Fauna	———
Ampelis cedrorum	Canad. Fauna	Florid. Fauna	Allegh. Fauna	———
Collurio ludovicianus . . .	Allegh. Fauna	Florid. Fauna	Louis. Fauna	———
Vireosylvia gilva	Canad. Fauna	Carol. Fauna	———	———
Lanivireo solitarius	Hudson Fauna	Allegh. Fauna	Florid. Fauna	———
Cistothorus palustris . . .	Hudson Fauna	Louis. Fauna	Louis. Fauna	———
Troglodytes aëdon . . .	Allegh. Fauna	Florid. Fauna	Louis. Fauna	———
? Sitta carolinensis	Canad. Fauna	Louis. Fauna	Canad Fauna	Florid. Fauna
Sitta canadensis	Canad. Fauna	Carol. Fauna	Canad. Fauna	Louis. Fauna ?
Parus atricapillus	Hudson. Fauna	Florid. Fauna	Canad. Fauna	Florid. Fauna
Carpodacus purpureus . . .	Canad. Fauna	Allegh. Fauna	Allegh. Fauna	Louis. Fauna
Chrysomitris tristis . . .	Canad. Fauna	Louis. Fauna	Allegh. Fauna	———
Passerculus savanna . . .	Hudson. Fauna	Allegh. Fauna	Carol. Fauna	———
Poocætes gramineus . . .	Canad. Fauna	Carol. Fauna	Carol. Fauna	———
Spizella socialis	Hudson. Fauna	Louis. Fauna	Carol. Fauna	———
? Melospiza melodia	Hudson. Fauna	Louis. Fauna	Carol. Fauna	Florid. Fauna
Melospiza Lincolnii	Hudson. Fauna	Allegh. Fauna	Carol. Fauna	———
Molothrus pecoris	Hudson. Fauna	Louis. Fauna	Carol. Fauna	———
Agelæus phœniceus	Hudson. Fauna	Florid. Fauna	Carol. Fauna	———
Sturnella ludoviciana . . .	Allegh. Fauna	Florid. Fauna	Carol. Fauna	———
Corvus americanus	Hudson. Fauna	Florid. Fauna ?	Canad. Fauna	———
Tyrannus carolinensis . . .	Canad. Fauna	Louis. Fauna	———	———
? Myiarchus crinitus	Canad. Fauna	Louis. Fauna	———	———
? Sayornis fuscus	Hudson. Fauna	Loûis. Fauna	Louis. Fauna	———
Contopus borealis	Hudson. Fauna	Allegh. Fauna	———	———
? Contopus virens	Canad. Fauna	Louis. Fauna	———	———
? Empidonax minimus . . .	Hudson. Fauna	Allegh. Fauna	———	———
? Empidonax acadicus . . .	Canad. Fauna	Carol. Fauna	———	———
Empidonax flaviventris . .	Canad. Fauna	Carol. Fauna	———	———
Picus villosus	Hudson. Fauna	Louis. Fauna	Canad. Fauna ?	Florid. Fauna
Picus pubescens	Hudson. Fauna	Louis. Fauna	Canad. Fauna ?	Florid. Fauna
Hylotomus pileatus	Hudson. Fauna	Florid. Fauna	Canad. Fauna ?	Florid. Fauna
Chordeiles popetue	Hudson. Fauna	——— *	———	———
Ceryle alcyon	Arctic Coast	Florid. Fauna	Carol. Fauna	———
Accipiter Cooperi	Canad. Fauna	———	Carol. Fauna	———
Accipiter fuscus	Canad. Fauna	———	Carol. Fauna	———
Buteo borealis	Hudson. Fauna	———	Canad. Fauna	———
Buteo lineatus	Canad. Fauna	Florid. Fauna	Allegh. Fauna	Florid. Fauna
Buteo pennsylvanicus . . .	Canad. Fauna	———	Allegh. Fauna	———
Haliaëtus leucocephalus .	Arctic Coast	Florid Fauna	Hudson. Fauna?	Florid. Fauna
Scops asio	Canad. Fauna	———	Canad. Fauna	———
Zenædura carolinensis . . .	Allegh. Fauna	Florid. Fauna	Carol Fauna	Florid. Fauna
Meleagris gallopavo	Allegh. Fauna	Florid. Fauna	Allegh Fauna	Florid. Fauna
Bonasa umbellus	Hudson. Fauna	Carol. Fauna	Hudson. Fauna?	Carol. Fauna
Botaurus lentiginosus . . .	Hudson. Fauna	Carol. Fauna	Louis. Fauna	———
Ardetta exilis	Canad. Fauna	Florid. Fauna	Louis. Fauna	———
Gallinago Wilsoni	Hudson. Fauna	Allegh Fauna	Louis. Fauna	———
Rhyacophilus solitarius . .	Hudson. Fauna	Allegh. Fauna	Carol. Fauna	———
Tringoides macularius . . .	Hudson. Fauna	Louis. Fauna	Louis. Fauna	———
Limosa fedoa	Arctic Coast ?	Hudson. Fauna	Louis. Fauna	———
Rallus virginianus	Allegh. Fauna	Louis. Fauna	Louis. Fauna	———
Porzana carolina	Canad. Fauna	Carol. Fauna	?	———
Fulica americana	Canad. Fauna	Carol. Fauna	Louis. Fauna	———
Nettion carolinensis . . .	Hudson. Fauna	Canad. Fauna	Carol. Fauna	———

* Blanks in the third column indicate that the species ranges southward in the breeding season into the Tropical Realm. Blanks in the fourth column that the species retires wholly within the Tropical Realm in winter; in the fifth column, that the southern limit in winter is within the Tropical Realm.

2. *List of Species which breed throughout the greater Part of the Cold-temperate Portions of the North American Region, with Indications of their Boreal and Austral Limits of Distribution in the Eastern Province.*

Species.	Boreal Limit in Breeding Season.	Austral Limit in Breeding Season.	Boreal Limit in Winter.	Austral Limit in Winter.
Turdus Pallasi	Hudson. Fauna	Canad. Fauna	Carol. Fauna	——— *
Turdus Swainsoni	Hudson. Fauna	Canad. Fauna	Florid. Fauna	———
Regulus calendula	Hudson. Fauna	Canad. Fauna	Carol. Fauna	———
Regulus satrapa	Hudson Fauna	Canad. Fauna	Allegh. Fauna	———
Anthus ludovicianus	Arctic Coast	Hudson. Fauna	Carol. Fauna	———
Helminthophaga celata	Hudson. Fauna	Canad. Fauna	Carol. Fauna	———
Wilsonia pusilla	Hudson. Fauna	Allegh. Fauna	——— *	———
Collurio borealis	Hudson. Fauna	Canad. Fauna	Canad. Fauna	Carol. Fauna
Troglodytes hyemalis	Hudson. Fauna	Canad. Fauna	Allegh. Fauna	———
Chrysomitris pinus	Hudson. Fauna	Canad. Fauna	Canad. Fauna	Florid. Fauna
Curvirostra americana	Hudson Fauna	Canad Fauna	Hudson. Fauna	Allegh. Fauna
Zonotrichia leucophrys	Hudson. Fauna	Canad. Fauna	Carol. Fauna	Louis. Fauna
Scolecophagus ferrugineus	Hudson. Fauna	Canad. Fauna	Carol. Fauna	———
Perisoreus canadensis	Hudson. Fauna	Canad. Fauna	Hudson. Fauna	Canad. Fauna
Picoides arcticus	Hudson. Fauna	Canad. Fauna	Hudson. Fauna	Canad. Fauna
Picoides hirsutus	Hudson. Fauna	Carol. Fauna	Hudson. Fauna	Canad. Fauna
Falco columbarius	Hudson. Fauna	Carol. Fauna	Canad. Fauna	Louis. Fauna
Astur atricapillus	Hudson. Fauna	Canad. Fauna	Hudson. Fauna	Allegh. Fauna
Ægialitis semipalmatus	Arctic Coast	Hudson. Fauna	Louis. Fauna	———
Phalaropus Wilsoni	Arctic Coast	Canad. Fauna	Louis. Fauna	———
Ereunetes pusillus	Arctic Coast	Canad. Fauna	Carol. Fauna	———
Gambetta melanoleuca	Hudson. Fauna	Carol. Fauna	Louis. Fauna	———
Gambetta flavipes	Hudson. Fauna	Canad. Fauna	Louis. Fauna	———
Numenius longirostris	Hudson. Fauna	Canad. Fauna	Louis. Fauna	———
Numenius hudsonius	Hudson. Fauna	Canad. Fauna	Louis. Fauna	———
Cygnus americanus	Arctic Coast	Canad. Fauna	Allegh. Fauna	———
Bernicla canadensis	Arctic Coast	Canad. Fauna	Allegh. Fauna	———
Mareca americana	Arctic Coast	Canad. Fauna	Allegh. Fauna	———
Fulix collaris	Arctic Coast	Canad. Fauna	Canad. Fauna	———
Aythya americana	Arctic Coast	Hudson. Fauna	Carol. Fauna	———
Aythya vallisneria	Arctic Coast	Hudson. Fauna	Carol. Fauna	———
Bucephala albeola	Arctic Coast	Hudson. Fauna	Allegh. Fauna	———
Erismatura rubida	Arctic Coast	Hudson. Fauna	Allegh. Fauna	———
Lophodytes cucullatus	Arctic Coast	Canad. Fauna	Allegh. Fauna	———
Pelecanus erythrorhynchus	Arctic Coast	Canad. Fauna	Allegh. Fauna	———
Graculus dilophus	Arctic Coast	Canad. Fauna	Canad. Fauna	———
Larus delawarensis	Arctic Coast	Hudson. Fauna	Allegh. Fauna	———
Chrœcocephalus philadephia	Arctic Coast	Hudson. Fauna	Canad. Fauna	———

3. *List of Species which breed only in the Warm-temperate Portions of the North American Temperate Region, and range Southward in the Breeding Season into the Tropical American Realm.*

Species.	Boreal Limit in Breeding Season.	Austral Limit in Breeding Season.	Boreal Limit in Winter.	Austral Limit in Winter.
Mimus polyglottus	Carolin. Fauna	——— *	Louis. Fauna	——— *
Polioptila cærulea	Carolin. Fauna	———	Louis. Fauna	———
Icteria virens	Carolin. Fauna	———	———	———
Stelgidopteryx serripennis	Carolin. Fauna	———	Louis. Fauna ?	———
Thryothorus Bewicki	Carolin. Fauna	———	Louis. Fauna	———
? Sitta pusilla	Louis. Fauna	———	———	———
Guiraca cærulea	Carolin. Fauna	———	Louis. Fauna	———
? Quiscalus major	Louis. Fauna	———	———	———
Elanus leucurus	Louis. Fauna	———	Florid. Fauna	———
Polyborus tharus	Florid. Fauna	———	———	———
Craxirex unicinctus	Louis. Fauna	———	Louis. Fauna	———
Chamapælia passerina	Louis. Fauna	———	Louis. Fauna	———
Tantalus loculator	Louis. Fauna	———	Louis. Fauna	———
Garzetta candidissima	Carolin. Fauna	———	Louis. Fauna	———
Herodias egretta	Carolin. Fauna	———	———	———
Himantopus nigricollis	Carolin. Fauna ?	———	Louis. Fauna	———
Rallus elegans	Carolin. Fauna	———	Florid. Fauna ?	———
Phœnicopterus ruber	Florid. Fauna	———	Florid. Fauna	———
Querquedula cyanoptera	?	———	Louis. Fauna	———
Pelecanus fuscus	Carolin Fauna	———	———	———
Tachypetes aquila	Florid Fauna	———	Florid. Fauna	———
Plotus anhinga	Louis. Fauna	———		———

* Within the Tropical Realm

4. *List of Species whose Breeding Range extends throughout the greater Part of the North American Realm, and Southward into the Tropical Realm, with Indications of their Boreal and Austral Limits in the Eastern Province.*

Species.	Boreal Limit in the Breeding Season.	Austral Limit in the Breeding Season.	Boreal Limit in Winter.	Austral Limit in Winter.
Dendrœca æstiva	Hudson. Fauna	——— *	———	——— *
Grus canadensis	Arctic Coast	———	Carolin. Fauna	———
Butorides virescens	Allegh. Fauna	———	Louis. Fauna	———
Ardea herodias	Hudson. Fauna	———	Carolin. Fauna	———
Hæmatopus palliatus . . .	Arctic Coast	———	Louis. Fauna	———
Ægialitis vociferus	Allegh. Fauna	———	Louis. Fauna	———
Recurvirostra americana . .	Hudson. Fauna	Louis. Fauna ?	Louis. Fauna	———
Symphemia semipalmata . .	Canad. Fauna	———	Louis. Fauna	———
Aix sponsa	Canad. Fauna	———	Carolin. Fauna	———
Podilymbus podiceps . . .	Hudson. Fauna	———	Carolin. Fauna	———

* Within the Tropical Realm.

5. *List of Species whose Breeding Habitat includes the greater Part of both North and South America, with Indications of their Boreal Limit, both in the Breeding Season and in Winter.*

Species.	Boreal Limit in the Breeding Season.	Austral Limit in the Breeding Season.	Boreal Limit in Winter.	Austral Limit in Winter.
Cathartes aura	Allegh. Fauna	———	Carolin. Fauna	———
Cathartes atratus	Louis. Fauna	———	Louis. Fauna	———
* Pandion haliaëtus	Arctic Coast	———	Louis. Fauna	———
Falco sparverius	Hudson. Fauna	———	Hudson. Fauna	———
* Falco peregrinus	Hudson. Fauna	———	Canad. Fauna	———
Bubo virginianus	Hudson. Fauna	———	Hudson. Fauna	———
* Strix flammea	Carolin. Fauna	———	Carolin. Fauna	———
* Otus vulgaris	Hudson. Fauna	———	Canad. Fauna ?	———
* Otus brachyotus	Arctic Coast	———	Hudson. Fauna?	———

* Also circumpolar species.

Summary of the Preceding Five Tables. — The total number of species given in the above lists of the species characteristic (mainly exclusively so) of the North American Temperate Region is 135. Of these 38 are restricted in the breeding season in their austral range to the Cold-temperate District; about one third of them, chiefly natatorial species, reach the Arctic coast; 61 are similarly mainly limited to the Middle-temperate District, but two or three reach the Arctic coast, and nearly one third range into the Hudsonian Fauna; 21 are limited in their boreal range to the Warm-temperate District, the greater part of which, even in the breeding season, range southward into the tropics. Of the whole number, 90 are land birds, 23 being raptorial species. Of the remaining 45 water birds, 7 are herons, 20 are *Grallæ*, and 18 are *Natatores*, 12 of the latter being *Anatidæ*.

In the list of those whose breeding habitat is the cold-temperate portions of the continent (Table 1), 20 of the species are aquatic and

18 terrestrial; of those breeding throughout the greater part of the continent, 10 only are aquatic and 51 are terrestrial; of those breeding in the warm-temperate portions of the continent, 9 are aquatic and 9 terrestrial; of the 10 wide-ranging species, whose breeding habitats embrace not only nearly the whole of temperate North America, but extend also into the tropics, 1 only is a land bird, 3 being *Herodiones*, 4 *Grallæ*, and 2 *Natatores*. Of the 9 species given in the Fifth Table, which range in the breeding season throughout both the North American and South American continents, none are aquatic; 4 are owls, 3 hawks, and 2 vultures. The most numerously represented family, and one of those almost exclusively characteristic of the North American Temperate Region (the *Sylvicolidæ*), has but three species which range across the continent, and only one of these is a typical representative of the family.

IV. Species limited in Longitude to the Eastern Province of the North American Temperate Region.

1. *List of Species restricted in the breeding Season to the Cold-temperate Portion of the Eastern Province, with Indications of their Boreal and Austral Limits.*

Species.	Boreal Limit in Breeding Season.	Austral Limit in Breeding Season.	Boreal Limit in Winter.	Austral Limit in Winter.
Mniotilta varia	Hudson. Fauna	Carolin. Fauna	——— *	——— *
Parula americana	Canad. Fauna	Carolin. Fauna	Florid. Fauna	———
Geothlypis philadelphia	Allegh. Fauna	?	———	———
Oporornis agilis	Canad. Fauna	Allegh. Fauna?	———	———
Helminthophaga chrysoptera	Canad. Fauna	Carolin. Fauna	———	———
? Helminthophaga peregrina	Canad. Fauna	Carolin. Fauna	———	———
Helminthophaga ruficapilla	Hudson. Fauna	Carolin. Fauna	———	———
Dendrœca coronata	Hudson. Fauna	Canad. Fauna	Carolin. Fauna	———
Dendrœca castanea	Hudson. Fauna?	Canad. Fauna	———	———
Dendrœca striata	Hudson. Fauna	Canad. Fauna	———	———
Dendrœca Blackburniæ	Canad. Fauna	Allegh. Fauna	———	———
Dendrœca cærulescens	Canad. Fauna?	Canad: Fauna	———	———
Dendrœca maculosa	Canad. Fauna?	Canad. Fauna	———	———
Dendrœca virens	Hudson. Fauna	Allegh. Fauna	———	———
Dendrœca palmarum	Hudson. Fauna	Canad. Fauna	Louis. Fauna	———
Perissoglossa tigrina	Canad. Fauna?	Allegh. Fauna	———	———
Euthlypis canadensis	Hudson. Fauna	Canad. Fauna	———	———
Setophaga ruticilla	Hudson. Fauna	Allegh. Fauna	———	———
Vireosylvia olivacea	Hudson. Fauna	Louis. Fauna	Florid. Fauna	———
Vireosylvia philadelphica	Hudson. Fauna	Allegh. Fauna	———	———
Cistothorus stellaris	Hudson. Fauna	Carolin. Fauna	Louis. Fauna	———
Parus hudsonicus	Hudson. Fauna	Canad. Fauna	Hudson. Fauna	Canad. Fauna
Zonotrichia albicollis	Hudson. Fauna	Canad. Fauna	Louis. Fauna	———
Junco hyemalis	Hudson. Fauna	Canad. Fauna	Allegh. Fauna	Louis. Fauna
Spizella monticola	Hudson. Fauna	Canad. Fauna	Allegh. Fauna	Louis. Fauna
Passerella iliaca	Hudson. Fauna	Canad. Fauna	Carolin Fauna	Louis. Fauna
Dolichonyx oryzivora	Allegh. Fauna	Allegh. Fauna	———	———
Tetrao canadensis	Arctic Coast	Canad. Fauna	Hudson. Fauna?	Canad. Fauna
Ægialitis melodus	Canad. Fauna	Carolin. Fauna	Louis. Fauna	———
Limosa hudsonica	Arctic Coast?	Hudson. Fauna	Louis. Fauna	———
Numenius borealis	Arctic Coast?	Hudson. Fauna	Louis. Fauna	———
Porzana noveboracensis	Hudson. Fauna	Carolin. Fauna	Louis. Fauna	———
Anas obscura	Hudson. Fauna	Allegh. Fauna	Allegh. Fauna	Florid. Fauna
Querquedula discors	Hudson. Fauna	Canad. Fauna	Allegh Fauna	———
Camptolæmus labradorius	Arctic Coast	Hudson. Fauna	———	Canad. Fauna

* The blanks in the fourth and fifth columns indicate that the limit in question is within the Tropical Realm.

2. *List of Species which breed throughout the Middle-temperate Portions of the Eastern Province, with Indications of their Boreal and Austral Limits in the Eastern Province.**

Species.	Boreal Limit in the Breeding Season.	Austral Limit in the Breeding Season.	Boreal Limit in Winter.	Austral Limit in Winter.
Turdus mustelinus	Allegh. Fauna	Louis. Fauna ?	———	———
Galeoscoptes carolinensis . .	Allegh. Fauna	Florid. Fauna ?	Louis. Fauna	———
Harporhynchus rufus . . .	Allegh. Fauna	Florid. Fauna ?	Louis. Fauna	———
Sialia sialis	Allegh. Fauna	Louis. Fauna	Carolin. Fauna	———
Dendrœca pennsylvanica . .	Canad. Fauna	Allegh. Fauna	———	———
Dendrœca discolor	Allegh. Fauna	———	———	———
Dendrœca cærulea	Allegh. Fauna	Carolin. Fauna	———	———
Wilsonia mitrata	Allegh. Fauna	Louis. Fauna	———	———
Pyranga rubra	Allegh. Fauna	Louis. Fauna	———	———
Vireo noveboracensis . . .	Allegh. Fauna	Florid. Fauna ?	Louis. Fauna	———
Lanivireo flavifrons	Allegh. Fauna	———	———	———
Lophophanes bicolor . . .	Carolin. Fauna	Louis. Fauna ?	Carolin. Fauna	Florid. Fauna
Coturniculus passerinus . .	Allegh. Fauna	Louis. Fauna	Carolin. Fauna ?	———
Coturniculus Henslowi . . .	Allegh. Fauna	Louis. Fauna	Carolin. Fauna ?	———
Ammodromus caudacutus . .	Allegh. Fauna	Louis. Fauna	Louis. Fauna	———
Ammodromus maritimus . .	Carolin. Fauna	Louis. Fauna	Louis. Fauna	———
Spizella pusilla	Allegh. Fauna	Louis. Fauna ?	Carolin. Fauna	———
Euspiza americana	Carolin. Fauna	———	———	———
Hedymeles ludovicianus . .	Allegh. Fauna	Carolin Fauna	———	———
Cyanospiza cyanea	Allegh. Fauna	———	———	———
Cardinalis virginianus . . .	Carolin. Fauna	———	Carolin. Fauna	———
Pipilo erythrophthalmus . .	Allegh. Fauna	Florid. Fauna	Louis. Fauna	———
Icterus spurius	Allegh. Fauna	Louis. Fauna	———	———
Icterus baltimore	Allegh. Fauna	Louis. Fauna	———	———
Corvus ossifragus	Carolin. Fauna	Florid. Fauna ?	Louis. Fauna	———
Centurus carolinus	Carolin. Fauna	———	Carolin. Fauna	———
Melanerpes erythrocephalus .	Allegh. Fauna	Florid. Fauna ?	Louis. Fauna	———
Antrostomus vociferus . . .	Allegh. Fauna	Florid. Fauna	Louis. Fauna	———
Coccygus americanus . . .	Allegh. Fauna	———	———	———
Coccygus erythrophthalmus .	Allegh. Fauna	———	———	———
Ortyx virginianus	Allegh. Fauna	———	Allegh. Fauna	———
Cupidonia cupido	Allegh. Fauna	Carolin. Fauna ?	Allegh. Fauna	Louis. Fauna

* The blanks in this and the following tables have the same significance as in the last preceding table.

3. *List of Species which breed throughout the Temperate Portions of the Eastern Province, with Indications of their Boreal and Austral Limits.*

Species.	Boreal Limit in the Breeding Season.	Austral Limit in the Breeding Season.	Boreal Limit in Winter.	Austral Limit in Winter.
Seiurus aurocapillus . . .	Canad. Fauna	Carolin. Fauna	Florid Fauna	———
Seiurus noveboracensis . . .	Hudson. Fauna	Louis. Fauna	———	———
Cyanura cristata	Canad. Fauna	Florid. Fauna	Allegh. Fauna	Florid. Fauna
Sphyrapicus varius	Canad. Fauna ?	Florid. Fauna ?	Carolin. Fauna	———
Colaptes auratus	Hudson. Fauna	———	Carolin. Fauna	———
Trochilus colubris	Canad. Fauna	———	Florid. Fauna ?	———
Chætura pelasgia	Canad. Fauna	Louis. Fauna ?	———	———
Grus americanus	Canad. Fauna	Florid. Fauna	Louis. Fauna	———
Actiturus Bartramius . . .	Canad. Fauna	———	Louis. Fauna	———
Sterna antillarum	Canad. Fauna	———	———	———
Dendrœca pinus	Hudson. Fauna ?	Louis. Fauna ?	Carolin. Fauna	———
Quiscalus purpureus . . .	Hudson. Fauna	———	Carolin. Fauna	———

4. *List of Species which breed in the Eastern Province only within the Warm-temperate and Subtropical Districts.*

Species.	Boreal Limit in the Breeding Season.	Austral Limit in the Breeding Season.	Boreal Limit in Winter.	Austral Limit in Winter.
Prothonotaria citrea	Carolin. Fauna	——	——	——
Oporornis formosus	Allegh. Fauna	Louis. Fauna ?	——	——
Helmitherus vermivorus	Carolin. Fauna	Florid. Fauna?	——	——
Helmitherus Swainsoni	Louis. Fauna	——	——	——
Helminthophaga pinus	Carolin. Fauna?	——	——	——
Helminthophaga Bachmani	Louis. Fauna	——	——	——
Dendrœca dominica	Louis. Fauna	——	——	——
Vireosylvia barbatula	Florid. Fauna	——	——	——
Pyranga æstiva	Carolin. Fauna	——	——	——
Thryothorus ludovicianus	Carolin. Fauna	——	Louis. Fauna	——
Peucæa æstivalis	Louis. Fauna	——	Florid. Fauna	——
Cyanospiza ciris	Louis. Fauna	——	——	——
Tyrannus dominicensis	Florid. Fauna	——	——	——
Campephilus principalis	Carolin. Fauna	——	Louis. Fauna	——
Picus borealis	Louis. Fauna	——	Louis. Fauna	——
Conurus carolinensis	Louis. Fauna	——	Louis. Fauna	——
Crotophaga ani	Florid. Fauna	——	——	——
Antrostomus carolinensis	Louis. Fauna	——	Louis. Fauna	——
Nauclerus furcatus	Carolin. Fauna	——	——	——
Ictinia mississippiensis	Louis. Fauna	——	——	——
Rosthramus sociabilis	Florid. Fauna	——	——	——
Florida cærulea	Carolin. Fauna	——	Louis. Fauna	——
Nyctherodius violaceus	Carolin. Fauna	——	——	——
Demiegretta Pealei	Florid. Fauna	——	Florid. Fauna	——
Demiegretta ludoviciana	Carolin. Fauna	——	Louis. Fauna	——
Ibis alba	Carolin. Fauna	——	Louis. Fauna	——
Platalea ajaja	Louis. Fauna	——	——	——
Ægialitis Wilsonius	Carolin. Fauna	——	Louis. Fauna	——
Aramus giganteus	Florid. Fauna	——	Florid. Fauna	——
Rallus crepitans	Carolin. Fauna	——	Louis. Fauna	——
Porzana jamaicensis	Carolin. Fauna	——	Florid. Fauna	——
Gallinula galeata	Carolin. Fauna	——	Louis. Fauna	——
Gallinula martinica	Carolin. Fauna	——	Louis. Fauna	——
Sula fiber	Louis. Fauna	——	Florid. Fauna	——
Graculus floridanus	Louis. Fauna	——	Florid. Fauna	——
Plotus anhinga	Carolin. Fauna	——	Florid. Fauna	——
Chrœcocephalus atricilla	Allegh. Fauna	——	Florid. Fauna	——
Thalasseus acuflavidus	Louis. Fauna	——	——	——
Anoüs stolidus	Florid. Fauna	——	——	——
Haliplana fuliginosa	Florid. Fauna	——	——	——
Rhynchops nigra	Carolin. Fauna	——	Florid. Fauna	——

Summary of the Four Preceding Tables. — About one hundred and twenty species occur in the Eastern Province of the North American Temperate Region that do not appear as regular residents in the Western Province of the same region, of which a small proportion are in part tropical. Of these one hundred and twenty, thirty-five are northern, or range in the breeding season only over the cold-temperate portions of the Eastern Province; twenty-eight of the latter being land birds, and only seven aquatic. Eighteen species of the land birds belong to the single family of the *Sylvicolidæ.* About one fourth of the Eastern Province species (thirty-two), *all land birds,* range in the breeding season over only the middle-temperate part of the province. Of these only three belong to the family *Sylvicolidæ,* and only one is a typical representative of that group; ten belong to the family *Fringillidæ,* three to the *Turdidæ*, and two each to the *Icteridæ*, *Picidæ*, *Cuculidæ*, and *Tetraonidæ;* several other families have one representative each. The total absence of any species of *Falconidæ*, *Strigidæ*, *Herodiones*, *Grallæ*, and *Natatores* is one of the most striking features in the list of the species restricted to the Eastern Province.

Twelve of the Eastern Province species breed throughout the greater part of the province, three of which are *Sylvicolidæ*, two are *Picidæ*, one is a humming-bird, one a wader, and one a tern.

Forty-one of the one hundred and twenty species restricted in longitudinal range to the Eastern Province extend so far into the Tropical American Realm in the breeding season as to be essentially tropical species, exclusive of a considerable number that appear only in the Floridian Fauna. Twenty-one of these are land birds and twenty aquatic; the latter embracing six *Herodiones*, six *Grallæ*, and eight *Natatores*, five of which are terns. The land species embrace three hawks, two species of *Fringillidæ*, seven of *Sylvicolidæ*, two of *Picidæ*, and one each of seven other families.

General Remarks on the Distribution and Migration of the Birds of the Eastern Province.

The preceding tables, illustrative of the geographical distribution of the birds of the Eastern Province of the North American Temperate Region, and the summary remarks already given respecting them, indicate a number of interesting general facts.

I. The species which have the greatest longitudinal range in the breeding season are the hawks, owls, and vultures, the swallows, the *Turdinæ* or typical thrushes, the woodpeckers and flycatchers, and the water birds; among the latter, especially the *Scolopacidæ*, the *Charadriidæ* and their allies, the *Anatidæ*, and the *Laridæ;* in fact, nearly all the *Natatores*. All the land birds ranging widely in longitude are hence species which possess highly developed powers of flight, and have also a wide latitudinal range. The few circumboreal *Natatores*, which have only moderate or greatly reduced powers of flight, possess great power of locomotion in the water. Their habitat is, moreover, not only generally the sea-shores, but the boreal shores of the converging continents of the northern hemisphere. Hence all the species having a wide geographical range — as the circumpolar and continental — are either pre-eminently strong fliers or powerful swimmers. It also explains the occurrence of the large proportion of long-winged birds, and especially of the preponderance of the water birds, in the three first primary divisions given above of the birds of the Eastern Province, namely, the cosmopolitan, the circumpolar, and the continental, and the small proportion of such species among those restricted in their longitudinal range to the Eastern Province. Most of the circumpolar species are also boreal ones.

II. The aquatic species, while forming only about four tenths of the birds found in the Eastern Province, greatly predominate over the land species in the boreal regions, in the Arctic Realm they outnumbering the land birds in the proportion of five to one, or form eight tenths of the whole. In the Cold-temperate District of the North American Region the water birds form about six tenths of the whole; in the Middle-temperate Districts, between four and five tenths; in the Warm-temperate District, rather less than four tenths. Farther southward, although a few groups (as the *Rallidæ*, *Herodiones*, and *Sterninæ*) are more numerously represented, the relative proportion of water birds to the terrestrial seems scarcely to increase. In the breeding season, however, a numerical comparison of the land and water birds yields very different results, in respect to the proportion characteristic of localities situated under different parallels of latitude. Passing from the extreme boreal regions southward, the number of *Grallæ* (exclusive of the *Paludicolæ*), *Anatidæ*, *Larinæ*, and *Lestridinæ* decreases rapidly, so that the number of the *Grallæ* (exclusive of the rails and their allies) is reduced in the breeding season, in the warm-temperate parts of the Eastern Province, to only seven or eight species, the *Anatidæ* to one (*Aix sponsa*), the *Larinæ* to one (*Chræcocephalus atricilla*), and the *Lestridinæ* disappear entirely.

III. A large proportion of the accessions to the land birds near the tropics being species of a comparatively low grade of structure, the prevalence of the water birds in the arctic and subarctic faunæ, the comparative absence of water birds in the temperate latitudes, and the great development here of the higher groups of the land birds, give to the temperate regions the maximum proportion of high forms of avine life, — a fact as true in respect to mammalian life as it is of birds.

IV. In respect to the distribution and relative development of particular families, the *Sittidæ* (*Sittæ*), the *Paridæ*, and the *Alcidæ* are alone restricted to the North Temperate Realm. The species of these groups (except the *Alcidæ*) are also sedentary throughout nearly their whole range. In the Eastern Province, *Larus*, *Stercorarius*, and their allies, as well as *Colymbus*, are restricted in the breeding season to its northern half, as are also, as already remarked, most of the water birds, except the *Rallidæ* and the *Herodiones*, which are chiefly southern. The representatives of the *Troglodytidæ* and *Icteridæ* increase rapidly in number towards the tropics, while the *Miminæ* and several genera of the more brightly colored *Fringillidæ* are confined to the southern half of the province. The *Sylvicolidæ*, the most exclusively distinctive family of the North American temperate region, reaches its maximum development in the Middle and Cold-temperate Districts. The section *Sylvicoleæ*, and especially the genus *Dendrœca*, is more numerously represented in the Eastern Province than in the Western, and the greater part breed in the colder latitudes, their "metropolis" during the breeding season being the Canadian Fauna of the Eastern Province and the corresponding fauna of the Western. The species of the section *Vermivoreæ* (genera *Helmitherus* and *Helminthophaga*) have a wider longitudinal distribution than the *Dendrœcæ*, the species of which genus are mainly restricted either to the Eastern or to the Western Province. The two species of *Helmitherus* are southern in their distribution, while four or five of the six *Helminthophagæ* are northern.

V. At the extreme north, or from the Arctic coast southward to the Canadian Fauna, nearly all the birds are migratory, owing to the extreme severity of the winter season; they also spend a smaller portion of the year at their breeding stations than do the species which breed farther to the southward. Even as far south as the Alleghanian and Carolinian Faunæ, the greater proportion of the species are to a greater or less extent migratory. In the Carolinian and Louisianian Faunæ a much greater proportion are resident, even including many species whose boreal limit of distribution is the Carolinian Fauna. From the Hudsonian Fauna southward many species are found the whole year at the same localities, and are hence termed "resident," though the individuals representing such species are migratory, there being a general movement of the winter habitat southward, but too limited to carry the wave of migration entirely beyond the southern limit of the summer habitat of these species.

VI. The representatives of the various groups differ from each other widely, as is well known, in respect to the extent of their migrations. Those of a few families (as the *Tetraonidæ*, the *Picidæ*, *Sittidæ*, *Corvidæ*, and *Strigidæ*) are nearly sedentary, the nature of their food being such that the supply is almost equally sure at all seasons. The insectivorous species have the most extended migratory range; the piscivorous, the graminivorous, and the raptorial the least of the non-sedentary species, the migrations of the latter being mainly governed by those of their prey. Hence the wood warblers (*Sylvicolidæ*), the flycatchers (*Tyrannidæ*), and the swallows make the longest journeys, and leave their breeding stations the earliest. Requiring apparently a temperature in winter similar to that of their summer habitats, as well as a constant supply of insect food, they begin their southward journeys almost before the close of the short northern summer, proceeding gradually southward during the autumn months to pass the winter in the tropics. The *Grallæ* have also to seek districts almost wholly beyond the reach of severe frosts, their food being only accessible to them where the ground continues unfrozen. The *Natatores* also necessarily migrate to localities where the streams and estuaries are nearly free from ice; the strictly littoral and pelagic species hence making shorter journeys than the inland species.

The migratory insessorial birds that pass the winter wholly or in part within the Eastern Province are principally fringilline species. The others are a single flycatcher (*Sayornis fuscus*), the northern members of the *Icteridæ*, two species of *Turdus*, three of *Miminæ*, three wrens, and three or four sparrows. All of these species are resident the whole year in those sections to which the northern members of these species mainly resort in winter. In these species there is hence only a partial recession southward in winter from the northern portions of their respective summer habitats. Most of the *Fringillidæ*, however, which pass the summer within or to the northward of the Alleghanian Fauna, remove wholly in winter from their summer stations. While some in winter barely abandon their summer stations, as shown in the preceding tables, of distribution, others pass entirely over one fauna, throughout which they occur only as spring and autumn passengers; others in a similar way pass over two faunæ in reaching their winter quarters. Hence some which breed in the Canadian Fauna and farther northward pass only into the Alleghanian and Carolinian Faunæ in winter, while others pass over the Alleghanian into the Carolinian, and others over both the Alleghanian and the Carolinian into the Louisianian.

Some species which in general breed far to the northward of the tropics, to which they retire in winter, are also known to breed on the mountains within the tropics (as *Dendrœca coronata* and *Perissoglossa tigrina*), and doubtless many others will be found to do so when the mountain faunæ of these regions become fully known; it being already well ascertained that there is a migration from the plains and lowlands to the mountains (more especially in the Tropical Realms) as well as (in the northern hemisphere) from the south northward.

VII. The lack of suitable food and the low temperature in winter in northern latitudes are evidently the causes which impel so many species to leave their breeding stations at that season to seek a warmer zone. While in most cases a degree of cold sufficient to cut off the supply of food of any species, especially of the insectivorous ones, would of itself also prove fatal to the birds themselves, it is by no means the case with the baccivorous and graminivorous species, their winter migrations appearing to be primarily and principally controlled by the accessibility of their food. This is evidently indicated by the irregular dispersion in winter of such species near the boreal limit of their range at that season, they being numerous where their food abounds and entirely absent in the immediately adjoining districts.*

VIII. The breeding range, as well as the migratory range, differs greatly not only among the species of different families (nearly all the species of some families having a wide range, while nearly all the species of other families have a quite restricted range, as in the *Corvidæ* and *Picidæ*, for instance, as compared with the *Sylvicolidæ*), but also among those of the same family and even of the same genus. The two extremes are well illustrated by the osprey or fish-hawk and

* These remarks are illustrated by the winter distribution of the robin and the cedar-bird in the Alleghanian Fauna, and by the sudden southward incursion of the snow-buntings and other northern sparrows when deep snows suddenly render their food more than usually difficult to procure in their usual winter resorts. The early return of birds to their breeding stations, — their real homes, — as soon as the causes that impelled their winter migration are removed, is further corroborative of the same view. Most of even the insectivorous species visit regions in winter whose average winter temperature differs but little from that of their breeding stations, and when the excessive heats of spring and summer arrive in the southern latitudes, they gradually retire again to their northern breeding stations, keeping pace in their migration with the northward advance of the summer warmth.

the bobolink, the one having an almost cosmopolitan breeding range, while the breeding range of the other is nearly or quite restricted to the Alleghanian Fauna. Several of the *Sylvicolidæ* have a breeding range as restricted as the bobolink, while a few other species of the same family breed throughout nearly the whole of North America. One of the species of *Dendrœca* (*D. æstiva*) has this wide breeding range, while other species of the same genus appear to breed only in the Canadian Fauna.

IX. Species which have a wide breeding range usually present a greater or less number of easily distinguishable local forms, which merge generally the one into the other in the regions lying between the localities at which these several forms are most fully developed. A part of these local forms have received distinctive names, and have of late been quite commonly regarded as distinct species, while many are as yet not so regarded. Every year additional races of this character are discovered, and doubtless many still remain unknown. Much time will probably elapse before naturalists will generally agree as to their true character and relations, though evidence indicative of their being the result of general and uniformly acting laws of geographical variation is apparently by no means wanting. The difference in color, size, form of the bill, length of the tail, etc., that appear almost universally to obtain between the northern and southern representatives of the same species, have already been sufficiently dwelt upon in the preceding pages; but the insertion of a few species in the list of those alleged above (Class IV of the preceding tables) to range across the North American continent calls for an additional word in respect to the differences which have led to the specific separation of the western representatives of these species from their eastern representatives, or to suggestions that they might prove to be specifically distinct. Most of the cases of this kind have been distinguished in the tables under Class IV by the prefix of a [?] before their names. In all these cases almost the sole difference alleged for the separation of the western forms is that of either the darker or brighter or, in other words, the more intense colors of those from the Pacific coast; this character being always the one most strongly urged as distinguishing them, and not unfrequently the only one, especially in those species that breed wholly to the northward of the latitude of San Francisco. The frequency of this difference seems to be a strong reason for regarding it as the result of a general law, and to remove it from the category of genuine specific distinctions.*

X. The number of species which breed in the American Arctic Fauna appears to be not far from sixty. In the Hudsonian Fauna the number is increased to upwards of one hundred and fifty, in the Canadian to probably about one hundred and sixty. In the Alleghanian the number is nearly one hundred and forty; in the Carolinian about one hundred and thirty-five; in the Louisianian about one hundred and thirty. The Hudsonian and Canadian Faunæ hence have a greater number of species, in areas of the same extent, and probably a far greater number of individuals, than the Carolinian and Louisianian Faunæ. In respect to the number of individuals, it is evident that this must result, in consequence of the hordes of wading and swimming birds, of thrushes, sparrows, and *Sylvicolidæ* that pass through the southern and middle districts of the Eastern Province to breed in its boreal portions; few of the species that breed at the southward being there as numerously represented as are scores of species that breed exclusively at the northward. Taking the whole number of species found at particular localities during the year, there is a constant increase in number to the southward, the increase, however, being less rapid from the southern boundary of the Canadian Fauna southward than from that point northward.† There is also a steady

* In addition to the list of examples of this variation already cited in Part III of this paper, the following may be added as marked instances: *Regulus satrapa*, Pacific coast specimens forming the variety *olivaceus* Baird; *Troglodytes hyemalis*, Pacific coast specimens forming the variety *pacificus* Baird; *Cistothorus palustris*, Pacific coast specimens forming the variety *paludicola* Baird; *Helminthophaga celata*, Pacific coast specimens being, according to Professor Baird, "much brighter and clearer yellowish beneath and olivaceous above," than those from the interior of North America; *Myiodioctes* (= *Wilsonia*) *pusilla*; *Hirundo bicolor*; *Collurio excubitoroides* (= *C. ludovicianus*), Pacific coast specimens being darker than those from the Mississippi Valley, and much darker than those from the Plains (*Baird*); *Carpodacus purpureus*, Pacific coast specimens being darker and forming the *C. californicus* of Baird; *Melospiza melodia*, the darker Pacific coast specimens forming the *M. Heermanni* Baird, etc.

† There seems to have been no exhaustive list published of the birds occurring at any locality north of the Alleghanian Fauna. Dr. Richardson's list is the largest, and gives two hundred and thirty-eight as the whole number known in 1831 to inhabit British North America north of the Canadas and east of the Rocky Mountains. Probably this number, and even more, may occur at a single locality on the Saskatchewan; but probably not more than two hundred and sixty or seventy. Three hundred and ten have been detected in Massachusetts, including those of irregular and very rare

increase southwards in the number and even in the proportion of species which are resident at the same locality the whole year. But from the absence of exhaustive lists of the species occurring at numerous localities, differing in latitude, it is difficult to make at present a wholly satisfactory numerical comparison of the different ornithological faunæ.*

occurrence ; three hundred and twenty-seven in the vicinity of New York City (*Lawrence*), and three hundred and forty-three in New Jersey (*Turnbull*). The number given by Ross as observed by him in the "Mackenzie's River District" is one hundred and ninety-two. The greatest number I have seen recorded from any restricted locality within the American tropical Realm is five hundred and forty, the number given from Costa Rica by Messrs. Lawrence and Salvin.

* Dr. Richardson in 1831, found that the number of species "known to rear their young on the banks of the Saskatchewan" amounted to one hundred and forty-one. At least twenty species more may now be safely added. Bonaparte, in 1827, estimated the number of species breeding at Philadelphia to be one hundred and four. Messrs. Coues and Prentiss in their list of the birds of the District of Columbia, published in 1861, mention forty-four species as being permanent residents, and fifty-nine others as summering, making one hundred and three that probably breed in the District, — one less than the number given by Bonaparte as breeding at Philadelphia. Messrs. S. F. and W. M. Baird gave, in 1844, one hundred and nine species as breeding at Carlisle, Pennsylvania. The three latter being inland localities, they may properly be compared with the Saskatchewan district. The numerous lakes at the latter locality, however, afford favorite breeding places for numerous water birds, while few such localities are afforded by the other localities mentioned; but since few water birds breed so far south as these localities, the difference in this respect is a fact of small importance. Dr. Turnbull, in 1869, gave the number of permanently resident species in "East Pennsylvania and New Jersey" as fifty-two, and the number of summer visitors as one hundred and fourteen, making a total of one hundred and sixty-six species that occur there in summer; but the area included in this list is more extended, and embraces a greater variety of surface than in the other cases, and includes several strictly coast species. Farther than this, an examination of his list shows that at least thirty of the one hundred and sixty-six are either wholly of accidental or of very rare occurrence, and hence do not regularly (many of them never) breed at the locality named. The number of resident species in Massachusetts is not far from thirty, of summer visitors one hundred and six, making one hundred and thirty-six that are more or less frequent in summer, — a number considerably less than undoubtedly occur in an equal area on the Saskatchewan. Mr. C. J. Maynard, in his careful analysis of the birds of Eastern Massachusetts (Naturalist's Guide, Part II, pp. 162 – 164, 1870), gives only one hundred and fifteen as being known with certainty to breed in the eastern half of that State, one or two of which cannot be considered as breeding there regularly. While this somewhat exceeds the number generally given as breeding at localities more to the southward, it is far less than the number given by Dr. Richardson as breeding on the Saskatchewan, and much less than the number now well known to be found there in summer. Dr. Coues, in his "Synopsis of the Birds of South Carolina," indicates only about one hundred and thirty-five as being known to breed regularly in that State.

APPENDIX TO PART V.

LIST OF AUTHORITIES.

In the following list are given the titles of general works and special papers that may be profitably consulted in a study of the geographical distribution of the birds of North America. An attempt has been made to cite all the papers of much importance bearing upon this subject that have appeared prior to the beginning of the year 1870, those published in this country having been also brought down to the present date (April, 1871). Some of those published in the transactions of foreign societies during 1869 have been necessarily omitted, as also a large proportion of those published in 1870, since most of these publications are usually several months in reaching this country.

In compiling the present list I have been greatly aided by the "List of Authorities" published by Professor Baird in the Appendix of his Report on the Birds of North America in 1858; Agassiz's "Bibliographia Zoölogiæ"; Carus and Engleman's "Bibliotheca Zoölogica," ending with the year 1860; by Dr. G. Hartlaub's "Bericht über die Leistungen in der Naturgeschichte der Vögel," in Wiegmann's Archiv für Naturgeschichte, and by Professor Alfred Newton's admirable ornithological record in Dr. Günther's "Zoölogical Record." * I am also indebted to Dr. Elliott Coues for the addition of the titles of a considerable number of articles to the proof-sheets, which he has had the kindness to carefully examine. The few titles enclosed in brackets indicate those papers I have not myself consulted.

In aiming at brevity I may have excluded from the list a few papers that might well have been added. Usually papers mentioning less than half a dozen species have been excluded, including announcements of the capture of species at localities beyond their usual range. To cite all such notices would nearly double the length of the list, without materially adding to its value, at least for general purposes. Papers in which new species were described are frequently omitted where the geographical data given in them have been subsequently incorporated in other papers published by the same author.

* The volume for 1869 I regret to find has not yet appeared.

The geographical arrangement of the papers serves to show at a glance what portions of the continent are tolerably well known, so far as regards the birds occurring there, as well as to indicate the considerable areas that are still almost unknown, and the amount of information possessed respecting the regions partially known. In order to indicate to some extent the character of the papers mentioned, the number of species given in each is usually stated, as well as the number of pages the papers embrace.

Occasionally valuable notes on the distribution of our birds, and sometimes nearly complete local lists, are to be found in the various agricultural periodicals, in the transactions or reports of agricultural societies, and in the various State agricultural reports. Although a number of such have been entered in the following list, others may have escaped notice; and information of such omissions, or of the omission of any local list, would be thankfully received by the writer.

GENERAL WORKS.

AUDUBON, J. J. Facts and Observations connected with the permanent Residence of Swallows in the United States. Ann. N. Y. Lyc. Nat. Hist., Vol. I, pp. 166 – 168, 1824.

AUDUBON, J. J. Ornithological Biography. 5 vols. 8vo. Edinburgh, 1831 – 1839.

AUDUBON, J. J. A Synopsis of the Birds of North America. 8vo. Edinburgh, 1839.

AUDUBON, J. J. Birds of America. 7 vols. 8vo. Philadelphia, 1840 – 1844. (506 species.)

BACHMAN, JOHN. On the Migration of the Birds of North America. Am. Journ. Sci. and Arts, 1st Ser., Vol. XXX, pp. 81 – 100, 1836.

BAIRD, S. F. Review of American Birds, in the Museum of the Smithsonian Institution, Part I, North and Middle America. 8vo. Washington, pp. 1 – 176, 1864; pp. 177 – 320, 1865; pp. 321 – 450, 1866.

BAIRD, S. F. The Distribution and Migration of North American Birds. Am. Journ. Sci. and Arts, 2d Ser., Vol. XLI, pp. 78 – 90, 184 – 192, 337 – 347, 1866.

BAIRD, S. F., CASSIN, JOHN, and LAWRENCE, G. N. Birds of North America: a Report on the Birds collected during the Explorations and Surveys for a Railroad Route from the Mississippi River to the Pacific Ocean. Pacific R. R. Rep. of Expl. and Surv., Vol. IX, 4to, 1858. (Republished under the title of "Birds of North America," with an Atlas of Plates. — 738 species, 22 of them extralimital.)

Bannister, H. M. A sketch of the Classification of the American *Anseres*. Proc. Acad. Nat. Sci. Philad., 1870, pp. 130 – 132. (12 North American species.)

Bonaparte, C. L. Additions to the Ornithology of the United States. Journ. Acad. Nat. Sci. Philad., Vol. V, pp. 28 – 31, 1825. (7 species.)

Bonaparte, C. L. Specchio comparativo della ornitologia di Roma e Filadelfia. 8vo. Pisa, 1827. Supplement to the same, 1832.

Bonaparte, C. L. Saggio di una distribuzione metodica degli Animali vertebrati etc. 8vo. Roma, 1831 – 1832; Isis, 1832, pp. 283 – 319; 1833, pp. 1183 – 1230.

Bonaparte, C. L. A Geographical and Comparative List of the Birds of Europe and North America. 8vo, London, 1838. (471 American species.) See a review of the same in Ann. & Mag. Nat. Hist., Vol. I, p. 318, 1839.

Bonaparte, C. L. American Ornithology, or the Natural History of the Birds inhabiting the United States, not given by Wilson. 4 vols. 4to. New York, 1825 – 1833.

Bonaparte, C. L., and Schlegel, H. Monographie des Loxiens. 4to. Leiden et Düsseldorf, 1850. (38 species.)

Brandt, Johann Friedrich. Rapport sur une Monographie de la famille des Alcadées. Bull. Sc. Acad. Imp. St. Petersb., Tome II, pp. 344 – 349, 1837.

Brandt, Johann Friedrich. Ergänzungen und Berichtigungen zur Naturgeschichte der Familie der Alciden. Bull. Sc. Acad. Imp. St. Petersb, Tome VII, pp. 199 – 256, 1869. (21 species.)

Brandt, Johann Friedrich. Bemerkungen über die Classification, die Entwickelungsstufen und die Verbreitung der Alciden. Bull. Sc. Acad. Imp. St. Petersb., Tome VII, pp. 256 – 268, 1869.

Brewer, Thomas M. North American Oölogy, Part I, *Raptores* and *Fissirostres*, June, 1857. Smithsonian Contributions, Vol. XI.

Brewer, Thomas M. A List of Birds common to Europe and America, with others not identical but confounded together from close resemblance. Proc. Bost. Soc. Nat. Hist., Vol. IV, pp. 324 – 328, 1853. (115 species.)

Brewer, Thomas M. A Synopsis of the Birds of North America. Published as an Appendix to Brewer's edition of Jardine's Wilson's Ornithology, pp. 682 – 746, 1840. (491 species).

Bryant, Henry. Remarks on the Variations of Plumage of *Buteo borealis* auct. and *Buteo Harlani* Aud. Proc. Bost. Soc. Nat. Hist., Vol. VIII, pp. 107 – 119, 1861.

Bryant, Henry. Monograph of the Genus *Cataractes* Moehring. Proc. Bost. Soc. Nat. Hist., Vol. VIII, pp. 134 – 144, 1861. (4 species.)

Cabot, S., Jr. On the Distribution of *Tetrao cupido*, and of other Grouse in New England. Proc. Bost. Soc. Nat. Hist., Vol. V, p. 154, 1855.

CASSIN, JOHN. Illustrations of the Birds of California, Texas, Oregon, British and Russian America, etc. 8vo. Philadelphia, 1853–1856. (Contains also Synopses of the *Parinæ*, *Raptores*, and *Fissirostres* of North America).

CASSIN, JOHN. Notes on North American *Falconidæ*, etc. Proc. Acad. Nat. Sci. Philad., 1855, pp. 277–284. (25 species.)

CASSIN, JOHN. Ornithology of the United States and British and Russian America. United States Mag., Vol. III, pp. 18, 109, 205, 481, 1856; Vol. IV, p. 10, 1857. (*Raptores* only.)

CASSIN, JOHN. Studies of the *Icteridæ*. Proc. Acad. Nat. Sci. Philad., 1866, pp. 10–25, 403–417; 1867, pp. 45–74.

CASSIN, JOHN. Notes on the *Picidæ*, etc. Proc. Acad. Nat. Sci. Philad., 1863, pp. 194–204, 322–328.

CASSIN, JOHN. Notes of an Examination of the Birds of the Subfamily *Cœrebinæ*. Proc. Acad. Nat. Sci. Philad., 1864, pp. 265–275.

COPE, E. D. The Birds of Palestine and Panama compared. Am. Nat., Vol. II. pp. 351–359, 1868.

COUES, ELLIOTT. A Monograph of the *Alcidæ*. Proc. Acad. Nat. Sci. Philad., 1868, pp. 2–81. (33 species.)

COUES, ELLIOTT. A Critical Review of the Family *Procellariidæ*. Proc. Acad. Nat. Sci. Philad., 1864, pp. 72–91, 116–144; 1866, pp. 25–33, 134–197. (92 species.)

COUES, ELLIOTT. A Critical Review of the Sub-family *Lestridinæ*. Proc. Acad. Nat. Sci. Philad., 1863, pp. 121–138. (7 species.)

COUES, ELLIOTT. Review of the Gulls of North America, etc. Proc. Acad. Nat. Sci. Philad., 1862, pp. 291–312. (25 species.)

COUES, ELLIOTT. Synopsis of the North American Forms of *Colymbidæ* and *Podicipidæ*. Proc. Acad. Nat. Sci. Philad., 1862, pp. 226–233.

COUES, ELLIOTT. A Revision of the Terns of North America, etc. Proc. Acad. Nat. Sci. Philad., 1862, pp. 535–559. (14 species.)

COUES, ELLIOTT. A Monograph of the *Tringeæ* of North America. Proc. Acad. Nat. Sci. Philad., 1861, pp. 170–205. (12 species.)

COUES, ELLIOTT. On Variation in the Genus *Ægiothus*. Proc. Acad. Nat. Sc. Philad., 1869, pp. 180–189. (Special reference to the geographical races.)

[COUES, ELLIOTT. Key to North American Birds (in press). 1 vol. imp. 8vo. Salem.]

DRESSER, H. E., and SHARPE, R. B. Notes on *Lanius excubitor* and its Allies. Proc. Zoöl. Soc. Lond., 1870, pp. 590–600. (9 species.)

ELLIOT, D. G. Remarks on the Species composing the Genus *Pediœcetes*. Proc. Acad. Nat. Sci. Philad., 1862, pp. 402–404.

ELLIOT, D. G. Monograph of the *Tetraonidæ*. Folio.

ELLIOT, D. G. The New and heretofore Unfigured Birds of North America. 2 vols. folio. New York, 1869.

ELLIOT, D. G.. A Monograph of the Genus *Pelecanus*. Proc. Zoöl. Soc. Lond., 1869, pp. 571 – 591. (9 species.)

[FINSCH, OTTO. Die Papageien, monographische bearbeitet. 2 vols. royal 8vo. Leiden, 1867 – 1868.]

GAMBEL, WM. Contributions to American Ornithology. Proc. Acad. Nat. Sci. Philad., Vol. IV, pp. 126 – 129, 1848. (Remarks on 6 species.)

GOULD, JOHN. Monograph of the *Trochilidæ*, or Humming-Birds. Folio, 1850, *et seq.*

HEERMANN, A. L. Additions to North American Ornithology. Proc. Acad. Nat. Sci. Philad., Vol. VII, pp. 177 – 180, 1854. (5 species.)

HERBERT, WILLIAM HENRY. Frank Forester's Field Sports of the United States and British Provinces of North America. 2 vols. 8vo. New York, 1848. (Contains many valuable observations on the game birds of North America.)

HUXLEY, T. H. On the Classification and Distribution of the *Alectoromorphæ* and *Heteromorphæ*. Proc. Zoöl. Soc. Lond., 1868, pp. 294 – 319. With a map.

LAWRENCE, G. N. Additions to the Ornithology of the United States. Ann. N. Y. Lyc. Nat. Hist., Vol. V, pp. 112 – 124, 1851. (22 species.)

LESSON, R. P. Distribution géographique de quelques Oiseaux marins. Ann. Sci. Nat., Tome VI, pp. 88 – 102, 1825.

LESSON, R. P. Essai sur la dispersion des Oiseaux sur la surface du globe. Ferrus. Bull. Sc. Nat., Tome XXVII, pp. 92 – 100, 1831.

LEWIS, E. J. American Sportsman. 8vo. Philadelphia, 1863. (Contains valuable information respecting the *Rasores*, *Grallæ*, and *Anatidæ*.)

MALHERBE, A. Monographie des Picidées. 2 vols. folio, avec 125 pl. coloriées, Metz, 1859 – 1862.

[MÜLLER, J. W. VON. Reisen in den Vereinigten Staaten, Canada und Mexico. 3 vols. royal 8vo. Leipzic, 1864 – 1865. Vol. III, pp. 551 – 594. (Gives a list of 611 species of North American birds.)]

NUTTALL, THOMAS. A Manual of the Ornithology of the United States and Canada. 2 vols. 12mo. Boston. Vol. I, Land Birds, 1832; 2d ed., with additions, 1840; Vol. II, Water Birds, 1834.

ORD, GEORGE. List of North American Birds, Guthrie's Geog., 2d Am. ed., 2 vols. 8vo. Philad., 1815.

PENNANT, THOMAS. Arctic Zoölogy. 3 vols. 4to, 1784 – 1787; 2d ed., 2 vols. 4to, 1792.

RIDGWAY, ROBERT. Notices of certain obscurely known Species of American Birds [of the Genera *Turdus* (*Hylocichla*), *Pyranga* and *Quiscalus*.] Proc. Acad. Nat. Sci. Philad., 1869, pp. 125 – 135.

RIDGWAY, ROBERT. A New Classification of the North American *Falconidæ*. Proc. Acad. Nat. Sci. Philad., 1870, pp. 138-150.

ROOSEVELT, R. B. The Game Birds of the Northern States of America. 12mo. New York, 1866, pp. 338.

SALVIN, O. On the Genus *Cinclus*. Ibis, 2d Ser., Vol. III, pp. 109-122, 1867. (Contains remarks on the geographical distribution of 13 species.)

SCHMARDA, L. K. Die geographische Verbreitung der Thiere, 3 Bd. Wien, 1853.

SCLATER, P. L. Catalogue of a Collection of American Birds, belonging to Philip Lutley Sclater, M. A., etc., etc. 8vo. London, 1862.

SCLATER, P. L. On the general Geographical Distribution of the Members of the Class Aves. Journ. of Proc. of Linnæan Society of London. Vol. II, Zoölogy, pp. 130-145, 1858.

SCLATER, P. L. Notes on Krueper's Nuthatch and on the other known Species of the Genus *Sitta*. Ibis, 1865.

SCLATER, P. L. Notes on the Genera and Species of *Cypselidæ*. Proc. Zoöl. Soc. Lond., 1865, pp. 593-617.

SCLATER, P. L. Notes upon the American *Caprimulgidæ*. Proc. Zoöl. Soc. Lond., 1866, pp. 123-145.

SCLATER, P. L., and SALVIN, O. Notes on the Species of the Genus *Asturina*. Proc. Zoöl. Soc. Lond., 1869, pp. 129-134. (7 species.)

SCLATER, P. L., and SALVIN, O. Synopsis of the American Rails (*Rallidæ*.) Proc. Zoöl. Soc. Lond., 1868, pp. 442-470. (48 species.)

SCLATER, P. L., and SALVIN, O. Exotic Ornithology. 4to. London, 1866 *et seq.* (Publishing in parts.)

SELYS-LONGCHAMPS, EDM. DE. Sur les Oiseaux américains admis dans la Faune européenne. Mém. de la Soc. R. de Liège, Vol. IV, pp. 35-50, 1849.

[SHARPE, R. B. A Monograph of the *Alcedinidæ*, or Family of Kingfishers. Imp. 8vo. London, 1870. Introduction published separately. (4to, London, 1871.)]

SUNDEVALL, C. J. Conspectus Avium *Picinarum*. 1 vol. 8vo. pp. 116. Stockholm, 1866. (254 species.)

SUNDEVALL, C. J. Conspectus Generis *Dendrœcæ*. Öfvers. Kongl. Vetensk. Acad. Forhandl., 1869, pp. 605-618. (30 species.)

TAYLOR, G. C. Birds observed during two Voyages across the North Atlantic. Ibis, 1869, p. 388.

VERRILL, A. E. On the supposed Eastward Migration of the Cliff Swallow (*Hirundo lunifrons* Say). Proc. Bost. Soc. Nat. Hist., Vol. IX, pp. 276-278, 1863.

VERRILL, A. E. Report of some Investigations upon the Geographical Distribution of North American Birds, made with Reference to the Physical Causes that determine their Limit in Latitude. Proc. Bost. Soc. Nat. Hist., Vol. X, pp. 259-262, 1866.

WALLACE, A. R. Letter concerning the Geographical Distribution of Birds. Ibis, Vol. I, pp. 449-454, 1859.

WIED, MAX PRINZ VON. Verzeichniss der Vögel welche auf seiner Reise in Nord-Amerika beobachtet wurden. Cabanis's Journ. für Ornithol., VI Jahrgang, pp. 1-29, 97-124, 177-204, 257-284, 337-354, 417-444, 1858; VII Jahrgang, pp. 81-96, 161-180, 241-260, 1859.

WILSON, ALEXANDER. American Ornithology, or the Natural History of the Birds of the United States. 9 vols. 4to. Philadelphia, 1808-1814. (283 species.)

GREENLAND.

FABRICIUS, O. Fauna Grœnlandiæ, etc. 8vo. Hafniæ et Lipsiæ, 1780. (54 species.)

HOLBÖLL, CARL. Ornithologiske Bidrag til den grönlandske Fauna. Kröyer's naturhist. Tidskr., Bd. IV, pp. 361-457, 1843; Isis, 1845, pp. 739-792. (88 species.)

REINHARDT, J. List of Birds hitherto observed in Greenland. Ibis, 1st Ser., Vol. III, pp. 1-19, 1861. (118 species.)

[RICHARDSON, J. Birds and Mammalia collected during the last Arctic Land Expedition under Sir John Franklin. Proc. of the Com. of Sci. and Corres. of Zoöl. Soc. Lond., Vol. I, p. 132, 1831.]

SABINE, EDWARD. A Memoir on the Birds of Greenland; with Descriptions and Notes on the Species observed in the late Voyage of Discovery in Davis's Straits and Baffin's Bay. Transact. Linn. Soc. of London, Vol. XII, pp. 527-559, 1819. (54 species.)

BRITISH NORTH AMERICA EAST OF THE ROCKY MOUNTAINS.

Labrador and Anticosti.

COUES, ELLIOTT. Notes on the Ornithology of Labrador. Proc. Acad. Nat. Sci. Philad., 1861, pp. 215-257. (83 species.)

VERREAUX, JULES. Der in Labrador befindlichen Landsäugethiere und Vögel. Bull. Akad. München, 1844, pp. 122-126, 129-133.

VERRILL, A. E. Notes on the Natural History of Anticosti. Proc. Bost. Soc. Nat. Hist., Vol. IX, pp. 137-145. (Contains a list of 62 species of birds.)

WEIS, SAMUEL. List of Vertebrates observed at Okak, Labrador, by Rev. Samuel Weis, with Annotations by Dr. A. S. Packard, Jr. Proc. Bost. Soc. Nat. Hist., Vol. X, pp. 264-277, 1866. (Birds, pp. 267-269; 85 species.)

Interior of British North America.

[ARMSTRONG, A. A personal Narrative of the Discovery of the Northwest Passage, etc. 8vo. London, 1857. (31 species from the Arctic Coast.)]

BARNSTON, GEORGE. Recollections of the Swans and Geese of Hudson's Bay. Can. Nat. and Geol., Vol. VI, pp. 337 – 344, 1861.

[BLAKISTON, THOMAS. [Ornithological] Scraps from the Far West. Zoölogist, 1859, pp. 6318 – 6325, 6373 – 6376.]

BLAKISTON, CAPT. On Birds collected and observed in the Interior of British North America. Part I, Ibis, 1st Ser., Vol. III, pp. 314 – 320, 1861; Part II, Ibid., Vol. IV, pp. 3 – 10, 1862.

BLAKISTON, CAPT. Notes on the Interior of British North America. Ibis, 1st Ser., Vol. V, pp. 39 – 87, 1863. (Devoted principally to remarks on the birds.)

CASSIN, JOHN. Remarks on Birds from the Arctic Regions, presented by Dr. Kane. Proc. Acad. Nat. Sci. Philad., 1852, p. 107.

EDWARDS, GEORGE. Natural History of Birds, etc. 4 vols. 4to. London, 1743 – 1751. (Contains the first descriptions of 38 species of birds from Hudson's Bay.)

GUNN, DONALD. Notes of an Egging Expedition to Shoal Lake, west of Lake Winnipeg. Smithsonian Report, 1867, pp. 427 – 432.

FORSTER, J. R. An Account of the Birds sent from Hudson's Bay, etc. Phil. Trans., Vol. LXII, pp. 382 – 422, 1772. (58 species.)

LEACH, W. E. Zoölogical Appendix to Ross's Voyage to Baffin's Bay. Birds, App. No. 2, pp. xlviii – lx, 1819. (20 species.)

LORD, J. K. List of Birds collected and presented by the British North American Boundary Commission to the Royal Artillery Institution. Proc. Roy. Artill. Instit., 1864, pp. 110 – 126. (87 species.)

LORD, J. K. Catalogue of the Birds, Nests, and Eggs collected by him when Naturalist to the North American Boundary Commission. Proc. Roy. Art. Inst., 1865, pp. 337 – 339.

MURRAY, ANDREW. Contributions to the Natural History of the Hudson's Bay Company's Territory. Part III, Birds. Edinburgh New Phil. Journ., 2d Ser., Vol. IX, pp. 221 – 231, 1857. (82 species.)

RICHARDSON, J., and SWAINSON, W. Fauna Boreali-Americana. Vol. II, Birds. 4to, 1831. (238 species.)

RICHARDSON, JOHN. Geographical Distribution of some American Birds. Ann. Nat. Hist., Vol. XI, pp. 484, 1843. (A list of 12 species collected at Fort Simpson, Mackenzie's River.)

ROSS, BERNARD R. Mammals and Birds of the Arctic Regions. Edinb. Phil. Journ., 2d Ser., Vol. XIII, 1861, pp. 161 – 164. (A nominal list of 94 species.)

ROSS, BERNARD R. List of Species of Mammals and Birds collected in Mackenzie's River District during 1860 – 1861 (June, 1860, to April, 1861). Can. Nat. and Geol., Vol. VI, pp. 441 – 444, 1862. (80 species of birds mentioned.)

ROSS, BERNARD R. List of Mammals, Birds and Eggs, observed in the Mackenzie's River District, with Notices. Nat. Hist. Rev., Vol. II, pp. 276 – 290, 1862. (192 species.)

SABINE, EDWARD. Supplement to the Appendix to Captain Parry's First Voyage. London, 4to, 1824. (73 species.)

SABINE, JOSEPH. Birds of Sir John Franklin's [First] Journey. Franklin's Journey to the Shore, of the Polar Sea. 4to. London, 1823. Append., pp. 669 – 703. (94 species.)

Nova Scotia, New Brunswick, and Newfoundland.

AMBROSE, JOHN. Observations on the Sea-Birds frequenting St. Margaret's Bay, N. S. Proc. Nov. Scot. Inst. Nat. Sci., Vol. II, Part II, pp. 51 – 59. (19 species.)

BREWER, THOMAS M. A few Ornithological Facts gathered in a hasty Trip through Portions of New Brunswick and Nova Scotia, in June, 1850. Journ. Bost. Soc. Nat. Hist., Vol. VI, pp. 297 – 308, 1852. (7 species.)

BRYANT, HENRY. List of Birds observed at Grand Menan and at Yarmouth, N. S., from June 16 to July 8, 1856. Proc. Bost. Soc. Nat. Hist., Vol. VI, pp. 114 – 123, 1857. (55 species.)

BRYANT, HENRY. Remarks on some Birds that breed in the Gulf of St. Lawrence. Proc. Bost. Soc. Nat. Hist., Vol. VIII, pp. 65 – 75, 1860; Can. Nat. and Geol., Vol. VI, 1861, pp. 255 – 267. (14 species.)

DOWNS, A. On the Land Birds of Nova Scotia. Proc. Nov. Scot. Inst. Nat. Sci., Vol. II, pp. 38 – 51, 1865. (61 species.)

JONES, J. M. On some of the Rarer Birds of Nova Scotia. Proc. Nov. Scot. Inst. Nat. Sci., Vol. II, Part II, pp. 70 – 73, 1868.

REEKS, HENRY. Notes on the Birds of Newfoundland. Can. Nat., Vol. V, pp. 38 – 47, 151 – 159, 1860.

WILLIS, J. R. List of the Birds of Nova Scotia, compiled from Notes by Lieut. Blakiston and Lieut. Bland. Smithsonian Report, 1858, pp. 280 – 286. (207 species.)

The Canadas.

BELL, ROBERT. Catalogue of Birds collected around Lakes Superior and Huron in 1860. Can. Nat. and Geol., Vol. VI, pp. 270 – 275, 1861. (77 species.)

CABOT, J. E. Report of the Birds collected and observed at Lake Superior. Agassiz's Lake Superior, pp. 383 – 385, 1850. (69 species.)

COUPER, WM. List of the Birds found in the Vicinity of Quebec, Lower Canada. Samuels's Birds of New England, pp. 80, 81, 368 – 372, 396, 477 – 478, 571 – 573, 1868. (154 species.)

D'URBAN, W. S. M. Notes on the Land Birds observed around Montreal during the Winter of 1856-57. Can. Nat. and Geol., Vol. II, pp. 138-145, 1857. (18 species.)

[Hadfield, Henry. Birds of Canada observed near Kingston during the Spring of 1858. Zoölogist, 1859, pp. 6701-6709, 6744-6752, 6751-6787.]

Hall, Archibald. On the Mammals and Birds of the District of Montreal. Part II, Birds. Canadian Nat. and Geol., Vol. VII, pp. 44-54, 1862. (199 species.)

King, W. Ross. The Sportsman and Naturalist in Canada; or, Notes on the Natural History of the Game, Game Birds, and Fish of that Country. Imp. 8vo, London, 1866.

Le Moine, J. M. Notes on Land and Sea Birds observed around Quebec. Can. Nat. and Geol., Vol. IV, pp. 411-414, 1859.

Le Moine, J. M. Ornithologie du Canada. Oiseaux de Proie et Palmipèdes. 8vo, pp. 96, Quebec, 1860.

McIlwraith, Thomas. List of Birds observed near Hamilton, Canada West. Proc. Essex Institute, Vol. V, pp. 79-96, 1866. (241 species.)

Vennor, H. G. A short Review of the *Sylviadæ*, or Wood-Warblers, found in the Vicinity of Montreal. Can. Nat. and Geol., VI, pp. 349-362, 1861.

Vennor, H. G., Jr. Notes on Birds wintering in and around Montreal. Can. Nat. and Geol., Vol. V, pp. 425-430, 1860. (27 species.)

W., J. F. Notes on Canadian Birds. Can. Nat. and Geol., Vol. V, pp. 230, 231, 1870. (The capture of 9 rare species near Quebec mentioned.)

NEW ENGLAND STATES.

General.

Brewer, T. M. Seaside Ornithology. Am. Nat., Vol. III, pp. 225-235, 1869.

Coues, Elliott. A List of the Birds of New England. Proc. Essex Institute, Vol. V, pp. 249-314, 1868. (335 species.)

Samuels, E. A. Birds of New England and Adjacent States, 5th ed. (of "Ornithology and Oölogy of New England"). Boston, 1870.

Maine.

Boardman, G. A. Catalogue of the Birds found in the Vicinity of Calais, Maine, and about the Islands at the Mouth of the Bay of Fundy. Proc. Bost. Soc. Nat. Hist., Vol. IX, pp. 122-132, 1862. (231 species.)

Cabot, S., Jr. Notes of the Breeding of *Mergus serrator*, *M. cucullatus*, *Anas sponsa*, *A. obscura*, and *Fuligula clangula*, on the Sources of the Androscoggin River, Maine. Proc. Bost. Soc. Nat. Hist., Vol. II, p. 55.

Hamlin, C. E. Catalogue of Birds found in the Vicinity of Waterville, Kennebec County, Maine. Rep. of Sec'y Maine Board of Agric., 1865, pp. 168-173. (135 species.)

HOLMES, EZEKIEL. Birds of Maine. Agr. Maine, 6th Ann. Rep., 1861. pp. 113–122. (A nominal list of 193 species.) Addenda to the same. 2d Ann. Rep. Nat. Hist. and Geol. of Maine, 1862, p. 118. (13 species.)

VERRILL, A. E. Catalogue of Birds found at Norway, Oxford County, Maine. Proc. Essex Institute, Vol. III, pp. 136–158, 1863. (159 species.)

VERRILL, A. E. Additions to the Catalogue of Birds found in the Vicinity of Calais, Maine, and about the Bay of Fundy. Proc. Bost. Soc. Nat. Hist., Vol. IX, pp. 233, 234, 1863. (12 species.)

Vermont.

THOMPSON, ZADOC. Birds of Vermont. History of Vermont, Natural, Civil, and Statistical, 8vo. Burlington, 1842. Birds, pp. 56–112.

Massachusetts.

ALLEN, J. A. Catalogue of the Birds found at Springfield, Massachusetts, with a List of those found in the State not yet observed at Springfield, etc. Proc. Essex Institute, Vol. IV, pp. 48–98, 1864. (296 species.)

ALLEN, J. A. Winter Notes of an Ornithologist. Amer. Nat., Vol. I, pp. 38–48, 1867. (On the Winter Birds of Massachusetts.)

ALLEN, J. A. The Birds of Spring. Amer. Nat., Vol. I, pp. 141–144, 1867.

ALLEN, J. A. Notes on some of the Rarer Birds of Massachusetts. Amer. Nat., Vol. III, pp. 505–519, 568–585, 631–648, 1869–70. (92 species.)

ANONYMOUS. Notes on the Ducks found on the Coast of Massachusetts in Winter. Amer. Nat., Vol., IV, p. 49. (8 species.)

BREWER, THOMAS M. Some Additions to the Catalogue of the Birds of Massachusetts in Prof. Hitchcock's Report, etc. Jour. Bost. Soc. Nat. Hist., Vol. I, pp. 435–439, 1837. (45 species.)

BREWER, T. M. Observations on the Appearance of the Cliff Swallow (*Hirundo lunifrons*), giving Data of its Appearance in New England. Proc. Bost. Soc. Nat. Hist., Vol. IV, p. 270, 1852.

BRYANT, H. Notice of *Mareca Penelope* and *Querquedula crecca* taken in Eastern Massachusetts. Proc. Bost. Soc. Nat. Hist., Vol. V, p. 195, 1855.

EMMONS, E. Catalogue of the Birds of Massachusetts. Hitchcock's Report on the Geology of Mass., Amherst, 1833. (165 species.)

MAYNARD, C. J. Catalogue of the Birds of Eastern Massachusetts. Naturalist's Guide, Part II, Boston, 1870. (299 species.)

NUTTALL, THOMAS. Remarks and Inquiries concerning the Birds of Massachusetts. Mem. Amer. Acad. Arts and Sciences, 2d Ser., Vol. I, pp. 91–106, 1833. (45 species remarked upon.)

PEABODY, W. B. O. A Report on the Ornithology of Massachusetts. Rep on the Fishes, Reptiles, and Birds of Mass., 8vo, 1839, pp. 256–404;

Ibid., Jour. Bost. Soc. Nat. Hist., Vol. III, pp. 65 – 246, 1840. (284 species).

PUTNAM, FRED. W. Catalogue of the Birds of Essex County, Mass. Proc. Essex Institute, Vol. I, pp. 201 – 231, 1856. (245 species.)

SAMUELS, E. A. Descriptive Catalogue of the Birds of Massachusetts. Agr. of Mass., Sec'y's Rep., 1863, App. pp. xvii – xxix, 1864. (267 species.)

Connecticut.

LINSLEY, J. H. A Catalogue of the Birds of Connecticut, etc. Am. Jour. of Sci. and Arts, 1st Ser., Vol. XLIV, pp. 249 – 273, 1843. (302 species.)

WOOD, WM. Rapacious Birds of Connecticut. *Hartford Times*, March 14 to August 8, 1861.

MIDDLE ATLANTIC STATES.

New York.

DE KAY, JAMES E. Zoölogy of New York, or the New York Fauna. 4to. Albany, 1844. Part II, Birds. (317 species.)

GIRAUD, J. P. The Birds of Long Island, 8vo. New York, 1843 – 1844. (276 species.)

HOPKINS, WM. List of Sea-Birds obtained at Auburn, New York. Proc. Bost. Soc. Nat. Hist., Vol. V, p. 13, 1854. (8 species.)

LAWRENCE, GEO. N. Catalogue of Birds observed on New York, Long and Staten Islands. Ann. New York Lyc. Nat. Hist., Vol. VIII, pp. 279 – 300, 1867. (327 species.)

New Jersey.

ABBOTT, C. C. Catalogue of the Birds of New Jersey. Geol. of New Jersey, Appendix, pp. 761 – 798, 1868. (301 species.)

ABBOTT, C. C. Notes on certain Inland Birds of New Jersey. Amer. Nat., Vol. IV, pp. 536 – 550, 1870. (26 species.)

TURNBULL, WM. P. The Birds of East Pennsylvania and New Jersey. 8vo pamphlet. Philadelphia, 1869. (343 species.)

Pennsylvania.

BAIRD, WM. M. and S. F. List of the Birds found in the Vicinity of Carlisle, Cumberland County, Penn. Am. Jour. Sci. and Arts, 1st Ser., Vol. XLVI, pp. 261 – 273, 1844. (201 species.)

BAIRD, S. F. Catalogue of the Birds found in the Neighborhood of Carlisle, Cumberland County, Penn. Literary Rec. and Journ. Linn. Ass. Pennsyl. College, Vol. I, p. 249, 1845.

BARNARD, VINCENT. A Catalogue of the Birds of Chester County, Pennsylvania, etc. Smiths. Rep., 1860, pp. 434 – 438. (191 species.)

BARTON, BENJ. S. Fragments of the Natural History of Pennsylvania.

Folio. Philadelphia, 1799. (Contains a list of the birds of Pennsylvania, with dates of migration.)

HALDEMANN, S. S. Facts in Ornithology. Proc. Acad. Nat. Sci. Philad., Vol. I, p. 54, 1841. (*Falco peregrinus* stated to breed "on the cliffs along the Susquehanna.")

TAYLOR, R. C. On the Geology and Natural History of the Northeastern Extremity of the Alleghany Mountain Range in Pennsylvania, U. S. Loudon's Mag. Nat. Hist., Vol. VIII, pp. 539–540. (Contains a nominal list of 56 species of birds.)

TAYLOR, R. C. A Supplement to the List of Species of Birds occurring at the Northeastern Extremity of the Alleghany Mountains. Loudon's Mag. Nat. Hist., Vol. IX, pp. 72–74, 1836. (36 species.)

See also TURNBULL, under *New Jersey*.

SOUTH ATLANTIC STATES.

General.

BARTRAM, JOHN. Travels through North and South Carolina, Georgia, East and West Florida, etc., 8vo. Philadelphia, 1791. Contains a list of the birds of the United States, pp. 289–296. (216 species.)

CATESBY, MARK. The Natural History of Carolina, Florida, and the Bahama Islands. 2 vols. folio, 3d ed. London, 1771.

Virginia and the District of Columbia.

[CLAYTON, JOHN. On the Birds of Virginia. Phil. Trans., Vol. XVII, p. 988, 1693.]

COUES, ELLIOTT, and PRENTISS, D. WEBSTER. List of the Birds ascertained to inhabit the District of Columbia, with the Times of Arrival and Departure of such as are Non-residents, etc. Smithsonian Report, 1861, pp. 399–421. (226 species.)

North Carolina.

COPE, E. D. Observations on the Fauna of the Southern Alleghanies. Amer. Nat., Vol. IV, pp. 392–401, 1870. (Madison County, North Carolina, and Giles County, East Virginia.)

[COUES, ELLIOTT. Notes on the Natural History of Fort Macon, N. C. Birds, Part I. Proc. Acad. Nat. Sci. Philad., 1871 (in press).]

South Carolina.

BACHMAN, JOHN. On the Migration of the Birds of North America. Am. Journ. Sci. and Arts, 1st Ser., Vol. XXX, pp. 81–100, 1836. (Contains numerous observations on the birds of South Carolina.)

BURNETT, W. I. Notes on the Fauna of the Pine Barrens of Upper South Carolina. Proc. Bost. Soc. Nat. Hist., Vol. IV, pp. 115–118, 1851. (69 species of birds mentioned.)

COUES, ELLIOTT. Synopsis of the Birds of South Carolina. Proc. Bost. Soc. Nat. Hist., Vol. XII, pp. 104 – 127, 1868. (294 species.)

GIBBS, LEWIS R. Catalogue of the Birds of South Carolina. Tuomey's Geology of South Carolina, Appendix, pp. iii – viii, 1848. (271 species.)

Georgia.

GERHARDT, ALEX. Über die Lebensweise der Vögel Nord-Amerikas (Georgia). Naumannia, 1855, pp. 380 – 397, 458 – 469; 1856, pp. 1 – 18.

GERHARDT, ALEX. Verzeichniss der Vögel des Staats Georgia, nach White. Naumannia, 1855, pp. 382 – 383. (A nominal list of 270 species.)

Florida.

ALLEN, J. A. List of the Winter Birds of East Florida, with Annotations. Bulletin Mus. Comp. Zoöl., Vol. II, pp. 250 – 368, 1871. (183 species.)

BRYANT, H. Observations on some of the Birds of East Florida. Proc. Bost. Soc. Nat. Hist., Vol. VII, pp. 5 – 21, 1859. (26 species.)

TAYLOR, GEORGE CAVENDISH. Five Weeks in the Peninsula of Florida during the Spring of 1861, with Notes on the Birds observed there. Ibis, 1st Ser., Vol. IV, pp. 127, 197, 1862.

GULF STATES.

Mississippi.

WAILES, B. L. C. Birds of Mississippi. Rep. on Agr. and Geol. of Mississippi, pp. 317 – 327, 1854. (91 species; mainly a nominal list.)

CENTRAL STATES

Ohio.

KIRKPATRICK, JOHN. The Rapacious Birds of Ohio. *Ohio Farmer* for 1858 – 59; Ohio Agricult. Rep., 1858, pp. 340 – 383, 1859.

KIRTLAND, J. P. Fragments of Natural History. II, Ornithology. Am. Journ. Sci. and Arts, 1st Ser., Vol. XL, pp. 19 – 24, 1841. (Contains notes on 16 species observed near Cleveland.)

KIRTLAND, J. P. Catalogue of the Birds of Ohio. Second Ann. Report of the Geol. Survey of Ohio, pp. 177 – 187, 1838. (223 species.)

READ, M. C. Catalogue of the Birds of Northern Ohio. Proc. Acad. Nat. Sci. Philad., 1853, pp. 395 – 402. (145 species.)

WHEATON, J. M. Catalogue of the Birds of Ohio. Ohio Agri. Rep., 1860, pp. 359 – 380, 480. (285 species.)

Indiana.

ALLEN, J. A. List of Birds observed in Richmond, Wayne Co., Indiana, in June, 1867. Mem. Bost. Soc. Nat. Hist., Vol. I, pp. 522 – 526, 1868. (72 species.)

HAYMOND, RUFUS. Birds of Southeastern Indiana. Proc. Acad. Nat. Sci. Philad., 1856, pp. 286 – 299. (147 species.)

HAYMOND, RUFUS. Birds of Franklin County, Indiana. Cox's Geol. Survey of Ind., 1869, pp. 209 – 235. (164 species.)

Illinois.

ALLEN, J. A. Notes on Birds observed in Northern Illinois, in June, 1867. Mem. Bost. Soc. Nat. Hist., Vol. I, pp. 502 – 522, 1868. (94 species.)

BRENDEL, F. Vögel der Umgegend Peoria's in Illinois. Giebel's Zeitsch. für Naturk., 1857, p. 420.

KENNICOTT, ROBERT. Catalogue of the Animals of Cook County, Illinois. [Birds.] Transact. Ill. State Agric. Soc., I, 1853 – 54, pp. 580 – 591. (187 species.)

KENNICOTT, R. Notes of the Occurrence of *Plotus anhinga* and *Tantalus loculator* in Southern Illinois. Proc. Bost. Soc. Nat. Hist., Vol. V, p. 391, 1856.

PRATTEN, HENRY. Catalogue of the Birds of Southern Illinois (Wayne and Edwards Counties). Transact. Ill. State Agric. Soc., I, pp. 596 – 609, 1853 – 54. (184 species.)

Michigan.

HUGHES, D. DARWIN. The Game Birds of Michigan. *Detroit Free Press*, 1870.

KNEELAND, S., JR. On the Birds of Keweenaw Point, Lake Superior. Proc. Bost. Soc. Nat. Hist., Vol. VI, pp. 231 – 240, 1858. (147 species.)

Minnesota.

HEAD, J. F. Some Remarks on the Natural History of the Country about Fort Ripley, Minnesota. Smithsonian Report, 1854, pp. 291 – 293. (60 species of birds mentioned.)

[TRIPPE, T. MARTIN. Notes on the Birds of Minnesota (in press). Proc. Essex Institute, Vol. VI. (138 species.)]

Wisconsin.

BARRY, A. CONSTANTINE. Ornithological Fauna of Wisconsin. Proc. Bost. Soc. Nat. Hist., Vol. V, pp. 1 – 13, 1854. (193 species.)

HOY, P. R. Notes on the Ornithology of Wisconsin. Proc. Acad. Nat. Sci. Philad., 1853, pp. 304 – 313, 381 – 385, 425 – 429. Ibid., Transact. Wisconsin Agric. Soc., 1852, pp. 341 – 364. (283 species.)

Iowa.

ALLEN, J. A. Notes on Birds observed in Western Iowa, July 1 to Sept. 20, 1867. Mem. Bost. Soc. Nat. Hist., I, pp. 491 – 502, 1868. (108 species.)

ALLEN, J. A. Catalogue of the Birds of Iowa. White's Geol. of Iowa, Vol. II, Appendix, pp. 419 – 427, 1870. (283 species.)

Missouri.

Hoy, P. R. Journal of an Exploration of Western Missouri in 1854, etc. Smithsonian Rep., 1864, pp. 431–438. (Includes a nominal list of the birds observed, — 153 species.)

Texas.

Butcher, H. B. List of Birds collected at Laredo, Texas, in 1866 and 1867. Proc. Acad. Nat. Sci. Philad., 1868, pp. 148–150. (89 species.)

Dresser, H. E. Notes on the Birds of Southern Texas. Ibis, 2d Ser., Vol. I, pp. 312–330, 466–495; Vol. II, pp. 23–46, 1866.

McCall, Geo. A. Remarks on the Habits, etc., of Birds met with in Western Texas, between San Antonio and the Rio Grande, and in New Mexico, with Descriptions of several Species believed to have been hitherto undescribed. Proc. Acad. Nat. Sci. Philad., 1850–51, pp. 213–224. (68 species.)

McCall, Geo. A. [Ornithological] Facts and Observations from Notes when in Texas. Ann. N. Y. Lyc. Nat. Hist., Vol. VI, pp. 9–14. (18 species.)

Roemer, Ferdinand. Texas: mit besonderer Rücksicht auf deutsche Auswanderung und die physichen Verhältnisse des Landes, nach eigener Beobachtung geschildert. 8vo. Bonn, 1849. (Contains a list of birds, — 40 species.)

ROCKY MOUNTAINS AND ADJACENT PLAINS.

General.

Baird, S. F. [Nominal] List of Birds inhabiting America, west of the Mississippi, not described in Audubon's Ornithology. Stansbury's Expl. and Survey of the Valley of the Great Salt Lake, pp. 327–335, 1852. (153 species.)

Baird, S. F. Descriptions of New Birds, collected between Abuquerque, New Mexico, and San Francisco, during the Winter of 1853–54, by Dr. C. B. R. Kennerly and H. B. Möllhausen, etc. Proc. Acad. Nat. Sci. Philad., Vol. VII, p. 118–120, 1854. (8 species.)

Baird, S. F. Report on the Birds collected by the United States and Mexican Boundary Survey, with Notes by J. H. Clark, D. N. Couch, and C. B. R. Kennerly. U. S. and Mex. Bound. Sur., Vol. II, Part II, No. 2, pp. 1–32, 1859. (236 species.)

Baird, S. F. Report on the Birds collected on the Route near the 38th and 39th Parallels, explored by Capt. J. W. Gunnison, and near the 41st Parallel, explored by Lieut. E. G. Beckwith. Pacific R. R. Rep. of Expl. and Surv., Vol. X, Part IV, No. 2, pp. 11–16, 1859. (25 species.)

COOPER, J. G., and SUCKLEY, G. Report on the Birds collected on the Route near the 47th and 49th Parallels, explored by I. I. Stevens, Governor of Washington Territory. Pacific R. R. Rep. of Expl. and Surveys, Vol. XII, Part II, No. 3, pp. 140–291, 1859. (Republished, "with additions and corrections," in 1860, forming the work entitled "Natural History of Washington Territory.") (212 species.)

COOPER, J. G., and SUCKLEY, G. List of Birds heretofore reported as found in the Northwestern Part of America, but of which no Specimens have been reported by recent Explorers. Pacific R. R. Rep. of Expl. and Surv., Vol. XII, Part II, Zoölogy, pp. 288–290, 1859. (120 species.)

COUES, ELLIOTT. Ornithology of a Prairie Journey, and Notes on the Birds of Arizona. Ibis, 1st Ser., Vol. V, April, 1865.

HAYDEN, F. V. List of the Birds of the Upper Missouri. Geol. and Nat. Hist. of the Upper Missouri, pp. 157–176, 4to. Philadelphia, 1862. (191 species.)

HEERMANN, A. L. Report on the Birds collected on the Route near the 32d Parallel, from the Rio Grande to the Pimas Villages, explored by Lieut. J. G. Parke. Pacific R. R. Rep. of Expl. and Surveys, Vol. X, Part —, No. 1, pp. 9–20, 1859. (25 species.)

KENNERLY, C. B. R. Report upon the Birds of the Route near the 35th Parallel, explored by Lieut. A. W. Whipple. Pacific R. R. Rep. of Expl. and Surveys, Vol. X, Part VI, No. 3, pp. 19–35, 1859. (88 species.)

NEWBERRY, J. S. Report upon the Zoölogy of Lieut. R. S. Williamson's Explorations in Oregon and California. Pacific R. R. Expl. and Surveys, Vol. VI, Part IV, No. 2, chap. ii, Birds, pp. 73–110, 1857. (175 species.)

SAY, THOMAS. The Ornithological Articles in Long's Expedition (from Pittsburg to the Rocky Mountains). 2 vols. 8vo. Philadelphia, 1823.

TOWNSEND, JOHN K. List of the Birds inhabiting the Regions of th Rocky Mountains, the Territory of Oregon, and the Northwest Coast of North America. Journ. Acad. Nat. Sci. Philad., 1st Ser., Vol. VIII, pp. 151–158, 1839. (231 species.)

TOWNSEND, JOHN K. Description of twelve new Species of Birds, chiefly from the Vicinity of the Columbia River, with an Appendix containing a [nominal] List of the Birds of the Columbia River. Journ. Acad. Nat. Sci. Philad., 1st Ser., Vol. VII, pp. 187–193, 1837. (72 species.)

WOODHOUSE, S. W. Report on the Birds of Capt. L. Sitgreaves's Expedition down the Zuñi and Colorado Rivers. Sitgreaves's Report, pp. 58–105, 1853. (247 species.)

Montana.

COOPER, J. G. The Fauna of Montana Territory. Part II, Birds. Amer. Nat., Vol. II, pp. 596–600; Vol. III, pp. 31–35, 73–84, 1869. (107 species.)

Utah.

BAIRD, S. F. Exploration and Survey of the Valley of the Great Salt Lake, etc., by H. Stansbury, T. E., U. S. A. 8vo. Philad., 1852, Appendix C, Birds, pp. 314–324. (31 species.)

BAIRD, S. F. Birds found at Fort Bridger. Gen. Rep. on Birds of N. Amer., Append., pp. 926, 927; P. R. R. Rep. of Expl. and Surv., Vol. IX, 1858. (104 species.)

Arizona.

COUES, ELLIOTT. List of Birds collected in Southern Arizona by Dr. E. Palmer, with Remarks. Proc. Acad. Nat. Sci. Philad., 1868, pp. 81–85 (53 species.)

COUES, ELLIOTT. List of the Birds of Fort Whipple, Arizona, with which are incorporated all other Species ascertained to inhabit the Territory, etc. Proc. Acad. Nat. Sci. Philad., 1866, pp. 39–100. (245 species.) — Republished under the title, "Prodrome of a Work on the Ornithology of Arizona Territory." *

New Mexico

BAIRD, S. F. On Birds collected in New Mexico by Lieut. Abert. Stansbury's Exped., pp. 325–327. (9 species.)

HENRY, T. CHARLTON. Notes derived from Observations made on the Birds of New Mexico, during the Years 1853 and 1854. Proc. Acad Nat. Sci. Philad., 1855, pp. 306–317. (170 species.)

HENRY, T. CHARLTON. Catalogue of the Birds of New Mexico, as compiled from Notes and Observations made while in that Territory, during a Residence of six Years. Proc. Acad. Nat. Sci. Philad., 1859, pp. 104–109. (198 species.)

PACIFIC COAST.

General.

CASSIN, JOHN. Catalogue of Birds collected by the North Pacific Surveying and Exploring Expedition, in Command of Capt. Rogers, U. S. N., etc. Proc. Acad. Nat. Sci. Philad., 1862, pp. 312–328. (Contains many species from the West Coast of North America.)

PEALE, TITIAN R. Report on the Mammalia and Birds of the U. S. Exploring Expedition under Capt. Wilkes. 4to. Philadelphia, 1848. (Includes many species from the West Coast of North America.)

VIGORS, N. A. Ornithology of Captain Beechey's Voyage. Zoöl. of Beechey's Voyage, pp. 13–40, 1839.

* I regret to learn from Dr. Coues that this work will not be published.

Alaska.

Baird, S. F. On Additions to the Bird-Fauna of North America, made by the Scientific Corps of the Russo-American Telegraph Expedition. Transact. Chicago Acad. Sci., Vol. I, pp. 311–337, 1870. (16 species.)

Dall, W. H., and Bannister, H. M. List of the Birds of Alaska, with Biographical Notes. Transact. Chicago Acad. Science, Vol. I, pp. 267–310, 1870. (212 species.)

Dall, W. H. Alaska and its Resources. 8vo. Boston, 1870, pp. 580–586. (Contains a nominal list of 212 species.)

Vancouver's Island.

Brown, Robert. Synopsis of the Birds of Vancouver's Island. Ibis, 1868, pp. 414–428. (153 species.)

[Lord, John K. The Naturalist in Vancouver's Island and British Columbia. 2 vols. 8vo. London, 1866. Vol. II, pp. 291–301, contains a list of 226 species of birds collected by Mr. Lord.]

Sclater, P. L. On a Collection of Birds from Vancouver's Island. Proc. Zoöl. Soc. Lond., 1859, pp. 235–237. (35 species.)

California.

Brewer, T. M. List and Descriptions of Eggs obtained in California by Mr. E. Samuels. Proc. Bost. Soc. Nat. Hist., Vol. VI, pp. 145–149, 1857. (16 species.)

Bridges, Thomas. Notes on California Birds, with Remarks by P. L. Sclater. Proc. Zoöl. Soc. Lond., 1858, pp. 1–31, 1858. (11 species.)

Cooper, J. G. The Fauna of California and its Geographical Distribution. Proc. Cal. Acad. Sci., Vol. IV, pp. 61–81, 1869.

Cooper, J. G. Species of Animals new to the State of California. Proc. Cal. Acad. Sci., Vol. II, pp. 118–123, 1861. (18 species of birds.)

Cooper, J. G. Some recent Additions to the Fauna of California. Proc. Cal. Acad. Sci., Vol. IV, pp. 3–13, 1868. (33 species of birds added.)

Cooper, J. G. The Naturalist in California. Amer. Nat., Vol. III, pp. 182–189, 470–481, 1869.

Cooper, J. G. Monterey in the dry Season. Amer. Nat., Vol. IV, pp. 756–758, 1871. (About 30 species of birds mentioned as summer residents.)

Cooper, J. G., and Baird, S. F. Ornithology of California. Vol. I, [Land Birds,] royal 8vo, 1870.

Coues, Elliott. From Arizona to the Pacific. Ibis, 2d Ser., Vol. I. pp. 260–275, 1866.

Feilner, John. [Ornithological] Explorations in Upper California in 1860. Smith. Rep., 1864, pp. 421–430.

GAMBEL, WM. Remarks on the Birds observed in Upper California, with Descriptions of New Species. Proc. Acad. Nat. Sci. Philad., 1846-47, pp. 44-48, 110-115, 154-158, 200-205. (84 species.) Ibid., Journ. Acad. Nat. Sci. Philad., 2d Ser., Vol. I, pp. 25-56, 215-229, 1847-1849. (176 species.)

HEERMANN, A. L. Notes on the Birds of California observed during a Residence of three Years in that Country. Jour. Acad. Nat. Sci. Phil., 2d Ser., Vol. II, pp. 259-272, 1853. (103 species.)

HEERMANN, A. L. Report on the Birds collected during Explorations in California for Railroad Routes to connect with the Routes near the 35th and 32d Parallels of North Latitude, made by Lieut. R. G. Williamson. Pacific R. R. Rep. of Expl. and Surveys, Vol. X, Part IV, No. 2, pp. 29-80. (221 species.)

LICHTENSTEIN, H. Beiträge zur ornithologischen Fauna von Californien, etc. Abhandl. der Berlin Akad., Phys. Kl., 1838, pp. 417-451.

SCLATER, P. L. List of Birds collected by Mr. Thomas Bridges, in the Valley of San José, California. Proc. Zoöl. Soc. Lond., 1857, pp. 125-127. (33 species.)

VIGORS, N. A. On a Collection of Skins of Birds from California. Proc. Zoöl. Soc. Lond., Vol. I, p. 65, 1833. (Only a paragraph of ten lines.)

XANTUS, JOHN. List of Birds collected at Fort Tejon, California. Proc. Acad. Nat. Sci. Philad., 1859, pp. 189-193. (144 species.)

MEXICO.

BAIRD, S. F. Notes on a Collection of Birds made by John Xantus, Esq., at Cape St. Lucas, Lower California. Proc. Acad. Nat. Sci. Philad., 1859, pp. 299-306. (42 species.)

BONAPARTE, C. L. Catalogue d'Oiseaux du Méxique et du Pérou. Rev. Zool., 1840, p. 19.

CABOT, SAMUEL. Descriptions and Habits of some of the Birds of Yucatan. Journ. Bost. Soc. Nat. Hist., Vol. IV, pp. 460-467, 1843. (4 species.)

CASSIN, JOHN. Catalogue of Birds collected by Mr. Wm. S. [H.] Pease during the March of the Army of the United States from Vera Cruz to the City of Mexico. Proc. Acad. Nat. Sci. Philad., 1848-49, pp. 87-91. (45 species.)

LAWRENCE, G. N. List of a Collection of Birds from Northern Yucatan. Ann. N. Y. Lyc. Nat. Hist., Vol. IX, pp. 198-210, 1869. (103 species.)

MONTES-DE-OCA, RAFAEL. The Mexican Humming-Birds. Proc. Acad. Nat. Sci. Philad., 1860, pp. 47, 48, 80, 81, 551-553. (4 species.)

[PARZUDAKI, E. Catalogue des Oiseaux du Méxique qui composent les collections de M. Sallé. (432 species.)]

SCLATER, P. L. On a Collection of Birds made by Mr. H. S. le Strange, near the City of Mexico. Proc. Zoöl. Soc. Lond., 1869, pp. 361 – 364. (Of the 262 species said to have been contained in the collection only 15 are particularly mentioned.)

SCLATER, P. L. List of a Collection of Birds procured by Mr. G. H. White in the Vicinity of the City of Mexico. Proc. Zoöl. Soc. Lond., 1864, pp. 172 – 179. (156 species.)

SCLATER, P. L. Notes on some Birds from Southern Mexico. Proc. Zoöl. Soc. Lond., 1858, pp. 95 – 99. (14 species.)

SCLATER, P. L. On a Collection of Birds received by Mr. Sallé from Southern Mexico. Proc. Zoöl. Soc. Lond., 1857, pp. 226 – 230. (29 species.)

SCLATER, P. L. Catalogue of the Birds collected by M. Aug. Sallé in Southern Mexico, etc. Proc. Zoöl. Soc. Lond., 1856, pp. 283 – 311. (233 species.)

SCLATER, P. L. On a Collection of Birds made by Sign. Matteo Botteri in the Vicinity of Orizaba, Southern Mexico. Proc. Zoöl. Soc. Lond., 1857, pp. 210 – 215. (38 species.)

SCLATER, P. L. Notes on a Collection of Birds from the Vicinity of Orizaba and neighboring Parts of Southern Mexico. Proc. Zoöl. Soc. Lond., 1860, pp. 250 – 254. (44 species.)

SCLATER, P. L. List of Birds collected by Mr. A. Boucard in the State of Oaxaca in Southern Mexico. Proc. Zoöl. Soc. Lond., 1859, pp. 369 – 393. (236 species.)

SCLATER, P. L. List of Additional Species of Mexican Birds obtained by M. Sallé, from the Environs of Jalapa and San Andres Tuxtla. Proc. Zoöl. Soc. Lond., pp. 201 – 207, 1857. (62 species.)

SCLATER, P. L. On a Series of Birds collected in the Vicinity of Jalapa, Mexico. Proc. Zoöl. Soc. Lond., 1859, pp. 362 – 369. (226 species.)

SCLATER, P. L. On some Birds to be added to the Avifauna of Mexico. Proc. Zoöl. Soc. Lond., 1862, pp. 368, 369. (8 species.)

SUMICHRAST, F. The Geographical Distribution of the Native Birds of the Department of Vera Cruz, with a List of the Migratory Species, translated from the French, by Dr. T. M. Brewer. Mem. Bost Soc. Nat. Hist., Vol. I, pp. 542 – 563, 1869. (225 species.)

SWAINSON, WM. A Synopsis of the Birds discovered in Mexico by W. Bullock, Sen. and Jun. Phil. Mag., 2d Ser., Vol. I, pp. 364 – 369, 433 – 442, 1827. (101 species.)

CENTRAL AMERICA.

General.

BONAPARTE, C L. Notes sur les collections rapportées en 1853, par M. A. Delattre de son voyage en California et dans le Nicaragua. Comptes

Rendus, Tome XXXVIII, pp. 827–835, 913–927, 1853; Tome XXXVIII, pp. 1–11, 53–66; 258–266, 365–389, 533–542, 650–665, 1854.

COUES, ELLIOTT. Notes on Certain Central American *Laridæ*. Ibis, 1864.

MOORE, T. J. List of Mammals and Birds collected by Mr. Jos. Leland in Honduras, Belize, and Guatemala. Proc. Zoöl. Soc. Lond., 1859, pp. 50–65. (129 species.)

LAWRENCE, GEO. N. Descriptions of New Species of Birds of the Families *Zanagradæ*, etc., etc. Ann. Lyc. Nat. Hist. N. Y., Vol. VIII, pp. 126–135. (12 species.)

LAWRENCE, GEO. N. Descriptions of Six New Species of Birds from Central America. Ann. Lyc. Nat. Hist. N. Y., Vol. VIII, pp. 170–173.

SALVIN, OSBERT. A List of Species to be added to the Ornithology of Central America. Ibis, 1861.

SALVIN, O., and SCLATER, P. L. Descriptions of New Species of Birds of the Families *Dendrocolaptidæ*, *Strigidæ*, and *Columbidæ*. Proc. Zoöl. Soc. Lond., 1868, pp. 53–60.

SALVIN, OSBERT. On the *Psittacidæ* of Central America. Ibis, 1871, pp. 86–100. (27 species.)

SCLATER, P. L., and SALVIN, O. On the Ornithology of Central America. Ibis, I, pp. 1–22, 117–138, 213–234, 1859. (382 species.)

SCLATER, P. L., and SALVIN, O. Descriptions of New and Little-known American Birds, etc. Proc. Zoöl. Soc. Lond., 1868, pp. 322–329.

SCLATER, P. L., and SALVIN, O. Descriptions of Six New Species of American Birds, etc. Proc. Zoöl. Soc. Lond., 1869, pp. 416–420.

SUNDEVALL, C. J. Conspectus genris *Certhiolæ*. Öfvers. Kongl. Vetensk. Acad. Forhand., 1869, pp. 619–625. (20 species.)

Guatemala.

SALVIN, OSBERT, and SCLATER, P. L. Contributions to the Ornithology of Guatemala. Ibis, 1860, pp. 28–45, 272–278. (119 species.)

SALVIN, OSBERT. Notes on the Humming-Birds of Guatemala. Ibis, 1860, pp. 259–272. (23 species.)

SALVIN, OSBERT The Sea-Birds and Waders of the Pacific Coast of Guatemala. Ibis, 1865, pp. 187–199.

SALVIN, OSBERT. A further Contribution to the Ornithology of Guatemala. Ibis, 1866, pp. 188–206. (81 species.)

SALVIN, OSBERT. On some Collections of Birds from Veragua. Proc. Zoöl. Soc. Lond., 1867, pp. 129–161. (220 species.)

SALVIN, OSBERT. Descriptions of Eight New Species of Birds from Veragua. Proc. Zoöl. Soc. Lond., 1866, pp. 67–76.

SCLATER, P. L., and SALVIN, O. Descriptions of Four New Species of Birds from Veragua. Proc Zoöl. Soc. Lond., 1868, pp 387–390.

SCLATER, P. L., and SALVIN, O. Descriptions of Three New Species of Tanagers from Veragua. Proc. Zoöl. Soc. Lond., 1869, pp. 439-440.

Honduras.

LAWRENCE, G. N. Catalogue of a Collection of Birds in the Museum of the Smithsonian Institution, made by Mr. H. E. Holland at Greytown, Nicaragua, etc. Ann. Lyc. Nat. Hist. N. York, 1865, pp. 179-185. (61 species.)

SALVIN, OSBERT. A Fortnight amongst the Sea-Birds of British Honduras. Ibis, 1864, pp. 372-387.

SCLATER, P. L. List of Birds collected by Geo. Cavendish Taylor in the Republic of Honduras. Proc. Zoöl. Soc., 1858, pp. 356-360. (39 species.)

SCLATER, P. L., and SALVIN, OSBERT. List of Birds collected on the Bluefields River, Mosquito Coast, by Mr. Henry Wickham. Proc. Zoöl. Soc. Lond., 1867, pp. 278-280. (39 species.)

TAYLOR, GEO. CAVENDISH. On Birds collected in the Republic of Honduras, etc. Ibis, 1860, pp. 10-24, 110-122, 222-228, 311-317. (96 species.)

Costa Rica.

CABANIS, J. Uebersicht der im Berliner Museum befindlichen Vögel von Costa Rica. Journ. für Orn., VIII Jahrgang, pp. 321-336, 401-416, 1860; IX Jahrgang, pp. 1-14, 96, 241-256, 1861. (113 species.)

CASSIN, JOHN. On some Conirostral Birds from Costa Rica, etc. Proc. Acad. Nat. Sci. Philad., 1865, pp. 169-172. (22 species.)

[FRANZIUS, V. Ueber die geographische Verbreitung der Vögel Costa-Rica's. Caban. Journ. für Ornith., 1869, pp. 195, 290.]

LAWRENCE, G. N. A Catalogue of the Birds of Costa Rica. Ann. Lyc. Nat. Hist. New York, 1868, pp. 86-149. (511 species.)

SALVADORI, TOMMASO. Intorno ad alcuni Uccelli di Costa Rica. Atti della R. Accad. Sc. di Torino, 1868, pp. 170-185. (23 species.)

SALVIN, OSBERT. Descriptions of Seventeen New Species of Birds from Costa Rica. Proc. Zoöl. Soc. Lond., 1864, pp. 579-586.

SALVIN, OSBERT. Notes on Mr. Lawrence's List of Costa Rica Birds. Ibis, 1869, pp. 310-319; 1870, pp. 107-116. (29 species added to Mr. Lawrence's List.)

New Granada.

LAWRENCE, GEO. N. Catalogue of a collection of Birds made in New Granada, etc. Ann. Lyc. Nat. Hist. New York, Vol. VII, pp. 288-302, 315-334; 461-479, 1861; Vol. VIII, pp. 1-13, 1862. (415 species.)

LAWRENCE, GEO. N. List of Birds from near David, Chiriqui, New Granada, etc. Ann. Lyc. Nat. Hist. New York, 1865, pp. 175-179. (39 species.)

THE BERMUDA AND WEST INDIA ISLANDS.

Bermudas.

BLAND, LIEUT. List of the Birds of Bermuda. Smithsonian Report, 1858, pp. 286 - 289. (114 species.)

WEDDERBURN, J. W., and HURDIS, J. L. Notes and Observations on the Birds of Bermuda, etc. Jones's The Naturalist in Bermuda. London, 1859, pp. 23 - 97.

Bahamas.

BRYANT, HENRY. A List of Birds seen at the Bahamas, from January 20 to May 14, 1859, etc. Proc. Bost. Soc. Nat. Hist., Vol. VII, pp. 102 - 134, 1860. (87 species.)

BRYANT, HENRY. Additions to a List of the Birds seen at the Bahamas. Proc. Bost. Soc. Nat. Hist., Vol. XI, pp. 63 - 70, 1866. (31 species.)

Cuba.

BREWER, T. M. [Nominal] List of the Birds of Cuba, compiled from two Lists furnished by Dr. John Gundlach. Proc. Bost. Soc. Nat. Hist., Vol. VII, pp. 305 - 308, 1860. (251 species.)

GUNDLACH, JOHN. Description of Five new Species of Birds, and other Ornithological Notes of Cuban Species. Journ. Bost. Soc. Nat. Hist., Vol. VI, pp. 313 - 319, 1852. (18 species in all.)

GUNDLACH, J. Beiträge zur Ornithologie Cuba's. Journ. für Orn., II Jahrgang, pp. —, 1854; III Jahrgang, 465, 1855; IV Jahrgang, pp. 1, 97, 337, 417, 1856; V Jahrgang, pp. 225 - 242, 1857. (251 species.)

GUNDLACH, J. Tabellarische Uebersicht aller bisher auf Cuba beobachteten Vögel. Journ. für Orn., IX Jahrgang, pp. 321 - 349, 1861. (249 species.)

GUNDLACH, J. Zusätze und Berichtigungen zu den "Beiträgen zur Ornithologie Cuba's." Journ. für Orn., IX Jahrgang, pp. 401 - 416; X Jahrgang, pp. 81 - 96. (249 species.)

GUNDLACH, JUAN. Revista y Catálogo de las Aves cubanas. Repert. fisico-nat. de Cuba, Vol. I, pp. 165 - 180, 221 - 242, 281 - 302, 347 - 363, 386 - 403, 1865, 1866. (257 species.)

GUNDLACH, J. Briefliches von Cuba. Cab. Journ. für Orn., 1866, pp. 352 - 353.

LAWRENCE, Geo. N. Notes on some Cuban Birds, etc. Ann. Lyc. Nat. Hist. New York, Vol. VII, p. 247, 1860. (27 species.)

LEMBEYE, J. Aves de la Isle de Cuba, 8vo. Habana, 1850. (222 species.)

RAMON DE LA SAGRA. Histoire physique, politique et naturelle de l'ile de Cuba, etc. Ornithologie par A. d'Orbigny. 8vo, 1839. (129 species.)

VIGORS, N. A. On some Birds from Cuba. Zoöl. Journ., Vol. III, pp. 432 - 448, 1828. (45 species.)

Jamaica.

GOSSE, P. H. The Birds of Jamaica. 8vo. London, 1847.

MARCH, W. T. Notes on the Birds of Jamaica, with Remarks by S. F. Baird. Proc. Acad. Nat. Sci. Philad., 1863, pp. 150 – 154, 283 – 304; 1864, pp. 62 – 72. (109 species.)

[OSBURN, W. Notes on the Mountain Birds of Jamaica. Zoölogist, pp. 6709, 6761.]

SCLATER, P. L. A List of a Collection of Birds made by the late Mr. W. Osburn in Jamaica, with Notes. Proc. Zoöl. Soc. Lond., 1861, pp. 69 – 82. (92 species.)

St. Domingo, St. Thomas, Porto Rico, etc.

BRYANT, HENRY. A List of the Birds of St. Domingo, etc. Proc. Bost. Soc. Nat. Hist., Vol. XI, pp. 89 – 98, 1866. (78 species.)

BRYANT, H. A List of Birds from Porto Rico, etc. Proc. Bost. Soc. Nat. Hist., Vol. X, pp. 248 – 257, 1866. (41 species.)

CASSIN, JOHN. Catalogue of Birds from the Island of St. Thomas, West Indies, collected and presented to the Academy of Natural Science by Mr. Robert Swift. Proc. Acad. Nat. Sci., 1860, pp. 374 – 379. (27 species.)

LAWRENCE, G. N., and JULIEN, A. A. Catalogue of Birds collected at the Island of Sombrero, W. I. Ann. Lyc. Nat. Hist. New York., VIII, pp. 93 – 107, 1864. (34 species.)

NEWTON, A. and E. Observations on the Birds of St. Croix, West Indies, made between February 20 and August 6, 1857, and March 4 and September 28, 1858. Ibis, I, pp. 59 – 69, 138 – 150, 253 – 264, 365 – 379, 1859. (64 species.)

SALLÉ, AUG. Liste des Oiseaux rapportés et observés dans la République Dominicaine, pendant son voyage de 1849 – 1851. Proc. Zoöl. Soc. Lond., 1857, pp. 230 – 237. (61 species.)

SUNDEVALL, C. J. Aves Insulæ Portorico. Öfvers. Kongl. Vetensk. Akad. Forhand., 1869, pp. 593 – 603. (90 species.)

SUNDEVALL, C. J. Aves Insulæ Sancti Barthelemy. Öfvers. Kongl. Vetensk. Akad. Forhand., 1869, pp. 579 – 592. (47 species.)

CAMBRIDGE, April, 1871.

N. B. — The Museum of Comparative Zoölogy would gladly accept books not yet upon the shelves of its library in exchange for its publications or zoölogical specimens.

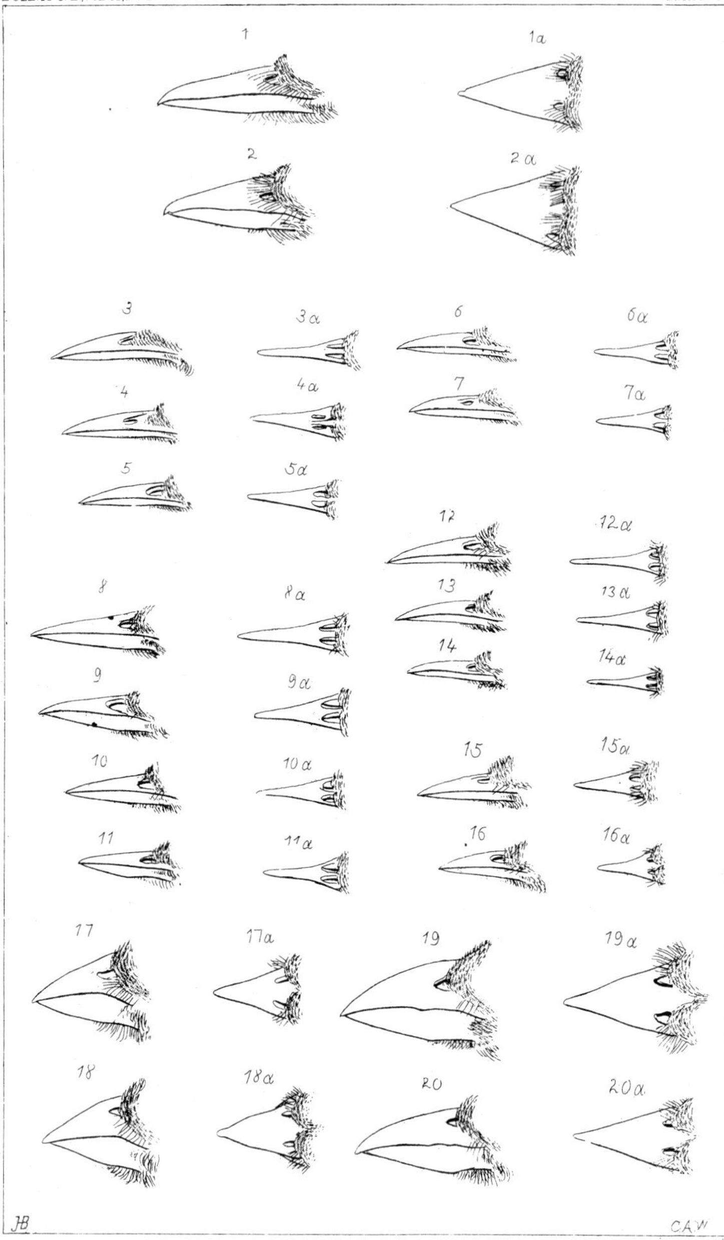

1-2, Tyrannus carolinensis; 3-7, Troglodytes ædon; 8-11, Seiurus noveboracensis; 12-14, Mniotilta varia; 15-16, Dendræca striata; 17-18, Pipilo erythrophthalmus; 19-20, Pyranga rubra.

Plate IV.*

Fig. 1 – 1*a*. *Tyrannus carolinensis*, specimen No. 6942, from Eastern Massachusetts.

" 2 – 2*a*. *Tyrannus carolinensis*, specimen No. 6945, from Eastern Massachusetts.

" 3 – 3*a*. *Troglodytes aëdon*, specimen No. 10931, ♀, from Jacksonville, Fla.

" 4 – 4*a*. *Troglodytes aëdon*, specimen No. 10684, ♀, from Dummitt's, Fla.

" 5 – 5*a*. *Troglodytes aëdon*, specimen No. 10683, from Dummitt's, Fla.

" 6 – 6*a*. *Troglodytes aëdon*, specimen No. 5212, from Welaka, St. John's River, Fla.

" 7 – 7*a*. *Troglodytes aëdon*, specimen No. 10930, from Jacksonville, Fla.

" 8 – 8*a*. *Seiurus noveboracensis*, specimen No. 5447, from Mount Tom, Mass.

" 9 – 9*a*. *Seiurus noveboracensis*, specimen No. 1442, from Weston, Mass.

" 10 – 10*a*. *Seiurus noveboracensis*, specimen No. 6794, from Brookline, Mass.

" 11 – 11*a*. *Seiurus noveboracensis*, specimen No. 4246, from Waterville, Maine.

" 12 – 12*a*. *Mniotilla varia*, specimen No. 5148, ♂, from Jacksonville, Fla.

" 13 – 13*a*. *Mniotilla varia*, specimen No. 6806, ♂, from Brookline, Mass.

" 14 – 14*a*. *Mniotilla varia*, specimen No. 8216, ♂, from Hudson, Mass.

" 15 – 15*a*. *Dendrœca striata*, specimen No. 5052, ♂, from Watertown, Mass.

" 16 – 16*a*. *Dendrœca striata*, specimen No. 4367, ♂, from Newtonville, Mass.

" 17 – 17*a*. *Pipilo erythrophthalmus*, specimen No. 4727, ♂, from Weston, Mass.

" 18 – 18*a*. *Pipilo erythrophthalmus*, specimen No. 10721, ♂, from Dummitt's, Fla.

" 19 – 19*a*. *Pyranga æstiva*, specimen No. 10629, ♂, from Jacksonville, Fla.

" 20 – 20*a*. *Pyranga æstiva*, specimen No. 5431, ♂, from Jacksonville, Fla.

* At bottom of Plate IV, last line, for " Pyranga *rubra* " read " Pyranga *æstiva*."

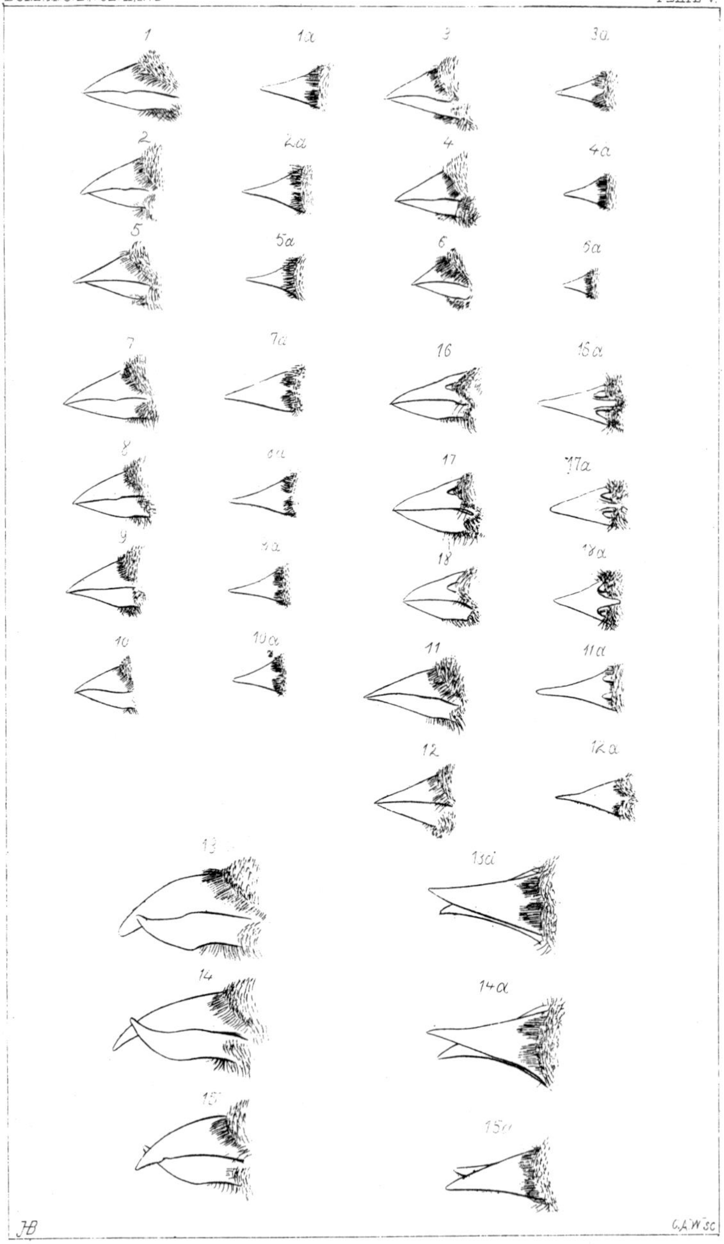

1-6, Ægiothus linarius; 7-10, Chrysomitris tristis; 11-12, C. pinus; 13-15, Curvirostra americana; 16-18, Passerculus savanna

Plate V.

Fig. 1 – 1*a*. *Ægiothus linaria*, specimen No. 10859, ♂, from Newton, Mass.
" – 2*a*. *Ægiothus linaria*, specimen No. 10860, ♂, from Newton, Mass.
" 3 – 3*a*. *Ægiothus linaria*, specimen No. 6392, ♂, from Fort Anderson, British America. (An original specimen of *Æ. fuscescens* Coues. — Smith. Inst., No. 43386.)
" 4 – 4*a*. *Ægiothus linaria*, specimen No. 4943, from Newton, Mass.
" 5 – 5*a*. *Ægiothus linaria*, specimen No. 3229, from Southern Maine.
" 6 – 6*a*. *Ægiothus linaria*, specimen No. 6489, ♀, from Fort Simpson, British America. (An original specimen of *Æ. exilipes* Coues. — Smith. Inst., No. 27431.)
" 7 – 7*a*. *Chrysomitris tristis*, specimen No. 6453, ♂, from Rocky Mountains, west of Denver, Colorado.
" 8 – 8*a*. *Chrysomitris tristis*, specimen No. 8125, ♂, from Springfield, Mass.
" 9 – 9*a*. *Chrysomitris tristis*, specimen No. 4925, from Newtonville, Mass.
" 10 – 10*a*. *Chrysomitris tristis*, specimen No. 4631, from Newtonville, Mass.
" 11 – 11*a*. *Chrysomitris pinus*, specimen No. 9556, from Waterville, Maine.
" 12 – 12*a*. *Chrysomitris pinus*, specimen No. 10875, from Gorham, New Hampshire.
" 13 – 13*a*. *Curvirostra americana*, specimen No. 4639, ♂, from Newton, Mass.
" 14 – 14*a*. *Curvirostra americana*, specimen No. 4638, ♂, from Newton, Mass.
" 15 – 15*a*. *Curvirostra americana*, specimen No. 4637, ♂, from Newton, Mass.
" 16 – 16*a*. *Passerculus savanna*, specimen No. 5084, ♂, from Ipswich, Mass.
" 17 – 17*a*. *Passerculus savanna*, specimen No. 5175, ♂, from Hibernia, St. John's River, Florida.
" 18 – 18*a*. *Passerculus savanna*, specimen No. 7119, ♂, from Henley Harbor, Labrador.

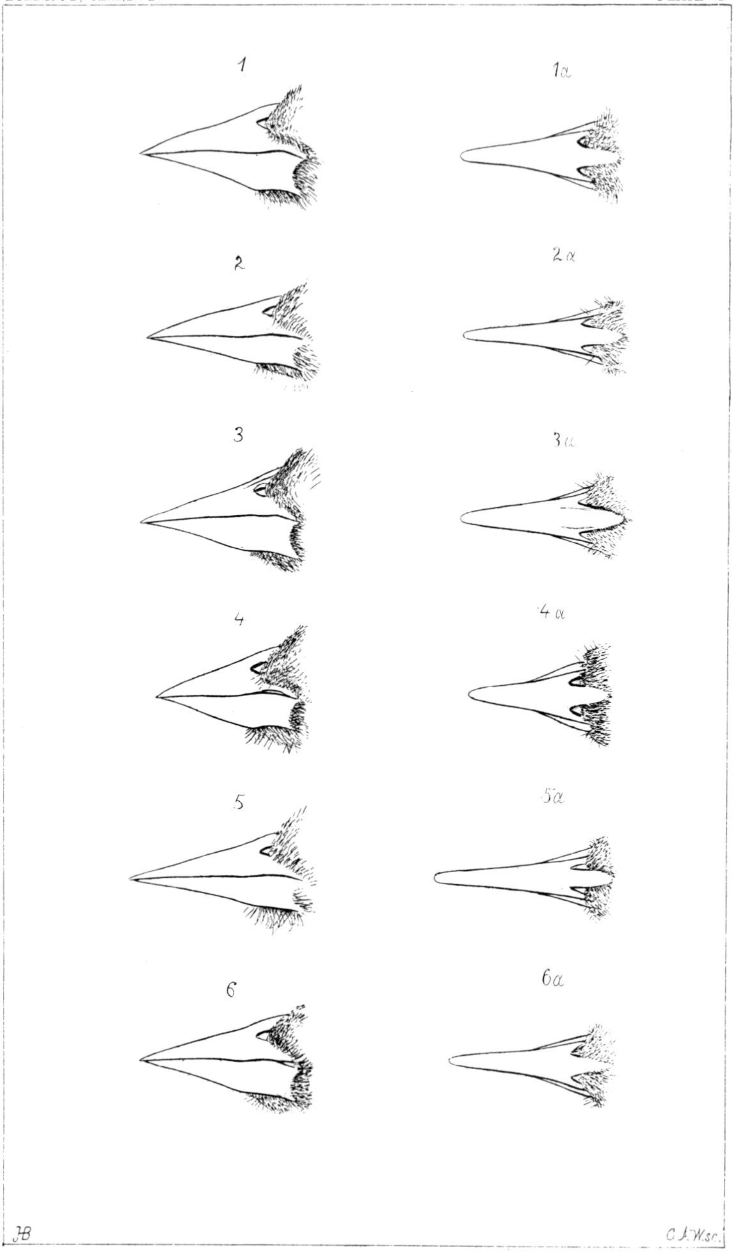

Agelæus phœniceus.

Plate VI.

Fig. 1 – 1*a*. *Agelæus phœniceus*, specimen No. 8071, ♂, from Orleans, Mass.

" 2 – 2*a*. *Agelæus phœniceus*, specimen No. 10572, ♂, from Dummitt's, Fla.

" 3 – 3*a*. *Agelæus phœniceus*, specimen No. 4589, ♂, from Newtonville, Mass.

" 4 – 4*a*. *Agelæus phœniceus*, specimen No. 8068, ♂, from Orleans, Mass.

" 5 – 5*a*. *Agelæus phœniceus*, specimen No. 10569, ♂, from Dummitt's, Fla.

" 6 – 6*a*. *Agelæus phœniceus*, specimen No. 10576, ♂, from Dummitt's, Fla.

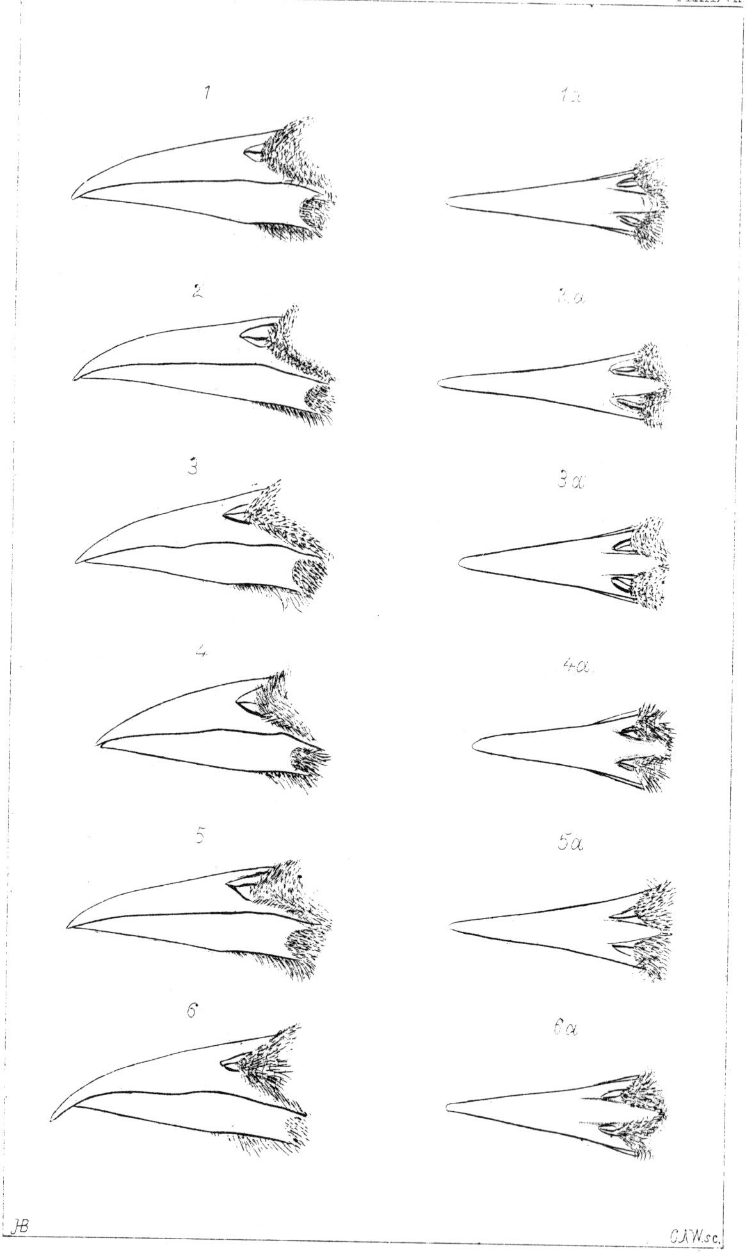

Quiscalus purpureus.

Plate VII.

Fig. 1 – 1*a*. *Quiscalus purpureus*, specimen No. 4601, ♂, from Newtonville, Mass.

" 2 – 2*a*. *Quiscalus purpureus*, specimen No. 5201, ♂, from Welaka, St. John's River.

" 3 – 3*a*. *Quiscalus purpureus*, specimen No. 4603, ♂, from Newtonville, Mass.

" 4 – 4*a*. *Quiscalus purpureus*, specimen No. 8072, ♂, from Orleans, Mass.

" 5 – 5*a*. *Quiscalus purpureus*, specimen No. 6834, ♂, from Phillipsburg New Jersey,

" 6 – 6*a*. *Quiscalus purpureus*, specimen No. 6848, ♂, from Cape Florida.

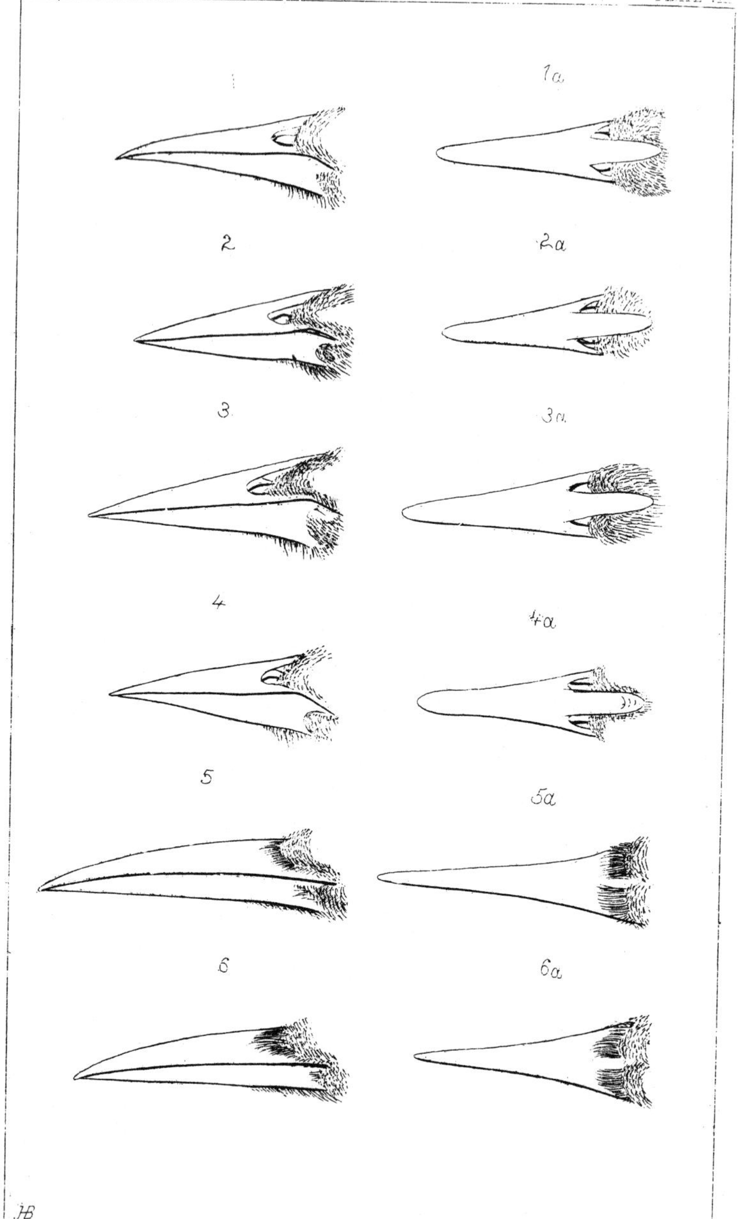

1-4. Sturnella ludoviciana. 5-6. Colaptes auratus

Plate VIII.

Fig. 1–1*a*. *Sturnella ludoviciana*, specimen No. 5370, ♀, from Hawkinsville, Fla.

" 2–2*a*. *Sturnella ludoviciana*, specimen No. 5372, ♀, from Hawkinsville, Fla.

" 3–3*a*. *Sturnella ludoviciana*, specimen No. 5339, ♂, from Enterprise, Fla.

" 4–4*a*. *Sturnella ludoviciana*, specimen No. 5340, ♂, from Enterprise, Fla.

" 5–5*a*. *Colaptes auratus*, specimen No. 4881, ♀, from Newton, Mass.

" 6–6*a*. *Colaptes auratus*, specimen No. 5464, ♀, from Newton, Mass.

Errata. At the upper left-hand corner of all the plates, for "Bull. M. C. Z., Vol. II, No. 2," read "Bull. M. C. Z., Vol. II, No. 3."

HUMBOLDT SCHOLARSHIP.

AT a meeting of the Faculty of the Museum of Comparative Zoölogy, held May 6, 1871,

THE HUMBOLDT SCHOLARSHIP

was awarded

TO J. A. ALLEN,

in consideration of his paper upon the Mammals and Winter Birds of East Florida, and the proceeds of the Humboldt Fund, for one year, granted to him in aid of his exploration of the Fauna of the Rocky Mountains.

INFLUENCE OF ELEVATION AND LATITUDE UPON THE DISTRIBUTION OF SPECIES

INFLUENCE OF ELEVATION AND LATITUDE UPON THE DISTRIBUTION OF SPECIES.—It is surprising to what extent observers still overlook the fact of the influence of elevation upon the distribution of animal and vegetable life;—that they should still regard parallels of latitude, instead of isothermal lines, as bounding the habitats of species. That such is the case, however, is sufficiently apparent from such notices as that in the December NATURALIST, respecting the summer distribution of the chestnut-sided warbler (*Dendroica Pensylvanica*), which is but a sample of such remarks as frequently occur in reference to the distribution of our birds and mammals. The merest tyro in the study of the geographical distribution of animals knows that their range is not only determined by climatic influences, but that these influences depend largely upon the character of the surface of the country, as, for instance, whether it is a level plain or is broken by mountain ranges, and that increase in elevation is climatically equivalent to an increase in latitude. If authors would use isothermal lines in giving the distribution of a species, instead of arbitrary political divisions, they would be able to speak with much greater precision in such matters than is customary at present. The isotherms are now so well established, and a knowledge of them may be so easily acquired by means of our meteorological maps, that it seems quite time to adopt them in speaking of the distribution of species.

While southern New England may, generally speaking, form the southern limit of the breeding range of a bird, or of the distribution of a mammal, reptile or plant, the same species may, and generally does, exist in the highlands of the Alleghanies as far south as northern Georgia; and even species which occur only to the northward of southern Maine, in the lowlands, not only occur in the highlands of Berkshire county, Massachusetts, but also southward in the Alleghanies to North Carolina.

A great point will be gained in precision when naturalists come to use natural faunal areas, instead of arbitrary political divisions, in speaking of the distribution of species. If Dr. Brewer had said, in speaking of the chestnut-sided warbler, "*not known to breed south of the Alleghanian Fauna*," instead of "not known to breed farther south than Massachusetts," he would not only have expressed the fact with precision, so far as our present knowledge extends, but would have saved himself the exposure to such criticism as that made by Mr. Stark (Am. Nat., VIII, p. 756, Dec.,

1874). I mention the present case merely by way of illustration, and not for the purpose of making any special strictures upon my friend Dr. Brewer, who is by no means in this respect an exceptional transgressor. If it is urged that the people would not understand such expressions as the "Alleghanian Fauna," and the like, it may be said that the time has come when they should be familiar with them. Most intelligent people know that isothermal lines vary in direction with the elevation and contour of the land over which they pass, sweeping, in our own country, far to the southward in leaving the lowlands of the Atlantic coast; that they pass southward of the Appalachian Highlands, and then bend abruptly northward again along their western base. It is time they knew, also, that the different zones of animal life follow the flexures of the isotherms, and that there are natural faunal belts, sufficiently distinct to be capable of recognition, whose boundaries coincide very nearly with certain of these isotherms. Furthermore, that throughout eastern North America, at least, these faunal belts are already well known to specialists of the subject, and that there already exist definite expressions for such cases as the one that has furnished the text for the present note. I will add, also, that so much is already known of the laws of the geographical distribution of animal life, that one could have safely assumed, from our present knowledge of the general range of the chestnut-sided warbler, that from its being a rather common summer resident in southern New England, it would also be found to breed in the mountainous districts as far south even as northern Georgia.—J. A. Allen.

THE INFLUENCE OF PHYSICAL CONDITIONS IN THE GENESIS OF SPECIES.

THE INFLUENCE OF PHYSICAL CONDITIONS IN THE GENESIS OF SPECIES.

AMONG biologists who accept the modern theory of evolution as the only reasonable hypothesis available for the explanation of the diversity of structure among organized beings, there is a wide difference of opinion as to what are the leading causes of differentiation. The doctrine of natural selection, or the survival of the fittest, has recently been brought prominently forward as the key to this complex problem, and is upheld by a large class of enthusiastic adherents, who accept it as the full solution of the whole question. By others the conditions of environment are believed to be far more influential in effecting a certain class of modifications, at least, than the necessarily precarious influence of natural selection, which must take its origin in isolated instances of variation in favorable directions, and depend for its continuance upon these fortuitous advantages being inherited by the descendants of the favored individuals in which they originate. The modifying influence of conditions resulting from geographic or climatic causes, was long since noticed, and for nearly a century has been considered by many writers as explanatory of much of the diversity existing not only in the human race, but among animals. It has, however, remained, until recently, vaguely grounded, being based more in conjecture than on observed facts. Scarcely, indeed, have two decades passed since the real nature and extent of geographical variation among animals, and even as yet among only a few species, began to receive careful attention, while only within the last fifteen years has any attempt been made to correlate the observed differences with the climatic or geographical conditions of habitat. Only within recent years have the differences in general size, and in the relative size of different parts, been ascertained by careful measurement, and the differences in the character of the tegumentary covering (as the pelage in mammals) and in color, in individuals of the same species inhabiting distant portions of a common habitat, been duly recorded. In the work of registering these instructive data, it has fallen to Americans to take a leading part; large credit in the matter being due not only to the activity of our professional biologists, but to the liberality of the general Government in attaching competent natural-history observers and collectors to the numerous surveying parties it has sent out during the last twenty years to explore the, till then, practically unknown geography and productions of our Western Territories. The combined fruits of their labors, together with those of the agents and correspondents of the Smithsonian Institution, have resulted in the accumulation of an amount of material far exceeding that elsewhere accessible to single investigators; representing, as it does, at least two of the vertebrate classes of animals from the whole North American continent so fully, that generalizations may be made from their study which could not otherwise have been reached for many years, and for which no similar facilities for any other equal area as yet exist. The recent investigations of American mammalogists and ornithologists have been, in consequence, largely directed to the subject of geographical variation; and their publications teem with tabulated measurements and records of variations in form and color that accompany differences in the climatic or geographical conditions of habitat. Among the results that have followed are the discovery of numerous interesting geographical varieties or sub-species, and the demonstration of the complete intergradation of many forms, often quite widely diverse in color, size, and proportion of parts, formerly regarded (and properly so as then known) as unquestionably distinct species; which discoveries have of course necessitated a large reduction in the number of recognized "specific" or non-intergrading forms. But most important of all has been the correlation of local variations with the conditions of environment, and the deduction therefrom of certain laws of geographical variation. Upon these have been based hypotheses that go far toward explaining many of the phenomena of intergradation and differentiation observed among existing animals. In the present paper will be given not only a summary of the results thus far attained, but enough of the details of the subject to show the nature of the evidence on which rest the conclusions already reached. These results, it is claimed, show that other influences than natural selection opérate powerfully in the differentiation of specific forms, and that geographical causes share more largely in the work than naturalists have heretofore been prepared to admit,—at least to consider as proven.

As is well known, animals vary greatly in respect to the extent of the areas they inhabit. While a few species are nearly or quite cosmopolitan, many others are restricted to single small islands, or to limited portions of a continent. Not a few range over the greater part of whole hemispheres, while by far the larger number are confined within comparatively narrow limits. Of the numerous species of mammals and birds inhabiting North America, none are equally common throughout the whole extent of the continent. The habitats of a few only extend from the Barren Grounds of the Arctic regions to Mexico, and from the Atlantic coast westward to the Pacific; one or two only among the mammals range over the whole continent from Alaska to Central America, while some occupy merely the extreme boreal parts of the continent. The latter, in many cases, range also over the Arctic and sub-Arctic regions of the Old World. Others extend from Arctic America southward to the United States. Still others occupy only the middle or more temperate latitudes, being unrepresented in the extreme north or the extreme south. Others, again, first appear in the middle or more southerly parts, and range thence southward far into the tropics. A large number are restricted to the region east of the Rocky Mountains; others are confined to a narrow belt along the Pacific coast; and others still to limited areas of the great Rocky Mountain Plateau. In general, their distribution accords with climatal regions or zones, their respective ranges being limited in part by latitude and in part by geographical barriers, as treeless, arid plains or high mountain ranges. The northern and southern boundaries of the habitat of a species are found to agree, not generally with the arbitrary parallels of the geographer, but with isothermal lines, these being more or less different for each species. The geographical distribution of a species is thus mainly determined by climatic or other physical causes, though in part, doubtless, by its organic constitution. In most cases, species that are wide-ranging are the most variable, as would naturally follow from their being subjected, in the different portions of their habitats, to widely different environing circumstances. Hence such species are often found to run into numerous local races, some of them greatly differing from others, but still inseparably connected by individuals inhabiting the intervening regions. Over districts slightly diversified, even if of large extent, species generally preserve comparative constancy of character, while, conversely, local races are of frequent occurrence in regions of alternating valleys, mountain ranges, and table lands; and more especially is this true if the highly diversified region be situated in the warmer latitudes. Small islands, remotely situated from other lands, have usually many species peculiar to themselves, their differentiation being proportionate to the geologic antiquity of the islands and their remoteness from larger land areas. In islands of recent origin, and not widely separated from continental lands, the ancestral stock of the species is still often clearly apparent, the forms thus differentiated through insular influences not having passed beyond the varietal stage; in other cases they are specifically different from their nearest continental allies, or may even have advanced far toward generic distinctness, while their origin may still remain tolerably apparent.

Plasticity, or susceptibility to the influences of physical surroundings, often differs even among quite closely allied species, as those of the same family or even genus; and different species are evidently affected differently by the same circumstances. Variability in color may, or may not, accompany variability in size, or in the character of particular organs. Generally, however, a species which varies greatly in one feature, varies to a similar degree in many others. Species having a wide geographical range not only commonly run into a greater or less number of local races, but they generally present more than the average amount of strictly individual variation, as though species ranging widely in space were originally more plastic than those having more circumscribed habitats, and were thus able more easily to adapt themselves to their surroundings; they are also more persistent, their fossil remains being far more frequently met with in the quaternary deposits than are those of the more local and generally more specialized forms.

Geographical variation, as exhibited by the mammals and birds of North America, may be summarized under the following heads: namely, (1) variation in general size, (2) in the size of peripheral parts, and (3) in color; the latter being subdivisible into (*a*) variation in color with latitude, and (*b*) with longitude. As a rule, the mammals and birds of North America increase in size from the south northward. This is true not only of the individual representatives of each species, but generally the largest species of each genus and family are northern. There are, however, some strongly marked exceptions, in which the increase in size is in the opposite direction, or southward. There is for this an obvious explanation, as will be presently shown; the increase being found to be almost invariably toward the region where the type or group to which the species belongs receives its greatest numerical development, and where the species attain the largest size, and are also most specialized. Hence the representatives of a given species increase in size toward its hypothetical centre of distribution, which is in most cases doubtless also its original centre of dispersal. Consequently there is frequently a double decadence in size within specific groups, and both in size and numerically in the case of species, when the centre of development of the group to which they belong is in the warm-temperate or tropical regions. This may be illustrated by reference to the distribution of the higher classes of vertebrates in North America. Among the species occurring north of Mexico, there are very few that may not be supposed to have had a northern origin; and the fact that some are circumpolar in their distribution, while most of the others (especially among the mammals) have congeneric Old World allies, further strengthens the theory of their northern origin. Not only do individuals of the same species increase in size toward the north, but the same is true of the species of the different genera. Again, in the exceptional cases of increase in size southward, the species belong to southern types, or, more correctly, to types having their centre of development within or near the intertropical regions, where occur not only the greatest number of the specific representatives of the type, but also the largest.

For more detailed illustration we may take three families of the North American Carnivora; namely, the Canidæ (wolves and foxes), the Felidæ (lynxes and wild cats), and the Procyonidæ (raccoons). The first two are to some extent cosmopolitan, while the third is strictly American. The Canidæ have their largest specific representatives, the world over, in the temperate or colder latitudes. In North America the family is represented by six species,[1] the smallest of which (speaking generally) are southern, and the largest northern. Four of them are among the most widely-distributed of North American mammals, two (the gray wolf and the common fox) being circumpolar species; another (the arctic fox) is also circumpolar, but is confined to high latitudes. The three widest-ranging species (the gray wolf, the common fox, and the gray fox) are those which present the most marked variation in size. Taking the skull as the basis of comparison, it is found that the common wolf is fully one-fifth larger in the northern parts of British America and Alaska than it is in Northern Mexico, where it finds the southern limit of its habitat. Between the largest northern skull and the largest southern skull there is a difference of *about thirty-five per cent. of the mean size!* Specimens from the intermediate region show a gradual intergradation between these extremes, although many of the examples from the upper Missouri country are nearly as large as those from the extreme north.

The common fox, though occurring as far north as the wolf, is much more restricted in its southward range, especially along the Atlantic coast, and presents a correspondingly smaller amount of variation in size. The Alaskan animal, however, averages about one-tenth larger than the average size of specimens from New England. In the gray fox, whose habitat extends from Pennsylvania southward to Yucatan, the average length of the skull decreases from about five inches in Pennsylvania to considerably less than four in Central America,—a difference equal to about thirty per cent. of the mean size for the species.

[1] The gray wolf (*Canis lupus*), the prairie wolf (*C. latrans*), the Arctic fox (*Vulpes lagopus*), the common fox (*V. alopex*), the kit fox (*V. velox*), and the gray fox (*Urocyon virginianus*).

The Felidæ, unlike the Canidæ, reach their greatest development, as respects both the number and the size of the species, in the intertropical regions. This family has but a single typical representative—the panther (*Felis concolor*)—north of Mexico, and this ranges only to about the northern boundary of the United States. The other North American representatives of the family are the lynxes, which, in some of their varieties, range from Alaska to Mexico. They form, however, the most northern, as well as the most specialized or "aberrant" type of the family. While they vary greatly in color, as well as in the length and texture of the pelage, at different localities, they afford a most remarkable exception to all laws of variation in size with locality; for a large series of skulls, representing localities as widely separated as Louisiana, Northern Mexico, and California on the one hand, and Alaska and the Mackenzie River District on the other, as well as various intermediate localities, reveals no appreciable difference in size throughout this wide area. The true cats, however, as the panther and the ocelots, are found to greatly increase in size southward, or toward the metropolis of the family. The panther ranges from the Northern States southward over most of South America. Skulls from the Adirondack region of New York have an average length of about seven and a half inches, the length increasing to eight and three-quarters in Louisiana and Texas, from beyond which points there is lack of data. The ocelot (*Felis pardalis*) finds its northern limit near the Rio Grande of Texas, and ranges thence southward far into South America. The average size of Costa Rican examples is about one-fifth greater than that of specimens from the Rio Grande.

The Procyonidæ are chiefly represented in tropical America, a single species—the common raccoon (*Procyon lotor*)—being found in the United States, and thence northward to Alaska. Here again the increase in size is southward, or toward the metropolis of the family,—Pennsylvania specimens averaging about one-tenth smaller than Costa Rican examples.

The common otter (*Lutra canadensis*) affords another example of increase in size southward among our Carnivora, although belonging to a family essentially northern in its distribution. The otters, however, form a distinct sub-family, which attains its greatest number of species in the warmer regions of the earth; and hence offers, not an exception to, but a confirmation of, the law of increase toward the centre of distribution of the group to which it belongs.

Instances of increase in size northward among the Carnivora of North America are so generally the rule that further space need not be taken in recounting examples in detail. It may suffice to state that the badger (*Taxidea americana*), the marten (*Mustela americana*), the fisher (*M. pennanti*), the wolverine (*Gulo luscus*), and the ermine (*Putorius ermineus*),—all northern types,—afford examples of variation in size strictly parallel with that already noticed as occurring in the foxes and wolves.

To refer briefly to other groups, it may be stated that the Cervidæ (deer family) are mainly rather northern in their distribution; that the largest species occur in the colder zones, and that individuals of the same species increase rapidly in size toward the north. Some of the species, in fact, afford some of the most striking instances of northward increase in size; among which are the common Virginia deer and its several representatives in the interior of the continent and on the Pacific slope. It is also noteworthy that the most obviously distinctive characteristic of the group—the large, annually deciduous antlers—reaches its greatest development at the northward. Thus all the northern species, as the moose, the elk, and the caribou, have branching antlers of immense size, while the antlers are relatively much smaller in the species inhabiting the middle region of the continent, and are reduced to a rudimentary condition—a simple slender sharp spike, or a small and singly forked one—in the tropical species; the antlers declining in size much more rapidly than the general size of the animal. This is true in individuals of the same species as well as of the species collectively.

The Rodentia (the squirrels, marmots, spermophiles, mice, and their affines) offer the same illustrations in respect to the law of increase in size as the species already mentioned, the size sometimes increasing to the southward, but more generally to the northward, since the greater number of the species belong decidedly to northern types. There is no more striking instance known among mammals of variation in size with locality than that afforded by the flying squirrels, in which the northern race is more than one-half larger than the southern; yet the two extremes are found to pass so gradually the one into the other, that it is hardly possible to define even a southern and a northern geographical race except on the almost wholly arbitrary ground of difference in size. The species, moreover, is one of the most widely distributed, ranging from the Arctic regions (the northern limit of forests) to Central America.

Among birds the local differences in size are almost as strongly marked as among mammals, and, in the main, follow the same general law. A decided increase in size southward, however, or toward the warmer latitudes, occurs more rarely than in mammals, although several well-marked instances are known. The increase is generally northward, and is often very strongly marked. The greatest difference between northern and southern races occurs, as in mammals, in the species whose breeding stations embrace a wide range of latitude. In species which breed from northern New England to Florida, the southern forms are not only smaller, but are also quite different in color and in other features. This is eminently the case in the common quail (*Ortyx virginianus*), the meadow lark (*Sturnella magna*), the purple grackle (*Quiscalus purpureus*), the red-winged blackbird (*Agelæus phœniceus*), the golden-winged woodpecker (*Colaptes auratus*), the towhee (*Pipilo erythrophthalmus*), the Carolina dove (*Zenædura carolinensis*), and in numerous other species; and is quite appreciable in the blue jay (*Cyanurus cristatus*), the crow (*Corvus americanus*), in most of the woodpeckers, in the titmice, numerous sparrows, and several thrushes and warblers, the variation often amounting to from ten to fifteen per cent. of the average size of the species.

As a general rule, certain parts of the organism vary more than does general size, there being a marked tendency to enlargement of peripheral parts under high temperature, or toward the tropics,—hence southward in North America. This is more readily seen in birds than in mammals, in consequence, mainly, of their peculiar type of structure. In mammals it is manifested occasionally in the size of the ears and feet, and in the horns of bovines, but especially and more generally in the pelage. At the northward, in individuals of the same species, the hairs are longer and softer, the under fur more abundant, and the ears and the soles of the feet better clothed. This is not only true of individuals of the same species, but of northern species collectively as compared with their nearest southern allies. Southern individuals retain permanently, in many cases, the sparsely clothed ears and the naked soles that characterize northern individuals only in summer, as is notably the case among the different squirrels and spermophiles.

In mammals which have the external ear largely developed,—as the wolves, foxes, some of the deer, and especially the hares,—the larger size of this organ in southern as compared with northern individuals of the same species is often strikingly apparent. It is more especially marked, however, in species inhabiting extensive open plains and semi-desert regions. The little wood hare, or gray "rabbit" (*Lepus sylvaticus*), affords a case in point. This species is represented, in some of its varieties, across the whole breadth of the continent, and from the northern border of the United States southward to Central America, but in different regions by different geographical races or sub-species. In addition to certain differences of color and general size, the ears vary still more strongly. In the form inhabiting the Great Plains, commonly known as the little sage-brush hare (*L. sylvaticus, nuttalli*), the ears are considerably longer than in the eastern variety, and increase in size from the north southward, reaching their greatest development in Western Arizona and the desert region further westward and southward, where the variety of the plains proper passes into still another variety characterized mainly by the large size of its ears, which are in this race nearly twice the size they attain in the eastern variety. In the large long-eared "jackass" hares of the Plains, the ear likewise increases in size to the southward. In *Lepus callotis*, for example, which ranges from Wyoming southward far into Mexico, the ear is about one-fourth to one-third larger in the southern examples than in the northern. The little brown hare of the Pacific coast (*L. trowbridgei*) presents a similar increase in the size of the ear southward, as does, to a less extent, the prairie hare (*L. campestris*). Not only are all of the long-eared species of American hares confined to the open plains of the arid interior of the continent, but over this same region is the tendency to an enlargement of the ear southward stronger than elsewhere. It is also of interest in this connection that the largest-eared hares of the Old World occur over similar open, half-desert regions, as do also the largest-eared foxes. On our western plains, the deer are represented by a large-eared species. Among the domestic races of cattle, those of the warm temperate and intertropical regions have much larger and longer horns than those of northern countries; as is shown by a comparison of the Texan, Mexican, and South American breeds, with the northern stock, or those of the south of Europe with the more northern races. In the wild species of the Old World, the southern or sub-tropical are remarkable for the large size of their horns. The horns of the American pronghorn (*Antilocapra americana*) are also much larger at southern than at northern localities.[1] Naturalists have also recorded the existence of larger feet in many of the smaller North American mammalia at the southward than at the northward, among individuals of the same species, especially among the wild mice, in some of the squirrels, the opossum and raccoon, as well as in other species.

In birds, the enlargement of peripheral parts, especially of the bill, claws, and tail, is far more obvious and more general than in mammals. The bill is particularly susceptible to variation in this regard,—in many instances being very much larger, among individuals of unquestionably the same species, at the southward than at the northward. This accords with the general fact that all the ornithic types in which the bill is remarkably enlarged occur in the intertropical regions. The southward enlargement of the bill within specific groups may be illustrated by reference to almost any group of North American birds, or to those of any portion of the continent. As in other features of geographical variation, the greatest differences in the size of the bill are met with among species having the widest distribution in latitude. Among the species inhabiting eastern North America we find several striking examples of this enlargement among the sparrows, black-birds, thrushes, crows, wrens, and warblers; in the quail, the meadow lark, the golden-winged woodpecker, etc. Generally the bill, in the slender-billed forms, becomes longer,

[1] The deer tribe, in which the antlers increase in size toward the north, offer an apparent exception to the rule of increase in size of peripheral parts toward the tropics. The antlers of the deer, however, are merely seasonal appendages, being annually cast and renewed, and are thus entirely different physiologically from the horns of bovines, which retain a high degree of vitality throughout the life of the animal.

more attenuated, and more decurved (in individuals specifically the same) in passing from the New England States southward to Florida, while in those which have a short, thick, conical bill there is a general increase in its size so that the southern representatives of a species, as a rule, have thicker and longer bills than their northern relatives, though the birds themselves are smaller. There is thus not only generally a relative, but often an absolute, increase in the size of the bill in the southern races. The species of the Pacific coast and of the interior afford similar illustrations, in some cases more marked even than in any of the eastern species. More rarely, but still quite frequently, is there a similar increase in the size of the feet and claws.

The tail, also, affords an equally striking example of the enlargement of peripheral parts southward. Referring again to the birds of the Atlantic coast, many of the above-named species have the tail absolutely longer at southern localities than at northern, and quite often relatively longer. Thus while the general size decreases, the length of the tail is wholly maintained, or decreases less than the general size; but, in some cases, while the general size is one tenth or more smaller at the south, the tail is ten to fifteen per cent. longer than in the larger northern birds. Some western species are even more remarkable in this respect; and in consequence mainly of this fact the southern types have been varietally separated from the shorter-tailed northern forms of the same species.

Variations in color with locality are still more obvious, particularly among birds, in which the colors are more positive, the contrasts of tint greater, and the markings consequently better defined than is usually the case in mammals. The soft, finely-divided covering of the latter is poorly fitted for the display of the delicate pencilings and the lustrous, prismatic hues that so often characterize birds. Mammals, however, present many striking instances of geographical variation in color.

As already stated, geographical variation in color may be conveniently considered under two heads. While the variation with latitude consists mainly in a nearly uniform increase in one direction, the variation observed in passing from the Atlantic coast westward is more complex. In either case, however, the variation results primarily from nearly the same causes, which are obviously climatic, and depend mainly upon the relative humidity, or the hygrometric conditions of the different climatal areas of the continent. In respect to the first, or latitudinal variation, the tendency is always toward an increase in intensity of coloration southward. Not only do the primary colors become deepened in this direction, but dusky and blackish tints become stronger or more intense, iridescent hues become more lustrous, and dark markings, as spots and streaks or transverse bars, acquire greater area. Conversely, white or light markings become more restricted. In passing westward a general and gradual blanching of the colors is met with on leaving the wooded regions east of the Mississippi, the loss of color increasing with the increasing aridity of the climate and the absence of forests, the greatest pallor occurring over the almost rainless and semi-desert regions of the Great Basin and Colorado Desert. On the Pacific slope, north of California, the color again increases, with a tendency to heavy, sombre tints over the rainy, heavily-wooded region of the north-west coast.

Geographical variation in color among mammals, for reasons already stated, is generally, but not always, manifested merely through the varying intensity or depth of the tints. It is, however, often strongly marked. The common chickaree, or red squirrel (*Sciurus hudsonius*), for example, which ranges from high northern latitudes southward over the northern portion of the United States, shows an increase in the color over the middle of the dorsal surface from pale yellowish or fulvous to rufous. The fox squirrel of the Mississippi Valley (*Sciurus niger, ludovicianus*), which ranges from Dakota southward to the Gulf of Mexico, has the lower parts, at the northward, very pale yellowish-white, which tint gradually increases in intensity southward, till in Louisiana it becomes deep reddish-orange, the dorsal surface also becoming at the same time somewhat darker. Excepting the fox squirrels, and a Pacific coast variety of the chickaree, all the squirrels living north of Mexico have the lower parts white, while those inhabiting tropical America have the lower parts fulvous, deep golden, orange, or even dark brownish-red; specimens with the belly white being exceptional, though occasionally occurring in several of the species.

Mammals tend strongly to run into melanitic phases, which are especially developed at particular localities or over limited regions, but whether or not the result of geographical influences is not clearly evident. The whitening of the pelage in winter at the north in a considerable number of species of mammals, and in one genus of birds, and not elsewhere, is, on the contrary, a strictly geographical phenomenon, but seems to be the result of other than the ordinary causes of geographical variation in color. Its occurrence in some species, and its absence in others closely allied to them, is a fact not readily explained. It shows, however, how differently different animals are effected by the same influences. The change to a white winter livery is more complete in the higher latitudes, where the whiteness pervades the pelage to a greater depth, and continues for a longer period, the change being only partial in the southern representatives of species that exhibit this seasonal change of color.

In respect to southward increase in color among birds, a few examples only, out of the many almost equally striking, can be here given. These will be chosen from widely different groups, and will represent localities remotely separated, as well as very diverse styles of coloration. In comparing, for instance, New England examples of the common quail with others from southern Florida, the colors are found to be so much stronger and darker in the southern birds as to give the appearance of their being entirely distinct species; particularly when the smaller size and larger bills of the southern race are also considered. While in the northern birds the color of the dorsal surface is gray and rufous, slightly varied with black, the gray is wholly wanting in the southern type, the rufous is much stronger, and the black markings are very much broader. The lower surface is varied by transverse bars of black and white, but while in the northern birds the white bars are twice, or more than twice, the width of the black ones, in the southern birds they are often of equal width; or the black bars may be the broader, with much more black bordering the white throat-patch, giving, on the whole, a very much darker aspect to this region of the body. Yet, when a series is brought together from many intermediate localities, there is found to be a complete intergradation between the most extreme phases. In the common towhee the style of coloration is entirely different from that seen in the quail, the colors being chiefly massed in large areas, with white markings on the wings, and large white spots at the ends of the outer tail-feathers. In this species, southern specimens differ from northern ones in the black of the upper parts and the chestnut of the sides being more intense, while the white markings on the wings and tail are greatly reduced in area. In the northern bird four of the outer pairs of tail-feathers have a large white spot near the end, while in the southern form only three pairs are thus marked.

In the purple grackle the plumage (in the males) is everywhere black, with, at the north, greenish or bronzy reflections; in the southern or Floridan form the black is more intense, and the reflections are steel-blue and purple, with iridescent bars across the middle and lower parts of the back. In the northern form the female is dull brownish-black, with little or no iridescence, while in the southern form the female is nearly as lustrous as the northern male. The two types differ so widely, not only in color, but, as previously noticed, in size and in the form of the bill, that, without the connecting specimens from intermediate localities, no ornithologist would hesitate to regard them as entirely distinct species; and they were, indeed, at one time so regarded. The red-winged black-bird has, excepting its red wing-patches, also a lustrous black plumage throughout, and presents a similar range of variation in general color with the preceding; while the red of the wing-patch becomes much darker at the southward, and its creamy-white border seen in the northern form changes to yellowish-orange in the southern.

The common blue jay, and the long-crested jays of the Rocky Mountain region, may be cited as illustrations of southward increase in brilliancy or intensity of coloring where the prevailing tint is blue; the green Mexican and Rio Grande jays of a passage from yellowish-green tints into bright yellow; the yellow-throated warblers (genus *Geothlypis*), several of the fly-catchers (genera *Myiarchus* and *Tyrannus*), and the meadow lark, as examples of increase in the area and intensity of yellow; several of the woodpeckers (genera *Centurus* and *Sphyrapicus*), the cardinal finches (genus *Cardinalis*), and some of the tanagers (genus *Pyranga*), of a similar increase of red; the goldfinches (genus *Chrysomitris*), and most of the species above-named, of increase in extent and purity of black areas. The Rocky Mountain jays have, at the northward, a large portion of the plumage rather dark ashen, which further southward becomes bluish ash, and still further south culminates, in the Central American States, in blue. In the genus *Geothlypis*, the Maryland yellow-throat (*G. triches*), which ranges over the whole United States, and thence far southward, has at the northward the abdomen whitish; more to the southward, yellowish; and, in the West Indies, Mexico, and northern South America, runs into races in which the abdomen is bright yellow. At the same time the black markings about the head increase in extent and purity, and the general size becomes larger, the group having its metropolis in the tropical regions. In consequence of these variations in color and size, this species at the southward becomes differentiated into several more or less well-marked sub-species (formally accorded full specific rank), which are connected by an unbroken series of intergradations.

In the great-crested flycatcher (*Myiarchus crinitus*) of the United States, the yellow of the abdominal region is much the stronger in the southern birds: while the same is true of several of the western species of the same genus, which at the southward also pass into several recognizable sub-species.

The western goldfinch (*Chrysomtris psaltria*) affords a well-known instance of increase of black. This species is found in the western half of North America from about the parallel of 40° southward to Ecuador. The northern form has the black of the upper parts mainly restricted to the head, wing, and tail, the rest of the dorsal surface being olive-green. In northern Mexico the back begins to be more or less clouded with black, which tint increases in extent in Central America till it wholly replaces the olive-green, while in northern South America it becomes more intense and lustrous. In northern specimens the tail is marked with white spots, which either decrease greatly in size, or become wholly obsolete, in the southern races. The extremes, as may well be imagined, are widely diverse in their coloration, and, though formally regarded as entirely distinct species, have been found so thoroughly to intergrade that it is impossible to draw any lines of separation between the several races. Lawrences' flycatcher (*Myiarchus lawrencci*) affords also a striking example of southward increase in the area and intensity of black. At the northward this species has a grayish-black crown, which gradually passes southward into a form with the crown wholly deep black. With the increase southward of the area and intensity of black markings, there is also in this, as in other species, a general increase in the intensity or depth of the other accompanying tints.

The red-bellied, or Carolina woodpecker (*Centurus carolinus*), a common bird of the United States, shows a strong increase of red on the head and lower surface of the body at the southward, in which this tint is not only much brighter, but also much more extended in the south-Florida birds than in those from the Northern States. At the same time it presents, in common with other species of the same family, a marked southward decrease in the size of the white transverse bars and spots of the dorsal plumage.

In the southern portion of the Mississippi Valley, the variation is in a tropical direction, and is merely due to the more northward extension there of tropical influences. In passing to the Plains and the Great Basin west of the Rocky Mountains, however, an entirely different phase of color variation is met with. Here, as a general rule, there is a loss of color, this region being characterized by the presence of subdued or faded tints in the representatives of species having a nearly continental range. The transition, however, is as gradual as is that of the climatic conditions, the paleness beginning near the eastern border of the great plains, and, increasing westward, reaching its extreme phase in the arid wastes of the almost wholly rainless districts of the far south-west,—South Nevada, Arizona, and the contiguous region westward and southward. In respect to this part of the subject, it is hardly necessary to say more than that the representatives of continental species found here are uniformly much paler than those inhabiting the adjoining regions; that in many cases the paler forms were originally described as distinct species, and are commonly recognized as varietally distinct, though found to inseparably intergrade with the neighboring darker forms. In addition to the general paleness, there is often an increase in the areas of white, and in some cases an accession of new ones.

The wooded, mountainous districts embraced in this region also give rise to peculiar local phases of color variation, to give a detailed account of which would too greatly extend the present paper. The tendency is mainly toward the development of more or less well-marked rufous or fulvous phases of coloration, with sometimes an accession of red, while not a few species have more than the usual amount of black. A most striking instance of increase of red at western localities is seen in the yellow-bellied woodpecker (*Sphyrapicus varius*), which, in some of its forms, ranges in the breeding season over the more northern and elevated wooded portions of the continent. In its eastern form the male has merely the chin, throat, and crown red, while in the female the red is restricted to the crown. In rare instances there is a trace also of a narrow red nuchal band. In the Rocky Mountain form, however, there is always a red nuchal band, the red on the throat is more extended in the male, and a small area of red appears also on the throat of the female. In the form met with in the Cascade Range the red begins to spread over contiguous portions of the plumage, while in the form occurring along the Pacific coast the red overspreads the whole head, neck, and breast, through which, however, the markings of the eastern birds can generally be readily traced. Here we have, at one end of the series, the red confined to a few distinct patches about the head, while at the other it extends over the whole anterior half of the body. Yet the intergradation between the two has been so fully traced that these diverse forms are now held, by competent authorities, as merely local races of a single species!

Another case of the increase of red over the same region is afforded by the golden-winged and red-shafted woodpeckers (genus *Colaptes*), in which yellow in the eastern form is replaced by red in the other; in the middle region of the continent the species being largely represented by individuals in which are variously combined the special characteristics of the two forms. In the present case the black cheek-patch of the eastern form is replaced by a red one in the western. Traces of the characteristics of the western type occasionally appear in the most eastern representatives of the eastern type, and, conversely, features of the eastern bird appear in the most western representatives of the western, showing at least their close affinity and probable community of origin.

The Pacific coast region from California northward is characterized by a great accession of color, all the continental species being here represented by forms much darker even than on the Atlantic coast. Here the coloration is duller than at the southward, though perhaps equally strong, the general tendency being to fuscous or dusky tints. We consequently find among the mammals and birds of the United States three strongly marked phases of color-differentiation among representatives of the same species, characterizing the three most strongly marked climatal regions,—a bright, strongly-colored form east of the Great Plains, a pallid form over the dry central region, and a deeply-colored fuscous form over the rainy, heavily-wooded region of the north-west coast. Examples of this differentiation are afforded by apparently all the species whose habitats extend entirely across the continent, the several local forms being in some species only more strongly marked than in others. Among mammals illustrations are afforded by different species of squirrels, hares, mice, lynxes, deer, etc.; and among birds by six or eight species of sparrows, a number of woodpeckers, several flycatchers, thrushes, and warblers, the meadow lark, various hawks, owls, etc. Generally these several geographical forms were originally described as distinct species, and many of them are still thought worthy of recognition by varietal names. As intermediate links began to be discovered, they were at first looked upon as the result of hybridity between the supposed distinct species whose characters they respectively combined; but eventually such links were found to be too frequent, and too general over the areas where the habitats of the several forms come together, to render such a supposition longer tenable, it finally appearing evident that they were only the connecting forms between merely local races or incipient species.

The local races of any given region, as compared collectively with those of contiguous regions, and the manner of their mutual intergradation, point plainly to some general or widely acting cause of differentiation. This is indicated by the constancy of the results, so many species, belonging to numerous and widely distinct groups, being similarly affected. Will the fortuitous, spontaneous results of natural selection yield a satisfactory explanation of these phenomena, or must we seek some more uniform and definitely acting cause? To briefly summarize the results above detailed, we have a somewhat uniform increase of size in some given direction affecting many species simultaneously and similarly over the same areas. We have a frequent enlargement of peripheral parts, affecting not a few but many species, and all in a similar manner, though in varying degrees. We have a very general increase in the depth or intensity of colors southward, a general loss of color in approaching the central, arid portions of the continent, and again an excessive increment of color under still different climatic conditions and over a different area. We find the increase of size among the individuals of any given species to be quite uniformly in the direction of the centre of distribution of the group to which the species belongs, this being especially well-marked in mammals. We find the increase in the size of peripheral parts,—as the external ear and the length of the pelage in mammals, and the size of the bill and length of the tail in birds,—to be in the direction of the regions where these parts meet respectively their greatest development,—the increase in color (especially among birds) toward the region where are developed the richest and most lustrous tints, the loss of color in the direction of the region where the greatest general pallor prevails. We find again that the enlargement of peripheral parts correlates with increase of temperature; the southward increase of color with an increase of atmospheric humidity and temperature, and consequently with the protective influences of luxuriant arboreal vegetation and clouds; and, conversely, the loss of color accompanying excessive aridity, a scanty vegetation, and an almost cloudless sky,—the conditions, in short, of all others the most powerfully effective in the blanching of color; and again the sombre, dusky tints of the north-west coast accompanying the most humid conditions of climate and the conditions generally most favorable for the protection or preservation of color. Are these merely accidental coincidences, or are they the evident results of cause and effect? Because the white winter livery of some of the northern species is more protective against cold than darker tints would be, or aids in concealing them in some cases (as in the hares and ptarmigans) from their enemies, or in other cases (as in the ermines and the Arctic fox) tends to aid them in stealing unperceived upon their prey, are they to be regarded as unquestionably the beneficial results of the working of natural selection? Because the dull gray tints of species inhabiting the semi-desert regions of the interior harmonize well with the general gray aspect of their surroundings, is this concordance the result again of the operation of the law of natural selection, the less favorably colored having been weeded out in the struggle for existence? Are the heavy, dull colors of the humid region of the north-west the result, again, of the necessary influence of natural selection in perpetuating only the individuals whose colors best accord with their sombre conditions of environment? Has the same action brought about the bright, rich coloration of birds, insects, and other animals under the warm humid conditions of the hotter parts of the earth, preserving the ratio of brilliancy of coloration with that of the conditions that everywhere most favor such differentiation? Finally, is the exact correlation of the changes in coloration with the gradual change of climatic conditions in passing from one geographical region to another the result in like manner of the long-continued weeding out of the less-favored? Or are these modifications severally due to the *direct* action of the conditions of environment?

In answering these questions it may be well to glance first at the nature of the theoretical origin of differentiation through the influence of natural selection as expounded by the leading advocates of the theory. As is well known, all the individuals of a species found at the same locality (differences resulting from sex and age aside) are not all cast in the same mould, but differ constantly, the average range of purely individual variation in general size and in the size of different parts ranging (in birds and mammals) from eight to fifteen or twenty per cent. of the average size for the species, with a corresponding amount of variation in color. These variations are found to tend in every conceivable direction, and it of course follows that some of them must be in directions exceptionally favorable to the species. The theory of modification by the action of natural selection only supposes that the stronger or otherwise more favored individuals transmit their favorable qualities to their offspring, and that the latter, in consequence of their inherited advantages, multiply more rapidly than their less favored relatives; that these favorable deviations from the parental stock become in subsequent generations more pronounced, and that the original form is eventually overpowered and supplanted by its modified descendants. From the same original stock may be conceived to arise, even simultaneously, other forms diverging in different, though still favorable, directions, these in turn giving rise to several lines of descent, occupying perhaps different portions of the habitat of the original species, where they also multiply and become dominant, and eventually pass on from the stage of incipient species to more or less widely differentiated types. These premises being admitted,—and they are certainly within the bounds of reasonable conception,—only the element of time, apparently, is requisite for the development of an endless variety of unstable forms, constantly increasing in number and following divergent lines of development, and thus capable, apparently, of giving rise to all the diversity of organisms at present peopling the earth.

But there are many adverse circumstances with which the favored forms have in the outset to contend, and to which, in the majority of instances, they must succumb. These are, first, the minuteness of the first favorable divergence, the isolation of the individuals in which it appears, and consequently the impossibility of such individuals pairing with others similarly favored, and the consequent tendency of the offspring to possess the favorable characters in a less rather than in a greater degree than the parent, and to be absorbed into the original stock. Secondly, in case the incipient advantages are perpetuated, as it is necessary to suppose, the new offshoot must originate from a single point, and spread thence gradually to contiguous regions as its representatives slowly multiply.

But it is supposed, again, that new forms are not always thus gradually evolved from minute beginnings, but sometimes—perhaps not unfrequently—arise by a *saltus;* that individuals may be born widely different from their parents, differing so widely and persistently as not to be so readily absorbed by the parental stock. In proof of this, instances are cited of new species apparently appearing suddenly, and of varieties thus originating under artificial conditions resulting from domestication.

Granting that new forms may thus arise, although as yet few facts have been adduced in its support, they are necessarily at first local, and in no way accord with the observed geographical differences that characterize particular regions, and which affect similarly many species belonging to widely different groups.

The direct influence of climatic or geographical conditions upon animals is, in the main, ignored by the leading exponents of the doctrine of natural selection. To quote Mr. Darwin's own words on this point:—

"The action of climate seems at first sight to be quite independent of the struggle for existence; but, in so far as climate chiefly acts in reducing food, it brings on the most severe struggle between the individuals, whether of the same or of distinct species, which subsist on the same kind of food. Even when climate, for instance extreme cold, acts directly, it will be the least vigorous, or those which have got the least food through the advancing winter, which will suffer most. When we travel from south to north, or from a damp region to a dry, we invariably see some species gradually getting rarer and rarer, and finally disappearing; and the change of climate being conspicuous, we are tempted to attribute the whole effect to its direct action. But this is a false view: we forget that each species, even where it most abounds, is constantly suffering enormous destruction at some period of its life from enemies, or from competitors for the same place and food; and, if these enemies or competitors be in the least degree favored by any slight change of climate, they will increase in numbers, and, as each area is already fully stocked with inhabitants, the other species will decrease. When we travel southward and see a species decreasing in numbers, we may feel sure that the cause lies quite as much in other species being favored, as in this one being hurt. So it is when we travel northward, but in a somewhat lesser degree, for the number of species of all kinds, and therefore of competitors, decreases northwards; hence, in going northward, or in ascending a mountain, we far oftener meet with stunted forms, due to the *directly* injurious action of climate, than we do in proceeding southward or in descending a mountain. When we reach the Arctic regions, or snow-capped summits, or absolute deserts, the struggle for life is almost exclusively with the elements. That climate acts in main part indirectly by favoring other species, we may clearly see in the prodigious number of plants in our gardens which can perfectly well endure our climate, but which never become naturalized, for they cannot compete with our native plants, nor resist destruction by our native animals." [1]

While there is perhaps little reason to question the general correctness of the above-quoted generalizations, they have little bearing upon the question of the modification of species by the direct action of climatic conditions, but relate mainly to such unfavorable climatic influences as tend toward the extinction of species, or to the circumscription of their ranges. Indeed, the

---

[1] "Origin of Species," 5th ed., pp. 80, 81.

phenomena of variation detailed in the foregoing pages were almost wholly unknown at the time the earlier editions of the "Origin of Species" were published, and have hardly as yet become the common property of naturalists. Gradual decrease in size southward in hundreds of species inhabiting the same continent, or a gradual increase or decrease in color in given directions on a similarly grand scale, are facts but recently made known, and have not as yet been very fully discussed by evolutionists of the purely Darwinian school. Mr. Darwin, indeed, in referring to the "effects of changed conditions" upon animals, alludes to facts of a similar character,—as the alleged brighter colors of European shells near their southern limit of distribution and when living in shallow water, and the more sombre tints of birds that live on islands or near the coast under overcast skies, as compared with those of the same species living more in the interior, etc.; but is in doubt as to how much should be attributed, even in such cases, "to the accumulative action of natural selection, and how much to the definite action of the conditions of life." "Thus," he says, "it is well known to furriers that animals of the same species have thicker and better fur the farther north they live; but who can tell how much of this difference may be due to the warmest-clad individuals having been favored and preserved during many generations, and how much to the action of the severe climate? for it would appear that climate has some direct action on the hair of our domestic quadrupeds."[1] Since, however, it happens that some species do not vary at all, although living under the most opposite climates, he is thereby inclined "not to lay much weight on the direct and definite action of the conditions of life," though he fully admits "that strong arguments of a general nature may be advanced on the other side." "In one sense," he adds, "the conditions of life may be said, not only to cause variability, but likewise to include natural selection; for the conditions determine whether this or that variety shall survive."[2] But he says again: "I believe that natural selection generally acts slowly in effecting changes, *at long intervals of time, and only on a few of the inhabitants of the same region.*" In a latter work, however, he

[1] "Origin of Species," pp. 166, 167. [2] Ibid., p. 168.

refers to the variation in color with locality seen in many species of birds in the United States, and says explicitly, in reference to northern and southern localities, "this seems to be the direct result of the difference in temperature, light, etc., between the two regions." [1]

There is, however, a vast amount of unquestionable proof of the direct and constant action of climate and other conditions of life upon animals, and that such geographical variations as the thicker and softer fur of mammals inhabiting cold regions, smaller size and brighter colors at the southward, etc., etc., do not require the action of natural selection, in its strict and proper sense, for their explanation. It is well known, for instance, that a flock of fine-wooled sheep, when taken to a hot climate, rapidly acquire a coarser and coarser fleece, till, in a few generations, it nearly loses its character of proper wool, and becomes simply hair; that the change affects simultaneously the whole flock, and is not brought about by one or two individuals acquiring a coarser fleece and through their descendants modifying the character of the herd. Furthermore, in the case of sheep, it is well known that certain countries are very favorable to the production of a fine fleece, and that fine-wooled breeds, even by man's aid, cannot be perpetuated in other regions. Again, it is a fact of common observation that in birds and mammals colors become more or less faded toward the moulting season simply by the direct action of the elements,—the tints of the fresh and the long-worn plumage or pelage being more or less strikingly different in the same individuals,—and that this contrast at different seasons is more marked in arid than in moist regions, through the greater bleaching effect of a dry heated atmosphere and the more intense dazzling sunlight of regions that are not only cloudless, but lack the protection afforded by abundant vegetation.

While so much is claimed by the writer as due to the direct action of climatic causes, it is admitted also that habits and food, and other conditions of life than those resulting from climate, have a marked effect in determining modifications of form and color among animals. A scarcity of a favorite kind of food

---

[1] "The Descent of Man," 2d ed., p. 225.

will undoubtedly force species to subsist upon the next best that offers, which may be so different as to modify certain characters and fit the species to live upon the less desired food. A change of food may lead to modification of dentition, the muscles of mastication, and the organs of digestion, and, correlatively, of other organs or parts of the body; the modification, however, arising simultaneously among all the descendants of the individuals thus driven to a change of diet, instead of appearing first in a single individual and becoming perpetuated in its descendants alone. Entomologists have found that, among insects of the same species, the forced or voluntary use of different food-plants gives rise to modifications of color and structure, and hence result in what have been termed phytophagic varieties or sub-species, and that man can also effect such changes at will by simply changing the food of the species. Again, the geological character of a country is well known to have a marked effect upon the size and color of animals inhabiting it, as is strikingly illustrated among molluscous animals, whose abundance, and even presence, is largely dependent upon the constituents of the soil. Over regions of the United States, for example, where the underlying rock is non-calcareous, the species are both few in number and sparsely represented, while in other regions, where limestone abounds, but which are in other respects essentially the same, the species are far more numerous and far more abundantly represented. In respect to the fresh-water mussels, those of the same species from different streams are easily distinguishable by differences in the thickness of the shell, in color, shape, and ornamentation, so that the character of the shells themselves affords a clew to the locality of their origin. At some localities the species tend to become tuberculous or spinous, this being particularly the case toward the southward; at other localities they acquire a very much thickened shell, or different colors, the same characteristics appearing simultaneously in quite diverse species, and thus becoming distinctive of particular localities. In regard to mammals, measurements of large series of the skulls of minks, martens, squirrels, and other native species, show that the representatives of these species living in northern New England and north-eastern New York are smaller than the representatives of the same species occurring in the limestone districts of

Pennsylvania and the States more to the westward, and the same is true of the different kinds of domestic cattle. This is in opposition to the law of decrease in size southward that elsewhere and generally characterizes these same species, and seems obviously related to the geological character of the country at these localities; small size, in opposition to a general law, occurring over northern non-calcareous districts, and larger size more to the southward where the underlying rock is limestone. In this case the difference obviously results from the direct action of the conditions of habitat upon every individual rather than from "slowly effected changes" originating in "only a few of the inhabitants" of these respective districts.

Use and disuse of organs, through changes of habit resulting from changed conditions of environment, must result in some modification of the organ involved. As an example may doubtless be cited the passerine birds of some of the smaller, remotely-situated islands, as the Guadelupe and Galapagos groups, where recent investigations have shown that most of the species differ similarly in several features from their nearest allies of the mainland, and of which they are unquestionably insular forms. These differences consist in the greater size of the bill, shorter wings, longer tails, and darker colors. The sedentary life necessitated by the confined habitats of species thus situated would naturally act more or less strongly on the organs of flight, and a reduction in the size of the wing would follow;—not necessarily through the round-about process of natural selection, through the modification originally of a single individual, but by the direct action on all the individuals alike of the changed conditions of life.

It is doubtless unnecessary to further multiply examples of the modification of animals by the direct action of the conditions of life. The subject is one that can be but imperfectly treated at best in a short paper like the present. The illustrations have here been drawn from a limited geographical field, and mainly from among the two higher classes of vertebrates. There are, however, abundant indications that other fields and other classes would yield results equally confirmatory of the direct action of physical conditions in the evolution of specific forms among animals and plants. Changes in environing conditions will, however, go but a short way toward explaining the origin of the great diversity of structure among existing organisms; the character of the food, habit, or the increased use or the disuse of particular organs, may explain many of the modifications, leaving a large share of the work to as yet unknown causes. Natural selection, as sometimes defined, is made to cover all causes of differentiation, it being stated by Mr. Darwin himself that, if organic beings undergo modification through changes in their conditions of life, "uniformity of character can be given to their modified offspring solely by natural selection preserving similar favorable variations." In its strict sense, variation by natural selection results only through favorable differences appearing at first in single isolated individuals, which transmit these favorable qualities to their offspring, in virtue of which they multiply till they outnumber, crowd out, and finally destroy the less favored form from which they originated.

It is hardly conceivable, for example, how the peculiar structure seen in the woodpecker, the kingfisher, the swift, the heron, or the duck, or the peculiar dentition and correlated characters of the rodents, the ruminants, or the shrews and moles, as compared with the Carnivora, can have been initiated by the direct action of climatic conditions, however much other conditions of environment may have favored the development of these diverse types.

Having thus far mainly detailed merely facts and coincidences relating to the subject of variation with locality, it may be well in conclusion to consider more fully some of the possible or probable causes of purely geographical variation. In regard to geographical variation in color, it seems evident that high temperature, conjoined with moisture, favors increase of color and especially the acquisition of lustrous tints, while moisture alone favors simply increase in depth or the production of dull, heavy, and especially fuscous phases of coloration; on the other hand, that aridity and exposure favor the loss of color. The latter is due apparently no less to the influence of a dry and often intensely heated atmosphere than to the direct action of light intensified by the reflection of the sun's rays from almost verdureless sands. That the latter conditions act powerfully in blanching color there is most abundant proof. Hence we have the necessary correlation of increase of bright rich tints of coloration with the increase of atmospheric humidity. In respect to the enlargement of peripheral parts at the southward, it is obvious that a high temperature favors the more rapid circulation of the blood in these parts, while, as is well known, a low temperature produces the opposite effect, and necessarily retards their development.

With the decrease in size among birds, there has been observed a decrease of vivacity and deterioration of song, which may reasonably be attributed to the enervating influence of a high temperature. Since the northern types of animals reach their highest physical development toward the northward, it seems fair to suppose that decrease in size southward may be directly due to the enfeebling influences of increase of temperature, since certainly it cannot be attributed, in the majority of cases at least, to greater scarcity of food, for, in many instances, just the reverse obtains. This supposition is in accordance with the known effects of similar climatic conditions upon the northern races of man, which reach their greatest vigor and highest intellectual status under temperate conditions of climate, and deteriorate, both physically and mentally, on removing to intertropical regions. Again, the mammals and birds of the United States reach their maximum size within the United States under the stimulating climate of the region drained by the Upper Mississippi and Upper Missouri Rivers, being, as a rule, larger here than in corresponding latitudes more to the eastward. The decrease in size toward both the northern and southern borders of the habitat of a given species or genus, of which there are many marked instances, further shows that size varies with the varying conditions of habitat, and reaches the maximum only where the conditions are most favorable to the life of the species.

Much has been written respecting the influence of climate on man, and many speculations have been indulged in in relation to the part the conditions of life have taken in bringing about the diversity at present existing among the different races. A striking parallelism is often observable between the leading features of geographical variation among animals and the physical differences that obtain among nations or races of men inhabiting the same areas and subjected to the same influences. While civilized man is, in a measure, less the subject of such influences than the lower animals, he is not wholly above them. Certain regions more favor both physical and intellectual development than others; and these prove to be, as would be expected, the milder temperate portions of the globe, where the struggle for a mere vegetative existence is reduced to a minimum.

The influence of different climatic conditions upon members of the same nationality find exemplification in different parts of our own country, and are so obvious as to be the subject of frequent observation and comment. The same original stock is found to gradually develop certain peculiar physical and mental characteristics when placed under diverse conditions of climate, certain localities more favoring intellectual growth and activity than others; just as certain regions are characterized by the frequent occurrence of particular diseases, which in other regions are exceptional. While humidity and a high temperature, when combined, are found to be enervating and deteriorating, a clear dry atmosphere favors vigor of both mind and body. But the subject of the influence of climatic conditions upon man is too vast to be entered upon in detail in the present connection. The study of man from a geographical standpoint, or with special reference to conditions of environment, offers a most important and fruitful field of research, which, it is to be hoped, will soon receive a more careful attention than has as yet been given it.

In conclusion, a few words seem called for concerning the question, What is a species? as well as in respect to the bearing of the general facts above detailed upon the evolution of specific forms.

As is well known, the belief that species were distinct and immutable creations was long the prevailing one among naturalists. Yet the question of what constitutes a species is one about which endless discussions have arisen, and one respecting which the most discordant opinions have been held by naturalists equally eminent in their respective fields of research. The amount and kind of difference necessary to characterize a species has been variously defined; forms that some have considered as specific others have regarded as merely varieties, and the reverse. In certain groups of organisms intermediate forms have been constantly met with, constituting steps of easy intergradation between quite diverse types. Such forms have been, and still are, held by some writers as varieties of a single species, and by others as constituting a group (genus or sub-genus) of distinct, but nearly related, species. Through the frequent discovery of such intergradations, however, the instability of so-called "species" has been made manifest, and the contrary doctrine of the stability or fixity of species refuted. Indeed, naturalists now generally agree that the terms variety, species, genus, sub-genus, family, sub-family, super-family, and the like, are but conventional and more or less arbitrary designations for different degrees of differentiation,—convenient formulæ for the expression of general facts in biology. Not a few high authorities even maintain that the differences which characterize these several groups are of the same nature, differing only in degree, in opposition to others who hold that they are based on different categories of structure, or on differences of *kind* rather than of degree. The falsity of the latter view is shown more and more clearly with the increase of our knowledge of the structure and affinities of animals.

While formerly species were considered as necessarily characterized either by differences of a particular kind, or by a certain amount of difference, the present tendency is to regard neither as a sufficient criterion, the test of specific diversity being merely absence of intergradation,—in other words, breaks in the continuity of closely allied beings. Local races, or geographical forms, are thrown together under one specific designation whenever they are found to intergrade, however diverse may be their extreme phases of differentiation. The term species is now made to cover groups which were, not many years since, frequently regarded as sub-genera, or even genera,—the forms then supposed, in numberless instances, to be "good species" now ranking merely as sub-species. The reduction in the number of species has necessarily entailed a considerable reduction in the number of currently accepted genera, which in turn are limited by hiati rather than by any given amount or particular kind of difference. It was formerly urged against the theory of evolution that its advocates could point to no instance of the gradual change of one species into another, and that, until this was done, the theory was untenable. Among the species of North American vertebrates recognized as valid ten years ago, hundreds of instances can now be cited of thoroughly proven intergradation; forms then regarded as unquestionable species being found to be but connected phases of one and the same specific type, which assumes, at remote localities, under the evident action of climatic agencies, phases widely diverse, which gradually merge the one into the other through the individuals inhabiting the intervening districts. So long as species are based on the absence of intergradation,—and biologists have found no other satisfactory criterion for their limitation,—there can of course be no passage of one species into another. Let, however, some of the connecting links become extinct, and these now intergrading forms would be resolved into distinct species. In this way insular and other local forms are passing beyond the so-called varietal stage, and species are similarly tending to generic distinctness. That varieties may and do arise by the action of climatic influences, and pass on to become species, and that species become, in like manner, differentiated into genera, is abundantly indicated by the facts of geographical distribution and the obvious relation of local forms to the conditions of environment. The present more or less unstable condition of the circumstances surrounding organic beings, together with the known mutations of climate our planet has undergone in past geological ages, points clearly to the agency of physical conditions as one of the chief factors in the evolution of new forms of life. So long as the environing conditions remain stable, just so long will permanency of character be maintained; but let changes occur, however gradual or minute, and differentiation begins. If too sudden or too great, extinction of many forms must result, giving rise to breaks in the chain of genetically connected organisms. In the deep abysses of the sea, where the temperature is low and stable, where the conditions of life must have remained almost unvaried since the early geological periods, the same low organisms still exist that were the prevailing forms of life when life first dawned upon the earth. The recent explorations of the depths of the sea have gone far to prove that stability of organic forms is in direct ratio to the stability of the conditions of existence, while the facts of geographical distribution show that change of structure and diversity of life are directly related to the physical conditions of habitat.

JOEL A. ALLEN.

PROGRESS OF ORNITHOLOGY IN THE UNITED STATES DURING THE LAST CENTURY.

PROGRESS OF ORNITHOLOGY IN THE UNITED STATES DURING THE LAST CENTURY.

BY J. A. ALLEN.

EARLY PAPERS.

PRIOR to the year 1808, when the first volume of Alexander Wilson's great work was published, little had been written on American birds by Americans. A few lists of the birds of limited portions of the United States that appeared during the last fifteen years of the eighteenth century constituted our whole ornithological literature at that date. The first of these was a list of about one hundred and twenty species, published by Thomas Jefferson in 1787, in his celebrated Notes on Virginia. This is a catalogue of the species described by Catesby, with the addition, in parallel columns, of the Linnæan and common names, and of the popular names of a few species not described by Catesby, — merely a nominal list of no special importance. This was followed, in 1791, by William Bartram's enumeration of the birds known to this excellent naturalist as inhabitants of the eastern portion of the United States, published in his Travels in North and South Carolina, Georgia, East and West Florida, etc.; this list embraced about two hundred and fifteen species, and, though mainly a nominal one, was accompanied by typographical signs indicating the range of the species at the different seasons of the year, while a few were described at some length in other parts of the same work. In many cases Linnæan names were used for the designation of the species; but to the larger number he gave new names, which, being generally unaccompanied by descriptions, are now to a very large extent undeterminable. This paper is especially noteworthy as forming the first important contribution by an American to American ornithology.

The next paper in order of time was by Jeremy Belknap, who in his History of New Hampshire (vol. iii.), published in 1792, devoted ten pages to birds, giving a list of about one hundred and sixty species. The current Linnæan names were used, to which a few new names were added. Although no descriptions are given, the names generally clearly indicate the species meant. Though not commonly referred to by scientific writers, it forms a highly important paper and one worthy of attention. Samuel Williams, in his Natural and Civil History of Vermont (published in 1794), devoted also about the same number of pages to the birds of Vermont. Scientific names of about fifty species are given in full, and the generic names of a dozen others. The paper also embraces a number of pages of valuable notes. These two last-named articles show that several species were then common in New Hampshire and Vermont that long since became extinct there, and are hence papers of considerable historic interest. In 1799, Benjamin S. Barton published his Fragments of the Natural History of Pennsylvania; this rare folio contains a list of the birds of Pennsylvania, with notes on their migrations.

Although John Clayton published a valuable paper on the birds of Virginia as early as 1693 (Philosophical Transactions, vol. xviii.), and although many enumerations of a few of the birds of many portions of the Atlantic States, from Florida to Maine, were made by different writers during the seventeenth and eighteenth centuries, some of which possess especial value in a historic point of view, the four papers already mentioned constitute the bulk of the ornithological literature written by Americans prior to the time of Alexander Wilson.

GENERAL WORKS.

Alexander Wilson, a Scotchman by birth, came to America in 1794, and some ten years later conceived the idea of writing a history of the birds of his adopted country. Receiving hearty encouragement from his kind friend, Mr. William Bartram, he entered seriously upon his great work in 1805, to which he devoted almost his whole time and energy till his untimely death in 1813. The first volume appeared in 1808, followed by eight others, the last two of which were published after his death, under the editorship of Mr. George Ord. Of these nine folio volumes, accompanied by colored plates, Bonaparte wrote in 1825 as follows: "We may add, without hesitation, that such a work as he [Wilson] has published in a new country is still a desideratum in any part of Europe." His earlier figures were sometimes stiff and awkward, but they were generally of a high grade of excellence for that time, and his descriptions were concise and exact. His accounts of the habits of many of the species have rarely been surpassed in point of truthfulness or felicity of expression. Never extended by irrelevant matter, some of them are models of descriptive writing, evincing a poet's love and appreciation of nature. Although adopting Pennant's system of classification, by far the best then extant, the species were not systematically grouped, but taken in the order in which they most conveniently came to his hand. Had he lived to complete his work, many others would doubtless have been added, and the whole rearranged in accordance with his own ideas of their affinities.

The work so well begun by Wilson was continued by Bonaparte, whose American Ornithology, or the Natural History of the Birds inhabiting the United States, not given by Wilson, was extended to four volumes, similar in style to those of Wilson. The first appeared in 1825, the second and third in 1828, and the fourth in 1833.

Several editions of Wilson's work were subsequently issued, either separately or combined with Bonaparte's continuation. The first American edition was that published by George Ord, in 1828–29, in three octavo volumes of text and a folio volume of plates. In this work the species are arranged systematically, but the editor adhered to the original text, correcting merely a few erroneous references and verbal inaccuracies. In 1831, Jameson published in Edinburgh an 18mo edition of Wilson's and Bonaparte's works. The succeeding year this was followed by Jardine's edition (London and Edinburgh, 1832), in three octavo volumes, of the same authors. This last was reissued in this country in 1840, in one volume, with the addition of a Synopsis of the Birds of North America, by Dr. T. M. Brewer. Of this there were subsequently several reprints from the same stereotyped plates.

In 1827, John James Audubon began the publication of his celebrated work on North American birds, which was not completed till 1839. The whole work forms five octavo volumes of text, with an elephant folio atlas in four volumes of four hundred and thirty-five plates. The text was published in Edinburgh, with the title of Ornithological Biography; the plates in London, as Birds of America. This magnificent work remains as yet unequaled in respect to its illustrations, which are unrivaled in point of accuracy and life-like character, the birds being all represented of the size of life. In his animated descriptions there is at times a tendency to exaggeration and redundancy of personal incidents. The species are arranged according to the convenience of the author, a systematic arrangement being in such a work obviously impracticable. In 1839, however, on its completion, the author published his Synopsis of the Birds of North America. In this work the nomenclature is revised and greatly changed, principally through the adoption of many of the then recently introduced generic designations. This Synopsis (one volume, octavo, Edinburgh) was a methodical catalogue of all the species at that time known to inhabit North America north of Mexico, and was intended to serve as a systematic index to his Ornithological Biography, and Birds of America. The work, however, is much more than this, giving, as it does, the characters of the families, genera, and species, the range of each species, and numerous bibliographical references. It includes also a few species not given in his larger works.

Subsequently Audubon republished his Ornithological Biography, and Birds of America in a single work, under the title of Birds of America. This is simply a reissue of the text of the former, with the species systematically arranged under the names employed in the Synopsis, and the addition of the plates of the large folio work, reduced by the camera lucida. This work was published at Philadelphia (1840 to 1844), in seven imperial octavo volumes. It has also been since republished in New York, with chromo-lithographic illustrations of a character far inferior to those of the original work. The appendix to the seventh volume includes a few species previously unpublished by this author.

In 1832 appeared the first volume of Thomas Nuttall's Manual of the Ornithology of the United States and of Canada, embracing the land-birds. This was followed in 1834 by the second volume, devoted to the water-birds. A new edition of the land-birds was published in 1840, including all the species discovered in the mean time by Townsend and Audubon. These two volumes (12mo, Boston), illustrated with numerous small wood-cuts, contain a succinct history of the birds of North America known at the time of their publication, and, being written in a pleasing style, form a work that has been deservedly popular, despite many inaccuracies, and has recently been republished.

The next general work on the birds of North America was John Cassin's Illustrations of the Birds of California, Texas, Oregon, British and Russian America. This was published in parts (Philadelphia, 1853–55), and was intended to contain descriptions and figures of all North American birds unfigured by previous authors, and a general synopsis of North American ornithology. It was issued in ten parts, with fifty plates. The author's plan of continuing the work was probably superseded by the preparation soon after of a far more important work, in the labor of which Mr. Cassin shared.

Dr. T. M. Brewer's North American Oölogy (Smithsonian Contributions, vol. xi.), published in 1857, is the only American work devoted exclusively to the eggs and breeding habits of North American birds. It is, however, as yet uncompleted, Part I. (quarto, pp. 140) embracing the *Raptores* and *Fissirostres*, being the only portion yet published. The five colored plates give figures of the eggs of all the species of these groups at that time accessible, while the text contains a very full account of the distribution and breeding habits of the species embraced.

In 1858 appeared the well-known General Report on the Birds of North America, forming volume viii. of the Pacific Railroad Reports of Explorations and Surveys. This was the joint work of Spencer F. Baird, John Cassin, and George N. Lawrence, the parts devoted to the *Raptores*, *Grallæ*, and *Alcidæ* being by Mr. Cassin, while Mr. Lawrence wrote that relating to the *Longipennes*, *Totipalmes*, and *Colymbidæ*, the remainder being by Professor Baird, the whole work making a quarto volume of over one thousand pages. The abundant material at the command of these eminent authors, and the elaborate and critical manner in which it was treated, render this work by far the most complete and important, in a scientific point of view, that had at that time appeared relating to North American ornithology, or, in fact, to any similar area anywhere. It is, however, a strictly technical treatment of the subject. The special reports on the ornithology of the different routes explored, contained in other volumes of this series of reports, supplemented this general work with much biographical matter, in connection with which appeared thirty-eight finely executed colored plates of species described in the general and special reports, but previously unfigured by American authors. This was followed in 1859 by Professor Baird's report on the Birds of the United States and Mexican Boundary Survey, with twenty-five admirable plates of previously unfigured species. In 1870, the text of the General Report was re-issued, with some additional matter, under the title of Birds of North America, with an accompanying volume of one hundred plates, including those above-mentioned as accompanying the special reports of the Pacific Railroad and Mexican Boundary Surveys.

In 1864 Professor Baird began the publication of his Review of American Birds (Part I. North and Middle America, Smithsonian Miscellaneous Publications, No. 181). Installments of the work were published at intervals during the next two years, the last signature bearing the date of "June 9, 1866." The work, so far as published, forms an octavo volume of 450 pages, taking the *Oscines* as far as the genus *Collurio.* The very thorough character of this much-needed work renders it a source of sincere regret that its busy author has not found time to carry it forward to completion.

In 1869 appeared D. G. Elliot's work (two volumes in one, folio, New York), entitled New and Unfigured Species of the Birds of North America, containing sixty-four colored plates, and short critical notices of one hundred and fourteen species. This was followed in 1870 by Cooper's Birds of California, edited by Professor Baird "from the manuscript and notes of J. G. Cooper." We have as yet of this important work only the first volume (forming volume i. of the Zoölogical Reports of the Geological Survey of California, Professor J. D. Whitney, director), embracing the land-birds. This is a quarto of six hundred pages with life-size colored figures of the heads of each species and small full-length figures of some species of each genus, *inserted in the text.* This method of illustration was novel and advantageous to the student, the work being intended as a manual of the ornithology of the whole region west of the Rocky Mountains north of Mexico.

The next general work on North American birds is Dr. Elliott Coues's Key to North American Birds (one volume, quarto, with upwards of 250 wood-cuts and six steel plates), published in October, 1872. This highly useful and deservedly popular work is unique in ornithological literature, being a manual of the birds of North America, designed especially for the use of students, in which are introduced analytical tables for the determination of the species, similar in character to those so successfully adopted in botanical text-books. This was followed, in 1874, by the same author's Field Ornithology and Check-List of North American Birds, intended as a supplement to the Key, the first part being a "manual of instruction for procuring, preparing, and preserving birds," and the latter, as its title implies, a "check-list" of the species.

By far the most important recent work on North American ornithology, however, is the joint work of Prof. S. F. Baird, Dr. T. M. Brewer, and Mr. Robert Ridgway, published under the title of Birds of North America. The first three volumes of this indispensable work, embracing the land-birds, appeared in 1874, the portion relating to the water-birds being still in course of preparation. The more technical portion is the joint work of Professor Baird and Mr. Ridgway, while the biographical portion is written by Dr. Brewer. This work, it is needless to say, represents the most advanced views of our best authors, and must long remain the leading authority on the subject of North American ornithology. It is illustrated with five hundred and ninety-three wood-cuts, devoted largely to the external anatomy, and sixty-four colored plates giving life-size figures of the heads of the greater part of the species and varieties. Later works possessing a general character are Dr. Elliott Coues's Birds of the Northwest (1874), and Mr. H. W. Henshaw's recent Report (1875) upon the ornithological work of the Survey of the Territories West of the One Hundredth Meridian. The first of these is an octavo of nearly eight hundred pages, forming No. 2 of the Miscellaneous Publications of Dr. F. V. Hayden's Geological Survey of the Territories. The work is intended as a hand-book of the region drained by the Missouri River and its tributaries, and hence embraces a wide field. Giving as it does a summary of the ornithology of this extensive region, combining much new matter with much previously published in scattered papers, and the fullest tables of bibliographical references for the larger part of the birds of North America that has yet appeared, it is a work of the highest value to students of American ornithology.

Mr. Henshaw's work (Chapter III. of vol. v. of the Reports of the Geological and Geographical Surveys West of the One Hundredth Meridian, under Lieut. G. M. Wheeler) is limited to the actual work of the Wheeler Survey, of which it presents a general systematic summary, based on the collections and field-notes made chiefly by the author and Drs. Yarrow and Rothrock, and Messrs. Bischoff and Aiken, and is hence made up wholly of original matter, adding largely to our knowledge of the ornithology of this previously little-known region. It is accompanied by fifteen chromo-lithographic plates of previously unfigured species and varieties, and embraces about five hundred quarto pages of text.

WORKS AND PAPERS OF A SPECIAL OR LOCAL CHARACTER.

The long list of general works by no means comprises all the important contributions made by Americans to North American ornithology. The special papers, many of them of a high scientific value, are too numerous to be mentioned even by name within the limits of the present paper. These number several hundred, varying in length from a few pages to hundreds each. While a considerable proportion are limited to the descriptions of a few new or little-known species, or to the enumeration of the species found at particular localities, others are exhaustive monographs of genera or families, or are devoted to a discussion of general questions of nomenclature, of the geographical distribution of the species, or of laws of geographical variation.

Of papers of a strictly local character may be mentioned the Rev. W. B. O. Peabody's Report on the Birds of Massachusetts (Fishes, Reptiles, and Birds of Massachusetts, 1839); Zadock Thompson's chapter on the Birds of Vermont (Natural, Civil, and Statistical History of Vermont, 1842); J. P. Giraud's Birds of Long Island (8vo, 1844); Dr. J. E. Dekay's Report on the Birds of New York (New York Zoölogy, Part II., one vol. 4to, with colored plates, 1844); E. A. Samuels' Birds of New England (8vo, pp. 600, 1867); C. J. Maynard's Birds of Florida (now publishing in parts); and T. G. Gentry's Life-Histories of the Birds of Eastern Pennsylvania (vol. i., 1876), the two latter still in course of publication. Linsley's Catalogue of the Birds of Connecticut (American Journal of Science and Arts, first series, vol. xliv., 1843), and Wm. M. and S. F. Baird's List of the Birds of Carlisle, Pennsylvania (ibid., vol. xlvi., 1844), are noteworthy, as being the first of a long series of papers of a strictly faunal character that have contributed so much to our knowledge of the distribution of our birds, their periods of migration, and their seasonal ranges throughout the eastern half of North America from Labrador to Florida. For notes on the avifauna of Labrador and Anticosti, we are indebted to Coues, Verrill, and Packard; Drs. Brewer and Bryant have written on the birds of New Brunswick and Nova Scotia; Cabot, Holmes, Boardman, Verrill, and Hamlin on those of Maine, and Maynard and Brewster on those of Maine and New Hampshire; Emmons, Brewer, Putnam, Samuels, Allen, and Maynard on those of Massachusetts; Linsley and Wood, and others, on those of Connecticut; Coues and Brewer on those of New England in general; Lawrence and Fowler on those of New York; Abbott and Turnbull on those of New Jersey; Taylor, the Bairds, and Barnard on those of Pennsylvania; Coues and Webster and Jouy on those of the District of Columbia; Scott and Brewster on those of West Virginia; Burnett, Gibbs, Cope, and Coues on those of North and South Carolina; Bryant and Allen on those of Florida; Kirtland, Read, Kirkpatrick, and Wheaton on those of Ohio; Haymond and Allen on those of Indiana; Kennicott, Pratten, Allen, Ridgway, and Nelson on those of Illinois; Kneeland, Hughs, and Covert on those of Michigan; Head, Trippe, and Hatch on those of Minnesota; Hoy and Barry on those of Wisconsin; Allen, Parker, and Trippe on those of Iowa; Hoy on those of Missouri; Roemer, McCall, and Butcher on those of Texas; Coues, Allen, Snow, and others on those of Kansas; Allen, Aiken, Holden, Trippe, and Henshaw on those of Colorado and Wyoming; Hayden, Cooper, Allen, and Coues on those of the Upper Missouri country; Townsend, Cooper, Suckley and Bendire on those of Oregon; Baird, Allen, Ridgway, and Henshaw on those of Utah; Henry, Baird, and Henshaw on those of New Mexico; Coues, Henshaw, and Yarrow on those of Arizona; Ridgway, Henshaw, and Yarrow on those of Nevada; Xantus, Feilner, Brewer, Gambel, Heerman, Cooper, Ridgway, Nelson, and Henshaw on those of California; and Dall, Bannister, H. W. Elliott, and Coues on those of Alaska. Other writers on the birds of the far West are Say, Woodhouse, Kennerly, Newberry, Suckley, Stevenson, and Merriam.

Among the numerous papers of a critical or monographic character may be mentioned Bonaparte's Observations on the Nomenclature of Wilson's Ornithology, published in 1824–26 (Journal of the Academy of Natural Science, Philadelphia, vols. iii., iv., v.), in which were first introduced into the annals of North American ornithology a large proportion of the generic names now in current use, a few only of which were first proposed in this essay. It created, however, a revolution in the Wilsonian nomenclature. Audubon's Synopsis (1839) again brought down the nomenclature to the date of its publication, through the adoption of the changes made necessary by the further increase of knowledge. The Audubonian nomenclature was generally in use in this country till the appearance of Baird, Cassin, and Lawrence's great work in 1858, when a number of new generic and subgeneric names were proposed, and the nomenclature again modernized. Comparatively few generic names have since been introduced, but revisionary work of a somewhat different character (to be noticed later) has already greatly modified the nomenclature of 1858.

Among revisionary papers of a minor but more general character are numerous critical essays by Cassin, especially on the *Icteridæ* and *Picidæ*, the *Raptores*, the *Cœrebidæ*, *Caprimulgidæ*, etc.; by Baird, in his Review of American Birds; by Coues among the *Laridæ*, *Procellariidæ*, *Grallæ*, *Colymbidæ*, *Spheniscidæ*, etc.; by Ridgway among the *Raptores* and by Bannister among the *Anseres*, etc. Of separate monographic papers may be mentioned those of Coues on the genus *Ægiothus*, the *Tringaæ*, *Alcidæ*, etc., Bryant on the genus *Cattaractes*, and Ridgway on various groups of the *Raptores* and *Oscines*, the genus *Leucosticte*, etc.

The anatomy and embryology of our birds have not received the attention these subjects deserve, but a number of papers have appeared treating more or less in detail of particular points. Dr. Coues has published an account of the myology of *Colymbus torquatus* and various notes on the structure of the *Spheniscidæ;* Streets has written on the characters of the skull in birds, and Mr. Ridgway has called attention to various points in the osteology of the *Raptores.* Agassiz and Wyman have noticed a few points in the development of birds, while Prof. E. S. Morse has carefully studied the development of the tarsus and carpus with results confirmatory of the previously suspected affinities existing between birds and reptiles.

Not only has the geographical distribution of North American birds been studied in connection with the general history of each species, but attempts have been made to discover and limit the different faunal areas of the continent, special attention having been given to this department of the subject by Baird, Verrill, Allen, and Ridgway, and incidentally to a less extent by others. The subject of geographical variation has also been thoroughly investigated, with results of high importance, which will later be referred to more fully. Among the prominent workers in this field are Baird, Allen, Coues, Ridgway, and Henshaw, while the subject has received much attention from others.

Within the last ten years also a field previously wholly unexplored has been opened up, that of palæontological ornithology, mainly through the labors of Professors O. C. Marsh and E. D. Cope, the former of whom has alone made known, mostly in preliminary papers, the remains of upwards of thirty species, found in the Cretaceous green-sands of New Jersey, and the Cretaceous and Tertiary formations of Kansas, Colorado, and Wyoming. Those described by Professor Cope are likewise mainly from the same localities.

Much has also been written by American ornithologists on exotic birds, especially on those of Central and South America. In 1838, J. P. Giraud described "sixteen new species" of Mexican birds, and Dr. S. Cabot published (1842 to 1844), several papers on the birds of Yucatan, describing a number of new species. Mr. George N. Lawrence has published numerous papers on the birds of Mexico, the West Indies, and Central and South America, and Dr. Bryant on the birds of some of the Lesser Antilles. Cassin published not only on the birds of Central and South America, but also on those of Africa and other distant countries, including reports on the collections made by the United States Exploring Expeditions under Commodores Rogers and Wilkes. Baird's Review of American Birds treats largely of the birds of tropical America, and some of the monographic and revisionary papers of Coues and Ridgway have taken a wide range. D. G. Elliot has published elegantly illustrated monographs of the *Pittidæ*, the *Phasianidæ*, and the *Tetraoninæ*, and many papers on different genera of the humming-birds and other exotic groups, mostly, however, prepared and published abroad. Professor Alpheus Hyatt has published a monograph of the *Spheniscidæ*, and Drs. Kidder and Coues reports on the birds of Kerguelen Island; the present writer reports on collections of birds made at Lake Titicaca, and at Santarem, on the River Amazons. Prof. James Orton has also published on the birds of Ecuador, and the Rev. J. H. Bruce on those of India.

Among the many aids to a better knowledge of American birds should be mentioned the explorations of Kennicott, Dall, H. W. Elliott, Bannister, and Bischoff, in British North America and Alaska, conducted under the auspices of the Smithsonian Institution, several of whom have forwarded to Washington immense collections of the birds of the Northwest. The surveys of our western Territories made under the direction of the War Department should be especially mentioned as among the auxiliaries of ornithological science in this country, the various government expeditions having been usually accompanied by competent naturalists. The recent surveys (still fortunately in progress) under the Department of the Interior have likewise been productive of important results. The numerous correspondents of the Smithsonian Institution, both in the boreal and in the tropical portions of the continent, though not always citizens of the United States, have added greatly to our store of ornithological material and knowledge. The several naval exploring expeditions have also contributed, from distant foreign shores, rich stores of specimens and facts. Among more private enterprises should not be forgotten the Thayer and Hassler expeditions from the Museum of Comparative Zoölogy to South America, both of which returned with rich ornithological collections, numbering in the aggregate thousands of specimens. The expeditions of Professor Orton, Mr. Alexander Agassiz, and Mr. Linden have also been fruitful in ornithological results. Of the explorations of Dr. Bryant in the West Indies, and of numerous other ornithological explorers elsewhere, the want of space forbids a further notice.

SUMMARY OF PROGRESS.

When Alexander Wilson, the "Father of American Ornithology," began his great work, less than three fourths of a century ago, probably not a dozen species of American birds had been scientifically described by American writers, and almost nothing had been published relating to their distribution or habits. On the completion of Bonaparte's continuation of Wilson's work in 1833, about four hundred species had been described by these two authors, of which colored figures had also been published, with a more or less full account of the habits and range of each species. In 1833, Bonaparte (in his Geographical and Comparative List) gave the number of North American species as four hundred and seventy-one, while in 1839 the number described by Audubon (in his Synopsis) was four hundred and ninety-one. Nearly all of these had been re-figured by Audubon, the figures being all of life size, and as yet unsurpassed in fidelity or artistic effect. Audubon had likewise largely increased our knowledge of their distribution and habits, while the greater part had, moreover, found a valuable biographer in Thomas Nuttall. The publication of numerous minor papers had also contributed largely to a better knowledge of many of the species.

In 1844 the number of species had increased to five hundred and six, the number given by Audubon in his Birds of North America. In 1858 the number had risen to seven hundred and sixteen, all of which had been ably elaborated by the authors of the General Report, their affinities thoroughly discussed, and their nomenclature carefully revised. In the mean time numerous special papers had appeared relating to their habits and distribution. Yet the nests and eggs, and even the winter and summer resorts, of many of the species still remained unknown.

From this time onward information respecting our birds increased more rapidly than before; new workers came prominently into the field, and a rapid advance marked every year. At the present time the number of generally accepted species entitled to recognition as birds of that portion of North America north of Mexico is not less than six hundred and fifty, with, in addition, about one hundred and fifty commonly recognized subspecies, or about eight hundred recognized forms. The nests, eggs, and general habits of nearly all are now well known, particularly of those which occur east of the Rocky Mountains. Among the recently discovered extinct species have been found entirely new types representing what is believed to be a new sub-class, they having true teeth and other characters assimilating them to reptiles, between which and true birds they undoubtedly form connecting links.

Another phase of progress that should not pass unnoticed in this connection is the attention that has been paid to the geographical distribution of the species, with especial reference to the determination of the different faunal areas in North America, many of which are already known with a tolerable degree of definiteness; also the tendency to study the various sub-specific and specific forms from a geographical and evolutionary standpoint. Formerly the study of our birds was pursued wholly analytically, and forms from distant, little-known localities which differed slightly from their near affines of neighboring regions were looked upon as distinct "species." Later, as the material for a better knowledge of the subject accumulated, specimens of an intermediate character came to light, which, so long as they were few, were naturally looked upon as probably hybrids between the forms whose characters they seemed to combine. Still later, however, it was found that certain strains of deviation from pronounced types occurred in a large number of species belonging to widely different families inhabiting the same areas. This led to the discovery of laws of geographical variation, connecting particular phases of local differentiation with the topographical and climatic peculiarities of the regions where they so uniformly occur. Many of the isolated facts bearing on this subject had been observed and placed on record prior even to 1860, but their full import was not realized till after the lapse of another decade, during which our stores of material had become vastly increased. In 1871 the "new departure" was for the first time fairly entered upon, which in three years revolutionized the nomenclature of North American ornithology, adding an important chapter to philosophical zoölogy, and exerting great influence in many other departments of North American zoölogy. Naturally, a view that threatened either to assign fully one sixth of the previously recognized species to the limbos of synonomy or to lower them to the grade of geographical races was not rashly espoused by those to whom belonged the credit of the recognition and description of these previously supposed specific forms; but so overwhelming were the facts in its favor found to be, that one after another of our leading writers soon gave it their indorsement, so that probably a greater degree of unanimity of opinion respecting any problems in ornithology never obtained, than now exists among our ornithologists respecting the subject of geographical variation among our birds and the sub-specific relationship of many forms which when first made known seemed unquestionably of specific rank.

The next step, and apparently a wholly logical one in the revolution, will doubtless be the general adoption of a trinomial system of nomenclature for the more convenient expression of the relationship of what are conventionally termed "sub-species," so that we may write, for instance, *Falco communis anatum* in place of the more cumbersome *Falco communis* subsp. *anatum*. This system is already, in fact, to some extent in use here, though looked upon with strong disfavor by our transatlantic fellow-workers, who seem as yet not fully to understand the nature of the recent rapid advance ornithology has made in this country, or to appreciate the thoroughly substantial nature of the evidence on which it is based. The constant and energetic exploration of the great North and Northwest, of the vast trans-Mississippian region, and of our sub-tropical borders, during the last two decades, by scores of indefatigable collectors and observers, has certainly not been in vain, as witness the hundreds and often thousands of specimens of single species, representing the gradually varying phases presented at hundreds of localities, that have passed through the hands of our specialists.

While the field of North American ornithology is far from an exhausted one, the progress made during little more than a half century is certainly creditable to American enterprise and to American students, though to Americans alone, of course, belongs only a share of the credit of the marked advancement.

In a short article like the present, devoted exclusively to what Americans have accomplished, justice can hardly be done to all, nor is there room to more than allude to the fact that much has been done in aid of the general advance by numerous foreign writers. By no means have all the names of even Americans that are deserving of recognition here, been mentioned in the present article, nor have all articles been cited that are entitled to a high degree of prominence; the omissions, however, are those of limitation and not of choice. Neither is there space to notice the several important ornithological collections that have been gathered, to which alone many pages might be profitably devoted.

ON THE DECREASE OF BIRDS IN THE UNITED STATES.

ON THE DECREASE OF BIRDS IN THE UNITED STATES.

IN the accounts the early settlers of the United States have left us of their experiences, they often refer to the scarcity of the smaller birds in the undisturbed forests, and note the fact that no sooner is a clearing made, a house erected, and the ground broken for agriculture, than many kinds of the smaller birds, before scarcely observed, gather about and take up their residences in the immediate vicinity of the new settlement. A few species that are rarely seen so long as the country remains in its primitive condition, forsake their wilder haunts with man's appearance on the scene, and become his associates. Notably is this the case with the robin, the blue-bird, the house-wren, and several kinds of sparrows and swallows, while others resort to the fields in large numbers with the first attempts at their cultivation. Doubtless the species in question do not thus actually increase suddenly in abundance, but simply leave their former resorts for surroundings they at once recognize as more congenial to them. Many species, however, possessing in other respects quite similar habits, seem to ever look upon man with distrust, and immediately retire from his presence. Hence the settlement of a region previously unoccupied by civilized man, rapidly entails a general revolution in the haunts of many of the species as well as in their relative abundance. While many of the smaller birds take up their abode within the precincts of civilization, the larger soon find in man an enemy whose presence they quickly learn to shun. Many of the larger species are eagerly pursued by him for food, often to such an extent as to materially reduce their number. At first easily approached, they soon lose their confidence in him as a harmless invader of their haunts, and through bitter experience learn his true character. Other species whose natural resorts are the deep forests, are soon forced to seek new homes, through the necessary removal of the forests to make way for agriculture. Other species prove obnoxious from their frequent inroads upon the fruits of the husbandman's industry, and hence become forever proscribed races. Through these and other very natural causes, great changes have resulted in the relative abundance of the different species of birds within the settled portions of the United States since these areas came under the dominion of the white race of man. So scarce indeed have many species become, and so differently limited in their ranges, that their former abundance over districts from which some have become wholly extirpated, and over which many others have become a thousand-fold reduced in numbers, would seem hardly creditable were it not a matter of historic record.

When Europeans first visited what is now the United States, they found all the bays and inlets of the Atlantic Coast, and all the rivers and lakes of the interior, literally swarming with water-fowl, their countless numbers at times "darkening the air," while the noise of their wings resounded like the "rumbling of distant thunder." While few of the species have become totally extinct, none are nearly so numerous as formerly. From Florida to Maine, the early settlers had no trouble in shooting as many of these birds as they wished, almost from their very doors, where now the sportsman may range the country for many a weary mile without even a view of the coveted game. The representatives of this large class of game-birds occur in any tolerable degree of abundance at only a few distant points, and when found they are obtainable by only the utmost vigilance and caution. The terrestrial game-birds have also similarly decreased, some of the more important species having become wholly exterminated throughout vast areas. The rapacious species, so numerous once as to be an absolute annoyance, are now so scarce as to be looked upon as trophies when captured by the youthful sportsman, and as rare acquisitions when obtained by the ornithological collector. This has resulted in part through the persecution to which they have been so long subjected, a premium having been at times set by law upon their heads; and in part from the deforestation of so large a portion of the older States. The removal of the forests has also doubtless greatly diminished the numbers of all of the strictly forest species, but in none is it more noteworthy than among the different species of woodpeckers, amounting in a few instances to partial extirpation. A few of the smaller occupants of the more open country have also suffered a notable diminution in numbers, particularly observable among the black-birds and the crows, their extirpation having been fostered by the payment of premiums by the local governments for their destruction. Many other species, wholly unobnoxious, have been greatly affected by the settlement of the country and its subjugation to agriculture, while some have simply fallen a prey to the ruthless spirit of destruction that kills simply for the pleasure of taking life. To refer to the destruction of our birds more in detail is to expose, in many cases, reprehensible acts of cruelty which offer often few or no ameliorating features.

When Massachusetts was first settled, Cape Cod was still, in all probability, the abode of the great flightless auk (*Alca impennis*), as was the more northerly coast for many years after. There are, at least, references to the existence of birds called "pengwins," by the early writers, as far south as the coast of Massachusetts. What these so-called "pengwins" were, we distinctly learn from Richard Whitbourn's account of his voyage to Newfoundland in 1618. He says: "These Penguins are as bigge as Geese, and flie not, for they have but little short wings, and they multiply so infinitely upon a certaine flat Iland, that men drive them from thence upon a boord into their Boates by hundreds at a time; as if God had made the innocencie of so poore a creature to become an admirable instrument for the sustentation of man."[1] Their "innocencie" and man's cupidity very soon, however, effected their total extirpation south of Nova Scotia.

The turkey formerly existed throughout all of the region south of the latitude of the Great Lakes, from central New England to Florida and the Gulf of Mexico, but in the wild state it now exists in numbers only in the least inhabited portions of the Alleghanies and the South Atlantic and Gulf States. The pinnated grouse, or prairie hen (*Cupidonia cupido*), likewise inhabited portions of New England, New York, New Jersey, and Pennsylvania as well as the western prairies; but it has not only disappeared from all of the more easterly States, but has also become considerably restricted in its range, even in the prairie States, where its extirpation, unless prevented by timely legislation, will be one of the near events of the future.

---

[1] Purchas, Vol. IV., p. 1886.

The noisy, chattering Carolina parroquet—the sole representative in the United States of the large and interesting order to which it belongs—which in early times ranged over most of the eastern half of the United States, from the Lake States southward, and was abundant from the Ohio Valley and Maryland southward to the Gulf, will soon doubtless be numbered among the totally extinct birds of the United States. Its destructive proclivities certainly rendered it somewhat obnoxious wherever it was abundant, and formed one of the chief causes that led to its rapid restriction. From the novelty of its form, plumage and habits, it has ever had to run the gauntlet of numberless shot-guns wherever it has been a casual visitor. At present it is hardly anywhere abundant, and is frequent only over portions of the South Atlantic and Gulf States.

The pileated woodpecker (*Hylotomus pileatus*), or log-cock of the woodsman, was formerly a numerous inhabitant of the forests of the United States, from Maine to Florida and Texas. The destruction of the forests began quite early to affect its range, and for many years its harsh notes and loud tapping have been heard only in the most unsettled parts of the States east of the Mississippi, it being already nearly or quite extinct over two-thirds of the area it formerly inhabited east of the Mississippi River. Most other species of the woodpecker family are doubtless far less numerous then formerly, owing mainly to the deforestation of the country. One species, however, the well-known red-headed woodpecker (*Melanerpes erythrocephalus*) has almost entirely disappeared from portions of its former habitat. Formerly as abundant in New England as it now is in the Middle and Western States, it has, in comparatively recent times, nearly abandoned all of the country east of the Hudson River.

None of the swimming and wading birds, except the great auk, have as yet become wholly extinct, but all have become greatly reduced in numbers, the greater part probably being not a hundredth part as numerous now as when the country was first visited by Europeans. During the seventeenth century, the sand-hill crane, and probably also the whooping crane, were common species all along the Atlantic coast eastward to Maine, though for the last hundred years they have been rarely met with on the Atlantic slope north of North Carolina, and are of rather rare occurrence now between the Ohio river and the Great Lakes. They have, in fact, never been recognized as birds of the New England fauna, and appear in the local lists of the birds of the Atlantic States, north of Georgia, only as occasional or accidental visitors. Captain Philip Amadas, however, speaks of meeting, in 1584, with immense flocks of these birds on "Wokokon" Island, on the coast of "Virginia." He says that his party having discharged their "harquebuz-shot, such a flocke of Cranes (the most part white) arose under us, with such a cry, redoubled by many ecchoes, as if an armie of men had showted all together."[2] The crane also figures in most of the early accounts of the natural productions of the Atlantic coast region from Virginia northward. Thomas Morton speaks of their abundance in Massachusetts as late as 1630. Of "Cranes," he says, "there are greate store, that ever more came at S. David's day, and not before: that day they never misse. These sometimes eate our corne, and doe pay for their presumption well enough; and serveth there in powther, with turnips to supply the place of powthered beefe, and is a goodly bird in a dishe, and no discommodity."[3]

The swan is also mentioned by the same writer as a bird of New England, although its occurrence there is almost unrecognized in the annals of ornithology. He says, in his enumeration of the birds: "And first of the Swanne, because shee is the biggest of all the fowles of that Country. There are of them in Merrimack River, and in other parts of the country, greate store at the seasons of the yeare. The flesh is not much desired of the inhabitants, but the skinnes may be accompted a commodity, fitt for divers uses, both for fethers, and quiles."[4] The swan is also mentioned by other early writers as a common bird of the whole Atlantic coast, although for many years few have been seen north of New Jersey, and it has, in comparatively recent years, greatly declined in abundance throughout the region south of the Great Lakes.

The white pelican is also mentioned by several writers as a former inhabitant of New England, as well as of the region more to the southward, but of late it has occurred north of the Potomac only as a straggler or so called "accidental" visitor; yet from its present known range in the remote interior it seems reasonable to suppose that it may have been formerly a common bird of the Atlantic coast as far north at least as Maine. The snow goose was

---

[2] Hakluyt's Voyages, new edition, vol. III., p. 302.

[3] New English Canaan, p. 69.

[4] Ibid. p. 67.

also a common winter visitor southward to the Middle Atlantic States, but is now, even in New England, of rather uncertain occurrence, while the other geese and ducks were so abundant that the early colonists had no trouble in supplying their tables by visiting the nearest pond, river or inlet.

The herons, nearly useless as food, have suffered an immense decrease in number, mostly through very natural causes, but often through wholly reprehensible acts of wantonness. Many have of late been destroyed for their feathers where, in Florida especially, the havoc made with these poor defenseless birds is a subject of painful contemplation and a disgrace to the age. The poor birds are attacked at their breeding grounds, and hundreds are slain in a few hours by single parties, whose only use of them is to secure the beautiful plumes with which nature has unfortunately adorned them. In this way colony after colony is broken up, the greater part of the birds being actually killed on the spot, often leaving nestlings to suffer a lingering death by starvation. The few old birds that survive usually abandon the locality where for generations their progenitors had lived and reared their young undisturbed, only to be attacked at some new point the following year. The effect of the wholesale destruction that for the last few years has prevailed in Florida and other portions of the Gulf States, is already apparent in the rapid decrease there of these beautiful birds. The habit most of the species of herons have of breeding together in communities renders their destruction during nesting time an easy matter, their strong parental affection leading them to be neglectful of their own safety when their young are in danger. Disgraceful and inhuman as the act may seem, many a heronry of the qua bird, or night heron, is annually destroyed in mere wantonness in order that the perpetrator may boast of the "cart load" of birds he shot in a single day.

The terns and gulls that form such graceful objects as they course over our bays and harbors or along the sea-coast, have long been subject to wicked and needless, if not wanton, persecution at nearly all their breeding grounds. Nesting on the ground, in communities, on low, barren, sandy islands, they are readily preyed upon by the people of the vicinity, who as "eggers" regularly visit the islands to rob the poor birds of their eggs. First breaking all the eggs found on the first visit, they afterwards daily frequent the breeding grounds and secure the fresh eggs subsequently laid, leaving perhaps a few nests undisturbed toward the close of the season, from which late and probably enfeebled broods are usually reared. This results in a great decrease in numbers among the species so persecuted, and often in their abandonment of favorite breeding grounds for those whose remoteness from man insures them greater security. Some species, as the large herring-gull, have wholly left such exposed breeding grounds, and resort to dense swampy forests, where they place their nests in trees instead of on the ground; a strange departure from their natural preferences, indicative of a high degree of intelligence. While these are among the changes that have resulted from the increase of population, they are brought about by the needless and reprehensible acts of heartless men, discreditable to any civilized community or age. The results of this needless spoliation add nothing to our general wealth or comfort. The birds thus persecuted are harmless species, whose beautiful forms and graceful evolutions one never tires of watching, and whose destruction should be amenable to severe legal penalties. Unless in some way protected, they must soon cease to be summer residents of our coast.

Reference has already been made to the decrease of all kinds of game-birds, but a few words further on this point may not be out of place. In the absence of definite statistics respecting their former abundance, with which to compare their present number, a few historical references may suffice to indicate how great has been the depletion. While the bays and rivers of the Middle States have ever been the great winter resorts of the ducks and geese, these birds were in early times far more numerous at more northern and less favorable localities than they are at the present time at their most populous resort. References to the occurrence of flocks so large as to darken the air, deafening the ears of the observer with the sound of their wings as they rose from the water, are too frequent in the early records to be considered as merely figures of speech. The eastern portion of Massachusetts, from its natural configuration, was never pre-eminently fitted for the resort of waterfowl, yet in the early days of the Plymouth Colony, one writer says that the planters lived during the winter on the roast meat of the native fowls they killed, and that every man had his own duck before him on his trencher. The quaint old Thomas Morton, in speaking of geese, says, "I have often had one thousand before the mouth of my gun," and adds that "the fethers of the Geese that I have killed in a short time, have paid for all the powther and shott I have spent in a year;" and that he had plenty of ducks and teal in the ponds and rivers about his house. The numerous species of plovers, snipes and sandpipers have undoubtedly decreased proportionately with the ducks and geese. Of these birds, says Morton, referring to them as "sanderlings," "I was much delighted to feede on them, because they were fatt, and easie to come by, because I went but a step or two for them; and I have killed betweene foure and five dozen at a shoot, which would load me home." Josselyn also says of these birds, "I have known twelve score and above kill'd at two shots"[5] The contrast in this respect of these early colonial times with the present requires no further comment.

Reference has also already been made to the extirpation of the turkey over large portions of its former habitat. There is little now to remind one that it formerly ranged in New England from Maine, southern New Hampshire, and southern Vermont, thence southward, except the tradition of its former occurrence at certain localities. Morton, however, who is more explicit in these matters than most of the other early historians, thus speaks of its former abundance in Eastern Massachusetts: "Turkies there are, which divers times in great flocks have sallied by our doores; and then a gunne (being commonly in a redinesse,) salutes them with such a courtesie, as makes them take a turne in the Cooke roome. They daunce by the doore so well ....I had a Salvage who hath taken out his boy in a morning, and they have brought home their loades about noone. I have asked them what number they found in the woods, who answered *Neent Metawna*, which is a thousand that day; the plenty of them is such in those parts."[6] Josselyn thus refers to their early decrease: "I have also seen threescore broods of young Turkies on the side of a Marsh, sunning themselves betimes, but this was thirty years since, the English and the Indian having now [1672] destroyed the breed, so that 'tis very rare to meet with a wild Turkie in the Woods."[7]

Our markets are now periodically so well supplied with wild pigeons, and such numbers of them are occasionally reported at

---

[5] Voyages to New England, p. 102.

[6] New English Canaan, pp. 69, 70.

[7] New England's Rarities, p. 9.

their different roosting places, that it would seem that their extirpation is far distant, if indeed they can as yet have at all diminished in number. Yet their abundance is evidently far less now than formerly, and a moment's reflection on the wholesale destruction they have annually suffered for the last two centuries leads one to wonder that so many still exist. Doubtless no other bird of their size was ever so numerous in the United States as was this species in early times. The accounts given by Wilson and Audubon of the numbers formerly seen in Kentucky and other of the interior States has to some degree made us familiar with their former astonishing abundance at certain localities. Yet the reports given by these naturalists are eclipsed by the accounts given by writers of the seventeenth and eighteenth centuries of their numbers over districts where they are now of comparatively infrequent occurrence. Strachey, speaking of their abundance in Virginia in 1610, says: "A kind of wood-pidgeon we see in the winter-time, and of them such nombers as I should drawe..........the creditt of my relation concerning all the other in question, yf I should express what extended flocks, and how manie thousands in one flock, I have seen in one day, wondering (I must confess,) at their flight, when, like so manie thickened clowds, they (having fed to the norward in the day tyme) retourne againe more sowardly towards night to their roust; but there be manie hundred witnesses, who may confirm this my report yf herein yt testifieth an untruth."[8]

It appears from the testimony of other writers that at about this date they occurred in similar numbers at various points from Virginia to Massachusetts. Morton, writing a few years later of New England, refers to "millions of turtle-doves on the green boughs which sate pecking of the full ripe pleasant grapes that were supported by the lusty trees." Watson relates that in early times such immense flocks of pigeons passed over Philadelphia that they "obscured the sun for two or three hours," and were killed in hundreds with sticks by people standing on the tops of the houses, and that they were "brought into the city by cart loads." Makins, in a poetic description of Pennsylvania, written about 1729, says:

> "Here, in the fall, large flocks of pigeons fly
> So numerous that they darken all the sky."

[8] Historic of Travaile into Virginia, Hakluyt Soc., p. 126.

Watson also states that "Hector St. John, of Carlisle," refers to pigeons being so numerous in 1782 as to darken the sun, and says that as many were sold for a penny as a man could carry home. They are also reported to have appeared about Philadelphia again in 1793 in similar abundance.

Lewis states that soon after the settlement of Lynn, Massachusetts, "the wild pigeons are represented to have been so numerous that they passed in flocks so large as 'to obscure the light,' and continued flying 'for foure or five houres together,' to such an extent that a person could see neither 'beginning nor ending, length or breadth of these millions.' When they alighted in the woods they frequently broke down large limbs of the trees with their weight, and the crashing was heard at a great distance. A single family has been known to have killed more than one hundred dozen in one night with poles and other weapons; and they were often taken in such numbers that they were thrown into piles and kept to feed the swine."[9] The same writer states that as late as 1829 they "continued to make their annual visits, but their flocks appear smaller every year. Instead of darkening the air, as formerly, by their numbers, the sportsman can seldom find enough to compensate his time."[10]

Josselyn says, in his account of some of the birds of New England, published in 1675: "The *Pidgeon*, of which there are millions of millions. I have seen a flight of *Pidgeons* in the spring, and at *Michaelmas*, when they return back to the southward for four or five miles, that to my thinking had neither beginning nor ending, length nor breadth, and was so thick that I could see no Sun, they join Nest to Nest, and Tree to Tree by their Nests many miles together in *Pine*-Trees. But of late they are much diminished, the English taking them with Nets. I have bought at *Boston* a dozen *Pidgeons* ready pull'd and garbidged for three pence."[11]

Williams, in his history of Vermont, thus speaks of the former abundance of the pigeon in that state: "The surveyor, Richard Hazen, who ran the line which divides Massachusetts from Vermont, in 1741, gave this account of the appearances which he met with to the westward of the Connecticut River: 'For three miles together

---

[9] Alonzo Lewis's History of Lynn, p. 22.

[10] Ibid, p. 241.

[11] Voyage to New England, p. 99.

the pigeons' nests were so thick that five hundred might have been told on the beech trees at one time; and could they have been counted on the hemlocks, as well, I doubt not but five thousand at one turn round.' The remark of the first settlers of Vermont fully confirm this account. The following relation was given me, by one of the earliest settlers at Clarendon,: 'The number of pigeons was immense. Twenty-five nests were frequently to be found in one beech tree. The earth was covered with these trees, and with hemlocks, thus loaded with the nests of pigeons. For an hundred acres together, the ground was covered with dung to the depth of two inches. Their noise in the evening was extremely troublesome, and so great that the traveler could not get any sleep where their nests were thick. About an hour after sunrise, they rose in such numbers as to darken the air. When the young pigeons were grown to a considerable bigness, before they could readily fly, it was common for the settlers to cut down the trees, and gather a horse-load in a few moments.' The settlement of the country," continues Williams, "has since set bounds to this luxuriance of animal life,"[12] they having in a great measure disappeared from that section of the country.

Less than a generation since the destruction of pigeons by means of nets was very general throughout the North eastern States, the sale of the birds thus captured forming quite a source of income to the few in each neighborhood who engaged in it. The capture of several hundred dozens in a season by a single person was not unusual. Of late, however, they have rarely appeared here in numbers, the greater part seen in the Eastern markets being shipped thither from the West.

While the transformation of the country from its primitive to its present condition has been necessarily disastrous to the bird-life that formerly occupied its forests and interval lands, or peopled its numerous lakes, rivers and maritime waters, the unrestrained destructive proclivities of our people have added greatly to the needless sacrifice of our feathered associates. In many cases the early settlers were compelled, in self-protection, to wage an exterminating warfare upon many of the larger rapacious beasts, and eagles, hawks and owls were commonly included in the doomed list of animals,

---

[12] The Natural and Civil History of Vermont, p. 114.

the destruction of which not only met with popular approval, but was encouraged by rewards offered by the lòcal governments. To these were commonly added crows, jays and the different species of blackbirds, but in some cases also other species of the smaller song birds. The robin and other kinds of thrushes have often been on the list of proscribed species. The records of the different towns show that large sums were often paid in premiums for the destruction of crows and blackbirds. In early times these birds were so numerous that they proved a great annoyance, taking to themselves a large share of the maize crop. For this reason they were currently known in many sections as "maize-thieves." In 1697, on March 8, the town of Lynn, Massachusetts, voted "that every householder in town should, sometime before the fifteenth day of May next, kill or cause to be killed, and bring the heads of them, at or before the time aforesaid, twelve blackbirds" to certain persons named in the act, a refusal or neglect of which entailed a fine of three pence for each black-bird less than the number specified.[13] In Norwich, Connecticut, early in the eighteenth century, a half-penny a head was paid for every blackbird or crow killed in the town, and, though the amount offered was so small, it is stated that it formed a considerable item of expense to the town.[14] Kalm states that in Pennsylvania and New Jersey a premium of three pence a dozen was settled by law for "dead-maize thieves." The desired reduction in their numbers seems to have been speedily effected in New England with results little anticipated by the inhabitants. Kalm says that he was told by Dr. Franklin in 1750 that "by means of the premiums which have been settled for killing them in New England, they have been so extirpated that they are very rarely seen, and in a few places only." In the summer of 1749, however, an immense quantity of worms appeared in the meadows and did great damage, so that the people gladly abated their enmity of the maize-thieves, believing that these birds lived chiefly upon the worms before the maize was ripe, and thus kept the worms in check. The hay crop was that year almost wholly destroyed, so that hay had to be obtained from Pennsylvania, and was even imported from England.[15] The present number of these birds, including both the purple

[13] Lewis's History of Lynn, p. 186.

[14] Miss F. M. Caulkins's History of Norwich, p. 130.

[15] Kalm's Travels, Forster's translation, Vol. II., pp. 78, 79.

grackle (*Quiscalus purpureus*) and the red-winged blackbird (*Agelæus phœniceus*), is far less — propably many hundred-fold less—throughout a large part of the United States than it was two centuries ago, when in fall and spring they assembled in such immense flocks as "to darken the air."

The periodical competitive squirrel and bird hunts, once so popular and so indiscriminately destructive of every thing that wore either fur or feathers, are fortunately now, in many localities at least, things of the past. A favorable reaction has in recent years set in, resulting in the passage of laws by most of the States and Territories for the better protection of game and for the protection of the so-called "beneficial" birds. In the main these enactments are well adapted to the objects in view, though often in their attempted discrimination between "beneficial" and "injurious" species, betoken a lamentable lack of any proper knowledge on the part of the legislators of the subject respecting which they are called upon to legislate. In too many cases, however, they fail of their object from the want of an intelligent public opinion to ensure their enforcement. In the immediate vicinity of the larger cities it may be unsafe for pot-hunters and ambitious youngsters to openly indulge in the indiscriminate slaughter of the smaller song-birds, or to shoot grouse or woodcock out of season, but in the rural districts the case is generally far otherwise. Here the country lads and idlers destroy at all times such birds as their fancy or their love of "sporting" may dictate. The petty inroads upon the garden or grain-field made by a few species of otherwise highly beneficial song-birds is considered ample cause for a murderous onslaught upon a dozen or two of our most useful and ornamental species of birds, and the numbers annually killed in this way throughout the country is undoubtedly rapidly diminishing the representatives of a considerable number of our practically harmless birds. The indiscriminate destruction of the nests and eggs of all the birds found in the suburbs of our larger towns by the irrepressible urchins of the lower classes, merely in wanton cruelty, is, however, an evil of no small magnitude and one not easy to remedy. Add to this the inroads made by thousands of worthless feline pets upon both the young and the old birds that nest in the gardens and cultivated grounds in our larger villages and towns, and it becomes almost a matter of surprise that our feathered neighbors survive at all in such localities, and that the increase of noxious insects is not far more rapid than it is. Not only are all of our smaller birds worthy of the most stringent protection from an esthetic point of view alone, but especially from their being the natural enemies of the increasing hordes of noxious insects that make such sad havoc with every species of vegetation.

The custom that has prevailed in the rural districts ever since the settlement of the country of shooting certain kinds of thrushes, jays, sparrows, and blackbirds, and even some of our most useful species of flycatchers, because they now and then steal a few berries or cherries from the garden, or destroy a few young blades of corn or other grain, or now and then seize upon a luckless bee in procuring their varied insect diet, has undoubtedly caused a needless and harmful decrease in the number of the feathered inhabitants of our orchards, gardens, fields and pasture-lands, far more, doubtless, than has resulted from the changes naturally attending the agricultural development of the country. The numerous game protective associations that have originated during the last few years are doing much for the proper limitation of the destruction of game and fish, but the decrease of the smaller song-birds does not so strongly appeal to their self-interest. Societies should be formed whose express object should be the protection throughout the country of not only these practically innocent and pleasure-giving species, but also the totally innoxious herons, terns and gulls, whose extirpation is progressing with needless and fearful rapidity. Perhaps no more legitimate or appropriate work than this could engage the attention of the associations for the "Prevention of Cruelty to Animals." Unless something is soon done to awaken public opinion in this direction, and to enlist the sympathies of the people in behalf of our persecuted birds, the close of the next half-century will witness a large increase in the list of the wholly extirpated species, and a great decrease in numbers of others that are now comparatively abundant. While science proves capable of easily restocking our exhausted streams and ponds with fish, it is far otherwise with birds and mammals, which it is beyond the power of man to replace when once exterminated over a large area.

J. A. Allen.

Geographical Variation in North American Birds.

Geographical Variation in North American Birds.

By J. A. Allen.

Probably the birds of no equal area of the earth's surface are better known than those of North America north of Mexico, or of the whole continent southward even to the Isthmus of Panama. No museums in the world, probably, possess so large suites of specimens of single species as there are of North American birds in the Museum of the Smithsonian Institution and in the Museum of Comparative Zoology, nor from so many localities. In many instances single species are represented by hundreds of specimens collected at frequent intervals throughout their known range. Those contained in the Smithsonian Institution have been most carefully elaborated by Prof. Baird and others, whose reports upon them have justly acquired a world-wide reputation for their thoroughness and accuracy. Those in the Museum of Comparative Zoology have also been carefully studied.

Briefly, then, what are the facts and the general results that have followed the investigation of this exceptionally large amount of material? What are the allowable inferences, and what general principles have been apparently established? To answer these questions as briefly as may be is the object of the present remarks,—premising, however, that the formerly current opinions respecting the rank of a certain class of forms heretofore generally regarded as specific have been radically modified. Intergradation has been frequently traced between widely different forms, a gradual coalescence in scores of instances having been positively established, and rendered extremely probable in a large number of others.

In North America geographical variation exhibits two marked phases:—(1) a differentiation with differences of latitude and elevation, and (2) differentiation with differences of longitude; which, for convenience, may be termed respectively latitudinal and longitudinal variation.[1] In respect to both, differentiation occurs in different degrees in different groups, in accordance with their general tendency to variation, or, as it were, in proportion to their normal degree of plasticity. In regard to variation with latitude the modifications are apparently more general than in what I have termed longitudinal variation. In latitudinal variation the differentiation affects not merely color, but size and the details of structural parts, whereas

[1] See Bull. Mus. Comp. Zool., Vol. II, pp. 229-247, *et seq.*, April, 1871.

color appears to be the main element affected by longitudinal variation. The fact of variation in size has been conceded as a general law by the majority of at least American ornithologists and mammalogists since it was so fully established by Prof. S. F. Baird in 1857 and 1858, in his admirable reports on the mammals and birds of North America, published in the series of Government Reports on the explorations and surveys of the various Pacific Railroad Routes. Prof. Baird then and subsequently[1] called attention to the fact of the greater length of the tail in several species of birds at certain localities, and cites instances of the larger size of the bill at southern points, and the paler color of the plumage of the birds of the Plains and the arid peninsular of Lower California. All his subsequent works have furnished numerous citations of similar variation with locality, but instead of insisting upon any common tie connecting these phenomena as the result of general laws, they were viewed as evidences of specific differentiation. The differences are, indeed, so great between many of the forms now known to intergrade that it is not surprising that they were regarded as different species when known from only a few examples, apparently unconnected by intermediate forms. Subsequently, however, it has been found that they are not trenchantly separated, intermediate forms so linking them together that they can be only vaguely diagnosed. These connecting links, inhabiting — at least in the breeding season — localities intermediate in geographical position and in climatic conditions to those frequented by the more extreme forms, suggest an intimate genetic relationship and a differentiation mainly or wholly through climatic influence, or the diverse conditions of environment.

Latitudinal variation presents the following phenomena, which are of such general occurrence that even the exceptions, if such there really be, are exceedingly few.

1. *As regards Size.* There is a general reduction in the size of the individual from the north southward, amounting not unfrequently to as high as ten to fifteen per cent. of the maximum size of the species. The reduction is much greater in some species, and in some groups of species, than in others, but is almost invariably considerable and easily recognizable.

2. *In respect to the Bill.* The variation of the bill is somewhat inverse to that of the general size, as a rule the southern forms having generally relatively, and often absolutely, larger bills than north-

[1] Amer. Journ. Sci. and Arts, Vol. XLI, 1866.

ern ones, the increased size taking different proportions in different species and different styles of bill. Those of a stout, thick, conical form generally increase in general size, but especially in thickness. Those of a slender, attenuate form become slenderer and relatively longer at the southward, with a decidedly greater tendency to curvature.

3. *In respect to the Claws.* A similar increase in size is apparent in the claws, especially in that of the hallux, at southern localities, perhaps less marked and less general than the increase of the bill, with which it evidently correlates.

4. *In respect to the Tail.* A marked elongation of the tail at the southward has been noticed in many cases, both in Cape St. Lucas birds (*Baird*) and in those of Florida.

5. *In respect to Color.* The differences in color are especially obvious, and may be reduced to two phases of modification:—(*a*) a general increase in intensity at the southward, and (*b*) an increase in the extent of dusky or black markings at the expense of the intervening lighter or white ones; or, conversely, the reduction in size of white spots and bars. Under the general increase in intensity the iridescence of lustrous species becomes greater, and fuscous, plumbeous, rufous, yellow and olivaceous tints are heightened in species with the color continuous in masses. Under the repression of light colors the white or yellowish edgings and spots on the wings and tail become more or less reduced, and frequently to a great degree, in species barred transversely with light and dark colors; the dark bars widen at the southward at the expense of the white or lighter ones, sometimes to such an extent as to greatly change the general aspect of the species, as is the case in the *Ortyx virginianus* of the Atlantic States, and in other well known species. Also under the tendency to the increase of dark colors, longitudinal streaks and blotches on a light ground increase in extent and intensity of color.

In respect to longitudinal variation, the differences appear to be mainly those of color, and to hold a direct relationship to the humidity of the climate. On the arid plains of the middle and western portions of the continent the annual rainfall is less than half that of the eastern half of the continent, while a rainy belt occurs on the Pacific coast, stretching northward from near the mouth of the Columbia River to Alaska, over which the annual rainfall is double that of any portion of the eastern half of the continent. Taking the species that present a nearly continental range, we find that almost invariably they pass gradually into the pallid forms of the interior at the eastern edge of the arid plains, the greatest pallor being developed in the driest regions, as the peninsular of Lower California and the almost rainless belt along the Colorado River, and northward along the eastern base of the Sierra Nevada Mountains; that on the Pacific slope they again reassume nearly the tints of the eastern form, but more to the northward, over the above-mentioned rainy region, they acquire a depth of color far in excess of what the species presents in the Atlantic region. This coincidence of bright and pale tints, with the relative humidity of the locality is certainly suggestive, if not demonstrative, of the relation of cause and effect between these two phenomena, since the same rule is traceable over large portions, at least, of the Old World; the Scandinavian forms, for instance, being darker colored than the conspecific races of Central Europe, and these again darker than those of Northern Africa and the adjacent regions. Humidity alone, or in conjunction with greater intensity of light, seems equally well to account for the increase of color to the southward. Yet, from the well known bleaching effect of sunlight, intensified by reflection, upon the colors of animals living upon sandy islands, and seabeaches, and desert interior regions, it seems doubtful whether the larger share of modification in intensity of color in birds may not be due to humidity alone, or to humidity and a high temperature together, rather than to intensity of light.

In regard to the enlargement of peripheral parts at the southward, it seems not unreasonable to suppose that, the increase of temperature in stimulating the circulation in these exposed members may have something to do with it, especially in view of the evidence afforded by mammals, which in general present climatic modifications parallel with those of birds.

Whatever may be the cause of the above modifications of structure and color at different localities, we certainly find the following coincidences: I. In accordance with the increase in the intensity of color in individuals of the same species from the north southward, in the northern hemisphere, the brighter colored species in genera represented in both the temperate and tropical regions occur, as a general rule, at the southward; the same fact holding good also for sub-families. In cosmopolitan genera, families, etc., the tropical species are almost always brighter colored than the extra-tropical ones. All the most gorgeously colored families of birds are either exclusively tropical or semi-tropical, with generally the outlying species more plainly colored than the average for the family. II. In accordance with the increase in the size of the bill at the southward, all the species that have this member enormously developed are tropical or semi-tropical, not only such families as have the beak at its maximum of development, as the toucans and hornbills, but in all groups in which it is unusually large, the extreme development is reached in the intertropical regions. III. In respect to the tail, with very few exceptions, all long-tailed forms reach their highest development within or near the equatorial regions.

The facts indicated above in respect to the inosculation of forms formerly regarded as specifically differentiated, will evidently require modifications of the hitherto accepted nomenclature. Evidently many of these forms are so strongly marked that they should be in some manner recognized in nomenclature, though admittedly of less than specific rank. Most naturalists now practically recognize as species such groups of individuals as are not known to graduate by nearly imperceptible stages into any other similar group; and as varieties, such groups of individuals as occur at certain localities, or over certain areas, which differ more or less from other groups inhabiting other (generally contiguous) localities, with which there is evidence that they do, more or less fully, intergrade. Convenience seems to demand such a course, in order to enable the naturalist to specify what particular variety or race of a species inhabits a given section of country—a method, in fact, already more or less generally practiced.

Finally, what is the bearing of these facts of geographical variation upon the question of origin of genera and species? Having approached the subject from a geographical standpoint, my own impression of the importance of the conditions of environment in modifying the characteristics of animals may have unduly impressed me; yet that they exercise a greater influence than is currently recognized I think must be admitted. How, for instance, can natural or sexual selection satisfactorily account for the occurence of pallid forms in arid, semi-desert regions, and of brighter colored forms in contiguous humid districts, or the generally increased intensity of color southward, and its maximum development only toward and within the tropical regions? In many cases, it is true, the change in color may be protective, as it doubtless is in the assimilation of the pale tints of birds and other animals inhabiting arid plains to the generally gray color of the vegetation and the earth itself in such localities; yet, as the resemblance of the birds of these arid districts when young or in fresh plumage to those of the adjoining regions at the same season is much greater, as a general rule, than at the end of the breeding season, we have thus palpable evidence of the direct modification of color by environing conditions. Again, it is hard to see how the intenser and darker shades of the iridescence of the *Quiscali* in the South Atlantic and Gulf States, or their slenderer and more decurved bill, or the greater breadth of the transverse black bars on the breast of the southern form of *Ortyx virginianus* can be in the one case any more "protective," or in the other give greater facility in obtaining food, than the different colors and the differently proportioned beaks of the northern forms of these species; or of what advantage the large claws and long tails can be at southern localities rather than at northern. The variation in color is not apparently any better explained by sexual selection than are the other modifications by natural selection, for it is hardly supposable that sexual selection should act in so uniformly an accelerated degree toward the southward, or so generally from arid regions toward moister ones. On the contrary, it is just this gradual and general modification over wide areas that apparently points to climatic influence as the differentiating cause. There is, further, frequently a closer assimilation of the sexes at the southward, as among the *Icteridæ*, through the greater increased brilliancy of the female as compared with the male, which is rather the reverse than otherwise of what is commonly supposed to be the result of sexual selection.

Freely admitting, however, that both natural selection and sexual selection are causes of modification in the gradual differentiation of animals, I am led to regard them as secondary rather than primary elements, and that climate and other environing conditions take a larger share in the work than the majority of evolutionists seem willing to admit. Evidently no single law will explain all the phases of modification by descent, and in addition to those above alluded to, doubtless what Hyatt and Cope, among American zoologists, have termed the laws of acceleration and retardation are among the other causes of the modification. In birds, even, phenomena are apparent that cannot be strictly admitted into the category of geographical or climatic variations, but seem to singularly combine some evident features of this character with a retention of a few embryonic characteristics, especially in respect to coloration, of allied intergrading forms, as occurs in some of the birds of the middle portion of the

North American continent as compared with those of the eastern portion. Again, in respect to insular regions, while the above mentioned general laws of climatic variation are there evident, certain other exceptional modifications obtain, that seem to specially characterize those regions.

A word, in conclusion, respecting hybridity:—When comparatively few instances were known, in which specimens combined in various degrees the characters of two quite distinct forms, their synthetic character was generally explained by the theory of hybridity; but the irrefragibility of the evidence now at hand in proof of the intergradation of such forms over large areas,—the transition being so gradual as to occupy hundreds of miles in the passage, — and also coincident with a similarly gradual change in the conditions of environment, together with the demonstrable evidence of the power of climatic influence, seems to furnish a far more satisfactory explanation of these perplexing phenomena. But an advocate of the theory of hybridity might still assume that this gradual transition over a wide area is no objection to the theory, since the gradual fading out of the impression of contact in either direction from the line of junction of the respective habitats of two forms is just the result that would be anticipated from such a sexual intermingling of the forms in question. But the real objection to the theory — granting the possibility of hybridization on such a gigantic scale, which seems really improbable — is, that widely different forms occur also at different points in latitude, between which each successive stage of gradual differentiation can be readily traced, where hybridity can scarcely be supposed to account for the gradual change. Furthermore, gradual differentiation is now known in so many cases that it amounts to the demonstration of climatic variation as a general law, by means of which a species may be safely predicted to take on a given character under certain specific climatic conditions. If the theory of hybridity be urged to account for the intergradation of forms occurring at localities differently situated in respect to latitude, as has sometimes been done, it evidently falls under the weight it has to support; and yet there seems to be little better evidence in its behalf in cases where the intergrading forms happen to be differently situated in respect to longitude.

To describe in detail, or even to give illustrations, of geographical modification would require more space than would be proper to use in this connection, especially since a preliminary exposition of the facts upon which the preceding generalizations have been based, has already been presented in two papers in the Bulletin of the Museum of Comparative Zoology (Vol. II, No. 3, April, 1871, and Vol. III, No. 6, June, 1872).

The Committee on the Walker Prizes offered their report on the essays presented on the "Darwinian Question," which was accepted. A first prize of one hundred dollars was awarded to Prof. E. D. Cope; and a second prize, fifty dollars, to B. F. Ferris.

The thanks of the Society were voted to Mr. C. M. Tracy, for two fine Ornithorhynchus skins.

TO WHAT EXTENT IS IT PROFITABLE TO RECOGNIZE GEOGRAPHICAL FORMS AMONG NORTH AMERICAN BIRDS?*

THE AUK:

A QUARTERLY JOURNAL OF

ORNITHOLOGY.

VOL. VII. JANUARY, 1890. No. 1.

TO WHAT EXTENT IS IT PROFITABLE TO RECOGNIZE GEOGRAPHICAL FORMS AMONG NORTH AMERICAN BIRDS?*

BY J. A. ALLEN.

IN THE early days of natural history minute differences of structure, size or color received little attention, and the groups looked upon in early times as species now take, in not a few instances, the rank of a genus, the one wide-ranging species of the early authors having been found to include several species, each with its own circumscribed habitat. Buffon, for example, considered the exotic forms of life which closely resembled European types either as being degenerate forms of the latter, or as slight modifications of them, due to climatic influences, differences of food, etc. Even the species of Linnæus, and of his contemporaries and immediate followers, were often groups of a highly composite character. It was not till much later that the importance of nicer discriminations became apparent.

By the middle of the present century the smallest appreciable deviations became of specific import, and even a difference of habitat was not unfrequently thought to be sufficient ground for the presumption of specific diversity. Consequently individual variations were unwittingly made the basis of specific distinctions.

* Read at the Seventh Congress of the American Ornithologists' Union, Nov. 15, 1889.

In the progress of descriptive zoölogy the quest for new species and new genera became rampant. The distinction of imposing a new name, to be followed by *nobis* or *mihi*, led always to a search for differences. It was natural, in such a scramble, that resemblances should be overlooked, and that the study of individual variation should receive little attention. The synonyms that form such an array in our modern systematic treatises on zoölogy attest the result.

But not all descriptive naturalists erred equally or in the same way. There always have been, of course, both splitters and lumpers, the one class more or less at arms with the other, the one holding the other more or less in contempt. But aside from this individual diversity among naturalists there have been periodic phases of change in the matter of excessive splitting and the reverse, which may be likened to the oscillations of a pendulum. This is especially true in respect to our own country, and notably in regard to North American ornithology during the last half century. Within this period there have been at least three well-defined oscillations of the ornithological pendulum. The first, in the direction of excessive subdivision, reached its culmination about the year 1870. A reverse oscillation immediately followed, sweeping in character, and of strong and sudden impetus. It gradually spent itself during the latter part of the following decade. A third oscillation, in the direction of the first, feebly set in soon after, and, slowly acquiring momentum, now seems to have reached a degree of force that challenges candid consideration.

In this connection a brief review of the progress of our science during the last fifty years may not be out of place.

Fifty years ago there were no large collections of birds, either in public museums or in private cabinets, anywhere in America About this time the Government instituted a series of transcontinental surveys, extending in nearly parallel lines across the continent from about the 90th meridian westward to the Pacific coast, these lines being run at rather wide intervals from the Mexican border on the south to the British boundary on the north. These several military surveying parties were usually accompanied by good field naturalists and collectors. Also at about the same time the Smithsonian Institution, through the influence and direction of the late Professor Baird, secured correspondents and collectors at many points in British America. All of the collections thus made were deposited in the National Museum, then commonly known as 'The Smithsonian.' In a few years a large amount of new material was thus brought together from the far West and the remote North. These were the first collections of any magnitude ever received from trans-Mississippian North America. Material from east of the Mississippi River was still scanty, and hence the proper means for careful comparison of eastern and western forms was often lacking.

During the elaboration of this material, derived from hundreds of isolated localities, the discovery of new forms, both generic and specific, was the paramount incentive in the investigation. Not till some years later did the subject of climatic influences upon animals, in other words the evolution of species by environment, receive much consideration. Nor could it have been otherwise. We have first to gather our facts before we can generalize. By these seeming strictures no discredit is intended to the naturalists of that period, nor any condemnation of their methods. They were the pioneers, conducting a careful reconnaissance, and preparing the way for the occupation of the field by the grand army of workers who followed later.

Variations, even when slight, are facts of great import, whether they be individual, climatic, or specific, in the usually accepted sense of this last term; and when, except in the case of purely individual variation, they are made the basis of a name their availability in the discussion of profounder questions is thereby greatly enhanced. Down to as late as 1865, a collection of individuals indicated by a binomial name was a species, theoretically, at least, distinct and definable from any other similar group. An occasional intermediate, when found, was a great stumbling block, usually to be accounted for on the theory of hybridization. Varieties, in the geographical sense of today, were rare indeed, and the term subspecies was almost unknown in zoölogy.

Between the years 1865 and 1870 the question of what was a species had already become a very troublesome one. Many of the then currently recognized species were found to be as elusive as the proverbial flea. The increase of material and better opportunities for investigation showed that the range of individual variation had been underrated, and that many of the alleged differences on which species had been founded were of little or no value, while in many other cases closely allied so-called species were merely local, intergrading, geographical forms, correlated with special features or conditions of environment.

During the three years including and following the year 1871 ornithological opinion on the subject of species and subspecies, or respecting the status of a large proportion of the birds of North America, experienced a radical disruption. The first great wave, which for generations had been increasing in volume and force, met a barrier against which it recoiled and subsided with phenomenal suddenness, giving place to a strong and sweeping counter current. The key-note to the situation had been struck, and after a brief period of wavering a happy medium course was hit upon, which seemed to solve most of the difficulties that had beset the general subject of species. This was no less than the reduction of numerous so-called species to the rank of geographical races or subspecies, namely such as were found to intergrade with other forms, or which it seemed probable, on general principles, might so intergrade. Immediately inveterate splitters became bold lumpers, and the ornithological pendulum swung quickly back with a momentum sufficient to carry it somewhat beyond the vertical. The term 'var' interposed between the name of the original species (in the sense of the earliest described species) and its various local offshoots was the magic link which was to connect and duly correlate the discordant bird elements of our North American fauna. This, of course, was the origin and first phase of our present trinomial system of nomenclature, which ten years later was formally endorsed and adopted by the American Ornithologists' Union.

While this great step — little less than a revolution in its results — was in the main in the right direction, it led to some rash conclusions, theoretical reasoning now and then overstepping the hard line of facts. Consequently in a few instances species were unduly merged, and it has been necessary to reconsider these hasty rulings. The oscillation in the direction of unwarrantable lumping, however, soon reached its extreme limit; the pendulum settled back, and for a time remained at what we may consider as very near its normal point of equilibrium. For nearly a decade, dating from 1875, the deflections were slight and variable, now to one side, and now slightly to the other. This period of comparative stability includes the work of the A. O. U. Committee, in the years 1883-84, on the status of the described forms of North

American birds. As some hard-and-fast line in respect to subspecies was necessary it was very properly agreed by the Committee that no species should be reduced to a subspecies except on proof of intergradation. This kind of proof is sometimes lacking where its existence is a theoretical certainty. Yet it seemed better to leave such questions open to be settled by a later increase of material.

The work of the A. O. U. Committee on the momenclature and status of North American birds yielded as satisfactory results as could have been anticipated, nearly nine-tenths of its decisions being reached by practically unanimous consent. The outcome of its deliberations, as embodied in the A. O. U. Code and Check-List, has well stood the test of time, there having been found thus far very few cases where subsequent discoveries have necessitated any revision of the Committee's decisions. A few forms then considered as ineligible to subspecific rank, and therefore rejected, have since been admitted, in consequence of the acquisition of material from new localities rendering their proper status more evident; while others have been described and added to the list. Some of these later discoveries have proved, indeed, little less than startling. These facts, and the very great increase of material during the last five years, have had possibly a too stimulating effect; they have unquestionably started the pendulum again in the direction of finer discriminations and excessive splitting. The majority of the old A. O. U. Committee, the authors of the Check-List, will doubtless now admit forms to subspecific rank they would not have admitted in 1884, had they been then placed before them. They certainly know much more about North American birds at present than they did five years ago, but is their judgment as sound and are their tendencies as rationally conservative?

If any be without sin, let him cast the first stone. Conscious of my own changed tendencies, it has seemed to me well to raise the above question for brief consideration, since it can do no harm to survey the field calmly and take note of the present drift in respect to a very important subject.

Recent investigations have taken me over fields I worked, with some care, ten to fifteen years ago. In the meantime material has greatly increased; series of specimens have been obtained from localities then unknown; thus I find myself looking at things in a new light, but from, I trust, a more advanced position. My former tendencies, in common with those of others at that time were in the direction of reducing doubtful forms to synonyms, and closely related species to geographical forms. Now, with much additional experience, some increase of knowledge in respect to particular points at issue, and much more abundant material, some of my former conclusions seem open to revision, as I now realize that the resources then at command were far less adequate for the settlement of questions at issue than I then supposed them to be.

The discoveries made during the last five years show that the subject of North American ornithology is far from exhausted, even in respect to the cataloguing of its numerous forms of bird life, and especially as regards their distribution. To show how much we did not know five years ago of the birds of even our long-settled southeastern States, I have but to instance five or six species—namely, Swainson's Warbler, Bachman's Warbler, Leconte's Sparrow, the Raven, and the Seaside Finches. Add to this the new forms recently brought to light in this supposed well-known area, and we must conclude that we are still only on the threshold of a thorough knowledge of the birds of our South Atlantic and Gulf States. What do we as yet know of the distribution of many of the southern subspecies of this area, and of their lines of inosculation with the northern forms? Nothing, with exactness. What do we yet know of the breeding ranges of the summer birds south of the Ohio Valley? Practically nothing. To how slight an extent are we able to unravel the many perplexing problems of the bird fauna of the great State of Texas, so peculiarly situated in relation to the East and the West, the North and the South, as regards North America at large. The great Southwest and the great Northwest, with their opposite extremes of climatic conditions and peculiarities of environment, as compared with the region to the eastward, still present to us many perplexing problems.

Under such a condition it is no wonder that the pendulum again tends in the direction of refined subdivision. We are alert for differences, with our wits sharpened to recognize slight variations in size, in form, and in tones of color. Our material is constantly becoming more ample, and the meaning of slight variations is thus more apparent than it otherwise would be. When large series of specimens of any species from distant points are compared, in cases where the environment is more or less diverse, we are accustomed to find appreciable differences—in some cases slight, in others so well-marked as to be obvious at a glance. In many instances, however, the differences are apparent only when large series are available for comparison; the differences being merely average differences; a greater or less proportion of the specimens of the two series are practically indistinguishable, the range of individual variation in either series overlapping the difference characterizing the two slightly differentiated forms. In other cases the occurrence of specimens that cannot be easily referred, without knowing their origin, to one or to the other, is exceptional.

These being the general facts in the case we are at once confronted with a serious question and a grave danger. The splitters of an earlier time regarded every form, however slightly differentiated, as a species. We arbitrarily define a species as a group of individuals standing out distinct and disconnected from any similar group, within which, though occupying different parts of a common habitat, we recognize other forms characteristic of, and restricted to particular areas. These reach a maximum degree of differentiation at some point in the habitat, and thence gradually shade into other conspecific forms geographically contiguous.

The distinction we thus make between species and subspecies, though a purely conventional one, forms an indispensable basis for the convenient recognition of the various minor stages in the evolution of organized beings. The serious question is where to draw the line in recognizing local forms in nomenclature. While it is important to discover, and in some way record, even the very slight differences due to peculiarities of environment, there is obviously a reasonable limit to the naming of such forms by the use of the trinominal system of nomenclature. How well-marked then, must be a set of intergrades to entitle them to recognition? On this point no arbitrary hard-and-fast line can be laid down. Much, at least for the present, must be left to the discretion of the investigator. We are still groping in the dark; our steps are, in the main, tentative and provisional. We cannot act decisively in respect to the bird life of North America, or of any large area, till we know thoroughly the phases of variation throughout every nook and corner of the area in question. At present new forms are coming to light, often where least expected; every considerable series of specimens from any locality previously known only superficially presents us with, if not new nameable forms, at least a new set of puzzling intergrades, tending to unsettle opinions we thought were safely grounded, and showing that every question touching the status of species and subspecies is still more or less open to revision.

Unquestionably the tendency at present is to name forms which six years ago would have been considered too slightly differentiated to require such recognition. On the other hand, differences long since noticed, have but recently come to be properly understood. In the light of new material they prove to have a significance previously unappreciated, owing to the absence of the requisite data.

It is still evident, however, that great caution should be exercised in bestowing trinomials, in order to guard against drawing too fine distinctions. Very little is gained by naming races distinguishable only by experts, aided by a large amount of material, or where the differentiation is largely a matter of a slight average difference between forms contiguous in habitat—forms which nine out of ten ornithologists of average acuteness and experience, and with only ordinary resources, will be more or less unable to satisfactorily distinguish. In fact, a form based on a certain series of specimens may seem to any investigator of this same material well founded, but when judged by other material not unfrequently loses much of the distinctness it seemed to present when tested by the first set of specimens.

There are necessarily, in cases of wide-ranging species which run into well-marked geographical forms, numerous connectent series, made up of intergrades of all degrees of relationship to the more extreme phases. Some of these intergrades may seem different enough from either extreme to warrant recognition as an additional subspecies. By such a course what do we gain? We bridge the difficulty by doubling it; we get rid of one set of of troublesome intergrades by creating two others; leading the way to further subdivision of like character, and increasing the difficulties. Obviously the situation is not in this way improved. Yet the tendency to this sort of division is evidently increasing, each step in this direction making the next one easier. Only the exercise of due discretion can prevent the reduction of "our beneficent system of trinomials" to an absurdity. It is much easier to name a dozen new species or subspecies than to get rid of one, though erected on a false basis. Let us then weave our ornithological net so open-meshed that the undesirable small fry of incipient local forms may readily pass through and escape till further differentiation, in future ages, shall render them desirable captures.*

THE GEOGRAPHICAL ORIGIN AND DISTRIBUTION OF NORTH AMERICAN BIRDS, CONSIDERED IN RELATION TO FAUNAL AREAS OF NORTH AMERICA

THE AUK:

A QUARTERLY JOURNAL OF ORNITHOLOGY.

VOL. X. APRIL, 1893. NO. 2.

THE GEOGRAPHICAL ORIGIN AND DISTRIBUTION OF NORTH AMERICAN BIRDS, CONSIDERED IN RELATION TO FAUNAL AREAS OF NORTH AMERICA.[1]

BY J. A. ALLEN.

THE PRESENT paper consists essentially of two parts. The first treats of the probable geographical origin and present distribution of the genera of birds represented in North America, regardless of any inferences that may be drawn from this general presentation of the facts of the case; the second relates to the subdivision of the continent into faunal areas of various grades, with reference to their relationships, classification and nomenclature. No very novel views, nor many new facts are presented, the paper being in great measure a restatement of generally known facts in a new sequence, with a view to giving them greater emphasis in their bearing upon the special subject in hand. The leading ideas here embodied have already been set forth by the writer in other connections,[2] but the evidence is here for the first time presented

[1] Read before the Tenth Congress of the American Ornithologists' Union, held in Washington, D. C., Nov. 15-17, 1892.

[2] See the author's recent paper, The Geographical Distribution of North American Mammals (Bull. Am. Mus. Nat. Hist., Vol. IV, 1892, pp. 199-243, pll. v-viii), and the earlier papers there cited.

in detail from the standpoint of birds. As an introduction to Part I a few pages are devoted to questions having a more or less direct bearing upon the general subject.

I.—The Geographical Origin and Distribution of North American Birds.

It has long been recognized by nearly all writers on geographical biology that the two leading factors governing the distribution and dispersal of life over the globe are the land connections which now exist, or have formerly existed, between the principal land masses, and climate; and that by far the most potent of the climatic influences is temperature. In considering the faunal relations of North America to the Old World, it is important, therefore, to keep in mind the present slight separation of northern North America from Eurasia, and that, as currently believed by many geologists, the shallow basin now forming Bering Sea was at one time dry land, and thus formed a broad land connection between northwestern North America and northeastern Asia, during at least a portion of the Tertiary.

As is well known, a large proportion of the genera, and many of the species, of both animals and plants occurring in North America have a circumboreal distribution, even in many cases where their present habitats do not extend quite to the Arctic regions. Furthermore, that many genera, particularly of birds, which are at present limited to the warm temperate and tropical latitudes, are common to both the Old World and the two American continents. It is, in certain cases, hard to see how their present dispersion could have been brought about under the geographic and climatic conditions now existing. Geology here comes to our assistance, furnishing evidence that in earlier times the climate of the globe was not only more uniform, but also much warmer over the regions now buried half the year under snow and ice. It is well known that in Miocene times a warm temperate flora prevailed over the present Arctic regions, and that subtropical plants flourished in Central Europe and in corresponding latitudes in North America. Also that many types of mammals, now represented only in the tropics, formerly ranged over the greater part of the northern hemisphere, as shown by their fossil remains, long since buried in the Tertiary deposits of both North America and Eurasia. Furthermore, there is abundant evidence of a considerable interchange of life between the two northern continents at a time when there was no climatic barrier, as now, to the northward extension of subtropic types; and that Eurasia was in part populated by types originating in North America; and also that North America has derived a portion of its life from the Old World. There is no reason to suppose that birds were then any less restricted in their means of dispersion than now, in comparison, for example, with mammals, insects, land mollusks and plants.

The past history of birds, while so defective in comparison with that of some other groups, affords proof of the former much wider dispersion of certain types than obtains at the present time. While for this class the geological record is so imperfect, it suffices to show that what are now strictly tropical genera, for example, formerly reached southern and central Europe. It also affords evidence that birds in Tertiary times were not so very different from the birds of to-day. The leading genera of the raptorial, gallinaceous, wading and swimming birds were much the same as now[1]; with them existed other genera which have since become extinct, while many of the now prevalent forms were absent, and have doubtless come later into existence. The

---

[1] Among the existing genera of birds found in the Tertiary of Europe and North America are the following, most of which date back to the Miocene, and some to the Eocene.

Struthio	Olor	Columba
Colymbus	Phœnicopterus	Buteo
Podiceps	Leptoptilus	Aquila
Podilymbus	Ibis	Falco
Uria	Ardea	Milvus
Larus	Grus	Strix
Sterna	Rallus	Bubo
Puffinus	Fulica	Psittacus
Phaëthon	Phalaropus	Picus
Sula	Scolopax	Leptosoma
Phalacrocorax	Tringa	Trogon
Pelecanus	Totanus	Callocalia
Merganser	Numenius	Corvus
Anas	Charadrius	Scolecophagus
Spatula	Coturnix	Passer
Anser	Phasianus	Sitta
Branta	Meleagris	Motacilla

fragile nature of most bird skeletons, and the small size of the greater part of the species, and the fact that only those of more or less aquatic habits would be liable to rapid entombment under conditions favorable for their preservation, are circumstances which render hopeless any expectation of the future discovery of the ancestral lines of the great bulk of our present avian fauna.

Two hypotheses, however, may be hazarded regarding the present limited distribution of many groups now restricted within comparatively small areas. First, that they had not only a local origin, but that for some not very obvious reason they have always had a local distribution, as for example the Todies in the West Indies, and various South American and tropical Old World types. Second, that they have become specialized since the close of the Tertiary, with adaptations to a tropical or semi-tropical environment.

There is evidence that towards the close of the Tertiary a marked change in the earth's climate took place, culminating in the Glacial Period, during which the whole northern half of the northern hemisphere became covered with a heavy ice cap, lasting for possibly thousands of centuries, and extending its chilling influence nearly to the northern tropic. The rise of the glacial period was of course gradual, and the southward progress of the great ice cap drove before it all forms of life capable of any considerable power of locomotion, while those unable thus to escape must have perished from cold. There was hence a great crowding together of exiles from the north into the more favored regions to the southward, leading to an intense struggle for existence, and a weeding out on a grand scale of forms least fitted for the contest. This period must thus have been one of great activity in the evolution of new types. Opportunity was given for the gradual adaptation of many forms to a lower temperature than that to which they had been accustomed, and to an enforced change of food. The recession of the ice fields was accompanied by the gradual extension northward of habitable land. A broader area becoming available in summer than in winter an annual migration for a greater or less distance on the part of the pioneer life became a necessity. Finally the ice receded to its present limits and the whole north, under radically altered climatic conditions, became again available for occupation by the more or less modified descendants of the pre-glacial exiles. To some of these a comparatively low temperature had doubtless become congenial, and from them may have been derived the distinctively arctic and subarctic birds of the present age. They belong mainly to our present circumboreal genera, and are unrestricted by climatic conditions in their dispersal throughout the arctic and subarctic regions.

Other forms proved less flexible, and remained in latitudes more nearly corresponding to the climatic conditions of preglacial times. They had, however, before the beginning of the Glacial Period, become broadly dispersed, and now are found in widely disconnected areas. We have thus a reasonable explanation of the disconnected distribution of congeneric species in such groups as the Tree Ducks, Egrets and Herons, Spoonbills, Flamingoes, Snakebirds, Gannets, Gallinules, Barn and Pigmy Owls, Kites (genus *Elanus*), Trogons, Barbets, Kingfishers, Swifts, Goatsuckers, Piculets, and a few Passerine birds. On the other hand, doubtless many of the peculiar tropical types of land birds were already restricted to somewhat near their present limits, and that they have never had a much wider dispersion than they have at the present day. Many of them are also possibly of comparatively modern origin. It is only on this supposition that we can account for the numerous peculiar types that at present characterize the subtropical and tropical areas of the several continents.

It is not probable, for example, that such exclusively tropical American families as the Formicariidæ, the Dendrocolaptidæ, the Galbulidæ, the Todidæ, the Toucans, the Motmots, the Cotingas, etc., have ever had a much wider range than now. It seems also probable that such distinctively American types as the Hummingbirds, the Icteridæ, the Tyrant Flycatchers, the Tanagers, the Vireos and the Mniotiltidæ, which for the most part have their centers of abundance in the tropics, with merely outlying members in temperate North America, have never had a wider range than at present, and that most of their outlying genera and species have, under the intense struggle for existence in the overcrowded tropics, become gradually somewhat modified to adapt them to slightly more boreal conditions, thus in course of time more or less extending the general habitat of the families to which they respectively belong. At the far north they are still cut off from further extension by an insuperable climatic barrier, and have thus been unable to reach any part of the Eurasiatic continent.

The Glacial Period and the change of climate it produced has not only had a profound and permanent influence upon the distribution of life in the northern hemisphere, but proved a powerful agent in the evolution of new types, and in the extinction of old ones; it also left its impress upon the majority of northern birds through the development in them of the habit of migration: for it is now generally conceded that this habit must have originated in consequence of the great reduction of temperature beginning about the close of the Tertiary. Prior to this period, owing to the comparatively uniform climatic conditions then obtaining almost everywhere, there could have been no necessity for extended periodic movements. Later a nearly uniform temperature throughout the year gave place in northern latitudes to warm summers followed by winters of great severity. During the waning of the ice period the area offering a congenial summer home to a great multitude of birds became greatly extended, from which, however, they were driven by semi-arctic winters to seek favorable winter haunts further southward.

We have here what seems a natural and reasonable hypothesis for the origin of migration, and one now currently accepted by ornithologists. In this way, it is believed, the habit of migration not only originated but has become established as an irresistible hereditary impulse, as inherent and mandatory as the 'instinct' of reproduction.

In this connection, a few words may not be out of place respecting the question, Why do birds migrate? In considering this question, it should be borne in mind that there is everywhere, and has ever been, a constant and intense struggle for existence—that throughout nature the birth rate is far above the possibilities of permanent increase. Hence, in the bird world, as elsewhere, every station affording favorable conditions for existence must be occupied; there can be no unutilized areas. Many birds are organized to subsist only upon insects or soft fruits, or upon both combined; such food abounds in summer far to the northward of where it can be obtained in winter. Thus some of our Swifts, Swallows and Flycatchers range in summer almost to the borders of the Arctic Circle, where for a few weeks in summer they find abundance of food and a congenial temperature. Here they bring forth their young, but are forced to retire at the approach of autumn, retreating gradually before the southward advance of colder weather, passing through the middle latitudes in September, and reaching the tropics in October or November, the time varying more or less with different species. Here they remain till the increased warmth of March or April, or the stimulus of the approaching breeding season admonishes them of the necessity of revisiting their breeding stations, when they begin to retrace the journey toward their summer haunts, keeping pace so exactly with the advance of the season as not to lose even a day of the brief interval available for their sojourn in their far northern home. They are thus wanderers—exiles, as it were,—for three fourths of the year. Evidently our northern-breeding insectivorous and berry-eating birds could not survive a winter at their breeding grounds. They might perhaps be able to pass the whole year in the tropics, or in the lower temperate latitudes,—at the risk, however, of over-crowding the regular occupants, and also of leaving a habitable area unoccupied. As a matter of fact, nature not only 'abhors a vacuum,' using the phrase in a strict sense, but allows no waste places; living space is always at a premium. Near relatives of our subarctic and cold temperate insectivorous birds are found in the tropics and throughout the temperate latitudes; while the tropical forms are non-migratory, those breeding in the temperate latitudes are less so than their more northern kin; they remain, owing to the longer summer, for a greater length of time at their breeding stations and have a shorter journey to reach their winter haunts.

This may serve as a general illustration, showing that the absence of proper food in the high north forces the summer insectivorous birds to leave these regions for warmer latitudes, where can be found the food their peculiar organization renders necessary. In short, as our knowledge of the habits and migrations of birds increases, it becomes more and more evident that the cause of the autumnal migration is failure of proper food at the breeding station.

Why migratory birds ever leave their winter haunts seems at first sight less obvious, since in most instances it can hardly be due to failure of the food supply, nor to any absolute incompatibility of climate. The return in spring is often attributed to strong home love, evidence of the existence of which is shown by the return of birds to the same locality—even to the same nesting site—for many successive years. There is so much indubitable proof of this, that it is commonly assumed as the rule in most species. It is certainly beyond question that birds do not select their breeding stations in any hap-hazard way, journeying north along a vague course and stopping to nest wherever the proper conditions of season and other surroundings happen to prove favorable. Hence the impulse that governs their spring movements has been loosely termed the 'home instinct.'

If we consider that migration consists really of two movements—that is from the breeding station to the winter quarters and then back again—and that the one movement is the necessary complement of the other, it is hardly necessary to seek for a separate cause for the two movements; the two together constitute migration in a complete sense, which, as already explained, is an inherited habit,—an inherent, irresistible impulse, closely blended with the function of reproduction. The promptings which lead to the migratory movement, respectively in fall and spring, have unquestionably a different origin; the autumnal movement being doubtless prompted by a reduction of temperature and a failing food supply, while the spring movement is incited by the periodic activity of the reproductive organs, resulting in the necessity for the return of the species to the peculiar conditions and surroundings to which for long ages it has been undergoing special adaptation—in other words, to its home.

In the present paper, North America is considered in its faunal rather than its geographical sense, and in the synopsis here following the area covered by the A. O. U. Check-List and its Supplements is the region mainly considered, namely, North America north of Mexico, but including the peninsula of Lower California. The number of species now recognized in the A. O. U. Check-List as occurring within this area is about 795, with nearly 270 additional subspecies, making a total of about 1065 species and subspecies. These are arranged under 303 genera, with about 65 additional subgenera. Of these, however, 75 species and 25 genera occur merely as accidental stragglers from the Old World, the West Indies and Mexico. Deducting these as not valid components of the North American fauna, leaves about 720 species and 278 genera as legitimately North American, under the present restriction of the term. In the following synopsis, subgenera will be in the main ignored, as their consideration is found to rarely modify the results derived from a consideration of the genera. Unless otherwise stated, the breeding range alone is taken into account.

The Pygopodes, as recognized in the A. O. U. Check-List, consist of three families, Podicipidæ, Urinatoridæ, and Alcidæ. The Podicipidæ constitute a small, well circumscribed group, of almost universal distribution, and of great antiquity, dating at least from the Miocene, with a closely related ancestral form in the Eocene. The number of genera recognized varies from two to half a dozen or more, according to the views of different systematists. Most of these so-called genera, however, are little more than subgenera, and as such are geographically limited to particular regions, some being tropical or austral, while others are confined to the temperate portions of the northern hemisphere, where they have a circumpolar distribution, as *Colymbus* and *Dytes* of the North American fauna. *Æchmophorus*, scarcely more than a well-defined subgenus, is confined to western temperate North America, though allied to a connectent form between *Æchmophorus* and *Colymbus* in South America. It is thus a distinctively American type. *Podilymbus*, consisting of a single species common to a large part of both North and South America, is quite sharply separated from the other members of the family, as a type peculiarly American, fossil remains of which, according to Dr. Shufeldt, occur in the Pliocene deposits of Oregon.

The Loons, forming the family Urinatoridæ, are distinctively northern, and to a large degree arctic or subarctic, though some of the species range in the breeding season throughout the cold temperate latitudes. It is a compact, sharply defined group, consisting of a single genus and less than half a dozen species; nearly all of them are circumboreal, none being distinctively American. The Loons form one of the early types, having survived with apparently little change from the close of the Eocene.

The Alcidæ, more numerous in genera and species than either the Grebes or Loons, are, like the Loons, a strictly northern type several of the genera being among the most arctic of birds. With a superficial resemblance to the Penguins of the antarctic seas, but no close kinship, they may be regarded as their arctic representatives in the rôle they fill in nature's economy. The Auks number 13 genera and about 24 species, displaying considerable diversity in size and form, and especially in the size and structure of the bill, and in the character of the nuptial ornaments. Considering their high northern range, one is struck with the greatly circumscribed distribution of many of the genera, only five out of the 13 being circumpolar, three being confined to the shores of the North Atlantic and Eastern Arctic Oceans, and five to the shores of the North Pacific and Bering Sea. Their limited distribution and diversity of structure indicate a plastic group of comparatively modern development. Being maritime species, none of even the non-circumpolar species can be claimed as distinctively either North American or Eurasiatic. In favor of their recent origin is the fact that, while almost arctic in habitat, few of the genera have acquired a circumpolar distribution, as they would in all probability have done had they existed in the same diversity in preglacial times. The locally restricted forms, moreover, occupy a region where food is exceptionally abundant.

Of the three families of Longipennes, — Stercorariidæ, Laridæ and Rynchopidæ, — the first is arctic, the second of world-wide distribution, and the third essentially tropical. Even the genera of the Longipennes are for the most part either circumpolar, cosmopolitan or tropicopolitan. Not a single genus is distinctively North American, and many of the species are either circumpolar or nearly cosmopolitan.

The Tubinares are so largely pelagic, so wide-ranging, and for the most part so little known as regards their breeding stations, that they will be dismissed without further consideration.

Of the six families of Steganopodes four — Phaëthontidæ, Sulidæ, Anhingidæ and Fregatidæ — are essentially tropical; another, Pelecanidæ, is semi-cosmopolitan, being found almost everywhere outside of the arctic and subarctic regions. The remaining family, the Cormorants (Phalacrocoracidæ), is universally distributed, though the species are largely maritime rather than inland, where they frequent only the larger rivers and lakes. Not a single genus of this whole order is distinctively North American. They are all birds of strong flight, and are apparently old types which have long had a wide distribution, the remains of Gannets, Cormorants and Pelicans occurring in deposits of Miocene age in both North America and Europe.

The Anseres are also a cosmopolitan group, containing many genera of very wide distribution. Particular styles of the Anserine type, however, have been developed over particular regions, as in South America, Australia, and in the Old World tropics. Of the 23 North American genera four only are peculiar to this continent, namely, *Lophodytes*, *Charitonetta*, *Camptolaimus*, and *Chen*, closely allied respectively to the circumpolar genera *Merganser*, *Glaucionetta*, *Somateria* and *Anser*, and are thus obviously derived from northern stock. *Histrionicus* is also almost wholly confined to the northern half of North America, while *Arctonetta* and *Philacte* occur only along the arctic coasts of Alaska and Siberia, like several of the genera of the Alcidæ. *Dendrocygna* is tropicopolitan, barely reaching the southern border of the area here under consideration. *Anas*, *Dafila*, *Spatula*, and *Erismatura* are nearly cosmopolitan. The remaining 14 genera of Ducks, Geese, and Swans are circumpolar, and most of them pass the breeding season in the cold temperate or subarctic regions. Thus of the Anseres only about one sixth of the genera are distinctively North American, and in every case are only slight modifications of circumpolar types, and hence of boreal origin, of probably comparatively recent date.

The Odontoglossæ, or the Flamingoes, are tropical,—a small, structurally widely isolated group, dating at least from the Miocene. A single species barely reaches our southern border, though in Pliocene and Post-Pliocene times Flamingoes ranged much farther north than now, both in Europe and North America, and were doubtless more numerous in species, the Flamingoes being a waning group.

Of the Herodiones, the Spoonbills, the Ibises, and the Storks are for the most part tropical, and hence do not enter prominently into the North American fauna. *Ajaja* and *Guara* are tropical American; *Plegadis* is subcosmopolitan and slightly more northern; *Tantalus* is also essentially tropical American, but has closely related forms in the Old World tropics. Our Herons all belong to nearly cosmopolitan genera and subgenera, many of the species being closely representative of Old World forms of corresponding latitudes. The North American Herodiones thus present no forms distinctively American.

The Paludicolæ afford us one form, *Aramus*, peculiar to America, but it barely reaches our southern border. Of the other genera, *Grus* is common to the whole northern hemisphere; *Rallus* and *Porzana* are almost cosmopolitan, as are likewise *Gallinula* and *Fulica*, leaving only *Ionornis* as peculiar. This, however, is tropical American rather than North American, and has closely allied forms in the Old World tropics. The group is thus not only widely dispersed but of known great antiquity, the fossil remains of *Grus*, *Rallus* and *Fulica* having been found in the Eocene and Miocene deposits of Europe and in the Pliocene of North America.

The six families of the Limicolæ also afford very few peculiar North American genera, and these occur where, at first thought, they would be least expected, namely, among the Snipes and Sandpipers. Of the Phalaropes *Steganopus*, a subgenus of *Phalaropus*, is exclusively North American and the most southern in distribution of the three known species of the group, the others being both arctic and circumboreal. It is also of interest to note that, according to Dr. Shufeldt, its remains occur in the Post-Pliocene deposits of southeastern Oregon, rendering it certain that North America was within the pre-glacial range of the group.

The Stilts and Avocets, on the other hand, are of nearly world-wide distribution, being absent from only the cold temperate and arctic portions of the globe. The same is true also of the Oyster-catchers. The Surf-bird, genus *Aphriza*, is an isolated type of extended range on the Pacific coast of both North and South America during migration; its breeding haunts have as yet escaped discovery.

The two remaining families are the Scolopacidæ and the Charadriidæ. Of the former fifteen genera are represented in North America. One only, *Gallinago*, may be termed cosmopolitan. Of the other fourteen, six are circumpolar, and *eight* are distinctively North American, the North American genera being *Philohela*, *Macrorhamphus*, *Micropalama*, *Ereunetes*, *Symphemia*, *Bartramia* and *Tryngites*. Three of them breed from the cold temperate southward. Three of them are also restricted to eastern North America, while the other two—*Ereunetes* and *Macrorhamphus*, both monotypic—have each developed an eastern and a western subspecies. *Heteractitis* has a representative in Alaska and another in Eastern Siberia. Of the two genera of Charadriidæ, one, *Charadrius*, is arctic and circumboreal; the other, *Ægialitis*, is cosmopolitan.

We have now passed rapidly in review the Swimmers, Waders, and Shore-birds, with the following results: Total number of genera, 93; of which 74, or 80 per cent., are either cosmopolitan, tropicopolitan, circumpolar, or subcircumpolar, leaving 19, or 20 per cent., as American, of which only 11, or about 12 per cent., are distinctively North American; one half of these belong to the single family Scolopacidæ.

The North American Gallinæ number 12 genera, of which one, *Ortalis*, is tropical, while *Meleagris* is partly so. The remaining ten genera fall into the two subfamilies, Perdicinæ and Tetraoninæ. The four genera of the former are exclusively American; none reach the cold temperate zone, and all range to the southward of the United States; three of them are also exclusively western. Their metropolis is Mexico, where all occur, and where they have their greatest numerical representation. They have no representatives in South America, and no near relatives in any part of the Old World.

The Tetraoninæ are as emphatically northern as the Perdicinæ are southern. Of the six North American genera, three are circumboreal, one (*Tympanuchus*) is practically eastern, though formerly, in Post-Pliocene times, according to Dr. Shufeldt, ranging to Oregon; the remaining two, *Pediocætes* and *Centrocercus*, are western.

Of the eight genera of Pigeons, one, *Columba*, is subcosmopolitan; two, *Ectopistes* and *Zenaidura*, are North American, though the breeding range of the latter extends to the tropics; the remaining five are tropical American, of which four barely cross our southern border. The Columbine element in our fauna is thus obviously of tropical American origin.

Of the Birds of Prey, the three genera of Vultures are of course tropical American in origin, and still largely so in distribution, reaching only the warmer parts of North America. Of the sixteen genera of the Falconidæ, nine are either circumboreal or subcosmopolitan, and these all date back, with about the same distribution as now, to the Miocene, while some are known from the Eocene. The remaining seven genera must be ranked as tropical, five of them extending but a short distance into the United States. Hence not a single genus of this large family can be classified as distinctively North American. The same remarks are nearly true of the Owls, four of the twelve genera being circumboreal and mainly subarctic; six others are subcosmopolitan, while two only are American, namely, *Micropallas* and *Speotyto*, both of which have a rather restricted North American range. The latter doubtless reached North America from two sources, the West Indies, from which the Florida stock came, and Mexico, from which the western stock was derived.

Conurus, the only representative of the great order Psittaci, is of course only a northern outlier of the numerous group of tropical American Conures.

Our Cuckoos are all of tropical American origin, only the single genus *Coccyzus* having an extended North American distribution. *Geococcyx* is doubtless of Mexican origin. Our single genus of Kingfishers is subcosmopolitan.

Of our eight genera of Woodpeckers two, *Picoides* and *Dryobates*, are circumpolar, one being subarctic and the other of wide distribution; three are distinctively North American, while three others range at large over most of the two Americas.

Of the three genera of Swifts one, *Chætura*, is nearly cosmopolitan, one is exclusively North American, and one ranges over a large part of both the American continents.

All of the five genera of Goatsuckers, if *Antrostomus* be recognized as separable from *Caprimulgus*, are exclusively American; one only, *Phalænoptilus*, is restricted to North America, while *Nyctidromus* does not pass much beyond the Rio Grande.

Seven of the ten genera of Hummingbirds are essentially tropical, barely extending within our limits, while only one ranges over a very large part of the continent. Of the fourteen species, only one is eastern.

The Tyrannidæ, though preëminently a group of the American tropics, is represented by ten genera in North America; five of them, however, are scarcely entitled to rank as North American, and of the remaining five, only two, *Contopus* and *Empidonax*, reach the cold temperate latitudes.

The only American representative of the numerous Old World family Alaudidæ is the genus *Otocoris*, which ranges throughout the greater part of the northern hemisphere.

The Corvidæ are represented by eight genera, three of which are cosmopolitan and one, *Xanthoura*, is tropical. The other four are not only almost exclusively North American, but two of them are limited to the western half of the continent.

The Icteridæ are of course all American. Of the nine genera represented in North America, three (*Dolichonyx*, *Xanthocephalus*, and *Scolecophagus*) are exclusively North American. The six remaining genera range far into the tropics, two of them being mainly tropical; one of them (*Quiscalus*) is also mainly limited in North America to the eastern half of the continent.

The large cosmopolitan family Fringillidæ is represented in North America by thirty genera, twenty-one of which are strictly American and nine circumpolar. Of the twenty-one American, or twenty-three if we give generic rank to *Passerculus*, *Ammodramus* and *Coturniculus*, as I believe should be done, seventeen are exclusively North American; three others are mainly tropical, and three are shared in common by both North and South America. The Fringillidæ, although a cosmopolitan family, give a high ratio of forms peculiarly North American, amounting to fully 60 per cent. Furthermore, three of the North American genera are strictly western, and two strictly eastern, while five of the nine circumboreal genera are either arctic or subarctic.

The tropical American family Tanagridæ is represented by only two genera, namely, *Euphonia* and *Piranga*, only the latter extending over even the warm temperate portion of the continent, while the former barely reaches our border.

Of the six genera of Swallows none is essentially North American. *Tachycineta*, *Progne* and *Stelgidopteryx* range over both Americas; *Chelidon*, *Petrochelidon* and *Clivicola* are old World, each with a single species in North America.

Of the family Ampelidæ, the genus *Ampelis* is common to the colder parts of the northern hemisphere, but is most likely of American origin; the other genus, *Phainopepla*, is western and southern.

Lanius is the single representative of the very numerous and otherwise exclusively Old World family Laniidæ.

The Vireos range over both Americas, to which they are limited, with their metropolis in Middle America.

The Mniotiltidæ constitute the most characteristic family of North American birds, 70 per cent. of the twenty genera occurring in North America. Considering the family as a whole, three genera are exclusively West Indian; *Basileuterus*, which scarcely reaches our border, is continental and tropical; *Compsothlypis* and *Geothlypis* range throughout the warmer parts of both continents, being tropical rather than North American, and nearly the same may be said of *Setophaga*. *Granatellus*, *Cardellina* and *Ergaticus* are Mexican and Central American, the two latter barely reaching our Mexican frontier. *Dendroica* is represented by about forty species, of which twenty-four, or about 60 per cent., may be considered as distinctively North American, while eleven, or 80 per cent., of the remaining species are West Indian; two are Central American and one occurs in the Galapagos Islands. Thus, in general terms, about one fourth of the species are West Indian and three fourths North American. The remaining eight genera are strictly North American, while three of them, *Helmitherus*, *Helinaia* and *Protonotaria*, are restricted to the eastern half of the United States, as is also *Compsothlypis*, so far as its United States distribution is concerned. *Mniotilta* is also eastern.

The large and widely dispersed Old World family Motacillidæ has only two genera in North America—*Budytes*, barely reaching Alaska, and the nearly cosmopolitan genus *Anthus*.

Cinclus is a mountain type, common to most of the higher mountain ranges of America and Eurasia.

The Troglodytidæ is almost exclusively an American family, represented in Eurasia by the subgenera *Anorthura* and *Elachura*. A few East-Indian genera are sometimes placed here, as *Sphenocichla*, *Pneopyga*, etc., but I think erroneously. The metropolis of the true Wrens is tropical America, where are found more than nine tenths of all the species of the group. The genera *Cistothorus*, *Thryothorus*, and *Campylorhynchus* extend into the warmer parts of the United States. *Salpinctes* and *Catherpes* are peculiar forms of the West and Southwest, of probably Mexican origin.

The subfamily Miminæ, of late associated with the Wrens, is exclusively American, and four out of the five North American genera doubtless originated near where they are now found. One of them, *Galeoscoptes*, is essentially eastern and one, *Oroscoptes*, is western. *Mimus* is tropical, with a single outlying species in North America.

Of the Paridæ, *Sitta* and *Parus* are found throughout the greater part of the northern hemisphere; *Chamæa*, *Psaltriparus*, and *Auriparus* are mainly limited to the northern border of Mexico and the adjoining tier of States to the northward, a single species of *Psaltriparus* extending southward to the higher mountains of Guatemala, and another northward along the Pacific coast to Washington.

Certhia and *Regulus* belong to the northern hemisphere, ranging over its northern half; *Polioptila* is tropical American, with outlying species extending across the warm temperate.

The cosmopolitan family Turdidæ is represented by three genera—*Merula*, *Turdus* and *Saxicola*—of wide dispersion, and by two—*Hesperocichla* and *Sialia*—which may be considered as autocthonous; the sixth genus, *Myiadestes*, is tropical American, with a single species in the western United States.

In concluding our review of the land birds, the results may be summarized as follows: Total number of genera, 181; of these 55, or 30 per cent., are circumboreal or otherwise wide-ranging Old World forms; 126 genera, or 70 per cent., are American, of which 35, or 28 per cent., are essentially tropical, leaving 91 genera, or about 50 per cent., as distinctively North American.

Separating the Land Birds into the two categories of Passerine and non-Passerine, we find that of the 75 non-Passerine genera, 36 per cent. are wide ranging Old World forms, leaving 64 per cent. as American, of which latter 40 per cent. are tropical American, leaving 30 per cent. of the whole as strictly North American. Of the 106 genera of Passeres, only 25 per cent. are circumboreal or otherwise widely distributed in the Old World, leaving 75 per cent. as American, of which four fifths are strictly North American, or 60 per cent. of the Passerine genera.

The total number of genera represented in North America (as geographically restricted in the A. O. U. Check-List) of which account is here taken, is 274; of these 129, or 44 per cent., are either circumboreal or widely dispersed over the Old World, leaving 145, or 56 per cent., as American. Of the latter 102 genera, or about 40 per cent of the whole, are distinctively North American, and 43 genera, or about 12 per cent. of the whole, may be classed as tropical American.

Of the distinctively North American genera 57 per cent. have a general distribution over the continent, while of the remaining 43 per cent., 17 per cent. are eastern and 26 per cent. are western.

In order to bring still more closely into relief the extent and manner of the relationship of the North American avifauna to that of the rest of the world, we will divide the non-tropical part of the continent into three belts, namely: (1) An Arctic, including that part of the continent and its adjacent islands north of about the limit of forest vegetation. (2) The Cold Temperate, limited on the north by the Arctic, and on the south by what is commonly recognized as the Canadian Fauna (see Pl. III). Its southern border thus coincides in a general way with the northern limit of the successful cultivation of the staple grains and fruits of the temperate zone. (3) The Warm Temperate, extending from the southern border of the Cold Temperate to about the edge of the palm belt, or to the hot lowlands of Mexico.

The Arctic belt is inhabited during the breeding season by about 65 genera of birds, of which only 5 are exclusively American; the remaining 60 have either a general circumarctic distribution or occur in Eastern Siberia or on the Arctic coast of Europe as well as in North America. None of the 5 American genera is strictly Arctic; they merely extend into the Arctic from the Cold Temperate, while quite a proportion of the circumpolar genera is strictly Arctic in their breeding range.

In the Cold Temperate belt 120 genera are represented, of which 98 are circumpolar and 22 American. Of these 46, or nearly half of the circumpolar genera, range also into the Warm Temperate belt, as do also 14 of the strictly American genera.

In the Warm Temperate belt 95 genera occur which do not range into the Cold Temperate, of which only 12 are Old World, 83 being exclusively American. In addition 60 genera are common to both the Cold Temperate and Warm Temperate, of which 46 are Old World and 14 American. This gives a total of 155 genera in the Warm Temperate, of which 58 are Old World and 97 exclusively American. Besides these, 50 essentially tropical genera reach or extend somewhat into the Warm Temperate, of which 43 are American and 7 tropicopolitan, increasing the total number of genera occurring in the Warm Temperate to about 205.

These statistics illustrate a number of important points: (1) the rapid increase of bird life in North America from the Arctic regions southward, notwithstanding the fact that the continent steadily and greatly decreases in breadth from the north southward, the number of genera in the Arctic belt being 65, in the

Cold Temperate 120, and in the Warm Temperate 205. (2) The decrease of Old World forms from the north southward, in the Arctic belt 91 per cent. of the genera being circumpolar, 82 per cent. in the Cold Temperate, and 37 to 23 per cent. in the Warm Temperate, according to whether the northern or the southern edge of the belt be considered. (3) As the distinctively American genera increase in passing southward, they became more or less differentiated into eastern and western types. Taking land birds alone, to the exclusion of pelagic and strictly maritime, the number of peculiarly eastern genera increases from two or three in the Cold Temperate to 17 in the Warm Temperate, and in the western from three or four in the Cold Temperate to 26 in the Warm Temperate, exclusive in each case of intrusive, essentially tropical forms which if taken into account would still further differentiate the two regions. Besides, as is well known, many genera which range across the continent are represented by different species on the two sides, while most of the continental species are differentiated into eastern and western subspecies, and frequently into a considerable number of local races.

We may now inquire, Whence was derived the present avifauna of North America? It is evident from the facts already presented that it is made up of four prominent elements. The first consists of types common to a large part of the northern hemisphere, which more and more prevail as we proceed from the south northward, till in subarctic and arctic America we meet with little else. The second consists of a rather strong infusion at the southward of types of almost universal distribution over the warmer temperate and intertropical latitudes. Third, a very prominent tropical American element, developed to the maximum just along our southern border and fading out gradually to the northward, little being left of it after we enter the Cold Temperate. Fourth, a very conspicuous autocthonous element, reaching its maximum in the Warm Temperate, and continuing prominent far into the Cold Temperate. This is made up partly of strongly modified Old World types, but mainly of peculiar genera, or at least peculiar species, derived obviously from tropical American stock, which entered our borders partly by way of the West Indies, but mainly by way of Central America and Mexico.

A large part of the distinctively American element in the North American avifauna seems almost obviously of tropical American origin; for example our Vultures, several genera of our Hawks and Owls, our Cuckoos, most of our Woodpeckers, our Nighthawks, Whippoorwills, Swifts, and all of our Humming-birds; all of our Flycatchers, Orioles and Blackbirds, and our Vireos and Tanagers; many of our Sparrows and Grosbeaks; all of our Gnatcatchers, and the Mockingbirds, some of our Wrens, and a few of our more southern genera of Warblers, as the Yellowthroats and Redstarts.

It seems probable that another portion originated in Mexico, mainly on the great Mexican Plateau, as the Quails, the Turkeys, and some of our Pigeons; a number of our Woodpeckers; the greater part of our Jays, the Pipilos, the various species of *Peucæa*, *Amphispiza* and *Passerina;* the Titmice of the genera *Psaltriparus*, *Auriparus*, and *Chamæa;* our Wrens of the genera *Catherpes* and *Salpinctes;* the Solitaire, most of the Thrashers, and the Bluebirds.

Doubtless we may properly recognize as autochthonous or indigenous a half dozen genera of Sandpipers, our Prairie, Sharp-tailed and Sage Grouse, our Woodpeckers of the genera *Ceophlœus*, *Xenopicus* and *Sphyrapicus;* the Jays of the genera *Cyanocephalus* and *Cyanocitta* and probably *Aphelocoma;* the genera *Scolecophagus*, *Xanthocephalus*, and *Dolichonyx* among the Icteridæ, and among our Sparrows such genera and subgenera as *Passerella*, *Passerculus*, *Ammodramus*, *Spizella*, *Zonotrichia*, *Melospiza*, *Poocætes*, *Spiza*, and *Calamospiza;* of the Warblers the genera *Mniotilta*, *Protonotaria*, *Helinaia*, *Helmitherus*, *Helminthophila*, *Sylvania*, and *Seiurus*, and many of the species of *Dendroica;* also *Galeoscoptes*, *Oroscoptes*, *Hesperocichla*, and *Hylocichla*.

As modifications of Old World or circumpolar stock may be mentioned *Camptolaimus*, *Lophodytes*, and *Chen* among Ducks and Geese, and *Picicorvus*, *Rhynchophanes*, *Hesperiphona*, and *Leucosticte* among Passeres, with of course the hundred or more species and subspecies which belong to circumpolar genera.

Probably three fourths of the distinctively North American genera and species have reached their present habitats by way of Mexico, and perhaps one fourth or less by way of the West Indies. Many of the local genera and species, as those restricted respectively to the eastern or western sides of the continent, have originated in some part of their present habitats, at a somewhat remote period, but most of them doubtless since the retreat of the glacial ice. In earlier times the barrier separating the ranges of species occupying respectively the eastern and western parts of the United States must have been much stronger than it is at present; and it will be less doubtless a few decades hence than it is now. There is good evidence that the eastern species are gradually extending their range westward, as I have already pointed out in the case of *Colaptes auratus*, and that western species are, in some cases at least, extending their range eastward. This is doubtless in great part due to, and is certainly aided by, the westward extension of agriculture, which is so rapidly transforming regions not long ago thought to be almost irreclaimable deserts. Irrigation and tree-planting, and the general pursuits of agriculture, cannot fail to extend westward the ranges of many eastern species, as the Bobwhite and various Sparrows, and probably of many of the smaller Passeres.

PART II.—THE FAUNAL SUBDIVISIONS OF NORTH AMERICA, CONSIDERED WITH REFERENCE TO THEIR RELATIONSHIPS, CLASSIFICATION AND NOMENCLATURE.[1]

1.—*Faunal Areas.*

In zoögeography it is necessary to recognize faunal areas varying in grade and importance, just as in zoölogy it is necessary to divide animals into groups differing in rank, as *classes*, *orders*, *families*, *genera* and *species*. The terms employed in zoögeography, however, have not been used with the same precision as the practically corresponding terms in zoölogy. Identical designations have sometimes been used in

[1] The classification and nomenclature here adopted is the same as in my recent paper on 'The Geographical Distribution of North American Mammals' (Bull. Am. Mus. Nat. Hist., Vol. IV, pp. 199–243, Dec. 1892). Also much of the matter, as regards the generalities of the subject, is the same in substance and often in phraseology. The accompanying maps are also reproduced, by permission, from the paper just cited. As that paper is not likely to be readily accessible to the majority of the readers of 'The Auk,' it is hoped that the reproduction of some matter previously published in another connection may not be considered reprehensible.

widely different senses, in accordance with the individual preferences of different writers. The terms *realm*, *region*, *kingdom*, *province*, *district*, and *fauna* have sometimes been applied by different writers to nearly the same geographical area, and each of them to divisions considered as of primary rank, while most of them have also been used for regions of secondary or still lower grade. In many cases they have been used merely in a quasi-technical sense, and there is hence difficulty in determining their claims to recognition by any rule of priority, or by frequency of use in any particular sense; so we are forced to choose mainly on grounds of appropriateness and convenience, the main thing being to secure uniformity of application. In view of this unfortunate state of affairs I attempted, in a paper on the geographical distribution of North American birds, published in 1871,[1] to devise a system of terms that should be not only appropriate, but in as close conformity as possible with previous usage. Finding that *realm* had been used by Agassiz and others for regions of the first rank, and that *region* and *province* had generally been employed for areas of secondary or lower rank, I proposed a scheme of nomenclature, which to some extent has been followed by others, but of course not fully or rigidly, since little attention has been paid to exact terminology in geographical biology. The scheme then proposed is as follows:—

For divisions of the first rank,					*Realm.*
"	"	"	second rank,		*Region.*
"	"	"	third	"	*Province.*
"	"	"	fourth	"	*Subprovince or District.*
"	"	"	fifth	"	*Fauna.*

Their grade and order of sequence may be indicated by a comparison with the leading groups in zoölogy; thus *realm* would correspond in rank with *class; region* with *order; province* with *family; district* with *genus; fauna* (or *flora*, as the case may be) with *species*. It sometimes becomes convenient to recognize other divisions intermediate to these—as in zoölogy we have suborder, subfamily, subgenus, and sub-

[1] Bull. Mus. Comp. Zoöl., II, No. 3, April, 1871.

species, so we may have in zoögeography *subregions*, *subprovinces*, and even *subfaunæ*. In addition to and independent of these divisions we may also recognize *zones*. To illustrate by an example, temperate North America forms a *region* of a *realm* and includes two *subregions*, one of which is divisible into *provinces* and *subprovinces*, and some of these latter into *districts*, while each may contain several *faunæ*. It also proves desirable to subdivide the continent into a number of transcontinental belts or *zones*, which in a way coincide with the climatic zones of the physiographer.

The term *fauna* is unfortunately in current use both in a general sense and also in a restricted, taxonomic sense. In the first case it is employed to designate the animal life of a given area, geographical or political, varying in extent from a township to a continent, from a transient pool to a lake or an ocean, or in geology from a definite stratigraphic horizon to a geologic age, as the qualifying words may chance to indicate; in the second case it is a definite taxonomic term for the ultimate subdivisions of a realm, as is the term *species* for the ultimate divisions of a *class*. The term *flora* has also the same double use; and it would perhaps be futile to attempt to displace either by some specially coined substitute, to take the place of these terms when used in a specific sense for an ultimate faunal or floral area. Such an attempt, however, is perhaps unnecessary, since the qualifying prefix will rarely, if ever, leave one in doubt as to whether the term is used in a general or in a taxonomic sense.

Realms are sometimes characterized by the presence of certain orders, and usually by the presence of certain families which give to them a particular impress, and by the absence of others which in a similar way characterize other realms. Regions are usually characterized by the prevalence over them of certain genera, or even by entire families; provinces by the prevalence of particular genera or by the presence or absence of prominent species. Faunæ, on the other hand, are seldom characterized by having restricted to them any particular genera or species, but by the combination or overlapping of a number of genera and species not found elsewhere associated,—in other words, by a peculiar commingling of a considerable number of genera and species.

The transition between faunæ, between provinces, or between adjoining divisions of any grade, is rarely abrupt; it is impossible to give them absolute boundaries; yet they may be limited, in a general way, with considerable definiteness. They depend upon climatic conditions, which are in a measure determined or modified by features of topography; in other words, they are determined by the same conditions that govern the distribution of species. Hence they can seldom be bounded by geographical meridians, or by parallels of latitude, or by political boundaries; they do, however, closely coincide with certain isothermal lines, which are generally those of the season of reproduction. As temperature is influenced by altitude as well as by latitude, elevated table-lands and mountain ranges deflect the isotherms, in the northern hemisphere, far to the southward of their position over the contiguous low country, and furnish congenial homes for northern forms of animals and plants under comparatively low latitudes. Thus in the Appalachian Highlands we find northern types far to the southward of their range in the low country on either side; and the same is more strikingly apparent in the case of the Rocky Mountains, where northern types extend far down into Mexico, and in the Cascade and Sierra Nevada chain where northern types extend to southern California; isolated patches of northern life also occur on the summits of detached peaks and outlying ranges throughout the Plateau and Great Basin regions of the West. Northern and southern forms of life thus interdigitate according to elevation of the land.

While the life of the middle temperate latitudes, at ordinary levels, is so different from that of the high north on the one hand, and from that of the tropics on the other, the change is effected by a very gradual transition between the two extremes. If we suppose the life of either of the northern continents to be represented by a cube, and this cube to be composed of two blunt-edged wedges, and the two wedges so placed that one thin edge and one thick edge shall form respectively the base and the top of the cube, the two wedges would represent the fading out of the northern life southward and of the southern life northward; except that in nature the two elements are diffused through the mass instead of being segregated as in our supposed cube.

Furthermore, this northern life is largely circumpolar, so that as we proceed northward, as in North America for example, the genera become in increasing ratio more and more those of wide distribution, till in the extreme north we meet with few that are not circumpolar. How gradually and completely this transition is made has already been shown (*antea*, p. 114), 47 per cent of the non-pelagic birds of North America occurring in corresponding latitudes in the Old World. Extra-tropical North America has, in fact, so much in common with extra-tropical Eurasia, that the two areas constitute merely two regions of a single realm.

To indicate my views in respect to the faunal relations of North America to the world at large, I will recapitulate briefly what I consider may be properly regarded as the primary life areas of the earth's surface, namely:—

(1) An *Arctic Realm*, occupying the region northward from about the limit of forest vegetation, or from about the isotherm of 32° F. It is characterized by its paucity of life and its homogeneousness, nearly all its forms of both animal and vegetable life ranging throughout its whole extent.

(2) A *North Temperate Realm*, extending from the northern limit of forest vegetation to the northern border of the palm belt, or between the annual isotherms of 32° and 70° F.

(3) An *American Tropical Realm*, consisting, as the name implies, of tropical America.

(4) An *Indo-African Realm*, consisting of Africa (except the northern border), and tropical Asia and its outlying islands.

(5) A *South American Temperate Realm*, embracing extra-tropical South America.

(6) An *Australian Realm*, including not only Australia, but New Guinea, New Zealand, and the various groups of islands to the northward and eastward.

(7) A *Lemurian Realm*, consisting of Madagascar.

An eighth or *Antarctic Realm* is also sometimes recognized, as the Antarctic counterpart of the Arctic realm, though perhaps less well characterized, its fauna consisting almost exclusively of maritime and pelagic species.

The North Temperate Realm is divisible into two regions, (1) a *North American Region*, consisting of temperate North America, and (2) a *Eurasiatic Region*, consisting of temperate Eurasia.

Before proceeding to consider the North American Region and its faunal subdivisions, a few words may be devoted to the *American Arctic*, which is here set off from the rest of North America as a part of the Arctic Realm. If North America were entirely isolated from the rest of the world, it would be quite proper to treat the American Arctic as merely a subdivision of the North American Region; but in view of the fact that it is in reality a part of a homogeneous hyperborean fauna of circumpolar distribution, it seems more in accordance with general facts to consider it as forming part of an Arctic Realm. The propriety of this seems especially emphasized when we consider that (to quote the words of Dr. Merriam) "the animals and plants inhabiting the Arctic regions are usually specifically identical throughout Arctic America, Greenland, and the polar parts of Eurasia and outlying islands," "the types inhabiting the Arctic Zone being few in number and uniform in character throughout their distribution."[1] The fauna of this Arctic Zone is thus no more American than it is Eurasiatic, and differs far more from that of the adjoining region to the southward, both in North America and Eurasia, than does the American Arctic from the Eurasian Arctic. The Arctic Realm possesses only a small number of peculiar types in proportion to its area or in comparison with the other realms, yet its ratio of peculiar types, in comparison with its meagre fauna, is by no means low. It seems an eminently natural division from the fact that its southern boundary marks the termination of forest vegetation, with which necessarily stop all the mammals, birds and insects which depend upon forests for food, shelter and a congenial home. Of the 65 genera of birds occurring in the American Arctic 60 are, as already shown, circumpolar, and 5 are American water birds that reach it for a short stay during the breeding season.

The American Arctic may be divided into two areas which may take the rank of faunæ, namely: (1) *Barren Ground*, (2) *Alaskan-Arctic*. The last has been characterized by Mr. E. W. Nelson,[2] and the first by me in a recently published paper on 'The Distribution of North American Mammals' (l. c. p. 220).

---

[1] Proc. Biol. Soc. Wash., VII, 1892, pp. 39, 40.

[2] Rep. Nat. Hist. Coll. in Alaska, 1887, pp. 26-32.

The exact limits of the breeding ranges of our arctic and subarctic birds is still very imperfectly known. Of the following list of 80 species and subspecies, a few doubtless range in the breeding season somewhat below the southern boundary of the Arctic, but for the most part they may be considered as distinctively characteristic of the Arctic fauna. A large proportion of them are either strictly circumpolar, or are represented in Eurasia by closely allied species or subspecies.

Distinctively Arctic.

(Those marked with a * are apparently confined mainly or wholly to the Eastern Arctic or Barren Ground Fauna; those marked with a † are Western or Alaskan Arctic; those with no sign prefixed are of general distribution, by far the greater part being circumpolar.)

† Urinator adamsi
Urinator arcticus
† Urinator pacificus
Fratercula arctica glacialis
Cepphus mandtii
Stercorarius pomarinus
Stercorarius parasiticus
Stercorarius longicaudus
Gavia alba
† Rissa tridactyla pollicaris
† Rissa brevirostris
* Larus glaucus
† Larus barrovianus
Larus leucopterus
† Larus glaucescens
* Larus kumlieni
† Larus nelsoni
* Larus marinus
† Larus schistisagus
Rhodostethia rosea
Xema sabinii
† Sterna aleutica
Fulmarus glacialis
† Fulmarus g. glupischa
† Phalacrocorax pelagicus
† Phalacrocorax urile
† ? Eniconetta stelleri
† Arctonetta fischeri
* Somateria borealis
† Somateria v-nigra
Somateria spectabilis
Oidemia americana
Chen hyperborea
Chen rossii
Anser albifrons gambeli
† Philacte canagica
Olor columbianus
Crymophilus fulicarius
Phalaropus lobatus
Tringa canutus
* Tringa maritima
† Tringa couesi
† Tringa ptilocnemis
† Tringa acuminata
Tringa maculata
* Tringa fuscicollis
† Tringa bairdii
Tringa alpina pacifica
* Ereunetes pusillus
† Ereunetes occidentalis
Calidris arenaria
Limosa hæmastica
† Limosa lapponica baueri
† Heteractitis incanus
Numenius hudsonicus
Numenius borealis
Charadrius squatarola
Charadrius dominicus

† Charadrius d. fulvus
Ægialitis semipalmata
* Ægialitis hiaticula
Arenaria interpres
Arenaria melanocephalus
Lagopus lagopus
Lagopus rupestris
* Lagopus r. reinhardti
Falco islandus
Falco rusticolus gyrfalco
* Falco rusticolus obsoletus
Otocoris alpestris
Acanthis hornemanni exilipes
* Acanthis linaria rostrata
Acanthis linaria holbœlli
Plectrophenax nivalis
† Plectrophenax hyperboreus
Calcarius lapponicus
Calcarius pictus
Anthus pensilvanicus
Saxicola œnanthe

The following are arctic faunally, if not geographically, breeding mostly above timberline in the Rocky Mountains:—

Lagopus leucurus
Leucosticte tephrocotis
Leucosticte atrata
Leucosticte australis

Many other species, more properly Cold Temperate than Arctic, range into the Arctic so that the above list is by no means a complete enumeration of the American Arctic avifauna.

The North American Region (see Pl. III) consists of two *Subregions*, namely, (1) a *Cold Temperate Subregion* and (2) a *Warm Temperate Subregion*. The Cold Temperate Sub region extends across the continent from about the mean latitude of 43° northward to the limit of forests, with also a narrow prolongation southward along the Appalachian Highlands as far as northern Georgia; another in the interior along the main chain of the Rocky Mountains and its principal outliers south into Mexico; and a third along the Cascade and Sierra Nevada ranges. Its southern border also sweeps to the northward so as to exclude the great Saskatchewan Plains. In other words, the Cold Temperate coincides exactly with Dr. Merriam's 'Boreal Province' as laid down on his 'Provisional Biological Map of North America'.[1] As compared with the Arctic, it has 120 genera instead of 65, of which about 70 per cent. are circumpolar, and 30 per cent. American, showing the close connection of the life of the whole northern half of the northern hemisphere.

The following list of 213 species and subspecies includes only such as may be properly considered as distinctively characteristic of the Cold Temperate Subregion, as contrasted on the one hand

[1] N. Am. Fauna, No. 3, 1891, map 5.

with the American Arctic, and on the other with the Warm Temperate Subregion. Not all of them, however, are strictly limited to the Cold Temperate, a few ranging into the southern border of the Arctic, and a few others extending southward into the northern edge of the Warm Temperate; but their ranges, generally speaking, may be regarded as limited to the Cold Temperate. In scanning the list it should be borne in mind that this subregion extends very far southward along the principal mountain ranges of the continent.

Distinctively Cold Temperate.

Colymbus auritus
Colymbus holbœlli
Urinator imber
Urinator lumme
Fratercula arctica
Cyclorrhynchus psittaculus
Simorhynchus cristatellus
Simorhynchus pygmæus
Simorhynchus pusillus
Synthliborhamphus antiquus
Cepphus grylle
Alca torda
Plautus impennis
Larus argentatus smithsonianus
Larus delawarensis
Larus vegæ
Larus brachyrhynchus
Larus franklini
Larus philadelphia
Sterna paradisæa
Puffinus tenuirostris
Oceanodroma furcata
Oceanodroma leucorrhoa
Sula bassana
Phalacrocorax carbo
Phalacrocorax dilophus
Phalacrocorax d. cincinnatus
Phalacrocorax pelagicus robustus
Merganser americanus
Merganser serrator
Anas carolinensis
Anas discors
Dafila acuta
Aythya americana
Aythya vallisneria
Aythya marila nearctica
Aythya affinis
Aythya collaris
Glaucionetta clangula americana
Glaucionetta islandica
Charitonetta albeola
Histrionicus histrionicus
Camptolaimus labradorius
Oidemia deglandi
Oidemia perspicillata
Branta canadensis
Branta c. occidentalis
Branta bernicla
Olor buccinator
Porzana carolina
Porzana noveboracensis
Gallinago delicata
Macrorhamphus griseus
Macrorhamphus scolopaceus
Micropalama himantopus
Tringa minutilla
Limosa fedoa
Totanus melanoleucus
Totanus flavipes
Totanus solitarius
Totanus s. cinnamomeus
Tryngites subruficollis
Dendragapus obscurus
Dendragapus obscurus fuliginosus
Dendragapus obscurus richardsoni
Dendragapus canadensis

Dendragapus franklini
Bonasa umbellus togata
Bonasa u. umbelloides
Bonasa u. sabini
Lagopus welchi
Lagopus lagopus alleni
Lagopus rupestris nelsoni
Lagopus r. atkhensis
Pediocætes phasianellus
Accipiter atricapillus
Accipiter a. striatulus
Archibuteo sancti-johannis
Falco columbarius
Falco richardsoni
Scotiaptex cinerea
Bubo virginianus arcticus
Bubo virginianus saturatus
Nyctala tengmalmi richardsoni
Nyctea nyctea
Surnia ulula caparoch
Dryobates villosus leucomelas
Dryobates v. hyloscopus
Dryobates pubescens gairdneri
Dryobates pubescens oreœcus
Picoides arcticus
Picoides americanus
Picoides a. dorsalis
Picoides a. alascensis
Sphyrapicus varius
Sphyrapicus v. nuchalis
Sphyrapicus thyroideus
Contopus borealis
Empidonax flaviventris
Empidonax difficilis
Empidonax pusillus
Empidonax minimus
Empidonax hammondi
Otocoris alpestris leucolæma
Pica pica hudsonius
Cyanocitta stelleri
Cyanocitta s. frontalis
Cyanocitta s. macrolopha
Cyanocitta s. annectens
Perisoreus canadensis
Perisoreus c. capitalis
Perisoreus c. fumifrons
Perisoreus c. nigricapillus
Perisoreus obscurus
Picicorvus columbianus
Cyanocephalus cyanocephalus
Corvus corax principalis
Scolecophagus carolinus
Coccothraustes vespertinus
Coccothraustes v. montanus
Pinicola enucleator
Leucosticte griseonucha
Carpodacus purpureus
Carpodacus p. californicus
Carpodacus cassini
Loxia leucoptera
Spinus pinus
Ammodramus princeps
Ammodramus sandwichensis
Ammodramus s. savanna
Ammodramus s. alaudinus
Zonotrichia querula
Zonotrichia leucophrys
Zonotrichia l. intermedia
Zonotrichia l. gambeli
Zonotrichia coronata
Zonotrichia albicollis
Spizella monticola
Spizella m. ochracea
Junco aikeni
Junco hyemalis
Junco h. oregonus
Junco h. shufeldti
Junco h. carolinensis
Junco h. thurberi
Junco annectens
Junco cinereus dorsalis
Junco cinereus palliatus
Melospiza cinerea
Melospiza lincolni
Passerella iliaca
Passerella iliaca unalaschcensis
Passerella iliaca megarhyncha
Passerella iliaca schistacea
Tachycineta bicolor
Tachycineta thalassina
Ampelis garrulus
Lanius borealis

Vireo philadelphicus
Helminthophila ruficapilla
Helminthophila r. gutturalis
Helminthophila celata
Helminthophila c. lutescens
Helminthophila peregrina
Dendroica tigrina
Dendroica cærulescens
Dendroica coronata
Dendroica auduboni
Dendroica maculosa
Dendroica castanea
Dendroica striata
Dendroica blackburniæ
Dendroica virens
Dendroica townsendi
Dendroica palmarum
Dendroica p. hypochrysea
Seiurus noveboracensis
Seiurus n. notabilis
Geothlypis agilis
Geothlypis philadelphia
Geothlypis macgillivrayi
Sylvania pusilla
Sylvania p. pileolata
Sylvania canadensis
Cinclus mexicanus
Troglodytes hiemalis
Troglodytes h. pacificus
Troglodytes alascensis
Certhia familiaris americana
Certhia familiaris montana
Certhia familiaris occidentalis
Sitta canadensis
Sitta pygmæa
Parus atricapillus
Parus a. septentrionalis
Parus a. occidentalis
Parus gambeli
Parus cinctus obtectus
Parus hudsonicus
Parus h. stoneyi
Parus rufescens
Regulus satrapa
Regulus s. olivaceus
Regulus calendula
Turdus aliciæ
Turdus aliciæ bicknelli
Turdus ustulatus
Turdus u. swainsonii
Turdus aonalaschkæ
Turdus a. auduboni
Turdus a. pallasii
Hesperocichla nævia
Sialia arctica

Of the above list of 213 species and subspecies, about 20 are maritime, about 60 range across the continent, and about 130 have a more limited range. Of these a small number is limited to the immediate vicinity of the Northwest Coast, a few to the Upper Missouri region, and a large number to the Rocky Mountain plateau and the various mountain ranges of the western half of the continent.

The Cold Temperate Subregion has been divided into four *faunæ*, as follows: (1) *Hudsonian*, (2) *Canadian*, (3) *Aleutian*, (4) *Sitkan*. The first two have been long recognized, and their boundaries are well known. They were first mapped by Dr. Merriam[1] in 1890, and more recently by the

[1] North American Fauna, No. 3, map 5.

present writer,[1] on practically the same lines. The last two were first defined by Mr. E. W. Nelson[2] as respectively the 'Sitkan District' and the 'Aleutian District,' and re-defined and mapped in my recent paper on the 'Geographical Distribution of North American Mammals.'[3] As noted later in the present paper (p. 139), there remain to be defined, when our knowledge of the subject becomes sufficiently detailed, a series of local mountain faunæ in the Rocky Mountains and other principal mountain chains in the West.

The *Warm Temperate Subregion* occupies middle North America, extending from the southern boundary of the Cold Temperate Subregion, as defined above, to the northern edge of the American Tropical Realm (see Pl. III). It is thus geographically the same as Dr. Merriam's 'Sonoran Province.' It includes the greater part of the United States, Lower California, and the Mexican tableland. It is cut into along the principal mountain systems by the southern prolongations of the Cold Temperate Subregion, and also extends northward over the Saskatchewan Plains. The extreme southern parts of the peninsulas of Florida and Lower California, however, are excluded, as also the lower coast region of Texas, these excluded districts, though of comparatively small extent, belonging to the Tropical Realm.

The Warm Temperate Subregion contrasts strongly with the Cold Temperate in respect to its topographic and climatic features, as well as in its faunal aspects. Aside from the very marked difference of temperature between the two, the Cold Temperate is a homogeneous region, covered almost continuously with principally coniferous forests; and aside from its southern extension along the mountain ranges, presents little diversity in topography. The Warm Temperate, on the other hand, is highly diversified topographically, and consequently in climate, especially in respect to rainfall, which of course greatly modifies the distribution of forests, and of plants in general, which in turn exerts a marked influence upon the distribution of animal life. We have hence conditions favorable for the development of

---

[1] Bull. Am. Mus. Nat. Hist., IV, 1892, pl. viii.
[2] Rep. Nat. Hist. Coll. made in Alaska, 1887, pp. 24--27.
[3] Bull. Am. Mus. Nat. Hist., IV, 1892, pp. 223, 224, pl. viii.

locally restricted types, including families as well as genera and species, to say nothing of merely geographical races or subspecies. Consequently it is divisible into a larger number of faunal areas, varying to a much greater extent as regards their relative taxonomic rank, and much more strongly contrasted in respect to their faunal characteristics.

The Warm Temperate Subregion may be considered as primarily divisible along a nearly north and south line into two *Provinces*, namely (1) an eastern or *Humid Province* and (2) a western or *Arid Province.* The Humid Province extends from the Atlantic coast westward to about the 100th meridian, or to the eastern border of the Great Plains, exclusive, of course, of the Appalachian Highlands; the Arid Province extends thence westward to the Pacific coast, exclusive, of course, of the southward extensions of the Cold Temperate along the mountain systems. These regions thus correspond respectively with Dr. Merriam's 'Humid Sonoran' and his 'Arid Sonoran,' as he originally defined them. The Humid Province, as the name implies, has a rather moist climate and is (or was a century ago) in most parts heavily forested; the Arid Province is, on the other hand, a region of open plains and deserts, with, in general, an excessively arid climate.

These two provinces thus coincide with the two strongly marked regions of the middle portion of North America in respect to climate as modified by rainfall. The line of separation is thus meridional instead of transcontinental as are the faunal boundaries determined by temperature. In this case temperature, as a controling influence in the distribution of animals and plants, is subordinated to the usually less powerful agent, humidity.

The transition between the Humid and Arid Provinces is nowhere abrupt; they gradually merge into each other everywhere along their line of junction, as the prairies of the Mississippi Valley gradually become more arid and take on the characteristic aspect of the plains. There is thus here the usual 'transition' belt occurring between contiguous faunal areas. It is, however, rather broader than between regions where temperature is primarily the limiting influence, as in the case of boundaries trending in a nearly east and west direction, the transition being first from a forested region to one of fertile prairies, and thence to arid plains and deserts. The dividing line may be considered as coincident with the isohyetal curve marking an annual rainfall of 20 inches or less, as shown on the rainfall charts of the United States, — in other words, as already stated, near the 100th meridian.

The contrasting climatic conditions of these two regions are due to peculiar physiographic and geographic conditions which extend back to a comparatively remote period — so remote as to have given time for the development of many generic forms limited almost entirely to one or the other of these two regions. Thus, in respect to birds, we have 17 genera in the Humid Province which do not range to any great extent into the Arid Province, and 28 genera in the Arid Province which are not found to any extent in the Humid. The distinctively eastern or Humid Province genera are:—

Philohela	Ictinia	Protonotaria
Bartramia	Dolichonyx	Helinaia
Colinus	Quiscalus	Helmitherus
Tympanuchus	Ammodramus[1]	Compsothlypis
Catharista	Spiza	Galeoscoptes
Elanoides	Mniotilta	

A few of these genera range into the eastern border of the Plains, and sometimes, as in the case of *Galeoscoptes*, occur locally westward to the Pacific coast. They are nevertheless to be regarded as distinctively characteristic of the eastern side of the continent. *Colinus* is more difficult to locate, as southern forms of the genus occur in the Sonoran Subprovince of the Arid, while its northern form, *C. virginianus*, is distinctively an eastern form characteristic of the Humid Province.

On the other hand, the following genera may in like manner be considered as distinctively characteristic of the Arid Province, although some of them range northward into the Cold Temperate:—

Æchmophorus	Pediocætes	Micropallas
Oreortyx	Centrocercus	Geococcyx
Callipepla	Pseudogryphus	Xenopicus
Cyrtonyx	Glaucidium	Phalænoptilus

[1] Subgenus.

Cypseloides	Amphispiza	Chamæa
Aëronautes	Calamospiza	Psaltriparus
Selasphorus	Phainopepla	Auriparus
Xanthocephalus	Salpinctes	Myiadestes
Calcarius	Catherpes	
Rhyncophanes	Oroscoptes	

Were we to include in this connection some 38 additional tropical genera which range only a short distance into the Warm Temperate, as here defined, we should have to add 7 genera to the eastern list and about 30 to the western list. In other words, out of a total of about 150 genera distinctive of the Warm Temperate,[1] about 65, or 43 per cent., have a transcontinental distribution, and about 85, or 57 per cent., are either eastern or western.

The higher ratio of peculiar types in the Arid Province as compared with the Humid Province is obviously due to geographic conditions, the Arid Province adjoining at the southward a broad tropical land area, while the southern boundary of the Humid Province is formed by the Gulf of Mexico. A large proportion of the genera peculiar to either the Humid or Arid Provinces range across the whole north and south breadth of the Warm Temperate Subregion.

The northern half of the Warm Temperate, however, also differs faunally quite markedly from the southern half, in consequence of the extension southward of a few northern genera over most of its northern half, and the extension northward of many tropical genera over a portion of its southern half. As regards birds, while a considerable list of species would fall respectively into one or the other of these two categories, the number of genera is small, owing to the fact that if we take them on the basis of their transcontinental range their representation is often very different in the two provinces, as regards both their geo graphical distribution and the species which represent them. The northern list would include perhaps such genera as *Botaurus* (excluding *Butorides*), *Bonasa*, *Passerculus* (subgenus), and perhaps *Tachycineta* and *Ampelis*, and in the East *Carpodacus* and *Merula;* although most of these are really Cold

[1] Cosmopolitan and other wide-ranging genera are not included in this connection, but only such as are distinctively American.

Temperate rather than Warm Temperate forms. The southern list would include a large number of tropical genera which range into the southern part of the Warm Temperate—such for instance as *Columbigallina*, *Scardafella*, *Urubitinga*, *Asturina*, *Geococcyx*, *Conurus*, *Nyctidromus*, various genera of Hummingbirds, *Milvulus*, *Pyrocephalus*, *Campylorhynchus*, *Catherpes*, etc.; but they are nearly all limited to the Arid division of the Warm Temperate, and thus serve to emphasize the differences distinguishing the Arid from the Humid Province quite as much as a northern and southern division along a transcontinental line. The differentiation of the Warm Temperate into two transcontinental belts—a northern and a southern—is thus due to the increase of tropical forms near the southern border, in accordance with the law of general increase of the forms of life from the north southward.

In respect to species and subspecies, the following lists may serve to indicate the forms characteristic respectively of the Humid and Arid Provinces. As the transition between the two is gradual, with many interdigitations, through the extension of forest or tree-loving species up the partially wooded river valleys into the Plains, the lists are to some degree open to the criticism that many species classified as eastern are not sharply limited at the eastern edge of the Plains, and that a few of the properly Arid Province species range eastward to the prairies east of the Mississippi River, yet in general terms the distribution is as implied in the headings of the two lists. The straggling of Catbirds and Kingbirds to the Great Basin, and even to the Pacific coast, does not weigh heavily against the propriety of including them in the Humid Province list since they are here such characteristic elements of the fauna. The lists, furthermore, are not intended to be complete; to save space subspecies are in many instances omitted; the more tropical forms, and generally also those of local distribution, as well as the maritime water birds, are excluded.

Humid Province.

Ajaja ajaja
Guara alba
Ardea rufescens
Ardea tricolor ruficollis
Ardea cœrulea
Nycticorax violaceus
Rallus elegans
Ionornis martinica

Philohela minor
Symphemia semipalmata
Ægialitis meloda
Colinus virginianus
Tympanuchus cupido
Tympanuchus americanus
Meleagris gallopavo
Columbigallina passerina
Catharista atrata
Elanoides forficatus
Ictinia mississippiensis
Buteo lineatus
Buteo latissimus
Syrnium nebulosum
Megascops asio
Conurus carolinensis
Dryobates villosus
Dryobates borealis
Melanerpes erythrocephalus
Melanerpes carolinus
Antrostomus vociferus
Antrostomus carolinensis
Tyrannus tyrannus
Tyrannus dominicensis
Myiarchus crinitus
Sayornis phœbe
Contopus virens
Empidonax acadicus
Otocoris alpestris praticola
Cyanocitta cristata
Aphelocoma floridana
Corvus ossifragus
Dolichonyx oryzivorus
Molothrus ater
Sturnella magna
Agelaius phœniceus
Quiscalus major
Quiscalus quiscula
Ammodramus savannarum
Chondestes grammacus
Spizella pusilla
Peucæa æstivalis
Melospiza fasciata
Pipilo erythrophthalmus
Cardinalis cardinalis
Guiraca cærulea
Habia ludoviciana
Passerina cyanea
Passerina ciris
Piranga rubra
Piranga erythromelas
Lanius ludovicianus
Vireo flavifrons
Vireo solitarius
Vireo noveboracensis
Protonotaria citrea
Helinaia swainsonii
Helmitherus vermivorus
Helminthophila pinus
Helminthophila chrysoptera
Helminthophila bachmani
Dendroica cærulea
Dendroica pensylvanica
Dendroica dominica
Dendroica kirtlandi
Dendroica vigorsii
Dendroica discolor
Seiurus aurocapillus
Seiurus motacilla
Geothlypis trichas
Geothlypis formosa
Icteria virens
Sylvania mitrata
Galeoscoptes carolinensis
Harporhynchus rufus
Thryothorus ludovicianus
Thryothorus bewickii
Troglodytes aëdon
Cistothorus palustris
Cistothorus stellaris
Sitta carolinensis
Sitta pusilla
Parus bicolor
Parus carolinensis
Polioptila cærulea
Turdus mustelinus
Turdus fuscescens
Sialia sialis

Arid Province.

Anas cyanoptera
Dendrocygna fulva
Dendrocygna autumnalis
Plegadis guarauna
Symphemia s. inornata
Ægialitis meloda circumcincta
Ægialitis nivosa
Ægialitis montana
Callipepla squamata
Callipepla gambeli
Callipepla californica
Oreortyx pictus
Colinus virginianus texensis
Pediocætes phas. columbianus
Pediocætes phas. campestris
Tympanuchus pallidicinctus
Meleagris gallopavo ellioti
Meleagris gallopavo mexicanus
Columba fasciata
Columbigallina p. pallescens
Pseudogryphus californianus
Buteo borealis calurus
Buteo borealis harlani
Buteo lineatus elegans
Archibuteo ferrugineus
Falco mexicanus
Syrnium occidentale
Megascops asio (subsps.)
Megascops flammeolus
Speotyto cunicularia hypogæa
Glaucidium gnoma
Micropallas whitneyi
Geococcyx californianus
Dryobates villosus (subsps.)
Dryobates nuttalli
Dryobates scalaris
Dryobates arizonæ
Xenopicus albolarvatus
Melanerpes torquatus
Melanerpes formicivorus bairdi
Melanerpes uropygialis
Colaptes cafer
Colaptes chrysoides
Phalænoptilus nuttalli
Antrostomus vociferus arizonæ
Chordeiles virginianus henryi
Chordeiles acutipennis texensis
Cypseloides niger
Chætura vauxi
Aëronautes melanoleucus
Trochilus alexandri
Selasphorus rufus
Selasphorus alleni
Selasphorus platycercus
Calypte anna
Tyrannus verticalis
Tyrannus vociferans
Myiarchus mexicanus
Myiarchus cinerascens
Sayornis saya
Sayornis nigricans
Contopus richardsonii
Contopus pertinax
Empidonax wrightii
Empidonax fulvifrons
Otocoris alpestris (subsps.)
Aphelocoma woodhousei
Aphelocoma californica
Aphelocoma sieberi arizonæ
Corvus caurinus
Molothrus ater obscurus
Xanthocephalus xanthocephalus
Sturnella magna neglecta
Agelaius phœniceus sonoriensis
Agelaius gubernator
Agelaius tricolor
Icterus bullocki
Icterus cucullatus (subsp.)
Icterus parisorum
Scolecophagus cyanocephalus
Carpodacus mexicanus frontalis
Spinus psaltria
Calcarius ornatus
Rhyncophanes mccownii
Poocætes gramineus confinis
Ammodramus s. perpallidus
Ammodramus h. occidentalis
Chondestes grammacus strigatus
Spizella pusilla arenacea
Spizella pallida

Spizella breweri
Spizella atrigularis
Amphispiza bilineata
Amphispiza belli
Peucæa cassini
Peucæa carpalis
Peucæa ruficeps
Melospiza fasciata (subsps.)
Pipilo maculatus (subsps.)
Pipilo fuscus
Pipilo aberti
Pipilo chlorurus
Cardinalis cardinalis (subsps.)
Pyrrhuloxia sinuata
Guiraca cærulea eurhyncha
Habia melanocephala
Passerina amœna
Passerina versicolor
Calamospiza melanocorys
Piranga rubra cooperi
Piranga ludoviciana
Phainopepla nitens
Lanius ludovicianus (subsps.)
Vireo solitarius (subsps.)
Vireo atricapillus
Vireo belli
Vireo huttoni
Vireo vicinior
Helminthophila luciæ
Helminthophila virginiæ
Dendroica graciæ
Dendroica chrysoparia
Dendroica nigrescens
Dendroica occidentalis
Geothlypis trichas occidentalis
Icteria virens longicauda
Oroscoptes montanus
Harporhynchus longirostris
Harporhynchus curvirostris
Harporhynchus bendirei
Harporhynchus lecontei
Harporhynchus crissalis
Salpinctes obsoletus
Catherpes mexicanus
Campylorhynchus brunneicapillus
Thryothorus ludovicianus (subsp.)
Thryothorus bewickii (subsps.)
Troglodytes aëdon (subsps.)
Cistothorus palustris paludicola
Sitta carolinensis aculeata
Sitta pygmæa
Parus bicolor texensis
Parus inornatus
Parus atricristatus
Parus wollweberi
Parus carolinensis agilis
Chamæa fasciata
Psaltriparus lloydi
Psaltriparus minimus
Psaltriparus plumbeus
Psaltriparus santaritæ
Auriparus flaviceps
Polioptila cærulea obscura
Polioptila plumbea
Polioptila californica
Myiadestes townsendi
Turdus fuscescens (subsp.)
Sialia mexicana

A careful tabulation of the thousand or more species and subspecies of North American birds shows that about 400 occur in the Warm Temperate Subregion that do not extend much to the northward of its northern border, and which may be hence termed distinctively Warm Temperate species. This excludes about 75 essentially tropical species and subspecies which range into the southern border of the United States, and also about 86 wide-ranging species whose habitats either broadly overlap both subregions or have even a much greater distribution and are thus not properly distinctive of either the Cold Temperate or the Warm Temperate.

Of the 400 essentially Warm Temperate forms only about 6 per cent. have a transcontinental range; about 25 per cent. may be considered as having a more or less general dispersion over the Humid Province, to which they are restricted, and about 44 per cent. range more or less at large over the Arid Province, to which they are practically confined. The remaining 25 per cent. are of more or less local distribution, part being water birds confined to certain portions of the Atlantic, Pacific or Gulf coasts; part are more or less maritime land birds, while a few are limited to particular areas of comparatively small extent in the interior, or to a narrow belt along the Pacific coast. As will be noted later, some 28 species and 24 subspecies are confined to Lower California and its outlying islands.

The northern and southern differentiation of the Warm Temperate noted above serves, however, as a basis for primary subdivisions (secondary divisions of the Warm Temperate) of both the Humid and Arid provinces, each being separable into two *Subprovinces* (see Pl. IV). Thus the Humid Province is divisible into (1) an *Appalachian Subprovince*, consisting of the long recognized Alleghanian and Carolinian Faunæ, and (2) an *Austroriparian Subprovince*, consisting of the Louisianian Fauna, as commonly recognized. The Austroriparian Subprovince is Dr. Merriam's 'Louisianian or Austroriparian subregion' of his 'Sonoran' region; the Appalachian Subprovince is the northeastern part of his 'Sonoran,' left by him as an unnamed remainder after setting off and naming as 'subregions' all the other parts of his 'Sonoran Region.' The line separating these two subprovinces marks the southern limit of several northern genera and many northern species, and the northern limit of a still greater number of southern genera and species.

The Humid Province comprises three *faunæ*, as follows: (1) *Alleghanian*, (2) *Carolinian*, (3) *Louisianian*. These faunæ have been so long recognized, and have recently been so well mapped by Dr. Merriam,[1] that a detailed account of them may be here omitted.

The Arid Province not only extends, as already stated, from the eastern edge of the Great Plains to the Pacific coast, but also northward over the Saskatchewan Plains, the Plains of the Colum-

[1] N. Am. Fauna, No. 3, map 5.

bia, and thence northward into southern British Columbia. It thus includes the so-called 'Central' and 'Pacific' Provinces of Baird and most subsequent writers, excepting of course the more elevated parts of the Rocky Mountain plateau. It is thus coëxtensive with Dr. Merriam's 'Arid Sonoran.' While it is true that a narrow belt along the Pacific coast, from southern California northward to the Alaskan Peninsula, possesses a few peculiar types, and lacks a few of those occurring in the region immediately to the eastward, the differentiation is on the whole too slight to give the Pacific coast district the rank of a region coördinate in grade with the formerly so-called Middle and Eastern Provinces. These differences serve at best merely to mark off from the interior region at large a tier of narrow coast faunæ of the same grade as those bordering the Atlantic coast, although the latter, owing to the widely different physiography of the eastern and western borders of the continent, have a much greater east and west extent.

The Arid Province, like the Humid, is divisible into two *subprovinces*, namely, (1) a northern or *Campestrian Subprovince*, and (2) a southern or *Sonoran Subprovince* (see Pl. IV). These two regions correspond respectively with Dr. Merriam's 'Arid Upper Sonoran' and his 'Arid Lower Sonoran'; just as the two subprovinces of the Eastern Province correspond with his 'Humid Upper Sonoran' and 'Humid Lower Sonoran,' as laid down on his 'Second Provisional Bio-geographic Map of North America,' except that the 'humid' and 'arid' portions of his 'Transition Zone'[1] are also included respectively in the Alleghanian and Campestrian Subprovinces. The Sonoran Subprovince is equal to Dr. Merriam's restricted 'Arid or Sonoran subregion proper' plus his 'Lower Californian subregion,' while the Campestrian Sub-province includes his 'Great Basin subregion' and his 'Great Plains subregion.'[2] The name 'Campestrian' has reference to the fact that this subprovince is largely made up of plains, including as it does the greater part of the Great Plains, the Plains of the Saskatchewan, and the Plains of the Columbia and Snake Rivers.

Many species are limited in their southward distribution by the southern border of the Campestrian Subprovince, but few genera

[1] *Cf.* Proc. Biol. Soc. Washington, VII, 1892, pp. 26-33, and accompanying map.

[2] *Cf.* N. Am. Fauna, No. 3, 1890, p. 25.

appear to be thus restricted. This boundary also forms about the northern limit of many species and genera of the Sonoran Subprovince. These two subprovinces are hence characterized mainly by the presence of a large number of forms found in the Sonoran which are absent from the Campestrian, and are thus distinguished, like many northern divisions when compared with adjoining southern ones of coördinate rank, from the Arctic southward, by what they lack rather than by the posession of any peculiar types.

The Campestrian Subprovince itself may be divided into three areas which may be termed *districts*, namely (1) the *Great Plains District*, (2) the *Great Basin District*, and (3) the *Pacific Coast District*. (See Pl. IV.) The first two are respectively the 'Great Plains subregion' and the 'Great Basin subregion' of Dr. Merriam's first 'Biological Map of North America.'[1] Although these two districts are separated by the main chain of the Rocky Mountains, they are faunally but slightly differentiated. But few genera occur in the one that do not occur in the other, a few species, and a larger number of subspecies are restricted to one or the other during the breeding season, but they often become more or less mixed during winter, when they meet on common ground in the Sonoran Subprovince. Thus the Rocky Mountains, while forming so imposing a feature in the configuration of the continent, fail to be by any means an impassable barrier to the dispersal of species, owing to their numerous comparatively low depressions, and to their meridional trend.

The Pacific Coast District consists of a narrow belt situated mainly west of the Sierra Nevada and Cascade Ranges, and is characterized by the presence of a few species and a considerable number of subspecies mainly restricted to it.

The Sonoran Subprovince consists of Dr. Merriam's restricted 'Sonoran subregion,'[1] with the addition of Lower California, which Dr. Merriam gave the rank of a 'subregion.' A careful synopsis of the bird life of Lower California and its outlying islands shows that it has 28 species and 24 subspecies which are not found in the United States. Of the 29 species 10 are confined exclusively to the outlying islands, and 9 others are either strictly

[1] N. Am. Fauna, No. 3, 1890, p. 25, and map 5.

maritime or pelagic. Of the 10 peculiar land birds, all but one are congeneric with and for the most part closely allied to North American species. Of the 24 subspecies, 23 are merely local races of North American species. A number of these peculiar species and subspecies are confined to the subtropical portions of the region below La Paz, which belongs to the American Tropical Realm rather than to the Temperate Realm, thus leaving very little of distinctive importance for the non-tropical portion of the Peninsula.

It is at present impracticable to attempt to define in detail the numerous faunæ of the Arid Province. While certain portions might be thus treated, our knowledge of the region as a whole is too defective to warrant even a provisional subdivision into faunæ. The same is true also of the southern prolongations of the Cold Temperate along the mountain ranges of the western half of the continent. It is evident, without going into a detailed analysis, that many of the wide-ranging species that prevail over and give character to the Arctic, Hudsonian, and Canadian Faunæ, respectively, are absent from the Rocky, the Cascade, and the Sierra Nevada ranges and their outlying spurs; also that many eastern forms are here replaced by more or less closely related western forms, and that quite a number of peculiar mountain types are superimposed upon this new combination. The case is obviously quite different from the conditions met with in the Alleghanies, where merely a small percentage of Canadian forms occur far to the southward of where they are found in the contiguous lowlands. To give due expression to the faunal conditions met with, for example, in the Rocky Mountains south of the Canadian Pacific Railroad, it evidently will be necessary to recognize a succession of mountain faunæ as we go southward, as well as in descending from timber-line at any given point to the foot-hills. Dr. Merriam, in defining the life zones at high elevations in Arizona, Idaho, and California, has made an excellent beginning in this comparatively new line of work, and we may confidently look forward to still more important results from the great mass of unpublished data he has so industriously brought together. The proper collation of our mountain areas, from the British boundary across the United States into Mexico, will be a work of exceptional interest and importance, and will require much additional field research.

As already noted, the extreme southern portions of North America belong to the American Tropical Realm, which consists of tropical America at large. It thus includes not only a large part of South America, Central America and the West Indies, but the lowlands of Mexico, including the low eastern coast region to some distance north of the mouth of the Rio Grande, and the low western coast to some distance north of Mazatlan. To the Tropical Realm belong also the extreme southern portion of the peninsula of Lower California, and the extreme southern portion of the peninsula of Florida. There are thus three small portions of 'North America,' as defined in the A. O. U. Check-List, which belong with the Tropical rather than to the North Temperate Realm.[1]

The fauna of neither of the tropical areas within the United States is typically tropical, but the infusion of tropical elements is so great as to render them tropical rather than temperate. They have also little in common with each other, as would be naturally anticipated from their wide geographical separation through the interposition of the Gulf of Mexico, thus preventing a tropical land connection. Consequently the Floridian area, or the *Floridian Fauna*, as it has long been technically known,[2] belongs to the Antillean Region of the American Tropical, while the Texan area is an outlying arm of the Central American Region of the American Tropical. The tropical portion of Lower California also belongs to the Central American Region.

The Floridian Fauna has recently been treated in much detail by Dr. Merriam,[3] and hence need not be considered at length here. The following birds, however, may be mentioned as among those distinctively characteristic of this limited area, though having generally a very extended range into tropical America.

---

[1] See maps, plates III and IV, where the uncolored portions to the south of the colored areas belong to the Tropical Realm. The uncolored portion at the top of plate III may be taken as representing that portion of the continent belonging to the Arctic.

[2] *Cf.* Bull. Mus. Comp. Zoöl., II, 1871, p. 391.—The general provisional northern limit here given—"near the latitude of Lake George"—proves to have been carried a little too far north, its limits as now recognized being Cape Malabar on the east coast and Tampa Bay on the west coast. (*Cf. Merriam*, Proc. Biol. Soc. Washington, VII, 1892, p. 33.)

[3] *Cf.* Merriam, Proc. Biol. Soc, Washington, VII, 1892, pp. 52-54.

Sula sula
Fregata aquila
Phœnicopterus ruber
Ardea occidentalis
Aramus giganteus
Columba leucocephalus
Zenaida zenaida
Rostrhamus sociabilis
Buteo brachyurus
Polyborus cheriway
Speotyto cunicularia floridana
Crotophaga ani
Coccyzus minor maynardi
Euetheia bicolor
Ammodramus nigrescens
Vireo altiloquus barbatulus
Cœreba bahamensis

The semi-tropical area occupying the extreme lower portion of the Rio Grande Valley and extending as a narrow belt thence northward for a short distance along the Texas coast, and also southward along the Mexican coast to the neighborhood of Tampico, I have recently designated[1] as the *Tamaulipan Fauna.* Among the birds that extend northward from the tropics to find here about their northern limit are the following:—

Podiceps dominicus
Ortalis vetula maccalli
Engyptila albifrons
Buteo albicaudatus
Crotophaga sulcirostris
Trogon ambiguus
Nyctidromus albicollis merrilli
Amazilia fuscicaudata
Amazilia cerviniventris
Pitangus derbianus
Myiozetetes texensis
Xanthoura luxuosa
Embernagra rufivirgata
Sporophila morelleti sharpei
Euphonia elegantissima
Vireo flavoviridis
Compsothlypis nigrilora
Geothlypis poliocephala palpebralis
Basileuterus culicivorus
Harporhynchus longir. sennetti

The Tamaulipan Fauna has fewer distinctively tropical types than would be expected from its low altitude and geographical position. This is doubtless due to the extreme aridity of the country, since in the forest regions further inland under the same parallels Trogons, Motmots and Parrots occur to a much greater extent than in the arid, nearly treeless coast region.

The extreme southern portion of Lower California, from about La Paz southward, may be considered as fairly subtropical, but contains, on the whole, comparatively few distinctively tropical types of birds. Among such may perhaps be placed the following:—

[1] Bull. Am. Mus. Nat. Hist. IV, p. 241, Dec. 1892.

Glaucidium hoskinsi
Crotophaga sulcirostris
Dryobates scalaris lucasanus
Basilinna xantusi
Empidonax cineritius
Empidonax griseus
Dendroica bryanti
Geothlypis beldingi
Harporhynchus cinereus
Campylorhynchus affinis
Merula confinis

A few additional subspecies of northern forms also characterize the subtropical portion of the peninsula, which may perhaps well bear the name *Saint Lucas Fauna*.

Independently of the subdivisions already enumerated, and in addition to them, North America may be divided into a number of transcontinental belts or *Zones* in accordance with the climatic conditions prevailing over the different parts of the continent. Several of these zones have been repeatedly recognized by various writers on the distribution of animals and plants. A division of extra-tropical North America into about seven zones will greatly facilitate the correlation of the faunæ of different regions. These are, (1) the Arctic or Hyperborean Zone, equivalent to the American portion of the Arctic Realm; (2) the Subarctic or Hudsonian Zone; (3) the Cold Temperate or Canadian Zone; (4) the North Warm Temperate or Alleghanian Zone; (5) the Middle Warm Temperate or Carolinian Zone; (6) the South Warm Temperate or Louisianian Zone; (7) the Subtropical or Floridian Zone.[1] Dr. Merriam has already in his recent papers on the distribution of North American mammals distinguished these zones more or less definitely in treating of the life zones of the San Francisco Mountain region in Arizona and in Idaho. He has done much also toward correlating the life zones of mountain faunas with the long recognized faunæ of the Atlantic Coast. Much further information is required before these zones can be geographically defined over the western half of the continent.

In selecting names for their designation several alternatives present themselves, as for example, names derived from the climatic zones, or from some leading characteristic, as 'Spruce

---

[1] This is a modification of my recently published classification (Bull. Am. Mus. Nat. Hist., IV, p. 240), rendered necessary from the evident desirability of reserving the term 'Subtropical Zone' for the designation of the most northern belt of the Tropical Realm, instead of applying it to the lowest or most southern belt of the Temperate Realm, as is done when it is used for the Louisianian Zone.

Zone' for the Hudsonian, 'Arctic-alpine' for treeless mountain summits equivalent in character to the Arctic, etc., or from those of the Atlantic coast faunæ. This latter method has the merit of at once suggesting a well-known standard of comparison when applied to belts in the interior or on the Pacific coast, representative of the commonly recognized faunæ of the Atlantic coast.

Dr. Merriam has already recognized the equivalents of the Arctic, Hudsonian, Canadian, Alleghanian, etc., in portions of the mountainous districts of the West, and has presented in substance the following correlations.[1]

Zone.	*Fauna.*
Alpine	} Arctic.
Subalpine or Timber-line . . .	} Arctic.
Hudsonian or Spruce	Hudsonian.
Canadian or Fir	Canadian.
Neutral or Pine	Alleghanian.
Piñon or Cedar	[Carolinian].
Desert	[Louisianian].

2. *Classification and Nomenclature.*

A few words in regard to the names chosen for the several major divisions of the North American Region, and the choice of names in general in bio-geography. It is natural that the influences controlling the geographical distribution of life, namely, climate, and hence the principal climatic zones, should suggest the names of many of the larger ontological regions; and we find that to a large extent such names have been chosen, as by Dana, in 1852, in discussing the distribution of marine life, and by botanists generally, and notably by German writers. For the lesser regions geographical names, as Hudsonian, etc., are admirably appropriate when suggestive of some characteristic portion of the region in question. Whenever feasible, names first given should be retained in preference to later names.

In concluding this paper a few words of explanation are necessary in relation to various points of nomenclature and classification. In comparing the present scheme of faunal areas of North

---

[1] N. Am. Fauna, No. 3, 1890, pp. 7-34, and maps 1-4; *ibid*, No. 5, 1891, pp. 9-12, 21-25.

America with those employed by Dr. Merriam in his recent well-known admirable papers on the geographical distribution of North American mammals, so frequently cited in the preceding pages, it will be noticed that there is a striking agreement in their number and boundaries, although a few new minor divisions have here been introduced; yet the terms employed for their designation are to a great extent different. As already intimated, the present system of classification and nomenclature is a further development of that first instituted by me in 1871, and used later in 1878, and now carried out in greater detail and extended to the whole North American Continent. The present revision of the subject is therefore not to be looked upon as unfriendly criticism of Dr. Merriam's classification and nomenclature, which he evidently adopted provisionally,[1] selecting such terms as would suffice to clearly indicate the areas under discussion; his attention was given mainly and most successfully to an elucidation of the facts of distribution; a detailed consideration of the nomenclature of the subject was outside of his special field.

In attempting to establish a consistent scheme of classification and terminology, the aim is to assign definite terms for areas of similar taxonomic grade. Many of the terms in more or less current use have been employed so loosely, and used in so many different senses by different writers, that, as already said, a strict 'rule of priority' cannot be enforced, at least without leading in many instances to very unsatisfactory and inharmonious results. As already explained, the system here adopted is analogous to the schemes followed in systematic biology and stratigraphic geology. In the selection of names for the higher divisions, reference has been had to the influences controling the geographic distribution of life, namely, *climate*, and the climatic zones have been allowed to suggest the names of many of the major ontological areas. Indeed, such names have been employed before in nearly the same sense, not only by physical geographers, but by many botanists and some zoölogists. Thus 'Humid' and 'Arid' become appropriate and suggestive designations for the eastern and western subdivisions of the North American Warm Temperate Subregion. For the lesser regions geographical

---

[1] This is not only apparent from his papers, but I am informed by him that this was intentionally the case.

names, as 'Hudsonian,' etc., are admirably appropriate when suggestive of some characteristic portion of the region in question. Terms designating grade should, of course, be used with the same strictness as the corresponding terms,—order, family, genus, etc., expressive of rank,—in biology.

Realm is employed as a designation for primary regions, taking the world at large, and Fauna for the ultimate subdivisions. Region, the term selected for divisions of the second rank, has been used by different writers for areas of various grades, but it is proposed to limit its application in a technical sense to the primary divisions of Realms. Below this, in successively descending order, we have Subregions, Provinces, Subprovinces, Districts and Faunæ; faunæ being subdivisions of districts (in cases when it seems desirable to recognize districts), while districts are the primary subdivisions of the subprovinces. In the Arctic Realm the only subdivisions it seems necessary to recognize are faunæ; in the Cold Temperate, possibly both districts and faunæ; in the Warm Temperate, at least so far as North America is concerned, it seems desirable to recognize (1) provinces, (2) subprovinces, (3) districts, and (4) faunæ.

As early as 1878[1] I separated the 'North American Region' into two *Subregions*, namely, a *Cold Temperate Subregion* and a *Warm Temperate Subregion*, as is done in the present paper, using these terms as headings in tables giving the distribution of the genera of North American mammals. Baird's 'Eastern,' 'Middle,' and 'Western' *Provinces* were recognized as "natural regions," with the designation of 'Provinces,' but with the Eastern Province modified so as to restrict it to the Warm Temperate Subregion, and all three reduced in grade to regions of the third rank[2] instead of the second rank as regarded by Professor Baird.

In 1883 Dr. Packard[3] substantially adopted this classification in treating of the faunal regions of North America, with, however, a change of name for the 'Cold Temperate Subregion,' he

---

[1] Bull. U. S. Geol. and Geogr. Survey (Hayden), IV, 1878, pp. 338–344.

[2] That is, of the North American Region; really of fourth rank, considered from the basis of the world as a whole.

[3] Twelth Ann. Rep. U. S. Geol. and Geogr. Survey (Hayden), pt. I, 1883, pp. 362–370, and map; the latter republished in the Third Rep. U. S. Entomol. Comm., 1883, map iv.

adopting for it that of 'Boreal Province' — an unfortunate suggestion of my own made later in the paper above cited (l. c., p. 376, where, in some unaccountable way my former division of the 'North Temperate Realm' into 'Subregions' was wholly overlooked!). Dr. Packard, in his otherwise excellent 'Zoögeographical Map of North America,' failed, however, to recognize the southward extension of the Cold Temperate Subregion along the principal mountain systems of the continent.

Dr. Merriam in 1890[1] again set off the Cold Temperate Subregion, under the name 'Boreal Province,' and mapped in detail its southern prolongations into the mountainous parts of the Warm Temperate. The Warm Temperate Subregion was also recognized as a contrasting region of coördinate rank, under the designation 'Sonoran Province,' while the old 'Eastern,' 'Middle,' and 'Western' Provinces were properly repudiated as having no basis in nature. Particularly is this the case in respect to the Central Province, of which Dr. Merriam observes: "The region almost universally recognized by recent writers as the 'Central Province' is made up of the Great Plains, the Rocky Mountains and the Great Basin. A critical study of the life of the Rocky Mountains has shown it to consist of a southward extension of the Boreal Province, with an admixture of southern forms resulting from an intrusion or overlapping of representatives of the Sonoran Province, some of which, from long residence in the region, have undergone enough modification to be recognized as distinct subspecies or even species. A similar analysis of the Great Plains and Great Basin has shown them to consist of northward extensions of the Sonoran Province, somewhat mixed with the southernmost fauna and flora of the Boreal Province. Thus the whole of the so-called 'Great Central Province' disappears.

"This explains a multitude of facts that are utterly incomprehensible under the commonly-accepted zoölogical divisions of the country. These facts relate particularly to the distribution of species about the northern boundaries of the supposed Central and Pacific Provinces, and to the dilemma we find ourselves in when attempting to account for the origin of so many primary

---

[1] N. Am. Fauna, No. 3, Sept. 1890, pp. 24–26, and map 5; see also Proc. Biol. Soc. Washington, VII, 1892, pp. 21–40, and accompanying map.

life areas in a country where there are no impassable physical barriers to prevent the diffusion of animals and plants." [1]

Dr. Merriam's generalizations respecting the Central Province of authors mark an important advance in the study of North American bio-geography. Taking this region with its original boundaries and significance it is a highly artificial division, embracing within its area very unlike faunal elements. Eliminating from it, however, the broad central arm of the 'Boreal' or Cold Temperate Subregion, which occupies so much of the great central plateau, relieves it of an extraneous element, and reduces it to a more natural and geographically quite different region.

The first discrepancy between Dr. Merriam's classification and my own that requires notice is in respect to the primary divisions of the North American Region, which he first termed 'Provinces'[2] and later 'Regions,'[3] with the prefixes 'Boreal' and 'Sonoran' respectively for the 'Cold Temperate' and 'Warm Temperate' *Subregions* of the present writer. The use of Boreal, however, as shown above, was not an innovation; but the term 'Sonoran' was used in a new and greatly extended sense, the term Sonoran being applied to a region identical in geographical extent with the Warm Temperate,—a designation previously used for the same area,—and hence including the region east of the Mississippi (as well as that west of it), from the Great Lakes and southern New England south to Florida and the Gulf Coast. The terms 'Sonoran' and 'Sonoran Province' were used as early as 1866 by Prof. Cope,[4] and also later by Cope, Heilprin, and others, for a region of comparatively small extent, consisting of Sonora and adjoining portions of Arizona and New Mexico. In 1887 Heilprin[5] extended the region to include "the peninsula of Lower California, the State of Sonora in Mexico, New Mexico, Arizona, and parts, not yet absolutely defined, of Nevada, California, Texas, and Florida," and modified its title by calling it the 'Sonoran Transition Region.' The Sonoran Province or Region of these authors is thus not at all the 'Sonoran Region' of Merriam, which is an area of much greater extent and of higher rank.

---

[1] N. Am. Fauna, No. 3, pp. 22, 23.
[2] N. Am. Fauna, No. 3, p. 19, 20.
[3] Proc. Biol. Soc. Washington, VII, 1892, pp. 22, 26.
[4] Proc. Acad. Nat. Sci. Phila., 1866, p. 300.
[5] The Geogr. and Geol. Distrib. of Anim., p. 106.

The term Sonoran, used in this extended sense, seems at least inappropriate if not misleading, as there are few if any strictly 'Sonoran' types represented in that portion of the United States situated to the eastward of the Mississippi River. The more descriptive and appropriate designation of 'Warm Temperate' is therefore preferred for the region in question, since it not only has priority, but is in harmony with the terms Arctic, Cold Temperate, and Tropical, used currently for other coödinate areas of the continent.

Another, and perhaps the only other, important discrepancy between Dr. Merriam and myself is in respect to the primary subdivisions of the Warm Temperate or 'Sonoran' Subregion. Here the difference is in respect to classification, Dr. Merriam dividing the Warm Temperate into two transcontinental divisions which he terms respectively 'Upper' and 'Lower Sonoran'; while according to my best judgment the primary division is in a meridional line into an eastern and a western division, which I have termed respectively Humid and Arid Provinces, borrowing the terms from Dr. Merriam, who has used them in the same geographical sense but not in the same nomenclatural relation, as already shown in preceding pages. As the evidence, pro and con, has already been submitted (see *antea*, pp. 128–131, and Bull. Am. Mus. Nat. Hist., IV, pp. 230–232), it is unnecessary to rediscuss the matter here.

In respect to the subdivisions respectively of the Humid and Arid Provinces, my subprovinces correspond to his secondary divisions of the 'Sonoran' (with slight modifications, as already explained), my Appalachian Subprovince being essentially his 'Humid Upper Sonoran,' my Austroriparian Subprovince his 'Humid Lower Sonoran,' my Campestrian Subprovince his 'Arid Upper Sonoran,' and my Sonoran Subprovince his 'Arid Lower Sonoran.' My division of the Campestrian Subprovince into Districts conforms nearly with Dr Merriam's division of the same geographical area,[1] and I have adopted for these areas the same descriptive appellations.

Taking Dr. Merriam's latest paper on the faunal areas of North America and accompanying map,[2] one other point of difference calls for notice, namely, his recognition of a 'Transition

[1] N. Am. Fauna, No. 3, p. 25.
[2] Proc. Biol. Soc. Wash., VII, pp. 1–64, April. 1892.

Zone,' interposed between and separating the Cold Temperate, or 'Boreal,' from the Warm Temperate, or 'Sonoran' (l. c. pp. 30–33). This is equivalent to what is termed in this paper the Alleghanian Zone, and forms the northern transcontinental belt of the Warm Temperate. This zone was first recognized by Dr. Merriam in the West as the 'Neutral or Pine Zone,'[1] and correlated later with the Alleghanian Fauna of the East, as the 'Neutral or Transition Zone.'

The Alleghanian Zone is beyond question a transition belt, being necessarily so from its geographical position; its affinities, however, are decidedly with the Warm Temperate division of the continent rather than with the Cold Temperate, as the case was first interpreted by Dr. Merriam,[2] since its northern boundary coincides closely with the northern limit of distribution of a large number of southern genera of both plants and animals, including most of the staple grains and fruits of the Warm Temperate Zone.

As is well known, there is always a belt of neutral territory along the common boundary line of two adjoining areas, varying in breadth with the rank of the two areas; and the present case of the Alleghanian Zone is thus not exceptional. All things considered it therefore seems best to regard it as the northern transcontinental belt of the Warm Temperate, rather than to give it the anomalous position of a minor faunal area interposed between and completely separating two areas of a higher grade.[3] Besides, the term 'transition,' for reasons already given, is not a distinctive designation for a faunal area of any grade, although it has been repeatedly used in this way by different writers; its proper function is that of a descriptive term—not a designation to be used in a taxonomic sense.

The various faunal areas recognized in the present paper may be tabulated as follows:—

---

[1] N. Am. Fauna, No. 3, 1890, p. 11, map 5.

[2] N. Am. Fauna, No. 3, p. 20, and *ibid.*, No. 5, p. 21.

[3] In biology 'aberrant,' 'transition or intermediate genera are frequently met with, and in some cases it is difficult to refer them to one of the two subfamilies to which they are allied rather than to the other. Yet we feel compelled to refer them to one or the other, or else to make a new subfamily for the aberrant genus, in case it shows sufficient differentiation, rather than to leave it as an isolated genus, with the rank of a genus, to be interposed between two subfamilies, or families, as the case may be.

Tabular Synopsis of the Faunal Areas of North America.

REALMS	Arctic.	
	North Temperate.	
	American Tropical.	
REGIONS	North American	< *North Temperate Realm.*
	Central American	< *American Tropical Realm.*
	Antillean	
SUBREGIONS	Cold Temperate	= *North American Region.*
	Warm Temperate	
PROVINCES	Humid	= *Warm Temperate Subregion.*
	Arid	
SUBPROVINCES	Appalachian	= *Humid Province.*
	Austroriparian	
	Campestrian	= *Arid Province.*
	Sonoran	
DISTRICTS	Great Plains	= *Campestrian Subprovince.*
	Great Basin	
	Pacific Coast	
FAUNÆ[1]	Barren Ground	*Arctic.*
	Alaskan-Arctic	
	Aleutian	*Cold Temperate.*
	Hudsonian	
	Canadian	
	Sitkan	
	Alleghanian	*Humid Warm Temperate.*
	Carolinian	
	Louisianian	
	Floridian	*Tropical.*
	Tamaulipan	
	Saint Lucas	
ZONES	Arctic or Hyperborean.	
	Subarctic or Hudsonian.	
	Cold Temperate or Canadian.	
	North Warm Temperate or Alleghanian.	
	Middle Warm Temperate or Carolinian.	
	South Warm Temperate or Louisianian.	
	Subtropical or Floridian.	

[1] Mainly those of Eastern North America only.

GEOGRAPHICAL VARIATION AMONG NORTH AMERICAN MAMMALS, ESPECIALLY IN RESPECT TO SIZE.

GEOGRAPHICAL VARIATION AMONG NORTH AMERICAN MAMMALS, ESPECIALLY IN RESPECT TO SIZE.

By J. A. Allen.

FERÆ (Suborder FISSIPEDIA).

Having recently had an opportunity (through the kindness of Professor Baird) of studying with some care the magnificent series of skulls of the North American *Mammalia* belonging to the National Museum (amounting often to eighty or a hundred specimens of a single species), I have been strongly impressed with the different degrees of variability exhibited by the representatives of the species and genera of even the same family. The variation in size, for instance, with latitude, in the Wolves and Foxes is surprisingly great, amounting in some species (as will be shown later) to 25 per cent. of the average size of the species, while in other species of the *Feræ* it is almost *nil*. Contrary to the general supposition, the variation in size among representatives of the same species is not always a decrease with the decrease of the latitude of the locality, but is in some cases exactly the reverse, in some species there being a very considerable and indisputable *increase southward*. This, for instance, is very markedly true of some species of *Felis* and in *Procyon lotor*. Consequently, the very generally-received impression that in North America the species of *Mammalia* diminish in size southward, or with the decrease in the latitude (and altitude) of the locality, requires modification. While such is generally the case, the reverse of this too often occurs, with occasional instances also of a total absence of variation in size with locality, to be considered as forming "the exceptions" necessary to "prove the rule".

That there are such exceptions, both among Birds and Mammals, I have been long aware, and long since noticed that where there is an actual increase in size to the southward it occurs in species that belong to families or genera that are mainly developed within the tropics, there reaching their maximum development, both in respect to the number of their specific representatives, and in respect to the size to which some of the species attain. This fact seems also to have been observed by others.*

Most of the Mammals of North America belong to families, subfamilies, or genera which have their greatest development in the temperate or colder portions of the northern hemisphere, as the *Cervidæ*, the *Canidæ*, the *Mustelidæ*, the *Sciuridæ* (especially the subfamily *Arctomy-*

*I find that Mr. Robert Ridgway, some two years since, thus referred to this point. In alluding to the smaller size of Mexican specimens of *Catharpes mexicanus* as compared with specimens from Colorado, (*C. mexicanus* var. *conspusus*) he says: "As we find this peculiarity exactly paralleled in the *Thryotharus ludovicianus* of the Atlantic States, may not these facts point out a law to the effect that in genera and species in the temperate zone the *increase in size with latitude is toward the region of the highest development of the group?*"—Baird, Brewer, and Ridgway's Birds of North America, Vol. III, App., p. 503, 1874.

inæ), the *Leporidæ*, the *Castoridæ*, the *Arvicolinæ* among the *Muridæ*, the *Saccomyidæ*, *Geomyidæ*, etc. These rarely present an exception to the general law of decrease in size southward, though the variation is less (in fact, occasionally almost *nil*) in some species than in the others. The more marked exceptions, or those in which there is an actual increase in size southward, occur in those families hat reach their highest development with the tropics, as the *Felidæ* and *Procyonidæ*.

In some species (as I have elsewhere noticed), there probably exists a double decadence in size, the individual reaching its maximum dimensions where the conditions of environment are most favorable for the existence of the species, diminishing in size toward the northern (through scarcity of food and severity of climate) as well as toward the southern (in consequence of the enervating influence of tropical or semi-tropical conditions) limit of its distribution.

In a general way, the correlation of size with geographical distribution may be formulated in the following propositions:

(1) *The maximum physical development of the individual is attained where the conditions of environment are most favorable to the life of the species.* Species being primarily limited in their distribution by climatic conditions, their representatives living at or near either of ther respective latitudinal boundaries are more or less unfavorably affected by the influences that finally limit the range of the species. These influences may be the direct effects of too high or too low a temperature, too little or too much humidity, or their indirect effects acting upon the plants or other sources of food. Hence the size of the individual generally correlates with the abundance or scarcity of food. Different species being constitutionally fitted for different climatic conditions, surroundings favorable to one may be very unfavorable to others, even of the same family or genus. Hence

(2) *The largest species of a group* (genus, subfamily, or family, as the case may be) *are found where the group to which they severally belong reaches its highest development, or where it has what may be termed its center of distribution.* In other words, species of a given group attain their maximum size where the conditions of existence for the group in question are the most favorable, just as the largest representatives of a species are found where the conditions are most favorable for the existence of the species.

(3) *The most "typical" or most generalized representatives of a group are found also near its center of distribution, outlying forms being generally more or less "aberrant" or specialized.* Thus the *Cervidæ*, though nearly cosmopolitan in their distribution, attain their greatest development, both as respects the *size* and the *number* of the species, in the temperate portions of the northern hemisphere. The tropical species of this group are the smallest of its representatives. Those of the temperate and cold-temperate regions are the largest, where, too, the species are the most numerous. Most of the species of this family also have a wide geographical range, and their representatives respectively present great differences in size with locality; namely, a very marked decrease in size to the southward. The possession of large, branching, deciduous antlers forms one of the marked features of the family. These appendages attain their greatest development in the northern species, the tropical forms having them reduced almost to mere spikes, which in some species never pass beyond a rudimentary state. Beginning at the northward, we have first, in the subarctic and cold-temperate regions, the Alcine and Saugerine forms, species of the largest size, with heavy, large antlers. Next, in the colder-temperate regions, come the Elaphine species, also of very large size, with nearly the largest antlers of any of the *Cervidæ*. We

next meet, in the temperate and warmer regions generally, the smaller Capreoline and Rusine forms, decreasing in size southward, with a rapid reduction also in the size of the antlers. Finally, in the subtropical and tropical portions of the Old World, we meet with antlerless forms, that constitute the smallest species known among the *Cervidæ* and their allies.

The decrease in the size of the antlers southward among the different genera and species is also well marked among individuals of the same species, especially among the Cariacine deer of North America.

The *Canidæ* form another family, which, while having a nearly cosmopolitan distribution, is most numerously represented in the temperate regions of the northern hemisphere, where also occur nearly all of the larger species, and where are exclusively found the true Wolves and Foxes. In respect to the latter, the larger species of each occur only at the northward, and the smaller at the southward. Thus, in North America, the large Gray Wolf ranges from the arctic regions to Florida and Mexico, while the Coyote is not found much to the northward of the great campestrian region of the interior. The Common Fox ranges also from the subarctic districts southward to the Gulf of Mexico, while the smaller Gray Fox finds its northern limit near the parallel of 42°, while a third still smaller species is confined within the warmer-temperate latitudes. At the extreme northward, we find, however, a smaller arctic form, on the extreme northern confines of the habitat of the family. In the Wolves and Foxes, decrease in size to the southward is strongly marked, being probably not exceeded in any other group, though perhaps nearly equaled in some of the Cariacine Deer.

The *Ursidæ*, while having a wide geographical range, are confined mainly to the north hemisphere, throughout which they have representatives. Here again the larger species are northern, while all the warm-temperate and subtropical forms are small. There is also a corresponding decrease in size southward among the representatives of the several species. (See later portions of the paper for a somewhat detailed discussion of the North American species.)

The *Mustelidæ*, while mainly confined to the northern hemisphere, have also representatives south of the equator. Of the *Mustelinæ* proper, all the larger species are boreal, though some of the smaller extend also to the arctic regions. The Wolverine, the largest of the group, is the most boreal; the Fisher and the Marten, the next in size, are mainly confined to the subarctic and cold-temperate regions; the Mink, next in size, extends farther southward; the Weasels range also into the middle-temperate latitudes, with a single species occurring (only at considerable altitudes) under the tropics. *Galictis* is its single tropical representative, and is also the most specialized (though not the smallest) type of the group. The *Melinæ* and *Enhydrinæ*, each with a single American representative, and both boreal, are also among the largest representatives of the family. The *Mephitinæ*, of medium or rather small size, are strictly a warm-temperate and tropical group, with representatives extending from the northern parts of the United States southward to the southern parts of South America. The *Lutrinæ* have a wider range, being found throughout the tropics as well as in the temperate and colder regions, and apparently present not a very great range of geographical variation.

The *Felidæ*, while possessing an almost cosmopolitan range, have their greatest development within the tropics, where they attain their maximum size and number of species. The single boreal genus found in America is one of the most specialized forms of the family. As will be shown later, the American representatives of this family present a notable exception to the general law of decrease in size toward the south, and confirm the law of increase in size toward the geographical center of the group to which they belong.

The *Procyonidæ* are essentially a tropical family, in which regions are found the largest species and the greatest variety of forms. The single North American species presents a marked *increase in size southward*, as will be fully shown later.

The *Glires*, or *Rodentia*, are found throughout the greater part of the world, but are represented by special groups in different regions. Being strictly herbivorous, they are most numerously developed in the temperate and warmer latitudes. The largest known species are tropical, but others of large size are more or less boreal. In the northern hemisphere, the largest species is the Beaver, which formerly ranged throughout the temperate latitudes. Of the *Muridæ*, the larger species are southern, the smaller northern; and there is a tendency (among some of the species, at least) to an increase in size southward, as in some of the varieties of *Hesperomys leucopus*. The *Arvicolinæ*, on the other hand, are subarctic and temperate in their distribution, and markedly increase in size to the northward. Here, likewise, the largest species of the group are met with.

The *Sciuridæ* are also a nearly cosmopolitan group, with different genera and subfamilies specially characteristic of different regions. The *Sciurinæ* are most numerously represented in the warm-temperate and subtropical latitudes, where also occur the largest species. Yet some of those of the more northern districts show a decided tendency to diminution in size southward, while in others the decrease in this direction is less marked. The *Arctomyinæ* are temperate and subarctic, and the largest species occur at the northward. Parry's Marmot is the most boreal and much the largest. Franklin's Spermophile next succeeds, and is one of the largest of the group. *Spermophilus grammurus* (with its varieties *Beecheyi* and *Douglassi*), of about the same dimension, occupies the elevated interior and the Pacific slope, extending, however, quite far southward. The smallest of the group, *S. Harrisii*, *S. spilosoma*, and *S. mexicana*, have a more southern range. In all of these species, there is a marked decrease in size to the southward in their respective representatives, as there is among the species themselves. *Arctomys* and *Sciuropterus* are boreal genera, with their larger species and varieties occurring at the northward, and a northward increase in size in the representatives of their several forms.

The *Leporidæ* of America are mainly restricted to the northern continent, their center of development as respects the number of species, being the United States. Here occur also nearly all of the larger forms. The Polar Hare, one of the largest, is strictly arctic; three or four others of nearly equal size find their northern limit, with one exception, south of the fo[rty-]ninth parallel. The most remarkable trait of the family is the rat[her] small degree of geographical variation its representatives presen[t, bo]th as respects size and coloration. The difference in size between the largest and smallest species is less than is often found in any co-ordinate group having the same number of species, and the species themselves present great constancy of character. There is generally a slight decrease in size southward among individuals of the same species, but sometimes the difference is scarcely perceptible. In the most northerly but one* of the species (*Lepus americanus*), there is apparently a very slight decrease (certainly no increase) in size northward.

---

* The material at hand is too scanty to afford grounds for any satisfactory generalization respecting the Polar Hare.

With these general remarks, we will pass now to a more special examination of geographical variation in size in several of the more common species of the North American *Feræ*, based on the abundant material in the National Museum.

CANIS LUPUS.

The common Gray Wolf of the northern hemisphere presents a range of individual variation in color exceeded by but few known species of Mammals; gray, white, and black individuals, with various intermediate stages of coloration, occurring with greater or less frequency wherever the species abounds, several of these varieties sometimes occurring in the same litter. Black and white wolves seem to occur more frequently at some localities than others, but gray is generally nearly everywhere the prevailing color. Cream-colored and rufous varieties are also said to have a wide prevalence over some parts of the great plains of the interior. To what extent these variations in color are to be considered as geographic is not yet well established.* With such an evident tendency to variability, it is not surprising that geographical variation in size is displayed in this species to a marked degree. The variation in this respect constitutes a pretty uniform decrease in size southward, as shown (see the subjoined table) by the size of the skull, only fully adult skulls being here taken. The largest are from Fort Simpson and other localities in or near the Mackenzie River district, six of which, out of a series of nine specimens, exceed 10.25 inches in length (one reaching 11.50!), and the other three average above 9.50, the whole averaging 10.38. The next in size are from the region about Puget Sound, a series of three (the only ones in the collection), averaging nearly 10.50. Of sixteen specimens from Forts Benton, Union, and Randall, on the Upper Missouri, the average is 9.45, the extremes being 10.50 and 8.50. Nine specimens from Forts Kearney and Harker (chiefly from Fort Kearney, and all pretty old) average a little larger than the Upper Missouri specimens, the extremes being 10.15 and 9.35. A single specimen from the mountains of New Mexico reaches 10.00, while the three most southern (from the Rio Grande and Sonora, Mexico) average only 8.37, being the smallest of the whole series, and averaging 2.00 shorter than the series of nine from the Mackenzie River region. This difference is fully 25 per cent. of the average size of a series of upward of eighty specimens; while the difference between the smallest (from Saltillo, Mexico) and the largest (from Fort Simpson) is 3.75, or nearly 40 per cent. of the average size of the whole series!

---

* See further on color variation in this species, Bull. Mus. Comp. Zoöl., vol. i, pp. 154–158.

Measurements of forty-five North American skulls of CANUS LUPUS.*

Catalogue-number.	Original number.	Locality.	Sex.	Total length.	Greatest width.	Distance between orbits.	Nasal bones, length.	Upper incisors from front to hinder margin of palate.	Upper incisors, width between external edges.	Lower jaw, length.	Lower jaw, height.	Remarks.
9001	......	Fort Simpson, Hudson's Bay Territory	......	11.50	6.20	2.08	4.45	5.50	2.10	8.00	3.34	
9003	......	......do	......	10.75	5.85	2.07	3.94	5.00	2.05	7.75	3.25	
9000	......	......do	......	10.60	5.65	1.85	4.20	4.95	1.95	7.40	3.00	
9002	......	......do	......	10.35	......	1.87	4.12	......	......	......	......	"White." Skull imperfect.
6217	1054	100 miles northwest of Fort Yukon	♂	10.35	5.50	1.90	3.87	5.00	1.95	7.00	3.10	
7154	2	Fort Yukon	♀	10.25	5.20	1.80	4.00	5.05	1.85	7.65	3.00	"Black."
6218	......	Pool's River	......	9.50	5.35	1.70	3.65	4.65	1.73	7.00	2.75	
6256	......	Fort Rae	......	9.20	4.50	1.45	3.57	4.60	1.65	6.85	2.50	
6257	......	......do	......	9.80	4.90	1.67	3.50	4.50	1.75	7.00	2.82	
1194	......	Minnesota	......	9.95	5.15	1.65	3.87	4.90	1.85	7.45	2.80	Very aged.
1804	......	Adirondack Mountains, New York	......	9.20	4.95	1.73	3.50	4.35	1.55	6.85	2.70	Rather young.
3438	......	Puget Sound	......	10.55	5.67	2.07	4.07	5.12	1.90	......	......	
4114	......	Oregon	......	9.50	5.00	1.80	3.43	4.65	1.75	7.10	2.86	Marked "*L. gigas*".
	......	......do	......	10.70	5.90	......	......	......	......	......	......	"*Lupus gigas* Townsend." Measurements from Baird.
13949	......	Little Chief Mountain, forty-ninth parallel	......	9.15	5.00	1.75	3.55	4.60	1.75	6.80	2.65	
13965	......	Fort Benton	......	9.80	5.35	1.80	3.70	4.75	1.83	7.20	2.92	
13966	......	......do	......	9.40	5.20	2.15	3.50	4.45	1.95	7.00	2.92	
13967	......	......do	......	9.60	5.35	1.77	3.50	4.70	1.70	7.00	3.00	
2972	......	Fort Union	......	10.50	......	1.80	3.88	4.85	2.05	7.20	2.90	
3037	......	......do	......	9.90	5.15	1.84	3.70	4.75	1.78	6.92	2.85	
3031	......	......do	......	9.50	5.00	1.85	3.33	4.70	2.00	6.85	2.70	
4124	......	......do	......	8.85	4.75	1.50	3.40	4.35	1.55	6.30	2.50	
3020	......	......do	......	9.15	......	1.53	3.35	4.60	1.62	6.80	2.68	
3005	......	......do	......	8.50	4.50	1.30	3.12	4.27	1.45	6.00	2.33	
2954	......	......do	......	9.70	5.50	2.05	3.70	4.85	1.88	7.10	2.85	A short, very broad skull.
2984	......	......do	......	9.40	5.25	1.98	3.20	4.48	1.85	......	......	Six others in the same lot of similar dimensions.
2994	......	......do	......	9.50	......	1.92	3.45	4.75	1.87	6.75	2.75	Very old.
12906	......	Fort Randall, Dak	......	9.00	4.85	1.70	3.20	4.37	1.75	6.65	2.58	
12907	......	......do	......	9.42	4.75	1.50	3.53	4.53	1.68	6.80	2.75	
12302	......	......do	......	9.75	5.25	1.72	3.62	4.57	1.78	7.00	2.95	
2611	......	Fort Kearney, Nebr	......	9.35	5.00	1.50	3.50	4.57	1.62	6.70	2.62	
884	......	......do	......	10.15	5.40	1.85	3.80	4.98	1.87	7.35	3.20	
1314	......	......do	......	9.65	5.35	1.83	3.50	4.75	2.00	6.90	2.85	
3521	......	......do	......	9.85	4.87	1.68	3.75	4.80	1.80	7.05	2.90	
887	......	......do	......	9.80	5.35	1.85	3.65	4.80	1.90	......	......	

3686	......	......do	......	10.00	5.50	1.87	3.80	4.80	1.97	7.18	3.00	
1312	......	......do	♂	9.75	5.25	1.78	3.67	5.00	1.80	7.88	3.15	"Gray."
11592	......	Fort Harker, Kans	......	9.50	5.06	1.72	3.50	4.65	1.80	......	......	
11591	......	......do	......	10.15	4.85	1.62	3.80	5.00	1.85	7.20	3.00	
3341	......	New Mexico	......	10.00	4.92	1.65	3.65	4.90	1.75	7.35	2.93	
8098	......	Fort Cobb, Ind. Ter	......	9.25	4.85	1.72	3.50	4.50	1.65	6.60	2.65	
1380	......	Matamoras, Mexico	......	8.75	4.62	1.65	3.18	4.25	1.53	6.35	2.63	
2193	......	Santa Cruz (Sonora), Mexico	......	8.60	4.55	1.55	3.07	4.20	1.57	6.30	2.63	
1379	......	Saltillo, Mexico	......	7.75	4.05	......	......	......	......	......	......	Measurements from Baird.

* The measurements in this and the following tables are all taken either with calipers or dividers, and are expressed in hundredths of the English inch. They hence always indicate the length in a straight line between the points indicated.

CANIS LATRANS.

The Coyote, or Prairie Wolf, the nearest American affine of the Gray Wolf, is as remarkable for its constancy of character, especially in respect to color, as the latter is for its variability. The individual variations in the color of *Canis latrans* consist generally in the depth or intensity of the shadings of black or rufous that more or less pervade the pelage of certain parts of the head and body. Although considerable variations have been noticed in respect to the form of the skull, they are small in proportion to those presented by *Canis lupus*. It is also much less influenced apparently by locality. The species has, however, a less extended range than *Canis lupus*, and the specimens at command represent localities less widely separated than do the series of skulls of *Canis lupus*.

Measurements of forty skulls are given below, mainly from Nebraska, Dakota, and Wyoming. The most distant localities are Columbia River and Fort Tejon, California, Southern Texas, and Fort Union, Montana. Of this series of forty skulls, the average is 7.40; only two attain a length of 8.00, one of which (measuring 8.00 in length) is from Fort Union, and the other (8.05 inches in length) is from Fort Massachusetts, New Mexico. Only two fall below 6.95, one of which measures 6.65 and the other 6.50; the smaller being from the Coppermine River, New Mexico, and the other from Fort Randall, Dakota. Of thirteen specimens from Fort Randall, the largest measures 7.60 in length and the smallest 6.65, the majority (more than three-fourths) falling between 7.00 and 7.50, thus presenting a remarkable uniformity in size. Ten others from Fort Kearney average fully as large, the extremes being 6.95 and 7.60, while four-fifths of them fall between 7.00 and 7.50. Three specimens from Fort Tejon, California, measure respectively 7.95, 7.60, and 7.45, or above the average of those from Dakota and Nebraska! Four specimens from Wyoming Territory, however, measure each 7.80. A single San Diego specimen measures 7.75, and two specimens from Southern Texas respectively 6.95 and 7.00, or but little below the average of northern specimens. Of four specimens from New Mexico, three attain or exceed 7.40, one reaching 8.05 and forming the largest of the series; the other, with a length of only 6.50, forms the smallest of the series, both the largest and the smallest being from New Mexico. It thus appears that in *Canis latrans* there is comparatively little decrease in size southward, instead of the southern averaging fully 25 per cent. smaller, as is the case in *Canis lupus*. The difference between the extremes is only 1.55, or about 20 per cent., against twice that amount in *Canis lupus*. Throwing out the two skulls that fall below 6.95 would reduce the difference between the extremes to 1.10, and the variation to only 15 per cent. of the average! In both *Canis latrans* and *Canis lupus*, the width of the skull averages about one-half the length, ranging in *Canis latrans* from 0.49 to 0.52, while in *Canis lupus* the range in this proportion is from 0.48 to 0.56.

A glance at the table shows that while the Upper Missouri specimens are rather younger than those from Fort Kearney, they rather exceed them in size, and the difference would be somewhat greater if they were of strictly corresponding ages. The single very large skull from New Mexico is also that of a very old individual.

Measurements of forty skulls of CANIS LATRANS.

Catalogue-number.	Locality.	Sex.	Total length.	Greatest width.	Distance between orbits.	Nasal bones, length.	Upper incisors from front to hinder margin of palate.	Upper incisors, width between external edges.	Lower jaw, length.	Lower jaw, height.	Remarks.
3025	Fort Union	.....	8. 00	3. 95	1. 30	2. 82	4. 10	1. 48	.....	.....	
12902	Fort Randall, Dak	♂	7. 40	3. 43	1. 13	2. 80	3. 60	1. 15	5. 25	1. 95	Rather young.
12893	......do	♂	7. 50	3. 65	1. 18	2. 82	3. 67	1. 15	5. 50	2. 06	Do.
12905	......do	♂	7. 50	3. 77	1. 18	2. 80	3. 67	1. 20	5. 50	2. 16	Do.
12899	......do	♂	7. 60	3. 80	1. 25	2. 87	3. 75	1. 20	5. 53	2. 25	Do.
12897	......do	♂	7. 40	3. 80	1. 15	2. 80	3. 60	1. 19	5. 30	2. 00	Do.
12896	......do	.....	7. 25	3. 70	1. 15	2. 75	3. 60	1. 19	5. 32	1. 95	Do.
12888	......do	.....	7. 15	3. 52	1. 20	2. 70	3. 53	1. 10	5. 27	1. 85	Do.
12903	......do	.....	7. 10	3. 45	1. 18	2. 88	3. 50	1. 15	5. 05	1. 85	Quite young.
12904	......do	♀	7. 00	3. 56	1. 15	2. 50	3. 53	1. 09	5. 12	1. 95	
12892	......do	♀	6. 95	3. 52	1. 14	2. 40	3. 27	1. 12	5. 02	2. 10	
12901	......do	♀	6. 65	3. 38	1. 12	2. 37	3. 38	1. 04	4. 95	1. 85	
12887	......do	♀	7. 25	3. 70	1. 16	2. 70	3. 58	1. 12	5. 25	2. 08	Aged.
1327	Fort Kearney, Nebr	♀	7. 00	3. 65	1. 26	2. 65	3. 38	1. 15	5. 10	1. 90	"Eight to ten years old."
1335	......do	♀	6. 95	3. 75	1. 25	2. 60	3. 45	1. 15	5. 15	2. 00	"Ten or eleven years old."
1336	......do	♀	7. 40	3. 80	1. 25	2. 90	3. 93	1. 13	5. 27	2. 00	"Four years old."
1338	......do	♀	7. 12	3. 62	1. 20	2. 62	3. 63	1. 20	5. 15	1. 88	"Two to four years old."
1322	......do	♀	7. 35	3. 62	1. 23	2. 75	3. 63	1. 13	5. 30	1. 90	"Eight years old."
1332	......do	♀	7. 10	3. 62	1. 20	2. 58	3. 74	1. 18	5. 37	2. 00	"One to two years old."
1328	......do	♀	7. 50	3. 62	1. 05	2. 95	3. 65	1. 18	5. 40	2. 18	"One and a half to three years old."
1341	......do	♂	7. 60	3. 62	1. 30	2. 70	3. 75	1. 18	5. 38	2. 00	
1323	......do	♂	7. 25	3. 50	1. 20	2. 70	3. 57	1. 12	5. 15	1. 97	"Eight to twelve years old."
1339	......do	♂	7. 30	3. 68	1. 20	2. 88	3. 65	1. 15	5. 40	2. 00	
11552	Wyoming Territory		7. 80	.....	.....	.....	.....	.....	.....	.....	
11554	......do	.....	7. 80	3. 95	.....	.....	.....	.....	.....	.....	
11559	......do	.....	7. 80	3. 82	.....	.....	.....	.....	.....	.....	
11558	......do	.....	7. 80	3. 95	.....	.....	.....	.....	.....	.....	
1916	Columbia River	.....	7. 80	3. 78	.....	.....	.....	.....	.....	.....	
4113	Oregon	.....	7. 25	3. 75	.....	.....	.....	.....	.....	.....	
3323	Nebraska	♂	7. 80	3. 80	.....	.....	.....	.....	.....	.....	
3332	Fort Massachusetts, N. Mex	.....	8. 05	3. 94	.....	.....	.....	.....	.....	.....	
1000	Coppermine River, N. Mex	.....	7. 40	.....	.....	.....	.....	.....	.....	.....	Very aged.
1001	......do	.....	6. 50	3. 28	.....	.....	.....	.....	.....	.....	
1019	Texas	♂	6. 95	3. 53	.....	.....	.....	.....	.....	.....	Rather aged.
975	Frontera, Tex	.....	7. 00	3. 62	.....	.....	.....	.....	.....	.....	Very aged.
2198	San Francisco Mountains, N. Mex	.....	7. 60	3. 62	.....	.....	.....	.....	.....	.....	Do.
1218	San Diego, Cal	.....	7. 75	3. 77	.....	.....	.....	.....	.....	.....	Do.
3577	Fort Tejon, Cal	.....	7. 95	4. 00	.....	.....	.....	.....	.....	.....	Do.
3616	......do	.....	7. 60	3. 65	.....	.....	.....	.....	.....	.....	Adult.
3617	......do	.....	7. 45	3. 48	.....	.....	.....	.....	.....	.....	Rather young.

VULPES ALOPEX ET VARS.

In the Common Fox of North America, we meet with a range of color-variation, irrespective of locality, somewhat akin to that seen in *Canis lupus*. The prevalent tendency, however, is toward melanism, which tendency is much more strongly developed in the colder than in the warmer latitudes. Frequently, individuals of the melanistic type occur in litters of the common variety. The varying degrees of melanism occurring in this species have given rise to several commercial varieties, which have received at the hands of naturalists systematic designations, and been regarded more or less generally as valid species. Generally, these melanistic varieties are more fully furred, with larger and heavier tails, than the common form. The difference in the fineness and softness of the fur is recognized to such an extent by furriers as to greatly affect the price of the skins, the so-called "Silver" and "Cross" furs being considered far more valuable than the fulvous type.

The so-called "Cross Fox" (*Vulpes "decussatus"*) is more or less frequent as far south as Northern New England and Northern New York, and throughout the more elevated portions of the great Rocky Mountain plateau, where it constitutes a large proportion of the representatives of the so-called *Vulpes "macrurus"*. More rarely, the Black or so-called "Silver Fox" (*Vulpes "argentatus"*) is met with over the same regions, becoming frequent in the higher parts of the Rocky Mountains* and northward. The fulvous form seems, however, to be generally the more prevalent form throughout the range of the species. To the southward, it is the form exclusively met with; but near timber-line in the Rocky Mountains, and throughout the "fur countries", it seems to be not much more frequent than the melanistic forms.

With this tendency to great variability in color, we meet, as usual in such cases, a great variation in size. In the present case, the variation in color may be properly regarded as geographical, through an increasing tendency to melanism northward. The variation in size is also chiefly of the same character, the size uniformly increasing toward the north, as shown by the subjoined table of measurements. A glance at this table shows at once the nature of this variation. The largest specimens come from the Aleutian Islands and Alaska; the smallest from Essex County, New York, which is the most southerly locality well represented in the collection.

A series of nine skulls from Alaskan localities range in length from 5.70 to 6.20, five out of the nine having a length of 6.00 to 6.20 (two 6.15 and two 6.20), and give an average of 5.98. In another series of eighteen from the Mackenzie River district (mainly from Fort Anderson), the range is from 5.55 to 6.10. Only one, however, exceeds 6.00, and three only reach this size, the average being 5.80. These series consist about equally of the so-called "Silver" and common fulvous varieties, and, as may be seen from the table, there is no material difference in size between the two so-called varieties.

A third series of nine skulls, of the so-called "*macrurus*", chiefly from the Upper Missouri country (including two, however, from the Pacific slope), ranges from 5.40 to 6.00, with an average of 5.75. Two only reach 6.00, and two only fall as low as 5.50. Hence the series forms a third appreciable step in the southward decrease in size. Though the latitude is much less, the elevation of the region is much greater than that of the localities more to the northward. With a similar altitude, the decrease would have been more marked, as is proven by the series

*See Bulletin Essex Institute, vol. vi, p. 54.

next to be considered. A fourth series of twelve specimens, from the Adirondack region of New York, ranges from 5.20 to 5.68, with an average of 5.40. Only three specimens range above 5.50, while four fall below 5.30. A fifth series of five skulls, from European localities, ranges from 5.50 to 5.70, with an average of 5.58.

In the Alaskan series, the width ranges from 2.90 to 3.32, averaging 3.20; in the Mackenzie River district series, from 2.87 to 3.28, averaging 3.02; in the "*macrurus*" series, from 2.70 to 3.20, averaging 2.90; in the Adirondack series, the width ranges from 2.70 to 2.95, averaging 2.80; in the European series, from 3.05 to 3.15, averaging 3.08. Hence—

	Average length.	Average width.
Alaskan series	5.98	3.20, or 0.535 of length.
Mackenzie River District series	5.80	3.02, or 0.521 of length.
"*Macrurus*" series	5.75	2.90, or 0.504 of length.
Adirondack series	5.40	2.80, or 0.518 of length.
European series	5.58	3.08, or 0.552 of length.

It thus appears that in the American specimens there is not only a well-marked southward decrease in size, but also a decrease in the relative breadth of the skull, through the greater elongation of the facial portion; also that the relative breadth is quite appreciably greater in the European form, as noticed long since by Professor Baird.*

While the European *Vulpes vulgaris* may be considered as subspecifically distinct from the American (*Vulpes vulgaris* subsp. *fulvus*), through its wider skull, less pointed and shorter muzzle, harsher and more reddish fur, etc., the different so-called American "species" or "varieties" (*fulvus*, "*decussatus*", "*argentatus*", and "*macrurus*") do not have the same claim to subspecific recognition. The Foxes of the colder regions, it is true, have a fuller and softer pelage, a greater tendency to melanism, shorter muzzles, and are larger, yet these differences are so inconstant, especially the differences of color, and so insensibly intergrade, that any attempt at their subspecific recognition seems impracticable, the most diverse varieties in color occurring at the same localities and even among individuals of the same litter.†

*Mam. N. Amer., pp. 126, 130.
†On this point see Bulletin Mus. Comp. Zoöl., vol. i, pp. 159, 160.

Measurements of fifty-three skulls of VULPES ALOPEX *et vars.*

Catalogue-number.	Locality.	Sex.	Length.	Width.	Remarks.
9481	Kinai, Alaska	------	6.20	3.03	
8039	Kodiak, Alaska	------	6.20	3.32	
8037	......do	------	5.90	3.00	
8417	Aleutian Islands	------	6.15	3.25	
6037	Yukon River	♂	6.15	3.00	Silver.
6039	......do	♀	6.00	3.15	Do.
6040	......do	♀	5.70	3.05	Do.
6034	......do	------	5.80	3.00	Fulvous.
6035	......do	------	5.70	2.90	Do.
4300	Fort Liard	------	5.95	3.14	Silver.
6221	Peel River	------	5.87	3.10	Fulvous.
6038	......do	------	5.95	2.88	Silver.
7175	Fort Good Hope	♂	6.00	3.00	Do.
6221	......do	------	5.90	3.07	Fulvous.
7179	......do	♂	5.75	2.94	Silver.
6264	Fort Anderson	♂	6.10	3.28	Black.
1485	......do	------	6.00	3.25	Fulvous.
6262	......do	♂	5.95	3.00	Silver.
7186	......do	------	5.80	2.90	Fulvous.
1484	......do	------	5.75	2.95	Do.
6225	......do	♀	5.75	3.00	Silver.
1492	......do	♂	5.75	3.10	Do.
6263	......do	♂	5.66	2.87	Do.
1519	......do	------	5.65	3.00	Fulvous.
6259	......do	♂	5.60	3.00	Silver.
6258	......do	♂	5.55	2.90	Do.
1518	......do	------	5.55	------	Fulvous.
11573	Wyoming Territory	------	6.00	3.20	"*Macrurus.*
11297	......do	------	6.00	3.00	Do.
11296	......do	------	5.82	------	Do.
11218	......do	------	5.50	2.78	Do.
7855	Fort Berthold	------	5.90	3.03	Do.
12909	Fort Randall, Dak	------	5.65	2.85	Do.
1325	Fort Kearney, Nebr	------	5.77	2.85	Do.
2014	Fort Dalles, Oreg	♂	5.70	2.83	Do.
4191	Fort Crook, Cal	------	5.40	2.70	Do.
3078	Essex County, New York	------	5.65	2.87	
3073	......do	------	5.68	2.85	
3067	......do	------	5.60	2.85	
3719	......do	------	5.50	2.95	
3695	......do	------	5.50	2.87	
3698	......do	------	5.30	2.70	
3709	......do	------	5.30	2.70	
3716	......do	------	5.37	2.75	
3701	......do	------	5.27	2.80	
3715	......do	------	5.25	2.75	
3074	......do	------	5.25	2.73	
3075	......do	------	5.20	2.78	
1038	Sweden	------	5.70	3.10	
790	England	------	5.65	3.04	
868	Germany	------	5.62	3.15	
869	......do	------	5.62	3.05	
870	......do	------	5.50	3.07	

UROCYON VIRGINIANUS.

Measurements of a series of fifteen skulls of this species (all of the available material) form a series grading by slight differences in length from 3.73, the smallest, to 4.77, the largest. The largest specimens are from Pennsylvania, Washington, and Virginia; all these exceed 4.60 in length. The next in size are from Southern Texas and Southern California, which range in length from 4.63 down to 4.50. Next come three specimens from Tehuantepec, Southern Mexico, which range from 4.40 to 4.15. Between these and the next—a series of three "*littoralis*" skulls from the islands off Southern California—is an interval of three-tenths of an inch, the three "*littoralis*" skulls ranging from 3.85 to 3.75. The smallest of all, however, is a single well-matured skull from Merida, Yucatan, 3.73 in length, and hence smaller even than the smallest "*littoralis*" skull, its breadth being only 1.98 against a breadth of 2.05 in the narrowest "*littoralis*" specimen. The localities represented are few and widely separated; there being no specimens from points between Virginia and Southern Texas, and none between Texas and Tehuantepec, Mexico, nor between these two last-named localities and Fort Tejon, Cal. The small insular race known as "*littoralis*", from the islands off the coast of Southern California, come in between the Tehuantepec specimens and the example from Merida. While there are no very considerable breaks in the chain, the gradation would be more complete if specimens could be included from other intermediate localities. The specimens at hand are sufficient to show a very great but still very gradual decrease in size southward, amounting to over 25 per cent. of the mean size. The mean of the two extremes is 4.25, with a difference of 1.04; while, with a single exception, there is a gap at no point of more than 0.08.

With this rapid decrease in size may be noticed a considerable range of variation in breadth in specimens of nearly the same length, indicating the existence of an unusual amount of individual variation, the ratio of width to length varying from 0.54 to 0.59.

Measurements of fifteen skulls of UROCYON VIRGINIANUS.

Catalogue-number.	Locality.	Sex.	Length.	Width.	Remarks.
4729	Pennsylvania	......	4.77	2.70	
968	Washington, D. C	♀	4.70	2.56	
671	White Sulphur Springs, Va	......	4.62	2.65	
7491	Washington County, Texas	......	4.60	2.70	
1175	Eagle Pass, Texas	......	4.50	2.58	
3543	Fort Tejon, Cal	......	4.63	2.53	
3545	......do	......	4.55	2.65	
4140	Cape Saint Lucas, Lower California	......	4.50	2.43	
8659	Tehuantepec, Mexico	......	4.40	2.35	
8662	......do	......	4.27	2.37	
13851	......do	......	4.15	2.25	
2275	San Miguel Island, California	......	3.85	2.23	Var. *littoralis*.
2154	San Nicolas Island, California	......	3.80	2.05	Do.
6323	......do	......	3.75	2.10	Do.
13477	Merida, Yucatan	......	3.73	1.98	

FELIS CONCOLOR.

The amount of material available for the study of variation in size with locality in the present species is too small to yield very satisfactory results. In the eight specimens of which measurements are given below, it will be noticed that there is a decided increase in size southward. Between the three skulls from northern localities (one each from Northern New York and Washington and Oregon Territories) and the three (mature) skulls from southern localities (Louisiana and the Rio Grande, Texas), the average difference is fully an inch, or about one-eighth of the mean size.

Measurements of eight skulls of FELIS CONCOLOR.

Catalogue-number.	Locality.	Sex.	Length.	Width.	Remarks.
3811	Essex County, New York	......	7.40	5.25	Old.
8597	Puget Sound	......	7.50	5.05	
3267	Oregon	......	7.80	5.15	
1148	Eagle Pass, Texas	......	7.75	4.95	Quite immature.
1356	Rio Grande, Texas	......	7.50	5.00	Do.
1355	......do	......	8.40	5.35	
1895	Brazos River, Texas	......	8.50	5.60	
1158	Prairie Mer Rouge, Louisiana	......	8.75	5.50	

FELIS PARDALIS.

Fourteen skulls of *Felis pardalis* show a most decided southward increase in size. A series of five skulls from the Lower Rio Grande average about an inch shorter than another series of nine from Southern Mexico and Central America. The largest of the Rio Grande skulls has a length of 5.25, while the smallest of the Mexican and Central American series (excluding one rather young specimen) has a length of 5.20, and the largest a length of 6.20. The three largest (6.00 to 6.25) are from Costa Rica, while one other from Panama and another from Surinam are but little smaller. The smallest of the Rio Grande series (a rather young specimen) is but 4.50 in length; the smallest of the tropical series (a specimen of corresponding age) 5.35.

The difference in size with locality is thus as great in this species and in *Felis concolor* as it is in the Wolves and Foxes; but the increase is in the opposite direction,—to the *northward* in the former and to the *southward* in the latter; the one group being a northern type, the other a tropical.

Measurements of fourteen skulls of FELIS PARDALIS.

Catalogue-number.	Locality.	Sex.	Length.	Width.	Remarks.
1363	Matamoras, Mexico	......	4.50	3.05	Mature but not very old.
1362	......do	......	4.90	3.35	
1361	......do	......	5.05	3.35	
1359	......do	......	5.20	3.50	
1358	......do	......	5.25	3.40	
6023	Panama	......	5.60	3.75	
7080	Mirador, Mexico	......	5.70	3.55	
13852	Tehuantepec, Mexico	......	5.50	3.70	
11743	Isthmus of Darien	......	5.85	3.80	
14182	Costa Rica	......	6.00	3.73	Very old.
14179	......do	......	6.00	3.94	Do.
14180	......do	......	6.20	4.19	Do.
14178	......do	......	5.35	3.60	Adult but not very old.
13005	Surinam	......	5.80	3.83	Very old.

LYNX RUFUS ET LYNX CANADENSIS.

In the subjoined table are given measurements of thirty-four skulls of North American Lynxes, namely, seven of *L. fasciatus*, ten of *L. rufus*, eight of *L. maculatus*, and nine of *L. canadensis*, representing localities as distant from each other as Alaska and Northern Mexico on the one hand, and New York and Fort Tejon, Cal., on the other. Yet the extremes of variation met with at single localities are as great as those from the most widely separated of the above-named localities; in other words, no geographical variation in size is perceptible. The largest northern specimen (*canadensis*), from Peel River, Arctic America, with a length of 5.30, a little exceeds in size the largest specimens from any locality south of the latitude of 40°; but it in turn is slightly smaller than a specimen (*fasciatus*) from Fort Townsend, Wash., which has a length of 5.50, and by another of the same dimensions (*rufus*) from the Big Sioux River. Eight specimens of the most southern type (*L. maculatus*), all from Texas and the Mexican side of the Lower Rio Grande, differ in the average from nine specimens of the most northern type (*L. canadensis*), all from Arctic or sub-Arctic America, almost inappreciably, the *canadensis* series having an average length of 5.01 and the *maculatus* series of 5.00! The difference in breadth is also only about one-tenth of an inch, which the addition of a single specimen to either series might cancel. This is certainly a surprising result when it is remembered that one of the chief alleged distinctive characters of *L. canadensis* has been its supposed larger size!

The average dimensions of these several series are as follows:

Species.	Number of specimens.	Length.	Breadth.
L. canadensis	9	5.01	3.52
L. fasciatus	7	5.03	3.56
L. maculatus	8	5.00	3.40
**L. rufus*	10	4.91	3.41
Mean of all	34	4.98	3.47

* The specimens placed under *rufus* are those that are so marked in the collection, being the specimens so identified by Professor Baird.

The *fasciatus* series is the largest, but this series happens to include more very old specimens than the others, and hence its higher average. Such a constancy of size as is here shown to prevail over an area embracing more than 40 degrees of latitude is probably without a parallel in any other conspecific group of North American Mammals.

The difference between these heretofore commonly-recognized "species" of the genus *Lynx* must hence be sought elsewhere than in size. The specific distinctness of *L. canadensis*, the most northern type, has been heretofore scarcely questioned, in consequence of its supposed larger size, larger limbs, longer, softer pelage, longer ear-tufts, more indistinct markings, and generally lighter or grayer color. The longer ear-tufts correlate with the longer, softer pelage, that always characterizes the boreal representatives of species having a wide latitudinal range. The difference in coloration is not greater than, or even so great as, that which obtains between *fasciatus* and *rufus*, or between *fasciatus* and *maculatus*, which forms naturalists now seem disposed to refer to one and the same species under the name *L. rufus*. *Maculatus*, the most southern form, differs from the "typical" or eastern *rufus* in its shorter, coarser fur, more reddish tints, and more distinct markings. Its reputed range extends from the Lower Rio Grande westward across the continent to Southern California; but in the National Museum collection are also specimens marked *rufus* from many points within this area, including a considerable series from Fort Tejon. The gradation from the "typical" *rufus* type into *maculatus* is complete and by almost insensible stages.

The *L. fasciatus* or Columbia River race differs from *rufus* in its more uniform and darker (chestnut rather than reddish) coloration, by the markings on the dorsal surface and sides of the body being nearly obsolete, and the fuller, softer fur, which is about as heavy and soft as in *canadensis*. We have hence, in this form, only another instance of the duller, darker, and more uniform coloration that characterizes the greater part of the Mammals (and many Birds also) from the humid, heavily-wooded Columbia River region, as compared with their conspecific allies of the other portions of the continent.

L. canadensis differs from these several southern races mainly as the northern representatives of a given species usually differ from its southern representatives, namely, in its softer and longer pelage, more heavily-clothed feet, longer ear-tufts, paler or grayer general color, and more indistinct markings, and especially in a tendency to entire obsolescence of the markings on the lower surface of the body and inner side of the legs. The tail has a shorter area of black at the end, and lacks the white on the lower surface at the extreme tip, so constantly seen in the other forms. The tail is but little, if any, shorter, although the greater length and thickness of the fur give it that appearance. There is, however, a tendency to a greater length of tail to the southward. Its supposed greater size and larger limbs are also due almost wholly to the greater fullness and length of the pelage, the fresh carcass (in a specimen from Houlton, Me.) with the skin removed giving the same measurements as in *L. rufus* (a specimen from Colorado).

The prior name for the group of American Lynxes is undoubtedly *rufus* of Guldenstädt (1776), which antedates by about forty years Rafinesque's names of *canadensis*, *montanus*, and *floridanus* (1817). The *L. maculatus* of Horsfield and Vigors (1829), which was admitted as a valid species by Baird, but regarded as merely a variety of *rufus* by Audubon and Bachman, is evidently subspecifically indistinguishable from the true *rufus* of authors. *L. fasciatus* of Rafinesque (based on the "Tiger Cat" of Lewis and Clarke, from the Columbia River region) is far more tangible, sufficiently so to be properly recognizable as a subspecies (*Lynx rufus* subsp. *fasciatus*). The *L. canadensis* of authors seems to have even still stronger claims for nominal recognition, though the differences are still clearly such as characterize geographical races. We hence believe its relationship to the rest of the group is better indicated by a name (*L. rufus* subsp. *canadensis*) indicating subspecific rather than specific rank.

A single adult skull (from Sweden) of the large Lynx of the northern parts of the Old World (*Lynx borealis*) exceeds in size by an inch the largest specimens of the American Lynxes, and hence seems to indicate an animal fully one-fifth larger than even exceptionally large specimens of *L. rufus*.

Measurements of thirty-four skulls of LYNX RUFUS *et vars.*

Catalogue-number.	Locality.	Sex.	Length.	Width.	Remarks.
8599	Puget Sound, Wash	------	4.80	3.45	"*fasciatus.*"
8600	......do	------	4.65	3.30	Do.
3426	Steilacoom, Wash	♀	5.20	3.60	Do. Very old.
3197	......do	------	5.45	3.95	Do.
3427	Fort Umpqua, Wash	------	------	3.32	Do.
3426	......do	------	4.90	3.50	Do.
2032	Shoalwater Bay, Wash	------	4.75	------	Do.
3147	Fort Townsend, Wash	------	5.50	3.82	Do. Very old.
2883	Big Sioux River	♀	5.50	3.82	"*rufus.*" Very old.
3775	Mississippi	------	4.95	3.38	Do.
7465	Arizona	------	5.10	3.55	Do.
3120	Florida	------	4.85	3.22	Do. Quite young.
2391	Louisiana	------	4.90	3.27	Do.
3574	Fort Tejon, Cal	♂	4.80	3.32	Do.
3542	......do	------	4.80	3.50	Do.
3541	......do	♂	4.93	3.38	Do.
3570	......do	------	4.65	3.37	Do.
3576	......do	------	4.65	3.37	Do.
1887	Fort Belknap, Tex	------	5.12	3.72	"*maculatus.*"
1109	Eagle Pass, Texas	------	5.27	3.51	Do.
7493	Washington County, Texas	------	4.72	3.25	Do.
1376	Matamoras, Mexico	------	4.55	3.10	Do. Rather young.
1367	......do	------	5.10	3.40	Do.
1368	......do	------	4.80	3.25	Do.
1006	Texas	------	5.15	3.57	Do.
1159	Prairie Mer Rouge, Louisiana	------	4.80	3.28	Do.
9478	Kinai, Alaska	------	4.85	3.35	"*canadensis.*
6031	Yukon, Alaska	------	4.95	3.53	Do.
6030	......do	♀	4.75	3.35	Do.
6216	Peel River	♀	4.95	3.53	Do.
6211	......do	♂	5.30	3.70	Do.
4468	Fort Simpson	------	5.15	3.60	Do.
4296	Liard River	------	5.00	3.45	Do.
3279	Red River Settlement	------	5.00	3.52	Do.
2570	Medicine Bow Creek, Wyoming	------	5.15	3.60	Do.

PROCYON LOTOR.

The present species presents another well-marked case of gradual increase in size southward. In a series of fifteen skulls from the Atlantic States (New York, Pennsylvania, and Georgia), only a single specimen exceeds 4.38 in length or 3.00 in width, the largest specimen being from Saint Simon's Island, Georgia. Three from Essex County, New York, average 4.28; five from Pennsylvania average 4.29; seven from Saint Simon's Island, Georgia, average 4.26 (or 4.29, excluding one very small one). Six specimens from the interior (Nebraska, Missouri, Indian Territory, and the Lower Rio Grande) average 4.49, two only falling below 4.50, and the largest (Rio Grande) 4.70. Three from California ("*hernandezi*") average 4.63, the largest reaching 4.78, with a width of 3.38. Six from Southern Mexico average 4.58, the largest reaching 4.73 in length, with a width of 3.42. Three from Costa Rica average 4.69, the largest reaching 4.85.

In addition to the above, there is a single very aged specimen from Detroit, Mich., which has a length of 4.35, and two others from Alaska (one middle-aged, the other rather young) which measure, respectively, 4.25 and 4.05 in length, the latter being the smallest of the whole series, although it contains others equally young.

Between the three specimens from Essex County, New York, and the three from Costa Rica, specimens of corresponding ages and constituting the two extremes, the average difference is nearly six-tenths (0.57) of an inch, or about one-seventh of the size of the northern examples.

Besides the difference in size, there is also a considerable range of variation in respect to the general form of the skull in the ratio of width to length, in the shape, degree of concavity of the palate, in specimens from the same locality, the ratio of width to length varying from 0.65 to 0.73.

In addition to the increase in size southward, there is a tendency to an increase in the intensity of the colors in the same direction, with a stronger contrast between the light and dark markings. These differences, taken collectively, have given rise to several nominal species, of which the *P. hernandezi* of Wagler and *P. psora* of Gray have become the most prominent. The species normally presents a considerable range of color-variation, tending on the one hand more or less to melanism and on the other to albinism. On these extreme phases of coloration have also been based other nominal species, as the *P. obscurus* of Wiegmann and the *P. nivea* of Gray. All these names have been already placed by Gray, in his later notices of the group, under the head of *P. lotor*, but separated as being varietally distinct. It seems doubtful, however, whether even the large southern form, usually called *hernandezi*, is really entitled to subspecific recognition.

Measurements of thirty-six skulls of PROCYON LOTOR.

Catalogue number.	Locality.	Sex.	Length.	Width.	Remarks.
8690	Alaska	......	4.25	2.87	Middle-aged.
8693	......do	......	4.05	2.92	Young.
1068	Detroit, Mich	......	4.35	......	Very old.
3723	Essex County, New York	......	4.25	2.87	Do.
3722	......do	......	4.27	2.70	Middle-aged.
3079	......do	......	4.32	3.00	Very old.
898	Pennsylvania	......	4.35	3.00	Do.
6025	......do	......	4.12	2.51	Do.
4817	......do	......	4.38	2.64	Do.
575	......do	......	4.25	2.88	Do.
766	......do	......	4.35	2.93	Do.
2443	Saint Simon's Island, Georgia	......	4.25	2.62	Middle-aged.
2437	......do	......	4.38	3.03	Very old.
2447	......do	......	4.06	2.65	Do.
2450	......do	......	4.30	2.87	Do.
2444	......do	......	4.12	......	Do.
2446	......do	......	4.12	2.90	Do.
2202	......do	......	4.57	3.00	Do.
9649	Nebraska	......	4.50	3.07	Do.
8085	Fort Cobb, Indian Ter	......	4.50	3.03	Do.
3325	Independence, Mo	......	4.23	2.78	Young.
7739	Long Point, Tex	......	4.32	2.98	Old.
1386	Lower Rio Grande	......	4.52	2.90	Young.
1387	......do	......	4.70	3.15	Old.
3224	San Francisco, Cal	......	4.42	......	Very old; "*hernandezi.*"
3933	Sacramento, Cal	......	4.78	3.38	Do.
13312	California	......	4.70	3.12	Very old.
70[illegible]1	Mirador, Mexico	......	4.75	3.15	Middle-aged.
6119	Colima, Mexico	......	4.50	3.33	Old.
6481	......do	......	4.46	3.15	Middle-aged.
9706	Tehuantepec, Mexico	......	4.50	2.83	Young.
13853	......do	......	4.52	......	Old.
13854	......do	......	4.73	3.42	Very old.
13300	Costa Rica	......	4.55	3.03	Middle-aged.
14190	......do	......	4.78	3.32	Old.
14191	......do	......	4.85	3.00	Do.

PUTORIUS VISON.

Eighteen skulls from the northern parts of the continent, mainly from Alaska, average 2.66 in length and 1.58 in width, the extremes being, length, 3.02 and 2.30; width, 1.90 and 1.40. Thirteen skulls from the highlands of Northeastern New York average 2.40 in length and 1.34 in width, the extremes being, length, 2.60 and 2.17. Three skulls from Pennsylvania (undoubtedly males) average 2.49 in length and 1.48 in width. In the northern series, the sex of the skull is given by the collector, whence it appears that the twelve males have an average length of 2.81, and the six females an average length of 2.48, showing a considerable sexual variation in size. Yet the smallest males (2.64 and 2.63) fall below the largest female (2.68), if the skulls are all correctly marked. None of the other females, however, exceed 2.55, and only three of the males fall below 2.70. In the New York series, the sex is not indicated; but, judging from the proportion of the small to the large skulls, the sexes are about equally represented in the two series, but in the New York series there is a very gradual decline from the largest to the smallest. The northern series of eighteen is selected from a series of twenty-three; the New York series of thirteen from a series of thirty. In each case only very old skulls were chosen, the immature specimens in each case being thrown out in order to have a fair basis for comparison. The immature and middle-aged specimens greatly predominate in the New York series, owing, doubtless, to the species being more closely hunted there than in the more unsettled districts of the far north.

Taking these two series as a basis for a general comparison, there is indicated a considerable decrease in size from the north southward, amounting to 0.26 in length and 0.24 in width, or about one-tenth of the average size of the New York series. A single specimen, marked "Brookhaven, Miss.", and another marked "Tuscaloosa, Ala.", however, have a length respectively of 2.60 and 2.80, the former equaling the largest New York specimens, and the latter nearly equaling the average size of the males of the northern series, while a single male skull from Fort Randall, D. T., 2.90 in length, is the second in size of the whole series; one Fort Yukon specimen only being larger! Other specimens from the Upper Missouri region, however, are much smaller, as are other specimens from Prairie Mer Rouge, La., indicating that the specimens above mentioned are much above the average for their respective localities.

Measurements of thirty-seven skulls of PUTORIUS VISON.

Catalogue-number.	Locality.	Sex.	Length.	Width.	Remarks.
6330	Fort Yukon, Alaska	♂	3.02	1.90	Very old.
8709	Alaska	♂	2.82	1.64	Do.
8797	......do	♂	2.83	1.62	Do.
8796	......do	♂	2.75	1.61	Do.
8707	......do	♂	2.73	1.62	Do.
8703	......do	♂	2.75	1.57	Do.
8702	......do	♂	2.68	1.62	Do.
8798	......do	♂	2.64	1.55	Do.
8648	Alaska (Kadiak)	♂	2.63	1.52	Do.
8708	Alaska	♀	2.68	1.58	Do.
6531	......do	♀	2.55	1.50	Do.
8704	......do	♀	2.45	1.45	Do.
8706	......do	♀	2.32	1.40	Do.
8705	......do	♀	2.30	1.40	Do.
3284	Nelson River	♂	2.86	1.62	Do.
4309	Fort Simpson	♂	2.70	1.51	Do.
8132	......do	♂	2.90	1.78	Do.
4305	......do	♀	2.55	1.46	Do.
12915	Fort Randall	......	2.90	1.61	Do.
3730	Essex County, New York	......	2.60	1.48	Old.
3824	......do	......	2.60	1.38	Do.
1169	......do	......	2.40	1.32	Old. *P.* "*nigrescens*" A. & B.
3085	......do	......	2.40	1.38	Do.
3084	......do	......	2.40	1.31	Do.
3823	......do	......	2.32	1.32	Old.
3822	......do	......	2.30	1.23	Do.
2242	Saranac Lake, New York	......	2.47	1.37	Do.
2243	......do	......	2.40	1.30	Do.
2241	......do	......	2.35	1.31	Do.
2244	......do	......	2.20	1.18	Do.
2250	......do	......	2.40	1.48	Do.
2267	......do	......	2.17	1.20	Do.
1847	Pennsylvania	......	2.50	1.48	Do.
4834	......do	......	2.50	1.48	Do.
4835	......do	......	2.47	1.48	Do.
1894	Tuscaloosa, Ala	......	2.80	1.61	Do.
11315	Brookhaven, Miss	......	2.60	1.50	Do.

MUSTELA AMERICANA.

The forty-six male skulls of this species, of which measurements are given below, are mainly from four or five localities differing widely in latitude. A comparison of the average size of a considerable number from each shows a well-marked decrease in size southward. Four skulls from Peel River, the largest, and also from the most northerly locality, have an average length of 3.39, and an average width of 2.07, the extremes being 3.50 and 3.35 in length and 2.12 and 2.02 in width. Nine skulls from the Yukon (probably mostly from near Fort Yukon) give an average length of 3.34 and an average width of 1.98, the extremes being 3.55 and 3.00 in length and 2.15 and 1.73 in width. Five skulls from Fort Good Hope give an average length of 3.24 and an average width of 1.95, the extremes in length being 3.37 and 3.15 and in width 2.05 and 1.73. Ten skulls from the northern shore of Lake Superior average 3.14 in length and 1.76 in width, the extremes in length being 2.23 and 3.02 and in width 1.89 and 165. Eight skulls from the vicinity of Umbagog Lake, Maine (Coll. Mus. Comp. Zoöl.), average 2.96 in length and 1.72 in width, the extremes in length being 3.10 and 2.73, and in width 1.85 and 1.50. Five skulls from Northeastern New York average 3.02 in length and 1.61 in width, the extremes being in length 3.10 and 2.92 and in width 1.68 and 1.50. There is thus a gradual descent in the average length from 3.39 to 3.02, and in width from 2.07 to 1.61. The largest and the smallest of the series are respectively 3.55 and 2.92 in length. Several fall as low as 3.00, and an equal number attain 3.50. The difference between the largest and the smallest, excluding the most extreme examples, is one-sixth of the dimensions of the smaller and one-seventh of the size of the larger.

The sexes differ considerably in size, relatively about the same as in *Putorius vison*; but the above generalizations are based wholly on males, and in each case on those of practically the same age, only specimens indicating mature or advanced age being used.

The series of fully one hundred skulls of this species contained in the National Museum presents a considerable range of variation in details of structure, involving the general form of the skull, the relative size of different parts, and the dentition, especially the form and relative size of the last molar. In a former paper,* I had occasion to notice somewhat in detail the variations in color our American Martens present, and the difficulty of finding any features of coloration that seemed to indicate more than a single American species, or that would serve to distinguish this even from the Martens of the Old World. Dr. J. E. Gray, it is true, had already called attention to the small size of the last molar in the American Martens as compared with the size of the same tooth in the Old World Martens; but, as his observation was apparently based on a single American skull, and as I was at the time strongly impressed with the wide range of individual variation I had found in allied groups, even in dental characters, and also with the great frequency of Dr. Gray's characters failing to be distinctive, I was misled into supposing all the Martens might belong to a single circumpolar species, with several more or less strongly-marked geographical races. My friend Dr. Coues some months since kindly called my attention to the validity of Dr. Gray's alleged difference in respect to the size and form of the last molar, which I have since had opportunity of testing. This character alone, however, fails to distinguish *Mustela foina* from *Mustela americana*, in which the last molar is alike, or so nearly so that it fails to furnish distinctive differences. The size and general form of the skull in the two are also the same, the shape of the skull and the form of the last upper molar failing to be diagnostic. The second lower true molar, however, in *Mustela foina* presents a character (shared by all the Old World Martens) which serves to distinguish it from *Mustela americana*, namely, the presence of an inner cusp not found in the latter. In *Mustela flavigula*, the last molar is relatively *smaller* than even in *Mustela americana*, and of the same form. *Mustela martes* differs in its more massive dentition and in the heavier structure of the skull, but especially in the large size of the last molar and the very great development of its inner portion. Hence, while the size and shape of the last upper molar serves to distinguish *Mustela martes* from *Mustela americana*, it fails as a valid distinction between *Mustela americana* and *Mustela flavigula* and *Mustela foina*. As already remarked, however, *Mustela americana* lacks the inner cusp of the second lower molar, which is present in the Old World Martens, or at least possesses it only in a very rudimentary condition.

* "Mammals of Massachusetts", Bull. Mus. Comp. Zoöl., vol. i, pp. 161–167, Oct., 1869

Measurements of forty-six skulls of MUSTELA AMERICANA.

Catalogue number.	Locality.	Sex.	Length.	Width.	Remarks.
6043	Yukon River	♂	3.55	2.15	
6049	......do	♂	3.50	1.85	
6085	......do	♂	3.45	1.83	
6047	......do	♂	3.37	1.82	
6044	......do	♂	3.30	1.85	
6051	......do	♂	3.00	1.73	
6048	......do	♂	3.28	......	Imperfect.
6046	......do	♂	3.28	1.82	
9099	Kenai, Alaska	♂	3.30	2.03	
7159	Fort Good Hope	♂	3.37	2.05	
7167	......do	♂	3.25	1.98	
7168	......do	♂	3.25	1.93	
7164	......do	♂	3.25	1.76	
7163	......do	♂	3.15	1.73	
6061	Peel River	♂	3.50	2.02	
6080	......do	♂	3.37	2.12	
6063	......do	♂	3.35	......	Imperfect.
6059	......do	♂	3.35	......	Do.
3285	Red River	♂	3.40	1.94	
4670	Lake Superior (north shore)	♂	3.23	1.75	
4668	......do	♂	3.18	1.65	
4664	......do	♂	3.15	1.65	
4663	......do	♂	3.16	1.65	
4666	......do	♂	3.15	1.87	
4675	......do	♂	3.15	1.83	
4674	......do	♂	3.15	1.85	
4667	......do	♂	3.10	1.89	
4672	......do	♂	3.12	1.65	
4681	......do	♂	3.02	1.83	
	Washington Territory	♂	3.23	1.90	
	......do	......	3.15	1.72	Rather young.
	......do	......	3.03	1.55	Do.
	......do	......	3.00	......	Do.
1668	Essex County, New York	♂	3.10	1.57	
1163	......do	♂	3.03	1.63	
3819	......do	♂	3.00	1.68	
3818	......do	♂	2.92	1.50	
2245	Saranac Lake, New York	♂	3.03	1.68	
541	Umbagog Lake, Maine	♂	3.10	1.85	
550	......do	♂	3.00	1.70	
542	......do	♂	3.00	1.72	
552	......do	♂	3.00	1.72	
553	......do	♂	3.00	1.78	
543	......do	♂	2.90	1.78	
545	......do	♂	2.92	1.68	
544	......do	♂	2.73	1.50	

TAXIDEA AMERICANA.

The subjoined measurements of eleven skulls of this species (embracing all at present available) show also a well-marked southward decrease in size. A fuller series would be more satisfactory, but would doubtless only confirm what is here indicated. Six of the specimens are from rather northern localities and five from rather southern localities, the region represented extending from the Upper Missouri southward to the Lower Rio Grande. The specimens composing the two series are of very nearly corresponding ages. The northern series (four from different points on the Upper Missouri, one from Iowa, and one from Oregon) average 5.00 in length and 3.18 in width, the extremes being, in length, 5.22 and 4.92 (4.76 if we include one rather young example), the width ranging from 3.50 to 2.97. The southern series (including two or three from the vicinity of Matamoras, Mexico, and one each from New Mexico and California) averages 4.62 in length and 2.92 in width, the extremes being, in length, 4.75 and 4.50, and in width, 3.07 and 2.80.

The skulls, and especially the molar teeth, in the American Badgers, vary considerably in different individuals, as long since pointed out by

Professor Baird.* Southern specimens differ from northern ones not only in being smaller, but somewhat in color, so that the *T. berlandieri* of Professor Baird may perhaps be entitled to subspecific rank (*T. americana* subsp. *berlandieri*), though the material at hand indicates that the two forms will be found to thoroughly intergrade. The chief differences in coloration consist in the more reddish-gray tint of the southern form, with a decided tendency to a continuous light dorsal stripe, instead of this stripe being restricted to the head.

Measurements of eleven skulls of TAXIDEA AMERICANA.

Catalogue-number.	Locality.	Sex.	Length.	Width.	Remarks.
11505	Upper Missouri	......	5.22	3.50	
1178	......do	......	5.12	3.12	
2148	......do	......	4.75	3.07	Rather young.
2078	Quisquaton, Iowa	......	5.06	......	Imperfect.
12908	Fort Randall, Dak	......	4.95	3.25	
2033	Upper Des Chutes, Oreg	......	4.92	2.97	
4196	Fort Crook, Cal	......	4.60	3.07	
3767	New Mexico	......	4.50	2.80	Rather young; *berlandieri.*
1390	Matamoras, Mexico	......	4.75	2.94	*berlandieri.*
	......do	......	4.66	2.85	Do.
4135	Texas	......	4.57	2.94	Do.

LUTRA CANADENSIS.

Specimens of this species from northern and southern localities do not differ materially in size; skulls from Newfoundland, Maine, Lake Superior, Washington, and Georgia agreeing very closely in dimensions. In a series of eighteen (mainly from northern localities), nine attain or exceed a length of 4.25, and three reach 4.50, while two only fall as low as 4.00. Seven specimens from the vicinity of Lake Umbagog, Maine, (in Mus. Comp. Zoöl.) average 4.28 in length and 2.93 in width; two of these reach 4.50 in length and two fall slightly below 4.00 (3.96 and 3.97). Two specimens from Washington, D. C., have a length respectively of 4.45 and 4.50; one specimen from Saint Simon's Island, Georgia, is nearly as large (4.32), while a Fort Cobb specimen has a length of 4.22. These four are the only ones from very southerly points. Four other specimens, from as many localities, range from 4.05 to 4.15; while three specimens from Newfoundland range from 4.03 to 4.25. While these specimens are too few to warrant positive conclusions as to geographical variations, they seem to point to a great constancy of size throughout a wide range of latitude.

*U. S. and Mex. Bound. Survey, Zoöl., p. 21.

Measurements of eighteen skulls of LUTRA CANADENSIS.

Catalogue-number.	Locality.	Sex.	Length.	Width.	Remarks.
501	Newfoundland	......	4.20	2.75	
498	......do.	......	4.03	2.53	
500	......do.	......	4.15	2.57	
490	......do.	......	4.35	2.90	
555	Umbagog Lake, Maine	......	4.40	3.00	
556	......do.	......	4.27	2.85	
557	......do.	......	4.50	2.90	
559	......do.	......	3.97	2.70	
558	......do.	......	3.96	2.70	
489	......do.	......	4.50	3.00	
4446	Lake Superior	......	4.15	2.85	
11839	Fort Berthold, Dak.	......	4.25	2.82	
2247	Saranac Lake, New York	......	4.05	2.57	
13671	Bayfield, Wis.	......	4.06	2.82	
8097	Fort Cobb, Indian Ter.	......	4.22	2.87	
	Washington, D. C	......	4.50	2.95	
433	......do.	......	4.45	......	Imperfect.
3142	Saint Simon's Island, Georgia	......	4.32	2.75	

MEPHITIS MEPHITICA.

The twenty-nine skulls of this species of which measurements are given below show a wide range of variation in size, and a decided decrease southward. The localities embrace such distant points as California and the Atlantic seaboard on the one hand, and Maine and Texas on the other; but, with one or two exceptions, the specimens from any single locality are unsatisfactorily few. The specimens range in length from 2.60 to 3.50, and in width from 1.60 to 2.25! Yet there is not a specimen included in the series that is not so old as to have all the cranial sutures obliterated. A portion of the difference is doubtless sexual, but the specimens, unfortunately, have not the sex indicated. Ten of the specimens may be considered as western, coming mainly from Utah and California; ten others are from Maine and Massachusetts, and one from Northeastern New York; three are from Pennsylvania; and of the remaining five, four are from Texas, and one from Louisiana. The western series of ten average 3.10 in length and 1.95 in width, ranging in length from 2.85 to 3.50 and in width from 1.70 to 2.25. The New England series of ten average 2.88 in length and 1.72 in width, ranging in length from 2.70 to 3.25 and in width from 1.53 to 1.85. The single New York specimen scarcely varies from the average of the New England series, while the Pennsylvania specimens fall a little below. The five southern specimens average 2.73 in length, or a little below the New England series, ranging in length from 2.60 to 290.*

It thus appears that the western specimens are decidedly the largest of all, and that the northern are somewhat larger than the southern, the specimens compared being of corresponding ages, though of unknown sex, but doubtless comparable in this respect also.

The difference in size amounts to above one-fourth the size of the largest specimen and above one-third the size of the smallest. Between the western and southern series, the average difference amounts to one-third of the average size of the larger series! The western series includes the so-called *Mephitis occidentalis* of Baird, based on California specimens, and whose chief difference is merely that of larger size; yet the four specimens from Ogden, Utah (Coll. Mus. Comp. Zoöl.), considerably

* The range in width is not fairly indicated, owing to two of the smaller specimens being imperfect.

excelled in size the three from California. The southern series represents the so-called *M. varians* of Gray and Baird.

The unsatisfactory character of the several species of North American Skunks of the *mephitica* group, and the wide range of color-variation among individuals from the same locality, I have previously had occasion to notice,* and a re-examination of the subject confirms the conclusions then announced, which, I am happy to find, have recently received the support of Dr. Coues, who has lately made a study of this group.† As Dr. Coues has remarked, and as the subjoined measurements show, few species of animals vary so much in size and in cranial characters as the present, independently even of sex and age. Some specimens are not only more than one-fourth larger than others, but "there is a corresponding range of variation in contour. Compared with an ordinary ratio of osteological variability," says Dr. Coues, "the discrepancies are almost on a par with those exhibited by the coloration of the animal when set over against the more constant markings of most animals." In view of this great degree of variability, however, Dr. Coues has ventured to describe a "new species" (*M. frontata*), based on a fossil skull from one of the bone-caves of Pennsylvania, as it seems to me, unadvisedly. The specimen, though that of a very aged individual, is scarcely larger (see subjoined table) than the average of specimens from the Eastern States, its chief difference from the average skull consisting in an abnormal tumidity of the frontal region, arising evidently from disease. It is a feature by no means confined to the present example, but is merely an extreme enlargement of the sinuses of the frontal region often seen in specimens of the existing animal, evidently resulting from disease. In No. 917 (Albany, N. Y.), No. 8099 (Fort Cobb, Ind. T.), No. 1878 (Calcasieu Pass, La.), and No. 1620 (Indianola, Tex.), the same tendency is strongly marked, which, in some of these specimens, had they attained equal age, must have resulted in a malformation nearly or quite as great as is seen in the fossil skull in question.

In this connection, I may add that a pretty careful examination of the fossil remains of *Carnivora*, collected by Professor Baird many years since from the bone-caves of Pennsylvania (of which this fossil skull of the Skunk forms a part), has failed to show any of them to be specifically different from the species now or recently living in the same region. Many of them are remains of individuals of large size, but not exceeding the dimensions of specimens of the recent animal from the same or contiguous regions. These remains include, among others, the following species:—*Lynx rufus*, *Urocyon virginianus*, *Mustela pennanti*, *Mustela americana*, *Putorius vison*, *Lutra canadensis*, *Mephitis mephitica* (other specimens than the "*frontata*" skull), *Procyon lotor*, *Ursus americanus*, etc.

---

* See Bull. Mus. Comp. Zoöl., vol. i, pp. 178–181, Oct., 1869.

† Bull. U. S. Geol. and Geog. Surv. of the Territories, vol. i, No. 1, pp. 7–15, 1875.

Measurements of twenty-nine skulls of MEPHITIS MEPHITICA.

Catalogue-number.	Locality.	Sex.	Length.	Width.	Remarks.
2617	Petaluma, Cal	......	3.30	2.07	
3271	......do	......	3.08	2.04	
2434	Port Townsend, Oreg	......	2.93	1.70	
4195	Fort Crook, Cal	......	2.85	......	
417	Ogden, Utah	......	3.12	1.87	
419	......do	......	3.50	2.25	Very old.
416	......do	......	3.10	1.90	
418	......do	......	2.98	1.85	
10008	Wyoming Territory	......	3.15	2.05	
3327	Fort Laramie	......	2.96	1.78	
575	Upton, Me	......	3.25	......	Very old.
580	......do	......	3.00	1.85	
577	......do	......	2.87	1.75	
574	......do	......	2.85	1.73	
583	Norway, Me	......	2.90	1.75	
578	......do	......	2.70	1.70	
569	......do	......	2.87	1.78	
567	Massachusetts	......	2.70	1.53	
568	......do	......	2.75	......	
576	......do	......	2.72	1.70	
3816	Essex County, New York	......	2.88	1.78	
2232	Bone-caves, Pennsylvania	......	2.90	......	Fossil; *M. frontata* Coues.
610	Carlisle, Pa	......	2.87	......	Imperfect.
4833	Chester County, Pennsylvania	......	2.60	1.65	
1620	Indianola, Tex	......	2.80	1.78	
1004	Eagle Pass, Tex	......	2.60	......	Imperfect.
1113	......do	......	2.68	1.60	
1395	Matamoras, Tex	......	2.90	1.90	
1878	Calcasieu, La	......	2.68	......	Imperfect.

URSUS ARCTOS, ET VARS.

In a series of seventeen rather aged skulls of *Ursus arctos* and its varieties (all but one of the specimens being American), the largest specimens are from California, the great metropolis of the "Grizzlies". Of the eight skulls from this State, five attain a length of 14.50 or more, three exceeding 15.00, and one reaching 15.60, while the smallest falls as low as 13.25. Of five specimens from different localities in the Rocky Mountains, three reach or exceed 14.40, the extremes being 14.75 and 13.25. Of three specimens from the Arctic coast, one has a length of 13.40, and the others respectively 12.40 and 12.35. A single specimen from Russia has a length of 13.75. These I regard as being all unquestionably conspecific, though perhaps referable to two or three subspecies. Whether strictly so or not, we have the fact of the culmination in size in the region where the Grizzlies are most abundant, namely, in California; these two facts, greatest abundance and largest size, seeming to indicate this region as presenting the most favorable conditions for the existence of these animals. The Rocky Mountain specimens average considerably smaller than the Californian; and though the species is pretty frequent here it is far less abundant than on the Pacific slope, especially in California and Oregon. The Franklin Bay specimens, representing the so-called "Barren Ground Bear", and indistinguishable from the true *arctos* of the Old World, are smaller even than the specimens from the Rocky Mountains.

Measurements of seventeen skulls of URSUS ARCTOS (*chiefly subsp.* HORRIBILIS).

Catalogue-number.	Locality.	Sex.	Length.	Width.	Remarks.
3837	Sacramento, Cal	......	15.60	9.05	Very old.
1218	Monterey, Cal	......	15.75	......	Do.
7401	......do	......	14.05	7.75	
3630	......do	......	16.00	8.50	Very old.
6905	...A.do	♂	15.40	8.10	Do.
3538	Fort Tejon, Cal	♀	13.25	7.45	
3537	......do	♂	14.75	8.90	
3536	......do	♀	14.50	9.20	
2086	Los Nogales, Sonora	......	14.40	8.00	Var. "*horriæus*" Baird.
990	Copperminea, N. Mex	♂	14.50	8.25	Do.
3818	Medicine Bow Mountains (eastern slope)	......	14.75	8.50	
13245	Big Porcupine Creek, Mont	♀	13.25	7.40	
14785	Nebraska	......	13.45	6.90	
7146	Franklin Bay, Arctic Sea	♂	12.35	7.30	"Barren Ground Bear."
6551	......do	......	13.40	8.65	Do.
6548	......do	......	12.45	7.25	Do.
4441	Russia	......	13.75	7.53	

The question of the relationship of the large Bears of North America to those of the Old World has long been a vexed one, and is, of course, one not easily settled. In the present collection are thirty-three skulls, representing various ages, but the greater part are adult. These include two only from the Old World, six from the Arctic coast, eleven from California, and fourteen from various localities in the Rocky Mountains, from Idaho Territory to Arizona.

Among the American specimens are two rather easily distinguishable forms, one of which is the large Grizzly, or *U. horribilis* of authors, from the western parts of the United States; the other, the smaller so-called Barren Ground Bear of Arctic America; both being undoubtedly specifically distinct from the *Ursus americanus*. The Barren Ground form* differs from the more southern Grizzly not only in its smaller size, but in its strong tendency to a depression of the frontal region of the skull, where the simple flattening of this region in the Grizzly is here often carried so far as to form a well-marked concavity as in the true *arctos* of the Old World. Sometimes, however, *U. horribilis* also presents a considerable depression between the postorbital processes, as great even as in average specimens of *U. arctos*, as is the case in No. 7401 from Monterey, Cal. The Barren Ground Bear's skull generally presents a more dog-like aspect, in consequence of the thickening superiorly of the postorbital border of the frontals, than is seen in *U. horribilis*, it approaching in this respect to the form seen in *Ursus spelæus*, where this feature attains its highest development, resulting in the very strong frontal depression so characteristic of the skulls of that species.

The dentition of *U. arctos*, *U. richardsoni*,† and *U. horribilis* presents no important differences, the chief difference being the relatively rather smaller size of the teeth in the latter. The form of the last upper molar is almost precisely the same in the two first named, and the differences presented by *U. horribilis* are both slight and inconstant. In *U. richardsoni*, this tooth narrows gradually, and about equally, on each side posteriorly, almost exactly as in *U. arctos*, it being widest at or near its extreme anterior border. While this is sometimes the case in *U. horribilis*, its greatest breadth is generally one-fifth the length of the tooth behind the anterior border, and the tooth is relatively broader posteriorly

* Named by Captain Mayne Reid, in one of his stories, "*Ursus Richardsoni*"!
† The Barren Ground Bear skulls in the collection are labeled with this name.

than in the others. Specimens of *U. horribilis*, from the same locality, however, differ more among themselves in this respect than the average difference between *U. horribilis* and *U. arctos.* The teeth, however, in *U. arctos* are relatively larger than in *U. horribilis*, the difference being quite appreciable. The teeth of the Franklin Bay specimens (*U. richardsoni*), on the other hand, are of the same relative size as in the Old World examples of *U. arctos.*

After a careful consideration of the subject, I believe the Barren Ground Bear of Richardson (*U. richardsoni* of Mayne Reid) to be not even subspecifically distinct from the true *U. arctos* of the Old World. The Grizzly, from its larger size, widely different geographical distribution, apparently larger claws, slight differences in the dentition and in the form of the frontal region of the skull, may be so regarded (*U. arctos* subsp. *horribilis*), as it can hardly be doubted that it gradually passes into the Barren Ground form.

The subjoined table of detailed measurements of the skulls of *U. arctos horribilis* indicates the wide range of individual variation that may be looked for among skulls from the same locality. These variations not only affect the ratio of width to length, through the greater or less elongation of the facial portions of the skull as compared to the rest, but also all the other proportions are more or less variable, including even the teeth themselves. Thus, two specimens from California, of practically the same length (15.60 and 15.40), vary in breadth from 8.10 to 9.05, while two others vary still more, one, with a breadth of 9.20, having a length of only 14.50, while another, with a breadth of 8.50, has a length of 16.00! In these last, the ratio of width to length varies from 0.53 to 0.63. In two California specimens of practically the same length (15.60 and 15.75), the length of the last molar varies from 1.43 to 1.58. In the series of California specimens alone, the length of the last molar varies from 1.35 to 1.66, and the width of the same from 0.67 to 0.80, the widest tooth being, furthermore, *not* the longest. As already stated, the last upper molar attains its greatest width near the anterior border, but in several specimens the width of the anterior third is nowhere greater than the width of the tooth at its middle; and the same is also sometimes true in *U. richardsoni.*

***Measurements of seventeen skulls of* URSUS ARCTOS (*belonging chiefly to subsp.* HORRIBILIS).**

Catalogue number.	Original number.	Locality.	Sex.	Total length.	End of intermaxillaries to occipital condyles.	Greatest width.	Distance between orbits.	Expanse of orbital processes.	Nasal bones, length.	Nasal bones, width before.	Least width of muzzle behind the canines.	Upper incisors from front to molars.	Upper incisors from front to hinder margin of palate.	Distance between external edges of canines.	Upper molars, length taken together.	Upper molars, distance between.	Lower jaw, length.	Lower jaw, height.	Anterior end of intermaxillaries to anterior end of nasals.	Anterior end of intermaxillaries to the orbit.	Anterior end of intermaxillaries to postorbital processes.	Anterior end of nasals to postorbital processes.	Length of crown of last molar.	Nature of specimen.
3837	....	Sacramento, Cal	....	15.60	14.45	9.05	3.25	4.65	....	1.48	3.28	3.30	7.50	3.32	2.92	2.10	10.50	4.65	2.80	5.65	7.85	5.00	1.43	Very old.
1218	....	Monterey, Cal	....	15.75	14.50	....	3.40	5.20	....	1.45	3.05	3.52	7.85	3.65	3.15	1.92	10.55	4.70	3.17	6.05	8.05	5.15	1.58	Do.
7401	....	......do	....	14.05	13.25	7.75	3.00	4.15	3.70	1.35	2.70	3.00	7.40	2.90	3.00	1.75	9.60	4.25	2.90	5.35	7.35	4.65	1.52	
3630	....	......do	....	16.00	14.75	8.50	3.25	4.70	....	1.58	3.27	3.48	8.15	3.50	3.26	1.95	10.90	4.78	3.20	6.20	8.52	5.50	1.57	Very old.
6905	....	......do	♂	15.40	14.20	8.10	3.25	4.58	4.05	1.50	3.25	3.40	7.40	3.40	3.08	2.13	10.45	4.43	2.90	5.75	7.75	5.15	1.53	
3538	1322/134	Fort Tejon, Cal	♀	13.25	12.40	7.45	2.97	4.40	3.70	1.32	2.75	2.87	6.60	2.87	2.75	1.92	.....	....	2.57	5.05	7.25	4.72	1.35	
3537	1333/1332	......do	♂	14.75	13.[illegible]5	8.90	2.93	4.65	3.78	1.45	3.00	3.22	7.40	3.37	3.08	1.87	10.15	4.40	2.90	5.50	7.75	4.90	1.55	
3536	1985/763	......do	♀	14.50	13.65	9.20	3.20	5.00	4.12	1.52	3.12	3.15	7.75	3.60	3.10	2.10	10.00	4.60	2.93	5.60	7.65	4.80	1.50	
2086	....	Los Nogales, Sonora	....	14.40	13.65	8.00	2.95	4.50	3.35	1.48	2.77	3.00	6.80	2.95	2.87	1.85	9.45	4.07	3.15	5.25	7.45	4.38	1.42	Variety "*horriæus*" Baird.
990	....	Copper Mines, N. Mex	♂	14.50	13.05	8.25	3.15	4.72	....	1.45	2.76	3.12	6.90	3.15	2.83	1.80	9.45	4.30	3.10	5.65	7.60	4.67	1.37	"*horriæus*" (type).
3818	....	Medicine Bow Mts., east slope	....	14.75	13.55	8.50	3.10	4.45	....	1.35	2.90	3.10	7.15	3.00	2.63	1.73	10.65	4.45	2.94	5.50	7.45	4.58	1.35	
13245	....	Big Porcupine Creek, Mont	♀	13.25	12.60	7.40	2.75	3.75	3.55	1.33	2.75	2.85	6.40	2.80	2.85	1.80	9.00	3.85	2.82	4.90	6.80	4.20	1.47	Very old.
14785	....	"Nebraska"	....	13.45	12.80	6.90	2.57	3.73	3.90	1.35	2.50	3.08	6.90	2.57	2.87	1.60	9.25	3.75	2.62	5.32	7.25	4.67	1.43	Middle-aged.
4441	....	Russia	....	13.75	13.20	7.53	2.78	4.20	3.50	1.33	2.83	2.98	6.85	2.97	3.00	1.80	9.40	3.75	3.08	5.62	7.50	4.62	1.47	Do.
7146	....	Franklin Bay, Arctic Sea	♂	12.35	11.85	7.30	2.88	4.00	3.32	1.16	2.80	2.60	6.25	2.70	2.78	1.82	8.43	3.65	2.20	4.53	6.25	4.23	1.40	Rather young.
6551	....	......do	....	13.40	12.70	8.65	3.08	4.50	3.40	1.38	3.05	2.80	6.70	3.08	2.90	2.25	9.25	4.12	2.90	4.87	6.65	3.90	1.33	Middle-aged.
6548	....	......do	....	12.45	.....	7.25	2.75	4.05	3.30	1.20	2.55	2.72	6.05	2.55	2.72	1.82	8.20	3.60	2.63	4.65	6.42	3.93	1.37	Do.

Measurements of the molar teeth of URSUS ARCTOS *et var.*

Catalogue number.	Locality.	Sex.	Upper first molar.		Upper second molar.		Upper third molar.		Remarks.
			Length.	Width.	Length.	Width.	Length.	Width.	
13945	Big Porcupine Creek, Mont.	......	0.60	0.52	0.91	0.63	1.40	0.65	Subsp. *horribilis.*
3318	Medicine Bow Mountains, Wash.	......	0.57	0.48	0.74	0.60	1.36	0.67	Do.
990	Copper Mines, N. Mex.	......	0.60	0.54	0.92	0.70	1.35	0.70	Do.
7401	Monterey, Cal.	......	0.67	0.52	0.91	0.65	1.50	0.74	Do.
3630	......do.	......	0.72	0.65	0.97	0.75	1.66	0.75	Do.
6905	......do.	......	0.67	0.53	0.93	0.64	1.56	0.80	Do.
3537	Fort Tejon, Cal.	......	0.65	0.50	0.93	0.69	1.52	0.75	Do.
3536	......do.	......	0.65	0.51	0.93	0.67	1.43	0.72	Do.
3538	......do.	......	0.55	0.45	0.87	0.66	1.35	0.67	Do.
6557	Arctic coast.	......	0.66	0.51	0.95	0.70	1.37	0.70	"*richardsoni.*"
6548	......do.	......	0.57	0.45	0.88	0.64	1.41	0.68	Do.
7146	......do.	......	0.62	0.48	0.92	0.65	1.40	0.71	Do.
4441	Russia	......	0.68	0.57	0.94	0.72	1.40	0.75	*arctos.*
1033	Northern Sweden	......	0.63	0.40	0.90	0.65	1.27	0.67	Do.

URSUS AMERICANUS.

Seventeen skulls of this species, embracing all the aged ones in the collection, seem to indicate a slight increase in size to the southward. Four aged skulls from Louisiana and Florida range in length from 12.50 to 13.10, and three others, more or less immature, would doubtless have attained an equal size had they lived to be as old. A Georgia specimen, also not full-grown, has a length of 11.15, and in old age would probably have considerably exceeded 12.00. The other specimens, all full-grown and some of them very old, range from 9.90 to 12.15, most of them falling between 10.25 and 11.75. The largest (12.15) is from Puget Sound. A New York specimen comes next in size (11.90); New Mexican specimens next, the Alaskan being the smallest. This certainly points to a southward increase in size; but a much larger series would, of course, be necessary in order to establish positively whether the increase is in this direction. It would seem natural to expect it to be so, since the Bear is a hibernating animal, and is active for a much shorter period in northern than in southern localities.

It seems worthy of remark that only a small proportion of the skulls of Bears, and even of other *Carnivora*, including the Minks, Otters, and Martens, seen in collections, are specimens of mature age. The proportion of fully adult and very aged specimens is much greater among those from the unsettled parts of the continent than among those from the older States, owing, doubtless, to these animals being so closely hunted in the more settled districts that they rarely live to a very great age.

Measurements of seventeen skulls of URSUS AMERICANUS.

Catalogue number.	Locality.	Sex.	Length.	Width.	Remarks.
3834	Key Biscayne, Fla	----	13.10	......	Very old.
1155	Prairie Mer Rouge, La	----	12.90	7.40	Do.
1156	......do	----	12.70	7.45	Do.
987	......do	----	12.50	7.35	Do.
1154	......do	----	11.10	6.10	Middle-aged.
988	......do	----	10.60	5.95	Rather young.
3894	Georgia	----	11.15	6.10	Middle-aged.
3798	New York	----	11.80	7.35	
2250	......do	----	11.00	7.55	
994	Copper Mines, N. Mex	----	9.90	6.07	Very old; brown.
992	......do	----	11.35	7.05	Do.
991	......do	----	11.75	6.85	Very old; black.
12398	Henry's Lake, Wyo	♂	11.40	7.40	
3650	Puget Sound	----	12.15	7.40	
6949	......do	----	10.20	6.00	
8695	Alaska	----	10.25	6.30	
9477	......do	----	10.07	5.15	

The range of variation not dependent upon locality is more fully indicated in the table of detailed measurements of these skulls given below, but certain of the most prominent points of variation are not well shown by any series of measurements. Especially is this the case in respect to the amount of convexity different specimens present, in which individual variation is strongly marked. One of the most prominent distinctions of *U. americanus* as compared with *U. arctos* and its varieties is the great convexity of the upper outline of the skull, both antero-posteriorly and transversely. Another feature is the constriction of the facial portion, giving a concave outline to the nasals when viewed in profile. But there are exceptions, even to the first of these distinctions, one or two specimens occurring (especially No. 2250 from New York) in which the flattening of the frontal region is as marked as in average skulls of *U. horribilis*. This flattening is also well marked in Nos. 1155 and 1156, from Louisiana. The greatest convexity is reached in No. 3484, from Key Biscayne, Fla.; this and No. 2250 (New York) presenting the two extremes in respect to convexity. No. 3894, from Georgia, has about the same degree of convexity as the Florida specimen. No. 2250 is also remarkable for the shortness of the facial portion of the skull, thereby imparting to it a greater than the usual ratio of width to length. In this specimen (mentioned by Professor Baird as remarkable for its width*), the width is 0.69 of the length. In another, from Louisiana (No. 1155), it falls as low as 0.54! The average ratio of width to length is about 0.56 to 0.60.

The teeth of *U. americanus* seem, in looking at them, to be relatively much smaller than in *U. arctos*, but, upon careful measurement, the difference is quite small, while they are of the same relative size as those of *U. horribilis*. In *U. americanus*, the temporal ridges pass more abruptly inward toward the medial line of the skull than in either *U. horribilis* or *U. arctos*.

The most important distinction presented by *U. americanus* is the form of the last upper molar. In *U. americanus*, the crown is widest at the middle, narrowing both anteriorly and posteriorly, but most rapidly posteriorly. The inner border is nearly straight; the outer has a prominent medial convexity, while in *U. horribilis* and *U. arctos* both outlines are nearly straight and generally about equally convergent. In *U. americanus*, the anterior third of the last molar is generally narrower

*Mam. N. Amer., p. 227.

than the middle third, though sometimes equaling it; but it is never wider, as it almost invariably is in *U. horribilis* and *U. arctos.* The Puget Sound specimens have the anterior third the narrowest; in Alaskan specimens, it reaches its extreme width, while New York and Louisiana examples present the medium phase.

The skulls of *U. cinnamomeus* do not seem to be in any way distinguishable from average skulls of *U. americanus*, the distinction between them being one of color only and inconstant as characterizing any particular locality or region.

The upper molar teeth of *U. americanus*, as shown by the subjoined measurements, differ considerably in size in fully adult specimens. The first molars range in length of crown from 0.40 to 0.52, and in the width of the same from 0.27 to 0.42. The second ranges in length from 0.67 to 0.78; the third from 0.94 to 1.22, and in width from 0.51 to 0.67! In two specimens, with the first 0.44 in length, the third in one has a length of only 0.94 and the other 1.07! In another, the length of the first molar is 0.41 and the third 1.11. In still another, with the length of the first molar 0.43, the length of the third is 0.96. In two others, while the length of the first molar is 0.50 in each, the third molar in one has a length of 1.22 and in the other 1.15.

The largest skulls of *U. americanus* nearly equal in size the smaller skulls of *U. arctos horribilis*, and actually overlap the series from Franklin Bay and the measurements given by authors of the true *arctos* of the Old World. In view of this fact, and of the great range of individual variation in size, cranial and dental characters, and the unreliability of color as a specific character, I too hastily, in former papers,* referred all the American land-bears, including the *U. americanus*, to the *U. arctos*, which I am now convinced was a mistake; *U. americanus* being, I now believe, unquestionably specifically distinct, and the Grizzly subspecifically separable from the *U. arctos* of the Old World.

* Bulletin Mus. Comp. Zoöl., vol. i, pp. 184-192, Oct., 1869; Bulletin Essex Institute, vol. vi, pp. 46, 54, 59, 63, 1874.

Measurements of seventeen skulls of URSUS AMERICANUS.

Catalogue-number.	Locality.	Sex.	Total length.	Anterior end of intermaxillaries to occipital condyles.	Greatest width.	Distance between orbits.	Expanse of postorbital processes.	Nasal bones, length.	Nasal bones, width before.	Least width of muzzle behind the canines.	Upper incisors, from front to molars.	Upper incisors, from front to hinder margin of palate.	Distance between external edges of canines.	Upper molars, length taken together.	Upper molars, distance between.	Lower jaw, length.	Lower jaw, height.	Anterior end of intermaxillaries to anterior end of nasals.	Anterior end of intermaxillaries to the orbit.	Anterior end of intermaxillaries to postorbital processes.	Anterior end of nasals to postorbital processes.	Length of crown of last molar.	Remarks.
3650	Puget Sound	....	12. 15	11. 35	7. 40	2. 80	4. 00	3. 00	1. 10	2. 32	2. 85	5. 82	2. 53	2. 14	1. 64	8. 03	3. 18	2. 55	4. 52	6. 15	3. 80	1. 08	Very old.
6949	......do	....	10. 20	9. 80	6. 00	2. 48	3. 45	2. 54	1. 12	2. 18	2. 35	4. 93	2. 15	2. 07	1. 35	7. 10	2. 74	2. 22	3. 75	5. 32	3. 25	1. 00	Do.
8695	Alaska	....	......	9. 75	6. 30	2. 50	3. 60	2. 70	1. 10	2. 08	2. 43	5. 30	2. 15	2. 00	1. 25	7. 15	3. 03	2. 07	4. 00	5. 50	3. 47	0. 98	Do.
9477	......do	....	10. 07	9. 25	5. 15	2. 33	3. 07	2. 40	0. 90	1. 95	2. 33	4. 85	1. 95	2. 00	1. 37	6. 57	2. 50	1. 98	3. 68	5. 00	3. 10	0. 93	
3798	New York	....	11. 80	11. 10	7. 35	2. 57	3. 80	3. 13	1. 15	2. 35	2. 74	5. 53	2. 63	2. 22	1. 55	7. 50	3. 30	2. 60	4. 35	6. 00	3. 70	1. 00	
2250	......do	....	11. 00	10. 50	7. 55	2. 77	4. 12	2. 55	1. 18	2. 25	2. 50	5. 33	2. 43	1. 85	1. 75	7. 50	3. 33	2. 12	3. 78	5. 34	3. 35	0. 90	
3484	Key Biscayne, Fla	....	13. 10	11. 75	.....	2. 70	4. 25	.....	1. 20	2. 53	2. 75	5. 85	2. 62	2. 22	1. 72	8. 00	4. 03	2. 80	4. 57	6. 35	3. 85	1. 00	
1155	Prairie Mer Rouge, La	....	12. 90	12. 10	7. 40	2. 65	3. 80	2. 55	1. 02	2. 68	2. 57	6. 20	2. 78	2. 48	1. 66	8. 80	3. 73	2. 70	4. 57	6. 20	3. 50	1. 21	
1156	......do	....	12. 70	11. 60	7. 45	2. 65	4. 00	2. 90	1. 19	2. 37	2. 50	5. 85	2. 70	2. 38	1. 78	8. 05	3. 75	2. 80	4. 30	5. 90	3. 35	1. 18	
987	......do	....	12. 50	11. 75	7. 35	2. 85	3. 80	3. 12	1. 03	2. 50	2. 70	6. 10	2. 65	2. 36	1. 62	8. 25	3. 70	2. 30	4. 45	6. 20	4. 00	1. 15	
1154	......do	....	11. 10	10. 30	6. 10	2. 44	3. 35	2. 83	1. 10	2. 06	2. 30	5. 30	2. 07	2. 00	1. 35	7. 05	3. 15	2. 07	3. 90	5. 45	3. 50	0. 96	Middle-aged.
988	......do	....	10. 60	10. 15	5. 95	2. 32	3. 25	2. 73	1. 00	2. 23	2. 25	5. 45	2. 20	2. 25	1. 57	7. 20	3. 05	2. 03	3. 62	5. 20	3. 27	1. 11	Rather young.
3894	Georgia	....	11. 15	10. 50	6. 10	2. 35	3. 27	2. 80	0. 96	2. 03	2. 46	5. 60	2. 10	2. 20	1. 35	7. 30	3. 15	2. 20	4. 05	5. 62	3. 50	1. 08	Middle-aged.
994	Copper Mines, N. Mex	....	9. 90	9. 20	6. 07	2. 25	3. 30	2. 80	1. 12	1. 95	2. 30	5. 00	2. 10	2. 03	1. 55	6. 75	2. 73	2. 00	3. 60	5. 25	3. 40	1. 03	Very old; brown.
992	......do	....	11. 35	10. 60	7. 05	2. 45	3. 60	2. 92	1. 08	2. 20	2. 55	5. 60	2. 57	2. 07	1. 54	7. 75	3. 00	2. 50	4. 26	6. 00	3. 50	1. 02	Do.
991	......do	....	11. 75	10. 60	6. 85	2. 56	3. 70	2. 87	1. 12	2. 30	2. 40	5. 75	2. 56	2. 05	1. 56	7. 75	3. 35	2. 30	4. 00	5. 65	3. 65	1. 07	Very old; black.
12398	Henry's Lake, Wyo	♂	11. 40	10. 60	7. 40	2. 55	3. 82	2. 75	1. 19	2. 35	2. 60	5. 70	2. 45	2. 00	1. 70	7. 80	3. 35	2. 30	3. 95	5. 50	3. 35	0. 95	

Measurements of the molar teeth of URSUS AMERICANUS.

Catalogue-number.	Locality.	Sex.	Upper first molar.		Upper second molar.		Upper third molar.	
			Length.	Width.	Length.	Width.	Length.	Width.
897	Saint Lawrence County, New York	......	0.50	0.42	0.78	0.60	1.22	0.64
3857	Pennsylvania	......	0.52	0.35	0.75	0.54	1.17	0.60
8161	Nulato, Alaska	......	0.50	0.33	0.76	0.54	1.15	0.64
9477	Alaska	......	0.45	0.38	0.72	0.48	1.00	0.52
8695	......do	......	0.44	0.27	0.67	0.46	0.94	0.51
1155	Prairie Mer Rouge, La	......	0.42	0.33	0.69	0.47	1.00	0.53
1156	......do	......	0.50	0.39	0.76	0.55	1.18	0.67
987	......do	......	0.50	0.38	0.76	0.58	1.15	0.65
1154	......do	......	0.43	0.35	0.70	0.52	0.96	0.57
988	......do	......	0.41	0.35	0.75	0.57	1.11	0.63
3894	Georgia	......	0.40	0.32	0.71	0.51	1.08	0.57
991	Copper Mines, N. Mex	......	0.44	0.33	0.68	0.50	1.07	0.57
994	......do	......	0.43	0.32	0.70	0.50	1.03	0.54

TABULAR SUMMARY.

Species.	Locality.	Number of specimens.	Sex.		Length.	Width.	Remarks.
Canis lupus	Forts Simpson, Yukon, and Rae.	9	....	Average	10.38	5.40	
Do	Forts Benton and Union	16	....	....do	9.45	5.07	
Do	Forts Kearney and Harker	9	....	....do	9.69	5.18	
Do	Rio Grande and Sonora	3	....	....do	8.37	4.31	
Do	Forts Simpson, Yukon, and Rae.	9	....	Maximum	11.50	6.20	
Do	Forts Benton and Union	16	....	....do	10.50	5.50	
Do	Forts Kearney and Harker	9	....	....do	10.15	5.50	
Do	Rio Grande and Sonora	3	....	....do	8.75	4.62	
Do	Forts Simpson, Yukon, and Rae.	9	....	Minimum	9.20	4.50	
Do	Forts Benton and Union	16	....	....do	8.75	4.50	
Do	Forts Kearney and Harker	3	....	....do	9.35	4.85	
Do	Rio Grande and Sonora	9	....	....do	7.75	4.05	
Canis latrans	Fort Randall	12	....	Average	7.23	3.61	
Do	Fort Kearney	10	....	....do	7.25	3.65	
Do	California	4	....	....do	7.69	3.75	
Do	Fort Randall	12	....	Maximum	7.60	3.80	
Do	Fort Kearney	10	....	....do	7.95	4.00	
Do	California	4	....	....do	7.60	3.80	
Do	Fort Randall	12	....	Minimum	6.65	3.38	
Do	Fort Kearney	10	....	....do	6.95	3.50	
Do	California	4	....	....do	7.45	3.48	
Vulpes alopex	Alaska	9	....	Average	5.98	3.20	Var. *fulvus*.
Do	Mackenzie River District	18	....	....do	5.80	3.02	Do.
Do	Upper Missouri	9	....	....do	5.78	2.90	"*macrurus*."
Do	Essex County, New York	12	....	....do	5.40	2.80	Var. *fulvus*.
Do	Europe	5	....	....do	5.58	3.08	Var. *alopex*.
Do	Alaska	9	....	Maximum	6.20	3.32	Var. *fulvus*.
Do	Mackenzie River District	18	....	....do	6.10	3.28	Do.
Do	Upper Missouri	9	....	....do	6.00	3.20	"*macrurus*."
Do	Esex County, New York	12	....	....do	5.68	2.95	Var. *fulvus*.
Do	Europe	5	....	....do	5.70	3.15	Var. *alopex*.
Do	Alaska	9	....	Minimum	5.70	2.90	Var. *fulvus*.
Do	Mackenzie River District	18	....	....do	5.55	2.87	Do.
Do	Upper Missouri	9	....	....do	5.40	2.78	"*macrurus*."
Do	Esex County, New York	12	....	....do	5.20	2.70	Var. *fulvus*.
Do	Europe	5	....	....do	5.50	3.04	Var. *alopex*.
Urocyon virginianus	Pennsylvania, Washington, and Virginia.	3	....	Average	4.97	2.64	Var. *virginianus*.
Do	Texas	2	....	....do	4.56	2.64	Do.
Do	Southern California	3	....	....do	4.56	2.54	Do.
Do	Tehuantepec, Mexico	3	....	....do	4.20	2.32	Do.
Do	Islands off California	3	....	....do	3.80	2.03	Var. *littoralis*.
Do	Merida, Yucatan	1	....		3.75	1.98	Var. *virginianus*.
Do	Pennsylvania, Washington, and Virginia.	3	....	Maximum	4.70	2.70	Do.
Do	Texas	2	....	....do	4.60	2.70	Do.

TABULAR SUMMARY—Continued.

Species.	Locality.	Number of specimens.	Sex.		Length.	Width.	Remarks.
Urocyon virginianus.	Southern California	3	....	Maximum.	4.63	2.65	Var. *virginianus*.
Do	Tehuantepec, Mexico	3	....	....do	4.40	2.37	Do.
Do	Islands off California	3	....	....do	3.85	2.23	Var. *littoralis*.
Do	Pennsylvania, Washington, and Virginia.	3	....	Minimum.	4.62	2.56	Var. *virginianus*.
Do	Texas	2	....	....do	4.50	2.58	Do.
Do	Southern California	3	....	....do	4.50	2.43	Do.
Do	Tehuantepec, Mexico	3	....	....do	4.15	2.25	Do.
Do	Islands off California	3	....	....do	3.75	2.05	Var. *littoralis*.
Felis concolor	New York and Oregon	3	....	Average	7.57	5.15	
Do	Texas and Louisiana	3	....	....do	8.72	5.48	
Do	New York and Oregon	3	....	Maximum.	7.80	5.25	
Do	Texas and Louisiana	3	....	....do	8.75	5.60	
Do	New York and Oregon	3	....	Minimum	7.40	5.05	
Do	Texas and Louisiana	3	....	....do	8.40	5.35	
Felis pardalis	Matamoras, Mexico	5	....	Average	4.93	3.33	
Do	Costa Rica	4	....	....do	5.89	3.86	
Do	Southern Mexico and Central America	8	....	....do	5.78	3.74	Includes the Costa Rican series.
Do	Matamoras, Mexico	5	....	Maximum.	5.25	3.50	
Do	Costa Rica	4	....	....do	6.20	4.19	
Do	Southern Mexico and Central America.	8	....	....do	6.20	4.19	
Do	Matamoras, Mex	5	....	Minimum	4.50	3.05	
Do	Costa Rica	4	....	....do	5.35	3.60	Includes the Costa Rican series.
Do	Southern Mexico and Central America.	8	....	....do	5.35	3.60	
Lynx rufus	British North America	9	....	Average	5.01	3.52	"*canadensis*."
Do	Washington and Oregon Territories.	7	....	....do	5.03	3.56	"*fasciatus*."
Do	Texas and Matamoras, Mex	8	....	....do	5.00	3.40	"*maculatus*."
Do	United States (mainly Ft. Tejon, Cal.)	10	....	....do	4.91	3.47	"*rufus*."
Do	British North America	9	....	Maximum	5.30	3.70	"*canadensis*."
Do	Washington and Oregon Territories.	7	....	....do	5.50	3.95	"*fasciatus*."
Do	Texas and Matamoras, Mex	8	....	....do	5.27	3.72	"*maculatus*."
Do	United States (mainly Ft. Tejon, Cal.)	10	....	....do	5.50	3.82	"*rufus*."
Procyon lotor	New York, Pennsylvania, and Georgia.	15	....	Average	4.27	2.91	
Do	Southern Texas and California.	6	....	....do	4.57	3.11	
Do	Southern Mexico and Costa Rica.	9	....	....do	4.60	3.15	
Do	New York, Pennsylvania, and Georgia.	15	....	Maximum.	4.57	3.03	
Do	Southern Texas and California.	6	....	....do	4.73	3.38	
Do	Southern Mexico and Costa Riea.	9	....	....do	4.85	3.42	
Putorius vison	Alaska (chiefly)	18	♂ ♀	Average	2.63	1.53	
Do	New York	13	♂ ♀	....do	2.40	1.34	
Do	Alaska (chiefly)	12	♂	....do	2.81	1.63	
Do	....do	6	♀	....do	2.42	1.46	
Do	....do	12	♂ ♀	Maximum.	3.02	1.90	Male.
Do	New York	6	♂ ♀	....do	2.60	1.48	Do.
Do	Alaska (chiefly)	2	♂ ♀	Minimum	2.30	1.40	Female.
Do	New York	4	♂ ♀	....do	2.17	1.13	Do.
Mustela americana	Peel River	4	♂	Average	3.39	2.07	
Do	Yukon River	9	♂	....do	3.34	1.98	
Do	Fort Good Hope	5	♂	....do	3.24	1.95	
Do	Lake Superior	10	♂	....do	3.14	1.76	
Do	Umbagog Lake, Maine	8	♂	....do	2.96	1.72	
Do	Northern New York	5	♂	....do	3.02	1.61	
Do	Peel River	4	♂	Maximum.	3.50	2.12	
Do	Yukon River	9	♂	....do	3.55	2.15	
Do	Fort Good Hope	5	♂	....do	3.37	2.05	
Do	Lake Superior	10	♂	....do	3.23	1.89	
Do	Umbagog Lake	8	♂	....do	3.10	1.85	
Do	Northern New York	5	♂	....do	3.10	1.68	
Do	Peel River	4	♂	Minimum	3.35	2.02	
Do	Yukon River	9	♂	....do	3.00	1.73	
Do	Fort Good Hope	5	♂	....do	3.15	1.73	
Do	Lake Superior	10	♂	....do	3.02	1.63	

TABULAR SUMMARY—Continued.

Species.	Locality.	Number of specimens.	Sex.		Length.	Width.	Remarks.
Mustela americana ..	Umbagog Lake............	8	♂	Minimum .	2.73	1.50	
Do...............	Northern New York	5	♂	....do	2.92	1.50	
Taxidea americana ..	Northern localities........	5	....	Average ..	5.00	3.18	
Do...............	Southern localities	5	....	....do	4.62	2.92	
Do...............	Northern localities	5	....	Maximum.	5.22	3.50	
Do...............	Southern localities	5	....	....do	4.75	3.07	
Do...............	Northern localities........	5	....	Minimum .	4.92	2.97	
Do...............	Southern localities	5	....	....do	4.50	2.80	
Lutra canadensis	Newfoundland and Umbagog Lake, Maine.	10	....	Average ..	4.24	2.79	
Do...............	Southern localities	4	....	....do	4.37	2.86	
Do...............	Newfoundland and Umbagog Lake, Maine.	10	....	Maximum.	4.50	3.00	
Do...............	Southern localities	4	....	....do	4.50	2.95	
Do...............	Newfoundland and Umbagog Lake, Maine.	10	....	Minimum .	3.96	2.53	
Do...............	Southern localities	4	....	....do	4.22	2.75	
Mephitis mephitica..	Western localities.........	10	....	Average ..	3.10	1.95	
Do...............	New England	10	....	....do	2.88	1.72	
Do...............	Southern localities	5	....	....do	2.73	......	
Do...............	Western localities.........	10	....	Maximum.	3.50	2.25	
Do...............	New England	10	....	....do	3.25	1.85	
Do...............	Southern localities	5	....	....do	2.90	......	
Do...............	Western localities.........	10	....	Minimum .	2.85	1.70	
Do...............	New England	10	....	....do	2.70	1.53	
Do...............	Southern localities	5	....	....do	2.60	......	
Ursus arctos.........	California	8	....	Average ..	14.81	8.42	Subsp. *horribilis*.
Do...............	Rocky Mountains..........	5	....	....do	12.07	7.81	Do.
Do...............	Arctic coast...............	3	....	....do	12.77	7.73	Subsp. *arctos*.
Do...............	California	8	....	Maximum.	16.00	9.20	Subsp. *horribilis*.
Do...............	Rocky Mountains	5	....	....do	14.75	8.50	Do.
Do...............	Arctic coast...............	3	....	....do	13.40	8.65	Subsp. *arctos*.
Do...............	California	8	....	Minimum .	13.25	7.45	Subsp. *horribilis*.
Do...............	Rocky Mountains	5	....	....do	13.25	6.90	Do.
Do...............	Arctic coast...............	3	....	....do	12.45	7.25	Subsp. *arctos*.
Ursus americanus ...	Georgia, Florida, and Louisiana.	7	....	Average ..	12.01	6.72	
Do...............	New York	2	....	....do	11.40	7.45	
Do...............	New Mexico	3	....	....do	11.00	6.66	
Do...............	Puget Sound and Alaska..	4	....	....do	10.67	6.21	
Do...............	Georgia, Florida, and Louisiana.	7	....	Maximum.	13.10	7.40	
Do...............	New York	2	....	....do	11.80	7.55	
Do...............	New Mexico	3	....	....do	11.75	7.05	
Do...............	Puget Sound and Alaska..	4	....	....do	12.15	7.40	
Do...............	Georgia, Florida, and Louisiana.	7	....	Minimum .	10.60	5.95	
Do...............	New Mexico	3	....	....do	9.90	6.07	
Do...............	Puget Sound and Alaska..	4	....	....do	10.07	5.15	

GEOGRAPHICAL DISTRIBUTION OF THE MAMMALIA

ART. XV.—THE GEOGRAPHICAL DISTRIBUTION OF THE MAMMALIA, CONSIDERED IN RELATION TO THE PRINCIPAL ONTOLOGICAL REGIONS OF THE EARTH, AND THE LAWS THAT GOVERN THE DISTRIBUTION OF ANIMAL LIFE.

By Joel Asaph Allen.

I.—DISTRIBUTION OF MAMMALIAN LIFE IN THE NORTHERN HEMISPHERE, CONSIDERED IN RELATION TO LAWS OF GEOGRAPHICAL DISTRIBUTION.

When, in 1871, I published* a few preliminary remarks concerning the general subject of geographical zoölogy, it was my intention soon to present more fully the facts whereon were based the few general principles then stated. In this paper I claimed, in accordance with the views of Humboldt, Wagner, Dana, Agassiz, De Candolle, and others, that life is distributed in circumpolar zones, which conform with the climatic zones, though not always with the parallels of the geographer. Subsequent study of the subject has confirmed the convictions then expressed. These are directly antagonistic to the scheme of division of the earth's surface into the life-regions proposed by Dr. Sclater in 1857, based on the distribution of birds, and since so generally adopted. Their wide acceptation, it seems to me, has resulted simply from the fact that so few have taken the trouble to sift the facts bearing upon the subject, or to carefully examine the basis on which Dr. Sclater's divisions are founded. The recent appearance of Mr. Wallace's laborious and in many respects excellent and praiseworthy work† has now rendered a critical presentation of the subject more necessary than before, since, instead of seeking in the facts of geographical zoölogy a basis for a natural scheme of division, he has unhesitatingly accepted Dr. Sclater's ontological regions and marshalled his facts and arranged his work wholly in conformity with this, as I shall presently attempt to show, grossly misleading scheme. The source of error, as I hope to make evident, lies in method of treatment. Assuming apparently that the larger or continental land-areas are necessarily coincident with natural ontological regions, divisions of the earth's surface wholly incompara-

---

* On the Geographical Distribution of the Birds of Eastern North America, with special reference to the Number and Circumscription of the Ornithological Faunæ. <Bull. Mus. Comp. Zoöl., vol. ii, No. 3, pp. 375–450. April, 1871.

† The Geographical Distribution of Animals. With a Study of Living and Extinct Faunas as Elucidating the Past Changes of the Earth's Surface. By Alfred Russel Wallace. Two vols. 8°. With maps and illustrations. London, 1876.

ble have been contrasted, and erroneous deductions have been the result. In the division of the northern hemisphere into two primary regions, the so-called "Nearctic" and "Palæarctic", no account has been taken of the almost homogeneous character of life throughout the Arctic and Sub-Arctic regions, and the equally important principle of temperature as a powerful limiting agent, nor of the facts of the rapid increase of organic forms and the consequent differentiation of life from the Arctic regions toward the Equatorial in an ever increasing ratio in proportion to the extent and divergence of the principal land-areas. At the northward, this method of division separates, into primary life-regions, areas of the closest ontological resemblances, while at the southward these divisions each embrace faunæ so unlike those of their northern portions respectively that the two extremes of either region have little in common, scarcely more than have the southern portions of these two regions as compared with each other. It is the neglect of the above-stated fundamental facts and principles that forms the fatal weakness of the scheme of life-regions proposed by Dr. Sclater, and so widely and thoughtlessly accepted. That the facts and principles above alluded to are fundamental,—in other words, that life is distributed in circumpolar zones under the controlling influence of climate and mainly of temperature,—I propose to show by a tabular presentation of the facts of distribution of mammalian life in the northern hemisphere.

One of the reasons given by Mr. Wallace for adopting Dr. Sclater's regions is that "it is a positive, and by no means an unimportant advantage to have our named regions approximately equal in size, and with easily defined, and therefore easily remembered, boundaries", providing that "we do not violate any clear affinities or produce any glaring irregularities". It is further claimed that "all elaborate definitions of interpenetrating frontiers, as well as regions extending over three fourths of the land surface of the globe, and including places which are the antipodes of each other, would be most inconvenient, even if there were not such difference of opinion about them".*

These arguments can be scarcely characterized as otherwise than trivial, since they imply that truth, at least to a certain degree, should be regarded as secondary to convenience. They further show that the author of these propositions has not worked out in detail the distribution of life, species by species, over a diversified area of considerable extent, like, for instance, that of Eastern North America, where an interdigitation of the lesser faunal areas is one of the marked features of the region, as it is elsewhere wherever there is a varied topography and consequent inequality of climate under the same parallels of latitude. Again, Mr. Wallace says,—"On two main points every system yet proposed, or that probably can be proposed, is open to objection; they are,—1stly, *that the several regions are not of equal rank;*—2ndly, that they are not equally applicable to all classes of animals. As to the first objection, it will be found impossible to form any three

*Geogr. Dist. Anim., vol. i, pp. 63, 64.

or more regions, each of which differs from the rest in an equal degree or in the same manner. One will surpass all others in the possession of peculiar families; another will have many characteristic genera; while a third will be mainly distinguished by negative characters. There will also be found many intermediate districts, which possess some of the characteristics of two well-marked regions, and a few special features of their own, or perhaps with none; and it will be a difficult question to decide in all cases which region should possess the doubtful territory, or whether it should be formed into a primary region by itself."*

In geographical zoölogy, as in the genetic relation of animals, we find, as a rule, no strongly marked boundary-lines, and in the life-regions, especially those of lesser rank, the boundaries can be given only approximately, owing to the intergradation of contiguous faunæ and floræ, contingent upon the gradual modification of climatic conditions; yet it is not hard to find boundary-lines that shall be, if not sharply definable, at least easy of recognition. This at least proves to be the case wherever the distribution of specific forms is thoroughly known. The first objection, "that the several regions are not of equal rank," forms to my mind no objection at all, since it matters little whether they are equal or unequal if they correctly indicate the distribution of life.

The second objection Mr. Wallace has himself satisfactorily answered, in discussing the question "*Which class of animals is of most importance in determining Zoölogical Regions.*" As Mr. Wallace here points out, and as must become apparent to every careful investigator of this question, the mammalia are pre-eminently of the greatest importance in determining zoölogical regions. To summarize Mr. Wallace's argument on this point,† their dispersal is less dependent on fortuitous circumstances than that of the representatives of other classes; from their high organization they are less dependent upon "other groups of animals", and have so much power of adaptation that they are "able to exist in one form or another over the whole globe", as is certainly not the case with two of the lower classes of vertebrates, the reptilia and amphibia. Their distribution and dispersal are dependent on the distribution of the land-areas, and are modified by such physical conditions as mountain barriers, areas of forest, and grassy or desert plateaus. Furthermore, their geological history, as well as their geographical range, is better known than that of most other classes, and there is also a greater unanimity of opinion respecting their natural affinities and the limitation of families and genera in this class than in most others. "We should therefore", says Mr. Wallace (and I heartily agree with the remark), "construct our typical or standard Zoölogical Regions in the first place, from a consideration of the distribution of mammalia, only bringing to our aid the distribution of other groups to determine doubtful points. Regions so established will be *most closely in accordance with*

* Geogr. Dist. Anim., vol. i, p. 53.

† See Geogr. Distr. Anim., vol. i, pp. 56–58.

*those long-enduring features of physical geography, on which the distribution of all forms of life fundamentally depends;** and all discrepancies in the distribution of other classes of animals must be capable of being explained, either by their exceptional means of dispersion or by special conditions affecting their perpetuation and increase in each locality." "If these considerations are well founded," he continues, "the objections of those who study insects or molluscs, for example,—that our regions are not true for their departments of nature—cannot be maintained. For they will find, that a careful consideration of the exceptional means of dispersal and conditions of existence of each group, will explain most of the divergences from the normal distribution of higher animals." †

In the present paper I shall consequently, in my discussion of the zoölogical regions of the northern hemisphere, confine myself primarily to mammals. Throwing aside, for the moment, all theoretical considerations, I shall endeavor first to present the facts of the case, and then consider what generalizations may be legitimately drawn from them.

A word, however, first in respect to the conformation and distribution of the land-areas. In reference to this part of the subject I can hardly do better than to again quote the words of Mr. Wallace, who has thus forcibly presented the subject:—"One great peculiarity of the distribution of land lies in its freedom from complete isolation . . . The continents, indeed, resembling as they do a huge creeping plant, with roots at the North Pole, and the matted stems and branches of which cover a large part of the northern hemisphere and send three great offshoots toward the South Pole, offer great facilities for the transmission of varied forms of animal life. There is evidence to prove that during the greater part of the Tertiary period the relative positions of our conti-

---

* The italicizing is my own.

† The question, Which class of animals is best fitted to form the basis of a division of the earth's surface into life-regions? has a wider bearing than might be at first supposed, since the same power of adaptation to diverse climatic conditions that results in a wide distribution in some cases and a limited range in others would also impart different degrees of ability to resist the influence of geological changes, and is hence related to the question, Which class forms the best index for marking geological time? The relative importance of different groups as geological indices is necessarily connected with their power to resist unfavorable influences, and hence groups that succumb most readily would give the best clue to such changes in the past. Among vertebrates the mammalia are undoubtedly, as a *class*, the best able to survive a wide range of climatic conditions. Birds are to so great a degree migratory that they are in great measure able to avoid seasonal extremes of climate by a change of habitat. Extremes that mammals readily survive prove quickly fatal to reptiles and amphibians.

Climate, though in itself a powerful geological agent, is, of course, subject to profound modification due to geological causes. Any great amount of upheaval or subsidence of the earth's crust, or the gradual uplifting of mountain chains, must necessarily induce changes in the climate of the regions where such disturbances occur, the effect of which must extend over an area far greater than that of the disturbed district. A comparatively slight change of climate, either in respect to temperature or humidity, has a most marked influence upon vegetation, and especially upon the distribution of forests. The presence or absence of particular species of plants is well known to determine the presence or absence of many species of insects, while the distribution of whole families of the latter is determined wholly by the character of the vegetation;

nents and oceans did not greatly differ from their present form, and the former, back to the time of the Devonian formation, were never so completely submerged as to be replaced by oceans comparable in depth with our Atlantic and Pacific."* "This curious fact," he says again, "of the almost perfect continuity of all the great masses of land, notwithstanding their extremely irregular shape and distribution, is no doubt dependent on the [geological] circumstances just alluded to; that the great depth of the oceans and the slowness of the process of upheaval, has almost always produced the new lands close to, or actually connected with, pre-existing lands; and this has necessarily led to a much greater uniformity in the distribution of organic forms, than would have prevailed had the continents been more completely isolated from each other. . . . the whole land is almost continuous. It consists essentially of only three masses: the American, the Asia-African, and the Australian. The two former are only separated by thirty-six miles of shallow sea at Behring's Straits, so that it is possible to go from Cape Horn to Singapore or the Cape of Good Hope without ever being out of sight of land; and owing to the intervention of the numerous islands of the Malay Archipelago the journey might be continued under the same conditions as far as Melbourne and Hobart Town."† The close proximity of the great land-masses in the Arctic regions is a fact to be kept in mind in any discussion of the distribution of life in the northern hemisphere, and also the fact that in Tertiary times the connection was almost indisputably more intimate than it is now.

---

and even mammals and birds are greatly affected, and even some are mainly controlled, in their range by the presence or absence of forests, the distribution of which is so intimately connected with climate. The reptiles, unlike mammals and birds, are quickly influenced by changes of temperature, and are unable to exist in the colder parts of the earth. Amphibians also require a moderately warm, or at least temperate, climate, and though ranging beyond the true reptiles become reduced to a few types in the cold-temperate latitudes, beyond which they wholly disappear. Fluviatile and terrestrial mollusks are also exceedingly susceptible to changes in the conditions of life that affect but slightly either insects or vertebrates, especially the two higher classes of the latter, even the geological character of a country having a powerful influence upon their distribution, as well as affecting their size and the thickness of their calcareous covering. While the mammalia are able to survive changes that would exterminate reptiles and amphibians, and are somewhat independent of the influences that govern the existence of many insects and mollusks, their fossil remains must give, for this reason, a less minute record of past geological and climatic changes than either the lower classes of vertebrates, the mollusca, or the insects, and afford a far less detailed record than plants. Among mammals sometimes the same species, and often the same genus, has a range extending from the Arctic regions to the warm-temperate or subtropical latitudes, thus showing an adaptability to varied conditions of existence not exhibited by the lower vertebrates, or by mollusks or plants. While their lack of exceptional means of dispersal and their superiority to forces of restriction that limit many groups of animals render them highly useful as a standard of reference in respect to present life-regions, the latter necessarily detracts from their importance as a medium of geological record, so far at least as regards the minuter details.

* Report of a Lecture before the Royal Geographical Society, in Geogr. Mag., vol. iv, August, 1877, p. 221.

† Geogr. Dist. Anim., vol. i, p. 37.

As is well known, and almost universally admitted, the animal and plant life of the Arctic lands is nearly everywhere the same, many of the species having a circumpolar range, while the genera are mainly, and the families almost entirely, the same throughout. Especially is this the case with mammals. To show how gradual is the change from almost absolute uniformity in the Arctic regions to the ultimate diversity met with in the intertropical latitudes it is only necessary to divide latitudinally the so-called "Nearctic" and "Palæarctic" regions into several minor areas, and to tabulate and compare the genera found in each. Adopting as our first division the region approximately bounded southward by the isotherm of 36° F., and hence embracing the Arctic, Sub-Arctic, and Cold Temperate lands of the northern hemisphere, we find that of the fifty-four commonly recognized genera of non-pelagic mammals occurring north of this boundary, five are subcosmopolitan; twenty-seven, or more than one-half, are strictly circumpolar, being represented throughout the greater part of the region north of this boundary; that five more are found on both shores of the Atlantic, and that five others are common to both shores of the Pacific. This leaves only twelve—less than one-fourth—that are peculiar to either the northern portion of North America or to the corresponding portion of the Old World, of which eight are restricted to America and four to the Europæo-Asiatic continent. These genera and their distribution are approximately shown in the subjoined table.

Genera of mammals of the Arctic and Cold Temperate portions of the northern hemisphere (the region north of the mean annual of 36° F.).

Circumpolar.			
Lynx.	Ursus.	Ovis.	Castor.
Canis.	Rosmarus.	*Vesperugo.†	*Sciurus.
Vulpes.	Phoca.	*Vespertilio.	Sciuropterus.
Mustela.	Alces.	Sorex.	Tamias.
Putorius.	Tarandus.	Arvicola.	Spermophilus.
Gulo.	Cervus.	Evotomys.	Arctomys.
*Lutra.	Bison.	Myodes.	Lagomys.
Thalassarctos.	Ovibos.	Cuniculus.	*Lepus.
American.	**American and Asiatic.**	**Europæo-Asiatic.**	**American and European.**
Mephitis.	Enhydris.	Meles.	Pagomys.
Taxidea.	Eumetopias.	Bos.	Pagophilus.
Procyon.	Zalophus.	Mus.	Erignathus.
Aplocerus.	Callorhinus.	Sminthus.	Halichærus.
Zapus.	Rhytina.		Cystophora.
Hesperomys.			
Fiber.			
Erethizon.			

* Subcosmopolitan. † *Scotophilus* of American authors, not of Dobson.

Summary.

Total number of genera	54
Subcosmopolitan	5
Circumpolar	27
Shores of North America and Asia	5
Shores of North America and Europe	5
Exclusively either American or Europæo-Asiatic	12
Peculiar to America	8
Peculiar to the Europæo-Asiatic continent	4

The above-given statistics show most clearly that the mammals of the northern third of the northern hemisphere present few generic or subgeneric forms that are peculiar to either North America or to the Europæo-Asiatic continent. In many cases, these are closely representative forms; in other cases, the peculiar genera extend but a short distance into the region, being temperate forms rather than hyperboreal.

The close relationship of the mammalian life of the northern lands, as compared with the diversity met with between that of the northern and southern portions of the two northern continents, is further shown by a tabulation of the genera met with in the region intervening between the cold-temperate and sub-tropical zones of life, the northern and southern boundaries of which may be considered respectively as the isotherms of 36° and 68° to 70° F. Rather more than one-half of the above-enumerated genera extend also over a large portion of this more southern belt, and impart thereby a general similarity to the *facies* of the mammalian faunæ of the two regions. In addition to these, however, we find in North America thirty-one genera and seven subgenera that are not found much, if any, to the northward of the isotherm of 36° F., and about the same proportion of new generic and subgeneric types make their appearance in the corresponding region of the Old World. Turning first to North America, we find that of these added forms one has so wide a distribution that it may be properly considered as subcosmopolitan, being found in the corresponding region of the Europæo-Asiatic continent as well as far to the southward of the region under notice. One other occurs also in Eastern Asia and six more belong rather to Tropical America than to Temperate North America. Excluding these, leaves about thirty as strictly American and twenty-two that are almost wholly restricted to Temperate North America; there is, hence, twice as great a difference between the mammalian faunæ of the middle temperate region of North America and the colder portion of the same continent as there is between those of the colder parts of the two northern continents, or the northern portions of the so-called "Nearctic" and "Palæarctic Regions". But we get in Temperate North America not only twenty-two generic and subgeneric forms peculiar to this region, but a differentiation of this region into three well-marked faunal areas, differing more from each other than do the boreal parts of the New World ("Nearctic Region") from the boreal parts of the Old World ("Palæarctic Region"). While thirteen of the genera, or about one-third, have a general distribution throughout Temperate North America, there are four genera and one subgenus peculiar to the so-called Eastern Province, five genera and one subgenus mainly restricted to the Middle Province, and five genera and two subgenera almost wholly limited to the geographically much smaller Western Province. In addition to this, there are five other genera and one subgenus common to the greater part of the Middle and Western Provinces that are not found in the Eastern.* The genera that may be regarded as characterizing the middle temperate region of North America and their relative distribution is shown in the subjoined table.

Terrestrial genera and subgenera of Middle North America (between the mean annuals of 36° and 68° F.), not found in the Arctic and Cold Temperate latitudes.

[NOTE.—Subgenera are enclosed in parentheses.]

Of general distribution.		Limited to the Eastern Province.	Limited to the Middle Province.	Limited to the Western Province.	Common to the Middle and Western Provinces.
Felis.	*Sigmodon.	Nycticejus.	*Bassaris.	*Macrotus.	(Otospermophilus.)
Urocyon.	Ochetodon.	*Synotus.	Antilocapra.	Antrozous.	(Pedomys)
Cariacus.	(Pitymys.)	Condylura.	*Dicotyles.	*Nyctinomus.	Perognathus.
Lasiurus.	Geomys.	Blarina.	*(Notiosorex.)	†Urotrichus.	Cricetodipus.
Scapanus.	Didelphys.	(Oryzomys.)	Synaptomys.	(Onychomys.)	Dipodomys.
Scalops.			Cynomys.	(Chilotus.)	Thomomys.
Neosorex.				Haplodon.	
Neotoma.					

* Chiefly tropical. † Occurs also in Asia.

Summary.

Total number of genera (plus 7 subgenera)	38
Of general distribution	13
Peculiar to the Eastern Province	5
Peculiar to the Middle Province	6
Peculiar to the Western Province	7
Common to the Western and Middle Provinces, but not found in the Eastern	6
Mainly tropical or subtropical	8

* Mr. Wallace, in his late work (Geogr. Dist. Anim., vol. i, p. 6), refers to the Rocky Mountains as forming a barrier to species, "almost all the mammals, birds, and insects" belonging to different species on the two sides of the Rocky Mountains. Nothing, so far as mammals and birds are concerned (and I am informed by good authorities that the same is true of insects), could well be further from the truth. Only in rare instances do the Rocky Mountains form such a barrier, the division between the Eastern and Middle Provinces being more than six hundred miles to the eastward of this range, while the boundary between the Middle and Western Provinces is formed by the Sierra Nevada chain. The same species, as a rule, range over the greater part of the great elevated interior plateau, of which the Rocky Mountains constitute the axis. So far as the distribution of both birds and mammals is concerned, the presence or absence of forests, and the accompanying diverse climatic conditions, have far more to do with the limitation of habitat than the commonly so-called "Rocky Mountain barrier". This is obviously due to the longitudinal direction of this supposed barrier, which, if trending in a latitudinal direction, would certainly form an impassable obstacle to very many species.

Between the warm-temperate belt we have been considering and the zone next to the southward—the subtropical—the faunal differences are far greater than between the warm-temperate and colder zones. Aside from the few subcosmopolitan genera still present, and the few essentially tropical genera that range northward into the warmer temperate zone, there is little in common to the mammalian faunæ of these two regions. At or near this boundary (the isotherm of about 68° F.—say 68° to 70° F.) several strictly tropical families first make their appearance, and tropical genera begin largely to replace those of the colder region to the northward.

In respect to the Europæo-Asiatic continent, we have already seen how small a proportion of the genera of mammals met with north of the thirty-sixth isotherm are really peculiar to this region, the number being less than twelve per cent., the remainder being circumpolar. Passing, however, to the warm-temperate division of this Europæo-Asiatic continent, or that portion between the isotherms of 36° and 68° to 70° F., and we meet with many genera not found to the northward. While many circumpolar genera still prevail, at least three-fourths of the whole number are here first met with. A considerable proportion (about one-fifth) are properly southern or subtropical, and extend far to the southward of the warm-temperate zone. About one-half, however, are peculiar to this zone, and belong to groups (families of subfamilies) especially characteristic of the North Temperate Realm. In adopting the isotherm of 70° F. as its southern boundary, we include not only the Mediterranean Province (and hence Northern Africa), but all of Asia north of the great Himalayan chain, together with Northern China and the Persian Peninsula. Hence quite a number of such southern forms occur as *Macacus*, *Herpestes*, *Genetta*, *Hyæna*, *Hystrix*, etc., that are more properly members of the intertropical fauna. Owing to the great extent of this region, we meet with many genera peculiar to special districts, giving a higher proportion of peculiar forms than is met with in the corresponding portion (but far more limited in area) of North America. Of about fifty genera met with here that do not occur to the northward, about one-fourth may be thrown out as more properly tropical, since they in most cases barely enter the southern border.

Of the remainder, fully one-half are restricted in their range wholly or almost wholly to this region, the rest extending far into or throughout the Old World tropics. There is thus more than thrice as great a difference between the mammalian fauna of the boreal parts of the Europæo-Asiatic continent and that of the warmer parts of the same continent as between the fauna of the boreal parts of the Europæo-Asiatic continent and the corresponding region of North America. The differentiation is here again, as in North America, from the north southward, not through the rapid increase of land-area and diversity of physical structure, but purely from climatic conditions,—through the multiplication of life in consequence of increase of temperature and means of subsistence. This is still more strikingly shown by a comparison of the fauna of the middle portion of the so-called "Palæarctic Region" with that of its southern border, at which point the truly tropical forms begin to appear. The genera of a zone, say two degrees in width, at these two points would be not only in large part different, but those of the southern belt would be far more numerous.

Genera of mammals of the warm-temperate portions of the eastern hemisphere (between the isotherms of 36° and 68° to 70° F.), not occurring to the northward of the 36th isotherm.

* Macacus.	† *Moschus.*	Rhinolophus.	† *Nectogale.*
Felis.	† *Hydropotes.*	* Plecotus.	Spalax.
* Genetta.	† *Poëphagus.*	* Synotus.	Rhizomys.
* Herpestes.	* *Addax.*	Scotophilus.	† *Siphneus.*
* Hyæna.	* Oryx.	Miniopterus.	Meriones.
† *Nyctereutes.*	Damalis.	* Nyctinomus.	† *Cricetulus.*
† *Lutronectes.*	† *Procapra.*	† *Scaptochirus.*	† *Alactaga.*
† *Ælurus.*	† *Saiga.*	† *Scaptonyx.*	* Gerbillus.
* Equus.	† *Pantholops.*	† *Anusorex.*	* Dipus.
† Camelus.	† *Budorcas.*	† *Mygale.*	Muscardinus.
† *Dama.*	† *Rupicapra.*	Urotrichus.	Eliomys.
† *Elaphodus.*	Nemorhædus.	† *Uropsilus.*	* Hystrix.
† *Lophotragus.*	Capra.	Crocidura.	

Summary.

Total number	51
Occurring in southern portions only	13
Peculiar to the region, and generally restricted to a limited range	24
Of rather wide range southward	14

A comparison of the families represented in different portions of the northern hemisphere north of the isotherm of 70° F. brings into prominence some of the points already stated, without the confusion of detail incident to a comparison on the basis of genera, and gives also a more convenient standard for the next stage of comparison, namely, a comparison of the faunæ of the temperate zones with those of the tropical, as well as with the faunæ of the two great land-areas of the northern hemisphere. Of thirty-three families of non-pelagic mammals found north of about the isotherm of 70° F. (68° to 70°), thirteen have a nearly cosmopolitan distribution, and six others are common to both the Old World and the New, leaving fourteen, or about one-third, peculiar to either North America or to Europe and Asia. Three of these are essentially subtropicopolitan or tropicopolitan, having merely straggling representatives north of the 68th isotherm, and five others are represented each by only a single species. Seven of these fourteen families (four only according to many systematists‡) are North American and seven European

* Occurring in southern portions only; chiefly tropical.

† Peculiar to the region and mostly of restricted range.

‡ I here admit to family rank *Antilocapridæ*, *Zapodidæ*, and *Geomyidæ*, the two former of which are treated by Mr. Wallace as subfamilies of subcosmopolitan families, while the other is not commonly recognized as distinct from *Saccomyidæ*. On the other hand, I refer the *Cercolabidæ* to the *Hystricidæ*.

and Asiatic. One or two others barely touch, or possibly overlap slightly, the above-given boundary. North of the isotherm of 36° F. not more than two or three families are met with that are not cosmopolitan, and two of these have each but a single species north of this line.

The following is a list of the families referred to above, with approximate indications of their distribution.

Families of non-pelagic mammals occurring north of the mean annual of 70° F.

Subcosmopolitan.		Circumpolar.	North American.	Europæo-Asiatic.
Felidæ.	Bovidæ.	Rosmaridæ.	†Procyonidæ.	Erinaceidæ.
Canidæ.	Vespertilionidæ.	*Rhytinidæ.	‡Antilocapridæ.	‡Suidæ.
Mustelidæ.	Muridæ.	Talpidæ.	‡Zapodidæ.	Equidæ.
Ursidæ.	Sciuridæ.	Soricidæ.	Geomyidæ.	Myoxidæ.
Otariidæ.	Hystricidæ.	Castoridæ.	Saccomyidæ.	Spalacidæ.
Phocidæ.	Leporidæ.	Lagomyidæ.	‡Haplodontidæ.	Dipodidæ.
Cervidæ.			†Didelphyidæ.	Rhinolophidæ.

* Formerly occurring on the shores of the North Pacific only, but now extinct.
† Tropical; one species only found north of 70th isotherm. ‡ Represented by a single species.

Summary.

Whole number	33
Subcosmopolitan	13
Circumpolar (arctopolitan)	6
American (exclusively)*	7
Europæo-Asiatic (or exclusively Old World) †	7

In regard to the southern extension of these thirty-three families, thirteen range far into, and most of them over, the greater part of Intertropical America, and eighteen far into, and most of them over, the greater part of the intertropical portion of the Old World.

In Intertropical America, only thirty families are represented. Of these, thirteen occur over much of Temperate North America, while eleven are subcosmopolitan, and the same number are peculiar to the region, while one-half of the whole do not range much beyond the northern tropic. Seven are semitropicopolitan, or occur also in the warmer parts of the Old World; but of these, three are *Chiroptera* and another is marine. The approximate range of the families represented in Intertropical America is indicated in the annexed table.

* Five only are exclusively North American.
† Two only are exclusively "Palæarctic".

Families of non-pelagic mammals occurring in Intertropical America (between the northern and southern isotherms of 70° F.).

[NOTE.—The names of families peculiar to the region are printed in italics.]

Cebidæ.	Otariidæ.	Soricidæ.	Hystricidæ.
Mididæ.	Cervidæ.	*Centetidæ.	Leporidæ.
Felidæ.	*Trichechidæ.‡	Sciuridæ.	*Brachypodidæ.*
Canidæ.	*Tapiridæ.	Muridæ.	*Dasypodidæ.*
Mustelidæ.	*Dicotylidæ.*	*Octodontidæ.	*Myrmecophagidæ.*
†Procyonidæ.	Phyllostomidæ.	*Dinomyidæ.*	†*Didelphyidæ.*
Bassarididæ.	Emballonuridæ.	*Caviidæ.*	
Cercoleptidæ.	Vespertilionidæ.	*Dasyproctidæ.*	

Summary.

Total number	30
Peculiar to the region	12
Not found in temperate parts of North America	16
Subcosmopolitan	11
Occurring in the warmer parts (only) of the Old World	5
Occurring in North America (at large)	13

Fifty families are represented in the intertropical portions of Asia and Africa. Of these nearly thirty do not range much beyond the Northern Tropic, of which about twenty-three are limited to this region. Of the thirty-two families occurring in the north-temperate zone (of which only six or seven are exclusively Europæo-Asiatic), nearly one-half range over most of the Indo-African tropics. The following is a list of the families represented in the Old World tropics, exclusive of those limited to Madagascar and the Australian Realm.

Families of non-pelagic mammals occurring in the Indo-African Tropics (between the northern and southern isotherms of 70° F.)

[NOTE.—The names of families not occurring northward of the region are printed in italics.]

Simiidæ.	*Giraffidæ.*	§*Trichechidæ.*	*Tupayidæ.*
Cynopithecidæ.	Bovidæ.	*Pteropidæ.*	*Lophiomyidæ.*
Lemuridæ.	Cervidæ.	Rhinolophidæ.	Dipodidæ.
Tarsiidæ.	Tragulidæ.	*Nycteridæ.*	Muridæ.
Felidæ.	*Hippopotamidæ.*	Vespertilionidæ.	Myoxidæ.
Protelidæ.	*Phacochœridæ.*	§*Emballonuridæ.*	Sciuridæ.
Hyænidæ.	Suidæ.	*Galeopithecidæ.*	*Anomaluridæ.*
Viverridæ.	*Equidæ.*	Talpidæ.	*Hystricidæ.*
Canidæ.	*Rhinocerotidæ.*	Soricidæ.	§*Octodontidæ.*
Mustelidæ.	§*Tapiridæ.*	Erinaceidæ.	Leporidæ.
Ursidæ.	*Hyracidæ.*	*Potamogalidæ.*	*Manididæ.*
Æluridæ.	*Elephantidæ.*	*Macroscelidæ.*	*Orycteropodidæ.*
Camelidæ.	*Halicoridæ.*		

*Occurring in the Old World Tropics.
†Occurring also in Extratropical America.
‡*Manatidæ* of most authors.
§ Also represented in Intertropical America

Summary.

Total number	50
Peculiar (or almost wholly restricted) to the region	22
Subcosmopolitan	13
Represented in the American tropics (only)	4
Occurring in the Old World north of the tropics	23
Tropical	29

It thus appears that only about three-fifths as many families of mammals occur in the intertropical parts of the New World as in the corresponding parts of the Old World. The disproportion in the same direction in respect to genera and species is still greater. This is obviously due to the difference in size and configuration of the two areas. The Old World intertropical land surface is not only several times greater than the American (embracing thrice as great a breadth longitudinally), but is differentiated into one continental (Africa), two large peninsular (India and China) areas, and a group of large, highly differentiated islands (Malay Archipelago), while the intertropical region of America forms a single unindented region, with a single narrow isthmic prolongation. In the one case (America) we have a striking uniformity of mammalian life throughout, corresponding with the general uniformity of the climatic conditions characteristic of this area, contrasting with well-marked subdivisions in the other, and a much greater diversity of environing circumstances, originating geologically far back in the history of these several land-masses. As Mr. Wallace has remarked,—"To those who accept the theory of development as worked out by Mr. Darwin, and the views as to the general permanence and immense antiquity of the great continents and oceans so ably developed by Sir Charles Lyell, it ceases to be a matter of surprise that the tropics of Africa, Asia, and America should differ in their productions, but rather that they should have anything in common. Their similarity, not their diversity, is the fact that most frequently puzzles us."*

In the foregoing remarks, no reference has been made to Madagascar or to Australia, for the reason that they belong to distinct primary life-regions having little in common with the great Europæo-Asiatic land-area (of which Africa, on the other hand, is an inseparable appendage), which, with America, form the regions to which the discussion has thus far been intentionally limited. As will be more fully considered later, the intertropical Old World area is divisible into secondary regions, which for the present need not enter into the questions immediately at issue. These are, first, Does that portion of the northern hemisphere north of the northern subtropical zone admit of division into two primary life-regions, conforming in their boundaries to the configuration of the two great northern land-areas? And, secondly, In accordance with what principle does the life of the northern hemisphere become differentiated from the homogeneity characteristic of the northern regions

* Geogr. Dist. Anim., vol. i, p. 51.

to the great diversity met with under tropical latitudes? The fundamental question which underlies the whole subject is, Is, or is not, the life of the globe distributed in circumpolar zones? The second is, How and under what influences does it become differentiated?

To the first of these questions, I ventured some six years since,* to give an affirmative answer, in accordance not only with the views of numerous high authorities on the subject of the geographical distribution of life, but with what seemed to me to be incontrovertibly the facts in the case. While this view has since received the support of other high authorities, it has been altogether ignored by the advocates of Dr. Sclater's division of the earth's surface. Mr. Wallace, who faithfully reflects the views of the Sclaterian school, in referring to this subject says:—"Mr. Allen's system of 'realms' founded on climatic zones . . . calls for a few remarks. The author continually refers to the '*law of the distribution of life in circumpolar zones*', as if it were one generally accepted and that admits of no dispute. But this supposed 'law' only applies to the smallest details of distribution—to the range and increasing or decreasing numbers of *species* as we pass from north to south, or the reverse; while it has little bearing on the great features of zoölogical geography—the limitation of groups of *genera* and *families* to certain areas. It is analogous to the '*law of adaptation*' in the organization of animals, by which members of various groups are suited for an aerial, an aquatic, a desert, or an arboreal life; are herbivorous, carnivorous, or insectivorous; are fitted to live underground, or in fresh waters, or on polar ice. It was once thought that these adaptive peculiarities were suitable foundations for a classification,—that whales were fishes, and bats birds; and even to this day there are naturalists who cannot recognize the essential diversity of structure in such groups as swifts and swallows, sun-birds and humming-birds, under the superficial disguise caused by adaptation to a similar mode of life. The application of Mr. Allen's principle leads to equally erroneous results, as may be well seen by considering his separation of 'the southern third of Australia' to unite it with New Zealand as one of his secondary zoölogical divisions."†

Leaving Mr. Wallace's last-quoted objection for notice in another connection (see a foot-note beyond, under the sub-heading "Australian Realm"), I unblushingly claim, in answer to the main point, that the geographical distribution of life is *by necessity* in accordance with *a "law of adaptation"*, namely, of *climatic adaptation;* that such a law is legitimate in this connection, and that the reference to the "superficial disguise" adapting essentially widely different organisms to similar modes of life is wholly irrelevant to the point at issue,—a comparison of things that are in any true sense incomparable; furthermore, that the "law of distribution of life in circumpolar zones" does apply as well in a general sense as to details—"to groups of genera and families" as well as

---

*Bull. Mus. Comp. Zoöl., vol. ii, p. 376, 1871.

†Geogr. Dist. Anim., vol. i, p. 67.

to species. In the foregoing remarks I have had little to say respecting the range of *species*, and have tabulated merely genera and families. These tables clearly show that a large proportion of the mammalian genera and families of the northern hemisphere have a circumpolar range, the same genera and families occupying the Arctic and Sub-Arctic lands in both the Old World and the New, and that only a small per cent. of the whole number found here are peculiar to either of the northern land-areas; that a large part of the genera and families met with in the temperate and warmer latitudes occur on the eastern continent as well as on the western; that again a considerable proportion of the genera and families met with in the warmer parts of the earth occur also both in the Old World and the New, while many others are well known to have been common to the two during the Tertiary period. It has been further shown that there is a greater diversity of life between contiguous climatic belts of the same continent than between corresponding belts of the two continents, especially north of the forty-fifth parallel of latitude, and that any marked faunal differentiation of the two continents begins only in the warm-temperate and subtropical latitudes. On each continent, the arctic, temperate, and tropical zones are each marked in their general *facies* respectively by corresponding phases of life. So obvious is this that we have in current use the expressions "arctic life", "temperate life", and "tropical life", in recognition of certain common features of resemblance by which each of these regions is distinguished as a region from the others. This is in accordance with a law I have termed the law "of differentiation from the north southward",* or in accordance with increase of temperature and the conditions resulting therefrom favorable to increased abundance of life.

In this connection it may be well to recall certain general facts previously referred to respecting the geographical relations of the lands of the northern hemisphere and their past history. Of first importance is their present close connection about the northern pole and their former still closer union at a comparatively recent date in their geological history; furthermore, that at this time of former, more intimate relationship, the climatic conditions of the globe were far more uniform than at present, a mild or warm-temperate climate prevailing where now are regions of perpetual ice, and that many groups of animals whose existing representatives are found now only in tropical or semitropical regions lived formerly along our present Arctic coasts. We have, hence, an easy explanation of the present distribution of such groups as Tapirs, Manatees, many genera of Bats, etc., in the tropics of the two hemispheres, on the wholly tenable assumption of a southward migration from a common wide-spread northern habitat, to say nothing of the numerous existing arctopolitan and semi-cosmopolitan genera. The former greater community of life in the northern hemisphere in preglacial times is further evinced by the wide-spread occurrence there of the remains of Camels,

---

* Bull. Mus. Comp. Zoöl., vol. ii, p. 379.

Elephants, Mastodons, Rhinoceroses, and Horses, which, though extinct in America, have living representatives in the tropics of the so-called "Old World", to say nothing of the evidence afforded by the remains of still earlier types of arctopolitan range. The succeeding epochs of cold caused extensive migrations of some groups and the extinction of others; with the diverse climatic conditions subsequently characterizing high and low latitudes came the more pronounced differentiation of faunæ, and the development, doubtless, of many new types adapted to the changed conditions of life—the development of boreal types from a warm-temperate or semi-tropical stock. The accepted theories respecting the modification of type with change in conditions of environment—changes necessarily due mainly to climatic influences—render it certain that if animals are so far under the control of circumstances dependent upon climate, and emphatically upon temperature, as to be either exterminated or greatly modified by them, the same influences must govern their geographical distribution.

Recent discoveries respecting the mammalia inhabiting North America during the Tertiary period have shown that many of the leading types of mammals—including not only those above named, but also many others—now found only in the eastern hemisphere, originated in North America, and migrated thence to Asia, Europe, and even Africa, either as somewhat generalized types, or after they had nearly reached their present degree of differentiation; in short, so far as mammalian life is concerned, that America is *the* "Old World" from which the so-called "Old World" has been mainly peopled. The present genetic convergence of life about the northern pole seems to show that not only has there been here a comparatively free intercommunication, but that the mammalian life now existing there has lived there for a long period under similar conditions of environment; and that these conditions are unfavorable, in consequence of a comparatively low temperature, to rapid change of form or structure.

This is shown not only by the great diversity of life met with in the intertropical regions, as compared with the uniformity met with in the semi-frigid regions (equal areas being, of course, compared), but by the coincident occurrence of a simple, homogeneous arctic marine fauna, with the low temperature over the sea-floor far to the southward of where such forms occur in the warmer surface and shore-waters. The intimate relation between temperature and the distribution of life is most forcibly shown by the existence under the same parallel of latitude of diverse faunæ not only at different elevations above the sea on mountain-slopes, but at different depths beneath the surface of the ocean, where the several faunæ are characterized not only by the presence of different species, but by the prevalence of different genera, and even families. In fact, it is to me a matter of surprise that, with our present knowledge of the subject, any naturalist of note should assume that temperature has nothing to do with the circumscription of faunæ, or that any law based on it can have "little bearing on the great features of zoölogical geography—the limitation of groups of genera and families to certain areas"

II.—MAMMALIAN REGIONS OF THE GLOBE.

The influence of temperature as a limiting agent in the distribution of life, as well the "*law* of the distribution of life in *circumpolar zones*", was fully recognized by Humboldt nearly three-fourths of a century ago, and later, practically if not explicitly, by Ritter, De Candolle, Agassiz, Wagner, Forbes, Dana, Günther, Meyen, Middendorff, and many other leading zoölogists and botanists. While this law must incontrovertibly underlie every philosophic scheme of lief-regions, the number of zones to be recognized, as well as their boundaries, must in a measure be open to diversity of opinion. Professor Dana, in 1852, recognized five primary zones for marine animals, namely, a torrid, a north and a south temperate, and a north and a south frigid. The torrid and temperate were subdivided, the first into three, the others each into five sub-zones, the two frigid being left undivided. Mr. A. Agassiz, in treating of the distribution of the *Echini*,* recognizes also five zones, a torrid, two temperate, and two frigid. These five primary zones prove to be applicable also to the mammalia, and even their subdivisions may be readily traced, but are rather too detailed for practical use. Owing to the irregular surface of the land-areas, occasioned by elevated plateaus and mountain-chains, these zones of distribution have of course a less regular breadth and trend than they preserve over the oceans. Their boundaries, however, approximate to the courses of the isotherms, by certain of which they may be considered as in a general way limited.

In recognition of these zones, and also of the law of differentiation of life with the relative isolation of the principal land-areas, I proposed in a former paper (*l. c.*, p. 380) a division of the land-areas into eight "Realms", namely: I, Arctic; II, North Temperate; III, American Tropical; IV, Indo-African; V, South American Temperate; VI, African Temperate; VII, Antarctic; VIII, Australian. A subdivision of most of these primary regions was provisionally suggested, but only the North American was treated with any degree of detail, and this mainly with reference to the birds, and more especially those of its eastern portion. Subsequent study of the distribution of mammalian life over the globe has led me to modify some of the views then expressed, especially in relation to the divisions of the Australian Realm, and to unite the South African Temperate with the Indo-African, as a division of the latter, and also to recognize Madagascar and the Mascarene Islands as forming together an independent primary region, in accordance with the views of Sclater, Wallace, and others. Whether or not the Arctic and Antarctic Regions should stand as primary divisions seems also open to question. While perhaps tenable on general

* Illustr. Cat. Mus. Comp. Zoöl., No. vii, 1872, pls. A–F.

grounds, they are hardly required for the elucidation of the distribution of the mammalia, since they must be mainly characterized negatively.

Beginning with the Arctic Region, we meet, as already shown, and as is almost universally admitted, a continuous homogeneous fauna, of considerable geographical area, but mainly characterized by what it lacks. Its southern boundary may be considered as the northern limit of forest vegetation. Continuing southward, few other than arctopolitan genera of mammals are met with north of the mean annual of 36° F. This considerable belt hence includes what may be termed the cold-temperate zone. The American and Europæo-Asiatic portions of this zone are only to a slight degree differentiated, while each is essentially homogeneous.

Below this, non-arctopolitan genera, or those restricted to more or less limited areas, become more frequent, and, indeed, form a considerable proportion of the genera represented. This belt occupies the remainder of the north-temperate zone, extending to about the mean isotherm of 70° F., and may be termed the warm-temperate zone. Unlike the cold-temperate zone, it is divisible on each continent into several well-marked minor regions, which are, however, more strongly differentiated, *inter se*, in the Old World than in the New.

The tropical zone embraces, of course, in its fullest extension, a much greater latitudinal breadth than the temperate, but its southern land-border is very irregular, its only considerable development south of the equator being in South America and Africa. It is also so much diversified in many parts by mountain-chains that subdivision into secondary zones seems less feasible than in the case with the north-temperate zone. A central torrid and a north and a south sub-torrid zones might, however, be readily made, but such a division has not been attempted in the present connection. A northern sub-torrid division may indeed be very conveniently recognized, extending from about the annual isotherm of 67° to that of about 74° F., and including a transitional region consisting of the extreme southern border of what has been above defined as the warm-temperate zone and the northern border of the tropical.

In like manner, the distribution of life seems to warrant the recognition, in Africa and South America, of a corresponding transitional belt between the two torrid and the southern warm-temperate zones. Aside from these divisions, the Torrid Zone admits of others of a more practical or useful character. These become at once obvious, since they result from the position and configuration of its component land elements. The first is a primary separation into two "realms", an American and an Indo-African. Each of these is again divisible into several minor portions or "provinces"; but the Indo-African admits also of division into two "regions", an African and an Indian, which are divisions of secondary rank, each having several "provinces".

The South Temperate Zone has a very limited land-surface, consisting of the southern third of South America, a small portion of Southern Africa, and the greater portion of Australia. Extra-tropical South Africa is all comprised within the Warm Temperate Zone, and is so small in area and so intimately related, both geographically and faunally, with Tropical Africa, that its formal separation, while, perhaps, warranted in the abstract, is hardly practically necessary. Temperate South America is exceedingly irregular in its northern outline, owing to peculiarities of configuration, resulting from the presence of the great Andean Plateau, by means of which it extends along the western border of South America far northward of the southern tropic. Temperate Australia is clearly separable from the tropical portion of the Australian Realm. The South Temperate Zone hence consists of three comparatively small land-areas, widely separated from each other, and consequently, as would be supposed, have little in common.

The Antarctic Region has a very limited amount of land-surface, and the few species that compose its fauna are almost wholly either marine or pelagic. As previously stated, as a mammalian region it has little significance.

This hasty sketch shows that the differentiation of the land-surface of the earth into realms, regions, and minor divisions has relation not only to climate, but to the divergence and isolation of the different principal land-areas; that at the northward, where the lands converge, there is no partitioning in conformity with continental areas, the temperate and colder portions of the northern hemisphere all falling into a single primary division, and that only the southern half is susceptible of divisions of the second rank. Within the tropics, on the other hand, the lands of the eastern and western hemispheres fall at once into different primary regions, and one of these is again divisible into regions of second rank. Beyond the tropics, the land-surfaces are of small extent, widely separated, and faunally have almost nothing in common.

With these preliminary remarks, we may now pass to a detailed consideration of the several primary regions and their subdivisions.

I.—ARCTIC REALM.

Whether or not an Arctic Region should be recognized as a division of the first rank is a question not easy to satisfactorily answer. Naturalists who have made the distribution of animal life in the boreal regions a subject of special study very generally agree in the recognition of a hyperboreal or circumpolar fauna, extending in some cases far southward over the Temperate Zone. The Arctic portion of this hyperborean region has been frequently set off as a secondary division, or subregion,* and generally recognized as possessing many features not

* It forms Mr. Blyth's "Arctic Subregion" (Nature, vol. iii, p. 427, March 30, 1871), Mr. Brown's "Circumpolar" division (Proc. Zool. Soc., Lond., 1868, p. 337), and Dr. von Middendorff's "Zirkumpolar-Fauna" (Sibirische Reise, Bd. iv, p. 910, 1867). It also accords very nearly with Agassiz's "Arctic Realm" (Nott and Gliddon's Types of Mankind, 1854, p. lx and map).

shared by the contiguous region to the southward. For the present I prefer to still retain it as a division of the first rank. It is characterized mainly by the paucity of its life, as compared with every region except the Antarctic, and by what it has not rather than by the possession of peculiar species or groups. It wholly lacks both Amphibian and Reptilian life, is almost exclusively the summer home of many birds, and forms the habitat of the Esquimaux, the Arctic Fox, the Polar Bear, the Musk Ox, the Polar Hare, the Lemmings, the Walruses, the Narwhal, and the White Whale, which are confined within it. It has no *Chiroptera* nor *Insectivora*, two or three species of Shrews, however, barely reaching its southern border. It shares with the cold-temperate belt the presence of the Moose and the Reindeer, several Pinnipeds, a number of boreal species of *Glires*, several fur-bearing *Carnivora*, and a considerable number of birds. Its southern boundary may be considered as coinciding very nearly with the northern limit of arboreal vegetation, and hence approximately with the isotherm of 32° F. Its more characteristic terrestrial forms range throughout its extent, none being restricted to either the North American or Europæo-Asiatic continent. Hence it is indivisible into regions of the second and third grades (regions and provinces), and may be considered as embracing a single hyperborean assemblage of life.

II.—NORTH-TEMPERATE REALM.

Very few writers on zoölogical geography have failed to recognize the striking resemblance the fauna of Temperate North America bears to that of the corresponding portion of the Old World. The resemblance is less in the Avian class than among mammals, but is generally acknowledged as obtaining even there. Dr. Sclater, while admitting a strong resemblance between these areas, considered them as separable into two primary regions, in which view of the case he has been followed, among prominent writers on the subject, by Dr. Günther, Mr. Wallace, Mr. Murray, and Professor Cope. Dr. Günther, while provisionally accepting Dr. Sclater's "Nearctic" and "Palæarctic" regions, refers pointedly to the disagreement of the distribution of Batrachians with these divisions; for in discussing the distribution of this class he says,—"Dissimilarity and similarity of the Batracho-fauna depend upon zones. Palæarctic and Nearctic regions resemble each other more than any other third; the same is the case with Australia and South America; the Ethiopian region exhibits similarity with South America, as well as with the East Indies, but more especially with the latter."* Mr. Murray admits that "the boreal extremity of North America is tinged with a Europeo-Asiatic admixture", which he regards as "an extraneous element grafted upon the genuine stock, and easily eliminated from it".† But in his map of "Great Mammalian Regions" the boreal parts of

* Proc. Zoöl. Soc. Lond., 1858, p. 390.

† Geogr. Dist. Mam., p. 312.

both continents are similarly colored, the same color, however, extending only to about the forty-ninth degree of north latitude in North America, while in Africa it descends to north latitude 18°, and in Asia ranges from north latitude 30° to 25°! His divisions as recognized in the text are still more arbitrary and unphilosophic.

Mr. Wallace, in his discussion of zoölogical regions, says,—"The distinction between the characteristic forms of life in tropical and cold countries is, on the whole, very strongly marked in the northern hemisphere; and to refuse to recognize this in a subdivision of the earth which is established for the very purpose of expressing such contrasts more clearly and concisely than by ordinary geographical terminology, would be both illogical and inconvenient. The one question then remains, whether the Nearctic region should be kept separate or whether it should form part of the Palæarctic or of the Neotropical. Professor Huxley and Mr. Blyth advocate the former course; Mr. Andrew Murray (for mammalia) and Professor Newton (for birds) think the latter would be more natural. No doubt," Mr. Wallace adds, "much is to be said for both views," but decides in favor of the separation of the two regions in accordance with Dr. Sclater's scheme.*

While Mr. Blyth includes North America in his "Boreal Region" (as "2. Neo-septentrional Sub-region"), he adds also Central America and the Antilles (as "3. Neo-meridional Sub-region"), and, still more strangely, the Andean Region, with Chili, Patagonia, and the Fuegian and Falkland Archipelagos (as "4. Andesian Sub-region").†

Professor Huxley, in writing of the primary ontological regions of the globe, thus observes:—"In a well-known and very valuable essay on the Geographical Distribution of Birds, Dr. Sclater divides the surface of the globe primarily into an eastern and a western area, which he terms respectively *Palæogæa* and *Neogæa*. However, if we take into consideration not merely the minor differences on which the species and genera of birds and mammals are often based, but weigh the morphological value of groups, I think it becomes clear that the Nearctic province is really far more closely allied with the Palæarctic than with the Neotropical region, and that the inhabitants of the Indian and Ethiopian regions are much more nearly connected with one another and with those of the Palæarctic region than they are with those of Australia. And if the frontier line is latitudinal rather than longitudinal, and divides a north world from a south world, we must speak of *Arctogæa* and *Notogæa* rather than of Neogæa and Palæogæa as the primary distributional areæ. The secondary divisions, or geographical provinces, proposed by Dr. Sclater, answer, in great measure, to those which are suggested by the distribution of the *Alectoromorphæ*—except that, in common with many other naturalists, I think it would be convenient to recognize a *circumpolar* province, as distinct from the Nearctic and

* Geogr. Dist. Anim., vol. i, pp. 65, 66.

† Nature, vol. iii, p. 427, March 30, 1871.

Palæarctic regions."* Professor Huxley thus emphatically recognizes a region equivalent to my North Temperate Realm.

Mr. Robert Brown, in writing of the distribution of the mammals of Greenland, also recognizes a North Temperate Region, which he divides into a European Temperate Province and a North American Temperate Province, from which he separates a Circumpolar Region, equivalent to the Arctic Realm above characterized.†

Dr. Gill, in regard to fishes, recognizes an "Arctogæan" region, "embracing Europe, Northern Asia, and Northern America", as distinct on the one hand from the American Tropical and Transtropical Region, and on the other from Tropical Asia and Africa.‡

Dr. Packard, in discussing the distribution of the Phalænid Moths, recognizes both an Arctic Realm and a North Temperate Realm, as here characterized. Referring to a previously given table of subalpine and circumpolar species, he says,—"This table indicates how wide are the limits of distribution of these species, and it will be seen how important it is to follow circumpolar and north-temperate insect-faunæ around the globe, from continent to continent. It will be then seen how inadequate must be our views regarding the geographical distribution of the animals and plants of our own continent, without specimens from similar regions in the same zones in the Old World. It will be found that for the study of the insect-fauna of the Rocky Mountains and Pacific coast we must have ample collections from the Ural and Altai Mountains and surrounding plateaus," etc.§

Dr. August von Pelzeln also recognized a circumboreal region ("arktische Region"), and considers the "Nearctic" and "Palæarctic" as forming inseparable parts of a single region. He says:—"Die paläarktische Region scheint mir von der nearktischen nicht trennbar zu sein, sondern beide dürften ein Ganzes bilden, welches man als arktische Region bezeichnen könnte. Ihre Zusammengehörigkeit tritt mit voller Evidenz in den hochnordischen Ländern des alten und neuen Continentes hervor und erst in niedereren Breiten macht sich die Differenzirung geltend.

---

* Proc. Zoöl. Soc. Lond., 1868, pp. 314, 315.

† Proc. Zoöl. Soc. Lond., 1868, pp. 337, 338.

‡ Says Dr. Gill:—"In fine, dividing the earth into regions distinguished by general ichthyological peculiarities, several primary combinations may be recognized, viz.:—1, an *Arctogæan*, embracing Europe, Northern Asia, and Northern America; 2, an *Asiatic*, embracing the tropical portions of the continent; 3, *African*, limited to the region south and east of the Desert; 4, an *American* (embracing the America *par excellence* dedicated to Amerigo Vespucci), including the tropical and transtropical portions; and, 5, an *Australasian*. Further, of these (*a*) the first two [Arctogæan and Asiatic] have intimate relations to each other, and (*b*) the last three others among themselves; and some weighty arguments may be adduced to support a division of the faunas of the globe into two primary regions coinciding with the two combinations alluded to—(*a*) a *Cænogæa* and (*b*) an *Eogæa*, which might represent areas of derivation or gain from more or less distant geological epochs."—*Ann. and Mag. Nat. Hist.*, 4th ser., vol. xv, 1875, pp. 254, 255.

§ Monograph of Geometrid Moths, or Phalænidæ, of the United States, pp. 567, 586, 1876.

Die Vergleichung der Thierwelt beider Continente zeigt nämlich, dass die circumpolare Fauna in beiden dieselbe ist, dass in der Hochgebirgsfauna noch bedeutende Uebereinstimmung herrscht, dass in der übrigen paläo- und neoborealen Thierbevölkerung sowohl identische Arten als gemeinsam eigenthümliche Gattungen sich finden, endlich dass selbst jene Typen, welche jedem Continente eigenthümlich sind, doch eine gewisse Uebereinstimmung hinsichtlich des Charakters der Fauna an sich tragen, so dass sie einander näher stehen als Angehörigen anderer Regionen. In der neuen Welt ist eine Modification der Fauna auch durch das Eindringen neotropischer Formen gegeben."* He further also calls attention to the similarity of life which prevailed throughout this circumpolar region during the Quaternary period.

It is unnecessary to cite further, from the abundant material at hand, the opinions of specialists in reference to the propriety of recognizing a North Temperate Realm, as distinguished from the tropical regions of the globe, and in contradistinction from a north and south line of division of the North Temperate Zone into two primary ("Palæarctic" and "Nearctic") regions.

The chief differences between Dr. Sclater's division of the northern hemisphere and the present consist in setting off at the northward an Arctic Realm, the union of the so-called Nearctic and Palæarctic Regions into one circumpolar belt, and in the adoption for the same of a more northern limit than that proposed as the boundary of the two above-named Sclaterian regions. As will be shown later, the subdivisions of the North Temperate Realm or ("*Arctogœa*") as here defined agree in the main with the "subregions" of Sclater and Wallace. The more northward location of the southern boundary of the North Temperate Realm in North America results in the elimination of several characteristic tropical types, which extend a short way only into Dr. Sclater's Nearctic and Palæarctic Regions, and which, when considered as members of these regions, give false or misleading results when the two regions are contrasted on a numerical basis, grounded on the proportion of peculiar types,—numerous forms being thus reckoned as components of the Nearctic and Palæarctic regions which are properly tropical.

In North America, the division between characteristic temperate and tropical forms of life approximately coincides with the isotherm of 68° F., or somewhere between 68° and 70° F. This line begins on the Atlantic coast a little below the northern boundary of Florida, and runs thence westward along the Gulf coast to Southern Texas, and thence farther westward to the Pacific, not far from the international boundary between the United States and Mexico, swerving more or less northward or southward in accordance with the configuration and elevation of the land-surface. It thus leaves the greater part of the peninsula of Florida within the American Tropical Realm, to which the fauna of its

* Verhandl. der K. K. Zool.-Bot. Gesell. in Wien, Bd. xxv, 1876, pp. 50, 51; see also p. 62.

southern half is certainly closely allied. A portion of the Mexican highlands are undoubtedly to be included in the North Temperate Realm, but their fauna is too little known to admit of the boundary being at present definitely drawn.

On the other hand, the lower portion of the Great Colorado Valley and the coast region of Southern California are, perhaps, better referable to the American Tropical Realm than to the North Temperate. At the junction of the two realms, there must be a belt of debatable or doubtful ground. The approximate boundary I would place near the northern limit of distribution of such mammalian forms as *Nasua*, *Dicotyles*, *Manatus*, *Dasypus*, and the tropical species of *Felis* (as, *F. onca*, *F. pardalis*, *F. eyra*, and *F. yaguarundi*). This boundary also coincides quite nearly with the southern limit of distribution of the Lynxes, the Gray and Prairie Wolves, the Common Fox, the Mink, the Black and Grizzly Bears, the Wapati and Virginian Deer, the Bison, the Pronghorn, the Beaver, Prairie Dogs, Muskrat, the *Arvicolæ*, and the Moles (*Scalops* and *Condylura*). *Bassaris* is properly tropical, although straggling considerably farther northward than the other above-mentioned forms. Florida, for convenience, might be allowed to stand as a portion of the North Temperate Realm, although, as I have previously shown, it forms a distinct fauna, with strongly tropical affinities,* it having not less than twelve characteristically tropical genera of birds, several tropical genera of mammals (notably the Manatee and several Bats), and also several tropical genera of Reptiles and Batrachians, none of which range much, if any, to the northward of its southern half.

The southern boundary of the North Temperate Realm in the Old World may be doubtless approximately drawn near the same isotherm (about the mean annuals of 68° to 70° F.). This coincides closely with the southern boundary of the so-called Palæarctic Region. There is, however, here a broader belt of debatable or transitional ground than in the New World, into which so many tropical forms extend that it becomes almost a question whether the boundary between Tropical and Temperate life should not be carried considerably more to the northward, so as to leave Mr. Wallace's "sub-regions" 2 and 4 (Mediterranean and Manchurian) in the Tropical Realm rather than in the North Temperate. Despite, however, the presence of a considerable number of tropical genera in these regions, the North Temperate forms still greatly predominate. In the Western or "Mediterranean" district, for instance, we have species of *Macacus*, one of which even reaches the Spanish Peninsula. *Herpestes* has a similar northward extension. *Hyæna* and *Hystrix* range not only over most of this district, but also over the greater part of the Manchurian, where we again find a species of *Macacus*, and meet with *Semnopithecus*, while *Hyrax* just enters the Mediterranean from the southward. On the western border of the Manchurian we get also Pteropine Bats, and species of *Equidæ*, straggling remnants of the more

* Bull. Mus. Zoöl., vol. ii, pp. 391, 392.

northward extension of tropical life which inhabited this region during the middle and later portions of the Tertiary Period and in the Quaternary.

Divisions of the North Temperate Realm.—The North Temperate Realm is primarily divisible in two directions, giving in each two regions, namely, (1) by a longitudinal division into (*a*) a *North American Region* and (*b*) a *Europæo-Asiatic Region;* and (2) latitudinally, into (*a*) a *Cold Temperate* and (*b*) a *Warm Temperate Region.* The *Cold Temperate*, if limited on both continents by the isotherm of 36° F., presents a nearly uniform fauna throughout, its southern limit in both corresponding with the natural (that is, before modified by human agency) southern limit of distribution of *Tarandus* and *Alces.* While there is at this point in North America a well-marked transition in the fauna, the change in Europe and Asia appears to be less marked, the first important transition in the Old World being much farther southward, even as low almost as the isotherm of 60° F. Hence the divisions of the Temperate Realm in the Old World partake of the nature of *temperate* and *subtropical* rather than cold-temperate and warm-temperate. Here, in consequence of the great elevation and extent of the Himalayan Plateau, the northern or temperate division is greatly narrowed in Central Asia, where it becomes, according to Mr. Wallace, almost wholly separated into two quite widely detached regions, namely, the "Mediterranean" and "Manchurian Subregions".

As thus divided, the temperate and subtropical divisions of the Old World are very strongly marked. The latter consists mainly of Northern Africa, Asia Minor, Persia, Afghanistan and Beloochistan, Northern China, and Manchuria, with barely a narrow belt along the Mediterranean coast of Europe and the Spanish Peninsula. As already stated, it is strongly tinged with tropical forms. While there is a general prevalence of temperate types, we meet also with the large and essentially tropical forms of *Felis*, several Monkeys, several species of *Viverridæ*, *Hyæna*, *Hystrix*, *Equus*, and other distinctively tropical or subtropical types. The northern or temperate division of the Europæo-Asiatic Region seems to constitute two well-marked *provinces*, the one Eastern or European, the other Western or Asiatic. The former corresponds with Mr. Wallace's "European Subregion", exclusive of its northern third; the latter with his "Siberian Subregion", exclusive likewise of its boreal portion. For the southern or subtropical division I adopt the subdivisions proposed by Mr. Wallace, with, for the present, the boundaries he has assigned them,—namely, a Western or Mediterranean Province and an Eastern or Manchurian Province. These two provinces, as already noted, are quite widely separated, in consequence of the southward extension of the cold-temperate fauna over the Thibetan plateau to the Himalayas. The fauna of the Thibetan plateau is said by Mr. Blandford to be "essentially Boreal, Alpine and even Arctic types prevailing, the country having in many parts a climate scarcely equalled elsewhere for intensity of cold out of the Arctic Regions. This high barren tableland extends from Afghanistan to Yunan; it comprises the drainage-areas of the Upper Indus and the Sanpú, and is bounded on the north in its western portion by the Kuenluen range, but it is less defined and its boundaries less accurately known to the eastward, although much light has been thrown upon the subject by Prejewalski's explorations".* In the "List of Mammalia known to inhabit the Thibetan Plateau", given by Mr. Blandford, the only distinctively southern genus is *Equus*. The only peculiar genus is *Poëphagus*, but the list is evidently quite incomplete, the only Bat given being a species of *Plecotus*, and the only Insectivore a species of "*Crocidura*". *Budorcas*, usually attributed to Thibet, is excluded, and several other genera, as *Nectogale*, *Uropsilus*, and *Æluropus*, currently given as peculiar to the Thibet plateau, are not mentioned. While the Thibetan plains belong certainly to the colder division, so many types mainly restricted to this region occur that the question arises whether it may not be proper to recognize the region as a *Thibetian Province* of the Temperate Subregion.

North American Region.—The North American Region has been divided by Professor Baird into three "provinces", termed respectively "Eastern", "Middle", and "Western". Though not co-ordinate in point of differentiation with the divisions of the Europæo-Asiatic Region above recognized as provinces, they nevertheless possess distinctive features and form natural regions. They are of course far smaller in area, and possess a much smaller number of genera, but have about the same proportion of peculiar generic and subgeneric types.

In the subjoined tables an attempt is made to give lists of the genera of the two primary divisions of the North Temperate Realm, with approximate indications of their distribution in the various subdivisions of the two regions.†

* Proc. Zoöl. Soc. Lond., 1876, pp. 632, 633.

† In these lists, as elsewhere in the tabulated lists given in this paper, it is not assumed that the groups adopted as "genera" are always of co-ordinate value. The equation attempted is doubtless open in many cases to criticism. While the attempt is made to assume an intermediate position between undue conservatism and excessive multiplication in respect to groups assumed by different writers as "generic", the lists can of course be considered only as provisional. Again, it is occasionally difficult to decide whether certain genera should be assigned, even in a general way, to one of the faunal divisions rather than to another. However defective the result, the intent has of course been to give a fair presentation of the facts of distribution.

Genera of the North American Region.

[NOTE.—The names of circumpolar genera are in *italics;* those of genera peculiar to the region, in SMALL CAPITALS.]

Genera.	Subregions.		Provinces.		
	Cold Temperate.	Warm Temperate.	Eastern.	Middle.	Western.
Felis	−	+	+	+	+
Lynx	+	+	+	+	+
Canis	+	+	+	+	+
Vulpes	+	+	+	+	+
UROCYON	−	+	+	+	+
Gulo	+	−	+	−	−
Mustela	+	?	+	+	+
Putorius	+	+	+	+	+
MEPHITIS	+	+	+	+	+
SPILOGALE	−	+	+	+	+
TAXIDEA	+	+	+	+	+
Lutra	+	+	+	+	+
Enhydris	+	−	−	−	+
Procyon	+	+	+	+	+
Ursus	+	+	+	+	+
Phoca	+	−	+	−	+
Pagomys	+	−	+	−	−
Pagophilus	+	−	+	−	−
Erignathus	+	−	+	−	−
Halichœrus	+	−	+	−	−
Cystophora	+	−	+	−	−
Eumetopias	+	+ ?	−	−	+
Zalophus	+	+ ?	−	−	+
Callorhinus	+	+ ?	−	−	+
Alces	+	−	+	−	−
Rangifer	+	−	+	−	−
Cervus	+	+	+	+	+
CARIACUS	−	+	+	+	+
MAZAMA	+	−	−	+	+
Ovis	+	−	−	+	+
Bison	+	+	+	+	−
ANTILOCAPRA	−	+	−	+	−
Nyctinomus	−	+	?	+	+ ?
Nycticejus	−	+	+	?	−
Lasiurus	+	+	+	+	+
Vesperugo	+	+	+	+	+
Vespertilio	+	+	+	+	+
Synotus	−	+	+	+	− ?
Antrozous	−	+	−	+	+
SCALOPS	−	+	+	+	+
SCAPANUS	−	+	+	+	+
Urotrichus	−	+	−	−	+
NEOSOREX	−	+	+	+	+
Sorex	+	+	+	+	+
BLARINA	−	+	+	+ ?	−
Sciuropterus	+	+	+	+	+

Genera of the North American Region—Continued.

Genera.	Subregions.		Provinces.		
	Cold Temperate.	Warm Temperate.	Eastern.	Middle.	Western.
Sciurus	+	+	+	+	+
Tamias	+	+	+	+	+
Spermophilus	+	+	—	+	+
Cynomys	—	+	—	+	+
Arctomys	+	+	+	+	+
Haplodon	+	—	—	—	+
Neotoma	—	+	+	+	+
Sigmodon	—	+	+	—	—
Ochetodon	—	+	+	+	+
Hesperomys	+	+	+	+	+
Arvicola	+	+	+	+	+
Evotomys	+	—	+	—	—
Synaptomys	+	+	+	+	+
Fiber	+	+	+	+	+
Zapus	+	—	+	—	—
Perognathus	—	+	—	+	+
Cricetodipus	—	+	—	+	+
Dipodomys	—	+	—	+	+
Geomys	—	+	+	+	—
Thomomys	—	+	—	+	+
Castor	+	+	+	+	+
Erethizon	+	— ?	+	+	+
Lepus	+	+	+	+	+
Lagomys	+	—	— ?	+	+
Didelphys	—	+	+	+	+

Summary.

Whole number of genera	72
Peculiar to the region	23
Circumpolar	32
Of general distribution throughout the region	26
Occurring in the Cold Temperate Subregion	47
Occurring in the Warm Temperate Subregion	53–56
Land genera represented in the Eastern Province	47
Genera represented in the Middle Province	51
Land genera represented in the Western Province	48
Land genera restricted to the Eastern Province	*6
Genera common to the Middle and Western Provinces not represented in the Eastern Province	8
Genera restricted to the Middle Province	2
Land genera restricted to the Western Province	†3
Maritime genera restricted to the Eastern Province	5
Maritime genera restricted to the Western Province	5
Maritime genera occurring in both Eastern and Western Provinces	1

* Plus 5 maritime = 11.

† Plus 5 maritime = 7.

Europæo-Asiatic Region.—The Europæo-Asiatic Region embraces a far greater (about four times greater) area than the North American, and is physically much more highly diversified. It is similarly divisible into a Cold Temperate Subregion and a Warm Temperate Subregion, and is further differentiated into a number of well-marked provinces, two of which belong to the Cold Temperate Subregion, and three or more to the Warm Temperate Subregion.*

Genera of the Europæo-Asiatic Region.

[NOTE.—A few almost exclusively tropical genera, which barely reach or doubtfully extend a short distance over the southern boundary of the region, are omitted as being not properly faunal elements of the region.

The names of circumpolar genera are in *italics;* those of genera peculiar to the region in SMALL CAPITALS.]

Genera.	Subregions.		Provinces.			
	Cold Temperate.	Warm Temperate.	Western Temperate.	Eastern Temperate.	Mediterranean.	Manchurian.
Macacus	−	+	−	−	+	+
Felis	−	+	−	−	+	+
Lynx	+	+	+	+	+	+
Cynælurus	−	+	−	+	+	+
Genetta	−	+	−	−	+	−
Herpestes	−	+	−	−	+	+
Hyæna	−	+	−	−	+	+
Canis	+	+	+	+	+	+
Cuon	+	+	−	+	−	+
Vulpes	+	+	+	+	+	+
Arctonyx	−	+	−	+	−	+
NYCTEREUTES	−	+	−	−	−	+
Gulo	+	−	+	+	−	−
Mustela	+	−	+	+	−	−
Putorius	+	+	+	+	+	+
Lutra	+	+	+	+	+	+
LUTRONECTES	−	+	−	−	−	+
Enhydris	+	−	−	+	−	−
Mellivora	−	+	−	+	−	+
MELES	+	−	+	+	−	−
Ælurus	−	+	−	−	−	+
ÆLUROPUS	−	+	−	−	−	+

* I am far from sure that what is here recognized as the "Mediterranean Province" should not be subdivided, and the Eastern or Persian division recognized as a "Persian Province". If the Eastern, Middle, and Western divisions of the North American Region are to be accorded the rank of "Provinces", it may be necessary to admit, on similar grounds, a "Japanese Province"; but I am not at present prepared to adopt these divisions as "Provinces". To make the Provinces of the North American and Europæo-Asiatic Regions more nearly co-ordinate, I should prefer to unite the Middle and Western Provinces of the North American Region as forming a single Province. In fact, it seems doubtful whether the North American Region is differentiated into primary divisions that should be regarded as having co-ordinate rank with the Mediterranean and Manchurian divisions of the Europæo-Asiatic Region.

Genera of the Europæo-Asiatic Region—Continued.

Genera.	Subregions.		Provinces.			
	Cold Temperate.	Warm Temperate.	Western Temperate.	Eastern Temperate.	Mediterranean.	Manchurian.
Ursus	+	+	+	+	+	+
Callorhinus	+	−	−	+	−	−
Zalophus	+	+	−	+	+	+
Eumetopias	+	+	−	+	+	+
Phoca	+	−	+	+	−	−
Pagomys	+	−	+	+	−	−
Pagophilus	+	−	+	+	−	−
Halichœrus	+	−	+	−	−	−
Erignathus	+	−	+	−	+	−
Pelagius	−	+	−	−	+	−
Cystophora	+	−	+	−	−	−
Camelus	−	+	−	−	+	+
Alces	+	−	+	+	−	−
Rangifer	+	−	+	+	−	−
Cervus	+	+	+	+	+	+
Dama	−	+	−	−	+	−
Elaphodus	−	+	−	−	−	+
Lophotragus	−	+	−	−	−	+
Capreolus	+	+	+	−	+	+
Moschus	−	+	−	−	+?	+
Hydropotes	−	+	−	−	−	+
Bison	+	−	+	−	−	−
Poëphagus	−	+	−	−	−	+
Addax	−	+	−	−	+	−
Oryx	−	+	−	−	+	−
Gazella	−	+	−	−	+	−?
Procapra	−	+	−	+	+	+
Saiga	−	+	−	−	+?	+?
Panthalops	−	+	−	+	−	−
Budorcas	−	+	−	+	−	+
Rupicapra	−	+	−	−	+?	+?
Nemorhædus	−	+	−	−	−	+
Capra	−	+	−	−	+	−?
Ovis	+	+	−	+	+	+
Ammotragus	−	+	+	−	+	−
Sus	+	+	+	+	+	+
Asinus	−	+	−	−	+	+
Rhinolophus	−	+	−	−	+	+
Plecotus	−	+	−	−	+	−
Synotus	−	+	−	−	+	+
Vesperugo	+	+	+	+	+	+
Vespertilio	+	+	+	+	+	+
Miniopterus	−	+	−	−	+	+
Taphozous	−	+	−	−	+	+
Rhinopoma	−	+	−	−	+	+?
Nyctinomus	−	+	−	−	+	+
Erinaceus	+	+	+	+	+	+
Talpa	+	+	+	+	+	+

Genera of the Europæo-Asiatic Region—Continued.

Genera.	Subregions.		Provinces.			
	Cold Temperate.	Warm Temperate.	Western Temperate.	Eastern Temperate.	Mediterranean.	Manchurian.
SCAPTOCHIRUS	−	+	−	−	−	+
ANUROSOREX	−	+	−	−	−	+
MYOGALE	−	+	+?	−	+	−
NECTOGALE	+	−	−	+	−	−
Urotrichus	−	+	−	−	−	+
UROPSILUS	+?	+?	−	+?	−	+?
Sorex	+	+	+	+	+	+
CROSSOPUS	+	−	+	−	−	−
Crocidura	−	+	−	−	+?	+?
Mus	+	+	+	+	+	+
CRICETUS	+	+	+	+	+	+
CRICETULUS	−	+	−	−	−	+
Meriones	−	+	−	−	+	+?
RHOMBOMYS	−	+	−	−	+	+?
PSAMMOMYS	−	+	−	−	+	−
SMINTHUS	+	−	+	+	−	−
Arvicola	+	+	+	+	+	+
Evotomys	+	−	+	+	−	−
MYOSPALAX	−	+	−	−	−	+
ELLOBIUS	+	−	+	+?	−	−
SIPHNEUS	+	−	+	−	−	−
SPALAX	+	+	+	+	−	−
Dipus	−	+	−	−	+	+
ALACTAGA	+	+	+	+	+	−
MYOXUS	+	+	+	−	+	−
Muscardinus	+	+	+	+	+	+
Eliomys	−	+	+	−	+	−
Castor	+	−	+	+	−	−
Sciurus	+	+	+	+	+	+
Tamias	+	−	+	+	−	−
Sciuropterus	+	+	+	+	+	+
Pteromys	−	+	−	−	−	+
Spermophilus	+	−	+	+	−	+
Arctomys	+	−	+	+	−	−
Hystrix	−	+	−	−	+	−
Lagomys	+	−	+	+	−	−
Lepus	+	+	+	+	+	+

Summary.

Whole number of genera	107
Peculiar to the region	36
Circumpolar	22
Of general distribution throughout the region	15
Occurring in the Cold Temperate Subregion	54
Occurring in the Warm Temperate Subregion	80
Genera occurring in the Western Temperate (European) Province	48

Genera occurring in the Eastern Temperate (Asiatic) Province	46
Genera of the Mediterranean Province	60
Genera of the Manchurian Province	65
Genera common to the Eastern and Western Temperate Province	38
Genera common to the Mediterranean and Manchurian Province	50
Maritime genera of the Asiatic coast	8
Maritime genera of the European coast	6
Maritime genera common to both European and Asiatic coasts	3

In comparing the North American Region with the Europæo-Asiatic Region, the following resemblances and differences become apparent:—1. The number of genera in the Europæo-Asiatic Region is rather more than one-fourth greater than in the North American Region, with consequently a smaller proportion of circumpolar genera. 2. But this difference results almost wholly from the greater preponderance of peculiar types in the Southern Subregion, due evidently to the immensely greater extent and greater physical diversity of this portion of the Europæo-Asiatic Region as compared with the corresponding portion of the North American Region. 3. While the colder portions of the two regions have each about the same number of genera, which are in great part (*nearly two-thirds*) common to the two regions, the Warm Temperate (really Subtropical) Subregion of the Europæo-Asiatic Region has a far greater number of genera that do not extend to the northward of it than has the Warm Temperate Subregion of the North American Region, while a small proportion only (chiefly arctopolitan and subtropicopolitan) are common to the two subregions. Hence, 4. The two regions (Europæo-Asiatic and North American) are mainly differentiated (as already noticed) through the presence of genera limited to their southern subregions.

III.—AMERICAN TROPICAL REALM.

The American Tropical Realm is approximately bounded by the northern and southern mean annuals of 70° F. Its northern boundary has been already indicated in defining the southern limit of the North Temperate Realm, it being concurrent with the southern boundary of the North American Temperate Region. The southern boundary of the American Tropical Realm leaves the Atlantic coast near the thirtieth degree of south latitude, or near the southern extremity of Brazil, but in passing from the coast sweeps rapidly northward till it nearly or quite reaches the Tropic of Capricorn in Northeastern Buenos Ayres; it then bends to the southward and continues westward to the eastern base of the Andes. The Andean chain forms its western limit thence northward to Ecuador, where it crosses the Andean highlands and is again deflected southward, thus including a narrow belt of the coast region west of the Andes in Northwestern Peru.

As thus defined, the southern border of the American Tropical Realm is nearly coincident with the southern boundary of the "Brazilian

Region" as mapped by Mr. Wallace,* Brazil, nearly all of Paraguay, and Bolivia east of the Andes being included within this realm.

Its characteristic genera include all of the American Quadrumanes (families *Cebidæ* and *Mididæ*,=*Hapalidæ* of most authors), all the American Edentates, and nine-tenths of the American Marsupials. It is also the home of nearly all the American *Felidæ*, except the Lynxes.. It also has many peculiar genera of *Glires* and *Chiroptera*, while it almost altogether lacks the characteristic forms of mammalian life found in the northern temperate regions. Among the characteristic North American types unrepresented in the American Tropical Realm are, among Carnivores, not only the Lynxes, but the true Wolves and Foxes, the Martens, Wolverenes, Badgers, and Bears; among Ungulates, the Prong-horn, the Bison, Mountain Sheep, and Mountain Goat, and several important genera of the *Cervidæ;* among Rodents, the Spermophiles, Marmots, Muskrat, Beaver, Pouched Rats, "Gophers" (*Geomys* and *Thomomys*), the numerous species of *Arvicola*, etc.,—in short almost all of the prominent and characteristic genera of the order except the almost cosmopolitan genera *Lepus* and *Sciurus;* among Insectivores, all the Moles and Shrews, except a few forms of the latter, which extend over most of the Central American Region.

The American Tropical Realm is divisible into three regions,—the Antillean, the Central American, and the Brazilian. The Antillean Region includes only the West Indies and the southern extremity of Florida. The Central American Region embraces Mexico (exclusive of the elevated tablelands), the whole of Central America, and the extreme northern parts of South America (Venezuela north of the Orinoco Basin, Northern and Western New Granada, and most or all of that portion of Ecuador west of the Andes). The Brazilian Region comprises all the intertropical parts of South America not embraced in the Central American Region, including the whole area east of the Andes southward to the boundary already given.

Central American Region.—Of the genera occurring in the Central American Region (see subjoined table), only about one-ninth can be considered as peculiar to the region; about one-sixth are either sub-cosmopolitan or tropicopolitan; about three-fifths range also over the Brazilian Region, and a few over nearly all of South America; about one-half extend far into North America, among which are several that are also common to the greater part of the North Temperate Realm, while about one-eleventh are also found over most of both North America and South America. Aside from the few peculiar genera, the fauna is composed largely of genera common also to the Brazilian Region, which find their northern limit of distribution within the Central American Region, plus a very large proportion that extend southward from the North American Temperate Region, and which find their southern limit of distribution within the region under consideration. Its distinctive

* Geogr. Dist. Anim., vol. ii, map facing page 3.

feature is hence an approximately equal blending of temperate and tropical forms, whose respective habitats here overlap. Many of the northern forms do not quite reach the southern limit of the region, just as many of the southern forms do not quite reach its northern limit. It is distinguished from the North American Temperate Region by the preponderance of tropical life, and from the Brazilian Region by the copious intermingling therewith of northern forms, an element wholly lacking in the Brazilian Region.

Genera of the Central American Region.

Mainly or wholly restricted to the region.	Ranging also over much of the Brazilian Region.			
Bassariscyon.	Aluatta.	Cercoleptes.	Phyllostoma.	Arctopithecus.
Bassaris.	Cebus.	Nasua.	Arctibeus.	Tatusia.
Elasmognathus.	Sapajou.	Coassus.	Stenoderma.	Tamandua.
Macrotus.	Nyctipithecus.	Notophorus.	Centurio.	Cyclothurus.
Myxomys.	Callithrix.	Dicotyles.	Desmodus.	†Didelphys.
Heteromys.	Saimiris.	*Manatus.	Reithrodon.	Chironectes.
	Galictis.	Noctilio.	Cercolabes.	
	Grisonia.	Mormops.	Dasyprocta.	
	Conepatus.	Vampyrus.	Cœlogenys.	

Ranging also over much of the North American Temperate Region.		Tropicopolitan.	Subcosmopolitan.	
Urocyon.	§Spermophilus.	Nyctinomus.	Felis.	Lepus.
‡Procyon.	Neotoma.	Molossus.	Putorius.	
Cariacus.	Sigmodon.		Lutra.	
‡Lasiurus.	Ochetodon.		Vesperugo.	
Nycticejus.	Hesperomys.		Vespertilio.	
Blarina.	§Arvicola.		Sorex.	
§Sciuropterus.	Perognathus.		Sciurus.	

* Also West African.
† Also warmer parts of North America.
‡ Also nearly all of both North and South America.
§ Arctopolitan.

Summary.

Whole number of genera	63
Peculiar or mainly limited to the region	6
Occurring also over most of the Brazilian Region	40
Occurring also over much of the North American Region	24
Occurring also over most of both North and South America, but not in the Old World	5
Subcosmopolitan	8
Tropicopolitan	2

Antillean Region.—The Antillean Region differs from both the Central American and Brazilian most strongly in negative characters—through what it lacks rather than in what it has—although it possesses a number of peculiar genera. The *Chiroptera* form two-thirds of the genera and not less than five-sixths of the species. Of the eight peculiar genera, five are Bats, the others being *Solenodon* (the only Insectivore), *Capromys*, and the closely allied *Plagiodonta*, which together constitute a family peculiar to the region. Two orders—*Primates* and *Bruta*—highly characteristic of the Central American and Brazilian regions, are wholly absent. There are also no Ungulates, very few Carnivores, and very few Rodents; the latter, however, are of mostly peculiar species, as are many of the Bats. The single Insectivore is of a remarkable type, which finds its nearest ally in the remote island of Madagascar, the ordinary Insectivores of the neighboring Central American and North American Regions being wholly unrepresented.

Genera of the Antillean Region.

Peculiar to the region.	Tropical American.		Wide-ranging.
Nycticellus.	Nasua.	Mormops.	Lutra.
Lonchorhina.	Cercoleptes.	Macrotus.	Procyon.
Phyllodia.	Manatus.	Vampyrus.	Lasiurus.
Brachyphylla.	Natalus.	Arctibeus.	Vesperugo.
Phyllonicteris.	Thyroptera.	Stenoderma.	Vespertilio.
Solenodon.	Noctilio.	Heteromys.	* Hesperomys
Capromys.	Molossus.	Dasyprocta.	
Plagiodonta.	Nyctinomus.	Didelphys.	
	Chilomycteris.		

Brazilian Region.—Of about ninety commonly recognized genera, a little less than one-third may be considered as either wholly or mainly restricted to the region; a little less than another third range to the northward over much of the Central American Region, and may be considered as characteristic of the American Tropical Realm at large rather than of the Brazilian Region. About one-tenth of the remaining genera occur also over a large part of the Central American Region, while the remainder are divided about equally between tropicopolitan and cosmopolitan genera, and those that range southward over the South American Temperate Realm. One genus is also East Indian and another African, while quite a number range throughout the temperate and tropical parts of both Americas, and a few others over Temperate South America.

It is eminently characterized by its dozen genera of Monkeys, which, excepting a few that range into the Central American Region, are restricted wholly to this region; also by twelve to fifteen genera of Bats, which are scarcely found beyond its borders; nearly as many genera of Rodents, and quite a number of peculiar genera of other groups. Negatively it is characterized by the absence of Insectivores, the great bulk of the northern types of Carnivores, Ungulates, and Rodents. Its sole affinity with the life of the North Temperate Realm consists in the presence of a few such wide-ranging (cosmopolitan) genera as *Felis*, *Sciurus*, *Lepus*, *Vespertilio*, etc., and two other genera (*Procyon* and *Didelphys*) that range far into North America.

It is susceptible of division into several provinces, upon the detailed

* Dr. Coues gives *Hesperomys* (*Oryzomys*) *palustris* as Jamaican.—*Mon. N. Am. Rod.*, 116, foot-note.

consideration of which it is not proposed at present to enter. These are the Upper Amazonian Province, embracing the region drained by the Upper Amazon and its principal tributaries (Western Brazil and those portions of Peru and Bolivia east of the Andes); the Lower Amazonian Province, embracing the Lower Amazonian and Orinoco Basins; and the Southeast Brazilian Province, embracing Southeastern Brazil and Paraguay. They are characterized by the occurrence of numerous peculiar species rather than by peculiar genera. The genus *Lagothrix* appears to be confined, however, to the Upper Amazonian Province, *Chrysothrix* to the Lower Amazonian, and *Brachyteles* to the Southeast Brazilian, where occur also *Icticyon*, *Thous*, *Lycolepex*, etc., not found in the other regions, but ranging thence southward to Patagonia.

Genera of the Brazilian Region.

Mainly confined to the Brazilian Region.			
Lagothrix.	Pteronura.	Oxymicterus.	Bradypus.
Eriodes.	*Tapirus.	Dactylomys.	Prionodontes.
Pithecia.	Macrophyllum.	Cercomys.	Xenurus.
Brachyurus.	Vampyrus.	Mesomys.	Tolypeutes.
Nyctipithecus.	Saccopteryx.	Echimys.	Myrmecophaga.
Cheropotes.	Diphylla.	Loncheres.	Hyracodon.
Midas.	Habrothrix.	Chætomys.	Chironectes.
Icticyon.	Holochilus.	Hydrochœrus.	
Tropical America generally.			
Aluatta.	§Procyon.	Arctibeus.	Dasyprocta.
Cebus.	†Manatus.	Stenoderma.	Cœlogenys.
Sapajou.	Coassus.	Natalus.	Arctopithecus.
Callithrix.	Dicotyles.	Furripterus.	Chœlopus.
Saimiris.	Notophorus.	Thryoptera.	Tatusia.
Hapale.	Desmodus.	Noctilio.	Tamandua.
Galictis.	Schizostoma.	§Nycticejus.	Cyclothurus
Grisonia.	Centurio.	§Lasiurus.	§Didelphys.
Conepatus.	Sturnira.	‡Chilomycteris.	
Cercoleptes.	Phyllostoma.	‡Calomys.	
Nasua.	Glossophaga.	Cercolabes.	
Extending also over Temperate South America.		**Subcosmopolitan and tropicopolitan.**	
Chrysocyon.	Cavia.	Felis.	Vesperugo.
Lycalopex.	Kerodon.	Lutra.	Vespertilio.
Pseudalopex.	Myopotamus.	Nyctinomus.	Sciurus.
Blastocerus.	Dasypus.	Molossus.	Lepus.
Ctenomys.		Dysopes.	

* Also East Indian.
† Also West African.
‡ Also Temperate South America.
§ Also North American.

Summary.

Whole number of genera	90
Mainly restricted to the region	31
Of general distribution throughout the American Tropical Realm	41
Occurring also over much of the South American Temperate Realm	9
Occurring also in the warmer parts of the North Temperate Region	6
Tropicopolitan	3
Cosmopolitan	6

IV.—SOUTH AMERICAN TEMPERATE REALM.

What is here termed the South American Temperate Realm embraces all that portion of the South American continent and adjacent islands not included in the American Tropical Realm as already defined. It coincides very nearly with Mr. Wallace's "South Temperate America or Chilian Subregion".* Its northern limit on the Atlantic coast is near the thirtieth parallel. On leaving the Atlantic coast, the northern boundary passes obliquely northwestward, rising in the region of the Chaco Desert, to, or possibly a little beyond, the Tropic of Capricorn. Again descending to about the twenty-fifth parallel, it turns abruptly northward and eastward, along the eastern border of the Andean chain, nearly to the fifth degree of south latitude, near which point it strikes the Pacific coast. It thus embraces a large part of the great Andean plateau, with the neighboring coast region to the westward, nearly all the La Plata plains, and the region thence southward to Tierra del Fuego, which belongs also to this region.

As contrasted with the Tropical Realm to the northward, it is characterized, in respect to mammals, by the absence of all *Quadrumana* and the paucity of Edentates and Marsupials, there being neither Sloths nor Anteaters, while only two or three species of Opossums barely extend over its borders; the absence of all genera of Leaf-nosed Bats, and of not less than a dozen important genera of Rodents, the Coatis, the Kinkajou, the Tapirs, and many other genera characteristic of the American tropics.† As noted by Mr. Wallace, it is further characterized by the possession of the entire family of the *Chinchillidæ*, the genera *Auchenia*, *Habrocomus*, *Spalacopus*, *Actodon*, *Ctenomys*, *Dolichotis*, *Myopotamus*, *Chlamadophorus*, to which may be added the marine genera *Otaria*, *Arctocephalus*, *Morunga*, *Lobodon*, and *Stenorhynchus*, very few of which range beyond the northern border of this region. The Spectacled Bear is also confined to it, and here are also most largely developed the Murine genera *Calomys*, *Acodon*, and *Reithrodon*.

Although one of the smallest of the primary regions, it is apparently divisible into two more or less well-marked provinces, which may be

* Geog. Distr. Animals, vol. ii, p. 36, and map of the "Neotropical Region".

†Among the genera of the Brazilian Region here unrepresented are, aside from the *Quadrumana*, *Cercoleptes*, *Nasua*, *Tapirus*, *Bradypus*, *Chœlopus*, *Myrmecophaga*, *Tamandua*, *Cyclothurus*, *Phyllostoma*, *Glossophaga*, *Arctibeus*, *Dysopes* (and other genera of *Chiroptera*), *Hydrochœrus*, *Cercomys*, *Dactylomys*, *Loncheres*, *Echimys*, *Cœlogenys*, *Dasyprocta*, *Chœtomys*, *Cercolabes*, *Lepus*, *Sciurus*, *Habrothrix*, *Oxymycterus*, *Holochilus*, etc., =27+.

respectively termed the Andean and Pampean. The Andean Province is principally characterized by the presence of *Ursus* (*Tremarctus*) *ornatus*, the genera *Pudu*, *Furcifer*, *Tolypeutes*, *Chlamydophorus*, *Chinchilla*, *Lagidium*, *Spalacopus*, *Habrocomus*, and *Octodon*. *Auchenia* and several genera of Rodents range from the Andean Province southward over the plains of Patagonia to Tierra del Fuego. The Patagonian plains share largely in the general *facies* of the Andean fauna. A few genera only are restricted to the Pampean Province, these being mainly *Ctenomys*, *Lagostomus*, and *Dolichotis*. The differences between these two provinces relate mainly to species rather than to genera. The Pampean Province is much the smaller, embracing only the comparatively level pampa district bordering the La Plata and Lower Parana Rivers. So little is definitely known respecting the range of the mammals of this general region that it is scarcely practicable to attempt at present a definition of the boundaries between the Pampean and Andean divisions.

The relation of the South Temperate American to the Tropical American Realm is of course far closer than to any other, there being as usual a gradual transition between the two along their line of junction, through the extension of a few forms characteristic of the one for a short distance into the other, just as has been observed to be the case between the North Temperate and Tropical American Realms. It has, however, nothing in common with the North Temperate American Realm beyond the presence of a few cosmopolite types that extend across the intermediate Tropical Realm. So far as land mammals are concerned, it has no genera common to the South Temperate portions of the Old World, except a few that are almost cosmopolite. The case is different, however, with the marine species. Of the half dozen or more genera of Pinnipeds (the only marine forms we are here called upon to consider), none are peculiar to the shores of Temperate South America but are common to South Temperate and Antarctic shores generally. None of them, however, occur north of the tropics,* and it is hence only through these that there is any closer affinity between the mammalian life of this region and the South Temperate Zone generally than between it and that of north-temperate latitudes.

Of the thirty-four land genera below enumerated as occurring in the South American Temperate Realm, rather more than one-half (eighteen) are nearly or wholly confined to it. Most of the remainder extend far to the northward into Tropical America, and others reach North America, while five are almost cosmopolitan.

---

* *Otaria* alone reaches the Galapagos, which, although situated under the equator, are still within the influence of the cold Peruvian current, and appear to constitute an outlying element of the South American Temperate Realm.

Genera of the South American Temperate Realm.

Mainly or wholly limited to the region.		Marine, and mainly Antarctic.	Wide-ranging.	
Tremarctus.	Ctenomys.	Otaria.	Felis.	Cavia.
Furcifer.	Drymomys.	Arctocephalus.	Pseudalopex.	Myopotamus.
Pudu.	Dinomys.	Lobodon.	Lycolopex.	Calomys.
Auchenia.	Chinchilla.	Stenorhynchus.	Chrysocyon.	Habothrix.
Lophostoma.	Lagidium.	Morunga.	Putorius.	Tatusia.
Octodon.	Lagostomus.		Lutra.	Didelphys.
Spalacopus.	Dolichotis.		Conepatus.	
Habrocoma.	Chlamydophorus.		Vespertilio.	
Reithrodon.			Vesperugo.	
Acodon.			Kerodon.	

V.—THE INDO-AFRICAN REALM.

The Indo-African Realm consists mainly of Intertropical Africa and Intertropical Asia, to which it seems proper to add Extratropical South Africa. The small portion of Africa south of the Southern Tropic lies wholly within the warm-temperate zone. Its small extent and broad connection with Tropical Africa render its separation as a distinct realm (as I at one time rather hastily considered it) almost inadmissible, since it is especially open to the influence of the great intertropical African fauna, as is shown by the extension of many tropical forms down to within a few degrees of its southern extremity. The area really possessing a temperate climate is restricted to its extreme southern border, where alone appear the few generic and family types that do not have a very general range over the tropical portions of the continent. This area is many times smaller than the temperate portion of South America, but, though so small, has quite a number of peculiar genera, which impart to it quite distinctive features. It yet seems better to regard it as an appendage of the great Indo-African Realm rather than as a distinct primary region. Madagascar, with the Mascarene Islands, on the other hand, while perhaps possessing a closer affinity with Africa than with any other continental region, has yet a fauna made up so largely of peculiar types that it seems more in accordance with the facts of distribution to regard it as a separate primary region.

The Indo-African Realm, as thus restricted, forms a highly natural division. Although its two principal areas are quite widely separated, being in fact geographically almost wholly disassociated, they possess a wonderful degree of similarity. Of the fifty commonly recognized families of mammalia occurring within its limits, three-fifths are distributed throughout almost its whole extent. Of the remainder, one-half are confined to Africa, and one is African and American, leaving only nine in India that are unrepresented in Africa; three only of these latter are, however, peculiar to the Indian Region; all extend beyond it to the northward, five of them even occurring over the greater part of the northern hemisphere. Thus the African Region is the more specialized division, only a small portion of the tropical element in the Indian Region, through which it is differentiated from the great Europæo-Asiatic Temperate Region, being unrepresented in the African, while the African has three times as many peculiar families as the Indian.* As shown by the subjoined table, thirty of the fifty Indo-African families have a wide extralimital distribution, not less than one-fourth being emphatically cosmopolitan.

Families of Mammals represented in the Indo-African Realm, arranged to show (approximately) their distribution.

Occurring in the Indian Region, but not in the African.	Peculiar to the African Region.[1]	Common to both regions.		Common to both Regions, and also of wide extra-limital range.
* TARSIIDÆ.	Protelidæ.	Simiidæ.	Nycteridæ.	Felidæ.
† Æluridæ.	Hippopotamidæ.	Semnopithecidæ.	Erinaceidæ.	Canidæ.
‡ Ursidæ.	Phacochœridæ.	Cynopithecidæ.	Myoxidæ.	Mustelidæ.
‡ Cervidæ.	Giraffidæ.	Lemuridæ.	Spalacidæ.	Bovidæ.
† Camelidæ.	Hyracidæ.	Viverridæ.	Dipodidæ.	Pteropodidæ.
§ Tapiridæ.	Chrysochloridæ.	Hyænidæ.	Manididæ.	Rhinolophidæ.
* GALEOPITHECIDÆ.	Macroscelidæ.	Tragulidæ.		Vespertilionidæ.
‡ Talpidæ.	Potamogalidæ.	Equidæ.		Soricidæ.
* TUPAYIDÆ.	Lophyiomyidæ.	Suidæ.		Octodontidæ.
	Orycteropodidæ.	Rhinocerotidæ.		Sciuridæ.
		Elephantidæ.		Hystricidæ.
		Halicoridæ.		Leporidæ.

[1] The *Trichechidæ* (= *Manatidæ*) occur in Africa but not in India, but are found also in the warmer parts of America.

* Wholly restricted to the Indian Region.
† Mainly restricted to the Indian Region.
‡ Of wide extralimital range.
§ Found also in Intertropical America.
|| Chiefly African.

Summary.

Whole number	50
Of general distribution throughout the realm	30
Peculiar to the African Region	10
Peculiar to the Indian Region	3
Occurring in the Indian Region, but not in the African	6
Of wide extralimital range	16

African Region.—The African Region, as here recognized, is nearly equivalent to Mr. Wallace's "Ethiopian Region", with the exclusion

* Mr. Wallace has arrived at rather different conclusions respecting the specialization of the African Region, since he considers its specialization due wholly to the peculiar forms developed in Madagascar. Deducting these—for he considers Madagascar and its neighboring islands as forming a "subregion" merely of the "Palæotropical"—he believes would leave, in respect to specialization, the African and Indian Regions "nearly equal". In this comparison, however, I wholly exclude the Madagascan or "Lemurian" fauna, and still find Africa a considerably more specialized region.

of his "Lemurian Subregion". Its northern boundary will be provisionally considered as the northern mean annual of 70° F.

As thus limited, the greater part of the Arabian Peninsula and the southern portion of the Great Sahara belong to it. But just how much of the latter belongs here, and how much to the Mediterranean Region, cannot at present be readily determined. As already noticed, it consists largely of transitional ground, and is as yet quite imperfectly known. It is to some extent, doubtless, also a barrier region; but that it is by no means an impassable obstacle is sufficiently shown by the large number of generic types of mammals that extend from the Indian Region as far southward even as the Cape of Good Hope. Even if it were an insurmountable barrier, the comparatively humid and fertile eastern coast border would afford a sufficient highway of intercommunication between Tropical Asia and Tropical Africa, and the community of life of the two regions shows that for long ages there has been this open way of interchange.

The African Region, considering its great extent and its tropical climate, is to a great degree zoölogically a unit, yet it is by no means homogeneous. At least, three subdivisions may be recognized, each of which is characterized by many peculiar genera. These subregions have already been characterized by Mr. Wallace under the names of Eastern, Western, and Southern. The Western (West African Province) consists of the humid, heavily wooded region of the west coast, extending to a considerable, but at present not definitely determinable, distance into the interior, but probably with boundaries nearly as drawn by Mr. Wallace.* The Eastern (East African Province) includes the remainder of Intertropical Africa, while to the Southern (South African Province) belongs the southern extratropical portion of the continent.

Of these divisions, the Eastern contains the greatest number of genera, as it likewise contains by far the greatest area; but it is the least specialized, only *two-fifteenths* of its genera being peculiar to it, while of the genera of each of the other regions about *one-fourth* are peculiar. Nearly one-half (about forty-four per cent.) of the genera of the Eastern Province have a more or less general distribution over the whole African Region, while only a little more than a third (thirty-three to thirty-eight per cent.) of the genera of the other province have a similarly wide range.

A much larger proportion of Indian genera are represented in the Eastern and Southern Provinces than in the Western. This difference is due to obvious conditions, the fertile belt of the Nile district and adjoining coast forming an easy way of intercommunication between the

* The conclusions and details here presented were worked out independently and *de novo* by the present writer. That they agree so closely with the views and results attained by Mr. Wallace, so far as Africa south of the Great Desert is concerned, is to me a source of gratification. In order to avoid unconscious bias I purposely avoided a detailed study of Mr. Wallace's writings on this subject till my own results were written out, and on then comparing my own conclusions with those reached by Mr. Wallace, became for the first time aware of their close agreement.

two former not equally open to the Western Province. The Eastern and Southern Provinces further resemble each other in consisting largely of grassy plains, and in being, *par excellence*, the land of Antelopes. On the other hand, the Western Province, in consequence of its moist climate and dense forests, is the metropolis of the African Quadrumanes, to which region no less than six genera are restricted, and where all but one are represented, while only four occur in the Eastern, and merely a few outlying species reach the Southern. Hence the Eastern and Southern Provinces are far more closely allied than is either with the Western.

Eastern Province.—The East African Province or "Subregion" includes, as claimed by Mr. Wallace, not only East Africa proper, but also a considerable portion of the Great Sahara and the whole of the northern portion of Tropical South Africa, thus bounding the Western Province on three sides. In other words, it not only includes East Africa and Southern Arabia, but all of Tropical Africa, except the western portion, situated (speaking generally) between latitude 15° north and latitude about 22° south. As is well known, it consists mainly of a moderately elevated plateau, rising, in Abyssinia, into lofty mountains. It is generally an open region, "covered with a vegetation of high grasses or thorny shrubs, with scattered trees and isolated patches of forest in favorable situations. The only parts where extensive continuous forests occur are on the eastern and western slopes of the great Abyssinian plateau, and on the Mozambique coast from Zanzibar to Sofala."* It is worthy of note that the species peculiar to the province occur almost exclusively in Mozambique, or in Abyssinia and adjoining portions of Northeast Africa, a few extending into the Arabian Peninsula.

Of the ninety genera occurring in this province, ten, which are almost cosmopolite, may be considered as having too wide a range to possess any special significance. Of the remaining eighty, about one-fourth are found also in the Indian Region, leaving three-fourths (thirty-nine) as peculiarly African. Of these, twelve only are restricted to the Eastern Province, sixteen being common to the Southern Province, and ten to the Western. The subjoined tabular list indicates approximately the distribution of the genera of the Eastern Province.

---

* Wallace, Geogr. Dist. Anim., vol. i, p. 259.

Genera of the East African Province.

Restricted to the province.	Exclusively African, but occurring also in the other provinces.		
*Theropithecus.	Colobus.	Giraffa.	Phacochœrus.
*Galerella.	Guerza.	Oreas.	Rhinaster.
Rhinogale.	Cercopithecus.	Tragelaphus.	Hyrax.
*Neotragus.	Cynocephalus.	Oreotragus.	Dendrohyrax.
Nesotragus.	Galago.	Æpyceros.	Epomophorus.
† Petrodromus.	Athylax.	Kobus.	Macroscelides.
† Rhynchocyon.	Ichneumea.	Nanotragus.	Cricetomys.
† Saccostomus.	Bdeogale.	Cephalophus.	Steatomys.
† Peleomys.	Helogale.	Ægocerus.	Otomys.
*Lophiomys.	Mungos.	Alcelaphus.	Georychus.
† Heliophobius.	Crossarchus.	Connochetes.	Xerus.
Pectinator.	Lycaon.	Hippopotamus.	Aulacodus.
	Zorilla.	Potamochœrus.	Orycteropus.

Occurring also in the Indian Region.			Wide-ranging.
Viverra.	Elephas.	Scotophilus.	Felis.
Genetta.	‡ Sus.	Miniopterus.	Canis.
Herpestes.	Asinus.	Taphozous.	Vespertilio.
Calogale.	Halicore.	Rhinopoma.	Vesperugo.
Mellivora.	Cynonycteris.	Nyctinomus.	Erinaceus.
Aonyx.	Cynopterus.	Crocidura.	Mus.
Hyæna.	Rhinolophus.	Acanthomys.	Dipus.
Bubalus.	Phillorhina.	Rhizomys.	Meriones.
Oryx.	Megaderma.	Hystrix.	Sciurus.
Gazella.	Nycteris	Manis.	Lepus.

* Restricted to Abyssinia and Northeast Africa. † Restricted to Mozambique.
‡ See Rolleston, Trans. Linn. Soc. Lond., 2d ser., Zoöl., vol. i, pp. 256, 257, 1877.

The Southern Province.—The South African Province consists of only that small portion of the continent lying south of the Southern Tropic, and is hence situated wholly within the southern warm-temperate zone. In consequence of its configuration, its limited extension, and its geographical position in relation to Intertropical Africa, it could scarcely be expected to form more than an appendage of the inter-tropical zone, and such it proves really to be. Its area is equal to only about one-tenth of that of the Eastern Province, yet it has eight-ninths as many genera, fully two-thirds of which are common to the two. It hence presents to only a limited degree the features of a strictly temperate fauna, and these become prominent only over the narrow belt of country south of the mountain ranges forming the northern boundary of Cape Colony and Caffraria; but here even there is a strong invasion of essentially tropical forms.

In general *facies* it differs little zoölogically from the Eastern Province, of which it is merely a somewhat modified continuation. From its semi-temperate character it is less rich in Quadrumanes, but many other properly tropical types range nearly or quite to its southern border. It has, however, about one-fourth more peculiar genera, divided about equally, and mainly between Carnivores and Rodents, four only being Antelopes, and one only (*Chrysochloris*) an Insectivore. Of the twenty-four genera common also to the Indian Region, one-third are Chiropters. The remaining genera are, with very few exceptions, such as occur also in the Eastern Province, only three or four being common to the Southern and Western Provinces that do not also occur in the Eastern.

Of the eighty-two genera below enumerated as occurring in the Southern Province, a considerable portion are restricted to its southern half, while many others extend only over its northern portions. A few others, while mainly restricted to this region, and eminently characteristic of it, also extend somewhat into the Eastern Province.

Genera of the South African Province.

Restricted to the province.	Ranging into Tropical Africa.		Occurring also in the Indian Region.	
Ariela.	Galago.	Cephalophus.	Cyanælurus.	Miniopteris.
Cynictis.	Athylax.	Eleotragus.	Genetta.	Scotophilus.
Suricata.	Ichneumia.	Ægocerus.	Herpestes.	Taphozous.
Proteles.	Helogale.	Rhinaster.	Calogale.	Rhinopoma.
Megalotis.	Mungos.	Hyrax.	Mellivora.	Crocidura.
Hydrogale.	Lycaon.	Dendrohyrax.	Aonyx.	Hystrix.
Strepsiceros.	Zorilla.	Macroscelides.	Hyæna.	Manis.
Antidorcas.	Phacochœrus.	Steatomys.	Oryx.	
Scopophorus.	Giraffa.	Otomys.	Bubalus.	
Pelea.	Oreas.	Georychus.	Gazella.	
Chrysochloris.	Tragelaphus.	Graphiurus.	Asinus.	
Dendromys.	Damalis.	Xerus.	Elephas.	
Malacothrix.	Alcelaphus.	Aulacodus.	Cynonycteris.	
Mystromys.	Connochetes.	Orycteropus.	Cynopterus.	
Bathyergus.	Kobus.		Rhinolophus.	
Pedetes.	Æpyceros.		Phillorhina.	
Petromys.	Calotragus.		Nycteris.	
Wide-ranging.				
Felis.	Vespertilio.	Erinaceus.	Meriones.	Sciurus.
Canis.	Vesperugo.	Mus.	Dipus.	Lepus.

The Western Province.—As already stated, the Western Province differs greatly in respect to its physical characteristics from either of the other provinces of the African Region, and has, in consequence, a correspondingly specialized mammalian fauna. It resembles the Indian Region in its hot, damp climate and dense forests. And its fauna, though distinguished by many peculier genera, is also, in respect to its *general facies*, more like that of the Indian Region than is the fauna of any other portion of the African Region. It is similarly rich in the higher Quadrumanes and poor in Antelopes, while it shares with the

Indian Region the possession of the *Tragulidæ*. Its peculiar genera consist largely of Anthropoid Apes, found elsewhere only in India, but also includes several each of Carnivores Bats, and Rodents. It is pre-eminently the tropical province of the African Region. While it contains a smaller number of genera than either of the others, it has relatively a much larger number restricted to it, having eighteen peculiar genera out of a total number of seventy-five, while the Eastern Province, with ninety-one genera, has only twelve that are peculiar, and the Southern seventeen out of eighty-two.

Genera of the West African Province.

Restricted to the province.		Restricted to the African Region.		
Gorilla.	Cephalopus.	Cercopithecus.	Tragelaphus.	Rhinaster.
Mimetes.	Hyomoschus.	Cynocephalus.	Kobus.	Hyrax.
Miopithecus.	Chœropus.	Colobus.	Nanotragus.	Dendrohyrax.
Cercocebus.	*Manatus.	Guereza.	Ægocerus.	Epomophorus.
Arctocebus.	Hypsignathus.	Galago.	Damalis.	Xerus.
Perodicticus.	Potamogale.	Aythlax.	Oreas.	Aulacodus.
Poiana.	Lasiomys.	Mungos.	Hippopotamus.	Cricetomys.
Nandinia.	Anomalurus.	Zorilla.	Potamochœrus.	Graphiurus.
Adenota.		Crossarchus.	Phacochœrus.	
Occurring also in the Indian Region.				**Wide-ranging.**
Viverra.	Zorilla.	Phillorhina.	Crocidura.	Felis.
Herpestes.	Bubalus.	Nycteris.	Atherura.	Canis
Genetta.	Antilope.	Nyctinomus.	Manis.	Vespertilio.
Calogale.	Elephas.	Miniopterus.		Vesperugo.
Hyæna.	Cynonycteris.	Scotophilus.		Mus.
Aonyx.	Cynopterus.	Taphozous.		Sciurus.
Mellivora.	Rhinolophus.	Rhinopoma.		Lepus.

* Also American.

General Summary.

The number of genera represented in the African Region, and their range, is approximately as follows :—

	Eastern Province.	Southern Province.	Western Province.
Peculiar genera	12	17	18
Restricted to the African Region, but occurring more or less generally over two or more of the provinces	39	31	26
Occurring also in the Indian Region	30	24	24
Wide-ranging	10	10	7
Whole number	91	82	75

Indian Region.—The Indian Region may be defined, in general terms, as consisting of Intertropical Asia. It hence embraces Continental India from the Lower Indus to the Formosa Straits, the islands of the Indian Archipelago, as well as Formosa, the Philippines, Celebes, and all of the Sunda Islands. As far as the mammalia are concerned, only two primary subdivisions, or provinces, seem to be recognizable, the one a Northern, or Continental, the other a Southern, or Insular ("Malayan"). The former, or Continental, includes nearly all of the Hindostan and Indo-Chinese Peninsulas, excepting the extreme southern border of the latter and Malacca. These areas belong to the Insular Province, which comprises not only Borneo, Sumatra, and Java, but all of the above-named smaller islands to the eastward, except Formosa, which pertains to the Continental Province.

The long, narrow Malaccan Peninsula is almost insular in position and character, and agrees far better, climatologically, and in its productions, with Borneo and Sumatra, than with the mainland to the northward, as does, in fact, the extreme coast border of the mainland, embracing Lower Cochin China, Cambodia, etc. The small outlying islands to the eastward have nothing in common with the Australian Realm (if we exclude the wide-ranging *Chiroptera* and a few marine forms, which are, of all mammals, of least importance in a zoögeographical point of view), except the single Marsupial genus *Cuscus* occurring in Timor and Celebes, while no placental mammals except *Sus*, a few Murine genera, the Dugong, and *Chiroptera*, reach any portion of the Australian Realm. Malacca, Borneo, and Sumatra form the central and typical portion of the Insular or Malayan Province, being, from their larger area and closer proximity to each other and to the tropical mainland, far richer in genera and species than the smaller and more remote islands to the southward and eastward. Even Java has a less varied mammalian fauna than either Borneo or Sumatra, and thus differs from them negatively rather than by the possession of peculiar types. Thence eastward, throughout the Sunda Islands, the differences are almost wholly such as result from the small size and isolated position of these insular areas, through a gradual disappearance of many types present in the larger islands. The Philippines, for similar reasons, lack a large proportion of the genera found in the central portion of the province, while those they do possess, with few exceptions, are such as are common to the larger areas. The few that are peculiar are Indian, rather than Australian, in their affinities.

Celebes and Timor contain one strictly Australian genus (*Cuscus*, represented by several species), but the few other mammals found there are either Indian or possess strictly Indian or Indo-African affinities. Hence I fail to see any good reason for assigning Celebes and all the smaller Sunda Islands to the Papuan Province, as Mr. Wallace and others have done, but abundant evidence that such is not their real affinity. Even Mr. Wallace's own tables of distribution show at a glance the wide disassociation of these islands from the Papuan fauna, and their much nearer relation to the Indian, there being but one typically Australian or Papuan form represented in any of them, while none of the placental land mammals (excepting several subtropicopolitan genera of Bats and a few Muriform Rodents) are common to these islands and the Papuan-Australian division. The genera peculiar to the Philippines and Celebes (except *Cuscus* in the latter) have little if any more significance than the occurrence in Borneo and Sumatra of a few genera wholly restricted to one or the other of these last-named islands.

Ceylon and the adjoining low-coast portions of the Hindostan Peninsula are more tropical in character than the plateau region to the northward. While a few genera are restricted to this small area, and many more species occur here that are not found to the northward, the differentiation seems hardly great enough to warrant the separation of these areas as a region of co-ordinate rank with the "Malayan". It hence seems to me that Mr. Wallace has too emphatically recognized this comparatively unimportant difference in making it the basis of a distinct subregion (termed by him the "Ceylonese Subregion"). The only mammalian genera peculiar to this division are a genus of Lemurs (*Loris*), three genera (or subgenera) of *Herpestinæ* (*Calictis*, *Tæniogale*, *Onychogale*), and a genus of Mice (*Platacanthomys*), each represented by a single species, and, so far as known, of limited distribution.

Continental Province.—As already intimated, the Continental Province includes nearly all of Hindostan and Indo-China, or the whole of the tropical portion of the Asiatic continent excepting Malacca and the southern portions of Tenasserim, Siam, Cambodia, and Cochin China. It also extends into Southern China somewhat beyond the tropic (probably to the divide between the Li-kiang and Yang-tse-kiang Rivers), and also to the southern slope of the Himalayas.*

The plains of the Upper Indus appear, however, to belong to the Temperate Region to the northward, as does probably most of the country northwest of Delhi. The greater part of the interior of the Hindostan Peninsula has a less tropical character and a less varied fauna than Bengal, Assam, and Burmah, situated under the same parallels. I cannot agree, however, with Messrs. Blyth, Blandford, and von Pelzeln,†

* "On the southern slope of the Himalayas there is everywhere, until it has been cleared, luxuriant forest up to at least 12,000 feet above the sea, inhabited by a fauna which extends, without any great change of generic forms, throughout the Malay Peninsula and into the hill tracts of some at least of the Malay Islands."—BLANDFORD, *Proc. Zoöl. Soc. Lond.*, 1876, p. 632.

† Mr. Blyth makes "Hindostân proper, or the plains of Upper India east and south of the North West desert; Dukhun, or tableland of the Peninsula of India, and the intervening territory, inclusive of the Vindhaian ghâts; Coromandel Coast and low northern half of Ceylon" a subregion of his "Ethiopian Region" (Nature, vol. iii, p. 428). Mr. Blandford holds that the "hills of Southern India with the Malabar Coast and Southern Ceylon form a province of the Malay region, *whilst the greater* portion of the Indian peninsula is African in its affinities" (Proc. Zoöl. Soc. Lond., 1876, p. 632). Von Pelzeln considers India proper, from the Lower Brahmaputra River westward, a distinct primary region, which he calls the "hindostanische Region". His "malayische Region" hence consists of Warm-temperate and Tropical Asia, minus the Hindostan Peninsula, to which he adds the Philippines, Borneo, Bali, Java, and Sumatra. It includes China as far as the Yang-tse-kiang River, and the Himalayan plateau from

that the larger part of Hindostan should be joined to the African Region rather than the Indian, since only a very few African genera occur here that do not also range far to the eastward, or almost throughout the Indian Region. According to von Pelzeln,* about one-third of the genera of the "hindostanischen Fauna" are peculiar to it, while it shares almost another third with Indo-China. The remaining third (fourteen genera) are common to the African Region, *but all except four of them occur also more or less generally over the Indian Region.* Of these, two (*Hyæna* and "Ratelus" = *Mellivora*) scarcely reach the limits of the Indian Region as here defined. Among the genera given by him as peculiar are, however, several that range beyond the Indian Peninsula.

There is more reason for Mr. Wallace's separation of the Hindostan Peninsula from the Indo-Chinese portion of the Indian Region, and its subdivision into two "subregions"—a northern "Hindostan Subregion" and a southern "Ceylonese Subregion". As already shown, the latter has a number of peculiar forms, while three or four genera are also peculiar to the Hindostan Peninsula at large. But the scale of division that would make the Hindostan Peninsula separable into two subregions would also require a somewhat similar subdivision of Indo-China, making four divisions of what I here term the Continental Province. While these divisions would have some natural basis, they are too detailed to come into the category of divisions for which I adopt the term "province".

Continental Province.—The Continental Province, with the limitations here assumed, is nearly equivalent to Mr. Wallace's three "subregions", termed respectively "Hindostan", "Ceylonese", and "Indo-Chinese". Of about ninety-four genera represented in it, about two-thirds have a pretty general range throughout the province, while only about one-eighth are limited to the Hindostanese portion, including those already named as almost peculiar to Ceylon and the low coast region east of the Eastern Ghâts. Excluding about a dozen that range over at least half the surface of the globe, one-third of the remainder (more than one-fourth of the whole) are common to the African Region; more than one-half (almost one-half of the whole) are restricted to the Indian Region and a little more than one-fifth (about one-eighth of all) are peculiar to the province. This shows, as already noted in discussing the fauna

Burmah, Assam, and Bengal to the Kuenluen Mountains, thus embracing Nepal, Butan, and Thibet. It is divided into five subregions, the two northernmost of which belong mainly to the North Temperate Realm. (Festschrift z. Feier des fünfundzwanzigjährigen Bestehens d. K.-K. Zool.-Bot. Gesells. in Wien, 1876, pp. 53–74 u. Karte.) The fauna of the Thibetan plateau, as claimed by Mr. Blandford, being boreal and alpine, and having almost nothing in common with the tropical region to the southward, the artificial character of von Pelzeln's "subregions" is shown by his assuming the Yang-tse-kiang River to be a natural boundary between two primary regions, and his separation of Malacca from Sumatra and Borneo to form a part of his "hinter-indische Unterabtheilung", which thus consists of the whole of the Indo-Chinese Peninsula down to the very southern extremity of Malacca!

* Verhandl. d. K.-K. Zool.-Bot. Gesells. in Wien, xxv. Bd., p. 57, 1875.

of the Indo-African Realm, how strong an affinity exists between the African and Indian Regions, two-fifths of all the genera of the Indian Region which have an extralimital range occurring also in the African Region. The close affinity of the two provinces of the Indian Region is shown by the fact that two-thirds of the peculiar Indian genera found in the Northern or Continental division range also into the Southern or Insular. As will be shown later, the Insular Province is the more highly specialized of the two divisions.

Genera of the Continental Province.

Restricted to the Indian Region.				
Restricted to the province.		Occurring also in the Insular Province.		
Rhinopithecus.	[3]Tetracerus.	Hylobates.	Gymnopus.	Tragulus.
[1]Loris.	Porcula.	Semnopithecus.	Helictis.	Rhinoceros.
[2]Urva.	Eonycteris.	Macacus.	Mydaus.	Pteropus.
[1]Tæniogale.	Cœlops.	Nycticebus.	Cuon.	Macroglossus.
[1]Calictis.	Nesokia.	Viverricula.	Bibos.	Harpiocephalus.
[1]Onychogale.	Neodon.	Arctitis.	Rusa.	Tupaia.
Melursus.	[1]Platycanthomys.	Prionodon.	Rucervus.	Pteromys.
[3]Tragops.		Paguma.	Axis.	Spalacopus.
[3]Portax.		Paradoxurus.	Cervulus.	Acanthion.

Of wide extralimital range.				
Ranging into the African Region.		Ranging into the Europæo-Asiatic Region.		Subcosmopolite.
Viverra.	[4]Miniopterus.	[5]Putorius.	[6]Sus.	Felis.
Herpestes.	Taphozous.	[2]Ælurus.	[6]Rhinolophus.	Canis.
Aonyx.	Rhinopoma.	Arctonyx.	[6]Erinaceus.	Vulpes.
Bubalus.	[4]Nyctinomus.	[5]Ursus.	Talpa.	Mustela.
Halicore.	Crocidura.	Bos.	[5]Sorex.	Lutra.
Elephas.	Rhizomys.	Nemorhædus.	[5]Sciuropterus.	Vespertilio.
Cynopterus.	Gerbillus.	Capra.		Vesperugo.
Cynonycteris.	Meriones.	Gazella.		Sciurus.
Phillorhina.	Acanthomys.	Procapra.		Mus.
[4]Megaderma.	Hystrix.	[5]Cervus.		Lepus.
Scotophilus.	Atherura.	Hydropotes.		
Kerivoula.	Manis.	Moschus.		

[1]Restricted to Ceylon and Southern Hindostan.
[2]Restricted to the northern part.
[3]Hindostan generally.
[4]Also tropics of America.
[5]Whole northern hemisphere.
[6]Also African.

Summary.

Whole number	94
Restricted to the Indian Region	43
Restricted (almost wholly) to the province	16
Other genera ranging over most of the Indian Region and restricted to it	27
Common to the African Region	28
Common to portions of the Europæo-Asiatic Region	34
Ranging over most of the northern hemisphere	17
Nearly cosmopolite	10
Restricted to Southern Hindostan and Ceylon	5

Insular or "Malayan" Province.—The northern boundary of the Insular Province is not at present easily determinable, but it is quite evident that, as already stated, the southern maritime portions of Indo-China belong here rather than with the northern division of the Indian Region. To the southward and eastward it embraces, as already explained, the Sunda Islands, the Philippines, and Celebes. Of the eighty-three genera occurring in it, twenty-five, or nearly one fourth, are peculiar, while twenty-seven others do not range beyond the Indian Province. Twenty of the remainder are properly Indo-African genera, while about a dozen others have a wide extralimital range, and about the same number have a very local range, the larger islands having each one or two peculiar genera. Aside from several tropicopolitan genera of Bats, and the wide-ranging genera *Sus* and *Mus*, only one genus is properly Australian, and this is a straggler that merely reaches Timor and Celebes. As would be expected, the larger central islands, together with Malacca and the mainland belt, possess the richest and most varied fauna, the smaller outlying islands presenting a paucity of types proportionate to their size and isolation.

Timor, considering its close proximity to Australia, is remarkably free from Australian forms, presenting, in common with Celebes, the single Marsupial genus *Cuscus*. The distribution of the genera of this province is roughly indicated in the subjoined table. Notwithstanding its much smaller land-area, and the fact that it has ten less genera than the Continental Province, it has, as would be naturally expected, many more peculiar genera,* the ratio of peculiar genera in the one being as 16 to 94, and in the other as 25 to 83.

* Four, however, are peculiar only in regard to the Indian Region, they being simply wide-ranging tropical forms that are unrepresented in the Continental Province.

Genera of the Insular Province.

Restricted to the Indian Region.			
Restricted to the province.		Ranging over much of the Continental Province.	
Simia.	[6] Megærops.	Hylobates.	Rusa.
Simianga.	[8] Harpyia.	Semnopithecus.	Rucervus.
[1] Nasalis.	[4] Phyllotis.	Macacus.	Axis.
[3] Cynopithecus.	Chiromeles.	Nycticebus.	Cervulus.
Tarsius.	[8] Emballonura.	Viverricula.	Tragulus.
Hemigalea.	Hylomys.	Arctitis.	Rhinoceros.
Arctogale.	[1] Ptilocerus.	Prionodon.	[8] Pteropus.
[2] Cynogale.	Gymnura.	Paguma.	[8] Macroglossus.
[6] Barangia.	[4] Phlæomys.	Paradoxurus.	[8] Harpiocephalus.
Helarctos.	Rhinosciurus.	Gymnopus.	Tupaia.
[5] Anoa.	[5] Cuscus.	Helictis.	Pteromys.
[5] Babirusa.		Mydaus.	Spalacopus.
[9] Tapirus.		Cuon.	Acanthion.
Galeopithecus.		Bibos.	

Ranging into Africa and elsewhere.		
African.		Wide-ranging.
Viverra.	[7] Nycteris.	Felis.
Herpestes.	[8] Scotophilus.	Canis.
Aonyx.	[8] Kerivoula.	Mustela.
Bubalus.	[8] Miniopterus.	Lutra.
Sus.	[8] Taphozous.	Rhinolophus.
Elephas.	[8] Nyctinomus.[9]	Vespertilio.
Halicore.	Crocidura.	Vesperugo.
[8] Cynopterus.	Rhizomys.	Sorex.
[8] Cynonycteris.	Manis.	Mus.
[8] Phyllorhina.		Sciurus.
Megaderma.		Sciuropterus.

[1] Borneo only.
[2] Borneo and Sumatra.
[3] Philippines and Celebes.
[4] Philippines only.
[5] Celebes; *Cuscus* also in Timor and the Papuan region.
[6] Sumatra only.
[7] Java only.
[8] Tropics of the Old World generally.
[9] Also American tropics.

Summary.

Total number of genera	83
Restricted to the province*	25
Restricted to the Indian Region	52
Found outside of the Indian Region in the African only	20
Common to the African and Indian Regions	29
Wide-ranging (exclusive of tropicopolitan)	12
Of local distribution	12
Restricted to Borneo	2
Restricted to Borneo and Sumatra	1

* Exclusive of several tropicopolitan genera not occurring elsewhere in the Indian Region.

Restricted to Sumatra	2
Restricted to Java	1
Restricted to the Philippines	2
Restricted to the Philippines and Celebes	1
Occurring only in Celebes	2
Non-placental genera	1

VI.—AUSTRALIAN REALM.

The Australian Realm will be here restricted so as to embrace none of the islands situated to the westward of the Moluccas. The Molucca Group forms a transitional link between the Indo-African and the Australian Realm, but they are faunally more closely allied to the latter than to the former. These islands embrace, excluding *Chiroptera* and species probably or known to have been introduced by man,* only a single genus (*Sorex*) of Placental Mammals, while two genera of Papuan Marsupials (*Cuscus* and *Belideus*) are abundantly represented.

The Australian Realm, considered as a whole, is made up of very heterogeneous elements, its land-surface consisting of islands, many of them of small size and widely scattered. The mammals are almost wholly limited to its three larger constituents,—Australia, Tasmania, and New Guinea,—and a few of the larger islands in close proximity to them. Among the prominent types very generally represented throughout all of these areas are several wide-ranging (almost tropicopolitan) genera of Bats, which, in consequence of their wide geographical range, wholly fail to be distinctive, and may hence be safely ignored in the following general analysis of the region. The marine species (the Dugong and various species of Seals) are likewise of small importance in the present connection, since they are all wide-ranging species, not properly characteristic of the region. After these eliminations, we have left a few genera of *Muridæ* and the distinctively characteristic implacental mammalia. The latter, with the exception of a single family (*Didelphidæ*, occurring now only in the warmer parts of the two Americas), are found nowhere else, and hence give to the region an exceptional distinctness as a primary zoögeographical region. The numerous groups of small, widely scattered islands, usually considered as collectively forming the Polynesian Region, being destitute of mammalia, need not be here further considered.

New Zealand, situated more than a thousand miles to the southeastward of Australia (its nearest large land-area), is also wholly deficient in characteristic forms of mammalia; the only representatives of this class, aside from Seals and Bats, being a Rodent, supposed, rather than certainly known, to be found there. The Seals are wide-ranging species, and of the two species of Bats, one has Australian and the other South

* These include, besides the common domestic species, *Cynopithecus nigrescens*, *Viverra tangalunga*, *Babirusa alfurus*, and *Cervus hippelaphus* var. *moluccensis*, considered by Mr. Wallace as "probably" or "almost certainly" introduced by man, since they are species "habitually domesticated and kept in confinement by the Malays".—*Geogr. Dist. Anim.*, vol. i, p. 417.

American affinities. Judged by other classes of animals, the fauna of New Zealand is Australian (or Australian and Polynesian), but is yet so specialized that the New Zealand islands must be recognized as forming a distinct and highly differentiated region (New Zealand Region) of the Australian Realm.

As regards mammalia (and the same is true of the fauna and flora considered collectively), Tasmania, Australia, and New Guinea have many features in common, fully one-half of the genera (seven out of fourteen) of mammals occurring in Tasmania being represented not only throughout the greater part of Australia, *but also in New Guinea.*

Tasmania and New Guinea are less rich in mammalia than Australia, but this is obviously due to their insular character and small area. Tasmania is scarcely more closely related to Southern Australia than New Guinea is to Northern Australia. Formerly, New Guinea was thought to be very distinct from Australia, but the recent exploration of the interior of New Guinea by MM. Beccari, d'Albertis, and Laglaize, has brought to light the existence there of many forms before supposed to be restricted to Australia and Tasmania. M. Alphonse Milne-Edwards, in a recent communication to the French Academy respecting some new species of mammalia discovered in New Guinea by M. Laglaize, in referring to the close relationship existing between the faunæ of New Guinea and Australia, thus observes:—"Plus on étudie la faune de la Nouvelle-Guinée, plus on lui trouve de ressemblance avec celle de l'Australie, et les indications fournies par la répartition des espèces animales permet d'affirmer qu'autrefois ces terres ne formaient qu'un seul grand continent. Déjà les résultats dès voyages de circumnavigation entrepris dans la première moitié de ce siècle . . . avaient permis de soupçonner cette conformité d'origine; mais elle a été principalement mise en lumière à la suite des explorations de M. Wallace, de M. Beccari et de M. d'Albertis. Enfin les collections qui M. Laglaize a formées dans ces régions, ainsi que celles qui lui ont été remises par M. Bruijn et qui viennent d'arriver en France, fournissent des faits nouveaux qui accentuent encore les ressemblances entrevues."*

Formerly the Monotremes were supposed to be restricted to the southern half of Australia and Tasmania, but within the last two or three years the existence of *Tachyglossus* in North Australia (latitude 21°) has been established, and an allied species has been discovered in the mountains of New Guinea. M. A. Milne Edwards has also just described a species of *Dromicia* from New Guinea, and also a species of *Hapalotis*, and Dr. Peters has recently added species of *Phalangista*, *Chætocercus*, and *Hydromys*, making six genera recently discovered in New Guinea that were previously known only from Australia and Tasmania.

So far as at present known, only three or four genera (*Uromys*, *Dendrolagus*, *Dorcopis*, and *Myœctis*) of mammals are peculiar to New Guinea and the small islands situated between New Guinea and Australia, and

*Compte-rendu, tom. lxxxv, 1079, déc. 3, 1877.

probably some of these will yet be found in Australia. One of these (*Myœctis*) has been thus far reported only from the Aru Islands. As Tasmania has two peculiar genera (*Thylacinus* and *Sarcophilus*), New Guinea, in view of its four or five times greater area, is in reality scarcely more specialized than is Tasmania, and is hence faunally as much a part of Australia as is the latter. As will be shown later, nearly as many of the genera occurring in Southern Australia have been found in New Guinea as in Tasmania. Scarcely two years ago Mr. Wallace stated that "as yet no other [referring to the genus *Sus*] non-marsupial terrestrial mammal has been discovered [in "Papua, or the New Guinea Group"] except a Rat, described by Dr. Gray as *Uromys aruensis*, but about the locality of which there seems some doubt."* This genus has not only now been established as occurring there, but four additional species of it have been described by Dr. Peters, who has also added a species of *Hydromys*, and Mr. Alston has added a species of *Mus* and M. A. Milne-Edwards a species of *Hapalotis*, in all seven species, belonging either to Australian genera or having decided Australian affinities.

Regions of the Australian Realm.—Accepting the Polynesian Islands as forming one region (the Polynesian), and New Zealand as constituting another (the New Zealand), we have left for detailed consideration only the larger land-masses, consisting of Tasmania, Australia, and New Guinea with its associated islands, forming the third or Australian. The close zoölogical affinity of Tasmania and Australia no one questions, and it has been already shown that New Guinea and Australia are almost equally inseparable. Although many genera range from Tasmania across Australia into New Guinea, this large area, embracing as it does nearly fifty degrees of latitude, falls naturally into two well-marked subdivisions, the one tropical the other temperate.† These

* Geogr. Distr. Anim., vol. i, pp. 409, 410.

† In 1871, in referring to the Australian Realm (Bull. Mus. Comp. Zoöl., vol. ii, p. 381), I said:—"It is divisible into a Temperate and a Tropical Region, the former embracing New Zealand and Australia." The latter portion of this statement was of course made without due consideration. As already stated, New Zealand has no intimate relationship with Australia, and should be treated as a separate and independent region of the Australian Realm. Mr. Wallace, in stating his "Objections to the System of Circumpolar Zones" (Geogr. Distr. Anim., vol. i, p. 67), has very naturally taken notice of this unfortunate slip, and cites it as evidence of the "erroneous results" that follow from the adoption of the principle of the "distribution of life in circumpolar zones". My "separation of New Zealand to unite it with the southern third of Australia" was certainly most thoroughly erroneous; but while, as Mr. Wallace says, the fauna of Australia, taken as a whole, is exceptionally homogeneous, I cannot agree with him that New Guinea, so far at least as its mammalian fauna is concerned, is "as sharply differentiated from Australia as any adjacent parts of the same primary zoölogical region can possibly be"—in other words, that it can be only arbitrarily joined with the northern portion of Australia. I freely admit that I was not only in error as regards New Zealand, but also in respect to my division of the Australian continent, and I accept this portion of Mr. Wallace's criticism as fairly made. That the error was not one of "principle", but merely a wrong application of a principle, I think the text

I consider, so closely are they related, rather as provinces than regions, and may be termed respectively the Papuan Province and the Australian Province. The former is situated almost wholly between the equator and the twentieth degree of south latitude. The latter embraces that portion of Australia south of this line, together with Tasmania. The boundary between the two regions can of course be drawn only approximately, but may be provisionally assumed as the vicinity of the isotherm of 70° F.* The reason for uniting the northern portion of Australia with New Guinea as a part of the Papuan Province lies in the fact that not only so many of the mammalian genera are common to the two, but that these genera are absent from the more southern portions of Australia, where they are replaced by others wholly restricted to South Australia and Tasmania. Three-fourths of all the genera of Marsupials (excluding, of course, the American family *Didelphidæ*) are, so far as at present known, restricted to the Australian Province, as are several genera of *Muridæ* and the *Ornithorhynchus.* Of the remaining Marsupial genera, six only are limited to the Papuan Province.

The Papuan Province.—The Papuan Province embraces not only New Guinea, but the Molucca and Aru Islands on the west and the Solomon

---

here following sufficiently shows. The principle I still hold as applying to Australia with the same force as elsewhere, only I make the division more to the northward, as a little more care would have led me to do originally. The York Peninsula, and most probably the whole northern coast region north of 20° S. lat. (except the high arid interior), has certainly closer affinities, as regards mammals, with New Guinea than it has with any portion of South Australia. Of the strictly Papuan genera, only two out of nine are restricted to New Guinea, the rest being common to both North Australia and Papua. Of the other North Australian genera, about one-half occur generally throughout the continent, but the remainder are essentially South Australian, represented by only stragglers in Northern Australia. On the other hand, more than twenty genera occurring in Southern Australia and Tasmania, are wholly unrepresented in the portion of Australia I here assign to the Papuan Region. In other words, we get the same wide faunal differences between the tropical and temperate portions of the Australian Realm that we get elsewhere under similar climatic conditions.

In the same connection, Mr. Wallace cites my separation of Temperate South Africa as a primary region as another instance of the misleading nature of the principle of the distribution of life in zones. This I have also seen fit to abandon (see *anteà*, p. 351) on a détailed re-examination of the subject, not because the principle is erroneous, but in consequence of certain peculiar geographical conditions, namely, the comparatively small area subject to a temperate climate and to its limited extension into the temperate region. It is, in fact, wholly within the warm-temperate belt, and widens rapidly northward to abut very broadly against the tropical zone. Only a very small portion really comes under the influence of temperate conditions. Here we get, as usual, a temperate aspect in the fauna, and I still maintain my separation of South Africa as a faunal division, simply lowering its grade from a primary region to a "province" of the great Indo-African Realm, simply from the fact that the smallness of its area and warm-temperate, rather than temperate, conditions have prevented, as would be naturally expected, any great amount of differentiation.

* Mr. E. Blyth, in a paper (Nature, vol. iii, p. 428, issue of March 30, 1871) published almost simultaneously with my own cited in the last foot-note, included a portion of Northern Australia in his "Papuan Sub-region", namely, "York Peninsula and eastern half of Queensland (as far as the dividing range), on the main land of Australia".

Group on the east, as well as the most northerly portion of Australia, including the York Peninsula, and probably the whole northern coast region, or that portion of Australia north of the Southern Tropic, except the elevated arid interior. Of the twenty-seven genera (exclusive of *Chiroptera* and marine species) represented in the Papuan Province, ten are not found elsewhere in the Australian Realm. Three of these (*Sus*, *Sorex*, found only in the Moluccas, and *Mus*) have a wide Indo-African range; four (*Uromys*, *Dendrolagus*, *Dorcopsis*, and *Myœctis*) are found only in New Guinea and the Aru Islands; and one (*Dactylospila*) in the Aru Islands and the York Peninsula.

The seventeen remaining genera belong more properly to the Australian Province, or perhaps to Australia at large. Many of them, while numerous in species, have here (like *Halmaturus*, *Antechinus*, *Podabrus*, *Mus*, *Hapalotis*, etc.) only straggling representatives, but are numerously represented in the temperate region to the southward. The distribution of the genera is approximately indicated in the subjoined table.

Genera of the Papuan Province.

[NOTE.—The New Guinea representatives of the genera *Hapalotis*, *Phalangista*, and *Tachyglossus* have recently been separated from their Australian affines as distinct subgenera. *Babirusa* is also reported from Bouru, but as probably introduced from Celebes.]

Restricted to New Guinea and neighboring islands.	Restricted to New Guinea and North Australia.	Also ranging over most of the Australian Region.		
Sus.[1]	Acanthomys.[3]	*Mus.	Halmaturus.[6]	Petrogale.
Sorex.[2]	Phascogale.	*Hapalotis.[6]	*Perameles.[6]	*Phalangista.[6]
Uromys.	Cuscus.[4]	*Hydromys.[6]	Macropus.[6]	*Belideus.[6]
Dendrolagus.	Dactylopsila.[5]	Dasyurus.[6]	Osphranter.	*Dromicia.[6]
Dorcopsis.		*Antechinus.[6]	Onychogalea.	*Tachyglossus.
Myœctis.		*Chætocercus.	Largorchestes.	

[1] New Guinea only.
[2] Moluccas only.
[3] North Australia only.
[4] Also Celebes, Timor, and Moluccas.
[5] Aru Islands, New Guinea (*Peters*), and York Peninsula (*Krefft*).
[6] Mainly large South Australian genera, sparingly represented in North Australia and New Guinea.
*Occurring in New Guinea.

Summary.

Total number of genera	27
Restricted to the region (including, however, two Indo-African genera)	10
Represented in New Guinea	18
Ranging also over the Australian Region	16
Restricted to New Guinea and neighboring islands (exclusive of two Indo-African genera)	4
Common to only New Guinea and North Australia	4
Genera properly belonging to the Australian Region, but sparingly represented in the Papuan Region	10
Distinctively characteristic of the Papuan Region, about	15

Australian Province.—The Australian Province, embracing Tasmania and all of Australia south of about the southern isotherm of 70° F., contains not less than fifteen to eighteen genera, out of a total number of thirty-four that are restricted to this region, while of the remainder much more than one-half have their chief development here. One-third of the whole are represented in Tasmania, and nearly one-fourth range into New Guinea. Two only are peculiar to Tasmania. The distribution of the genera is shown somewhat in detail in the subjoined table.

In this connection it may be added that the close affinity of the Papuan fauna with that of Australia is sufficiently evinced by the fact that of the thirty-four genera represented in South Australia nine range into New Guinea—nearly as many as occur in Tasmania!

Genera of the Australian Province.

Restricted to Temperate Australia and Tasmania.		Occurring also in the Papuan Region.	
Pseudomys.	Æpyprymnus.	*Mus.	[3]Halmaturus.[2]
Echiothrix.	Bettongia.[2]	*Hapalotis.[3]	Petrogale.
*Antechinus.[2]	Hypsiprymnus.[2]	*Hydromys.[3]	Onychogale.
Antechinomys.	Phalascolarctos.	*Chætocercus.	Largorchestes.
Sarcophilus.[1]	Petaurista.	*Dasyurus.[3]	*[3]Phalangista.[2]
Thalacinus.[1]	Acrobata.	*[3]Perameles.[2]	*[3]Dromicia [2]
Podabrus.	Tarsipes.	[3]Macropus.[2]	*Belideus.[3]
Myrmecobius.	Phascolomys.[2]	Osphranter.	*Tachyglossus.[2]
Chæropus.	Ornithorhynchus.[2]		

[1]Restricted to Tasmania. [2]Represented in Tasmania. [3]Mainly restricted to the Papuan Region. *Occurring in New Guinea.

Summary.

Total number of genera	34
Restricted to the Australian Region	18
Occurring also in the Papuan Region	16
Represented in Tasmania	12
Represented in New Guinea	10
Restricted to Tasmania	2

VII.—LEMURIAN REALM.

As was long since claimed by Dr. Sclater,* Madagascar is faunally so distinct from every other ontological division of the globe as to be entitled to the rank of a primary zoögeographical region. With it, as is generally admitted, should be associated the Mascarene Islands. The very few mammals indigenous to these islands are decidedly Madagascarene in their affinities, as are the birds and other land animals. While the Lemurian fauna shows decided African affinities, it is second only to the Australian in its degree of specialization. It departs most strikingly from all other regions in what it lacks, through the absence of all Carnivores save one peculiar family (*Cryptoproctidæ*),

* Quarterly Journ. Sci., vol. i, April, 1864, pp. 213–219.

represented by a single species, and four peculiar genera of the family *Viverridæ;* of all Ruminants and Proboscidians; all Pachyderms except a single African genus of *Suidæ;* and all Rodents except a few species of *Muridæ.* The Insectivores are almost wholly represented by one or two species of *Crocidura*, and a family, embracing several genera, not found elsewhere, save a single genus in the West Indies. Four families of Bats occur, but are represented, with one exception, each by a single species. They belong to groups of semi-cosmopolitan range, and owing also to the exceptional means of dispersal possessed by the *Chiroptera*, have little weight in determining the affinities of the fauna. The Quadrumanes are represented only by the *Prosimiæ*, of which three-fourths of all the species occur here, while about four-fifths of the remainder are African. The remains of an extinct species of *Hippopotamus* have been found, a type existing at present only in Africa. Although the Indian genus *Viverricula* has recently been established as occurring in Madagascar, the few types that connect the Lemurian mammalian fauna with the faunæ of other parts of the world are preponderatingly African.

With the exception of the Bats, which, for reasons already given, are scarcely entitled to consideration in the present connection, the mammalia of "Lemuria" are, generally speaking, the lowest existing representatives of their respective orders. The most prominent type, embracing, in fact, about three-fifths of all the species (excluding the half dozen species of *Chiroptera*), belong to the *Prosimiæ*, the lowest of the Quadrumanes, which in early Tertiary times had representatives over a large part of the northern hemisphere, and perhaps had at that time a nearly cosmopolitan distribution. The Carnivores are likewise allied to early types of the *Viverridæ*, which formerly had a much wider range than at present; and the Insectivores are also of low forms, and allied to early types. These facts seem, at first sight, to lend support to the hypothesis, first advanced by Dr. Sclater, that Madagascar and the Mascarene Islands are but remnants of a former extensive land-area that possibly had connection with America as well as India, and embraced portions of Africa. The supposed former relationship with America is indicated perhaps not so much by the presence of *Solenodon* in the West Indies, and of American forms of Serpents, Lizards, and Insects in Madagascar, as by the abundant occurrence of Lemuroid remains in the North American Eocene. Since, however, these early Lemuroid forms appear not to have been true Lemurs, but a more generalized type, having affinities also with the Carnivores and Insectivores, and since they occurred also in Europe, and probably in Asia (for recent palæontological discoveries in our American Tertiaries show that much may be expected from future explorations elsewhere), it is possible that the explanation of the present distribution of the *Prosimiæ* needs not the supposition of the existence of any very extensive land-area that has since disappeared: in other words, that the African and Madagascarene *Lemuridæ* may have reached their present homes by migration from the northward (leaving a remnant in India), at a time when North America and Asia formed a continuous land-area, just as there is good reason for believing that the greater part of the present faunæ of India, Southern Europe, and Africa are a comparatively recent immigration from the northward; that Madagascar derived, at a comparatively early period, its existing fauna from Africa, as Mr. Wallace believes to have been the fact; and, finally, that at a time antedating the appearance of the present African fauna, Madagascar was actually united to the African continent.* America is now not only currently considered to be the "Old World" geologically, but it seems probable, as has recently been suggested,† that the Equine, Tapiroid, Rhinoceroid, Cameloid, Suilline, and Cervine forms, the *Prosimiæ*, and possibly the Proboscidians, Marsupials, and Edentates, were either first developed in America, or had their origin there in early generalized forms, and have since spread to the more recently formed continents of the eastern hemisphere. Many of them, as well as other early, generalized types, are known to have had a nearly contemporaneous existence during the early part of the Tertiary period both in America and Europe. This certainly lends probability to Mr. Wallace's hypothesis respecting the origin of the present Lemurian fauna.

The families and genera represented in "Lemuria", their faunal alliances, and areas of chief distribution, are as follows:—

LEMURIDÆ.—Chiefly developed in Madagascar, but occurring in Tropical Africa, Southern India, and the Malay Archipelago. Represented by about twelve genera and about fifty species, three-fifths of which are peculiar to Madagascar, and three-fourths of the remainder to Africa. Genera:—*Indris, Propithecus, Lemur, Hapalemur, Microcebus, Lepilemur, Chirogaleus.*

DAUBENTONIIDÆ.—Peculiar to Madagascar and represented by a single species—*Daubentonia* (=*Chiromys*) *madagascariensis.*

CRYPTOPROCTIDÆ.—One species (*Cryptoprocta ferox*), found only in Madagascar.

VIVERRIDÆ.—Warmer parts of Asia, the Malayan Islands, and Africa. Represented in Madagascar by several peculiar genera and the Indian genus *Viverricula*. Genera:—*Fossa, Galidia, Galidictis, Viverricula.* Species of the African genus *Herpestes* also reported.

EUPLERIDÆ.—Peculiar to Madagascar, and embracing the single genus *Eupleres.*

SUIDÆ.—Eastern hemisphere generally. Represented in Madagascar by species of the African genus *Potamochœrus.*

HIPPOPOTAMIDÆ.—African. Represented in Madagascar by the remains of a species believed to have but recently become extinct.

PTEROPIDÆ.—The tropics everywhere, except Tropical America. Represented in Madagascar and the Mascarene Islands by two species of the Indian and Australian genus *Pteropus.*

RHINOLOPHIDÆ.—Warmer parts of the eastern hemisphere. Represented in "Lemuria" by species of *Rhinolophus.*

---

* Geogr. Distr. Anim., vol. i, p. 273; Nature, vol. xvi (Oct. 25, 1877), p. 548.

† See especially Prof. O. C. Marsh's address on "the Introduction and Succession of Vertebrate Life in America", delivered before the Nashville meeting of the American Association for the Advancement of Science, Aug. 30, 1877.

VESPERTILIONIDÆ.—Cosmopolitan. Represented by the cosmopolite genus *Vespertilio.*

EMBALLONURIDÆ.—Warmer parts of the world. Represented by the genus *Taphozous.*

CENTETIDÆ.—Confined to Madagascar except one genus (*Solenodon*) in the West Indies. Represented in Madagascar by nearly a dozen species. Genera:—*Centetes, Hemicentetes, Ericulus, Oryzorictes, Echinops.*

SORICIDÆ.—The whole world, except South America and Australia. Represented in Madagascar by one or two species of *Crocidura*, a genus found in Africa, and the warmer parts of the eastern hemisphere generally.

MURIDÆ.—Cosmopolitan. Represented by several genera of African affinities, namely, *Nesomys, Brachytarsomys, Hypogeomys.*

VIII.—ANTARCTIC REALM.

The Antarctic Realm is geographically almost wholly oceanic, and its fauna hence consists almost exclusively of marine or pelagic species. It necessarily embraces not only the Antarctic Zone, but a large part of the cold south-temperate, since very few of its characteristic species are wholly restricted to the Antarctic waters. It will hence include not only the few small groups of Antarctic Islands, but also Tierra del Fuego and the Falkland Islands, and perhaps also the extreme southern shores of South America, while some of its characteristic forms also extend to New Zealand, and even Australia and the Cape of Good Hope. The only mammals that can be considered as strictly characteristic of this region are Pinnipeds and Cetaceans, of which several genera of each are almost wholly restricted to it. A "South Frigid", "Antarctic", or "South Circumpolar" "Zone", "Region", or "Realm", has been recognized by various writers for the marine invertebrates, and, by von Pelzeln for birds, with limitations much as here assigned. While the number of species peculiar to it is small, it is large relatively to the whole number represented, especially in the colder latitudes. There is, of course, a broad belt along its northern border of a transitional character, where Antarctic types overlap the range of groups characteristic of south-temperate latitudes.

One of the most important features of the South Circumpolar or Antarctic Realm is the resemblance of its life to the marine life of the Arctic or North Circumpolar Realm. While perhaps in no case are the species identical, the genera are frequently the same, not only among the mammalia, but among invertebrates. This is especially significant as regards the mammalia, since the terrestrial mammals of the extreme north and extreme south present no such parallelism, but the utmost divergence. Among Pinnipeds, most of the genera are peculiar to either the northern or southern waters, but in several instances the genera of the two regions are strictly representative. Thus, *Otaria* and *Arctocephalus* of the Southern Seas are represented in the Northern by *Eumetopias* and *Callorhinus*. *Zalophus* and *Macrorhinus* are both Northern and Southern. *Stenorhynchus*, *Lobodon*, *Leptonyx*, and *Ommatophoca* are strictly Southern, while *Phoca*, *Halichœrus*, *Erignathus*, *Cystophora*, *Monachus*, and one or two others, are strictly Northern, as are also the Walruses. The *Mysticete*, or Baleen Whales, among Cetaceans, have a somewhat similar distribution. While a few genera are restricted respectively to the Northern and Southern waters, the larger number are common to both, though represented by different species in the two regions, while they are (in some cases at least) absent from the intervening tropical seas. A large proportion of the *Denticete*, or Toothed Whales (Dolphins, Porpoises, Rorquals, etc)., are either limited to the warmer seas or have there their chief development, quite a number of genera being peculiar to the tropics. Others, however, like *Monodon*, are eminently boreal, while others, like *Beluga*, are common to the colder waters both north and south of the tropics. In most cases, however, we know as yet too little respecting the range of the different species and genera of *Cetacea* to be able to make much use of them in determining questions in geographical zoölogy.

This similarity between the marine life of the Arctic and Antarctic Regions evidently indicates that the forms common to the two had a common origin, and, at some former period, a continuous, probably circumtropical, distribution, and that on the increase of temperature in the intertropical regions, through well-known geological causes, they sought the more compatible cooler waters toward the poles. The similarity of the Arctic and Antarctic marine life is also a feature that sharply differentiates the fauna of the South Circumpolar Realm from that of the South Temperate and Tropical Zones.

III.—GENERAL SUMMARY.

As stated at the beginning of the present paper, one of the chief topics here proposed for discussion was the influences and laws which govern the distribution of life,—whether it is or is not co-ordinated with climatic zones, and governed in a large degree by climatic conditions, and especially by temperature. In fact, so generally is temperature recognized by the leading writers on the distribution of marine life that it seems superfluous to reiterate or emphasize this principle. That the zones of life should be perhaps a little less obvious over the land-areas,—in consequence of the diversity of contour resulting from differences of elevation, and the interruptions and exceptional conditions due to mountain chains and high plateaus,—than over the oceanic expanses, is naturally to be expected. That there is, however, a similar correspondence between climatic belts and the zones of life seems to me abundantly evident. As has been already shown, the broader or primary zones are, first, an Arctic or North Circumpolar Zone, embracing the arctic, subarctic, and colder temperate latitudes of the northern hemisphere, throughout the whole of which area there is a marked homogeneity of mammalian life, as well as of animal and vegetable life in general; secondly, that below this there is a broad belt of life, which, in its general *facies*, is distinctive of the temperate and warm-temperate latitudes, and that these two zones of life are far more closely related *inter se* than with the life of the intertropical regions, with which regions they may be collectively contrasted, and together receive the appropriate name of "*Arctogæa*";

thirdly, it has been shown, so far as the northern hemisphere is concerned, that the life of the tropical and temperate regions of the same continent is more widely different than is the life of corresponding portions of the temperate and colder parts of the (so-called) Old World and the New; fourthly, that the life of Tropical America has very little in common with that of the tropical portions of Asia and Africa; fifthly, that the life of the South Temperate Zone presents a *facies* distinct from that of the tropics, and has still less in common with that of the North Temperate Zone; sixthly, that Australasia is so highly differentiated as to form a distinct primary region, having little in common with other lands, even with those of contiguous regions, or those having a similar geographical position; seventhly, that Madagascar and its contiguous islands, while to some extent African in affinity, form also a highly specialized region; lastly, that the antarctic and cold south-temperate oceanic regions are recognizable as a primary region, characterized by a peculiar general *facies* of life that more strongly recalls that of the corresponding portions of the northern hemisphere than of any other portion of the earth. It has been further shown that the Australian Realm is divisible into temperate and tropical portions, and also that the land surface is separable into zones of even still narrower limits, corresponding in a general way with those recognized by Dana for marine life.

The almost total absence of identical genera, or even of families, excepting such as are essentially cosmopolitan, in the American and Old World tropics, as well as the distinctness of the Lemurian Realm, and the almost total isolation of the Australian Realm, evidently require for their explanation other causes than merely the existing climates. The geological history of these land-areas and their faunæ must be of course considered in order to understand their present relationships. As the northern hemisphere at present most clearly shows, nearly continuous land surface and similarity of climatic conditions implies identity of fauna, while isolation, especially when joined with diverse climatic conditions, implies diversity of life, and a differentiation proportionate to the degree of isolation, and the length of time such isolation has existed; in other words, that the present want of affinity between the life of the Lemurian and Australian Realms and that of the rest of the world is due rather to their long geographical isolation than to present climatic conditions, and that we here find, for reasons perhaps not wholly apparent, the remnants of a somewhat primitive or early fauna that was formerly shared more largely by other areas than at present,—that these regions became isolated before the development of many of the higher and now prevalent types of the larger and more diversified land-areas, and that here differentiation has proceeded less rapidly and along fewer and narrower lines than elsewhere; furthermore, that the present highly diversified fauna of the chief tropical areas, in comparison with the fauna of the north-circumpolar lands, is due in part to the southward migration, near the close of the Tertiary period, of forms adapted to a high temperature, and in part to the high rate of differentiation favored by tropical conditions of climate. Hence, given: 1. Arctic and cold-temperate conditions of climate, and we have a fauna only slightly or moderately diversified; 2. A moderate increase of temperature, giving warm-temperate conditions of climate, and we have the addition of many new types of life; 3. A high increase of temperature, giving tropical conditions of climate, and we have a rapid multiplication of new forms and a maximum of differentiation. Again, given: 1. A long-continued continuity of land surface, and we have an essential identity of fauna; 2. A divergence and partial isolation of land-areas, and we find a moderate but decided differentiation of faunæ; 3. A total isolation of land-areas, and we have a thorough and radical differentiation of faunæ, proportioned to the length of time the isolation has continued. Hence, the present diversity of life is correlated with two fundamental conditions: 1. Continuity or isolation, past as well as present, of land surface; and, 2. Climatic conditions, as determined mainly by temperature.*

In accordance with these principles, which rest on incontrovertible facts of distribution, it follows that the nearly united lands of the North present a continuous, almost homogeneous, arctopolitan fauna; that farther southward, in the warmer temperate latitudes, we begin to find a marked differentiation on the two continents; that this differentiation is still further developed in the tropical continuations of these same land-areas, till an almost total want of resemblance is reached, except that there is what may be termed, in contrast with the more northern regions, a "tropical *facies*" common to the two. The small amount of land surface belonging to these primary land regions south of the tropics have no more in common (a few marine species excepted) than have these two tropical areas, but it is hardly possible for them to have much less. The Antarctic (mainly oceanic) region has a fauna strongly recalling the marine fauna of the Arctic, but has no resemblance to that of the intervening area.

The northern circumpolar lands may be looked upon as the base or centre from which have spread all the more recently developed forms of mammalian life, as it is still the bond that unites the whole. Of the few cosmopolitan types that in a manner bind together and connect the whole mammalian fauna of the globe (the Lemurian and Australian Realms in part excepted), nearly all have either their true home or belong to groups that are mainly developed in the northern lands. A few

* In illustration of the above, it may be added that the circumpolar lands north of the mean annual of 36° F., or, in general terms, north of the fiftieth parallel, with approximately an area of about 12,500,000 square miles, have representatives of about fifty-four genera of mammals; Tropical America, with an approximate area of about 5,000,000 square miles, has about ninety genera; the Indo-African Realm, with an approximate area of about 15,000,000 square miles, has about two hundred and fifty genera. Hence the tropical lands are four to five times richer in genera, in proportion to area, than those of the Cold-temperate and Arctic regions.

have been pressed a little to the southward by the extreme rigor of an Arctic climate, but are still characteristic elements of all boreal faunas. The very few truly tropicopolitan mammalia are either *Chiroptera*, or marine, or at least aquatic, and have thus exceptional means of dispersal.

The primary regions and their subdivisions, recognized in the preceding pages, are enumerated in the subjoined schedule.

1.—*Primary divisions, or "Realms"*.

I. An ARCTIC, or NORTH CIRCUMPOLAR.
II. A NORTH TEMPERATE, divided into two regions and eight provinces.
III. An AMERICAN TROPICAL, with three regions. (Provinces not characterized.)
IV. An INDO-AFRICAN, with two regions and five provinces.
V. A SOUTH AMERICAN TEMPERATE, with two provinces.
VI. An AUSTRALIAN, with three regions and two provinces.
VII. A LEMURIAN.
VIII. An ANTARCTIC or SOUTH CIRCUMPOLAR.

2.—*Secondary divisions, or "Regions"*.

II. North Temperate Realm: 1, American; 2, Europæo-Asiatic.
III. American Tropical Realm: 1, Antillean; 2, Central American; 3, Brazilian.
IV. Indo-African Realm: 1, African; 2, Indian.
VI. Australian Realm: 1, Australian (Australia, Tasmania, and New Guinea); 2, Polynesian; 3, New Zealand.

3.—*Divisions of third rank, or "Provinces"*.

II, 1. American Region: *a*, Boreal*; *b*, Eastern; *c*, Middle; *d*, Western.
II, 2. Europæo-Asiatic Region: *a*, European; *b*, Siberian; *c*, Mediterranean; *d*, Manchurian.
IV, 1. African Region: *a*, Eastern; *b*, Western; *c*, Southern.
IV, 2. Indian Region: *a*, Continental; *b*, Insular.
V. South American Temperate Realm: *a*, Andean; *b*, Pampean.
VI, 1. Australian Region: *a*, Australian; *b*, Papuan.

* A "Boreal" province has not been distinctly recognized in the preceding pages as a division belonging to the same category as the other so-called or commonly recognized provinces, and is not at all recognized in the table of distribution given at p. 339. It is nearly equivalent to what is there implied by "Cold Temperate". I hope soon to be able, in a paper to be devoted especially to a consideration of the geographical distribution of North American mammals, to define and characterize it more definitely.

The relation of the different primary regions and their subdivions may be approximately indicated diagrammatically as follows:—

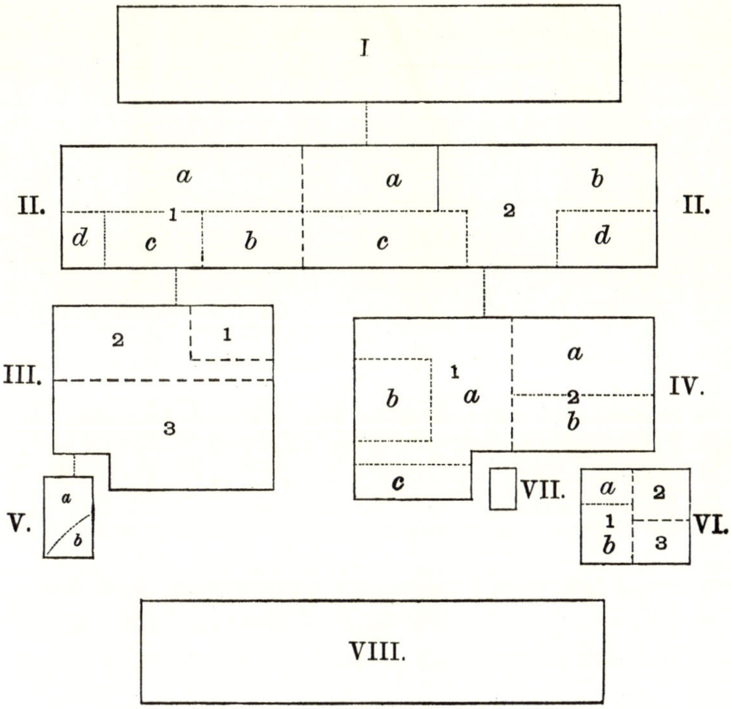

THE GEOGRAPHICAL DISTRIBUTION OF NORTH AMERICAN MAMMALS.

Article XIV.—THE GEOGRAPHICAL DISTRIBUTION OF NORTH AMERICAN MAMMALS.[1]

By Joel Asaph Allen.

(*With Four Maps, forming Plates V–VIII.*)

Influences determining the Geographical Distribution of Life.

It has long been recognized that the influences determining the distribution of life over the earth's surface are climate and the interrelation of the principal land areas. It is questionable which of the two, all things considered, has been the more important factor in bringing about the present distribution of life, since the climate of any given area depends largely upon the relative distribution of land and water, and is further greatly modified by the topography of the principal land areas, the presence or absence of lofty mountain chains greatly modifying the climate of an entire continent.

Climate.—Of strictly climatic influences, temperature is by far the most important, although moisture plays an influential part. Where a low temperature prevails life, both animal and vegetable, is represented by comparatively few forms ; under a high temperature it is characterized by great diversity and luxuriance. Within the Arctic Circle the species of both plants and animals are not only few but they are widely distributed, being for the most part everywhere the same. Under the tropics they are a hundred fold more numerous and of comparatively restricted distribution, the general facies, as regards both the fauna and the flora, changing within short distances, with few elements in common when widely separated areas are compared.

The influence of temperature is perhaps most strikingly displayed in the distribution of life upon the slopes of a high mountain, especially if situated near the tropics. While its base may be clothed with palms and luxuriant tropical vegetation its sum-

[1] Read before the New York Academy of Sciences, January 26, 1891. In revising for publication some new matter has been added, chiefly in the last third of the paper and in the footnotes.

mit may be snow-capped and barren, or scantily covered with only a small variety of hardy alpine shrubs and plants. The animal life becomes likewise correspondingly changed, tropical forms of mammals, birds and insects of the lower slopes gradually giving place to such as are characteristic of arctic latitudes.[1]

The influence of moisture is most strikingly shown in the distribution of forest trees, where on the same continent, under corresponding parallels of latitude, and at nearly the same elevation, the country may be either heavily or only sparsely wooded, or even wholly devoid of a forest growth, in accordance with the abundance or scarcity of the rain-fall. We may thus have, on the same continent, immense areas of forest alternating with vast stretches of open prairies and plains, or even almost verdureless deserts. The effect of humidity upon plant life is thus obvious; but it is equally potent, though less evident, upon animal life. Many animals—mammals and birds as well as insects—are so fitted for a forest life, as regards both food and shelter, that their very existence depends upon such surroundings. Others are equally specially adapted for life on the open plains, or even in arid deserts. Thus moisture alone may determine the character of life over extensive regions, regardless of temperature, which under ordinary conditions is the ascendant controlling influence.

Interrelation of Land Areas.—The relation of the principal land areas to each other, in respect to continuity on the one hand and isolation on the other, is coëval and perhaps more than coördinate with climate in its influence upon the distribution of life, as it is also in the evolution of life. Palæontology teaches us that the present characteristics of the faunas and floras of the principal land areas are the results of a long period of evolution, during which there have been no very sudden transitions, but, in general, a gradual development from ancestral forms having about the same geographical distribution as their descendents. While some of the continents have unquestionably derived part of their life from neighboring continents, and islands from the mainland to which they are contiguous, these

[1] An excellent illustration of the influence of temperature upon the distribution of life on mountain slopes is afforded by Dr. C. Hart Merriam's thorough survey of the San Francisco Mountain region in Arizona, the life-zones of which he has defined and illustrated in detail. (See N. Am. Fauna, No. 3, Sept., 1890, pp. 5-20, pll. 1, 2, and maps 1-4.)

migrations are in the main of great antiquity, dating back at least nearly to the Miocene. At that early time, and also previously, there was not only a greater uniformity of climate, but a more uniform distribution of life. With the coming in of the Glacial Period a great change was wrought in respect to both, and the former equilibrium in neither has been restored. While it would lead me too far from the subject especially in hand to discuss at any length the probable geographical origin of the leading types of even the present mammalian life of North America, it seems well to recall, in the present connection, certain facts of general import. In the first place, the so-called Old World is admittedly the most advanced and the most specialized of the several continents, as regards both its physical features and its biology ; and that probably many of its present leading types of mammalian life are of American origin ; that North America is behind Eurasia in development, and South America behind North America ; and that in reality Australia is the old continent in the sense of being behind all the others in its development, and thus the lowest, the least specialized, the most primitive.

In regard to the present distribution of the mammals of North America and their faunal relationship to the mammalia of the rest of the world, it is important to recall the present close proximity of North America at the northward to the northern portion of the Old World. Alaska is separated from Siberia by a shallow strait of less than forty miles in width, while in Tertiary times it is supposed they may have been united. Three-fourths of the land area is not only situated north of the equator, but is mainly massed about the northern pole, the only extensive stretch of sea being the few hundred miles between Iceland and Norway. If now we bear in mind the close similarity in climate and general physical conditions of the northern half of the northern hemisphere, its comparatively low temperature and meagre fauna and flora as compared with the tropics, and that continuity of land area tends to uniformity of life, and divergence and isolation tend to diversity of life, and increase of temperature to abundance and variety, the generalizations about to follow respecting the life regions of North America will, I trust, be recognized as resting on a sound basis.

IMPORTANCE OF MAMMALS AS A BASIS FOR THE CLASSIFICATION OF LIFE AREAS.

First, however, a word in reference to the class Mammalia as a basis for the distribution of the earth's surface into ontological divisions. On this point I cannot do better than to summarize the argument made by Mr. A. R. Wallace in his great work on 'The Geographical Distribution of Animals,' simply premising that his presentation of the case has my hearty approval.

The mammalia, he affirms, are pre-eminently of the greatest importance in determining zoölogical regions. Their dispersal is less dependent on fortuitous circumstances than that of the representatives of other classes; from their high organization they are less dependent upon other groups of animals, and have so much power of adaptation that they are able to exist in one form or another over the whole globe, as is certainly not the case with two of the lower classes of vertebrates, the reptilia and amphibia. Their distribution and dispersal are dependent on the distribution of the land areas, and are modified by such physical conditions as mountain barriers, areas of forest, and grassy or desert plateaus. Furthermore, their geological history, as well as their geographical range, is better known than that of most other classes, and there is also a greater unanimity of opinion respecting their natural affinities and the limitation of families and genera. "We should therefore," says Mr. Wallace, "construct our typical or standard Zoölogical Regions in the first place, from a consideration of the distribution of mammalia, only bringing to our aid the distribution of other groups to determine doubtful points. Regions so established will be most closely in accordance with these long-enduring features of physical geography, on which the distribution of all forms of life fundamentally depend; and all discrepancies in the distribution of other classes of animals must be capable of being explained, either by their exceptional means of dispersion or by special conditions affecting their perpetuation and increase in each locality. If these considerations are well founded, the objections of those who study insects or molluscs, for example,—that our regions are not true for their departments of nature—cannot be maintained. For they will find, that a careful consideration of the exceptional means of dispersal and conditions of existence of each group, will explain most of the divergences from the normal distribution of higher animals."[1]

While the divisions of North America, as set forth in the present paper, are presented from the standpoint of mammals, it may be premised that many of them were first outlined on the basis of the distribution of birds, and that they have been found equally applicable to reptiles and bratracians, and also in a measure to insects and plants.[2]

Systematic Classification of Life Areas.

In zoö-geography it is customary to recognize faunal areas belonging to several different categories, as regards their grade and extent, just as in zoölogy we divide animals into classes, orders, families, genera, and species. Unfortunately, however, the terms employed for their designation have not been used with the same precision as in zoölogical terminology. Identical terms have sometimes been used in diametrically opposite senses, in accordance with each writer's individual preferences, regardless of their prior use in a different sense by other authors. In view of this unfortunate state of affairs I attempted, in a paper on the distribution of North American birds, published in 1871,[3] to devise a system of terms that, while appropriate, should at the same time be in as close conformity as possible with current usage. Previously the terms *zone*, *realm*, *region*, *kingdom*, *province*, and even *fauna*, had been used more or less interchangeably for even the primary subdivisions, while some of these terms were also frequently employed in a narrower and more special sense. The scheme then proposed is as follows:

For divisions of the	first	rank	*Realm.*
" "	second	"	*Region.*
" "	third	"	*Province.*
" "	fourth	"	*Subprovince* or *District.*
" "	fifth	"	*Fauna*, *Flora.*

[1] Geogr. Dist. Anim., Vol. I, pp. 56–58.

[2] Since this was written Dr. C. Hart Merriam, in his admirable presidential address, entitled 'The Geographic Distribution of Life in North America with Special Reference to the Mammalia,' read before the Biological Society of Washington, February 6, 1892 (Proc. Biol. Soc. Wash., Vol. VII, April, 1892, pp. 1-64), has given a 'Historical Synopsis of Faunal and Floral Divisions Proposed for North America' (l. c., pp. 6–21) which may be profitably consulted in the present connection.

[3] On the Mammals and Winter Birds of East Florida, with....a Sketch of the Bird Faunæ of Eastern North America. Bull. Mus. Comp. Zoöl., Vol. II, No. 3, April, 1871.

Their grade and order of sequence may be indicated by a comparison with the leading groups in zoölogy: thus *realm* would correspond in rank with *class; region* with *order; province* with *family; district* with *genus;* and *fauna* (or *flora*, as the case may be) with *species.* It sometimes becomes convenient, as we shall see later, to recognize other divisions intermediate to those above named—as in zoölogy we have suborder, subfamily, subgenus, etc., so we may have here subregions, subprovinces, and even subfaunæ. There may also be a subdivision of a continental or other large area into zones.

Temperate North America forms a *region* of the *North Temperate Realm*, and includes two *subregions*, one of which is divisible into two *provinces;* each *province* is separable into two *subprovinces* and these again into several lesser well-marked areas termed *faunæ*, as will be presently shown in detail. Realms are sometimes characterized by the presence of certain orders, commonly by the presence of certain families, which give to the region a particular impress, and by the absence of others which in a similar way characterize other realms. Regions are usually characterized by the prevalence over them of certain genera, or even by entire families; provinces by the presence or absence of prominent generic types. Faunæ, on the other hand, are seldom characterized by the presence or absence of particular genera or species but by the association, through the overlapping of their habitats, of a number of genera and species not elsewhere found together.

The transition between faunæ, between provinces, or between adjoining divisions of any grade is rarely abrupt; it is impossible to draw a hard-and-fast line between any of them; yet in a general way they may be limited with considerable definiteness. They depend upon climatic conditions, which, in a measure, are determined or modified by features of topography. They are of course limited and determined by the same conditions that govern the distribution of species. Hence we can seldom bound our faunal areas by geographical meridians or by parallels of latitude, and very rarely are they found to agree with any political boundaries; they do, however, closely coincide with certain isothermal lines, which are generally those of the season of reproduction, or, in the northern hemisphere, those for the months of May, June and July. As temperature is influenced by altitude as well as by latitude, elevated plateaus and mountain ranges deflect the isotherms in the northern hemisphere far to the southward of their position over the contiguous low country, and furnish congenial habitats for northern forms of life under comparatively low latitudes. Thus in eastern North America the Appalachian Highlands carry the fauna of northern New England southward along the higher parts of the Alleghanies as far as northern Georgia. In the Rocky Mountains boreal types extend far down into Mexico, and along the Cascade and Sierra Nevada chain to southern California, with insular patches of northern life on the summits of detached peaks and ranges throughout the Plateau and Great Basin regions of the West. There is thus an interdigitation of the northern and southern life areas throughout the middle and southern portions of the North American continent.

The life of the globe is everywhere closely linked together. While the relationship at the northward is obviously intimate, and while in general the life of the intertropical zone is very different from that of the arctic or even of the temperate zones, and that of tropical America is very unlike that of tropical Africa or India, yet certain types are common to the whole. For example, the Cat family is represented throughout all countries except Australia, but there is a great difference between the Cats of the colder zones and those of the equatorial regions; yet the genus *Felis*, taken in a broad sense, is almost cosmopolitan. The Dog family, embracing the Wolves and Foxes, has an even wider distribution. The family Mustelidæ, embracing the Otters, Badgers, Skunks, Sables, Weasels, Minks and Martens, is more restricted, some of its leading forms, however, having a much wider and a very different distribution from others. Thus the Otters (subfamily Lutrinæ) are nearly cosmopolitan, while the true fur-bearing animals (subfamily Mustelinæ) are distinctively northern and circumpolar; the Skunks (subfamily Mephitinæ), on the other hand, are exclusively American, and range over the temperate and tropical portions of both continents. The order Insectivora has no representatives in either South America or Australia; the Hedghogs, forming a family of the order, are exclusively an Old

World type ; the Shrews, forming a second family, have a circumpolar distribution, the genus *Sorex* ranging on both continents from the Arctic regions to within the tropics. The Moles, forming a third family of this order, are confined to the north temperate latitudes, and are represented by different and peculiar genera in Europe, Asia, Africa and North America, and in the latter continent are confined mainly to the United States east of the Great Plains. Among Rodentia the genus *Sciurus*, consisting of the true or arboreal Squirrels, the genus *Sciuropterus*, or the Flying Squirrels, and the genus *Lepus*, embracing the Hares and Rabbits, are other examples of wide-ranging genera, the arboreal Squirrels and the Hares occurring everywhere except in Australia, although both are sparingly represented in America south of the Isthmus of Panama, beyond which the Flying Squirrels wholly cease to exist. But the species are generally of local distribution, and to some extent different styles of Squirrels and Hares characterize different areas of the common habitat of the group. Northern North America has several genera of Field Mice which are also common to the northern parts of the Old World, but the great bulk of our native Rats and Mice belong to genera peculiar to America. Also our Jumping Mouse, the numerous species of Kangaroo Rats and Pocket Mice, all the Pocket Gophers and the Prairie Dogs are distinctively North American.

These few illustrations, from the many that might be given, will serve to indicate, in a general way, the basis on which the life areas of North America, and of the world at large, are founded.

Primary Life Regions.

Seven primary life regions, speaking of the world as a whole, may be recognized, as follows :

1. An *Arctic Realm*, occupying the region north of the isotherm of 32° F., its southern boundary conforming very closely to the northern limit of trees. Its more characteristic terrestrial forms of both animal and vegetable life range nearly throughout its extent. It is thus so homogeneous in its ontological characters as not to require subdivision into regions and provinces, though embracing several slightly-marked areas of the rank of faunæ.

2. A *North Temperate Realm*, embracing the whole of that portion of the northern hemisphere embraced between the annual isotherms of 32° and 70° F.

3. An *American Tropical Realm*, consisting, as the name implies, of tropical America.

4. An *Indo-African Realm*, consisting of Africa (except the northern border), and tropical Asia and its outlying tropical islands.

5. A *South American Temperate Realm*, embracing extra-tropical South America.

6. An *Australian Realm*, including not only the continent of Australia but New Guinea, New Zealand, and the various groups of islands to the northward and eastward.

7. A *Lemurian Realm*, consisting of Madagascar.

An eighth or *Antarctic Realm* is also often recognized. It is almost wholly oceanic, and its fauna hence consists almost exclusively of marine or pelagic species, and is of course the Antarctic counterpart of the Arctic Realm, though perhaps less well characterized.

North Temperate Realm.

The North Temperate Realm is divisible primarily into two Regions, namely, (1) a *North American Region*, occupying the whole of North America from the beginning of forest vegetation southward to about the northern limit of palms, or the area between the annual isotherms of 32° and 70° F.; and (2) an *Eurasiatic Region*, consisting of the corresponding portion of the Old World.

This region, and the Old World in general, lying outside of the special scope of the present paper, we will now pass to a detailed consideration of the mammalian fauna of North America, and the principal faunal subdivisions of the North American Continent, their distinctive characteristics, their relation to each other, and to the Eurasiatic Region.

Attention has already been directed to the intimate geographic relation of northern North America to northern Eurasia, and the

Genera of Land Mammals of the North Temperate Realm.

[NOTE.—The names of circumpolar genera are in *italics*; those of genera peculiar respectively to the North American and Eurasiatic Regions are in SMALL CAPS.]

NORTH AMERICAN REGION.			EURASIATIC REGION.		
	Subregions.			Subregions.	
Genera.	Cold Temp.	Warm Temp.	Genera.	Cold Temp.	Warm Temp.
Didelphys	—	+	Sus	+	+
Dicotyles	—	+	CAMELUS	—	+
Cariacus	—	+	MOSCHUS	+	+
Cervus	+	—	*Cervus*	+	+
Alces	+	—	*Alces*	+	—
Rangifer	+	—	*Rangifer*	+	—
ANTILOCAPRA	—	+	DAMA	—	+
Bison	+	+	CAPREOLUS	+	+
Ovis	+	—	ELAPHODUS	—	+
MAZAMA	+	—	HYDROPOTES	—	+
Arctomys	+	+	*Bison**	+	—
CYNOMYS	—	+	POËPHAGUS	—	+
Tamias	+	+	ADDAX	—	+
Spermophilus	+	+	Oryx	—	+
Sciurus	+	+	Gazella	—	+
Sciuropterus	+	+	SAIGA	—	+
APLODONTIA	+	—	PANTHOLOPS	—	+
Castor	+	+	RUPICAPRA	+	—
FIBER	+	+	BUDORCAS	—	+?
Cuniculus	+	—	Nemorhœdus	+	+
Myodes	+	—	Capra	+	+
SYNAPTOMYS	+	+	*Ovis*	+	+
Evotomys	+	—	Equus	—	+
Arvicola	+	+	*Castor*	+	—
PHENACOMYS	+	—	*Sciurus*	+	+
ONYCHOMYS	—	+	*Sciuropterus*	+	+
Sitomys	+	+	Pteromys	—	+
Reithrodontomys	—	+	*Tamias*	+	—
Sigmodon	—	+	*Spermophilus*	+	—
Oryzomys	—	+	*Arctomys*	+	—
Neotoma	—	+	MYOXUS	—	+
THOMOMYS	—	+	ELIOMYS	+	+
GEOMYS	—	+	MUSCARDINUS	—	+
PEROGNATHUS	—	+	Mus	+	+
MICRODIPODOPS	—	+	Gerbillus	+	+
DIPODOMYS	—	+	CRICETUS	+	+
PERODIPUS	—	+	Meriones	—	+
ZAPUS	+	+	SMINTHUS	+	—
ERETHIZON	+	—	*Arvicola*	+	+
Lagomys	+	—	*Evotomys*	+	—
Lepus	+	+	*Cuniculus*	+	—
ANTROZOUS	—	+	*Myodes*	+	—
Vesperugo	+	+	ELLOBIUS	+	—
NYCTICEJUS	—	+	SIPHNEUS	+	—

North American Region.

Genera.	Subregions. Cold Temp.	Warm Temp.
Atalapha	+	+
Plecotus	—	+
Vespertilio	+	+
EUDERMA	—	+
Molossus	—	+
Nyctinomus	—	+
Otopterus	—	+
Sorex	+	+
NOTIOSOREX	?	+
BLARINA	—	+
SCALOPS	—	+
SCAPANUS	—	+
CONDYLURA	+	+
Urotrichus	—	+
Ursus	+	+
Procyon	+	+
BASSARISCUS	—	+
Lutra	+	+
MEPHITIS	+	+
Conepatus	—	+
SPILOGALE	—	+
TAXIDEA	+	+
Gulo	+	—
Lutreola	+	+
Putorius	+	+
Mustela	+	—
UROCYON	—	+
Vulpes	+	+
Canis	+	+
Lynx	+	+
Felis	—	+

Eurasiatic Region.

Genera	Subregions. Cold. Temp.	Warm Temp.
SPALAX	+	—
Rhizomys	—	+
Dipus	—	+
Alactaga	+	+
PLATYCERCOMYS	+	—
Hystrix	—	+
Lagomys	+	—
Lepus	+	+
Rhinolophus	—	+
TRIÆNOPS	—	+
SYNOTUS	—	+
Plecotus	—	+
Vesperugo	+	+
Harpiocephalus	—	+
Vespertilio	+	+
Miniopterus	—	+
RHINOPOMA	—	+
Nyctinomus	—	+
Erinaceus	+	+
Sorex	+	+
CROSSOPUS	+	—
Crocidura	—	+
ANUSOREX	—	+
DIPLOMESODON	—	+
CHIMARROGALE	—	+
NECTOGALE	—	+
MYOGALE	–	+
Urotrichus	—	+
UROPSILUS	—	+
SCAPTONYX	—	+
TALPA	—	+
SCAPTOCHIRUS	—	+
Ursus	+	+
ÆLUROPUS	—	+
ÆLURUS	—	+
MELES	+	+
MELLIVORA	—	+
ARCTONYX	—	+
Lutra	+	+
Gulo	+	—
Lutreola	+	+
Putorius	+	+
Mustela	+	—
NYCTEREUTES	—	+
Vulpes	+	+
CYON	—	+
Canis	+	+
Hyæna	—	+
Genetta	—	+
Herpestes	—	+
Lynx	+	+
Felis	+	+
Macacus	—	+

similarity of the climatic conditions of the two regions; and to the corresponding similarity in their mammalian life, as well as of their general faunal and floral facies.

In this connection it will be instructive to compare somewhat in detail the land mammals of the North American and Eurasiatic Regions, that is, omitting the Seals and the Sea Otter. We will also exclude the strictly Arctic genera—*Ovibos* and *Thalassarctos*. With these restrictions we have 75 genera for the North American Region and 97 for the Eurasiatic Region. These are enumerated in the accompanying table, where they are divided into four categories, as follows: (1) North American, subdivided into (*a*) Cold Temperate, (*b*) Warm Temperate; (2) Eurasiatic, subdivided into (*a*) Cold Temperate, and (*b*) Warm Temperate. By the use of distinctive type, the circumpolar genera and the genera peculiar respectively to the two Regions are distinguished from those having a more or less wide distribution to the southward of the North Temperate Realm. (See pp. 208, 209.)

The total number of genera tabulated is 140, a number of genera which barely enter the southern portion of the area under consideration being excluded as not properly pertaining to it. Of these 140 genera 75 are found in North America and 97 in Eurasia, the Eurasiatic Region, as would be expected from its much larger and more diversified area, having considerably the larger number. Of these 32 are circumpolar, and thus are common to the two regions, while quite a number of others are closely-allied representative genera. Thus nearly one-half of the North American genera are either identical with or closely related to Eurasiatic genera, while only a little more than one-third of the total number are respectively peculiar to one or the other of the two regions, namely: for the North American 27 out of 75, and for the Eurasiatic 40 out of 97, or 67 out of a total of 140.

If we compare, however, the Cold Temperate Subregions of the two Regions we find that out of 43 genera characteristic of the Cold Temperate Subregion in North America 32, or about three-fourths, are circumpolar, and that of the 49 genera of the Cold Temperate Subregion of the Eurasiatic Region 32 are also of course circumpolar. We find further that in the North American Region 17 out of the 27 peculiar genera do not extend north of the Warm Temperate Subregion ; that 6 are common to both subregions, while 4 are peculiar to the Cold Temperate Subregion. Also that in the Eurasiatic Region 30 of the 40 peculiar genera are confined to the Warm Temperate Subregion, that 4 are common to both subregions, while 6 only are peculiar to the Cold Temperate. Consequently the peculiar genera of the two Cold Temperate Subregions, taken together, number only 10 out of a total of 140. This shows that the chief difference between the Eurasiatic and North American Regions is confined respectively to their warm temperate subdivisions, less than one-third of their peculiar or distinctive genera occurring in their cold temperate subdivisions. It is also of interest to note that the peculiar genera of the North American Region belong mainly to two or three families of Rodents—particularly the Heteromyidæ and the Geomyidæ—while many of the peculiar Eurasiatic genera belong to the Talpidæ, and in each case are restricted to comparatively limited areas.

The Sclaterian System.

While these and similar facts have been given due weight by the majority of writers on zoö-geography, there has been one notable exception to which it may not be out of place in the present connection to pointedly call attention. In 1858 Dr. P. L. Sclater, the eminent ornithologist and Secretary of the London Zoölogical Society, published a memoir on the geographical distribution of birds, in which he divided the earth's surface into two primary and four secondary zoölogical regions nearly in accordance with the principal land areas. His two primary regions are equivalent respectively to the eastern and the western hemispheres; or the Old World, termed 'Palæogæa,' and the New World, termed 'Neogæa.' These primary areas were divided on the same principle into (1) a Palæarctic Region, (2) a Nearctic Region, (3) a Palæotropical Region, (4) a Neotropical Region; thus entirely ignoring the close similarity of life throughout the cold temperate and arctic regions of the globe. These divisions, as has been urged recently in their favor,[1] are *convenient* and *easy to remember*, since they are approximately equal in size, are easily defined, and

[1] Wallace, Geogr. Distrib. Anim., Vol. I, pp. 63, 64.

avoid complicated boundaries. The names chosen for them have a classical appearance, are euphonius, and hence captivating. Moreover, this scheme of classification was based on a class of animals respecting which the proposer of the scheme is recognized as an eminent authority. At this time, and even for many years later, there were few special students in the field of zoölogical geography. Hence it was natural that the classification here laid down should meet with wide acceptance, particularly among English writers. Later its fallacies were exposed, and even several eminent English naturalists proposed much more rational schemes—as Huxley on the basis of birds, Günther of the British Museum on the basis of reptiles, and Blyth on general grounds, etc. Yet in 1876, Mr. A. R. Wallace, in his very useful and in many ways admirable work on the 'Geographical Distribution of Animals,' gave new life to the scheme by adopting it as the basis of his own classification, and attempting its defense. We may recognize this as a system based on continental areas, regardless of the actual distribution of life; and also as the Sclaterian method, in opposition to nearly all other systems, whether of botanists or zoölogists, who in general recognize that the distribution of life is in accordance with the climatic zones, in virtue of climatic influences, which the Sclaterian school consider as superficial and misleading. Like so many other misnomers, the terms 'Palæarctic' and 'Nearctic,' 'Palæotropical' and 'Neotropical,' have apparently become ineradicable, their convenience for the designation of particular geographic areas contributing to their adoption even by authors who protest against their use in their original sense.[1]

---

[1] Dr. Packard, in writing of the 'American Arctic Province' in 1883, speaks emphatically on this point as follows: "We reject the term 'Nearctic' proposed by Mr. P. L. Sclater, and adopted by Mr. A. R. Wallace, for America north of Central America, for the reason that it seems to us an unnatural and artificial term. The fauna is essentially American north temperate, while the Arctic regions of America and Europe-Asia form a realm by itself, of much less importance, it is true, than the north temperate realm (American and Europæo-Asiatic regions), when we consider the land plants and animals, but of nearly as much importance as regards marine life. To apply the term *Nearctic* to so vast a region as the American involves the idea that the region covers an area essentially arctic in its features. It is to be hoped that the term will not be adopted by American writers, as it is not by German and French writers, and we heartily endorse Mr. J. A. Allen's protest against the use of the term by American writers on this subject. The circumpolar or Arctic realm is a realm by itself, limited by the low degree of temperature and mainly bounded by the isothermal of 32°, and the adoption of this term will conduce, it appears to us, to clearer and more concise ideas of the geographical distribution of life on our continent."—*Twelfth Ann. Rep. U. S. Geol. and Geog. Surv.* (*Hayden*), pt. I, p. 363.

THE MAMMALS OF NORTH AMERICA CONSIDERED IN RELATION TO THE NORTH AMERICAN REGION AND ITS SUBDIVISIONS.

Having now compared the North American Region with the Eurasiatic Region, we may proceed to an analysis of the North American Region itself. As shown by the table of distribution already given (pp. 208, 209), the North American Region is divisible into two Subregions, namely, a Cold Temperate and a Warm Temperate.[1] The most natural boundary for separating the two Subregions seems to be, in a general way, the northern limit of the successful cultivation of wheat, rye, barley, maize, peas, beans, hops, tobacco, potatoes and tomatoes, and the apple, peach and plum, or about the isotheral line of 65° F. This is approximately the boundary line separating the Alleghanian and Canadian Faunæ, as commonly recognized. The mammalian life of one of these two subregions differs vastly more from that of the other than does the mammalian life of the boreal parts of North America from that of the corresponding portions of Eurasia. The transition is, however, somewhat gradual. New elements appear near the southern boundary of the Cold Temperate, and increase in a rapidly progressive ratio as we proceed southward, while northern types fade out, and the general aspect eventually becomes radically changed.

The mammalian fauna of the Warm Temperate subdivision is found to consist of three pretty distinct elements: first, a generally diffused and more or less modified northern element, forming about one-fourth of the whole; second, a southern element, forming about another fourth; and third, an indigenous element, comprising about the remaining half. This of course is exclusive of the southward extension of purely northern forms along the various mountain ranges, which in a measure masks or obscures the general character of the fauna throughout the Rocky Mountain Plateau region.[2]

[1] These subdivisions were recognized by me in 1878, but rather informally, and chiefly in a tabular way. See Bull. U. S. Geol. and Geogr. Surv., IV, pp. 339–343, and *passim* in the text at p. 337.

[2] Dr. Merriam has already called attention to the mixed origin of the fauna and flora of the middle temperate portion of North America is his discussion of the fauna and flora of San Francisco Mountain and the Painted Desert in Arizona (North American Fauna, No. 3, Sept., 1890, pp. 20–22).

To review briefly the leading characteristics of the region south of about the latitude of the Great Lakes, as compared with the area to the northward, we have as new elements among the Carnivores the Panther or Mountain Lion, the Bay Lynx instead of the Canada Lynx, the Gray Fox, the little Kit Fox and the Prairie Wolf, the Black-footed Ferret, the Raccoon, and the whole Skunk tribe represented by two genera and numerous species. Among the Ruminants the Prong-horned Antelope and the Bison; while the Caribou and the Moose of the north are replaced by an entirely different genus of the Deer tribe, represented by three species. Among the Rodents, we have an entirely new set of both Squirrels and Hares, and a greatly increased number of species of each; many new Field Mice, a ten-fold increase in the ground Squirrels and Spermophiles; also entirely new and numerously represented genera among the Field Mice, and three new families of other rodents, including some 20 to 30 species of Pocket Gophers, Kangaroo Rats and Pocket Mice. Among the Insectivores, in place of the single genus of small Shrews characterizing the northern region, we have not only new species but additional genera, and also the whole family of Moles, comprising several genera. In Bats, in place of a few straggling species which barely reach the milder parts of the high north, we have a very great increase in the number of both genera and species. We have further in the Opossum a distinctly tropical type.

Passing now down to near the southern border of this region we find still fewer northern types, but meet in their place a decidedly tropical element. In Texas, and along our southern border thence westward, we have the Armadillo, the Peccary, the Coati, the so-called Texas Civet-cat or Cacomistle, the Jaguar, the Ocelot, and leaf-nosed and big-eared Bats of several genera. Although the width of the continent along the 60th parallel is three times as great as it is along the 30th parallel, the number of species of mammals is probably ten times greater along the 30th parallel than it is along the 60th, and the life, so far as mammals are concerned, has become almost entirely changed, under the influence of the greatly altered climatic conditions.

As shown by the table of distribution already given (pp. 208, 209), 14 genera occur in the Cold Temperate which do not range to any extent into the Warm Temperate, namely:

Cervus,	Aplodontia,	Erethizon,
Alces,	Cuniculus,	Lagomys,
Rangifer,	Myodes,	Gulo,
Mazama,	Phenacomys,	Mustela.
Ovis,	Evotomys,	

On the other hand, 33 genera found in the Warm Temperate do not occur in the Cold Temperate, namely:

Didelphys,	Geomys,	Otopterus,
Dicotyles,	Perognathus,	Notiosorex,
Cariacus,	Microdipodops,	Blarina,
Antilocapra,	Dipodomys,	Scalops,
Cynomys,	Perodipus,	Scapanus,
Onychomys,	Antrozous,	Urotrichus,
Reithrodontomys,	Nycticejus,	Bassariscus,
Sigmodon,	Plecotus,	Conepatus,
Oryzomys,	Euderma,	Spilogale,
Neotoma,	Molossus,	Urocyon,
Thomonys,	Nyctinomus,	Felis.

The remaining 27 genera are to a greater or less extent common to both the Cold Temperate and the Warm Temperate. These genera are as follows:

Bison,	Arvicola,	Procyon,
Arctomys,	Sitomys,	Lutra,
Tamias,	Lepus,	Mephitis,
Spermophilus,	Vesperugo,	Taxidea,
Sciurus,	Atalapha,	Lutreola,
Sciuropterus,	Vespertilio,	Putorius,
Castor,	Sorex,	Vulpes,
Fiber,	Condylura,	Canis,
Synaptomys,	Ursus,	Lynx.

Hence while the two Subregions have much in common, and are thus thoroughly bound together, their differential elements are strongly marked. The 42 genera occurring in the Cold Temperate are either obviously of boreal origin, or find their nearest relationships with boreal types. Of the 62 genera occurring in the Warm Temperate Subregion, about 14 are wide ranging southern or subcosmopolitan types (some of them disappear before reaching the southern third of the subregion), 24 may be regarded as indigenous, and about 13 as of southern (tropical or subtropical) origin.

Continuing the analysis in further detail, we find that the following 28 genera of the North American Region are of either circumpolar or subcosmopolitan distribution and hence not distinctively North American, namely:

Cervus,	Castor,	Gulo,
Alces,	Evotomys,	Lutreola,
Rangifer,	Arvicola,	Putorius,
Bison,	Lagomys,	Mustela,
Ovis,	Lepus,	Vulpes,
Arctomys,	Vesperugo,	Canis,
Tamias,	Vespertilio,	Lynx,
Spermophilus,	Sorex,	Felis.
Sciurus,	Ursus,	
Sciuropterus,	Lutra,	

Only the remaining genera can therefore be considered as distinctively North American. These may be divided, according to their distribution, as follows :

Of general distribution, and thus distinctive of the North American region as a whole rather than of any particular subdivision :

Fiber,	Atalapha,
Synaptomys,	Mephitis,
Sitomys,	Taxidea.
Zapus,	

They are not, however, all evenly distributed throughout the region, *Taxidea*, for example, being absent from the southeastern States, while some of the others do not apparently extend northward beyond the middle of the Cold Temperate Subregion. On the other hand, *Mazama*, *Aplodontia*, *Latax*, and the subgenus *Neürotrichus* are of such local and peculiar distribution as not to be diagnostic of any of the lesser divisions, the first two being mountain forms and another (*Latax*) strictly littoral. *Erethizon* and *Phenacomys* are distinctive of the northern subregion ; *Synaptomys* and the subgenus *Tamiasciurus* occupy a middle position, though mainly northern ; *Condylura*, though of more loçal distribution, is not distinctively either northern or southern.

The remaining genera (with their subgenera, which it will be convenient to use in the present connection) do not range north of the Warm Temperate Subregion, though some of them extend far beyond it to the southward. Of these the following range over the subregion at large, and are consequently distinctive of the Warm Temperate Subregion as a whole :

Cariacus,	Scapanus,
Neotoma,	Urocyon.
Neosciurus,	

The remaining genera and subgenera are of comparatively limited distribution, and may be conveniently divided into four categories. First, however, may be eliminated a number of intrusive southern forms which are properly tropical and extend only a short distance over the southern border of the Warm Temperate Subregion. These are *Tatusia*, *Dicotyles*, *Heteromys*, *Molossus*, *Nyctinomus*, and *Nasua* ; the first and the last (*Tatusia* and *Nasua*), and perhaps also *Heteromys*, though occurring north of the Rio Grande on the coast of Texas, appear to be really confined to the narrow northward extension of the American Tropical Realm along the lower coast of Texas, and thus really form no part of the proper fauna of the Warm Temperate Subregion. The four categories into which the remaining genera and subgenera may be divided are (1) *northern*, (2) *southern*, (3) *eastern*, and (4) *western*. The first two include the few types that range nearly across the continent from ocean to ocean as follows :

1. *Northern.*—Taken in a strict sense, the northern half of the Warm Temperate, has not a single genus, among those peculiar to the region, which ranges across the continent. By taking into consideration wide-ranging types, which reach this region (mainly) from the northward, we have the following as coming into the present category, namely : *Tamias*, *Lutreola*, *Zapus* and *Putorius*. *Tamias* (subgenus *Eutamias*), however, extends far southward in the West.

2. *Southern.*—The genera and subgenera that fall strictly into this category are few, even if we include besides those peculiar to the region, also those barely entering the region from the southward, as follows :

Didelphys,	Sigmodon,	Otopterus,
Dicotyles,	Oryzomys,	Corynorhinus,
Reithrodontomys,	Nyctinomus,	Spilogale.

Thus the northern and southern divisions of this subregion are distinguished mainly, so far as genera having a transcontinental distribution are concerned, by the presence of a number of types in the southern which do not reach the northern, only six of which, however, are strictly transcontinental.

3. *Eastern.*—The distinctively eastern genera and subgenera are the following:

Neofiber,	Nycticejus,
Blarina,	Parasciurus,
Scalops,	Tamias.

The first of these is local and southern; the others have a more general range.

4. *Western.*—The following list of 20 genera and subgenera are distinctively western or southwestern, only one of them (*Ictidomys*) occurring east of the Mississippi River.

Antilocapra,	Perodipus,	Otospermophilus,
Cynomys,	Onychomys,	Ictidomys,
Thomomys,	Antrozous,	Xerospermophilus,
Perognathus,	Euderma,	Ammospermophilus,
Chætodipus,	Molossus,	Bassariscus,
Microdipodops,	Notiosorex,	Conepatus.
Dipodomys,	Eutamias,	

All but two of these (*Molossus* and *Conepatus*) are indigenous to the region under consideration. *Geomys* may also be best placed here, though limited in its United States range to the plains and prairies east of the Rocky Mountains and to the coast region of the South Atlantic and Gulf States.

From the foregoing it is evident that the Cold Temperate and Warm Temperate Subregions differ greatly in respect to homogeneousness. No part of the Cold Temperate is very strongly differentiated from the rest of the subregion, whereas different parts of the Warm Temperate are found to be very unlike, even though situated under the same parallels of latitude. Hence the Warm Temperate admits of separation into two quite unlike parts, a western and an eastern, nearly on the line of the 100th meridian, while the Cold Temperate admits of no such subdivision.

The Major Faunal Areas of the North American Continent.

We are thus led to adopt the following scheme for the division of the North American continent into major faunal areas, as illustrated in the accompanying maps. First, the Arctic portion of the continent, or the region beyond the limit of arboreal vegetation, is to be assigned to the *Arctic Realm.* Second, the region south of the Mexican tableland, and also the low eastern coast region of Mexico north into Texas (about to Corpus Christi), and the low western coast region of Mexico north about to Mazatlan, may be assigned to the *American Tropical Realm*, with which also belong the extreme southern part of the Peninsulas of Lower California and Florida. Third, the remaining and by far the greater part of the continent belongs to the *North Temperate Realm*, of which it constitutes the *North American Region.*.

The North American Region is divisible into two *Subregions*, namely, a *Cold Temperate* and a *Warm Temperate*. The latter comprises two *Provinces*, a *Humid* or Eastern, and an *Arid* or Western. The Humid Province is divisible into two *Subprovinces*, namely, an *Appalachian* or Northern, and an *Austroriparian* or Southern. The Arid Province is also divisible into two Subprovinces, namely, a *Campestrian* or Northern, and a *Sonoran* or Southern. The Campestrian Subprovince is susceptible of division into two or three *Districts*, as the *Great Plains District*, the *Great Basin District*, and a *Pacific Coast District*. Each of these areas usually consists of two or more minor divisions or *Faunæ*. (See accompanying maps, plates V–VIII.)

We will now pass in more formal review the faunal subdivisions of the North American Continent, from Region to Fauna, so far as the latter are at present clearly determinable.

The American Arctic.

If North America were isolated from the rest of the world it would be quite proper to treat the American portion of the Arctic Realm as merely a subdivision of the North American *Region*; but in view of the fact that it is really a part of a homogeneous hyperborean fauna of circumpolar distribution it seems more in accordance with general facts to consider it as a part of the Arctic Realm. The propriety of this seems especially emphasized when we consider that the "animals and plants inhabiting the Arctic regions are usually specifically identical throughout Arctic America, Greenland, and the polar parts of Eurasia and outlying islands," the "types inhabiting the Arctic Zone being few in number and uniform in character throughout their distribution."[1] The fauna of this Arctic Zone is thus no more American than it is Europæo-Asiatic, and differs far more from that of the adjoining region to the southward, both in North America and Eurasia, than does the American arctic from the Eurasian arctic. The Arctic Realm possesses only a small number of peculiar types in proportion to its area or in comparison with the other realms situated under more favorable conditions for the development of a diversified abundance of life; yet its peculiar types are quite numerous when considered in relation to the general meagreness of the fauna in these inclement latitudes. It is especially characterized by its poverty of life, and consequently largely by negative characters—by what it lacks rather than by a high ratio of peculiar forms.

The American Arctic[2] may be divided into two areas, which may take the rank of faunæ, or perhaps more properly of subfaunæ, in view of their slight inter-differentiation, namely: (1) the *Barren Ground Fauna*, and (2) the *Alaskan Arctic Fauna.*[3]

The Arctic American as a whole is characterized by being the home of the Eskimo, the Polar Bear, the Arctic Fox, the Arctic Hare, Parry's Marmot, the White Lemming, the Musk Ox, the Barren-ground Caribou, the Walrusses, and various species of Seals. Its southern boundary also forms the northern limit of nearly all of the characteristic species of the adjoining region southward, the greater part of which find their northern limit very near where the forest vegetation gives place to that of the Barren-grounds. The two faunæ into which this area appears separable are principally characterized by each having certain marine mammals along its coast not common to the other. The Alaskan Arctic has the Fur Seal, the Sea Lion, the Banded Seal, the Pacific Walrus, and the Sea Otter, neither of which occur in the region designated as the Barren Ground Fauna, which on the other hand has a few species, including the Musk Ox, not found in the Alaskan Arctic.

---

[1] Merriam, Proc. Biol. Soc. Wash., VII, pp. 39, 40.

[2] The American Arctic was ranked by me in 1871 (Bull. Mus. Comp. Zoöl., II, No. 2, p. 403) as a 'fauna' of the Arctic Realm, and termed the 'American Arctic Fauna.'

[3] The Alaskan Arctic Fauna is Mr. Edward W. Nelson's Alaskan Arctic District, characterized by him in his 'Report upon Natural History Collections made in Alaska,' 1887, pp. 27–32.

THE NORTH AMERICAN REGION.

The North American Region forms two *Subregions*, namely : (1) a *Cold Temperate Subregion*, extending southward to about the mean latitude of the Great Lakes, with outlying portions extending further southward along the principal mountain systems of the continent ; and (2) a *Warm Temperate Subregion*, occupying the remainder of the North American Region. The differential features of the two subregions have already been shown in the analysis of the mammalian fauna of North America (see pp. 213–215).

I. COLD TEMPERATE SUBREGION.—The Cold Temperate Subregion extends across the continent from the Atlantic to the Pacific, and from about latitude 43° northward to the limit of forest vegetation, with, however, a narrow prolongation southward along the Appalachian Highlands as far as northern Georgia, another in the interior along the main chain of the Rocky Mountains south into Mexico, and a third along the Cascade and Sierra Nevada ranges. Its southern border also sweeps to the northward, in the region of the Great Plains, so as to exclude the plains of the Saskatchewan, which belong to the Warm Temperate. It is subdivisible into two transcontinental *zones*, termed respectively the *Hudsonian* and *Canadian Zones*,[1] named from the two principal faunæ of which this subregion mainly consists. These zones correspond respectively with the Subarctic and Cold Temperate Zones of physical geographers and botanists.

The *faunæ* of the Cold Temperate Subregion are the *Hudsonian* and *Carolinian* in the east, and the *Aleutian* and *Sitkan* on the northwest coast, with a series of closely-related mountain faunæ or subfaunæ in the Rocky Mountains and the Sierra Nevada, as yet not well defined.

1. *Hudsonian Fauna.*—The Hudsonian Fauna occupies a belt at the northern border of the Cold Temperate Subregion, extending from Newfoundland to and across Alaska nearly to Bering Sea. It is thus bounded on the north by the Barren Ground Fauna, and west and northwest by the Alaskan Arctic Fauna. Its southern limit may be tentatively given as the isotheral of 57° F.

[1] See Merriam, N. Am. Fauna, No. 5, 1891, pp. 10-12, and *ibid.*, Proc. Biol. Soc. Washington, VII, 1892, p. 24.

The Hudsonian, as thus defined, forms the northern limit of a comparatively large number of mammals dependent upon forests for food and shelter. These are apparently the Red Fox, the Timber Wolf, the Black Bear, the Canada Lynx, the Weasels, Mink, Marten, Fisher, Wolverine, and the Otter; also of the Red Squirrel, the Flying Squirrel, the Jumping and other Mice, and the Varying Hare. These all enter the Hudsonian Fauna from the southward, and range northward to about its northern border. The northern limit of these various species is not in each case coincident, some ranging more or less beyond certain others, the exact limit being determined by the peculiar needs and habits of the species, and somewhat with local conditions. Hence here, as at the boundaries of faunæ generally, the species limited by a given fauna do not all stop along one abrupt line, but gradually fade out, one after another, within, however, comparatively narrow and locally varying limits.

The Hudsonian Fauna is further characterized by forming the southern limit of several species of Seals, as the Gray, Hooded and Harp, the Barren-ground Caribou, the Tawny Lemming, and the Yellow-cheeked Meadow Mouse, which range southward over this fauna from the Arctic Realm. The Hudsonian Fauna is thus distinctively characterized by the assemblage of species just enumerated, which occur together nowhere else.

2. *Canadian Fauna.*—The Canadian Fauna, like the Hudsonian, forms a nearly transcontinental belt, with the southern border of the Hudsonian for its northern boundary. It apparently includes a small portion of southwestern Newfoundland, nearly all of the provinces of New Brunswick and Quebec, northern Ontario, the northern half of New England, northern Michigan, the northern border of Wisconsin, and northeastern Minnesota. West of Lake Superior it stretches northwestward in a broad belt to the Peace River and northern British Columbia districts of Canada. It thus extends across the Rocky Mountains nearly to the Pacific coast. Its southern border trends at first southward, reaching its most southern point in the interior in northern Michigan and north-eastern Wisconsin; it then runs northwest along the border of the plains, till it nearly or quite reaches the 55th parallel, when it then descends, west of the Saskatchewan Plains, as a narrow belt

along the eastern base of the Rocky Mountains, as far as northeastern Montana, with a further southward extension in the mountains proper. West of the Rocky Mountains it appears not to descend below the southern boundary of British Columbia, but runs southward in the mountains to an as yet undetermined limit. To the westward it is cut off from the Pacific coast by a distinctively coast belt. It has outlying insular areas in the Adirondack region of New York and on the higher crests of the Appalachian Highlands; it also forms a zone in the Rocky Mountain chain and its outlying ranges.[1]

The Canadian Fauna forms the northern limit of the common Skunk, the Star-nosed Mole, the Hoary, Red and Brown Bats, the northern form of the common Striped Squirrel, the Woodchuck, and the Virginia Deer. It also forms the southern limit of the Canada Lynx, the Wolverene, Pine Marten, Moose, Caribou, Canada Porcupine, and various species of Short-tailed Meadow Mice of the genus *Phenacomys*, etc.

3. *Sitkan Fauna.*—This is the 'Sitkan District,' as defined by Mr. E. W. Nelson.[2] It is of limited area, being confined to a narrow strip along the Pacific coast of Alaska, extending from Bristol Bay to about the southern end of the Territory of Alaska. So far as mammals go it has very little to distinguish it from the Canadian Fauna, of which it is practically little more than a littoral district. It is characterized by a few local forms of wide-ranging species, particularly of birds, and probably of animal life in general. It is a region of heavy precipitation and overcast skies, the climatic effect of which is strongly shown in the great intensification of color which marks most of the species coming within its influence. For this reason it seems entitled to recognition as a distinct faunal area of low rank.

4. *Aleutian Fauna.*—This is the 'Aleutian District' of Mr. E. W. Nelson,[3] previously termed by me the 'Aleutian Fauna.'[4]

---

1 On the southern boundary of the Canadian Fauna in the East, see Verrill, Proc. Bost. Soc. Nat. Hist., Vol. X, p. 260, and Allen, Bull. Mus. Comp. Zoöl., Vol. II, p. 398. See also Dr. Merriam's Map 5 in 'North American Fauna' No. 3, where the southern boundary of his Boreal Province may be taken as the southern boundary of the Canadian Fauna.

2 Rep. Nat. Hist. Coll. made in Alaska, 1887, pp. 24-26.

3 *Ibid.*, 1887, p. 26.

4 Bull. Mus. Comp. Zoöl., II, 1871, p. 401.

It includes not only the Aleutian chain of islands, but also the western and southern portions of the Alaskan Peninsula and probably the island of Kadiak, although this may more properly belong to the Sitkan Fauna. So far as mammals are concerned there is very little to distinguish the Aleutian Fauna, but it is well characterized by its bird life.

Both the Hudsonian and the Canadian Faunæ are represented in the higher parts of the Cascades, and in the more northerly parts of the Rocky Mountains in the United States, including their outlying spurs in both Idaho and Montana. More to the southward, in Utah and Colorado, and thence further southward in New Mexico, Arizona, and in Mexico, and also in the mountains of California, a few new elements come in, the differentiation being progressive toward the southward. Our knowledge of the mammalian life of these subalpine regions is, however, still too limited to render practicable any attempt in the present connection to define or characterize these southern extensions of the Hudsonian and Canadian life zones. We are thus far indebted to Dr. C. Hart Merriam, Chief of the Division of Mammalogy and Ornithology of the Department of Agriculture, for most of the exact knowledge we at present possess on the subject, and for a very successful attempt to correlate the life zones of some of our western mountain areas with those of the northern part of the continent. In the San Francisco Mountain region in Arizona he has very clearly traced[1] seven life zones and in part correlated them with the corresponding life areas of the continent at large. He subsequently extended[2] his careful methods of field work to the mountains of South-Central Idaho, with like praiseworthy and satisfactory results. The zones recognized in the San Francisco Mountain region and their correlations may be indicated as follows:

Zone.	*Fauna.*
Alpine	Arctic.
Subalpine or Timber-line	
Hudsonian or Spruce	Hudsonian.
Canadian or Fir	Canadian.
Neutral or Pine	Alleghanian.
Piñon or Cedar	[Carolinian].
Desert	[Louisianian].

[1] N. Am. Fauna, No. 3, 1890, pp. 7–34 and maps 1–4.
[2] *Ibid.*, No. 5, 1891, pp. 9–12, 21–25.

II. Warm Temperate Subregion.—The Warm Temperate Subregion occupies middle North America, extending from the Cold Temperate Subregion on the north to the American Tropical Realm, as already defined, on the south. It thus includes the greater part of the United States, Lower California, and the Mexican tableland. It is cut into along the principal mountain systems by the southern prolongations of the Cold Temperate Subregion, and also extends northward over the Saskatchewan Plains. The extreme southern parts of the peninsulas of Florida and Lower California, however, are excluded, as also the lower coast region of Texas, these excluded districts, though of comparatively small extent, belonging to the Tropical Realm.

The Warm Temperate Subregion, regarded as a whole, is very unlike the comparatively homogeneous Cold Temperate Subregion, as already shown (pp. 213–218). It is vastly more varied in its physical features, is situated for the most part under climatic conditions more favorable to abundance and diversity of life, and thus presents a greater proportion of peculiar types, and also a larger number of more sharply contrasted faunal areas. Its life is largely indigenous, with, however, a strong infusion of both northern and southern elements. The indigenous elements appear to have had their origin in the Mexican plateau region, and are thus properly designated as Sonoran. The following list of mammalian genera and subgenera may be safely placed in this category:

Cariacus,	Thomomys,	Spilogale,
Antilocapra,	Dipodomys,	Notiosorex,
Cynomys,	Perodipus,	Scalops,
Xerospermophilus,	Microdipodops,	Scapanus,
Ammospermophilus,	Perognathus,	Corynorhinus,
Reithrodontomys,	Chætodipus,	Euderma,
Onychomys,	Bassariscus,	Antrozous.
Geomys,	Conepatus,	

The genera of tropical origin are:

Didelphys,	Sigmodon,	Molossus,
Dicotyles,	Neotoma,	Nyctinomus,
Oryzomys,	Procyon,	Otopterus.

Only two of these, *Neotoma* and *Procyon*, extend very far northward.

Besides the above, several genera of wide distribution beyond North America also occur, as

Lepus,	Spermophilus,	Vulpes,
Castor,	Lutra,	Lynx,
Sciurus,	Mustela,	Felis,
Sciuropterus,	Putorius,	Vesperugo,
Tamias,	Canis,	Vespertilio, etc.

Several American genera of rather extended range to the northward also occur, as *Sitomys*, *Fiber*, *Atalapha*, *Mephitis*, *Taxidea*, etc.

Nomenclature and History.—Before passing to a detailed consideration of the subdivisions of the Warm Temperate Subregion, it may be proper to refer briefly to its nomenclature and history, as treated by previous authors. This region, as here defined, consists of the southern portion of Baird's Eastern Province, together with his Middle and Western Provinces. As early as 1878[1] I separated the 'North American Region' into two *Subregions*, namely, a *Cold Temperate Subregion* and a *Warm Temperate Subregion*, as is done in the present paper, using these terms as headings in tables giving the distribution of the genera of North American Mammals. Baird's 'Eastern,' 'Middle,' and 'Western' *Provinces* were recognized as "natural regions," with the designation of 'Provinces,' but with the Eastern Province modified so as to restrict it to the Warm Temperate Subregion, and all three reduced in grade to regions of the third rank[2] instead of the second rank, as regarded by Professor Baird.

In 1883 Dr. Packard[3] substantially adopted this classification in treating of the faunal regions of North America, with, however, a change of name for the 'Cold Temperate Subregion,' he adopting for it that of 'Boreal Province'—an unfortunate suggestion of my own made later in the paper above cited (l. c., p. 376, where, in some unaccountable way my former division of the 'North Temperate Realm' into 'Subregions' was wholly overlooked!). Dr. Packard, in his otherwise excellent 'Zoö-geographical Map of North America,' failed, however, to recognize the southward

[1] Bull. U. S. Geol. and Geogr. Survey (Hayden), IV, 1878, pp. 338–344.

[2] That is, of the North American Region; really of fourth rank, considered from the basis of the world as a whole.

[3] Twelth Ann. Rep. U. S. Geol. and Geogr. Survey (Hayden), pt. I, 1883, pp. 362–370, and map; the latter republished in the Third Rep. U. S. Entomol. Comm., 1883, map iv.

extension of the Cold Temperate Subregion along the principal mountain systems of the continent.

Dr. Merriam in 1890[1] again set off the Cold Temperate Subregion, under the name 'Boreal Province,' and mapped in detail its southern prolongations into the mountainous parts of the Warm Temperate. The Warm Temperate Subregion was also recognized as a contrasting region of coördinate rank, under the designation 'Sonoran Province,' while the old 'Eastern,' 'Middle,' and 'Western' Provinces were properly repudiated as having no basis in nature. Particularly is this the case in respect to the Central Province, of which Dr. Merriam observes: "The region almost universally recognized by recent writers as the 'Central Province' is made up of the Great Plains, the Rocky Mountains and the Great Basin. A critical study of the life of the Rocky Mountains has shown it to consist of a southward extension of the Boreal Province, with an admixture of southern forms resulting from an intrusion or overlapping of representatives of the Sonoran Province, some of which, from long residence in the region, have undergone enough modification to be recognized as distinct subspecies or even species. A similar analysis of the Great Plains and Great Basin has shown them to consist of northward extensions of the Sonoran Province, somewhat mixed with the southernmost fauna and flora of the Boreal Province. Thus the whole of the so-called 'Great Central Province' disappears.

"This explains a multitude of facts that are utterly incomprehensible under the commonly-accepted zoölogical divisions of the country. These facts relate particularly to the distribution of species about the northern boundaries of the supposed Central and Pacific Provinces, and to the dilemma we find ourselves in when attempting to account for the origin of so many primary life areas in a country where there are no impassable physical barriers to prevent the diffusion of animals and plants."[2]

Dr. Merriam's generalizations respecting the Central Province of authors mark an important advance in the study of North American bio-geography. Taking this region with its original boundaries and significance it is a highly artificial division, em-

[1] N. Am. Fauna, No. 3, Sept. 1890, pp. 24–26, and map 5; see also Proc. Biol. Soc. Washington, VII, 1892, pp. 21–40, and accompanying map.

[2] N. Am. Fauna, No. 3, pp. 22, 23.

bracing within its area very unlike faunal elements. Eliminating from it, however, the broad central arm of the 'Boreal' or Cold Temperate Subregion, which occupies so much of the great central plateau, relieves it of an extraneous element, and reduces it to a more natural and geographically quite different region.

As already seen, Dr. Merriam selected for his two primary divisions of the North American Region the terms 'Boreal' and 'Sonoran' *Provinces.* These regions, both as to grade and nomenclature, were at first apparently adopted provisionally, as he says, in speaking of the United States: "Indeed, the present investigation demonstrates that there are but two primary life provinces in this country: a northern, which may be termed *Boreal*, and a southern, which, for our purposes, may be termed *Sonoran*, since it comes to us from Mexico through Sonora."[1] Later, however, he has termed these divisions 'Regions' instead of 'Provinces,' but has continued the use of the terms 'Boreal' and 'Sonoran.' The term Sonoran is thus applied to a region identical in geographical extent with the Warm Temperate—a designation previously used for the same area—and hence includes the whole region east of the Mississippi (as well as that west of it), from the Great Lakes and southern New England south to Florida and the Gulf Coast. The terms 'Sonoran' and 'Sonoran Province' were used, however, as early as 1866 by Prof. Cope,[2] and also later by Cope, Heilprin, and others, for a region of comparatively small extent, consisting of Sonora and adjoining portions of Arizona and New Mexico. In 1887 Heilprin[3] extended the region to include "the peninsula of Lower California, the State of Sonora in Mexico, New Mexico, Arizona, and parts, not yet absolutely defined, of Nevada, California, Texas, and Florida," and modified its title by calling it the 'Sonoran Transition Region.' The Sonoran Province or Region of these authors is thus not at all the 'Sonoran Region' of Merriam, which is an area of much greater extent and of higher rank. The term Sonoran, used in this extended sense, seems at least inappropriate if not misleading, as there are few if any strictly 'Sonoran' types represented in that portion of the United States

[1] N. Am. Fauna, No. 3, p. 19.
[2] Proc. Acad. Nat. Sci. Phila., 1866, p. 300.
[3] The Geogr. and Geol. Distrib. of Anim., p. 106.

situated to the eastward of the Mississippi River. The more descriptive and appropriate designation of 'Warm Temperate' is therefore preferred for the region in question, since it not only has priority but is in harmony with the terms Arctic, Cold Temperate, and Tropical, used currently for other coördinate areas of the continent.

Dr. Merriam in his important contributions to North American bio-geography has evidently not attempted to devise a systematic scheme of terminology and classification for the various grades of faunal areas, but, at first at least, simply employed provisionally such terms as would suffice to clearly indicate the regions under consideration, his attention being mainly and most successfully given to an elucidation of the facts of distribution. In following out, in the present connection, a consistent scheme of nomenclature, first attempted many years ago, the aim is to fix definitely designations for areas of different grades, and to combine the whole into a consistent system of classification. The nomenclature of the subject has ever been in such a chaotic and inharmonious state that a strict 'rule of priority' cannot be enforced, the same terms having been used in widely different senses, while not unfrequently a number of different names have been given to the same area. As already explained (p. 204) the system here proposed in respect to the *rank* of areas of different grades, as from Realm to Fauna, is analagous to the systematic schemes of classification in biology, and also in stratigraphic geology. The selection of distinctive names for divisions of the higher grades has relation to the influences controlling the geographic distribution of life, namely, *climate*, and hence it is natural that the climatic zones and their principal subdivisions should suggest the names of many of the major ontological areas. This indeed has been the custom to a large extent with both botanists and physiographers, and has often been practiced by zoölogists. Thus 'Humid' and 'Arid' become appropriate and suggestive designations for the eastern and western subdivisions of the North American Warm Temperate Subregion. For the lesser regions geographical names, as 'Hudsonian,' etc., are admirably appropriate when suggestive of some characteristic portion of the region in question. Whenever feasible, names first given should of course be retained in preference to later names. Furthermore, the terms indicative of grade should be used with the same uniformity and strictness as are the terms order, family, genus, etc., expressive of rank, in biology.

PROVINCES OF THE WARM TEMPERATE.—The Warm Temperate Subregion, considered as a whole, is primarily divisible into two *Provinces*, namely, (1) a *Humid Province*, extending from the Atlantic Coast to the vicinity of the 100th meridian, and (2) an *Arid Province*, extending from the western border of the Humid Province to the Pacific coast, excluding of course all the higher mountain ranges, which are more or less wooded and constitute southern extensions of the Cold Temperate Subregion.

The Humid Province, as the name implies, has a humid climate and is in general heavily forested; the Arid Province is a dry region, some of it excessively arid, and consists mainly of open plains and deserts. It is highly diversified in respect to its physiographic features, and presents consequently a more varied fauna than is met with in the Humid Province. As already shown (p. 218) these two provinces are distinguished by the occurrence in each of a few peculiar types not possessed by the other, the Arid Province, however, having by far the greater number of peculiar types, owing to the large continuous land area extending from its southern boundary southward, from which direction much of its life has been derived, while the Humid Province has for its southern boundary a wide expanse of sea, namely, the Gulf of Mexico.

These two provinces thus coincide with the two strongly-marked climatic divisions of the middle, or United States, portion of North America in respect to rain-fall. Unlike the major divisions heretofore characterized (realms, regions and subregions), and also unlike the transcontinental zones, they are not separated by isothermal lines, trending in an east and west direction, but by a north and south line determined by the amount of rain-fall. Thus, in the present instance, temperature as a climatic influence governing the distribution of animals and plants is subordinated to the other leading climatic influence, humidity, which varies greatly in these two contrasting regions, in consequence of the long-continued peculiar physiographic and geographic conditions of the two regions. The many peculiar or indigenous types characterizing the Arid Province as compared with the Humid owe their existence, as already intimated, to the adjoining broad land area stretching far to the southward, whence they have been in large part derived, as contrasted with the absence of such a land area adjoining the Humid Province on its southern border. A large proportion of these peculiar types extend northward to the northern border of the Arid Province, or across the whole breadth of the Warm Temperate Subregion. The Warm Temperate Subregion is further subdivisible in a transcontinental direction into two zones, which might be termed a North Warm Temperate and a South Warm Temperate, as shown later on, but neither would be characterized by any considerable number of transcontinental genera. The northern belt would be characterized by less than half-a-dozen Cold Temperate genera which range a little way into the Warm Temperate, and the southern belt by about the same number of semi-tropical genera which extend into it from the southward. The remaining genera of the Warm Temperate are either wide-ranging transcontinental genera common to both belts, or else genera peculiar to either the Arid or the Humid Province.

The transition between the Humid and Arid Provinces is nowhere abrupt; they gradually merge into each other everywhere along their line of junction, as the prairies of the Mississippi Valley gradually become more arid and take on the characteristic aspect of the more arid plains. There is thus here the usual 'transition' belt occurring between contiguous faunal areas. It is, however, rather broader than between regions where temperature is primarily the limiting influence, as in the case of boundaries trending in a nearly east and west direction, the transition being first from a forested region to one of fertile prairies, and thence to arid plains and deserts. The dividing line may be considered as coincident with the isohyetal curve marking an annual rain-fall of 20 inches or less, as shown on rain-fall charts of the United States—hence, as above said, near the 100th meridian.

As already stated, the regions here designated as Humid and Arid 'Provinces' coincide with the 'humid' and 'arid' portions respectively of Dr. Merriam's 'Transition,' 'Upper Sonoran,' and

'Lower Sonoran' *Zones*;[1] he thus recognizing, but in a somewhat different way, the Humid and Arid areas here classified as *Provinces*.

HUMID PROVINCE.—The Humid Province corresponds to the warm temperate part of Prof. Baird's 'Eastern Province,' and is exactly coincident with Dr. Merriam's 'Humid Sonoran' and 'Humid Transition.'[2] It comprises the United States east of the Great Plains, including also southeastern Ontario and the upper St. Lawrence Valley, but excluding northern New England, a portion of northern New York, northern Michigan and northern Wisconsin, and the higher crests of the Alleghanies. It contains a few genera and subgenera which do not occur in the Arid Province, as *Nycticejus*, *Blarina*, *Scalops*, *Condylura*, *Neofiber*, *Parasciurus* and *Tamias* (restricted subgenus). It lacks about 20 genera and subgenera that are confined in their eastward range to the Arid Province.

The Humid Province is separable into two *subprovinces*, namely, (1) an *Appalachian Subprovince*, and (2) an *Austroriparian Subprovince*.

Appalachian Subprovince.—The Appalachian Subprovince consists of the Alleghanian and Carolinian Faunæ, with the boundaries as long recognized by ornithologists and mammalogists, and as recently revised and mapped by Dr. Merriam.[3] It is characterized by the presence of a number of somewhat northern genera and subgenera which do not extend south of the southern boundary of the Carolinian Fauna, namely:

Arctomys,	Mynomes,	Synaptomys,
Tamias,	Pitymys,	Condylura.
Tamiasciurus,		

It is further characterized by the absence of a considerably greater number of southern genera and subgenera which do not pass north of its southern border, as enumerated in the next paragraph.

[1] Proc. Biol. Soc. Washington, VII, pp. 27–31, and accompanying map.
[2] *Ibid.*, pp. 27 and 30, and accompanying map.
[3] North Am. Fauna, No. 3, Map 5.

Austroriparian Subprovince.—The Austroriparian Subprovince consists of the long-recognized Louisianian Fauna, or 'Austroriparian' Fauna, as sometimes termed. Its fauna differs so much from that of the Appalachian Subprovince that they form two strongly-contrasted faunal areas. It is reached by few northern types, southern forms prevailing, to which are added genera of wide general distribution, like *Lutra*, *Canis*, *Vulpes*, *Sciurus*, *Sciuropterus*, *Lepus*, etc. It is characterized by the absence of the genera tabulated in the preceding paragraph, and by the presence of a larger number of others which do not extend north of its northern boundary, namely :

Reithrodontomys,	Neotoma,	Spilogale,
Oryzomys,	Neofiber,	Corynorhinus,
Sigmodon,	Geomys,[1]	Nyctinomus.

FAUNÆ OF THE HUMID PROVINCE.—The Humid Province, as here limited, has long been divided into three *Faunæ*, as follows : (1) *Alleghanian Fauna ;* (2) *Carolinian Fauna ;* (3) *Louisianian Fauna*. They have been so long recognized, and of late so clearly defined, that their boundaries have become well known.[2]

1. *Alleghanian Fauna.*—The northern border of the Alleghanian Fauna forms about the northern limit of the Panther, the Bay Lynx, the Raccoon, the Mole Shrew, the common and Brewer's Moles, the Gray Squirrel, and the Wood Hare. Its southern border forms about the southern limit of the Fisher, the Ermines, the Harbor Seal, the Elk (in former times), the Northern Striped Squirrel, several species of Field Mice (genera *Evotomys* and *Synaptomys*), the varying Hare, etc. It also forms the northern limit of the Cat and Fox Squirrels (subgenus *Parasciurus*), the Opossum, and various other species soon to be mentioned in characterizing the Carolinian Fauna. The Alleghanian Fauna is thus characterized by the overlapping and commingling of a particular set of species not found elsewhere associated. It is bounded on the north by the Canadian Fauna, on the south by the Carolinian Fauna, and extends westward to the edge of the Great Plains.

[1] So far as its distribution east of the Mississippi River is concerned.

[2] See Allen, Bull. Mus. Comp. Zoöl., II, pp. 395-397; Merriam, North. Am. Fauna, No. 3, map No. 5. Respecting especially the southern boundary of the Alleghanian Fauna, see Bicknell, Bull. Nutt. Orn. Club, III, p. 128; Allen, *ibid.*, p. 149; Chapman, Auk, VI, p. 179.

2. *Carolinian Fauna.*—The northern boundary of this fauna forms in a general way the northern limit of the Gray Fox, the Northern Fox Squirrels, the Pine Mouse, the Opossum, and a Bat of the genus *Nycticejus.* It also forms the southern limit of the Star-nosed Mole, the common Red Squirrel, the Southern Chipmunk, the Woodchuck, the Muskrat, and the common Meadow Mouse (*Arvicola riparius*).

3. *Louisianian Fauna.*—The Louisianian Fauna joins the Carolinian Fauna on the south, and occupies the rest of the Eastern United States to the southward, excepting the extreme southern portion of the peninsula of Florida, which has long been recognized as a *Floridian Fauna*, and as belonging to the Antillean Subregion of the American Tropical Realm. The Louisianian Fauna is characterized by the possession of a number of both genera and species not found north of its limits, although most of the Louisianian genera have a wide distribution southward in Mexico, and westward and northwestward, where, however, they are commonly represented by different species. The characteristic elements of the Louisianian Fauna are the Wood Rat, Cotton Rat, Cotton Mouse, Golden Mouse, Rice-field Mouse, Harvest Mouse, Pocket Gopher, the Southern Gray and Southern Fox Squirrels, the Marsh Hare, Swamp Hare, Little Striped Skunk, and a species each of Big-eared and Leaf-nosed Bats. These represent eight genera not found in the Carolinian Fauna, as follows: *Neotoma*, *Sigmodon*, *Oryzomys*, *Reithrodontomys*, *Geomys*, *Spilogale*, *Corynorhinus* and *Nyctinomus.*

Several interesting facts may be here noted in relation to the faunæ of the Humid Province of the Warm Temperate Subregion as compared with the eastern tier of faunæ in the Cold Temperate Subregion. While the Alleghanian, Carolinian and Louisianian Faunæ terminate at the eastern border of the Great Plains, and hence extend over less than half the width of the continent, the Canadian and Hudsonian sweep across nearly its whole breadth, from the Atlantic coast nearly to the Pacific, and this too at a point where the continent presents its greatest breadth. The mammals and birds found from about the latitude of 43° northward extend as a rule uninterruptedly northwestward from the eastern seaboard over the comparatively low, generally forested interior to the Rocky Mountains and across Alaska almost to the Pacific coast. They also pass over the depressed portions of the Rockies in about latitude 57° to 59°, and follow the Peace and Liard Rivers to the sources of the Yukon, spreading thence northward and westward to the coast ranges of southern Alaska. In fact, the Rocky Mountains, as is well known, present too many points of depression to form much of a barrier to the dispersion of species, so that from southwestern British Columbia northward, except along the coast, the fauna is nearly identical with that of eastern Canada and northern New England. The arid, treeless plains of the interior thus form a greater barrier to the extension westward of eastern forms than do the Rocky Mountains themselves.

Arid Province.—The Arid Province extends from the eastern border of the Great Plains to the Pacific, and northward over the Saskatchewan Plains, the Plains of the Columbia, and thence north into southern British Columbia. It thus includes the so-called 'Central' and 'Pacific' Provinces of Baird and most subsequent writers, excepting of course the more elevated parts of the Rocky Mountain plateau. It is thus coëxtensive with Dr. Merriam's 'Arid Sonoran.' While it is true that a narrow belt along the Pacific coast from Southern California northward to the Alaskan Peninsula possesses a few peculiar types, and lacks a few of those occurring in the region immediately to the eastward, the differentiation is on the whole too slight to give to this Pacific coast district the rank of a region coördinate in grade with Baird's so-called Middle and Eastern Provinces.[1] These differences serve at best merely to mark off from the interior region at large a tier of narrow coast faunæ of the same grade as those bordering the Atlantic coast,

[1] In writing in 1871 of the 'Natural Provinces of the North American Temperate Region,' from the standpoint of Ornithology, I adopted the present classification, as shown by the following: "Within this Region may be recognized two Provinces—an Eastern and a Western—quite distinct from each other in their general features as well as in many special characteristics. The Eastern Province is characterized by the uniformity of its geographical and climatic features, and by a corresponding uniformity in its faunal and floral aspects. The Western Province, on the other hand, is characterized by the diversity of its geographical and climatic features—different areas situated under the same parallels differing greatly in these respects—and by the number and small extent of its zoölogical and botanical areas, and its comparatively numerous restricted floræ and faunæ....The Western Region [*lege* Province] commences at the western border of the Eastern and extends thence to the Pacific coast."—*Bull. Mus. Comp. Zoöl.*, Vol. II, pp. 384, 385.

although the latter, owing to the widely different physiography of the eastern and western borders of the continent, have a much greater east and west extent.

The Arid Province is characterized by the presence of about 20 genera not found in the Humid Province, and by lacking a few occurring in the latter. The 20 genera and subgenera of the Arid Province not found in the Humid Province are the following:

Antilocapra,	Perodipus,	Otospermophilus,
Cynomys,	Onychomys,	Ictidomys,
Thomomys,	Antrozous,	Xerospermophilus,
Perognathus,	Euderma,	Ammospermophilus,
Chætodipus,	Molossus,	Bassariscus,
Microdipodops,	Notiosorex,	Conepatus.
Dipodomys,	Eutamias,	

Among its characteristic mammals are the Badger, Coyote, Kit Fox, Black-footed Ferret, Texas Civet Cat, the Mule and Black-tailed Deer, the Prong-horned Antelope, all of the numerous species and subspecies of the genus *Thomomys* and nearly all the species of *Geomys*, four genera and some thirty species of Kangaroo Rats and Pocket Mice, all of the several species of Prairie Dog, eight or ten species of Spermophiles, a dozen species of Ground Squirrels, including the whole subgenus *Eutamias*, all of the several species of Jackass Hares, all of the Grasshopper Mice (*Onychomys*), and numerous species of Bats, Shrews and Arboreal Squirrels.

Subprovinces of the Arid Province.—The Arid Province, like the Humid, is divisible into two subprovinces, namely, (1) a northern or *Campestrian Subprovince*, and (2) a southern or *Sonoran Subprovince*. These two regions correspond respectively with Dr. Merriam's 'Arid Upper Sonoran' and 'Arid Lower Sonoran'; just as the two subprovinces of the Eastern Province correspond with his 'Humid Upper Sonoran' and 'Humid Lower Sonoran,' as laid down on his 'Second Provisional Bio-geographic Map of North America,' except that the 'humid' and 'arid' portions of his 'Transition Zone' are also included respectively in the Alleghanian and Campestrian Subprovinces. The Sonoran Subprovince is equal to Dr. Merriam's restricted "Arid or Sonoran subregion proper" plus his "Lower Californian subregion," while the Campestrian Subprovince includes his "Great Basin subregion" and his "Great Plains subregion."[1] The name 'Campestrian' has reference to the fact that this subprovince is largely made up of plains, including as it does the greater part of the Great Plains, the Plains of the Saskatchewan, and the Plains of the Columbia and Snake Rivers. (See map, pl. VII.)

Many species are limited in their southward distribution by the southern border of the Campestrian Subprovince, but few genera appear to be thus restricted. This boundary also forms about the northern limit of many species and genera of the Sonoran Subprovince. These two subprovinces are hence characterized mainly by the presence of a large number of forms found in the Sonoran which are absent from the Campestrian, which is thus characterized, like many northern divisions when compared with adjoining southern ones of coördinate rank, from the Arctic southward, by what it lacks rather than by the possession of any peculiar types.

Districts of the Campestrian Subprovince.—The Arid Province is further divisible into a number of areas intermediate in rank between faunæ and subprovinces, which may be called *Districts.* Thus the Campestrian Subprovince is separated by the main chain of the Rocky Mountains into two areas each of which includes two or more faunæ, equivalent to and representative of the Alleghanian and Carolinian Faunæ of the Atlantic coast. These two areas are (1) the *Great Plains District*, consisting of the Great Plains region east of the Rocky Mountains from northern Texas to and including the Saskatchewan Plains; and (2) the *Great Basin District*, including the Great Basin region at large, from southern New Mexico, southern Utah, and southern Nevada north to the Plains of the Columbia and Snake Rivers, and thence northward over the more open and arid portions of eastern British Columbia. It encloses outlying spurs and insular areas of the Cold Temperate Subregion. These two districts possess few distinctive genera and but few distinctive species, though physiographically so well separated. A narrow Pacific Coast belt, situated mainly west of the Sierra Nevada and Cascade Ranges, may be recognized as (3) a *Pacific Coast District*,

[1] *Cf.* N. Am. Fauna, No. 3, p. 25.

characterized by the presence of a few species and a considerable number of subspecies mainly restricted to it. (See map, pl. VII.)

The *Sonoran Subprovince* is apparently not so distinctly separable into Districts of very marked distinctness, even the Peninsula of Lower California presenting few peculiar forms of higher grade than subspecies. Apparently, however, the main continental divide serves to separate a well-marked eastern from a well-marked western subdivision, each characterized by many species and subspecies not found in the other.

The ultimate faunal areas, or the *Faunæ*, of the Arid Province have not as yet been outlined, and their detailed treatment is beyond the scope of the present paper.

Zones of the Warm Temperate Subregion.—In addition to the subdivisions of the Warm Temperate already recognized, and independent of them, this subregion may be divided also into several *zones* or *belts* of transcontinental extent, namely, (1) *Alleghanian Zone*, (2) *Carolinian Zone*, and (3) *Louisianian Zone*—these names being based respectively on those of the long-known Atlantic coast faunæ, of which they respectively in part consist, just as 'Canadian' and 'Hudsonian' have been adopted for the transcontinental zones of the Cold Temperate.

The *Alleghanian Zone*, east of the Great Plains, consists of the Alleghanian Fauna; from the eastern edge of the Plains westward it consists of a succession of faunæ—one in the Great Plains District, another in the Great Basin District, and a third in the Pacific Coast District, as yet not clearly defined—equivalent in faunal character to and representative of the Alleghanian Fauna of the East. This zone has already been traced across the continent and mapped by Dr. Merriam under the name '*Neutral* or Transition Zone.'[1] It was first recognized by him in his exploration of the San Francisco Mountain region in Arizona, under the name of 'Neutral or Pine Zone,'[2] and later

[1] Proc. Biol. Soc. Washington, VII, 1892, pp. 30–33, and accompanying map.

[2] N. Am. Fauna, No. 3, 1890, p. 11.

in Idaho.[1] In his later treatment of the subject he has strangely separated this zone as an independent region of minor grade, interposed between his two primary divisions of the continent! He says: "Interposed between the Boreal and Sonoran Regions throughout their numerous windings and interdigitations, is the Neutral or Transition Zone. The humid division of this zone, known as the Alleghanian Fauna, covers the greater part of New England....and extends westerly over the greater part of New York, southern Ontario and Pennsylvania, and sends an arm south along the Alleghanies all the way across the Virginias, Carolinas, and eastern Tennessee, to northern Georgia and Alabama. In the Great Lake region this zone continues westerly across southern Michigan and Wisconsin, and then curves northward over the prairie region of Minnesota, covering the greater parts of North Dakota, Manitoba, and the plains of the Saskatchewan; thence bending abruptly south it crosses eastern Montana and Wyoming, including parts of western South Dakota and Nebraska, and forms a belt along the eastern base of the Rocky Mountains in Colorado and northern New Mexico, here as elsewhere occupying the interval between the Upper Sonoran and Boreal Zones."[2]

The Alleghanian Zone is beyond question a transition belt, being necessarily so from its geographical position; its affinities, however, are decidedly with the Warm Temperate division of the continent rather than with the Cold Temperate, as the case was first interpreted by Dr. Merriam,[3] since its northern boundary coincides closely with the northern limit of distribution of a large number of southern genera of both plants and animals, including most of the staple grains and fruits of the Warm Temperate Zone.

As is well known, there is always a belt of neutral territory along the common boundary line of two adjoining areas, varying in breadth with the rank of the two areas; and the present case of the Alleghanian Zone is thus not exceptional. All things considered it therefore seems best to regard it as the northern transcontinental belt of the Warm Temperate rather than to give it the

[1] N. Am. Fauna, No. 5, 1891, p. 24.
[2] Proc. Biol. Soc. Washington, VII, 1892, pp. 30, 31.
[3] N. Am. Fauna, No. 3, p. 20, and *ibid.*, No. 5, pp. 21 and 25.

anomalous position of a minor faunal area interposed between and completely separating two areas of a higher grade.[1]

The *Carolinian Zone* consists of the Carolinian Fauna, with its several western equivalents. It correlates with the 'Piñon Zone' of Dr. Merriam, as recognized by him in the San Francisco Mountain Region,[2] and later in Idaho, under the designation of 'Arid Upper Sonoran Zone.'[3] Taken as a whole it corresponds to what he has denominated Upper Sonoran.

The *Louisianian Zone* includes the Louisianian Fauna and its equivalent faunæ in the West. It may be correlated with Dr. Merriam's 'Desert Zone or Area' in Arizona,[4] which became later[5] his 'Arid Lower Sonoran Zone,' and is, as a whole, the same as his 'Lower Sonoran.'

It thus appears that extra-tropical North America may be separated into about six transcontinental belts or zones, for the purpose of conveniently correlating the numerous faunæ of the continent, as follows:

(1) An Arctic or Hyperborean Zone, coëxtensive with the American portion of the Arctic Realm.

(2) A Subarctic, Hudsonian or 'Spruce' Zone.

(3) A Cold Temperate, Canadian, or 'Douglass Fir' Zone.

(4) A Temperate or Alleghanian Zone.

(5) A Warm Temperate or Carolianian Zone.

(6) A Subtropical or Louisianian Zone.

Tropical North America.

It has long been recognized that the extreme southern portion of the Peninsula of Florida, the lower portion of the Rio Grande Valley, and a narrow belt extending thence northward for a

[1] In biology 'aberrant,' 'transition' or intermediate genera are frequently met with, and in some cases it is difficult to refer them to one of the two subfamilies to which they are allied rather than to the other. Yet we feel compelled to refer them to one or the other, or else to make a new subfamily for the aberrant genus, in case it shows sufficient differentiation, rathei than to leave it as an isolated genus, with the rank of a genus, to be interposed between two subfamilies, or families, as the case may be.

[2] N. Am. Fauna, No. 3, pp. 12 and 20.

[3] *Ibid.*, No. 5, p. 25, and Proc. Biol. Soc. Washington, VII, p. 27.

[4] *Ibid.*, No. 3, pp. 13 and 20.

[5] Proc. Biol. Soc. Washington, VII, p. 28.

N Y ENG & PTG CO

REALMS.

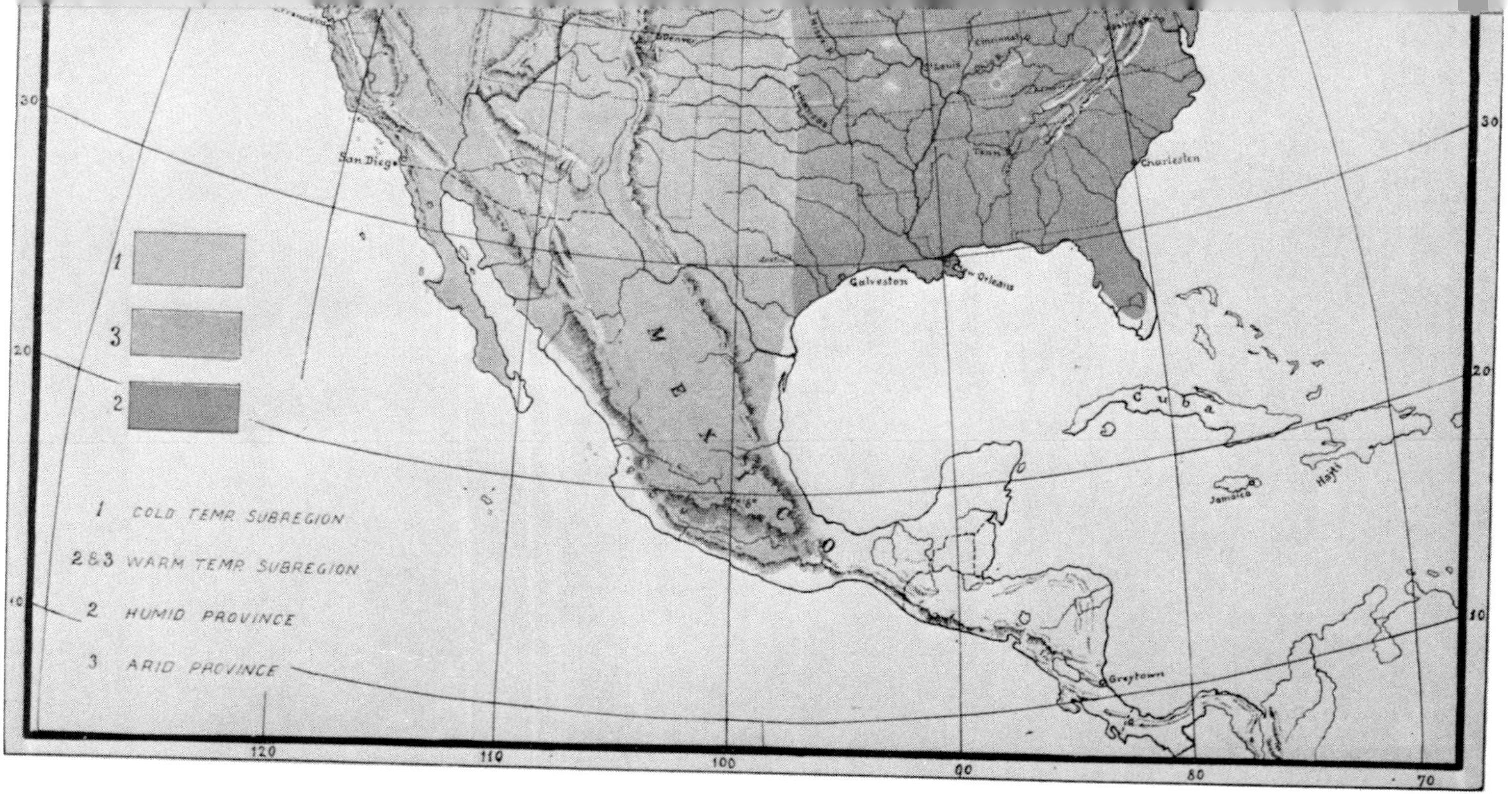

NORTH AMERICAN REGION.

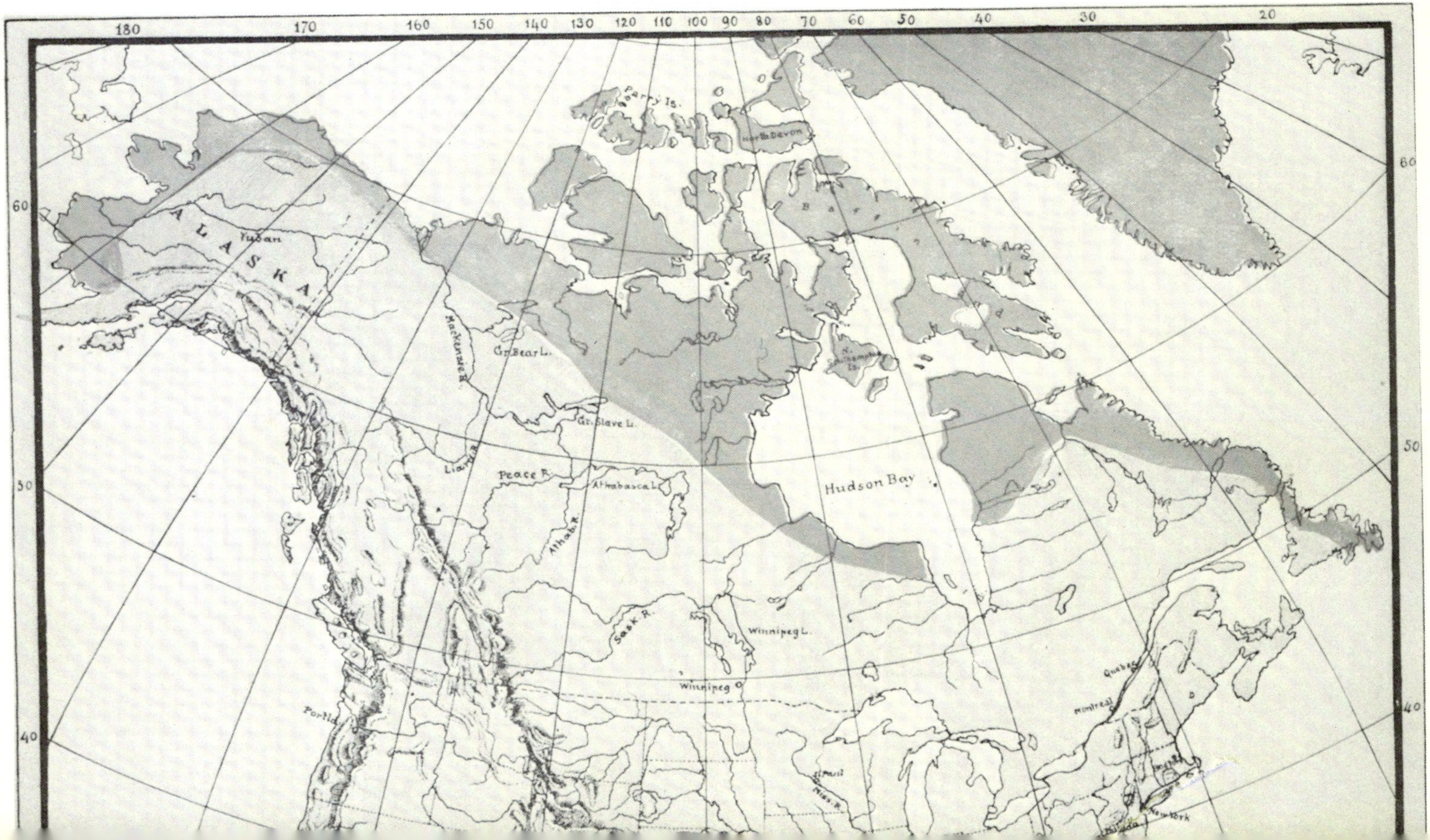
180
170
160
150
140
130
120
110
100
90
80
70
60
50
40
30
20
60
50
40
Parry Is.
North Devon
ALASKA
Yukon
Mackenzie R.
Gr. Bear L.
Gr. Slave L.
Liard R.
Peace R.
Athabasca L.
Athab. R.
Sask. R.
Winnipeg L.
Winnipeg
Hudson Bay
Quebec
Montreal
New York
Portld.
Miss. R.

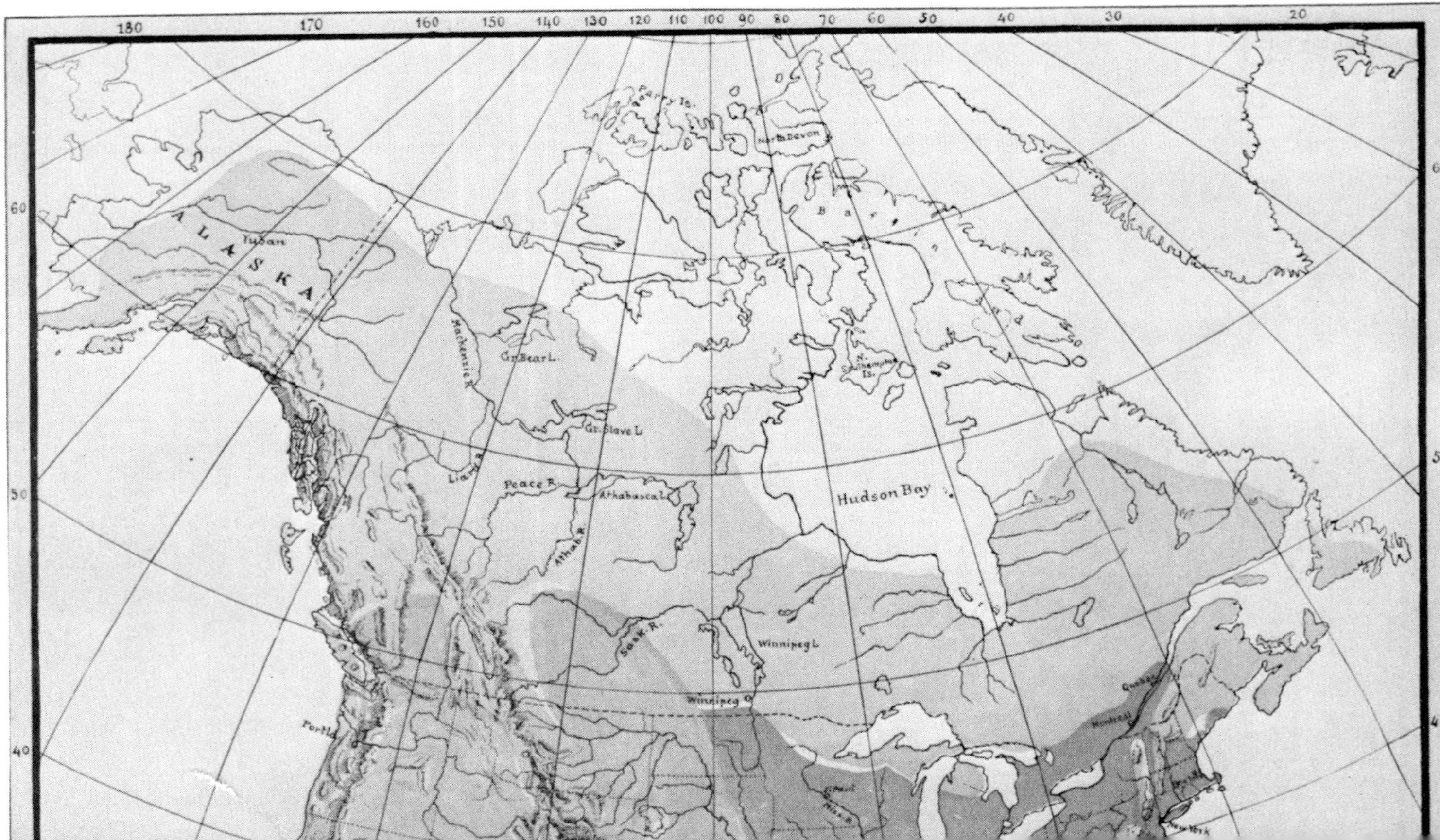
180
170
160
150
140
130
120
110
100
90
80
70
60
50
40
30
20
ALASKA
Yukon
Mackenzie R.
Gr. Bear L.
Gr. Slave L.
Liard R.
Peace R.
Athabasca L.
Athab. R.
Sask. R.
Winnipeg L.
Winnipeg
Hudson Bay
Parry Is.
North Devon
N. Southampton Is.
Portland
Quebec
Montreal
New York

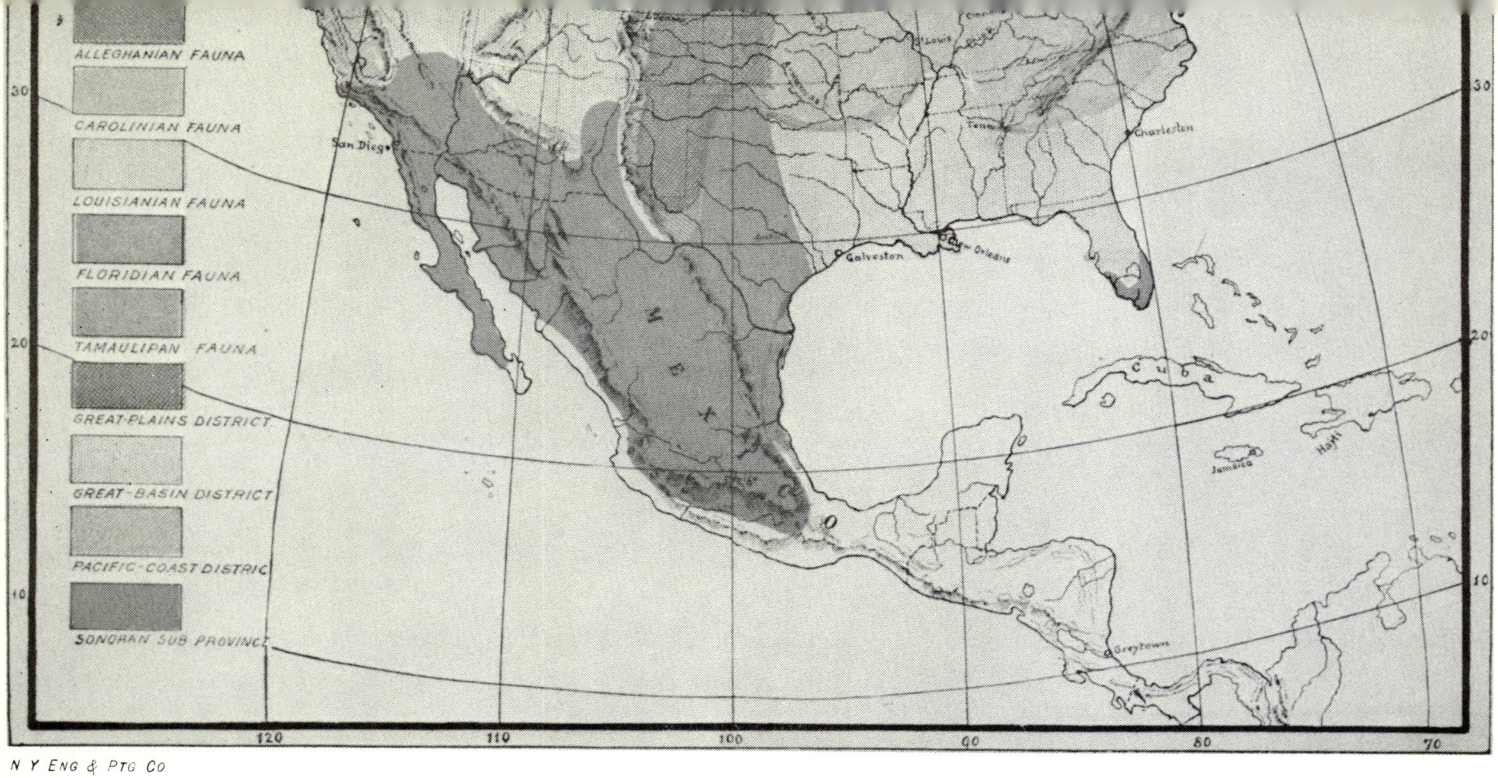

DISTRICTS AND FAUNAS.

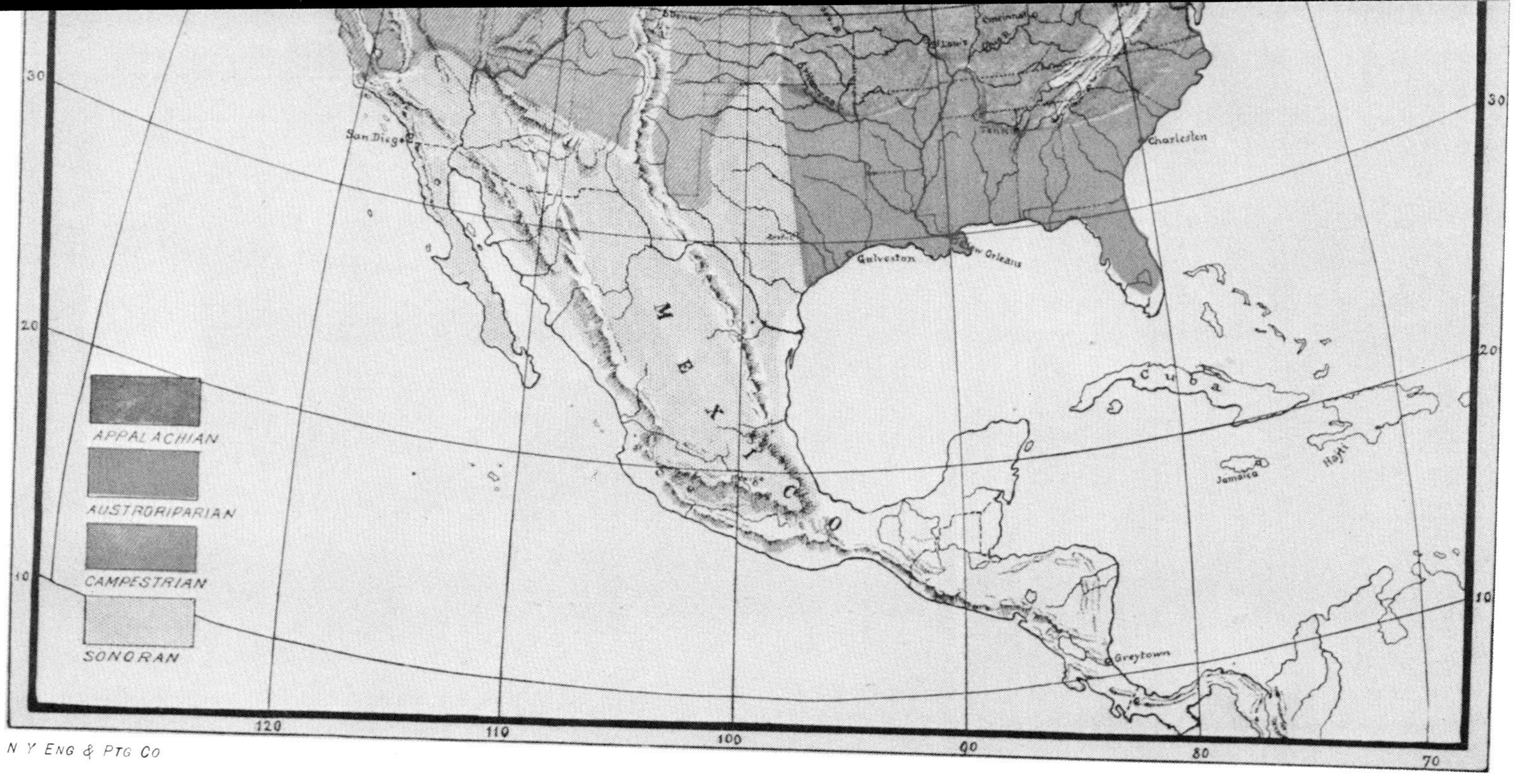

SUBPROVINCES.

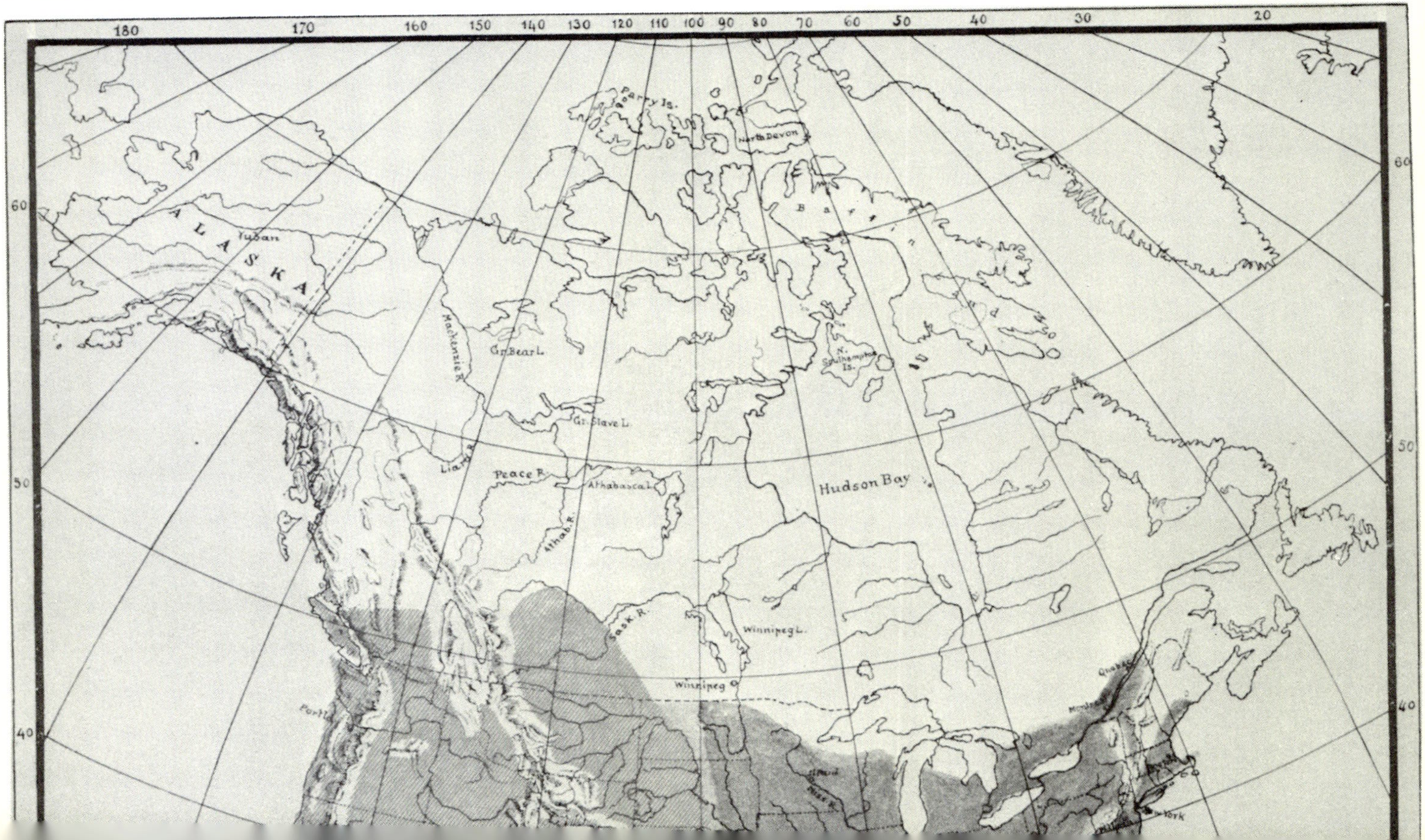
180
170
160
150
140
130
120
110
100
90
80
70
60
50
40
30
20
60
50
40
Parry Is.
North Devon
Baffin
ALASKA
Yukon
Mackenzie R.
Gr. Bear L.
N. Southampton Is.
Gr. Slave L.
Liard R.
Peace R.
Athabasca L.
Hudson Bay
Athab. R.
Sask. R.
Winnipeg L.
Winnipeg
Quebec
Montreal
York

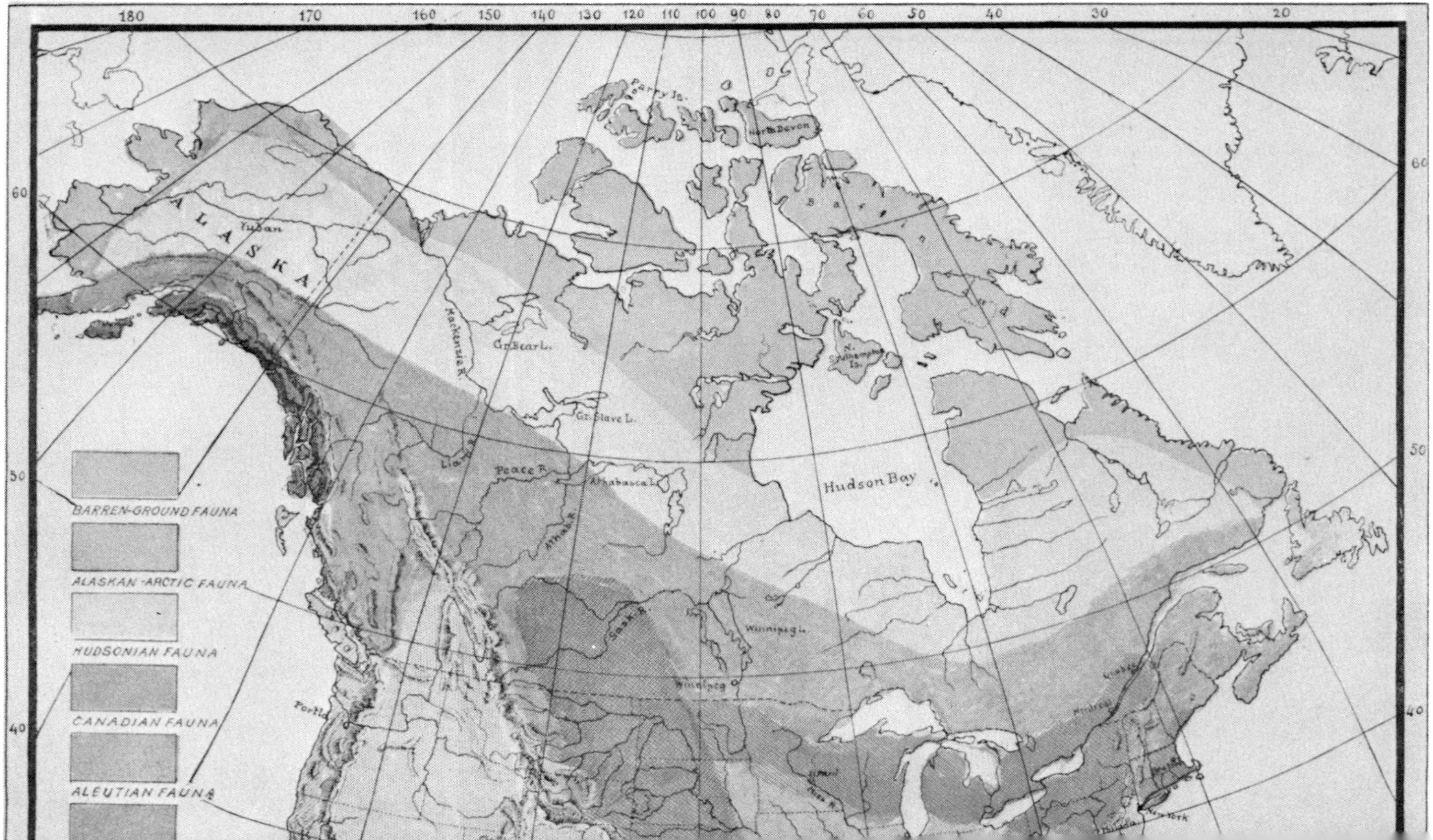
BARREN-GROUND FAUNA
ALASKAN-ARCTIC FAUNA
HUDSONIAN FAUNA
CANADIAN FAUNA
ALEUTIAN FAUNA
ALASKA
Yukon
Mackenzie R.
Gr. Bear L.
Gr. Slave L.
Peace R.
Athabasca L.
Athab. R.
Sask. R.
Winnipeg L.
Winnipeg
Parry Is.
North Devon
Hudson Bay
Portland
New York
180
170
160
150
140
130
120
110
100
90
80
70
60
50
40
30
20

short distance along the Texas coast contain such a strong infusion of tropical types as to render these limited areas properly referable to the Tropical Realm, to which belong also all of the lowland of Mexico, including a narrow coast belt extending from the Rio Grande southward on the eastern coast and from Mazatlan southward on the western coast.

The fauna of neither of the tropical areas within the United States is typically tropical, but the infusion of tropical elements is so great as to render them tropical rather than temperate. They have also little in common with each other, as would be naturally anticipated from their wide geographical separation through the interposition of the Gulf of Mexico, thus preventing a tropical land connection. Consequently the Floridian area, or the *Floridian Fauna*, as it has long been technically known, belongs to the Antillean Region of the American Tropical, while the Texan area is an outlying arm of the Central American Region of the American Tropical.

Floridian Fauna.—So far as mammals are concerned the Floridian Fauna has few distinctive elements. It has, however, a peculiar subspecies of the Wood Hare, and also of the Cotton Rat, and several species of Field Mice of the genus *Sitomys;* and the Manatee is a characteristic animal of the coast lagoons and rivers. From the standpoint of birds, insects, mollusks and plants, the Floridian Fauna is strongly characterized, not less than a dozen distinctly tropical genera of birds being represented, with a much larger infusion of tropical insects and mollusks.[2] Among the tropical birds which occur more or less regularly may be mentioned the Man-o'-War Bird, the Flamingo, the Florida Burrowing Owl, four West Indian species of Pigeons, the Ani, the Mangrove Cuckoo, the Dusky Seaside Sparrow, the Grassquit, the Black-whiskered Vireo, the Bahaman Honey Creeper, and various peculiar subspecies of northern birds.

Tamaulipan Fauna.—This semitropical area occupies the extreme lower portion of the Rio Grande Valley, probably not extend-

[1] *Cf.* Bull. Mus. Comp. Zoöl., II, 1871, p. 391.—The general provisional northern limit here given—"near the latitude of Lake George"—proves to have been carried a little too far north, its limits as now recognized being Cape Malabar on the east coast and Tampa Bay on the west coast. (*Cf. Merriam*, Proc. Biol. Soc. Washington, VII, 1892, p. 33.)

[2] *Cf.* Merriam, Proc. Biol. Soc. Washington, VII, 1892, pp. 52–54.

ing much above Hidalgo, and thus limited to within about the 100-foot contour line. It extends along the coast, mainly within the same contour line, northward to about the mouth of the Nuesces River, thence gradually fading out northward, a few tropical forms extending as far north as the mouth of the Colorado River, where a number of Louisianian forms gradually disappear. South of the Rio Grande it occupies the low coast region of Mexico southward nearly to Tampico. It thus includes the greater part of the State of Tamaulipas, and the southeastern part of the State of Nuevo Leon. Among mammals the following distinctively tropical forms are either limited in their northward range by the Tamaulipan Fauna, or extend but a little way beyond it:

Tatusia novemcinctus,	Nasua narica,
Heteromys alleni,	? Putorius brasiliensis frenatus,
Dipodomys phillipsii,	Felis eyra,
Sigmodon hispidus texianus,	" yaguarandi,
Oryzomys aquaticus,	" onca,
Sitomys mearnsii,[1]	" pardalis.
Atalapha intermedia,	

The last two have been reported from as far north along the Texas coast as the Brazos River, but they are tropical rather than warm temperate species. *Dicotyles tajacu* also properly belongs here, though ranging a little further to the northward than do the others.

Among tropical birds that here reach their northern limit are the following:

Podiceps dominicus,	Myiozetetes texensis,
Ortalis vetula maccalli,	Xanthoura luxuosa,
Engyptila albifrons,	Embernagra rufivirgata,
Buteo albicaudatus,	Sporophila morelleti sharpei,
Crotophaga sulcirostris,	Euphonia elegantissima,
Trogon ambiguus,	Vireo flavoviridis,
Nyctidromus albicollis merrilli,	Compsothlypis nigrilora,
Amazilia fuscicaudata,	Geothlypis poliocephala palpebralis,
" cerviniventris,	Basileuterus culicivorus,
Pitangus derbianus,	Harporhynchus longirostris sennetti.

The Tamaulipan Fauna has fewer distinctively tropical types than would be expected from its low altitude and geographical position. This is doubtless due to the extreme aridity of the

[1] = *Vesperimus mearnsii* Allen, Bull. Am. Mus. Nat. Hist., III, p. 300.

country, since in the forest regions further inland under the same parallels Trogons, Motmots and Parrots occur to a much greater extent than in the arid, nearly treeless coast region.

Tabular Synopsis of the Faunal Areas of North America.

REALMS.	Arctic.	
	North Temperate.	
	American Tropical.	
REGIONS.	North American.	<*North Temperate Realm.*
	Central American.	<*American Tropical Realm.*
	Antillean.	
SUBREGIONS.	Cold Temperate.	=*North American Region.*
	Warm Temperate.	
PROVINCES.	Humid.	= *Warm Temperate Subregion.*
	Arid.	
SUBPROVINCES.	Appalachian.	=*Humid Province.*
	Austroriparian.	
	Campestrian.	=*Arid Province.*
	Sonoran.	
DISTRICTS.	Great Plains.	=*Campestrian Subprovince.*
	Great Basin.	
	Pacific Coast.	
FAUNÆ.	Barren Ground.	*Arctic.*
	Alaskan-Arctic.	
	Aleutian.	*Cold Temperate.*
	Hudsonian.	
	Canadian.	
	Sitkan.	
	Alleghanian.	*Humid Warm Temperate.*
	Carolinian.	
	Louisianian.	
	Floridian.	*Tropical.*
	Tamaulipan.	

RECENT WORK IN NORTH AMERICAN MAMMALOGY

PROF. ALLEN announced that a new species of thrush had been discovered in Dominica, an island of the Lesser Antilles.

The following paper was read, entitled

RECENT WORK IN NORTH AMERICAN MAMMALOGY.

BY J. A. ALLEN, PH.D.

(Abstract.)

The present review of recent progress in North American mammalogy relates especially to the period since the appearance of Baird's great work on the mammals of North America, published in 1857, forming Volume VIII. of the Government Reports on Explorations and Surveys for a Railroad Route from the Mississippi River to the Pacific Ocean, made during the years 1853–56. While that work is taken as the basis of departure, it is of interest to note briefly some of the more important of the earlier works relating to North American mammals.

A review of the history of North American natural history necessarily begins with Catesby, whose great work, "The Natural History of Carolina, Florida," etc. (two volumes folio), was published in parts from 1730 to 1748. In this were figured and described for the first time a number of our mammals.

Pennant's "Arctic Zoölogy" (three volumes quarto, 1784–87) includes notices of many North American species.

Besides numerous more or less important incidental contributions made by explorers and travellers, and the descriptions of new species based on specimens collected by Lewis and Clarke in the years 1804–6, and by Major Long in 1819–20, and by various arctic explorers, as Sabine, Scoresby, and Richardson, there is little to note till we reach the year 1825, when Richard Harlan published his "Fauna Americana," an octavo volume of about 300 pages, devoted exclusively to North American mammals. It is now mainly of interest as the first general treatise on the subject, although for its time a creditable production. He described 127 species as "inhabiting North America north of Mexico." Of this number 11 were based on the fossil remains of extinct species, reducing the number of living species to 116.

This work was soon followed by a much more extended and elaborate general treatise by John D. Godman, published under the title "American Natural History: Part I., Mastology," an octavo work in three volumes, the first two published in 1826, the third in 1828. The work was illustrated with numerous plates engraved on copper, giving very good figures of nearly all the species. For the date of publication the work is one of high merit and may yet be consulted with much profit.

In 1829 appeared an exceedingly important work on North American mammals, namely, the first volume of Richardson's "Fauna Boreali-Americana," a quarto of 300 pages and 24 plates, on the mammals of British North America. It is still a standard work, and is the chief source of information on the habits and distribution of the mammals of the region north of the United States.

Among later important works of a general or faunal character may be mentioned Emmons' "Report on the Quadrupeds of Massachusetts," published in 1840; and DeKay's "Report on the Mammals of the State of New York," a quarto volume with thirty-three colored plates, published in 1842. This work includes references to all of the then known species of North American mammals, and served for many years as a general reference work, though it failed to take very high rank as an authority. Linsley's list of the mammals of Connecticut [1] was published in 1842; Thompson published a valuable fully annotated list of the mammals of Vermont [2] in 1853; and Kennicott published an important paper on the mammals of Illinois in 1857.[3]

The great work of Audubon and Bachman, entitled "The Viviparous Quadrupeds of North America," published in three volumes, royal octavo, 1846–54, with colored plates, marks a new era in North American mammalogy. About 200 species are described and 160 figured. The work, however, does not include either the bats or the marine species. It is devoted especially to the life histories of the mammals, and deals only to a limited extent with technical matters. Strange as it may seem, this forms the latest general work on North American mammals which treats of the subject from its popular or non-technical side, although one or two later works treat certain groups from a general standpoint, dealing with both the popular and the systematic phases of the subject. Audubon drew most of the plates of the great work with which his name is associated, and jointly with Bachman contributed largely to the text; but the strictly systematic portion is well known to be the work of Bachman, the father of systematic mammalogy in America, and, prior to Baird, the leading authority on the subject. His monographic revision of the genera *Sorex*, *Scalops*, *Sciurus*, and *Lepus*, form-

---

[1] American Journal of Science and Arts, XLIII., 1842, pp. 345–354.

[2] Natural History of Vermont, 1853, Mammals, pp. 28–56, and App. pp. 11–20.

[3] Report Commissioner of Patents, Agriculture, 1856, pp. 52–110, pll. v.–xiv

ing a series of special papers published 1837–42, were contributions of the highest importance, and the first careful, systematic work on any special groups of American mammals by an American writer.

Incidentally important contributions had been made by numerous foreign authors, but it would carry us far beyond our present scope to consider them in detail.

Baird began to write on North American mammals as early as 1852, his work culminating in 1857 in his great work already mentioned—a work which gave him pre-eminent standing as an authority, and which still remains the basis from which present investigators take their departure.

Baird's volume included only the land mammals, exclusive of the bats, which, with the seals and Cetacea, were not treated. With these several groups omitted, Baird recognized about 220 species as entitled to a place in the list of North American mammals,—nearly twice the number given by Harlan in 1825. Thirty-six others are mentioned as having been attributed to North America, which he was unable to recognize. A few of these are extralimital, but the greater part are either purely nominal, or so poorly described as to render their recognition impossible.

Baird treated the subject exclusively from the systematic side, there being nothing in his work relating to the habits of the species. Classification, and all questions of synonymy and nomenclature, were treated exhaustively; specific and generic characters were presented in detail and with great discrimination. As a technical treatise the work remains a monument to the industry and sagacity of its talented author, and, considering its extended scope, has not been excelled by any work, in any country, on systematic mammalogy. It was illustrated by sixty lithographic plates and nearly forty woodcuts, relating mainly to cranial, dental, and other structural characters. In 1860 the text was reissued, with a large number of additional plates, some of them new, but mostly from other volumes of the Pacific Railroad and Mexican Boundary Survey reports.

Professor Baird brought to his work not only a mind well trained in modern methods of research, but he was favored with resources in the way of material far beyond what had fallen to the lot of any of his predecessors.

In March, 1853, Congress appropriated $150,000 for the prosecution of surveys by the War Department of various routes across the continent, from the Mississippi River to the Pacific coast, for the purpose of determining practicable lines for the construction of railroads. Six well-equipped parties were placed at once in the field, and six others were soon added, each accompanied by a competent naturalist. Large collections were at the same time made by other officers of the army stationed at various posts throughout the distant West, and also by the United States and Mexican Boundary Survey. Although the collections made by the Boundary Survey were made the subject of a special report, they formed also a part of the material on which was based Baird's memorable "Eighth Volume," as this work is familiarly known among mammalogists. These collections were, by act of Congress, all transmitted to the Smithsonian Institution, and were thus brought together as a single collection, to be elaborated, fortunately, by one of the most competent naturalists America has yet produced.

But the volume under notice was not the sole outcome of these various Government surveys. Other volumes of the Pacific Railroad Reports contain the special field reports of the various naturalists of the different routes surveyed, accompanied by numerous plates of new or previously unfigured species. These reports embody the field notes of the collectors, and thus admirably supplement with biographical matter Baird's exhaustive systematic treatise. Hence to Gambel, Woodhouse, Kennerly, Cooper, Gibbs, Suckley, Heermann, Newberry, Trowbridge, and Gunnison—names well ingrained in the literature of many departments of North American natural history—we are deeply indebted for much of our information respecting the habits and distribution of the mammals of Western North America.

Since the publication of Professor Baird's great work in 1857, several monographs have appeared treating of particular orders or families. In 1864 the Smithsonian Institution published Dr. Harrison Allen's "Monograph of the Bats of North America," an octavo of about 100 pages, illustrated with numerous woodcuts. This was the first general systematic treatise on North American bats since the works of Godman and Harlan, published nearly forty years before, and it remains still our standard work on this group, and the only special treatise on the North American species of the order. Twenty species are recognized, and many others referred to as unidentifiable.

In 1866 Dr. Theodore N. Gill published a "Prodrome of a Monograph of the Pinnipedes," giving a systematic synopsis of the families, genera, and species of the marine Carnivora—the seals and their allies.

In 1870 I published a monograph of the family Otariidæ, or Eared Seals,[1] a paper of 108 pages, with three plates and a number of woodcuts, giving a detailed account of their osteology and

---

[1] Bull. Mus. Comp. Zoöl., II., No. 1, 1870, pp. 1-108, pll. I.-III.

external characters, supplemented by an extended account of their habits by Capt. Charles Bryant, formerly a Government agent at the Fur Seal Islands. Later very elaborate reports by Mr. Henry W. Elliott, on the habits of the species, have been published by the Treasury Department of the Government.

In 1874 appeared Capt. C. M. Scammon's work, "The Marine Mammals of the Northwestern Coast of North America, described and illustrated, together with an account of the American Whale-Fishery," a quarto volume of about 325 pages and 27 lithographic plates. The appearance of this work marked an era in the history of our marine mammalia, though restricted to those of the Pacific coast. It covered a new field, and furnished an inexhaustible fund of information, to which we are still almost wholly limited, as far as the life histories of the species are concerned. Cope, Gill, and Dall have contributed also various important papers on the Cetacea, treating the subject from the systematic side, and largely relating to the species of the Pacific coast.

More recently Mr. F. W. True, Curator of Mammals in the United States National Museum, has entered the field, and, besides numerous minor papers on various species of the Atlantic coast, has recently published "A Review of the Family Delphinidæ," an octavo memoir of 200 pages, with 47 plates. This is a monograph of the porpoises and dolphins of the world, and is a most welcome and able contribution to the subject. Sixty-two species are described, and nearly all are figured.

J. D. Caton has published numerous papers on our deer, and in 1877 an octavo volume of nearly 500 pages, with numerous woodcuts, entitled "The Antelope and Deer of America." It deals principally with their habits, affinities, and susceptibility to domestication, and is in its way an excellent treatise, and the only one relating exclusively to these animals.

Our most important ruminant, the American bison, was monographed by myself in 1876 in a quarto memoir of about 260 pages and 12 plates. Mr. Hornaday, in an octavo paper of nearly 200 pages, with many illustrations, entitled "The Extermination of the American Bison," published in 1889, has admirably brought the subject down to date.

In 1877 Dr. Coues published his "Fur-Bearing Animals: a Monograph of North American Mustelidæ." It is an octavo of about 350 pages and 20 plates, and forms a most important contribution to the literature of North American mammals.

In 1877, also, Dr. Coues and myself published a series of monographs of the various families of North American rodents, forming a quarto volume of 1,100 pages, entitled "Monographs of North American Rodentia." While the two authors shared the labor about equally, the joint authorship implied in the title relates merely to the fact that their separate memoirs are bound together under one general title, the different monographs being each by a single author. The work consists of eleven separate and distinct memoirs. In two cases, as in the monographs of the squirrels and hares, the scope is greater than is indicated by the general title of the work, all of the species found in Mexico, South and Central America, being included, as well as those from North America. An important feature of this work is the bibliographical appendix, by Gill and Coues, consisting of about 130 pages, and bringing the bibliography of North American mammals down to the year 1874, in a very thorough manner, and including most of the titles down to as late as 1876.[1]

In 1880 I published an octavo volume of about 800 pages, with numerous illustrations in the text, entitled "History of North American Pinnipeds: a Monograph of the Walruses, Sea-Lions, Sea-Bears, and Seals of North America," treating the subject at length in all its bearings.

In addition to the special works already enumerated, numberless papers, some of them of great importance, have appeared in Government reports and in the transactions of scientific societies. Some of these are simply more or less fully annotated faunal lists; others are faunal lists, including incidentally much technical matter; others are more or less elaborate monographs of particular genera or families; others still are brief papers describing new species or recording rare ones. Baird published practically nothing on mammals after 1859, since which time the most prominent names in the field have been Harrison Allen, J. A. Allen, Caton, Cooper, Cope, Coues, Dall Gill, Kennicott, Mearns, Merriam, Scammon, and True.

Besides these, Alston, of England, has published on the mammals of Mexico, in Godman and Salvin's "Biologia Centrali-Americana"; De Saussure has also written important papers on the same subject; while Brown and Ross have published valuable memoirs on the mammals of Arctic America.

---

[1] The titles of works and papers briefly mentioned in the present paper will be found there given in full. The intended scope of the work was to include "(*a*) all works and papers on mammalia at large published in North America, (*b*) all works and papers, wherever published, relating to North American mammals, and (*c*) such general works or collateral special papers as bear more or less directly upon the subject." The titles are arranged chronologically under a few general heads. Though not put forth as absolutely complete, it still remains the most extended compilation on the subject yet published.

Two orders of our mammals have received very little attention for nearly thirty years, and in respect to them we are at present entirely at sea as regards the number of species, their relationships, or their correct names. These are the bats, which have not been monographed since 1864, and the moles and shrews, which have not been thoroughly revised since 1857.[1] We are happy to say, however, in respect to the former, that Dr. Harrison Allen has a work in press, under the auspices of the Smithsonian Institution, relating to the bats, which will give us a new and most welcome point of departure for this group. The outlook regarding the moles and shrews is less encouraging, there being unfortunately no similar work under way treating of these groups from the American standpoint. Some years since, however, Dr. Dobson, of England, undertook a monographic revision of the Insectivora of the world, but the third and last part, which is to treat of the shrews, has not yet appeared. In 1874 this same author published a monograph of the bats of the world, including, of course, North American. This work was a boon to workers in this field, and a great help respecting our own bats, though not altogether satisfactory.

We have now said all that our limited time will permit concerning the progress of North American mammalogy from the bibliographic standpoint; its progress from the scientific standpoint remains to be considered.

Prior to the Government explorations which yielded the material on which Baird's work was based, we had only the most superficial knowledge of the mammalogy of North America at large, or beyond the more settled parts of the eastern portion of the United States. No museum, public or private, had a good series of specimens of even our most common mammals. Species without number had been described, generally in the most imperfect manner, from either single or comparatively few specimens, and the types had been preserved only in rare instances. Almost nothing was known of either seasonal or individual variation, nor was the material extant for the study of either of these very important phases of the subject. Even Professor Baird, in working up his comparatively abundant material from the West, found himself often baffled in his comparisons by the absence of specimens from the East. Pro-

[1] In 1877 Coues published an important preliminary paper on the North American Insectivora, entitled "Precursory Notes on American Insectivorous Mammals, with Descriptions of New Species" (Bull. U. S. Geol. Surv., III., 1877, pp. 631-658). He intended to follow this with an elaborate monograph of the group, which, however, unfortunately was never completed.

fessor Baird was thus confronted at the beginning of his work with a chaos of names and meagre descriptions, the legacy of his predecessors. Even not a few of the species of Audubon and Bachman were based on specimens from unknown localities, not always correctly supposed to be North American. Leconte, DeKay, Say, Ord, Rafinesque, and Harlan, and various foreign authors, had each contributed to the list of nominal species, many of which will forever remain undeterminable, while others have been duly relegated to the limbos of synonymy. The work of separating the chaff from the wheat, the false from the true, was a perplexing task, an intricate riddle, which Baird attacked with wonderful acumen and success, leaving a comparatively smooth and easy path for his successors.

This was, however, the period of excessive subdivision, the smallest recognizable differences serving as the basis of specific separation. Baird, as was natural, influenced by the methods of his time, contributed his share to the list of synonyms; yet, considering the inadequateness of his material, his lapses were few, and more than atoned for by the accuracy and detail of his descriptions. If he failed to substantiate tangible differences between closely allied forms, it was because they did not exist, not from any lack of care in his work. Sometimes features of seasonal or individual variation were mistaken for specific difference, but, as a rule, any forms recognized by Baird as species will bear the closest scrutiny, and are rarely found to be without some basis in nature—that is, if not "good species," they will be found to be what we now call geographical forms, or "subspecies."

The history of North American mammalogy and the history of North American ornithology run in closely parallel lines, the prominent workers being practically the same in each. In both of these departments there have been periods of excessive splitting, followed by undue lumping. On a former occasion I have referred at length to "the oscillations of the ornithological pendulum" during the last thirty years.[1] In mammals, as in birds, a period of "lumping" set in about 1870, following a long period during which the tendency had been to excessive subdivision, and extending into the present decade. The period from 1870 to 1880 was thus not only a lumping period, but a transition period, and also one of unequalled activity. Work was carried on also from a basis in some respects the exact reverse from that of the immediately preceding epoch. While the outcome was on the whole a marked advance on what had preceded, it is obviously open to material revision as seen from the standpoint of

[1] The Auk, Vol. VII., Jan., 1890, pp. 1–9.

to-day. A third period, differing much from either of the two preceding, has now opened; its culmination and final results cannot as yet be foreseen.

During the Bairdian period of North American mammalogy, the discovery of new forms was the paramount incentive to investigation. The method of study was hence intensely analytic; differences, rather than resemblances, filled the mind of the investigator. At this period the subject of variation under climatic influences, the evolution of species by environment, had attracted little attention. But the facts of evolution were being minutely recorded; a harvest was being reaped which was to form the basis of important generalizations, as then unsuspected even by the most industrious of the reapers. But they did their work well, and it is not to their discredit that they failed to see at first some of the fundamental principles of the evolution of life under the influences of environment. We have always first to gather our facts before we can generalize. Professor Baird himself, however, was the first to perceive and formulate some of the fundamental laws of geographical variation, as regards both North American birds and mammals, in the study of which, be it said to the credit of American naturalists, such principles were first recognized and established. Baird early perceived that, as a rule, individuals of the same species decreased in size from the north southward; that the animals from the arid plains were paler in color than their nearest relatives of the wooded region to the eastward; and that over the heavy rainfall belt of the Northwest they assumed a depth of coloring met with nowhere else on the continent. He also recognized that, in some instances, the size of peripheral parts, especially in birds, increased in size from the north southward, while the general size of the individual decreased. Later it was also noticed that, as a rule, individuals of the same species became brighter colored at the southward as compared with their northern relatives—that black bars and streaks were broadened at the expense of the intervening lighter spaces; that birds with metallic tints became more iridescent, and that the duller colors, of both birds and mammals, became deepened and intensified.[1]

In subsequent years, as material increased and large series of specimens of the same species and from the same locality were brought together, it was found that the range of purely individual variation, in respect to every feature, was far greater than had previously been suspected. This, of course, seemed to invalidate certain characters on which species had previously often

---

[1] See Allen, "Mammals and Winter Birds of East Florida," Bull. Mus. Comp. Zoöl., II., No. 8, 1871.

been based. Again, the increase of material from new localities showed that certain forms described as species gradually varied in the direction of other species, through specimens from intermediate points of the common habitat. In other words, we began to be troubled with a lot of specimens that were variously intermediate between two or more previously supposed wholly distinct species. This occurred not merely in a few isolated cases, but frequently over most of the continent. Under these circumstances, and in the light already thrown upon the general subject, it was natural that many species which had been held for years in good standing—or while known from only a few specimens—should be reduced either to pure synonyms or to the rank of geographical forms or subspecies. It was further natural that in some instances it should be presumed, on general grounds, that certain of the previously recognized species were either purely nominal or else only local forms, even where the evidence was far from conclusive. For a time, under the new order of things, it was not strange that undue lumping should result, as later researches have proved was the case.

"Intermediates" were at first explained on the ground of hybridity, and it was customary for a time to treat the few refractory specimens then known as hybrids. But the hybrids soon became troublesomely numerous; furthermore, it was noticed that over certain areas, and along certain geographical lines, the variations were in general of a similar character. As already said, this gave rise to the recognition of certain laws of geographical variation. A crisis resulted respecting methods of treatment, culminating about 1870. Down to about this date a collection of individuals known as a species was, theoretically at least, a distinct and definable group. Varieties, in a geographical sense, were rarely recognized, and "subspecies" was an almost unknown term. Many species to which we had long pinned our faith were turning out badly; instead of being stable and well marked, they were found to intergrade with others from which they were supposed to be distinct. This, of course, was always in the case of closely allied species; between many other species no such intergradation was apparent. Thus we had, as it were, two kinds of species, some which intergraded and some which did not intergrade. How were such diverse elements to be treated? This was the question of the hour. Should these strongly marked geographical forms be lumped together under a common specific name and thus wholly ignored in nomenclature, or should we continue to recognize under binomial names all strongly marked forms characterizing particular physiographic areas—in other words, forms having approximately definable habitats—although known to run into something else; or was there a still better way of expressing such important biologic facts? Here were evidently recognizable stages in the process of evolution by environment. How were they to be recognized so as to be made use of in philosophical biology; in what way could they be chronicled in descriptive biology? That many of these intergrading forms possessed noteworthy differences is evident from the fact that they formerly held the rank of unquestioned species.

For a time a wavering course was followed, some writers lumping as one species all forms that seemed to intergrade, while others continued to recognize all of the more prominent phases as species. Soon after it became the fashion to distinguish them as varieties, under what may be termed a qualified quadrinomial system of nomenclature. Thus our Eastern red squirrel, or chickaree, was called *Sciurus hudsonius;* its Rocky Mountain representative, or Fremont's chickaree, became *Sciurus hudsonius*, var. *fremonti;* another Rocky Mountain form, known as Richardson's chickaree, became *Sciurus hudsonius*, var. *richardsoni;* a Northwest coast form, known as Douglass' chickaree, became *Sciurus hudsonius*, var. *douglassi;* and so on in all similar cases. This proved a cumbersome method, and soon the term "var." was dropped, resulting in a trinomial name, pure and simple, for the designation of *subspecies*. This in reality is the binomial system of Linné in spirit, modified slightly in the letter to fit what was an ideal system of nomenclature at the time of its inception, to the requirements of modern biology. Hence American mammalogists adopted, in common with ornithologists, trinomial designations for nascent species, reserving the binomial distinctively for species in full standing. This system became generally current about 1883.

The several geographic, intergrading forms of a wide-ranging species often differ more from each other in their extreme phases than do, in other cases, perfectly distinct species. Hence, through the use of trinomials, two important things are accomplished: first, a means of distinguishing two kinds of relationship among congeneric forms, the use of a binomial name indicating entire distinctness, without regard to kind or degree of difference, while the use of a trinomial implies known intergradation, notwithstanding the very wide difference which frequently exists between the extreme phases of a group of intergrading or conspecific forms; secondly, provision for a convenient and tangible way of giving expression, by means of a definite formula, to some of the most suggestive facts in the evolution of life. It is not claimed that this is a perfect system of nomenclature, but it is the best and most convenient at present available. It has its objections, which, theoretically, seem insuperable, but which, in practice, prove insignificant.

At first, under this method, lumping flourished to a remarkable degree. Forms that were known to intergrade were not only reduced to subspecies, but not infrequently such reductions were made on theoretical grounds. Certain forms, judging from what had been established as occurring in allied groups, and in accordance with what might be presumed to occur as the result of certain laws of geographical variation, were presumptively merged with others, sometimes correctly, but often improperly.

It was during this phase of affairs that the "Monographs of North American Rodentia" appeared, as also Coues' "Fur-Bearing Animals" and my "North American Pinnipeds." It was also from this point of view that True prepared his "Check-list of North American Mammals," published in 1884. Incidentally both Coues and myself did some revisionary work in other groups, on nearly the same lines.

To throw into strong contrast the work of the three leading periods in the recent history of North American mammalogy, we will compare briefly two or three groups as left by Baird in 1857, by Coues and Allen in 1877, and as they appear from the standpoint of to-day. For this purpose we will select the Muridæ, or the field rats and mice; the Saccomyidæ, or the pouched rats and mice; the Leporidæ, or the hares; and the Sciuridæ, or the squirrels, spermophiles, and marmots.

In 1857 Baird recognized forty-eight species of field rats and mice; of these Coues reduced eighteen to synonyms and three others to subspecies. There has been no recent revision of the group as a whole, nor in fact of any of the genera. An entirely new genus, however, of meadowmice (*Phenacomys*), with four species, has been recently described by Dr. Merriam,[1] of which not a single specimen had been recognized by naturalists three years ago, although the genus is now known to range across North America from British Columbia to Labrador, and southward along the Rocky Mountains into the United States. Six different writers have described new species of the genus *Hesperomys*, or white-footed mice, three of them being remarkably unlike anything previously known. Several of Baird's species have also been revived as perfectly tenable subspecies. Other genera of this large family present a similar history. In short, not less than thirty species and subspecies have been

[1] "North American Fauna," No. 2, Oct., 1889, pp. 27-35, pll. ii.-vii.

added during the last six years, nearly doubling the number recognized in 1877.

Of the squirrels, Baird admitted thirty-nine species — an enormous reduction, by the way, from the number recognized by Audubon and Bachman. In 1877 I confirmed Baird's previous reductions, and still further reduced the number to thirty-three species and subspecies, adding, however, eight to those recognized by Baird. Since 1877 a large number of species and subspecies has been added, mainly among the ground squirrels and spermophiles. These additions are based on almost wholly new materials, and include species as distinct from their nearest allies as any previously known. That is, they are not the result of a new way of looking at things, but are actual discoveries of forms not at all closely related to anything known before.

Of the Saccomyidæ, Baird in 1857 recognized nine species; Coues, in 1877, admitted the same number. This group during the last four years has been thoroughly revolutionized, mainly by Dr. Merriam.[1] The names of several of the old species have been shown to have been strangely misapplied; one new genus and a new subgenus, and some twenty or more new species, have been added, besides numerous subspecies. Among the new forms are some strikingly different from any known to either Baird or Coues, while others differ in only comparatively slight particulars. These animals are peculiar to the arid plains and deserts of the West; are nocturnal and somewhat subterranean in their habits, and seem, moreover, especially susceptible to the influence of conditions of environment. Owing to their peculiar habits, and to the fact that the area they inhabit has been, till lately, little explored, very few specimens were extant in collections up to a very recent date.

Of the Leporidæ, or hares, Baird in 1857 was able to recognize but thirteen species as North American. All but one of these were admitted by me in 1877 and two others were added. Within the last year five others have been recognized, three of which are very distinct from any previously known, while two are old ones revived as subspecies.

To what, then, is all this change due? To several very obvious causes. First, to a vast increase of material; second, to its greatly improved quality; third, to more thorough and intelligent work in the field, conducted on a tenfold greater scale than ever before. Thousands of specimens are taken now where formerly only a few scores were obtained. Collecting and all

[1] "North American Fauna," No. 1, Oct., 1889; ibid., No. 3, Sept., 1890, pp. 71-75; and ibid., No. 4, Oct., 1890, pp. 41-49.

that relates to field work in mammals has been brought to an almost ideal state of proficiency. And all this is due, primarily, to the efforts and example of one man—Dr. C. Hart Merriam, formerly of Locust Grove, Lewis County, New York, now Chief of the Division of Ornithology and Mammalogy of the United States Department of Agriculture. He, realizing the importance of well-prepared skins, in large series from as many localities as possible, and of accurate data as to time and place of capture, and of measurements taken from the specimen before skinning, began, many years since, to form a collection for his own cabinet, which eventually became one of the most extensive and valuable in the country. An expert himself in all that relates to the capture and preservation of specimens, he trained collectors and sent them to remote parts of the country, and soon gathered a rich harvest, replete with novelties, some of them of startling character, so widely did they differ from anything previously known. Later, under the auspices of the Department of Agriculture, he has pushed the work with still greater activity, resulting in the recent establishment of what practically amounts to the inception of a Biological Survey, on a grand scale, of the United States, which, in the interest of science and in the economic interests of the country, it is to be hoped will be long liberally sustained.

The number of skilled collectors is rapidly increasing, with the result that our leading museums are now receiving much more and far better material than was ever before available for study. To contrast once more the old with the new, it may be stated that more specimens of the smaller mammals of North America were collected during the year 1890 than all that Professor Baird had at his command in 1857 for the preparation of his great work on North American mammals—specimens, too, incomparably better as regards their quality and utility for scientific investigation.

While, ten or twelve years ago, it was commonly supposed that comparatively little remained to be learned respecting the mammals of North America, beyond a few details regarding their distribution and habits, we are now little less than awed by the evident extent of our ignorance of the subject, as shown by the astonishing discoveries of the last four or five years, and recognize the obvious necessity of a careful revision of the whole field.

Recent Progress in the Study of North American Mammals.

Recent Progress in the Study of North American Mammals.

By J. A. Allen.

Three years since I presented a paper to the New York Academy of Sciences on "Recent Work in North American Mammalogy,"[1] giving a brief review of the principal works relating to North American Mammals, and dwelling in some detail upon the history of the subject from the year 1852 to the year 1890, comparing and contrasting the methods, the results, and the resources of the three periods into which these four decades may be divided. The concluding paragraph of this paper may be here quoted as a fitting introduction to the present article, as follows: "While, ten or twelve years ago it was commonly supposed that comparatively little remained to be learned respecting the mammals of North America, beyond a few details regarding their distribution and habits, we are now little less than awed by the evident extent of our ignorance of the subject, as shown by the astonishing discoveries of the last four or five years, and recognize the obvious necessity of a careful revision of the whole field" (l. c., p. 84). The results of the work of the last three years show that this statement, though a surprise to those unfamiliar with the then recent developments, was none too strong.

While this paper will relate mainly to the work of the last five years, it is desirable, for a proper understanding of

[1] Trans. N. Y. Acad. Sci., Vol. X., 1891, pp. 71-85.

the subject, to extend the field of view somewhat further into the past, summarizing briefly some portions of the paper already cited. Beginning then with what may be termed the Bairdian period we may trace in outline the history of the subject as regards (1) methods, (2) resources, and (3) results.

(1.) METHODS.—From 1850 to about 1865 the chief incentive to research in this field was apparently the discovery of new forms. The subjects of individual and seasonal variation were to a large extent necessarily neglected; their importance even had not come to be duly appreciated, there being rarely available for study a series of specimens of any species of sufficient extent to throw much light upon either of these questions. Neither was much attention paid to the equally important subject of geographic variation, for the very good reason that adequate material for its investigation did not exist. Accordingly very slight differences, especially if accompanied by a difference of habitat in the specimens presenting them, were regarded as of specific importance. The word subspecies, in its modern sense, was an unknown term in biologic terminology.

Following Baird's work, done mainly between 1852 and 1858, was a considerable interval of almost complete inactivity. There was, it is true, a gradual increase of material in a few of the principal museums, notably that under the direction of the Smithsonian Institution, but little was done toward its elaboration between the years 1860 and 1870.

At about this later date new workers entered the field, and with the greatly increased material at their disposal it became possible to take up, in the case of a few species, the neglected subjects of individual, seasonal, and geographic variation. It was found that many differences—as of size, color, and even in the relative size of different parts—which had previously been depended upon as of specific importance, were open to suspicion. These discoveries, while unquestionably important, unfortunately sometimes led to erroneous conclusions. They formed, however, a phase of progress ; although they gave the pendulum a swing in the right direction, the impetus was too great ; the tendency to excessive subdivision gave place to a brief period of undue lumping.

Later, with the rapid increase of material, a healthy reaction followed, although there came with it, for a time and in certain quarters, a tendency to excessive subdivision along the line of subspecies. But the continued rapid accumulation of fresh material, it is to be hoped, will supply the required antidote, especially in the case of investigators having access to large collections.

(2.) RESOURCES.—Professor Baird had as the basis for his great work on North American Mammals what was then looked upon as a large collection. Compared with what previous workers had possessed, it was indeed enormous ; compared with the collections of to-day it was very small, and the wonder is that his work based thereon so well stands the test of time. Doubtless more specimens have been gathered during any one of the last four years than Professor Baird had for the entire basis of his great monograph, which will ever remain a monument to his sagacity and accurate and painstaking methods.

But the *amount* of material available for study in recent years, in comparison with that available twenty to forty years ago, is only one of the points to be considered, the difference in *quality* being a far more important factor in the case than the difference in quantity. In the earlier days it was a rare thing to have any considerable number of specimens of the same species from any one locality, even the larger collections consisting of what might be termed sporadic material,—a specimen or two from one locality, and a few more from other localities, separated by perhaps hundreds of miles. Furthermore they were often without precise data as to either place or time of collection. The skull, as a rule, was left in the skin, and hence could be examined only by great labor and serious injury to the skin. The skins themselves were often left flat, or rolled up, or greatly over-stuffed, or otherwise distorted, with the ears crumpled, the toes and feet bent at all angles, and the tail shrunken or broken, so that through the faults of careless taxidermy accurate measurements were almost impossible, and even approximate ones were difficult to obtain. Well-prepared and carefully labeled specimens were the exception. Large numbers of the smaller mammals, it is true, were preserved in alcohol, and were thus available for measurement and for anatomical examination, but such material is almost useless for the study of color characters, and very inconvenient for comparative study in respect to specific and subspecific differences. Specimens of mammals preserved in spirits or other solutions are also in this respect very untrustworthy, from the fact that the coloration of the pelage is so liable to undergo great change, especially if wood alcohol or alum solutions happen to be the preservative employed.

Very different indeed from all this is the present method of forming collections of the smaller mammals. Now series of specimens of the same species, from the same locality, numbering from ten to fifty or more, are almost the rule; and it is considered essential, wherever possible, to have the series duplicated at different seasons of the year, in order to show the seasonal variation. Of course this has not yet been done for all species, but it is the aim to collect as far as possible on this plan. The skull is removed from the skin, carefully cleaned and preserved separately, and cross-reference numbers refer to both skin and skull, so that they may be studied together. Several measurements of the animal are taken by the collector, who has special instructions in this regard, before skinning, and recorded on the label; the skin is then carefully filled to the natural size as indicated by the measurements, the tail vertebræ being replaced by a wire; then, in pinning the specimen out to dry, the tail and the feet are carefully extended in straight lines, and the ears given a natural set, with the result of rendering all of the external characters readily available for study. Such a specimen is also a pleasing object to the eye, in comparison with the commonly more or less distorted and unattractive specimens of earlier days. Furthermore, not only is the sex, date, and place of collection given on the label, but the altitude of the locality, if in a mountainous district, is also recorded.

With the old-time material it was often difficult to determine satisfactorily even the color characters of a specimen, to say nothing of size and proportions, owing to its faulty preservation; while in that of to-day all of the external features can be utilized, in addition to the measurements taken by the collector from the animal before skinning. Thus in respect to resources the worker of to-day has advantages immensely superior to those of his predecessors prior to a very recent date.

As is well known to mammalogists generally, and as I have before stated,[1] this great improvement in the amount and character of the material now available for investigation is due primarily to the enthusiastic and well-directed efforts of Dr. C. Hart Merriam, Chief of the Divison of Ornithology and Mammalogy of the United States Department of Agriculture.

(3.) RESULTS.—As already indicated, methods of research have undergone radical change since 1857, the date of Baird's great work on North American mammals. While new forms are still looked for with considerable avidity, it is not by any means so much the end and main purpose of investigation as was the case thirty to forty years ago. Then the idea of evolution by environment had scarcely been suggested and formed no part of the working hypothesis of the naturalist. Twenty years ago it had become fairly established. At the present time the relation of forms to each other, geographically and phylogenetically, and to their environment, is the one interest-

[1] Trans. N. Y. Acad. Sci., X., p. 84.

ing problem underlying the whole subject. There is no more fascinating or profitable work for the student of the present mammalian fauna of North America than the tracing out of the habitats, and determining the intergradation or non-intergradation of such forms as compose many of our leading genera of mammals. We have as yet scarcely reached the point where this can be done with entire certainty for any group, but here and there boundaries have been established, and we can begin to foresee in some instances what will be the final results. Should the activity of the last five years be continued for the next decade, it will doubtless be possible at the end of that time to map the distribution of most of our mammals with considerable accuracy; to know what forms intergrade, and over what areas and under what conditions the intergradation occurs; also what are sharply isolated and localized, though closely related to others; and in many instances to determine the lines of evolution and of closest genetic relationship among the congeneric forms of several of the leading families.

A few statistics will throw into strong light some at least of the results of work during the last ten years. In 1884 Mr. Frederick W. True, Curator of Mammals in the United States National Museum, published "A Provisional List of the Mammals of North and Central America, and the West Indies,"[1] which fairly reflected the status of the subject as then understood. Mr. True's List contained 378 species and 45 subspecies—a total of 423 recognized forms. At the present time the number, as nearly as can conveniently be determined, is 732 species and 131 subspecies, or a total of 863—an increase exceeding 100 per cent., mainly within the last five years. These additions include 8 new genera and about 12 new subgenera, while 10 groups rated as subgenera in 1884 have been raised to the rank of genera. On the other hand, however, several genera have

[1] Proc. U. S. Nat. Mus., VII., 1884, pp. 587–611.

been reduced to subgeneric rank, and several others entirely canceled.[1]

In this connection we will consider North America not in its continental or geographic sense but as commonly faunally limited, or as defined in the A. O. U. Check-List of North American Birds, namely, North America north of Mexico but including Lower California. In the following statistical comparison we will also exclude the marine mammals, as the Whales and Porpoises. Taking the species by ordinal groups we have the following results :

Marsupials (Opossums): 1884, 1 species; 1894, 1 species and 1 subspecies.

Edentates (Armadillos, etc.): 1884, 1 species; 1894, 1 species.

Ungulates (Deer, Antelope, Sheep, etc.); 1884, 12 species and 3 subspecies; 1894, 12 species and 4 subspecies.

Rodents (Squirrels, Spermophiles, Gophers, Mice, Rats, Pocket Mice, Kangaroo Rats, Porcupines, Hares, etc.): 1884, 80 species and 34 subspecies; 1894, 231 species and 105 subspecies,—an increase of 151 species and 71 subspecies, or nearly 200 per cent.

Bats: 1884, 15 species; 1894, 25 species,—an increase of 60 per cent.

Insectivores (Moles and Shrews): 1884, 19 species and 1 subspecies; 1894, 30 species and 3 subspecies,—an increase of 65 per cent.

Carnivores (Bears, Wolves, Cats, Skunks, Weasels, etc.): 1884, 53 species and 4 subspecies; 1894, 69 species and 12 subspecies,—an increase of 42 per cent.

As would be expected, the increase proves to be greatest among the smaller nocturnal and burrowing species, as the Pocket Mice, Kangaroo Rats, Gophers, Spermophiles, Voles and other Field Mice, Shrews and Moles, where the increase in different groups ranges from 60 to 200 per cent. as against 15 to 40 per cent. among the Ungulates and

---

[1] In October, 1892, Mr. Walter E. Bryant published a very useful paper entitled "Recent Additions to the North American Mammal Fauna" (Zoe, III., pp. 201–223), giving a list of additions made between 1884 and October, 1892, with notes on the changes that had occurred in nomenclature. Although restricted to that part of North America north of Mexico, the changes in generic names number 17, mainly due to the revival of older names for those current in 1884, or through the raising of subgeneric names to generic rank, or the relegation of generic names to subgeneric rank, with about three subgenera and five generic names proposed for new groups or for pre-occupied names. The paper records 190 species and subspecies as actual additions to the list of 1884.

Carnivores. In other words, the larger and diurnal species, as the Deer and larger Carnivores, were relatively better known in 1884 than the smaller species that burrow in the ground and are for the most part abroad only at night.

These statistics, however, very inadequately express the results of the work of the last decade,—or more correctly of the last seven years, for little advance was made prior to 1887. Since this date have been made not only the additions and other changes indicated above, but many radical changes in the nomenclature of species have been found necessary, in consequence of the former misapplication of names. Also many forms formerly ranked as subspecies have been found to be entitled to specific rank; many names that had been reduced to synonyms have been revived for forms which prove to be tenable as subspecies, and to which they were evidently intended to apply, though originally very inadequately characterized. Again, forms originally characterized as species, and known only from limited areas and few specimens, have proved separable into several well-marked subspecies, and in some instances into species, and their known geographic range greatly extended.

In order to emphasize some of these points it will be necessary to pass in review a few special groups. Although little monographic work has as yet been attempted, owing to lack of material and time for final revisionary work, yet the results that have been reached in the few instances in which such work has been undertaken have stood the test of re-examination much better than might have been expected. It is almost too soon yet for final work in any group, but we may hope to reach the point ere long when the material will be sufficiently abundant to warrant the attempt, in the case at least of some of the better known generic groups.

The following genera will be taken in illustration of the subject: *Lepus*, *Heteromys*, *Perognathus*, *Dipodomys*, *Perodipus*, *Thomomys*, *Geomys*, *Phenacomys*, *Evotomys*, *Synap-*

tomys, *Arvicola*, *Neotoma*, *Sigmodon*, *Oryzomys*, *Onychomys*, *Sitomys*, *Sciurus*, *Tamias*, *Spermophilus*, and *Spilogale;* or, practically the families Leporidæ, Heteromyidæ, Geomyidæ, Muridæ, and Sciuridæ.[1]

In 1884 the genus *Lepus*, or the Hares, was recognized in Mr. True's List as consisting of 11 species and 7 additional subspecies; it now numbers 24 species and 8 subspecies. Of the new forms 9 have been added from the United States[2] and 5 from Mexico.

---

[1] In the following review, and in the lists given in the footnotes, North America is taken in its geographic sense, and the subject is brought down to June, 1894. The writer assumes no responsibility as an endorser of the species and subspecies recorded in the lists, which are in no sense revisionary, but merely intended to reflect the present status of the subject as it stands in the literature of this date. To give the lists consistency, and to facilitate comparison with the 1884 List, a few changes have been made in the generic allocation of some of the recently described forms, followed always, however, by the name employed by the original describer.

[2] The United States list of additions is as follows:

Lepus sylvaticus floridanus Allen, Bull. Am. Mus. Nat. Hist., III., p. 160, Oct., 1890. Brevard Co , Fla.

Lepus sylvaticus bachmani (Waterhouse). Revived by Allen, Bull. Am. Mus. Nat. Hist., VI., p. 170, May, 1894. Texas.

Lepus sylvaticus mearnsii Allen, *ibid.*, p. 171, May, 1894. Minnesota and adjoining portions of the Upper Mississippi region.

Lepus idahoënsis Merriam, N. Am. Fauna, No. 5, p. 75, July, 1891. Idaho and northern Nevada.

Lepus cinerascens Allen, Bull. Am. Mus. Nat. Hist., III., p. 159, Oct., 1890. Southern California.

Lepus insularis Bryant, Proc. Cal. Acad. Sci., 2d Ser., III., p. 92, April, 1891. Espiritu Santo Island, Lower California.

Lepus alleni Mearns, Bull. Am. Mus Nat. Hist., II., p. 294, Feb., 1890. Arizona.

Lepus melanotis Mearns, *ibid.*, p. 297., Feb., 1890. Kansas, Indian Territory, and western Texas.

Lepus paludicola Miller and Bangs, Proc. Biol. Soc. Wash., IX, p. 105, June 9, 1894. Western Florida.

Mexican and Central American species recently added are the following:

Lepus sylvaticus aztecus Allen, Bull. Am. Mus. Nat. Hist., III., p. 188, Dec., 1890. Isthmus of Tehuantepec, Mexico.

Lepus orizabæ Merriam, Proc. Biol. Soc. Wash., VIII., 1893, p. 143. Mt. Orizaba, Mexico.

The genus *Heteromys*, which is developed mainly in Mexico and Central America, one species barely reaching the lower Rio Grande valley in the United States, has been added to the United States fauna,[1] and the number of species raised from 3 in 1884 to 9, with an additional subspecies, in 1894.[2]

The genus *Perognathus* has been increased from the 6 forms recognized in 1884 to 29.[3] The range of characters,

---

Lepus veræcrucis Thomas, P. Z. S., 1890, p. 74, pl. vi. Jalapa, Mexico.

Lepus insolitus Allen, Bull. Am. Mus. Nat. Hist., III., p. 189, Dec. 1890. Plains of Colima, Mexico.

Lepus truei Allen, *ibid.*, p. 192. Mirador, near Vera Cruz, Mexico.

[1] Allen, Bull. Am. Mus. Nat. Hist., III., pp. 268–272, June, 1891. See also Allen and Chapman, *ibid.*, V., pp. 218–220, Sept., 1893.

[2] Cf. Thomas, Ann. and Mag. Nat. Hist. (6), XI., 1893, pp. 329-332, and *ibid.*, XII., pp. 233, 234.

The forms of this genus recently recognized by Mr. Oldfield Thomas appear to be as follows:

Heteromys alleni Coues. Lower Rio Grande Valley. (Cf. Allen, as cited above.)

Heteromys bulleri Thomas, Ann. and Mag. (6), XI., p. 330, April, 1893. La Laguna, Sierra de Juanacatlan, Jalisco, Mexico.

Heteromys salvini Thomas, *ibid.*, p. 331. Dueñas, Guatemala.

Heteromys salvini nigrescens Thomas, *ibid.*, XII., p. 234, Sept., 1893. Costa Rica.

Heteromys pictus Thomas, *ibid.*, p. 233, Sept., 1893. Mineral San Sebastion, Jalisco, Mexico.

Heteromys longicaudus Gray, P. Z. S., 1868, p. 204. "Hondurus" (= Venezuela *apud* Alston, Biol. Cent. Am. Mam., p. 167.)

Heteromys irroratus Gray, *ibid.*, p. 205. Oaxaca, Mexico.

Heteromys albolimbatus Gray, *ibid.*, p. 205. La Parada, Mexico.

Add also, to complete the list of the recorded species:

Heteromys longicaudatus Gray, *ibid.*, p. 204, Mexico.

Heteromys desmarestianus Gray, P. Z. S., 1843, p. 79, and 1868, p. 204. Coban, Guatemala.

Heteromys adspersus Peters, Monatsb. Ak. Berlin, 1874, p. 357. Panama.

Heteromys anomalus (Thompson). Trinidad.

[3] Cf. Merriam, N. Am. Fauna, No. 1, pp. 1–29, pl. i–iv. Oct., 1889.

The following were here recognized by Dr. Merrian, of which 15 were described as new:

Perognathus fasciatus Wied. Upper Missouri, near its junction with the Yellowstone.

Perognathus fasciatus flavescens Merriam. Kennedy, Neb.

both external and cranial, in this genus is strikingly great, yet the transition between the extreme phases of the group is so gradual that it is difficult to separate it into more than two subgenera, and these are by no means trenchantly limited.

In Mr. True's List the genus *Dipodomys* contained I species with I subspecies The group has since been

---

Perognathus flavus Baird. Western Texas.
Perognathus bimaculatus Merriam. Arizona.
Perognathus longimembris (Coues). Southern California.
Perognathus apache Merriam. Apache Co., Arizona.
Perognathus inornatus Merriam. Fresno, Cal.
Perognathus olivaceus Merriam. Kelton, Utah.
Perognathus olivaceus amœnus Merriam. Nephi, Utah.
Perognathus monticola Baird. Montana.
Perognathus formosus Merriam. St. George, Utah.
Perognathus intermedius Merriam. Mud Spring, Arizona.
Perognathus fallax Merriam. Southern California.
Perognathus obscurus Merriam. Grant Co., N. Mex.
Perognathus spinatus Merriam. Lower Colorado River, Cal.
Perognathus penicillatus Woodhouse. San Francisco Mountain, Arizona.
Perognathus hispidus Baird. Charco Escondido, Mexico.
Perognathus paradoxus Merriam. Trego Co., Kansas.
Perognathus paradoxus spilotus Merriam. Cook Co., Texas.
Perognathus californicus Merriam. Berkely, Cal.
Perognathus armatus Merriam. Mount Diablo, Cal.

Two other "undetermined species" were also mentioned, namely, *Perognathus lordi* (Gray) and *Perognathus mollipilosus* Coues. There have since been added :

Perognathus fuliginosus Merriam, N. Am. Fauna, No. 3, p 74, Sept., 1890. San Francisco Mountain, Arizona.

Perognathus femoralis Allen, Bull. Am. Mus. Nat. Hist. III., p. 281, June, 1891. San Diego Co., Cal.

Perognathus merriami Allen, *ibid.*, IV., p. 45, March, 1892. Southeastern Texas.

Perognathus infraluteus Thomas, Larimer Co., Col.

Perognathus lordi (Gray). British Columbia. Revived by Rhoads, Proc. Acad. Nat. Sci. Phila., 1893, p. 405, Jan. 27, 1894.

Perognathus copei Rhoads, *ibid.*, p. 404. Staked Plains, Texas.

Perognathus alticolus Rhoads,. *ibid*, p. 412. San Bernardino Mts., Cal.

Perognathus latirostris Rhoads, Am. Nat., XXVIII., Feb., 1894, p. 185. "Rocky Mountains."

divided into two genera,[1] *Dipodomys* and *Perodipus*,[2] with

---

[1] Cf. Merriam, N. Am. Fauna, No. 3, p. 72, Sept., 1890. *Dipodops*, gen. nov., later changed to *Perodipus* Fitzinger, an earlier name based on the same type. (Cf. Merriam, Proc. Biol. Soc. Wash., VII., p. 26, April, 1892.)

[2] Following are the recorded species and subspecies of these two genera:

Dipodomys phillipsi Gray. Valley of Mexico and adjacent mountain slopes and plains. (On this species see an important paper by Merriam, Proc. Biol. Soc. Wash., VIII., pp. 83–96, July 18, 1893.)

Dipodomys ornatus Merriam, Proc. Biol. Soc. Wash., IX, p. 110, June 21, 1894. Berriozabal, Zacatecas, Mexico.

Dipodomys perotensis Merriam, *ibid.*, p 111. Perote, Vera Cruz, Mexico.

Dipodomys elator Merriam, *ibid.*, p. 109. Henrietta, Clay Co., Texas.

Dipodomys deserti Stephens, Am. Nat., XXI., Jan., 1887, p. 42, pl. v. Deserts of Southeastern California.

Dipodomys merriami Mearns, Bull. Am. Mus. Nat. Hist., II., p. 290, Feb., 1890. Arizona.

Dipodomys merriami melanurus Merriam, Proc. Cal. Acad. Sci. (2), III., p. 345, June, 1893. San José del Cabo, Lower California.

Dipodomys merriami nevadensis Merriam, Proc. Biol., IX., p. 111, June 21 1894, Pyramid Lake, Nevada.

Dipodomys merriami nitratus Merriam, *ibid.*, p. 112. Keeler, east side of Owens Lake, California.

Dipodomys merriami nitratoides Merriam, *ibid.*, p. 112. San Joaquin Valley, California.

Dipodomys merriami exilis Merriam, *ibid.*, p. 113. Fresno, San Joaquin Valley, California.

Dipodomys merriami atronasus Merriam, *ibid.*, p. 113. Hacienda La Parada, San Luis Potosi, Mexico.

Dipodomys ambiguus Merriam, N. Am. Fauna, No. 4, p. 42, Oct., 1890. El Paso, Texas.

Dipodomys spectabilis Merriam, *ibid.*, p. 46. Cochise Co., Arizona.

Dipodomys californicus Merriam, *ibid.*, p. 49. Mendocino Co., Cal.

Dipodomys simiolus Rhoads, Proc. Acad. Nat. Sci. Phila., 1893, p. 410, Jan. 27, 1894. Agua Caliente, California.

Dipodomys similis Rhoads, *ibid.*, p. 411. San Diego Co., California.

Dipodomys parvus Rhoads, Am. Nat., XXVIII., Jan., 1894, p. 70. San Bernardino Valley, California.

Perodipus agilis (Gambel). Southern California.

Perodipus ordii (Woodhouse). Western Texas.

Perodipus ordii palmeri (Allen), Bull. Am. Mus. Nat. Hist., III., p. 276, June, 1891. San Luis Potosi, Mexico.

Perodipus ordii columbianus Merriam, Proc. Biol. Soc. Wash., IX, p. 115, June 21, 1894. Umatilla, Plains of Columbia, Oregon.

Perodipus streatori Merriam, *ibid.*, p. 113. Carbondale, Mariposa Co., Cal.

Perodipus panamintinus Merriam, *ibid.*, p. 14. Panamint Mountains, California.

12 species and 6 subspecies in the former, and 9 species and 2 subspecies in the latter, or a total increase of from 2 to 29 forms, with a considerable extension of the known range of the group. Besides these additions, an allied genus *Microdipodops* has been described from Halleck, Nevada,[1] which combines the external characters of some of the species of *Perognathus* with many of the cranial characters of *Perodipus*.

The Pocket Gophers form a group of burrowing Rodents, restricted to North America, and constituting two strongly marked genera, *Thomomys* and *Geomys*. They live almost wholly underground, and can be secured only by trapping them in their burrows. In Mr. True's List *Thomomys* is credited with two species (one of them then known only from the type specimen) and two additional subspecies. This genus has now 16 species and 1 subspecies,[2] and its

---

Perodipus richardsoni (Allen), Bull. Am. Mus. Nat. Hist., III, p. 277. Oklahoma Territory.

Perodipus sennetti (Allen), *ibid.*, p. 226, April, 1891. Cameron Co., Texas.

Perodipus compactus (True), Proc. U. S. Nat. Mus., XI., p. 160. Jan. 5, 1889. Padre Island, Texas.

Perodipus chapmani (Mearns), Bull. Am. Mus. Nat. Hist., II., p. 291., Feb., 1890. Fort Verde, Arizona.

Perodipus longipes (Merriam), N. Am. Fauna, No. 3, p. 71, Sept., 1890. Painted Desert, Arizona.

[1] Merriam, N. Am. Fauna, No. 5, p. 115, July 30, 1891. Type, *M. megacephalus*, sp. nov.

[2] For a recent preliminary revision of the genus see Allen, Bull. Am. Mus. Nat. Hist., V., 1893, pp. 47–68. On *T. bulbivorus* see an important paper by Gerrit S. Miller, Jr., Proc. Biol. Soc. Wash., VIII., pp. 113–116., Aug., 1893.

Following is a list of the United States species:

Thomomys bulbivorus (Rich.).
Thomomys laticeps Baird.
Thomomys bottæ (Eyd. and Gerv).
Thomomys townsendii (Rich.).
Thomomys monticolus Allen.
Thomomys douglasii (Rich.).
Thomomys douglasii fuscus (Merr.).
Thomomys clusius Coues.
Thomomys talpoides (Rich.).
Thomomys fulvus (Woodh.).
Thomomys perpallidus Merr.
Thomomys aureus Allen.
Thomomys fossor Allen.
Thomomys toltecus Allen.

known geographical range has been extended far to the southward in Mexico. This genus has the further curious history that of the 12 forms recently added 5 of them were described some thirty years ago and had lapsed into synonymy. While many of the species closely resemble each other externally they are easily separated by cranial differences.

In 1884 *Geomys* was credited with 5 species, three of which were known only from the United States, and the others, one each respectively, from Mexico and Central America. The recognized United States forms have been increased to 6, and the Mexican and Central American forms to 9, or a total increase of from 5 to 15.[1] A monographic revision of the genus has been undertaken by Dr. Merriam and is nearly ready for publication, with numerous figures of cranial characters and distribution maps of the species.

While the increase in our knowledge of the two families Heteromyidæ[2] and Geomyidæ, or the Pouched Rats and Pocket Gophers, has been so great, it has scarcely been less so in respect to the Muridæ. Here the ratio of increase in the number of recognized forms is not only very high,

---

Recently described Mexican species are:

Thomomys orizabæ Merriam, Proc. Biol. Soc. Wash., VIII., p. 145; Dec. 1893. Mt. Orizaba, Mexico.

Thomomys peregrinus Merriam, *ibid.*, p. 146. Salazar, State of Mexico.

[1] The present status of the group is as follows, the United States species being:

Geomys tuza (Ord).
Geomys bursarius (Shaw).
Geomys bursarius lutescens Merriam.
Geomys castanops (Baird).
Geomys personatus True.
Geomys clarkii Baird.

The Mexican and Central American species are:

Geomys mexicanus (Licht.).
Geomys gymnurus Merriam.
Geomys nelsoni Merriam.
Geomys merriami Thomas.
Geomys grandis Thomas.
Geomys scalops Thomas.
Geomys hispidus Le Conte.
Geomys cherriei Allen.
Geomys fumosus Merriam.

[2] Saccomyidæ of Baird, Lilljeborg, Coues, etc. As *Saccomys* is a synonym of *Heteromys* it becomes untenable as the basis of the family name. Heteromyidæ is equal to Heteromyinæ of Alston, 1876.

but there has also been a notable increase in the number of genera and subgenera, and changes in the status of several of those previously recognized, as well as in the nomenclature of some of the older species. The old genus *Hesperomys* has not only been dismembered by the raising of the United States subgenera *Vesperimus* (Coues=*Sitomys* Fitzinger, of earlier date), *Onychomys* and *Oryzomys*, and the Mexican and Central American *Tylomys*, *Rhipodomys*, and *Abrothrix* to full generic rank, but a new subgenus *Baiomys*[1] has been proposed, and the name *Hesperomys* discarded as untenable, at least so far as the North American Muridæ are concerned.[2] The genus *Phenacomys* was described by Dr. Merriam[3] in 1889, with 4 species, to which others have since been added.[4]

This is a boreal type which barely reaches the mountainous portions of the United States, north of which it ranges across the continent from ocean to ocean; yet prior to 1889 not a specimen appears to have been seen, or at least critically examined, by any naturalist.

The Red-backed Meadow Mice, forming the genus *Evotomys*, known from 1 species and 1 additional subspecies in 1884, now numbers 9 species with 3 additional subspecies,

---

[1] True, Proc. U. S. Nat. Mus. XVI., p. 758, Feb., 1894. Type, *Hesperomys taylori* Thomas.

[2] Cf. Allen, Bull. Am. Mus. Nat. Hist., III., pp. 291-294, June, 1891.

[3] N. Am. Fauna, No. 2, pp. 27-35, Oct., 1889.

[4] *Phenacomys intermedius* Merriam, N. Am. Fauna, No. 2, p. 32, Oct., 1889. Kamloops, British Columbia.

Phenacomys celatus Merriam, *ibid.*, p. 33. Godbout, P. Q., Canada.

Phenacomys latimanus Merriam, *ibid.*, p. 34. Fort Chimo, Ungava, Labrador.

Phenacomys ungava Merriam, *ibid.*, p. 35, Fort Chimo, Ungava, Labrador.

Phenacomys longicaudus True, Proc. U. S. Nat. Mus., XIII., p. 303, Nov., 1890. Coos Co., Oregon.

Phenacomys orophilus Merriam, N. Am. Fauna, No. 5, p. 65, July, 1891. Idaho.

some of the new forms differing radically in coloration from those earlier known.[1]

Mr. True has recently described a new genus *Myctomys*[2] from Fort Chimo, Ungava, Labrador, which appears to combine somewhat the characters of the genera *Phenacomys* and *Synapatomys*.

The Lemming Mouse, forming the still monotypic genus *Synaptomys*, has had its known range remarkably extended, and furnishes a cogent commentary upon our lack of knowledge of the mammalian fauna of even the long-settled parts of the United States. A few years since this species (*S. cooperi*) was positively known only from southern Illinois, Indiana and Ohio, east of the Mississippi River, though the type of the species was reputed to have come from either New York or New England, or possibly from New Jersey. Recently Dr. Merriam reported its occurrence in Maryland, in Virginia, in the mountains of North Carolina, and at Alfred Centre, N. Y.[3] It has since

---

[1] The following have been recorded:

Evotomys rutilus (Pallas). "Circumpolar regions."

Evotomys gapperi (Vigors). Northern border of eastern United States.

Evotomys gapperi ochraceus Miller, Proc. Boston Soc. Nat. Hist., XXVI., p. 193, March 24, 1894. White Mts., N. H.

Evotomys gapperi rhoadsi Stone, Am. Nat., Jan., 1893, p. 55. May's Landing, N. J.

Evotomys gapperi brevicaudus Merriam, No. 5, p. 66, July, 1891. Black Hills, South Dakota.

Evotomys carolinensis Merriam, Am. Journ. Sci., XXXVI., Dec., 1888, p. 460. Mountains of North Carolina.

Evotomys dawsoni Merriam, Am. Nat., July, 1889, p. 649. Finlayson River, N. W. T.

Evotomys galei Merriam, N. Am. Fauna, No. 4, p. 23, Oct., 1890. Boulder County, Colorado.

Evotomys occidentalis Merriam, *ibid.*, p. 26. Chehalis Co., Washington.

Evotomys californicus Merriam, *ibid.*, p. 26. Humboldt Co., California.

Evotomys idahoënsis Merriam, *ibid.*, No. 5, p. 66, July, 1891. Salmon River Mountains, Idaho.

Evotomys fuscodorsalis Allen, Bull. Am. Mus. Nat. Hist., VI., p. 103, April, 1894. New Brunswick.

[2] Proc. U. S. Nat. Mus., XVII., p. 2 (of advance sheet), April 26, 1894, Type *M. innuitus*, sp. nov.

[3] Merriam, Proc. Biol. Soc. Wash., VII., pp. 175–177, Dec., 1892.

been taken in Southern New Jersey,[1] and in Wareham, Massachusetts.[2] As Dr. Coues,[3] in 1877, reported it from Kansas, Oregon, and Alaska, the genus has probably a wide distribution, and may be found to embrace several quite distinct forms. It evidently occurs in the East over portions of the country where it has escaped all collectors for half a century, since, as compared with other field mice, it proves to be a singularly difficult species to trap. Indeed, the Virginia and New York records rest on skulls found in pellets from stomachs of owls, and the Maryland record on a specimen taken from the stomach of a hawk.

A Meadow Mouse, also with grooved incisors, like *Synaptomys*, has been recently described by Mr. Rhoads from Kittitas Co., Washington, as the type of a new genus *Aulacomys*.[4]

The genus *Arvicola*, embracing the common Meadow Mice, has been increased in the last ten years from 8 species and 2 subspecies to 15 species and 5 subspecies.[5]

The Wood Rats (*Neotoma*) and Cotton Rats (*Sigmodon*) have undergone corresponding revision and increase. Not only has *Neotoma micropus* of Baird been revived, but the number of forms of *Neotoma* has been raised from 4 to

---

[1] S. N. Rhoads, Am. Nat., Jan., 1893 p 53. Described as *Synaptomys stonei*, sp. nov.

[2] Outram Bangs, Proc. Biol. Soc. Wash., IX., pp. 99–104, April, 1894.

[3] Mon. N. Am. Roden., pp. 235, 236.

[4] Am. Nat., Feb. 1894, p. 182. Type *A. arvicoloides*, sp. nov.

[5] Additions subsequent to Bryant's List of 1892 are as follows:

Arvicola edax Baird (Revived; cf. Allen, Bull. Am. Mus. Nat. Hist., V., 1893, p. 184.)

Arvicola aztecus Allen, *ibid.*, p. 73, April, 1893. Aztec and La Plata, New Mexico.

Arvicola operarius Nelson, Proc. Biol. Soc. Wash., VIII., p. 139, Dec., 1893. St. Michaels, Norton Sound, Alaska.

Arvicola phæus Merriam, *ibid.*, VII., p. 171, Sept., 1892. Sierra Nevada de Colima, Jalisco, Mexico.

Arvicola chrotorrhinus Miller, Proc. Bost. Soc. Nat. Hist., XXVI., p. 190 March, 1894. White Mts., N. H.

40. Four forms have been revived, and 22 species and 10 subspecies[1] have been described as new!

---

[1] *Neotoma cinerea* (Ord).

Neotoma cinerea occidentalis (Baird). Revived by Allen, Bull. Am. Mus. Nat. Hist., III., p. 287, June, 1891.

Neotoma cinerea drummondii (Rich.). Revived by Merriam, Proc. Biol. Soc. Wash., VII., p. 25, April, 1892.

Neotoma occidentalis fusca True, Proc. U. S. Nat. Mus., XVII., No. 1,006, p. —, (advance sheet, p. 2, June 27, 1894). Fort Umpqua, Oregon.

Neotoma lepida Thomas, Ann. and Mag. Nat. Hist. (6), XII., p. 235, Sept., 1893. Utah.

Neotoma arizonæ Merriam, Proc. Biol. Soc. Wash., VIII., p. 110, July, 1893. Eastern Arizona and New Mexico.

Neotoma floridana (Ord).

Neotoma pennsylvanica Stone, Proc. Acad. Nat. Sci. Phila., 1893, p. 16. Cumberland Co., Pa.

Neotoma mexicana Baird. Revived by Merriam, N. Am. Fauna, No. 3, p. 67, Sept. 1890.

Neotoma mexicana bullata Merriam, Proc. Biol. Soc. Wash., IX., p. 122, July 2, 1894. Santa Catalina Mts., Arizona.

Neotoma leucodon Merriam, *ibid.*, p. 120. San Luis Potosi, Mexico.

Netoma latifrons Merriam, *ibid.*, p. 121. Querendaro, Michoacan, Mexico.

Neotoma fulviventer Merriam, *ibid.*, p. 121. Toluca Valley, Mexico.

Neotoma orizabæ Merriam, *ibid.*, p. 122. Mt. Orizaba, Puebla, Mexico.

Neotoma baileyi Merriam, *ibid.*, p. 123. Valentine, Nebraska.

Neotoma fallax, Merriam, *ibid.*, p. 123. Gold Hill, Boulder Co., Colorado.

Neotoma desertorum Merriam, *ibid.*, p. 125. Furnace Creek, Death Valley, California.

Neotoma desertorum sola Merriam, *ibid.*, p. 126. San Emigdio, Kern Co., California.

Neotoma orolestes Merriam, *ibid.*, p. 128. Saugache Valley, Colorado.

Neotoma albigula Hartley, Proc. Cal. Acad. Sci. (2), IV., p. 157, pl. xii., skull, May 9, 1894. Fort Lowell, Arizona.

Neotoma micropus Baird. Revived by Allen, Bull. Am. Mus. Nat. Hist., III., p. 282., June, 1891.

Neotoma micropus canescens Allen, *ibid.*, p. 285. Oklahoma Territory.

Neotoma fuscipes Baird.

Neotoma fuscipes streatori Merriam, Proc. Biol. Soc. Wash., IX., p. 124, July 2, 1894. Carbondale, Amador Co., California.

Neotoma fuscipes dispar Merriam, *ibid.*, p. 124. Lone Pine, Owens Valley California.

Neotoma monochroura Rhoads, Am. Nat., XXVIII., Jan., 1894, p, 67 Josephine Co., Oregon.

Dr. Merriam has also described a new genus, from the State of Colima, Mexico, allied to *Neotoma*, under the name *Xenomys*.[1]

The Cotton Rats (genus *Sigmodon*) have been increased from 1 species to 2 species and 4 subspecies, the genus now consisting of two quite distinct groups, represented respectively by *S. hispidus* and *S. fulviventer*.[2]

Neotoma californica Price, Proc. Cal. Acad. Sci. (2), IV., p. 154, pl. xi., skull, May 9, 1894. Bear Valley, San Benito Co., Cal.

Neotoma splendens True, Proc. U. S. Nat. Mus., XVII., No. 1,006, p. — (advance sheet, p. 1, June 27, 1894). Marin Co., California.

Neotoma venusta True, *ibid.*, p. 2. Carrizo Creek, California.

Neotoma macrotis Thomas, Ann. and Mag. Nat. Hist. (6), XII., p. 234, Sept., 1893. San Diego, Cal.

Neotoma macrotis simplex True, Proc. U. S. Nat. Mus., XVII., No. 1,006, p. —, (advance sheet, p. 2, June 27, 1894. Fort Tejon, California.

Neotoma intermedia Rhoads, Am. Nat., XXVIII., Jan., 1894, p. 69. San Diego Co., Cal.

Neotoma intermedia gilva Rhoads, *ibid.*, p. 70. Banning and San Bernardino, Cal.

Neotoma intermedia melanura Merriam, Proc. Biol. Soc. Wash., IX., p. 126, July 2, 1894. Ortiz, Sonora, Mexico.

Neotoma intermedia angusticeps Merriam, *ibid.*, p. 127. Southwest corner of Grant Co., New Mexico.

Neotoma pinetorum Merriam, Proc. Biol. Soc. Wash., VIII., p. 111, July, 1893. San Francisco Mountains, Arizona.

Neotoma ferruginea Tomes.

Neotoma torquata Ward, Am. Nat., Feb., 1891, p. 160. State of Morelos, Mexico.

Neotoma alleni Merriam, Proc. Biol. Soc. Wash., VII., p. 168, Sept., 1892. Manzanillo, Colima, Mexico.

Neotoma tenuicauda Merriam, *ibid.*, p. 169. Sierra Nevada de Colima, Jalisco, Mexico.

1 Proc. Biol. Soc. Wash., VII,, pp. 159–163, Sept., 1892. Type *X. nelsoni*, sp. nov.

2 *Sigmodon hispidus* Say & Ord. Southeastern United States.

Sigmodon hispidus littoralis Chapman, Bull. Am. Mus. Nat. Hist., II., p. 118, June, 1889. Southern Florida.

Sigmodon hispidus texianus (Aud. & Bach.). Revived by Allen, *ibid.*, III, p. 287, June, 1890. Texas.

In 1884 the genus *Oryzomys* was ranked as a subgenus of "*Hesperomys*," with 2 species, one of which was the Rice-field Mouse (*O. palustris*) of the Southern States, and the other a Mexican and Central American species (*O. couesi*). The number has since been increased by the addition of one species and two subspecies from the United States, and 6 species from Mexico and Central America.[1]

The Grasshopper Mice, genus *Onychomys*, likewise merged in *Hesperomys* in 1884, and numbering only 2 species, now stands as a full genus with 8 species and 1 subspecies.[2]

Sigmodon hispidus toltecus (De Saussure). Revived by Allen, *ibid.*, p. 207, Southern Mexico, South to Costa Rica.

Sigmodon hispidus arizonæ Mearns, *ibid.*, II., p. 287, Feb., 1890. Fort Verde, Arizona.

Sigmodon fulviventer Allen, *ibid.*, II., p. 180, Oct., 1889. Zacatecas, Mexico.

[1] *Oryzomys palustris* (Harlan). South Atlantic States.

Oryzomys palustris natator Chapman, Bull. Am. Mus. Nat. Hist., V., p. 44, March, 1893. Florida and Gulf Coast.

Oryzomys palustris texensis Allen, *ibid.*, VI., p. 177, May, 1894. Coast of Texas.

Oryzomys aquaticus Allen, *ibid.*, III., p. 289, June, 1891. Cameron Co. Texas.

Oryzomys talamancæ Allen, Proc. U. S. Nat. Mus., XIV., 1891, p. 193. Costa Rica.

Oryzomys alfaroi Allen, Bull. Am. Mus. Nat. Hist., III., p. 214, April, 1891. Costa Rica.

Oryzomys costaricensis Allen, *ibid.*, V., p. 239, Sept., 1893, Costa Rica.

Oryzomys couesi (Alston). See Thomas, Ann. and Mag. Nat. Hist. (6), XI., 1893, p. 403.

Oryzomys fulgens Thomas, *ibid.*, p. 403, Mexico.

Oryzomys melanotis Thomas, *ibid.*, p. 404. Mineral San Sebastian, Jalisco, Mexico.

[2] *Onychomys leucogaster* (Wied). Upper Missouri.

Onychomys leucogaster brevicaudus Merriam, N. Am. Fauna, No. 5, p. 52 July, 1891. Idaho.

Onychomys longipes Merriam, *ibid.*, No. 2, p. 1, Oct., 1889, Texas.

Onychomys longicaudus Merriam, *ibid.*, p, 2. St. George, Utah.

Onychomys melanophrys Merriam, *ibid.*, p. 2, and No. 3, pp. 61, 62. Kanab, Utah.

The very numerous group of Deer Mice, or White-footed Mice, forming the restricted genus *Sitomys* (= subgenus *Vesperimus* Coues), has of late received a large share of attention, both from collectors and systematists. In 1884 probably the specimens in the museums did not exceed 700 or 800, of which only a small part would now be considered as available, or at least desirable, for scientific use, while now probably 7,000 to 10,000 could be brought together if the specimens in the various public and private collections in this country should be combined. The number of forms recognized in this particular group of mice in 1884 was 9—6 species and 3 subspecies. The number now recorded is 37—21 species and 16 subspecies; an increase of about 300 per cent. in seven years.[1] I am more

---

Onychomys melanophrys pallescens Merriam, *ibid.*, No. 3, p. 61, Sept., 1890. Apache Co., Arizona.

Onychomys fuliginosus Merriam, *ibid.*, p. 59, Oct., 1890. Arizona.

Onychomys ramona Rhoads, Am. Nat., Sept., 1893, p. 833. San Bernardino Valley, Cal.

Onychomys torridus (Coues). Arizona.

[1] The United States additions made prior to October, 1892, may be found recorded in detail by Bryant in Zoe, III., pp. 212–214. Mexican and later additions are included in the following full list:

a. UNITED STATES.

Sitomys americanus (Kerr).

Sitomys americanus canadensis Miller, Proc. Biol. Soc. Wash., VIII., p. 55, June, 1893. Northern New England, northern New York, and northward.

Sitomys americanus gossypinus (Le Conte).

Sitomys americanus arcticus (Mearns).

Sitomys americanus nebracensis (Baird).

Sitomys americanus texanus (Woodhouse).

Sitomys americanus rufinus (Merriam).

Sitomys americanus austerus (Baird).

Sitomys americanus gambeli (Baird). Revived by Allen, Bull. Am. Mus. Nat. Hist., V., 1893, p. 190.

Sitomys americanus deserticolus (Mearns).

Sitomys americanus sonoriensis (Le Conte).

Sitomys americanus thurberi Allen, Bull. Am. Mus. Nat. Hist., V., p. 185, August, 1893. Northern Lower California.

or less familiar with all but three or four, and consider that very few of them are not well founded ; while I am cogni-

Sitomys mearnsii (Allen).

Sitomys boylii (Baird).

Sitomys auripectus Allen, Bull. Am. Mus. Nat. Hist., V., p. 75, April, 1893. San Juan region of Utah, Colorado, New Mexico, and Arizona.

Sitomys rowleyi Allen, *ibid.*, p. 76. Southeastern Utah and adjoining portions of Colorado, New Mexico, and Arizona.

Sitomys rowleyi pinalis Miller, *ibid.*, p. 331, Dec., 1893. Grant County, New Mexico, and Pinal County, Arizona.

Sitomys fraterculus Miller.

Sitomys eremicus (Baird).

Sitomys major Rhoads, Am. Nat., Sept., 1893, p. 831. San Bernardino Co., Cal.

Sitomys herroni Rhoads, *ibid.*, p. 832. San Bernardino Valley, Cal.

Sitomys californicus (Gambel).

Sitomys gilberti Allen, Bull. Am. Mus. Nat. Hist., V., p. 188, Aug., 1893. San Benito Co., Cal.

Sitomys martirensis Allen, *ibid.*, p. 187. San Pedro Martir Mountains, Lower California.

Sitomys robustus Allen, *ibid.*, p. 335, Dec., 1893. Lake Co., California.

Sitomys floridanus (Chapman).

Sitomys macropus (Merriam).

Sitomys truei (Shufeldt).

Sitomys truei megalotis (Merriam).

Sitomys truei nasutus (Allen).

Sitomys truei crinitus (Merriam).

Sitomys aureolus (Aud. & Bach.).

Sitomys michiganensis (Aud. & Bach.).

Sitomys anthonyi (Merriam).

Sitomys taylori (Thomas).

Sitomys niveiventris (Chapman).

Sitomys niveiventris subgriseus Chapman, Bull. Am. Mus. Nat. Hist., V., p. 340, Dec., 1893. Florida.

b. Mexico and Central America.

Sitomys aztecus (De Saussure).

Sitomys melanophrys (Coues).

Sitomys difficilis (Allen) = *Vesperimus difficilis* Allen, Bull. Am. Mus. Nat. Hist., III., p. 298, June, 1891. Sierra de Valparaiso, Zacatecas, Mexico.

Sitomys musculus Merriam, Proc. Biol. Soc. Wash., VII., p. 170, Sept., 1892. Colima and Jalisco, Mexico.

Sitomys cherriei Allen = *Hesperomys* (*Vesperimus*) *cherrii* Allen, Bull. Am. Mus. Nat. Hist., III., p. 211, April, 1891. Costa Rica.

Sitomys nudipes Allen = *Hesperomys* (*Vesperimus*) *nudipes* Allen, *ibid.*, p. 213, and V., 1893, p. 239. Costa Rica.

zant of quite a number of still undescribed forms entitled to recognition in nomenclature. Five admitted by Baird in 1857, but later reduced to synonyms, have been reinstated as subspecies. The group is divisible into several sections or subgenera, one of which (as already noticed) has already been characterized by Mr. True (see *antea*, p. 31). Neither this nor some of the other leading types of this group were represented in our leading museums by a single specimen prior to 1885. They were thus not overlooked by previous workers, but are genuine discoveries of the present decade.

The Sciuridæ, or the Squirrels, Spermophiles and their allies, present a nearly parallel case with the Muridæ, but only the genera *Sciurus*, *Tamias* and *Spermophilus* can be noticed in the present connection.

Of the genus *Sciurus* there were recognized in 1884, 13 species and 7 subspecies; in 1894, 18 species and 16 subspecies.[1]

---

[1] The additions from the United States are :

Sciurus hudsonius mogollonensis Mearns.

Sciurus hudsonius vancouverensis Allen.

Sciurus hudsonius californicus Allen. (Perhaps = *Sciurus bottæ* Less., 1832.)

Sciurus carolinensis hypophæus Merriam.

Sciurus niger limitis (Baird). Revived by Allen, Bull. Am. Mus Nat. Hist., VI., p. 183, May, 1894.

Sciurus fossor nigripes Bryant.

Sciurus aberti concolor True, Proc. U. S. Nat. Mus., XVII., No. 999, (advance sheet), April 26, 1894. Larimer Co., Colorado.

Mexican and Central American additions are :

Sciurus niger melanonotus Thomas, P. Z. S., 1890, p. 73, pl. vi. Jalapa, Mexico.

Sciurus apache Allen, Bull. Am. Mus. Nat Hist., V., p. 29, March, 1893. Northern Chihuahua, Mexico.

Sciurus aberti durangi Thomas, Ann. & Mag. Nat. Hist., (6) XI., p. 50, Jan., 1893. Cuidad, Durango, Mexico.

Sciurus nayaritensis Allen, Bull. Am. Mus. Nat. Hist., III., p. vii., Feb., 1890 = *S. alstoni* Allen (nec Anderson). *ibid.*, II., p. 167, Oct., 1889. Zacatecas Mexico.

Sciurus cervicalis Allen, *ibid.*, p. 183. Sierra Nevada de Colima, Jalisco, Mexico.

In 1884 the genus *Tamias*, as recorded in Mr. True's List, contained 4 species and 4 additional subspecies; in 1894 (taking *Tamias* as limited in 1884), the recognized forms number 21 species and 12 subspecies.

The species in the 1884 list were (1) the common Eastern Chipmunk (*Tamias striatus*) which has since been separated into three subspecies; (2) Say's Chipmunk (*T. lateralis*), ranging from the Rocky Mountains westward, in suitable localities, to the Pacific coast. This now consists of a group of three species and several additional subspecies. (3) Harris's Chipmunk (*T. harrissi*), more southern in distribution than the last, now stands as a group of 4 species and 2 subspecies. (4) The Four-lined or Rocky Mountain Chipmunk (*T. quadrivittatus*), with 4 subspecies, now stands as a group of 23 species and 16 subspecies, under several subgeneric subdivisions.[1]

Sciurus leucops (Gray). Revived by Allen, *ibid.*, p. 166, and III., p. 182.

Sciurus nelsoni Merriam, Proc. Biol. Soc. Wash., VIII., p. 144, Dec., 1893. Morelos, Mexico.

[1] Following is the list as at present recognized (for full reference to additions made prior to October, 1892, see Bryant, as before):

Tamias striatus (Linn.).
Tamias striatus lysteri (Rich.).
Tamias striatus griseus Mearns.
Tamias quadrivittatus (Say).
Tamias quadrivittatus neglectus Allen.
Tamias quadrivittatus gracilis Allen.
Tamias quadrivittatus luteiventris Allen.
Tamias quadrivittatus affinis Allen.
Tamias quadrivittatus borealis Allen.
Tamias merriami Allen.
Tamias obscurus Allen (ex Townsend MS.).
Tamias senex Allen.
Tamias frater Allen.
Tamias minimus (Bach.).
Tamias minimus consobrinus Allen.
Tamias minimus pictus Allen.
Tamias amœnus Allen.
Tamias cinereicollis Allen.
Tamias umbrinus Allen.
Tamias speciosus Merriam.
Tamias townsendii Bachm.
Tamias townsendii hindsii (Gray).
Tamias quadrimaculatus (Gray).
Tamias macrorhabdotes Merriam.

Tamias panamintinus Merriam, Proc. Biol. Soc., VIII., p. 134, Dec., 1893.

Tamias callipeplus Merriam, *ibid.*, p. 136.

Tamias alpinus Merriam, *ibid.*, p. 137.

Tamias bulleri Allen, Bull. Am. Mus. Nat. Hist., III., p. 92 = *T. asiaticus bulleri* Allen, *ibid.*, II., p. 173.

Tamias dorsalis Baird.

Tamias lateralis (Say).

The genus *Spermophilus* was credited in the 1884 List with 11 species and 6 subspecies; it now contains 20 species and 13 subspecies.[1]

Tamias castanurus Merriam, N Am. Fauna, No. 4, p. 19, Oct., 1890.

Tamias cinerascens Merriam, *ibid*, p. 20.

Tamias chrysodeirus Merriam, *ibid.*, p. 19.

Tamias chrysodeirus brevicaudus (Merriam) = *Spermophilus chrysodeirus brevicaudus* Merriam, Proc. Biol. Soc. Wash., VIII., p. 134, Dec., 1893.

Tamias harrisii (Aud. & Bach.).

Tamias leucurus Merriam, N. Am. Fauna, No. 2, p. 20, Oct., 1889.

Tamias leucurus cinnamomeus Merriam, *ibid.*, No. 3, p. 51, Sept., 1890.

Tamias leucurus peninsulæ Allen, Bull. Am Mus. Nat. Hist., V., p 197, Aug., 1893.

Tamias nelsoni (Merriam) = *Spermophilus nelsoni* Merriam, Proc. Biol. Soc. Wash, VIII., p. 129, Dec., 1893.

[1] *Spermophilus grammurus* (Say).

Spermophilus grammurus douglasii (Rich.).

Spermophilus grammurus beecheyi (Rich).

Spermophilus grammurus atricapillus Bryant, Proc. Cal. Acad. Sci., (2), II., p. 26, June 20, 1889.

Spermophilus grammurus macrourus (Bennett). Revived by Allen, Bull. Am. Mus. Nat. Hist., II., 1889, p. 170. Zapotlan, Jalisco, Mexico.

Spermophilus grammurus fisheri (Merriam) = *Spermophilus beecheyi fisheri* Merriam, Proc. Biol. Soc. Wash, VIII, p. 133, Dec., 1893.

Spermophilus empetra (Pallas).

Spermophilus empetra kodiacensis Allen.

Spermophilus empetra columbianus (Ord) = *Spermophilus columbianus* Merriam, N. Am. Fauna, No. 5, p. 39, July, 1891.

Spermophilus richardsoni (Sabine).

Spermophilus townsendi Bachman.

Spermophilus elegans Kennicott.

Spermophilus armatus Kennicott.

Spermophilus beldingi Merriam.

Spermophilus perotensis Merriam, Proc. Biol. Soc. Wash., VIII., p. 131, Dec., 1893.

Spermophilus obsoletus Kennicott.

Spermophilus spilosoma Bennett.

Spermophilus spilosoma macrospilotus Merriam.

Spermophilus spilosoma major Merriam.

Spermophilus spilosoma pratensis Merriam.

Spermophilus spilosoma obsidianus Merriam.

Spermophilus spilosoma annectens Merriam, Proc. Biol. Soc. Wash., VIII, p. 132, Dec., 1893.

Spermophilus cryptospilotus Merriam.

Besides the addtions above indicated to the genera *Sciurus*, *Tamias*, and *Spermophilus*, each of these genera is considered as separable into several subgenera, most of which, however, were proposed prior to 1884. Thus Trouessart,[1] in 1880, proposed to divide the North American species of *Sciurus* into (1) *Neosciurus* (type, *S. carolinensis*), (2) *Parasciurus* (type, *S. niger*), and (3) *Tamiasciurus* (type, *S. hudsonius*). *Tamias* was divided by the same author into (1) *Tamias* (type, *T. striatus*), and (2) *Eutamias* (type, *T. asiaticus*), to which Dr. Merriam[2] has added *Ammospermophilus* (type, *T. lucurus*), and has transferred this group to the genus *Spermophilus*. To the four previously recognized subgenera of *Spermophilus* Dr. Merriam has added[2] also *Xerospermophilus* (type, *S. mohavensis*).

In respect to the other genera of the Sciuridæ, additions have been made also to both *Arctomys* and *Cynomys*, with radical changes in the nomenclature of some of the species of the latter.

There have likewise been numerous additions among the Bats and Shrews, and also among the Carnivores, particularly in the case of the Skunks of the genus *Spilogale*. This group stood in 1884 as a subgenus of *Mephitis*, with 1 species; it now ranks as a genus,[3] with 8 species and 2 additional subspecies.

Spermophilus canescens Merriam.

Spermophilus sonoriensis Ward, Am. Nat., Feb., 1891, p 158. Hermosillo, Sonora, Mexico.

Spermophilus tereticaudus Baird.

Spermophilus mohavensis Merriam.

Spermophilus neglectus Merriam.

Spermophilus mexicanus (Licht.).

Spermophilus tridecemlineatus (Mitchell).

Spermophilus tridecemlineatus pallidus Allen.

Spermophilus franklini (Sabine).

Spermophilus annulatus (Aud. & Bach.).

[1] Bull. Soc. d'Études Scientif. d'Angers, 1888, pp. 76-81 and 86.

[2] Proc. Biol. Soc. Wash., VII., p. 27, April, 1892.

[3] Cf. Merriam, N. Am. Fauna, No. 4, pp. 1-15, Oct., 1890.

As already said, the work here under notice is recognized as tentative. As additional material is acquired, and the various groups are taken up monographically, probably some of the recently described forms will be found to have been too hastily recognized in nomenclature, and that others now ranked as species will have to take the status of subspecies. On the other hand, it is quite certain that there still remain many new-forms to be described, even among some of the groups to which so many additions have already been made.

In many groups of our mammals little has been done as yet beyond the description of a few new forms. Especially is this the case with nearly all of the genera of the Carnivores, the Shrews and Moles. Some preliminary work has recently been done on the latter by Mr. True, who has in hand a monograph of the group.[1] The Shrews present a most difficult and interesting group, where at present it is almost impossible to properly allocate the names already given, or to form even a surmise as to the number of species and their relationships. The Bats have fortunately been brought into comparative order through the labors of Dr. Harrison Allen, whose recent monograph of the subject[2] is a valuable contribution to our knowledge of their anatomy, relationships, and proper nomenclature. Thus

---

[1] To the Moles have recently been added 1 new genus, 3 new species, and 1 new subspecies, as follows:

Parascalops, gen. nov., True, Proc. U. S Nat. Mus., XVII., 1894, (advance sheet, April 26, 1894). Type, *Scalops breweri* Bachman.

Scapanus dilatus True, *ibid.* Fort Klamath, Oregon.

Scapanus anthonyi Allen, Bull Am. Mus. Nat. Hist., V., p. 200, Aug., 1893. San Padro Martir, Lower California.

Scalops texanus Allen, *ibid.*, III., 1891, p. 221, and V., p. 200. Texas.

Scalops aquaticus australis Chapman, *ibid.*, V., p. 339, Dec., 1893. Florida.

Scalops parvus Rhoads, Proc. Acad. Nat. Sci. Phila., 1894, p 157. Tarpon Springs, Florida.

[2] A Monograph of the Bats of North America. By Harrison Allen, M.D.,= Bulletin of the U. S. Nat. Mus., No. 43. 8vo., pp. i-ix., 1-198, pll. i-xxxviii., 1893 (March, 1894).

although so much has been done during the last five years, the published results form but a beginning toward a proper knowledge of the mammalian fauna of North America.

Thus far in this paper reference has been made mainly to the discovery of new forms as evidence of the recent advances in our knowledge of North American mammals, and little in respect to increase in knowledge along other lines, or of particular geographical areas. Most of the new forms have of course come from the little known parts of the West, and from countries south of the United States, where collecting had been exceedingly superficial and sporadic, the material previously gathered being such as chance threw in the way of comparatively untrained collectors. Less than a decade ago the more unsettled parts of the West, and especially the extensive arid regions, were practically virgin ground, so far as the systematic trapping of small mammals was concerned. Now there are few areas of any great extent that have not been visited by trained collectors with the best modern devises for capturing small mammals; while considerable portions of the country have been methodically explored, with a view not merely to securing large series of specimens from certain selected localities for comparative study, but especially for the purpose of determining the geographical distribution of particular forms or groups of forms, as regards both their vertical and horizontal range. In the western and southwestern portions of the United States many thousands of square miles have been systematically surveyed biologically under the direction of Dr. C. Hart Merriam,[1] Chief of the Division of Ornithology and Mammalogy, United States Department of Agriculture, who for the last five years has been able to keep in the field a large corps of skilled collectors. Under the same auspices much collecting has been done also in the East and South,

---

[1] See N. Am. Fauna, Nos. 3, 4, 5 and 7, for reports on his explorations in Arizona, Idaho, and southwestern California and contiguous regions.

and in northern and central Mexico. Thousands upon thousands of specimens have also been secured by independent collectors, which have found their way either into various private collections or into public museums.[1]

This material has of course all been subjected to preliminary examination, and most of the novelties made known, but for the important results that must follow from its careful elaboration we must still wait. Various faunal papers—far too numerous for enumeration in the present connection—have also been published. New workers are also every year entering the field; so that the outlook for the further progress of North American mammalogy is full of encouragement.

[1] For example, during the last four or five years some 6,000 specimens of North American Mammals have been acquired by the American Museum of Natural History in New York City.

AUTOBIOGRAPHICAL NOTES

AND

A BIBLIOGRAPHY

OF THE

SCIENTIFIC PUBLICATIONS

J. A. Allen

June, 1885

AUTOBIOGRAPHICAL NOTES

AND

A BIBLIOGRAPHY

OF THE

SCIENTIFIC PUBLICATIONS

OF

JOEL ASAPH ALLEN

PUBLISHED BY THE MUSEUM

NEW YORK

1916

Dedicated to
Henry Fairfield Osborn,
President of the American Museum of Natural History,
and to the Memory of my revered Teacher,
Louis Agassiz.

FOREWORD.

The life and writings of Joel Asaph Allen have exerted so great an influence on the progress of ornithology and mammalogy in America that all who have the interests of these branches of science at heart, both in this country and abroad, will welcome this biographical and bibliographical volume. It is issued as an expression of the appreciation of Doctor Allen's life work by the Trustees of The American Museum of Natural History and his devoted colleagues on its Scientific Staff. The biographical sketch was especially desired in connection with the Bibliography because it sets forth so clearly the broad groundwork of travel, of field observation and of field record which has established a model for all modern American work on the birds and the mammals. At this time we are sure that naturalists in all parts of the world will unite with Dr. Allen's colleagues in felicitating him on the great work which he has accomplished and in wishing him many more years of strength and activity.

HENRY FAIRFIELD OSBORN.

November 9, 1916.

PREFACE.

The present publication was prepared by request of Henry Fairfield Osborn, President of the American Museum of Natural History, to whom the author is indebted for courtesies extending over many years of the most pleasant official relationship.

It is a pleasure, realized by few, to be able to supervise the publication of a list of one's own technical papers, extending over more than half a century, and the opportunity for it was duly welcomed.

The biographical sketch, preceding the list of papers, however, was written with great reluctance and with many misgivings, as being, in the author's opinion, inopportune in the present connection. If such notes are worthy of record, it seems more fitting that their publication should be posthumous. Their present appearance is due to solicitations difficult to disregard.

J. A. Allen.

August 20, 1916.

CONTENTS.

AUTOBIOGRAPHICAL NOTES.

AUTOBIOGRAPHICAL NOTES.

EARLY SURROUNDINGS AND TRAINING (1838–1858).

I was born in Springfield, Massachusetts, July 19, 1838, the eldest son of Joel and Harriet (Trumbull) Allen, both of early New England stock. My father was a descendant, in the seventh generation, of Samuel Allen, who settled in Windsor, Connecticut, in 1640. The date of his birth is not positively known, nor has a definite record been found of when he emigrated from England. He is supposed to have come over with the Dorchester Company in 1630, in the ship 'Mary and John.'[1] The Windsor land records show that he received a grant of land from the plantation at Windsor, Connecticut, January 27, 1640. He died at Windsor in 1648. The present public library building at Windsor marks the site of the Samuel Allen homestead. Joel Allen, my father, was born in Springfield, Massachusetts, December 11, 1810.

On the maternal side the descent is from John Trumbull, great-grandfather of Governor Jonathan Trumbull (said to have been the original "Brother Jonathan" and familiar friend of Washington), who was born in Newcastle-on-Tyne, England, and settled in Roxbury, Massachusetts, in 1639.

My immediate progenitors were farmers. My father, however, learned the carpenter's trade and was a house-builder in his earlier days, but later bought a farm on which he spent the greater part of his life. He was respected by his neighbors as a man of excellent judgment and sterling integrity, whose advice was often solicited in the affairs of the neighborhood. In politics he was a Free-soiler, and later a Republican. He died January 9, 1886, at the age of 75 years and one month, survived by my mother, three sons and a daughter. My mother died June 2, 1892, at the age of 80 years and six months. My mother taught school for several years before her marriage. She was a true helpmeet, practical, conscientious, and in every way lovable and inspiring. My father had little appreciation for my natural history tastes, but was kind and generous, offering to share his farm with me if I would remain with him on the old homestead. My mother, on the other hand, was much in sympathy with my yearnings, and often used her influence in my favor.

[1] The first leaves of the Dorchester records are stated to be missing.

Our family numbered five children, one girl and four boys, one of whom (a boy, Edwin) died in infancy. My sister, Harriet Emma, taught school for a number of years and later married a farmer. She died suddenly of pneumonia, contracted in attending a sick brother, April 2, 1900, in her 54th year. My two brothers, Edgar, a moulder by trade, and Irving, a farmer, are still (August, 1916), living, the latter at the old homestead in Springfield.

My early training was rigidly puritanical. My parents were both members of the Congregational church, and strict in their religious observances. Family prayers invariably followed breakfast, and also closed the routine of Sunday, all the religious requirements of the day being strictly observed.

My earliest recollections are naturally associated with the surroundings of my birthplace on the old farm, situated on a hill about a mile and a half east of the then thickly settled part of Springfield, known as the Watershops, where the United States Government has for more than a century carried on the manufacture of fire-arms. The family home was a large two-story square-roofed house, at that time innocent of paint and unshaded by trees. One of the pleasantest remembrances of my younger days is of helping my father plant the row of maples and elms which long since became the prominent feature of the road frontage of the farm, and in recovering and painting the house. We were not crowded by neighbors, the nearest residence on the west being half a mile away, and there was only one house within half a mile to the eastward. Subsequently others much nearer were built in both directions, the lonely country road has become Allen Street, and a trolley car line has been projected to connect the rapidly extending suburbs with the business portion of the city.

Dandelions and daisies and other wild flowers were early attractions, the profuse gathering of which at an early age led my elders, and particularly my mother, to predict that when the toddling youngster grew up he would favor the profession of medicine, and I was often facetiously dubbed "Dr. Sykes," in allusion to our then family physician, an herb-doctor of local reputation. In due time I was assigned a share in the household chores, and trained to perform the allotted tasks with promptness and care.

The nearest schoolhouse was a mile distant, of the conventional red type, situated as usual on the crest of a hill. In summer the school was taught by a schoolmistress, while the winter session was conducted by a schoolmaster selected for his ability to keep the larger boys in order as well as to teach the "three Rs." Some years later a schoolmistress was employed for both the winter and summer sessions. In these days the services of boys of even six and seven years were considered too valuable for farm-work to be sacrificed in summer for school purposes, so that to them only the winter session of the school year was available.

Despite hard work and long hours, the farm proved attractive and satisfying for a time, but at about the age of fourteen the love inspired by this free contact with natural surroundings developed a desire to know more of the animal and plant life, the soil and the rocks, and the ever changing phenomena of sky and air, than could be gained merely by association.

At the age of thirteen, after much pleading on my part, to my great delight, my father presented me with a gun. At first it merely afforded the pleasure all boys experience in being able to shoot something, either as game or on the pretext that certain birds and animals are destructive to crops, and that it is desirable to reduce their numbers. But very soon the destructive instinct gave place to a desire to possess specimens for study, particularly of birds, which I found were so numerous in kinds that comparatively few of them were known by name to any of the people, either of town or country, whom I met. Warblers, vireos, kinglets, sparrows and many other kinds of birds were shot, measured, weighed, described and given provisional names in my notebooks, so that I might again recognize them when met with, long before I knew that books had been written about them and that they all had names, Latin as well as English. I even made attempts to draw and color them, but entire lack of instruction in the work led only to failure and disappointment. A little later, however, I made the acquaintance of Bradford Horsford, a teacher of drawing, who was also an amateur ornithologist and taxidermist, with a good knowledge of all the commoner birds. From him I borrowed a copy of the Brewer edition of Wilson's 'American Ornithology,' which, to my unspeakable delight, he later sold to me; Nuttall's and Audubon's works on North American birds were also found in the public library of Springfield, and a new world was opened to me!

A little later I made the acquaintance of a man of broader education than I had ever before met, who taught our district school for several winter terms, and became a resident of the neighborhood. As he was a nature-lover himself he could appreciate my aspirations, and most generously presented me with a copy of Blythe's 'Cuvier's Animal Kingdom,' a work of which I previously had never heard. Thus equipped, and with the resources of a public library now at my command, acquaintance with not only the local birds, mammals, reptiles and fishes, but with many of the insects, became a delightful experience. Interest in farm work as an occupation as rapidly declined, but a filial desire to share fully in the family burdens led to no neglect of duties but often to excessive effort in manual labor to demonstrate an interest otherwise unfelt.

In winter I usually attended the district school, but was so far in advance of all the other pupils that I was in a class by myself in most of the branches taught, and received little aid from the teacher. A large part of my last term in the district school was devoted to solving the "miscellaneous examples" at the end of Greenleaf's higher arithmetic, then commonly in use in the schools of the region, and in such other school arithmetics as I could borrow from friends who had long since finished their school days. One winter was spent entirely at home (I was then probably in my fifteenth year), in the study of such natural history books as I could command, and of a borrowed copy of Webster's unabridged dictionary, from which I copied a large part of the natural history definitions.

At Wilbraham Academy and Home Studies (1858–1862).

Then followed several winter terms at the Wilbraham Academy where I selected my own studies, which included at first physiology, natural philosophy, astronomy, and chemistry, in addition to English grammar and arithmetic; later rhetoric, algebra, Latin, French and German were substituted for the natural sciences. Humboldt's 'Cosmos' (in English), and works of similar character constituted my favorite reading outside of my school work.

My summers were still spent on the farm, where, equipped with elementary works on meteorology, geology and mineralogy, I had ample food for thought. Lyell's 'Principles of Geology,' Dana's 'Mineralogy,' the 'Proceedings' of the American Association for the Advancement of Science (the full series as far as then published), and various State geological and natural history reports were studied with avidity.

During this period my ever present ambition was to write a history of the "Birds of New England" that should be as complete and exhaustive as possible, and based on original observation, including the necessary explorations in northern New England where so many of the migratory species were supposed to pass the breeding season. Next to this I looked upon editorial work as an enviable goal. Yet at the time these aspirations began to develop composition was a slow and difficult task, and to acquire facility in writing I forced myself to keep a daily journal, in which I recorded not only the current weather conditions in detail but every incident of my daily experiences that seemed to offer a subject for comment.

In Professor Oliver Marcy, teacher of the natural sciences at Wilbraham Academy and later professor and finally dean of the faculty at Northwestern University, I found a most sympathetic friend and counsellor. It was he who first secured my introduction to the public through the 'New England Farmer.' It was my custom at the Academy to offer, when composition day came round, some of the results of my natural history observations as a composition. One of my first offerings of this sort was a summary of my weather journal for the previous seasonal period of three months. When the time came for the return of compositions mine was withheld with the request from Professor Marcy, who had charge of my division, that he might retain it for a short time. A few days later he handed me a copy of the 'New England Farmer' containing my article, printed in full and without change, with a very complimentary introduction by the editor. It was to me an exceedingly great surprise, and sufficient explanation of why my composition had not been returned to me at the usual time. I was at this time about eighteen or nineteen years of age.

With this cordial welcome from the editor of the 'Farmer,' I mustered courage to offer, soon after, other contributions of a similar character, my weather reports thereafter appearing at regular intervals till circumstances prevented the continuation of the observations. These articles were followed (in 1859) by others on quite different subjects, including a series of twenty-five on New England birds; also a paper on the moon's alleged influence upon the weather, in reply to previous articles on the subject by other contributors. My article was based on meteorological observations kept at the United States Armory at Springfield, Mass., for a long period of years, latterly by Mr. Joseph Weatherhead, who kindly placed them at my disposal for study. It is needless to say that I found no evidence of lunar influence on the character of the weather, and no basis for sundry other assumptions regarding the moon's influence on mundane life, or for popular beliefs respecting the proper time in the moon for weaning pigs, colts and calves or planting crops.

During these years I spent much time studying botany, adding a collection of dried plants to my other natural history gatherings, nearly all of which I was able to identify without other aid than Gray's well-known 'Manual of the Botany of the Northern United States,' my copy of which eventually became well annotated with my field notes. At this time I had not the acquaintance of any one who had any knowledge of the scientific names of plants, or who, so far as I knew, cared to know them. I still recall the exquisite pleasure it gave me to discover the name of some wild plant I had long admired in total ignorance of its affinities or name.

During the years 1859–1861, I collected and mounted (as attested by my catalogue, still extant) some 300 birds, representing nearly 100 species, and also such native mammals as I could find near my home, and I preserved in jars of alcohol specimens of all the reptiles, amphibians and fishes;

such mollusks as were available were also gathered and several hundred insects. Best of all, I knew the technical names of nearly all except the insects, of which, however, I knew many. The local minerals and rocks found place on the shelves of my little museum, for which a small room was kindly provided by my parents, and which I equipped with shelves and a flat table case for insects. In addition there were rows of bottles containing the products of my boyish experiments with such cheap chemicals as I could afford to purchase at the neighboring drug stores, each duly labeled with its proper chemical formula. The whole was amateurish in the extreme, and represented merely a superficial acquaintance with a wide range of subjects, but enough to add immensely to the pleasure of living, giving, as it did, the sense of being in touch with the plant and animal life and the geological features of my immediate environment. My notebooks contained pages of descriptions of unusual atmospheric phenomena, from the prismatic tints of fleecy clouds floating past the midday sun, haloes, unusual storm conditions, auroral displays, and the August and November shooting-star periods, to the varied forms of the snow crystals of a winter storm — things for the most part unobserved by my friends and neighbors, and which hence gave them no added joy to living.

It is needless to say that my interest in every day practical affairs was limited to a conscientious and cheerful discharge of the obligations natural to my position as a helper to my father in the routine of farm work. Every spare moment of the day when in the house was spent in my room poring over books or specimens or jotting down things seen out of doors in the corn or hay field. These constant disappearances when off duty were naturally an annoyance to my father, who could not appreciate my absorption in such unpractical affairs. To the oft-made inquiry of my Father, "Where's Asaph?" was Mother's gentle response, "upstairs," and the contemptuous paternal rejoinder: "Upstairs; he's always 'upstairs.'" Although unappreciative of his son's "foolish notions," he was not harsh or unkind, as an agreement, lasting for several seasons, granting one day a week for the prosecution of my hobbies is ample evidence. For these foibles my mother had always a degree of sympathy, which increased as years passed to active influence in their behalf.

To demonstrate my hearty interest in forwarding the farm work, I often as I afterwards found, exerted myself beyond my proper physical endurance, which with the absorption in my natural history work told heavily on my health. It was often necessary in the busy season for my father to employ day laborers and it was always my ambition to "lead the field," which I was always able to do except in the heavier work, even when a young boy just entering the teens. My evening task, before retiring, was to write in my journal the notes of the day and to change the dryers in my extemporized botanical press, consisting of several pieces of thick board, cut the proper length, a lot of old newspapers, and a heavy, smoothly waterworn stone for the top of the pile to afford the requisite pressure. Many, many a time this bedtime task found me almost too exhausted by the day's labor to accomplish. These long periods of overwork undoubtedly laid the foundation for much of the semi-invalidism of many later years.

There was, of course, respite in seasons when farm work was not pressing, when much spare time could be found for collecting trips and the preparation of specimens. My excursions were generally limited to the radius of a few miles from the farm, adjoining which, in my early days, were a few small patches of virgin forest. But the ax had already begun to thin their ranks, and before my collecting days had fairly begun, these grand old remnants of former forest conditions were swept away and the land devoted to farm purposes or allowed to grow up in a shrubby second-growth.

My excursions extended to the famous new-red sandstone quarries in East Longmeadow, my first visit to which is still vividly remembered. Other trips were made to Mount Tom, with boon companions — sportsmen who had become amateur taxidermists and enthusiastic general collectors of birds, beasts, fishes, reptiles and insects, for the Springfield Museum of Natural History, under the auspices of the City Library Association.[1] One memorable trip of several weeks duration was made to visit relatives in Windsor County, Vermont. The physical character of the country was in strong contrast with that of my home surroundings, and the fauna and the flora were perceptibly different. Most interesting of all the many excursions in this new field was the ascent of Mount Ascutney, and many samples of the rocks of the general region visited were taken home for my cabinet. A second trip to this region, made in August of a later year, included a visit to the gold washings at Springfield (Vermont), where I spent a night in camp and gained much valued information respecting the methods there employed in securing the precious gold dust and diminutive nuggets that barely paid for the time and labor expended.

My several terms at Wilbraham Academy prior to the winter of 1861–62 had been a serious draft upon my father's limited resources, with interest still to be met on farm mortgages and a considerable family to support, and, eager as I was for such advantages, I could not consent to accept his

[1] These men were Caleb W. Bennett, a house and sign painter, who will receive later mention in this narrative; Solomon Stebbins, a paper hanger; and Charles Emery, a draughtsman at the U. S. Armory at Springfield. Their leisure time was holidays and Sundays, which they intelligently and enthusiastically devoted to natural history collecting, and their evenings to the study and preparation of specimens for exhibition in the Springfield Museum.

further aid. I had my little museum, so dear to me that it had seemed impossible to part with it under any circumstances. In this emergency it occurred to me to offer to sell the collection to Wilbraham Academy, and use the proceeds for my school expenses. To my surprise as well as delight, the offer of sale was accepted, and to this extent my way was clear.

Special Student under Louis Agassiz (1862–1871).

On again entering the Academy I found a congenial spirit, whose tastes and aspirations were similar to my own, but under better guidance, as my new friend [1] was a nephew of my loved teacher, Professor Marcy. He was planning to enter the Lawrence Scientific School at Cambridge, to become a pupil of the great Agassiz. Why should not I do the same? The balance still due me for my collection, if paid to me in cash would enable me to establish myself at Cambridge, on our proposed plan of hiring an inexpensive room and boarding ourselves. The necessary preliminaries having been arranged, we arrived in Cambridge early in February, 1862. I well remember plodding through the knee-deep snow of Divinity Avenue to reach the Agassiz Museum, in exceedingly inclement weather, and our cordial greeting by the great scientist. Plans for laboratory work were at once arranged, and also for attending certain of the courses of lectures at the Lawrence Scientific School that were to form a part of our curriculum. Besides those of Agassiz himself we were to attend the course by Jeffries Wyman on comparative anatomy, a course on physics by Joseph Lovering, a course on chemistry by Josiah P. Cook, and the course on botany by Asa Gray, all eminent specialists of world-wide renown.

I had naturally chosen as my specialty the study of birds, and was not a little disappointed at having assigned to me the same task as that set for Mr. Niles. We were both given collections of corals of several genera and requested to find out their methods of growth and laws of development. Not a hint was given us as to what details we were to look for, and no books of reference were suggested. The first lesson, we were told, was to learn to observe, to use our eyes. Equipped with hand lenses, we proceeded to our task. After a few hours of application we were asked "Well, what have you seen?" and the same query was daily repeated. We reported what we thought we had discovered, and if we had seen aright we were encouraged with a few words of approval; if we were mistaken

---

[1] William Harmon Niles, of Worthington, Mass., later for many years professor of Geology and Physical Geography at the Boston Institute of Technology.

the reply was "You are wrong; you must look again; you must learn to see."

We were not left, however, without friendly words of intercourse about other matters. The great master was engaged with his own work in the same room, which was in the southeast corner of the first floor of the first section of the Museum, of which this was at that time the only part constructed. Agassiz was then engaged in studying the great DeKonink and other large collections of fossil mollusks, at that time recently received and not fully unpacked. From time to time he gave expression to his delight over some new discovery, and to the trays containing the specimens and their labels were frequently added new labels of his own, on which the name of a new genus or a new species was written, as the case might be. His workroom, or laboratory, was almost daily the resort of eminent scientists: Jules Marcou was almost a daily visitor, when the conversation was always in French; Jeffries Wyman was a frequent caller, as was also Benjamin Pierce, the eminent mathematician; on the days when Agassiz gave his lectures many distinguished notables from Boston were present, among whom was frequently Oliver Wendell Holmes.

Besides Agassiz and his French artist Bourkhart and ourselves, the room was occupied by the osteological preparator Guggenheim, with whom conversation was always in German. For many years this old German worked at preparing fish skeletons for the Museum, whose eccentricities of dress and habits will be long remembered by the Agassiz students of that early day. He was an inveterate smoker and snuff-taker, wore a wig, and made his midday coffee amid the débris of his work.

Time at first passed slowly with the two new students. They had been given a difficult problem, and weeks passed with little progress in its solution. It was a trial of persistence, of character, as well as of keenness of observation, a test to show whether the interest of the student was real or imagined. While our efforts were at times disheartening in their results, perseverance and continued application won the day. Gradually we saw the light and were able to show, each in his particular problem, the successive stages of growth in the young coral, and discovered the law governing the multiplication of plates in the successive stages of growth. We were then given other subjects to study, Mr. Niles taking up crinoids, while I was given a large miscellaneous collection of fossil gastropods to assort and arrange according to their relationships. We were both happy, having been assured by our great teacher that we were making good progress in "learning to observe."

Soon, however, I contracted a bad case of measles, which ended in serious impairment of my eyes, from which I suffered for years after. For some six months it was impossible for me to read ordinary print, and for several years I could not use my eyes for evening study.

I returned to my Springfield home early in May, equipped with tanks of alcohol (the well known 'Agassiz tanks') for the preservation of specimens. Although in wretched health, suffering from chronic indigestion as well as from weak eyes, I collected over four hundred specimens of animals, chiefly vertebrates, and largely birds, besides taking a share in the farm work.

On my return to the Museum in the autumn I ventured to express my strong desire to study birds. The condition of my eyes, however, was a great handicap. The mammals and birds were in charge of A. E. Verrill, then a student at the Museum (later professor of zoölogy for many years at Yale University), who was requested to assign me material for study. The collection of birds consisted at that time of several hundred skins (possibly a thousand or two, all North American), and several thousand in alcohol, nearly all uncatalogued and the alcoholics unidentified. I began with the identification and cataloguing of the alcoholics, which occupied most of my time for many months, and really extended over years, as new collections came in. I soon acquired facility in recognizing birds in obscure plumages after long storage in alcohol, their feathers wet and colors disguised. It was, therefore, good training in the art of "learning to see." My weak eyes, however, prevented study at night, and even the little writing required in entering specimens in the catalogue often entailed much suffering. I attended numerous courses of lectures, but my progress in acquiring further knowledge of languages, so essential to a scientific student, was for a considerable time at a standstill.

At this time, by direction of Professor Agassiz, I took up the study of Pterylography, with the aid of C. L. Nitzsch's classic 'System der Pterylographie' (Halle, 1840), and made many preparations in illustration of the subject from both fresh and alcoholic material, with a view to the publication of an extended illustrated monograph of the Pterylography of the Owls. Illness, however, retarded the work, and other interests and duties intervening, it was never completed. But the investigation proved of great use to me in subsequent work in ornithology.[1]

Thus the time passed during the next two years and a half, my vacations being spent at home collecting specimens for the Agassiz Museum. For

[1] Professor Agassiz thus refers to the subject in the Annual Report of the Museum of Comparative Zoölogy for the year 1863, p. 17: "Ill health has also taken away Mr. Allen from his field of labors. I regret it the more since he has made excellent progress in Ornithology and promised to become a valuable assistant in the arrangement of the specimens of birds. He has left unfinished a very interesting investigation upon the structure and arrangement of the feathers of birds."

assistance rendered in cataloguing and in other routine work at the Museum, a monthly allowance was received sufficient for my necessary living expenses, together with a furnished room in the Museum dormitory, known as 'Zoölogical Hall,' at that time the home of the student assistants of the Museum.

During the summer of 1864 I made a trip to Sodus, Wayne County, N. Y., where several weeks were spent with relatives. The region, bordering Lake Ontario, was originally heavily forested, and some of the virgin forest still remained. The type of country was new to me and intensely interesting. A large collection of the land and fresh-water mollusks was gathered and eagerly studied, and a few insects, mammals and birds were collected, including a few species of both mammals and birds I had not before seen in life.

With the Agassiz Expedition in Brazil (1865).

During the winter of 1864–65, Professor Agassiz's usual vigorous health began to give way under the strain of years of incessant work and care, so that a change of scene and some degree of relaxation became imperative. A generous Boston friend of means (Nathaniel Thayer) suggested his making a journey to Brazil, and offered to provide the means not only for his own expenses but also for a small corps of assistants. This kind offer he accepted. It was then early in March, 1865, and the expedition was to start as soon as the necessary preparations could be made. On March 12 he asked me if I would like to join the expedition. Naturally I accepted the unexpected invitation, and was instructed to join three other Museum assistants who had also been invited to accompany the expedition, and meet him, with them, at a designated place in Boston to arrange for our passports. Thus suddenly opportunity opened to us for exploration in a distant land. We left Boston for New York on the evening of March 26, and sailed for Rio de Janeiro on April 2, on the steamship 'Colorado,' of the Pacific Mail Steamship Company. The 'Colorado' was a new vessel, of 2000 tons, then regarded as remarkably large and luxuriously furnished. This was her maiden voyage, and Professor Agassiz and his party of sixteen persons were the only passengers.

As we passed down the coast of Virginia, on April 3, heavy masses of smoke were seen over the western horizon, which the Captain of the 'Colorado' believed to indicate a distant battlefield. But it was not till May 16 that news of the fall of Richmond reached us in Brazil and confirmed the Captain's surmise that the smoke we saw on April 3 was from a great battle — the closing contest of the civil war.

Early in the voyage Professor Agassiz began a series of lectures, which was continued almost daily till we reached Rio de Janeiro, dealing in turn with the Gulf Stream, the Sargosso Sea, embryology, and the classification of the animal kingdom. We arrived at Rio April 22, after an exceptionally pleasant voyage.

Save for a few excursions, by rail or boat, to near points, the whole party remained at Rio for several weeks, but work was organized immediately on our arrival and collections of specimens were gathered daily in the surrounding suburbs. At the same time preparations were under way for the various projected expeditions into the interior that were to occupy us the rest of the year. Besides the six trained assistants from the Museum, as many other young men were attached to the expedition as volunteers; all were duly assigned positions on the different exploring parties.[1]

Following several short excursions from Rio, including a two weeks' trip to Mendez and the Barra do Parahyba, on the Dom Pedro II railroad, some 60 miles from Rio, I was assigned to one of the parties that was to visit the northern Provinces of Brazil. We left Rio de Janeiro June 9 for Petropolis, a beautiful mountain suburb of Rio, where we spent the night.[2] The following morning we resumed our journey by a mule-drawn stage for Juiz da Fora, over a perfect road, making the 100 miles in ten hours, with relays of mules every ten miles. The road traversed a highly picturesque section of country, parts of it clothed with verdure of almost Amazonian luxuriance, while flocks of toucans and large parrots frequently attracted attention.

At this point we were to outfit for our long journey into the wilderness; but instead of finding our equipment of mules and camarados awaiting us, as we had expected (it having been ordered in advance), our animals were still at distant ranches and nothing was in readiness. It required five days to assemble the mules and their drivers. It was June 15 when our little cavalcade of 14 mules, 4 horses, and 4 camarados left Juiz da Fora for the headwaters of the Rio das Velhas, some 300 miles distant.

We reached Barbacena on the third day, where Mr. Ward with his share

---

[1] The personnel of the Expedition, in addition to Professor Agassiz, was as follows: Assistants from the Museum staff: John G. Anthony, conchologist; Charles Frederick Hartt, geologist; Orestes H. St. John, geologist; George Sceva, osteologist; J. Burkhardt, artist. Volunteers: William James, later the eminent psychologist of Harvard University; Edward Copeland, Newton Dexter, James Hunnewell, S. V. R. Thayer, and Thomas W. Ward. D. Bourget, a French naturalist, joined the Expedition at Rio de Janeiro.

[2] Our party consisted of Orestes H. St. John (geologist and chief), George Sceva, Thomas W. Ward (mineralogist), and myself. We were to travel together, however, for only a short distance, as Mr. Ward was to leave the party at Barbacena and Mr. Sceva at Lagoa Santa.

of the cavalcade left us for Diamantina. We were delayed here several days while Mr. Sceva made a fruitless side-trip to some limestone caves at San José to prospect for fossil bones. We resumed the journey on the 24th, reaching Morro Velho July 7 and Lagoa Santa July 13, much of the way by obscure and difficult roads. Here we spent several days, collecting specimens and visiting the famous bone caverns[1] made classic by the Danish naturalist, Dr. P. W. Lund, who was still resident at Lagoa Santa and whom we had the pleasure of meeting. Six days later we reached Trahiras, on the Rio das Velhas, where we disposed of our animals and outfitted with a large canoe and boatmen for the voyage down the river. This involved another delay of eight days although our river equipment had been ordered in advance and was supposed to be awaiting our arrival. But the delay had its advantages, for it enabled us to repack the collections we had made *en route* for shipment to the United States and gave opportunity for making a large collection of fishes, besides obtaining many birds and mammals. We received here also our first letters from home, letters that had taken three months to reach us.

We were able to secure a boat well adapted to our needs, a canoe 50 feet in length and four in width, dug out of a single tree. This we rigged with outriders and covered with canvas to protect us from the sun by day and afford shelter at night. We named her the 'Gequitibá,' an Indian name borne not only by the tree from which it was made, but by the fazenda and village from which we outfitted. We also supplied ourselves with a smaller canoe to serve as a tender. Our crew consisted of four boatmen, a cook, and an English boy we had previously secured at Morro Velha, with St. John and myself, making eight persons in all.

Our journey from Juiz da Fora (distant about 120 miles from Rio) by pack train to Gequitibá (or Trahiras), on the upper Rio das Velhas, was made in twenty days of actual travel, the distance by our circuitous route being about 300 miles, but the time occupied in outfitting for the two journeys, and in various delays *en route* was much more, so that the time between our arrival at Juiz da Fora and our departure down the river was about 50 days (June 9–July 31).

On July 31 we left Trahiras to descend the river, reaching its junction (Barro do Rio das Velhas) with the Rio São Francisco August 28. Our progress was greatly delayed by the low stage of water rendering the numerous rapids difficult, and in some instances dangerous, to pass. Nearly every day we lost several hours in exploring for channels that would permit the

---

[1] Mr. Sceva was left here to make further exploration, but his search for fossils was poorly rewarded.

passage of our boat, and had several narrow escapes from shipwreck. These delays gave opportunity to explore the adjacent country and to make collections, particularly of birds, frogs, fishes and mollusks. The river is very tortuous, particularly in its upper course, where it occasionally returns on itself in great loops. The distance by land between Barro do Rio do Gequitibá, our point of departure, and Barro do Rio das Velhas is said to be 55 leagues or about 220 miles, and about 350 miles (estimated) by the river. Our journey down the river, we were told, was one of the quickest and most successful ever made at such a low stage of water, it being accomplished in twenty-one days, without having to unload the boat or losing any baggage. We were fortunate in our choice of boat, as with an ajojo (two boats lashed together) we should certainly have been wrecked in passing rapids, and with a barca (house boat) have been repeatedly forced to unload our freight. We had also an excellent crew of boatmen, who often took to the water and guided the boat by hand through dangerous passes.

We remained three days at the Barra, where we were obliged to get a relay of boatmen, and left September 1 for Januaria, arriving there September 9. The Rio São Francisco is about twice the size of the Rio das Velhas and unobstructed by rapids; we were hence able to make good progress, though at times delayed by strong head winds. While at the Barra we made very satisfactory collections, but spent much time in unsuccessful efforts to obtain barrels and spirits for our specimens, which we had intended to send from this point to Rio Janeiro for shipment to Cambridge. Consequently we were obliged to take them with us, in the hope of being able to send them to the coast for shipment home via Bahia or Pernambuco.

During the three months since we left Rio de Janeiro it had become evident that my physical condition rendered it impracticable for me to attempt to carry out the journey as originally planned, namely, to descend the Rio São Francisco to the Barra do Rio Grande and then cross the divide to the headwaters of the Rio Parahyba do Norte and thence reach the coast in Ceará. After long consultation we decided to divide our party, Mr. St. John to attempt to complete the journey as originally planned, while I was to remain a short time at Januaria and then continue down the São Francisco to a point from which it would prove practicable to make an overland journey to Bahia.

In accordance with this plan Mr. St. John left Januaria in our small canoe, with one man, September 14, with the alternative in mind that should it prove impracticable to make the Ceará portion of the journey he was to join me again at Barra do Rio Negro for the overland trip to Bahia. Pluck and perseverance enabled him to carry out the original schedule of the expedition, though short of funds and finally seriously ill with fever.

I remained at Januaria till the 19th (ten days in all), having in the meantime sold our big canoe and hired a barca for the descent of the São Francisco, discharged our old boatmen and engaged others, repacked our collections for the journey to Bahia, and made several excursions into the neighboring country, which resulted in important accessions to our natural history collection and many pages to my field notes.

The journey down the Rio São Francisco was made comfortably, save for excessive heat (maximum daily temperature 80–98° Fahr.), and without untoward incident beyond frequent delays by heavy head winds and by illness of the boatmen. I decided to terminate the river journey at the village of Chique-Chique, which point we reached in the afternoon of October 8, the descent from Januaria to this point having occupied 18 days. Chique-Chique is 530 miles below Januaria and about 580 miles from Bahia, by the overland route.

The region about Chique-Chique is excessively arid and the heat intense. It was near the beginning of the rainy season, which had already set in further in the interior, as shown by the slight daily rise of the river during my descent from Januaria. In fact, nearly the whole journey from Lagoa Santa to this point was made just in advance of the rainy season, occasional showers at different points indicating its near approach.

It was necessary for me to remain several weeks at Chique-Chique for an opportunity to join a pack-train bound for the coast, there being considerable traffic between this point and Bahia. The caravans, however, depart infrequently and at uncertain intervals. The route was by way of Jacobina, Arraial do Riacho, Jacuhipe, Villa da Feira da Sta. Anna, and Cachoeira.

The country between Chique-Chique and the coast at Bahia presents three natural regions, which are plateaus, differing widely from each other in their geological features. The first is excessively arid and extends from the Rio São Francisco to Jacobina, a distance of rather more than 200 miles. It is a vast limestone plain, practically without inhabitants, so that it was necessary to take with us not only food for the animals as well as the men, but also drinking water, which was carried in large leather water bags, as in parts of the Far East. This plain rises gradually toward the east, culminating in the Taboleira de Jacobina. To the eastward the descent to the Jacobina valley is abrupt, through a narrow precipitous defile known as the Tombador (the tumble-down). Almost vertical walls of rock, nearly a thousand feet in height, surround the head of the valley.

The second or middle plateau likewise has a breadth of some 200 miles, extending from Jacobina eastward to the Sierra da Terra Dura, a point midway between Jacobina and Cachoeira, at the head of Bahia Bay. The underlying rock is gneiss, with vast exposures of bare rock, and with nowhere more than a very thin covering of soil. Slight knolls and shallow basins alternate, with rarely a difference in altitude of more than twenty or thirty feet. Many of the hollows form shallow lakes in the rainy season, from which most of the water evaporates in the dry season. At frequent intervals there are circular holes in the rock, called by the natives caldeirãos, which I found on examination to be genuine pot-holes, some of them of great size (the largest one measured was 20 feet in diameter).

A considerable descent is made in passing from the middle to the eastern or coast plateau, which is characterized — at least along our route of travel — by the general absence of rock exposures and heavy superficial deposits of sand, doubtless from the abraded plains to the westward. The whole region between the São Francisco and the sea is covered, generally speaking, with low open forest, or catinga, except for a narrow belt along the coast, where a moister atmosphere permits a more luxuriant growth. The whole area presents a barren aspect, the vegetation being dwarfed and scanty and the aridity excessive. The greatest aridity and the highest temperature occurs in the limestone district, where little or no moisture is precipitated for nine months in the year, and where all the herbaceous vegetation annually withers. Cacti occur in great variety, including arboreal forms of gigantic proportions. With the exception of a few species, the trees are leafless throughout the long dry season, and the streams become dry or merely form chains of brackish pools.[1] The convolvuli and other vines clinging to the arboreal vegetation, though dead at this season, indicated the presence of a wealth of flowers and foliage during the short rainy period.

Our train comprised several score of pack mules and a considerable number of riding horses. Eight animals were necessary for the transportation of my collections. The journey occupied several weeks of rather tedious travel, Bahia not being reached till about the end of November. I soon found storage in a warehouse for my collections until I could arrange for their shipment to Boston. They included several cases of bird and mammal skins, mollusks, and geological specimens, besides some six or eight barrels of fishes, reptiles and other vertebrates in alcohol. These required repacking for shipment, new casks being necessary for part of the alcoholics. Some days later, I found that my fine work in repacking was only temporary, for at the custom house an export duty of $8\frac{1}{2}$ per cent was levied on natural history specimens, which involved the opening of my cases for inspection!

As I knew Professor Agassiz had an agent, who was also a warm personal

---

[1] I found the water at midday in some of the trickling streams we crossed had a temperature of 102° F.

friend of his, at Bahia, I had anticipated little trouble in arranging for my passage home and the forwarding of my collections. It unfortunately happened that he was absent at this time from the country, and also that no provision had been made for the arrival of any of the field parties at this point, and consequently that no arrangements had been made for sending home any members of the Expedition or their collections from Bahia. I was thus received as a stranger, and as my funds had become exhausted, and I was also seriously ill, the outlook was not exhilarating. It was, however, my great good fortune to find at my hotel Captain Joseph Swift, of Gloucester, Mass., a retired shipmaster now acting as a super-cargo for a Boston firm engaged in the Brazil trade. I revealed to him my straits and he immediately took me under his care, even advancing money for my expenses. He also secured passage for me on his vessel soon to sail for Boston. I shipped the collections by the steamship 'North America' of the United States and Brazilian Steamship Company.

On December 15 I left Bahia on the Hanoverian brigantine 'Gehardina,' of 300 tons, chartered here for Boston and loaded with sugar and hides by Captain Swift, who with myself were the only passengers, payment for my passage to be made on my arrival in Boston. The accommodations were primitive — a berth in the Captain's cabin under the companion-way, the fare conforming in character to what would be expected in a small sailing vessel in the commercial trade. The officers and crew were Germans, but some of the former could speak a little English. During the three weeks spent in Bahia I made the acquaintance of a number of foreigners resident in the city, among them several who were much interested in scientific pursuits, notably Dr. O. Wucherèr and Charles Williams.

The 'Gerhadina' was favored for a month with fine weather and made a quick voyage as far as Cape Hatteras, with a fair prospect of reaching Boston in less than forty-five days from Bahia. But January 21, in passing the Cape, we were suddenly struck by a furious gale, which rapidly drove us off our course to the eastward. We were constantly under close-reefed sails for fourteen days, the wind shifting about every twelve hours from northwest to northeast, and back again, with tremendously heavy cross seas. At the end of this period the ship had become strained and was leaking, the food supply was getting short, and as a last resource Captain Swift and Captain Aalderks decided to try to make Bermuda, then not far away. But as we had not seen the sun for two weeks, it was dangerous to trust to dead reckoning in attempting to make so difficult a port in stormy weather. So the 'Gerhadina' was soon headed southward for St. Thomas, the harbor of last resort for North Atlantic seamen in distress. In a short time we passed into smoother water and in due time reached the harbor of St. Thomas, which we found filled with shipping driven thither, like ourselves, by stress of weather, most of the vessels with broken masts or otherwise seriously disabled. Not a few, as we learned later, caught in this memorable storm were never again heard from, and our ship was among those reported as overdue and supposed to have been lost.

A week at St. Thomas sufficed for repairs and the taking on of fresh supplies, and gave me time to gather a small general collection of natural history specimens, and we were again on our way north, with fine weather and full sails as far as the dreaded Cape, when we again encountered heavy gales and were driven from our course. It was overcast and stormy with high winds for the rest of the voyage; we hoped we still held enough westing to make our port, but were in doubt, as we were again dependent on dead reckoning. It was therefore a crucial time when at sunset of a boisterous March day the sailors were sent aloft to shorten sail, preparatory to laying-to for the night, presumably near the eastern end of Long Island, but with doubt as to whether we should not be so far off our course as to render necessary another long voyage to recover our lost westing. Shortly, however, the sailors from the topmast announced a light, and soon after a second light, thus making us sure of the position of the ship and the course to lay. The next morning the 'Gerhadina' was anchored in the harbor of Woods Hole, and a few days later ended her ninety days' voyage from Bahia to Boston.

When I left Bahia I was rather glad that the fates had determined that I was to make the voyage home in a sailing vessel, and on completing it I was still glad I had had the opportunity of such a varied experience. I had seen the ocean for weeks in its most amiable moods, and I had seen it again for weeks in its most angry paroxysms. The trade-wind belts, the doldrums with their huge, slowly rolling swells, the North Atlantic in its February anger, were all precious memories. The Sargosso Sea had added specimens of the surface life of the sea to my collections — minute mollusks and a varied assortment of crustaceans gathered from the gulf weed — and many observations on sea-birds, flying-fishes, and the atmospheric phenomena of mid-ocean.

Collecting Trip to the Middle West (1867).

During the last two months of the voyage my health naturally improved, but I had suffered so long from chronic indigestion and intestinal troubles, and especially during my journey in Brazil, that I finally felt it best to try and abandon my ideal of a life devoted to natural history research, and accordingly resigned my position at the Museum of Comparative Zoölogy and returned to the farm. With this change of occupation my health improved, and with it my old love strengthened. During the winter of 1866–1867, the point of non-resistance was reached, and I planned a natural history collecting trip to the Middle West, in parts of which several of my mother's brothers and sisters had settled as pioneers, thus affording me convenient bases from which to prosecute my work. I was successful in securing orders in advance for specimens of both animals and plants, and so successful in filling the orders that after thousands of miles of travel and the acquisition of a most valuable experience I returned home free of debt, and therefore perfectly satisfied with the results. This trip furnished material for several faunal papers, published in the 'Proceedings' and 'Memoirs' of the Boston Society of Natural History.

The month of May was spent near Sodus Bay, on Lake Ontario. During June the Chicago Academy of Sciences was the point from which excursions were made into the adjoining country, including a trip to Richmond, Indiana, where as a guest of a fellow naturalist, I passed one of the most enjoyable weeks of my life, collecting fossils as well as birds, insects and shells. Several weeks were spent in north-central Illinois, and also in the vicinity of Des Moines, Iowa. During most of July and August I was a guest of the Iowa Geological Survey, through the kindness of the State Geologist, Dr. Charles A. White, and of his field assistant, my friend Orestes H. St. John, who had been a fellow-student with me at the Cambridge Museum and my companion on the Brazil expedition. I accompanied him during his reconnaissance of an area about sixty miles square in south-west-central Iowa (mainly the nine counties of Dallas, Guthrie, Boone, Greene, Carroll, Crawford, Sac, Calhoun and Audubon), the greater part of which region, and thence westward to the Missouri River, was still practically an unsettled wilderness. Our camp wagon was our sole shelter and our immediate source of supply, and our teamster-cook our only human associate for many days together. Here I made general collections, but for the most part found plants and insects the most profitable, securing during the trip several species of each that proved to be new to science. On my return I spent a few weeks in southern Michigan, and made another short stay at Sodus, N. Y.

This out of door life proved of great benefit to my health, and while still in the West I wrote to Professor Agassiz that I felt again equal to resuming my duties at the Museum of Comparative Zoölogy. His response was a most cordial invitation to return, and in October I resumed my post, where I remained uninterruptedly for the next eighteen years, except when in the field on Museum expeditions, as noted in the following narrative. I had already been placed in charge of the departments of mammals and birds when I accompanied the Agassiz Expedition to Brazil, in 1865, and in 1871 was promoted to the official status of "assistant in ornithology," and continued as acting curator of mammals and birds till my resignation in 1885 to assume similar duties at the American Museum of Natural History in New York.

East Florida Expedition (1868–1869).

In the winter of 1868–1869 I conducted a three month's exploration of the country bordering the St. John's River in eastern Florida, between Jacksonville and Enterprise, which furnished the basis of my paper 'On the Mammals and Winter Birds of East Florida,' etc., published in the 'Bulletin' of the Museum of Comparative Zoölogy (April, 1871). This journey was made, with two volunteer assistants, in a ship's yawl fitted with a large sail. As the country was then only slightly settled above Pilatka, our boat was our home and base of supplies, but at times we occupied rude huts that had been deserted by their former occupants. Parakeets were still abundant, and alligators had almost undisputed possession of the bayous and river banks. At certain points of our journey no human beings were seen for many days together except the members of our own party.

Assistant at Museum of Comparative Zoölogy (1871–1885).

Great Plains and Rocky Mountain Expedition (1871–1872).

In April, 1871, I started on a nine months' collecting trip to the Great Plains and the Rocky Mountains, in the interest of the Museum of Comparative Zoölogy, with two assistants, Richard Bliss, a fellow-student at the Agassiz Museum, and Caleb W. Bennett, of Springfield, Massachusetts. Mr. Bliss returned to Cambridge about the beginning of August, Mr. Bennett remaining with me, as hunter and taxidermist, till the end of the expedition in January, 1872. General collections were made at intervals from the Missouri River to Great Salt Lake, Utah. As western Kansas was then subject to raids by hostile Indians, we were provided with letters of introduction from Major-General Pope, Commander of the Department of the Missouri, to the commandants of the military posts near our proposed route in Kansas and Wyoming, requesting them to aid the expedition by providing escorts and transportation, should such assistance prove necessary. Accordingly Fort Hays was selected as the most convenient base from which to hunt buffalo, where a vacant suite of unoccupied officer's quarters was assigned us for use during the six weeks we spent at this point, making collections of the most varied and comprehensive character.

We began work at Leavenworth, where we spent the first ten days of

May, and then went to Topeka, remaining there about two weeks collecting, chiefly birds, fishes, reptiles, and insects. We arrived at Fort Hays May 26, where we remained till July 3, becoming well acquainted with the surrounding country while waiting for the military escort we were assured was necessary for our safety in going out to the buffalo range, some thirty miles northwest of the post. The cavalry, however, was absent on escort duty with government supply trains and the date of their return was tantalizingly indefinite. While waiting for several weeks for this alleged essential to our buffalo work, we made the acquaintance of Charles Clarkson, a professional buffalo hunter who for several years had hunted buffalo in winter for shipment to the eastern cities. This intelligent, enterprising and level-headed New Englander, who lived in a dugout on the outskirts of the military reservation, who owned a good pair of mules and a wagon, and was familiar with the ways of the Sioux Indians, assured us it would be quite safe to go on a buffalo hunt without the encumbrance of a military escort. As he was willing to act as our scout and hunter and furnish us with the necessary transportation for a reasonable consideration, we deemed it safe to trust to his judgment regarding the risks entailed.

We left Fort Hays on our buffalo hunt June 21, returning four days later with a wagon load of buffalo skeletons and skulls, besides leaving on the open prairie five skeletons we had prepared that we were unable to bring with us for lack of room. As we still needed an old cow, and had secured no calves, Clarkson and I returned to the buffalo range the next day, and at the end of another four days had completed our desiderata, having not only secured a fine old cow skeleton and a number of young calves, but had also retrieved the skeletons left on our former trip, which, however, we found had been somewhat damaged by coyotes. In all our spoils numbered 14 complete skeletons and several additional skulls, representing both sexes and various ages, from yearlings to old bulls and cows; also the skins as well as skeletons of five young calves. The time thus occupied was eight days, involving about thirty-six hours of travel. We saved no skins, except those of the calves, as at this season the old coat had been shed (except of course on the shoulders and head) and the new coat was so short that it barely concealed the skin.

The experience was one long to be remembered, as we took no camp outfit but our blankets, a little flour and canned fruits, depending naturally upon buffalo meat for our main subsistence, buffalo chips supplying us with fuel. Our blankets were our only shelter at night, and our wagon was the only available screen from the hot midday sun, under the shade of which we crept to eat our dinner.[1] We saw no Indians, but the landscape was every-

[1] Temperature 105° F. in the shade on one occasion, and usually over 100° at midday.

where dotted with small bands of buffalos; which were so numerous on one occasion that they darkened the plains to the west of us as far as the eye could reach. They were given no peace by the skin hunters, several parties of whom we met on the range. The skins then sold for two dollars each at the nearest railroad station. Once a small band of buffalo, stampeded by these hunters, sent us to our wagon for safety, the herd passing on both sides of us almost within arm's reach.

On returning from the buffalo hunt we packed the collections we had accumulated at Fort Hays and shipped them East, part to Rochester for preparation and part direct to the museum at Cambridge. July 3 we took the train for Denver, and at daylight the next morning obtained our first view of the Rocky Mountains, the snowfields on the higher peaks coming clearly into view with the rising of the sun. We spent two days at Denver, outfitting for a wagon trip into the mountains, and left for South Park on the afternoon of July 6, camping the first night on Turkey Creek, just behind the first range of foothills. While in Denver we had noticed a band of about 300 Ute Indians, encamped on the outskirts of the city, who had come in to receive their annual supplies from the United States Government, and to barter their robes and other furs with the people of Denver. They swarmed through the town on horseback, men, women and children, clad in their native costumes, their faces painted, and their persons decorated with beads and brass and tin trinkets. They were nearly all bare-headed, their coarse jet black hair reaching usually to their shoulders. Some wore hats, or hats and coats, and a few full suits of white man's clothes, but usually a deer skin or a woolen blanket wrapped about them was the principal article of clothing. It was our first sight of the red man in his native splendor and naturally these Indians greatly interested us. But great was our disgust to find that they had left Denver for the mountains just ahead of us and were encamped only a few hundred yards beyond the place we had chosen for our own camp. While they might be 'honest Indians' our guide deemed it best for us to keep a sharp lookout for our horses and other belongings. We decided to stay in camp in the morning till they passed on. They broke camp early and we did not anticipate meeting them again. In this we were disappointed, for we found they had gone only a few miles and again encamped. Thus we passed and repassed each other's camps for several days, until we decided to remain for some time at a point that offered good collecting, in the hope that they would continue on their journey and we should be rid of their visits. But this plan failed, as we again overtook them and they later became once more a nuisance. They were friendly, indeed quite too familiar, and inveterate beggars, particularly for tobacco and matches.

Our journey to South Park was of great interest, not only for its scenic features but for the transitions in the fauna and flora as we advanced to higher altitudes. We reached the base of Mount Lincoln, at the northern end of the Park, July 19, where we camped for a week, extending our collecting trip to above timber line. On the top of Mount Lincoln we found the pipit (*Anthus rubescens*) breeding, and collected specimens of a new species of rosy finch (*Leucosticte australis*), and at timber line, in the taluses obtained about thirty specimens of a new pica (*Ochotona saxatilis* Bangs, described in 1899 from the specimens we collected in 1871). We returned from South Park by way of Pike's Peak, Manitou Springs and the Garden of the Gods.[1] We remained here nearly a week collecting and thence proceeded along the outer base of the foothills northward to Denver, where we arrived August 18, the trip having occupied a little more than six weeks.

Our next collecting point was Cheyenne, where we spent ten days, and then Bennett and I (Mr. Bliss having returned to Cambridge) went direct to Ogden, Utah. This was our base for the next seven weeks, from which we made excursions to the northern end of Great Salt Lake. Here we found avocets, stilts, phalaropes and other marsh and shore birds plentiful, and an abundance of ducks, terns and gulls. We also obtained many fishes, mollusks and crayfishes.

October 9 we took the train east to Green River, where we remained till the 17th. Here we also secured many fishes from pools and streams, and made a considerable collection of fossil fishes from the famous Green River shales, but high winds, ice and snow rendered collecting difficult.

From Green River we went to Fort Fred Steele, arriving at 3 A. M., October 18. This was the point we had selected from which to make our departure into the mountains for big game. We soon found, however, that this plan was impracticable, owing to the expense and the uncertainty of the results. Fortunately at this juncture I met Messrs. Ferris and Hunt, professional hunters, who were supplying game to the eastern markets as a business. I soon made an agreement with them to furnish a specified number of the various kinds of such big game as could be obtained in this region, for a certain price per head, to be delivered to us at Percy, a station on the Union Pacific Railroad about thirty miles east of Fort Steele.

After packing and shipping the collections made at Fort Steele, which included among other things several boxes of fossils, we went to Percy, where we remained from October 20 till December 18. Percy consisted at this time of the railway station, a small hotel, and one or two occupied log-cabins and many vacant ones,— all that was left of what a few years

[1] There were only a few log-cabins and a post-office at the present site of Colorado Springs, and Manitou was merely a name.

before, while the Union Pacific Railroad was under construction at this point, was a booming railroad town. Some of the houses were in decay and others had been removed, but the evidences of its former activities were shown by those remaining and the characteristic debris of tin-cans, empty bottles and other refuse. We hired for our workshop a log cabin that had previously seen service as a saloon, as it was large and fairly well lighted. It served our purpose admirably. Here for two months we worked six days a week in preparing the big game delivered to us at frequent intervals by Ferris and Hunt, as per contract. About one day in seven we devoted to hunting, and added thus to our spoils several antelopes and coyotes, jack-rabbits and cottontails, besides many birds, including a large series of sage grouse, then so abundant that Bennett and I, on one occasion, shot thirty in an hour — all we could carry to camp, and could have killed as many more the next hour had we needed them. As no hay or similar material could be obtained at Percy, we had to substitute dry grass for filling the skins of the mammals, and to obtain this we had to tramp to a moist ravine a mile and a half away, cut it with our hunting knives and carry it home on our backs. A journey of a hundred miles by rail to Laramie was necessary to obtain material for packing cases, namely, discarded dry-goods boxes, which we dismantled and shipped as freight to Percy and remade to suit our needs. Roughing out skeletons and preparing skins of deer, elk, mountain sheep and antelope occupied our time and kept us confined to our laboratory for the greater part of these eight weeks, but our enthusiasm was well sustained by the results, and now and then a day's tramp through the sage brush and snow relieved the monotony.

Our shipment of 17 large cases from Percy included skins and skeletons of 8 elk, 12 black-tailed deer, 1 white-tailed deer, 25 prong-horned antelopes, and 11 bighorn sheep; also 35 skulls of antelope and a fine series of the skulls of elk and black-tailed deer, besides small game (coyotes, foxes, porcupines, beaver, rabbits, etc.) and birds. It nearly filled a freight car, and was shipped on December 17, but, as will be explained later, did not reach its destination for several months.

On the 19th we took an east-bound train for Omaha, already twelve hours late when it reached Percy. A mile and a half east of Percy we ran into a snowbank and the train was partly derailed and delayed for another two hours. Between Carbon and Medicine Bow we were again stalled in the snow for twenty-two hours, where we had to await the arrival of a wrecking train, and two engines to replace our disabled ones. We got our first square meal in two days at Sidney. We struck another heavy snowstorm at North Platte, and at Elkhorn, near Omaha, it was necessary to divide the train to overcome the heavy grade. When we reached Omaha at 9 A. M. of the 22nd, the train was nearly three days overdue. It was also the last train that made the run over the Union Pacific Railroad that winter, owing to a snow-blockade which lasted for several months.

As I had already made an arrangement by letter with Clarkson of Hays City for a winter buffalo hunt, we left Omaha for Kansas City at 4 P. M. of the same day, reaching Kansas City at 4 A. M. of the 23rd. Here I obtained free passes to Ellis and return for Mr. Bennett and myself, and transportation for our specimens on the return trip to Kansas City at half rates, and thence east to Boston by the Merchant's Union Express Company on the same terms. At 10.45 P. M. we were again on a train bound west over the Kansas Pacific. At 11 A. M. the next day we were stalled in a snow drift near Bunkerhill. At Bunkerhill we found five freight trains, eleven engines and a snowplow stalled in the snow. At 3 P. M. orders were received for all trains to remain here till the storm abated. At 8.30 A. M. the next day (December 25) our train, equipped with a snowplow, started on its journey (temperature 12° F. below zero), but in an hour we were again stuck fast in the snow in a cut, but were finally pushed through by two freight engines attached to the rear end of the train. A delay of an hour at Fossil, for switching, gave the passengers, at 11 A. M., an opportunity to procure a luncheon of crackers and cheese,— our Christmas dinner and the first food we had had since the previous evening.

We reached Hays, thirty miles west of Fossil, that evening. Within sight of the station were five stalled trains, each with two locomotives. Behind us were two freight trains with 'double-headers,' and two snowplows, one propelled by three engines and the other by four. During the following night the wind refilled the cuts behind us between Walker and Bunkerhill, causing another blockade.

On the 26th I went to Clarkson's ranch and found he was at Coyote, fifty miles further west; but there was no train west till the evening of the 28th, when we reached Coyote, at 9.30 P. M. Fortunately the weather had now moderated, and on the 29th we went to Clarkson's camp and made arrangements with him for another buffalo hunt. At this time we were in quest of skins for mounting; our summer hunt, as already stated, was primarily for skeletons. Two weeks earlier buffalo had been abundant as far east as Coyote, but they had been so relentlessly persecuted by hunters that they had moved west and were now massed chiefly between Sheridan and Wallace, about one hundred miles to the westward. Between one hundred and one hundred and fifty hunters were said to be in constant pursuit of them. It was roughly estimated that at least 15,000 buffalo had been killed along the Kansas Pacific Railroad during the year (1871).

We were prepared to start on our hunt on the morning of December 31, but the weather turned severely cold and delayed our departure till the following day, January 1, 1872, when we drove south with Clarkson's outfit for nine miles and then returned, finding no buffalo, and being assured by hunting parties we met that there were no buffalos in that direction. We lunched at Coyote and then drove north nearly to the Saline River, camping for the night six miles north of Coyote, and the following night on the South Fork of the Solomon, 30 miles from Coyote. The next morning we went northwest, along the divide between the Saline and the Solomon, to a point opposite Buffalo Station, thence turning north and camping on the North Fork of the Solomon. We saw no buffalo, but were informed by returning hunters that there was a large herd six or eight miles to the northward, where a band of Omaha Indians were hunting them. The next night we camped on a tributary of the Solomon, ten miles northeast of Grinnell. During the day's drive we saw a few buffalo about 2 P. M., and were in sight of small bands for the rest of the day, but they were too wild to permit of a near approach. The next day, January 4, we came up with the first old bulls about five miles from camp, but they were wary, and the ground was unfavorable for stalking them. The weather became suddenly threatening and the wind keen and piercing from the north. Clarkson and his partner, Alden, deemed it imprudent to go further from the railroad, as if a storm should overtake us we would be far from shelter and without wood. They decided to return toward the Saline, from which point we could easily reach shelter should a storm render it necessary. We turned southward and in a few miles came upon a small herd of buffalo, from which Clarkson killed five, Alden three, and I got an old bull. The weather continuing cold and the sky overcast, with signs of a storm, we put up our large Sibley tent, which protected us from the wind and served as our base for the next three days, during which, although buffalo were scarce, our hunters secured their loads of meat, and we obtained nearly the desired number of skins. On the return journey we secured two more, reaching Buffalo Station on the evening of the 7th. The station consisted of the section house and two freight cars, one of them fitted up with sleeping bunks and the other serving as kitchen, and a water tank and two or three dugouts. The owner of one of the latter, a buffalo hunter, happened to be absent and Mr. Bennett and I slept in his dugout. Although the entrance was ankle deep with water, and the floor also covered with water, we passed a comfortable night, such shelter as this being far preferable to sleeping out of doors in the chilling south wind then prevailing. We learned here that the buffalo had left the whole line of the railroad, presumably driven off by the 600 or more Omaha Indians we had seen encamped on the Solomon. So we felt satisfied that we had done as well as was possible under the conditions.

We drove the next day to Coyote, where we finished preparing our specimens and boxed and shipped them. They included eight skins and three heads of buffalo, and skins and skeletons of two lynxes, several coyotes, wolves and jack-rabbits, and a few birds. We settled with Clarkson for $50, which sum covered our transportation and board for a week, and his promise to get for us three more skins of buffalo — two cows and a yearling bull, which were still lacking to complete our series.

We left Fort Hays for the East at 3 P. M. on January 12, and reached Cambridge on the 22d, stopping by the way at my old home in Springfield for a couple of days to visit the home folks, well satisfied with the results of our nine months' work in the field. We had collected and sent to Cambridge 200 skins, 60 skeletons and 240 additional skulls of mammals (mostly large species); 1500 bird skins, over 100 birds in alcohol, and a large number of nests and eggs; a considerable number of fishes, both fossil and recent, and a few mollusks, and many insects and crustaceans.

The following year I remained at the Cambridge Museum, cataloguing and labeling the rapidly increasing collections of birds and mammals.

Yellowstone Expedition (1873).

In May, 1873, Professor Baird, Assistant Secretary of the Smithsonian Institution, saw an opportunity to advance our knowledge of the natural history of a then little known portion of the West through utilizing opportunities afforded by the large military expedition the War Department was to provide as an escort for the railroad surveyors who were to locate the Northern Pacific Railroad westward from Bismark. A number of prominent specialists were invited to coöperate in the natural history work, some of whom accepted but later found it impracticable to join the expedition. I was invited to take charge of the work in vertebrate zoölogy, and the authorities of the Museum of Comparative Zoölogy kindly granted me leave of absence to accompany the expedition, the Museum to have a share of the duplicates obtained. The large scientific staff contemplated at the outset dwindled to a few persons, mostly with little or no experience in scientific field work, so that I was installed as chief of the party. I secured the services of Mr. C. W. Bennett, who accompanied me on my 1871 expedition, as my personal assistant. The other members of the party were a geologist, a photographer, and an artist, the latter being Mr. Konopicky of the Museum of Comparative Zoölogy. The photographer's wagon unfortunately did not arrive until the expedition had reached the Yellowstone, and he was accordingly greatly handicapped in his work.

Mr. Bennett and I left Boston for Bismark, on the Upper Missouri, May 27, via New York, Chicago, St. Paul, Duluth and Fargo, the other members of our party joining us at New York. Passenger trains were then running over the Northern Pacific only as far as Fargo, North Dakota. Construction trains were in use beyond this point, running "on orders" at irregular and uncertain intervals. On reaching the 17th siding, some twelve miles or more east of Bismark, we had orders to wait over till the next day for a train, or return to the 16th siding, signifying at least a day's delay. The terrain here was a wet grassy prairie, the home of myriads of mosquitoes; the shelter for the night was a box car, without seats, blankets, food, or even potable water. It seemed better to complete the journey on foot, and with two of my party and a sergeant from Fort Lincoln, with the mail for Forts Lincoln, Rice and Buford, as guide, we reached Bismark at midnight. The distance from Fargo to Bismark is 200 miles, which required nearly three days to make.

Reported the next day at Fort Lincoln, opposite Bismark, and made arrangements for transportation to Fort Rice, some twenty miles south of Fort Lincoln, and the rendezvous of the Yellowstone Expedition, where we arrived on the 8th of June. Here our party was most cordially welcomed by Gen. D. S. Stanley, commander of the Expedition, who assured us of every assistance possible in our work. Captain H. M. Lazelle, Company H, 8th Infantry, U. S. A., an officer greatly interested in several lines of scientific research, was detailed to take charge of the Scientific Corps, which was assigned to the Headquarters Division of the Expedition. In due time we were provided with the necessary equipment, comprising two (later five) six-mule baggage wagons, five tents, five saddle horses, and a detail of a sergeant and three soldiers to look after our comfort and belongings, one of them serving as cook.

The previous year the survey of the Northern Pacific Railroad route was seriously impeded in the Yellowstone region by hostile Indians. Hence a heavy military escort was this year provided for the protection of the engineer corps. The present escort comprised 1400 troops and 400 civilian employees, accompanied by a train of 300 wagons, each drawn by six mules, and 1200 beeves on the foot to furnish fresh beef en route. The escort comprised the famous Seventh Cavalry, with Gen. George A. Custer in command, and parts of the Eighth and Twenty-second Infantry, and a company of Indian scouts.

The military escort broke camp at Fort Rice on June 20, following the engineers who had taken the field some ten days before. Our course was nearly due west, and the first objective point was the Great Bend of Heart River; a later landmark was the crossing of Big Muddy Creek, where we were detained several days by heavy rains and high water. The Big Muddy was finally crossed on a pontoon bridge, constructed of wagon bodies filled with empty water casks. It required the greater part of two days to make the crossing after the construction of the bridge. We reached the Little Missouri July 11, via the Bad Lands and Davis Creek. The descent of the latter was a tedious day's march of twelve hours, to gain the same number of miles on our course. The stream was tortuous and the valley narrow, so that it was necessary to build ten corduroy bridges in a distance of eight miles. The infantry was deployed in single file in the bluffs on each side of the creek, to prevent a surprise attack by Indians. The ford of the Little Missouri was over a treacherous bottom, and before attempting it with the wagon train it was found necessary to consolidate the quicksands by treading with the cavalry and the herd of beeves. It was also necessary for the pioneers to construct a roadway up the gorge on the west side in order to reach the plateau on the western side of the river.

The Yellowstone River was reached July 15, and the crossing made a few days later by means of the steamboat 'Josephine,' which met us here for the purpose of ferrying the expedition across. We remained in camp near this point for two weeks, waiting for the engineers to connect their preliminary line with their work of the previous year. A stockade was built here and named Camp Thorne, at which a large quantity of supplies was left, with a strong guard of troops, including two squadrons of cavalry and two companies of infantry. The site of Camp Thorne is now the town of Glendive. The distance from Fort Rice to Camp Thorne was 310 miles.

On July 28 we began our march up the west side of the Yellowstone. Owing to the bad lands bordering the river our line of travel was some distance to the westward, and very devious, it being necessary at times to make long detours in order to get around deep ravines, a long, hard day's march of ten or twelve hours resulting in some instances in a gain of not more than two or three miles on our course. We passed the mouth of Powder River July 30, the mouth of Tongue River August 4, the mouth of the Rosebud August 7, the mouth of the Big Horn August 12, reaching Pompey's Pillar about 8 A. M. August 15, 190 miles above Camp Thorne. This was the end of our reconnaissance on the Yellowstone, the line above Pompey's Pillar having been surveyed the previous year. On the 17th of August we turned west toward the Musselshell River, which we struck at a point 50 miles northwest from Pompey's Pillar on the 19th. This point marked the end of the outward journey.

The next day we crossed to the left bank of the Musselshell and followed down this river, to the Big Bend of the Musselshell, which we reached

August 27. We then struck across to the Yellowstone along the divide between Big Porcupine and Little Porcupine Creeks, reaching the Yellowstone River at the mouth of the Little Porcupine, a little above the mouth of Tongue River. Here we struck our outward trail, which we retraced to Camp Thorne, where we arrived September 7, and later, over the same trail proceeded to Fort Rice, our starting-point.

The daily routine of the march was usually as follows: Reveille at 3 A. M.; breakfast at 3.30; tents struck at 5; marching at 5.30; with variations to earlier or later according to special exigencies. The head of the column usually made camp at from 12.30 to 3 P. M.; when delayed by bridging streams or road-making, camping was delayed till 4.30 or later — once till 6.40 and once till 7.30, after a day's march of ten, twelve, or even fourteen hours. The daily distance ranged from eight to twenty miles, according to the amount of delay incident to bad roads.

Field observations and collecting were usually restricted to the line of march, and the preparation of specimens to the daylight hours in camp. The weeks spent in camp at Fort Rice and Camp Thorne, and the delays at the crossing of the Big Muddy, the Little Missouri, and at a few other points, were utilized to the utmost by Mr. Bennett and myself for the increase of our collections. But parts of the journey were less favorable, owing to Indian trouble. Indians were first seen watching us from neighboring bluffs near the mouth of Powder River; they soon became bolder and were seen daily, when orders were given forbidding straying from the line of march, or the use of firearms without permission from the commanding officer. This compelled us to abandon bird collecting and side excursions for several weeks. The first attack upon the Expedition by the Indians was an attempt by them to ambush the advance guard of cavalry, while resting at midday some miles in advance of the main column, about twenty miles above Tongue River. Here General Custer, with seventy men, had a four hour battle with about 300 Indians on August 3, finally charging and dispersing them. But while the battle was in progress a part of the hostile band, hovering along the left flank of our main command, succeeded in killing several of our stragglers, including the veterinary surgeon, the sutler, and a cavalryman. During the fight General Custer had one private wounded and several horses killed and others wounded. The extent of the casualties of the Indians was not ascertained, but at least two were killed and several wounded. Two days later an Indian alarm caused General Stanley to halt and park the train, in preparation for an attack, but the precaution proved to have been unnecessary. Indians were seen, however, in the bluffs on the east bank of the river, and a few shots were fired at them from our six-pounders, which effectively dispersed them.

On the 8th General Custer received orders to advance and chastise the Indians, whose hastily abandoned camps we had been passing for several days, as village after village joined the main band. He set out at 8 P. M., with 450 cavalrymen and our Indian scouts, and by a forced march overtook them about daylight on the morning of the 11th near Pompey's Pillar. The Indians were on the east bank of the river, but at once proceeded to cross and, 500 strong, opened an attack upon him from all sides. The fighting for a time was very sharp, but finally the cavalry were ordered to charge, and the Indians soon scattered in every direction. Custer had three men killed, including his orderly, and Lieut. Brayden was severely wounded in the thigh. For the next twelve hours our camp presented a decidedly warlike aspect, with our battery mounted for action and five companies of the 22d Infantry doing picket duty. An attack seemed probable, but our foes evidently decided we were too well prepared, and slowly retired.

These incidents gave proof that our heavy military escort was not a needless demonstration. It was only three years later, and about sixty miles south of Pompey's Pillar, on the Little Big Horn, that General Custer and his whole command were massacred in a fight with this same band of Sioux Indians.

The opportunities for natural history collecting and field research on this expedition were far from ideal, but we did not return empty handed nor without well-filled notebooks. Very little of the large area traversed had previously been visited by a naturalist, and was still unrepresented by specimens in the National Museum. Much valuable information was gathered respecting the general character of the country and its biology, a portion of which was promptly published.[1] Besides the birds, their nests and eggs, and mammals, small collections of reptiles, fishes, insects and plants were made, and also of invertebrate fossils, which were everywhere scarce. The badlands were searched at every opportunity for vertebrate remains, but always, to our great disappointment, in vain. Mr. Konopicky, the artist, made colored drawings from life of the few fishes obtained, and many excellent sketches of geological exposures and striking topographic features in the bad lands. The photographer also secured many characteristic views of the country traversed. To me it was an experience of great value from the naturalist's point of view, and one I have never ceased to recall with much pleasure for its personal associations and its dash of military flavor.

---

[1] *Cf.* Notes on the Natural History of portions of Dakota and Montana Territories, being the substance of a report to the Secretary of War on the collections made by the North Pacific Railroad Expedition of 1873, Gen. D. S. Stanley, Commander. By J. A. Allen, Naturalist of the Expedition. *Proc. Boston Society of Natural History*, Vol. XVII, pp. 33–91, June, 1874. Also separately, repaged.

Special Collaborator, U. S. Geological and Geographical Survey of the Territories (1876–1882).

From 1874 to 1882 I gave much the larger part of my time to original research, the results of which appeared in numerous minor papers and in several monographs. The first of the latter was 'The American Bisons, Living and Extinct,' published in 1876, under the joint auspices of the Museum of Comparative Zoölogy and the Geological Survey of Kentucky (N. S. Shaler, Geologist-in-Charge). In 1876, with the approval of the director of the Agassiz Museum, I divided my time between this institution and the United States Geological and Geographical Survey, of which for the following six years I was a "special collaborator." At the end of this period the United States Geological Survey was reorganized and the scope of its work greatly restricted. Also my health became so seriously impaired that I was obliged to suspend almost entirely my research work for a considerable period (as explained later in this narrative).

My work for the Geological Survey comprised a number of minor papers published in the 'Bulletin' of the Survey and monographs of various families of North American rodents, in coöperation with the late Dr. Elliott Coues, published in 1877 as 'Monographs of North American Rodentia,' forming volume XI of the 'Memoirs' of the Survey. In 1880 was published my 'History of North American Pinnipeds, a Monograph of the Walruses, Sea-Lions, Sea-Bears and Seals of North America,' as No. 12 of the 'Miscellaneous Publications' of the Survey (8vo, pp. xvi + 785, 60 text figures). This was to have been followed by a volume of the 'Memoir' series on the North American species of the mammalian orders Cete and Sirenia. To this undertaking I had devoted about three years of intense application, and when overtaken by illness had written a large part of the text, and most of the plates had been lithographed. The only parts ever published were a portion (about one-third) of the bibliography,[1] and the account (with reduced half-tone copies of the plates) of the Atlantic Right Whale,[2] more than a quarter of a century after it was written. The printing of the bibliography was left uncompleted in consequence of the inability of the author to revise the proofsheets.

A serious attack of pleurisy interrupted my work in December, 1881.

---

[1] Preliminary List of Works and Papers relating to the Mammalian Orders Cete and Sirenia. *Bull. U. S. Geol. and Geogr. Survey*, VI, pp. 399–562, August 30, 1882 (1013 titles, carrying the subject from the year 1495 to the end of the year 1840).

[2] The North American Right Whale and its Allies. *Bull. Amer. Mus. Nat. Hist.*, XXIV, pp. 277–329, pll. xix–xxiv, April 8, 1908. As here published the article is changed only by the addition of new matter to bring the subject down to the date of publication.

In March, 1882, my physicians were so dissatisfied with my condition that they strongly urged a change of climate, and I was sent to Colorado Springs, Colorado. I went with the fullest confidence that the change of scene and air would soon completely restore my strength, and accordingly took with me a collecting outfit for ornithological field work. My good friend William Brewster, of Cambridge, joined me in a few weeks. I did not regain strength, however, and found myself quite exhausted by a half mile walk.

During the six weeks Mr. Brewster remained with me, we made many collecting trips, by carriage, to the adjacent cañons, forming a considerable collection of bird skins and acquiring much greatly valued information regarding the spring migration of birds at Colorado Springs. When he returned, two months later, to the East I was in worse condition than when I left home, and was compelled to spend the summer at a neighboring ranch in the foothills of the Rockies, thoroughly mystified at my lack of recuperative power, for the organic trouble that led to my being sent to Colorado had quite disappeared.

On returning to Cambridge in September I learned that my invalidism was due to nervous breakdown. Recovery was exceedingly slow; after a few months I regained strength for part time work at the Museum, and for a small amount of literary work, slowly increasing it, month by month, but for years afterward my physical condition was a serious handicap.

Early in 1885, the financial resources of the Cambridge Museum became much reduced, leading to the discharge of several of the assistants and the prospective dismissal of others, including myself. Through the kindness of Mr. Alexander Agassiz, three alternatives were open to me — to remain at the Museum at the risk of its being soon closed; to accept a position open to me on the United States Geological Survey; or to accept a curatorship offered me by the American Museum of Natural History in New York. For family reasons (I had a motherless three-year-old boy whom it was necessary to leave with friends in Massachusetts), field work with the Geological Survey did not appeal to me, and led to my acceptance of the New York curatorship, which I have ever since felt was a wise decision.

Curator at American Museum of Natural History (since 1885).

On May 1, 1885, I entered upon my duties as curator of the 'Department of Ornithology and Mammalogy' at the American Museum of Natural History in New York. The collection of mammals then consisted of about 1000 mounted skins and 300 mounted skeletons, all on exhibition in the exhibition halls. There was not even a nucleus of a study collection. The collection of birds numbered about 10,000 mounted skins and several hundred mounted skeletons, all on exhibition, and about 3000 unmounted skins, forming the beginning of a study collection.

The exhibition collections were rich and varied, the specimens having been selected with excellent discrimination, mainly by the late Dr. D. G. Elliot, and included the famous Maximilian collection of mammals and birds. The exhibition collections of this department thus compared favorably with the best in this country. The North American collection of birds was essentially complete, nearly every procurable species being represented, and the specimens were all correctly determined and neatly labeled.[1] But no part of either collection had been catalogued, except the European birds, which had recently been carefully registered by Dr. Edgar A. Mearns. In the case of the exotic species, of both mammals and birds, the labels bore the names under which they were originally purchased, and hence, while in most cases correctly identified, the technical nomenclature was antiquated.

The first task was therefore to catalogue the collections, thus securing a permanent record of their history.[2] Fortunately, the original labels had in most cases been preserved by pasting or tacking them to the bottom of the stands. The next step was to renew the labels, in a uniform style, giving the currently accepted technical names of the specimens, with their localities, in conspicuous type.

At the end of the first year the mammals had been catalogued and relabeled. At the end of the third year most of the birds had been catalogued and provided in large part with new labels.

In the curator's first annual report,[3] on the condition and extent of the collections, the importance of fine exhibition collections was not only fully recognized, but the formation of adequate study collections, to serve as the basis for scientific research, was strenuously insisted upon in order to bring the department to a proper standard of efficiency. As often as opportunity arose for securing such material, urgent appeals were made to the

[1] The determination was made by Robert Ridgway of the U. S. National Museum shortly before the collection came under my care.

[2] "The museum assistant in charge of a special department must naturally, if the purpose for which large collections are brought together is carried out, spend the greater part of his time in preparing them for the specialist who is at some future time to avail himself of the treasures brought together for his benefit. There is, therefore, the same danger that an eminent specialist, after his appointment to the curatorship of a great museum, will find his museum duties so arduous as to prevent him, as his colleague in the professional chair has been prevented, by official work, from doing original work."— *A. Agassiz*, in the Ann. Report of the Curator of the Museum of Comparative Zoölogy at Harvard College for 1883–84, p. 7.

[3] Annual Report of the Trustees for the year 1885 (1886), pp. 9–12.

President for their purchase,— for the first two years usually without avail, owing to the lack of funds.[1]

The year 1887, however, proved to be epochal in the history of the department, not only through the addition of scientific material but in other lines of development. It was in this year that the Trustees purchased the George N. Lawrence collection of birds, a strictly research collection of immense historic importance, numbering some 12,000 specimens of (mainly) tropical American birds; and also the Herbert H. Smith collection of 4000 birds from southern Brazil. The D. G. Elliot collection of 2000 humming birds (including many types) was presented by Mr. Elliot, and a collection of about 2250 Arizona birds was presented by Dr. Edgar A. Mearns, U. S. A. Further accession, comprising some 500 North American birds, were also added, partly by purchase and partly through the Elliot-Richardson Expedition to Montana (noteworthy as the first Museum expedition from this department), making the grand total of 21,000 birds added in this notable year, besides a large number of nests and eggs and some osteological material. The previous year a taxidermist, Mr. Jenness Richardson, had been added to the personnel of the department, through which means 18 bird groups were added to the exhibition collection, these forming the beginning of the Museum's magnificent series of 'Habitat Groups.'

During this year the efficiency of the department was further greatly enhanced by the purchase of the D. G. Elliot ornithological library, consisting of about 1000 carefully selected volumes, thus providing the Museum with an ornithological library surpassed by few similar libraries in this country. A large number of moth-proof tin cans were also provided for the safe storage of the rapidly increasing research collections.

The reception and care of these large accessions naturally checked progress in cataloging and labeling the exhibition collections, but the bird collection as a whole had suddenly been transformed from merely a show collection to one of impressive scientific importance.

Thus far the curator had had to depend upon his own efforts in cataloging, labeling, and caring for the collections, except for a little aid now and then from interested volunteer assistants; but in 1888, Mr. Frank M. Chap-

[1] From the Treasurer's Report for the year 1885, it appears that the contribution by the City toward maintenance was $15,000, and that the income from invested funds was $300. The rest of the $30,509 expended for running expenses was met by personal contributions of the Trustees. The additional $6,054 expended for collections and for books for the library came also from private sources.

It is of interest to compare, in this connection, the income and expenditures in 1915 with those of 1885, as indicative of the growth of the Museum in resources during the thirty-year interval. The financial statement for 1915, as given in the Forty-seventh Annual Report of the Trustees, shows the contribution from the City to have been $200,000, leaving a deficit of $38,500 to be met by personal contributions of the Trustees. The net receipts from endowment are given as $265,275.16.

man was secured as a permanent assistant, to whom twenty years later (in 1907) was assigned the curatorship of ornithology. He proved from the first an enthusiastic and efficient helper, at all times in fullest sympathy with the policies and work of the department. His various expeditions to Florida, Trinidad, Cuba, and Mexico in the earlier years of his connection with the Museum added important accessions to the research collections of mammals and birds, and later he took in hand the construction of the Habitat Groups of birds, which soon became an impressive feature of the Museum exhibits. This has been followed in recent years by his well known series of expeditions to South America, in the interests of this department.

In 1887, on the death of Dr. J. B. Holder, who for many years had been curator of the Department of Invertebrate Zoölogy, Fishes and Reptiles, this department was also placed in my charge,[1] and the collections of fishes and reptiles remained in my care till 1901.

Following the bounteous year 1887, were a number of lean years, as regards accessions to the collections; there was, however, a slow but steady increase of material from both North and South America, and in later years also important additions from Asia and Africa. These accessions afforded not only material for exhibition but for scientific research, the results of which were published in the Museum 'Bulletin.' With the gradual increase of the collections the scientific staff of the Department also increased, from a single assistant in 1888 to six assistants in 1915, besides stenographic and other non-staff office assistants and four field assistants.

As previously noted, Assistant Curator Chapman was placed in charge of the ornithological division of the department in 1907, with W. DeW. Miller as assistant; Roy C. Andrews became assistant in the mammalogical division in 1908, and assistant curator in 1911; H. E. Anthony, became

---

[1] In the list of Officers and Committees in the Annual Report for 1888, the entry for my department reads: "Prof. J. A. Allen, Curator of the Department of Ornithology, Mammalogy, Fishes and Reptiles. Also temporarily in charge of the Department of Invertebrate Zoölogy." In 1890 the Department of Invertebrate Zoölogy was transferred to the Department of Geology, Mineralogy and Conchology (Prof. R. P. Whitfield, curator), and the name of my department changed to "Department of Mammalogy, Ornithology, Herpetology, and Ichthyology." In 1898 it was renamed "Department of Vertebrate Zoölogy," and this was its official designation till 1901, when Fishes and Reptiles were assigned elsewhere and it became "Department of Mammalogy and Ornithology," the designation it has since retained.

There having been, during these years, no available exhibition space for the fishes and reptiles, I was responsible merely for their safe storage. I catalogued, however, the accessions of reptiles and batrachians, in which classes I had formerly taken much interest, and of which for a time I was curator at the Boston Society of Natural History.

A department of taxidermy was established, as stated above, early in 1886, and, with a chief taxidermist and a staff of several assistants, remained an adjunct of the Department of Mammals and Birds till 1903, when it was separated as a distinct department and broadened to cover a much wider field than mammals and birds.

assistant in the same division in 1914. In 1915 Herbert Lang and James P. Chapin, on their return from the American Museum Congo Expedition, were made, respectively, assistants in mammalogy and ornithology.

Thus the curator was gradually relieved of much routine work and able to devote his time mainly to research and to the editorial supervision of the Museum 'Bulletin' and the zoölogical portion of the 'Memoirs.'

While these pages are not intended to comprise a detailed history of the department, it seems desirable to mention here some of the larger increments of the collection, on account of their bearing upon my later scientific work. One of the primary duties of a curator and his scientific staff is the identification of accessions and the prompt publication of the results of their elaboration, in the interest not only of science but of the institution with which they are connected. This obligation gives direction to their scientific activities, forcing them from general to more special lines. In this way their researches are in a measure restricted within certain boundaries.

As already noted, in 1885 there was not even the beginning of a research collection in mammalogy; in 1915 the unmounted mammals exceeded 40,000 specimens, and included a fair amount of osteological material and spirit specimens in addition to skins and skulls; while the old mounted specimens had been supplemented, and to a large extent superseded, by elaborate group exhibits, illustrating the life histories of many prominent types of North American and exotic mammals. During recent years this work and the installation of a synoptic mammal collection has been under the direct supervision of Director Lucas and Assistant Curator Andrews.

The bird collection in 1885 included a small number of skins, mostly North American, in addition to the 10,000 mounted specimens on exhibition; in 1915 it comprised not only the most extensive and elaborate series of 'Habitat Groups' in any museum in the world, but more than 130,000 study skins. Further available as research material was the collection of Dr. Jonathan Dwight, of chiefly North American birds, numbering 45,000 specimens; Dr. L. C. Sanford's collection of North American water birds, numbering about 8000 specimens; and the Brewster-Sanford collection of about 6000 specimens, mostly seabirds in large series from the South Atlantic and South Pacific oceans and the southern coasts of South America. Thus the American Museum in 1915 contained a total of about 190,000 birds available for scientific research, exclusive of mounted birds and groups. Collectively they doubtless formed the largest and by far the most valuable collection of American birds yet assembled in any single museum.

The first notable accessions of birds were received, as already related, in 1887, and comprised two important authoritative and historic collections, the Lawrence collection of American birds (largely from tropical America and containing nearly all of Lawrence's types), and the Elliot collection of Hummingbirds, at that time one of the largest and most authoritative collections in the world of this numerous and most interesting suborder of birds. In the same year were also received a notable collection of birds from Arizona for the North American study series, and our first freshly collected and wholly unworked collection of exotic birds, the Herbert Smith collection of some 4000 specimens from the Province of Matto Grosso, Brazil. This became the basis of a series of papers by the curator published in the Museum 'Bulletin' in 1889–1892. In 1889 small collections of birds were received from Ecuador and Bolivia, aggregating about 600 specimens (200 species), which formed also the basis of papers in the 'Bulletin.'

During the years 1890, 1891, 1892, important accessions of mammals and birds were received through purchases of small collections from Mexico and Central America, and collections obtained through Museum expeditions to Florida, Texas, Colorado, and Washington, upon which reports were published in the 'Bulletin,' by the curator and the assistant curator. In 1893, and again in 1894, Mr. Chapman made collecting trips to the Island of Trinidad, B. W. I., resulting in the addition of 550 mammals and about 690 birds, which formed the basis of extended papers on the mammal and bird faunas of the island.

Among later collections may be mentioned the Streator collection of birds and mammals made in British Columbia in 1899; collections of birds and mammals made in Alaska and northern British Columbia by the A. J. Stone expeditions in 1897–1903; the H. H. Smith collections of birds and mammals from the Santa Marta region of Colombia, received in 1898–1901; the Peary collections of birds and mammals from Greenland and Ellsmere Land (1898–1909); the collections of Siberian birds and mammals made by N. G. Buxton on the Jesup North Pacific Expedition in 1901; the Batty collections of birds and mammals from Chiriqui and Mexico received in 1901–1906; the Nicaragua collections of birds and mammals made by W. B. Richardson in 1907–1908; the E. T. Seton collection of mammals from the Athabasca-Mackenzie region of Canada, 1908; the Tjäder East African collection of mammals, 1907; the Rainsford collection of East African mammals in 1913; the Congo collections of birds and mammals, each numbering some 6000 specimens, made by Herbert Lang and James P. Chapin, received in 1915; and the constant inflow of birds and mammals from South America, from 1911 to date, through expeditions directed by Curator Chapman and collected mainly by Leo E. Miller, W. B. Richardson, and George K. Cherrie; and the birds and mammals collected by Cherrie and Miller on the Roosevelt Brazilian Expedition; also birds and mammals collected by H. E. Anthony in Panama, and in Oregon and Washington.

The Asiatic material received, in addition to the Buxton Siberian collection already mentioned, has been mainly collections of mammals from the Island of Hainan, China, received by purchase, 1906–1908, and a few small lots of mammals received from western and northern China (by purchase); and small collections of birds and mammals collected in the Dutch East Indies by Roy C. Andrews in 1909 and in Korea in 1912. Especially should also be mentioned the important collection of cetacean material collected by Mr. Andrews in the North Pacific in 1910–1912.

This material has all been carefully studied and reported upon in the 'Bulletin' and 'Memoirs' of the American Museum, except the recently received mammals and birds of the Congo Expedition, and some of the South American collections of birds, the Cetacean material having been the basis of monographs by Mr. Andrews, while Mr. Anthony has published on the mammals collected by him in Panama and Oregon. The other mammal collection have all been reported upon in numerous papers by the writer, including a memoir on Muskoxen, recent and extinct (1913), and a revision of the South American Sciuridæ (1915). My last faunal papers on birds were reports on the Smith collection from Santa Marta, Colombia (1900–1904) and the Buxton collection from Siberia (1905). Two papers on the types of the North American genera of birds (1907) are also among my later papers on birds.

The collections of birds received prior to 1905 have been the basis of several faunal papers by Curator Chapman and Assistant Curator W. DeW. Miller. Chapman has published important preliminary papers on the recently received South American collections of birds, describing many new forms, and has now in press a report on the birds of Colombia, which forms Volume XXXVI of the American Museum 'Bulletin.' It is thus a pleasure to record that the recent field activities of my department of the Museum, so extensive and so varied, have been adequately supplemented by laboratory research and prompt publication of results.

Affiliations with the American Ornithologists' Union and other Scientific Societies (since 1883).

As regards general activities, my interests have been confined within narrow limits, partly from an inborn shrinking from functions that necessitate appearance in public positions and partly through lack of the physical endurance to meet such demands. The exception has been my connection with the American Ornithologists' Union, the founding of which was due in part to my instigation,[1] and with whose activities I became from the first

[1] The call for the meeting of ornithologists which led to its organization was issued by Elliott Coues, William Brewster and the writer, in 1883.

intimately associated. Although unable to attend the meeting of ornithologists at which the Union was founded it happened that I was selected as its first president, and, contrary to my expressed wish, was annually reëlected for the following six years. This embraced of course the formative period of the society's history, when its work was planned and carried forward with the energy and enthusiasm that so strongly marked its early activities. Committees were appointed at its first meeting to take in hand various important subjects, as the migration and geographic distribution of North American birds, which investigation was prosecuted so successfully that in a few years it outgrew the financial resources of the society and led to the establishing of a Division of the United States Department of Agriculture to carry on the work, which ultimately became the present Bureau of Biological Survey, for many years under the direction of C. Hart Merriam, the chairman of this committee of the American Ornithologists' Union.

Another important committee was at the same time established to prepare a standard Check-List of North American birds, to displace the discordant check-lists then in vogue, with of course undesirable results. To accomplish this it was found necessary at the outset to have as the basis of the list a consistent and carefully prepared code of rules of nomenclature in place of then existing inharmonious codes and go-as-you-please methods of nomenclature. This led to intense research on the part of several members of the committee, and eventually to the adoption and publication of a new code, departing widely in some important respects from any of its predecessors. This A. O. U. Code later became the basis of the International Code of Zoölogical Nomenclature, framed on essentially the same lines and departing from it in no essential respect, except in point of brevity, through omission of adequate illustrations of the rules, and thereby rendering necessary the issuance of official 'Opinions' to clear up obscure points. A draft of the A. O. U. Code was made by two members of its Nomenclature Committee, to serve as a working basis for the Committee, which after long and careful consideration by the full Committee was adopted essentially as written, not only as to form and phraseology but also (with the exception of the addition of a single provision, since abandoned) as to its rulings.[1]

The work of the A. O. U. Nomenclature Committee resulted not only in a new check-list of North American birds, standardizing their nomenclature, but also a new and elaborate Code of Zoölogical Nomenclature,

---

[1] It may be pardonable in this connection to state that the preliminary draft was written in part by Elliott Coues (mainly the part comprising the 'General Principles') and the rest, including the historical part of the 'Introduction' and most of the 'Canons' and their explanatory 'remarks', by the author of these 'Notes.'

which has had vast influence in standardizing the rules of nomenclature the world over.[1]

Still another important A. O. U. Committee had its inception at the first meeting of the Union in 1883, namely, a Committee on the Protection of North American Birds, which carried on the work of bird protection for many years with great energy and very important results. From it originated later the so-called Audubon Societies movement, from which, through the special efforts of one of its enthusiastic members, William Dutcher, was developed the present powerful organization known as the National Association of Audubon Societies. From this committee also emanated the A. O. U. "Model Bird Law," which has been enacted by most of the States in this country and the Provinces of Canada, in the essential phraseology of the committee's draft. In this work I shared actively, furnishing in 1886 most of the matter for the 'Bulletins' of the committee, issued first in 'Science' and afterwards separately reprinted in large editions.

For many years my interest in the activities and welfare of the American Ornithologists' Union was absorbing. For seven years I was its president, for twenty-seven years the editor of its journal 'The Auk' and its other publications; and a member of its Committee on the Classification and Nomenclature of North American Birds from 1883 to date. For twenty-eight years I maintained an unbroken record of attendance at its annual congresses and council meetings.

Before leaving Cambridge I had begun to take a more or less active part in the work of the Boston Society of Natural History, having served as acting secretary and editor of its publications for a number of months, to fill an interim, and for a number of years had been a member of its Council.

On my arrival in New York I was warmly welcomed by the New York Academy of Sciences, and almost immediately given a place on its Council, and later for several years was a Vice-President; but I failed to identify myself heartily with its interests, due in part to limitations imposed by impaired health.

My affiliation with the Linnæan Society of New York was more intimate, and the office of president for a number of years never proved very strenuous, as the attendance was usually small and the meetings more or less informal and semisocial.

---

[1] A revised edition of the A. O. U. Code was published in 1910, in which the article adopted by the International Congress in 1909, newly defining the method for the determination of types of genera, was substituted for the corresponding matter in the original A. O. U. Code.

The editing and carrying through the press of the three editions of the Check-List and of fifteen of its sixteen supplements, and also the original and the revised editions of the Code, was a labor of love for the author of these 'Notes.'

Elliott Coues was chairman of the Nomenclature Committee from its inception in 1883 till his death in 1899. He was succeeded by C. Hart Merriam, who held this position till 1904, when he was succeeded by the writer, who resigned in 1912.

General Considerations and Home Life (since 1874).

As a fact of possibly some psychological interest I may here briefly revert to a handicap with which probably few called to public functions have had, at least in equal degree, to contend. From early boyhood I was painfully embarrassed in the presence of strangers. Later in life attempts to present papers verbally before scientific societies were always unsatisfactory and often failures, not from lack of familiarity with the details of the subject but from embarrassment. The same timidity prohibited my seriously considering teaching as a possible means of raising funds to aid in meeting the expenses of an education, or of giving public lectures for the same purpose, as many of my associates at the Agassiz Museum were doing, with both pleasure and profit. The ordeal of an examination for a degree at the Lawrence Scientific School at Harvard was sufficient to banish all aspiration for such honors. This in part, perhaps, led to a feeling of disrespect for this sort of a label, and to the belief that any worthy accomplishment would, sooner or later, receive due recognition. My wants were simple and inexpensive; all I aspired to was opportunity for scientific research, believing that diligence, singleness of purpose, and honest work would bring its own reward. I was content to follow my own lines of dominating interest to such limit as the circumstances of earning a living would permit. I never had any desire for money as such, nor any interest whatever in financial projects, nor any longing for honors beyond those my colleagues in science saw fit to impose. Therefore my election, in the course of time, to all of the leading academies of science in this country, and to honorary membership in foreign societies with which my lines of study were affiliated, have always come as exceedingly pleasant surprises. The greatest surprise of all, however, was the reception of a Ph.D. from Indiana University in 1886, the first intimation of which was the receipt from its president, David Starr Jordan, of the official notification that I had been thus honored.

In looking back to the beginning, it is difficult not to contrast my early surroundings with those of the average youth of today smitten with the nature study 'craze.' As stated in the early part of these notes, I was wholly isolated from everyone having even the slightest biologic interest, with only the inspirations and opportunities afforded by the little red schoolhouse on the hill, in the fifties of the last century, in a neighborhood of small farmers to whom the raising of potatoes and grain was the chief aim of life. My studies were not only selfchosen, but were carried on in odd hours without either advice or encouragement, in contrast with the systematised training of today, from the primary school to the post graduate course. My first systematic guidance was received at the Agassiz Museum in 1862; to the personality and influence of the 'Great Teacher,' LOUIS AGASSIZ, I can never adequately express my indebtedness.

I cannot close these 'Notes' without reference to the most important of the influences that have most happily and most strongly affected my career.

In 1874 it was my good fortune to have won the love of Mary Manning Cleveland, of Cambridge, Mass., daughter of the late Dr. Anthony Benezet Cleveland and Mary (Manning) Cleveland, sister of the late Professor William C. Cleveland of Cornell University and of Dr. Clement Cleveland, an eminent surgeon of New York. We were married October 6, 1874. She died April 17, 1879, leaving a son, Cleveland Allen, ten months old, now in business in New York. Never could marital affection nor mutual interest and sympathy be stronger than was our lot to enjoy. When the end came there seemed little worth while in life.

After seven years I had the equally good fortune to make the acquaintance of Susan Augusta Taft, daughter of the late Daniel and Emeline (Smith) Taft, of Cornwall-on-Hudson, N. Y. We were married April 27, 1886, and have enjoyed thirty-one years of happiness, unalloyed except by her several acute but temporary illnesses, when again the light of life seemed vanishing. My own health during this long period having been by no means strong, I owe to her deep love and sympathy, to her supreme optimism and constant watchfulness over my health, and to her inspiration, the greater part of the little I may have achieved in these last thirty years, and doubtless many years of activity beyond those I otherwise should have attained.

ADDENDA.

EXPEDITIONS.

1865. Zoölogical Assistant on the Thayer Expedition to Brazil, under Professor Louis Agassiz, April, 1865–March, 1866.

1867. Collecting Expedition to western New York, southeastern Indiana, northern Illinois, western Iowa, southern Michigan, May–October, 1867. (A private enterprise.)

1868–69. East Florida: Jacksonville, via St. Johns River, to the head of Lake George, December, 1868–April, 1869, with two volunteer assistants (Rev. Thomas Marcy and J. E. Brundage). Made under the auspices of the Museum of Comparative Zoölogy.

1871–72. Fort Leavenworth, Kansas, west to Northern Utah, May, 1871–February, 1872, with two assistants (Caleb W. Bennett and Richard Bliss, Jr., the latter from May to August). Made in the interest of the Museum of Comparative Zoölogy.

1873. Bismark, North Dakota, to the Yellowstone River, thence up the Yellowstone to Pompeys Pillar, north to the Musselshell River, and east to the Yellowstone and Bismark. Made under the auspices of the Smithsonian Institution, as chief of the party of Naturalists of the North Pacific Railroad Expedition of 1873, Gen. D. S. Stanley, Commander.

Positions Held.

Assistant in Ornithology and Curator of Birds and Mammals in the Museum of Comparative Zoölogy at Harvard College, 1870–1875.

Lecturer on Ornithology at Harvard College, 1871–1873.[1]

Curator of Reptiles, Boston Society of Natural History, 1868–1871.

Curator of Birds and Mammals, Boston Society of Natural History, 1870–1880.

Acting Secretary, Boston Society of Natural History, October, 1874–May, 1875.

Councilor, Boston Society of Natural History, 1881–1885.

Corresponding Secretary, Nuttall Ornithological Club, and Editor of its 'Bulletin,' 1876–1883.

President of the American Ornithologists' Union, August, 1883 – November, 1891. Editor of its journal 'The Auk,' and of its other publications, 1883–1912. Subcommittee (with E. Coues) to codify the rules of nomenclature for discussion by the full Committee on the Revision of the Nomenclature and Classification of North American Birds, 1883–1886. Chairman of the Committee on the Nomenclature and Classification of North American Birds, 1904–1912. Chairman of the Committee on Revision of the Code of Nomenclature, 1905–1908.

Curator of Birds and Mammals at the American Museum of Natural History, New York, since 1885. Editor of the 'Bulletin' and the zoölogical series of the 'Memoirs,' since 1889.

---

[1] A course of twelve lectures was planned but, if I remember rightly, only four or five were delivered. As they were intended to treat the subject seriously they were naturally technical rather than 'popular,' and the audience soon dwindled to a few students actually interested in the subject. While read from manuscript, they were elaborately illustrated by specimens with extemporized explanations, particularly as respects the osteology and pterylography of the subject. In my inexperience in such matters, an average attendance of half-a-dozen auditors seemed farcical, and I counted myself a failure as a lecturer and did not complete the course. On looking over the manuscript and the schedule of the course in later years I have felt chagrin that it was not completed, as the attendance was as large as should have been expected at a course so specialized in character. The following is the schedule of the course: I. The distinctive characteristics of birds in comparison with other vertebrates; birds as a modification of the vertebrate type for aërial locomotion; the osteology of birds. II. The muscular and nervous systems and the organs of sense. III. The digestive, circulatory and respiratory systems and vocal organs. IV. Reproductive organs and embryological development. V and VI. The tegumentary system, including the development of feathers, their arrangement, structure, modifications and functions. VII. Historical summary of the principal systems of classification. VIII, IX, and X. The classification of birds, with a general review of the leading groups. XI and XII. Geographical distribution of animals, with special reference to birds; geographical variation in birds.

In short, the lectures were planned to form a compendium of bird-lore, and had they been completed and published, with proper illustrations, they would have furnished in concise form a useful reference work for the bird student, the need for which was not supplied, even in part, till 1884, when Coues's 'General Ornithology' was added to the second edition of his invaluable 'Key to North American Birds.'

One of the five incorporators of the (first) Audubon Society for the Protection of Birds, New York, 1886.

A Founder and Director of the Audubon Society of the State of New York, 1897–1912.

Vice-President New York Academy of Sciences, 1891–1894. Member of Council, 1886–1896.

President of the Linnæan Society of New York, 1890–1897.

A Founder, Director, and member of the Executive Committee of the National Association of Audubon Societies, since 1905. Second Vice-President, 1908–1912.

Member of the Commission on Zoölogical Nomenclature of the International Congress of Zoölogy, since 1910.

HONORS CONFERRED.

Humboldt Scholarship, Lawrence Scientific School, Harvard College. 1871.
Honorary Ph.D., Indiana University. 1886.
Walker Grand Prize, Boston Society of Natural History. 1903.
Medal of the Linnæan Society of New York. 1916.

Honorary and Corresponding Memberships.

Fellow of the American Academy of Arts and Sciences, Boston. 1871.

Corresponding Member of the Lyceum of Natural History, New York. 1874.

Corresponding Member Sociedad Mexicana Historia Natural, Mexico. 1874.

Fellow of the American Association for the Advancement of Science. 1875.

Member of the National Academy of Sciences, Washington. 1876.

Corresponding Member of the Davenport Academy of Sciences, Iowa. 1877.

Fellow of the American Philosophical Society, Philadelphia. 1878.

Fellow of the American Ornithologists' Union. 1883. (A Founder; President, 1883–1891.)

Honorary Member of the Nuttall Ornithological Club, Cambridge. 1886.

Life Member of the Metropolitan Museum of Art, New York. (By kindness of Morris K. Jesup.) 1887.

Fellow of the New York Academy of Sciences. 1886.

Honorary Member of the New York Zoölogical Society. 1887.

Foreign Member of the British Ornithologists' Union. 1890. Honorary Member since 1907.

Foreign Member of the Zoölogical Society of London. 1891. Honorary Fellow since 1901.

Honorary Member of the California Academy of Sciences. 1892.

Corresponding Member of the Academy of Natural Sciences of Philadelphia. 1893.

Extraordinary and Foreign Member of the Deutsche Verein zum Schultze der Vogelwelt. 1900.

Honorary Member of the Australian Ornithologists' Union. 1904.

Charter Member of the American Association of Museums. 1905.

Honorary Member of the Deutsche Ornithologische Gesellschaft. 1907.
Honorary Member of the Hungarian Central Bureau of Ornithology. 1909.
Honorary Member of the South African Ornithologists' Union. 1909.
Life Member of the City Library Association, Springfield, Mass. 1910.

Active or corresponding membership is or has been held in the following Societies, mostly additional to those mentioned above: Boston Society of Natural History, 1862–1887 (corresponding member since 1887); Essex Institute, Salem, Mass., 1874–1887 (corresponding member since 1887); Torrey Botanical Club, New York, 1893–1909; Linnæan Society of New York, since 1885 (President, 1890–1897); American Society of Naturalists, since 1891; Biological Society of Washington, since 1882; Washington Academy of Sciences, since 1900; American Society of Geographers, since 1905.

BIBLIOGRAPHY.

BIBLIOGRAPHY.

THIS Bibliography does not include hundreds of minor reviews, particularly of ornithological papers published in the 'Bulletin' of the Nuttall Ornithological Club and 'The Auk,'[1] nor many of the unsigned editorial paragraphs in the 'Notes and News' department of the same journals, all of which, unless signed, are by the editor. It does, however, include the unsigned biographical notices, which, unless otherwise indicated, are by the editor. In the case of reviews, those omitted are usually brief announcements of technical papers, noting little beyond the fact of their publication and general import, or works or papers of a more or less popular character, hardly entitled to consideration as serious contributions to science.

The titles are arranged in nine sections: (1) Mammals, about 270 titles; (2) Birds, about 965 titles; (3) Reptiles, 5 titles; (4) Zoögeography, 9 titles; (5) Evolution, 22 titles; (6) Nomenclature, 27 titles; (7) Biography, 118 titles; (8) Miscellaneous, a few titles each on botanical, geological, and other subjects; (9) Editorial Work.

In the case of papers of a mixed character, like that on 'Mammals and Winter Birds of East Florida,' where four special subjects are formally treated under the same general title, or in the case of faunal papers which include mammals and birds, and in one case also reptiles and plants, titles are repeated under each section to which they in part formally relate; beyond these few special cases titles are not repeated under the different sectional headings, although subjects additional to that indicated by the title may be incidentally discussed, as in some reviews and faunal papers.

---

[1] Volumes I–XXVIII. My editorial connection with 'The Auk' closed with the issue for January, 1912, Volume XXXIX.

The general indexes to these two journals contain about 1400 entries under my name, of which about 1250 are indicated as reviews. Of this number less than 700 are included in this bibliography.

I.— MAMMALS.

1869.

1. Catalogue of the Mammals of Massachusetts, with a critical revision of the species. <*Bull. Mus. Comp. Zoöl.*, I, No. 8, pp. 143–252, Oct., 1869.

   68 species, with critical and other extended annotations. The revisional comment was unfortunately based largely on the literature of the subject, in the absence of actual specimens of many of the forms considered.

1870.

2. Notes on the Mammals of Iowa. <*Proc. Boston Soc. Nat. Hist.*, XIII, pp. 178–194, Feb., 1870.

   48 species, with brief field and critical notes.

3. On the Eared Seals (Otariadæ), with detailed descriptions of the North Pacific Species, by J. A. Allen. Together with an Account of the Habits of the Northern Fur Seal (*Callorhinus ursinus*), by Charles Bryant. <*Bull. Mus. Comp. Zoöl.*, II, No. 1, pp. 1–108, pll. i–iii, August, 1870.

   I. (1) Résumé of recent contributions to the natural history of the Otariadæ, pp. 4–19; (2) affinities, distinctive characters and synonymy, with remarks on sexual, age and individual variation, and a conspectus of the genera and species, pp. 19–45; (3) on the North Pacific species, pp. 45–89. II. Habits of the Northern Fur Seal, etc., by Charles Bryant, with notes by J. A. Allen, pp. 89–108.

   Oulophocinæ, p. 44, and *Trichiphocinæ*, p. 44, subfamm. nov. of Otariidæ; subsequently abandoned.

4. A Spike-horned Moose. <*Amer. Nat.*, IV, p. 443, with 1 fig., Sept., 1870.

   Captured in northern Maine, where 'spike-horns' are said to be well-known to moose hunters.

5. The Distribution of the Moose in New England. <*Amer. Nat.*, IV, pp. 505–506, Nov., 1870.

   Frequent in northern Maine, scarce in extreme northern parts of New Hampshire and Vermont, and in the Adirondacks in New York.

1871.

6. The Classification of the Eared Seals. <*Amer. Nat.*, V, pp. 37–42, March, 1871.

   Reply to criticisms by Theodore Gill of my paper on the Eared Seals (*supra*, No. 3), in *Amer. Nat.*, IV, pp. 675–684, Jan. 1871.

7. The Fauna of the Prairies. <*Amer. Nat.*, V, pp. 4–9, March, 1871.

   Mammals, *passim.*, pp. 4–6.

8. On the Mammals and Winter Birds of East Florida, with an examination of certain assumed specific characters in Birds, and a sketch of the Bird-faunæ of Eastern North America. <*Bull. Mus. Comp. Zoöl.*, II, No. 3, pp. 161–450, pll. 4–8, April, 1871.

   Annotated list of the mammals, 35 species, pp. 168–185. *Trichechus manatus* Linné employed as the name of the Manatee (p. 171).

1872.

9. "Spike-horned Muledeer." <*Amer. Nat.*, VI, pp. 692–693, Nov., 1872.

The "Spike-horned" mule deer recorded in a previous number of the *American Naturalist* (July, 1872) by E. D. Cope is shown to have been a young elk.

10. Description of a Specimen of *Balænopterus musculus.* <*Amer. Nat.*, VI, pp. 473–474, August, 1872.

Review of a memoir by Thomas Dwight, Jr., on this species (*Mem. Boston Soc. Nat. Hist.*, II, pp. 203–230, June, 1872).

11. The Habits of the Orca. <*Amer. Nat.*, VI, pp. 474–475, August, 1872.

Review of C. M. Scammon's paper 'The Orca' (*Overland Monthly*, July, 1872, pp. 52–57).

12. [Geographical Variation in Mammals and Birds.] <*Proc. Boston Soc. Nat. Hist.*, XV, pp. 156–159, Sept., 1872.

Mainly in reference to birds.

1874.

13. Laws of Geographical Variation in North American Mammals and Birds. <*Amer. Nat.*, VIII, pp. 227–229, April, 1874.

In reference to R. Ridgway's paper (*Amer. Nat.*, VII, pp. 548–555, Sept., 1873) on the same subject.

14. Notes on the Mammals of portions of Kansas, Colorado, Wyoming, and Utah. <*Bull. Essex Inst.*, VI, pp. 43–66, May, 1874.

Four distinct lists of mammals observed in the region named in the title, with notes on their habits, distribution, etc. I. Mammals of Middle and Western Kansas, 28 species, with three pages on *Cynomys ludovicianus*, pp. 45–52. II. Mammals of Park County, Colorado, 37 species, pp. 53–58. III. Mammals of Carbon County, Wyoming, 32 species, pp. 58–61. IV. Mammals of Great Salt Lake Valley, Utah, 42 species, pp. 61–66.

15. Notes on the Natural History of portions of Dakota and Montana Territories, being the substance of a report to the Secretary of War, on the collections made by the North Pacific Railroad Expedition of 1873, Gen. D. S. Stanley, commander. <*Proc. Boston Soc. Nat. Hist.*, XVII, pp. 33–91, June, 1874. (Also as a separate pamphlet of 61 pp., 1874.)

Mammals, pp. 36–45 (pp. 6–13 of the reprint), 31 species; notes on habits, distribution, etc.

16. On Geographical Variation in Color among North American Squirrels; with a list of the Species and Varieties of the American Sciuridæ occurring north of Mexico. <*Proc. Boston Soc. Nat. Hist.*, XVI, pp. 276–294, June, 1874.

Variation in color, pp. 276–286; list of species and varieties, pp. 286–294. Genera: *Sciurus*, *Sciuropterus*, *Tamias*, *Spermophilus*, *Cynomys*, *Arctomys*. Species, 25, with 19 additional varieties = 44 forms. Vars. nov.: (1) *Tamias quadrivittatus* var. *pallidus*, p. 289; (2) *Spermophilus tridecem-lineatus* var. *pallidus*, p. 291; (3) *Spermophilus parryi* var. *kodiacensis*, p. 292.

17. Scammon's 'Marine Mammals of the Northwest Coast and American Whale-fishery.' <*Amer. Nat.*, VIII, pp. 632–635, Oct., 1874.

An appreciative review of this important work (4to).

1875.

18. Synopsis of the American Leporidæ. <*Proc. Boston Soc. Nat. Hist.*, XVII, 1875, pp. 430–436.

10 species, and 5 varieties = 15 forms.

19. The Marine Mammals of Western North America. <*The Nation*, No. 501, Feb. 4, 1875.

Review of Scammon's 'Marine Mammalia' (*cf. supra*, No. 17).

20. Geographical Variation in Color among Squirrels. <*Amer. Nat.*, IX, pp. 504–509, Sept. 1875.

Extended abstract, with editorial comment, of No. 16, *supra*.

1876.

21. The North American Bison and its Extermination. <*Penn Monthly*, VII, No. 75, pp. 214–224, March, 1876.

A general summary of the subject.

22. Recent Memoirs on North American Mammals. <*Amer. Nat.*, X, pp. 423–425, July, 1876.

Synoptic review of Theodore Gill's 'Synopsis of Insectivorous Mammals' (*Bull. U. S. Geol. and Geogr. Surv. Terr.*, No. 2, May, 1875).

23. The Former Range of some New England Carnivorous Mammals. <*Amer. Nat.*, X, pp. 708–715, Dec., 1876.

Citations of many early records, showing their former occurrence far south of their present ranges, and their great decrease in numbers as well as range.

24. "Hare and Rabbit." <*Forest and Stream*, VI, No. 18, p. 284, June 8, 1876.

On the significance and use of the terms.

25. A Description of a new Generic Type (*Bassaricyon*) of Procyonidæ from Costa Rica. <*Proc. Acad. Nat. Sci. Philadelphia*, 1876, pp. 20–23, pl. i, April 18, 1876.

Bassaricyon gabbi, gen. et sp. nov.

26. List of Mammals and Birds collected at Lake Titicaca, Peru. With field-notes by Samuel Garman. <*Bull. Mus. Comp. Zoöl.*, III, No. 15, July, 1876, pp. 349–359.

Mammals, 10 species, pp. 349–352.

27. Description of some remains of an Extinct Species of Wolf and an Extinct Species of Deer from the Lead Region of the Upper Mississippi. <*Amer. Journ. Sci. and Arts*, 3d ser., XI, pp. 47–51, Jan., 1876.

Canis mississippiensis and *Cervus whitneyi*, spp. nov., p. 49.

28. Geographical Variation among North American Mammals, especially in respect to size. <*Bull. U. S. Geol. Surv. Terr.*, II, No. 4, pp. 309–344, July 1, 1876.

Relates exclusively to the fissiped Carnivora.

"In a general way, the correlation of size with geographical distribution may be formulated in the following propositions:

"(1) *The maximum physical development of the individual is attained where the conditions of environment are most favorable to the life of the species.* Species being primarily limited in their distribution by climatic conditions, their representatives living at or near either of their respective latitudinal boundaries are more or less unfavorably affected by the influences that finally limit the range of the species....

"(2) *The largest species of a group (genus, subfamily, or family, as the case may be) are found where the group to which they severally belong reaches its highest development, or where it has what may be termed its center of distribution.* In other words, species of a given group attain their maximum size where the conditions of existence for the group in question are the most favorable, just as the largest representatives of a species are found where the conditions are most favorable for the existence of the species.

"(3) *The most 'typical' or most generalized representatives of a group are found also near its center of distribution, outlying forms being generally more or less 'aberrant' or specialized* Thus the Cervidæ, though nearly cosmopolitan in their distribution, attain their greatest development, both as respects the size and the number of the species, in the temperate portions of the northern hemisphere. The tropical species of this group are the smallest of its representatives. Those of the temperate and cold temperate regions are the largest, where, too, the species are the most numerous....The possession of large, branching, deciduous antlers forms one of the marked features of the family. These appendages attain their greatest development in the northern species, the tropical forms having them reduced almost to mere spikes, which in some species never pass beyond a rudimentary state...." (p. 310).

29. Geographical Variation among North American Mammals, especially in respect to size. <*Amer. Nat.*, X, pp. 625–627, Oct., 1876.

Abstract of the preceding (No. 28).

30. Memoirs of the Geological Survey of Kentucky. | N. S. Shaler, director. | Vol. I, Part ii. | – | The American bisons, | living and extinct. | By J. A. Allen. | With twelve plates and map. | – | University press, Cambridge: | Welch, Bigelow, & Co. | 1876. Also:

Memoirs of the Museum of Comparative Zoölogy,| at Harvard College, Cambridge, Mass. | Vol. IV, No. 10. |— | The American bisons, living and extinct: | By J. A. Allen. | Published by permission of N. S. Shaler, director of the Kentucky | Geological Survey. With 12 plates and a map. | University press, Cambridge: | Welch, Bigelow, & Co. | 1876.

4°, pp. i–ix, 1–246, 1 col. map, 12 pll., 13ll., 2 woodcuts in text. Edition of 500 copies.

These two publications were simultaneous, and only differ in the titles. The following are the contents:—

PART I.

PART II.

Appendix I. Additional notes on the former range and early abundance of the bison east of the Mississippi River, pp. 223–231.

Appendix II. On the age of the bison in the Ohio Valley, by N. S. Shaler, pp. 232–236.

Index, pp. 237–246.

Map, colored, to illustrate the geographical distribution, past and present, of *Bison americanus*, and 12 plates, each with unpaged explanatory leaf: pl. i, horncores of *Bison latifrons;* pl. ii, atlas of *B. latifrons* and *B. americanus;* pl. iii, teeth of *B. antiquus* and *B. americanus;* pl. iv, skull of *B. antiquus;* pll. v and vi, individual variation in skulls of *B. americanus;* pl. vii, skulls of *B. americanus* and *B. bonasus*, and tarsal bones of *B. americanus* and *B. latifrons?;* pl. viii, horns of *B. americanus* and *B. antiquus;* pl. ix, milk dentition of *B. americanus;* pl. x, stages of attrition due to age in teeth of *B. americanus;* pl. ix, metacarpal bones of *B. americanus;* pl. xii, molar-premolar series in *B. americanus* and *Bos taurus*.

31. Recent Contributions to North American Mammalogy.<*Amer. Nat.*, X, pp. 362–365, June, 1876. (Anonymous.)

Review of Coues's 'Synopsis of the Muridæ (*Proc. Acad. Nat. Sci. Philadelphia*, 1874) and his 'Some Account. . . . of *Zapus hudsonius*' (*Bull. U. S. Geol. Surv. Torr.*, Ser. 2, No. 5, Jan., 1871.

32. Exploration of Lake Titicaca, by Alexander Agassiz and S. W. Garman. III. List of Mammals and Birds, with Field Notes by Mr. Garman.<*Bull. Mus. Comp. Zoöl.*, III, pp. 349–359, July, 1876.

Mammals, pp. 350–353; 10 species. [Accidental duplication of No. 26, *supra*.]

33. The Extirpation of the larger Indigenous Mammals in the United States. <*Penn. Monthly*, pp. 794–806, Oct., 1876.

1877.

34. History of the American Bison, *Bison americanus*.<Ninth Ann. Rep. U. S. Geol. Surv. Terr. for the year 1875, pp. 443–587, June, 1877. Edition of 7000 copies; also separate, 1000 copies.

A republication of the original memoir (see *supra*, No. 30) under Dr. Coues's editorship, exclusive of the map and plates, with the following changes: —

1. The omission of the illustrations, explanatory pages, and textual references.

2. The omission of the portion relating to the extinct species, the present reprint being confined to the one existing species, beginning at page 36 of the original.

3. The incorporation of the appendices in the body of the text.

4. The addition of much new matter by the author himself.

5. Various minor modifications with the slight alteration, chiefly verbal, of context incident thereto.

6. Alteration of the title and substitution of editorial preface for the preliminary matter of the original.

" No editorial abridgment or digest of any part of the memoir was made, the portions of the memoir here reproduced being according to copy furnished by author, who added much new matter and made some little changes *passim* in the context. A few editorial notes, chiefly explanatory of modifications of the text, are introduced in brackets." — *E. Coues*, editor.

35. Additional note on *Bassaricyon gabbi*.<*Proc. Acad. Nat. Sci. Philadelphia*, 1877, pp. 267–268, pl. ii.

By error, a description and figure of *Nasua narica*.

36. Monographs of North American Rodentia. By Elliott Coues and Joel Asaph Allen. = Report of the U. S. Geol. Survey of the Territories (F. V. Hayden,

U. S. Geol.-in-Charge), Vol. XI. Washington: Govern. Printing Office, August, 1877. 4°, pp. i–x, 1–1091, pll. i–v.

The following by J. A. Allen: —

Monograph II. Leporidæ, pp. 265–378. Mon. III. Hystricidæ, pp. 379–398. Mon. IV. Lagomyidæ, pp. 399–414. Mon. V. Castoroididæ, pp. 415–426. Mon. VI. Castoridæ, pp. 427–454. Mon. XI. Sciuridæ, pp. 631–940. Appendix A. Synoptical List of the Fossil Rodentia of North America, pp. 943–950.

Fam. nov. *Castoroididæ*, p. 419; *Lepus sylvaticus* var. *arizonæ*, var. nov., p. 332; *Lepus graysoni*, sp. nov., p. 347; *Lepus brasiliensis* var. *gabbi*, var. nov., p. 419; *Tamias asiaticus* var. *borealis*, var. nov., p. 793.

Respecting Mon. XI, Sciuridæ, see Allen, *Bull. Amer. Mus. Nat. Hist.*, XXXIV, p. 153, May 17, 1915 (*infra*, No. 259).

37. Material for Bibliography of North American Mammals. Prepared by Theodore Gill and Elliott Coues. Appendix B to Coues and Allen's 'Monographs of North American Rodentia,' pp. 951–1081, August, 1877.

"In the labor of preparing this matter for the press, which has devolved upon Dr. Coues, much assistance has been rendered by Mr. Allen, who has kindly revised the proofs, adding some titles, and making many valuable suggestions." — *Gill and Coues, l. c.*, p. 951.

"Q.— Addenda" to the foregoing, pp. 1074–1081. "For most of the following titles, about 250, the compilers are indebted to the kind attentions of Mr. J. A. Allen. They were received during the printing of the Bibliography, but too late for insertion under their proper heads, and represent, in particular, many important palæontological papers by Leidy, Marsh and Cope."— *Gill and Coues, l. c.*, p. 1074.

38. Northern Range of the Bison. <*Amer. Nat.*, XI, No. 10, Oct., 1877, p. 624.

Its presence on the Peace and Hay Rivers, near Great Slave Lake, recorded on information received from E. W. Nelson.

39. American Insectivorous Mammals. <*Amer. Nat.*, XI, No. 10, Oct., 1877, pp. 613–615. (Anonymous.)

Review of Coues's Precursory Notes on American Insectivorous Mammals,' etc. (*Bull. U. S. Geol. Surv. Terr.*, III, pp. 631–653, May, 1877).

40. North American Fur-bearing Animals. <*Amer. Nat.*, XI, No. 10, Oct., 1877, pp. 617–619. (Anonymous.)

Review of Coues's "Fur-bearing Animals," 1877.

41. Cope's Vertebrate Palæontology of New Mexico. <*Amer. Nat.*, XI, No. 12, Dec., 1877, pp. 750–753. (Anonymous.)

Review of Vol. IV, pt. ii, of Report U. S. Geogr. Surv. West of One Hundredth Meridian.

1878.

42. The Geographical Distribution of the Mammals, considered in relation to the principal ontological regions of the earth, and the laws that govern the distribution of animal life. <*Bull. U. S. Geol. and Geogr. Surv. Terr.*, IV, No. 2, pp. 313–377, May, 1878.

Contents: — I. Distribution of mammalian life in the Northern Hemisphere, considered in relation to laws of geographical distribution, pp. 313–329.— Historical résumé, with critical analysis of views of Sclater and Wallace.

II.— Mammalian regions of the Globe, pp. 329–373.— (1) Arctic Realm; (2) North Temperate Realm; (3) Tropical Realm; (4) South American Temperate Realm; (5) Indo-African Realm; (6) Australian Realm; (7) Lemurian Realm; (8) Antarctic Realm.

The realms are divided into regions and provinces.

III.— General Summary, pp. 373–377, with diagram of realms, regions and provinces.

43. Synonymatic list of the American Sciuri, or Arboreal Squirrels.<*Bull. U. S. Geol. and Geogr. Surv. Terr.*, IV, No. 4, pp. 877–887, 905, Dec., 1878.

Revised list of species, with nomenclatural changes from the Monograph of the Sciuridæ in Coues and Allen's 'Monographs of North American Rodentia,' 1877 (see *supra*, No. 36).

1879.

44. On the Coatis (genus *Nasua* Storr).<*Bull. U. S. Geol. and Geogr. Surv. Terr.*, V, No. 2, pp. 153–174, Sept., 1879.

Monographic revision of the genus *Nasua*.

45. On the Species of the Genus *Bassaris*.<*Bull. U. S. Geol. and Geogr. Surv. Terr.*, V, No. 3, pp. 331–340, Nov., 1879.

Monographic revision of the genus *Bassaris*.

1880.

46. The Fauna of Eastern Massachusetts: Forms brought in and expelled by Civilization.<Memorial History of Boston, edited by Justin Winsor, I, 1880, Chap. ii, pp. 9–16.

Relates to mammals only.

47. History of North American Pinnipeds: A Monograph of the Walruses, Sea-lions, Sea-bears and Seals of North America. Washington: Govern. Printing Office, 1880. = U. S. Geol. and Geogr. Surv. Territories (F. V. Hayden, U. S. Geol.-in-Charge), Miscel. Publ. No. 11. 8°, pp. i–xvi, 1–785, 60 woodcuts.

Family Odobænidæ, pp. 5–186. *Odobænus* Brisson (1762) the proper generic name for the walruses, and Odobænidæ the proper name for the family — not *Trichechus* (auct. non Linné), nor Trichechidæ, which are, respectively, the proper names of the Manatees;[1] two species of walrus established, *O. rosmarus* (Linn.) and *O. obesus* (Illiger), with figures of skulls, and full history of each species, including the nomenclature, osteology and dentition, geographical distribution, chase and commercial products.

Family Otariidæ, pp. 187–411. Technical and commercial history, with synopsis, characters, and geographic distribution of the extra-limital species, recent and fossil, as well as the North American.

Family Phocidæ, pp. 412–756. Technical history of the family, classification, synonymatic list of the genera and species, geographical distribution, fossil remains, milk-dentition, habits, migrations, locomotion on land, the seal-hunting industry and sealing-grounds, methods of capture, species hunted, products, decrease of seals from injudicious hunting, etc., pp. 412–557; systematic treatment of the North American species, pp. 557–756; Appendix: *a*, material examined (pp. 757–764); *b*, additions and corrections (pp. 765–773); index (pp. 775–785).

"A most valuable and complete history of these animals, especially of those found in North America, of their distribution and pursuit, with full synonymy, and copious tables and references. The history of the species of the group generally is also discussed, with remarks on their synonymy and distribution."— Zoöl. Rec. for 1880, Mamm. p. 2.

---

[1] I employed *Trichechus manatus* Linné for the Florida Manatee as early as 1871 (see *supra*, No. 8).

1881.

48. List of Mammals collected by Dr. Edward Palmer in Northeastern Mexico, with field-notes by the collector. <*Bull. Mus. Comp. Zoöl.*, VIII, No. 9, March, 1881, pp. 183–189.

Annotated list of 25 species. *Heteromys alleni* Coues, sp. nov., ined., p. 187.

1882.

49. Preliminary List of Works and Papers relating to the Mammalian Orders Cete and Sirenia. <*Bull. U. S. Geol. and Geogr. Surv.*, VI, No. 3, pp. 399–562, Aug. 30, 1882.

Covers the period from Albertus Magnus (1495) to the year 1840, and numbers 1013 annotated titles, the annotations in many cases amounting to a full statement of contents, so far as pertinent to the present subject, including names of species and genera and nature of treatment. All thus far published. The cause of the discontinuance of publication is explained in an insert, as follows:

"Owing to the illness of the author, which prevented his revision of the proof-sheets, it was necessary to stop the printing of the 'List' at the end of the year 1840. The present instalment comprises only a little more than one-third of the article; the remainder will be published as soon as the author's health renders it practicable.— J. A. Allen, Cambridge, Sept., 1882."

See further, a 'Personal Note' in *Bull. Amer. Mus. Nat. Hist.*, XXIV, 1908, pp. 279–280.

1883.

50. Capture of the Crested Seal on the Coast of Massachusetts. <*Science*, I, No. 19, pp. 542–543, June 15, 1883.

51. The Right Whale of the North Atlantic. <*Science*, I, No. 21, pp. 598–599, June 29, 1883.

Review of Dr. J. B. Holder's paper on this subject (*Bull. Amer. Mus. Nat. Hist.*, I, pp. 99–137, pll. x–xiii, May 30, 1883).

52. The Right Whale of the North Atlantic. <*Science*, II, No. 26, p. 134, Aug. 3, 1883.

Reply to Dr. Holder's rejoinder to above.

53. The Right Whale of the North Atlantic. <*Science*, II, No. 30, p. 267, Aug. 31, 1883.

Reply to a second rejoinder by Dr. Holder.

1884.

54. How long the Buffaloes remained in Illinois. <*American Field*, XXII, p. 128, Aug. 9, 1884.

Remained till as late as 1805, and were not wholly extirpated till 1808 or 1809; not wholly destroyed by a heavy snowfall in 1763, as stated by a previous correspondent.

55. The Mammals of the Adirondacks. <*Science*, IV, 1884, pp. 445–446.

Review of Dr. C. Hart Merriam's 'Vertebrates of the Adirondack Region, Northeastern New York,' Chap. II, Mammals (*Trans. Linn. Soc. New York*, I, 1882, pp. 27–106; II, 1884, pp. 9–214).

1885.

56. What is the Present Distribution of the American Bison? <*Forest and Stream*, XXIV, No. 8, p. 145, March 19, 1885.

A request for exact information as to its present numbers and distribution.

57. Prairie Dog Wells. <*Forest and Stream*, XXIV, No. 8, p. 145, March 19, 1885.

No evidence that prairie dogs ever burrow for water, but evidence to the contrary, and against considering "the prairie dog as a water witch," as believed by a previous correspondent.

58. On an Extinct Type of Dog from Ely Cave, Lee County, Virginia. <*Mem. Mus. Comp. Zoöl.*, X, No. 2, pp. 1–8, pll. i–iii, Dec. 1885.

Pachycyon robustus sp. nov., p. 4. With 'On the Age of the Ely Cave,' by N. S. Shaler.

1887.

59. The West Indian Seal (*Monachus tropicalis*). <*Bull. Amer. Mus. Nat. Hist.*, II, pp. 1–34, pll. i–iv, April 25, 1887.

Introduction, pp. 1–3; external characters, pp. 4–6; osteological characters, pp. 6–19; sexual differences, pp. 20–21; affinities of the genus *Monachus*, pp. 22–23; general history, pp. 23–26; geographical distribution, pp. 27–29; habits, pp. 29–34.

60. Note on Squalodont Remains from Charleston, S. C. <*Bull. Amer. Mus. Nat. Hist.*, II, pp. 34–39, pll. v–vi, April 25, 1887.

Squalodon tiedemani, sp. nov., p. 34. Based on the rostral portion of a skull, with teeth *in situ*.

61. The West Indian Seal. <*Science*, IX, No. 206, p. 35, Jan. 14, 1887.

Capture of specimens by Henry A. Ward at the Triangles, off Yucatan. Habits and distinguishing characters.

62. The West Indian Seal. <*Science*, IX, No. 207, p. 59, Jan. 21, 1887.

Notice of a paper on this species by F. W. True and F. A. Lucas, published in 1886, based on a female specimen taken near Havana, Cuba, in 1883.

1888.

63. List of Important Publications relating to Mammals. <Kingsley's Riverside Natural History, V, 1888, pp. 529–534.

About 130 titles.

1889.

64. Notes on a Collection of Mammals from Southern Mexico, with descriptions of new species of the genera *Sciurus*, *Tamias*, and *Sigmodon*. <*Bull. Amer. Mus. Nat. Hist.*, II, pp. 165–181, Oct. 21, 1889.

An annotated list of 16 species. Spp. et subspp. nov.: (1) *Sciurus alstoni*, p. 167 (name preoccupied and changed to *nayaritensis*, p. vii of same volume and Vol. III, p. 185); (2) *Tamias asiaticus bulleri*, p. 173; (3) *Tamias asiaticus merriami*, p. 176; (4) *Sigmodon fulviventer*, p. 180.

65. Former Range of the Buffalo in Virginia. <*New York Evening Post*, Dec. 10, 1889.

Not found in the immediate vicinity of Washington, as alleged by a previous correspondent, but on the upper James River.

1890.

66. On Seasonal Variations in Color in *Sciurus hudsonius*. <*Bull. Amer. Mus. Nat. Hist.*, III, pp. 41–44, June, 1890.

Based on a series of about 60 specimens, taken at Hastings, New York, and representing nearly every month of the year.

67. A Review of some of the North American ground Squirrels of the genus *Tamias*. <*Bull. Amer. Mus. Nat. Hist.*, III, pp. 45-116, June, 1890.

24 species and subspecies are recognized, of which 13 are here first described, as follows: (1) *Tamias obscurus*, p. 70; (2) *T. senex*, p. 83; (3) *T. speciosus* (Merriam, ined. MS.), p. 86; (4) *T. frater*, p. 88; (5) *T. amœnus*, p. 90; (6) *T. cinereicollis*, p. 94; (7) *T. umbrinus*, p. 96; (8) *T. quadrivittatus gracilis*, p. 99; (9) *T. q. luteiventris*, p. 101; (10) *T. q. affinis*, p. 103; (11) *T. q. neglectus*, p. 106; (12) *T. minimus consobrinus*, p. 112; (13) *T. m. pictus*, p. 115.

68. Notes on a small collection of West Indian Bats, with description of an apparently new species. <*Bull. Amer. Mus. Nat. Hist.*, III, pp. 169–173, Nov. 14, 1890.

An annotated list of 8 species, with critical comment on *Artibeus* "*perspicillatus* (Linn.)." *Artibeus coryi*, sp. nov., p. 173.

69. Notes on collections of Mammals made in Central and Southern Mexico, by Dr. Audley C. Buller, with descriptions of new species of the genera *Vespertilio*, *Sciurus*, and *Lepus*. <*Bull. Amer. Mus. Nat. Hist.*, III, pp. 175–194, Dec. 10, 1890.

Annotated list of 44 species, with additional notes on 3 species of Mexican Leporidæ. Spp. nov.: (1) *Vespertilio velifer*, p. 177; (2) *Sciurus cervicalis*, p. 183; (3) *Lepus sylvaticus aztecus*, p. 188; (4) *Lepus insolitus*, p. 189; (5) *Lepus truei*, p. 192.

70. Descriptions of a New Species and a New Subspecies of the Genus *Lepus*. <*Bull. Amer. Mus. Nat. Hist.*, III, pp. 159–160, Oct., 1890.

(1) *Lepus cinerascens*, p. 159; (2) *Lepus sylvaticus floridanus*, p. 160.

71. List of Mammals collected by Mr. Clark P. Streator in British Columbia, with descriptions of two new subspecies of *Sciurus*. <*Bull. Amer. Mus. Nat. Hist.*, III, pp. 161–168.

Annotated list of 20 species. Subspp. nov.: (1) *Sciurus hudsonius vancouverensis*, p. 165; (2) *S. h. californicus*, p. 165.

1891.

72. Description of a new species of Big-eared Bat, of the genus *Histiotus*, from Southern California. <*Bull. Amer. Mus. Nat. Hist.*, III, pp. 195–198, Feb. 20, 1891.

Histiotus maculatus, sp. nov., p. 195. Later the type of the genus *Euderma* H. Allen, 1892.

73. Notes on a Collection of Mammals from Costa Rica. <*Bull. Amer. Mus. Nat. Hist.*, III, pp. 203–218, April 17, 1891.

38 species, the following new: (1) *Blarina costaricensis*,[1] p. 205; (2) *Hesperomys (Vesperimus) cherriei*, p. 211; (3) *Hesperomys (Vesperimus) nudipes*, p. 213; (4) *Hesperomys (Oryzomys) alfaroi*, p. 214.

74. On a Collection of Mammals from Southern Texas and Northeastern Mexico. <*Bull. Amer. Mus. Nat. Hist.*, III, pp. 219–228, April 17, 1891.

An annotated list of 31 species, the following new: (1) *Scalops argentatus texanus*, p. 221; (2) *Dipodops sennetti*, p. 226.

75. Description of a new species of *Capromys*, from the Plana Keys, Bahamas. <*Bull. Amer. Mus. Nat. Hist.*, III, pp. 329–336, Aug. 31, 1891.

Capromys ingrahami, p. 329, sp. nov.

76. Notes on new or little-known North American Mammals, based on recent additions to the Collection of Mammals in the American Museum of Natural History. <*Bull. Amer. Mus. Nat. Hist.*, III, pp. 263–310, June 30, 1891.

Extended remarks on *Heteromys alleni* Coues, *Dipodomys phillipsii* Gray, *Neotoma micropus* Baird, *Hesperomys indianus* Wied (= *Mus musculus* Linn.), the generic name *Hesperomys* Waterhouse, and *Mus agrarius americanus* Kerr (1771) *vs. Hesperomys leucopus* auctorum (strictly, *vs. Mus sylvaticus noveboracensis* Fischer, 1829). *Mus agrarius americanus* shown later to be unavailable in this connection, through a previous *Mus americanus* Kerr.

Spp. et subspp. nov.: (1) *Dipodops ordii palmeri*, p. 276; (2) *Dipodops richardsoni*, p. 277; (3) *Perognathus (Chætodipus) femoralis*, p. 281; (4) *Neotoma micropus canescens*, p. 285; (5) *Oryzomys aquaticus* p. 289; (6) *Vesperimus difficilis*, p. 298; (7) *Vesperimus nasutus*, p. 299; (8) *Vesperimus mearnsii*, p. 300.

77. Recent work in North American Mammalogy. <*Trans. New York Acad. Sci.*, X, No. 5, 1891, pp. 71–85.

A review of "recent progress" (1857–1891) in North American mammalogy.

78. Descriptions of two supposed new species of Mice from Costa Rica and Mexico, with remarks on *Hesperomys melanophrys* of Coues. <*Proc. U. S. Nat. Mus.*, XIV, pp. 193–196, 1891.

Spp. nov.: (1) *Oryzomys talamancæ*, p. 193; (2) *Hesperomys (Vesperimus) affinis*, p. 195.

1892.

79. Description of a new species of *Perognathus* from Southeastern Texas. <*Bull. Amer. Mus. Nat. Hist.*, IV, pp. 45–46, pl. iii, March 25, 1892.

Perognathus merriami, p. 45.

80. On a small collection of Mammals from the Gallapagos Islands, collected by Dr. G. Baur. <*Bull. Amer. Mus. Nat. Hist.*, IV, pp. 47–50, March 25, 1892.

4 species, 2 new, as follows: (1) *Atalapha brachyotis*, p. 47; (2) *Oryzomys bauri*, p. 48.

81. The Geographical Distribution of North American Mammals. <*Bull. Amer. Mus. Nat. Hist.*, IV, pp. 199–244, pll. v–viii (maps).

Influences determining the geographic distribution of life (climatic), pp. 199–200; interrelation of land areas, pp. 200–201); mammals as the basis for the classification of life areas,

[1] Based on alcoholic specimens collected in all probability in the United States, wrongly attributed to Costa Rica, whence they were received by the author.

pp. 202–203; systematic classification of life areas, pp. 203–211; the Sclaterian system, pp. 211–212; the mammals of North America considered in relation to the North American Region and its subdivisions, pp. 213–240, with 3 maps.

82. A synopsis of the Pinnipeds, or Seals and Walruses, in relation to their commercial history and products. <Fur-seal Arbitration. Appendix to the case of the United States before the Tribunal of Arbitration, etc., I, 1892, pp. 367–391.

83. Fur-seal Hunting in the Southern Hemisphere. <*Ibid.*, pp. 393–404.

84. The Alaskan Fur-seal and Pelagic Sealing. <*Ibid.*, pp. 405–410.

85. Visitors' Guide to the Collection of Mammals in the American Museum of Natural History, Seventy-Seventh Street and Eighth Avenue, New York City. By J. A. Allen, Ph.D., Curator. New York: Printed for the Museum. 1892. 8°, pp. 1–89. Diagram of hall and 24 cuts in text.

1893.

86. Rabbits and Hares. <*Our Animal Friends*, XX, No. 11, July, 1893, pp. 248–251.

87. List of Mammals and Birds collected in Northeastern Sonora and Northwestern Chihuahua, Mexico, on the Lumholtz Archæological Expedition, 1890–92. <*Bull. Amer. Mus. Nat. Hist.*, V, pp. 27–42, March 16, 1893.

Mammals, 17 species, pp. 28–32. *Sciurus apache*, sp. nov., p. 29.

88. Descriptions of Four New Species of *Thomomys*, with remarks on other Species of the Genus. <*Bull. Amer. Mus. Nat. Hist.*, V, pp. 47–63, pl. i, April 28, 1893.

I. Descriptions of spp. nov.: (1) *Thomomys monticolus*, p. 48; (2) *T. aureus*, p. 49; (3) *T. fossor*, p. 51; (4) *T. toltecus*, p. 52. II. Questions of nomenclature, pp. 53–64. III. Cranial characters, pp. 64–65. IV. Species and subspecies of the genus *Thomomys*, pp. 66–67.

89. List of Mammals collected by Mr. Charles P. Rowley in the San Juan Region of Colorado, New Mexico and Utah, with descriptions of new species. <*Bull. Amer. Mus. Nat. Hist.*, V, pp. 64–84, April 28, 1893.

Annotated list of 34 species. Spp. nov.: (1) *Zapus princeps*, p. 71; (2) *Arvicola* (*Mynomes*) *aztecus*, p. 73; (3) *Reithrodontomys aztecus*, p. 79; *Sitomys auripectus*, p. 75; *S. rowleyi*, p. 76.

90. On a Collection of Mammals from the San Pedro Martir Region of Lower California, with notes on other species, particularly of the genus *Sitomys*. <*Bull. Amer. Mus. Nat. Hist.*, V, pp. 181–202, August 18, 1893.

Annotated list of 9 species. Spp. and subspp. nov.: (1) *Sitomys americanus thurberi*, p. 185; (2) *S. martirensis*, p. 187; (3) *S. gilberti*, p. 188; (4) *Tamias leucurus peninsulæ*, p. 197; (5) *Scapanus anthonyi*, p. 200.

Hesperomys gambelii Baird — type locality fixed as Monterey, Cal., p. 190; *Hesperomys boylei* Baird — type locality fixed as Middle Fork of the American River, El Dorado Co., Cal., p. 192; *Hesperomys austerus* Baird — type locality fixed as Fort Steilacoom, Puget Sound, p. 192.

91. On a Collection of Mammals from the Island of Trinidad, with descriptions of new species. <*Bull. Amer. Mus. Nat. Hist.*, V, pp. 203–234, Sept. 21, 1893. (With Frank M. Chapman.)

Annotated list of 34 species. Spp. nov.: (1) *Chœronycteris intermedia*, p. 207; (2) *Nectomys palmipes*, p. 209; (3) *Tylomys couesi*, p. 211; (4) *Oryzomys speciosus*, 212; (5) *O. trinitatis*, p. 213; (6) *O. velutinus*, p. 214; (7) *O. brevicauda*, p. 215; (8) *Loncheres castaneus*, p. 222; (9) *Echimys trinitatis*, p. 223.

List of known land mammals from Trinidad (65 species), pp. 231–234.

92. Further Notes on Costa Rican Mammals, with description of a new species of *Oryzomys*. <*Bull. Amer. Mus. Nat. Hist.*, V, pp. 237–240, Sept. 22, 1893.

Annotated list of 18 species. *O. costaricensis*, sp. nov., p. 239.

93. Description of a new species of Opossum, from the Isthmus of Tehuantepec, Mexico. <*Bull. Amer. Mus. Nat. Hist.*, V, pp. 235–236, Sept. 22, 1893.

Didelphys (Micoureus) canescens, sp. nov., p. 235.

94. Description of a new Mouse from Lake County, California. <*Bull. Amer. Mus. Nat. Hist.*, V, pp. 335, 336, Dec. 16, 1893.

Sitomys robustus, p. 335.

95. Description of a new species of *Geomys* from Costa Rica. <*Bull. Amer. Mus. Nat. Hist.*, V, pp. 337, 338, Dec. 16, 1893.

Geomys cherriei, p. 337.

1894.

96. Notes on Mammals from New Brunswick, with Description of a new Species of *Evotomys*. <*Bull. Amer. Mus. Nat. Hist.*, VI, pp. 99–106, April 14, 1894.

Annotated list of 33 species. *Evotomys fuscodorsalis* sp. nov., p. 103. Later found to be a melanistic phase of *E. gapperi* (see *infra*, No. 139).

97. On the Seasonal Change of Color in the Varying Hare (*Lepus americanus* Erxl.). <*Bull. Amer. Mus. Nat. Hist.*, VI, pp. 107–128, April 14, 1894.

Change due to moult, not to change of color of the hair. Based on the study of about 75 specimens collected in New Brunswick.

98. On the Mammals of Aransas County, Texas, with descriptions of new forms of *Lepus* and *Oryzomys*. <*Bull. Amer. Mus. Nat. Hist.*, VI, pp. 165–198, May 31, 1894.

Annotated list of 36 species. Subspp. nov.: (1) *Lepus sylvaticus mearnsii* p. 171 (footnote); (2) *Oryzomys palustris texensis*, p. 177.

Table of cranial measurements and ratios of 34 North American specimens of the genus *Mephitis* (insert, facing p. 190); note on variability of coloration in species of *Mephitis*, pp. 192–196.

99. Cranial Variations in *Neotoma micropus* due to Growth and Individual Differentiation. <*Bull. Amer. Mus. Nat. Hist.*, VI, 1894, pp. 233–246, pl. iv, August 3, 1894.

100. Remarks on Specimens of *Chilonycteris rubiginosus* from Western Mexico, and on the Color Phases of *Pteronotus davyi* Gray. <*Bull. Amer. Mus. Nat. Hist.*, VI, 1894, pp. 247–248, August 3, 1894.

101. Descriptions of ten new North American Mammals, and remarks on others. <*Bull. Amer. Mus. Nat. Hist.*, VI, 1894, pp. 317–333, Nov. 7, 1894.

Spp. et subspp. nov.: (1) *Perognathus pricei*, p. 318; (2) *Perognathus conditi*, p. 318; (3) *Reithrodontomys mexicanus fulvescens*, p. 319; (4) *Arvicola leucophæus*, p. 320; (5) *Sitomys americanus arizonæ*, p. 321; (6) *Neotoma campestris*, p. 322; (7) *Neotoma rupicola*, p. 323; (8) *Neotoma grangeri*, p. 324; (9) *Sciurus hudsonicus dakotensis*, p. 325; (10) *Phenacomys truei*, p. 331. Remarks on *Arvicola haydenii* Baird and *A. cinnamomeus* Baird.

102. Recent Progress in the Study of North American Mammals. <*Abst. Proc. Linnæan Soc. New York*, for the year ending March 27, 1894, pp. 17–45, July 20, 1894.

A review of recent methods and results, with lists of the North American species of *Lepus*, *Heteromys*, *Perognathus*, *Dipodomys*, *Perodipus*, *Thomomys*, *Geomys*, *Phenacomys*, *Evotomys*, *Arvicola*, *Neotoma*, *Sigmodon*, *Oryzomys*, *Sitomys*, *Sciurus*, *Tamias*, *Spermophilus*, etc., in illustration of the subject.

103. Descriptions of five new North American Mammals. <*Bull. Amer. Mus. Nat. Hist.*, VI, 1894, pp. 347–350, Dec. 7, 1894.

Spp. et subspp. nov.: (1) *Arvicola insperatus*, p. 347; (2) *Lepus texianus eremicus*, p. 347; (3) *Lepus sylvaticus pinetis*, p. 348; (4) *Sciurus arizonensis huachuca*, p. 349; (5) *Sciurus hudsonicus grahamensis*, p. 350.

104. Remarks on a Second Collection of Mammals from New Brunswick, and on the Re-discovery of the Genus *Neotoma* in New York State. <*Bull. Amer. Mus. Nat. Hist.*, VI, 1894, pp. 359–364, Dec. 22, 1894.

Notes on 7 species, pp. 359–361. Capture of *Neotoma pennsylvanicus* Stone on Storm King Mountain, Orange Co., N. Y.; and reference to its long previous record of occurrence at Piermont, Rockland Co., N. Y., pp. 362–364.

1895.

105. On the Species of the Genus *Reithrodontomys*. <*Bull. Amer. Mus. Nat. Hist.*, VII, 1895, pp. 107–143, May 21, 1895.

A monographic revision; 15 species and subspecies recognized, the following new: (1) *R. merriami*, p. 119; (2) *R. dychei*, p. 120; (3) *R. dychei nebrascensis*, p. 122; (3) *R. megalotis deserti*, p. 127; (4) *R. arizonensis*, p. 134; (5) *R. mexicanus intermedius*, p. 136; (6) *R. m. aurantius*, p. 137; (7) *R. costaricensis*, p. 139.

106. On the Names of Mammals given by Kerr in his 'Animal Kingdom,' published in 1792. <*Bull. Amer. Mus. Nat. Hist.*, VII, pp. 179–192, June 20, 1895.

Four generic and 16 specific names antedate current names taken from later authors; as follows: *Myocastor* = *Myopotamus* Geoffroy, 1805 (p. 182); *Cricetus* = *Cricetus* Cuvier, 1817 (p. 183); *Myotalpa* = *Siphneus* Brants, 1827 (p. 183); *Lynx* = *Lynx* Rafinesque, 1818 (p. 182). The necessary changes in specific names are:

Cercocebus torquatus (Kerr) = *C. collaris* Gray, 1843 (p. 185).
Nasalis nasuus (Kerr) = *N. nasalis* (Shaw), 1800 (p. 186).
Colobus badius (Kerr) = *C. ferruginea* (Shaw), 1800 (p. 186).
Cebus albulus (Kerr) = *C. hypoleucus* Humb., 1811 (p. 186).
Priodon maximus (Kerr) = *P. gigas* (Cuvier), 1817 (p. 187).
Mastodon americanus (Kerr) = *M. giganteus* Cuvier, 1817 (p. 187).
Canis australis Kerr = *C. antarcticus* Shaw, 1800 (p. 188).
Crocidura cærulea (Kerr) = *C. cærulescens* (Shaw), 1800 (p. 188).
Petaurus norfolcensis (Kerr) = *P. sciurea* (Shaw), 1794 (p. 190).
Alce fossilis (Kerr) [= *A. gigantea*[1] Blumenb., 1799 (p. 191).]

---

[1] By error "*Cervus giganteus* Goldfuss, 1821" is cited instead of *Alce giganteus* Blumenbach, 1799.

Mazama temama (Kerr) = *M. tema* Rafinesque,[1] 1817.
Ovis europæa (Kerr) = *O. musimon* Auct. ex Pallas, 1811 (p. 192).
Lynx texensis nom. nov. (p. 188) is given to replace *Felis maculata* Horfs. and Vigors (1829), preoccupied by *Felis (Lynx) vulgaris maculatus* Kerr, for the Texas lynx.

107. On a Collection of Mammals from Arizona and Mexico, made by Mr. W. W. Price, with Field Notes by the Collector. <*Bull. Amer. Mus. Nat. Hist.*, VII, pp. 193–258, fig. 1–17, June 29, 1895.

Annotated list of 70 species collected and 16 others observed but not collected, making a total of 86 species, with important field notes, extended measurements and technical comment.

Cervus hemionus Raf., 1817, replaces *C. macrotis* Say, 1823; *Ovis cervina* Desm., replaces *O. canadensis* Shaw. *Thomomys cervinus*, sp. nov., p. 203.

108. List of Mammals collected in the Black Hills Region of South Dakota and in Western Kansas by Mr. Walter W. Granger, with Field Notes by the Collector. <*Bull. Amer. Mus. Nat. Hist.*, VII, pp. 259–274, August 21, 1895.

Annotated list of 53 spp. *Lepus sylvaticus grangeri* subsp. nov., p. 264.

109. Descriptions of new American Mammals. <*Bull. Amer. Mus. Nat. Hist.*, VII, pp. 327–340, Nov. 8, 1895.

Spp. et subspp. nov.: (1) *Lepus aquaticus attwateri*, p. 327; (2) *Reithrodontomys australis*, p. 328; (3) *Oryzomys cherriei*, p. 329; (4) *Peromyscus attwateri*, p. 330; (5) *Neotoma cinnamomea*, p. 331; (6) *Sciurus* (*Microsciurus*, subgen. nov.) *alfari*, p. 333; (7) *Tamias pricei*, p. 333; (8) *Tamias wortmani*, p. 335; (9) *Spermophilus tridecemlineatus olivaceus*, p. 337; (10) *S. t. parvus*, p. 337; (11) *Blarina* (*Soriciscus*) *nigrescens*, p. 339; (12) *B.* (*S.*) *orophila*, p. 340.

110. Monographic Revision of the Pocket Gophers, Family Geomyidæ (exclusive of the genus *Thomomys*). <*Science*, N. S., I, pp. 241–243, March 1, 1895.

Review of Dr. C. Hart Merriam's paper of this title in *N. Am. Fauna*, No. 8, 1895.

111. The Pocket Gophers of the United States. <*Science*, N. S., I, pp. 689–690, June 21, 1895.

Review of paper by Vernon Bailey in *Bull. No. 5, U. S. Dept. Agric.*, 1895.

112. Those Pocket Gophers. Another Contribution about the Literature and Habits of the Geomyidæ. <*New York Sun* (newspaper), Feb. 26, 1895.

A humorous reply to abusive criticism of Dr. C. Hart Merriam's 'Monographic Revision of the Pocket Gophers' (*N. Amer. Fauna*, No. 8, Jan. 1895), under the pseudonym "Thomas Jones."

1896.

113. Note on *Macrogeomys cherriei* (Allen). <*Bull. Amer. Mus. Nat. Hist.*, VIII, pp. 45–46, pl. i, April 15, 1896.

Geomys cherriei further described and skull figured (*cf. supra*, No. 95).

114. On Mammals collected in Bexar County and Vicinity, Texas, by Mr. H. P. Attwater, with Field Notes by the Collector. <*Bull. Amer. Mus. Nat. Hist.*, VIII, pp. 47–80, April 22, 1896.

Annotated list of 53 spp.

[1] *Cervus temama* Kerr and *Mazama tema* Rafinesque I have recently rejected as indeterminable, accepting instead *Cervus sartorii* Saussure (*cf. Bull. Amer. Mus. Nat. Hist.*, XXXIV, p. 541, Nov. 2, 1915).

115. Descriptions of new North American Mammals. <*Bull. Amer. Mus. Nat. Hist.*, VIII, 1896, pp. 233–240, pll. x and xi, Nov. 21, 1896.

Spp. et subspp. nov.: (1) *Rangifer terrænovæ*, p. 233 (= *R. terrænovæ* Bangs of 11 days earlier date); (2) *Reithrodontomys laceyi*, p. 235; (3) *Perognathus mearnsi*, p. 237; (4) *Peromyscus michiganensis pallescens*, p. 238; (5) *Vespertilio incautus*, p. 239; (6) *Vespertilio chrysonotus*, p. 340.

116. List of Mammals collected by Mr. Walter W. Granger, in New Mexico, Utah, Wyoming and Nebraska, 1895–96, with Field Notes by the Collector. <*Bull. Amer. Mus. Nat. Hist.*, VIII, pp. 241–258, Nov. 25, 1896.

Annotated list of 48 spp.

117. On Mammals from the Santa Cruz Mountains, California. <*Bull. Amer. Mus. Nat. Hist.*, VIII, pp. 263–270, Dec. 4, 1896.

Annotated list of 16 species, with records of large series of measurement of various species of Muridæ and one shrew.

117*a*. North American Shrews. <*Science*, N. S., III, pp. 411–413, March 13, 1896.

Review of papers by C. Hart Merriam and Gerrit S. Miller, Jr., published in *N. Amer. Fauna*, No. 10, Dec. 31, 1895.

1897.

118. On Mammals from Yucatan, with Descriptions of New Species. <*Bull. Amer. Mus. Nat. Hist.*, IX, pp. 1–12, Feb. 23, 1897. With Frank M. Chapman.

Annotated list of 15 species, with 3 new, as follows: (1) *Peromyscus yucatanicus*, p. 8; (2) *Reithrodontomys mexicanus gracilis*, p. 9; (3) *Heteromys gaumeri*, p. 9.

119. On a Second Collection of Mammals from the Island of Trinidad, with descriptions of New Species, and a Note on some Mammals from the Island of Dominica, W. I. <*Bull. Amer. Mus. Nat. Hist.*, IX, pp. 13–30, Feb. 26, 1897. With Frank M. Chapman.

The annotated list of Trinidad species numbers 40, with the following new: (1) *Artibeus palmarum*, p. 16; (2) *Oryzomys delicatus*, p. 19; (3) *Akodon urichi*, p. 19; (4) *Akodon frustrator*, p. 20; (5) *Thylamys carri*, p. 27.

Five species recorded from Dominica.

120. Additional Notes on Costa Rica Mammals, with Descriptions of New Species. <*Bull. Amer. Mus. Nat. Hist.*, IX, pp. 31–43, pl. i, March 11, 1897.

Annotated list of 66 species, with 2 new genera and 4 new species, as follows: Genn. nov.: *Zygodontomys*, p. 38; *Sigmodontomys*, p. 38. Spp. nov.: (1) *Artibeus intermedius*, p. 33; (2) *Oryzomys chrysomelas*, p. 37; (3) *Sigmodontomys alfari*, p. 39; (4) *Sigmodon borucæ*, p. 40.

121. Further Notes on Mammals Collected in Mexico by Dr. Audley C. Buller, with Descriptions of New Species. <*Bull. Amer. Mus. Nat. Hist.*, IX, pp. 47–58, March 15, 1897.

An annotated list of 35 species, with the following as new: (1) *Peromyscus spicilegus*, p. 50; (2) *Peromyscus banderanus*, p. 51; (3) *Oryzomys mexicanus*, p. 52; (4) *Oryzomys bulleri*, p. 53; (5) *Sigmodon mascotensis*, p. 54; (6) *Sigmodon colimæ*, 55; (7) *Heteromys hispidus*, p. 56.

122. Preliminary Description of a new Mountain Sheep from the British Northwest Territory. <*Bull. Amer. Mus. Nat. Hist.*, IX, pp. 111–114, pll. ii and iii, April 8, 1897.

Ovis stonei, sp. nov.

123. On a Small Collection of Mammals from Peru, with Descriptions of New Species. <*Bull. Amer. Mus. Nat. Hist.*, IX, pp. 115–119, April 26, 1897.

Spp. 12, with 3 new: (1) *Sciurus (Microsciurus) peruanus*, p. 115; (2) *Oryzomys baroni*, p. 118; *Sigmodon peruanus*, p. 118.

124. Description of a new species of *Sigmodon*, from Bogotá, Colombia. <*Bull. Amer. Mus. Nat. Hist.*, IX, pp. 121–122, May 24, 1897.

Sigmodon bogotensis, sp. nov.

125. On a Collection of Mammals from Jalapa and Las Vigas, State of Vera Cruz, Mexico. <*Bull. Amer. Mus. Nat. Hist.*, IX, pp. 197–208, June 16, 1897. With Frank M. Chapman.

Annotated list of 19 species with 7 new: (1) *Mus musculus jalapæ*, p. 198; (2) *Reithrodontomys rafescens*, p. 199; (3) *Reithrodontomys saturatus*, p. 201; (4) *Peromyscus furvus*, p. 201; (5) *Peromyscus melanotis*, p. 203; (6) *Peromyscus musculus brunneus*, p. 203; (7) *Oryzomys jalapæ*, p. 206.

126. Description of a new Vespertilionine Bat from Yucatan. <*Bull. Amer. Mus. Nat. Hist.*, IX, pp. 231–232, Sept. 28, 1897.

Adelonycteris gaumeri, sp. nov.

1898.

127. Descriptions of new Mammals from Western Mexico and Lower California. <*Bull. Amer. Mus. Nat. Hist.*, X, pp. 143–158, April 12, 1898.

Spp. et subspp. nov.: (1) *Marmosa sinaloæ*, p. 143; (2) *Lepus peninsularis*, p. 144; (3) *Lepus cerrosensis*, p. 145; (4) *Lepus arizonæ confinis*, p. 146; (5) *Thomomys fulvus anitæ*, p. 146; (6) *Thomomys fulvus martirensis*, p. 147; (7) *Thomomys atrovarius*, p. 148; (8) *Perognathus pernix*, p. 149; (9) *Neotoma sinaloæ*, p. 149; (10) *Neotoma arenacea*, p. 150; (11) *Neotoma anthonyi*, p. 151; (12) *Peromyscus eremicus propinquus*, p. 154; (13) *Peromyscus cedrosensis*, p. 154; (14) *Peromyscus cineritius*, p. 155; (15) *Peromyscus geronimensis*, p. 156; (16) *Peromyscus exiguus*, p. 157; (17) *Peromyscus dubius*, p. 157.

Neotoma bryanti Merriam redescribed from a series of good specimens, pp. 152–153.

128. Revision of the Chickarees, or North American Red Squirrels (subgenus *Tamiasciurus*). <*Bull. Amer. Mus. Nat. Hist.*, X, pp. 249–298, August 31, 1898.

Monographic revision. Subspp. nov.: (1) *Sciurus hudsonicus baileyi*, p. 261; (2) *Sciurus hudsonicus ventorum*, p. 263; (3) *Sciurus hudsonicus streatori*, p. 267; (4) *Sciurus douglasii cascadensis*, p. 277; (5) *Sciurus fremonti neomexicanus*, p. 291.

Sciurus douglasii mollipilosus Aud. & Bachm. revived to replace *Sciurus hudsonicus orarius* Bangs, 1897, p. 277.

129. Nomenclatorial Notes on certain North American Mammals. <*Bull. Amer. Mus. Nat. Hist.*, X, pp. 449–461, Nov. 10, 1898.

Sciurus rubricatus Ord (1818), *Sciurus bottæ* Lesson (1831), names of too doubtful significance to be employed; *Sciurus californicus* Lesson (1847), of doubtful application, but renders *Sciurus hudsonicus californicus* Allen (1890) untenable, which is here replaced by *Sciurus douglasii albolimbatus* (nom. nov.), p. 453; *Sciurus griseus* Ord (1818) antedates *Sciurus fossor* Peale (1848) for the same species; *Spermophilus empetra* (Pallas, 1778) is again affirmed to = *S. parryi* Richardson; *Arctomys lewsii* Aud. & Bach. (1853), an earlier name for *Cynomys leucurus* Merriam (1890); *Glis canadensis* Erxleben (1777) = *Arctomys monax melanopus* Kuhl (1820); *Arctomys pruinosus* Gmelin (1788) = *A. caligatus* Eschscholtz (1829); *Hypudæus ochrogaster* Wagner (1843) = *Microtus austerus* LeConte (1853); *Lutra canadensis*

Schreber, 1776) the correct name of the common Otter of North America (vs. Rhoads); *Mustela pennanti* Erxleben (1777) the correct name for the Fisher (*vs.* Rhoads).

Sciurus wagneri nom. nov. for *Sciurus albipes* et *S. varius* Wagner, preoccupied (= *Sciurus poliopus* Fitzinger).

130. A Pocket Mouse in Confinement. <*Amer. Nat.*, XXXII, pp. 583–584, August, 1898.

A specimen of *Perognathus mearnsi* lived in a cage at the American Museum of Natural History for about four years in good health without partaking of water, though water was offered it, showing that water was not necessary to its welfare.

131. The Sea Otter. <*Amer. Nat.*, XXXII, pp. 356–358, May, 1898.

On their greatly reduced numbers, based on a report to U. S. Government by Capt. V. L. Hooper, Commanding Bering Sea Patrol Fleet, 1897. (Treas. Dept. Rep., Dec. 1897.)

132. The Mammals of Florida. <*Amer. Nat.*, XXXIII, pp. 433–436, June, 1898.

Review of Outram Bangs's paper 'The Land Mammals of Peninsular Florida and the Coast Region of Georgia' (*Proc. Boston Soc. Nat. Hist.*, XXVIII, 1898).

1899.

133. Trouessart's Catalogue of Mammals. <*Amer. Nat.*, XXXIII, pp. 69–70, Jan., 1899.

Review of second edition of Trouessart's 'Catalogus Mammalium tam Viventium quam Fossilium,' 1897.

134. Generic and Family Names of Rodents. <*Amer. Nat.*, XXXIII, pp. 70–72, Jan., 1899.

Review of Thomas's 'On the Genera of Rodents' (*Proc. Zool. Soc. London*, 1896, pp. 1012–1018) and Palmer's 'A List of the Generic Names of Rodents' (*Proc. Biol. Soc. Washington*, XI, pp. 241–270, Dec. 7, 1897).

135. The Fossil Bisons of North America. <*Amer. Nat.*, XXXIII, pp. 665–666, August, 1899.

Review of Lucas's paper (*Proc. U. S. Nat. Mus.*, XXI, pp. 755, 771, pls. lxv-lxxxiv, and text figures).

136. 'Wild Animals I have Known.' <*Amer. Nat.*, XXXIII, pp. 666–668, August, 1899.

Review of E. S. Thompson's book thus entitled (New York, 1898).

137. The North American Arboreal Squirrels. <*Amer. Nat.*, XXXIII, August, 1899, pp. 635–642.

A review of E. W. Nelson's 'Revision of the Squirrels of Mexico and Central America' (*Proc. Washington Acad. Sci.*, I, pp. 15–110, pll. i and ii, May, 1899), to which is here added a list of the 'Arboreal Squirrels found in North America north of Mexico,' with their geographic ranges.

138. Report of the Fur Seal Investigations, 1896–1897. <*Science*, N. S., X, pp. 885–890, Dec. 15, 1899.

Review of 'The Fur Seals and Fur Seal Islands of the North Pacific,' by David Starr Jordan, Commissioner in Charge of Fur Seal Investigations, 1896–1897.

139. On Mammals from the Northwest Territory collected by Mr. A. J. Stone. <*Bull. Amer. Mus. Nat. Hist.*, XII, pp. 1–9, March 4, 1899.

Annotated list of 17 species, the following new: (1) *Zapus saltator*, p. 3; (2) *Phenacomys constablei*, p. 4; (3) *Microtus stonei*, p. 5; (4) *Microtus vellerosus*, p. 7; (5) *Microtus cautus*, p. 7. Also note on *Evotomys fuscodorsalis* Allen = *E. gapperi* (Vigors).

140. Descriptions of new North American Rodents. <*Bull. Amer. Mus. Nat. Hist.*, XII, pp. 11–17, March 4, 1899.

Spp. et subspp. nov.: (1) *Lepus americanus phæonotus*, p. 11; (2) *Lepus bishopi*, p. 11; (3) *Lepus floridanus chapmani*, p. 13; (4) *Thomomys fulvus alticolus*, p. 13; (5) *Reithrodontomys tenuis*, p. 15; (6) *Peromyscus texanus subarcticus*, p. 15; (7) *Sciurus chapmani*, p. 16.

141. New Rodents from Colombia and Venezuela. <*Bull. Amer. Mus. Nat. Hist.*, XII, pp. 195–218, Dec. 20, 1899.

Spp. et subspp. nov.: (1) *Lepus* (*Sylvilagus*) *superciliaris*, p. 196; (2) *Isothrix rufodorsalis*, p. 197; (3) *Echimys mincæ*, p. 199; (4) *Echimys urichi*, p. 199; (5) *Echimys canicollis*, p. 200; (6) *Heteromys jesupi*, p. 201; (7) *Akodon venezuelensis*, p. 203; (8) *Akodon columbianus*, p. 203; (9) *Oryzomys maculiventer*, p. 204; (10) *Oryzomys trichurus*, p. 206; (11) *Oryzomys sanctæmartæ*, p. 207; (12) *Oryzomys mollipilosus*, p. 208; (13) *Oryzomys magdalenæ*, p. 209; (14) *Oryzomys villosus*, p. 210; (15) *Oryzomys palmarius*, p. 210; (16) *Oryzomys tenuicauda*, p. 211; (17) *Oryzomys modestus*, p. 212; (18) *Oryzomys fulviventer*, p. 212; (19) *Sciurus saltuensis bondæ*, p. 213; (20) *Sciurus* (*Guerlinguetus*) *quebradensis*, p. 217.

142. The Generic Names *Echimys* and *Loncheres*. <*Bull. Amer. Mus. Nat. Hist.*, XII, pp. 257–264, Dec. 26, 1899.

On *Echimys*, *Loncheres*, *Dactylomys*, *Nelomys* and *Mesomys*. *Loncheres* = *Echimys* of earlier date; *Proechimys* gen. nov., to replace *Echimys* auct., not of F. Cuvier, 1809.

1900.

143. The Mountain Caribou of Northern British Columbia. <*Bull. Amer. Mus. Nat. Hist.*, XIII, 1900, pp. 1–18, fig. 1–18, April 3, 1900.

Comparison with other forms; referred to *Rangifer montanus* Seton-Thompson, but later (see below, No. 179) became *Rangifer osborni* Allen. Skull and antlers figured.

144. Note on the Wood Bison. <*Bull. Amer. Mus. Nat. Hist.*, XIII, 1900, pp. 63–67, April 16, 1900.

Its range, former and present, and approximate present number of individuals living.

145. List of Bats collected by Mr. H. H. Smith in the Santa Marta region of Colombia, with descriptions of new species. <*Bull. Amer. Mus. Nat. Hist.*, XIII, pp. 87–94, May 12, 1900.

Annotated list of 22 species, the following new: (1) *Chiroderma jesupi*, p. 88; (2) *Micronycteris hypoleuca*, p. 90; (3) *Promops affinis*, p. 91; (4) *Promops milleri*, p. 92.

146. Note on the Generic Names *Didelphis* and *Philander*. <*Bull. Amer. Mus. Nat. Hist.*, XIII, pp. 185–190, Oct. 12, 1900.

Philander considered a synonym of *Didelphis*. *Caluromys* nom. nov. is proposed for *Philander* auct., with *Didelphis philander* designated as type.[1] *Caluromys alstoni* sp. nov., p. 189.

---

[1] Later, under the action of the principle of tautonymy, *Philander* became reinstated, leaving *Caluromys* a pure synonym of *Philander*.

147. Descriptions of new American Marsupials. <*Bull. Amer. Mus. Nat. Hist.*, XIII, 1900, pp. 191–199, Oct. 23, 1900.

Spp. et subspp. nov.: (1) *Didelphis pernigra*, p. 191; (2) *Didelphis karkinophaga caucæ*, p. 192; (3) *Didelphis karkinophaga colombica*, p. 193; (4) *Metachirus fuscogriseus*, p. 194; (5) *Metachirus tschudii*, p. 195; (6) *Metachirus nudicaudatus colombianus*, p. 196; (7) *Marmosa chapmani*, p. 197; (8) *Marmosa klagesi*, p. 198; (9) *Thylamys keaysi*, p. 198.

148. On Mammals collected in Southeastern Peru by H. H. Keays, with Descriptions of new species. <*Bull. Amer. Mus. Nat. Hist.*, XIII, pp. 219–227, Nov. 16, 1900.

Eighteen species, 5 new, as follows: (1) *Dactylomys peruanus*, p. 220; (2) *Oxymycterus juliacæ*, p. 223; (3) *Oxymycterus apicalis*. p. 224; (4) *Oryzomys keaysi*, p. 225; (5) *Oryzomys obtusirostris*, p. 226.

149. Some Results of a Natural History Journey to Northern British Columbia, Alaska and the Northwest Territory, in the interest of the American Museum of Natural History. By A. J. Stone. <*Bull. Amer. Mus. Nat. Hist.*, XIII, pp. 31–62, April 6, 1900.

Introductory Note and technical names by J. A. Allen.

150. The systematic name of the Cuban Red Bat. <*Proc. Biol. Soc. Washington*, XIII, 165, Oct. 31, 1900.

Lasiurus pfeifferi (Gundlach) — not *L. blossevillii* = *L. bonariensis* (Lesson & Garnot).

151. The proper name of the Viscacha. <*Proc. Biol. Soc. Washington*, XIII, 183, Nov. 30, 1900.

Vizcacia viscacia (Molina).

152. The North American Jumping Mice. <*Amer. Nat.*, XXXIV, pp. 199–202, March, 1900.

Review of E. A. Preble's 'Revision of the Jumping Mice of the Genus *Zapus*,' in *N. Amer. Fauna*, No. 15, 1899, with a list of the species and subspecies, giving their ranges.

153. Preliminary List of the Mammals of New York. <*Amer. Nat.*, XXXIV, pp. 316–318, April, 1900.

Review of G. S. Miller's 'Preliminary List of New York Mammals,' in *Bull. New York State Mus.*, VI, 1899, pp. 271–390.

1901.

154. The generic names *Myrmecophaga* and *Tamandua*, and the specific names of the Opossums of the genus *Didelphis*. <*Proc. Biol. Soc. Washington*, XIV, pp. 91–93, June 19, 1901.

Myrmecophaga tridactyla Linn.; *Tamanduas* F. Cuvier (1829) has priority over *Uroleptes* Wagler (1830); *Didelphis marsupialis* Linn. restricted to the Guiana form of the genus.

155. The proper generic names of the Viscacha, Chinchillas and their allies. <*Proc. Biol. Soc. Washington*, XIV, 181–182, Dec. 12, 1901.

Respectively *Viscacia* Schinz (1825), *Chinchilla* and *Lagidium*; *Callomys* is indeterminable.

156. Descriptions of two new species of South American Muridæ. <*Bull. Amer. Mus. Nat. Hist.*, XIV, pp. 39–40, Jan. 31, 1901.

Spp. nov., (1) *Zygodontomys thomasi*, p. 39; (2) *Sigmodon simonsi*, p. 40.

157. On a Further Collection of Mammals from Southeastern Peru, collected by Mr. H. H. Keays, with Descriptions of new species. <*Bull. Amer. Mus. Nat. Hist.*, XIV, pp. 41–46, Jan. 31, 1901.

An annotated list of 18 species. Spp. nov.: (1) *Rhipidomys ochrogaster*, p. 43; (2) *Phyllotis osilæ*, p. 44; (3) *Akodon lutescens*, p. 46.

158. The Musk-oxen of Arctic America and Greenland. <*Bull. Amer. Mus. Nat. Hist.*, XIV, pp. 69–86, pll. xiii–xvii, and 7 text figures, March 27, 1901.

Principally on the Ellesmere Land form, here referred to *Ovibos wardi* Lydekker (= *O. pearyi* Allen, Ms.).

159. Description of a new Caribou from Kenai Peninsula Alaska. <*Bull. Amer. Mus. Nat. Hist.*, XIV, pp. 143–148, and 4 text figures, May 28, 1901.

Rangifer stonei sp. nov.

160. A Preliminary Study of the North American Opossums of the Genus *Didelphis*. <*Bull. Amer. Mus. Nat. Hist.*, XIV, pp. 149–188, pll. xxii–xxv, June 15, 1901.

Spp. and subspp. nov.: (1) *Didelphis marsupialis texensis*, p. 172; (2) *Didelphis marsupialis tabascensis*, p. 173; (3) *Didelphis richmondi*, p. 175; (4) *Didelphis yucatanensis*, p. 178.

161. Descriptions of two new Opossums of the Genus *Metachirus*. <*Bull. Amer. Mus. Nat. Hist.*, XIV, pp. 213–218, July 3, 1901.

Spp. nov.: (1) *Metachirus fuscogriseus pallidus*, p. 215; (2) *Metachirus grisescens*, p. 217.

162. The generic names of the Mephitinæ. <*Bull. Amer. Mus. Nat. Hist.*, XIV, pp. 325–334, Nov. 12, 1901.

Chinca Lesson (1842) a synonym of *Mephtis* Cuvier (1800); *Spilogale* Gray (1865) is not a synonym of *Mephitis*, as construed by Howell (*N. Amer. Fauna*, No. 20, Aug. 31, 1901).

A list of the species and subspecies of *Mephitis* is given (pp. 333, 334).

163. New South American Muridæ and a new *Metachirus*. <*Bull. Amer. Mus. Nat. Hist.*, XIV, pp. 405–412, Nov. 30, 1901.

Spp. nov.: (1) *Oryzomys bolivaris*, p. 405; (2) *Oryzomys castaneus*, p. 406; (3) *Oryzomys perenensis*, p. 406; (4) *Oryzomys rivularis*, p. 407; (5) *Phyllotis chacoensis*, p. 408; (6) *Phyllotis cachinus*, p. 409; (7) *Eligmodontia morgani*, p. 409; (8) *Akodon tucumanensis*, p. 410; (9) *Metachirus nudicaudatus bolivianus*, p. 411.

164. Mammals of South Africa. <*Science*, N. S., XIII, pp. 105–106, Jan. 18, 1901.

Review of Volume I of W. L. Sclater's work of this title.

165. Miller's Key to the Land Mammals of Eastern North America. <*Amer. Nat.*, XXXV, pp. 59–61, Jan., 1901.

Review of G. S. Miller's paper thus entitled, with criticism of certain points of nomenclature, as the use of *Rosmarus* in place of the earlier *Odobenus*, etc.

166. Two Important Papers on North American Mammals. <*Amer. Nat.*, XXXV, pp. 221–224, March, 1901.

Bailey's 'Revision of American Voles of the Genus *Microtus*,' and Osgood's 'Revision of the Pocket Mice of the Genus *Perognathus*.'

1902.

167. Beddard's 'Mammalia.' <*Amer. Nat.*, XXXVI, pp. 911–914, Nov., 1902.

Review of this excellent manual (Cambridge Natural History, Vol. X, London, 1902).

168. Note on the Names of a few South American Mammals. <*Proc. Biol. Soc. Washington*, XIV, pp. 183–185, Dec. 12, 1901.

Several species names date from G. Fischer, 1814, which are usually attributed to later authors; several names attributed to Wied were first published by Schinz (1821), or by Kuhl, or by Temminck; Gray's names (1827) for certain genera of bats antedate later names now current, etc.

169. A Further Note on the Generic Names of the Mephitinæ. <*Proc. Biol. Soc. Washington*, XV, pp. 59–60, March 22, 1902.

Apropos of Howell's replacing *Mephitis* by *Chincha* and *Spilogale* by *Mephitis*.

170. A Further Note on the Name of the Argentine Viscacha. <*Proc. Biol. Soc. Washington*, XV, 196, Oct. 10, 1902.

Viscaccia dates from Oken, 1816, instead of from Schinz, 1825; the specific name *chilensis* Oken (1816) has priority over *maximus* Desmarest (1817) — hence *Viscaccia chilensis* Oken for the Argentine Viscacha.

171. North American Ruminants. <*Amer. Mus. Journal*, II, March, 1902, Supplement (= Guide Leaflet, No. 5), pp. 1–29. 18 half-tone illustrations.

172. The Caribou of British Columbia and Alaska. <*Outing*, 1902, pp. 555–561, 11 text figures.

Distribution, characters and habits of *Rangifer montanus*, *R. osborni*, *R. granti*, and *R. stonei*.

173. North American Deer. <*Amer. Nat.*, XXXVI, pp. 755–756, Sept., 1902.

Review of the work entitled 'The Deer Family,' by Theodore Roosevelt and others.

174. Zimmermann's 'Zoologiæ Geographicæ' and 'Geographische Geschichte' considered in their relation to Mammalian Nomenclature. <*Bull. Amer. Mus. Nat. Hist.*, XVI, pp. 13–22, Feb. 1, 1902.

Various names in the 'Geschichte' (1780) were previously employed in the 'Geographicæ' (1777) and thus take an earlier date; others antedate Erxleben (1777) to whom they are usually accredited. *Dama virginiana* Zimm., 1777, is considered tenable for the common Virginia Deer. *Dama lichtensteini* (p. 20) is a new name for *Cervus mexicanus* Lichtenstein, preoccupied by *Cervus mexicanus* Zimmermann and Gmelin, which is shown to have been based mainly on the Prong-horn (*Antilocapra americana*).

175. The Generic and Specific Names of some of the Otariidæ. <*Bull. Amer. Mus. Nat. Hist.*, XVI, pp. 111–118, March 15, 1902.

Eumetopias stelleri (Lesson, 1828) becomes *E. jubata* (Schreber, 1776); the first tenable specific name of the Southern Sea Lion is shown to be *byronia* Blainville (1820); the generic name of the northern Fur Seals is shown to be *Callotaria* Palmer, 1892, not *Otoes* Palmer, 1901 = *Otaria* Peron, 1816.

176. A new Caribou from the Alaska Peninsula. <*Bull. Amer. Mus. Nat. Hist.*, XVI, pp. 119–127, and 6 text figures, April 7, 1902.

Rangifer granti, sp. nov.

177. A new Bear from the Alaska Peninsula. <*Bull. Amer. Mus. Nat. Hist.*, XVI, pp. 141–143, pll. xxx, and xxxi, April 12, 1902.

Ursus merriami sp. nov. Skull and dentition figured.

178. A new Sheep from the Kenai Peninsula. <*Bull. Amer. Mus. Nat. Hist.*, XVI, pp. 145–148, and 2 text figures, April 23, 1902.

Ovis dalli kenaiensis subsp. nov.

179. Description of a new Caribou from Northern British Columbia, and remarks on *Rangifer montanus*. <*Bull. Amer. Mus. Nat. Hist.*, XVI, pp. 149–158, and 6 text figures. April 16, 1902.

Rangifer osborni, sp. nov.; skulls of true *Rangifer montanus* figured.

180. Nomenclatorial Notes on American Mammals. <*Bull. Amer. Mus. Nat. Hist.*, XVI, pp. 159–168, July 1, 1902.

Alce Blumenbach (1803 [*lege* 1799]) replaces *Megaceros* Owen (1844) for the extinct Irish Elk. *Alce* (non Blumenbach, *Alces* auct.) is changed to *Paralces*, nom. nov., p. 160. *Dama* vs. *Odocoileus* is again considered, anent comment on the case by Osgood. The names of the Peccaries, both generic and specific, are discussed, and a list of the species with their synonymy is given, the specific name *pecari* Fisher (1814) replacing *albirostris* Illiger (1815) and also *labiatus* Cuvier (1817). *Sciurus rufiventer* E. Geoffroy (1803) is adopted in place of *S. ludovicianus* Custis (1806).

181. List of Mammals collected in Alaska by the Andrew J. Stone Expedition of 1901. <*Bull. Amer. Mus. Nat. Hist.*, XVI, pp. 215–230, July 12, 1902.

An annotated list of 26 species, with field notes by J. D. Figgins, collector.

182. A Preliminary Study of the South American Opossums of the Genus *Didelphis*. <*Bull. Amer. Mus. Nat. Hist.*, XVI, pp. 249–279, Aug. 18, 1902.

A monographic revision, with copious tables of measurements. Subspp. nov.: (1) *Didelphis marsupialis insularis*, p. 259; (2) *D. m. etensis*, p. 262; (3) *D. paraguayensis andina*, p. 272; (4) *D. p. meridensis*, p. 274. *D. paraguayensis* Oken (1816) replaces *D. aurita* Temminck (1825).

183. Mammal Names proposed by Oken in his 'Lehrbuch der Zoologie.' <*Bull. Amer. Mus. Nat. Hist.*, XVI, pp. 373–379, Oct. 11, 1902.

Generic names considered available are *Citellus*, *Grison*, *Tayra*, *Thos*, *Panthera*, *Tigris*, *Leo*. The specific names considered available, or as rendering invalid later similar names, are *Lepus chilensis*, for the Argentine Viscacha; *Hystrix paraguayensis*, for *Spiggurus spinosa* F. Cuvier; *Felis* (*Lynx*) *brasiliensis*, unidentifiable with certainty, but precludes a later *Felis brasiliensis* (as of Schinz and F. Cuvier); *Felis* (*Panthera*) *mexicana* = *Felis mexicana* Desm. of even date.

184. A new Caribou from Ellesmere Land. <*Bull. Amer. Mus. Nat. Hist.*, XVI, pp. 409–412, with 2 text figures, Oct. 31, 1902.

Rangifer pearyi sp. nov.

185. The Hair Seals (Family Phocidæ) of the North Pacific Ocean and Bering Sea. <*Bull. Amer. Mus. Nat. Hist.*, XVI, pp. 459–499, with 10 text figures, Dec. 12, 1902.

Nomenclature; sexual differences in dentition; revision of the North Pacific species (11 species and subspecies are recognized), the following new: (1) *Phoca hispida gichigensis*, p. 488; (2) *Phoca ochotensis macrodens*, p. 483; (3) *Phoca stejnegeri*, p. 485; (4) *Phoca richardii pribilofensis;* (5) *Phoca richardii geronimensis*, p. 495.

Phoca nigra Pallas (p. 483 footnote) suggested as apparently available for *Callorhinus curilensis* (see *infra*, No. 192).

1903.

186. Description of a new species of *Sigmodon* from Ecuador.<*Bull. Amer. Mus. Nat. Hist.*, XIX, pp. 99–100, March 20, 1903.

Sigmodon puna sp. nov.

187. Report on the Mammals Collected in Northeastern Siberia by the Jesup North Pacific Expedition, with Itinerary and Field Notes by N. G. Buxton.<*Bull. Amer. Mus. Nat. Hist.*, XIX, pp. 101–184, 195, 196, May 9, 1903.

Descriptive, critical, and field notes on 35 species, the following new: (1) *Citellus buxtoni*, p. 139; (2) *Citellus stejnegeri*, p. 142; (3) *Evotomys* (*Craseomys*) *latastei*, p. 145; (4) *Evotomys jochelsoni*, p. 148; (5) *Lemmus obensis chrysogaster*, p. 153; (6) *Ochotona kolymensis*, p. 154; (7) *Lepus gichiganus*, p. 155; (8) *Vulpes anadyrensis*, p. 167; (9) *Putorius* (*Arctogale*) *pygmæus*, p. 176; (10) *Erinaceus orientalis*, p. 179; (11) *Sorex buxtoni*, p. 181.

188. Descriptions of new Rodents from Southern Patagonia, with a Note on the Genus *Euneomys* Coues, and an Addendum to Article [*supra*, No. 186] on Siberian Mammals.<*Bull. Amer. Mus. Nat. Hist.*, XIX, pp. 185–196, May 9, 1903.

(1) *Ctenomys robustus*, p. 185; (2) *Ctenomys sericeus*, p. 187; (3) *Ctenomys colburni*, p. 188; (4) *Oxymycterus microtis*, p. 189; (5) *Reithrodon cuniculoides obscurus*, p. 190, (6) *Reithrodon hatcheri*, p. 191; (7) *Euneomys petersoni*, p. 192, spp. et subspp. nov.
Euneomys Coues, a well-founded genus.

189. Mammals collected in Alaska and Northern British Columbia by the Andrew J. Stone Expedition of 1902.<*Bull. Amer. Mus. Nat. Hist.*, XIX, pp. 521–567, text figs. 1–9, Oct. 10, 1903.

Annotated list of 28 species, the following new: (1) *Citellus stonei*, p. 537; (2) *Synaptomys* (*Mictomys*) *chapmani*, p. 555; (3) *Synaptomys* (*Mictomys*) *andersoni*, p. 554; (4) *Erethizon epizanthus nigrescens*, p. 558; (5) *Putorius microtis*, p. 563.

Under the mistaken impression that *Alce* properly dated from Frisch (1775), thus antedating *Alce* Blumenbach (1799) and *Alces* Gray (1821), *Paralces* Allen (1902—*cf. supra*, No. 180) is here conceded to be untenable, but later investigation renders it evident that Frisch's work in which *Alce* was proposed is not to be accepted as nomenclaturally authoritative.

190. List of Mammals collected by J. H. Batty in New Mexico and Durango, with Descriptions of new Species and Subspecies.<*Bull. Amer. Mus. Nat. Hist.*, XIX, pp. 587–612, Nov. 13, 1903.

An annotated list of 34 species, the following new: (1) *Odocoileus battyi*, p. 591; (2) *Eutamias durangæ*, p. 594; (3) *Citellus* (*Otospermophilus*) *grammurus rupestris*, p. 595; (4) *Peromyscus paulus*, p. 598; (5) *Peromyscus texanus flaccidus*, p. 599; (6) *Sigmodon baileyi*, p. 601; (7) *Reithrodontomys megalotis sestinensis*, p. 602; (8) *Neotoma intermedia durangæ*, p. 602; (9) *Perodipus obscurus*, p. 603; (10) *Lepus* (*Macrotolagus*) *texianus micropus*, p. 605; (11) *Lepus* (*Macrotolagus*) *gaillardi battyi*, p. 607; (12) *Lepus* (*Sylvilagus*) *durangæ*, p. 609; (13) *Canis impavidus*, p. 609; (14) *Myotis californicus durangæ*, p. 612.

191. A new Deer and a new Lynx from the State of Sinaloa, Mexico.<*Bull. Amer. Mus. Nat. Hist.*, XIX, pp. 613–615, Nov. 14, 1903.

(1) *Odocoileus sinaloæ*, p. 613; (2) *Lynx ruffus escuinapæ*, p. 614.

192. Note on *Phoca nigra* Pallas.<*Proc. Biol. Soc. Washington*, XVI, p. 49, March 19, 1903.

The name being preoccupied cannot be used in place of *Callotaria curilensis* Jordan and Clark, as previously suggested (see *supra*, No. 185).

193. Note on *Sciurus mollipilosus* Audubon and Bachman. <*Proc. Biol. Soc. Washington*, XVI, p. 126, Sept. 30, 1903.

Defense of the use of *mollipilosus* in place of the later *orarius* Bangs.

1904.

194. New forms of the Mountain Goat (*Oreamnos*). <*Bull. Amer. Mus. Nat. Hist.*, XX, pp. 19–21, text fig. 1–18, Feb. 10, 1904.

(1) *Oreamnos montanus columbianus* and (2) *O. m. missoulæ* (p. 20), subspp. nov.

195. Mammals from Southern Mexico and Central and South America. <*Bull. Amer. Mus. Nat. Hist.*, XX, pp. 29–80, Feb. 29, 1904.

Annotated lists from Southern Vera Cruz, Mexico, 19 spp.; Central Costa Rica, 23 spp.; Chiriqui, Republic of Panama, 53 spp. Partial revisions of the genera *Tayra*, *Nasua*, and *Potos*.

New species and subspecies: (1) *Lepus (Sylvilagus) russatus*, p. 31; (2) *Lepus (Sylvilagus) parvulus*, p. 34; (3) *Tayra barbara irara*, p. 36; (4) *Akodon irazu*, p. 46; (5) *Felis carrikeri*, p. 47; (6) *Nasua narica bullata*, p. 48; (7) *Nasua narica panamensis*, p. 51; (8) *Nasua narica yucatanica*, p. 52; (9) *Nasua narica pallida*, p. 53; (10) *Sigmodon borucæ chiriquensis*, p. 68; (11) *Felis mearnsi* (nom. nov.), p. 71; (12) *Felis panamensis*, p. 71; (13) *Potos flavus chiriquensis*, p. 74; (14) *Potos flavus caucensis*, p. 75; (15) *Potos flavus chapadensis*, p. 76; (16) *Myotis chiriquensis*, p. 77.

196. The External Ear Bone in certain Rodents. <*Bull. Amer. Mus. Nat. Hist.*, XX, pp. 135–138, text fig., April 7, 1904.

External ear bone in *Heteromys*, here first described and figured.

197. Further Notes on Mammals from Northwestern Durango. <*Bull. Amer. Mus. Nat. Hist.*, XX, pp. 205–210, May 28, 1904.

An annotated list of 17 species, the following new: (1) *Sciurus aberti phæurus*, p. 205; (2) *Sciurus aberti barberi*, p. 207; (3) *Eutamias canescens*, p. 208.

198. New Bats from Tropical America, with note on Species of *Otopterus*. <*Bull. Amer. Mus. Nat. Hist.*, XX, pp. 227–237, June 29, 1904.

Spp. et subspp. nov.: (1) *Molossus coibensis*, p. 227; (2) *Molossus bondæ*, p. 228; (3) *Promops barbatus*, p. 228; (4) *Dermonotus suapurensis*, p. 229; (5) *Lonchophylla thomasi*, p. 230; (6) *Artibeus rusbyi*, p. 230; (7) *Artibeus insularis*, p. 231; (8) *Artibeus yucatanicus*, p. 232; (9) *Phyllostoma hastatus panamensis*, p. 233; (10) *Phyllostoma hastatus cauræ*, p. 234.

Two species of *Otopterus* in Mexico — *O. bocourtianus* (Dobson) and *O. mexicanus* (Saussure).

199. Mammals Collected in Alaska by the Andrew J. Stone Expedition of 1903. <*Bull. Amer. Mus. Nat. Hist.*, XX, pp. 273–292, fig. 1–19, Sept. 8, 1904.

An annotated list, with many measurements, of 28 species.

200. A new Sheep from Kamchatka. <*Bull. Amer. Mus. Nat. Hist.*, XX, pp. 293–298, fig. 1–6, Sept. 8, 1904.

Ovis storcki, sp. nov.

201. New Mammals from Venezuela and Colombia. <*Bull. Amer. Mus. Nat. Hist.*, XX, pp. 327–335, Oct. 8, 1904.

Spp. et subspp. nov.: (1) *Peramys brevicaudatus dorsalis*, p. 327; (2) *Oryzomys tenuipes*, p. 328; (3) *Akodon meridensis*, p. 329; (4) *Holochilus venezuelensis*, p. 330; (5) *Felis maripensis*, p. 331; (6) *Felis sanctæmartæ*, p. 332; (7) *Procyon proteus*, p. 333; (8) *Nasua phæocephala*. p. 334.

202. List of Mammals from Venezuela collected by Mr. Samuel M. Klages. <*Bull. Amer. Mus. Nat. Hist.*, XX, pp. 337–345, Oct. 8, 1904.

An annotated list of 46 species.

203. A Fossil Porcupine from Arizona. <*Bull. Amer. Mus. Nat. Hist.*, XX, pp. 383–384, Oct. 15, 1904.

Erethizon godfreyi, sp. nov.

204. The Tamandua Anteaters. <*Bull. Amer. Mus. Nat. Hist.*, XX, pp. 385–398, fig. 1–4, Oct. 29, 1904.

A review of the group, with respect to external and cranial characters. Subspp. nov.: (1) *Tamandua tetradactyla chapadensis*, p. 392; (2) *T. t. instabilis*, p. 392; (3) *T. t. tenuirostris*, p. 394; (4) *T. t. chiriquensis*, p. 395.

205. Report on Mammals from the District of Santa Marta, Colombia, collected by Mr. Herbert H. Smith, with Field Notes by Mr. Smith. <*Bull. Amer. Mus. Nat. Hist.*, XX, pp. 407–468, fig. 1–4, Nov. 28, 1904.

An annotated list of 78 species, the following new: (1) *Coendou sanctæmartæ*, p. 441; (2) *Lutra colombiana*, p. 452; (3) *Alouatta seniculus rubicunda*, p. 458; (4) *Alouatta seniculus caucensis*, p. 462.

Urocyon aquilus Bangs not referable to *Urocyon*.

206. Palmer's 'Index Generum Mammalium.' <*Science*, N. S., XIX, pp. 498–501, March 25, 1904.

A review of this notable work (*N. Amer. Fauna*, No. 23, Jan. 1904.)

1905.

207. Mammalia of Southern Patagonia. Reports of the Princeton University Expeditions to Patagonia, 1896–1899, Vol. III, 1905, Part I, pp. 1–120, pll. i–xxix.

Detailed treatment of 55 species, with special reference to nomenclature, that of the genera as well as the species discussed historically; full tables of references under the genera and species, and an annotated bibliography (pp. 192–210) of 65 titles.

Eunothocyon, gen. nov., p. 153 (in text); *Carcinocyon*, gen. nov., p. 153 (in text; *Ctenomys osgoodi*, nom. nov., to replace *C. robustus* Allen, preoccupied, p. 191; *Canis sclateri* nom. nov. (p. 153) to replace *Canis microtis* Sclater, preoccupied.

Species figured: *Zaedyus ciliatus*, pll. i–iii (animal, skeleton, and three skulls); *Kerodon australis* and *Ctenomys osgoodi*, pl. vii (skulls); *Ctenomys sericeus* and *C. colburni*, pl. viii (skulls); *Eligmodontia*, *Oryzomys* and *Oxymycterus*, pll. ix and x (skulls and dentition of various species); *Akodon*, pll. xi and xii (skulls and dentition of 6 species); *Phyllotis*, *Euneomys*, and *Reithrodon*, pll. xiii and xiv (skulls and dentition); *Arctocephalus australis* and *A. philippii*, pll. xv–xvii (skulls, three views of each); *Arctocephalus townsendi* pll. xviii–xx (skull, three views); *Otaria byronia*, pl. xxi (skeleton); *Conepatus humboldti*, pl. xxii (skulls and dentition); *Cerdocyon griseus*, pl. xxiii (skull, three views); *Lynchailurus pajeros crucina*, pl. xxiv (skull and dentition); *Puma pearsoni*, pll. xxv and xxvi (colored figures of animal, red and gray phases); *Puma pearsoni*, pll. xxvii–xxix (three views of skull).

208. The Andrew J. Stone Explorations in Arctic and Subarctic America. 4to, pp. i–xvi, 1–38, with 55 half-tone illustrations in text. American Museum of Natural History, 1905. Edition limited to 100 numbered copies. (Anonymous.)

The illustrations, from photographs, relate mostly to mammals, as does the greater part of the text.

209. Mammals from Beaver County, Utah, collected by the [Brooklyn Institute] Museum Expedition of 1904.<*Science Bulletin Mus. Brooklyn Inst. Arts and Sciences*, I, No. 6, pp. 117–122, March 31, 1905.

Annotated list of 16 species, the following new: (1) *Eutamias lectus*, p. 117; (2) *Eutamias adsitus*, p. 118; (3) *Cynomys parvidens*, p. 119; (4) *Marmota engelhardti*, p. 120; (5) *Ochotona cinnamomea*, p. 121.

1906.

210. Mammals from the States of Sinaloa and Jalisco, Mexico, collected by J. H. Batty during 1904 and 1905.<*Bull. Amer. Mus. Nat. Hist.*, XXII, pp. 191–262, pll. xx–xxxiii, and 3 text figures, July 25, 1906.

I. Mammals from Southern Sinaloa, pp. 192–237, 41 species; II. Mammals from the State of Jalisco, pp. 237–262, 60 species. Extended field notes, much critical comment, and many measurements, in text or tabulated. The deer, peccaries, and nasuas treated at length, with illustrations, in the case of *Nasua* with relation to sex and age; pathological conditions in skulls of *Tayassu* treated at length.

Spp. et subspp nov.: (1) *Heteromys pictus escuinapæ*, p. 211; (2) *Molossus sinaloæ*, p. 236; (3) *Sciurus poliopus tepicanus*, p. 243; (3) *Sigmodon vulcani*, p. 247; (4) *Heteromys jaliscensis*, p. 251.

211. Mammals from the Island of Hainan, China.<*Bull. Amer. Mus. Nat. Hist.*, XXII, pp. 463–490, pl. xix, Dec. 17, 1906.

An annotated list of 41 species, the following new: (1) *Manis pusilla*, p. 465; (2) *Atherurus hainanus*, p. 470; (3) *Ratufa gigantea hainana*, p. 472; (4) *Funambulus rindonensis*, p. 472; (5) *Sciurus erythræus insularis*, p. 473; (6) *Tamiops* (gen. nov.) *macclellandi hainanus*, p. 476; (7) *Tamiops macclellandi riudoni*, p. 477; (8) *Tupaia modesta*, p. 481; (9) *Rhinolophus hainanus*, p. 482; (10) *Hipposideros poutensis*, p. 483; (11) *Scotophilus kuhlii insularis*, p. 485; (12) *Scotophilus castaneus consobrinus*, p. 485; (13) *Pipistrellus portensis*, p. 487.

212. The proper name of the Mexican Tamandua.<*Proc. Biol. Soc. Washington*, XIX, p. 200, Dec. 31, 1906.

The specific name *mexicanus* Saussure, 1860, should replace *tenuirostris* Allen, 1904.

213. Vertebrata of the Land; Birds and Mammals.<*Science, N. S.*, XXIII, pp. 317–319, Feb. 28, 1906.

In report of a discussion before the American Society of Naturalists at Philadelphia, Dec. 27, 1895, on 'The Origin and Relations of the Floras and Faunas of the Antarctic and Adjacent Regions.'

1907.

214. The Families and Genera of Bats.<*Amer. Nat.*, XLI, pp. 671–672, Oct. 1907.

Review of Gerrit S. Miller's work of this title (*Bull. U. S. Nat. Mus.*, No. 57, June 29, 1907).

1908.

215. The North Atlantic Right Whale and its near Allies.<*Bull. Amer. Mus. Nat. Hist.*, XXIV, pp. 277–329, pll. xix–xxiv, and 1 text figure, April 8, 1908.

History, relationships, nomenclature, geographical distribution, and external and osteological characters of *Eubalæna glacialis* (Bonnaterre).

216. The Peary Caribou (*Rangifer pearyi* Allen).<*Bull. Amer. Mus. Nat. Hist.*, XXIV, pp. 487–493, fig. 1–12, May 22, 1908.

Detailed description, with measurements and figures of skulls and antlers, and comparison with allied forms.

217. Notes on *Solenodon paradoxurus* Brandt. <*Bull. Amer. Mus. Nat. Hist.*, XXIV, pp. 305–517, pll. xxviii–xxxiii, and fig. 1–9, June 8, 1908.

External and cranial characters described and figured, including a young skull with milk dentition, and comparison with *S. cubanus*.

218. Mammalogical Notes.—I–VI. <*Bull. Amer. Mus. Nat. Hist.*, XXIV, pp. 575–589, fig. 1–4, Sept. 11, 1908.

(1) Concrescence in premolars of a bat (*Artibeus quadrivittatus*); (2) bats from the Island of San Domingo (*Ardops haitiensis* and *Molossus verrilli*, spp. nov., p. 581); (3) note on the type of the genus *Sciuropterus;* (4) note on the type locality of *Rangifer arctica* (Richardson); (5) northward extension of range of Coyotes; (6) the generic name *Galera* Browne.

219. Mammals from Nicaragua. <*Bull. Amer. Mus. Nat. Hist.*, XXIV, pp. 647–670, text figs, 1–12, Oct. 13, 1908.

Annotated list of 59 species. Gen. nov., *Hoplomys*, p. 649; spp. et subspp. nov.: (1) *Lepus gabbi tumacus*, p. 649; (2) *Hoplomys truei*, p. 650; (3) *Heteromys vulcani*, p. 652; (4) *Heteromys fuscatus*, p. 652; (5) *Neotoma chrysomelas*, p. 653; (6) *Oryzomys alfaroi incertus*, p. 655; (7) *Oryzomys ochraceus*, p. 655; (8) *Oryzomys carrikeri*, p. 656; (9) *Sigmodon hispidus griseus*, p. 657; (10) *Ototylomys fumeus*, p. 658; (11) *Peromyscus nicaraguæ*, p. 658; (12) *Sciurus deppei matagalpæ*, p. 660; (13) *Lutra latidens*, p. 660; (14) *Tayra barbara inserta*, p. 662; (15) *Bassaricyon richardsoni*, p. 662; (16) *Blarina olivaceus*, p. 669; (17) *Artibeus jamaicensis richardsoni*, p. 669; (18) *Alouatta palliata matagalpæ*, p. 670.

1909.

220. Mammals from British East Africa, collected by the Tjädor Expedition of 1906. <*Bull. Amer. Mus. Nat. Hist.*, XXVI, pp. 147–175, fig. 1–10, March 19, 1901.

Annotated list of 56 species. Spp. et subspp. nov.: (1) *Tragelaphus tjaderi*, p. 148; (2) *Madoqua langi*, p. 153; (3) *Arvicanthis nairobæ*, p. 168; (4) *Mus kijabius*, p. 169; (5) *Crocidura kijabæ*, p. 173.

221. The White Bear of Southwestern British Columbia <*Bull. Amer. Mus. Nat. Hist.*, XXVI, pp. 233–238, figs. 1–4, April 17, 1909.

External and cranial characters of *Ursus kermodei* Hornaday.

222. Further Notes on Mammals from the Island of Hainan, China. <*Bull. Amer. Mus. Nat. Hist.*, XXVI, pp. 239–242, April 17, 1909.

Notes on 13 species, 2 new for the island. (1) *Paradoxurus* (*Paguma*) *larvatus hainanus* and (2) *Mungos rubrifrons*, p. 240, sp. et subsp. nov.

223. Mammals from Shen-si Province, China. <*Bull. Amer. Mus. Nat. Hist.*, XXVI, pp. 425–430, Oct. 21, 1909.

Annotated list of 15 species. Spp. nov.: (1) *Myotalpa rufescens*, p. 428; (2) *Sciurotamias owstoni*, p. 428; (3) *Eutamias albogularis*, p. 429.

224. Osgood's Revision of the Mice of the Genus *Peromyscus*. <*Amer. Nat.*, XLIII, pp. 633–639, Oct., 1909.

Review of W. H. Osgood's paper of this title (*N. Amer. Fauna*, No. 28, April 17, 1909), in part historical of the subject.

225. 'Life Histories of Northern Animals.' <*Science*, N. S., XXX, No. 782, pp. 924–927, Dec. 24, 1909.

Review of Ernest Thompson Seton's work of this title, 2 vols, roy. 8vo., New York, Charles Scribner's Sons, 1909.

1910.

226. The Black Bear of Labrador. <*Bull. Amer. Mus. Nat. Hist.*, XXVIII, pp. 1–6, Jan. 5, 1910.

Ursus americanus sornborgeri Bangs, not tenable. *Ursus americanus kenaiensis* (p. 6), subsp. nov.

227. Mammals from the Athabasca-Mackenzie Region of Canada. <*Bull. Amer. Mus. Nat. Hist.*, XXVIII, pp. 7–11, Jan. 5, 1910.

Annotated list of 25 species.

228. Mammals from Palawan Island, Philippine Islands. <*Bull. Amer. Mus. Nat. Hist.*, XXVIII, pp. 13–17, Jan. 5, 1910.

Annotated list of 13 species, the following new: (1) *Mus luteiventris*, p. 14; (2) *Arctictis whitei*, p. 15; (3) *Mungos palawanus*, p. 17.

229. Nelson's Monograph of the North American Leporidæ. <*Amer. Nat.*, XLIV, pp. 57–63, Jan., 1910.

Review of 'The Rabbits of North America,' by E. W. Nelson (*N. Amer. Fauna*, No. 29, August 31, 1909). *Lepus aquaticus littoralis* Nelson shown to be a synonym of *Lepus aquaticus aquaticus*.

230. Seton's 'Life Histories of North American Animals.' <*Amer. Nat.*, XLIV, pp. 124–127, Feb., 1910.

Review of Ernest Thompson Seton's work of this title.

231. Additional Mammals from Nicaragua. <*Bull. Amer. Mus. Nat. Hist.*, XXVIII, pp. 87–115, April 30, 1910.

Annotated list of 82 species represented by specimens in the American Museum of Natural History, and 50 additional species on the basis of previous records. Spp. nov.: (1) *Macrogeomys matagalpæ*, p. 97; (2) *Oryzomys richardsoni*, p. 99; (3) *O. nicaraguæ*, p. 100; (4) *Conepatus nicaraguæ*, p. 106. *Ursus americanus perniger*, p. 115, nom. nov. for *U. a. kenaiensis*, preoccupied.

232. Mammals from the Caura district of Venezuela, with description of a new species of *Chrotopterus*. <*Bull. Amer. Mus. Nat. Hist.*, XXVIII, pp. 145–149, May 27, 1910.

Annotated list of 14 species. *Chrotopterus carrikeri* (p. 147), sp. nov.

233. The Mammals of Colorado. <*Science*, N. S., XXXII, No. 814, pp. 178–179, Aug. 5, 1910.

Review of Edward Royal Warren's book of this title (16 mo, New York and London, 1910).

1911.

234. Mammals from Venezuela collected by Mr. M. A. Carriker, Jr., 1909–1911. <*Bull. Amer. Mus. Nat. Hist.*, XXX, p. 239–273, Dec. 2, 1911.

Annotated list of 66 spp. Spp. et subspp. nov.: (1) *Cavia porcella venezuelæ*, p. 250; (2) *Loncheres carrikeri*, p. 251; (3) *Urocyon cinereoargentea venezuelæ*, p. 259; (4) *Chilonycteris rubiginosa fusca*, p. 262.

235. Mammals collected in the Dutch East Indies by Mr. Roy C. Andrews, on the cruise of the 'Albatross' in 1909. <*Bull. Amer. Mus. Nat. Hist.*, XXX, pp. 335–339, Dec. 21, 1911.

Annotated list of 16 species. Spp. nov.: (1) *Mus andrewsi*, p. 336; (2) *M. buruensis*, p. 336; *Sciurus beebei*, p. 338; (4) *Tamiops sauteri*, p. 339.

236. The Okapi. <*Amer. Mus. Journ.*, XI, No. 2, March, 1911, pp. 73–75.

A general account of the Okapi with reference to specimens received from the Museum's Congo Expedition.

237. The Habitat Groups of Mammals and Birds in the American Museum of Natural History. <*Amer. Mus. Journ.*, XI, No. 7, pp. 248–249, Nov. 1911.

1912.

238. Sheldon's 'The Wilderness of the Upper Yukon.' <*Science*, N. S., XXXV, No. 890, pp. 105–106, Jan. 19, 1912.

Review of Charles Sheldon's 'The Wilderness of the Upper Yukon: A Hunter's Explorations for Wild Sheep in Sub-Arctic Mountains.'

239. Historical and Nomenclatorial Notes on North American Sheep. <*Bull. Amer. Mus. Nat. Hist.*, XXXI, pp. 1–29, text figs. 1–4, March 4, 1912.

McGillivray's original account of his discovery of Wild Sheep in Canada in 1800, and copy of his original figure (pp. 2–7); E. Geoffroy's 'Belier de Montagne' (p. 8); *Ovis cervina* Demarest and *Ovis canadensis* Shaw (pp. 9–11); collation of Shaw and Nodder's 'Naturalist's Miscellany,' with reference to the date of *Ovis canadensis* Shaw (pp. 11–15); other early references to the Mountain Sheep of North America (pp. 15–17); the Taye of "California," with a copy of Venega's figure (pp. 17–20); synonymic list of North American Sheep, with their type localities and ranges (pp. 22–29). Eleven forms recognized, including *Ovis californianus* Douglas.

240. Mammals from Western Colombia. <*Bull. Amer. Mus. Nat. Hist.*, XXXI, pp. 71–95, April 19, 1912.

Annotated list of 55 spp. Spp. et subspp. nov.: (1) *Sylvilagus* (*Tapeti*) *fulvescens*, p. 75; (2) *Heteromys lomitensis*, p. 77; (3) *Reithrodontomys milleri*, p. 77; (4) *Rhipidomys mollissimus*, p. 78; (5) *R. similis*, p. 79; (6) *R. cocalensis*, p. 79; (7) *Thomasomys cinereiventer*, p. 80; (8) *T. popayanus*, p. 81; (9) *Neacomys pusillus*, p. 81; (10) *Oryzomys palmiræ*, p. 83; (11) *O. pectoralis*, p. 83; (12) *O.* (*Oligoryzomys*) *munchiquensis*, p. 85; (13) *O.* (*O.*) *fulvirostris*, p. 86; (14) *O.* (*Melanomys*) *obscurior affinis*, p. 88; (15) *Æpeomys fuscatus*, p. 89; (16) *Microxus affinis*, p. 89; (17) *Sciurus milleri*, p. 91; (18) *Blarina* (*Cryptotis*) *squamipes*, p. 93.

241. A new Pika from Colorado. <*Bull. Amer. Mus. Nat. Hist.*, XXXI, pp. 103–104, Sept. 2, 1912.

Ochotona figginsi, p. 103.

242. The probable recent extinction of the Muskox in Alaska. <*Science*, N. S., XXXVI, No. 934, pp. 720–722, Nov. 22, 1912.

Evidence that Muskoxen lived in Alaska "as recently as fifty to sixty years ago."

243. Zoölogy of the Stephánsson-Anderson [Arctic] Expedition — a preliminary Estimate. <*Amer. Mus. Journ.*, XII, p. 237, Nov., 1912.

244. Miller's 'List of North American Land Mammals in the United States National Museum.' <*Science*, N. S., XXXVII, No. 951, pp. 453–454, March 21, 1913.

Review of the work (*Bull. 79, U. S. Nat. Mus.*, Dec. 31, 1912).

1913.

245. Mammals collected in Korea. <*Bull. Amer. Mus. Nat. Hist.*, XXXII, pp. 427–436, Sept. 2, 1913. (With Roy C. Andrews.)

Annotated list of 19 spp., with field notes by Mr. Andrews. Spp. nov.: (1) *Ochotona* (*Pika*) *coreanus*, p. 429; (2) *Meles melanogenys*, 433.

246. New Mammals from Colombia and Ecuador.<*Bull. Amer. Mus. Nat. Hist.*, XXXII, pp. 469–484, text figs. 1–18, Sept. 25, 1913.

Spp. et subspp. nov.: (1) *Cholœpus florenciæ*, p. 469; (2) *C. agustinus*, p. 470; (3) *C. andinus*, p. 472; (4) *C. capitalis*, p. 472; (5) *Tayassu niger*, p. 476; (6) *Sylvilagus (Tapeti) salentus*, p. 476; (7) *Myoprocta milleri*, p. 477; (8) *Coendu quichua richardsoni*, p. 478; (9) *Proechimys o'connelli*, p. 479; (10) *Sigmodon chonensis*, p. 479; (11) *Akodon tolimæ*, p. 480; (12) *Potos flavus tolimensis*, p. 481; (13) *Nasua olivacea lagunetæ*, p. 483; (14) *Tayra barbara senilis*, p. 484.

247. Revision of the *Melanomys* Group of American Muridæ.<*Bull. Amer. Mus. Nat. Hist.*, XXXII, p. 535–555, pl. lxviii, Nov. 17, 1913.

14 species and subspecies, the following new: (1) *Melanomys caliginosus oroensis*, p. 538; (2) *M. affinis monticola*, p. 540; (3) *M. phæopus vallicola*, p. 544; (4) *M. phæopus tolimensis*, p. 545; (5) *M. lomitensis*, p. 545; (6) *M. buenavistæ*, p. 547.

248. New South American Muridæ.<*Bull. Amer. Mus. Nat. Hist.*, XXXII, pp. 597–604, Dec. 3, 1913.

Spp. et subspp. nov.: (1) *Oryzomys helvolus*, p. 597; (2) *O. o'connelli*, p. 597; (3) *O. vincencianus*, p. 598; (4) *O. incertus*, p. 598; (5) *Zygodontomys griseus*, p. 599; (6) *Z. fraterculus*, p. 599; (7) *Akodon chapmani*, p. 600; (8) *Rhipidomys quindianus*, p. 600; (9) *R. caucensis*, p. 601; (10) *R. venezuelæ yuruanus*, p. 601; (11) *R. milleri*, p. 602; (12) *Œcomys mincæ*, p. 603; (13) *Œ. caicaræ*, p. 603.

249. Catalogue of Mammals of Western Europe.<*Science*, N. S., XXXVIII, No. 970, pp. 159–162, Aug. 1, 1913.

Review of Gerrit S. Miller's work of this title (8vo, British Museum, 1912).

250. Ontogenetic and other Variations in Muskoxen, with a systematic Review of the Muskox Group, recent and extinct.<*Mem. Amer. Mus. Nat. Hist.*, New Series, I, Pt. 4, 1913, pp. 101–226, pll. xi–xviii, 1 map, and 45 text figs., March, 1913.

Ontogenesis of the horns, teeth, skull and pelage, pp. 107–143; individual differentiation as indicated by the skull, pp. 143–157; systematic review, including historical summary, pp. 157–160; geographic distribution, past and present, pp. 160–164; classification and relationship, 164–171; *Ovibos*, characters, alleged species and subspecies, pp. 171–179; geographic variation, pp. 179–180; synopsis of species and subspecies, pp. 180–182; systematic description, habits, and distribution: *Ovibos moschatus moschatus*, pp. 183–189; *O. m. niphæcus*, pp. 189–191; *O. m. wardi*, pp. 191–201; *O. yukonensis* (extinct), pp. 201–203; *O. pallantis* (extinct), pp. 203–205; extermination, pp. 205–207; Muskoxen in Zoölogical Gardens, pp. 207–208; *Bootherium*, pp. 209–213; *Symbos*, pp. 213–215; *Liops*, p. 216; bibliography, pp. 221–226.

Plates xi–xv, *O. m. wardi* as follows: pl. xi, horncores; pl. xii, transverse sections of horncores; pl. xiii, longitudinal sections of horncores; pl. xiv, sections of horncores; pl. xv, maxillary toothrow at different ages; pl. xvi, mandibular toothrow at different ages. Plates xvii and xviii, skull of *Symbos cavifrons*.

Text figures 1–26, skulls and dentition of *O. m. wardi* from fœtal age to senescence; text fig. 27, map of distribution, present and recent, of Muskoxen in North America and Greenland; text figs. 28–31, skulls of *O. m. moschatus* and *O. m. wardi;* text figs. 32–36, mounted specimens of same; text figs. 38–44, photographs of calves of *O. m. wardi* in New York Zoölogical Park; text fig. 45, type skull of *Bootherium bombifrons*.

251. Shall the Walrus become Extinct?<*Amer. Mus. Journ.*, XIII, pp. 38–42, Jan., 1913. With illustrations.

Comment on its unrestricted destruction and need of international restriction for its preservation.

1914.

252. Review of the Genus *Microsciurus*. <*Bull. Amer. Mus. Nat. Hist.*, XXXIII, pp. 145–165, Feb. 26, 1914.

19 species and subspecies recognized. Spp. nov.: (1) *Microsciurus rubrirostris*, p. 163; (2) *M. florenciæ*, p. 164.

253. Two new Mammals from Ecuador. <*Bull. Amer. Mus. Nat. Hist.*, XXXIII, pp. 199–200, Feb. 28, 1914.

(1) *Sylvilagus daulensis*, p. 199; (2) *Thomasomys aureus altorum*, p. 200.

254. Mammals from British East Africa, collected on the Third African Expedition of the American Museum by William S. Rainsford. <*Bull. Amer. Mus. Nat. Hist.*, XXXIII, pp. 337–344, June 18, 1914.

Annotated list of 40 species.

255. New South American Bats and a new Octodent. <*Bull. Amer. Mus. Nat. Hist.*, XXXIII, pp. 381–389, pl. xxviii, July 9, 1914.

Spp. et subspp. nov.: (1) *Amorphochilus schnablii osgoodi*, p. 381; (2) *Eptesicus andinus*, p. 382; (3) *Dasypterus ega punensis*, p. 382; (4) *Myotis ruber keaysi*, p. 383; (5) *M. punensis*, p. 383; (6) *M. bondæ*, p. 384; (7) *M. maripensis*, p. 385; (8) *M. esmeraldæ*, p. 385; (9) *M. caucensis*, p. 386; (10) *Nyctinomus æquatorialis*, p. 386; (11) *Mormopterus peruanus*, p. 387; (12) *Thrinacodus apolinari*, p. 387, pl. xxxviii, skull, with skull of *Dactylomys dactylinus* for comparison.

256. New South American Sciuridæ. <*Bull. Amer. Mus. Nat. Hist.*, XXXIII, pp. 585–597, fig. 1–2, Oct. 8, 1914.

Notosciurus gen. nov., p. 585; spp. et subspp. nov.: (1) *N. rhoadsi*, p. 585, text figs. 1–2; (2) *Guerlinguetus pucheranii salentensis*, p. 587; (3) *G. hoffmanni quindianus*, 587; (4) *G. hoffmanni manavi*, p. 589; (5) *G. griseimembra*, p. 589; (6) *G. candelensis*, p. 590; (7) *Sciurus gerrardi salaquensis*, p. 592; (8) *S. gerrardi cucutæ*, p. 592; (9) *S. saltuensis magdalenæ*, 594; (10) *S. duida*, p. 594; (11) *S. igniventris zamoræ*, 594; (12) *S. langsdorffii urucumus*, p. 595; (13) *S. langsdorffii steinbachi*, p. 596; (14) *S. stramineus zarumæ*, 597.

257. New South American Monkeys. <*Bull. Amer. Mus. Nat. Hist.*, XXXIII, pp. 647–655, Dec. 14, 1914.

Spp. et subspp. nov.: (1) *Callicebus lugens duida*, p. 647; (2) *Alouatta seniculus bogotensis*, p. 648; (3) *A. seniculus caquetensis*, p. 650; (4) *Pithecia milleri*, p. 650; (5) *Cacajao roosevelti*, p. 651; (6) *Ateles longimembris*, p. 65; (7) *A. robustus*, p. 652; (8) *Cebus apella brunneus*, 653; (9) *C. æquatorialis*, p. 654.

258. The Generic Names *Speothos* and *Icticyon*. <*Proc. Biol. Soc. Washington*, XXVII, p. 147, July 10, 1914.

Both genera tenable, not synonyms as held by some authors.

258*a*. Individual Variation in Muskoxen. <IXe Congrés International de Zoologie tenu à Monaco, 1913, pp. 210–215. Separates received in June, 1914.

With reference to the bones of the skull.

1915.

259. Review of the South American Sciuridæ. <*Bull. Amer. Mus. Nat. Hist.*, XXXIV, pp. 147–309, pll. i–xiv, and 25 text figs., May 17, 1915.

Historical outline, pp. 151–158; general considerations, pp. 158–168; genera and subgenera of American squirrels, pp. 169–186; systematic review of the South American squirrels, pp. 187–296; geographic distribution and interrelationships of South American squirrels, pp. 297–307.

Gen. nov.: (1) *Leptosciurus*, p. 199; (2) *Mesosciurus*, p. 212; (3) *Hadrosciurus*, p. 265; (4) *Urosciurus*, p. 267; (5) *Simosciurus*, p. 280; (6) *Histriosciurus* (subgen. nov.), p. 236. Subspp. nov.: (1) *Guerlinguetus æstuans venustus*, p. 260; (2) *Mesosciurus gerrardi baudensis*, p. 308; (3) *M. gerrardi valdivia*, p. 309.

Plates i–v, skulls of North American species; vi, maxillary teeth of same; vii-xii, skulls of South American species; xiii and xiv, maxillary teeth of same; text figs. 1–16, skins of South American species, to show relative size; text figs. 17–19, skull of *Syntheosciurus brochus*, p. 176; text figs. 20–21, hind foot of *Notosciurus* and *Mesosciurus*, p. 211; text figs. 22–25, distribution maps of genera.

260. Notes on American Deer of the Genus *Mazama*. <*Bull. Amer. Mus. Nat. Hist.*, XXXIV, pp. 521–553.

24 forms recognized, the following new: (1) *Mazama trinitatis*, p. 532; (2) *M. americana tumatumari*, p. 536; (3) *M. americana juruana*, p. 537; (4) *M. gualea*, p. 545; (5) *M. fuscata*, p. 545; (6) *M. zamora*, p. 546; (7) *M. murelia*, p. 547; (8) *M. cita sanctæmartæ*, p. 550.

261. New South American Mammals. <*Bull. Amer. Mus. Nat. Hist.*, XXXIV, pp. 625–634, Dec. 30, 1915.

(1) *Dasyprocta fuliginosa candelensis*, p. 625; (2) *Dasyprocta variegata zamoræ*, p. 627; (3) *Dasyprocta variegata chocoensis*, p. 627; (4) *Dasyprocta variegata urucuma*, p. 634; (5) *Proechimys kermiti*, p. 629; (6) *Oryzomys mureliæ* (nom. nov., to replace *O. incertus* Allen, 1913, preoccupied), p. 630; (7) *Procyon* (*Euprocyon*) *æquatorialis*, p. 630; (8) *Margay tigrina elenæ*, p. 631; (9) *Margay caucensis*, p. 631; (10) *Oncoides pardalis tumatumari*, p. 632; (11) *Eptesicus chapmani*, p. 632.

262. Convenience versus Fitness. <*Science*, N. S., XLII, No. 1084, pp. 492–494, Oct. 8, 1915.

On the function of the genus in classification, in reply to objectors who oppose subdivision of the "good old genera" of fifty years ago. Certain groups of mammals cited in illustration.

1916.

263. The proper Generic Name of the Macaques. <*Bull. Amer. Mus. Nat. Hist.*, XXXV, pp. 49–52, Feb. 21, 1916.

Silenus Goldfuss, 1820. The subgeneric names of the macaques and their near allies also considered. Comparison (in parallel columns) of the generic and subgeneric names of macaques employed by Elliot (Primates, 1913) with those here adopted.

264. New South American Mammals. <*Bull. Amer. Mus. Nat. Hist.*, XXXV, pp. 83–87, April 28, 1916.

(1) *Metachirus nudicaudatus antioquiæ*, p. 83; (2) *Tamandua tetradactyla punensis*, p. 83; (3) *Sylvilagus boylei*, p. 84; (4) *Cavia* (*Cavia*) *anolaimæ*, p. 85; (6) *Glossophaga apolinari*, p. 86; (7) *Saimiri caquetensis*, p. 87.

265. The Neotropical Weasels. <*Bull. Amer. Mus. Nat. Hist.*, XXXV, pp. 89–111, April 26, 1916.

Historical résumé, pp. 90–91; color variations in the *Mustela tropicalis* group, pp. 91–98; the status of *Mustela affinis* Gray, pp. 98–99; systematic review of the species and subspecies, pp. 99–111.

8 forms recognized — 6 species and 2 additional subspecies. *Mustela tropicalis nicaraguæ*, p. 100, subsp. nov.

266. List of Mammals collected for the American Museum in Ecuador by William B. Richardson, 1912–1913. <*Bull. Amer. Mus. Nat. Hist.*, XXXV, pp. 113–125, May 11, 1916.

Annotated list of 70 species and subspecies, with locality records, etc.

267. List of Mammals collected in Colombia by the American Museum of Natural History Expeditions, 1910–1915. <*Bull. Amer. Mus. Nat. Hist.*, XXXV, pp. 191–238, May 31, 1916.

Annotated list of 185 species and subspecies, with locality records, and technical comment, particularly on some of the Muridæ, Carnivores and Primates. Several extralimital Primates included for purpose of record and comment. *Alouatta seniculus* group, pp. 228–231.

268. New Mammals collected on the Roosevelt Brazilian Expedition. <*Bull. Amer. Mus. Nat. Hist.*, XXXV, pp. 523–530, July 24, 1916.

Species et subspp. nov.: (1) *Proechimys boimensis*, p. 523; (2) *Œcomys milleri*, p. 523; (3) *Œ. florenciæ*, p. 524; (4) *Œ. emiliæ*, p. 525; (5) *Oryzomys* (*Oligoryzomys*) *microtis*, p. 525; (6) *O.* (*O.*) *utiaritensis*, p. 527; (7) *O.* (*O.*) *mattogrossæ*, p. 528; (8) *Zygodontomys tapirapoanus*, p. 528; (9) *Zygodontomys tapirapoanus*, p. 528; (10) *Molossus cherriei*, p. 529; (11) *Molossus daulensis*, p. 530.

269. Mammals collected on the Roosevelt Brazilian Expedition, with Field Notes by Leo E. Miller. <*Bull. Amer. Mus. Nat. Hist.*, XXXV, pp. 559–610, Aug. 9, 1916.

Introduction, pp. 559–562; systematic list, 97 species and subspecies, pp. 562–588, with much technical comment; field notes by Leo E. Miller, on 53 species, pp. 589–610.

Tapirus terrestris guianiæ (p. 566), subsp. nov.

270. An Extinct Octodont from the Island of Porto Rico, West Indies. <*Ann. New York Acad. Sci.*, XXVII, pp. 17–22, pll. i–v, Jan. 25, 1916.

Isolobodon portoricensis (gen. et sp. nov.), p. 19.

Note.— The above list of titles ends with August, 1916. They are numbered 1–270, plus 117*a* and 258*a*. No. 32 is an inadvertent duplication of No. 26. Actual number of titles, 271.

Families, Superfamilies, Subfamilies, Genera and Subgenera, and Species and Subspecies, described as new or renamed.[1]

Higher Groups.

Oulophocinæ,[2] Bull. Mus. Comp. Zoöl., II, 1870, p. 44.
Trichiphocinæ,[2] *ibid.*, p. 44.
Castoroididæ, Monographs N. Amer. Rodentia (Rep. U. S. Geolog. Surv. Terr. (Hayden), XI, 1877, p. 419.)
Odobænidæ, Hist. N. Amer. Pinnipeds, 1880, pp. 3, 5.
Gressigrada, *ibid.*, p. 3 = Odobænidæ + Otariidæ.
Reptigrada, *ibid.*, p. 4 = Phocidæ.

Genera and Subgenera.

Bassaricyon, Proc. Acad. Nat. Sci. Philadelphia, 1876, pp. 20–23, pl. i. Type, *Bassaricyon gabbi*, sp. nov.
Ictidomys, Monographs N. Amer. Rodentia, p. 821, August, 1877. Type, *Spermophilus tridecemlineatus* Mitchell.
Pachycyon, Mem. Mus. Comp. Zoöl., X, No. 2, pp. 4–8, pll. i–iii, Dec., 1885. Type, *Pachycyon robustus*, sp. nov. Extinct.

---

[1] Arranged chronologically.
[2] Tentatively proposed as subfamilies of the Otariidæ and subsequently abandoned.

Microsciurus (subgen. of *Sciurus*), Bull. Am. Mus. Nat. Hist., VII, p. 333, Nov. 8, 1895. Type, *Sciurus* (*M.*) *alfaroi.*

Zygodontomys, Bull. Am. Mus. Nat. Hist., IX, p. 38, March 11, 1897. Type, *Oryzomys cherriei* Allen.

Sigmodontomys, Bull. Am. Mus. Nat. Hist., IX, p. 38, March 11, 1897. Type, *S. alfari*, sp. nov. (= *Nectomys.*)

Proechimys, Bull. Am. Mus. Nat. Hist., XII, p. 264, Dec. 26, 1899. Type, *Echimys trinitatis* Allen and Chapman.

Caluromys, Bull. Am. Mus. Nat. Hist., XIII, p. 189, Oct. 12, 1900. Type, *C. alstoni.* To replace *Philander* Brisson, 1762, considered as properly a synonym of *Didelphis* Linné, 1758. Under the principle of tautonymy *Philander* later became reinstated for this group, with *Didelphis philander* Linné as type.

Paralces, Bull. Am. Mus. Nat. Hist., XVI, p. 160, July 1, 1902. Type, *Cervus alces.* A substitute name for *Alces* Gray, 1821, preoccupied by *Alce* Blumenbach, 1799, type *Alce gigantea* Blumenbach = *Megaceros hibernicus* Owen, 1844.

Eunothocyon, Mamm. Patagonia, in Princeton Univ. Exped. Patagonia, III, pt. i, 1905, p. 152. Type, *Canis sladeni* Thomas.

Carcinocyon, *ibid.*, p. 153. Type, *Canis thous* Linné. (*Cerdocyon* H. Smith, type *Canis azaræ* Wied, apud Thomas.)

Tamiops, Bull. Am. Mus. Nat. Hist., XXII, p. 477, July 25, 1906. Type, *Tamiops macclellandi riudoni*, subsp. nov.

Hoplomys, Bull. Am. Mus. Nat. Hist., XXIV, p. 650, Sept. 11, 1908. Type, *Hoplomys truei* sp. nov.

Notosciurus, Bull. Am. Mus. Nat. Hist., XXXIII, p. 585, Oct. 8, 1914. Type, *Notosciurus rhoadsi* sp. nov.

Leptosciurus, Bull. Amer. Mus. Nat. Hist., XXXIV, p. 199, May 17, 1915. Type, *Macroxus pucheranii* Fitzinger = *Sciurus rufoniger* Pucheran.

Mesosciurus, *ibid.*, p. 212. Type, *Sciurus æstuans* var. *hoffmanni* Peters.

Histriosciurus, *ibid.*, p. 213 (subgenus of *Mesosciurus*). Type, *Sciurus gerrardi* Gray.

Hadrosciurus, *ibid.*, p. 265. Type, *Sciurus flammifer* Thomas.

Urosciurus, *ibid.*, p. 267. Type, *Sciurus tricolor* Pœppig.

Simosciurus, *ibid.*, p. 280. Type, *Sciurus stramineus* Eydoux and Souleyet.

Isolobodon, Ann. New York Acad. Sci., XXXVII, p. 19, Jan. 26, 1916. Type, *Isolobodon portoricensis* sp. nov.

Species and Subspecies.

1874.

Tamias quadrivittatus var. *pallidus*, Proc. Boston Soc. Nat. Hist., XVI, Feb. 1874, p. 289. (*Eutamias.*) [1]

Spermophilus tridecemlineatus var. *pallidus*, *ibid.*, p. 291. (*Ictidomys.*)

Spermophilus parryi var. *kodiacensis*, *ibid.*, p. 291. (*Citellus.*)

1876.

Canis mississippiensis, Amer. Journ. Sci. and Arts, ser. 3, XI, Jan. 1876, p. 49. (Extinct.)

---

[1] Present current generic and specific names are given in parentheses when different from those originally employed.

Cervus whitneyi, ibid., p. 49. (Extinct.)
Bassaricyon gabbi, Proc. Acad. Nat. Sci. Philadelphia, 1876, p. 20.

1877.

Lepus sylvaticus var. *arizonæ*, Mon. N. Amer. Roden., p. 332, August, 1877. (*Sylvilagus audubonii arizonæ.*)
Lepus graysoni, ibid., p. 347. (*Sylvilagus.*)
Lepus brasiliensis var. *gabbi, ibid.*, p. 419. (*Sylvilagus.*)
Tamias asiaticus var. *borealis, ibid.*, p. 793. (*Eutamias quadrivittatus borealis.*)

1885.

Pachycyon robustus, Mem. Mus. Comp. Zoöl., X, pp. 4–8, pll. i–iii, 1885. (Extinct.)

1887.

Squalodon tiedemani, Bull. Amer. Mus. Nat. Hist., II, p. 34, April 25, 1887. (Extinct.)

1889.

Sciurus alstoni, Bull. Amer. Mus. Nat. Hist., II, p. 167, Oct. 21, 1889 (name preoccupied; replaced by *S. nayaritensis* Allen, 1889).
Tamias asiaticus bulleri, ibid., p. 173. (*Eutamias bulleri.*)
Tamias asiaticus merriami, ibid., p. 176. (*Eutamias merriami.*)
Sigmodon fulviventer, ibid., p. 180.

1890.

Tamias obscurus, Bull. Am. Mus. Nat. Hist., III, p. 70, June, 1890. (*Eutamias.*)
Tamias senex, ibid., p. 83. (*Eutamias.*)
Tamias frater, ibid., p. 88. (*Eutamias.*)
Tamias amœnus, ibid., p. 90. (*Eutamias.*)
Tamias cinereicollis, ibid., p. 94. (*Eutamias.*)
Tamias umbrinus, ibid., p. 96. (*Eutamias.*)
Tamias quadrivittatus gracilis, ibid., p. 99. (*Eutamias.*)
Tamias quadrivittatus luteiventris, ibid., p. 101. (*Eutamias.*)
Tamias quadrivittatus affinis, ibid., p. 103. (*Eutamias.*)
Tamias quadrivittatus neglectus, ibid., p. 106. (*Eutamias.*)
Tamias minimus consobrinus, ibid., p. 112. (*Eutamias.*)
Tamias minimus pictus, ibid., p. 115. (*Eutamias.*)
Lepus cinerascens, Bull. Am. Mus. Nat. Hist., III, p. 159, Oct., 1890. (*Sylvilagus.*)
Lepus sylvaticus floridanus, ibid., p. 160. (*Sylvilagus.*)
Sciurus hudsonius vancouverensis, Bull. Am. Mus. Nat. Hist., III, p. 165, Nov. 14, 1890.
Sciurus hudsonius californicus, ibid., p. 165 (name preoccupied; replaced by *S. albolimbatus* Allen, 1898).
Artibeus coryi, Bull. Am. Mus. Nat. Hist., III, p. 173, Nov. 14, 1890.
Vespertilio velifer, Bull. Am. Mus. Nat. Hist., III, p. 177, Dec. 10, 1890. (*Myotis.*)
Sciurus cervicalis, ibid., p. 183.
Lepus sylvaticus aztecus, ibid., p. 188. (*Sylvilagus floridanus aztecus.*)
Lepus insolitus, ibid., p. 189. (*Sylvilagus cunicularius insolitus.*)
Lepus truei, ibid., p. 192. (*Sylvilagus.*)

1891.

Histiotus maculatus, Bull. Am. Mus. Nat. Hist., III, p. 195, Feb. 20, 1891. (*Euderma maculatum.*)
Blarina costaricensis, Bull. Am. Mus. Nat. Hist., III, p. 205, April 17, 1891. (*? = Blarina brevicauda.*)
Hesperomys (*Vesperimus*) *cherriei*, *ibid.*, p. 211. (*Reithrodontomys.*)
Hesperomys (*Vesperimus*) *nudipes*, *ibid.*, p. 213. (*Peromyscus.*)
Hesperomys (*Oryzomys*) *alfaroi*, *ibid.*, p. 214. (*Oryzomys.*)
Scalops argentatus texanus, *ibid.*, p. 221. (*Scalopus.*)
Dipodops sennetti, *ibid.*, p. 226. (*Perodipus.*)
Dipodops ordii palmeri, Bull. Am. Mus. Nat. Hist., III, p. 276, June 30, 1891. (*Perodipus.*)
Dipodops richardsoni, *ibid.*, p. 277. (*Perodipus.*)
Perognathus (*Chætodipus*) *femoralis*, *ibid.*, p. 281.
Neotoma micropus canescens, *ibid.*, p. 285.
Oryzomys aquaticus, *ibid.*, p. 289.
Vesperimus difficilis, *ibid.*, p. 298. (*Peromyscus.*)
Vesperimus nasutus, *ibid.*, p. 299. (*Peromyscus.*)
Vesperimus mearnsi, *ibid.*, p. 300. (*Peromyscus.*)
Capromys ingrahami, Bull. Am. Mus. Nat. Hist., III, p. 329, Aug. 31, 1891.
Oryzomys talamancæ, Proc. U. S. Nat. Mus., XIV, 1891, p. 193.
Hesperomys (*Vesperimus*) *affinis*, *ibid.*, p. 195. (*Peromyscus.*)

1892.

Perognathus merriami, Bull. Am. Mus. Nat. Hist., IV, p. 45, pl. iii, March 25, 1892.
Atalapha brachyotis, *ibid.*, p. 47. (*Lasiurus.*)
Oryzomys bauri, *ibid.*, p. 48.

1893.

Sciurus apache, Bull. Am. Mus. Nat. Hist., V, p. 29, March 16, 1893.
Thomomys monticolus, Bull. Am. Mus. Nat. Hist., V, p. 48, April 28, 1893.
Thomomys aureus, *ibid.*, p. 49.
Thomomys fossor, *ibid.*, p. 51.
Thomomys toltecus, *ibid.*, p. 52.
Zapus princeps, Bull. Am. Mus. Nat. Hist., V, p. 71, April 28, 1893.
Arvicola (*Mynomes*) *aztecus*, *ibid.*, p. 73. (*Microtus.*)
Reithrodontomys aztecus, *ibid.*, p. 79.
Sitomys auripectus, *ibid.*, p. 75. (*Peromyscus.*)
Sitomys rowleyi, *ibid.*, p. 76. (*Peromyscus.*)
Sitomys americanus thurberi, Bull. Am. Mus. Nat. Hist., V, p. 185, August 18, 1893. (*Peromyscus.*)
Sitomys martirensis, *ibid.*, p. 187. (*Peromyscus.*)
Sitomys gilberti, *ibid.*, p. 188. (*Peromyscus.*)
Tamias leucurus peninsulæ, *ibid.*, p. 197. (*Ammospermophilus.*)
Scapanus anthonyi, *ibid.*, p. 200.
Chæronycteris intermedia, Bull. Am. Mus. Nat. Hist., V, p. 207, Sept. 21, 1893 (with F. M. Chapman).

Nectomys palmipes, ibid., p. 209.
Tylomys couesi, ibid., p. 211. (*Rhipidomys.*)
Oryzomys speciosus, ibid., p. 212.
Oryzomys trinitatis, ibid., p. 213.
Oryzomys velutinus, ibid., p. 214.
Oryzomys brevicauda, ibid., p. 215. (*Zygodontomys.*)
Loncheres castaneus, ibid., p. 222.
Echimys trinitatis, ibid., p. 223. (*Proechimys.*)
Oryzomys costaricensis, Bull. Am. Mus. Nat. Hist., V, p. 239, Sept. 22, 1893.
Didelphys (*Micoureus*) *canescens, ibid.*, p. 235. (*Marmosa.*)
Sitomys robustus, Bull. Am. Mus. Nat. Hist., V, p. 335, Dec. 16, 1893. (*Peromyscus.*)
Geomys cherriei, ibid., p. 337. (*Macrogeomys.*)

1894.

Evotomys fuscodorsalis, Bull. Am. Mus. Nat. Hist., VI, p. 103, April 14, 1894. (*Evotomys gapperi.*)
Lepus sylvaticus mearnsi, Bull. Am. Mus. Nat. Hist., VI, p. 171, May 31, 1894. (*Sylvilagus.*)
Oryzomys palustris texensis, ibid., p. 177.
Perognathus pricei, Bull. Am. Mus. Nat. Hist., VI, p. 318, Nov. 7, 1894.
Perognathus conditi, ibid., p. 318.
Reithrodontomys mexicanus fulvescens, ibid., p. 319.
Arvicola leucophæus, ibid., p. 321. (*Microtus.*)
Sitomys americanus arizonæ, ibid., p. 321. (*Peromyscus.*)
Neotoma campestris, ibid., p. 322.
Neotoma rupicola, ibid., p. 323.
Neotoma grangeri, ibid., p. 324.
Sciurus hudsonicus dakotensis, ibid., p. 325.
Phenacomys truei, ibid., p. 331.
Arvicola insperatus, Bull. Am. Mus. Nat. Hist., VI, p. 347, Dec. 7, 1894. (*Microtus.*)
Lepus texianus eremicus, ibid., p. 347.
Lepus sylvaticus pinetis, ibid., p. 348. (*Sylvilagus nuttalli pinetis.*)
Sciurus arizonensis huachuca, ibid., p. 349.
Sciurus hudsonicus grahamensis, ibid., p. 350.

1895.

Reithrodontomys merriami, Bull. Am. Mus. Nat. Hist., VII, p. 119, May 21, 1895.
Reithrodontomys dychei, ibid., p. 120.
Reithrodontomys dychei nebrascensis, ibid., p. 122.
Reithrodontomys megalotis deserti, ibid., p. 127.
Reithrodontomys arizonensis, ibid., p. 134.
Reithrodontomys mexicanus intermedius, ibid., p. 136.
Reithrodontomys mexicanus aurantius, ibid., p. 137.
Reithrodontomys costaricensis, ibid., p. 139.
Lynx texensis, Bull. Am. Mus. Nat. Hist., VII, p. 188, June 20, 1895.
Thomomys cervinus, Bull. Am. Mus. Nat. Hist., VII, p. 203, June 29, 1895.
Lepus sylvaticus grangeri, Bull. Am. Mus. Nat. Hist., VII, p. 264, August 21, 1895. (*Sylvilagus nuttalli grangeri.*)

Lepus aquaticus attwateri, Bull. Am. Mus. Nat. Hist., VII, p. 327, Nov. 8, 1896. (*Sylvilagus.*)
Reithrodontomys australis, *ibid.*, p. 328.
Oryzomys cherriei, *ibid.*, p. 329. (*Zygodontomys.*)
Peromyscus attwateri, *ibid.*, p. 330.
Neotoma cinnamomea, *ibid.*, p. 331.
Sciurus (*Microsciurus*) *alfari*, *ibid.*, p. 333. (*Microsciurus.*)
Tamias pricei, *ibid.*, p. 333. (*Eutamias.*)
Tamias wortmani, *ibid.*, p. 335. (*Callospermophilus.*)
Spermophilus tridecemlineatus olivaceus, *ibid.*, p. 337. (*Ictidomys.*)
Spermophilus tridecemlineatus parvus, *ibid.*, p. 337. (*Ictidomys.*)
Blarina (*Soriciscus*) *nigrescens*, *ibid.*, p. 339. (*Cryptotis.*)
Blarina (*Soriciscus*) *orophila*, *ibid.*, p. 340. (*Cryptotis.*)

1896.

Rangifer terrænovæ, Bull. Am. Mus. Nat. Hist., VIII, p. 233, Nov. 21, 1896.
Reithrodontomys laceyi, *ibid.*, p. 235.
Perognathus mearnsi, *ibid.*, p. 237.
Peromyscus michiganensis pallescens, *ibid.*, p. 238.
Vespertilio incautus, *ibid.*, p. 239. (*Myotis.*)
Vespertilio chrysonotus, *ibid.*, p. 340. (*Myotis.*)

1897.

Peromyscus yucatanicus, Bull. Am. Mus. Nat. Hist., IX, p. 8, Feb. 23, 1897 (with F. M. Chapman).
Reithrodontomys mexicanus gracilis, *ibid.*, p. 9.
Heteromys gaumeri, *ibid.*, p. 9.
Artibeus palmarum, Bull. Am. Mus. Nat. Hist., IX, p. 16, Feb. 26, 1897 (with F. M. Chapman).
Oryzomys delicatus, *ibid.*, p. 19.
Akodon urichi, *ibid.*, p. 19.
Akodon frustrator, *ibid.*, p. 20.
Thylamys carri, *ibid.*, p. 27.
Artibeus intermedius, Bull. Am. Mus. Nat. Hist., IX, p. 33, March 11, 1897.
Oryzomys chrysomelas, *ibid.*, p. 37. (*Melanomys.*)
Sigmodontomys alfari, *ibid.*, p. 39. (*Nectomys.*)
Sigmodon borucæ, *ibid.*, p. 40.
Peromyscus spicilegus, Bull. Am. Mus. Nat. Hist., IX, p. 50, March 15, 1897.
Peromyscus banderanus, *ibid.*, p. 51.
Oryzomys mexicanus, *ibid.*, p. 52.
Oryzomys bulleri, *ibid.*, p. 53.
Sigmodon mascotensis, *ibid.*, p. 54.
Sigmodon colimæ, *ibid.*, p. 55.
Heteromys hispidus, *ibid.*, p. 56. (*Liomys.*)
Ovis stonei, Bull. Am. Mus. Nat. Hist., IX, p. 111, April 7, 1897.
Sciurus (*Microsciurus*) *peruanus*, Bull. Am. Mus. Nat. Hist., IX, p. 115, April 26, 1897. (*Microsciurus.*)

Oryzomys baroni, *ibid.*, p. 118.
Sigmodon peruanus, *ibid.*, p. 118.
Sigmodon bogotensis, Bull. Am. Mus. Nat. Hist., IX, p. 121, May 24, 1897.
Mus musculus jalapæ, Bull. Am. Mus. Nat. Hist., IX, p. 198, June 16, 1897 (with F. M. Chapman).
Reithrodontomys rufescens, *ibid.*, p. 199.
Reithrodontomys saturatus, *ibid.*, p. 201.
Peromyscus furvus, *ibid.*, p. 201.
Peromyscus melanotis, *ibid.*, p. 203.
Peromyscus musculus brunneus, *ibid.*, p. 203.
Oryzomys jalapæ, *ibid.*, p. 206.
Adelonycteris gaumeri, Bull. Am. Mus. Nat. Hist., IX, p. 231, Sept. 28, 1897. (*Eptesicus.*)

1898.

Marmosa sinaloæ, Bull. Am. Mus. Nat. Hist., X, p. 143, April 12, 1898.
Lepus peninsularis, *ibid.*, p. 144. (*Sylvilagus.*)
Lepus cerrosensis, *ibid.*, p. 145. (*Sylvilagus.*)
Lepus arizonæ confinis, *ibid.*, p. 146. (*Sylvilagus.*)
Thomomys fulvus anitæ, *ibid.*, p. 146.
Thomomys fulvus martirensis, *ibid.*, p. 147.
Thomomys atrovarius, *ibid.*, p. 148.
Perognathus pernix, *ibid.*, p. 149.
Neotoma sinaloæ, *ibid.*, p. 149.
Neotoma arenacea, *ibid.*, p. 150.
Neotoma anthonyi, *ibid.*, p. 151.
Peromyscus eremicus propinquus, *ibid.*, p. 154.
Peromyscus cedrocensis, *ibid.*, p. 154.
Peromyscus cineritius, *ibid.*, p. 155.
Peromyscus geronimensis, *ibid.*, p. 156.
Peromyscus exiguus, *ibid.*, p. 157.
Peromyscus dubius, *ibid.*, p. 157.
Sciurus hudsonicus baileyi, Bull. Am. Mus. Nat. Hist., X, p. 261, August 31, 1898.
Sciurus hudsonicus ventorum, *ibid.*, p. 263.
Sciurus hudsonicus streatori, *ibid.*, p. 267.
Sciurus douglasii cascadensis, *ibid.*, p. 277.
Sciurus fremonti neomexicanus, *ibid.*, p. 291.
Sciurus wagneri, Bull. Am. Mus. Nat. Hist., X, p. 453, Nov. 1898 (nom. nov., given in error to replace *S. varius* Wagner, preoccupied).
Sciurus douglasii albolimbatus, *ibid.*, p. 453 (to replace *Sciurus hudsonius californicus* Allen, 1890, preoccupied).

1899.

Zapus saltator, Bull. Am. Mus. Nat. Hist., XII, p. 3, March 4, 1899.
Phenacomys constablei, *ibid.*, p. 4.
Microtus stonei, *ibid.*, p. 5. (*Microtus drummondii.*)
Microtus vellerosus, *ibid.*, p. 7. (*Microtus mordax.*)
Microtus cautus, *ibid.*, p. 7. (*Microtus mordax.*)

Lepus americanus phæonotus, ibid., p. 11.
Lepus bishopi, ibid., p. 11.
Lepus floridanus chapmani, ibid., p. 13. (*Sylvilagus.*)
Thomomys fulvus alticolus, ibid., p. 13.
Reithrodontomys tenuis, ibid., p. 15.
Peromyscus texanus subarcticus, ibid., p. 15. (*Peromyscus maniculatus artemisiæ.*)
Sciurus chapmani, ibid., p. 16. (*Mesosciurus.*)
Lepus (Sylvilagus) superciliaris, Bull. Am. Mus. Nat. Hist., XII, p. 196, Dec. 20, 1899.
Isothrix rufodorsalis, ibid., p. 197.
Echimys mincæ, ibid., p. 199. (*Proechimys.*)
Echimys urichi, ibid., p. 199. (*Proechimys.*)
Echimys canicollis, ibid., p. 200. (*Proechimys.*)
Heteromys jesupi, ibid., p. 201.
Akodon venezuelensis, ibid., p. 203.
Akodon colombianus, ibid., p. 203. (*Melanomys.*)
Oryzomys maculiventer, ibid., p. 204.
Oryzomys trichurus, ibid., p. 206. (*Œcomys.*)
Oryzomys sanctæmartæ, ibid., p. 207.
Oryzomys mollipilosus, ibid., p. 208.
Oryzomys magdalenæ, ibid., p. 209.
Oryzomys villosus, ibid., p. 210.
Oryzomys palmarius, ibid., p. 210.
Oryzomys tenuicauda, ibid., p. 211.
Oryzomys modestus, ibid., p. 212.
Oryzomys fulviventer, ibid., p. 212.
Sciurus saltuensis bondæ, ibid., p. 213.
Sciurus (Guerlinguetus) quebradensis, p. 217. (= *Mesosciurus chapmani.*)

1900.

Chiroderma jesupi, Bull. Am. Mus. Nat. Hist., XIII, p. 88, May 12, 1900.
Mycronycteris hypoleuca, ibid., p. 90.
Promops affinis, ibid., p. 91.
Promops milleri, ibid., p. 92.
Caluromys alstoni, Bull. Am. Mus. Nat. Hist., XIII, p. 189, Oct. 12, 1900. (*Philander.*)
Didelphis pernigra, Bull. Am. Mus. Nat. Hist., XIII, p. 191, Oct. 23, 1900.
Didelphis karkinophaga caucæ, ibid., p. 192.
Didelphis karkinophaga colombica, ibid., 193.
Metachirus fuscogriseus, ibid., p. 194.
Metachirus tschudii, ibid., p. 195.
Metachirus nudicaudatus colombianus, ibid., p. 196.
Marmosa chapmani, ibid., p. 197.
Marmosa klagesi, ibid., p. 198.
Thylamys keaysi, ibid., p. 198.
Dactylomys peruanus, Bull. Am. Mus. Nat. Hist., XIII, p. 220, Nov. 16, 1900.
Oxymycterus juliacæ, ibid., p. 223.

Oxymycterus apicalis, *ibid.*, p. 224. (*Lenoxus.*)
Oryzomys keaysi, *ibid.*, p. 225.
Oryzomys obtusirostris, *ibid.*, p. 226.

1901.

Zygodontomys thomasi, Bull. Am. Mus. Nat. Hist., XIV, p. 39, Jan. 31, 1901.
Sigmodon simonsi, *ibid.*, p. 40.
Rhipidomys ochrogaster, *ibid.*, p. 43.
Phyllotis osilæ, *ibid.*, p. 44.
Akodon lutescens, *ibid.*, p. 46.
Rangifer stonei, Bull. Am. Mus. Nat. Hist., XIV, p. 143, May 28, 1901.
Didelphis marsupialis texensis, Bull. Am. Mus. Nat. Hist., XIV, p. 172, June 15, 1901.
Didelphis marsupialis tabascensis, *ibid.*, p. 173.
Didelphis richmondi, *ibid.*, p. 175.
Didelphis yucatanensis, *ibid.*, p. 178.
Metachirus fuscogriseus pallidus, Bull. Am. Mus. Nat. Hist., XIV, p. 215, July 3, 1901.
Metachirus griscescens, *ibid.*, p. 217.
Orizomys bolivaris, Bull. Am. Mus. Nat. Hist., XIV, p. 405, Nov. 30, 1901.
Oryzomys castaneus, *ibid.*, p. 406.
Oryzomys perenensis, *ibid.*, p. 406.
Oryzomys rivularis, *ibid.*, p. 407.
Phyllotis chacoensis, *ibid.*, p. 408.
Phyllotis cachinus, *ibid.*, p. 409.
Eligmodontia morgani, *ibid.*, p. 409.
Akodon tucumanensis, *ibid.*, p. 410.
Metachirus nudicaudatus bolivianus, *ibid.*, p. 411.

1902.

Dama lichtensteini, Bull. Am. Mus. Nat. Hist., XVI, p. 20, Feb. 1, 1902 (nom. nov., to replace *Cervus mexicanus* Lichtenstein held to be preoccupied by *Cervus mexicanus* Gmelin, unidentifiable.) (*Odocoileus.*)
Rangifer granti, Bull. Am. Mus. Nat. Hist., XVI, p. 119, April 7, 1902.
Ursus merriami, Bull. Am. Mus. Nat. Hist., XVI, p. 141, April 12, 1902.
Rangifer osborni, Bull. Am. Mus. Nat. Hist., XVI, p. 149, April 16, 1902.
Ovis dalli kenaiensis, Bull. Am. Mus. Nat. Hist., XVI, p. 145, April 23, 1902.
Didelphis marsupialis insularis, Bull. Am. Mus. Nat. Hist., XVI, p. 259, August 18, 1902.
Didelphis marsupialis caucæ, *ibid.*, p. 261.
Didelphis marsupialis etensis, *ibid.*, p. 262.
Didelphis paraguayensis andina, *ibid.*, p. 272.
Didelphis paraguayensis meridensis, *ibid.*, p. 274.
Rangifer pearyi, Bull. Am. Mus. Nat. Hist., XVI, p. 409, Oct. 31, 1902.
Phoca hispida gichigensis, Bull. Am. Mus. Nat. Hist., XVI, p. 478, Dec. 12, 1902.
Phoca ochotensis macrodens, *ibid.*, p. 483.
Phoca stejnegeri, *ibid.*, p. 485.
Phoca richardii pribilofensis, *ibid.*, p. 495.
Phoca richardii geronimensis, *ibid.*, p. 495.

1903.

Sigmodon puna, Bull. Am. Mus. Nat. Hist., XIX, p. 99, March 20, 1903.
Citellus buxtoni, Bull. Am. Mus. Nat. Hist., XIX, p. 139, May 9, 1903.
Citellus stejnegeri, *ibid.*, p. 142.
Evotomys (*Craseomys*) *latastei*, *ibid.*, p. 145.
Evotomys jochelsoni, *ibid.*, p. 148.
Lemmus obensis chrysogaster, *ibid.*, p. 153.
Ochotona kolymensis, *ibid.*, p. 154.
Lepus gichiganus, *ibid.*, p. 155.
Vulpes anadyrensis, *ibid.*, p. 167.
Putorius (*Arctogale*) *pygmæus*, *ibid.*, p. 176. (*Mustela.*)
Erinaceus orientalis, *ibid.*, p. 179.
Sorex buxtoni, *ibid.*, p. 181.
Ctenomys robustus, Bull. Am. Mus. Nat. Hist., XIX, p. 185, May 9, 1903. (*C. osgoodi.*)
Ctenomys sericeus, *ibid.*, p. 187.
Ctenomys colborni, *ibid.*, p. 188.
Oxymycterus microtis, *ibid.*, p. 189.
Reithrodon cuniculoides obscurus, *ibid.*, p. 190.
Reithrodon hatcheri, *ibid.*, p. 191.
Euneomys petersoni, *ibid.*, p. 192.
Citellus stonei, Bull. Am. Mus. Nat. Hist., XIX, p. 537, Oct. 10, 1903.
Synaptomys (*Mictomys*) *andersoni*, *ibid.*, p. 554.
Synaptomys (*Mictomys*) *chapmani*, *ibid.*, p. 555.
Erethizon epizanthus nigrescens, *ibid.*, p. 558.
Putorius microtis, *ibid.*, p. 563. (*Mustela.*)
Odocoileus battyi, Bull. Am. Mus. Nat. Hist., XIX, p. 591, Nov. 13, 1903.
Eutamias durangæ, *ibid.*, p. 594.
Citellus (*Otospermophilus*) *grammurus rupestris*, *ibid.*, p. 595.
Peromyscus paulus, *ibid.*, p. 598.
Peromyscus texanus flaccidus, *ibid.*, p. 599. (*Peromyscus leucopus tornillo.*)
Sigmodon baileyi, *ibid.*, p. 601.
Reithrodontomys megalotis sestinensis, *ibid.*, p. 602.
Neotoma intermedia durangæ, *ibid.*, p. 602.
Perodipus obscurus, *ibid.*, p. 603.
Lepus (*Macrotolagus*) *texianus micropus*, *ibid.*, p. 605. (*Lepus californicus texianus.*)
Lepus (*Macrotolagus*) *gaillardi battyi*, *ibid.*, p. 607.
Lepus (*Sylvilagus*) *durangæ*, *ibid.*, p. 609. (*Sylvilagus floridanus holzneri.*)
Canis impavidus, *ibid.*, p. 609.
Myotis californicus durangæ, *ibid.*, p. 612.
Odocoileus sinaloæ, *ibid.*, p. 613.
Lynx ruffus escuinapæ, *ibid.*, p. 614.

1904.

Oreamnos montanus columbianus, Bull. Am. Mus. Nat. Hist., XX, p. 20, Feb. 10, 1904. (*O. americanus columbiæ.*)
Oreamnos montanus missoulæ, *ibid.*, p. 20. (*O. americanus missoulæ.*)
Lepus (*Sylvilagus*) *russatus*, Bull. Am. Mus. Nat. Hist., XX, p. 31, Feb. 29, 1904.

Lepus (Sylvilagus) parvulus, ibid., p. 34.
Tayra barbara irara, ibid., p. 36.
Akodon irazu, ibid., p. 46. (*Scoteomys.*)
Felis carrikeri, ibid., p. 47.
Nasua narica bullata, ibid., p. 48.
Nasua narica panamensis, ibid., p. 51.
Nasua narica yucatanica, ibid., p. 52.
Nasua narica pallida, ibid., p. 53.
Sigmodon borucæ chiriquensis, ibid., p. 68.
Felis mearnsi, ibid., p. 71 (nom. nov., to replace *F. costaricensis*, preoccupied).
Felis panamensis, ibid., p. 71.
Potos flavus chiriquensis, ibid., p. 74.
Potos flavus caucensis, ibid., p. 75.
Potos flavus chapadensis, ibid., p. 76.
Myotis chiriquensis, ibid., p. 77.
Sciurus aberti phæurus, Bull. Am. Mus. Nat. Hist., XX, p. 205, May 28, 1904.
Sciurus aberti barberi, ibid., p. 207.
Eutamias canescens, ibid., p. 208.
Molossus coibensis, Bull. Am. Mus. Nat. Hist., XX, p. 227, June 29, 1904.
Molossus bondæ, ibid., p. 228.
Promops barbatus, ibid., p. 228.
Dermonotus suarpurensis, ibid., p. 229.
Lonchophylla thomasi, ibid., p. 230.
Artibeus rusbyi, ibid., p. 230.
Artibeus insularis, ibid., p. 231.
Artibeus yucatanicus, ibid., p. 232.
Phyllostoma hastatus panamensis, ibid., p. 233.
Phyllostoma hastatus cauræ, ibid., p. 234.
Ovis storcki, Bull. Am. Mus. Nat. Hist., XX, p. 293, Sept. 8, 1904.
Peramys brevicaudatus dorsalis, Bull. Am. Mus. Nat. Hist., XX, p. 327, Oct. 8, 1904.
Oryzomys tenuipes, ibid., p. 328.
Akodon meridensis, ibid., p. 329.
Holochilus venezuelensis, ibid., p. 330.
Felis maripensis, ibid., p. 331.
Felis sanctæmartæ, ibid., p. 332.
Procyon proteus, ibid., p. 333.
Nasua phæocephala, ibid., p. 334.
Erethizon godfreyi, Bull. Am. Mus. Nat. Hist., XX, p. 383, Oct. 15, 1904. (Extinct.)
Tamandua tetradactyla chapadensis, Bull. Am. Mus. Nat. Hist., XX, p. 392, Oct. 29, 1904.
Tamandua tetradactyla instabilis, ibid., p. 392.
Tamandua tetradactyla tenuirostris (= *mexicana* Saussure, 1860), *ibid.*, p. 394.
Tamandua tetradactyla chiriquensis, ibid., p. 395.
Coendou sanctæmartæ, Bull. Am. Mus. Nat. Hist., XX, p. 441, Nov. 28, 1904.
Lutra colombiana, ibid., p. 452.
Alouatta seniculus rubicunda, ibid., p. 458.
Alouatta seniculus caucensis, ibid., p. 462.

1905.

Eutamias lectus, Science Bull. Brooklyn Inst. Arts and Sciences, I, No. 6, p. 117, March 31, 1905.
Eutamias adsitus, *ibid.*, p. 118.
Cynomys parvidens, *ibid.*, p. 119.
Marmota engelhardti, *ibid.*, p. 120.
Ochotona cinnamomea, *ibid.*, p. 121.
Ctenomys osgoodi, Mamm. Patagonia, in Princeton Univ. Exped. to Patagonia, III, pt. i, p. 191 (nom. nov., to replace *C. robustus* Allen, preoccupied).
Canis sclateri, *ibid.*, p. 153 (nom. nov., to replace *C. microtis* Sclater, preoccupied)

1906.

Heteromys pictus escuinapæ, Bull. Am. Mus. Nat. Hist., XXII, p. 211, July 25, 1906 (*Liomys.*)
Molossus sinaloæ, *ibid.*, p. 236.
Sciurus poliopus tepicanus, *ibid.*, p. 243.
Sigmodon vulcani, *ibid.*, p. 247.
Heteromys jaliscensis, *ibid.*, p. 251.
Manis pusilla, Bull. Am. Mus. Nat. Hist., XXII, p. 465, Dec. 17, 1906.
Atherurus hainanus, *ibid.*, p. 470.
Ratufa gigantea hainana, *ibid.*, p. 472.
Funambulus riudonensis, *ibid.*, p. 472.
Sciurus erythræus insularis, *ibid.*, p. 473.
Tamiops macclellandi hainanus, *ibid.*, p. 476.
Tamiops macclellandi riudoni, *ibid.*, 477.
Tupaia modesta, *ibid.*, p. 481.
Rhinolophus hainanus, *ibid.*, p. 482.
Hipposideros poutensis, *ibid.*, p. 483.
Scotophilus kuhlii insularis, *ibid.*, p. 485.
Scotophilus castaneus consobrinus, *ibid.*, p. 485.
Pipistrellus portensis, *ibid.*, p. 487.

1907 (None).

1908.

Ardops haitiensis, Bull. Am. Mus. Nat. Hist., XXIV, p. 581, Sept. 11, 1908.
Molossus verrilli, *ibid.*, p. 581.
Lepus gabbi tumacus, Bull. Am. Mus. Nat. Hist., XXIV, p. 649, Oct. 13, 1908. (*Sylvilagus.*)
Hoplomys truei, *ibid.*, p. 650.
Heteromys vulcani, *ibid.*, p. 652. (*Liomys.*)
Heteromys fuscatus, *ibid.*, p. 652.
Neotoma chrysomelas, *ibid.*, p. 653.
Oryzomys alfaroi incertus, *ibid.*, p. 655. (*O. alfaroi.*)
Oryzomys ochraceus, *ibid.*, p. 655.
Oryzomys carrikeri, *ibid.*, p. 656.

Sigmodon hispidus griseus, ibid., p. 657.
Ototylomys fumeus, ibid., p. 658.
Peromyscus nicaraguæ, ibid., p. 658. (*P. mexicanus saxatilis.*)
Sciurus deppei matagalpæ, ibid., p. 660.
Lutra latidens, ibid., p. 660.
Tayra barbara inserta, ibid., p. 662.
Bassaricyon richardsoni, ibid., p. 662.
Blarina olivaceus, ibid., p. 669. (*Cryptotis.*)
Artibeus jamaicensis richardsoni, ibid., p. 669.
Alouatta palliata matagalpæ, ibid., p. 670. (= *A. palliata palliata.*)

1909.

Tragelaphus tjaderi, Bull. Am. Mus. Nat. Hist., XXVI, p. 148, March 19, 1909.
Madoqua langi, ibid., p. 153.
Arvicanthis nairobæ, ibid., p. 168.
Mus kijabius, ibid., p. 169.
Crocidura kijabæ, ibid., p. 173.
Paradoxurus (Paguma) larvatus hainanus, Bull. Am. Mus. Nat. Hist., XXVI, p. 240, April 17, 1909.
Mungos rubrifrons, ibid., p. 240.
Myotalpa rufescens, Bull. Am. Mus. Nat. Hist., XXVI, p. 428, Oct. 21, 1909.
Sciurotamias owstoni, ibid., p. 428.
Eutamias albogularis, ibid., p. 429.

1910.

Ursus americanus kenaiensis, Bull. Am. Mus. Nat. Hist., XXVIII, p. 6, Jan. 5, 1910. (*U. americanus pernigra.*)
Mus luteiventris, ibid., p. 14.
Arctictis whitei, ibid., p. 17.
Mungos palawanus, ibid., p. 17.
Macrogeomys matagalpæ, Bull. Am. Mus. Nat. Hist., XXVIII, p. 97, April 30, 1910.
Oryzomys richardsoni, ibid., p. 99.
Oryzomys nicaraguæ, ibid., p. 100.
Conepatus nicaraguæ, ibid., p. 106.
Ursus americanus perniger, ibid., p. 115 (nom. nov., to replace *U. a. kenaiensis* Allen, preoccupied).
Chrotopterus carrikeri, Bull. Am. Mus. Nat. Hist., XXVIII, p. 147, May 27, 1910.

1911.

Cavia porcella venezuelæ, Bull. Am. Mus. Nat. Hist., XXX, p. 250, Dec. 2, 1911.
Loncheres carrikeri, ibid., p. 251.
Urocyon cinereoargentea venezuelæ, ibid., p. 259.
Chilonycteris rubiginosa fusca, ibid., p. 262.
Mus andrewsi, Bull. Am. Mus. Nat. Hist., XXX, p. 336, Dec. 21, 1911.
Mus buruensis, ibid., p. 336.
Sciurus beebei, ibid., p. 338.
Tamiops sauterei, ibid., p. 339.

1912.

Sylvilagus (*Tapeti*) *fulvescens*, Bull. Am. Mus. Nat. Hist., XXXI, p. 75, April 19, 1912.
Heteromys lomitensis, *ibid.*, p. 77. (= *H. australis* Thomas.)
Reithrodontomys milleri, *ibid.*, p. 77.
Rhipidomys mollissimus, *ibid.*, p. 78.
Rhipidomys similis, *ibid.*, p. 79.
Rhipidomys cocalensis, *ibid.*, p. 79.
Thomasomys cinereiventer, *ibid.*, p. 80.
Thomasomys popayanus, *ibid.*, p. 81.
Neacomys pusillus, *ibid.*, p. 81.
Oryzomys palmiræ, *ibid.*, p. 83.
Oryzomys pectoralis, *ibid.*, p. 83.
Oryzomys (*Oligoryzomys*) *munchiquensis*, *ibid.*, p. 85.
Oryzomys (*Oligoryzomys*) *fulvirostris*, *ibid.*, p. 86.
Oryzomys (*Melanomys*) *obscurior affinis*, *ibid.*, p. 88. (*Melanomys affinis.*)
Æpeomys fuscatus, *ibid.*, p. 89.
Microxus affinis, *ibid.*, p. 89.
Sciurus milleri, *ibid.*, p. 91. (*Mesosciurus gerrardi milleri.*)
Blarina (*Cryptotis*) *squamipes*, *ibid.*, p. 93.
Ochotona figginsi, Bull. Am. Mus. Nat. Hist., XXXI, p. 103, May 28, 1912.

1913.

Ochotona (*Pika*) *coreanus*, Bull. Am. Mus. Nat. Hist., XXXII, p. 429, Sept. 2, 1913.
Meles melanogenys, *ibid.*, p. 433.
Cholœpus florenciæ, Bull. Amer. Mus. Nat. Hist., XXXII, p. 469, Sept. 25, 1913.
Cholœpus agustinus, *ibid.*, p. 470.
Cholœpus andinus, *ibid.*, p. 472.
Cholœpus capitalis, *ibid.*, p. 472.
Tayassu niger, *ibid.*, p. 476.
Sylvilagus (*Tapeti*) *salentus*, *ibid.*, p. 476.
Myoprocta milleri, *ibid.*, p. 477.
Coendu quichua richardsoni, *ibid.*, p. 478.
Proechimys o'connelli, *ibid.*, p. 479.
Sigmodon chonensis, *ibid.*, p. 479.
Akodon tolimæ, *ibid.*, p. 480.
Potos flavus tolimensis, *ibid.*, p. 481.
Nasua olivacea lagunetæ, *ibid.*, p. 483.
Tayra barbara senilis, *ibid.*, p. 484.
Melanomys caliginosus oroensis, Bull. Am. Mus. Nat. Hist., XXXI, p. 538, Nov. 17, 1913.
Melanomys affinis monticola, *ibid.*, p. 540.
Melanomys phæopus vallicola, *ibid.*, p. 544.
Melanomys phæopus tolimensis, *ibid.*, p. 545.
Melanomys lomitensis, *ibid.*, p. 545.
Melanomys buenavistæ, *ibid.*, p. 547.

Oryzomys helvolus, Bull. Am. Mus. Nat. Hist., XXXII, p. 597, Dec. 3, 1913.
Oryzomys o'connelli, *ibid.*, p. 597.
Oryzomys vicencianus, *ibid.*, p. 598.
Oryzomys incertus, *ibid.*, p. 598. (*O. mureliæ.*)
Zygodontomys griseus, *ibid.*, p. 599.
Zygodontomys fraterculus, *ibid.*, p. 599.
Akodon chapmani, *ibid.*, p. 600.
Rhipidomys quindianus, *ibid.*, p. 600.
Rhipidomys caucensis, *ibid.*, p. 601.
Rhipidomys venezuelæ yuruanus, *ibid.*, p. 601.
Rhipidomys milleri, *ibid.*, p. 602.
Œcomys mincæ, *ibid.*, p. 603.
Œcomys caicaræ, *ibid.*, p. 603.

1914.

Microsciurus rubrirostris, Bull. Am. Mus. Nat. Hist., XXXIII, p. 163, Feb. 26, 1914.
Microsciurus florenciæ, *ibid.*, p. 164.
Sylvilagus daulensis, Bull. Am. Mus. Nat. Hist., XXXIII, p. 199, Feb. 28, 1914.
Thomasomys aureus altorum, *ibid.*, p. 200.
Amorphochilus schnablii osgoodi, Bull. Am. Mus. Nat. Hist., XXXIII, p. 381, July 9, 1914.
Eptesicus andinus, *ibid.*, p. 382.
Dasypterus ega punensis, *ibid.*, p. 382.
Myotis ruber keaysi, *ibid.*, p. 383.
Myotis punensis, *ibid.*, p. 383.
Myotis bondæ, *ibid.*, p. 384.
Myotis maripensis, *ibid.*, p. 385.
Myotis esmeraldæ, *ibid.*, p. 385.
Myotis caucensis, *ibid.*, p. 386.
Nyctinomus æquatorialis, *ibid.*, p. 386.
Mormopterus peruanus, *ibid.*, p. 387.
Thrinacodus apolinari, *ibid.*, p. 387.
Notosciurus rhoadsi, Bull. Am. Mus. Nat. Hist., XXXIII, p. 585, Oct. 8, 1914.
Guerlinguetus pucheranii salentensis, *ibid.*, p. 587. (*Leptosciurus.*)
Guerlinguetus hoffmanni quindianus, *ibid.*, p. 587. (*Mesosciurus.*)
Guerlinguetus hoffmanni manavi, *ibid.*, p. 589. (*Mesosciurus.*)
Guerlinguetus griseimembra, *ibid.*, p. 589. (*Mesosciurus.*)
Guerlinguetus candelensis, *ibid.*, p. 590. (*Mesosciurus.*)
Sciurus gerrardi salaquensis, *ibid.*, p. 592. (*Mesosciurus.*)
Sciurus gerrardi cucutæ, *ibid.*, p. 592. (*Mesosciurus.*)
Sciurus saltuensis magdalenæ, *ibid.*, p. 593. (*Mesosciurus.*)
Sciurus duida, *ibid.*, p. 594. (*Urosciurus.*)
Sciurus igniventris zamoræ, *ibid.*, p. 594. (*Urosciurus.*)
Sciurus langsdorffii urucumus, *ibid.*, p. 595. (*Urosciurus.*)
Sciurus langsdorffii steinbachi, *ibid.*, p. 596. (*Urosciurus.*)
Sciurus stramineus zarumæ, *ibid.*, p. 597. (*Simosciurus.*)
Callicebus lugens duida, Bull. Am. Mus. Nat. Hist., XXXIII, p. 647, Dec. 14, 1914.
Alouatta seniculus bogotensis, *ibid.*, p. 648.

Alouatta seniculus caquetensis, ibid., p. 650.
Pithecia milleri, ibid., p. 650.
Cacajao roosevelti, ibid., p. 651.
Ateles longimembris, ibid., p. 651.
Ateles robustus, ibid., p. 652.
Cebus apella brunneus, ibid., p. 653.
Cebus æquatorialis, ibid., p. 654.

1915.

Guerlinguetus æstuans venustus, Bull. Am. Mus. Nat. Hist., XXXIV, p. 260, May 17, 1915.
Mesosciurus gerrardi baudensis, ibid., p. 308.
Mesosciurus gerrardi valdiviæ, ibid., p. 309.
Mazama trinitatis, Bull. Am. Mus. Nat. Hist., XXXIV, p. 532, Nov. 2, 1915.
Mazama americana tumatumari, ibid., p. 536.
Mazama americana juruana, ibid., p. 537.
Mazama gualea, ibid., p. 545.
Mazama fuscata, ibid., p. 545.
Mazama zamora, ibid., p. 546.
Mazama murelia, ibid., p. 547.
Mazama cita sanctæmartæ, ibid., p. 550.
Dasyprocta fuliginosa candelensis, Bull. Am. Mus. Nat. Hist., XXXIV, p. 625, Dec. 30, 1915.
Dasyprocta variegata zamoræ, ibid., p. 627.
Dasyprocta variegata chocoensis, ibid., p. 627.
Dasyprocta variegata urucuma, ibid., p. 634.
Proechimys kermiti, ibid., p. 629.
Oryzomys mureliæ, ibid., p. 630 (to replace *Oryzomys incertus* Allen, 1913, preoccupied).
Procyon (Euprocyon) æquatorialis, ibid., 630.
Margay tigrina elenæ, ibid., p. 631.
Margay caucensis, ibid., p. 631.
Oncoides pardalis tumatumari, ibid., p. 632.
Eptesicus chapmani, ibid., p. 632.

1916.

Isolobodon portoricensis, Ann. New York Acad. Sci., XXVII, pp. 17–22, pll. i–iv, Jan. 25, 1916.
Metachirus nudicaudatus antioquiæ, Bull. Am. Mus. Nat. Hist., XXXV, p. 83, April 28, 1916.
Tamandua tetradactyla punensis, ibid., p. 83.
Sylvilagus boylei, ibid., p. 84.
Cavia (Cavia) anolaimæ, ibid., p. 85.
Oryzomys barbacoas, ibid., p. 85.
Glossophaga apolinari, ibid., p. 86.
Saimiri caquetensis, ibid., p. 87.
Mustela tropicalis nicaraguæ, Bull. Am. Mus. Nat. Hist., XXXV, p. 100, April 28, 1916.
Proechimys boimensis, Bull. Am. Mus. Nat. Hist., XXXV, p. 523, July 24, 1916.

Œcomys milleri, *ibid.*, p. 523.
Œcomys florenciæ, *ibid.*, p. 524.
Œcomys emiliæ, *ibid.*, p. 525.
Oryzomys (Oligoryzomys) microtis, *ibid.*, p. 525.
Oryzomys (Oligoryzomys) utiaritensis, *ibid.*, p. 527.
Oryzomys (Oligoryzomys) mattogrossæ, *ibid.*, p. 528.
Zygodontomys tapirapoanus, *ibid.*, p. 528.
Molossus cherriei, *ibid.*, p. 529.
Molossus daulensis, *ibid.*, p. 530.
Tapirus terrestris guianiæ, Bull. Am. Mus. Nat. Hist., XXXV, p. 566, Aug. 9, 1916.

Summary.

Higher Groups	6
Genera and Subgenera	21
Species and Subspecies	563
Nomina nova (to replace preoccupied names)	10

INDEX TO MAMMALS.

NOTE.— The first reference is to the Bibliography (pp. 50–83), the second to the list of species, subspecies and higher groups (pp. 83–99). The original designation of the form (species of subspecies) is given in full except where a subgeneric name was included, which latter has been omitted in indexing. *Under genera subspecies are alphabetized by the subspecific name.*

II. BIRDS.

1860–62.

1. Birds of New England. <*New England Farmer* (weekly), for Aug. 11, 25, Sept. 22, Oct. 6, 20, Nov. 3, 17, Dec. 1, 22, 1860; Jan. 5, 26, March 2, May 18, June 15, July 20, Sept. 21, Oct. 26, Nov. 23, 1861; April 26, July 26, Aug. 16, Sept. 13, Oct. 18, 25, Nov. 8, Dec. 6, 1862. Also, published in the same newspaper (monthly), large 8vo, for Sept., Oct., Nov., Dec., 1861, January, March, May, June, July, Sept., Nov., Dec., 1862, and Jan., 1863.

   "Twenty-five articles in all, taking the birds of New England in order from Accipitres to the middle of the Fringillidæ. Popular biographical accounts, written to interest farmers in the feathered life of their fields."— (*Coues*, 'Bibliography of North American Ornithology,' in 'Birds of the Colorado Valley,' 1878, p. 661.)

1864.

2. Catalogue of the Birds found at Springfield, Mass., with Notes on their Migrations, Habits, etc.; together with a List of those Birds found in the State not yet observed at Springfield. <*Proc. Essex Inst.*, IV, art. iv, pp. 48–98, August, 1864.

   "195 spp. at Springfield; 296 in Massachusetts, of which 131 breed, 28 are resident, 67 winter visitors, 75 migrants, 106 summer visitors, 35 stragglers."— (*Coues*, *l. c.*, p. 667.)

1865.

3. Notes on the Habits and Distribution of the Duck Hawk, or American Peregrine Falcon (*Falco peregrinus*), in the Breeding Season, and Description of the Eggs [from Mt. Tom, Mass.]. <*Proc. Essex Inst.*, IV, (for 1864) 1865, art. x, pp. 153–161. (Communicated Nov. 14, 1864.)

1867.

4. Winter Notes of an Ornithologist. <*Amer. Nat.*, I, pp. 38–48, March, 1867.

   Relates to the winter birds of Massachusetts — 55 to 60 spp., consisting mainly of permanent residents and visitors from the North. Comment on their manner of occurrence and relative abundance.

5. The Birds of Spring. <*Amer. Nat.*, I, pp. 141–144, May, 1867.

   Relates mainly to New England — 280 species there at the season named; contains a tabular summary.

6. Ornithological Calendars [for March, April, and May]. <*Amer. Nat.*, I, p. 54 [for March], p. 109 [for April], p. 160 [for May], March–May, 1867.

   Dates of usual appearance of migratory birds in Massachusetts in the spring months.

1868.

7. Notes on Birds observed in Western Iowa, in the months of July, August and September; also on birds observed in Northern Illinois in May and June, and

at Richmond, Wayne Co., Indiana, between June third and tenth.<*Mem. Boston Soc. Nat. Hist.*, I, pt. iv, art. xiii, pp. 488–526, Dec., 1868 (read June, 1868).

"Very full: includes some critical commentary on geographical distribution in general, and on relationships of certain disputed species. Iowa, 108 spp.; Illinois, *a*, Ogle Co., 84 spp.; *b*, Cook Co., 94 spp., with monographic account of certain Turdidæ; Indiana, 72; the annotations in each case chiefly field-notes.

"'Range in the breeding-season must form the basis for defining the limits of different ornithological districts. . . . Among migrants of the same species the examples which arrive in the spring the earliest are bigger and more brightly tinted than those which come later, and, conversely, on their return the examples last seen are bigger than the summer specimens.' Hence it would appear that the largest individuals are those which go furthest north in summer, and, he also adds, are those which live further north in winter. Some characteristics of the ornithological provinces of North America are next briefly mentioned; and then follow the lists of species observed, as stated in the title. In Iowa about 108 or 110 species were seen, of which at least 100 breed in the State. For Illinois two lists are given, one of 84 species in Ogle County, the other of 94 species in Cook County. At Richmond 72 species were observed by himself and Dr. Haymond. Some critical notes on supposed species (Turdidæ, Laridæ) are added in the course of the paper."—*Coues, l. c.*, p. 676, quoting A. Newton, Zool. Rec. for 1868, p. 676.

8. [Abstract of a paper on the Birds of Iowa and Illinois, with special reference to their migration.]<*Proc. Boston Soc. Nat. Hist.*, XII, 1868, p. 85.

See the full memoir cited above, No. 7.

9. Circular in reference to obtaining data concerning the distribution of North American birds in the breeding season.<Ann. Report Trustees of Mus. Comp. Zoölogy for 1868 (1869), pp. 26–27.

Originally issued June 4, 1868. Signed L. Agassiz, but instigated and written by J. A. Allen.

9*a*. Notes on the Red and Mottled Owls.<*Amer. Nat.*, II, pp. 327–329, August, 1868.

On the red and gray phases of *Scops asio*, and incidental reference to the question of whether there is more than a single species of the genus *Scops* in North America.

10. The "Dwarf Thrush" again.<*Amer. Nat.*, II, pp. 488–489, Nov., 1868.

With reference to a former article in this journal by E. A. Samuels on this subject.

11. 'Natural History of Birds.'<*Amer. Nat.*, II, pp. 554–555, Dec., 1868.

Review of Miss Grace Anna Lewis's work of this title (12mo, Philadelphia, 1868).

1869.

12. Philadelphia Vireo and Yellow-bellied Flycatcher.<*Amer. Nat.*, III, p. 504, Nov., 1869.

Brief history of these two species, in reply to inquiries of a correspondent (W. L. T., Minneapolis, Minn.).

1869–70.

13. Notes on some of the Rarer Birds of Massachusetts.<*Amer Nat.*, III, Dec., 1869, pp. 505–519; Jan. and Feb., 1870, pp. 568–585, 631–648.

"315 species (*cf.* footnote, p. 647) in Massachusetts. The paper is supplementary to the author's Massachusetts catalogue published five years before in *Proc. Essex Inst.*; it contains critical comments on the rarer or less generally known species, and discusses the cases of some doubtful ones. The supposed *Buteo 'cooperi'* proved to be *lineatus*."—(*Coues, l. c.*, p. 682.)

1870.

14. Occurrence of the Brown Pelican [*Pelecanus fuscus*] in Massachusetts. <*Amer. Nat.*, IV, p. 58, March, 1870.

15. What is the "Washington Eagle"? <*Amer. Nat.*, IV, pp. 524–527, Nov. 1870.

Identified as the young of *Haliaëtus leucocephalus*.

16. Catalogue of the Birds of Iowa. <White's Geolog. Survey of Iowa, Vol. II, 1870, Appendix B, pp. 419–427. (Des Moines, Iowa, 8vo, pp. viii, 443.)

"Nominal list of 283 spp., those actually observed to breed within the limits of the State being marked with asterisk. A few of the species are to be regarded rather as stragglers, chiefly winter visitors from the North; some, however, from the West and South."— (*Coues, l. c.*, p. 682.)

17. Summer Red Bird [*Pyranga æstiva*]. *Amer. Nat.*, IV, p. 56, March, 1870.

Its occurrence at Amherst, Mass.

18. The Ruby-crowned Kinglet. *Amer. Nat.*, IV, p. 54, March, 1870.

The red crown-patch absent in the female.

1871.

19. The Fauna of the Prairies [of the United States]. <*Amer. Nat.*, V, pp. 4–9, March, 1871.

Mention of various birds, p. 6.

20. Classification of Water Birds. <*Amer. Nat.*, IV, pp. 746–752, Feb., 1871.

An extended synoptical review of Dr. E. Coues's paper on this subject (*Proc. Acad. Nat. Sci. Philadelphia*, 1869, pp. 193–218).

21. The Migration of Hawks. <*Amer. Nat.*, V, p. 173, May, 1871.

Confirmation of Dr. William Wood's observations (*Amer. Nat.*, Feb. 1871, p. 759) on the migration of hawks, namely in loose straggling flocks.

22. On the Mammals and Winter Birds of East Florida, with an Examination of certain assumed Specific Characters in Birds, and a Sketch of the Bird-Faunæ of Eastern North America. <*Bull. Mus. Comp. Zoöl.*, II, No. 3, pp. 161–450, pll. iv–viii, April, 1871.

"Part I, The Topographical, Climatic and Faunal characteristics of East Florida. (Part II, On Mammals.) Part III, On Individual and Geographical Variation among Birds, considered in respect to its bearing upon the value of certain assumed specific characters — a highly important philosophic treatise upon the general subject, which is discussed at length with force and logical consistency; the author's broad views upon this subject had at once a marked influence upon ornithological thought. Variation in general size and proportion of parts, both individual and climatic, are illustrated with numerous tables of measurements. An essay on species and varieties follows. Part IV, List of the Winter Birds of East Florida, with annotations,— field-notes, measurements, and much synonymy and technical criticism. Part V, On the Geographical Distribution of the Birds of Eastern North America, with special reference to the number and circumscription of the Ornithological Faunæ. After general introductory remarks, the natural Provinces of the North American Temperate Region are discussed, and the Ornithological Faunæ of the Eastern Provinces are treated. The following Faunæ are laid down and characterized: — 1. Floridan. 2. Louisianian. 3. Carolinian. 4. Alleghanian. 5. Canadian. 6. Hudsonian. 7. American Arctic. Various tabular summaries follow, with general remarks on the distribution and migration of the birds of the Eastern Province. A copious bibliography of American Ornithological literature concludes. The plates illustrate the variation in the bill of many species. The article gained the Humboldt Scholarship, and is one of the most important of American ornithological works. *Cf. Ibis*, 1872, pp. 189–191; *Zool. Rec.* for 1871, pp. 24, 25; *Am. Nat.*, V, 1871, pp. 364–373."— (*Coues, l. c.*, p. 686.)

1872.

23. Birds of Kansas. <*Amer. Nat.*, VI, pp. 359–360, June, 1872.

Review of first edition of F. H. Snow's Birds of Kansas (8vo, 1872, pp. 8), calling attention to errors and adding 18 species to the list.

24. Prof. Snow's List of Kansas Birds. <*Amer. Nat.*, VI, pp. 483–484, August, 1872.

In reply to criticism by Dr. T. M. Brewer of the former notice of the first edition of Prof. Snow's list. (*Cf. supra*, No. 23.)

25. Birds of Kansas. <*Amer. Nat.*, VI, p. 765, Dec., 1872.

Notice of the *second edition* of F. H. Snow's List of Kansas Birds. (8vo, pp. 16, Oct. 1872). Commended as "a highly valuable and creditable list of the Birds of Kansas."

26. Geographical Variation in North American Birds. <*Proc. Boston Soc. Nat. Hist.*, XV, pp. 212–219, Dec., 1872.

In amplification of a previous verbal communication. (Reprinted in *Amer. Nat.*, VIII, pp. 534–541, Sept. 1874.)

27. Notes on the Natural History of Fort Macon, N. C., and Vicinity. <*Amer. Nat.*, VI, pp. 546–549, Sept., 1872.

A review of E. Coues's paper in *Proc. Acad. Nat. Sci. Philadelphia*, 1871, pp. 12–49, 120–148.

28. The Birds of the Tres Marias and Socorro Islands. <*Amer. Nat.*, VI, pp. 681–682, Nov., 1872.

A review of A. J. Grayson's paper (edited by G. N. Lawrence), in *Proc. Boston Soc. Nat. Hist.*, XIV, 1872, pp. 261–303.

29. Ornithological Works in Prospect. <*Amer. Nat.*, VI, pp. 478–482, Aug., 1872.

Anticipatory notices of E. Coues's 'Key to North American Birds,' 'Birds of North America,' by S. F. Baird, T. M. Brewer, and R. Ridgway, and of C. J. Maynard's 'Birds of Florida.'

30. Ornithological Blunders. <*Amer. Nat.*, VI, pp. 303–304, May, 1872. (Anonymous.)

Comment on the case of *Bonasa jobsii* Jaycox, suggesting palliating circumstances.

31. [Geographical Variation in North American Birds.] <*Proc. Boston Soc. Nat. Hist.*, XV, 1872, pp. 156–159. (An editorial abstract is given in *Amer. Nat.*, VIII, 1874, pp. 534–541).

32. Catalogue of the Penguins in the Museum of the Boston Society of Natural History. <*Amer. Nat.*, VI, pp. 545–546, Sept., 1872.

A review of A. Hyatt's paper (*Proc. Boston Soc. Nat. Hist.*, XIV, 1872, pp. 237–251).

33. Notes of an Ornithological Reconnaissance of Portions of Kansas, Colorado, Wyoming, and Utah. <*Bull. Mus. Comp. Zoöl.*, III, No. 6, pp. 113–183, July 10, 1872.

"After general introductory, the author gives: — 1. List of Birds observed at Fort Leavenworth and Topeka, Kansas, spring of 1871 (121 spp.); 2. At Fort Hays, Kansas, May–July, 1871 (61 spp.); 3. In Northwestern Kansas, Dec. 1871, Jan., 1872 (25 spp.); 4. At Cheyenne, Wyoming, Aug. 1871 (41 spp.); 5. In Colorado, July–August, 1871 (81 spp.); 6. In South Park, Colorado, July, 1871 (54 spp.); 7. On Mount Lincoln, Colorado, July, 1871 (36 spp.); 8. Ogden, Utah, Sept.–Oct., 1871 (137 spp.); 9. General Summary or Combination of all the Observations (228 spp.). Besides the extended field-notes, there is much critical annotation. The article is very complete, and highly interesting from its bearing on general questions."— (*Coues, l. c.*, pp. 689–690.)

34. Ornithological Notes from the West. <*Amer. Nat.*, May, June, July, 1872, pp. 263–275, 342–351, 394–404.

I, Notes on the Birds of Kansas, pp. 263–275; II, Notes on the Birds of Colorado, pp. 342–351; III, Notes on the Birds of the Great Salt Lake Valley, pp. 394–404. Field notes on the species observed, with brief sketches of the country traversed. (See above, No. 33, for the more formal report on the species observed.)

1873.

35. Monograph of the Spheniscidæ. <*Amer. Nat.*, VII, pp. 38–40, Jan., 1873.

Review of E. Coues's 'Material for a Monograph of the Spheniscidæ' (*Proc. Acad. Nat. Sci. Philadelphia*, 1872, pp. 170–212).

36. The Birds of Florida. <*Amer. Nat.*, VII, pp. 165–166, March, 1873.

A review of Part I, of C. J. Maynard's work of this title (4to, pp. 32, Salem, 1872).

37. A Text Book of North American Ornithology. <*Amer. Nat.*, VII, pp. 308–309, May, 1873.

A review of E. Coues's 'Key to North American Birds,' first edition.

38. Revision of the American or Tyrant Flycatchers. <*Amer. Nat.*, VII, pp. 35–38, Jan., 1873.

Extended review of E. Coues's 'Studies of the Tyrannidæ. Part I' (*Proc. Acad. Nat. Sci. Philadelphia*, 1872, pp. 58–81, July, 1872).

39. Crows and Ravens. <*Amer. Nat.*, VII, pp. 743–744, Dec., 1873.

In reference to their supposed antagonism. It is stated that the author had found both crows and ravens together over portions of Dakota and Montana, both frequent and breeding in the same forests.

40. Recent Contributions to American Geographical Ornithology. <*Amer. Nat.*, VII, pp. 361–364, June, 1873.

Review of Holden and Aiken's 'Notes on Birds of Wyoming and Colorado Territories' (*Proc. Boston Soc. Nat. Hist.*, XV, pp. 193–210, Dec., 1872), W. D. Scott's 'Partial List of the Summer Birds of Kenawha County, West Virginia' (*ibid.*, pp. 219–228, Jan., 1873), and T. Martin Trippe's 'Notes on the Birds of Southern Iowa' (*ibid.*, pp. 229–242, March, 1873).

41. The White-fronted Owl in Canada. <*Amer. Nat.*, VII, pp. 427–428, July, 1873.

On *Nyctale albifrons* Cassin (= *N. acadica* juv.); its rarity in comparison with the adult (*N. acadica*) noted, and evidence cited that this supposed species is rightly considered to be the young of *N. acadica*.

1874.

42. Notes on the Natural History of Portions of Dakota and Montana Territories, being the substance of a report to the Secretary of War on the Collections

made by the North Pacific Railroad Expedition of 1873, Gen. D. S. Stanley, Commander. <*Proc. Boston Soc. Nat. Hist.*, XVII, 1874, pp. 33–86. Also separately published, 8vo, Boston, 1874, pp. 1-61.

"III, Report on the Birds, pp. 44–68. 118 spp., fully annotated, the list preceded by general considerations of the avifauna of the region, and several partial local lists. An important contribution."— (*Coues, l. c.*, p. 699.)

43. Field Ornithology. <*Amer. Nat.*, VIII, pp. 418–420, July, 1874.

A review of E. Coues's work of this title (8vo, Salem and Boston, 1874).

44. Laws of Geographical Variation in North American Mammals and Birds. < *Amer. Nat.*, VIII, pp. 227–229, April, 1874.

In reference to R. Ridgway's article (*Amer. Nat.*, VII, 1873, pp. 548–555), and making definite claim for certain original generalizations on this subject. (See also Coues, *op. cit.*, pp. 415–421).

1875.

45. [Exhibition of a Specimen of the Sharp-tailed Finch from Illinois.] <*Proc. Boston Soc. Nat. Hist.*, XVII, 1875, pp. 292–295.

First description of *Ammodramus caudacutus* var. *nelsoni*, and the first new form of bird described by the author.

46. Influence of Elevation and Latitude upon the Distribution of Species. <*Amer. Nat.*, IX, pp. 181–182, March, 1875.

47. Coues's Birds of the Northwest. <*Atlantic Monthly*, XXXVI, pp. 365–368, Sept., 1875.

Review of the work (8vo, 1874, U. S. Geol. and Geogr. Surv. Terr., Miscel. Pub. No. 3).

48. Dr. Coues's 'Birds of the Northwest.' <*Amer. Nat.*, IX, pp. 466–468, Aug., 1875.

Review of the work.

49. 'Birds of the Northwest.' <*Rod and Gun*, VI, May 22, 1875, p. 119.

A letter of comment upon the criticisms of Dr. Elliott Coues's 'Birds of the Northwest,' by a previous contributor.

1876.

50. Exploration of Lake Titicaca, by Alexander Agassiz and S. W. Garman. III List of Mammals and Birds. By J. A. Allen, with Field-notes by Mr. Garman. <*Bull. Mus. Comp. Zoöl.*, III, pp. 349–359, July, 1876.

An annotated list of 69 spp. of birds, pp. 353–359. *Falcinellus ridgwayi*, *Gallinula garmani*, spp. nov.

51. Progress in Ornithology in the United States during the last Century. <*Amer. Nat.*, X, pp. 536–550, Sept., 1876.

An enumeration, with comment, of the principal works and papers of the previous century (1776–1876), followed by a 'Summary of Progress.'

52. The Availability of certain Bartramian Names in Ornithology. <*Amer. Nat.*, X, pp. 21–29, Jan., 1876.

In reference to Dr. E. Coues's attempted revival of sundry Bartramian names for birds in his 'Fasti Ornithologiæ Redivivi. No. I' (*Proc. Acad. Nat. Sci. Philadelphia*, 1875, pp. 338–358).

53. Bartramian Names Again: An Explanation. <*Amer. Nat.*, X, pp. 176, 177, March, 1876.

Explanation of a misconception of the position assumed by his critic in Coues's rejoinder (*Amer. Nat.*, X, Jan. 1876, pp. 98–102) to 'The Availability of Certain Bartramian Names.' (*Cf. supra*, No. 52).

54. Ornithological Calendar for March. <*Forest and Stream*, VI, p. 84, Mar. 16, 1876.

55. Calendar for Birds in Massachusetts, March 20–31. <*Forest and Stream*, VI, p. 116, March 30, 1876.

56. Calendar [for Birds in Massachusetts for April 1–20.] <*Forest and Stream*, VI, p. 132, April 6, 1876.

57. Send in the Reports. <*Forest and Stream*, VI, p. 115, March 30, 1876.

On the importance of securing data on the migrations and habits of birds, with suggestions to observers. (See also *supra*, No. 9, on this subject.)

58. The Extinction of the Great Auk at the Funk Islands. <*Amer. Nat.*, X, p. 48, Jan., 1876.

59. Geographical Variation in the Number and Size of the Eggs of certain North American Birds. <*Bull. Nutt. Ornith. Club*, I, pp. 74–75, Sept., 1876.

Southern representatives of northern species lay smaller eggs, and also a smaller number of eggs.

60. Decrease of Birds in Massachusetts. <*Bull. Nutt. Ornith. Club*, pp. 53–60, Sept., 1876.

Statement of the case, and discussion of the causes, special reference being made to the Great Auk, Pinnated Grouse, Wild Turkey, Wild Pigeon, Swans, Cranes, Geese, etc.

61. Breeding of the Canada Goose [*Bernicla canadensis*] in trees. <*Bull. Nutt. Ornith. Club*, p. 50, July, 1876.

62. *Anser* [*Chen*] *rossii* in Oregon. <*Bull. Nutt. Ornith. Club*, I, p. 52, July, 1876.

Secured at Camp Harney, Oregon, by Capt. Charles Bendire.

63. Occurrence of the Wood Ibis [*Tantalus loculator*] in Pennsylvania and New York. <*Bull. Nutt. Ornith. Club*, I, p. 96, Nov., 1876.

64. Field and Forest. <*Bull. Nutt. Ornith. Club*, I, p. 71, Sept., 1876.

Notice of Vol. II, No. 1, the only ornithological article it contains being on Wilson's Phalarope by L. Kumlien.

65. The Birds and Coming Storms. <*Forest and Stream*, VI, No. 9, p. 133, April 6, 1876.

66. Descriptions of New Species of American Birds. <*Bull. Nutt. Ornith. Club.*, I, p. 47, July, 1876.

Notice of two recent papers by G. N. Lawrence in *Ann. Lyc. Nat. Hist. New York*, XI, pp. 163–166, and *Ibis*, July, 1875, pp. 383–387.

67. Notes on the Breeding Habits of Clarke's Crow (*Picicorvus columbianus*), with an account of its Nest and Eggs. <*Bull. Nutt. Ornith. Club*, I, pp. 44–46, July, 1876.

Compiled from letters from Capt. Charles Bendire and published over his name, with explanatory and other comment.

68. The Birds of Kansas. <*Bull. Nutt. Ornith. Club*, I, pp. 47–48, July, 1876.

Notice of F. H. Snow's third edition of his 'A Catalogue of Kansas Birds' (8vo, pp. 14, Nov. 1875).

69. Extinct Birds with Teeth. <*Bull. Nutt. Ornith. Club*, I, p. 49, July, 1876.

Notice of O. C. Marsh's papers on *Ichthyornis*, *Apatornis*, *Hesperornis*, *Lestornis*, in *Amer. Journ. Sci.*, 1875 and 1876.

70. Life-Histories of the Birds of Eastern Pennsylvania. <*Bull. Nutt. Ornith. Club*, I, pp. 49–50, July, 1876.

Review of Vol. I of T. G. Gentry's work of this title (12mo, Philadelphia, 1876).

71. Decrease of Birds in the United States. <*Penn. Monthly*, pp. 931–944, Dec., 1876.

Their present numbers, particularly waterfowl and shore-birds, contrasted with the numbers recorded in the accounts of conditions in the 17th century.

72. The Birds of Ritchie County, Virginia. <*Bull. Nutt. Ornith. Club*, I, p. 72, Sept., 1876.

Brief notice of William Brewster's paper (*Ann. Lyc. Nat. Hist. New York*, XI, pp. 129–146).

73. Birds of Southwestern Mexico. <*Bull. Nutt. Ornith. Club*, I, p. 93, Nov., 1876.

Brief notice of G. N. Lawrence's paper on F. E. Sumichrast's collection (*Bull. U. S. Nat. Mus.*, No. 4, 1876).

74. Jordan's Manual of Vertebrate Animals. <*Bull. Nutt. Ornith. Club*, I, pp. 93–94, Nov., 1876.

Review of the first edition of this well-known work by David Starr Jordan (12mo, Chicago, 1876).

75. The Portland Tern [*Sterna portlandica*]. <*Bull. Nutt. Ornith. Club*, I, pp. 71–72, Sept., 1876.

Notice of William Brewster's paper on the "so-called *Sterna portlandica*," which he considers an unusual phase of *Sterna macroura* (*Ann. Lyc. Nat. Hist. New York*, XI, pp. 200–207, Feb. 18, 1876).

76. Sexual, Individual, and Geographical Variation in *Leucosticte tephrocotis*. < *Bull. U. S. Geol. and Geogr. Surv. Terr.*, II, No. 4, July, 1876, pp. 345–350.

A "considerable constant sexual variation in coloration and also. . . . in size," overlooked by R. Ridgway in his monograph on the genus *Leucosticte* (*Bull. U. S. Geol. and Geogr. Surv. Terr.*, No. 2, 2d ser., May, 1875).

77. Sexual Variation in the Genus *Leucosticte*. <*Field and Forest*, II, No. 5, pp. 76–79, Nov., 1876.

A rejoinder to Mr. Ridgway's reply to criticisms of his Monograph of the Genus *Leucosticte*. (*Cf. supra*, No. 76).

1877.

78. Recent Ornithological Papers. <*Amer. Nat.*, XI, pp. 615–617, Oct., 1877.

Reviews of Nelson's paper on Illinois Birds in *Bull. Essex Inst.*, VIII, 1877, pp. 90–155, and IX, pp. 32–65; McCauley's on Texas Birds, in *Bull. U. S. Geol. and Geogr. Surv. Terr.*, III, 1877, 655–695, and others.

79. Californian Ornithology. <*Bull. Nutt. Ornith. Club*, II, p. 76, July, 1877.

Review of Dr. J. G. Cooper's paper 'New Facts relating to Californian Ornithology' (*Proc. California Acad. Sci.*, 1876).

80. McCauley's Notes on Texan Ornithology. <*Bull. Nutt. Ornith. Club*, II, p. 76, July, 1877.

Review of Lieut. C. A. H. McCauley's paper (*Bull. U. S. Geol. and Geogr. Surv. Terr.*, III, No. 3, pp. 655–695, May 15, 1877).

81. Occurrence of the Western Nonpareil [*Cyanospiza versicolor*], and Berlandier's Wren [*Thryothorus ludovicianus* var. *berlandieri*] at Fort Brown, Texas. < *Bull. Nutt. Ornith. Club*, II, No. 4, pp. 109–110, Oct., 1877.

Based on information furnished by Dr. J. C. Merrill.

82. Ridgway's "Studies of the American Falconidæ." <*Bull. Nutt. Ornith. Club*, II, pp. 70–73, July, 1877.

Review of several of R. Ridgway's papers on this family, in *Bull. U. S. Geol. and Geogr. Surv. Terr.*, *Proc. Acad. Nat. Sci. Philadelphia*, and *Proc. Boston Soc. Nat. Hist.*

83. List of Birds collected by Mr. Charles Linden, near Santarem, Brazil. <*Bull. Essex Inst.*, VIII, No. 8, 1876, pp. 78–83. (Pub. Feb., 1877).

Annotated list of 128 spp.; *Coccygus lindeni*, sp. nov.

84. The Influence of Physical Conditions in the Genesis of Species. <*Radical Review*, I, pp. 108–140, May, 1877.

On the influence of environment in modifying forms of mammals and birds, and discussion of Darwin's theory of the origin of species by natural selection. Reprinted (by request), with slightly modernized nomenclature by the author, in Ann. Rep. Smiths. Institution for 1905 (1906).

1878.

85. Maynard's Birds of Florida. <*Bull. Nutt. Ornith. Club*, III, p. 145, July, 1878.

Notice of Part IV of C. J. Maynard's "Birds of Florida," etc. (4to, Newtonville, Mass., pp. 89–112, one pl.). The full title now employed is 'The Birds of Florida, with the Water and Game Birds of Eastern North America.'

86. Jordan's Manual of Vertebrated Animals. <*Bull. Nutt. Ornith. Club*, III, pp. 145–146, July, 1878.

Brief notice of the 2d ed., revised and enlarged (1878). (*Cf. supra*, No. 74.)

87. The Carolinian Fauna. <*Bull. Nutt. Ornith. Club*, III, July, 1878, pp. 149–150.

Apropos of Eugene Bicknell's paper (*Bull. Nutt. Ornith. Club*, III, pp. 128–132) and continuing the subject (northward extension into the lower Hudson River valley).

88. Brewer's Supplement to his Catalogue of New England Birds. <*Bull. Nutt. Ornith. Club*, III, p. 185, Oct., 1878.

Review of paper by Dr. T. M. Brewer (*Proc. Boston Soc. Nat. Hist.*, XIX, pp. 301–309), which adds 21 spp. to his 'Catalogue of the Birds of New England' published in 1875.

89. Elliot's Review of the Ibidinæ, or Ibises. <*Bull. Nutt. Ornith. Club*, III, p. 182, Oct., 1878.

Review of D. G. Elliot's paper of this title (*Proc. Zool. Soc. London*, 1877, pp. 477–510).

90. [The Glossy Ibis, *Ibis falcinellus*, in Massachusetts.] <*Bull. Nutt. Ornith. Club*, III, p. 152, July, 1878.

Note, following a record of this species in Massachusetts by C. B. Cory, on the occurrence of two other individuals on Cape Cod at about the same time. Also a foot-note on the correct generic name of the bird, which is stated to be *Plegadis* Kaup. (*Cf. Ibis*, 4th ser., II, Jan., 1878, p. 112).

91. A list of the Birds of Massachusetts, with Annotations. <*Bull. Essex Inst.*, X, pp. 3–37, 1878.

"This may be considered to supercede previous tracts on the same subject, both by the present and other authors, as it completely sums our knowledge of the subject. The paper opens with summary considerations, followed by a valuable historical résumé. 1. Species of authentic occurrence within the State, 317. 2. Extirpated, 4. 3. Of probable occurrence, 24. 4. Hypothetical and doubtful species, 3 (*Myiodioctes minutus*, *Empidonax pygmæus* of Minot, *Thaumatias linnæi*). 5. Introduced undomesticated species, 6.— 'Considered as fairly entitled to recognition as Massachusetts birds,' 340. Known to breed in the State, about 135. Extremely rare or accidental visitors, 90. North American species added since 1867, 35." — (*Coues, l. c.*, p. 736).

92. Sabine's Gull [*Xema sabinii*] in Maine. <*Bull. Nutt. Ornith. Club*, III, p. 195, Oct., 1878.

The only previous New England record for this species is Boston Harbor, Mass., Sept. 24, 1874 (Brewster, *Amer. Sportsman*, V, 1875, p. 370; Brewer, *Proc. Boston Soc. Nat. Hist.*, XVII, 1875, p. 449.

93. Occurrence of three Species of Sea-Ducks [*Œdemia americana*, *Œ. fusca*, *Œ. perspicillata*] at St. Louis, Missouri. <*Bull. Nutt. Ornith. Club*, III, pp. 148–149, July, 1878.

94. Description of a Fossil Passerine Bird from the Insect-bearing Shales of Colorado. <*Bull. U. S. Geol. and Geogr. Surv. Terr.*, IV, No. 2, May 3, 1878, art. xix, pp. 443–445, pl. 1, figg. 1, 2.

"*Palæospiza bella*, g. sp. n., p. 443 — the first fossil Passerine discovered in North America. It bears distinct impressions of feathers. N. B.— a few copies of this No. of the Bull. were in private circulation from about April 14, and some electros of the plate were sent out during this month. Thus, the cut appeared in a Californian newspaper [*Cf. infra*, No. 117] with a compiled account of the bird, April 27. The article was immediately reprinted in the *Am. Journ. Sci.*, for May, 1878."— (Coues, *l. c.*, p. 738). The type specimen, lost for many years was recovered in 1915.

95. Birds of the Vicinity of Cincinnati. <*Bull. Nutt. Ornith. Club*, III, p. 34, Jan., 1878.

Review of F. W. Langdon's 'Catalogue of the Birds of the Vicinity of Cincinnati' (8vo, Salem, 1877, pp. 18).

96. Rowley's 'The Pied Duck [*Camptolæmus labradorius*].' <*Bull. Nutt. Ornith. Club*, III, pp. 79–80, Apr., 1878.

Review of G. D. Rowley's Monograph (*Orn. Miscel.*, II, pt. vii, 1877).

97. Barrows's 'Catalogue of the Alcidæ.' <*Bull. Nutt. Ornith. Club*, III, p. 86, Apr., 1878.

Brief review of W. B. Barrows's his paper of this title (*Proc. Boston Soc. Nat. Hist.*, XIX, 1877, pp. 150–165).

98. Ridgway's 'Studies of the American Herodiones.' <*Bull. Nutt. Ornith. Club*, III, pp. 182–183, Oct., 1878.

Review of R. Ridgway's paper in *Bull. U. S. Geol. and Geogr. Surv. Terr.*, IV, pp.-219–251, Feb. 5, 1878.

99. Reichenow's Review of the Herons and their Allies. <*Bull. Nutt. Ornith. Club*, III, pp. 183–185, Oct., 1878.

Extended review of Dr. Anton Reichenow's 'Uebersicht der Schreitvögel,' in *Cabanis's Journ. für Ornith.*, Jahrg. XXV, Apr.–July, 1877, pp. 113–171, 225–278.

100. An Inadequate 'Theory of Birds' Nests.' <*Bull. Nutt. Ornith. Club*, III, pp. 23–32, Jan., 1878.

A criticism of A. R. Wallace's 'Theory of Birds' Nests.'

101. Late Capture of the Yellow-bellied Flycatcher [*Empidonax flaviventris*] in Massachusetts. <*Bull. Nutt. Ornith. Club*, III, pp. 101–102, Apr., 1878.

Nov. 29, and Dec. 1, 1876 — two records.

102. The Lark-Bunting [*Calamospiza bicolor*] in Massachusetts. <*Bull. Nutt. Ornith. Club*, III, p. 48, Jan., 1878.

First record of the capture of this species east of the Mississippi River.

103. Rufous-headed Sparrow [*Peucœa ruficeps*] in Texas. <*Bull. Nutt. Ornith. Club*, III, pp. 188–189, Oct., 1878.

Gainesville, Texas, April 24, 1878, collected by George H. Ragsdale.

104. The Snow-Bird [*Junco hyemalis*] in Summer on Mount Wachusett, Mass. < *Bull. Nutt. Ornith. Club*, III, p. 192, Oct., 1878.

On the authority of Bradford Torrey.

105. Early Nesting of the Shore Lark [*Eremophila alpestris*] near Indianapolis, Ind. <*Bull. Nutt. Ornith. Club*, III, p. 189, Oct., 1878.

Young birds half-grown April 24, 1878. Based on information communicated by David S. Jordan (*in lit.*).

106. Ridgway's Ornithology of the Fortieth Parallel. <*Amer. Nat.*, XII, p. 469, July, 1878.

Review of the work (Expl. 40th Parallel, IV, pt. iii, 4to, 1877).

107. Sharpe's 'Catalogue of the Birds in the British Museum.' <*Bull. Nutt. Ornith. Club*, III, pp. 77–79, Apr., 1878.

Review of Volume I, 1874 (Falconidæ), Volume II, 1875 (Strigidæ), and Vol. III, 1877 (the Coliomorphæ — Crows, Birds of Paradise, Orioles, etc.).

108. [*Pipilo erythrophthalmus* with white spots on the Scapulars.] <*Bull. Nutt. Ornith. Club*, III, p. 42 Jan., 1878.

Taken near Leavenworth, Kansas.

109. Summer Birds of the Adirondacks. <*Bull. Nutt. Ornith. Club*, III, p. 36, Jan., 1878.

Notice of T. Roosevelt and H. D. Minot's paper (8vo, 1877, pp. 4, privately printed.)

110. Birds of Central New York. <*Bull. Nutt. Ornith. Club*, III, pp. 34–35, Jan., 1878.

Review of F. B. Rathbun's List, in *Auburn Daily Advertiser*, Aug. 14, 1877.

111. Birds of Southern Illinois. <*Bull. Nutt. Ornith. Club*, III, p. 36, Jan., 1878.

Review of E. W. Nelson's paper (*Bull. Essex Inst.*, IX, pp. 32–65, June, 1877).

112. Gentry's 'Life-Histories of the Birds of Eastern Pennyslvania.' <*Bull. Nutt. Ornith. Club*, III, pp. 36–37, Jan., 1878.

Review of Vol. II, of this work (*cf. supra*, No. 70).

113. Street's Notes on the Birds of Lower California and the Hawaiian and Fanning Islands. <*Bull. Nutt. Ornith. Club*, III, pp. 80–81, April, 1878.

Review of Dr. Thomas H. Street's paper in Bull. U. S. Nat. Mus., No. 7 (birds, pp. 9–33, 1877).

114. Bendire's Notes on the Birds of Southeastern Oregon. <*Bull. Nutt. Ornith. Club*, III, p. 81, April, 1878.

Review of Capt. Charles Bendire's paper in *Proc. Boston Soc. Nat. Hist.*, XIX, 1877, pp. 109–149.

115. Ridgway's Report on the Ornithology of the Fortieth Parallel. <*Bull. Nutt. Ornith. Club*, III, pp. 81–83, April, 1878.

Review of R. Ridgway's 'Ornithology' in Vol. IV, Part iii, Explor. 40th Parallel, pp. 303–670, 1877. (*Cf. supra*, No. 106.)

116. Feilden's 'List of Birds Observed in Smith Sound,' etc. <*Bull. Nutt. Ornith. Club*, III, p. 86, April, 1878.

Review of H. W. Feilden's paper (*Ibis*, 4th ser., I, 1877, pp. 401–412).

117. A Remarkable Fossil Bird. [*Palæospiza bella* Allen] <*Pacific Rural Press* (newspaper), April 27, 1878. Vol. XV, No. 17, p. 257, fig.

"Anonymous, but contributed to the paper by the U. S. Geological Survey. It is an abstract of the orig. account, accompanied by an electro, and actually antedates the regular publication of the article by six days."— *Coues, l. c.*, p. 738. (*Cf. supra*, No. 94.)

118. The Nuttall Ornithological Club. <*Boston Journal*, Mar. 19, 1878.

"A dignified and well-considered defence of the Club from the attacks made in the article entitled 'History Repeating Itself,' and elsewhere. 'Neither the ideas, the similes, nor the phraseology of the article are new, having done service repeatedly in other Boston papers within a few weeks in a similar connection. Therefore the inference is natural, that they have emanated, either directly or indirectly, from a single source. The article in question is simply a presentation, in connected form, of the various anonymous squibs that have appeared repeatedly in other papers. . . . Unfortunate, indeed, is it if this purely practical and scientific question cannot be discussed on its merits, and that the defenders of the [English] sparrows must confess their weakness by a persistent resort to such unsafe weapons as misrepresentation and ridicule.'"— *Coues*, 'On Present Status of *Passer domesticus*,' etc., in *Bull. U. S. Geol. and Geogr. Surv. Terr.*, V, 1879, p. 185).

119. The Nuttall Ornithological Club. <*Evening Transcript*, Mar. 21, 1878.

"A defence of the Club, which has been referred to contemptuously as a body of "Cambridge juveniles," "precocious boys," "over-modest youths," etc., including a statement of its organization and operations, and the real character of its membership. T. M. Brewer is mentioned as a member, and a contributor to the Bulletin of the Club."— *Coues, l. c.*

120. The Nuttall Ornithological Club. Sketch of our Cambridge Ornithological Society. <*Cambridge Chronicle*, March 30, 1878.

121. Nuttall Ornithological Club. Sparrows [*Passer domesticus*].— The Nuttall Ornithological Club decides against them. <*Boston Daily Advertiser*, CXXXI, No. 47, Feb. 23, 1878. (Anonymous.)

"The inside history of this article is: Some time about January, 1878, Dr. Brewer was in my office in Washington, where some words on the subject grew a little heated toward the close of the interview, when I proposed that it would be well to ask the Nuttall Club to take up and sift the matter, that we might get at the facts, if possible. The reply was complimentary neither to the sincerity of my proposition nor to the ornithological ability of the Club, of which he was a member. I soon afterward addressed to the Club a communication inviting their attention to the matter, suggesting a full and fair discussion of the subject in open meeting, and representing that the Club was specially well qualified to come to just conclusions, consisting, as it did, of a number of working ornithologists of recognized ability and experience, who were perfectly familiar with the case as presented in Boston, Cambridge, and vicinity. The meeting was soon after held. Notice was sent to the resident members of the proposed consideration of the question, and corresponding members were also invited to take part in the discussion. Dr. Brewer for some reason did not attend the meeting. The report of the meeting, constituting the present article, indicates that Mr. J. A. Allen, Mr. William Brewster, Mr. H. A. Purdie, Mr. H. D. Minot, Mr. Ruthven Deane, and others, participated in the discussion. Mr. Allen's views are not here presented. All the testimony here reported is unfavorable to the Sparrows, but need not be here analyzed, as it is only what every competent and unprejudiced observer knows; it is explicit, emphatic, and irrefragable, substantiating every count that has been brought against the birds. The article includes an open letter to Dr. Brewer from John Dixwell, M. D., stating that in 39 individuals, taken at the height of the canker-worm pest, no trace of insect food could be found on dissecting; the publication of which letter gave offence to Dr. Brewer, who considered it an unwarrantable liberty to take with his name. At the close of the discussion a vote was taken on the question of whether or not, in the opinion of those present, the further increase of the house sparrow in this country was desirable. The result was a unanimous negative."— *Coues, l. c.*

122. Nuttall Ornithological Club. Sparrows [*Passer domesticus*] brought to Judgment. Discussion of the Nuttall Ornithological Club upon the Merits and Demerits of the English Sparrow in the United States. <*The Country* (newspaper of New York), pp. 245–246, Feb. 23, 1878.

"This is the full report of the meeting, communicated officially by the Club, occupying nearly two pages (5 columns). It gives much matter additional to that published in the Boston Daily Advertiser, and is especially important in presenting at length views of J. A. Allen, and in including communications from Mr. R. Ridgway and Dr. C. C. Abbott, not given in the Boston report of the proceedings. Allen's carefully considered testimony, though well guarded, is, emphatically and explicitly, against the Sparrows. 'Mr. Allen further stated, that every ornithologist of note throughout the country who has expressed himself upon the subject (and nearly all have done so) has, almost without exception, declared against the Sparrows. Not a few of them consider their rapid increase an alarming evil, which will soon call for legislative action to hold it in check.'"— *Coues, l. c.*

123. The Sparrows [*Passer domesticus*]. <*Evening Transcript*, Mar. 19, 1878. (By H. A. Purdie.)

"Covering a copy of The Country of February 23, 1878, which contained the full report of the Nuttall Club's proceedings, Mr. J. A. Allen's testimony having been omitted from the report as published in Boston. Mr. Allen's testimony follows."— *Coues, l. c.*

124. [Range of the Fish Crow in New York and New England]. <*Bull. Nutt. Ornith. Club*, II, p. 47, Jan., 1878.

125. Persistency in Nest-building by a Pair of City Robins [*Turdus migratorius*]. < *Bull. Nutt. Ornith. Club*, III, pp. 103–104, Apr., 1878.

Rebuilt their nest five times when it was removed by a human friend of the birds, owing to the unsafe nature of the site on account of cats.

1879.

126. Notes on the Sea-Birds of the Grand Banks. <*Bull. Nutt. Ornith. Club*, IV, pp. 127–128, April, 1879.

Based mainly on letters from Raymond L. Newcomb.

127. Lawrence and Ober on the Birds of Dominica and St. Vincent. <*Bull. Nutt. Ornith. Club*, IV, pp. 48–49, Jan., 1879.

Notices of five of G. N. Lawrence's papers on F. A. Ober's collections in the Lesser Antilles, in *Ann. New York Acad. Sci.*, and *Proc. U. S. Nat. Mus.*, 1878.

128. Elliot's Synopsis of the Trochilidæ. <*Bull. Nutt. Ornith. Club*, IV, pp. 230–232, Oct., 1879.

D. G. Elliot's 'A Classification and Synopsis of the Trochilidæ' (4to, Smiths. Inst., 1879).

129. Nest and Eggs of the Cerulean Warbler [*Dendrœca cærulea*]. <*Bull. Nutt. Ornith. Club*, IV, pp. 25–27, Jan., 1879.

130. Odd Behavior of a Robin [*Turdus migratorius*] and a Yellow Warbler (*Dendrœca æstiva*]. <*Bull. Nutt. Ornith. Club*, IV, pp. 178–182, July, 1879.

131. Brewer on the Nests and Eggs of *Empidonaces*. <*Bull. Nutt. Ornith. Club*, IV, p. 232, Oct., 1878.

Review of Dr. T. M. Brewer's paper (*Proc. U. S. Nat. Mus.*, Vol. I, 1879).

132. Belding and Ridgway's Birds of Central California. <*Bull. Nutt. Ornith. Club*, IV, July, 1879, pp. 167–171.

Review of L. Belding and R. Ridgway's 'A Partial List of the Birds of Central California' (in *Proc. U. S. Nat. Mus.*, I, April, 1879), with remarks on methods of designating "incipient species," including a history of the evolution and adoption of the trinomial system in the United States.

133. The Evening Grosbeak [*Hesperiphona vespertina*] in New Mexico. <*Bull. Nutt. Ornith. Club*, IV, p. 237, Oct., 1879.

134. Birds of the Colorado Valley. <*The Nation*, March 20, 1879. (Anonymous.)

Review of Dr. Coues's 'Birds of the Colorado Valley,' 8vo, 1878.

135. Birds of the Southwest. <*Science News*, I, pp. 81–84, Jan. 15.

Review of Dr. Coues's 'Birds of the Colorado Valley.'

1880.

136. Roberts on the Convolution of the Trachea in the Sandhill and Whooping Cranes [*Grus canadensis* and *G. americana*]. <*Bull. Nutt. Ornith. Club*, V, pp. 179–180, July, 1880.

Short notice of T. S. Roberts's paper (*Amer. Nat.*, XIV, 1880, pp. 108–114).

137. The Tennessee Warbler [*Helminthophaga peregrina*] destructive to Grapes. < *Bull. Nutt. Ornith. Club*, V, p. 48, Jan., 1880.

138. Eastward Range of the Western Meadow Lark [*Sturnella magna neglecta*]. < *Bull. Nutt. Ornith. Club*, V, pp. 53–54, Jan., 1880.

To central and western Iowa, southern Michigan, and northwestern Illinois.

139. [Note on *Chordeiles popetue minor* in Florida].<*Bull. Nutt. Ornith. Club,* V, p. 54, Jan., 1880.

Note to Mr. Greene Smith's record under this title.

140. The King Eider [*Somateria spectabilis*] at Buffalo, N. Y.<*Bull. Nutt. Ornith. Club,* V, pp. 62–63, Jan., 1880.

141. On recent additions to the Ornithological Fauna of North America.<*Bull. Nutt. Ornith. Club,* V, pp. 85–92, April, 1880.

The additions, formally listed, with references to places of original record, number 38, with an introductory 'Retrospective' of 4 pp.

142. A Crossbilled Horned Lark [*Eremophila alpestris.*]<*Bull. Nutt. Ornith. Club,* V, p. 115, April, 1880.

143. [Note on a Loggerhead Shrike (*Lanius ludovicianus*) taken in Northern New York.]<*Bull. Nutt. Ornith. Club,* V, p. 119, April, 1880.

144. Capture of Escaped Cage-birds having the Appearance of Wild Birds.<*Bull. Nutt. Ornith. Club,* V, pp. 119–121, April, 1880.

Records for various exotic species, with comment.

145. Note on the Little Brown Crane (*Grus fraterculus* Cassin).<*Bull. Nutt. Ornith. Club,* V, pp. 123–124, April, 1880.

Comparison of a specimen taken in eastern Mexico with specimens of *Grus canadensis.*

146. Destruction of Birds by Light-houses.<*Bull. Nutt. Ornith. Club,* V, pp. 131–138, July, 1880.

Statistics of migrating birds killed by lighthouses.

147. Origin of the Instinct of Migration in Birds.<*Bull. Nutt. Ornith. Club,* V, pp. 151–154, July, 1880.

Attributed to change of climate at the close of the Tertiary period. "What was at first a forced migration would soon become habitual, and through the heridity of habit give rise to that wonderful faculty we term the instinct of migration" (p. 153).

148. List of the Birds of the Island of Santa Lucia, West Indies.<*Bull. Nutt. Ornith. Club,* V, pp. 163–169, July, 1880.

Annotated list of 56 spp.; first record for the island of 16 species.

149. First Capture of the Blue Grosbeak [*Goniaphea cœrulea*] in Massachusetts.< *Bull. Nutt. Ornith. Club,* V, p. 184, July, 1880.

150. Capture of a South American Finch [*Gubernatrix cristatella*] near Providence, R. I.<*Bull. Nutt. Ornith. Club,* V, p. 240, Oct., 1880.

In perfect plumage, showing no sign of previous captivity.

151. Coues's Bibliography of American Ornithology.<*Bull. Nutt. Ornith. Club,* V, pp. 40–41, Jan., 1880.

Review of the Second Instalment (Faunal Publications relating to Central and South America), in *Bull. U. S. Geol. and Geogr. Surv. Terr.,* V, pp. 239–330, Sept., 1879. Also mention of the same author's paper 'On the Present Status of *Passer domesticus* in America,' etc., *op. cit.,* pp. 175–193.

152. Ridgway on the Species of the Genera *Scops* and *Tyrannus*, etc. <*Bull. Nutt. Ornith. Club*, V, pp. 41–42, Jan., 1880.

Review of four papers by R. Ridgway in *Proc. U. S. Nat. Mus.*, 1878.

153. McChesney's Notes on the Birds of Fort Sisseton, Dakota Territory. <*Bull. Nutt. Ornith. Club*, V, pp. 42–43, Jan., 1880.

Review of Dr. Chas. E. McChesney's paper of this title in *Bull. U. S. Geol. and Geogr. Surv. Terr.*, V, 1879, pp. 71–104.

154. Minor Ornithological Publications. <*Bull. Nutt. Ornith. Club*, V, Jan., 1880, pp. 43–46; April, 1880, pp. 112–115; July, 1880, pp. 180, 181.

Brief notices of 54 papers in *Forest and Stream*, *Field and Forest*, and *Forest and Stream* (Nos. 1–54). Continued in subsequent volumes of the *Nuttall Bulletin* and *The Auk* till 1887. The record begins with the close of that given by Dr. Coues in his 'List of Faunal Publications relating to North American Ornithology,' published in his 'Birds of the Colorado Valley,' or about July, 1878. (*Cf. infra*, Nos. 178, 181, 192, 200, 205, 214, 218.)

155. Henshaw's Report on Collections made in California, Nevada, and Oregon in 1877–78 <*Bull. Nutt. Ornith. Club*, V, pp. 105–107, April, 1880.

Review of H. W. Henshaw's report in Rep. U. S. Geogr. Surv. West of 100th Meridian, App. L, pp. 282–335, Feb. 1880.

156. Cory's Birds of the Bahamas Islands. <*Bull. Nutt. Ornith. Club*, V, p. 107, April, 1880.

Review of Chas. B. Cory's book of this title (4to, Boston, 1880).

157. Brewer's Additional Notes on New England Birds. <*Bull. Nutt. Ornith. Club*, V, pp. 108–109, April, 1880.

Review of Dr. T. M. Brewer's supplemental list, in *Proc. Boston Soc. Nat. Hist.*, XVII, pp. 436–454.

158. Kumlien's Contributions to the Natural History of Arctic America. <*Bull. Nutt. Ornith. Club*, V, pp. 109–110, April, 1880.

Review of the bird portion of Ludwig Kumlien's 'Contributions to the Natural History of Arctic America,' etc. (*Bull. U. S. Nat. Mus.*, No. 15, 1879; Birds, pp. 69–105).

159. Gibbs's List of the Birds of Michigan. <*Bull. Nutt. Ornith. Club*, V, p. 110, April, 1880.

Review of Dr. Morris Gibbs's 'Annotated List of the Birds of Michigan' (*Bull. U. S. Geol. and Geogr. Surv. Terr.*, V, No. 3, Nov. 1879).

160. Harvie-Brown on the Capercaillie in Scotland. <*Bull. Nutt. Ornith. Club*, V, pp. 110–111, April, 1880.

Review of J. A. Harvie-Brown's 'The Capercaillie in Scotland' (8vo, Edinburgh, 1879).

161. Sennett's 'Further Notes on the Birds of the Lower Rio Grande of Texas.' <*Bull. Nutt. Ornith. Club*, V, p. 111, April, 1880.

Review of George B. Sennett's second paper on Texas Birds (*Bull. U. S. Geol. and Geogr. Surv. Terr.*, V, 1879, pp. 371–440).

162. Minot's Diary of a Bird. <*Bull. Nutt. Ornith. Club*, V, p. 112, April, 1880.

Review of H. D. Minot's 'The Diary of a Bird' (8vo, Boston, 1880).

163. Freke on Birds common to Europe and North America. <*Bull. Nutt. Ornith. Club*, V, pp. 173–174, July, 1880.

Review of A Comparative Catalogue of Birds found in Europe and North America by Percy Evans Freke (8vo, Dublin, 1880).

164. Brayton's Catalogue of the Birds of Indiana. <*Bull. Nutt. Ornith. Club*, V, pp. 174–175, July, 1880.

Review of Dr. A. W. Brayton's paper on Indiana birds, in *Trans. Indiana Hort. Soc.* for 1879, pp. 89–166 (1880).

165. Mearns's Birds of the Hudson Highlands. <*Bull. Nutt. Ornith. Club*, V, p. 175, July, 1880.

Review of Dr. E. A. Mearns's 'List of the Birds of the Hudson Highlands,' in *Bull. Essex Inst.*, Vols. X, XI, 1879–80.

166. Harvie-Brown and Cordeaux's 'Report on the Migration of Birds in the Autumn of 1879.' <*Bull. Nutt. Ornith. Club*, V, pp. 175–176, July, 1880.

Review of their paper in *The Zoologist*, May, 1880, pp. 161–204.

167. Ridgway on the Nomenclature of North American Birds. <*Bull. Nutt. Ornith. Club*, V, pp. 177–178, July, 1880.

Review of his 'Revisions of Nomenclature of Certain North American Birds,' in *Bull. U. S. Nat. Mus.*, 1880, pp. 1–16.

168. Cooper on the Migration and Nesting Habits of West-Coast Birds. <*Bull. Nutt. Ornith. Club*, V, p. 232, Oct., 1880.

Review of Dr. J. G. Cooper's paper of this title in *Proc. U. S. Nat. Mus.*, 1879, pp. 241–251.

169. Langdon's Ornithological Field Notes. <*Bull. Nutt. Ornith. Club*, V, pp. 232–233, Oct., 1880.

Review of Frank W. Langdon's paper in *Journ. Cincinnati Soc. Nat. Hist.*, July, 1880, pp. 121–127.

170. Stearns's List of the Birds of Fishkill, New York. <*Bull. Nutt. Ornith. Club*, V, p. 233, Oct., 1880.

Review of a privately printed brochure (8vo, pp. 16, without date).

171. Harvie-Brown on the Effects of an unusually severe Winter upon Scottish Birds. <*Bull. Nutt. Ornith. Club*, V, pp. 233–234, Oct., 1880.

Review of a paper by J. A. Harvie-Brown, in *Proc. Nat. Hist. Soc. Glasgow*, 1879, pp. 123–190.

172. Frank R. Rathbun's 'Bright Feathers.' <*Bull. Nutt. Ornith. Club*, V, p. 234, Oct., 1880.

Review of Part I of this work (4to, Auburn, N. Y.).

173. C. B. Cory's 'Beautiful and Curious Birds of the World.' <*Bull. Nutt. Ornith. Club*, p. 236, Oct., 1880.

Review of Part I, of this work (elephant-folio, Boston, 1880).

1881.

174. New England Bird Life. <*The Nation*, No. 845, Sept. 8, 1881. (Anonymous.)

Review of Part I of Winifrid A. Stearns and Elliott Coues's 'New England Bird-Life' (8vo, Boston, 1881).

175. Insectivorous Birds in their Relation to Man.<*Bull. Nutt. Ornith. Club*, VI, pp. 22–27, Jan., 1881.

Summary of recent investigations in economic ornithology.

176. Coues's Third Instalment of American Ornithological Bibliography.<*Bull. Nutt. Ornith. Club*, VI, pp. 44–46, Jan., 1881.

Review of the Systematic Section, in *Bull. U. S. Geol. and Geogr. Surv. Terr.*, V, 1880, pp. 521–1066.

177. Coues's Fourth Instalment of Ornithological Bibliography.<*Bull. Nutt. Ornith. Club*, VI, p. 46, Jan., 1881.

Review of the section relating to British Birds, in *Proc. U. S. Nat. Mus.*, II, 1880, pp. 359–476.

178. Minor Ornithological Papers.<*Bull. Nutt. Ornith. Club*, VI, pp. 47–53, Jan., 1881.

Includes 86 titles (Nos. 55–140). *Cf. supra*, No. 154.

179. Some Observations on the Migration of Birds.<*Bull. Nutt. Ornith. Club*, VI, pp. 100, 188, April, 1881.

A half-page footnote to W. E. D. Scott's paper of this title, and a later correction of an erroneous statement in it. Mr. Scott's paper relates to observations made at night through a telescope at Princeton, N. J.

180. Shufeldt's Osteological Memoirs.<*Bull. Nutt. Ornith. Club*, VI, pp. 109–110, April, 1881.

Reviews of Dr. R. W. Shufeldt's papers on the Burrowing Owl and Horned Lark, in *Bull. U. S. Geol. and Geogr. Surv. Terr.*, VI, 1881.

181. Minor Ornithological Papers.<*Bull. Nutt. Ornith. Club*, VI, pp. 112–113, April, 1881.

Titles Nos. 141–160. (*Cf. supra*, Nos. 154 and 178.)

182. Supplementary List of Birds of the Island of Santa Lucia, W. I.<*Bull. Nutt. Ornith. Club*, VI, p. 128, April, 1881.

Adds 12 species to the previously published List (*cf. supra*, No. 148).

183. Winter Birds of Fort Walla Walla, W. T.<*Bull. Nutt. Ornith. Club*, VI, p. 128, April, 1881.

Based on letters from Capt. Charles Bendire.

184. Ridgway's Nomenclature of North American Birds.<*Bull. Nutt. Ornith. Club*, VI, pp. 164–171, July, 1881.

Review, with historical summary of the subject, of R Ridgway's paper under this title (*Bull. U. S. Nat. Mus.*, No. 21, 1881).

185. Ridgway's Revised Catalogue of the Birds of Illinois.<*Bull. Nutt. Ornith. Club*, VI, pp. 171–172, July, 1881.

Review of R. Ridgway's paper in Bull No. 4, Illinois State Lab. Nat. Hist., 1881.

186. Godman and Salvin's 'Biologia Centrali-Americana.'<*Bull. Nutt. Ornith. Club*, VI, pp. 174–176, July, 1881.

Review of the 'Aves' (Zoölogy, Parts I–X, p. 1–152, pll. i–x).

187. A Second Massachusetts Specimen of the Red-bellied Woodpecker (*Centurus carolinus*). <*Bull. Nutt. Ornith. Club,* VI, p. 183, July, 1881.

188. *Larus glaucus* in Texas. <*Bull. Nutt. Ornith. Club,* VI, No. 3, p. 187, July, 1881.

Record of a specimen taken in Clay County, Texas, Dec. 18, 1880.

189. The Migration of Birds. <*Scribner's Monthly Magazine,* XXII, pp. 932–938, Oct., 1881.

A general review of the subject from the viewpoint of the author, elaborating points stated in No. 147, *supra.*

1882.

190. Capture of *Plectrophenax nivalis* in Chester, South Carolina. <*Bull. Nutt. Ornith. Club,* VII, p. 54, Jan., 1882.

Given on authority of Leverett M. Loomis.

191. The Sharp-tailed Finch [*Ammodramus caudacutus nelsoni*] in Kansas. <*Bull. Nutt. Ornith. Club,* VII, p. 55, Jan., 1882.

192. Minor Ornithological Papers. <*Bull. Nutt. Ornith. Club,* VII, pp. 115–118, April, 1882.

Titles numbered 161–189. (*Cf. supra,* Nos. 154, 178, and 181.)

1883.

193. [Note on a Hybrid Sparrow (*Zonotrichia albicollis* + *Junco hyemalis*).] <*Bull. Nutt. Ornith. Club,* VIII, pp. 79–80, April, 1883. Description (*l. c.*) by Mr. Charles H. Townsend.

194. Note on Exceptions to the Law of Increase in Size Northward among North American Birds. <*Bull. Nutt. Ornith. Club,* VIII, pp. 80–82, April, 1883.

In certain genera of Oscines which are for the most part tropical in distribution.

195. The British Museum Catalogue of Birds. <*Bull. Nutt. Ornith. Club,* VIII, pp. 99–105, April, 1883.

Review of Vols. IV, V, VI.

196. Ridgway on the Tree-creepers. <*Bull. Nutt. Ornith. Club,* VIII, p. 113, April, 1883.

Review of R. Ridgway's paper on the group (in *Proc. U. S. Nat. Mus.,* 1882, pp. 111–116).

197. Ridgway's Review of the Genus *Centurus.* <*Bull. Nutt. Ornith. Club,* VIII, p. 114, April, 1883.

Notice of the paper (in *Proc. U. S. Nat. Mus.,* 1881, pp. 93–119).

198. Freke on North American Birds crossing the Atlantic. <*Bull. Nutt. Ornith. Club,* VIII, pp. 114–115, April, 1883.

Review of Percy Evans Freke's paper of this title, published in *Sci. Proc. Dublin Soc.,* III, 1881.

199. Freke on European Birds Observed in America. <*Bull. Nutt. Ornith. Club,* VIII, p. 115, April, 1883.

Review of Percy Evans Freke's paper of this title in *The Zoologist,* Sept., 1881.

200. Minor Ornithological Publications. <*Bull. Nutt. Ornith. Club*, VIII, pp. 116–119, April, 1883.

Continued, embracing titles numbered 190–235. (*Cf. supra*, Nos. 154, 178, 181, and 192.)

201. Capture of the Great Gray Owl in Massachusetts. <*Bull. Nutt. Ornith. Club*, VIII, p. 123, April, 1883.

Near Springfield, Feb., 1882.

202. List of Birds Observed in the Vicinity of Colorado Springs, Colorado, during March, April and May, 1882. <*Bull. Nutt. Ornith. Club*, VIII, pp. 151–161, 189–198, 1883.

With William Brewster. An Annotated list of 134 species.

203. Ridgway on New Species and Subspecies of Birds. <*Bull. Nutt. Ornith. Club*, VIII, pp. 168–169, July, 1883.

Brief notices of eight of R. Ridgway's recent papers.

204. Dubois on Geographical Variation in the Crossbills. <*Bull. Nutt. Ornith. Club*, VIII, p. 170, July, 1883.

Review of "De la Variabilité des Oiseaux du genre *Loxia*" (*Bull. du Musée royal d'Hist. Nat. de Belgique*, I, 1882).

205. Minor Ornithological Publications. <*Bull. Nutt. Ornith. Club*, VIII, pp. 171–178, July, 1883.

Continued, including titles numbered 236–364. (*Cf. supra*, Nos. 154, 178, 181, 192, and 200.)

206. The Wood Ibis in Massachusetts. <*Bull. Nutt. Ornith. Club*, VIII, p. 187, July, 1883.

Taken at Georgetown, Mass., June 19, 1880.

207. The American Ornithologists' Union. <*Bull. Nutt. Ornith. Club*, VIII, pp. 221–226, Oct., 1883.

Its origin and founding, etc.

208. Goss's Birds of Kansas. <*Bull. Nutt. Ornith. Club*, VIII, p. 227, Oct., 1883.

Review of 'A Catalogue of the Birds of Kansas' by N. S. Goss (8vo, Topeka, 1883).

209. Beckham's Birds of Nelson County, Kentucky. <*Bull. Nutt. Ornith. Club*, VIII, pp. 227–228, Oct., 1883.

Review of C. W. Beckham's paper in *Journ. Cincinnati Soc. Nat. Hist.*, VI, pp. 136–147, July, 1883.

210. Migration of Birds. <*Bull. Nutt. Ornith. Club*, VIII, pp. 228–231, Oct., 1883.

Notices of seven Migration Reports, of different countries, for the years 1880 and 1881.

211. Bean's Notes on Birds Collected in Alaska and Siberia. <*Bull. Nutt. Ornith. Club*, p. 231, Oct., 1883.

Review of Tarleton H. Bean's paper (*Proc. U. S. Nat. Mus.*, V, 1882, pp. 144–173)

212. Turner on *Lagopus mutus* and its Varieties. <*Bull. Nutt. Ornith. Club*, VIII, p. 232, Oct., 1883.

Review of Lucien M. Turner's paper (*Proc. U. S. Nat. Mus.*, V, 1882, pp. 225–233).

213. Reichenow's 'Die Vögel der Zoologischen Gärten.' <*Bull. Nutt. Ornith. Club*, VIII, p. 232, Oct., 1883.

Review of Part 1, of this work (8vo, Leipzig, 1882).

214. Minor Ornithological Publications. <*Bull. Nutt. Ornith. Club*, VIII, pp. 223–238, Oct., 1883.

Continued, including titles numbered 365–454. (*Cf. supra*, Nos. 154, 178, 181, 200, and 205.)

215. Intelligence of the Crow. <*Science*, N. S., I, No. 18, p. 513, June 8, 1883.

216. New England Bird Life. <*The Nation*, No. 939, p. 537, June 21, 1883. [Anon.]

Review of Part II of Stearns and Coues's 'New England Bird Life.'

1884.

217. Stejneger and Ridgway on Birds of the Commander Islands. <*Auk*, I, pp. 81–82, Jan., 1884.

Review of 'Contributions to the Natural History of the Commander Islands, No. 1' (*Proc. U. S. Nat. Mus.*, 1883, pp. 58–89).

218. 1884–87. Minor Ornithological Publications. <*Auk*, I, 1884, pp. 85–89, 185–190, 285–289; II, 1885, pp. 96–101, 209–214, 295–302; 372–375; III, 1886, pp. 270–273, 475–478; IV, 1887, pp. 66–69, 337–342.

An annotated list of 1199 titles of minor papers relating to ornithology, begun in 1880 in *Bulletin Nuttall Ornithological Club*, and closed with the last date here given. (Continued in later volumes of *The Auk* till 1894, by Charles F. Batchelder).

219. Coues's 'Key to North American Birds.' <*Science*, N. S., IV, pp. 86–87, July 25, 1884.

Review of the second edition.

220. Stejneger on the American Turdidæ. <*Auk*, I, pp. 181–182, April, 1884.

Review of Dr. L. Stejneger's 'Remarks on the Systematic Arrangement of the American Turdidæ' (*Proc. U. S. Nat. Mus.*, V, 1882, pp. 449–483).

221. Jeffries on the Epidermal System of Birds. <*Auk*, I, pp. 182–183, April, 1884.

Review of J. Amory Jeffries's 'The Epidermal System of Birds' (*Proc. Boston Soc. Nat. Hist.*, XXII, pp. 203–240, pll. iv–vi, Dec. 1883).

222. A Plea for the Metric System in Ornithology. <*Auk*, I, April, 1884, pp. 203–205.

Communication by Dr. C. Hart Merriam, with editorial comment in approval of the Plea.'

223. The British Museum Catalogue of Birds. <*Auk*, I, pp. 278–281, July, 1884.

Review of Vol. VII, by R. Bowdler Sharpe, and Vol. VIII, by Hans Gadow.

224. Coues's 'Key to North American Birds.' <*Auk*, I, pp. 282–283, July, 1884.

Review of second edition.

224. Key to North American Birds. <*The Nation*, XXXIX, No. 996, p. 99, July 31, 1888. (Anon.)

Review of second edition of Coues's work of this title.

226. Second Meeting of the American Ornithologists' Union. <*Auk*, I, pp 369–379, Oct., 1884.

227. Brewster on Birds in the Gulf of St. Lawrence. <*Auk*, I, pp. 379–380, Oct., 1884.

Review of William Brewster's 'Notes on the Birds of a Summer Cruise in the Gulf of St. Lawrence' (*Proc. Boston Soc. Nat. Hist.*, XXII, pp. 364–412).

228. Baird, Brewer, and Ridgway's 'Water Birds of North America.' <*Auk*, I, pp. 382–386, Oct., 1884.

Review of the work (2 vols., 4to, Boston, 1884).

1885.

229. A Naturalist's Rambles about Home. <*Auk*, II, pp. 86–88, June, 1885.

Review of Chas. C. Abbott's book of this title (8vo, New York, 1884).

230. Seebohm's History of British Birds. <*Auk*, II, pp. 88–91, Jan., 1885.

Review of the work entitled 'A History of British Birds, with colored illustrations of their Eggs' (2 vols., roy. 8vo, London, 1883).

231. Where Robin goes in Winter. <*Every Other Sunday*, p. 6, May, 1885.

232. Birds in the Bush. <*The Nation*, XLI, No. 1045, p. 39, July 9, 1885. (Anon.)

Review of Bradford Torrey's 'Birds in the Bush.'

233. Ridgway on New Species and Subspecies of American Birds, and on the Nomenclature of other Species. <*Auk*, II, pp. 290–293, July, 1885.

Brief notices of 19 papers by R. Ridgway in *Proc. U. S. Nat. Mus.*, Vols. VI, VII, VIII, 1883–1885.

234. Capture of Escaped Cage-birds. <*Auk*, II, p. 314, July, 1885.

An Australian Parakeet in California and a Java Sparrow in Massachusetts, with comment on the capture of wild exotic and even tropical cage-birds far from their natural environment.

235. [Annotations to] 'Early Spring Notes from the Mountains of Southern Arizona,' by W. E. D. Scott. <*Auk*, II, pp. 348–356, Oct., 1885.

Technical notes on *Turdus unalascæ auduboni* (p. 349), *Certhia familiaris mexicana* (p. 350), *Catherpes mexicanus conspersus* (p. 350), *Troglodytes aëdon marianæ* (p. 351), *Junco cinereus palliatus* (p. 354).

236. Sharpe's Catalogue of the Birds in the British Museum. <*Auk*, II, pp. 365–368, Oct., 1885.

Review of Vol. X (Fringilliformes, Part I), with critical comment on questions of nomenclature.

237. Turner's List of the Birds of Labrador. <*Auk*, II, pp. 368–369, Oct., 1885. See also *Auk*, III, p. 141.

Review of L. M. Turner's paper (*Proc. U. S. Nat. Mus.*, VIII, 1885, pp. 233–254).

238. Ridgway on New American Birds. <*Auk*, II, p. 370, Oct., 1885.

Review of five papers by R. Ridgway.

239. Ridgway's List of Emended Names of North American Birds. <*Auk*, II, pp. 371–372, Oct., 1885.

Comment on R. Ridgway's 'Some Emended Names of North American Birds' (*Proc. U. S. Nat. Mus.*, VIII, pp. 354, 355, Sept. 2, 1885).

240. A Hawk Owl (*Surnia ulula caparoch*) at Chatham, Mass. <*Auk*, II, p. 383, Oct., 1885.

241. A Crested Auk on the Massachusetts Coast. <*Auk*, II, p. 388, Oct., 1885.

Apparently *Simorhynchus cristátellus.*

1886.

242. Third Meeting of the American Ornithologists' Union. <*Auk*, III, pp. 115–122, Jan., 1886.

243. Murdoch on the Birds of Point Barrow, Alaska. <*Auk*, III, pp. 122–124, Jan., 1886.

Review of John Murdoch's Point Barrow Natural History Report.

244. Bird Protection by the A. O. U. <*Forest and Stream*, XXV, No. 26, Jan. 21, 1886. (Anon.)

Formation of a Committee on Bird Protection, etc.

245. Stejneger's Ornithological Explorations in Kamtschatka. <*Auk*, III, pp. 126–129, Jan., 1886.

Review of Stejneger's 'Results of Ornithological Explorations in the Commander Islands and Kamtschatka.' 8vo, 1895 (= *Bull. No. 29, U. S. Nat. Mus.*).

246. [John Burroughs on Ornithologists as Bird Enemies.] <*Auk*, III, p. 142, Jan., 1886.

Comment on his article in *The Century* in which ornithologists are classified as "among the worst enemies" of birds.

247. On the Avi-fauna of Pinal County, with Remarks on Some Birds of Pima and Gila Counties, Arizona. By W. E. D. Scott. With Annotations by J. A. Allen. <*Auk*, III, 1886, pp. 249–258, 383–389, 421–432; IV, 1887, pp. 16–24, 196–205; V, 1888, pp. 29–36, 159–168.

The annotations relate to the following species: *Rallus virginianus*, Vol. III, p. 385; *Porzana carolina*, p. 386; *Colinus ridgwayi*, p. 387; *Callipepla squamata*, p. 388; *Geococcyx californianus*, p. 425; *Dryobates stricklandi*, p. 426; *Melanerpes torquatus*, p. 428; *Iache latirostris*, p. 432; *Myiarchus cinerascens*, Vol. IV, p. 18; *Aphelocoma sieberi arizonæ*, p. 21; *Icterus parisorum*, p. 23; *I. cucullatus nelsoni*, p. 23; *Carpodacus purpureus californicus*, p. 196; *Loxia curvirostris stricklandi*, p. 197; *Spinus tristis*, p. 198; *S. pinus*, p. 199; *Spizella socialis arizonæ*, p. 200; *Junco hyemalis, oregonus, annectens, caniceps, dorsalis, palliatus*, pp. 201–202; *Amphispiza bilineata*, p. 202; *Peucæa ruficeps boucardi*, p. 203; *Pipilo fuscus mesoleucus*, p. 204; *Vireo gilvus swainsoni*, p. 32; *Vireo solitarius plumbeus* and *V. s. cassinii*, p. 32; *Dendroica æstiva morcomi*, p. 34; *Mimus polyglottus*, p. 160; *Catherpes mexicanus conspersus*, p. 162; *Troglodytes aëdon* group, pp. 163–165; *Turdus aonalashkæ auduboni*, p. 167.

248. The Present Wholesale Destruction of Bird-Life in the United States. <*Science*, VII, No. 160, pp. 191–195, Feb. 26, 1886. Also, *Bulletin No. 1*, of the A. O. U. Committee on Protection of Birds, pp. 1–5, March, 1886.

249. Destruction of Birds for Millinery Purposes. <*Science*, VII, No. 160, pp. 196–197, Feb. 26, 1886. Also, *Bull. No. 1*, A. O. U. Committee on Protection of Birds, pp. 6–7, March, 1886. (Anon.)

250. The Relation of Birds to Agriculture. <*Science*, VII, No. 160, pp. 201–202, Feb. 26, 1886. Also, *Bull. No. 1*, A. O. U. Committee on Protection of Birds, pp. 11–12, March, 1886. (Anon.)

251. Bird-Laws. <*Science*, Vol. VII, No. 160, pp. 202–204, Feb. 26, 1886. Also, *Bull. No. 1*, A. O. U. Committee on Protection of Birds, pp. 12–14. (Anon.)

252. An Appeal to the Women of the Country in Behalf of the Birds. <*Science*, VII, No. 160, pp. 204–205, Feb. 26, 1886. Also, *Bull. No. 1*, A. O. U. Committee on Protection of Birds, pp. 14–15. March, 1886.

253. Protection of Birds by Legislation. <*Forest and Stream*, XXVII, No. 16, pp. 304–305, Nov. 11, 1886. Also, *Bull. No. 2*, of the A. O. U. Committee on Protection of Birds, pp. 1–7, Nov., 1886.

Signed by the Committee, but written by J. A. A., and published with articles by other contributors.

254. Capture of a Pair of Wild Hybrid Ducks (Mallard + Muscovy) on Long Island. <*Auk*, III, pp. 274–275, April, 1886.

255. The Masked Bob-White (*Colinus ridgwayi*) in Arizona. <*Auk*, III, pp. 275–276, April, 1886.

256. [Note on an Abnormally Colored Bluebird (*Sialia sialis*).] <*Auk*, III, p. 282, footnote, April, 1886.

257. The A. O. U. Check-List of North American Birds. <*Auk*, III, pp. 397–398, July, 1886.

Review of the first edition. Descriptive account, following a critical review of the 'Code and Check-List' by David Starr Jordan.

258. Note on Two Specimens of the Snowy Plover from the Indian Territory. <*Auk*, III, p. 409, July, 1886.

259. 'Aptoso-Chromatism.' <*Auk*, III, pp. 413–414, July, 1886.

Criticism of Mr. Hoxie's theory respecting "a 'moultless color change' in the feathers of birds," designated 'Aptoso-Chromatism.'

260. 'The Standard Natural History' — 'Birds.' <*Auk*, III, pp. 473–474, Oct., 1886.

Review of Vol. IV of this work (imp. 8vo, Boston, 1885).

261. *Ægialitis meloda circumcincta* on the Atlantic Coast. <*Auk*, III, pp. 482–483, Oct., 1886.

In New Jersey and Maine.

262. A Revised List of the Birds of Massachusetts. <*Bull. Amer. Mus. Nat. Hist.*, I, No. 7, pp. 221–271, July, 1886.

An annotated list of 340 species, plus 4 species extirpated, 4 introduced, and 19 of probable occurrence. The third and last list of the birds of Massachusetts published by this author.

263. The Masked Bob-White (*Colinus ridgwayi*) of Arizona, and its Allies. <*Bull. Amer. Mus. Nat. Hist.*, I, No. 7, pp. 273–290, with col. pl., July, 1886.

264. The Type Specimen of *Colinus ridgwayi*. <*Auk*, III, p. 483, Oct., 1886.

Stated to be in the collection of G. Frean Morcom, of Chicago.

265. Three Interesting Birds in the American Museum of Natural History: *Ammodramus leconteii, Helinaia swainsoni*, and *Saxicola œnanthe*. <*Auk*, III, pp. 489–490, Oct., 1886.

The Leconte's Sparrow was taken by Prince Maximilian in 1832–34, some ten years before the discovery of the species by Audubon.

266. Bird Destruction. <*Science*, VIII, No. 183, pp. 118–119, Aug. 6, 1886.

1887.

267. Stejneger on Japanese Woodpeckers. <*Auk*, IV, pp. 63–64, Jan., 1887.

Notice of Dr. L. Stejneger's 'Review of Japanese Birds, I. The Woodpeckers' (*Proc. U. S. Nat. Mus.*, 1886, pp. 200–201).

268. Ferrari-Perez on the Birds of Mexico. <*Auk*, IV, p. 65, Jan., 1887.

Notice of Fernando Ferrari-Perez's 'Catalogue of Animals collected by the Geogr. and Expl. Comm. of the Republic of Mexico,' in *Proc. U. S. Nat. Mus.*, 1886, pp. 125–199 (birds, pp. 130–182).

269. Fourth Meeting of the American Ornithologists' Union. <*Auk*, IV, pp. 56–61, Jan., 1887.

270. A Further Note on *Colinus ridgwayi*. <*Auk*, IV, pp. 74–75, Jan., 1887.

Additional specimens noted, including two early examples in Mr. Henshaw's collection.

271. [Note on Florida Specimens of *Dendroica discolor*.] <*Auk*, IV, p. 135, footnote, April, 1887.

272. Sclater's Catalogue of the Cœrebidæ, Tanagridæ, and Icteridæ. <*Auk*, IV, pp. 149–150, April, 1887.

Review of Vol. XI of the British Museum Catalogue of Birds.

273. Ridgway's Nomenclature of Colors and Ornithologists' Compendium. <*Auk*, IV, pp. 151–152, April, 1887.

Review of R. Ridgway's work entitled 'A Nomenclature of Colors for Naturalists, and Compendium of Useful Knowledge for Ornithologists' (8vo, Boston, 1886).

274. [The Redpolls of Massachusetts.] <*Auk*, IV, p. 164, April, 1887.

A note of explanation, in reply to a communication by Mr. W. Brewster.

275. [The Antedating of Papers and Works on Natural History.] <*Auk*, IV, p. 176, April, 1887. See also *op. cit.*, p. 270.

276. [Additions in the Department of Ornithology at the American Museum of Natural History.] <*Auk*, IV, pp. 270–272, July, 1887.

They include the D. G. Elliot collection of Hummingbirds, the George N. Lawrence collection (with the types of about 300 species of American birds), and the H. H. Smith collection of birds from Chapada, Matto Grosso, Brazil; also an account of the beginning of the Museum's series of Bird Groups, which later came to form so striking a feature of the Museum's exhibition series of birds. The acquisition of the Elliot ornithological library is also noted.

277. The Pine Finch (*Spinus pinus*) Breeding at Cornwall-on-Hudson, N. Y. <*Auk*, IV, pp. 284–286, July, 1887.

The parent birds, with the nest and eggs, were used for a 'habitat group' in the American Museum of Natural History, where it is still (1916) on exhibition.

278. Ridgway's Manual of North American Birds. <*Auk*, IV, pp. 333–336, Oct., 1887.

Review of R. Ridgway's work of this title (roy. 8vo, Philadelphia, 1887).

279. Key to North American Birds. <*The New York Evening Post*, Dec. 21, 1887.

Review of Dr. Coues's 'Key,' not signed.

280. What is a Bird? <*Audubon Magazine*, I, pp. 31–33, March, 1887.

1888.

281. Fifth Meeting of the American Ornithologists' Union. <*Auk*, V, pp. 95–100, Jan., 1888.

282. Coues's 'Key to North American Birds.' <*Auk*, V, pp. 101–104, Jan., 1888.

Review of the third edition of E. Coues's work of this title (roy. 8vo, Boston, 1887).

283. [A Monument to Audubon.] <*Auk*, V, pp. 123–124, Jan., 1888.

Account of the movement to erect a monument in New York City to John James Audubon.

284. [The Bird Collection at the British Museum.] <*Auk*, V, pp. 124–125, Jan., 1888.

Summary of an article by R. B. Sharpe on 'Ornithology at South Kensington.'

285. [A 'Discovery' in the Mechanism of Birds' Wings.] <*Auk*, V, pp. 125–126, Jan., 1888. See also *op. cit.* p. 223.

Account of a discussion in the New York Academy of Sciences on an alleged "important discovery," etc. *Cf. infra*, No. 303.

286. Notes on the Summer Birds of Holderness, Bethlehem, and Franconia, N. H. (With Walter Faxon). <*Auk*, V, pp. 149–155, April, 1888.

Contains 'Notes on Birds Observed at Franconia and Bethlehem, in July and August, 1874,' by J. A. A., pp. 153–155, following two briefly annotated lists of birds of the same general region by Walter Faxon.

287. [On the Wrens of the *Troglodytes ædon* Group.] <*Auk*, V, pp. 163–165, April, 1888. (*Cf. supra*, No. 247.)

288. Seebohm's 'Geographical Distribution of the Charadriidæ.' <*Auk*, V, pp. 189–194, April, 1888.

Critical review of Henry Seebohm's work of this title (4to, London, 1888).

289. Ridgway on New or Little-known American Birds, etc. <*Auk*, V, pp. 194–195, April, 1888.

Brief notices of a number of R. Ridgway's papers.

290. Stejneger on Japanese Birds, etc. <*Auk*, V, pp. 195–197, April, 1888.

Reviews of several of L. Stejneger's recent papers on Japanese birds.

291. Chamberlain's 'Systematic Table of Canadian Birds.' <*Auk*, V, pp. 198–199, April, 1888.

Review of Montague Chamberlain's work of this title (folio, St. John, N. B., 1888, pp. 14).

292. Sclater and Hudson's 'Argentine Ornithology.' <*Auk*, V, pp. 199–200, April, 1888.

Review of Vol. I of this well-known work (8vo, London, 1888).

293. Descriptions of two New Subspecies of the Seaside Sparrow (*Ammodramus maritimus*). <*Auk*, V, pp. 284–287, July, 1888.

Ammodramus maritimus peninsulæ, p. 284; *A. m. sennetti*, p. 286.

294. Description of a New Species of the Genus *Tityra*, from Ecuador. <*Auk*, V, pp. 287–288, July, 1888.

Tityra nigriceps, p. 287.

295. Notes on Louisiana Birds. <*Auk*, V, pp. 324–325, July, 1888.

Brief notes on several species.

296. [Bird Destruction for Millinery Purposes.] <*Auk*, V, pp. 334–335, July, 1888.

Some statistics of the London sales of bird skins for this use.

297. Turner's Report of his Ornithological Observations in Alaska. <*Auk*, V, pp. 409–410, Oct., 1888.

Review of L. M. Turner's observations in Alaska in a 4to report (No. 11) of the U. S. Signal Service (Birds, pp. 115–196, with 10 col. plls.).

298. Sharpe's Catalogue of the Fringillidæ. <*Auk*, V, pp. 410–413, Oct., 1888.

Review of Vol. XII of the British Museum Catalogue of Birds.

299. Shufeldt on the Osteology of the Icteridæ and Corvidæ. <*Auk*, V, pp. 413–414, Oct., 1888.

Review of a paper by R. W. Shufeldt (*Journ. Anat. and Phys.*, XXII, pp. 309–350).

300. Note on the Correct Name of *Symphemia semipalmata inornata* Brewster. < <*Auk*, V, pp. 423–424, Oct., 1888.

Claimed to be *Symphemia semipalmata speculifera* (Cuv.).

301. Visitor's Guide to the Collection of Birds in the American Museum of Natural History, Seventy-seventh Street and Eighth Avenue, New York City.— J. A. Allen, Curator. New York: Printed for the Museum, 1888.— 8vo, pp. 62, + 3ll. unpaged, and 18 cuts in text.

302. List of Important Publications relating to Birds. <*Riverside Natural History*, Vol. IV, pp. 549–556, 1888.

188 titles.

1889.

303. On the Structure of Birds in Relation to Flight, with Special Reference to Recent Alleged Discoveries in the Mechanism of the Wing. <*Trans. New York Acad. Sci.*, pp. 89–100, 1888.

304. Sixth Congress of the American Ornithologists' Union. <*Auk*, VI, pp. 55–58, Jan., 1889.

305. Cooke's "Report on Bird Migration in the Mississippi Valley,' <*Auk*, VI, pp. 58–61, Jan., 1889.

Synoptic review of the work (U. S. Dept. Agric., Divis. of Economic Orn., Bull. No. 2).

306. Nelson's Report upon Natural History Collections made in Alaska. <*Auk*, VI, pp. 61–63, Jan., 1889.

Review of E. W. Nelson's 'Report' (4to, Washington, 1887–1888). Birds, pp. 19–230, with 12 col. plls.

307. Supplement to the A. O. U. Code of Nomenclature and Check-List of North American Birds and The Abridged A. O. U. Check-List. <*Auk*, VI, pp. 168–169, April, 1889.

Notice of the first Check-List supplement.

308. Ridgway on New or Little-known American Birds. <*Auk*, VI, pp. 171–173, April, 1889.

Review of eight papers by R. Ridgway.

309. Beckham on the Birds of Southwestern Texas. <*Auk*, VI, pp. 173–174, April, 1889.

Review of C. W. Beckham's 'Observations on the Birds of Southwestern Texas (*Proc. U. S. Nat. Mus.*, 1887, pp. 633–696.)

310. Note on the First Plumage of *Colinus ridgwayi*. <*Auk*, VI, p. 189, April, 1889.

311. Sclater's Catalogue of the Mesomyodian Passeres. <*Auk*, VI, pp. 266–268, July, 1889.

Review of Vol. XIV of the British Museum Catalogue of Birds.

312. Sclater and Hudson's Argentine Ornithology. <*Auk*, VI, pp. 268–269, July, 1889.

Review of Vol. II (London, 1889; for Vol. I *cf. supra*, No. 292). Also includes corrections of Barrows's 'Birds of the Lower Uruguay' (in *Auk*, I, 1884).

313. On *Cyclorhis viridis* (Vieill.) and its near Allies, with Remarks on other Species of the Genus *Cyclorhis*. <*Bull. Amer. Mus. Nat. Hist.*, II, pp. 123–135, June 17, 1889.

Cyclorhis flavipectus trinitatis, subsp. nov., p. 131.

314. Descriptions of New Species of South American Birds, with Remarks on various other little-known Species. <*Bull. Amer. Mus. Nat. Hist.*, II, pp. 137–151, June 28, 1889.

Spp. nov.: (1) *Thryothorus macrurus*, p. 137; (2) *Thryothorus longipes*, p. 138; (3) *Platyrhynchus bifasciatus*, p. 141; (4) *Platyrhynchus insularis*, p. 143; (5) *Sublegatus virescens*, p. 149; (6) *Empidonax lawrencei* (= *Octhoëca flaviventris* Lawr.), p. 150; (7) *Thamnophilus doliatus mexicanus*, p. 151.

315. [Note on] the Breeding of the Florida Gallinule in Vermont. <*Auk*, VI, p. 274, July, 1889.

Follows a communication on this subject by Jenness Richardson.

316. Report of the U. S. Ornithologist and Mammalogist for the Year 1888. <*Auk*, VI, p. 284, July, 1889.

Brief summary of report by C. Hart Merriam.

317. Barrows's Report on the English Sparrow in North America. <*Auk*, VI, pp. 326–328, Oct., 1889.

Review of W. B. Barrows's report on this subject, forming Bulletin I of the Division of Economic Ornithology and Mammalogy, U. S. Dept. Agric.

318. Gould's Ornithological Works. *New York Evening Post*, July 27, 1889. [Anonymous.]

Summary account of John Gould's principal ornithological works, with a brief biographical sketch of the author.

318*a*. Stone's Catalogue of the Muscicapidæ in the Collection of the Philadelphia Academy of Natural Sciences. <*Auk*, VI, pp. 330–331, Oct., 1889.

Review of Witmer Stone's paper of this title (*Proc. Acad. Nat. Sci. Philadelphia*, 1889, pp. 146–154).

319. Bergtold's List of the Birds of Buffalo and Vicinity. <*Auk*, VI, pp. 331–332, Oct., 1889.

Review of W. H. Bergtold's paper (in *Bull. Buffalo Naturalists' Field Club*, 1889).

320. Keyes and Williams's Preliminary Catalogue of the Birds of Iowa. <*Auk*, VI, p. 332, Oct., 1889.

Review of their 'Catalogue' (in *Proc. Davenport Acad. Nat. Sci.*, V, 1889).

321. Taylor's Catalogue of Nebraska Birds. <*Auk*, VI, pp. 332–333, Oct., 1889.

Review of W. Edgar Taylor's 'Catalogue' (in Ann. Rep. Nebraska State Board of Agric. for 1887, pp. 111–118, 1888).

322. Shufeldt's Recent Contributions to the Osteology of North American Birds. <*Auk*, VI, pp. 333–334, Oct., 1889.

Review of four papers by Dr. R. W. Shufeldt on different groups of birds.

323. Notes on a Collection of Birds from Quito. <*Bull. Amer. Mus. Nat. Hist.*, II, pp. 69–76, March 22, 1889.

An annotated list of 79 species. *Manacus* Brisson antedates *Chiromachæris* Cabanis (p. 73).

324. List of Birds Collected in Bolivia by Dr. H. H. Rusby, with Field Notes by the Collector. <*Bull. Amer. Mus. Nat. Hist.*, II, 77–112, March 22, 1889.

Annotated list of 267 species, the following new: (1) *Empidonax bolivianus*, p. 86; (2) *Chiroxiphia pareola boliviana*, p. 87; (3) *Enicornis striata*, p. 89; (4) *Leptasthenura fuscescens*, p. 90; (5) *Leptasthenura fuliginiceps boliviana*, p. 91; (6) *Synallaxis griseiventris*, p. 91; (7) *Anabazenops immaculatus*, p. 96; (8) *Picolaptes oblectus*, p. 94; (9) *Myrmochanes* (gen. nov.) *hypoleucus*, p. 95; (10) *Conopophaga rusbyi*, p. 96; (11) *Phlogopsis notata*, p. 97; (12) *Scytalopus bolivianus*, p. 96.

325. Remarks on Individual and Seasonal Variation in a large Series of *Elainea* from Chapada, Matto Grosso, Brazil, with a Revision of the Species of the restricted Genus *Elainea*. <*Bull. Amer. Mus. Nat. Hist.*, II, pp. 183–208, Oct. 31, 1889.

Critical comment on various alleged species of the genus.

326. On the Maximilian Types of South American Birds in the American Museum of Natural History. <*Bull. Amer. Mus. Nat. Hist.*, II, pp. 209–276, Dec., 1889.

About 160 species formally considered and incidental comment on 10 others. See also *infra*, No. 352.

1890.

327. To What Extent is it Profitable to Recognize Geographical Forms among North American Birds? <*Auk*, VII, pp. 1–9, Jan., 1890.

A protest against too fine splitting.

328. Seventh Congress of the American Ornithologists' Union. <*Auk*, VII, pp. 66–71, Jan., 1890.

329. Waterhouse's 'Index Generum Avium.' <*Auk*, VII, pp. 71–73, Jan., 1890.

Review, with critical comment, of F. H. Waterhouse's work of this title (8vo, London, 1889).

330. Blanchard on the Nomenclature of Organized Beings. <*Auk*, VII, pp. 73–74, Jan., 1890.

Review of Blanchard's Report 'De la Nomenclature des êtres organisés,' presented to the Congrès International de Zoologie, Paris, 1889.

331. Menzbier's Ornithology of Turkestan. <*Auk*, VII, pp. 78–79, Jan., 1890.

Review of the first livraison of the work.

332. Note on *Thryothorus ludovicianus miamensis*. <*Auk*, VII, pp. 115–117, April, 1890.

Inedited matter in a paper by W. E. D. Scott on Birds of the Gulf Coast of Florida.

333. Saunders's Manual of British Birds. <*Auk*, VII, pp. 195–196, April, 1890.

Review of 'An Illustrated Manual of British Birds,' by Howard Saunders (8vo, London, 1889).

334. Notes on Sport and Ornithology. <*Auk*, VII, pp. 196–197, April, 1890.

Review of a work of this title by the late Crown Prince Rudolf of Austria (8vo, London, 1889).

335. Ridgway on the Genus *Ziphocolaptes*. <*Auk*, VII, pp. 271–272, July, 1890.

Notice of R. Ridgway's 'A Review of the Genus *Xiphocolaptes* of Lesson' (*Proc. U. S. Nat. Mus.*, 1889, pp. 1–20).

336. Ridgway on the Genus *Sclerurus*. <*Auk*, VII, pp. 272–273, July, 1890.

Notice of R. Ridgway's 'A Review of the Genus *Sclerurus* of Swainson' (*Proc. U. S. Nat. Mus.*, 1889, pp. 21–31).

337. Ridgway on Birds from the Galapagos Islands, the Abrolhos, the Island of Santa Lucia, and the Straits of Magellan. <*Auk*, VII, pp. 273–274, July, 1890.

Notice of 'Scientific Results of Explorations by the U. S. Fish Commission Steamer Albatross. No. 1, Birds Collected on the Galapagos Islands in 1888,' and 'No. 2, Birds Collected on the Island of Santa Lucia,' etc. by R. Ridgway (*Proc. U. S. Nat. Mus.*, 1889, pp. 101–139).

338. Chapman on the Genus *Xiphorhynchus*. <*Auk*, p. 274, July, 1890.

Notice of 'A Review of the Genus *Xiphorhynchus* Swainson, with Descriptions of two new Species,' by F. M. Chapman (*Bull. Amer. Mus. Nat. Hist.*, II, pp. 153–162, July, 1889).

339. Stejneger and Lucas on Pallas's Cormorant. <*Auk*, VII, pp. 276–277, July, 1890.

Review of 'Contributions to the Natural History of the Commander Islands, X.' By L. Stejneger and F. A. Lucas (*Proc. U. S. Nat. Mus.*, XII, pp. 83–94, pll. ii–iv).

340. Lucas on the Osteology of the Thrushes and Wrens. <*Auk*, VII, p. 277, July, 1890.

Review of F. A. Lucas's 'Notes on the Osteology of the Thrushes, Miminæ, and Wrens' (*Proc. U. S. Nat. Mus.*, 1888, pp. 173–180).

341. Shufeldt on the Relationships of the Genus *Chamæa*. <*Auk*, VII, p. 278, July, 1890.

Review of R. W. Shufeldt's paper 'On the Position of *Chamæa* in the System' (*Journ. Morph.*, III, pp. 475–502).

342. Shufeldt's "Studies of the Macrochires." *Auk*, VII, pp. 278–279, July, 1890.

Review of his 'Studies,' etc., in *Journ. Linn. Soc., Zool.*, XX, pp. 299–394, pll. xvii–xxiv.

343. Shufeldt on the Osteology of the North American Passeres. <*Auk*, VII, pp. 279–280, July, 1890.

Review of his 'Contributions to the Comparative Osteology of Families of the North American Passeres,' in *Journ. Morph.*, III, 1889, pp. 81–112, pll. v, vi.

344. Maynard's 'Eggs of North American Birds.' <*Auk*, VII, p. 280, July, 1890.

Review of C. J. Maynard's work of this title (8vo, Boston, 1890).

345. Anthony on New Birds from Lower California. <*Auk*, VII, pp. 281–282, July, 1890.

Review of A. W. Anthony's paper of this title in *Proc. California Acad. Sci.*, 2d Ser., II, 1889, pp. 75–82.

346. Sennett on Bird Legislation. <*Auk*, VII, p. 282, July, 1890.

Review of an address on the subject by G. B. Sennett, published in Rep. Board of Agric. Penn. for 1889.

347. Description of a New Species of *Icterus* from Andros Island, Bahamas. <*Auk*, VII, pp. 343–346, Oct., 1890; and *ibid.*, VIII, 1891, pl. i (colored).

Icterus northropi sp. nov. Description and plate republished in 'A Naturalist in the Bahamas,' a memorial volume of the late John I. Northrop (8vo, New York, 1910).

348. Sclater's Catalogue of the Tracheophonæ. <*Auk*, VII, pp. 379–380, Oct., 1890.

Review of Vol. XV of the British Museum Catalogue of Birds.

349. Dr. Merriam's 'Biological Survey' of the United States. <*Auk*, VII, p. 414, Oct., 1890.

Inception of a 'Biological Survey' under Dr. C. Hart Merriam, and nature of the proposed work.

1891.

350. List of the Birds of Labrador, including Ungave, East Main, Moose, and Guld Districts of the Hudson Bay Company, together with the Island of Anticosti. By Lucien M. Turner. Reprinted by the author's permission from the Proceedings of the U. S. National Museum, 1885, pp. 233–254. Revised and brought down to date [1891], by J. A. Allen. <Packard's 'The Labrador Coast,' pp. 406–442, 1891.

351. The American Ornithologists' Union. A Seven Years' Retrospect. An Address delivered by the Retiring President at the Eighth Congress of the Union, Nov. 19, 1890. By J. A. Allen. Published by order of the Union, New York, January, 1891.— 8vo, pp. 19, and title page.

352. Further Notes on Maximilian Types of South American Birds. <*Bull. Amer. Mus. Nat. Hist.*, III, pp. 199–202, Feb. 20, 1891. (*Cf. supra*, No. 326.)

Rhopocichla, gen. nov., p. 199.

353. [Annotations to] 'The Birds of Andros Island, Bahamas.' By John I. Northrop. <*Auk*, VIII, pp. 64–80, Jan., 1891.

Notes on: *Mimus polyglottos*, p. 67; *M. gundlachi*, p. 67; *Seiurus aurocapillus*, p. 68; *Geothlypis rostrata*, p. 69; *Vireo crassirostris*, p. 70; *Myiarchus lucaysiensis*, p. 72; *Speotyto cunicularia dominicensis*, p. 75; *Rallus coryi*, p. 77.

354. Sharpe's 'Catalogue of the Sturniformes.' <*Auk*, VIII, pp. 90–92, Jan., 1891.

Review of Vol. XIII of the British Museum Catalogue of Birds.

355. Hargitt's Catalogue of the Woodpeckers. <*Auk*, VIII, pp. 92–95, Jan., 1891.

Review of Vol. XVIII of the British Museum Catalogue of Birds.

356. Merriam's 'Results of a Biological Survey of the San Francisco Mountain Region and Desert of the Little Colorado, Arizona.' *Auk*, VIII, pp. 95–98, Jan., 1891.

Review of C. Hart Merriam's 'Results,' etc., in *North Am. Fauna*, No. 3, Sept., 1891.

357. Dr. Merriam's Exploration of the Death Valley Region. <*Auk*, VIII, pp. 122–123, Jan., 1891.

Announcement of the Expedition.

358. Description of a New Species of *Mimocichla*, from the Island of Dominica, West Indies. <*Auk*, VIII, pp. 217–218, April, 1891.

Mimocichla verrillorum = *M. ardesiaca albiventris* Scl. of slightly earlier date (See *Auk*, VIII, 1891, p. 317.)

359. The Ornithology of the 'Century Dictionary.' <*Auk*, VIII, pp. 222–224, April, 1891.

Review of Dr. E. Coues's ornithological contributions to the 'Century Dictionary.'

360. Nicholson's Translation of Sundevall's 'Tentamen.' <*Auk*, VIII, pp. 227–228, April, 1891.

Review of the work (8vo, London, 1889).

361. Góss's 'History of the Birds of Kansas.' <*Auk*, VIII, pp. 228–230, April, 1891.

Review of N. S. Goss's work of this title (roy. 8vo, Topeka, 1891).

362. Gätke's 'Die Vogelwarte Helgoland.' <*Auk*, VIII, pp. 299–300, July, 1891.

Review of Heinrich Gätke's work of this title (roy. 8vo, Braunschweig, 1891).

363. Cory's 'Birds of the Bahama Islands.' <*Auk*, VIII, pp. 300–301, July, 1891.

Review of the revised edition, 1890.

364. Grant's 'Our Common Birds [and how to know them].' <*Auk*, VIII, p. 301, July, 1891.

Review of John B. Grant's work of this title (New York, 1891).

365. Thompson's 'Birds of Manitoba.' <*Auk*, VIII, pp. 301–302, July, 1889.

Review of E. E. Thompson's paper in *Proc. U. S. Nat. Mus.*, XIII, 1890, pp. 457–643, pl. xxxviii.

366. Canadian Bird Notes. <*Auk*, VIII, pp. 302–303, July, 1891.

Review of ornithological papers in *Trans. Canadian Inst.*, 1890.

367. Stone's List of 'Birds Collected in Yucatan and Southern Mexico' on the Heilprin Expedition. <*Auk*, VIII, p. 303, July, 1891.

Review of Witmer Stone's paper in *Proc. Acad. Nat. Sci. Philadelphia*, 1890, pp. 201–218.

368. Townsend on the Birds of the Coast and Islands of Upper and Lower California. <*Auk*, VIII, p. 305, July, 1891.

Review of Charles H. Townsend's 'Birds from the Coasts of Western North America and adjacent Islands' (*Proc. U. S. Nat. Mus.*, XIII, 1890, pp. 131–142).

369. Palmer on Birds observed during the Cruise of the 'Grampus.' <*Auk*, VIII, p. 305, July, 1891.

Review of William Palmer's paper in *Proc. U. S. Nat. Mus.*, XIII, 1890, pp. 249–262, on birds observed in Gulf of St. Lawrence, and at Funk Island, etc.

370. Lucas on the Anatomy and History of the Great Auk. <*Auk*, VIII, p. 306, July, 1891.

Review of F. A. Lucas's paper in Rep. U. S. Nat. Mus., 1887–88, pp. 493–529, pll. lxxi–lxxiii.

371. Capture of *Geothlypis poliocephala palpebralis* in Cameron County, Texas. <*Auk*, VIII, p. 316, July, 1891.

372. Note on *Mimocichla verrillorum*. <*Auk*, VIII, p. 317, July, 1891.

Name antedated by *Mimocichla albiventris* (Scl.). See *supra*, No. 358.

373. [Plates of *Otophanes mcleodii* Brewster, and of two species of *Megascops*]. <*Auk*, VIII, p. 320, Oct., 1891.

Explanatory note.

374. Sharpe's 'Review of Recent Attempts to Classify Birds.' <*Auk*, VIII, pp. 379–381, Oct., 1891.

Notice of Dr. R. B. Sharpe's address on this subject before Second Int. Orn. Congress, Budapest, 1891.

375. Hornaday's Handbook of Taxidermy and Zoölogical Collecting. <*Auk*, VIII, pp. 381–383, Oct., 1891.

Review of Wm. T. Hornaday's work of this title (8vo, New York, 1891).

376. Butler's Birds of Indiana. <*Auk*, VIII, pp. 383–384, Oct., 1891.

Review of Amos W. Butler's 'Birds of Indiana' (8vo, pp. 135).

377. Colburn and Morris's 'Birds of the Connecticut Valley in Massachusetts.' <*Auk*, VIII, pp. 384–385, Oct., 1891.

Review of their privately printed brochure of this title (16mo, pp. 24, Springfield, Mass.).

378. Merriam's List of Birds Observed in Idaho. <*Auk*, VIII, p. 385, Oct., 1891.

Review of list in *N. Am. Fauna*, No. 5.

379. Maynard's 'Contributions to Science.' <*Auk*, VIII, pp. 385–387, Oct., 1891.

Notice of 16 papers relating to ornithology contained in Vol. I of his work of this title.

380. On a Collection of Birds from Chapada, Matto Grosso, Brazil, made by Mr. Herbert H. Smith. Part I, Oscines. <*Bull. Amer. Mus. Nat. Hist.*, III, pp. 337–380, Sept. 29, 1891.

87 species here treated: *Callista margaritæ*, sp. nov., p. 351; *Zonotrichia capensis costaricensis*, subsp. nov., p. 374.

1892.

381. The New Nuttall. <*Auk*, IX, pp. 59–61, Jan., 1892.

Review of Montague Chamberlain's 'A Popular Handbook of the Ornithology of the United States and Canada, based on Nuttall's Manual.' 2 vols., 8vo, Boston, 1891.

382. Shufeldt's 'Myology of the Raven.' <*Auk*, IX, p. 62, Jan., 1892.

Review of R. W. Shufeldt's work of this title (8vo, London and New York, 1890).

383. Bolles's 'Land of the Lingering Snow.' <*Auk*, IX, p. 62, Jan., 1892.

Review of Frank Bolles's work of this title (12mo, Boston and New York, 1891).

384. Keyser's 'Bird-dom.' <*Auk*, IX, p. 63, Jan., 1892.

Review of Leander Keyser's work thus entitled (12mo, Boston, 1891).

385. Packard's 'The Labrador Coast.' <*Auk*, IX, p. 171, April, 1892.

Review of A. S. Packard's work thus entitled (8vo, New York, 1891).

386. Lucas on the Osteology of the Paridæ, *Sitta*, and *Chamæa*. <*Auk*, IX, p. 172, April, 1892.

Review of a paper by F. A. Lucas entitled 'Notes on the Osteology of the Paridæ, *Sitta*, and *Chamæa*, in *Proc. U. S. Nat. Mus.*, XIII, 1890, pp. 337–345, pl. xxvii.

387. G. K. Cherrie on Costa Rican Birds. <*Auk*, IX, p. 173, April, 1892.

Notice of his paper entitled 'Descriptions of New Genera, Species, and Subspecies of Birds from Costa Rica,' in *Proc. U. S. Nat. Mus.*, XIV, 1892, pp. 337–346.

388. Shufeldt on the Osteology of Arctic and Sub-Arctic Water Birds. <*Auk*, IX, pp. 173–174, April, 1892.

Review of Parts V–IX, of R. W. Shufeldt's 'Contributions to the Comparative Osteology,' etc., in Vols. XXIV and XXV, *Journ. Anat. and Phys.*, 1890–91.

389. Shufeldt on the Osteology and Classification of the North American Pigeons, Woodpeckers, and Kites. <*Auk*, IX, p. 174, April, 1892.

Notice of four papers by R. W. Shufeldt.

390. Ridgway on New or Little-known Central American and South American Birds. <*Auk*, IX, pp. 174–175, April, 1892.

Short notices of nine papers by R. Ridgway in *Proc. U. S. Nat. Mus.*, XIV, 1891.

391. Chapman on the 'Origin of the Avifauna of the Bahamas.' <*Auk*, IX, pp. 179–180, April, 1892.

Review of F. M. Chapman's paper of this title in *Amer. Nat.*, June, 1891, pp. 528–539.

392. Sclater on the Geographical Distribution of Birds. <*Auk*, IX, pp. 183–184, April, 1892.

Review of an address on this subject read before the Second Int. Orn. Congress, Budapest, 1891 (*Ibis*, 1891, pp. 514–527).

393. Sclater and Shelley on the Scansores and Coccyges. <*Auk*, IX, p. 184, April, 1892.

Review of Vol. XIX of British Museum Catalogue of Birds.

394. Hartert's Catalogue of the Birds in the Museum of the Senkenberg Natural History Society. <*Auk*, IX, p. 185, April, 1892.

Review of the work (8vo, Frankfort a. M., Jan., 1891).

395. Leverkuhn's 'Fremde Eier im Nest.' <*Auk*, IX, pp. 185–186, April, 1892.

Notice of Paul Leverkühn's work of this title (8vo, Berlin, 1891).

396. Jäckel's Birds of Bavaria. <*Auk*, IX, p. 186, April, 1892.

Notice of Andreas Johannes Jäckel's work on this subject (8vo, München und Leipzig, 1891).

397. The North American Species of *Colaptes*, considered with special reference to the Relationships of *C. auratus* and *C. cafer*. <*Bull. Amer. Mus. Nat. Hist.*, IV, 1892, pp. 21–44, and map, March 8, 1892.

Hybridization on a large scale shown to obtain between *C. auratus* and *C. cafer*.

398. The Blackfronted Warbler (*Dendroica nigrifrons* Brewst.). <*Auk*, IX, p. 207, April, 1892.

Explanatory note respecting a colored plate of this species published in the preceding number of *The Auk* (IX, pl. i, Jan., 1891).

399. Salvadori's Catalogue of the Parrots. <*Auk*, IX, pp. 277–279, July, 1892.

Review of Vol. XX of the British Museum Catalogue of Birds, with critical nomenclatural comment.

400. Oustalet on the Birds of Patagonia. <*Auk*, IX, p. 281, July, 1892.

Review of the 'Oiseaux' of the Mission Scientifique du Cap Horn (4to, pp. 341, pll. 6).

401. Stone on the Crows, Birds of Paradise, and Orioles in the Museum of the Philadelphia Academy of Natural Sciences. <*Auk*, IX, p. 282, July, 1892.

Review of Witmer Stone's paper in *Proc. Acad. Nat. Sci. Philadelphia*, 1891, pp. 441–450.

402. The Wanton Destruction of Bird Life in America. <*Our Animal Friends*, XIX, pp. 126–128, 150–152, Feb. and March, 1892.

403. Notice of some Venezuelan Birds Collected by Mrs. H. H. Smith. <*Bull. Amer. Mus. Nat. Hist.*, IV, 1892, pp. 51–59, April, 6, 1892.

48 species and subspecies, the following new: (1) *Ramphocœlus atrosericeus capitalis*, p. 51; (2) *Lophotriccus subcristatus*, p. 53; (3) *Picumnus obsoletus*, p. 55.

404. Description of a New Gallinule from Gough Island. <*Bull. Amer. Mus. Nat. Hist.*, IV, 1892, pp. 57–58, May 9, 1892.

Porphyriornis comeri, gen. et sp. nov.

405. Bendire's 'Life Histories of North American Birds.' <*Auk*, IX, pp. 375–376, Oct., 1892.

Review of Part I of this work (4to, Special Bulletin, U. S. Nat. Mus., 1892).

406. Ridgway's 'The Hummingbirds.' <*Auk*, IX, pp. 376–377, Oct., 1892.

Review of paper published in Rep. U. S. Nat. Mus. for 1890, pp. 255–383, pll. i–xliii.

406*a*. Merriam on the Life Areas of North America. <*Auk*, IX, pp. 377–382, Oct., 1892.

Review of Dr. Merriam's Presidential Address before the Biological Society of Washington, Feb. 6, 1892, entitled 'The Geographic Distribution of Life in North America with Special Reference to the Mammalia' (in *Proc. Biol. Soc. Washington*, VII, pp. 1–64).

407. Suchetet on Hybridity in Birds. <*Auk*, IX, pp. 382–383, Oct., 1892.

Review of 'Les Oiseaux Hybrides rencontrés à l'état sauvage par André Suchetet.' Troisieme Partie. Les Passereaux.

408. On a Collection of Birds from Chapada, Matto Grosso, Brazil, made by Mr. H. H. Smith, Part II, Tyrannidæ. <*Bull. Amer. Mus. Nat. Hist.*, IV, pp. 331–350, Dec., 1892.

1893.

409. British Museum Catalogue of the Picariæ. <*Auk*, X, pp. 66–69, Jan., 1893.

Review of Vols. XVI and XVII of Catalogue of Birds in the British Museum.

410. Cory's 'Catalogue of West Indian Birds.' <*Auk*, X, pp. 69–70, Jan., 1893.

Review of the 1st ed. (4to, Boston, 1892).

411. Dixon's 'The Migration of Birds.' <*Auk*, X, 70–73, Jan., 1893.

Review, with extended critical comment, of Charles Dixon's 'The Migration of Birds, an attempt to reduce Avian Season-Flight to Law (8vo, London, 1892).

412. [Pterylography — A Neglected Branch of Ornithology.] <*Auk*, X, pp. 94–95, Jan., 1893.

Remarks supplementary to an article of this title by Hubert Lyman Clark.

413. The Geographical Origin and Distribution of North American Birds, considered in Relation to Faunal Areas of North America. <*Auk*, X, pp. 97–150, pll. iii–v (maps), April, 1893.

414. Keeler on the 'Evolution of the Colors of North American Land Birds.' <*Auk*, X, pp. 189–195, April, 1893.

Review of Charles A. Keeler's work of this title (8vo, San Francisco, 1893), with much critical commentary.

415. Beddard's 'Animal Coloration.' <*Auk*, X, pp. 195–199, April, 1893.

Review of Frank E. Beddard's 'Animal Coloration, an Account of the Principal Facts and Theories relating to the Colors and Markings of Animals' (8vo, London, 1892).

416. Hawks and Owls in their Relation to Agriculture. <*Auk*, X, pp. 199–201, April, 1893.

Review of Dr. A. K. Fisher's work of this title, forming Bull. No. 3, Division of Orn. and Mam., U. S. Dept. Agriculture.

417. Bolles's 'Chronicles.' <*Auk*, X, pp. 201–202, April, 1893.

Review of Frank Bolles's 'At the North of Bear-camp Water: Chronicles of a Stroller in New England from July to December' (12mo, Boston and New York, 1893).

418. Foster's Bibliography of the Ornithological Writings of George N. Lawrence. < *Auk*, X, p. 202, April, 1893.

Review of L. S. Foster's work of this title (Bull. No. 40, U. S. National Museum, 1892).

419. Ornithology of the Death Valley Expedition. <*Auk*, X, pp. 285–288, July, 1893.

Review of Part II (Report on the Birds, by A. K. Fisher), of the Death Valley Expedition, a Biological Survey of Parts of California, Nevada, Arizona, and Utah (*North Amer. Fauna*, No. 7, May 31, 1893).

420. The Faunal Position of Lower California. <*Auk*, X, pp. 306–307, July, 1893.

Reply to remarks by Dr. C. Hart Merriam under the same title (*l. c.*, pp. 305, 306).

421. Shufeldt on Fossil Birds from Oregon. <*Auk*, X, pp. 343–345, Oct., 1893.

Review of 'A Study of the Fossil Avifauna of the Equus Beds of the Oregon Desert,' by Dr. R. W. Shufeldt (in *Journ. Acad. Nat. Sci. Philadelphia*, XI, pp. 389–425, pll. xv–xvii).

422. Shufeldt on *Ichthyornis*, and on the Classification of the Longipennes. <*Auk*, X, p. 345, Oct., 1893.

Notice of two papers by Dr. R. W. Shufeldt (*Journ. Anat. and Phys.*, XXVII, pp. 336–342, and *Amer. Nat.*, 1893, pp. 233–237).

423. The Affinities of Hummingbirds and Swifts. <*Auk*, X, pp. 345–346, Oct., 1893.

Notice of papers by R. Ridgway, R. W. Shufeldt, and F. A. Lucas on this subject.

424. Food Habits of Birds. <*Auk*, X, p. 347, Oct., 1893.

Notice of papers on this subject by W. B. Barrows and F. E. L. Beal in Dr. C. Hart Merriam's 'Report of the Ornithologist and Mammalogist' for 1892 (Rep. Secretary of Agriculture for 1892, pp. 181–200).

425. Hasbrouck on 'Evolution and Dichromatism in the Genus *Megascops*.' <*Auk*, X, pp. 347–351, Oct., 1893.

Critical review of E. M. Hasbrouck's paper on this subject in *Amer. Nat.*, 1893, pp. 521–533, 638–649.

426. Cook's 'Birds of Michigan.' <*Auk*, X, pp. 351–352, Oct., 1893.

Review of A. J. Cook's 'Birds of Michigan,' (Bull. 94, Michigan Agric. Station, State Agric. College).

427. Averill's List of the Birds of Bridgeport. <*Auk*, X, pp. 352–353, Oct., 1893.

Notice of a brochure by C. K. Averill, Jr. (8vo, pp. 19, 1892).

428. Summer Birds of Greene County, Pa. <*Auk*, X, p. 353, Oct., 1893.

Notice of a brochure of this title, by J. Warren Jacobs (8vo, pp. 15, 1893).

429. Nutting's Zoölogical Explorations on the Lower Saskatchewan River. <*Auk*, X, pp. 353–354, Oct., 1893.

Notice of C. C. Nutting's 'Report' on his Explorations, etc. (*Bull. Lab. Nat. Hist. Iowa State Univ.*, II, No. 3, pp. 235–293, Jan., 1893).

430. 'A New List of Chilian Birds.' <*Auk*, X, p. 354, Oct., 1893.

Notice of H. B. James and P. L. Sclater's work of this title (roy. 8vo, London, 1892).

431. Newton's 'A Dictionary of Birds.' Part I. <*Auk*, X, pp. 357–360, Oct., 1893.

Extended review of the first part (A-Ga; 8vo, London, 1893).

432. On a Collection of Birds from Chapada, Matto Grosso. Parts III and IV, Pipridæ to Rheidæ. <*Bull. Amer. Mus. Nat. Hist.*, V, 1893, pp. 107–158, July 19, 1893.

Spp. and subspp. nov: (1) *Pygmornis chapadensis*, pp. 122; (2) *Piaya cayana cabanisi*, p. 136; (3) *Buteo albicaudatus sennetti*, p. 144;

This concludes the paper, which comprises a list of 324 species, with annotations and comment on allied species in a number of groups. Part IV (pp. 152–158), 'Oological Notes,' contains notes on the nests and eggs of 27 species.

433. *Cœreba* versus *Certhiola*. <*Auk*, X, pp. 369–370, Oct., 1893.

Cœreba antedates *Certhiola* by 28 years, with *C. flaveola* as type by monotypy.

434. The Evolution of the Colors of North American Land Birds. <*Auk*, X, pp. 377–380, Oct., 1893.

Reply to Mr. Keeler's rejoinder (*l. c.*, pp. 375–377) to a review of his 'Evolution of Colors in North American Land Birds' (*Auk*. X, pp. 189–195; see *supra*, No. 414).

435. Nehrling's 'Our Native Birds of Song and Beauty.' <*Auk*, X, p. 387, Oct., 1893.

Brief notice of Volume I (4to).

436. List of Mammals and Birds Collected in Northeastern Sonora and Northwestern Chihuahua, Mexico, on the Lumholtz Archæological Expedition, 1890–92. <*Bull. Amer. Mus. Nat. Hist.*, V, 1893, pp. 27–42.

Annotated list of 162 species of birds.

1894.

437. Newton's 'A Dictionary of Birds.' Part II. <*Auk*, XI, pp. 56–60, Jan., 1894.

Review of Part II (Ga-Moa). (8vo, London, 1893.) (See *supra*, No. 431.)

438. Salvadori's Catalogue of the Pigeons. <*Auk*, XI, pp. 60–62, Jan., 1894.

Review of Vol. XXI of the British Museum Catalogue of Birds, with comment on nomenclatural points.

439. Elliot's Monograph of the Pittidæ. Parts I and II. <*Auk*, XI, pp. 62–63, Jan., 1894.

Review of D. G. Elliot's work of this title, 2d ed. (folio, London, April, 1893).

440. Sharpe on the Zoögraphical Areas of the World. <*Auk*, XI, pp. 63–65, Jan., 1894.

Review of R. Bowdler Sharpe's paper on this subject (*Nat. Sci.*, III, Aug., 1893, pp. 100–108). "But the recognition and definition of an Arctic Zone, or 'Realm,' as Mr. Allen calls it, is a fact which must henceforward be admitted by all ornithologists."

441. Non-significance of albinistic eggs of the Marsh Wren. <*Auk*, XI, p. 81, Jan., 1894.

Comment on theoretical suggestions of Dr. L. B. Bishop.

442. First Plumages. <*Auk*, XI, pp. 91–93, pl. ii, Jan., 1894.

Seiurus aurocapillus, ad. and juv.

443. Anthony on the Birds of San Pedro Martir, Lower California. <*Auk*, XI, pp. 167–168, April, 1894.

Review of A. W. Anthony's paper of this title (*Zoe*, IV, pp. 228–247).

444. Short's Birds of Western New York. <*Auk*, XI, p. 168, April, 1894.

Notice of Ernest H. Short's brochure of this title (8vo, Chili, N. Y., pp. 13, 1893).

445. Ridgway on the Genus *Myiarchus*, etc. <*Auk*, XI, pp. 168–170, April, 1894.

Notice of five papers on American birds, by R. Ridgway, in *Proc. U. S. Nat. Mus.*, XVI, pp. 605–614, 663–686.

446. Stejneger on Japanese Birds. <*Auk*, XI, p. 170, April, 1894.

Notice of Stejneger's 'Third Instalment of Japanese Birds,' etc. (*Proc. U. S. Nat. Mus.*, XVI, 1893, pp. 615–638).

447. Ogilvie-Grant's 'Catalogue of the Game Birds.' <*Auk*, XI, pp. 171–173, April, 1894.

Review of Vol. XXII of the British Museum Catalogue of Birds, 1893, with critical comment.

448. McIlwraith's 'Birds of Ontario.' <*Auk*, XI, pp. 240–241, July, 1894.

Review of Thomas McIlwraith's work of this title, 2d ed. (8vo, Toronto, 1894).

449. Sharpe's Catalogue of the Fulicariæ and Alectorides. <*Auk*, XI, pp. 242–243, July, 1894.

Review, with critical comment, of Vol. XXIII of the British Museum Catalogue of Birds (1894).

450. Elliot's Monograph of the Pittidæ. Part III. <*Auk*, XI, p. 243, July, 1894.

Notice of Part III, Feb., 1894. (*Cf. supra*, No. 439.)

451. Mrs. Wright's 'The Friendship of Nature.' <*Auk*, XI, p. 314, Oct., 1894.

Notice of Mrs. Mabel Osgood Wright's book of this title (8vo, New York and London, 1894).

452. Shufeldt's 'Comparative Oölogy of North American Birds.' <*Auk*, XI, pp. 314–315, Oct., 1894.

Critical notice of R. W. Shufeldt's paper of this title, in Report U. S. Nat. Mus. for 1892 (1894), pp. 461–493.

453. Stone on Old World Rallinæ. <*Auk*, XI, p. 317, Oct., 1894.

Notice of Witmer Stone's 'Review of Old World Rallinæ,' in *Proc. Acad. Nat. Sci. Philadelphia*, 1894, pp. 130–149.

454. Wallace on "Palæarctic" and "Nearctic." <*Auk*, XI, pp. 318–319, Oct., 1894.

Critical review of A. R. Wallace's 'Palæarctic and Nearctic Regions compared as regards the Families and Genera of their Mammalia and Birds' (*Nat. Sci.*, IV, 1894, pp. 435–445).

455. [Notice of George K. Cherrie's ornithological work in Costa Rica]. <*Auk*, XI, p. 340, Oct., 1894.

"Mr. Cherrie has made an enviable record for himself in Costa Rica, displaying an energy and a capacity for work rarely equalled."

456. Fashion Journals and Bird Destruction. <*Auk*, XI, p. 342, Oct., 1894.

Comment on an article in *Harper's Bazaar* of Aug. 18, 1894.

1895.

457. Elliot's Monograph of the Pittidæ. Part IV. <*Auk*, XII, pp. 65–66, Jan., 1895.

Review of Part IV, issued September, 1894. (*Cf. supra*, Nos. 439, 450.)

458. Two Popular Bird Books. <*Auk*, XII, pp. 66–67, Jan., 1895.

Review of 'In Bird Land,' by Leander S. Keyser, and 'The Birds' Calendar,' by H. E. Parkhurst.

459. Ridgway on New Birds from the Galapagos Islands. <*Auk*, XII, pp. 70–71, Jan., 1895.

Review of R. Ridgway's 'Descriptions of Twenty-two new Species of Birds from the Galapagos Islands' (*Proc. U. S. Nat. Mus.*, XVII, pp. 357–370).

460. Lucas on the Affinities of the Cœrebidæ. <*Auk*, XII, pp. 71–72, Jan., 1895.

Review of F. A. Lucas's paper 'Notes on the Anatomy and Affinities of the Cœrebidæ and other American Birds' (*Proc. U. S. Nat. Mus.*, XVII, pp. 299–312).

461. Rhoads's Reprint of Ord's North American Zoölogy. <*Auk*, XII, pp. 72–73, Jan., 1895.

Review of S. N. Rhoads's reprint of George Ord's 'Zoölogy' in Guthrie's Geography (8vo, Haddonfield, N. J., 1894).

462. Reichenow's Birds of German East Africa. <*Auk*, XII, pp. 73–74, Jan., 1905.

Review of Dr. A. Reichenow's 'Die Vögel Deutsch-Ost-Afrikas' (roy. 8vo, Berlin, 1894).

463. [Work of Dr. C. Hart Merriam, as Ornithologist and Mammalogist of U. S. Dept. of Agric. for 1893.] <*Auk*, XII, p. 98, Jan., 1895.

Notice of his 'Report' for the year 1893.

464. Shufeldt on the Osteology of Cranes and Rails. <*Auk*, XII, pp. 172–173, April, 1895.

Notice of R. W. Shufeldt's paper on this subject in *Journ. Anat. and Phys.*, XXIX, 1894, pp. 21–34.

465. Newton's 'A Dictionary of Birds.' Part III. <*Auk*, XII, pp. 169–170, April, 1895.

Notice of Part III (Moa-Sheathbill), 1894. (See *supra*, Nos. 431 and 437.)

466. Grundtvig on the Birds of Shiocton, Wisconsin. <*Auk*, XII, p. 173, April, 1895.

Notice of F. L. Grundtvig's paper in *Trans. Wisconsin Acad. Sci. Arts, and Letters*, X, 1894, pp. 73–184.

467. Bourns and Worcester on the Birds of the Philippines. <*Auk*, XII, pp. 173–174, Auk, 1895.

Review of their paper in *Occas. Papers of Minnesota Acad. Nat. Sci.*, I, No. 1, Dec. 1894.

468. Merriam's Laws of Temperature Control of the Distribution of Land Animals and Plants. <*Auk*, XII, pp. 172–173, April, 1895.

Review of Dr. C. Hart Merriam's paper of this title in *Nat. Geogr. Mag.*, VI, 1894, pp. 229–238, pll. xii–xiv.

469. Minot's Land-Birds and Game-Birds of New England. Second Edition. <*Auk*, XII, pp. 284–286, July, 1895.

Review of the second edition of H. D. Minot's work, edited by William Brewster (8vo, Boston and New York, 1895).

470. Clark on the Pterylography of North American Goatsuckers and Owls. <*Auk*, XII, pp. 287–288, July, 1895.

Review of Hubert Lyman Clark's paper (*Proc. U. S. Nat. Mus.*, XVII, pp. 551–572, June. 1895).

471. Sharpe and Wyatt's Monograph of the Swallows. <*Auk*, XII, pp. 373–375, Oct., 1895.

Review, with critical comment, of their 'A Monograph of the Hirundinidæ, or Family of Swallows' (2 vols., 4to, London, 1885–1894).

472. Food Habits of Woodpeckers. <*Auk*, XII, pp. 380–381, Oct., 1895.

Review of F. E. L. Beal's 'Preliminary Report on Woodpeckers,' and F. A. Lucas's 'The Tongues of Woodpeckers,' forming *Bull. No. 7*, U. S. Dept. Agric., 1895.

473. Barrows and Schwarz on the Food of the Common Crow. <*Auk*, XII, pp. 381–383, Oct., 1895.

Review of *Bull. No. 6*, Dept. of Agric., 1895.

474. Forbush on 'Birds as Protectors of Orchards.' <*Auk*, XII, pp. 383–384, Oct., 1895.

Review of E. H. Forbush's paper in Massachusetts Crop Report for July, 1895.

475. Suchetet on Hybridity in Birds. <*Auk*, XII, p. 384, Oct., 1895.

Review of Part V of André Suchetet's 'Les Oiseaux Hybrides rencontrés à l'état sauvage' (8vo, pp. 473–873, Lille, 1895). (See *supra*, No. 407.)

1896.

475*a*. Elliot's 'Monograph of the Pittidæ.' <*Auk*, XIII, pp. 60–61, Jan., 1896.

Review of the concluding Part V, with a summary notice of the completed work (folio, London, 1893–1895). (See *supra*, Nos. 439, 450, 457.)

477. A. O. U. Check-List of North American Birds, Second Edition. <*Auk*, XIII, pp. 69–70, Jan., 1896.

Comparison with the first edition (1886).

478. 'Progress in American Ornithology, 1886–95.' <*Science*, N. S., III, Nos. 73 and 75, pp. 777–779 and 842–843, May 22 and June 5, 1896.

Relates to the Second Edition of the A. O. U. Check-List of North American Birds, in reply to criticisms by Dr. R. W. Shufeldt in the *American Naturalist* (XXX, May, 1896, pp. 357–372), and *Science* (III, pp. 841–842, June 5, 1896).

479. Bendire on the Cowbirds. <*Auk*, XIII, pp. 71–72, Jan., 1896.

Review of Capt. Charles Bendire's 'The Cowbirds,' in Report U. S. Nat. Mus. for 1893 (1895), pp. 587–624, pll. i–iii.

480. Lucas on the Weapons and Wings of Birds. <*Auk*, XIII, p. 72, Jan., 1896.

Review of Frederic A. Lucas's paper in Report of U. S. Nat. Mus. 1893 (1895), pp. 653–663.

481. Fisher's Hawks and Owls from the Standpoint of the Farmer. <*Auk*, XIII, p. 73, Jan., 1896.

Review of Dr. A. K. Fisher's paper by this title, in Yearbook U. S. Dept. Agric. for 1894 (1895), pp. 215–232.

482. Beddard's 'Text-book of Zoögeography.' <*Auk*, XIII, pp. 73–75, Jan., 1896.

Review of Frank E. Beddard's work of this title (8vo, Cambridge, 1895), with critical comment.

483. [The Taxonomic Value of the Tongue in Birds]. <*Auk*, XIII, pp. 114–115, April, 1896.

Note to an article with this title (*l. c.*, pp. 109–114) by Frederic A. Lucas.

484. Gätke's 'Heligoland.' <*Auk*, XIII, pp. 137–153, April, 1896.

An extended critical review of the work, and of Gätke's views on bird migration, change of color in feathers without moult, etc.

485. The 'Birds' of 'The Royal Natural History.' <*Auk*, XIII, pp. 156–160, 252–253, April and July, 1896.

Review of the bird part (Vols. III and IV) of 'The Royal Natural History,' edited by Richard Lydekker (roy. 8vo, London, 1894–95).

486. Saunders and Salvin's Catalogue of the Gaviæ and Tubinares. <*Auk*, XIII, pp. 160–162, April, 1896.

Review, with comment on points of nomenclature, etc., of Vol. XXV, of the British Museum Catalogue of Birds.

487. Salvadori's Catalogue of the Chenomorphæ, Crypturi, and Ratitæ. <*Auk*, XIII, pp. 162–164, April, 1896.

Review, with technical comment, of Vol. XXVII of the British Museum Catalogue of Birds.

488. Loomis on California Water Birds. <*Auk*, XIII, p. 168, April, 1896.

Review of Leverett M. Loomis's 'California Water Birds, No. II,' in *Proc. California Acad. Sci.* (2), VI, 1896, pp. 1–30.

489. [Some Questions of Nomenclature]. <*Auk*, XIII, pp. 187–190, April, 1896.

In response, by request, to Mr. Witmer Stone's paper of this title (*l. c.*, pp. 183–187).

490. [The Seebohm Collection of Birds]. <*Auk*, XIII, pp. 195–196, April, 1896.

Notice of its extent and its transfer to the British Museum. Based on an article in *Nature* (Feb. 20, 1896).

491. Stone on Birds Collected in North Greenland. <*Auk*, XIII, p. 243, July, 1896.

Notice of Witmer Stone's paper in *Proc. Acad. Nat. Sci. Philadelphia*, 1895, pp. 502–505.

492. Schalow on a Collection of Birds from West Greenland. <*Auk*, XIII, pp. 243–244, July, 1896.

Review of a paper by Herman Schalow in *Journ. f. Orn.*, 1895, pp. 475–481.

493. Rotzell's Birds of Narberth, Pa., and Vicinity. <*Auk*, XIII, p. 244, July, 1896.

Review of Dr. W. E. Lotzell's brochure of 8 pp. (1895).

494. Rhoads's List of Tennessee Birds. <*Auk*, XIII, pp. 244–245, July, 1896.

Review of paper by S. N. Rhoads in *Proc. Acad. Nat. Sci. Philadelphia*, 1895, pp. 463–501.

495. Short's Birds of Western New York. <*Auk*, XIII, p. 245, July, 1896.

Review of Ernest H. Short's second edition of his 'Birds of Western New York' (8vo, pp. 20, 1896.) (See *supra*, No. 444.)

496. Cory's 'Hunting and Fishing in Florida,' with a 'Key to the Water Birds of the State.' <*Auk*, XIII, pp. 246–247, July, 1896.

Review of the work (sm. quarto, Boston, 1896).

497. Howe's 'Every Bird.' <*Auk*, XIII, p. 247, July, 1896.

Review of Reginald Heber Howe, Jr's., work of this title (sm. 8vo, Boston, 1896).

498. Winchell's 'Evolution of Bird Song.' <*Auk*, XIII, pp. 249–250, July, 1896.

Review of Charles A. Winchell's work of this title (8vo, London and New York).

499. Harvey-Brown and Buckley's 'A Vertebrate Fauna of the Moray Basin.' < *Auk*, XIII, pp. 251–252, July, 1896.

Review of the work (sm. 4to, Edinburgh, 1895).

500. The Origin and Relations of the Floras and Faunas of the Antarctic and adjacent Regions. Vertebrata of the Land; Birds and Mammals. <*Science* (2), III, No. 61, pp. 317–319, Feb. 28, 1896.

Birds offer no satisfactory evidence of a former "Antarctic Continent."

501. Alleged Changes of Color in the Feathers of Birds without Molting. <*Bull. Amer Mus. Nat. Hist.*, VIII, pp. 13–44, March 18, 1896.

"It is a summary and criticism of the work of some of the more important writers upon the subject of color changes in feathers without moult, and it deals unsparingly with those who have asserted as possible the complete rejuvenation of an abraded feather...."— J. Dwight, Jr., in *Auk*, XIII, pp. 166–167, April, 1896.

502. Schwann's Handbook of British Birds. <*Auk*, XIII, pp. 328–329, Oct., 1896.

Review of H. Kirke Schwann's 'Concise Handbook of British Birds' (16mo, London, 1896).

503. Ridgway and Lucas on a New Family [Procniatidæ] of Birds. <*Auk*, XIII, pp. 334–335, Oct., 1896.

Review of two papers on this subject by R. Ridgway and F. A. Lucas, in *Proc. U. S. Nat. Mus.*, XVIII, pp. 450, and 505–507.

504. Montgomery on Migration as a Check upon Geographic Variation. <*Auk*, XIII, p. 335, Oct., 1896.

Review of a paper by Thomas H. Montgomery, Jr., in *Amer. Nat.*, 1896, pp. 458–464.

505. Contributions to Economic Ornithology. <*Auk*, XIII, pp. 335–338, Oct., 1896.

Notice of papers by S. D. Judd, F. E. L. Beal, E. H. Forbush, and Miss F. A. Merriam.

1897.

506. Sharpe's Catalogue of the Limicolæ. <*Auk*, XIV, pp. 102–104, Jan., 1897.

Review of Vol. XXIV of the British Museum Catalogue of Birds, 1896.

507. Bendire's 'Life Histories of North American Birds.' <*Auk*, XIV, pp. 104–106, Jan., 1897.

Review of Part II of Capt. Charles E. Bendire's important work. (See *supra*, No. 405).

508. Miss Merriam's 'A-Birding on a Bronco.' <*Auk*, XIV, pp. 107–108, Jan., 1897.

Review of Miss Florence A. Merriam's work of this title (16mo, Boston and New York, 1896).

509. 'Papers Presented to the World's Congress of Ornithology.' <*Auk*, XIV, p. 108, Jan., 1897.

Review of a volume thus entitled, and edited by Mrs. E. Irene Rood (8vo, Chicago, 1896).

510. The Revised New Nuttall. <*Auk*, XIV, p. 109, Jan., 1897.

Review of Montague Chamberlain's second and revised ed. of Nuttall's Ornithology (2 vols., 8vo, Boston, 1896). (*Cf. supra*, No. 381.)

511. Wintle's 'Birds of Montreal.' <*Auk*, XIV, p. 112, Jan., 1897.

Review of Ernest D. Wintle's work of this title (8vo, Montreal, 1896).

512. Oberholser's Birds of Wayne County, Ohio. <*Auk*, XIV, pp. 112–113, Jan., 1897.

Review of H. C. Oberholser's 'A Preliminary List of the Birds of Wayne County, Ohio' (*Bull. Ohio Agric. Exper. Station*, tech. ser., I, No. 4, July, 1896, pp. 243–354).

513. Ridgway's 'Manual of North American Birds.' <*Auk*, XIV, pp. 232–233, April, 1897.

Review of the 2d ed. (roy. 8vo, Philadelphia, 1896).

514. Goode's 'The Published Writings of Philip Lutley Sclater.' <*Auk*, XIV, pp. 233–234, April, 1897.

Review of *Bull. No. 49*, U. S. Nat. Mus., 1896.

515. Bates's 'The Game Birds of North America.' <*Auk*, XIV, p. 244, April, 1897.

Review of the work (16mo, Boston, 1896).

516. Butler on 'A Century of Change in the Aspects of Nature in Indiana.' <*Auk*, XIV, p. 245, April, 1897.

Notice of A. W. Butler's paper of this title in *Proc. Indiana Acad. Sci.*, No. V, 1895, pp. 31–42.

517. Elliot's Catalogue of a Collection of Birds from Somali-Land. <*Auk*, XIV, p. 245, April, 1897.

Notice of D. G. Elliot's paper of this title in *Field Mus. Publ.*, Orn. Ser., I, No. 2, 1897, pp. 29–67).

518. Anderson's Birds of Winnebago and Hancock Counties, Iowa. <*Auk*, XIV, p. 246, April, 1897.

Notice of Rudolph M. Anderson's privately printed brochure, 16mo, pp. ii + 19, Forest City, Iowa, 1897.

519. Ricker's Notes on the Birds of Hull, Mass. <*Auk*, XIV, p. 246, April, 1897.

Brief notice of Everett W. Ricker's privately printed paper of this title (16mo, Newtonville, Mass., 1896).

520. Howe's Birds of Brookline, Mass. <*Auk*, XIV, p. 246, April, 1897.

Notice of paper of this title by Reginald Heber Howe, Jr. (privately printed, folio, 2 pp., Jan., 1897).

521. Tegetmeier's 'Pheasants.' <*Auk*, XIV, pp. 426–427, April, 1897.

Review of the third edition of W. B. Tegetmeier's work of this title (8vo, London, 1897).

522. Cory's List of the Birds of Eastern North America. <*Auk*, XIV, p. 248, April, 1897.

Review of C. B. Cory's work of this title (8vo, Boston, 1896).

523. Schalow's the Published Writings of Anton Reichenow. <*Auk*, XIV, p. 248, April, 1897.

Notice of the work (8vo, pp. 29, 1896).

523*a*. The Fate of some Carrier Pigeons. <*Forest and Stream*, XLVIII, p. 383, 1897.

524. The Proper Generic Name of the Loons. <*Auk*, XIV, p. 312, July, 1897.

Gavia Forster should replace *Urinator* Cuvier.

525. Ridgway's Birds of the Galapagos Archipelago. <*Auk*, XIV, pp. 329–330, July, 1897.

Review of R. Ridgway's monographic paper in *Proc. U. S. Nat. Mus.*, XIX, pp. 459–670.

526. Cooke's 'Birds of Colorado.' <*Auk*, XIV, pp. 331–332, July, 1897.

Review of W. W. Cooke's paper of this title in *Bull. No. 37*, Colorado State Agric. College, pp. 143, March 17, 1897.

527. Miller on Construction of Scientific Names. <*Auk*, XIV, pp. 332–333, July, 1897.

Review of Prof. Walter Miller's paper on this subject (*Proc. California Acad. Sci.*, 3d ser., I, No. 3, pp. 115–143).

528. The New York Zoölogical Society. <*Auk*, XIV, p. 344, July, 1897.

Announcement of its organization.

529. Cory's Shore Birds of North America. <*Auk*, XIV, pp. 418–419, Oct., 1897.

Notice of C. B. Cory's work of this title (sm. 4to, Boston, 1897).

530. Hartert on the Podargidæ, Caprimulgidæ and Macropterygidæ. <*Auk*, XIV, pp. 419–420, Oct., 1897.

Review of Ernest Hartert's monographs of these families (in 'Das Tierreich,' Lief. 1).

531. Papers on Economic Ornithology. <*Auk*, XIV, pp. 420–422, Oct., 1897.

Review of S. D. Judd's 'Methods in Economic Ornithology with Special reference to the Catbird'; F. E. L. Beal's 'The Blue Jay and its Food,' and his 'Some Common Birds in their Relation to Agriculture'; and Dr. T. S. Palmer's Extermination of Noxious Animals by Bounties.'

532. Whitlock's Review of Herr Gätke's Views on the Migration of Birds. <*Auk*, XIV, pp. 422–424, Oct., 1897.

Review of F. B. Whitlock's work on Herr Gätke's 'Views,' etc. (8vo, London, 1897), with excerps from his criticisms of Herr Gätke's theories and assumptions.

533. Suchetet on Hybrids among Wild Birds. <*Auk*, XIV, pp. 424–425, Oct., 1897.

Review of the completed work 'Des Hybrides à l'état sauvage,' by André Suchetet. (Large 8vo, Paris, 1897; see *supra*, Nos. 407, 475).

534. Birds and Millinery. <*New York Times*, Nov. 21, 1897.

Reply to an editorial on this subject in the *New York Times*, of Nov. 19, 1897.

535. An Ornithologist's Plea. <*New York Times*, Nov. 25, 1897.

Reply to criticism by 'A. B. C.' on the work of the Audubon Societies, in the *New York Times* of Nov. 22, 1897.

536. Heron Farming for Aigrettes. <*New York Times*, Nov. 29, 1897.

Exposure of the report of an alleged heron farm in Tunis, cited in the *New York Times* of Nov. 23 and 26, 1897.

1898.

537. Dixon's Migration of Birds. <*Auk*, XV, pp. 67–70, Jan., 1898.

Review of Charles Dixon's 'The Migration of Birds,' etc., "amended" edition (8vo, London, 1897). Critical notice of his theories of the origin and causes of migration. (*Cf. supra*, No. 411.)

538. Marsh on 'The Affinities of *Hesperornis*.' <*Auk*, XV, p. 70, Jan., 1898.

Critical notice of O. C. Marsh's paper in *Amer. Journ. Sci.*, III, pp. 347–348, April, 1897.

539. Stone on the Genus *Sturnella*. <*Auk*, XV, p. 70, Jan., 1898.

Notice of Witmer Stone's paper in *Proc. Acad. Nat. Sci. Philadelphia*, 1897, pp. 146–152.

540. The Proper Name of the Western Horned Owl. <*Auk*, XV, p. 71, Jan., 1898.

Review of Witmer Stone's paper on this subject in *Amer. Nat.*, March, 1897, p. 236.

541. Baskett's 'Story of the Birds.' <*Auk*, XV, pp. 71–72, Jan., 1898.

Review of James Newton Baskett's work of this title (12mo, New York, 1897).

542. Grinnell on the Birds of Santa Barbara, San Nicolas and San Clemente Islands, California. <*Auk*, XV, p. 73, Jan., 1898.

Notice of paper by J. Grinnell in *Publ. No. I, Pasadena Acad. Sci.*, pp. 26, Aug., 1897.

543. [New York Zoölogical Society.] <*Auk*, XV, pp. 79–80, Jan., 1898.

Its plans and progress.

544. 'Audubon and His Journals.' <*Auk*, XV, pp. 198–205, April, 1898.

Review of Miss Maria R. Audubon's work of this title (2 vols., 8vo, New York, 1897).

545. Miss Merriam's 'Birds of Village and Field.' <*Auk*, XV, p. 206, April, 1898.

Review of Miss Florence A. Merriam's work of this title (12mo, Boston and New York, 1898).

546. 'Hair and Feathers.' <*Auk*, XV, p. 207, April, 1898.

Review of paper by J. S. Kingsley of this title in *Amer. Nat.*, XXXI, pp. 767–777, figs. 1–14, Sept., 1897.

547. Bauer on the Birds of the Galapagos Archipelago. <*Auk*, XV, pp. 207–208, April, 1898.

Review of G. Bauer's paper on Galapagos birds in *Amer. Nat.*, XXXI, 1897, pp. 777–784.

548. Two New Popular Bird Books. <*Auk*, XV, pp. 275–278, July, 1898.

Review of W. E. D. Scott's 'Bird Studies' (4to, 1898), and of A. C. Apgar's 'Birds of the United States, east of the Rocky Mountains' (sm. 8vo, New York, 1898).

549. Cory's Ducks, Geese and Swans. <*Auk*, XV, pp. 278–279, July, 1898.

Review of C. B. Cory's 'How to know the Ducks, Geese and Swans of North America' (sm. 4to, Boston, 1897).

550. Worcester and Bourn's Contributions to Philippine Ornithology. <*Auk*, XV, p. 284, July, 1898.

Review of their paper on Philippine birds in *Proc. U. S. Nat. Mus.*, XX, No. 1143, pp. 549–625, 1898.

551. Butler's Birds of Indiana. <*Auk*, XV, pp. 335–336, Oct., 1898.

Review of A. W. Butler's descriptive catalogue, etc., in Rep. State Geologist of Indiana for 1897, pp. 515–1187.

552. Blanford's 'Birds of British India.' <*Auk*, XV, p. 336, Oct., 1898.

Review of Vol. IV of 'The Fauna of British India.'

553. Gurney's 'The Economy of the Cuckoo.' <*Auk*, XV, pp. 337–338, Oct., 1898.

Review of a paper by J. H. Gurney, published in *Norfolk and Norwich Nat. Soc.*, VI, pp. 365–384.

554. Eastman on Struthious Birds. <*Auk*, XV, pp. 338–339, Oct., 1898.

Review of paper by C. R. Eastman, in *Bull. Mus. Comp. Zoöl.*, XXXII, No. 7, pp. 127–144, 1898.

555. Bangs on Birds from Colombia. <*Auk*, XV, p. 339, Oct., 1898.

Notice of two papers by Outram Bangs, in *Proc. Biol. Soc. Washington*, XV, 1898, pp. 131–144, 157–160.

556. Nelson on New Birds from Mexico. <*Auk*, XV, pp. 339–340, Oct., 1898.

Notice of two papers by E. W. Nelson, in *Proc. Biol. Soc. Washington*, XV, 1898, pp. 5–11, 57–68.

1899.

557. Torrey's 'A World of Green Hills.' <*Auk*, XV, pp. 86–87, Jan., 1899.

Review of Bradford Torrey's work of this title (16mo, Boston and New York, 1898).

558. Mrs. Maynard's Birds of Washington. <*Auk*, XVI, pp. 87–88, Jan., 1899.

Review of Mrs. L. W. Maynard's 'Birds of Washington and Vicinity,' etc. (8vo, Washington, 1898).

559. Blanchan's 'Birds that Hunt and are Hunted.' <*Auk*, XVI, pp. 88–89, Jan., 1899.

Review of Neltjie Blanchan's work of this title (8vo, New York, 1898).

560. Huntington's 'In Brush, Sedge, and Stubble.' <*Auk*, XVI, pp. 89, 372–373, Jan. and July, 1899.

Notice of Parts I and II of Dwight W. Huntington's work thus entitled (folio, Cincinnati, 1898).

561. Oberholser on the Wrens of the Genus *Thryomanes*. <*Auk*, XVI, pp. 89–90, Jan., 1899.

Notice of H. C. Oberholser's paper in *Proc. U. S. Nat. Mus.*, XXI, No. 1153, pp. 421–450, Nov., 1898.

562. Bangs on Birds from Colombia. <*Auk*, XVI, pp. 90–91, Jan., 1899.

Notice of his third paper on Colombian birds (*Proc. Biol. Soc. Washington*, XII, 1898, pp. 171–192.

563. Proceedings of the Indiana Academy of Science. <*Auk*, XVI, p. 91, Jan., 1899.

Notice of the ornithological papers in the volume for the year 1897.

564. The Osprey. <*Auk*, XVI, pp. 95–213, Jan. and April, 1899.

The transfer of this journal from New York to Washington, changes in its management and editorial control, and comment (*l. c.*, p. 213) on Dr. Gill's views on the classification of birds.

565. [New York Zoölogical Park.] <*Auk*, XVI, p. 96, Jan., 1899.

Further account of progress.

566. Volume XXVI of the British Museum Catalogue of Birds. <*Auk*, XVI, pp. 198–203, April, 1899.

Critical review of the volume (by R. Bowdler Sharpe and W. R. Ogilvie-Grant), and a general survey of the whole series (27 volumes) of the 'Catalogue' (1874–1898).

567. [Chuck-will's-widow in Kansas.] <*Auk*, XVI, p. 187, footnote, April, 1899.

Second authentic record for Kansas. (See further, *Auk*, XVII, p. 175.)

568. Evans's 'Birds.' <*Auk*, XVI, p. 203, April, 1899.

Notice of the Bird volume (Vol. IX) by A. H. Evans, of the Cambridge Natural History Series.

569. 'Birds.' <*Science*, N. S., IX, pp. 647–648, May 5, 1899.

Review of the volume 'Birds,' by A. H. Evans, Cambridge Natural History Series, Vol. IX.

570. Von Ihering's Birds of São Paulo, Brazil. <*Auk*, XVI, pp. 203–204, April, 1899.

Review of Dr. Von Ihering's 'As Aves do Estado de S. Paulo,' in *Revista do Museu Paulista*, III, 1899, pp. 113–476.

571. Dearborn's Birds of Belknap and Merrimac Counties, New Hampshire. <*Auk*, XVI, p. 204, April, 1899.

Notice of Ned Dearborn's 'Preliminary List,' etc. (doctorial thesis, 8vo, Durham, New Hampshire College, 1898).

572. Nash's 'The Birds of Ontario in Relation to Agriculture.' <*Auk*, XVI, pp. 204–205, April, 1899.

Notice of Charles W. Nash's work of this title (8vo, Toronto, 1898).

573. Stejneger on the Birds of the Kurile Islands. <*Auk*, XVI, p. 205, April, 1899.

Notice of Dr. L. Stejneger's paper of this title (*Proc. U. S. Nat. Mus.*, XXI, pp. 269–296).

574. Clark on 'The Feather Tracts of North American Grouse and Quail.' <*Auk*, XVI, pp. 205–206, April, 1899.

Paper by Hubert Lyman Clark, in *Proc. U. S. Nat. Mus.*, XXI, pp. 641–654, pll. xlvii–xlix.

575. ['Bird-Lore.'] <*Auk*, XVI, pp. 212–213, April, 1899.

Notice of the inception, character and scope of this new ornithological journal.

576. Elliot's Wild Fowl of North America. <*Auk*, XVI, pp. 288–289, July, 1899.

Review of D. G. Elliot's 'Wild Fowl of the United States and British Possessions,' etc. (8vo, New York, F. P. Harper, 1898).

577. Thompson's 'Wild Animals I have known.' <*Auk*, XVI, pp. 289–290, July, 1899.

Review of his book of this title (8vo, New York, 1899).

578. Stone on the Types of Birds in the Collection of the Academy of Natural Sciences of Philadelphia. <*Auk*, XVI, p. 290, July, 1899.

Summary of Witmer Stone's paper in *Proc. Acad. Nat. Sci. Philadelphia*, 1899, pp. 5–62.

579. New North American Birds. <*Auk*, XVI, p. 291, July, 1899.

Notice of three papers by O. Bangs, one by W. Brewster, and one by W. H. Osgood.

580. Economic Relations of Birds to Agriculture. <*Auk*, XVI, pp. 294–295, July, 1899.

Notice of papers by F. E. L. Beal, S. D. Judd, and T. S. Palmer.

581. Genera and Subgenera of the A. O. U. Check-List. <*Auk*, XVI, p. 297, July, 1899.

General comment with reference to a paper on this subject by E. Coues in *The Osprey*, III, 1899, p. 144.

582. [A Sophism anent Bird Destruction.] <*Auk*, XVI, p. 304, July, 1899.

583. Republication of Descriptions of New Species and Subspecies of North American Birds. <*Auk*, XVI, pp. 338–350, July, 1899.

Includes those given in the 'Ninth Supplement' to the A. O. U. Check-List, and also those published subsequently, down to October, 1899.

584. Montgomery on the Food of Owls. <*Auk*, XVI, pp. 363–364, Oct., 1899.

Review and summary of Thomas H. Montgomery, Jr's., paper on this subject in *Amer. Nat.*, XXXIII, July, 1899, pp. 563–572.

585. D. E. Lantz's 'Review of Kansas Ornithology.' <*Auk*, XVI, pp. 364–365, Oct., 1899.

Review of his paper in *Trans. Kansas Acad. Sci.*, 1896–97, pp. 224–276 (July, 1899).

586. The Goss Collection of Mexican and Central American Birds. <*Auk*, XVI, pp. 365–366, Oct., 1899.

Review of D. E. Lantz's paper on Col. N. S. Goss's Collection (*Trans. Kansas Acad. Sci.*, 1896–97, pp. 218–224, July, 1899).

587. C. B. Cory's 'The Birds of Eastern North America. Water Birds, Part I.' <*Auk*, XVI, pp. 366–367, Oct., 1899.

Review of the work (sm. 4to, Field Museum, Chicago, 1899).

588. Edward Knobel's 'Field Key to the Land Birds.' <*Auk*, XVI, pp. 367–368, Oct., 1899.

Review of the work (12mo, Boston, 1899).

589. Mrs. Olive Thorne Miller's 'The First Book of Birds.' <*Auk*, XVI, pp. 368–369, Oct., 1899.

Review of the work (12mo, Boston, 1899).

590. Oberholser on Untenable Names in Ornithology. <*Auk*, XVI, p. 370, Oct., 1899.

Notice of H. C. Oberholser's paper of this title (*Proc. Acad. Nat. Sci. Philadelphia*, 1899, pp. 201–216).

591. Gurney and Gill on the Age to which Birds Live. <*Auk*, XVI, pp. 370–372, Oct., 1899.

Review of H. Gurney's (*Ibis*, Jan., 1899, pp. 19–42) and Theodore Gill's (*Osprey*, III, 1899, pp. 157–160) papers on this subject.

1900.

592. The Little Black Rail. <*Auk*, XVII, pp. 1–8, pl. i, Jan., 1900.

General history of *Porzana jamaicensis.*

593. Merriam's Biological Survey of Mount Shasta. <*Auk*, XVII, pp. 73–74, Jan., 1900.

Review of 'Results of a Biological Survey of Mount Shasta, California,' by C. Hart Merriam (*N. Amer. Fauna*, No. 16, Oct. 18, 1899).

594. Palmer's 'The Avifauna of Pribilof Islands.' <*Auk*, XVII, pp. 74–76, Jan., 1900.

Review of Wm. Palmer's paper of this title in Jordan's 'The Fur Seal Islands of the North Pacific' (Part III, 1899, pp. 355–431, pll. xxxviii–xli).

595. Howe and Sturtevant's 'Birds of Rhode Island.' <*Auk*, XVII, pp. 76–77, Jan., 1900.

Review of the work (8vo, Newport, 1899).

596. Cory's 'The Birds of Eastern North America. Part II, Land Birds.' <*Auk*, XVII, p. 78, Jan., 1900.

Review of C. B. Cory's work of this title (sm. 4to, Field Museum, Chicago, 1899).

597. 'Avium Generum Alphabeticus.' <*Auk*, XVII, pp. 78–79, Jan., 1900.

Review of the work, published by the British Ornithologists' Club, (*Bulletin*, Vol. IX, pp. 1–31, 1899).

598. Sharpe's 'Hand-List of the Genera and Species of Birds.' <*Auk*, XVII, pp. 79–81, Jan., 1900.

Review of Vol. I (8vo, British Museum, 1899).

599. Dubois's 'Synopsis Avium.' <*Auk*, XVII, p. 81, Jan., 1900.

Notice of fasc. i, of the work (4to, Bruxelles, 1899).

600. Salvadori and Festa on Birds of Ecuador. <*Auk*, XVII, pp. 81–82, Jan., 1900.

Review of their 'Viaggio del Dr. Enrico Festa nell' Ecuador' (in *Boll. Mus. Zool. ed Anat. comp. d. R. Univ. di Torino*, Vol. XV, pp. 1–31, 1899).

601. Kellogg's List of Biting Lice (Mallophaga) taken from North American Birds and Mammals. <*Auk*, XVII, pp. 82–84, Jan., 1900.

Review of Vernon L. Kellogg's paper in *Proc. U. S. Nat. Mus.*, XXII, No. 1183, 1889, pp. 39–100.

602. Lange's 'Our Native Birds, How to Protect them and Attract them to our Homes.' <*Auk*, XVII, pp. 85–86, Jan., 1900.

Review of D. Lange's work of this title (12mo, New York and London, 1899).

603. H. A. Macpherson's 'History of Fowling.' <*Auk*, XVII, pp. 85–86, Jan., 1900.

Review of the work (large 8vo, Edinburgh, 1899).

604. Bumpus on 'The Elimination of the Unfit.' <*Auk*, XVII, p. 87, Jan., 1900.

Review of Hermon C. Bumpus's paper entitled "The Elimination of the Unfit as illustrated by the Introduced Sparrow *Passer domesticus*" (in Biol. Lectures of the Marine Biol. Lab., Wood's Holl, Mass., Sess. of 1897–1898, pp. 209–226.

605. Whitman on 'Animal Behavior.' <*Auk,* XVII, pp. 87–88, Jan., 1900.

Review of C. O. Whitman's paper in Biol. Lect. of the Marine Biol. Lab., Wood's Holl, Mass., 1898 (1899), pp. 285–338.

606. [Bird Destruction for Millinery Purposes.] <*Auk,* XVII, pp. 94–96, Jan., 1900.

Extended comment on the subject.

607. Keeler's 'Bird Notes Afield.' <*Auk,* XVII, pp. 180–181, April, 1900.

Review of Charles A. Keeler's work of this title (8vo, San Francisco, 1899).

608. Russell on Birds of the Northwest Territory. <*Auk,* XVII, pp. 181–182, April, 1900.

Review of Frank Russell's 'Explorations in the Far North' (8vo, University of Iowa, 1899. Birds, pp. 253–270).

609. Loomis on California Water Birds. <*Auk,* XVII, pp. 182–183, April, 1900.

Review of Part IV of L. M. Loomis's series of papers under this general title (*Proc. California Acad. Sci.*, 3d Ser., II, pp. 277–322, Feb. 12, 1900).

610. Witmer Stone on 'The Summer Molting Plumage of Certain Ducks.' <*Auk,* XVII, pp. 183–184, April, 1900.

Review of his paper (*Proc. Acad. Nat. Sci. Philadelphia*, 1899, pp. 467–472).

611. Stark's Birds of South Africa. <*Auk,* XVII, pp. 189–190, April, 1900.

Review of Vol. I of Arthur C. Stark's 'Birds of South Africa' (8vo, London, 1900).

612. Rothschild and Hartert's 'Review of the Ornithology of the Galapagos Islands.' <*Auk,* XVII, pp. 300–303, July, 1900.

Review of their paper on this subject (*Novitates Zoologicæ*, VI, 1899, pp. 85–205).

613. Salvadori and Festa on Birds of Ecuador. <*Auk,* XVII, p. 303, July, 1900.

Review of Parts II and III of their paper on the birds of Ecuador. (See *supra*, No. 600.)

614. Oberholser on Birds from Central Asia, Madagascar, and Santa Barbara Islands, Cal. <*Auk,* XVII, pp. 304–305, July, 1900.

Review of three papers by H. C. Oberholser (*Proc. U. S. Nat. Mus.*, XXII, Nos. 1195–1197, pp. 205–234).

615. Dugmore's 'Bird Homes.' <*Auk,* XVII, pp. 306–307, July, 1900.

Review of A. R. Dugmore's work thus entitled (8vo, New York, 1900).

615*a*. 'Bird Homes.' <*New York Evening Post,* June 26, 1900.

Review of A. R. Dugmore's work of this title. (See *supra*, No. 615.)

616. Henry J. Pearson's 'Beyond Petsora Eastward.' <*Auk,* XVII, pp. 307–308, July, 1900.

Review of the work (royal 8vo, London, 1899).

617. Collett and Nansen's Birds of the Norwegian North Polar Expedition. <*Auk,* XVII, pp. 308–310, July, 1900.

Review of section IV, 'An account of the Birds,' of the Norwegian North Polar Expedition of 1893–1896 (4to, Christiania, London, etc., 1899).

618. Palmer on Legislation for the Protection of Birds. <*Auk*, XVII, pp. 314–315, July, 1900.

Résumé of Dr. T. S. Palmer's report on the subject, forming *Bull. No. 12* of the Division of the Biological Survey.

619. [The Outlook for Bird Protection in North America.] <*Auk*, XVII, pp. 323–324, July, 1900.

620. 'Aptosochromatism.' <*Auk*, XVII, pp. 327–336, Oct., 1900.

Apropos of papers by F. J. Bertwell, Dr. A. P. Chadbourne, and J. Lewis Bonhote on the alleged change of color in feathers without moult.

621. North American Birds Collected at Santa Marta, Colombia. <*Auk*, XVII, pp. 363–367, Oct., 1900.

43 species, 16 of them here recorded for the first time from Colombia.

621*a*. List of Birds collected in the District of Santa Marta, Colombia, by Mr. Herbert H. Smith. <*Bull. Amer. Mus. Nat. Hist.*, XIII, pp. 117–183, Aug. 25, 1900.

An annotated list of 388 species. Spp. nov.: (1) *Odontophorus atrifrons*, p. 127; (2) *Myiobis assimilis*, p. 144; (3) *Ochthœca jesupi*, p. 151; (4) *Ochthœca olivacea*, p. 152; (5) *Attila parvirostris*, p. 153; (6) *Attila rufipectus*, p. 153; (7) *Grallaria bangsi*, p. 159; (8) *Hylophilus brunneus*, p. 171.

622. Beyer's 'The Avifauna of Louisiana.' <*Auk*, XVII, pp. 392–393, Oct., 1900.

Review of George E. Beyer's annotated list of the Birds of Louisiana (*Proc. Louisiana Soc. Naturalists*, 1897–1899 [1900]. Birds, pp. 1–45 of reprint).

623. Burns's 'A Monograph of the Flicker.' <*Auk*, XVII, pp. 393–394, Oct., 1900.

Review of Frank L. Burns's paper in *Wilson Bulletin*, April, 1900, pp. 1–82.

624. John Macoun's 'Catalogue of Canadian Birds.' <*Auk*, XVII, pp. 394–395, Oct., 1900.

Review of Part I of this work (8vo, Ottawa, 1900).

625. Proceedings of the Delaware Valley Ornithological Club. <*Auk*, XVII, p. 395, Oct., 1900.

Synopsis of No. III, for the years 1898 and 1899 (1900).

626. Economic Ornithology. <*Auk*, XVII, pp. 396–399, Oct., 1900.

Review of various circulars and reports by Dr. T. S. Palmer, issued by the U. S. Department of Agriculture.

627. The Birds of Celebes and Neighboring Islands. <*Science*, N. S., XII, pp. 223–225, Aug. 10, 1900.

Review of A. B. Meyer and L. W. Wigglesworth's work of this title (2 vols., 4to, Berlin, 1898).

628. Meyer and Wigglesworth's 'Birds of Celebes.' <*Auk*, XVII, 1900, pp. 399–401.

(See *supra*, No. 627.)

1901.

629. Joseph Grinnell's 'Birds of the Kotzebue Sound Region.' <*Auk*, XVIII, pp. 119–120, Jan., 1901.

Review of the paper (*Pacific Coast Avifauna*, No. 1, pp. 1–80, and map).

630. Sharpe's 'Hand-List of the Genera and Species of Birds. <*Auk*, XVIII, pp. 120–121, Jan., 1901.

Review of Vol. II. (See *supra*, No. 598.)

631. Dubois's 'Synopsis Avium.' <*Auk*, XVIII, pp. 121–122, Jan., 1901.

Review of fascicles II–IV. (See *supra*, No. 599.)

632. G. E. Shelley's 'Birds of Africa.' <*Auk*, XVIII, p. 122, Jan., 1901.

Review of Vols. I and II (4to, London, 1896 and 1900).

633. Republication of Descriptions of New Species and Subspecies of North American Birds. No. 2. <*Auk*, XVIII, pp. 172–179, April, 1901.

Includes the forms described in 1900. (See *supra*, No. 583.)

634. Richard M. Barrington's 'The Migration of Birds at Irish Light Stations.' < *Auk*, XVIII, pp. 205–206, April, 1901.

Review of the work (8vo, London and Dublin, 1900).

635. Gätke's 'Heligoland.' Second German Edition. <*Auk*, XVIII, pp. 206–207, April, 1901.

Brief notice.

636. Collett on the Skull and Auricular Openings in North European Owls. <*Auk*, XVIII, p. 207, April, 1901.

Brief notice of Dr. R. W. Shufeldt's English translation in *Journal of Morphology*, XVII, 1900, pp. 119–176, pll. xv–xx.

637. Merriam and Preble on the Summer Birds of Western Maryland. <*Auk*, XVIII, p. 208, April, 1901.

Review of their paper in the Alleghany County Report of Maryland Geol. Surv., 1900, pp. 291–307, Nov. 1900.

638. Loomis on California Water Birds. <*Auk*, XVIII, pp. 208–209, April, 1901.

Review of No. V, of this series of papers (*Proc. California Acad. Sci.*, 3d ser., II, pp. 349–363, Nov. 24, 1900).

639. Eaton's 'Birds of Western New York.' <*Auk*, XVIII, pp. 212–213, April, 1901.

Review of Elon Howard Eaton's paper in *Proc. Rochester Acad. Sci.*, IV, 1901, pp. 1–64.

640. The Question of the Generic Name *Gavia*. <*Auk*, XVIII, pp. 270–271, April, 1901.

S. G. Gmelin's use of the name *Gavia* is merely a citation from Brisson, without nomenclatural significance, contrary to the opinion of Dr. Anton Reichenow.

641. Norton on Birds from Labrador. <*Auk*, XVIII, p. 277, April, 1901.

Review of Arthur H. Norton's paper in *Proc. Portland Soc. Nat. Hist.*, II, 1901, pp. 139–158, pl. ii.

642. Howe and Allen's 'Birds of Massachusetts.' <*Auk*, XVIII, p. 278, July, 1901.

Review of R. H. Howe and G. M. Allen's work (8vo, Cambridge, Mass., 1901).

643. R. M. Strong's 'Quantitative Study of Variation in the Smaller North American Shrikes.' <*Auk*, XVIII, pp. 283–284, July, 1901.

Review of his paper of this title in *Amer. Nat.*, XXXV, April, 1901, pp. 271–298.

644. Stone 'On Moult and Alleged Color-change in Birds.' <*Auk*, XVIII, p. 284, July, 1901.

Review of Witmer Stone's paper in *The Ibis*, April, 1901, pp. 177–183.

645. Selous's 'Bird Watching.' <*Auk*, XVIII, pp. 408–409, Oct., 1901.

Review of Edmund Selous's work (8vo, London, 1901).

646. The Woodpeckers. By Fannie Hardy Eckstorm. <*The Nation*, LXXII, No. 1867, April 11, 1901. Also *Auk*, XVIII, p. 210, April, 1901.

Review of the work (square 12mo, Boston, 1901).

1902.

647. 'Der Gesang der Vögel, seine anatomischen und biologischen Grundlagen.' < *Science*, N. S., XV, pp. 98–99, Jan., 1902.

Review of Dr. Valentin Häcker's work of this title (gr. 8vo, Jena, 1900).

648. The Birds of North and Middle America. <*Science*, N. S., XV, pp. 225–226, Feb. 7, 1902.

Review of Part I of R. Ridgway's work of this title (*Bulletin 50*, U. S. National Museum).

649. Ridgway's 'Birds of North and Middle America.' <*Auk*, XIX, pp. 97–102, Jan., 1902.

Review of Part I. (See *supra*, No. 648.)

650. Scott on the Song of Baltimore Orioles in Captivity. <*Auk*, XIX, pp. 102–104, Jan., 1902.

Review of 'Data on Bird Songs,' by W. E. D. Scott (*Science*, N. S., XIV, pp. 522–526, Oct. 4, 1901).

651. Sharpe's 'Hand-List of the Genera and Species of Birds.' <*Auk*, XIX, pp. 105–106, Jan., 1902.

Review of Vol. III. (See *supra*, Nos. 598 and 630.)

652. Stark's 'Birds of South Africa.' <*Auk*, XIX, pp. 106–107, Jan., 1902.

Review of Vol. II. (See *supra*, No. 611.)

653. Osgood's Contributions to the Natural History of the Queen Charlotte Islands and the Cook Inlet Region of Alaska. <*Auk*, XIX, pp. 108–110, Jan., 1902.

Review of W. H. Osgood's paper (*North Amer. Fauna*, No. 21, Sept. 1901).

654. Verrill's 'The Story of the Cahow.' <*Auk*, XIX, pp. 110–111, Jan., 1902.

Review of A. E. Verrill's paper in *Pop. Sci. Monthly*, LX, Nov., 1901, pp. 22–30.

655. Bonhote's 'On the Evolution of Pattern in Feathers.' <*Auk*, XIX, pp. 112–114, Jan., 1902.

Review of J. L. Bonhote's speculative essay (*Proc. Zool. Soc. London*, 1901, pp. 316–326).

656. Proceedings of the Nebraska Ornithologists' Union. <*Auk*, XIX, p. 215, April, 1902.

Review of the *Proceedings* of the second annual meeting, held in Omaha, Jan. 12, 1901.

657. The American and European Herring Gulls. <*Auk*, XIX, pp. 283–284, April, 1902.

Not separable even as subspecies.

658. The name of the Zenaida Dove. <*Auk*, XIX, pp. 286–287, April, 1902.

The *Columba meridionalis* Latham, 1801, is not the *Zenaida zenaida* Bonaparte, as claimed by Forbes and Robinson (*Bull. Liverpool Mus.*, I, 1898, p. 36).

659. Campbell's Nests and Eggs of Australian Birds. <*Auk*, XIX, pp. 301–302, July, 1902.

Review of Archibald James Campbell's work of this title (2 vols., Royal 8vo, Melbourn, 1900).

660. Woodcock's Birds of Oregon. <*Auk*, XIX, pp. 302–303, July, 1902.

Review of A. R. Woodcock's 'Annotated List,' etc. (*Bull. 68, Oregon Agric. Exper. Station*, Corvallis, Oregon, Jan., 1902).

661. Proceedings of the Delaware Valley Ornithological Club. <*Auk*, XIX, pp. 303–304, July, 1902.

Review of *Cassinia* for 1901.

662. Perkins and Howe's 'Preliminary List of the Birds of Vermont.' <*Auk*, XIX, pp. 304–306, July, 1902.

Review of two papers by, respectively, George H. Perkins and Reginald Heber Howe, the latter a critique of the former.

663. Upland Game Birds. <*Amer. Nat.*, XXXVI, p. 575, Sept., 1902.

Review of Edwyn Sandys and T. S. Van Dyke's work of this title (8vo, New York and London, 1902).

664. 'Upland Game Birds.' <*Auk*, XIX, pp. 306–307, July, 1902.

Review of the work. (See *supra*, No. 663.)

665. A. S. Packard's 'Lamarck, His Life and Work.' <*Auk*, XIX, p. 306, July, 1902.

Brief review of the work (8vo, London and New York, 1901).

666. Richmond's 'List of Generic Terms Proposed for Birds during 1890–1900.' < *Auk*, XIX, pp. 307–308, July, 1902.

Review of Dr. Charles W. Richmond's paper of this title (*Proc. U. S. Nat. Mus.*, XXIV, No. 1267, May, 1902, pp. 663–729).

667. Oberholser's Review of the Horned Larks. <*Auk*, XIX, pp. 308–309, July, 1902.

Review of H. C. Oberholser's 'A Review of the Larks of the genus *Otocoris*' (*Proc. U. S. Nat. Mus.*, XXIV, No. 1271, June, 1902, pp. 801–884).

668. Ogilvie-Grant on Recently Described American Gallinæ. <*Auk*, XIX, pp. 309–311, July, 1902.

Critical notice of W. R. Ogilvie-Grant's paper entitled 'Remarks on the Species of American Gallinæ,' etc. (*The Ibis*,' pp. 233–245, April, 1902).

669. Grinnell's 'Check-List of California Birds.' <*Auk*, XIX, pp. 405–407, Oct., 1902.

Critical review of Joseph Grinnell's paper of this title (*Pacific Avifauna*, No. 3, 1902, pp. 1–92, 2 col. maps).

670. Berlepsch and Hartert on the Birds of the Orinoco Region. <*Auk*, XIX, pp. 407–408, Oct., 1902.

Review of their paper in *Novitates Zoologicæ*, IX, 1902, pp. 1–134.

671. Dubois's 'Synopsis Avium.' <*Auk*, XIX, p. 409, Oct., 1902.

Review of Parts V–X, completing Vol. I, 1899–1902. (See *supra*, Nos. 599, 631.)

672. Chapman on Birds from Alaska. <*Auk*, XIX, p. 414, Oct., 1902.

Review of F. M. Chapman's paper (*Bull. Amer. Mus. Nat. Hist.*, XVI, pp. 231–247, Aug., 1902).

673. So-called Species and Subspecies. <*Science*, N. S., XVI, pp. 383–386, Sept. 5, 1902.

674. In re *Meleagris sylvestris* Vieillot. <*Auk*, XIX, pp. 419–420, Oct., 1902.

Comment on C. Davies Sherborn's letter to *The Auk* under this title, with reference to previous editorial criticism of Mr. Ogilvie-Grant (*Auk*, XIX, p. 311).

1903.

675. The A. O. U. Check-List — Its History and its Future. <*Auk*, XX, pp. 1–9, Jan., 1903.

676. Vernacular Names of Birds. <*Auk*, XX, pp. 70–73, Jan., 1903.

Mainly in reference to the use of the hyphen in compound names, with other comment in relation to Dr. Edwin Doran's article in the same issue of *The Auk* (pp. 38–42) proposing certain rules for the construction of vernacular names of birds.

677. Ridgway's 'Birds of North and Middle America.' <*Auk*, XIX, pp. 73–76, Jan., 1903.

Review of Part II. (See *supra*, Nos. 648 and 649.)

678. Mrs. Florence Merriam Bailey's 'Handbook of Birds of the Western United States.' <*Auk*, XX, pp. 76–78, Jan., 1903.

Review of the work (12mo, Boston and New York, 1902).

679. Brewster's 'Birds of the Cape Region of Lower California.' <*Auk*, XX, pp. 78–80, Jan., 1903.

Review of the work (*Bull. Mus. Comp. Zoöl.*, XLI, Sept., 1902, pp. 1–244, and map).

680. [H. W.] Henshaw's 'Birds of the Hawaiian Islands.' <*Auk*, XX, pp. 80–81, Jan., 1903.

Review of the work (8vo, Honolulu, 1902).

681. Snodgrass and Heller on the Birds of Clipperton and Cocos Islands. <*Auk*, XX, pp. 81–82, Jan., 1903.

Review of their paper in *Proc. Washington Acad. Sci.*, IV, Sept., 1902, pp. 501–520.

682. Strong on the Development of Color in Feathers. <*Auk*, XX, pp. 86–88, Jan., 1903.

Review of Dr. R. M. Strong's 'The Development of Color in the Definitive Feather' (*Bull. Mus. Comp. Zoöl.*, XL, 1902, pp. 146–186, pll. i–ix).

683. Note on *Psittacula modesta* Cabanis. <*Auk*, XX, p. 213, April, 1903.

Psittacula modesta and *P. m. sclateri* are considered as both entitled to recognition.

684. Note on *Sylvia cœrulea* Wilson. <*Auk*, XX, pp. 216–218, April, 1903.

Sylvia cærulea Wilson not preoccupied by *Sylvia cærulea* Latham, the latter being not the name of a new species but = *Motacilla cærulea* Linn. transferred to the genus *Sylvia*. The principle involved is discussed at length.

685. Ornithological Magazines. <*Auk*, XX, pp. 219–226, April, 1903.

Reviews of *The Condor*, *Bird-Lore*, and *Wilson Bulletin* for the year 1902.

686. Pycraft on 'The Significance of the Condition of Young Birds at Birth.' <*Auk*, XX, pp. 227–228, April, 1903.

Review of W. P. Pycraft's paper on this subject (*Pop. Sci. Monthly*, LXII, Dec., 1902, pp. 108–116).

687. "Some Suggestions." <*Auk*, XX, pp. 234–235, April, 1903.

Comment on a letter to the editors of *The Auk*, under the above title, suggesting its restricting the output of faunal lists in favor of articles of more popular interest.

688. [The American Museum of Natural History.] <*Auk*, XX, pp. 246–248, April, 1903.

Account of the 'Bird Group' exhibits in the Museum.

689. The California Meeting of the American Ornithologists' Union. <*Auk*, XX, pp. 299–302, July, 1903.

Held in San Francisco, May 15–16, 1903.

690. Winkenwerder on the Migration of Birds. <*Auk*, pp. 311–313, July, 1903.

Review of H. A. Winkenwerder's 'The Migration of Birds, with special Reference to Nocturnal Flights' (*Bull. Wisconsin Nat. Hist. Soc.*, II, No. 4, Oct. 1902, pp. 177–263).

691. North American Water-Fowl. <*Auk*, XX, pp. 313–314, July, 1903.

Review of 'The Water-Fowl Family,' by L. C. Sanford, L. B. Bishop and T. S. Van Dyke. (8vo, New York, 1903. American Sportsman's Library series).

692. The 'New' Edition of Nuttall. <*Auk*, XX, p. 314, July, 1903.

Review of Montague Chamberlain's, 'A Popular Handbook of the Birds of the United States and Canada. By Thomas Nuttall.' Criticism of the work.

693. Scott's 'The Story of a Bird Lover.' <*Auk*, XX, pp. 315–316, July, 1903.

Review of W. E. D. Scott's autobiographical work thus entitled (8vo, New York, 1903).

694. Snow's Catalogue of the Birds of Kansas. <*Auk*, XX, p. 317, July, 1903.

Review of the fifth edition of Professor Francis H. Snow's Catalogue of Kansas Birds (reprinted from *Trans. Kansas Acad. Sci.*, XVIII).

695. Bangs on New Subspecies of American Birds. <*Auk*, XX, p. 320, July, 1903.

Brief mention of eight papers by Outram Bangs, in *Proc. New England Zoöl. Club*, Vols. III and IV, 1902–1903.

696. North's 'Nests and Eggs of Australian Birds.' <*Auk*, XX, pp. 321–322, July, 1903.

Notice of the second edition of A. J. North's work of this title, issued in three parts, 1901–1903.

697. Seth-Smith's Handbook of Parrakeets. <*Auk*, XX, pp. 322–323, July, 1903, and XXI, p. 96, 1904.

Review of David Seth-Smith's 'Parrakeets, a Handbook to the imported Species' (8vo, London, 1903).

698. J. Macoun's Catalogue of Canadian Birds. <*Auk*, XX, p. 441, Oct., 1903.
Review of Part II. (See *supra*, No. 624.)

699. Dresser's 'A Manual of Palæarctic Birds.' <*Auk*, XX, pp. 441–442, Oct., 1903.
Review of H. E. Dresser's work thus entitled (8vo, London, Pt. I, 1902, Pt. II, 1903).

700. Huntington's 'Our Feathered Game.' <*Auk*, XX, p. 443, Oct., 1903.
Review of Dwight W. Huntington's book on North American Game Birds (crown 8vo, New York, 1903).

701. Degen on the 'Perennial Moult' in the Australian Crow. <*Auk*, XX, pp. 444–446, Oct., 1903.
Review of 'Ecdysis, as Morphological Evidence of the Tetradactyle Feathering of the Bird's Forelimb, based on the Perennial Moult in *Gymnorhina tibicen*, by Edward Degen (*Trans. Zool. Soc. London*, XVI, pt. viii, May, 1903, pp. 347–412, pll. xxxvi–xxxviii).

702. Weed's Bibliography of Economic Ornithology. <*Auk*, XX, p. 446, Oct., 1903.
A brief notice of Clarence M. Weed's 'A Partial Bibliography of the Economic Relations of some North American Birds.' (*Tech. Bull. No. 5*, New Hampshire College Agric. Exper. Station.)

703. Stone on Birds of Southern New Mexico and Western Texas. <*Auk*, XX, p. 447, Oct., 1903.
Review of a paper by Witmer Stone and James A. G. Rehn in *Proc. Acad. Nat. Sci. Philadelphia*, 1903, pp. 16–34.

704. Bonhote's List of Birds Collected in the Bahamas. <*Auk*, XX, pp. 447–448, Oct., 1903.
Review of J. Lewis Bonhote's paper in *The Ibis*, July, 1903, pp. 273–312.

705. Ornithological Magazines: 'The Osprey.' <*Auk*, XX, pp. 451–452, Oct., 1903.
Review of Vol. VI of this magazine.

706. F. M. Chapman's 'The Economic Value of Birds to the State.' <*Auk*, XX, p. 453, Oct., 1903.
Review of the work (4to, 1903, State of New York Forest, Fish and Game Commission).

707. The A. O. U. Model Law. <*Condor*, V, pp. 157–158, Nov., 1903.
Defense of the provision for ornithological collecting.

1904.

708. Mason A. Walton's 'A Hermit's Wild Friends.' <*Auk*, XXI, pp. 87–90, Jan., 1904.
Extended criticism of the work (8vo, Boston, 1893).

709. Fisher's 'Birds of Laysan.' <*Auk*, XXI, p. 90, Jan., 1904.
Review of Walter K. Fisher's paper on 'Birds of Laysan and the Leeward Islands, Hawaiian Group' (*U. S. Fish Comm. Bulletin* for 1903, pp. 4–15).

710. Jones's 'The Birds of Ohio.' <*Auk*, XXI, pp. 90–91, Jan., 1904.
Review of Lynds Jones's work of this title (*Ohio State Acad. Sci.*, Special Papers, No. 6, 1903, pp. 141, with map).

711. Sharpe's 'Hand List of the Genera and Species of Birds.' <*Auk*, XXI, pp. 92–93, Jan., 1904.

Review of Volume IV, 1903. (See *supra*, Nos. 598, 630 and 651.)

712. Ridgway on New American Birds. <*Auk*, XXI, p. 93, Jan., 1904.

Brief notice of two important papers (in *Proc. Biol. Soc. Washington*, XVI, pp. 105–113. 167–170).

713. Ernest Hartert's 'Der Vögel der Paläarktischen Fauna.' <*Auk*, XXI, pp. 94–95, Jan., 1904.

Review of Heft I of this important work (large 8vo, Berlin, 1903).

714. The Avicultural Magazine. <*Auk*, XXI, pp. 95–96, Jan., 1904.

Review of Vol. I, new series.

715. Coues's 'Key to North American Birds.' Fifth Edition. <*Auk*, XXI, pp. 292–296, April, 1904.

Review of this posthumous edition of a celebrated work (2 vols., roy. 8vo, Boston, 1903).

716. F. M. Chapman's 'Color Key to North American Birds.' <*Auk*, XXI, pp. 296–297, April, 1904.

Review of the work (8vo, New York, 1903).

717. Mrs. Wheelock's 'Birds of California.' <*Auk*, XXI, pp. 299–300, April, 1904.

Review of the work (sm. 8vo, Chicago, 1904).

718. Kumlien and Hollister's 'The Birds of Wisconsin.' <*Auk*, XXI, pp. 301–302, April, 1904.

Review of the work (*Bull. Wisconsin Nat. Hist. Soc.*, III, 1903).

719. Silloway's 'The Birds of Fergus County, Montana.' <*Auk*, XXI, pp. 302–303, April, 1904.

Review of P. M. Silloway's paper forming *Bull. No. 1*, Fergus County High School, Lewistown, Mont., 1903.

720. Oberholser's 'Review of the Wrens of the Genus *Troglodytes*.' <*Auk*, XXI, pp. 303–304, April, 1904.

Review of H. C. Oberholser's paper of this title (*Proc. U. S. Nat. Mus.*, XXVII, No. 1354, Feb., 1904, pp. 197–210, and map).

721. Oberholser on the American Great Horned Owls. <*Auk*, XXI, pp. 304–305, April, 1904.

Review of H. C. Oberholser's paper 'A Revision of the American Great Horned Owls' (*Proc. U. S. Nat. Mus.*, XXVII, No. 1352, pp. 177–192, Feb., 1904).

722. Snodgrass and Heller on the Birds of the Galapagos Archipelago. <*Auk*, XXI, pp. 305–306, April, 1904.

Review of paper by Robert Evans Snodgrass and Edmund Heller (*Proc. Washington Acad. Sci.*, V, 1904).

723. Shufeldt on the Osteology of the Halcyones and Limicolæ. <*Auk*, XXI, p. 306, April, 1904.

Review of two papers by R. W. Shufeldt (*Amer. Nat.*, XXXVII, pp. 697–724; *Ann. Carnegie Mus.*, II, pp. 15–70, pl. i and 27 text figs.).

724. Evans's 'Turner on Birds.' <*Auk*, XXI, pp. 306–307, April, 1904.

Review of A. H. Evans's annotated translation of this early work (8vo, Cambridge, 1903).

725. Recent Papers on Economic Ornithology. <*Auk*, XXI, pp. 307–308, April, 1904.

Review of papers by Sylvester D. Judd, Edward Howe Forbush, and H. P. Attwater.

726. Summary of Game Laws for 1903. <*Auk*, XXI, pp. 308–309, April, 1904.

Based on papers by Henry Oldys, T. S. Palmer, and R. W. Williams, Jr. (Dept. Agric. Washington).

727. The case of *Megalestris* vs. *Catharacta*. <*Auk*, XXI, pp. 345–348, July, 1904.

Catharacta Brünnich preoccupied by *Catarractes* Brisson; extended comment on the case.

728. Black-capped Petrel in New Hampshire. <*Auk*, XXI, p. 383, pl. xxii, July, 1904.

729. W. T. Hornaday's 'The American Natural History.' <*Auk*, XXI, pp. 394–395, July, 1904.

Review of the work (8vo, New York, 1904).

730. Boardman's 'The Naturalist of the St. Croix.' <*Auk*, XXI, pp. 397–398, July, 1904.

Review of Samuel Lane Boardman's 'Memoir of George A. Boardman' (8vo, Bangor, 1903).

731. Pearson's 'Three Summers among the Birds of Russian Lapland.' <*Auk*, XXI, pp. 398–399, July, 1904.

Review of the work (8vo, London, 1904).

732. Swarth on the 'Birds of the Huachuca Mountains, Arizona.' <*Auk*, XXI, pp. 401–402, July, 1904.

Review of Henry S. Swarth's paper of this title, forming *Pacific Avifauna*, No. 4, 1904 (70 pp.).

733. Nelson's 'Revision of the American Mainland Species of *Myiarchus*.' <*Auk*, XXI, pp. 403–404, July, 1904.

Review of E. W. Nelson's paper of this title in *Proc. Biol. Soc. Washington*, XVII, 1904, pp. 151–160.

734. Code of Botanical Nomenclature. <*Auk*, XXI, pp. 404–405, July, 1904.

Reviewed and compared with the A. O. U. Code (*Bull. Torrey Botan. Club*, XXXI, No. 5, May, 1904.)

735. Bird Groups in the American Museum of Natural History. <*Auk*, XXI, p. 408, July, 1904.

736. So-called Nature Books. <*Auk*, XXI, p. 409, July, 1904.

Comment on the works of William J. Long and Mason A. Walton.

737. The International Catalogue of Scientific Literature. <*Auk*, XXI, pp. 494–501, Oct., 1904.

A critical review of the section "Zoology" in the first annual issue.

738. Cooke's 'Some New Facts about Bird Migration.' <*Auk*, XXI, pp. 501–503, Oct., 1904.

Review of Wells W. Cooke's paper of this title in Yearbook of U. S. Dept. Agric. for 1903, pp. 371–386.

739. G. M. Allen's 'The Birds of New Hampshire.' <*Auk*, XXI, pp. 503–505, Oct., 1904.

Review of Glover M. Allen's paper in *Proc. Manchester* (*N. H.*) *Inst. Arts and Sciences*, IV, pt. I, 1902 (1903), pp. 23–222.

740. Hartert's 'Die Vögel der Paläarktischen Fauna.' <*Auk*, XXI, pp. 505–506, Oct., 1904.

Review of Part II. (See *supra*, No. 713.)

741. Kirtland's Warbler. <*Auk*, XXI, pp. 506–507, Oct., 1904.

Review of papers on this species by Charles C. Adams and Norman A. Wood.

742. Forbush on the Destruction of Birds by the Elements. <*Auk*, XXI, pp. 507–509, Oct., 1904.

Review of E. H. Forbush's 'Special Report' on this subject (Fifty-first Ann. Rep. Massachusetts State Board of Agric.).

1905.

743. Cooke's 'Distribution and Migration of North American Warblers.' <*Auk*, XXII, pp. 91–92, Jan., 1905.

Review of Wells W. Cooke's paper of this title in *Bulletin No. 18*, Division of Biological Survey, U. S. Dept. Agric., 1904.

744. Osgood on Birds of Alaska. <*Auk*, XXII, pp. 92–93, Jan., 1905.

Review of W. H. Osgood's 'A Biological Reconnaissance of the Base of the Alaska Peninsula' (*North Amer. Fauna*, No. 24, 1904).

745. Preliminary Review of the Birds of Nebraska. <*Auk*, XXII, pp. 94–95, Jan., 1905.

Review of a brochure of this title by Lawrence Bruner, Robert H. Walcott, Myron H. Swenk (8vo, Omaha, no date).

746. Scott on the Inheritance of Song in Passerine Birds. <*Auk*, XII, pp. 95–96, Jan., 1905.

Review of two papers on this subject by W. E. D. Scott, in *Science*, Vols. XIX and XX, 1904.

747. Scott's Ornithology of Patagonia. <*Auk*, XXII, pp. 96–97, Jan., 1905.

Review of W. E. D. Scott's Ornithology of Patagonia, Part I, Rheidæ-Sphenicidæ (Vol. II of the Reports of Princeton University Expeditions to Patagonia, pp. 1–112, 1896–1899).

748. Bryan's 'A Monograph of Marcus Island.' <*Auk*, XXII, pp. 98–99, Jan., 1905.

Review of Wm. Alanson Bryan's paper in *Occas. Papers of the Bernice Pauahi Mus.*, II, No. 1, 1903.

749. J. Macoun's 'Catalogue of Canadian Birds.' <*Auk*, XXII, pp. 99–100, Jan., 1905.

Review of Part III (8vo, Ottawa, 1904).

750. Riley on the Birds of Barbuda and Antigua. <*Auk*, XXII, p. 101, Jan., 1905.

Review of J. H. Riley's 'Catalogue of a Collection of Birds from Barbuda and Antigua, British West Indies.' (*Smithsonian Misc. Coll.*, quart. issue, XLVII, 1904, pp. 277–291).

751. Dubois's 'Synopsis Avium.' <*Auk*, XXII, p. 102, Jan., 1905.

Review of fasc. xiii–xvi, completing Alphonse Dubois's work of this title (roy. 8vo, Bruxelles, 1899–1902). (See *supra*, Nos. 599, 631, and 671.)

752. Report on the Birds Collected in Northeastern Siberia by the Jesup North Pacific Expedition, with Field Notes by the Collector. <*Bull. Amer. Mus. Nat. Hist.*, XXI, 1905, pp. 219–257, July 24, 1905.

An annotated list of 127 species. Spp. nov.: (1) *Alauda buxtoni*, p. 247; (2) *Anthus anadyrensis*, p. 254.

753. Madarázs's 'An Extraordinary Discovery in Ornithology.' <*Auk*, XXII, pp. 102–103, Jan., 1905.

Review of Julius V. Madarász's paper of this title (*Ann. Mus. Nat. Hist. Hungarici*, II, 1904, pp. 396–398).

754. Shalow on Arctic Birds. <*Auk*, XXII, pp. 103–104, Jan., 1905.

Review of Dr. H. Shalow's 'Die Vögel der Arktis' in Fauna Arctica, Vol. IV, pt. 1, pp. 81–288, 1904.

755. [Bird Groups in the American Museum of Natural History.] <*Auk*, XXII, pp. 107–109, Jan., 1905.

756. [Work of the A. O. U. Committee for Bird Protection.] <*Auk*, XXII, pp. 110–112, Jan., 1905.

757. The Loggerhead Shrike in Connecticut in Winter. <*Auk*, XXII, p. 211, April, 1905.

758. Ridgway's 'The Birds of North and Middle America.' <*Auk*, XXII, pp. 219–222, April, 1905.

Review of Part III.

759. Richmond on Birds described by Pallas in 1764. <*Auk*, XXII, pp. 222–223, April, 1905.

Review of Dr. Charles W. Richmond's "Notes on the Birds described by Pallas in the 'Adumbratiuncula' of Vroeg's Catalogue." (*Smithsonian Misc. Coll.*, quart. issue, XLVII, 1905, pp. 342–347).

760. Harvie-Brown and Macpherson's 'A Fauna of the Northwest Highlands and Skye.' <*Auk*, XXII, p. 223, April, 1905.

Review of the work (sm. 4to, Edinburgh, 1904).

761. Hagmann's Concordance of Brazilian Birds described by Spix, Wied, Burmeister, and Pelzeln. <*Auk*, XXII, p. 226, April, 1905.

Review of Dr. G. Hagmann's paper in *Boletim do Museu Goeldi*, IV, 1904.

762. Raine on the Eggs of the Solitary Sandpiper. <*Auk*, XXII, pp. 100–101, Jan., 1905.

Summary of Walter Raine's paper on the 'Discovery of the Eggs of the Solitary Sandpiper' (*Ottawa Nat.*, XVIII, pp. 135–138).

763. Shufeldt on the Families and Higher Groups of Birds. <*Auk*, XXII, pp. 227–228, April, 1905.

Review of R. W. Shufeldt's 'An Arrangement of the Families and Higher Groups of Birds' (*Amer. Nat.*, XXXVIII, 1904, pp. 833–857).

764. Shelley's 'Birds of Africa.' <*Auk*, XXII, pp. 228–229, April, 1905.

Review of Vol. IV, pt. 1 (London, 1905). (See *supra*, No. 632.)

765. Report of the Chief of the Biological Survey (Dr. C. Hart Merriam) for the Year ending June 30, 1904. <*Auk*, XXII, pp. 230–232, April, 1905.

Review of the Report (in Yearbook Dept. Agric. for 1904).

766. [Founding of the National Association of Audubon Societies.] <*Auk*, XXII, p. 232, April, 1905.

767. Townsend's 'The Birds of Essex County, Massachusetts.' <*Auk*, XXII, pp. 322–323, July, 1905.

Review of Dr. C. W. Townsend's work of this title (*Mem. Nuttall Orn. Club*, No. III, 1905).

768. Job's 'Wild Wings.' <*Auk*, XXII, pp. 324–325, July, 1905.

Review of the work (8vo, Boston and New York, 1905).

769. Sharpe on the Birds of the Antarctic Regions. <*Auk*, XXII, p. 325, July, 1905.

Review of Dr. R. B. Sharpe's Report on the Birds of the 'Southern Cross' Expedition (8vo, London, British Museum, 1904).

770. Butterfield on Bird Migration. <*Auk*, XXII, pp. 325–328, July, 1905.

Review of W. Ruskin Butterfield's 'Remarks upon Theories in regard to the Migration of Birds' (in *Novitates Zool.*, XII, pp. 15–20, 1905).

771. Nelson on the Names of Certain North American Birds. <*Auk*, XXII, pp. 330–331, July, 1905.

Review of two papers by E. W. Nelson in *Proc. Biol. Soc. Washington*, XVIII, 1905.

772. The Status of certain Swainsonian Genera of Birds. <*Auk*, XXII, pp. 400–407, Oct., 1905.

Discussion of the status of the genera *Xiphorhynchus*, *Vermivora*, *Tiaris*, and *Ammodramus*, first published incidentally in a paper preceding by a few weeks their formal publication with designation of a type species for each.

773. Stephens's 'Life Areas of California.' <*Auk*, XXII, pp. 424–246, Oct., 1905.

Review and synopsis of Frank Stephens's paper of this title (*Trans. San Diego, California, Soc. Nat. Hist.*, I, No. 1, 1905).

774. Chapman on the Life History of the American Flamingo. <*Auk*, XXII, p. 426, Oct., 1905.

Review of F. M. Chapman's paper in *Bull. Amer. Mus. Nat. Hist.*, XXI, 1905, pp. 53–77, with 15 text figs.

775. Hartert's 'Die Vögel der Paläarktischen Fauna.' <*Auk*, XXII, pp. 428–429, Oct., 1905.

Review of Heft III. (See *supra*, Nos. 713, 740.)

776. Clark on the Amount of Difference that should characterize Species and Subspecies. <*Auk*, XXII, pp. 429–434, Oct., 1905.

Review of Hubert Lyman Clark's 'The Limits of Difference in Specific and Subspecific Distinctions' (in *Fifth Ann. Rep. Michigan Acad. Sci.*, 1903), discussing at length his six "fundamental rules."

777. Oberholser on the Nomenclature of Certain Genera of Birds. <*Auk*, XXII, pp. 436–437, Oct., 1905.

Review of H. C. Oberholser's paper entitled 'Notes on the Nomenclature of Certain Genera of Birds' (in *Smithsonian Misc. Coll.*, quart. issue, XLVIII, 1905).

778. Forbush on the Decrease of Birds and Means for their Protection. <*Auk*, XXII, pp. 437–438, Oct., 1905.

Review of E. H. Forbush's 'Special Report' on this subject in 52d Ann. Rep. Massachusetts State Board of Agric., 1905.

779. Supplementary Notes on Birds Collected in the Santa Marta District, Colombia, by Herbert H. Smith, with Descriptions of Nests and Eggs. <*Bull. Amer. Mus. Nat. Hist.*, XXI, pp. 275–295, Oct. 4, 1905.

Additions and corrections, pp. 275–278; descriptions of nests and eggs, 42 species, pp. 279–295.

1906.

780. The Name of the Western Sandpiper. <*Auk*, XXIII, pp. 97–98, Jan., 1906.

Ereunetes occidentalis Lawrence (1864) should give place to *Ereunetes mauri* Gundlach 1856.

781. Stejneger's 'The Birds of the Genus *Cinclus* and their geographical Distribution.' <*Auk*, XXIII, pp. 111–112, Jan., 1906.

Review of Dr. L. Stejneger's paper of this title (in *Smithsonian Misc. Coll.*, quart. issue, XLVIII, 1905).

782. Clark's 'Birds of the Southern Lesser Antilles.' <*Auk*, XXIII, pp. 113–114, Jan., 1906.

Review of Austin E. Clark's paper of this title (in *Proc. Boston Soc. Nat. Hist.*, XXXII, No. 7, 1905).

783. Beebe's 'Two Bird Lovers in Mexico.' <*Auk*, XXIII, p. 115, Jan., 1906.

Review of C. William Beebe's book of this title (8vo, Boston and New York, 1905).

784. Hartzsch's Birds of Iceland. <*Auk*, XXIII, pp. 115–116, Jan., 1906.

Review of Bernhard Hantzsch's 'Beitrag zur Kenntniss der Vogelwelt Islands (8vo, Berlin, 1905).

785. Csörgey's Ornithological Fragments from the Manuscripts of Johann Salamon von Petényi. <*Auk*, XXIII, pp. 116–117, Jan., 1906.

Review of the work (8vo, Gera-Untermhaus, 1905).

786. Harvie-Brown's 'Travels of a Naturalist in Northern Europe.' <*Auk*, XXIII, pp. 117–118, Jan., 1906.

Review of J. A. Harvie-Brown's work of this title (2 vols., 8vo, London, 1905).

787. Economic Ornithology. <*Auk*, XXIII, pp. 119–120, Jan., 1906.

Review of three papers by Vernon Bailey, S. D. Judd, and W. L. McAtee (*Bull. Biol. Survey*, Nos. 21–23, 1905).

788. Subgenera and Other Matters. <*Auk*, XXIII, pp. 122–125, Jan., 1906.

Anent Dr. Ernst Hartert's protest against certain editorial criticisms of Heft III of his 'Die Vögel der Paläarktischen Fauna,' in *The Auk*, Vol. XXII, p. 428.

789. [First Annual Meeting of the National Association of Audubon Societies.] <*Auk*, XXIII, pp. 126–128, Jan., 1906.

General account of the proceedings.

790. Thayer and Bangs on Birds from Panama. <*Auk*, XXIII, pp. 234–235, April, 1906.

Review of four papers on Panama birds, chiefly by Outram Bangs.

791. Verrill on Birds of Dominica. <*Auk*, XXIII, pp. 235–237, April, 1906.

Review of A. Hyatt Verrill's paper on the 'Avifauna of Dominica' (privately printed, unpaged, and without date).

792. Whitaker's 'The Birds of Tunisia.' <*Auk*, XXIII, pp. 238–240, April, 1906.

Review of J. I. S. Whitaker's work of this title (2 vols., large 8vo, London, 1905).

793. Ralfe's 'The Birds of the Isle of Man.' <*Auk*, XXIII, p. 240, April, 1906.

Review of P. G. Ralfe's book of this title (8vo, Edinburgh, 1905).

794. Economic Ornithology. <*Auk*, XXIII, pp. 241–242, April, 1906.

Review of papers by McAtee, Judd, Palmer, and others, published by U. S. Dept. of Agric.

795. Buturlin's 'The Breeding-grounds of the Rosy Gull.' <*Auk*, XXIII, pp. 348–349, July, 1906.

Review of S. A. Buturlin's paper of this title in *The Ibis*, 1906, pp. 131–139, 333–337.

796. Clarke's 'Birds of the South Orkney Islands.' <*Auk*, XXIII, pp. 350–351, July, 1906.

Review of Wm. Eagle Clarke's paper of this title in *The Ibis*, Jan., 1906, pp. 145–187, pll. iii–xiii.

797. Menegaux and Hellmayr on the Passeres Tracheophones of the Paris Museum. <*Auk*, XXIII, pp. 351–363, July, 1906.

Review of three papers by these authors published in 1905 and 1906 in *Bulletins* and *Memoirs* of three different French scientific societies. See further on this matter, *Auk*, XXIII, pp. 480–483, Oct., 1906.

798. 'An Eccological Survey in Northern Michigan.' <*Auk*, XXIII, p. 354, July, 1906.

Review of a Report prepared under the direction of Prof. Chas. C. Adams, University of Michigan (8vo. Lansing, Mich., 1906).

799. Brewster's 'The Birds of the Cambridge Region.' <*Auk*, XXIII, pp. 466–470, Oct., 1906.

Review of William Brewster's work of this title (*Mem. Nuttall Orn. Club*, No. IV, 1906).

800. Report on the Immigration of Birds in England and Wales in the Spring of 1905. <*Auk*, XXIII, p. 472, Oct., 1906.

Review of the Report, by a Committee of the British Ornithologists' Club (*Bull. British Orn., Club*, XVII, 1906).

801 'A Hand-List of the Birds of the Philippine Islands.' <*Auk*, XXIII, pp. 472–473, Oct., 1906.

Review of Richard E. McGregor and Dean C. Worcester's work of this title (8vo, Manila, 1906).

802. New Names for North American Birds. <*Auk*, XXIII, pp. 474–475, Oct., 1906.

Review of three papers on this subject by H. C. Oberholser and of one by Outram Bangs.

803. Palmer on Federal Game Protection. <*Auk*, XXIII, pp. 475–477, Oct., 1906.

Review of Dr. T. S. Palmer's 'Federal Game Protection — A Five Years' Retrospect' (in Yearbook of U. S. Dept. of Agric., 1905, pp. 541–562).

1907.

804. The Rio Grande Seedeater, its Status and Technical History. <*Auk*, XXIV, pp. 26–30, Jan., 1907.

Sporophila morelleti sharpei (Lawrence) shown to be a well-marked subspecies.

805. Beebe's 'The Bird.' <*Auk*, XXIV, pp. 112–113, Jan., 1907.

Review of C. William Beebe's work of this title (large 8vo, New York, 1906).

806. Hellmayr on Spix's Types of Brazilian Birds. <*Auk*, XXIV, pp. 113–114, Jan., 1907.

Review of C. E. Hellmayr's 'Revision der Spix'schen Typen Brasilianische Vögel' (*Abhandl. der K. B. Akad. Wiss.*, XXII, pp. 563–726, May, 1906).

807. Lönnberg on the Birds of South Georgia. <*Auk*, XXIV, p. 115, Jan., 1907.

Review of Einar Lönnberg's paper on birds in 'Contributions to the Fauna of South Georgia' (8vo, = *Vet. Akad. Handl.*, XI, No. 5, 1906).

808. Harvie-Brown's 'A Fauna of the Tay Basin.' <*Auk*, XXIV, pp. 115–117, Jan., 1907.

Review of J. A. Harvie-Brown's book of this title (sm. 4to, Edinburgh, 1906, pp. 1–102, pll. i–xii and 7 text figs.).

809. Chapman's 'The Warblers of North America.' <*Auk*, XXIV, pp. 227–229, April, 1907.

Review of F. M. Chapman's work of this title (8vo, New York, 1907).

810. Alphéraky's 'The Geese of Europe and Asia.' <*Auk*, XXIV, pp. 229–230, April, 1907.

Review of Sergius Alphéraky's work (4to, London, 1905).

811. Schiebel's 'Phylogeny of the Species of *Lanius*.' <*Auk*, XXIV, p. 230, April, 1907.

Review of Dr. P. G. Schiebel's 'Die Phylogenese der *Lanius* Arten' (in *Journ. f. Orn.*, 1900, pp. 1–77, 161–219, pll. 8, 7 colored).

812. Hellmayr on the Types of Little-known Neotropical Birds. <*Auk*, XXIV, p. 231, April, 1907.

Review of C. E. Hellmayr's paper in *Novitates Zool.*, XIII, July, 1906, pp. 305–352.

813. Cooke's the 'Distribution and Migration of North American Ducks, Geese and Swans.' <*Auk*, XXIV, pp. 232–233, April, 1907.

Review of Wells W. Cooke's paper (*Bull. Biol. Surv.*, No. 26).

814. Forbush's 'Useful Birds and their Protection.' <*Auk*, XXIV, pp. 234–235, April, 1907.

Review of E. H. Forbush's work thus entitled (8vo, no date, Massachusetts State Board of Agriculture).

815. Dionne's Birds of the Province of Quebec. <*Auk*, XXIV, p. 236, April, 1907.

Review of C. E. Dionne's 'Les Oiseaux de la Province de Québec' (8vo, Quebec, 1906).

816. Cole on Birds from Yucatan. <*Auk*, XXIV, p. 237, April, 1907.

Review of Dr. Leon J. Cole's 'Aves' in 'Vertebrata from Yucatan' (*Bull. Mus. Comp. Zoöl.*, L, No. 5, 1906, pp. 100–159. Birds, pp. 109–146).

817. Salvin and Godman's 'Biologia Centrali-Americana, Aves.' <*Auk*, XXIV, pp. 350–352, July, 1907.

Review of the work (4 vols., 4to, 1879–1905).

818. Proceedings of the Fourth International Ornithological Congress. <*Auk*, XXIV, pp. 352–354, July, 1907.

Review of the volume (*Ornis*, Vol. XIV, 1907).

819. Newton's 'Ootheca Wolleyana.' <*Auk*, XXIV, pp. 354–355, July, 1907.

Review of Part IV, completing the work (London, 8vo, 1905).

820. Mershon's 'The Passenger Pigeon.' <*Auk*, XXIV, pp. 355–357, July, 1907.

Review of W. B. Mershon's work of this title (8vo, New York, 1907).

821. Forbes's 'An Ornithological Cross-section of Illinois in Autumn.' <*Auk*, XXIV, pp. 358–359, July, 1907.

Review of Prof. S. A. Forbes's paper of this title (*Bull. Illinois State Laboratory of Nat. Hist.*, VII, pp. 305–335, April, 1907).

822. Berlepsch 'On the Genus *Elænia*.' <*Auk*, XXIV, p. 360, July, 1907.

Review of paper in *Proc. IVth Internat. Orn. Congress*, 1905 (1906, pp. 372–448).

823. Hartert's 'Die Vögel der Paläarctischen Fauna.' <*Auk*, XXIV, p. 362, July, 1907.

Review of Heft IV, March, 1907.

824. Woodruff's 'The Birds of the Chicago Area.' <*Auk*, XXIV, p. 363, July, 1907.

Review of Frank M. Woodruff's paper of this title (*Chicago Acad. Nat. Sci., Bull. No. VI* of the Nat. Hist. Survey, 1907).

825. Fleming on Migrations of Brünnich's Murre. <*Auk*, XXIV, p. 364, July, 1907.

Review of J. H. Fleming's 'The Unusual Migration of Brünnich's Murre (*Uria lomvia*) in Eastern North America,' in *Proc. IVth Internat. Orn. Congress*, 1905 (1906), pp. 528–543, 2 maps.

826. The Types of North American Genera of Birds. <*Bull. Amer. Mus. Nat. Hist.*, XXIII, pp. 279–284, April 5, 1907.

On the basis of the principle of elimination in opposition to the "first species rule." The introductory matter deals with the 'first reviser' and 'Brissonian genera.' All of the original species of each genus are listed, and their subsequent generic disposition shown.

827. The *Bœlophus bicolor-atricristatus* Group. <*Bull. Amer. Mus. Nat. Hist.*, XXIII, pp. 467–481, June 12, 1907.

A study in geographic variation, etc. Mr. Ridgway's conclusion that *B. bicolor* and *B. atricristatus* freely hybridize in portions of Texas is confirmed.

828. Birds of North and Middle America. <*Amer. Nat.*, LXI, pp. 672–673, Oct., 1907.

Brief notice of R. Ridgway's work of this title, Parts I–IV.

829. Ridgway's 'The Birds of North and Middle America.' <*Auk*, XXIV, pp. 450–451, Oct., 1907.

Review of Part IV, 1907.

830. Townsend and Allen's 'Birds of Labrador.' <*Auk*, XXIV, pp. 451–452, Oct., 1907.

Review of Dr. Charles W. Townsend and Glover M. Allen's paper of this title (*Proc. Boston Soc. Nat. Hist.*, XXXIII, No. 7, 1907, pp. 277–428, with map.)

831. Townsend's 'Along the Labrador Coast.' <*Auk*, XXIV, pp. 452–453, Oct., 1907.

Review of Dr. Charles W. Townsend's work of this title (8vo, Boston, no date, [1907]).

832. Clark on New Birds from Eastern Asia and the Aleutian Islands. <*Auk*, XXIV, p. 453, Oct., 1907.

Review of paper by Austin H. Clark in *Proc. U. S. Nat. Mus.*, XXXII, pp. 467–475, 1907.

833. Clarke 'On the Birds of the Weddell and adjacent Seas.' <*Auk*, XXIV, pp. 454–455, 1907.

Review of William Eagle Clarke's report on the ornithological results of the Scottish National Antarctic Expedition (*Ibis*, 1907, pp. 325–349 and map.)

834. Goeldi's 'Album de Aves Amazonicas.' <*Auk*, XXIV, p. 455, Oct., 1907.

Review of Dr. Emilio A. Goeldi's work of this title (4to, 1900–1906).

835. Herman's 'The Protection of Birds in Hungary.' <*Auk*, XXIV, pp. 456–457, Oct., 1907.

Review of Otto Herman's Report to the Hungarian Government (8vo, Budapest, 1907).

836. Anderson's 'The Birds of Iowa.' <*Auk*, XXIV, pp. 458–459, Oct., 1907.

Review of Rudolph M. Anderson's paper of this title (*Proc. Davenport Acad. Sci.*, XI, pp. 125–417, March, 1907).

837. [The Seventh International Congress of Zoölogy.] <*Auk*, XXIV, pp. 462–466, Oct., 1907.

Short account of the Boston Congress (August 19–24, 1907), with a transcript of the new Article 30 of the International Code of Zoölogical Nomenclature adopted at this Congress, here first published.

838. A List of the Genera and Subgenera of North American Birds, with their Types, according to Article 30 of the International Code of Zoölogical Nomenclature. <*Bull. Amer. Mus. Nat. Hist.*, XXIV, pp. 1–50, Dec. 26, 1907.

Historical résumé, pp. 3–12, containing comment on 'Article 30,' and on 'G. R. Gray as a First Reviser.' The determination of types by Rule 30 gives practically the same results as by elimination, the two methods giving the same type in 192 out of 197 cases where the type depends on 'subsequent designation.'

1908.

839. The Generic Names *Mycteria* and *Tantalus* of Linnæus, 1758, <*Auk*, XXV, pp. 37–38, Jan., 1908.

Tantalus, a synonym of *Mycteria*, is replaced by *Jabiru* Hellmayr, 1906.

840. Widmann's 'Birds of Missouri.' <*Auk*, XXV, pp. 89–90, Jan., 1908.

Review of Otto Widmann's 'A Preliminary Catalogue of the Birds of Missouri' (*Trans. Acad. Sci. St. Louis*, XVII, No. 1, pp. 1–288, Nov., 1907).

841. Jones on 'The Development of Nestling Feathers.' <*Auk*, XXV, p. 90, Jan., 1908.

Review of Dr. Lynds Jones's paper of this title (*Laboratory Bull. No. 13*, Oberlin College, Ohio, 1907).

842. Beebe on Geographic Variation in Birds. <*Auk*, XXV, pp. 90–93, Jan., 1908.

Review of C. William Beebe's 'Geographic Variation in Birds with especial reference to the Effects of Humidity' (in *Zoologica*, I, No. 1, 1907, pp. 1–41, with 5 half-tone pll.).

843. Braislin's Birds of Long Island, New York. <*Auk*, XXV, p. 93, Jan., 1908.

Review of Dr. Wm. C. Braislin's 'A List of the Birds of Long Island, New York' (in *Abst. Proc. Linnæan Soc. New York* for 1907, pp. 31–123).

844. Finley's 'American Birds.' <*Auk*, XXV, pp. 93–94, Jan., 1908.

Review of William Lovell Finley's work of this title (8vo, New York, 1907).

845. Ihering's Birds of Brazil. <*Auk*, XXV, pp. 94–95, Jan., 1908.

Review of Dr. Hermann von Ihering's 'As Aves do Brazil' (8vo, São Paulo, Brazil, 1907).

846. Beal on 'Birds as Conservators of the Forest,' and on 'Birds of California in Relation to the Fruit Industry.' <*Auk*, XXV, pp. 95–97, Jan., 1908.

Review of two papers on these subjects by Prof. F. E. L. Beal, the first in Rep. New York Forest, Fish and Game Comm., 1902–03; the other in *Bull. Biolog. Survey*, No. 30, 1907).

847. Riddle's 'A Study of Fundamental Bars in Feathers.' <*Auk*, XXV, p. 98, Jan., 1908.

Review of Oscar Riddle's paper of this title (*Biol. Bull.*, XII, No. 3, 1907, pp. 165–174).

848. Riddle on the Cause of Production of 'Down' and other Down-like Structures in the Plumages of Birds. <*Amer. Nat.*, XLII, pp. 693–696, Oct., 1908.

Review of paper by Oscar Riddle (in *Biolog. Bull.*, XIV, pp. 163–176, Feb., 1908).

849. Hopkins on the Bony Semicircular Canals of Birds. <*Auk*, XXV, pp. 98–99, Jan., 1908.

Review of Miss May Agnes Hopkins's paper on this subject (*Biol. Bull.*, XI, No. 5, 1906, pp. 253–265).

850. A Twenty-five Year Index to 'The Auk' and 'Nuttall Bulletin.' <*Auk*, XXV, pp. 99–100, Jan., 1908.

Review of Dr. J. Dwight's 'Index,' etc., as above (8vo, Amer. Orn. Union, 1907).

851. Annual Report of the National Association of Audubon Societies for 1907. <*Auk*, XXV, pp. 101–102, Jan., 1908.

Review of President William Dutcher's Report as above (*Bird-Lore*, IX, pp. 285–372).

852. The Case of *Hortulanus* Vieillot. <*Auk*, XXV, pp. 223–224, April, 1908.

If recognized (under the rule of priority) would supercede *Zonotrichia* Swainson, 1832.

853. Shaw's 'The China or Denny Pheasant in Oregon.' <*Auk*, XXV, pp. 241–242, April, 1908.

Review of William T. Shaw's illustrated brochure of this title (Philadelphia, 1908).

854. Giglioli's 'Avifauna Italica.' <*Auk*, XXV, p. 242, April, 1908.

Review of Prof. E. H. Giglioli's work of this title (8vo, Firenze, 1907).

855. Bonhote's 'Birds of Britain.' <*Auk*, XXV, p. 243, April, 1908.

Review of J. Lewis Bonhote's illustrated work of this title (8vo, London, 1907).

856. 'British Birds.' <*Auk*, XXV, pp. 243–244, Jan., 1908.

Review of Vol. I of the magazine *British Birds*, London, 1907.

857. Godman's 'Monograph of the Petrels.' <*Auk*, XXV, p. 244, April, 1908.

Review of Part I of F. Du Cane Godman's 'Monograph' of the Tubinares (4to, London, Dec., 1907).

858. The Work of the Biological Survey. <*Auk*, XXV, pp. 246–247, April, 1908.

Review of 'Report on Work of the Biological Survey,' by James Wilson, Secretary U. S. Department of Agriculture (*Senate Doc. No. 132*, 60th Congress, 1st Session, 1907).

859. The Case of *Strix* vs. *Aluco*. <*Auk*, XXV, pp. 288–291, July, 1908.

Strix shown to belong to the *Strix aluco* group, while *Aluco* is the proper name for the Barn Owls.

860. *Columbina* vs. *Chæmepelia*. <*Auk*, XXV, pp. 301–306, July, 1908.

Columbina Spix (1825) has priority over *Chæmepelia* Swainson (1827), and under Rule 30 of the International Code is tenable in place of *Chæmepelia*.

861. Walter on Bird Migration. <*Auk*, XXV, pp. 329–333, July, 1908.

Review of 'Theories of Bird Migration' by Hubert Eugene Walter (in *School Science and Mathematics*, April–May, 1908), with extended comment on the general subject of bird migration.

862. 'Cassinia.' <*Auk*, XXV, p. 335, July, 1908.

Review of '*Cassinia, a Bird Annual*,' for 1907.

863. Beebe on the Seasonal Changes of Color in Birds. <*Auk*, XXV, pp. 335–336, July, 1908.

Review of C. William Beebe's 'Preliminary Report on an Investigation of the Seasonal Changes of Color in Birds' (*Amer. Nat.*, XLII, pp. 34–38, Jan., 1908).

864. Contributions to Philippine Ornithology. <*Auk*, XXV, p. 336, July, 1908.

Brief mention of 14 papers on Philippine birds by several different authors, in *Philippine Journ. Science*, II, No. 5, Oct., 1907.

865. Menegaux on the Birds of the French Antarctic Expedition. <*Auk*, XXV, p. 337, July, 1908.

Review of Dr. A. Menegaux's report on the Birds of the Expédition Antarctique Française, 1903–1905, under Dr. Jean Charcot (4to, no date).

866. Reichenow on Sea-Birds. <*Auk*, XXV, pp. 337–338, July, 1908.

Review of Dr. Anton Reichenow's 'Vögel des Weltmeers: Die Meeresvögel der ostlichen Erdhälfte,' in Deutsch Südpolar-Expedition 1901–1903, Vol. X, Zoologie I, pp. 435–467, with 6 pll. and map.

867. Godman's 'Monograph of the Petrels.' <*Auk*, XXV, pp. 338–339, July, 1908.

Review of Part II. *Puffinus godmani*, nom. nov., for "*Puffinus bailloni*" tentatively employed in this work for the Madeiran Shearwater.

868. Howard's 'The British Warblers.' <*Auk*, XXV, pp. 339–440, July, 1908.

Review of Parts I and II of H. Eliot Howard's work of this title (4to, London, Feb., 1907, and March, 1908).

869. Stone on 'Methods of Recording and Utilizing Bird-Migration Data.' <*Auk*, XXV, pp. 488–490, Oct., 1908.

Review of Witmer Stone's paper of this title (*Proc. Acad. Nat. Sci. Philadelphia*, 1908, pp. 128–156).

870. Forbes's Statistical Study of the Mid-Summer Bird Life of Illinois. <*Auk*, XXV, pp. 491–492, Oct., 1908.

Review of Prof. S. A. Forbes's 'The Mid-Summer Bird Life of Illinois: A Statistical Study' (in *Amer. Nat.*, XLII, August, 1908, pp. 505–519).

871. Membership Conditions in the A. O. U. <*Auk*, XXV, pp. 495–496, Oct., 1908.

Comment (jointly with F. M. Chapman) in reply to a letter to *The Auk* under this title by John Lewis Childs.

872. Riddle on the Genesis of Fault-bars and the Cause of Alternation of Light and Dark Bars in Feathers. <*Amer. Nat.*, XLII, pp. 550–552, August, 1908.

Review of Oscar Riddle's paper of this title (*Biol. Bull.*, XIV, No. 6, May, 1908, pp. 328–370, pll. xii–xv).

873. Pennant's 'Indian Zoology.' <*Bull. Amer. Mus. Nat. Hist.*, XXIV, pp. 111–116, Feb. 7, 1908).

Collation of the several editions, and list of species in the 1769 edition, with relation to the names employed for them by subsequent authors. *Otus bakkamoena* p. 3, pl. iii, 1769, is the first use of *Otus* in a generic sense (monotypic).

1909.

874. An American's Views of Bird Migration. <*British Birds*, III, pp. 12–19, June, 1909.

Written by request of the Editor of *British Birds*.

875. Chapman's 'Camps and Cruises of an Ornithologist.' <*Auk*, XXVI, pp. 89–90, Jan., 1909.

Review of F. M. Chapman's work of this title (8vo, New York, 1908).

876. Preble on the Birds of the Athabaska-Mackenzie Region. <*Auk*, XXVI, pp. 90–92, Jan., 1909.

Review of Edward A. Preble's 'A Biological Investigation of the Athabaska-Mackenzie Region' (*North Amer. Fauna*, No. 27, Oct., 1908).

877. MacFarlane on the Birds of Northwestern Canada. <*Auk*, XXVI, pp. 92–93, Jan., 1909.

Review of Roderick MacFarlane's 'Notes on the Mammals and Birds of Northern Canada,' in Charles Mair's 'Through the Mackenzie Basin,' etc. (8vo, Toronto, 1908).

878. Knight's 'The Birds of Maine.' <*Auk*, XXVI, pp. 93–95, Jan., 1909.

Review of Ora Willis Knight's work of this title (8vo, Bangor, 1908).

879. Godman's 'Monograph of the Petrels.' <*Auk*, XXVI, p. 95, Jan., 1909.

Review of Part III.

880. Gadow's 'Through Southern Mexico.' <*Auk*, XXVI, p. 95, Jan., 1909.

Review of Dr. Hans Gadow's work of this title (8vo, London, 1908).

881. Report on the Migration of Summer Residents in England and Wales, in the spring of 1907. <*Auk*, XXVI, p. 96, Jan., 1909.

Review of the report (= *Bull. British Orn. Club*, XX, 1908).

882. The Heath Hen. <*Auk*, XXVI, pp. 96–97, Jan., 1909.

Review of Dr. George W. Field's report of the 'Eastern Pinnated Grouse or Heath Hen, in 42d Ann. Rep. Massachusetts Commissioners on Fisheries and Game for 1907.

883. Taverner and Swales on the Birds of Point Pelee, Ontario. <*Auk*, XXVI, pp. 98–99, Jan., 1909.

Review of 'The Birds of Point Pelee,' by P. A. Taverner and B. H. Swales (in *Wilson Bulletin*, Vols. XIX and XX, 1907–08).

884. Bryan on the Birds of Molokai. <*Auk*, XXVI, pp. 99–100, Jan., 1909.

Review of Wm. Alanson Bryan's 'Some Birds of Molokai.' (*Occas. Papers, B. P. Bishop Mus.*, IV, No. 2, 1908, pp. 43–86).

885. Annual Report of the National Association of Audubon Societies. <*Auk*, XXVI, pp. 100–101, Jan., 1909.

Review of President William Dutcher's Annual Report for 1908 (in *Bird-Lore*, X, 1908, pp. 277–329).

886. Report of the Chief of the Bureau of Biological Survey for 1908. <*Auk*, XXVI, pp. 101–102, Jan., 1909.

Review of Dr. C. Hart Merriam's Report (in Ann. Reports of U. S. Dept. Agriculture, 1908).

887. Richmond's List of Generic Terms proposed for Birds during 1901–1905. <*Auk*, XXVI, pp. 102–104, Jan., 1909.

Review of Dr. Charles W. Richmond's paper in *Proc. U. S. Nat. Mus.*, XXXV, pp. 583–655, Dec., 1908.

888. The Habitat Groups of North American Birds in the American Museum of Natural History. <*Auk*, XXVI, pp. 165–175, pll. i–iv, April, 1909.

Inception and history of the construction and installation of the Bird Groups in this Museum with special mention of the large 'Habitat Groups' with illustrations from photographs of the groups.

889. Grinnell's 'The Biota of the San Bernardino Mountains.' <*Auk*, XXVI, pp. 202–203, April, 1909.

Review of Joseph Grinnell's paper of this title (in *University of California Publ., Zool.*, V, No. 1, 1908, pp. 1–170, pll. i–xxiv).

890. Grinnell on Birds of Southeastern Alaska. <*Auk*, XXVI, pp. 203–204, April, 1909.

Review of 'Birds and Mammals of the 1907 Alexander Expedition to Southeastern Alaska' (*University of California Publ., Zool.*, V, No. 2, 1909, pp. 171–264, ppl. xxv, xxvi, and 4 text figs.).

891. Grinnell on Birds Observed at Salton Sea. <*Auk*, XXVI, p. 204, April, 1909.

Review of J. Grinnell's 'Birds of a Voyage on Salton Sea' (*Condor*, X, 1908, pp. 185–191).

892. Chapman on the Life-Histories of the Booby and Man-o'-War Bird. <*Auk*, XXVI, p. 205, April, 1909.

Review of F. M. Chapman's 'A Contribution to the Life-Histories' etc., as above (in *Papers from the Tortugas Laboratory of the Carnegie Inst., Washington*, II, 1908 [1909], pp. 139–151, pll. i–vi).

893. Count von Berlepsch on the Birds of Cayenne. <*Auk*, XXVI, pp. 207–208, April, 1909.

Review of Hans Graf von Berlepsch's paper 'On the Birds of Cayenne' (*Novitates Zool.*, XV, 1908, pp. 103–164, 261–324).

894. 'Cassinia.' <*Auk*, XXVI, pp. 208–209, April, 1909.

Review of *Cassinia, a Bird Manual*, for 1908.

895. Stone's 'A Review of the Genus *Piaya* Lesson.' <*Auk*, XXVI, p. 209, April, 1909.

Review of Witmer Stone's paper of this title (*Proc. Acad. Nat. Sci. Philadelphia*, LX, Pt. 3, 1908, pp. 492–501).

896. Watson's 'The Behavior of Noddy and Sooty Terns.' <*Auk*, XXVI, pp. 209–214, April, 1909.

Review of John B. Watson's paper of this title (in *Papers from the Tortugas Laboratory of the Carnegie Inst., Washington*, II, 1908 (1909), pp. 187–225, pll. i–xi), with, incidentally, a protest against the pedantic jargon of many recent works on psychology and ecology.

897. [The Roosevelt Expedition to Africa.] <*Auk*, XXVI, p. 220, April, 1909.

898. Jubilee Meeting of the British Ornithologists' Union. <*Auk*, XXVI, pp. 317–319, July, 1909.

Based on the 'Jubilee Supplement' to *The Ibis*, 1909, Vol. II.

899. Thayer and Bangs on the Birds of Guadaloupe Island. <*Auk*, XXVI, pp. 319–320, July, 1909.

Review of 'The Present State of the Ornis of Guadaloupe Island' by John E. Thayer and Outram Bangs (*Condor*, X, No. 3, 1908, pp. 101–106).

900. New North American Birds. <*Auk*, XXVI, pp. 321–322, July, 1909.

Review of three papers, respectively by E. W. Nelson, Thayer and Bangs, and J. Grinnell, with comment on the status of *Melospiza melodia gouldi* Baird and *M. m. saltonis* Grinnell.

901. Widmann on 'The Summer Birds of Shaw's Garden.' <*Auk*, XXVI, p. 322, July, 1909.

Review of Otto Widmann's paper of this title in 20th Ann. Rep. Missouri Botanical Garden (pp. 41–80, col. pl.).

902. Godman's 'Monograph of the Petrels.' <*Auk*, XXVI, p. 323, July, 1909.

Brief notice of Part IV.

903. Howard's 'The British Warblers.' <*Auk*, XXVI, pp. 323–326, July, 1909.

Extended notice of Part III. (See *supra*, No. 868.)

904. Grinnell's 'A Bibliography of California Ornithology.' <*Auk*, XXVI, p. 326, July, 1909.

Review of Joseph Grinnell's paper of this title (*Pacific Coast Avifauna*, No. 5, 1909).

905. Dawson and Bowles's 'The Birds of Washington.' <*Auk*, XXVI, pp. 328–329, July, 1909.

Review of William Leon Dawson's work of this title (2 vols., 4to, Seattle, 1909).

906. [Scientific Ornithologists of America and Bird Protection.] <*Auk*, XXVI, p. 335–336, July, 1909.

The work of the American Ornithologists' Union, collectively and by individual members, for bird protection, in refutation of statements by Wm. T. Hornaday in *Bull. New York Zool. Soc.*, June, 1909.

907. Cory's 'The Birds of Illinois and Wisconsin.' <*Auk*, XXVI, pp. 442–443, Oct., 1909.

Review of C. B. Cory's work of this title (*Publ. Field Museum, Zool.*, Vol. IX, 1909).

908. Wright's 'Birds of the Boston Public Garden.' <*Auk*, XXVI, pp. 443–444, Oct., 1909.

Review of Horace W. Wright's book of this title (12mo, Boston and New York, 1909).

909. Hersey and Rockwell on the 'Birds of the Barr Lake District, Colorado.' < *Auk*, XXVI, p. 445, Oct., 1909.

Review of L. J. Hersey and R. B. Rockwell's paper of this title (*Condor*, XI, pp. 109–122, 1909).

910. Walcott's Analysis of the Bird Fauna of Nebraska. <*Auk*, XXVI, pp. 447–448, Oct., 1909.

Review of Robert H. Walcott's paper of this title (*Proc. Nebraska Orn. Union*, IV, Part 2, pp. 25–55, pll. i–vi).

1910.

911. Sharpe's Hand-List of Birds. <*Auk*, XXVII, pp. 93–95, Jan., 1910.

Review of the concluding volume (Vol. V) of R. Bowdler Sharpe's 'A Hand-List of the Genera and Species of Birds.' (See *supra*, Nos. 598, 630, 651, and 711.)

912. Sharpe's Hand-List of Birds. <*Science*, N. S., XXXI, No. 790, pp. 265–267, Feb. 18, 1910.

Review of R. Bowdler Sharpe's 'A Hand-List of the Genera and Species of Birds.'

913. Birds of New York. <*Science*, N. S., XXXII, No. 816, pp. 247–248, Aug. 19. 1910.

Review of Part I of Elon Howard Eaton's work of this title (4to, Albany, 1909).

914. Stone's 'Birds of New Jersey.' <*Auk*, XXVII, pp. 95–97, Jan., 1910.

Review of Witmer Stone's 'The Birds of New Jersey' in Ann. Report of New Jersey State Museum for 1908 (1909).

915. G. M. Allen's List of the Birds of New England. <*Auk*, XXVII, p. 97, Jan., 1910.

Review of 'Fauna of New England. II. List of Aves,' by Glover M. Allen (*Occasional Papers, Boston Soc. Nat. Hist.*, Vol. VII).

916. Osgood's 'Biological Investigations in Alaska and Yukon Territory.' <*Auk*, XXVII, pp. 97–98, Jan., 1910.

Review of Wilfred H. Osgood's paper of this title (*North Amer. Fauna*, No. 30).

917. Seton on the Mammals and Birds of Manitoba. <*Auk*, XXVII, p. 98, Jan., 1910.

Review of Ernest Thompson Seton's 'Fauna of Manitoba,' in British Association Handbook, Winnipeg, 1909 (12mo, pp. 47).

918. Job's 'The Sport of Bird Study.' <*Auk*, XXVII, p. 102, Jan., 1910.

Review of H. K. Job's book of this title (8vo, New York, 1908).

919. Report of the Chief of the Bureau of Biological Survey, 1909. <*Auk*, XXVII, pp. 104–106, Jan., 1910.

Review of the Annual Report of the Survey for 1909.

920. Macoun's 'Catalogue of Canadian Birds.' <*Auk*, XXVII, pp. 107–108, Jan., 1910.

Review of the second edition, published in 1909. (See *supra*, Nos. 624, 698, 749).

920*a*. The Bird Plume Bill. <*Forest and Stream*, Feb. 19, 1910.

921. Thayer on Concealing Coloration in Animals. <*Auk*, XXVII, pp. 222–225, April, 1910.

Review of Gerald H. Thayer's 'Concealing Coloration in the Animal Kingdom' (4to, New York, 1909 = Feb., 1910).

922. Reed and Wright on the Birds of Cayuga Lake Basin, New York. <*Auk*, XXVII, p. 226, April, 1910.

Review of 'The Vertebrates of the Cayuga Lake Basin, N. Y.,' by Hugh D. Reed and Albert W. Wright (in *Proc. Amer. Philos. Soc.*, XLVIII, No. 193, 1909).

923. L. H. Miller on California Fossil Birds. <*Auk*, XXVII, pp. 228–230, April, 1910.

Review of three papers by Loye Holmes Miller (in *The Condor*, and *Univ. California Publ., Geol.*, V, 1909).

924. Ticehurst's 'A History of the Birds of Kent.' <*Auk*, XXVII, pp. 230–232, April, 1910.

Review of the work (8vo, London, 1909).

925. J. Grinnell on New North American Birds. <*Auk*, XXVII, pp. 232–234, April, 1910.

Review of four papers by Joseph Grinnell, with critical comment (in *Univ. California Publ., Zool.*, V).

926. Godman's 'A Monograph of the Petrels.' <*Auk*, XXVII, p. 350, July, 1910.

Review of F. Du Cane Godman's concluding Part of the 'Monograph,' with comment on the work as a whole (large 4to, 106 col. pll., London, 1907–1910).

927. Coward's 'The Vertebrate Fauna of Cheshire.' <*Auk*, XXVII, pp. 351–353, July, 1910.

A review of a noteworthy monograph of an English County (2 vols., 8vo, London, 1910).

928. 'Our Search for a Wilderness.' <*Auk*, XXVII, pp. 353–354, July, 1910.

In Venezuela and British Guiana, by Mr. and Mrs. C. William Beebe (8vo, New York, 1910).

929. Festa's 'In Darien and in Ecuador.' <*Auk*, XXVII, p. 354, July, 1910.

Review of Dr. E. Festa's 'Nel Darien e nell' Ecuador' (8vo, Torino, 1909).

930. Clark on Birds Collected or Observed in the North Pacific Ocean and adjacent Seas. <*Auk*, XXVII, pp. 357–358, July, 1910.

Review of the paper (*Proc. U. S. Nat. Mus.*, No. 1727, April 30, 1910).

931. The A. O. U. Check-List of North American Birds. Third Edition. <*Auk*, XXVII, pp. 466–472, Oct., 1910. See also *ibid.*, XXVIII, p. 122.

A statistical and historical review of the third in comparison with previous editions of the A. O. U. Check-List. (8vo, pp. 430, New York, August, 1910.)

932. Wayne's 'Birds of South Carolina.' <*Auk*, XXVII, pp. 472–474, Oct., 1910.

Review of Arthur Trezevant Wayne's work of this title (8vo, Charleston Museum, 1910).

933. A Biography of William MacGillivray. <*Auk*, XXVII, pp. 474–476, Oct., 1910.

Review of William MacGillivray's 'Life of William MacGillivray,' the great Scottish ornithologist and professor of natural history at Aberdeen University (8vo, London, 1910).

934. Collation of Brisson's Genera of Birds with those of Linnæus. <*Bull. Amer. Mus. Nat. Hist.*, XXVIII, pp. 317–335, Nov. 11, 1910.

1911.

935. 'A Naturalist in the Bahamas.' *Auk*, XXVIII, p. 126, Jan., 1911.

A memorial volume, by Henry Fairfield Osborn, containing the collected papers and a biographical sketch of John I. Northrop (8vo, New York, 1910).

936. Cooke's 'Distribution and Migration of North American Shorebirds.' <*Auk*, XXVIII, pp. 126–128, Jan., 1911.

Review of Wells W. Cooke's paper of this title (*Biological Survey Bull.* No. 35).

937. Beck's 'Water Birds of the Vicinity of Point Pinos, California.' <*Auk*, XXVIII, pp. 128–129, Jan., 1911.

Review of Rollo Howard Beck's paper of this title (*Proc. California Acad. Sci.*, ser. 4, III, 1910).

938. Townsend's 'A Labrador Spring.' <*Auk*, XXVIII, p. 129, Jan., 1911.

Review of Charles W. Townsend's book of this title (12mo, Boston, 1909).

939. Hartert on the Proper Names of Various Species of British Birds. <*Auk*, XXVIII, p. 130, Jan., 1911.

Notice of Ernst Hartert's 'Notes on [the nomenclature of] Various Species of British Birds' (*British Birds*, IV, 1910, pp. 129–136).

940. Gladstone's 'The Birds of Dumfriesshire.' <*Auk*, XXVIII, p. 131, Jan., 1911.

Review of Hugh S. Gladstone's 'The Birds of Dumfriesshire, a Contribution to the Fauna of the Solway Area' (8vo, London, 1909).

941. Gunning and Haagner's 'A Check-List of the Birds of South Africa.' <*Auk*, XXVIII, pp. 133–134, Jan., 1911.

Review, with technical comment (*Ann. Transvaal Mus.*, II, July, 1910, suppl., 84 pp.).

942. [Gregory M.] Mathews's 'The Birds of Australia." <*Auk*, XXVIII, p. 135, Jan., 1911.

Review of Parts I and II (Roy. 4to, with text figs, and colored plates, London, 1910).

943. Swarth's 'Birds and Mammals of the 1909 Alexander Alaska Expedition.' <*Auk*, XXVIII, p. 277, April, 1911.

Review of Harry S. Swarth's paper of this title in *Univ. of California Publ., Zoology*, VII, No. 2, 1911).

944. Tracy's 'Significance of White Markings in Birds of the Order Passeriformes.' <*Auk*, XXVIII, pp. 278–279, April, 1911.

Summary of Henry Chester Tracy's paper of this title (*Univ. California Publ., Zoology*, VI, No. 13, 1910).

945. The Arizona 'Passenger Pigeons.' <*Science*, N. S., XXXII, No. 841, pp. 217–219, Feb. 10, 1911.

Critique of 'Notes on the Passenger Pigeon,' by W J McGee, in *Science*, N. S., XXXII, pp. 958–964, Dec. 30, 1910, in which he claimed Passenger Pigeons were seen by him in 1905 in the "Sierra Gila, seventy-five miles southwest of Yuma and near the Mexican boundary."

946. McGee's 'Notes on the Passenger Pigeon.' <*Auk*, XXVIII, p. 289, April, 1911.

Refers to the above-cited papers by McGee and Allen. (See *supra*, No. 945.)

947. Hellmayr's 'The Birds of the Rio Madeira.' <*Auk*, XXVIII, p. 376, July, 1911.

Review of C. E. Hellmayr's paper of this title, in *Novitates Zool.*, XVII, Dec., 1910.

948. Menegaux on Birds of Ecuador. <*Auk*, XXVIII, p. 376, July, 1911.

Notice of A. Menegaux's 'Étude des Oiseaux de l'Équateur rapportés par le Dr. Rivet' (in Mission du service géograph. de l'Armée pour la mesure d'un Arc de Méridian equatorial en Amér. du Sud, 1899–1906, t. IX).

949. 'Feathers and Facts.' <*Auk*, XXVIII, pp. 377–378, July, 1911.

Review of 'Feathers and Facts: A Reply to the Feather-Trade,' etc., printed for the Royal Society for the Protection of Birds, and of other (then) recent brochures dealing with the same subject.

950. Mathews's Proposed Changes in the Nomenclature of Birds. <*Auk*, XXVIII, pp. 495–498, Oct., 1911.

Critical review of three papers by Gregory M. Mathews containing numerous proposed changes in the nomenclature of birds, in opposition (in many cases) to the International Code of Zoölogical Nomenclature.

951. Roosevelt's 'Revealing and Concealing Coloration in Birds and Mammals.' <*Auk*, XXVIII, pp. 472–480, Oct., 1911.

Review of Theodore Roosevelt's paper in *Bull. Amer. Mus. Nat. Hist.*, XXX, pp. 119–231, Aug. 23, 1911, and incidental comment on the general topic.

1912.

952. Mathews's 'The Birds of Australia.' <*Auk*, XXIX, p. 124, Jan., 1912.

Review of Part 5 of this work, completing Vol. I, with critical comment on the nomenclature adopted in the case of certain genera of Penguins, etc. (See *supra*, No. 942.)

953. [Mathews's Notes on Nomenclature.] <*Auk*, XXIX, Jan., 1912, pp. 133–135.

An editorial rejoinder to Gregory M. Mathews's paper of this title (*ibid.*, pp. 131–133), with special reference to the genus (*Colymbus* Linné (1758) and *Colymbus* Brisson (1760).

954. *Columbina* vs. *Chæmepelia.* <*Science,* N. S., XXXIII, No. 844, pp. 336–337, March 3, 1912.

Columbina griseola Spix designated as type of *Columbina.*

955. [Ridgway's] 'Birds of North and Middle America.' <*Science,* N. S., XXXV, No. 900, pp. 499–501, March 29, 1912.

Review of Part V, issued Nov. 29, 1911.

1913.

956. Scharff's 'Distribution and Origin of Life in America.' <*Auk,* XXX, pp. 283–286, April, 1913.

Extended critical review of the work (8vo, New York, 1912).

1914.

957. Birds of New York. <*Science,* N. S., XL, No. 1036, p. 677, Nov. 6, 1914.

Review of Part 2 of Elon Howard Eaton's 'Birds of New York' (4to, Albany, 1914). (See *supra,* No. 912, for review of Part I.)

1915.

958. Roosevelt's 'Through the Brazilian Wilderness.' <*Amer. Mus. Journ.,* XV, pp. 64–65, Feb., 1915.

Short review of the work (8vo, New York, 1914).

959. 'The Auk' Index, 1901–1910 <*Auk,* XXXII, p. 242, April, 1915.

Notice of the 'Ten Year Index' to *The Auk* (Vols. XVIII–XXXVII, 1901–1910).

NOTE.— The above list contains 966 titles, numbered as follows: 1–959, plus 9*a*, 318*a*, 406*a*, 475*a*, 523*a*, 621*a*, 920*a*.

GENERA, SPECIES AND SUBSPECIES DESCRIBED AS NEW OR RENAMED.

New Genera.

Palæospiza, Bull. U. S. Geol. and Geogr. Surv. Terr., IV, No. 2, May 3, 1878, p. 443, pl. i.

Myrmochanes, Bull. Amer. Mus. Nat. Hist., II, p. 95, March, 1889.

Rhopocichla, Bull. Amer. Mus. Nat. Hist., III, p. 199, Feb. 20, 1891. (Preoccupied by *Rhopocichla* Oates, 1889).

Porphyriornis, Bull. Amer. Mus. Nat. Hist., IV, p. 57, May 9, 1892.

New Species and Subspecies.

1875.

Ammodramus caudacutus var. *nelsoni,* Proc. Boston Soc. Nat. Hist., XVII, March, 1875, p. 293.

1876.

Falcinellus ridgwayi, Bull. Mus. Comp. Zoöl., III, No. 15, July, 1876, p. 335.
Gallinula garmani, Bull. Mus. Comp. Zoöl., III, No. 15, July, 1876, p. 357.

1877.

Coccygus lindeni, Bull. Essex Inst., VIII, No. 8, 1876, p. 81, Feb., 1877.

1878.

Palæospiza bella, Bull. U. S. Geol. and Geogr. Surv. Terr., IV, No. 2, p. 443, pl. i, May 3, 1878.

1888.

Ammodramus maritimus peninsulæ, Auk, V, p. 284, July, 1888.
Ammodramus maritimus sennetti, Auk, V, p. 286, July, 1888.
Tityra nigriceps, Auk, V, p. 287, July, 1888.

1889.

Anabazanops immaculatus, Bull. Amer. Mus. Nat. Hist., II, p. 96, March 22, 1889
Picolaptes obtectus, *ibid.*, II, p. 94.
Myrmochanes hypoleucus, *ibid.*, II, p. 95.
Carpophaga rusbyi, *ibid.*, II, p. 95.
Phlogopsis notata, *ibid.*, II, p. 97.
Scytalopus bolivianus, *ibid.*, II, p. 96.
Empidonax bolivianus, *ibid.*, II, p. 86.
Chiroxiphia pareola boliviana, *ibid.*, II, p. 87.
Enicornis striata, *ibid.*, II, p. 89.
Leptasthenura fuliginiceps boliviana, *ibid.*, II, p. 91.
Synallaxis griseiventris, *ibid.*, II, p. 91.
Cyclorhis flavipectus trinitatis, Bull. Amer. Mus. Nat. Hist., II, p. 131, June 17, 1889.
Thryothorus longipes, Bull. Amer. Mus. Nat. Hist., II, p. 138, June 28, 1889.
Thryothorus macrurus, *ibid.*, II, p. 137.
Platyrhynchus bifasciatus, *ibid.*, II, p. 141.
Platyrhynchus insularis, *ibid.*, II, p. 143.
Euscarthmus ochropterus, *ibid.*, II, p. 143.
Sublegatus virescens, *ibid.*, II, p. 149.
Thamnophilus doliatus mexicanus, *ibid.*, II, p. 151.

1890.

Icterus northropi, Auk, VII, p. 343, Oct., 1890; *ibid.*, VIII, pl. i, colored.

1891.

Mimocichla verrillorum, Auk, VIII, p. 217, April, 1891.
Callista margaritæ, Bull. Amer. Mus. Nat. Hist., III, p. 351, Sept. 29, 1891.
Zonotrichia capensis costaricensis, *ibid.*, III, p. 374.

1892.

Ramphocœlus atrosericeus capitalis, Bull. Amer. Mus. Nat. Hist., IV, p. 51, April 6, 1892.

Lophotriccus substriatus, *ibid.*, IV, p. 53, May 9, 1892.

Picumnus obsoletus, *ibid.*, IV, p. 55.

Porphyriornis comeri, *ibid.*, IV, p. 57.

1893.

Pygmornis chapadensis, Bull. Amer. Mus. Nat. Hist., V, p. 122, July 19, 1893.

Piaya cayana cabanisi, *ibid.*, V, p. 136.

Buteo albicaudatus sennetti, *ibid.*, V, p. 144.

1900.

Odontophorus atrifrons, Bull. Amer. Mus. Nat. Hist., XIII, p. 127, Aug. 25, 1900.

Myiobis assimilis, *ibid.*, XIII, p. 144.

Ochthœca jesupi, *ibid.*, XIII, p. 151.

Ochthœca olivacea, *ibid.*, XIII, p. 152.

Attila parvirostris, *ibid.*, XIII, p. 153.

Attila rufipectus, *ibid.*, XIII, p. 153.

Grallaria bangsi, *ibid.*, XIII, p. 159.

Hylophilus brunneus, *ibid.*, XIII, p. 171.

1905.

Alauda buxtoni, Bull. Amer. Mus. Nat. Hist., XXI, p. 247, July 24, 1905.

Anthus anadyrensis, *ibid.*, XXI, p. 254.

1908.

Puffinus godmani, nom. nov. for "*Puffinus bailloni*" of Godman's 'Monograph of the Petrels,' Auk, p. 338, July, 1908.

III. REPTILES.

1868.

1. Catalogue of the Reptiles and Batrachians found in the Vicinity of Springfield, Mass., with notices of all other species known to inhabit the State. <*Proc. Boston Soc. Nat. Hist.*, XII, pp. 171–204, Dec., 1868.

   Reptilia, 24 spp.; Amphibia, 21 spp.

1869.

2. [Appendix to the Catalogue of Reptiles and Batrachians of Massachusetts.] <*Proc. Boston Soc. Nat. Hist.*, XII, pp. 248–250, Feb., 1869.

   Notes on 7 species.

1870.

3. Notes on Massachusetts Reptiles and Batrachians. <*Proc. Boston Soc. Nat. Hist.*, XIII, 1870, pp. 260–263, April, 1870.

   Notes on 6 species, and a list of Reptiles and Batrachians found in the vicinity of New Bedford, Mass., on information furnished by Mr. R. C. Ingraham. *Malacoclemmys palustris* Agassiz new to the Massachusetts list.

1874.

4. Notes on the Natural History of Montana and Dakota. <*Proc. Boston Soc. Nat. Hist.*, XVII, pp. 33–91, Nov., 1874.

   IV. Report on the Reptiles, pp. 68–69. An annotated list of 9 species. V. Report on the Batrachians, p. 70. An annotated list of 4 species.

1908.

5. Ruthven's Variations and Genetic Relationships of the Garter-snakes. <*Amer. Nat.*, XLII, pp. 552–559, August, 1908.

   Review of A. G. Ruthven's paper (*Bull. 61, U. S. Nat. Mus.*, 1908).

IV. ZOÖGEOGRAPHY.

1. On the Mammals and Winter Birds of East Florida....and a Sketch of the Bird Faunæ of Eastern North America. <*Bull. Mus. Comp. Zoöl.*, II, No. 3, April, 1871, pp. 161–450.

Part V. On the Geographical Distribution of the Birds of Eastern North America, with special reference to the Number and Circumscription of the Ornithological Faunæ, pp. 375–425; List of Authorities, pp. 426–450.

"In accordance with the facts stated above respecting the mode of the distribution of animals and plants over the earth's surface, and the zoölogical and botanical laws of the differentiation and mutual relations of the different regions, the following primary natural history divisions may be recognized: I, an Arctic Realm;[1] II, a North Temperate Realm; III, an American Tropical Realm; IV, an Indo-African Tropical Realm; V, a South American Temperate Realm; VI, an African Temperate Realm; VII, an Antarctic Realm; VIII, an Australian Realm" (p. 380).

For eastern North America are recognized the following seven faunæ: (1) Floridian, (2) Louisianian, (3) Carolinian, (4) Alleghanian, (5) Canadian, (6) Hudsonian, (7) American Arctic. Their boundaries and their characteristic species of birds are given (pp. 387–404), and they are further considered with reference to mammals and reptiles (pp. 404–406).

The species of North American birds are considered and tabulated with reference to their geographical ranges (pp. 407–418). General remarks on the distribution and migration of the birds of the Eastern Province (pp. 418–425). A Bibliography of ornithological works and papers, or "List of Authorities," relating to North America occupies pages 426–450, geographically arranged by States and countries, and numbering 346 titles — much the largest list of papers relating to North American ornithology that appeared prior to 1878.

1878.

2. The Geographical Distribution of the Mammalia, considered in relation to the principal Ontological Regions of the Earth, and the Laws that govern the Distribution of Animal Life. <*Bull. U. S. Geol. and Geogr. Surv. Terr.*, IV, No. 2, pp. 313–337, May 3, 1878.

I. General considerations, with criticism of the life-regions proposed by Dr. P. L. Sclater and supported by Mr. Alfred R. Wallace, pp. 313–329; II. Mammalian Regions of the Globe, pp. 329–373; III. General Summary, pp. 373–377.

The primary divisions are essentially as laid down in 1871 (see above), except that a South African Temperate Realm is admitted, and Madagascar is recognized as an additional Realm, designated as the Lemurian Realm. Under these are defined regions of secondary and tertiary rank where such subdivisions seemed to be required.

1891.

3. Merriam's 'Results of a Biological Survey of the San Francisco Mountain Region and Desert of the Little Colorado, Arizona.' <*Auk*, VIII, pp. 95–98, Jan., 1891.

Review of Dr. C. Hart Merriam's paper of this title (*N. Amer. Fauna*, No. 3, Sept. 11, 1898), mainly with reference to the portion relating to 'Generalizations concerning the Distribution of Life in North America.'

---

[1] Divisions I and II are essentially the same as *Triarctic* of Heilprin (1882), renamed *Holarctic* by Heilprin in 1883, by suggestion of Alfred Newton, and since commonly known as *Holarctic* by those who reject the Sclaterian system of zoögeographical divisions. *Arctogæa* of Huxley (1868) is more comprehensive.

1892.

4. Merriam on the Life Areas of North America. <*Auk*, IX, pp. 377–382, Oct., 1892.

Synoptic review of Dr. C. Hart Merriam's 'The Geographic Distribution of Life in North America with special reference to the Mammalia' (*Proc. Biol. Soc. Washington*, VII, pp. 1–64, and map, April, 1892).

5. The Geographical Distribution of North American Mammals. <*Bull. Amer. Mus. Nat. Hist.*, IV, pp. 199–243, pll. v–viii (colored maps), Dec. 29, 1892.

Influences determining the Geographical Distribution of Life, pp. 199–203; Systematic Classification of Life Areas, pp. 203–206; Primary Life Regions, pp. 206–207 (same as in No. 2 *supra*); North Temperate Realm, pp. 207–211; The Sclaterian System, pp. 211–212; The Mammals of North America considered in relation to the North American Region and its Subdivisions, pp. 213–240; Tropical North America, pp. 240–243; Tabular Synopsis, p. 243.

1893.

6. The Geographical Origin and Distribution of North American Birds, considered in relation to Faunal Areas of North America. <*Auk*, X, pp. 97–150, pll. iii, iv (colored maps), July, 1893.

I. The Geographical Origin and Distribution of North American Birds, pp. 98–117; II. The Faunal Subdivisions of North America, considered with reference to their Relationships, Classification and Nomenclature, pp. 117–150 (tabular synopsis, p. 150).

The classification adopted is essentially the same as that recognized in 1892 (see *supra*, No. 5).

7. The Faunal Position of Lower California. <*Auk*, X, pp. 306–307, July, 1893.

In reply to comment by Dr. Merriam under this title (*l. c.*, pp. 305–306).

1905.

8. Report on the Mammals collected in Northeastern Siberia by the Jesup North Pacific Expedition. <*Bull. Amer. Mus. Nat. Hist.*, XIX, pp. 101–184, March 31, 1903.

American affinities of certain East Siberian Mammals, pp. 182–184.

9. Stephens's 'Life Areas of California.' <*Auk*, XXII, pp. 424–426, Oct., 1905.

Synoptic summary of Frank Stephens's paper of this title (*Trans. San Diego, California, Soc. Nat. Hist.*, I, No. 1, 1905, pp. 1–8 and map), with comment.

V. EVOLUTION.[1]

1871.

1. Mammals and Winter Birds of East Florida, etc. <*Bull. Mus. Comp. Zoöl.*, II, No. 3, pp. 161–450, April, 1871.

   Part III. On Individual and Geographical Variation among Birds, considered in respect to its bearing upon the Value of certain assumed Specific Characters, pp. 186–250.

   Wide range of individual variation shown to occur in a considerable number of species, with extensive tables of measurements, pp. 186–226; correlation of variations in general size, size of bill, etc., and in coloration, with differences in climatic and geographic conditions, pp. 229–242; species, varieties, and geographical races, pp. 242–250. A presentation of facts, without discussion of any theories of evolution, which appeared later (see *infra*, No. 7).

1872.

2. Notes of an Ornithological Reconnaissance, etc. <*Bull. Mus. Comp. Zoöl.*, III, pp. 113–118, July, 1872.

   In the introduction the general subject of geographical variation is discussed and new facts presented.

3. Geographical Variation in Color among North American Mammals and Birds. <*Proc. Boston Soc. Nat. Hist.*, XV, 156–159, Sept., 1872.

   Abstract of a verbal communication, summarizing the general subject, with comment by other speakers.

4. Geographical Variation in North American Birds. <*Proc. Boston Soc. Nat. Hist.*, XV, pp. 212–219, Dec., 1872.

   A general résumé of the author's studies of the subject, to that date. (Republished in *Amer. Nat.*, VIII, pp. 534–541, Sept., 1874).

1874.

5. Geographical Variation in Color in North American Squirrels. <*Proc. Boston Soc. Nat. Hist.*, XVI, pp. 276–294, May, 1874.

   The introductory matter (pp. 276–286) relates to this subject. See especially the footnote to pp. 277, 278, giving references, etc., to previous discussions of the general subject of geographical variation in mammals and birds by the author and others.

1876.

6. Geographical Variation among North American Mammals. <*Bull. Geogr. and Geol. Surv. Terr.*, II, No. 4, July 1, 1876, pp. 309–344.

   The correlation of size with geographical variation is formulated (p. 310) under the three propositions:

   "(1) Maximum physical development of the individual is attained where the conditions of environment are most favorable to the life of the species....

[1] Exclusive of numerous incidental references in many general papers and reviews.

"(2) The largest species of a group (genus, subfamily, or family, as the case may be) are found where the group to which they severally belong reaches its highest development, or where it has what may be termed its center of distribution....

"(3) The most 'typical' or most generalized representatives of a group are found also near its center of distribution, outlying forms being generally more or less 'aberrant' or specialized...."

1877.

7. The Influence of Physical Conditions in the Genesis of Species. <*Radical Review*, I, No. 1, pp. 108–140, May, 1877. (Republished, by request, in the Ann. Report of the Smithsonian Institution for 1905 (1906), pp. 375–402).

"The doctrine of natural selection, or the survival of the fittest, has recently been brought forward as the key to this complex problem, and is upheld by a large class of enthusiastic adherents, who accept it as the full solution of the whole question. By others the conditions of environment are believed to be far more influential in effecting a certain class of modifications, at least, than the necessarily precarious influence of natural selection," etc.

The direct modifying influence of environment as a factor in evolution is regarded as more potent than natural selection taken in the narrow sense of the "survival of the fittest."

1880.

8. Origin of the Instinct of Migration. <*Bull. Nutt. Orn. Club*, V, pp. 151–154, July, 1880.

1883.

9. Note on Exceptions to the Law of Increase in Size Northward among North American Birds. <*Bull. Nutt. Orn. Club*, VIII, pp. 80–82, April, 1883.

In certain genera of Oscines, for the most part tropical in distribution.

1885.

10. Sexual Selection and the Nesting of Birds. <*Auk*, II, pp. 129–139, April, 1885.

In reference to Wallace's 'Theory of Birds' Nests' (*Intellectual Observer*, July, 1867), and Dixon's 'On the Protective Colour of Eggs' (in Seebohm's Hist. Brit. Birds, Introd., pp. x–xxxvii).

1892.

11. [Variations in Vertebrated Animals.] <*Amer. Nat.*, XXVI, pp. 87–89, Jan., 1892.

Abstract of a paper presented at the meeting of the American Society of Naturalists held in Philadelphia, Dec. 29–30, 1891, in a joint discussion of 'Definite *vs.* Fortuitous Variation.'

1893.

12. Keeler on the 'Evolution of the Colors of North American Birds.' <*Auk*, X, pp. 189–195, 377–380, April and Oct., 1893.

A review of Charles A. Keeler's work of this title (8vo, San Francisco, 1893), adverse to his theories.

13. Beddard's 'Animal Coloration.' <*Auk*, X, pp. 195–199, April, 1893.

A review of Frank E. Beddard's work of this title (8vo, London and New York, 1892), with approving comment of the author's attitude respecting "protective coloration," "warning colors," "sexual coloration," and "mimicry," for which he finds little in support of these popular theories.

1897.

14. The Discrimination of Species and Subspecies. <*Science*, N. S., V, pp. 877–879, June 4, 1897.

Apropos of Dr. C. H. Merriam's 'Suggestions for a New Method of Discriminating be-between Species and Subspecies,' published in a preceding number of *Science*.

1898.

15. 'A Precise Criterion of Species.' <*Science*, N. S., VII, pp. 801–802, June 10, 1898.

A criticism of Davenport and Blankinship's paper of this title in *Science* for May 20, 1898, on the quantitive study of variation in species.

1899.

16. Davenport's Statistical Methods. <*Amer. Nat.*, XXXIII, pp. 974–975.

Criticism of 'Statistical Methods with special reference to Biological Variation' by C. B. Davenport (12mo, New York, 1899).

1902.

17. 'So-called Species and Subspecies.' <*Science*, N. S., XVI, pp. 383–386, Sept. 5, 1902.

Discussion of the subject, apropos of an article under the same title in a previous number of *Science* by Hubert Lyman Clark.

1905.

18. The Probable Origin of Certain Birds. <*Science*, N. S., XXII, No. 562, pp. 431–434, Oct. 6, 1905.

Criticism of W. E. D. Scott's resort to the 'mutation theory' of deVries, in accounting for the origin of certain North American birds (*Science*, XXIII, pp. 272–282, Sept. 1, 1905).

19. The Evolution of Species through Climatic Conditions. <*Science*, N. S., XXII, No. 569, pp. 661–668, Nov. 24, 1905.

Apropos of a paper by D. S. Jordan on 'The Origin of Species through Isolation' (*Science*, XXII, pp. 545–552).

1906.

20. Heredity and Subspecies. <*Science*, N. S., XXIII, pp. 142–145, Jan. 26, 1906.

Discussed with reference to papers by Dr. D. S. Jordan on the same subject in previous issues of *Science*.

21. 'Barriers' and 'Bionomic Barriers'; or Isolation and Non-Isolation as Bionomic Factors. <*Science*, N. S., XXIII, pp. 310–312, Feb. 23, 1906.

In reference to papers on the same subject by Jordan and Evermann in previous issues of *Science*.

1907.

22. Mutations and the Geographic Distribution of nearly related Species in Plants and Animals. <*Amer. Nat.*, XLI, pp. 653–655, Oct., 1907.

VI. NOMENCLATURE.[1]

1883.

1. On Trinomial Nomenclature. <*Zoölogist,* 3d ser., VII, pp. 97–100, March, 1883.

1884.

2. 'Are Trinomials Necessary?' <*Auk,* I, pp. 102–104, pp. 200–202, Jan. and April, 1884.

   Replies to this question, raised by Montague Chamberlain (*ibid.,* pp. 101–102, 198–200).

3. 'Can We not have a Simpler System of Nomenclature?' <*Auk,* I, pp. 297–300, July, 1884.

   A letter of this title and editorial reply.

4. "A Lay View of 'Ornithophilologicalities.'" <*Auk,* I, pp. 300–304, July, 1884.

   An editorial reply to a letter of this title.

5. Zoölogical Nomenclature. <*Auk,* I, pp. 338–353, Oct., 1884.

   On trinomial nomenclature, with particular reference to the views and practices of English ornithologists, apropos of a discussion of the "Expediency, or otherwise, of adopting Trinomial Nomenclature in Zoölogy," held in the Lecture Room of the Natural History Museum, London, July 2, 1884.

6. Stejneger on Trinomials in American Ornithology. <*Auk,* I, pp. 381–382, Oct., 1884.

   Review of Dr. L. Stejneger's paper of this title (*Proc. U. S. Nat. Mus.,* VII, 1884, pp. 70–81, July 1, 1884), in which the history of trinomials is briefly given.

1887.

7. [The Antedating of papers and works on Natural History.] <*Auk,* IV, p. 176, April, 1887, and *ibid.,* p. 270, July, 1887.

   With special reference to the 'Proceedings' and other publications of the U. S. National Museum.

1890.

8. Waterhouse's 'Index Generum Avium.' <*Auk,* VII, pp. 71–73, Jan., 1890.

   Review of F. H. Waterhouse's work of this title (8vo, London, 1889), with criticisms of omissions and defects, as the omission of all generic names founded before 1766, except Brisson's, and the giving of vernacular French names the status of proper generic terms.

9. Blanchard on the Nomenclature of Organized Beings. <*Auk,* VII, pp. 73–74, Jan., 1890.

   Review of Dr. Raphaël Blanchard's report 'De la Nomenclature des êtres organizés, presented to the International Congress of Zoölogy (Reports to the Congress, pp. 87–157, July, 1889). Comment on some of the rulings in the report.

---

[1] With reference mainly to the general subject, and exclusive of special cases (Kerr, Oken, Pennant, etc.), and of incidental references in reviews and papers devoted chiefly to other subjects.

1896.

10. Sclater on Rules for Naming Animals. <*Auk*, XIII, pp. 325–328, Oct., 1896.

Review of a paper by Dr. P. L. Sclater entitled "Remarks on the Divergencies between the 'Rules for Naming Animals' of the German Zoölogical Society and the Stricklandian Code of Nomenclature" (*Proc. Zool. Soc. London*, 1896, pp. 306–319), with critical comment on many of his positions on important points.

11. A Question of Nomenclature. <*Auk*, XIII, pp. 191–192, April, 1896.

In reply, by request, to an article of this title by H. C. Oberholser (*ibid.*, pp. 190–191).

12. A reprehensible method of determining Priority of Publication. <*Science*, N. S., IV, pp. 691–693, Nov. 16, 1896.

A criticism of resolutions on this subject adopted by the Zoölogical Section of the American Association for the Advancement of Science, August, 1895.

13. The Date of Publication. <*Science*, IV, pp. 838, 839, Dec. 4, 1896.

Further discussion of the subject, in reply to Professor Cope's defense of the American Association resolution. (See *supra*, No. 12).

1897.

14. The Merton Rules. <*Science*, N. S., VI, pp. 16–19, July 2, 1897.

Comment on the rules compiled by Lord Walsingham and John Hartley Durant, "for regulating nomenclature with a view to secure a strict application of the Law of Priority in Entomological work."

1898.

15. A Defense of Canon XL of the A. O. U. Code. <*Auk*, XV, pp. 298–302, Oct., 1898.

A reply to criticism by D. G. Elliot of Canon XL of the A. O. U. Code (*ibid.*, pp. 294–298), in defense of taking names as originally spelled, and against the emendation of names.

16. 'Truth *versus* Error.' <*Auk*, XVI, Jan., 1899, pp. 46–51.

In reply to a second article, under this title, by D. G. Elliot (*ibid.*, pp. 38–45), a rejoinder to the reply made to his first paper on the A. O. U. Canon XL. (See *supra*, No. 15.)

1902.

17. A Method of Fixing the type in Certain Genera. <*Science*, N. S., XVI, pp. 114, 115, July 18, 1902.

A rule proposed and signed by Allen, Bangs, Evermann, Gill, Howell, Jordan, C. H. Merriam, G. S. Miller, Jr., Mary Rathbun, Oldfield Thomas — a formal recognition of the principle of tautonymy.

1903.

18 ntedated Publications. <*Science*, N. S., XVIII, pp. 631–632, Nov. 13, 1903.

Relates to publications of the Field Columbian Museum of Chicago.

1904.

of Botanical Nomenclature. <*Auk*, XXI, pp. 404–405, July, 1904.

eviewed and compared with the A. O. U. Code and their general agreement noted *Torrey Botanical Club*, XXXI, No. 5, May, 1904).

1905.

20. A New Code of Nomenclature. <*Science*, N. S., XXI, pp. 428–433.

Criticism of 'A Provisional Code' by Doctors Jordan, Evermann and Gilbert, published in *The Condor*, VII, pp. 28–30, Jan., 1905.

1906.

21. The 'Elimination' and 'First Species' Methods of Fixing the Types of Genera. <*Science*, N. S., XXIV, pp. 773–779, Dec. 14, 1906.

In reply to a paper by Witmer Stone on the same subject (*Science*, XXIV, pp. 560–565, Nov. 2, 1906).

22. Subgenera and other Matters. <*Auk*, XXIII, pp. 122–125, Jan., 1906.

Relates to the non-recognition of Brissonian genera by Hartert in his 'Die Vogel der paläarktischen Fauna.

23. The Determination of Types of Genera — a Correction. <*Science*, N. S., XXIV, p. 858, Dec. 28, 1906.

Respecting the type of *Vultur* Linn.

1907.

24. The First Species Rule for determining Types of Genera — How it works in Ornithology. <*Science*, N. S., XXV, pp. 546–554, April 5, 1907.

Fourth paper in the Stone-Allen discussion of the 'First Species Question.'

25. Another Word on the *Vultur* Case. <*Science*, N. S., XXV, p. 827, May 24, 1907.

Further discussion of the first species rule.

26. Article 30 of the International Code of Zoölogical Nomenclature. <*Science*, N. S., XXVI, pp. 719–723, Nov. 22, 1907.

In hearty approval of the "new Article 30," and criticism of statements in a previous article in *Science* by Witmer Stone.

1908.

27. Another Aspect of the Species Question. <*Amer. Nat.*, XLII, pp. 592–600, Sept., 1908.

With reference to a report of the symposium held by the Botanical Society of America in Chicago, Jan. 1, 1908, and published in the *Amer. Nat.* for April, 1908, pp. 217–281.

1909.

28. Genera without Species. <*Science*, N. S., XXIX, No. 754, pp. 934–936, June 11, 1909.

With reference to previous papers on this subject by Prof. T. D. A. Cockerell in the same volume of *Science*. The position is taken that such genera should each be treated on their merits, as against the proposed non-recognition of all such genera.

29. Genera without Species. <*Science*, N. S., XXX, No. 768, pp. 365–367, Sept. 17, 1909.

A continuation of the discussion (*supra*, No. 25), with special reference to an article by Mr. A. N. Caudell.

30. Recommendations regarding the treatment of Genera without Species, etc. <*Science*, N. S., XXX, No. 744, pp. 596–597, Oct. 29, 1909.

Covers two recommendations relating to genera without species, which are the outcome of the correspondence on this subject in *Science*, and directly with Prof. T. D. A. Cockerell.

1910.

31. The Determination of Genotypes in Zoölogy. Advance separate of a paper published in the *Proceedings* of the Seventh International Zoölogical Congress, Boston meeting, August 19–24, 1907 (8vo, pp. 7; separates received Jan. 24, 1910.)

32. Opinions rendered by the International Commission on Zoölogical Nomenclature. <*Science*, N. S., XXXII, No. 820, pp. 380–382, Sept. 10, 1910.

Summary of Opinions 1–25, with comment.

1912.

33. Mathews's Notes on Nomenclature. <*Auk*, XXIX, pp. 133–135, Jan., 1912.

Rejoinder to Gregory M. Mathews's paper of this title (*Auk*, *l. c.*, pp. 131–133), with special reference to the genus *Colymbus* Linné (1758) and *Colymbus* Brisson (1760), and to other Brissonian genera.

34. Suggested Amendments to the International Code of Zoölogical Nomenclature. Amendment Proposed by Dr. J. A. Allen. <*Zool. Anz.*, XLI, pp. 38–45, Nov. 26, 1912.

Amendment to replace the 'Recommendations' following Art. 36, placing generic names on a basis similar to that of specific names under the Gratz (1910) amendment to Art. 35.

1913.

35. Propositions by J. A. Allen and T. D. A. Cockerell for Amendments to the International Code of Zoölogical Nomenclature. <*Zool. Anz.*, XLI, pp. 426–427, March 11, 1913.

Species designated by only a vernacular name may, under certain conditions, be accepted as genotypes.

VII. BIOGRAPHY.[1]

1880.

1. Thomas Mayo Brewer.<*Bull. Nutt. Ornith. Club*, pp. 102–104, April, 1880; *Proc. Amer. Acad. Arts and Sciences*, VII, pp. 378–380, 1879–80; *Proc. Boston Soc. Nat. Hist.*, XX, pp. 413–415.

1884.

2. Wilson Flagg.<*Auk*, I, p. 306, July, 1884.
3. Henry George Vennor.<*Auk*, I, p. 306, July, 1884.

1885.

4. William Wood.<*Auk*, II, p. 391, Oct., 1885.
5. Harmon Albro Atkins.<*Auk*, II, p. 391, Oct., 1885.

1886.

6. Samuel Cabot.<*Auk*, III, p. 144, Jan., 1886.
7. Eugen von Boeck.<*Auk*, III, p. 496, Oct., 1886.

1887.

8. John Maynard Wheaton.<*Auk*, IV, p. 174, April, 1887.
9. Spencer Fullerton Baird.<*Auk*, IV, pp. 358–359, Oct., 1887.

1888.

10. Joseph Bassett Holder.<*Auk*, V, p. 220, April, 1888.
11. Charles Linden.<*Auk*, V, April, pp. 220–221, 1888
12. Philip Henry Gosse.<*Auk*, V, pp. 446–447, Oct., 1888.

1889.

13. Henry Stevenson.<*Auk*, VI, pp. 79–80, Jan., 1889.
14. August Friedrich Marschall.<*Auk*, VI, p. 80, Jan., 1889.
15. Nicolas Michaïlovitch Prjevalsky.<*Auk*, VI, pp. 80–81, Jan., 1889.

---

[1] The following biographical records relate almost exclusively to more or less prominent ornithologists, and were nearly all published in *The Auk* (1884–1911, inclusive) as editorial notices, without signature. They include not only American ornithologists but all the foreign ornithologists of note who died between the above-given dates. The references given here, it should be noted, include only a part of the formal biographies and obituaries published in *The Auk* during the period named, since the biographers of all deceased fellows of the A. O. U. are appointed by the President; also those of a number of the deceased Members and Associates have been prepared at the solicitation of the editor of *The Auk*, and not by him personally; such notices, of course, are not here included.

16. Thure Kumlien. <*Auk*, VI, pp. 204–205, April, 1889.
17. S. Lowell Elliott. <*Auk*, VI, p. 206, April, 1889.
18. Richard Spaulding Wray. <*Auk*, VI, p. 206, April, 1889.
19. Jerome Henry Kidder. <*Auk*, VI, p. 282, July, 1889.
20. Eugen Ferdinand von Homeyer. <*Auk*, VI, p. 341, Oct., 1889.

1890.

21. Ladislas Taczanowski. <*Auk*, VII, p. 218, April, 1890.
22. John Henry Gurney. <*Auk*, VII, pp. 299–300, July, 1890.
23. William Kitchen Parker. <*Auk*, VII, pp. 411–412, Oct., 1890.

1891.

24. Ferdinand Krauss. <*Auk*, VIII, p. 120, Jan., 1891.
25. Henry Davis Minot. <*Auk*, VIII, p. 121, Jan., 1891.
26. Nathaniel Stickney Goss. <*Auk*, VIII, pp. 245–247, April, 1891.
27. John C. Cahoon. <*Auk*, VIII, p. 320, July, 1891.
28. John I. Northrop. <*Auk*, VIII, p. 400, Oct., 1891.

1892.

29. Hermann Burmeister. <*Auk*, IX, pp. 399–400, Oct., 1892.
30. August von Pelzeln. <*Auk*, IX, pp. 74–75, Jan., 1892.
31. John Amory Jeffries. <*Auk*, IX, pp. 311–312, July, 1892.

1893.

32. John Strong Newberry. <*Auk*, X, p. 95, Jan., 1893.
33. Philo R. Hoy. <*Auk*, X, pp. 95–96, Jan., 1893.
34. Jenness Richardson. <*Auk*, X, pp. 307–308, July, 1893.
35. Austin F. Park. <*Auk*, X, pp. 384–385, Oct., 1893.
36. Benjamin Franklin Goss. <*Auk*, X, p. 385, Oct., 1893.
37. Charles F. Adams. <*Auk*, X, pp. 385–386, Oct., 1893.

1894.

38. Charles Glover Allen. <*Auk*, XI, pp. 86–87, Jan., 1894.
39. Samuel Lockwood. <*Auk*, XI, pp. 189–190, April, 1894.
40. William Cushman Avery. <*Auk*, XI, p. 263, July, 1894.
41. Alexander Theodore von Middendorf. <*Auk*, XI, p. 264, July, 1894.
42. Leopold von Schrenck. <*Auk*, XI, p. 264, July, 1894.

1895.

43. John Whipple Potter Jenks. <*Auk*, XII, p. 94, Jan., 1895.
44. Edward Baldamus. <*Auk*, XII, p. 94, Jan., 1895.
45. George Newbold Lawrence. <*Auk*, XII, pp. 198–199, April, 1895.
46. Frederick H. Hoadley. <*Auk*, XII, p. 199, April, 1895.

47. Edward Hargitt. <*Auk*, XII, p. 315, July, 1895.
48. John S. Cairns. <*Auk*, XII, p. 315, July, 1895.
49. Franklin Fairbanks. <*Auk*, XII, p. 315, July, 1895.
50. George Henry Ragsdale. <*Auk*, XII, p. 316, July, 1895.

1896.

51. Thomas Henry Huxley. <*Auk*, XIII, pp. 93–96, Jan., 1896.
52. Henry Seebohm. <*Auk*, XIII, pp. 96–97, Jan., 1896
53. Henry T. Wharton. <*Auk*, XIII, p. 97, Jan., 1896.
54. Juan Gundlach. <*Auk*, XIII, p. 267, July, 1896.
55. Thomas Lyttleton (Lord Lilford). <*Auk*, XIII, pp. 348–349, Oct., 1896.
56. Eugene Carleton Thurber. <*Auk*, p. 349, Oct., 1896.
57. George Brown Goode. <*Auk*, XIII, p. 349, Oct., 1896.

1897.

58. Howard Gardner Nichols. <*Auk*, XIV, p. 115, Jan., 1897.
59. Charles Emil Bendire. <*Auk*, XIV, p. 253, April, 1897.
60. Heinrich Gätke. <*Auk*, XIV, p. 254, April, 1897.
61. Platt Marvin Thorne. <*Auk*, XIV, pp. 254–255, April, 1897.
62. Robert Lee Lawrence. <*Auk*, XIV, p. 342, July, 1897
63. Edward Drinker Cope. <*Auk*, XIV, pp. 242–243, July, 1897.
64. Edward Newton. <*Auk*, XIV, p. 431, Oct., 1897.

1898.

65. George Baur. <*Auk*, XV, pp. 286–287, July, 1898.
66. Osbert Salvin. <*Auk*, XV, pp. 343–346, Oct., 1898.

1899.

67. Oliver Marcy. <*Auk*, XVI, p. 211, April, 1899.
68. Othniel Charles Marsh. <*Auk*, XVI, p. 211, April, 1899.
69. Joseph Wolf. <*Auk*, XVI, pp. 301–302, July, 1899.
70. John Cordeaux. <*Auk*, XVI, pp. 377–378, Oct., 1899.

1900.

71. Elliott Coues. <*Auk*, XVII, p. 91, Jan., 1900. (See also below Nos. **72** and **122**.)
72. Elliott Coues. <*Science*, N. S., XI, pp. 161–163, Feb. 2, 1900.
73. Daniel Webster Prentiss. <*Auk*, XVII, pp. 91–92, Jan., 1900.
74. William W. Colburn. <*Auk*, XVII, p. 92, Jan., 1900.
75. George B. Sennett. <*Auk*, XVII, p. 193, April, 1900. (See also below, No. **79**.)
76. William Edwin Brooks. <*Auk*, XVII, p. 194, April, 1900.
77. Alphonse Milne-Edwards. <*Auk*, XVII, pp. 320–321, Oct., 1900.
78. Edgar Leopold Layard. <*Auk*, XVII, pp. 321–322, Oct., 1900.

1901.

79. In Memoriam: George B. Sennett. <*Auk*, XVIII, pp. 11–23, Jan., 1901, with portrait.
80. Gustav Hartlaub. <*Auk*, XVIII, p. 219, April, 1901.
81. George Augustus Boardman. <*Auk*, XVIII, pp. 219–220, April, 1901.
82. Édmond de Sélys Longschamps. <*Auk*, XVIII, pp. 292–293, July, 1901.
83. William MacGillivray. <*Auk*, XVIII, pp. 293–294, July, 1901.

   Apropos of the unveiling of a memorial tablet to him at Marischal College, Aberdeen, Scotland, in 1900.

84. Newton Dexter. <*Auk*, XVIII, p. 413, Oct., 1901.

1902.

85. John Anderson. <*Auk*, XIX, p. 118, Jan., 1902.
86. Armand David. <*Auk*, XIX, p. 118, Jan., 1902.
87. Lionel William Wigglesworth. <*Auk*, XIX, p. 119, Jan., 1902.
88. James Graham Cooper. <*Auk*, XIX, pp. 421–422, Oct., 1902.

1903.

89. Emil Holub. <*Auk*, XX, p. 92, Jan., 1903.
90. Ludwig Kumlien. <*Auk*, XX, pp. 93–94, Jan., 1903.
91. Perry O. Simons. <*Auk*, XX, pp. 94–96, Jan., 1903.
92. John James Audubon. <*Auk*, XX, pp. 85–86, Jan., 1903.

   Review of John Burrough's biography of Audubon (12mo, Boston, 1903).

93. Thomas McIlwraith. <*Auk*, XX, p. 242, April, 1903.
94. John Nathaniel Clark. <*Auk*, XX, pp. 242–243, April, 1903.
95. Edward Stanley Waters. <*Auk*, XX, p. 243, April, 1903.
96. Mrs. E. S. Mogridge. <*Auk*, XX, pp. 327–328, July, 1903.
97. Gustav Ferdinand von Radde. <*Auk*, XX, pp. 458–459, Oct., 1903.

1904.

98. Lyman S. Foster. <*Auk*, XXI, p. 312, April, 1904.
99. John Fannin. <*Auk*, XXI, p. 510, Oct., 1904.
100. James Mortimer Southwick. <*Auk*, XXI, p. 511, Oct., 1904.

1905.

101. Adolph Boucard. <*Auk*, XXII, p. 333. July, 1905.

1906.

102. Jean Louis Cabanis. <*Auk*, XXIII, pp. 247–485, April, 1906
103. Paul Leverkuhn. <*Auk*, XXIII, p. 247, April, 1906.
104. Émile Oustalet. <*Auk*, XXIII, pp. 355–356, July, 1906.
105. George F. Breninger. <*Auk*, XXIII, p. 356, July, 1906.
106. LaRue Klingle Holmes. <*Auk*, XXIII, p. 356, July, 1906.

107. Joseph H. Batty. <*Auk*, XXIII, pp. 356–357, July, 1906.
108. Frank J. Thompson. <*Auk*, XXIII, pp. 357–358, July, 1906.
109. Henry Baker Tristram. <*Auk*, XXIII, p. 484, Oct., 1906.
110. Victor Fatio <*Auk*, XXIII, pp. 484–485, Oct., 1906.

1907.

111. William Thomas Blanford. <*Auk*, XXIV, p. 118, Jan., 1907.
112. Walter Lawry Buller. <*Auk*, XXIV, p. 119, Jan., 1907.
113. August Koch. <*Auk*, XXIV, pp. 238–239, April, 1907.
114. Alfred Newton. <*Auk*, XXIV, pp. 365–366, July, 1907.
115. Carl von Linné (Linnæus as a Zoölogist). <*Science*, N. S., XXV, pp. 953–961, June 21, 1907. Also *Ann. New York Acad. Sci.*, XVIII, pp. 9–19, pll. ii, 1908, portraits.

An address given at the celebration of the Bicentenary of the birth of Linnæus, in New York, May 23, 1907.

1908.

116. Howard Saunders. <*Auk*, XXV, pp. 103–104, Jan., 1908.
117. Rudolph Blasius. <*Auk*, XXV, p. 248, April, 1908.
118. Leslie Alexander Lee. <*Auk*, XXV, p. 340, July, 1908.
119. Edward Augustus Samuels. <*Auk*, XXV, p. 341, July, 1908.
120. J. V. Barboza du Bocage. <*Auk*, XXV, p. 498, Oct., 1908.
121. Francis Huntington Snow. <*Auk*, XXV, p. 498, Oct., 1908.

1909.

122. Elliott Coues. (Biographical Memoir.) <*National Acad. Sci., Biogr. Mem.*, VII, pp. 397–446, June, 1909.

With portrait, and lists of his principal works and papers (pp. 426–446).

123. Charles K. Worthen. <*Auk*, XXVI, p. 332, July, 1909, *ibid.*, XXVII, p. 112.
124. William H. Brownson. <*Auk*, XXVI, p. 453, Oct., 1909.

1910.

125. Richard Bowdler Sharpe. <*Auk*, XXVII, pp. 124–129, Jan., 1910, with portrait.
126. John Farwell Ferry. <*Auk*, XXVII, pp. 240–241, April, 1910.
127. Henry Hillyer Giglioli. <*Auk*, XXVII, pp. 484–485, Oct., 1910.
128. William Earl Dodge Scott. <*Auk*, XXVII, pp. 486–488, Oct., 1910.

1911.

129. Charles Otis Whitman. <*Auk*, XXVIII, p. 149, Jan., 1911.
130. Manly Hardy. <*Auk*, XXVIII, pp. 149–150, Jan., 1911; *ibid.*, p. 291, July, 1911.
131. George Ernest Shelley. <*Auk*, XXVIII, pp. 387–388, July, 1911.
132. Carl Parrot. <*Auk*, XXVIII, p. 388, July, 1911.
133. Adolf Bernhard Meyer. <*Auk*, XXVIII, p. 519, Oct., 1911.

1916.

134. Daniel Giraud Elliot. <*Science*, N. S., No. 1101, pp. 159–162, Feb. 4, 1916.

VIII. MISCELLANEOUS.

1865–1884.

1. Curatorial Reports on Mammals and Birds. <*Ann. Reports of the Trustees of the Mus. Comp. Zoöl.*, 1865–1884.

1866.

2. Notice of a Foray of a Colony of *Formica sanguinea* Latr. upon a Colony of a black species of *Formica*, for the purpose of making slaves of the latter. < *Proc. Essex Inst.*, V, pp. 14–16, April, 1866.

1870.

3. The Flora of the Prairies. <*Amer. Nat.*, IV, pp. 577–585, Dec., 1870.

   A running account of the more striking features, with particular mention of many species.

4. Notes on the Geological Character of the Country between Chique-Chique, on the Rio de São Francisco, and Bahia, Brazil. <*Hartt's Geology and Physical Geography of Brazil*, pp. 309–319. (8vo, Boston, 1870.)

   Also inedited geological and zoölogical notes, *passim*.

5. Our Weeds. <*Massachusetts Ploughman*, July 23–Oct. 8, 1870.

   A series of six articles, published as follows:

   Number 1, Introductory, July 23, 1870; No. II, The Crowfoots (Ranunculaceæ) and the Poppy Family (Papaveraceæ), July 30, 1870; No. III, The Cruciferæ, Cistaceæ or Rock Roses, and the Hypericaceæ (St. Johnswort Family), Aug. 13, 1870; No. IV, The Pink Family (Caryophyllaceæ), the Purselanes (Portulacaceæ), the Mallow Family (Malvaceæ), Aug. 27, 1870; No. V, The Pulse Family (Leguminosæ), Sept. 24, 1870; No. VI, The Rose Family (Rosaceæ), Evening Primrose Family (Onagraceæ), Orpine or Stone Crop Family (Crasulaceæ), Oct. 8, 1870.

1874.

6. Metamorphism produced by the burning of Lignite Beds in Dakota and Montana Territories. <*Proc. Boston Soc. Nat. Hist.*, XVI, 1873–1874, pp. 246–262, May, 1874.

7. Notes on the Natural History of Montana and Dakota. <*Proc. Boston Soc. Nat. Hist.*, XVII, 1874–1875, pp. 33–91, Nov., 1874.

   Report on the Plants, pp. 70–86 (pp. 40–56 of the separate issue). An annotated list of 224 species, based mainly on identifications by Dr. George Vasey, Botanist of the Department of Agriculture, to whom the collection of plants was referred for determination.

1874–75.

8. Boston Society of Natural History. Reports of Meetings. <*Boston Transcript* (newspaper), Nov. 15, 1874, to June 8, 1875.

Meeting of Oct. 7, 1874. *Ibid.*, Nov. 15, 1874.
Meeting of Oct. 21, 1874. *Ibid.*, Nov. 21, 1874.
Meeting of Nov. 4, 1874. *Ibid.*, Nov. 27, 1874.
Meeting of Nov. 18, 1874. *Ibid.*, Dec. 2, 1874.
Meeting of Dec. 2, 1874. *Ibid.*, Dec. 17, 1874.
Meeting of Dec. 16, 1874. *Ibid.*, Dec. 30, 1874.
Meeting of Jan. 6, 1875. *Ibid.*, Jan. 25, 1875.
Meeting of Jan. 20, 1875. *Ibid.*, Feb. 1, 1875.
Meeting of Feb. 3, 1875. *Ibid.*, Feb. –, 1875.
Meeting of Feb. 17, 1875. *Ibid.*, March 1, 1875.
Meeting of March 3, 1875. *Ibid.*, March –, 1875.
Meeting of March 17, 1875. *Ibid.*, April –, 1875.
Meeting of April 7, 1875. *Ibid.*, April –, 1875.
Meeting of April 21, 1875. *Ibid.*, May –, 1875.
Meeting of May 5, 1875. *Ibid.*, May –, 1875.
Meeting of May 19, 1875. *Ibid.*, June 8, 1875.

These popular reports of the meetings were furnished to the press by the author as a part of his duties as Acting Secretary of the Society

1876.

9. The Little Missouri "Bad Lands". <*Amer. Nat.*, X, pp. 207–216, April, 1876.

With reference to metamorphism due to the burning out of heavy beds of lignite.

10. Nuttall Ornithological Club. <*Bull. Nutt. Orn. Club*, I, pp. 29–32, July, 1876.

Historical sketch of the Club (1873–1876), giving an account of its origin and organization and a list of its publications prior to the establishment of its *Bulletin* in 1876.

1884.

11. Italics for Scientific Names. <*Science*, III, No. 51, p. 87, Jan. 25, 1884.

Refers to the advantage of using italic type for the technical names of genera and species, as an aid in finding the information sought.

1885.

12. Department of Birds and Mammals. <Ann. Rep. Trustees Amer. Mus. Nat. Hist. for the year 1885–6, pp. 9–12, March, 1886.

Report on the extent and condition of the collections; need of 'study' or research material urged; also of increased exhibition space.

1886.

13. [Establishment of the Division of Economic Ornithology and Mammalogy under the United States Department of Agriculture.] <*Auk*, I, p. 416, July, 1886.

1887.

14. Third Annual Report of the Curator of the Department of Mammalogy and Ornithology, 8vo, pp. 1–10.

A brochure privately printed by order of the President. Reprinted in part in Ann. Report Trustees Amer. Mus. Nat. Hist. for the year 1887–1888, pp. 12–16.

To a report on the accessions and activities of the year 1887 is added: 'Wants and Recommendations,' and letters to the President of date Nov. 7, 1887 and Jan. 21, 1888, in advocacy, respectively, of the purchase of an important collection of North American birds, and of sending expeditions to the West to secure specimens of bison, elk, caribou, pronghorn antelope, mountain sheep, and other forms of rapidly disappearing large North American mammals.

1888.

15. The Museum of Natural History. An Appeal for Aid in Carrying on its Work. <*New York Evening Post*, Feb. 28, 1888.

Of the $64,000 expended in 1877, all but $15,000 was furnished by the Trustees. The accessions received and the work accomplished by the Department of Mammalogy and Ornithology is recounted, and a plea made for financial assistance on the part of the public through aid by membership, etc., in return for free access to the exhibition halls of the Museum.

16. The Museum of Natural History. <*New York Tribune*, Feb. 24, 1888.

Published as editorial, about 500 words.

1890.

17. Zoe, A Biological Journal. <*Auk*, VII, p. 300, July, 1890.

Announcing the founding, objects and character of this excellent journal of which four volumes (1890–1893) were published.

1893.

18. The Natural History Museum of New York City. <*New York Evening Post*, Feb. 28, 1893.

Its scientific work as set forth in the *Bulletin* of the Museum.

19. Letter on alleged skinning of Kids and Birds while alive, to improve the quality of their Skins or Plumage for Commercial Purposes. <*Our Animal Friends*, XX, p. 268, No. 12, Aug., 1893.

20. Letter in reply to the question "How does the Study of Nature Pay?" <*The Hesperian* (newspaper), Gainesville, Texas, March 31, 1893.

Letter to Geo. H. Ragsdale.

Bibliography: Summary of Titles.

Mammals	271
Birds	966
Reptiles	5
Zoögeography	22
Nomenclature	35
Biography	134
Miscellaneous	20
	1453

IX. EDITORIAL WORK.

1874–1875.

The Proceedings of the Boston Society of Natural History, Vol. XVIII, for 1874–1875, and part of Vol. IX of the *American Naturalist*, in the temporary absence of their respective editors.

1876–1883.

Bulletin of the Nuttall Ornithological Club, Vols. I–VIII, Cambridge, Mass. Published by the Club, 8vo, 1876–1883.

The first number of this publication was issued in April, 1876, under the editorship of Mr. Charles J. Maynard, and under the title *Quarterly Bulletin of the Nuttall Ornithological Club*, Cambridge, Mass. With the second number (July, 1876) the title was changed to *Bulletin of the Nuttall Ornithological Club: A Quarterly Journal of Ornithology*, with, on the fourth page of the cover, a 'Prospectus,' which states: "The need in this country of a periodical exclusively devoted to Ornithology has been long apparent. . . . Each number will consist of not less than twenty-four pages, to be increased as soon as the receipts from subscriptions shall warrant the additional expense. Two more numbers will be issued during the present year, in order that the second volume may begin with January, 1877. . . . Professor Baird and Dr. Coues have kindly consented to act as Associate Editors. Its chief editorial management will be under the direction of Mr. J. A. Allen of the Museum of Comparative Zoölogy, Cambridge, Mass. . . ." No. 4 was issued November, 1876. Volume I consists of pp. i–iv, 1–100, with a colored plate, engraved on steel.

The volumes, under the same editorial staff, steadily increased in size to Vol. VII, 1882 (pp. i–vi, 1–275). Number 4 of Volume VIII (October, 1883) contained this "Important Notice: The First Series of the *Bulletin of the Nuttall Ornithological Club* closes with the present volume. A Second Series, with change of name, will begin with January, 1884, under the auspices of the American Ornithologists' Union, and under the Editorial Management of Mr. J. A. Allen, assisted by a Staff of Associate Editors. . . ."

The Associate Editors of the eight volumes of the *Bulletin* were Spencer F. Baird and Elliott Coues.

1884–1911.

The Auk: A Quarterly Journal of Ornithology, Vols. I–XXVIII, 1884–1911, and No. 1 of Vol. XXIX (Jan., 1912).

Title-page of Vol. I:

Old Series, Vol. IX	Continuation of the Bulletin of the Nuttall Ornithological Club	New Series, Vol. I

| — |

The Auk | A Quarterly Journal of Ornithology | – | Editor, | J. A. Allen | Associate Editors, | Elliott Coues, Robert Ridgway, William Brewster, | and Montague Chamberlain. [Cut of Great Auk, 4½ × 5 in. | Volume I | Published for the American Ornithologists' Union | Boston, Mass. | Estes and Lauriat | 1884.

The second page of the cover of the first number carries the following: "'The Auk,' published as the Organ of the American Ornithologists' Union, will be conducted as a *Magazine of General Ornithology*. In general character it will differ little from the late 'Bulletin of the Nuttall Ornithological Club,' of which it forms virtually a Second Series. Each number will contain about *one hundred pages*, and the illustrations will occasionally include *colored plates*."

The first volume of 'The Auk' consists of pp. i–vii, 1–419, and 1 colored plate. This is about the average size for the first 19 volumes, the last 7 averaging 500 pages, in each case exclusive of front-matter (contents and membership lists), which runs, in different volumes, from 8 to nearly 40 pages.

In *The Auk*, as in its predecessor, the *Nuttall Bulletin*, the department of 'Recent Literature' was intended to include notices (1) of all ornithological works and papers by American authors; (2) all papers, wherever published, relating especially to North American birds; (3) all of the more important monographs and works treating of special groups of birds or of birds in general. In 1880 was begun a series of papers entitle 'Minor Ornithological Papers,' giving the title, place of publication, and short notices or abstracts of the minor papers relating to North American birds, not otherwise noticed under 'Recent Literature,' subject however to certain stated restrictions as to their content. The record began "with the close of that given by Dr. Coues in his 'List of Faunal Publications relating to North American Ornithology' published in his 'Birds of the Colorado Valley,' or about July, 1878," and closed in 1894, after having been a feature of 'The Auk' for fourteen years. The last title of the series was No. 2697. Nos. 1–1199 were by the editor, the others (Nos. 1200–2697) by C. F. Batchelder, who from 1888 to 1893 (both inclusive) was Associate Editor of *The Auk* and greatly lightened for this period the duties of the senior editor.

The Associate Editors of *The Auk* for the first four volumes (1884–1887) were Elliott Coues, Robert Ridgway, William Brewster, and Montague Chamberlain. They were succeeded by Charles F. Batchelder (1888–1893) and Frank M. Chapman (1894–1915).

"Beginning with the initial volume of the Bulletin of the Nuttall Ornithological Club, and continuing to the present year, Dr. Allen has, without intermission guided the course of this journal and its successor The Auk; and the series of thirty-six volumes stands as a perpetual monument to his ability, and his painstaking devotion to the cause of ornithology and the interests of the American Ornithologists' Union. There have been few continuous editorships of equal length in the history of scientific periodicals."— Editorial statement by Witmer Stone, *Auk*, XXIX, p. 136, Jan., 1912.

1886.

The Code of Nomenclature | and | Check-List | of | North American Birds | adopted by the American Ornithologists' Union | being the Report of the Committee of the | Union on Classification and | Nomenclature — Zoölogical Nomenclature is a means, not an end, of Zoölogical Science | — | New York | American Ornithologists' Union | 1886. 8vo, pp. i–viii, 1–392.

Contains The Code of Nomenclature (pp. 1–69) and The Check-List of North American Birds, according to the Canons of Nomenclature of the American Ornithologists' Union (pp. 71–392).

1886–1916.

Bulletin of the American Museum of Natural History, Vols. I, No. 6, to XXXV, excluding Vols. XV, XVII and XVIII, devoted to anthropology. (8vo, New York.)

1889–1895.

Supplement to the Code of Nomenclature and Check-List of North American Birds, adopted by the American Ornithologists' Union. Prepared by a Committee of the Union. 8vo, pp. 23. New York, American Ornithologists' Union, 1889.

Check-List | of | North American Birds | prepared by a Committee | of the | American Ornithologists' Union | Second and revised edition | — | Zoölogical Nomenclature is a means, not an end, of Zoölogical Science | — |. New York | American Ornithologists' Union | 1895. 8vo, pp. i–xi, 1–372.

Supplements Nos. 2 to 15 to the Check-List were published at intervals in *The Auk* from 1890–1909, and also issued separately in an edition of 100 copies.

1893–1916.

Memoirs of the American Museum of Natural History. Vols. I and IX (Zoölogy and Palæontology), and general supervision of Vols. II, III, IV, and VI (Anthropology). Also, New Series, Vol. I, 1913–1916.

1908

The | Code of Nomenclature | adopted by the | American Ornithologists' Union | — | Zoölogical Nomenclature is a means, not an end, of Zoölogical Science | — | Revised edition | New York | American Ornithologists' Union | July, 1908. 8vo, pp. i–lxxxv.

1910.

Check-List | of | North American Birds | prepared by a Committee | of the | American Ornithologists' Union | Third Edition (Revised) | — | Zoölogical Nomenclature is a means, not an end, of Zoölogical Science | — | New York | American Ornithologists' Union | 1910. 8vo, pp. 1–430, 2 maps of North America, one of them colored to show faunal areas.

NATURAL SCIENCES IN AMERICA

An Arno Press Collection

Allen, J[oel] A[saph]. **The American Bisons,** Living and Extinct. 1876

Allen, Joel Asaph. **History of the North American Pinnipeds:** A Monograph of the Walruses, Sea-Lions, Sea-Bears and Seals of North America. 1880

American Natural History Studies: The Bairdian Period. 1974

American Ornithological Bibliography. 1974

Anker, Jean. **Bird Books and Bird Art.** 1938

Audubon, John James and John Bachman. **The Quadrupeds of North America.** Three vols. 1854

Baird, Spencer F[ullerton]. **Mammals of North America.** 1859

Baird, S[pencer] F[ullerton], T[homas] M. Brewer and R[obert] Ridgway. **A History of North American Birds:** Land Birds. Three vols., 1874

Baird, Spencer F[ullerton], John Cassin and George N. Lawrence. **The Birds of North America.** 1860. Two vols. in one.

Baird, S[pencer] F[ullerton], T[homas] M. Brewer, and R[obert] Ridgway. **The Water Birds of North America.** 1884. Two vols. in one.

Barton, Benjamin Smith. **Notes on the Animals of North America.** Edited, with an Introduction by Keir B. Sterling. 1792

Bendire, Charles [Emil]. **Life Histories of North American Birds** With Special Reference to Their Breeding Habits and Eggs. 1892/1895. Two vols. in one.

Bonaparte, Charles Lucian [Jules Laurent]. **American Ornithology:** Or The Natural History of Birds Inhabiting the United States, Not Given by Wilson. 1825/1828/1833. Four vols. in one.

Cameron, Jenks. **The Bureau of Biological Survey:** Its History, Activities, and Organization. 1929

Caton, John Dean. **The Antelope and Deer of America:** A Comprehensive Scientific Treatise Upon the Natural History, Including the Characteristics, Habits, Affinities, and Capacity for Domestication of the Antilocapra and Cervidae of North America. 1877

Contributions to American Systematics. 1974

Contributions to the Bibliographical Literature of American Mammals. 1974

Contributions to the History of American Natural History. 1974

Contributions to the History of American Ornithology. 1974

Cooper, J[ames] G[raham]. **Ornithology.** Volume I, Land Birds. 1870

Cope, E[dward] D[rinker]. **The Origin of the Fittest:** Essays on Evolution and **The Primary Factors of Organic Evolution.** 1887/1896. Two vols. in one.

Coues, Elliott. **Birds of the Colorado Valley.** 1878

Coues, Elliott. **Birds of the Northwest.** 1874

Coues, Elliott. **Key To North American Birds.** Two vols. 1903

Early Nineteenth-Century Studies and Surveys. 1974

Emmons, Ebenezer. **American Geology:** Containing a Statement of the Principles of the Science. 1855. Two vols. in one.

Fauna Americana. 1825-1826

Fisher, A[lbert] K[enrick]. **The Hawks and Owls of the United States in Their Relation to Agriculture.** 1893

Godman, John D. **American Natural History:** Part I — Mastology and **Rambles of a Naturalist.** 1826-28/1833. Three vols. in one.

Gregory, William King. **Evolution Emerging:** A Survey of Changing Patterns from Primeval Life to Man. Two vols. 1951

Hay, Oliver Perry. **Bibliography and Catalogue of the Fossil Vertebrata of North America.** 1902

Heilprin, Angelo. **The Geographical and Geological Distribution of Animals.** 1887

Hitchcock, Edward. **A Report on the Sandstone of the Connecticut Valley,** Especially Its Fossil Footmarks. 1858

Hubbs, Carl L., editor. **Zoogeography.** 1958

[Kessel, Edward L., editor]. **A Century of Progress in the Natural Sciences: 1853-1953.** 1955

Leidy, Joseph. **The Extinct Mammalian Fauna of Dakota and Nebraska,** Including an Account of Some Allied Forms from Other Localities, Together with a Synopsis of the Mammalian Remains of North America. 1869

Lyon, Marcus Ward, Jr. **Mammals of Indiana.** 1936

Matthew, W[illiam] D[iller]. **Climate and Evolution.** 1915

Mayr, Ernst, editor. **The Species Problem.** 1957

Mearns, Edgar Alexander. **Mammals of the Mexican Boundary of the United States.** Part I: Families Didelphiidae to Muridae. 1907

Merriam, Clinton Hart. **The Mammals of the Adirondack Region,** Northeastern New York. 1884

Nuttall, Thomas. **A Manual of the Ornithology of the United States and of Canada.** Two vols. 1832-1834

Nuttall Ornithological Club. **Bulletin of the Nuttall Ornithological Club:** A Quarterly Journal of Ornithology. 1876-1883. Eight vols. in three.

[Pennant, Thomas]. **Arctic Zoology. 1784-1787.** Two vols. in one.

Richardson, John. **Fauna Boreali-Americana;** Or the Zoology of the Northern Parts of British America, Containing Descriptions of the Objects of Natural History Collected on the Late Northern Land Expeditions Under Command of Captain Sir John Franklin, R. N. Part I: Quadrupeds. 1829

Richardson, John and William Swainson. **Fauna Boreali-Americana:** Or the Zoology of the Northern Parts of British America, Containing Descriptions of the Objects of Natural History Collected by the Late Northern Land Expeditions Under Command of Captain Sir John Franklin, R. N. Part II: The Birds. 1831

Ridgway, Robert. **Ornithology.** 1877

Selected Works By Eighteenth-Century Naturalists and Travellers. 1974

Selected Works in Nineteenth-Century North American Paleontology. 1974

Selected Works of Clinton Hart Merriam. 1974

Selected Works of Joel Asaph Allen. 1974

Selections From the Literature of American Biogeography. 1974

Seton, Ernest Thompson. **Life-Histories of Northern Animals: An Account of the Mammals of Manitoba.** Two vols. 1909

Sterling, Keir Brooks. **Last of the Naturalists:** The Career of C. Hart Merriam. 1974

Vieillot, L. P. **Histoire Naturelle Des Oiseaux de L'Amerique Septentrionale,** Contenant Un Grand Nombre D'Especes Decrites ou Figurees Pour La Premiere Fois. 1807. Two vols. in one.

Wilson, Scott B., assisted by A. H. Evans. **Aves Hawaiienses:** The Birds of the Sandwich Islands. 1890-99

Wood, Casey A., editor. **An Introduction to the Literature of Vertebrate Zoology.** 1931

Zimmer, John Todd. **Catalogue of the Edward E. Ayer Ornithological Library.** 1926